Lecture Notes in Civil Engineering   735

Series Editors

Marco di Prisco, *Politecnico di Milano, Milano, Italy*
Sheng-Hong Chen, *School of Water Resources and Hydropower Engineering, Wuhan University, Wuhan, China*
Ioannis Vayas, *Institute of Steel Structures, National Technical University of Athens, Athens, Greece*
Sanjay Kumar Shukla, *School of Engineering, Edith Cowan University, Joondalup, Australia*
Anuj Sharma, *Iowa State University, Ames, USA*
Nagesh Kumar, *Department of Civil Engineering, Indian Institute of Science Bangalore, Bengaluru, India*
Chien Ming Wang, *School of Civil Engineering, The University of Queensland, Brisbane, Australia*
Zhen-Dong Cui, *China University of Mining and Technology, Xuzhou, China*
Xinzheng Lu, *Department of Civil Engineering, Tsinghua University, Beijing, China*

Lecture Notes in Civil Engineering (LNCE) publishes the latest developments in Civil Engineering—quickly, informally and in top quality. Though original research reported in proceedings and post-proceedings represents the core of LNCE, edited volumes of exceptionally high quality and interest may also be considered for publication. Volumes published in LNCE embrace all aspects and subfields of, as well as new challenges in, Civil Engineering. Topics in the series include:

- Construction and Structural Mechanics
- Building Materials
- Concrete, Steel and Timber Structures
- Geotechnical Engineering
- Earthquake Engineering
- Coastal Engineering
- Ocean and Offshore Engineering; Ships and Floating Structures
- Hydraulics, Hydrology and Water Resources Engineering
- Environmental Engineering and Sustainability
- Structural Health and Monitoring
- Surveying and Geographical Information Systems
- Indoor Environments
- Transportation and Traffic
- Risk Analysis
- Safety and Security

To submit a proposal or request further information, please contact the appropriate Springer Editor:

- Pierpaolo Riva at pierpaolo.riva@springer.com (Europe and Americas);
- Swati Meherishi at swati.meherishi@springer.com (Asia—except China, Australia, and New Zealand);
- Wayne Hu at wayne.hu@springer.com (China).

All books in the series now indexed by Scopus and EI Compendex database!

Decheng Wan · Dezhi Ning · Hao Tian
Editors

Advanced and Emerging Marine Engineering Technologies

Proceedings of 2024 3rd International Joint Conference on Civil and Marine Engineering (JCCME 2024)

Editors
Decheng Wan
SJTU Computational Marine Hydrodynamics Lab
Shanghai Jiao Tong University
Shanghai, China

Dezhi Ning
State Key Laboratory of Coastal and Offshore Engineering
Dalian University of Technology
Liaoning, China

Hao Tian
Naval Architecture and Ocean Engineering College
Dalian Maritime University
Liaoning, China

ISSN 2366-2557   ISSN 2366-2565 (electronic)
Lecture Notes in Civil Engineering
ISBN 978-981-95-1486-1   ISBN 978-981-95-1487-8 (eBook)
https://doi.org/10.1007/978-981-95-1487-8

This work was supported by Shanghai Jiao Tong University.

© The Editor(s) (if applicable) and The Author(s) 2026. This book is an open access publication.

Open Access This book is licensed under the terms of the Creative Commons Attribution-NonCommercial-NoDerivatives 4.0 International License (http://creativecommons.org/licenses/by-nc-nd/4.0/), which permits any noncommercial use, sharing, distribution and reproduction in any medium or format, as long as you give appropriate credit to the original author(s) and the source, provide a link to the Creative Commons license and indicate if you modified the licensed material. You do not have permission under this license to share adapted material derived from this book or parts of it.

The images or other third party material in this book are included in the book's Creative Commons license, unless indicated otherwise in a credit line to the material. If material is not included in the book's Creative Commons license and your intended use is not permitted by statutory regulation or exceeds the permitted use, you will need to obtain permission directly from the copyright holder.

This work is subject to copyright. All commercial rights are reserved by the author(s), whether the whole or part of the material is concerned, specifically the rights of translation, reprinting, reuse of illustrations, recitation, broadcasting, reproduction on microfilms or in any other physical way, and transmission or information storage and retrieval, electronic adaptation, computer software, or by similar or dissimilar methodology now known or hereafter developed. Regarding these commercial rights a non-exclusive license has been granted to the publisher.

The use of general descriptive names, registered names, trademarks, service marks, etc. in this publication does not imply, even in the absence of a specific statement, that such names are exempt from the relevant protective laws and regulations and therefore free for general use.

The publisher, the authors and the editors are safe to assume that the advice and information in this book are believed to be true and accurate at the date of publication. Neither the publisher nor the authors or the editors give a warranty, expressed or implied, with respect to the material contained herein or for any errors or omissions that may have been made. The publisher remains neutral with regard to jurisdictional claims in published maps and institutional affiliations.

This Springer imprint is published by the registered company Springer Nature Singapore Pte Ltd.
The registered company address is: 152 Beach Road, #21-01/04 Gateway East, Singapore 189721, Singapore

If disposing of this product, please recycle the paper.

Contents

Thermal Performance Analysis of Polar Cruise Ship Coatings
under Extreme Cold Conditions 1
   Di Li, Ji Zeng, Yichao Bai, and Zicheng Wu

Hydrodynamic Analysis of Single Barge Unilateral Salvage of Wrecks
Under Mooring Conditions 15
   *Fei Wang, Guohong Zhou, Wei Zhang, JingJing Liu, Jinxing Li,
and ZiKai Xu*

Research on Safety Evaluation Technology of Ship's Mooring Ropes 29
   Nan Lu, Ji Zeng, and Qihang Wu

Hydrodynamic Analysis of Deep-Water Double Barges Lifting Wreck
Under Mooring Conditions 41
   ZiKai Xu, Yingjie Li, Yuantao Bi, and Wei Zhang

Efficiency of the Afterburner of the Solid Oxide Fuel Cell-Gas Turbine
Marine System 53
   Serhiy I. Serbin, Oleksii V. Patlaichuk, and Xianrui Zhao

Coastal Protection with Tensioned Viscoelastic Wave Barriers 62
   *Jianjun Chen, Chunxiao Li, Xin Liu, Jianjian Zhao, Cheng Bi,
Yong Zhang, Maquan Wang, Mingjun Nan, Hao Hu, Hewen Hu,
Ruichao Liu, and Yu Lei*

Design and Optimization for the Base of Gantry Used in Wreck Salvage 74
   *Shude Chen, Wei Deng, Yuantao Bi, Wei Zhang, Xinxin Wang,
and Yijian Han*

Research on Evaluation Methods for Lighting Layout of Manned
Submersibles 85
   Weizhe Xu, Jun Cao, Wenxin Qu, Dan Xing, and Hongtao Ji

An Analytical Study of the Multi-body Motion of the Installation
Operation between A Crane Ship and Gravity-based Wind Turbine
Foundation in Waves 95
   Pengfei Chen, Yu Lin, Bibo Li, Xiao Zhang, Jun Huang, and Yan Nie

Research on the Influencing Factors of Water Hammer Caused by Large
Spacing Water Transfer of Ship Based on Transient Flow Model 108
   Chao Fang, Bin Li, Weibo Pan, and Yu Guo

Development of a Pivoting Wave Energy Converter Concept 118
   Gianmaria Giannini, Victor Ramos, Paulo Rosa-Santos,
   Esmaeil Zavvar, Tomás Calheiros-Cabral, and Francisco Taveira-Pinto

Research on Hydrodynamic Characteristics of A SWATH Based on CFD 127
   Yilin Yang, Hao Hao, and Minmin Zheng

Influence of Stable Fins on SWATH Motion 135
   Yilin Yang, Minmin Zheng, and Hao Hao

Discussion on Active Fire Protection Systems for Offshore Substation
of Wind Power Station 143
   Song Li, Min Dai, Zhuoping Yuan, and Wei Jin

Risk Assessment of Ship Pilotage Based on WBS-RBS and Cloud Model:
A Case Study of Qinzhou Port 152
   Peng Liu, Yongtao Xi, Yi Su, and Haibo Huang

Fluid-Solid Coupling Analysis of Marine Stern Bearing Considering
Viscosity-Temperature and Cavitation Effect 171
   Xianjun Wu, Qianwen Huang, and Minghui Sheng

Analysis and Experimental Study on Water Yield of Seawater Desalination
Device by Light Distillation 179
   Lucai Wang, Tianhua Xie, and Zhilin Zheng

Analysis on the Application of Unmanned Equipment in Lifesaving
Missions at Sea 189
   Guanghui Yu, Lei Su, and Wei Li

An Improved Method of Constructing Mercator Chart 198
   Yong Zhang, Junting Xiong, and Qi Yang

Typhoon Prediction Analysis of Pangu Weather Model 207
   Peng Yuehua, Yao Libo, Song Jun, and Yuchen Zhang

Research on Intelligent Optimization of Deployment for High-Precision
Positioning of Submerged Buoys 215
   Leilei Han, Hairui Ma, and Minghai Li

Analyses on Consistency Between Level 1 and 2 Vulnerability Criterion for Dead Ship Stability 224
   Kaibo Yu, Feng Cai, and Jianjun Hou

Research on Evaluation of Personnel Evacuation Paths in Ship Fire 237
   Xiaogang Jiang, Zhijang Yuan, and Mingdong Lv

Crashworthiness Analysis of New Double-Layer Side Structure 244
   Ming-dong Lv, Xi-wei Wang, Zhi-jiang Yuan, Xiao-gang Jiang, Zhong-chang An, and Peng-yao Yu

Ship Navigation Risk Assessment in The Arctic Northeast Passage Via Stochastic Petri Nets Model 254
   Cuiwen Fang, Xiaoyu Du, Xinxin Zhang, and Shenping Hu

Preparation of the Super-sophobic Oil Surface on Steel Substrate by Composite Electrodeposition 266
   Hong Cheng, Dongliang Gu, and Yong Zhang

Assessment of Dynamic Maritime Navigation Risks During Cold Surges in Bohai Bay and Northern Yellow Sea: An Integration of Atmospheric and Wave Numerical Models 275
   Haoyu Chen and Chen Chen

Research on Significant Wave Height Forecast Based on Extreme Learning Machine 285
   Bo Yang, Xu Wang, and Xiao Wang

Research on the Recognition Method of Unsafe Action in Mooring Operations Based on Improved YOLOv7 297
   Zhen Fang, Wenjun Zhang, Xiangkun Meng, Xue Yang, and Xiangyu Zhou

A Framework of Predictive Maintenance for Maritime Autonomous Surface Ships Considering Component Degradation 303
   Hongqiang Li, Xiangkun Meng, Jing Liu, Wenjun Zhang, Xiang-Yu Zhou, and Xue Yang

Experimental Study on the Rudder Angle Optimization for a Twin Rudder Ship Resistance and Self-propulsion Performance 309
   Lei Xing, Li Zhang, Weimin Chen, and Yongyue Li

Hull Form Optimization of a Twin-Screw Full Ship Based on RANSE Method 318
   Li Zhang, Haikui Ren, and Jianting Chen

Effects of Mass Ratio on the Flow-Induced Vibration Characteristics
of Rigid Cylindrical Oscillator Supported by Maglev 330
    Guoqiang Lei, Xu Bai, Wen Zhang, Zhenbang Yang, Renwei Ji,
    and Jianjie Niu

Dynamic Risk Analysis for Escort Operations in Arctic Waters
Considering the Time Lag Feature of Risk Factors 339
    Jiaxuan Luo, Xiaofang Luo, Yingfei Zan, and Xu Bai

Risk Analysis of Container Ships Navigating in Deep-Water Channels
of Yangtze River Estuary in Severe Restricted Visibility 347
    Tingrong Qin, Pingping Luo, Idwimoh Daniel Ogooluwa,
    and Xiaojing Zhang

Enhancing Maritime Safety via a Hybrid TTT-LSTM Model for Vessel
Trajectory Prediction 360
    Xiangxing Zhou, Yuhao Li, Zhengchuan Qin, and Qing Yu

Study on Ice Resistance of Real Ship under Different Navigation
Environments and Ice Thickness 369
    Yujia Zhang, Bin Mei, Jiayun Ye, Congcong Zhao, Weifeng Li,
    and Guoyou Shi

Observations on Size Distribution Characteristics of Ice Floes
in Icebreaking Channel 377
    Congcong Zhao, Bin Mei, Yujia Zhang, Jiayun Ye, Weifeng Li,
    and Guoyou Shi

Brash Ice Contraction Phenomenon in Ice-Covered Channels 384
    Jiayun Ye, Bin Mei, Yujia Zhang, Congcong Zhao, Weifeng Li,
    and Guoyou Shi

Establishment of Multi-ship Joint Navigation Without Satellite Navigation
System Supporting 394
    Yundong Han, Fanjun Meng, Qian Huang, Guohui Lan, and Hairui Ma

Risk Analysis of LNG Ship Collision Accident in Arctic Waters Based
on Bayesian Network 402
    Zhuang Li, Haoran Jiang, Xiaoming Zhu, Hongbo Wang,
    and Kaixian Gao

Research Status of Resilience Governance of Transportation Safety
on the Maritime Silk Road 411
    Liu Zhu, Jianjun Wu, Shenping Hu, and Xiangqian Meng

Warship Formation Time Unification Based on High Precision Celestial
Observation 421
  Zhiyou Zhang, Fu Yu, and Yongxin Jiang

The Use of PNT System in the War and Enlightenment to Our Army's
Future Operations 430
  Zhenyu Zheng, Zhiyou Zhang, and Yulong Kong

Internal Wave Wake Characteristics of Submerged Vehicle in Different Fr
Numbers 443
  Hang Sun, Yongjie Zhang, Haoyu Yang, and Dawei Li

Simulation Research on Ship Drag Reduction Technology Based on Air
Blowing 452
  Xiao Wang, Haoyu Yan, Yongjie Zhang, and Dawei Li

Structural Study on High-Speed Catamaran in Regular Head Wave 460
  Yikang Chen, Zhipeng Deng, and Jinglei Yang

Research on Distortion Control Indexes of Large-Scale Temporary Hatch
Opening on Deck 476
  Zhiyuan Wei, Jianjun Hou, and Xiao Wang

Development and Advantages of the BeiDou Navigation Satellite System
Using in Maritime Search and Rescue 488
  Yulong Kong, Zhenyu Zheng, and Ming Wu

Study on the Effect of Free End on Flow-Induced Vibration and Energy
Harvesting of PTC Cylinders 496
  *Ruitao Tang, Sihan Liu, Yichen Ma, Ruipu Zhao, Fenglai Huang,
  and Chunhui Ma*

Weather Processes of Strong Wind over China's Coastal Waters in 2023 505
  Bin Zuo, Shipeng Su, and Yuepeng Yang

Prediction of Ship Motion Attitude Using an Improved Transformer Model 515
  Lingyi Hou, Hang Sun, Jianjun Hou, and Dawei Li

Analysis of Hydrometeorological Characteristics of Yellow Sea and Bohai
Sea during 1993-2023 527
  Yuepeng Yang, Hanjia Wang, and Peng Han

Calculation of the Existence Time of Bubbles in the Wave 537
  Zhijiang Yuan, Xiaogang Jiang, and Mingdong Lv

Research on Vibration Isolation Performance of Active Vibration Isolation
System Base Structure for Marine Power Equipment 551
  Jianlong Hu, Lin Lin, and Zhaowang Xia

Structural Strength Analysis of Winch for Marine Platforms 559
  Xin Ding, Lin Lin, and Zhaowang Xia

Research on Intelligent Monitoring of Ship Piping System Based
on Digital Twin 565
  Tingyu Jiao, Ji Zeng, Yu Zhang, Haoyun Gu, and Weiqi Ding

Research on Inland Waterway Remote Driving Based on Scaled Ship Model ... 572
  Tao Guo, Tao Li, Mao Zheng, Hao Wu, Yong Wu, and Chengguo Song

Image Object Detection and Tracking of Ships in Complex Inland Water
Areas 587
  Fang Jin, Yong Wu, Yongtao Wang, Xiao Liu, and Bing Han

Human Error Probability Analysis on Pilot's Operation in Changshu
Section of the Yangtze River 602
  *Haoxuan Yan, Cunlong Fan, Linglong Yan, Min Wang, Yongtao Xi,
  Shenping Hu, and Bing Han*

Research on Magnetic Field Distribution of Rotating Magnetic Target
Under Water 613
  Pengfei Lin, Ming Chang, Jun Li, Lei Xu, and Heda Zhao

An Optimization Model for Water Search and Rescue Sites Considering
Accident Black Spot 623
  Tingrong Qin, Xueli Chen, and Qinyou Hu

Three Dimensional Stress Analysis Method of Ring-Stiffened Cylindrical
Shell with New Design Features 633
  Shagu Chen, Yuan Gao, and Xiaozhong Xie

Numerical Simulation on Planar Motion Mechanism of a Bulk Carrier
with a Body-Force Propeller Model 650
  Hao Hao, Weimin Chen, and Minmin Zheng

Near-Field Low-Frequency Noise Prediction Analysis of Propeller Noise
in Shallow Water Environment 659
  He Yang, Chunyu Zhang, and Yun Liu

Research on Sea and Air Target Recognition Based on Otsu-Hough Sea
Antenna Detection and YOLOv8 Algorithm 666
    Wenjin Chen, Jingqiang Bi, and Xizhen Qiao

Design of Ship Air-Conditioning and Ventilation System in Low Latitude
and High Altitude Area 676
    Tianyu Tu, Liang Zhang, Dinghua Hu, and Lei Zhao

Investigation of Surrounding Rock Failure Modes in Marine Soft Soil
Strata for Shield Tunneling 687
    Wenbin Xu, Yindong Sun, Heng Zhang, Wu Ke, and Yajun Liu

Hydrodynamic Analysis of Unilateral Salvage of Wreck
by Semi-submersible 694
    Wei Zhang, Shaoshi Dai, Zikai Xu, and Jingjing Liu

Application and Discussion of Airbag in Salvage 711
    Yingchun Qi, Sihao Xing, Yingjie Li, and Yuantao Bi

Design and Optimization of Righting Hooks for Wreck Righting 724
    Fei Wang, Shude Chen, Yijian Han, Jingjing Liu, and Wei Zhang

Two-Stage Energy Management Strategy for Accommodation Area
of Multi-Energy Ship 734
    Lihao Wang, Jiahui Xue, and Can Cui

Cosine Wave Numerical Wave Tank Study and Its Force Analysis
on Small-Scale Structures 742
    *ZiFeng Sun, MingKai Li, XinYu Zhang, YanJun Guan, YuXiang Niu,
and JunSheng Zhang*

Research of Hydrodynamic and Energy Capture Characteristics of Wave
Energy Device with Eccentric Inner Rotor 756
    *Yuxiang Niu, Lujun Zhao, Yanjun Guan, Zifeng Sun, Junsheng Zhang,
and Wanqing Zhang*

Design and Analysis of Hexagonal Floating Photovoltaic Platform
and Mooring System 767
    Jiayang Sun, Yihou Wang, Tie Ren, and Taocui Yu

Application of Structural Response Prediction Methods for Real Time
Structural Health Monitoring of Multi-Linked Floating Offshore Structures 777
    *Kichan Sim, Byoung Wan Kim, Sa Young Hong, Hong Gun Sung,
and Kangsu Lee*

Two-Dimensional Diffraction Problem of a Floating Rectangular Box
Over a Sloping Seabed 785
   Ziqi Li and Bin Teng

Modal Decomposition Method for the Dynamic Response of a Submerged
Floating Tunnel Under Wave Action 793
   Mei Yu and Bin Teng

Experimental and Numerical Evaluation of Hydrodynamic Behavior
of 10 MW Floating Offshore Wind Turbine 802
   Byoung Wan Kim, Kichan Sim, Kangsu Lee, and Sa Young Hong

Numerical Investigation of Coastal Evolution and Protection Measures
Under Sea Level Rise 814
   *Renqiang Wen, Chen Gu, Yue Zheng, Tianxia Jia, Yang Hong,
and Zhipeng Qu*

Joint Statistical Analysis of Wave Height and Wave Period of Guangdong
Sea Area Based on Archimedean Copula 827
   Shiyun Wang, Jiangshan Zheng, Chen Gu, Yingying Li, and Bihong Zhu

Temporal and Spatial Distribution of Sea Surface Temperature in the East
China Sea Based on SODA Data 837
   Yingying Li, Chen Gu, Guoqu Cui, Shiyun Wang, Bihong Zhu, and Kan Yi

Research on Berth Allocation Problem for Jetty Ro-Ro Terminals Based
on Disruption Recovery 847
   Yu Lu, Zeyu Shi, Yi Sui, Shihao Wu, and Jiangang Jin

Numerical Study on Local Scour Propagation Below Two Intersecting
Pipelines on a Sandy Seabed Under Currents 862
   Zhuo Fang, Zhaolong Yin, Limeng Zhao, and Zhipeng Zang

Numerical Simulation of Cylindrical Gravity Anchor Penetrating
into Lakebed 868
   Yong Chen, Li Liang, Minghui Wang, Zan Fan, Lei Gao, and Song Liu

Cyclic Behaviour of Completely Overlapped Joint for Offshore Jacket
Substructures 877
   Ye Yang, Jiewen Wang, Jinkun Shi, Lei Gao, Wie-Min Gho, and Song Liu

Dynamic Anti-rolling Mechanism of Floating Wind Turbine Based
on Liquid-Sloshing Principles 884
   *Chongwei Zhang, Donghai Li, Shenghui Huang, Xunhao Zhu,
and Dezhi Ning*

Nonlinear Dynamics of Sloshing Liquid with Variable Mass 891
    Donghai Li, Chongwei Zhang, Xunhao Zhu, Shenghui Huang,
    and Dezhi Ning

Numerical Study of the Reeling Installation on China's First Reel-Lay
Vessel 897
    Aijin Xu, Junhan Yan, Weiping Xu, Lin Yuan, and Yan Chen

Study on the Influence of Tension in Deepwater REEL-Lay Process Based
on Finite Element Method 912
    Liwei Li, Zhibing Yu, Jing Hou, and Rundong Zhang

A New 6-DOF Maneuvering Model for Open Frame ROVs Considered
the Flow Memory Effect 921
    Ruinan Guo, Duanfeng Han, Yingfei Zan, Nan Sun, Binggang Yin,
    Fuxiang Huang, and Yaogang Sun

Modeling and FEM Analysis of the IEA15MW Wind Turbine Blade 933
    Ya Luo and Yan Gao

Peridynamic Analysis of Wind Turbine Tower Crack Propagation Under
Axial Force 942
    Yingying Chen and Yan Gao

Author Index 949

Thermal Performance Analysis of Polar Cruise Ship Coatings under Extreme Cold Conditions

Di Li[✉], Ji Zeng, Yichao Bai, and Zicheng Wu

College of Ocean Science and Engineering, Shanghai Maritime University, Shanghai, China
lidi0615@163.com

Abstract. Marine coating materials are crucial for protecting ship structures and equipment against harsh conditions. Composite coatings, known for their superior corrosion resistance, wear resistance, and environmental adaptability, have become the preferred choice for shielding ship base steel plates. With the burgeoning cruise ship industry's venture into polar expeditions, the need for marine composite coatings capable of withstanding extreme cold has intensified. This study leverages ANSIS finite element analysis software to investigate the adaptability and performance of composite coatings on polar cruise ships under severe cold conditions. Our research follows a threefold approach: (1) This paper focus on a high-performance composite coating with an epoxy resin base, employing nonlinear viscoelastic material theory to establish a model of the coating's behaviour. (2) This paper then derive a stress calculation equation for the composite coating, based on thermal analysis theory and the control equation's fixed solution conditions. (3) Utilizing ANSIS software, this paper simulates the thermal compatibility of the composite coating and steel plate substrate under extreme cold, assessing deformation, stress, and strain distribution. The evaluation, based on the Isai-Wu failure criterion, confirms the coating's cold resistance. This study not only provides a theoretical foundation for understanding composite coatings' environmental performance but also offers insights for future enhancements.

Keywords: Composite Coating · Thermal Performance · Polar Cruise Ship · Finite Element Analysis

1 Introduction

When polar cruise ships operate in extremely cold environments, as shown in Fig. 1, the durability and integrity of ship coatings are crucial for the safe and efficient operation of vessels in harsh conditions. As climate change opens new passages and extends navigable seasons, the cruise ship industry increasingly ventures into polar waters, facing unique challenges in material performance under severe cold. Marine coatings not only protect but also significantly enhance the operational lifespan and safety of the ship's structure and critical equipment [1].

Composite coatings, offering superior corrosion resistance, wear resistance, and environmental adaptability, have become essential for protecting marine structures. They

Fig. 1. Polar cruise ships sail in extreme cold conditions

are especially critical for ships in polar conditions, where traditional materials often fail. Low temperatures can reduce the mechanical flexibility of coatings, increase brittleness, and compromise vessel integrity.

Recent international research on marine coatings for polar cruise ships has been driven by the need to develop materials that can withstand extreme conditions, specifically under the rigorous demands of polar environments. The U.S. Department of Defense has been instrumental in this area with the establishment of the MIL-D-23003 standard [2, 3], which provides a comprehensive framework for the application and evaluation of non-skid coatings on ship decks, ensuring they can endure significant mechanical impacts and extreme temperatures. Researchers like F.J. Friederdorf have contributed by developing epoxy resin-based coatings with high friction coefficients that meet MIL-PRF-24667B specifications, capable of effective microwave radiation absorption, which may reduce radar signatures [4]. Additionally, D. Robinson's innovation in creating a dual-layer epoxy resin coating that replaces metallic fillers with thermosetting plastic particles reduces environmental impact and enhances durability, making it ideal for ship decks that experience frequent aircraft landings. These developments reflect a concerted effort to enhance the thermal and mechanical resilience of coatings to support modern naval operations in challenging polar conditions while promoting operational efficiency and sustainability.

2 Nonlinear Viscoelastic Intrinsic Modeling of Composite Coatings

In this study, a high-synthesis epoxy resin-based composite coating material was chosen for modeling. This material has a multi-layered structure, including a topcoat, an anti-slip layer, a primer, and a steel plate substrate. The steel and anti-slip layers are both 4 mm thick, while the topcoat and primer are thinner and negligible in physical layer analysis [5]. The coating material exhibits viscoelastic behavior, showing both viscous and elastic properties. The primer provides a strong adhesive bond, simplifying the model to a two-layer structure: the anti-slip layer and the steel plate [6].

For finite element analysis using ANSYS software, it is essential to define the performance parameters of the composite coating material and the stress-strain relationship curves. This section aims to derive a nonlinear viscoelastic model for the composite coating through mathematical formulations, providing necessary data for subsequent finite element analysis in ANSYS.

2.1 Nonlinear Viscoelastic Ontological Relationships for Composite Coating Materials

The general equation for stress and strain is:

$$\sigma(t) = E(t)\varepsilon \tag{1}$$

For a general viscoelastic material, the intrinsic relationship can be simply expressed by the following equation:

$$\sigma = f(\varepsilon, t) \tag{2}$$

This equation is used to account for nonlinear viscoelastic behaviour, where stress is a function of strain and time. The schematic diagrams of the elastic and viscous component are as shown in Fig. 2:

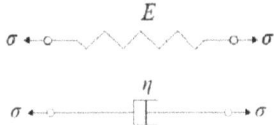

Fig. 2. Elastic and viscous components

where the elastic component is in accordance with the equation below:

$$\sigma = E\varepsilon \tag{3}$$

And the viscous component is in accordance with the equation below:

$$\sigma = \eta\varepsilon = \eta\frac{d\varepsilon}{dt} \tag{4}$$

where: E is the stiffness coefficient of the element, η is the viscosity, and $\frac{d\varepsilon}{dt}$ is the rate of strain onset. Typically, the combination of elastic and viscous components is changed to a viscoelastic model.

Maxwell models the series connection of the elastic and viscous components as shown below in Fig. 3.

Fig. 3. Maxwell model

The strain of the elastic component is denoted as ε_1 and the strain of the viscous component is denoted as ε_2, for the model the total strain can be expressed as ε. From the model the total strain can be the sum of the two strain components:

$$\varepsilon = \varepsilon_1 + \varepsilon_2 \tag{5}$$

Derive the above equation with respect to time and rewrite the collation:

$$\varepsilon(t) = \frac{\sigma_0}{\eta}t + c \tag{6}$$

Kelvin model is depicted as an analytical result of the parallel connection of two primitives, as shown in the Fig. 4 below:

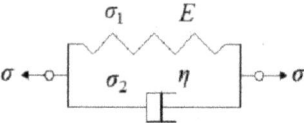

Fig. 4. Kelvin model

Kelvin's research model is like the Maxwell model mentioned above by expressing the stress in the elastic element as σ_1 and the stress in the viscous element as σ_2, so that its total stress is further expressed as σ. As guided by the deformation relationship, the total stress should be the sum of the two components as the following expression [7, 8].

$$\sigma = \sigma_1 + \sigma_2 \tag{7}$$

Substituting (4, 6) into the above equation yields the following equation:

$$\sigma = \frac{t}{\eta} - \frac{1}{E}\ln(\sigma - k\varepsilon) + c \tag{8}$$

where: c is a constant.

These are the constitutive equations of the Maxwell and Kelvin models [9, 10]. These are only the two most basic models, both of which have certain shortcomings and cannot accurately describe the creep and relaxation phenomena of materials. Therefore, Leaderman modified Boltzmann's superposition principle by replacing the linear theory with a nonlinear stress function to provide the constitutive equations for nonlinear viscoelastic materials by linking the current and instantaneous pressures through genetic integration in the study and analysis of viscoelastic materials [11]:

$$\sigma(t) = \int_{-\infty}^{t} \frac{E(t-\tau)}{E_0} d\sigma_e(\tau) = \frac{E(t)}{E_0}\sigma_e(0) + \int_{0}^{t} \frac{E(t-\tau)}{E_0} d\sigma_e(\tau) \tag{9}$$

$$\varepsilon(t) = \int_{-\infty}^{t} \frac{D_1(t-\tau)}{D_{10}} d\varepsilon_e(\tau) = \frac{D_1(t)}{D_{10}}\varepsilon_e(0) + \int_{0}^{t} \frac{D_1(t-\tau)}{D_{10}} d\varepsilon_e(\tau) \tag{10}$$

In the above equation: the first part represents the instantaneous viscoelastic stress term and the second part represents the genetic stress reflection effect, which also further reflects the effect of the reaction stress applied to τ on the stresses of T.

Where: E_0 is the initial elastic modulus, D_{10} is the initial creep flexure, $E(t)$ as well as $D_1(t)$ are the relaxation modulus and creep flexure, respectively.

2.2 Relaxation Modulus and Creep Compliance

Creep and stress relaxation are also fundamental properties of viscoelastic materials. Creep means that the stress does not change and the strain increases with time. Stress relaxation means that strain increases with time. The stress remains constant. The stress time history decreases with increasing stress time.

For standard creep compliance and relaxation modulus, the expressions are as follows:

$$D(t) = \frac{1}{E_0}\{1 + \beta[1 - \exp(-\lambda t)]\} = D_0\left\{1 + \left(\frac{D_\infty}{D_0} - 1\right)[1 - \exp(-\lambda t)]\right\} \quad (11)$$

$$E(t) = \frac{E_0}{1+\beta}\{1 + \beta \exp[-\lambda(1+\beta)t)]\} = E_0\left\{1 - \left(1 - \frac{E_\infty}{E_0}\right)[1 - \exp(-\lambda(1+\beta)t)]\right\} \quad (12)$$

where: D_0, E_0 represent for the initial creep compliance and relaxation modulus, D_∞, E_∞ represent for the equilibrium creep compliance and relaxation modulus, β, λ represent for the material parameters. Where β can be determined by equation $\beta = \frac{D_\infty}{D_0} - 1 = \frac{E_0}{E_\infty} - 1$, the other parameters are obtained from experimental and numerical calculations.

3 Stress theory for Composite Coatings and Steel Substrates

In this part of the article, the stress calculation theory of composite coating and matrix steel plate is studied and analyzed, and the corresponding calculation formula is derived.

3.1 Relaxation Modulus and Creep Compliance

In extreme weather conditions, the temperature field of the composite coating and the substrate steel plate will undergo significant changes in this stress state of the resulting strain increment there are five different forms based on the incremental form of elastic strain $\{\Delta\varepsilon_n^e\}$, the incremental form of temperature-generated strain $\{\Delta\varepsilon_n^T\}$, the incremental form of strain of creep stress $\{\Delta\varepsilon_n^c\}$, the self-generated volumetric strain mode $\{\Delta\varepsilon_n^0\}$, and the incremental form of strain of drying shrinkage $\{\Delta\varepsilon_n^s\}$, so that the total strain expression for the:

$$\{\Delta\varepsilon_n\} = \{\Delta\varepsilon_n^e\} + \{\Delta\varepsilon_n^c\} + \{\Delta\varepsilon_n^T\} + \{\Delta\varepsilon_n^s\} + \{\Delta\varepsilon_n^0\} \quad (13)$$

The elastic strain increment can be expressed as:

$$\{\Delta\varepsilon_n^e\} = \frac{1}{E\tau_n}[Q][\Delta\sigma_n] \quad (14)$$

where:

$$[Q] = \begin{bmatrix} 1 & -\mu & -\mu & 0 & 0 & 0 \\ & 1 & -\mu & 0 & 0 & 0 \\ & & 1 & 0 & 0 & 0 \\ & & & 2(\mu+1) & 0 & 0 \\ & & & & 2(\mu+1) & 0 \\ & & & & & 2(\mu+1) \end{bmatrix} \quad (15)$$

The temperature strain increment can be expressed as:

$$\{\Delta\varepsilon_n^T\} = \{\alpha_T \Delta T_n, \alpha_T \Delta T_n, \alpha_T \Delta T_n, 0, 0, 0\} \tag{16}$$

where: α_T is the coefficient of linear expansion, and ΔT_n indicates the amount of temperature change for the period.

The expression for creep strain increment is:

$$\{\Delta\varepsilon_n^c\} = \{\eta_n\} + C(t_n, \bar{\tau}_n)[Q][\Delta\sigma_n] \tag{17}$$

where:

$$\{\eta_n\} = \sum_i (1 - e^{-r_i \Delta\tau_n})\{\bar{\omega}_{sn}\} \tag{18}$$

The physical equation for the increment is obtained after collation as:

$$\{\Delta\varepsilon_n\} = [\overline{D}_n]\left(\{\Delta\varepsilon_n\} - \{\eta_n\} - \{\Delta\varepsilon_n^T\} - \{\Delta\varepsilon_n^s\} - \{\Delta\varepsilon_n^0\}\right) \tag{19}$$

where:

$$[\overline{D}_n] = \overline{E}_n[Q]^{-1} \tag{20}$$

$$[Q]^{-1} = \frac{1-\mu}{(1+\mu)(1-2\mu)} = \begin{bmatrix} 1 & \frac{\mu}{1-\mu} & \frac{\mu}{1-\mu} & 0 & 0 & 0 \\ & 1 & \frac{\mu}{1-\mu} & 0 & 0 & 0 \\ & & 1 & 0 & 0 & 0 \\ & & & \frac{1-2\mu}{2(1-\mu)} & 0 & 0 \\ & & & & \frac{1-2\mu}{2(1-\mu)} & 0 \\ & & & & & \frac{1-2\mu}{2(1-\mu)} \end{bmatrix} \tag{21}$$

$$\overline{E}_n = \frac{E(\bar{\tau}_n)}{1 + E(\bar{\tau}_n)C(t, \bar{\tau}_n)} \tag{22}$$

3.2 Finite Element Solution of Stress Field

We assume here that the composite coating material is a homogeneous and continuous viscoelastic structural body, which can be obtained from the geometric, physical and equilibrium equations to obtain the finite element equations for the entire region in each period (ΔT):

$$[K]\{\Delta\sigma_n\} = \{\Delta P_n\}^L + \{\Delta P_n\}^T + \{\Delta P_n\}^C + \{\Delta P_n\}^0 + \{\Delta P_n\}^s \tag{23}$$

where: the increment of nodal load $\{\Delta P_n\}^L$ is caused by the analyzed presence of external loads, $\{\Delta P_n\}^T$ is further caused by the temperature change of the environment, $\{\Delta P_n\}^C$ is caused by the corresponding creep, $\{\Delta P_n\}^0$ is caused by the self-generated volumetric deformation, $\{\Delta P_n\}^s$ is caused by the drying deformation, the deformation equations can be expressed by the following equation:

$$\Delta P_n = \sum_k \{\Delta P_n\}_k \tag{24}$$

This analysis mainly considers the deformation of the composite coating material at extreme temperatures, i.e. without external forces under working conditions, to obtain the following calculation formula:

$$\{\Delta P_n\}_k^C = \iiint [B]^T [\overline{D}_n]\{\eta_n\} dxdydz \tag{25}$$

$$\{\Delta P_n\}_k^T = \iiint [B]^T [\overline{D}_n]\{\eta_n^T\} dxdydz \tag{26}$$

$$\{\Delta P_n\}_k^0 = \iiint [B]^T [\overline{D}_n]\{\eta_e^0\} dxdydz \tag{27}$$

4 Finite Element Analysis of Thermal Compatibility of Composite Coatings and Substrate Steel Plates

This analysis is built using the ANSYS Workbench analysis module ACP (Pro) composite pre-processing module, transient thermodynamic analysis module, transient structural analysis module, and ACP (Post) composite post-processing module to meet the thermal matching performance analysis line of the coating and the matrix steel plate.

4.1 Modelling and Meshing of Composite Coatings and Substrate Steel Plates

Since the actual size of the matrix steel plate is very large in the application of finite element analysis is limited, this study applies a simplified model of the ship's matrix steel plate, whose dimensions are L = 1500 mm, W = 1500 mm, and thickness = 50 mm. modelling in the ACP Composites module is shown in Fig. 5.

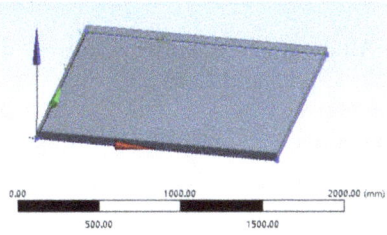

Fig. 5. Simplified 3D model of the ship's matrix steel plate

In the mesh division, considering the analysis accuracy and the results of composite coating analysis, the mesh division is a hexahedral mesh, set the mesh size of 20 mm, after the division of the mesh model as shown in Fig. 6. The number of mesh nodes is 91,808, and the total number of cells is 22,500.

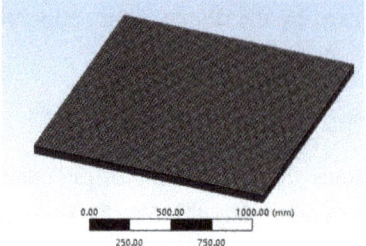

Fig. 6. Meshing of the model

4.2 Layer Setting of Composite Coating Materials

According to the actual composite coating laying conditions, the thickness of the coating is set to 4 mm, the laying angle of the composite coating is 90°, and its laying in the ACP is shown in Fig. 7, where a indicates the laying angle and t indicates the laying thickness. The transformed trend of Young's modulus (E) and shear modulus (G) observed in the polar axis of the composite coating material in the analysis is shown in Fig. 7, which also reflects that the composite coating belongs to a kind of nonlinear material.

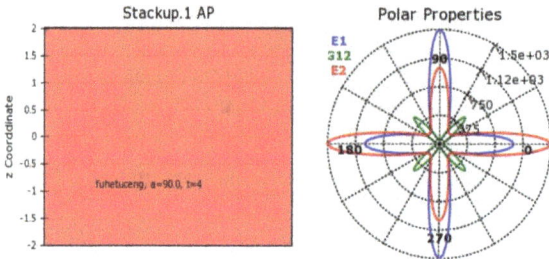

Fig. 7. Layer setting of composite coating materials

4.3 Transient Thermal Analysis of Composite Coatings and Substrate Steel Plates Under Extreme Cold Conditions

Working Conditions. This paper takes the ship Arctic navigation temperature environment as an example, the lowest temperature in the cold winter can reach − 40 °C ~ −50 °C. Currently, the ship's substrate steel plate is mainly affected by the ambient temperature, and is less affected by solar radiation. The room temperature in the cabin is kept around 25–30 °C, so the surface of the base steel plate forms a relatively high temperature difference currently, and the finite element analysis is carried out for the base steel plate of the ship under this working condition

Boundary Condition Setting for Working Conditions. In this analysis, the main consideration is the environmental impact on the substrate steel plate and the composite coating. The upper surface of the composite coating is exposed to air, resulting in heat exchange between the coating surface and the air. The lower surface of the steel plate is exposed to the ship's cabin, where the environment is stable, typically maintained at 22 °C. Due to poor air mobility, the lower layer of the substrate steel plate remains relatively closed. Under these conditions, the heat convection coefficient between the lower surface of the steel plate and the air is set at **5 $W/mm^2 \bullet °C$**, while the heat convection coefficient between the top surface of the composite coating and the air is **10 $W/mm^2 \bullet °C$**

Parameter Settings. Based on the computational results of the constitutive model, the parameter settings for the coating material are presented in Table **1**. Furthermore, in order to study the effect of temperature on the substrate steel plate and coating within a short period of time, in the analysis of the analysis time and the number of sub-steps for the analysis of the corresponding settings, set the results as shown in Table **2**: set the number of sub-steps for 1, the current number of steps for 1, the analysis time of 10 s. Set the step size: the initial step size is 10, the maximum step size is 100, the minimum step size is 10. As the composite coating material described in this paper belongs to a nonlinear material, and will produce large deformation when subjected to temperature changes, so in the analysis of large deformation will be opened to ensure that its analysis can converge

Table 1. Parameter settings for the coating material

Parameter	
α	1.25
β	0.35
$\beta 1$	2.1
γ	10.5
$D_0(Mpa^{-1})$	17.52
$E_0(MPa)$	16.2

Temperature Field Analysis of Composite Coatings and Substrate Steel Plates. The temperature fields of the matrix steel plate and the composite coating material were analyzed based on the settings, with results shown in Figs. 8 and 9 illustrates the temperature distribution across the composite coating on the matrix steel plate. Figure **8**(a) shows that the lowest temperature is on the upper surface of the composite coating, while the highest temperature is on the lower surface of the steel plate. Figure **8**(b) indicates a progressive temperature increase from the composite coating to the substrate steel plate. The temperature gradient shows that the entire substrate steel plate remains above 14 °C, while the composite coating stays below 6 °C. This distinct layering suggests effective thermal insulation by the coating

Table 2. Analysis setup

Step Controls	
Number of steps	1.
Current Step Number	1.
Step End Time	5.s
Auto Time Stepping	On
Define By	Substeps
Initial Substeps	10.
Minimum Substeps	10.
Maximum Substeps	100.
Initial Substeps	10.
Solver Controls	
Solver Type	Program Controlled
Weak Springs	Off
Large Deflection	On

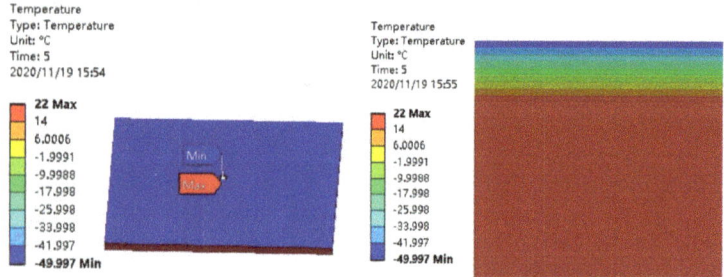

(a) Overall temperature of substrate steel plate composite coating(L).
(b) Enlarged view of the local temperature field(R).

Fig. 8. Temperature field distribution

Figure 9 presents additional findings. According to Fig. 9(a), the lowest temperature on the top surface of the composite coating is recorded during the 5-s analysis period, reaching its minimum in less than 1 s. Subsequently, the temperature exhibits minor fluctuations but stabilizes around $-50\,°C$. Conversely, Fig. 9(b) reports that the highest recorded temperature is $22\,°C$, a value that remains constant and stable over time. This stability correlates with the controlled conditions within the cabin, which is relatively sealed and exhibits limited air mobility. Additionally, the composite coating's poor heat transfer characteristics contribute to the minimal variability in the temperature field of the substrate steel plate.

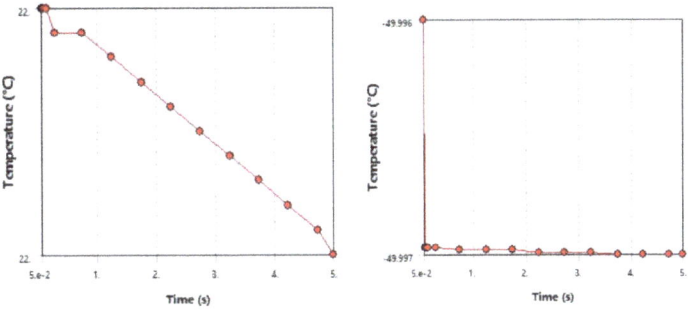

(a) Time history of the highest temperature on the coating (L).
(b) Time history of the lowest temperature under the deck (R).

Fig. 9. High and low temperature variation curve

Analysis of Strain and Stress Results for Composite Coatings and Substrate Steel Plates. Figure 10 illustrates the deformation of the analytical model comprising the composite coating and the matrix steel plate. The deformation increases progressively from the periphery towards the center of the model, where the maximum deformation occurs. This peak deformation is approximately 0.235 mm, observed in the central region, whereas the minimum deformation, approximately 0.0 mm, is located at the edges of the model. Additionally, Fig. **10** demonstrates that the deformations of the matrix steel plate and the composite coating are closely aligned, suggesting a high degree of concordance between these two components

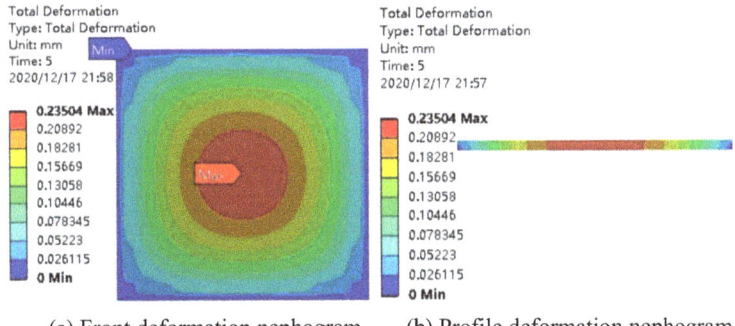

(a) Front deformation nephogram   (b) Profile deformation nephogram

Fig. 10. Deformation cloud of composite coating and substrate steel plate

Analysis of Strain Results for Composite Coatings and Substrate Steel Plates. According to Fig. **11**, the strain distribution within the composite coating and the matrix steel plate is depicted. Figure **11**(a) reveals that the strain progressively increases from the central region to the edge of the model. The maximum strain, approximately 0.00166, occurs at the model's edge, while the minimum strain, valued at 2.58×10^{-5}, is observed in the central region. This pronounced strain at the edges is attributed to greater temperature impacts on the composite coating, leading to increased strain in these areas. Furthermore, the time-dependent strain curves for the composite coating and the substrate steel plate can be analyzed from Fig. 11. As detailed in Fig. 11(b), the strain peaks near the value of 0.00166 at approximately 0.5 s and exhibits fluctuations over time. At 1 s, the strain reaches a peak, and beyond this point, both the maximum and minimum strains in the matrix steel plate and the composite coating stabilize, showing no significant fluctuations

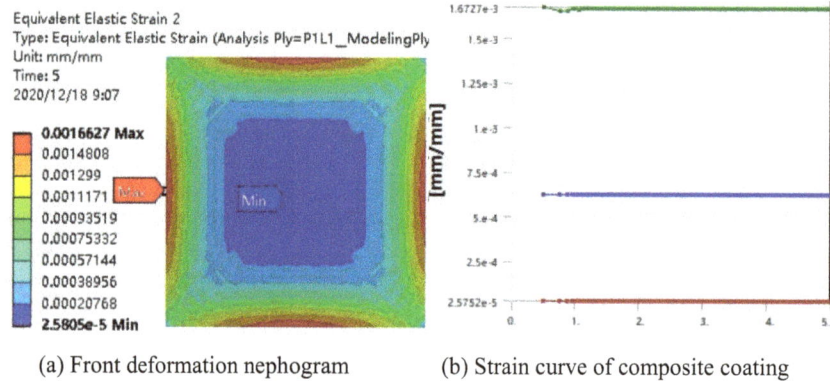

(a) Front deformation nephogram   (b) Strain curve of composite coating

Fig. 11. Strain cloud of composite coatings

Stress Result Analysis of Composite Coating and Substrate Steel Plate. According to Fig. 12, the strain distribution across the entire model is elucidated by the stress cloud diagrams of the composite coating and the substrate steel plate. Figure **12**(a) indicates that the stress distribution increases from the central region to the edge of the model, with the maximum stress occurring at the edge, approximately 0.295 MPa. Contrarily, the minimum stress is observed in the central region, valued at 225 MPa (noted as 2.25×10^3 MPa, which appears to be a typographical error). This discrepancy in stress

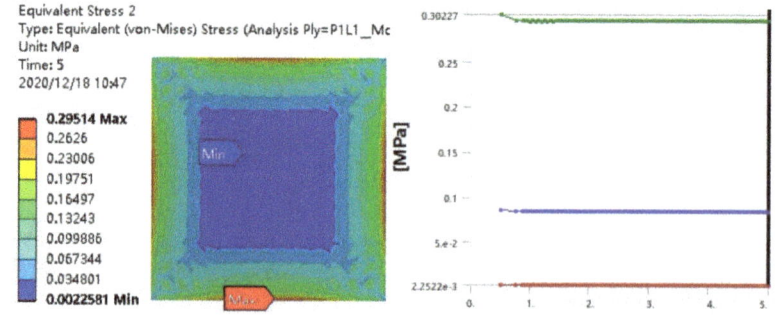

(a) Front deformation nephogram   (b) Stress curve of composite coating

Fig. 12. Front stress nephogram of composite coating

levels may be attributed to significant thermal impacts on the composite coating and constraints in displacement analysis, which amplify the stress at these points. Further insights are provided in Fig. **12**(b), which shows that the maximum stress nears 0.27 MPa at approximately 0.5 s into the analysis. Subsequently, the stress level increases, peaking at 1 s. Beyond 1 s, both the maximum and minimum stress values in the substrate steel plate and the composite coating stabilize, exhibiting no significant fluctuations

5 Conclusion

This paper comprehensively analyzes the thermal performance of high-performance composite coatings on polar cruise ships under extreme cold conditions. Using ANSYS finite element analysis software, we modeled the behavior of epoxy resin-based composite coatings and their interaction with steel plate substrates in polar environments.

Our findings show that the composite coatings exhibit excellent thermal compatibility and mechanical integrity in severe cold. Detailed simulations revealed minimal deformation and stress concentrations within safe limits per the Isai-Wu failure criterion, indicating robust resistance to cold-induced stresses and validating the coatings' effectiveness in protecting substrate steel plates against extreme temperatures and mechanical strains.

The theoretical framework established in this study enhances our understanding of composite coating behavior under polar conditions and contributes significantly to marine engineering by providing a reliable method for evaluating and improving the durability of marine coatings.

The implications of this research are profound, offering potential for future enhancements in coating formulations for greater resilience and operational efficiency. As the cruise ship industry expands into harsh polar regions, the findings from this study will aid in developing advanced materials that ensure the safety and efficiency of marine operations in extreme environments.

References

1. Friedersdorf FJ, Vestal C R, Garrett J T. High Friction Coating Formulations And Systems and Coated Articles Thereof Exhibiting Radar Signature Reduction and Methods of Providing the Same: US, 2009/0226673A1 [P] (2009). 2009-09-10
2. Robinson D, Cruz R S. Two-part Lightweight and Low Heavy Metal Content Epoxy Non-Skid Coating for a Deck or Floor: US 2008/0167401A1 [P] (2008). 2008-07-10
3. AASHTO: Report of the Joint Task Force on Rutting [R]. AASHTO Publishers (1989)
4. Staverman, A.J., Schwaral, F.: Nonlinear deformation behavior of high polymer. In: Stuart, H.A. (ed.) Die Physile der. Springer, Hochpolymeren, Berlin (1956)
5. Yang, T.: Viscoelastic Mechanics [M]. Huazhong University of Technology Press, Wuchang, China. (in Chinese)
6. Yi-he, L.: Study on Nonlinear Viscoelasticity Principal Structure Theory and its Solution Method [D]. Xiangtan University (2003) (in Chinese)
7. Schapery, R.A.: A theory of nonlinear thermoviscoelasticity based on irreversible thermodynamics. In: Proc. 5th U.S. Nat. Congr. Appl, pp. 511–530, Mech, ASME (1966)

8. Schapery, R.A.: On the characterization of nonlinear viscoelastic materials. Polym. Eng. Sci. **9**, 295–310 (1969)
9. Lemaitre, J.: A Course on Damage Mechanics. Spring -Verlag, Berlin (1992)
10. Deng Qi,Xu Jinwen,Gao Xinhua,et al. Status quo and development trend of anti-skid coating technology for ship substrate steel plate[J]. China Ship Research,2013.3(8)111–116. (in chinese)
11. Hansen G P, Rushing R A, Bulluck J W, et al. Coating with amine curing agent epoxide— containing toughener, epoxy resin and rubber toughener: US,7465477 [P] (2008). 2008-12-16

Open Access This chapter is licensed under the terms of the Creative Commons Attribution-NonCommercial-NoDerivatives 4.0 International License (http://creativecommons.org/licenses/by-nc-nd/4.0/), which permits any noncommercial use, sharing, distribution and reproduction in any medium or format, as long as you give appropriate credit to the original author(s) and the source, provide a link to the Creative Commons license and indicate if you modified the licensed material. You do not have permission under this license to share adapted material derived from this chapter or parts of it.

The images or other third party material in this chapter are included in the chapter's Creative Commons license, unless indicated otherwise in a credit line to the material. If material is not included in the chapter's Creative Commons license and your intended use is not permitted by statutory regulation or exceeds the permitted use, you will need to obtain permission directly from the copyright holder.

Hydrodynamic Analysis of Single Barge Unilateral Salvage of Wrecks Under Mooring Conditions

Fei Wang[1], Guohong Zhou[1], Wei Zhang[2], JingJing Liu[2(✉)], Jinxing Li[1], and ZiKai Xu[2]

[1] Offshore Engineering Department, China YanTai Salvage, Yantai, China
[2] Technology Center, China YanTai Salvage, Yantai, China
liujj@ytsalvage.com

Abstract. Single barge unilateral salvage method is more and more used in the overall salvage of large tonnage shipwrecks, which as a multi-body system has complex hydrodynamic characteristics, especially in the application of deep water needs further research. This paper is based on three-dimensional potential flow theory, with the help of hydrodynamic analysis software Moses simulation single barge unilateral salvage method in deep water during the overall salvage of large-tonnage shipwrecks, the barge and the shipwreck's response to the movement of different sea conditions, according to the results of the calculation of the analysis of the single barge unilateral salvage method applicable to the operating conditions at sea. The results of this study have certain reference value for the selection of operating conditions in the design of single barge unilateral salvage method.

Keywords: Lifting and Salvage · 3D Potential Flow Theory · Kinematic Response Prediction

1 Introduction

With the development of shipbuilding industry and shipping industry, the number of ships and accidents are increasing continuously. The rescue and salvage ability and technical level of the wreck is extremely important to ensure the safety of harbours, waterways and other strategic resources, and in special cases, the wreck needs to be salvaged as a whole in its original posture. At present, the commonly used wreck salvage methods [1] are: sealing and pumping salvage, float lifting salvage, ship lifting sled salvage and floating crane salvage and so on. In the traditional shipwreck salvage methods, float lifting salvage method is the most common application, but the method is usually applicable to small tonnage shipwreck salvage.

Applicable to large tonnage ship salvage methods, mainly double barge lifting the whole salvage and single barge unilateral salvage method. Single barge floating with tension equipment to lift the wreck salvage technology, typical cases are "Kursk" nuclear submarine salvage [2] and "K-129" nuclear submarine salvage [3]. 16 April 2014, the South Korea "Se Viet" passenger ship wrecked in the waters of Jindo County, South Jeolla

Province, Shanghai Salvage Bureau put forward the "steel beam lifting bottom" of the double barge lifting the overall salvage programme [4, 5], and successfully achieved the overall salvage of large tonnage shipwrecks in a water depth of 44 m. Zhang Fengrui [6] and other scholars studied the holding grasp assisted single barge salvage wreck hydraulic compensation analysis, Hou Joyi [7] and other scholars on the double barge salvage hydraulic compensation research, Zhu Ziyang [8] and other scholars carried out a double barge deep water salvage experimental research.

In this paper, for the 500 m water depth using a single barge unilateral salvage technology for the overall salvage of the wreck as a research case, based on the three-dimensional potential flow theory, with the help of hydrodynamic analysis software Moses were barge located in different sea conditions simulation, according to the results of the movement of the barge and the wreck of the response to analyse the salvage operation for the selection of the sea conditions of the actual project to provide a reference.

2 Design of Single Barge Unilateral Salvage of Wrecks Under Mooring Conditions

2.1 Demand Analysis

In the barge unilateral lifting of wrecks, wreck/substance lifting equipment is arranged on the deck of the operating barge, with a lower centre of gravity, and the ability to adapt to sea conditions is stronger in bad sea conditions, under the same roll angle of the operating mother ship, relative to the salvage process such as floating crane salvage, pontoon salvage, and double-barge lifting and floating salvage. Hydraulic pulling jack design adopts pulling rope type vector winch, compared with the current conventional use of strand type pulling jack, more friendly to deep water salvage scenarios, the operating water depth is only limited by the salvage wire rope strength and length. Multiple tension jacks can be used for simultaneous linkage lifting of the wreck, calculating and controlling the lifting strategy with higher controllability.

2.2 Single Barge Unilateral Salvage Wreck Design

Barge unilateral lifting of shipwreck is a salvage technology in which several groups of load-bearing steel wire ropes enter the water from the side side of the barge to connect the shipwreck, and the hydraulic pull jack arranged on the salvage barge drags the load-bearing steel wire ropes to lift the wreck. The hydraulic pull jack is a rope vector winch, which can be adapted to deep sea wreck/debris salvage operations. The barge for surface operation uses DP dynamic positioning barge, and the lifting device is arranged on the deck. In deep water operation, it is limited by the strength and length of the salvage wire rope, and has stronger adaptability to sea conditions. A number of tension jacks are used to jointly lift the wreck, calculate and control the lifting strategy, and the controllability is higher. The layout of the salvage system is shown in Fig. 1.

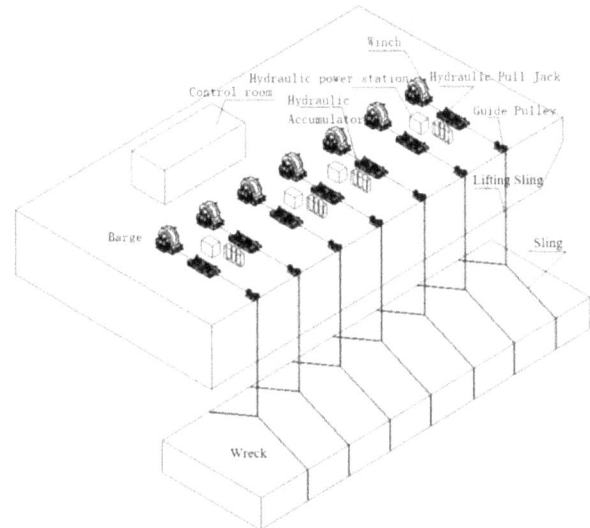

Fig. 1. Layout diagram of salvage system

2.3 Ship Parameters

Barge Parameters. In the simulation of this paper, DEBO2 ship is used as a barge, and the parameters of DEBO2 ship are shown in Table 1. the radius of inertia is obtained according to the empirical formula. The DEBO2 model is constructed in Moses according to the following parameters.

Table 1. Parameter list of barges

Parameter	Value
Length	159.6 m
Breadth	38.8 m
Depth	10.9 m
Draft	5.5 m
Centre of gravity	(73.38,0, 7.75)
Inertial radius	R_{xx} = 12.416 m, R_{yy} = 46.432 m, R_{zz} = 46.432 m
Weight	15096.3 t

Wreck Parameters. In this article, the length of the wreck is 90 m, the width of the wreck is 9 m, the depth of the wreck is 5 m and the weight is 2240 tons. In order to be able to determine the weight of each part of the wreck more accurately and to make the calculation results more precise, here, the weight distribution load calculation of the ship is applied to the wreck

Principles of Weight Distribution Load Strength Distribution. For wrecks, they are divided along the direction of the ship's length. The principle of static equivalence must be followed when approximating and idealising the distribution of each weight [9]. The weight of a certain item distributed with an arbitrary law within 2 standard theoretical station distances is P, and the distance of the centre of gravity from the 2 stations is α. Let the weights within the 2 theoretical station distances be P1 and P2 respectively, which can be obtained the formulae (1) (2) according to the principle of static equivalence

$$P_1 + P_2 = P \quad (1)$$

$$\frac{1}{2}P_1 \Delta L - \frac{1}{2}P_2 \Delta L = P\alpha \quad (2)$$

Weight Distributed Load Strength Calculation Method When the number of standard theoretical station distances within the theoretical station distance where the equipment is located is greater than or equal to 2, the specific calculation steps are as follows: the number of theoretical stations where the equipment is located is n pairs of points ($n \geq 2$), which is equivalent to 2 equivalent theoretical station distances, and the equivalent weight distribution P_1 and P_2 of the 2 equivalent theoretical station distances are calculated by the formulae (3) (4):

$$P_1 = P \times \left(0.5 + \frac{\alpha}{\Delta L}\right) \quad (3)$$

$$P_2 = P \times \left(0.5 - \frac{\alpha}{\Delta L}\right) \quad (4)$$

Where: P is the weight of the equipment, α is the distance between the point of action of the weight of this equipment and the geometric centre of the theoretical station range in which it is located; ΔL is the length of the equivalent theoretical station range after pairwise division. According to Table 2, the weight of each part of the wreck is estimated, the weight distribution curve of the wreck can be obtained, as shown in Fig. 2.

Table 2. Proportion of weight and space in each part of the wreck

	Weight(%)	Space(%)	Weight(t)
Structure	43	–	731
Main engine and auxiliary engine	35	56	595
Living and working equipment	4	11	68
Stock	1	5	17
Fixed ballast	8	–	136
Other	9	28	153

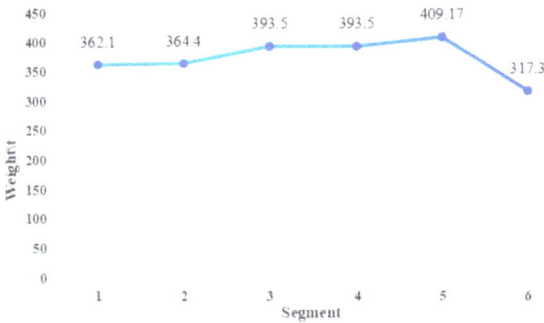

Fig. 2. Weight distribution map of shipwreck lifting

3 Numerical Forecasting Scheme

3.1 Mathematical Model

The basic assumptions are introduced here based on the theory of three-dimensional potential flow: the fluid is an incompressible rational fluid and the surface tension effect can be ignored; The fluid is irrotational, there is a velocity potential Φ = (x, y, z), and the gradient ∇Φ = (x, y, z) can obtain the velocity of the fluid particle. According to the linear potential flow theory, the total velocity potential of the flow field around the ship in waves can be divided into two parts: steady potential and unsteady potential.

$$\begin{cases} \Phi = (x, y, z) = [-Ux + \Phi_S = (x, y, z)] + \text{Re}\{\Phi_T = (x, y, z)e^{iwt}\} \\ \Phi_T = (x, y, z, t) = \varphi_I(x, y, z, t) + \varphi_D(x, y, z, t) + \varphi_R(x, y, z, t) \\ \varphi_I(x, y, z, t) = a\varphi_0(x, y, z, t) \\ \varphi_D(x, y, z, t) = a\varphi_7(x, y, z, t) \\ \varphi_R(x, y, z, t) = \sum_{j=1}^{6} iw\eta_i \cdot \varphi_j(x, y, z, t) \end{cases} \quad (5)$$

Where $[-Ux + \Phi_S = (x, y, z)]$ is the steady potential. $\text{Re}\{\Phi_T = (x, y, z)e^{iwt}\}$ is an unsteady potential; Φ_T is the constant wave potential generated by a ship in still water;The diffraction velocity potential is denoted by φ_D, and the incident velocity potential by φ_I; φ_R is the radiation velocity potential, φ_D and φ_R are the perturbation velocity potential; φ_7 is the diffraction potential per unit amplitude; φ_j is the J-mode motion radiation potential per unit complex velocity.

3.2 Calculation Model

The hydrodynamic analysis software Moses was used for numerical simulation. According to ship parameters in Chap. 1, barge and wreck models are established in Moses respectively. The wreck model is equivalent with a shell, and the weight of each part of the wreck is given in the software according to Sect. 2.3.2, as shown in Fig. 3 and Fig. 4 respectively.

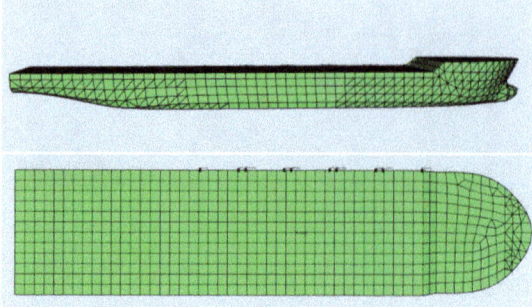

Fig. 3. Numerical modelling of barge

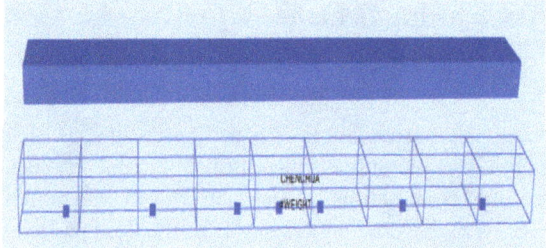

Fig. 4. Numerical modelling of shipwreck

According to the size of the wreck, the distribution of the crane arms needs to be allocated reasonably. According to the actual size of the wreck, therefore, the distribution of ten crane arms is adopted, and the crane arm is established on the port side of DEBO2, which is located on the deck for connecting the wreck, and the detailed distribution scheme is shown in Fig. 5.

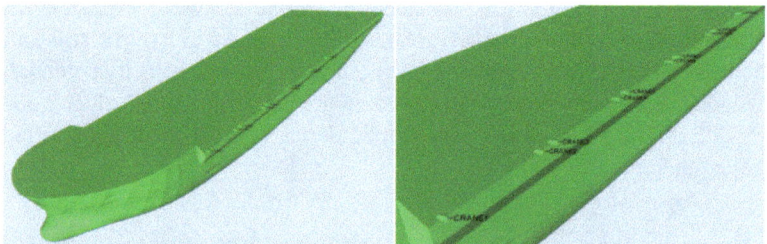

Fig. 5. Schematic diagram of the barge's port crane arm

The local water depth is 550 m, 8 anchor points are distributed on the barge, 8 mooring cables are defined to connect the seabed with the hull, the length of the mooring cable is 1000 m, the horizontal distance between the anchor point and the connection point of the upper end of the cable is 800 m, the distribution angles of the mooring cables are 30°, −30°, 45°, −45°135°, −135°, 150°, and − 150°, and the pretension of the mooring cables is 100 t. The schematic diagram of the barge salvage wreck is shown in Fig. 6.

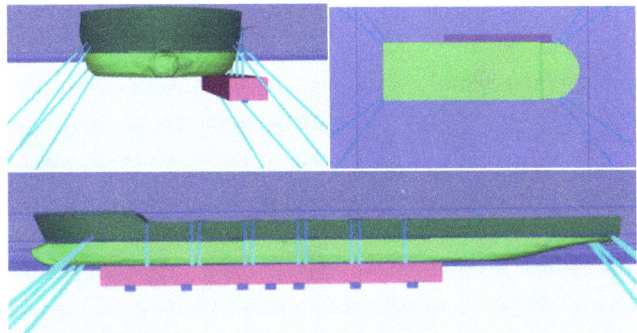

Fig. 6. Schematic diagram of single barge unilateral salvage of wrecks under mooring conditions in MOSES

3.3 Environmental Conditions

The simulation of the sea conditions in the south China sea as a reference, lifting target location depth of 500 m, by the long-term monitoring and statistics, and the march to September wave high under 1.2 m, the results are shown in Fig. 7, visible, the wave high below 1.2 m wave probability in 3–9 months above 50%, construction window appear probability is larger.

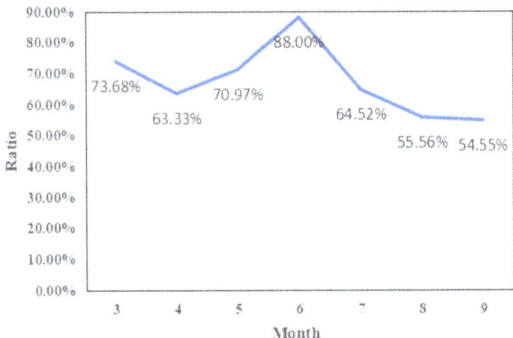

Fig. 7. Reproducibility of 1.2 m wave height at a point in the South China Sea

Therefore, this paper takes the JONSWAP wave spectrum with a meaningful wave height of 1.2 m as the incident wave to simulate the working conditions. In this paper, the main simulation conditions are the motion response of the surface ship and the wreck when the wreck is located at 1 m under the surface ship and when the wreck is located at a water depth of 500 m, and it is proposed to simulate the wave period of 6 s, 7 s, 8 s and 9 s, and the direction of the wave is 0°, 45°, 90°, 135°and 180°, respectively.

4 Motion Response Analysis

Under the conditions described above, the motion response of the surface ship and the wreck are 1 m below the barge and 500 m, and the numerical analysis is made according to the obtained motion response results.

4.1 Movement of Watercraft

Figure 8 shows the RAO curves for the six degrees of freedom of the barge when the wreck is located at 1 m below the bottom of the barge under sea conditions with wave directions of 0°, 45°, 90°, 135°, and 180°, and wave periods of 6 s, 7 s, 8 s, and 9 s, respectively

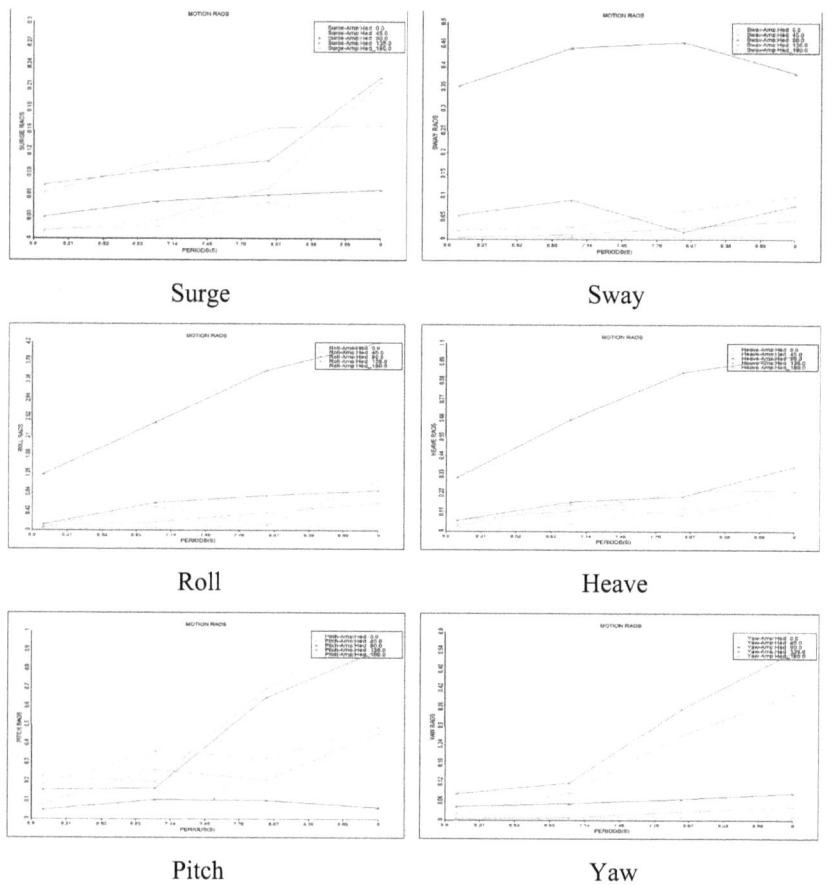

Fig. 8. RAO profile of the barge when the wreck is located 1 m below the bottom of the barge

Figure 9 shows the RAO curves for the six degrees of freedom of the barge when the wreck is located at a water depth of 500 m under sea conditions with wave directions of 0°, 45°, 90°, 135°, and 180°, and wave periods of 6 s, 7 s, 8 s, and 9 s, respectively.

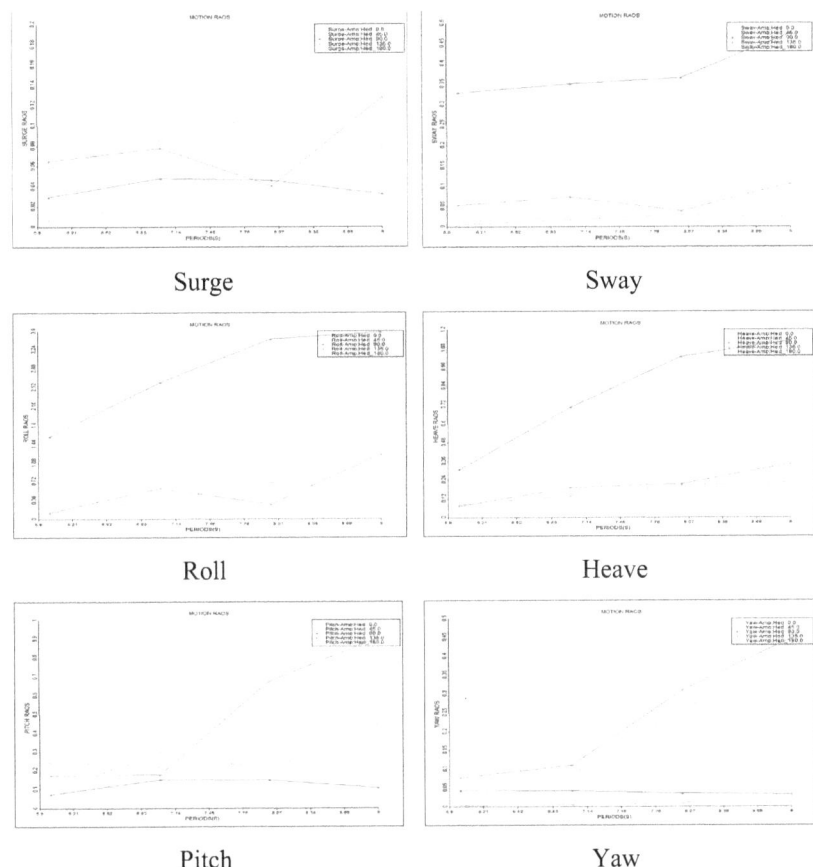

Surge   Sway

Roll   Heave

Pitch   Yaw

Fig. 9. RAO profile of the barge when the wreck is located at 500 m water depth

4.2 Movement of Wrecks

Figure 10 shows the RAO curves for the six degrees of freedom of the wreck when it is located at 1 m below the barge's bottom under sea conditions with wave directions of 0°, 45°, 90°, 135°, 180° and wave periods of 6 s, 7 s, 8 s and 9 s, respectively. Figure 11 show the RAO curves for the six degrees of freedom of the wreck when it is located at a water depth of 500 m under sea conditions with wave directions of 0°, 45°, 90°, 135°, 180° and wave periods of 6 s, 7 s, 8 s and 9 s, respectively.

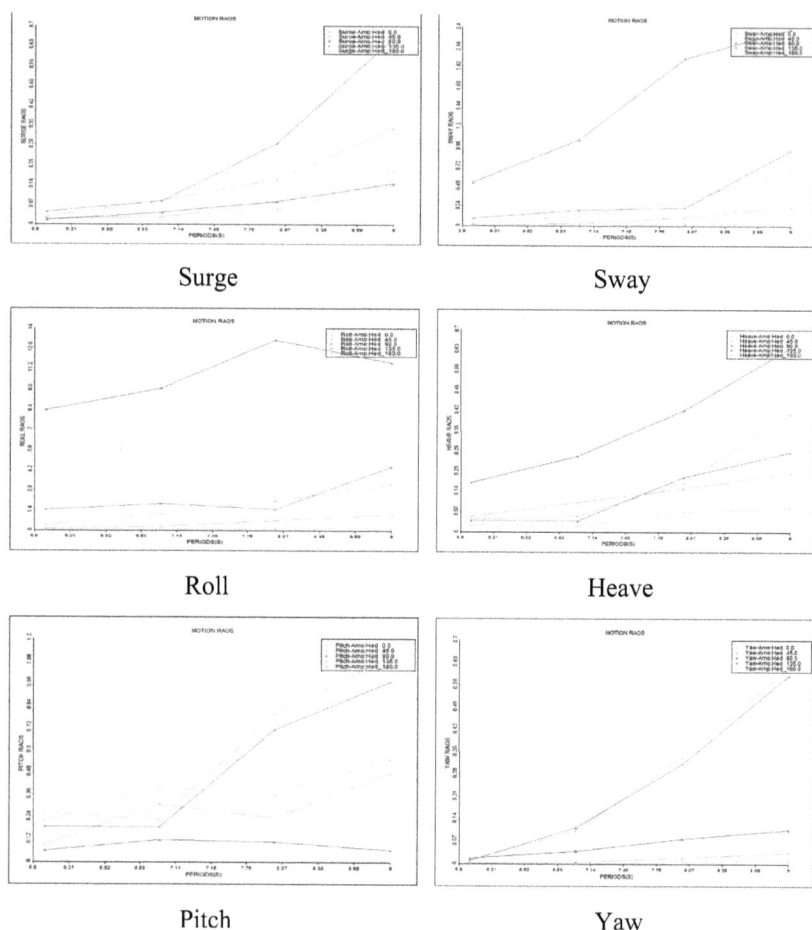

Fig. 10. RAO profile of the wreck when the wreck is located 1 m below the barge bottom.

Hydrodynamic Analysis of Single Barge Unilateral 25

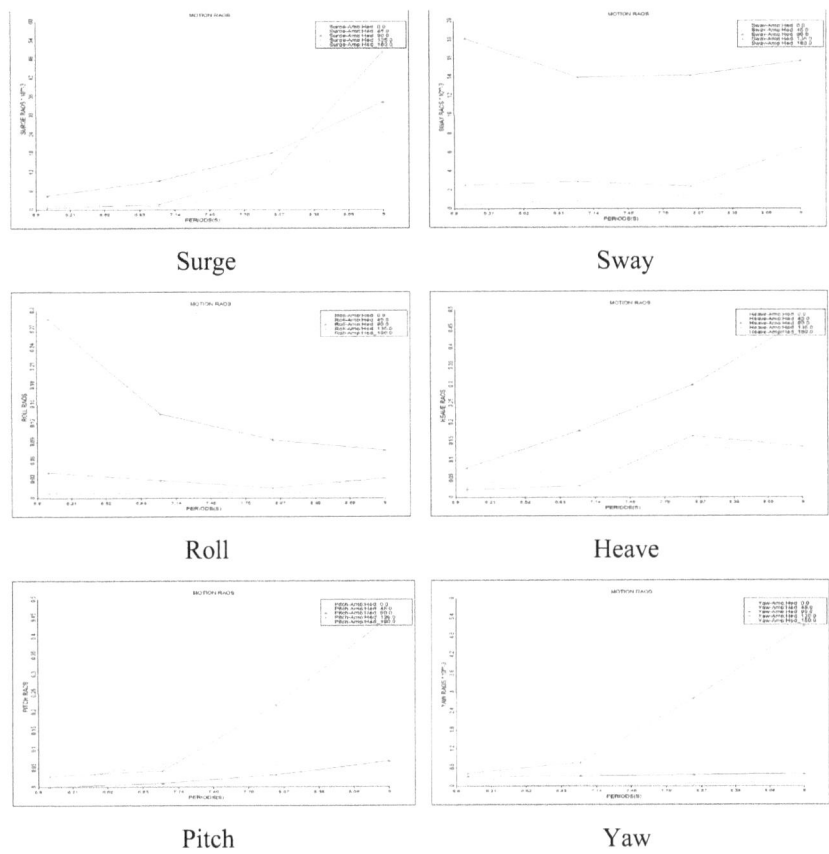

Fig. 11. RAO curve of the wreck at 500 m water depth

4.3 Numerical Analyses

Wreck Floated to 1 m Above Barge Bottom. When the wreck floats to a distance of 1 m from the bottom of the barge, according to the barge's motion response results, the maximum longitudinal oscillation amplitude of the barge is about 0.25 m under different sea conditions, all of which are less than 0.5 m; The lateral oscillation amplitude is larger than other sea conditions when the wave direction is 90°, but the maximum amplitude is about 0.45 m, both of which are less than 0.5 m; Similarly, when the wave direction is 90°, the amplitude of the heave is larger than other sea conditions, but the maximum amplitude is about 1 m. Moreover, when the wave period is 6 s, the amplitude of all downward waves is less than 0.5 m; The amplitude of the roll angle is also maximum when the wave direction is 90°, but the maximum amplitude reaches around 4.2° when the wave period is 9 s; The pitch angle's maximum amplitude is about 1°, and under sea conditions with wave periods of 6 and 7 s, the pitch amplitude of all waves downward is less than 0.5°; The yaw angle's maximum amplitude is around 0.6°, and under sea conditions with wave periods of 6 s, 7 s, and 8 s, the amplitude of all downward yaw angles is less than 0.5°.

When the wreck floats to a distance of 1 m from the bottom of the barge, according to the motion response results of the wreck, under different sea conditions, the longitudinal oscillation amplitude of the wreck is larger than other sea conditions when the wave direction is 135°. The maximum value occurs when the wave period is 9 s, which is about 0.7 m, and the longitudinal oscillation amplitude of other sea conditions is less than 0.5 m; The lateral oscillation amplitude is larger than other sea conditions when the wave direction is 90°, with a maximum amplitude of about 2.4 m; Similarly, when the wave direction is 90°, the amplitude of heave is larger than other sea conditions, with a maximum amplitude of about 0.6 m. Except for the sea conditions with a wave period of 9 s and a wave direction of 90°, the amplitude of heave in all other sea conditions is less than 0.5 m; The amplitude of the roll angle is also maximum when the wave direction is 90°, and the maximum amplitude angle reaches around 14° when the wave period is 9 s; The maximum amplitude of the pitch angle is about 1.2°, and under sea conditions with wave periods of 6 s and 7 s, the amplitude of all downward pitch angles of the waves is less than 0.5°; The yaw angle's maximum amplitude is around 0.6°, and under sea conditions with wave periods of 6 s, 7 s, and 8 s, the amplitude of all downward yaw angles is less than 0.5°.

The Wreck Is Located in A Water Depth of 500 m. When the wreck is located at a depth of 500 m, according to the motion response results of the barge, it can be seen that under different sea conditions, the maximum longitudinal oscillation amplitude of the barge is about 0.2 m, all of which are less than 0.5 m; The lateral oscillation amplitude is larger than other sea conditions when the wave direction is 90°, but the maximum amplitude is about 0.5 m; Similarly, when the wave direction is 90°, the amplitude of heave is larger than other sea conditions, but the maximum amplitude is about 1.1 m. Moreover, when the wave period is 6 s, the amplitude of all downward waves is less than 0.5 m; The amplitude of the roll angle is also maximum when the wave direction is 90°, but the maximum amplitude reaches around 3.6° when the wave period is 9 s; The pitch angle's maximum amplitude is about 1°, and under sea conditions with wave periods of 6 s and 7 s, the pitch angle amplitude of all waves downwards is less than 0.5°; The yaw angle's maximum amplitude is around 0.5°, and under other sea conditions, the amplitude of the yaw angle is less than 0.5°.

When the wreck is located at a depth of 500 m, according to the motion response results of the wreck, it can be seen that under different sea conditions, the maximum longitudinal amplitude of the wreck is about 0.06 m, all of which are less than 0.5 m; The lateral oscillation amplitude is larger than other sea conditions when the wave direction is 90°, but the maximum amplitude is about 0.018 m; Similarly, when the wave direction is 90°, the amplitude of heave is larger than other sea conditions, but the maximum amplitude is around 0.45 m, and the amplitude of heave in all sea conditions is less than 0.5 m; The maximum amplitude of the roll angle is also around 0.3° when the wave direction is 90°; The pitch angle's maximum amplitude is around 0.5°; The yaw angle's maximum amplitude is around 0.006°.

According to the calculation results and the above analysis, under sea conditions with a wave direction of 90°, the amplitude of oscillations that are vertical, longitudinal, and lateral is relatively large, but the amplitude of oscillations that are vertical, longitudinal, and lateral in most other sea conditions is less than 0.5 m, which belongs to good working sea conditions. After comprehensive analysis, the most suitable working conditions are sea conditions with wave periods of 6 s and wave directions of 0°, 45°, 135°, and 180°. Under these conditions, the six degree of freedom motion response amplitude of both surface and wrecks is in a very small state, making it a very good working sea condition.

5 Conclusion

This paper is based on three-dimensional potential flow theory, with the help of hydrodynamic analysis software Moses simulation of single barge unilateral salvage method in deep water overall salvage of large tonnage shipwrecks in the process of the operation, the barge and the shipwreck under different sea conditions of the motion response, according to the results of the calculation of the following conclusions: in the simulation of all the sea conditions, the most suitable for the single barge unilateral salvage operating conditions for the wave period of 6 s, the direction of the wave is 0°, 45°, 135°and 180°sea conditions, this condition, both the surface vessel or wreck six degrees of freedom motion response amplitude are in a very small state. Which is a very good operating condition, and at the same time, avoiding the operation under 90° wave direction.

References

1. Shumin, S., Yue, L.: Summary of salvage methods for steel wrecks. Guangdong Shipbuild. (1), 22–27 (2006)
2. MAMMOET (2021). Just One Way To Salvage A Sub And Beat The Winter. https://www.mammoet.com/cases/kursk/, 2001.
3. Ian H (2016). Project Azoria—Recovery of the Sunken Soviet Submarine K-129. https://www.thevintagenews.com/2016/12/19/project-azoria-recovery-of-the-sunken-soviet-submarine-k-129/
4. Yangqiao, Q.: When the salvage of the Sewol was about to end. China Notar. (11), 34–35 (2020)
5. Shihai, C., et al.: Design and implementation of Sewol salvage project. World Shipp. **40**(10), 6–23 (2017)
6. Zhang, F., Hou, J., Ning, D., et al.: performance analysis of the passive heave compensator for hydraulic shipwreck lifting systems in twin-barge salvaging. Ocean Engineering. **280**, 114469 (2023)
7. Hou Jiaoyi, et al (2018). Design of hydraulic heave compensation test platform based on double barge synchronous lifting technology. Mach. Tool Hydraul., 46(21):95–100, 158
8. Ziyang, Z., et al.: Experimental study on deep water salvage of large tonnage wreck using double barge lifting method. Ship Building of China. **63**(6), 245–254 (2022)
9. Kangkang, Y., Da, S.: Research on the calculation method of load strength for ship weight distribution. Ship Sci. Technol. **40**(19), 1–5 (2018)

Open Access This chapter is licensed under the terms of the Creative Commons Attribution-NonCommercial-NoDerivatives 4.0 International License (http://creativecommons.org/licenses/by-nc-nd/4.0/), which permits any noncommercial use, sharing, distribution and reproduction in any medium or format, as long as you give appropriate credit to the original author(s) and the source, provide a link to the Creative Commons license and indicate if you modified the licensed material. You do not have permission under this license to share adapted material derived from this chapter or parts of it.

The images or other third party material in this chapter are included in the chapter's Creative Commons license, unless indicated otherwise in a credit line to the material. If material is not included in the chapter's Creative Commons license and your intended use is not permitted by statutory regulation or exceeds the permitted use, you will need to obtain permission directly from the copyright holder.

Research on Safety Evaluation Technology of Ship's Mooring Ropes

Nan Lu(✉), Ji Zeng, and Qihang Wu

Shanghai Maritime University, Shanghai, China
15623018595@163.com

Abstract. Among the many safety problems in the field of ships, personal safety due to mooring rope accidents during mooring operations is particularly prominent. In the daily use of the mooring rope, there are many factors that affect the quality of the mooring rope and cause it to break. For mooring safety, the establishment of a scientific and reliable safety evaluation and management method for mooring ropes can prevent and resolve major mooring risks at a systematic level, safeguard the lives and health of crew members, and avoid unnecessary losses. Therefore, it is of great significance to systematically analyze the mooring rope breakage factors to prevent mooring rope accidents. The purpose of this paper is to develop a practical and versatile evaluation management system through ASP.NET technology and database technology on the basis of the research on the safety evaluation system of mooring ropes from different perspectives, which can effectively help the operators to evaluate the safety performance of mooring ropes, improve the management of mooring ropes by shipowners, shipyards or ship management companies, and reduce the maintenance cost.

Keywords: Ship's Mooring Ropes · Breakage Factors · Safety

1 Introduction

1.1 Background and Significance of the Study

Ships in the process of berthing and berthing involves towing, mooring, undoing and other operations, in these operations, the ship's mooring rope is subjected to a huge tensile force, in this case, the occurrence of accidents on the mooring rope will lead to very serious consequences [1], and if it can't be prevented in time, it will be a great threat to the life and health of the crew. The causes of accidents mainly include the following: casualties caused by the breakage of mooring ropes, casualties caused by the bouncing of mooring ropes in the course of operation, and casualties caused by the failure of equipment and machinery [2–4].

1.2 Research Status

In the theoretical study of mooring rope safety evaluation, related scholars and experts have put forward many valuable opinions. In recent years, with the rapid development of computer technology, the mooring rope safety evaluation method gradually shifts from qualitative method to comprehensive evaluation method based on mathematical model. Gao et al. analyzed the current safety and security situation of port ship's mooring operations based on the frequent ship's mooring breakage accidents around the world, and combined with the current domestic and international innovation and research and development application of mooring technology to develop a program to realize the intelligent mooring safety assistance system that serves as a cable and fender at the same time. Chen et al. analyzed the causes of ship's mooring breakage and proposed to reduce the risk caused by ship's mooring operation from the aspects of organization, personnel, and environment, and analyzed the factors of operators, special operation factors, and extreme environmental factors that affect the safety of the ship mooring [5, 6]. The purpose of this paper is to establish a generalized safety fuzzy comprehensive evaluation method for mooring ropes on the basis of previous work, such as mooring ropes of different material types shown in Fig. 1, and to develop a practical and multi-functional evaluation management system by means of ASP.NET technology and database technology, which can effectively help the staffs to evaluate the safety performance of mooring ropes and improve the management level of mooring ropes and reduce the maintenance cost of the shipowners, shipyards or ship management companies. It can effectively help the staff to evaluate the safety performance of mooring ropes, improve the management level of mooring ropes by shipowners, shipyards or ship management companies and reduce the maintenance cost.

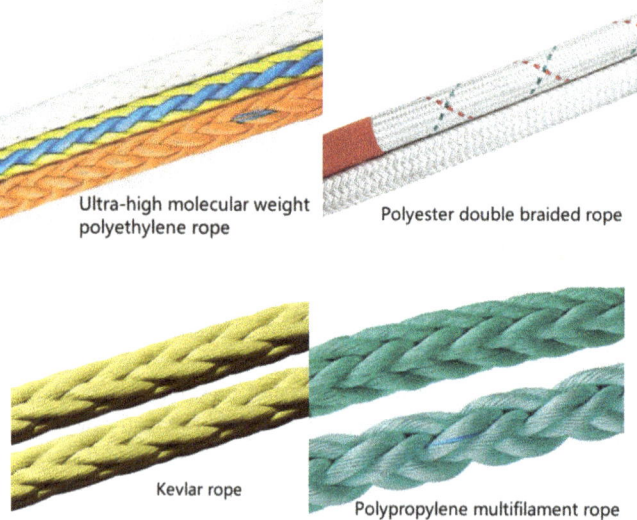

Fig. 1. Mooring ropes in various materials.

2 Research on the Safety Evaluation Method of Ship's Mooring Ropes

2.1 Safety Evaluation Theory and Examples

Fuzzy Comprehensive Evaluation Method, the Use of Model Objects in the Refinement of Relevant Influencing Factors, the Use of Fuzzy Theory to Make a General Evaluation of the Evaluation Object, the Use of Fig. 2 will be Evaluated in the Object of the Qualitative Issues into Quantitative Evaluation. In the actual processing, the complex nature of the qualitative method, such as: general, dangerous, very dangerous, and so on, and then solve a series of fuzzy problems. The method is characterized by strong operability, relatively simple evaluation process and intuitive evaluation results, and is suitable for the safety evaluation of complex systems.

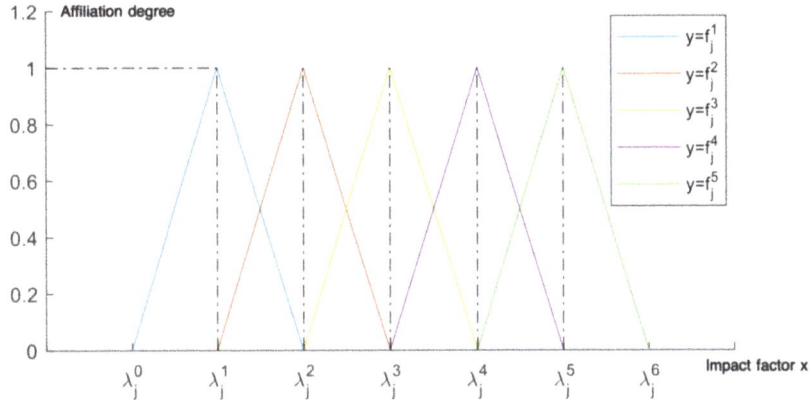

Fig. 2. Affiliation theory in centroid-based triangular whitening power fuzzy mathematics.

Compute the weight vector. When calculating the weight vector of the relative importance of a factor at a certain level to a factor at the previous level, it is first necessary to calculate the eigenvector W corresponding to the largest eigenvalue λ_{max} of the pairwise comparison matrix using Eqs. (1), (2).

$$|\lambda E - A| = 0 \tag{1}$$

$$AW = \lambda_{max} W \tag{2}$$

Afterwards, the weight vector W is normalized to obtain, where the components are normalized as shown in Eq. (3).

$$w'_j = \frac{w_j}{\sum_{i=1}^{n} w_i} \tag{3}$$

The final weight vector resulting from the calculation is Eq. (4):

$$W = (w'_1, w'_2, \ldots, w'_n)^T \tag{4}$$

In order to check the reasonableness of the calculations, it is necessary to perform a consistency test on the pairwise comparison matrix, which is calculated by the formula:

$$CR = \frac{CI}{RI} \tag{5}$$

$$CI = \frac{\lambda_{max} - n}{n - 1} \tag{6}$$

Among them, CR is the consistency ratio, the calculation formula is shown in formula Eq. 5, when CR < 0.10, that is to meet the consistency requirements, indicating that the results are reasonable and relatively reasonable; CI is the consistency index; RI is the stochastic consistency index, the value of which is determined by the dimensionality of the pairwise comparison matrix, as shown in Table 1.

Table 1. Stochastic consistency indicator values.

n	1	2	3	4	5	6	7	8	9	10	11
RI	0	0	0.58	0.90	1.12	1.25	1.32	1.41	1.45	1.49	1.51

Determination of evaluation indicator weights. According to the hierarchical analysis method to determine the "mooring rope safety evaluation index system", the criterion layer on the target layer of the weight W11, as well as the indicator layer on the criterion layer of the weight W12–W16. In order to further find the weight of the indicator layer on the target layer of the weight W, need to be in accordance with the Eq. (7) for the array multiplication of W11–W16 calculation.

$$W = \begin{pmatrix} w_1 \\ w_2 \\ \vdots \\ w_n \end{pmatrix} = W_{11} \cdot \begin{pmatrix} W_{12} \\ W_{13} \\ W_{14} \\ W_{15} \\ W_{16} \end{pmatrix} \tag{7}$$

A single indicator ui(i = 1,2,…,m) in the indicator set U, a single-indicator judgment is made, and for the indicator ui, its affiliation (degree of likelihood) rij is determined for the evaluation level vj(j = 1,2,…,n) the degree of affiliation (degree of likelihood) rij, so as to obtain the set of single-indicator judgments for the i factor ui ri = (ri1,ri2,…,rin). In addition, ri needs to be normalized so that it satisfies Eq. (8).

$$\sum_{j=1}^{n} r_{ij} = 1, (j = 1, 2, \ldots, m) \tag{8}$$

With Eq. (18), the evaluation set of m indicators constitutes a total evaluation matrix R:

$$R = \begin{bmatrix} r_{11} & r_{12} & \cdots & r_{1n} \\ r_{21} & r_{22} & \cdots & r_{2n} \\ \vdots & \vdots & \ddots & \vdots \\ r_{m1} & r_{m2} & \cdots & r_{mn} \end{bmatrix} \quad (9)$$

Determination of the degree of affiliation of the indicator. In this study, the trapezoidal type fuzzy distribution function is selected to construct the affiliation function of the indicator, combined with the characteristics of the indicator data. Trapezoidal distribution has the following three distribution types:

(1) Biased small distribution:

$$f(x) = \begin{cases} 1, x \leq a \\ \frac{b-x}{b-a}, a < x \leq b \\ 0, x > b \end{cases} \quad (10)$$

(2) Intermediate type distribution:

$$f(x) = \begin{cases} 0, x \leq a \\ \frac{x-a}{b-a}, a < x \leq b \\ 1, b < x \leq c \\ \frac{d-x}{d-c}, c < x \leq d \\ 0, x > d \end{cases} \quad (11)$$

(3) Skewed large distribution:

$$f(x) = \begin{cases} 0, x \leq a \\ \frac{x-a}{b-a}, a < x \leq b \\ 1, x > b \end{cases} \quad (12)$$

For the overall evaluation of "mooring line safety", it is necessary to take into account the weights and affiliations of the indicators, so the weighted average fuzzy operator is chosen for the final comprehensive evaluation operation. As shown in Eq. (13)

$$B = \sum (a_i \times r_{ij})(j = 1, 2, ..., m) \quad (13)$$

Constructing a Pairwise Comparison Matrix. Compare the elements of the indicators with each other, construct a pairwise comparison matrix A, as shown in Eq. (14), and then determine the weights between the factors at each level of the system.

$$A = \begin{bmatrix} a_{11} & a_{12} & \cdots & a_{1n} \\ a_{21} & a_{22} & \cdots & a_{2n} \\ \vdots & \vdots & \ddots & \vdots \\ a_{n1} & a_{n2} & \cdots & a_{nn} \end{bmatrix} \quad (14)$$

In the study, for the target "mooring line safety evaluation", there are five types of influencing factors in the criterion layer and two to five indicators in the indicator layer. In order to reflect the scientific rationality of the judgment of the importance of the influencing factors or indicators, five experts were firstly asked for their opinions on the evaluation of each factor or indicator by means of questionnaire survey. According to the formula (14), a pairwise comparison matrix is constructed for "criterion layer, ship factor, cable factor, equipment factor, environmental factor, personnel factor", as follows: A11, A12, A13, A14, A15, A16.

$$A_{11} = \begin{bmatrix} 1 & 1/7 & 1/2 & 1/4 & 1/3 \\ 7 & 1 & 1/4 & 3 & 3 \\ 2 & 1/6 & 1 & 3 & 1 \\ 4 & 1/3 & 3 & 1 & 2 \\ 3 & 1/3 & 1 & 1/2 & 1 \end{bmatrix} \tag{15}$$

$$A_{12} = \begin{bmatrix} 1 & 3/2 \\ 2/3 & 1 \end{bmatrix} \tag{16}$$

$$A_{13} = \begin{bmatrix} 1 & 3 & 3 & 4 & 4 \\ 1/3 & 1 & 3 & 3 & 4 \\ 1/3 & 1/2 & 1 & 2 & 1 \\ 1/4 & 1/3 & 1/2 & 1 & 1/2 \\ 1/4 & 1/3 & 1 & 2 & 1 \end{bmatrix} \tag{17}$$

$$A_{14} = \begin{bmatrix} 1 & 1 & 3 \\ 1 & 1 & 2 \\ 1/3 & 1/2 & 1 \end{bmatrix} \tag{18}$$

$$A_{15} = \begin{bmatrix} 1 & 2 & 4 & 3 \\ 1/2 & 1 & 3 & 1 \\ 1/4 & 1/3 & 1 & 1 \\ 1/3 & 1 & 1 & 1 \end{bmatrix} \tag{19}$$

$$A_{16} = \begin{bmatrix} 1 & 1/2 & 1/3 \\ 2 & 1 & 1/2 \\ 3 & 2 & 1 \end{bmatrix} \tag{20}$$

Through Matlab software, vector operations are performed on the pairwise comparison matrices A11 ~ A16 to solve the maximum eigenvalue and the corresponding eigenvector W, and the consistency test is carried out, in which the matrix A12 is a 2nd-order matrix, which satisfies the consistency by default. The corresponding CI and CR of A11 ~ A16 are calculated respectively according to Eqs. (5) (6), and the calculation results are shown in Table 2.

Table 2. Calculation results of matrices A11 ~ A16.

Matrices	Ordinal number	Maximum eigenvalue	RI	CI	CR	Consistency
A_{11}	5	5.0910	1.12	0.0228	0.0203	pass
A_{12}	2	2	/	/	/	pass
A_{13}	5	5.3329	1.12	0.0832	0.0743	pass
A_{14}	3	3.0183	0.58	0.0091	0.0176	pass
A_{15}	4	4.1031	0.90	0.0344	0.038	pass
A_{16}	3	3.0092	0.58	0.0046	0.0088	pass

According to the calculation results, it can be seen that the average random consistency indexes of A11 ~ A16 are all less than 0.1, so they all meet the consistency indexes. The eigenvectors corresponding to the largest eigenvalues of A11 ~ A16 are normalized according to Eqs. (3) (4) to obtain the weight coefficients of expert 1 on the scores of indicators at all levels:

Normative layer: w11 = {0.05,0.47,0.11,0.24,0.13};
Ship factors: w12 = {0.60,0.40};
Rope factors: w13 = {0.43,0.27,0.12,0.07,0.11};
Equipment factors: w14 = {0.44,0.39,0.17};
Environmental factors: w15 = {0.47,0.24,0.12,0.17};
Personnel factors: w16 = {0.16,0.30,0.54};

Similarly, the scoring data of the remaining four experts are processed, the pairwise comparison matrix is constructed, and the eigenvalues, eigenvectors, consistency indexes, etc. are calculated. Combining the results of the five experts, the weights of the indicators of the "mooring cable safety evaluation index system" are obtained, as shown in Table 3.

2.2 Demonstration of the Safety Status of the Rope According to the Calculation Results

In the study, the overall evaluation of "mooring line safety" needs to take into account the weights and degrees of affiliation of the indicators, so the weighted average fuzzy arithmetic is used for the final synthesis of the evaluation operation. As in Fig. 3, the cases of the ropes involved in the evaluation, as well as the internal structures.

In this section, the mooring situation of a container ship is taken as an example to study and evaluate the safety condition of the cable (FBR-POWERNEEMA® ROPE) provided by ZheJiang Four Brothers Rope Co., Ltd. Through the mooring statistics of this container ship for a period of 12 months, the relevant index data are organized and obtained as shown in Table 4.

Table 3. Indicator weights of the "mooring line safety evaluation index system".

Target Layers	Normative layer	Weights	Indicator layer	Weights
Mooring rope Safety Evaluation index system X	Ship factors x_1	0.07	Reasonable berthing rate x_{11}	0.57
			Reasonable departure rate x_{12}	0.43
	Rope factors x_2	0.32	Rope wear and tear x_{21}	0.18
			Rope maintenance rate x_{22}	0.27
			Rope inspection rate x_{23}	0.17
			Special operand x_{24}	0.20
			Number of ultra-safe loads x_{25}	0.18
	Equipment factors x_3	0.15	Maintenance rate of equipment x_{31}	0.52
			Equipment inspection rate x_{32}	0.33
			Rust degree of equipment x_{33}	0.15
	Environmental factors x_4	0.29	Meteorological scale x_{41}	0.44
			Wave height x_{42}	0.26
			Temperature difference x_{43}	0.10
			Humidity differences x_{44}	0.20
	Personnel factors x_5	0.17	Training completion rate x_{51}	0.12
			Proficiency x_{52}	0.34
			Compliance rate x_{53}	0.54

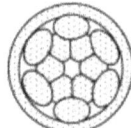

Fig. 3. Internal structure of the rope.

Table 4. Indicators statistics.

Indicators	Values	Indicators	Values
Reasonable berthing rate	0.87	Rust degree of equipment	0.11
Reasonable departure rate	0.95	Wind speed	3.25
Rope wear and tear	0.16	Wave height	1.12
Rope maintenance rate	0.75	Temperature difference	17.7
Rope inspection rate	0.86	Humidity differences	1.95
Special operand	6	Training completion rate	0.87
Number of ultra-safe loads	10	Proficiency	2.3
Maintenance rate of equipment	0.63	Compliance rate	0.89
Equipment inspection rate	0.59		

Combining the indicator statistics and the affiliation function for each indicator, the normalized affiliation matrix R is calculated according to Eqs. (7) (8) (9) (10) (11) (12), as follows:

$$R = (r_{ij}) = \begin{bmatrix} 0.85 & 0.15 & 0 & 0 & 0 \\ 1 & 0 & 0 & 0 & 0 \\ 0 & 0.40 & 0.60 & 0 & 0 \\ 0.25 & 0.75 & 0 & 0 & 0 \\ 0.80 & 0.20 & 0 & 0 & 0 \\ 0 & 0.80 & 0.20 & 0 & 0 \\ 0 & 0.50 & 0.50 & 0 & 0 \\ 0 & 0.65 & 0.35 & 0 & 0 \\ 0 & 0.45 & 0.55 & 0 & 0 \\ 0 & 0.90 & 0.10 & 0 & 0 \\ 0.375 & 0.625 & 0 & 0 & 0 \\ 0.317 & 0.683 & 0 & 0 & 0 \\ 0 & 0 & 0.46 & 0.54 & 0 \\ 0.525 & 0.475 & 0 & 0 & 0 \\ 0.85 & 0.15 & 0 & 0 & 0 \\ 0 & 0.65 & 0.35 & 0 & 0 \\ 0.95 & 0.05 & 0 & 0 & 0 \end{bmatrix} \quad (21)$$

R is fuzzy transformed according Eq. (13) to get the comprehensive evaluation vector B = (0.3389, 0.4812, 0.1645, 0.0153, 0). The maximum value of 0.4812 corresponds to the evaluation grade of "general", and the safety grade of the ship's mooring ropes is "general".

As a Web-based software system [7, 8], the mooring rope full life cycle safety management system is greatly convenient for the use of related personnel, as shown in Fig. 4. The safety of the system cable of the whole ship can be analyzed and checked.

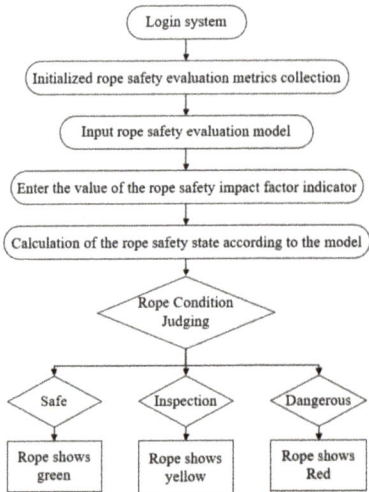

Fig. 4. FBR System Flowchart.

First enter the relevant data for the rope according to the indicators in Table 4. Evaluate and calculate the safety status of the cable according to different influencing factors, green if the safety status is safe, yellow if the safety status is to be checked, red if the safety status is dangerous, as shown in Fig. 5. Depending on the different values at different anchor points, different states of the ropes are displayed.

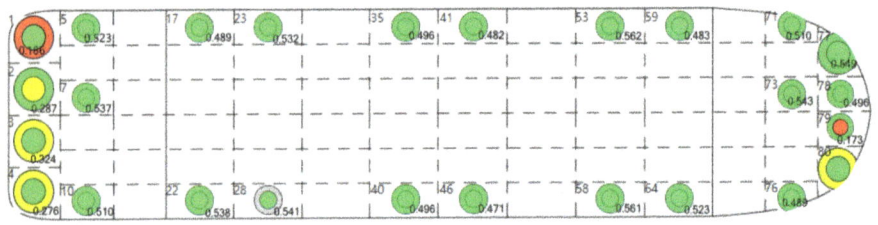

Fig. 5. Demonstration of the safety status evaluation of mooring ropes.

As shown in Fig. 6, a scatter plot of the safety status of the mooring cable is shown, recording the status of the cable when it is installed on top of the ship.

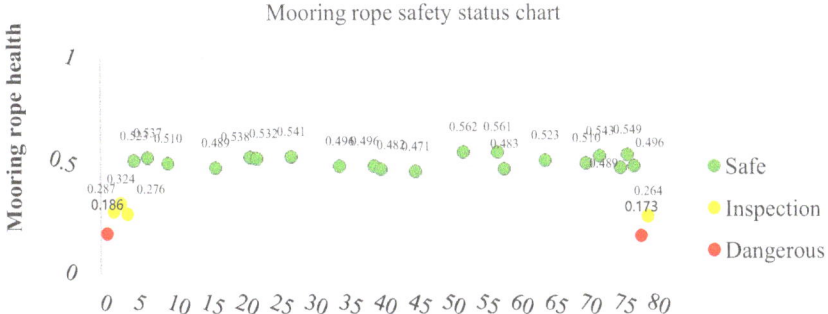

Fig. 6. Mooring rope safety status chart.

3 Conclusions

This reasearch develops the analysis and example calculations for the safety evaluation of mooring ropes. Firstly, the theory of safety evaluation is studied from the definition of safety evaluation and related evaluation methods. Secondly, on the basis of hierarchical analysis method, combined with the scoring of five experts, the weights of mooring cable safety evaluation indexes are calculated. Finally, the article proposes a safety evaluation model of mooring ropes based on fuzzy theory, and takes the mooring situation of a container ship as an example to study the safety condition of the ropes (FBR-POWERNEEMA® ROPE) supplied by ZheJiang Four Brothers Rope Co., Ltd. and the calculation results are in line with the actual situation, which in turn proves the validity of the model.

References

1. Valet, S., et al.: Accident caused by dynamic overloading of a ship mooring rope. Eng. Fail. Anal. **35**, 439–453 (2013)
2. Song, B., et al.: Current status of domestic and international applications and development of synthetic fiber cables for offshore engineering. Text. Herald **2021**(01), 76–79 (2021). (in chinese)
3. Wang, Y., et al.: Review of deep-sea polyester cables produced in China for ocean engineering. Synthetic Fiber **51**(10), 36–40 (2022). (in chinese)
4. Wei, C.: The necessity of establishing cable tension and fender pressure monitoring and warning system in large open wharf. Water Transport. Eng. **2007**(09), 115–118 (2007). (in chinese)
5. Chen, Y.: Safety and risk analysis of ship anchoring. China Water Transport. **2021**(04), 41–43 (2021). (in chinese)
6. Chen, Y.: Prevention of safety risks of marine cables. China Water Transport. (the second half of the month) **20**(06), 21–22 (2020). (in chinese)
7. Yang, X.: Development of ship oil monitoring information system based on ASP.NET. Jimei University (2017). (in chinese)

8. Yu, J.: Development of ship engine management information system based on B/S. Dalian Maritime University (2006). (in chinese)

Open Access This chapter is licensed under the terms of the Creative Commons Attribution-NonCommercial-NoDerivatives 4.0 International License (http://creativecommons.org/licenses/by-nc-nd/4.0/), which permits any noncommercial use, sharing, distribution and reproduction in any medium or format, as long as you give appropriate credit to the original author(s) and the source, provide a link to the Creative Commons license and indicate if you modified the licensed material. You do not have permission under this license to share adapted material derived from this chapter or parts of it.

The images or other third party material in this chapter are included in the chapter's Creative Commons license, unless indicated otherwise in a credit line to the material. If material is not included in the chapter's Creative Commons license and your intended use is not permitted by statutory regulation or exceeds the permitted use, you will need to obtain permission directly from the copyright holder.

Hydrodynamic Analysis of Deep-Water Double Barges Lifting Wreck Under Mooring Conditions

ZiKai Xu, Yingjie Li, Yuantao Bi, and Wei Zhang(✉)

Technology Center, YanTai Salvage, YanTai, China
`donsh328@163.com`

Abstract. The salvage method of lifting wreck with Twin-barges has the characteristics of large lifting capacity and high controllability, and its technology and process are relatively mature in shallow water salvage operations. However, there have been no operational cases of using this technology for deep water salvage above 100 m. This article is based on the salvage of a wreck at a depth of 150 m, using three-dimensional (3D) potential flow theory and Moses software for modeling. The hydrodynamic characteristics of the process of salvaging wreck with Twin-barges are analyzed, the motion response of barges and wreck has been studied, also the displacement amplitude of the barge under extreme sea conditions and the lifting force for salvaging the wreck are calculated. Unlike previous studies on Twin-barges for lifting wreck, this study uses two different barges as surface vessels for wreck lifting, which is similar to actual engineering vessels, can provide data support for this salvage project and reference for similar projects.

Keywords: Double Barges Lifting Wreck · 3D Potential Flow Theory · Hydrodynamic Analysis · Salvage Project

1 Introduction

The lifting and salvaging technology of installing tension jacks on double barges involves arranging two barges with tension jacks on both sides of the wreck, using multiple sets of steel wire ropes to hold the wreck. The tension jacks on both barges simultaneously apply force to drag the two ends of the salvaged steel wire ropes to carry out the wreck lifting and salvaging operation [1]. The lifting and salvaging technology of installing tension jacks on double barges involves arranging two barges with tension jacks on both sides of the wreck, using multiple sets of steel wire ropes to hold the wreck [2]. The tension jacks on both barges simultaneously apply force to drag the two ends of the salvaged steel wire ropes to carry out the wreck lifting and salvaging operation. In the salvage operation of the stern section of the "unblocked" ship, twin barges are used to cooperate with the steel strand tension jack, and the steel strand is connected with the bottom crossing wire rope of the wreck through a special connector to salvage the wreck. The maximum lifting force is more than 6000 T [3], and the single application of this technology that has the greatest impact is the Sewol wreck salvage project [4].

In recent years, researchers and scholars have continuously conducted in-depth research on double barge lifting technology to enhance its adaptability to high sea conditions. Scholars such as Zhu Ziyang [5] have conducted hydrodynamic analysis of double barge lifting wrecks, studied the motion response of ship groups, Fengrui Zhang [6] and others have studied the motion analysis model of double barge salvage operations, Zhang Zengmeng [7] and others have studied the calculation and analysis of double barge lifting salvage, Gao Dingquan et al. [7] have studied the coupled dynamic response of multiple ship groups in salvage operations, and Hou Jiaoyi et al. [8] have studied the motion response of single barge salvage ships.

This article takes the use of double barge salvage operations in a 150 m deep sea area as a case study to conduct hydrodynamic analysis of double barge salvage operations, and provides the motion response and salvage force of surface ships and underwater wrecks. The research results of existing scholars have all used Twin-barges to operate surface salvage ships. This article uses two completely different surface boats for hydrodynamic analysis of salvage operations, which can provide reference for related salvage operations.

2 Introduction to Ships

This article takes the environmental conditions of a 150 m deep sea area as the background, and uses a 100,000 ton semi submersible ship and a 25,000 ton load capacity barge as salvage operation ships to conduct hydrodynamic analysis of wreck salvage. The target sediment is 2000 tons, and the equivalent model is a rectangular prism with dimensions of L90m × B10m × D5m. The parameters of the semi submersible vessel and barge are shown in Table 1.

Table 1. Parameters of semi submersible ship and transport barge.

Ships	Length	Moulded Breadth	Moulded Depth	Design draft
Semi submersibles	290.0 m	76.0 m	15.0 m	11.0 m
Barges	159.0 m	38.8 m	10.9 m	5.0 m

3 Build Model

3.1 MOSES Coordinate System

The origin of coordinates is the projection point on the intersection line between the center plane and the base plane of the bow of the lifting-crane. (see Fig. 1).

The X-axis forward is the origin pointing to the stern.
The Y-axis forward is the origin pointing to the starboard.
The Z-axis forward is the origin straight up.

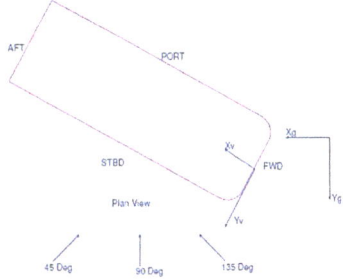

Fig. 1. The Coordinate System Definition and Environmental Condition Direction of Moses.

3.2 Calculation Model

Based on the three-dimensional potential flow theory, MOSES software is used to model the semi submersible and barge calculation models. The wreck is located between the two barges. To simplify the calculation, four suspension cables are designed for the connection between the barges and the wreck. The two operating vessels are anchored and positioned with 8 anchors, as shown in Fig. 2.

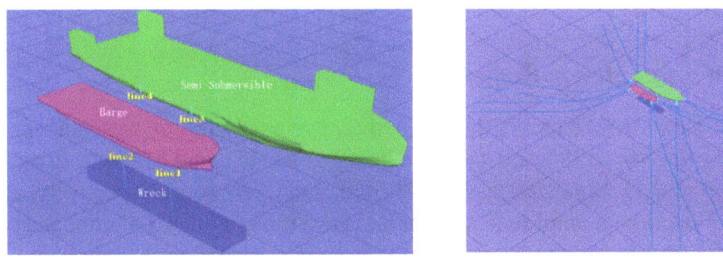

a. Connection model between surface shipwreck

b. Mooring model of working ship

Fig. 2. Calculation model for surface ships and wrecks.

4 Popper Function

Based on MOSES software, the three-dimensional potential flow theory is used to calculate the Response Amplitude Operator (RAO) at the set point. According to the official recommendation of MOSES, the ISSC spectrum is suitable for open sea areas [9] (recommended by the International Ship Structure Conference 1964, known as the ISSC spectrum). The spectrum function is

$$S_y(\omega) = |W(i\omega)|^2 S_r(\omega) \tag{1}$$

In the formula: Sy is the spectral density function of the response; W(iω) Is an operator; Sx(ω) Is the spectral density function of wave energy; ω Is the spectral frequency.

Taking into account the seakeeping of barges and semi submersibles, salvage operations are carried out under sea conditions with a wave height Hs of less than 1 m, a flow rate of 1 m/s, and a wind speed of 10 m/s. The ISSC wave spectrum is used as the incident wave to simulate the working conditions, with wave periods of 6 s, 7 s, 8 s, and 9 s, and wave directions of 0°, 45°, 90°, 135°, and 180°, respectively. The motion response of each structure is simulated.

5 Result

Based on the above conditions, numerical simulation analysis is conducted using Moses software to analyze various working conditions of the double ship lifting state. The motion response functions RAO of the wreck, semi submersible ship, and barge are analyzed, as well as the motion amplitudes of the wreck in the X, Y, and Z directions, and the calculation of the wreck's suspension force.

5.1 Motion Response Operator RAO

Organize the RAO data of wrecks, barges, and semi submersibles under sea conditions of 0°, 45°, 90°, 135°, and 180° with wave periods of 6–9 s, as shown in Figs. 3, 4, and 5.

5.2 Analysis of Barge Motion Response

According to the surge RAO curve of the barge, the extreme periods of surge under sea conditions of 0°, 45°, 90°, 135°, and 180° are 8 s, 9 s, 7 s, 9 s, and 8 s, with extreme values of 0.08 m/m, 0.11 m/m, 0.02 m/m, 0.08 m/m, and 0.09 m/m. Respectively, at a period of 6 s, the maximum amplitude at 0° in the wave direction is 0.06 m/m, and the minimum amplitude at 90° in the wave direction is 0.02 m/m; When the period is 7 s, the maximum amplitude at 135° in the wave direction is 0.07 m/m, and the minimum amplitude at 90° in the wave direction is 0.02 m/m; When the period is 8 s, the maximum amplitude at 180° in the wave direction is 0.09 m/m, and the minimum amplitude at 45° in the wave direction is 0.01 m/m; When the period is 9 s, the maximum amplitude at 45° in the wave direction is 0.11 m/m, and the minimum amplitude at 90° in the wave direction is 0.01 m/m.

According to the sway RAO curve of the barge, the extreme periods of sway under sea conditions of 0°, 45°, 90°, 135°, and 180° are 8 s, 9 s, 8 s, 9 s, and 8 s, with extreme values of 0.04 m/m, 0.12 m/m, 0.46 m/m, 0.16 m/m, and 0.04 m/m. When the period is 6–9 s, the maximum amplitudes in the 90° wave direction are 0.32 m/m, 0.4 m/m, 0.46 m/m, and 0.45 m/m, the minimum amplitudes in the 180° wave direction are 0 m/m, 0.01 m/m, 0.04 m/m, and 0.02 m/m, respectively.

According to the RAO curve of the barge's heave, the extreme periods of heave under sea conditions of 0°, 45°, 90°, 135°, and 180° are 8 s, 9 s, 9 s, 9 s, and 9 s, with extreme values of 0.1 m/m, 0.25 m/m, 1.08 m/m, 0.34 m/m, and 0.25 m/m, respectively. Among them, at a period of 6 s, the maximum amplitude at 90° wave direction is 0.23 m/m, and the minimum amplitude at 45° wave direction is 0.02 m/m; When the period is 7 s, the maximum amplitude at 90° in the wave direction is 0.6 m/m, and the minimum amplitude

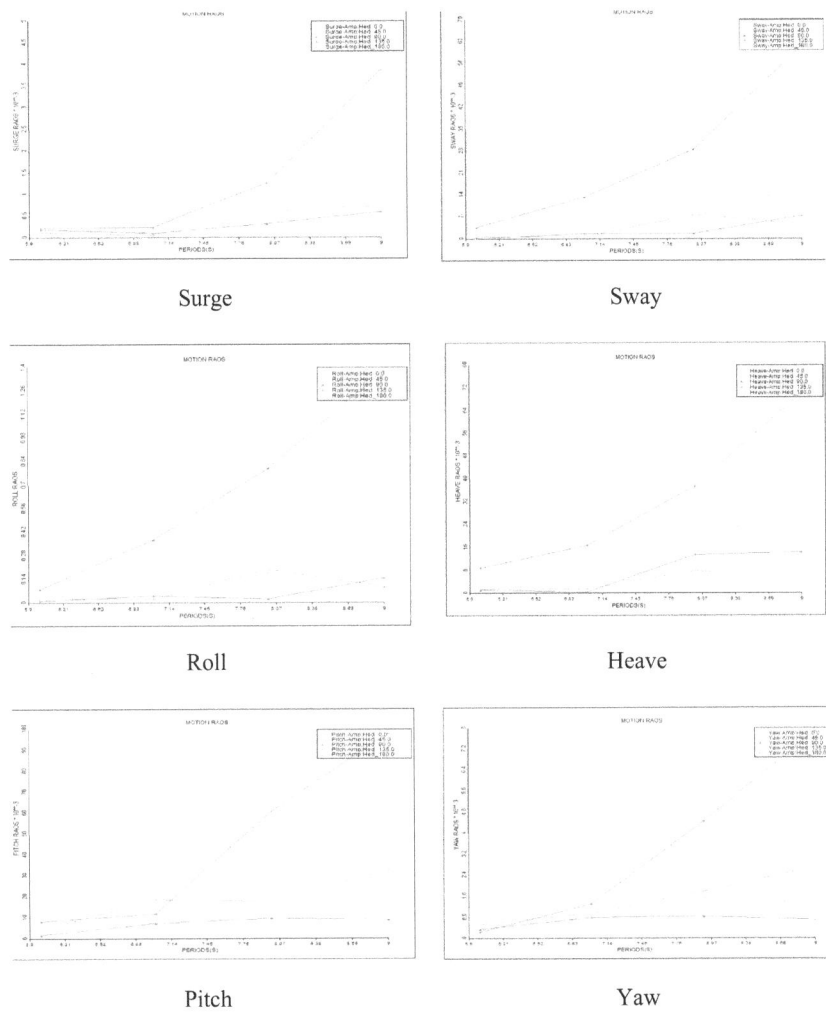

Fig. 3. RAO calculation curves for barges.

at 0° in the wave direction is 0.05 m/m; When the period is 8 s, the maximum amplitude at 90° wave direction is 0.92 m/m, and the minimum amplitude at 45° wave direction is 0.07 m/m; When the period is 9 s, the maximum amplitude at 90° in the wave direction is 1.08 m/m, and the minimum amplitude at 0° in the wave direction is 0.09 m/m.

From the roll RAO curve of the barge, it can be seen that under sea conditions of 0°, 45°, 90°, 135°, and 180°, the extreme roll periods are 8 s, 9 s, 9 s, and 8 s, with extreme values of 0.87°/m, 0.62°/m, 3.67°/m, 0.99°/m, and 0.67°/m, respectively. Among them, at a period of 6 s, the maximum amplitude at 90° in the wave direction is 1.03°/m, and the minimum amplitude at 135° in the wave direction is 0.01°/m; When the period is 7 s, the maximum amplitude at 90° in the wave direction is 1.42°/m, and the minimum amplitude at 180° in the wave direction is 0.12°/m; When the period is 8 s, the maximum

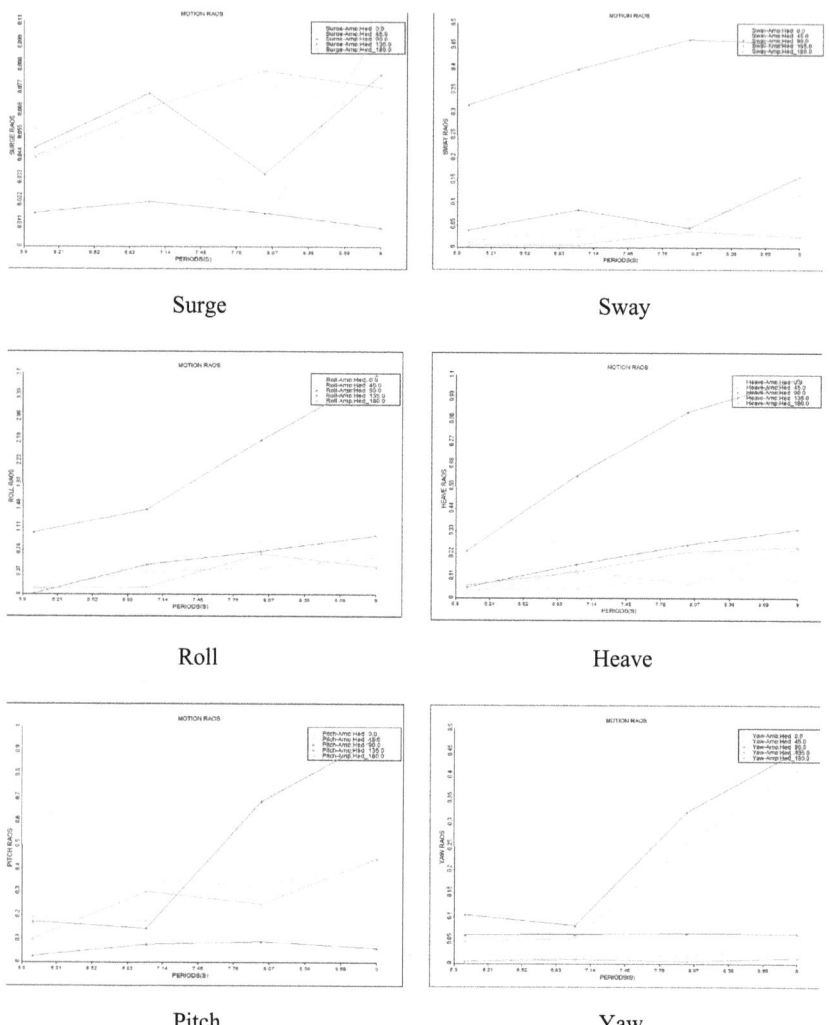

Fig. 4. RAO calculation curves for semi submersibles.

amplitude at 90° in the wave direction is 2.58°/m, and the minimum amplitude at 45° in the wave direction is 0.44°/m; When the period is 9 s, the maximum amplitude at 90° in the wave direction is 3.67°/m, and the minimum amplitude at 180° in the wave direction is 0.46°/m.

From the pitch RAO curve of barge, it can be seen that under sea conditions of 0°, 45°, 90°, 135°, and 180°, the extreme pitch periods are 9 s, 9 s, 8 s, 9 s, and 9 s, with extreme values of 0.44°/m, 0.96°/m, 0.09°/m, 0.95°/m, and 0.44°/m, respectively. Among them, at a period of 6 s, the maximum amplitude at 45° in the wave direction is 0.19°/m, and the minimum amplitude at 90° in the wave direction is 0.03°/m; When the period is 7 s, the maximum amplitude at 0° in the wave direction is 0.35°/m, and

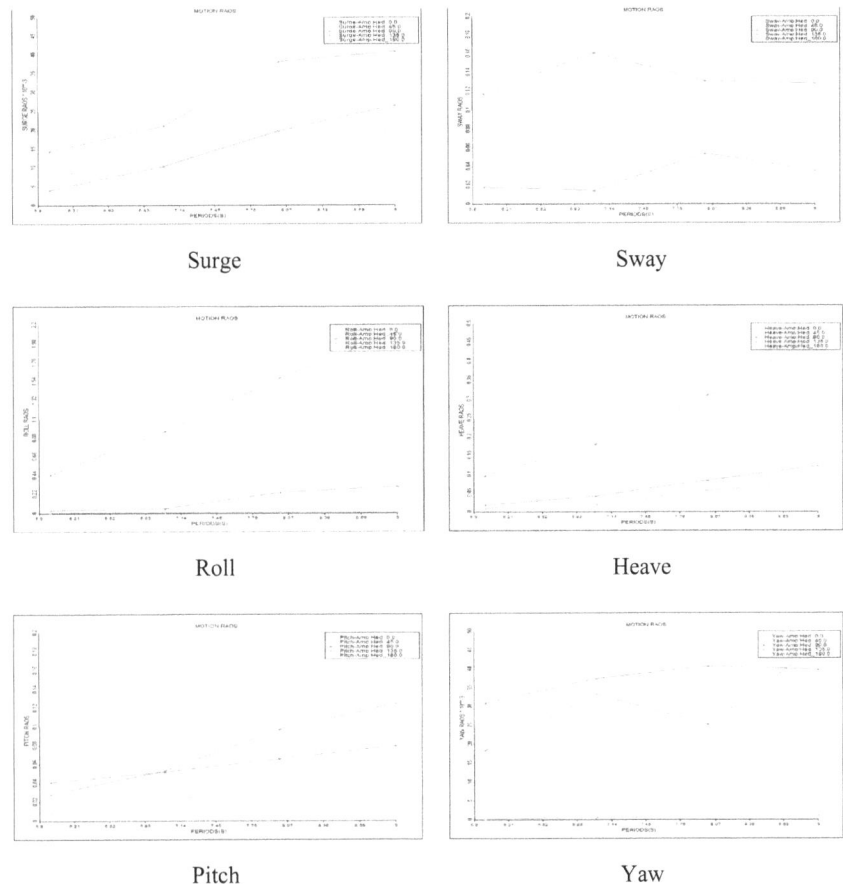

Fig. 5. RAO calculation curves for wrecks.

the minimum amplitude at 90° in the wave direction is 0.08°/m; When the period is 8 s, the maximum amplitude at 45° in the wave direction is 0.68°/m, and the minimum amplitude at 90° in the wave direction is 0.09°/m; When the period is 9 s, the maximum amplitude at 45° in the wave direction is 0.96°/m, and the minimum amplitude at 90° in the wave direction is 0.06°/m.

According to the yaw RAO curve of barge, under sea conditions of 0°, 45°, 90°, 135°, and 180°, the extreme periods of yaw are 9 s, 9 s, 9 s, and 9 s, with extreme values of 0.02°/m, 0.41°/m, 0.06°/m, 0.45°/m, and 0.01°/m, respectively. Among them, when the period is 6–9 s, the maximum amplitude at 135° in the wave direction is 0.1°/m, 0.08°/m, 0.32°/m, 0.45°/m, and the minimum amplitude at 180° in the wave direction is 0.01°/m.

The statistical summary of the extreme values of barge motion response is shown in Table 2. Comparison shows that the most unfavorable sea condition for barge salvage operations is 90° wave direction, with a wave period of 9 s. Although there are extreme values of heave, pitch, and yaw in the 45° and 135°, the RAO extreme values are much smaller compared to the 90°. The displacement fluctuation amplitudes of the barge along the X, Y, and Z directions under this sea condition are 0.6 m, 6.7 m, and 0.8 m, respectively.

Table 2. Statistical table of extreme RAO values for barge motion

Project	Angle	Period	Extreme
Surge	45°	9 s	0.11 m/m
Sway	90°	8 s	0.46 m/m
Heave	90°	9 s	1.08 m/m
Pitch	45°	9 s	0.96°/m
Roll	90°	9 s	3.67°/m
Yaw	135°	9 s	0.45°/m

Table 3. Statistical table of extreme RAO values for barge motion

Project	Angle	Period	Extreme
Surge	135°	9 s	0.04 m/m
Sway	90°	7 s	0.16 m/m
Heave	90°	9 s	0.47 m/m
Pitch	0°	9 s	0.17°/m
Roll	90°	9 s	2.11°/m
Yaw	135°	9 s	0.05°/m

5.3 Analysis of Motion Response of Semi Submersible

According to the surge RAO curve of the semi submersible, under sea conditions of 0°, 45°, 90°, 135°, and 180°, the extreme periods of surge are 9 s, 9 s, 9 s, and 8 s, with extreme values of 0.03 m/m, 0.02 m/m, 0.03 m/m, 0.04 m/m, and 0.02 m/m, respectively. Among them, at a period of 6 s, the maximum amplitude at 135° in the wave direction is 0.01 m/m, and the minimum amplitude at 45° in the wave direction is 0 m/m; When the period is 7 s, the maximum amplitude at 135° is 0.02 m/m, and the minimum amplitude at 0° is 0.01 m/m; When the period is 8 s, the maximum amplitude at 135° in the wave direction is 0.04 m/m, and the minimum amplitude at 0° in the wave direction is 0 m/m; When the period is 9 s, the maximum amplitude at 135° in the wave direction is 0.04 m/m, and the minimum amplitude at 45° in the wave direction is 0.02 m/m.

According to the sway RAO curve of the semi submersible, under sea conditions of 0°, 45°, 90°, 135°, and 180°, the extreme periods of sway are 9 s, 9 s, 7 s, 8 s, and 9 s, with extreme values of 0.01 m/m, 0.06 m/m, 0.16 m/m, 0.05 m/m, and 0 m/m, respectively.

Among them, when the period is 6–9 s, the maximum amplitudes in the 90° direction of the waves are 0.12 m/m, 0.16 m/m, 0.13 m/m, and 0.13 m/m, respectively. The minimum amplitudes in the 180° direction of the waves are all 0 m/m.

According to the heave RAO curve of the semi submersible, under sea conditions of 0°, 45°, 90°, 135°, and 180°, the extreme periods of heave are all 9 s, with extreme values of 0.08 m/m, 0.1 m/m, 0.47 m/m, 0.12 m/m, and 0.07 m/m, respectively. Among them, when the period is 6 s, the maximum amplitude at 90° in the wave direction is 0.1 m/m, and the minimum amplitude at 180° in the wave direction is 0.01 m/m; When the period is 7 s, the maximum amplitude at 90° in the wave direction is 0.18 m/m, and the minimum amplitude at 180° in the wave direction is 0.02 m/m; When the period is 8 s, the maximum amplitude at 90° wave direction is 0.31 m/m, and the minimum amplitude at 45° wave direction is 0.05 m/m; When the period is 9 s, the maximum amplitude in the 90° wave direction is 0.47 m/m, and the minimum amplitude in the 180° wave direction is 0.07 m/m.

From the roll RAO curve of semi submersible, it can be seen that under sea conditions of 0°, 45°, 90°, 135°, and 180°, the extreme roll periods are all 9 s, with extreme values of 0.07°/m, 0.43°/m, 2.11°/m, 0.31°/m, and 0.01°/m, respectively. Among them, during the period of 6–9 s, the maximum amplitudes at 90° in the wave direction are 0.45°/m, 0.96°/m, 1.58°/m, 2.11°/m, and the minimum amplitudes at 180° in the wave direction are 0°/m, 0°/m, 0.01°/m, and 0.01°/m.

From the pitch RAO curve of semi submersible, it can be seen that under sea conditions of 0°, 45°, 90°, 135°, and 180°, the extreme pitch periods are all 9 s, with extreme values of 0.17°/m, 0.14°/m, 0.08°/m, 0.12°/m, and 0.14°/m, respectively. Among them, at a period of 6 s, the maximum amplitude at 0° in the wave direction is 0.05°/m, and the minimum amplitude at 180° in the wave direction is 0.01°/m; When the period is 7 s, the maximum amplitude at 0° in the wave direction is 0.07°/m, and the minimum amplitude at 180° in the wave direction is 0.02°/m; When the period is 8 s, the maximum amplitude at 45° in the wave direction is 0.12°/m, and the minimum amplitude at 180° in the wave direction is 0.04°/m; When the period is 9 s, the maximum amplitude at 0° in the wave direction is 0.17°/m, and the minimum amplitude at 90° in the wave direction is 0.08°/m.

According to the yaw RAO curve of semi submersible, under sea conditions of 0°, 45°, 90°, 135°, and 180°, the extreme periods of yaw are 9 s, 9 s, 8 s, 9 s, and 9 s, with extreme values of 0°/m, 0.03°/m, 0.04°/m, 0.05°/m, and 0°/m. Among them, at a period of 6 s, the maximum amplitude at 90° in the wave direction is 0.03°/m, and the minimum amplitude at 180° in the wave direction is 0°/m; When the period is 7 s, the maximum amplitude at 90° in the wave direction is 0.04°/m, and the minimum amplitude at 180° in the wave direction is 0°/m; When the period is 8 s, the maximum amplitude at 90° in the wave direction is 0.04°/m, and the minimum amplitude at 180° in the wave direction is 0°/m; When the period is 9 s, the maximum amplitude at 135° in the wave direction is 0.05°/m, and the minimum amplitude at 180° in the wave direction is 0°/m.

The statistical summary of the extreme values of the motion response of the semi submersible is shown in Table 3. By comparison, it can be seen that the most unfavorable sea condition for the semi submersible salvage operation is the 90° wave direction, mainly reflected in the large heave and roll compared to other working conditions. Under this sea condition, the displacement fluctuation amplitude in the X, Y, and Z directions of the semi submersible is less than 0.9 m, 5.7 m, and 0.7 m, respectively.

5.4 Analysis of Motion Response of Wrecks

From the RAO curves of the surge and yaw of the wreck, it can be seen that the values under sea conditions with wave directions of 0°, 45°, 90°, 135°, and 180° are basically 0. From the sway RAO curve, it can be seen that under sea conditions with wave directions of 0°, 45°, 90°, 135°, and 180°, the sway extreme periods are 8 s, 9 s, 9 s, 9 s, and 8 s, with extreme values of 0.01 m/m, 0.02 m/m, 0.06 m/m, 0.01 m/m, and 0.01 m/m, respectively. From the RAO curve of the wreck's heave, it can be seen that under sea conditions of 0°, 45°, 90°, 135°, and 180°, the extreme periods of heave are 9 s, 9 s, 9 s, and 8 s, with their extreme values of 0.02 m/m, 0.01 m/m, 0.07 m/m, 0.01 m/m, and 0.01 m/m, respectively. From the roll RAO curve, it can be seen that under sea conditions with wave directions of 0°, 45°, 90°, 135°, and 180°, the roll extreme periods are 8 s, 9 s, 9 s, 9 s, and 8 s, with extreme values of 0.24°/m, 0.32°/m, 1.38°/m, 0.15°/m, and 0.2°/m, respectively. From the pitch RAO curve, it can be seen that under sea conditions with wave directions of 0°, 45°, 90°, 135°, and 180°, the pitch extreme periods are 9 s, 9 s, 8 s, 9 s, and 9 s, with extreme values of 0.05°/m, 0.09°/m, 0.01°/m, 0.1°/m, and 0.03°/m, respectively.

From the analysis results, it can be seen that the maximum motion response of the wreck occurs under the wave period of 9 s and the wave direction of 90° sea conditions, which is consistent with the extreme motion response of barges and semi submersible ships. The displacement fluctuations of the wreck in the X, Y, and Z directions under this sea condition are 1.2 m, 9.3 m, and 0.3 m, respectively.

5.5 Analysis of Cable Suspension Force

From the above analysis results, it can be seen that the extreme motion values of barges, semi submersibles, and wrecks occur at a wave direction of 90°, with a wave period of 9 s. Based on this, the lifting rope force is analyzed, and the calculation results are shown in Fig. 6.

From Fig. 4, it can be seen that the forces acting on the four mooring cables are not consistent, but the differences are small. Under the above working conditions, the maximum lifting force is line1, less than 2610 N, and the minimum lifting force is line4, greater than 2400 N. As the wave period increases in the same wave direction, the frequency of the fluctuation of each suspension cable force gradually decreases, but the amplitude and average value of the fluctuation of the suspension cable force remain basically unchanged. Under the same wave period conditions, the fluctuation amplitude and average value of the suspension cable force are basically symmetrical and similar to the 90° wave direction.

6 Conclusion

From the calculation results, it can be seen that for a sea condition of 150 m deep, using a double boat lifting salvage operation is not conducive to salvage under a 90° wave period of 9 s. Both surface operation ships and underwater wrecks have significant

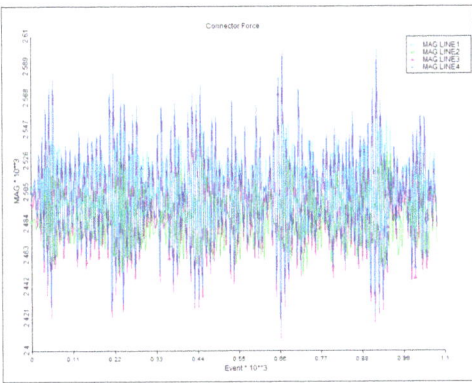

Fig. 6. Amplitude of cable force variation

motion responses. However, due to the possibility of the wreck being perpendicular to the wave direction after sinking, it is recommended to adjust the bow direction of the surface work ship shortly after the wreck leaves the seabed, in order to reduce the coupled motion response between the work ship and the underwater wreck.

References

1. Zhang, F., et al.: Research on the semi-active heave compensation of double barge and its simulation application in the salvage case of the "Shiyue" ship. Hydraul. Pneum. **45**(6), 41–49 (2021)
2. Lin, L., et al.: Design and test of wave compensation system based on flexible salvage. Mach. Tool Hydraul. **46**(19), 32–35 (2018)
3. Zhang, W., Man, Y.: Analysis of efficient wreck salvage technology. Hydraul. Pneum. Seal. **8**, 56–59 (2014)
4. Yao, Z., et al.: Dynamic mooring force calculation of twin barges during 'Sewol' lifted above seabed. Chin. J. Hydrodyn. **33**(1), 17–27 (2018)
5. Zhu, Z., et al.: Experimental study on deep water salvage of large tonnage wreck using double barge lifting method. Ship Build. China. **63**(6), 245–254 (2022)
6. Zhang, F., et al.: Performance analysis of the passive heave compensator for hydraulic shipwreck lifting systems in twin-barge salvaging. Ocean Eng. **280**, 114469 (2023)
7. Zhang, F., et al.: Semi-active heave compensation for a 600-meter hydraulic salvaging claw system with ship motion prediction via LSTM neural networks. J. Mar. Sci. Eng. **11**(5), 998 (2023). https://doi.org/10.3390/jmse11050998
8. Gao, D., et al.: Computational research on the coupling dynamic response of multi-body system for wreck salvage. Ship Eng. **44**(5), 59–64, 117 (2022)
9. Zhang, Z., et al.: Calculation and analysis of the hydraulic synchronizing lifting salvage based on GHS software. J. Dalian Marit. Univ. **43**(3), 25–30 (2017)

Open Access This chapter is licensed under the terms of the Creative Commons Attribution-NonCommercial-NoDerivatives 4.0 International License (http://creativecommons.org/licenses/by-nc-nd/4.0/), which permits any noncommercial use, sharing, distribution and reproduction in any medium or format, as long as you give appropriate credit to the original author(s) and the source, provide a link to the Creative Commons license and indicate if you modified the licensed material. You do not have permission under this license to share adapted material derived from this chapter or parts of it.

The images or other third party material in this chapter are included in the chapter's Creative Commons license, unless indicated otherwise in a credit line to the material. If material is not included in the chapter's Creative Commons license and your intended use is not permitted by statutory regulation or exceeds the permitted use, you will need to obtain permission directly from the copyright holder.

Efficiency of the Afterburner of the Solid Oxide Fuel Cell-Gas Turbine Marine System

Serhiy I. Serbin[1(✉)], Oleksii V. Patlaichuk[1], and Xianrui Zhao[2]

[1] Admiral Makarov National University of Shipbuilding, Mykolaiv, Ukraine
serhiy.serbin@nuos.edu.ua
[2] Jiangsu Maritime Institute, Nanjing, China

Abstract. This study focuses on evaluation of the effectiveness of an afterburner designed for solid oxide fuel cells (SOFC) located upstream of a gas turbine. The afterburner plays an important role in a hybrid marine power system, which includes solid oxide fuel cells and a gas turbine (SOFC-GT), aimed at maximizing the utilization of exhaust gas heat and enhancing overall system efficiency. The research involves theoretical investigations into the burning process of residual combustible components in the exhaust gases of fuel cells within an ejector-type afterburner. By conducting three-dimensional calculations using a model of turbulent reacting flows, the study identifies key directions for improving the afterburner's efficiency under conditions of low calorific value of the exhaust gases. The proposed design of the afterburner will ensure complete burnout of fuel residues after the fuel element stacks with minimal emissions of toxic components. The design is quite universal and does not require expensive catalysts. The calculations underscore the feasibility of utilizing the proposed combustion technology in marine SOFC-GT power systems. The results obtained pave the way for scientifically grounded strategies to improve the efficiency of future power systems integrating fuel cells and gas turbines, deepening our understanding of exhaust gas after burning processes. This knowledge will facilitate the development of highly efficient decarbonized power systems suitable for both marine and stationary applications.

Keywords: Afterburner · Ship Energy System · Gas Turbine

1 Introduction

The utilization of solid oxide fuel cells within hybrid marine power systems remains a pertinent topic. In such systems, primary energy conversions occur within the fuel cells, where electrochemical processes take place, while the heat from their exhaust gases is utilized in gas turbines. The operating temperature of solid oxide fuel cells is maintained within the range of 873–1373 K, allowing for the installation of a gas turbine at the outlet for additional electricity generation [1–3]. These cycles demonstrate competitiveness with conventional hybrid systems comprising a gas turbine and a waste heat boiler, as they achieve electrical efficiencies of approximately 70% or more.

The afterburner (combustion chamber) in the SOFC-GT hybrid system serves to incinerate any remaining fuel (typically, the fuel utilization factor in SOFC is approximately 80–85% [1, 2]). The afterburner must be specially designed for such a hybrid system. Conventional serial combustion chambers of gas turbine engines cannot be directly used in such systems without modifications because the fuel mixture after SOFC has a very low calorific value, and the air used as an oxidizer has a lower oxygen content compared to atmospheric air. Therefore, the combustion chamber needs to be specially designed (modified) for use in the SOFC-GT system. Another limitation is the necessity of stable operation of the chamber over a wide range of loads with low emissions of toxic components.

Previous research on low-emission fuel combustion technologies in power systems has identified key directions for improving their stability under conditions of low calorific values of gas-like fuels of various compositions and demonstrated the possibility of applying combustion technologies for partially mixed lean fuel-air mixtures in SOFC-GT systems, provided intensifiers of combustion are used.

In hybrid SOFC-GT systems, the gas turbine is thermodynamically coupled with the solid oxide fuel cell and operates by utilizing the heat of its exhaust gases. Based on the analysis of possibilities for improving systems with fuel cells and gas turbines, schemes for marine power plants have been developed in works [4, 5], utilizing stacks of solid oxide fuel cells in combination with a contact gas turbine unit that operates with over-expansion.

For the SOFC-GT system, a complex scheme of the gas turbine section operating with over-expansion has been adopted, in which a cycle with steam injection into the afterburner is implemented. Such a scheme allows for reducing the pressure in the SOFC stack housing to bring emissions to acceptable levels while substantially increasing the efficiency of the hybrid power plant.

A particular feature of the afterburner operation in such a system is the need to organize stable combustion of the exhaust gases of the solid oxide fuel cell under conditions of not only the low calorific value of these gases but also the additional injection of steam, which increases the likelihood of flame extinction without additional means of intensifying mixing and combustion processes.

Analysis of the available literature has shown that there are real opportunities for organizing efficient combustion of low-calorific gases in afterburners. Various constructive forms and means of intensification have been proposed [6–8]. CFD calculations of combustion processes of fuel cell exhaust gases in porous burners under conditions of premixing, in combustion systems similar to gas turbine combustion chambers, in catalytic heaters, etc., have been carried out. The accuracy of the modeling results was validated by comparison with experimental data [9–11]. However, serious problems persist related to the insufficient stability of afterburners under basic operating modes of the SOFC-GT system.

A review of recent studies and potential applications of afterburners for solid oxide fuel cell exhaust gases has highlighted several unresolved challenges. These include operational complexities within SOFC-GT systems and inefficiencies in the combustion chamber processes, which are influenced by the low calorific value of the exhaust gases and the significant steam injection into the combustion unit.

This work employs numerical modeling to conduct theoretical investigations into the efficiency of afterburning solid oxide fuel cell exhaust gases within a hybrid marine power system with an over-expansion contact gas turbine unit.

This study aims to investigate the efficiency of afterburning the exhaust gases of fuel cells in a hybrid marine power system consisting of solid oxide fuel cells and an over-expansion gas turbine. The research focuses on the operational processes in the ejector-type afterburner of the hybrid marine system for generating both electrical and thermal energy. The subject of the study encompasses the patterns of mixing and turbulent combustion of low-calorific exhaust gases from solid oxide fuel cells.

2 Mathematical Modeling

A crucial step towards the utilization of marine SOFC-GT power systems involves the modification of afterburners (combustion chambers) to combust thermochemical products formed within the stacks of solid oxide fuel cells. Employing 3D Computational Fluid Dynamics (CFD) analysis of operational processes is essential for determining rational afterburning technologies. In the development of a model for chemically reacting turbulent flows in the afterburner, it is imperative to incorporate a series of sub-models for the most critical physicochemical processes, such as the formation of the combustible mixture, turbulent transport of components, and their combustion in the presence of a significant amount of steam.

A mathematical model of the afterburner for solid oxide fuel cell exhaust gases has been developed based on solving a system of equations describing convective and diffusive concentration transport for each of the reacting mixture components. The proposed mathematical model is based on the equations of continuity, momentum conservation, energy conservation, transport of chemical components of the mixture, as well as equations of turbulence model: A detailed description of the model is provided in [12–15].

In the calculations, the combustion scheme of chemical reactants kee58 proposed in [16] was utilized. It comprises 58 chemical reactions involving 18 chemical components. The kinetic mechanism of exhaust gas combustion involves numerous chemical reactions, necessitating the use of a combustion model that can accurately simulate the chemical processes in the afterburner under turbulent mixing conditions. One such model is the Eddy-Dissipation-Concept (EDC) model, which accounts for the interplay between chemical kinetics and turbulence in the flow [17].

The mathematical model of nitrogen oxides (NO_x) emissions used in this work consists of a system of mass transport equations, which take into account convection, diffusion, as well as the formation and decomposition of nitrogen oxides and related compounds [18–21].

The object of investigation chosen for this study is the ejector-type syngas afterburner chamber [14]. The structural diagram of the afterburner is presented in Fig. 1. The oxidizer (cathode gas) enters receiver 2 through four inlet pipes 1, where it is divided into four streams. The first stream enters the ejector 3 through three channels 4. The exhaust gases from the solid oxide fuel cell (anode gases) enter through pipe 5, partially mixing with the oxidizer in the mixer 6 and then entering the combustion tube 7.

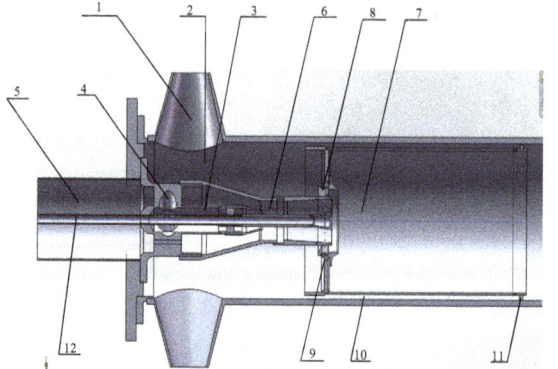

Fig. 1. Design of the afterburner: 1—inlet pipes; 2—receiver; 3—ejector; 4—oxidizing supply channel to the ejector; 5—nozzle for the supply of waste gases from the fuel cell; 6—mixer; 7—flame tube; 8—swirler; 9—oxidizing supply holes; 10—slot for cooling the heat pipe; 11—swirler; 12—additional fuel supply channel (if necessary).

The second stream of oxidizer is supplied to the swirl generator 8, spirals, and enters the combustion tube 7, forming a swirling combustion zone. The third stream of oxidizer enters the combustion tube through a series of holes 9. The fourth stream of oxidizer is directed into channel 10 to cool the wall of the combustion tube, and then after the swirl generator 11, it mixes with the combustion products to achieve the required temperature at the outlet of the afterburner. Natural gas can be supplied through channel 12 to maintain the stability of burning processes along the axis of the chamber.

The adequacy of the mathematical model used was verified by comparing the results of three-dimensional calculations with experimental data obtained during the investigation of the parameters of the afterburner chamber on a specialized test stand. The analysis of the obtained results [14] showed an acceptable level of agreement between the experimental and calculated values, indicate the possibility of using a similar model for calculations of ejector-type afterburners operating on low-calorie fuel.

3 Study of Afterburner Efficiency Using CFD Analysis

The calculations of the aerodynamic flow structure and combustion of the exhaust gases from the solid oxide fuel cell in the afterburner were carried out for the following fuel mixture composition (mass fractions): $H_2 = 1.18\%$, $H_2O = 38.88\%$, $CO = 7.06\%$, $CO_2 = 52.88\%$. The oxidizer composition (mass fractions) consisted: $O_2 = 16.53\%$, $H_2O = 3.84\%$, $CO_2 = 0.52\%$, $N_2 = 79.11\%$. The oxidizer flow rate after the cathode section of the fuel cell was 2.825 kg/s, with a temperature of 1193 K; the exhaust gas flow rate after the anode section of the fuel cell was assumed to be 0.41 kg/s, with a temperature of 1193 K. Three-dimensional calculations were performed using a finite difference computational grid consisting of 1.55 million tetrahedral elements.

As a result of conducting three-dimensional calculations using the Ansys Fluent software package [15] with the application of the kinetic mechanism [16], new data on the operating process of the afterburner for the exhaust gases from the solid oxide fuel cell were obtained.

Figure 2 shows the oxidizer streamlines, velocity vectors of the working fluid, and temperature distribution across the sections of the afterburner. In this design, due to the interaction of oxidizer streams passing through the swirl generator, ejector, and radial holes in the primary combustion device zone, an intensive recirculation zone is formed (Fig. 2, a, b) in the main part of the afterburner. This central recirculation zone is sufficiently developed and serves to effectively stabilize the flame front (Fig. 2, c)

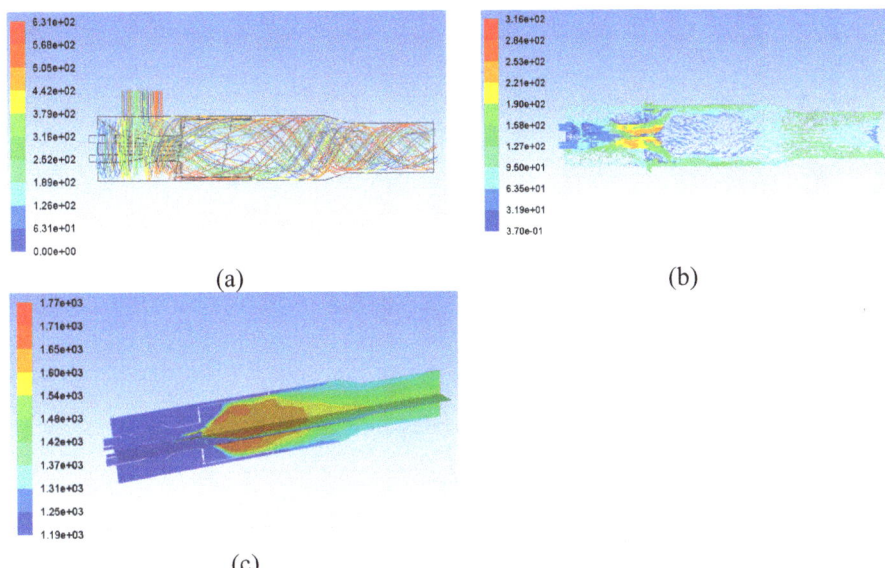

Fig. 2. Track lines of the oxidizer (a), velocity vectors of the working fluid, m/s (b), and temperature distribution, K (c) along the length of the afterburner

In Fig. 3, the distribution of flow velocity, mass fractions of combustible components: carbon monoxide (CO) and hydrogen (H_2), as well as molecular oxygen (O_2) along the length of the afterburner is presented. Due to the presence of an extensive recirculation zone (Fig. 3, a), where the combustion products with relatively high temperatures are located, combustible components (CO and H_2) intensively combust (Fig. 3, b, c) in the main part of the afterburner when interacting with oxygen (Fig. 3, d).

The variation of mass fractions of fully combusted products (H_2O, CO_2), active radical (H), and nitrogen oxide (NO) across the sections of the afterburner is depicted in Fig. 4.

It can be observed that the amount of fully combusted products is determined not only by the processes of burning the combustible components of the anode gases but also by their initial content in the fuel mixture. Among the active intermediate compounds, hydroxyl (OH) has the highest concentration. It is noteworthy that unstable compounds (radicals) start to form due to the high temperature of the solid oxide fuel cell exhaust gases even in the internal sections of the ejector, initiating the onset of afterburning

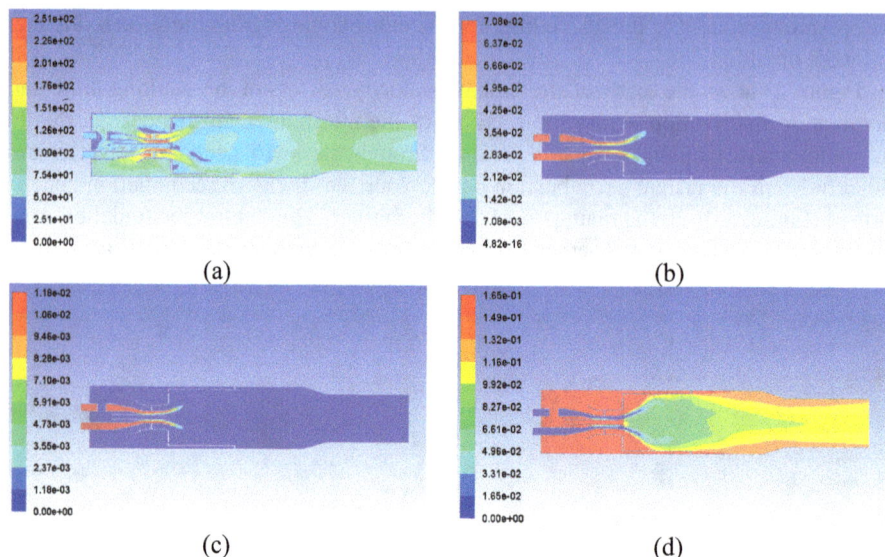

Fig. 3. Distribution of velocity, m/s (a), mass fractions: CO (b), H_2 (c), O_2 (d) along the length of the afterburner.

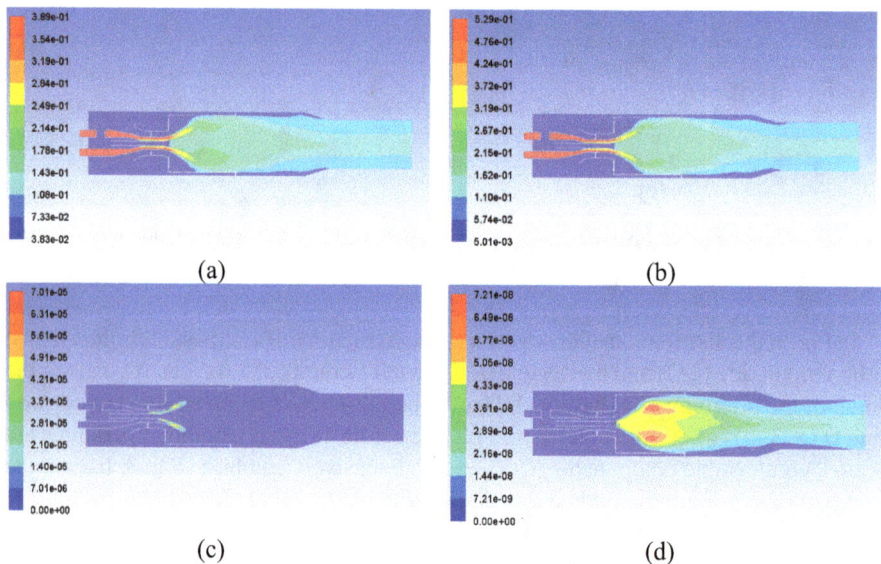

Fig. 4. Distribution of mass fractions: H_2O (a), CO_2 (b), H (c), NO (d) along the length of the afterburner.

processes inside the ejector. This may lead to overheating of the afterburner's structural elements and requires additional research to eliminate this phenomenon.

Nitrogen oxide (NO) is mainly formed in areas of highest temperatures. However, since the maximum combustion temperatures inside the afterburner do not exceed 1770 K, the concentrations of nitrogen oxide across its sections and at the outlet are minimal.

The molar fractions of the main components at the outlet of the afterburner are presented in Table 1.

Table 1. Mole fractions of components at the outlet of the afterburner

H_2	CO	H_2O	CO_2	O_2	NO
$2.332 \cdot 10^{-06}$	$2.384 \cdot 10^{-07}$	0.148	0.054	0.111	$9.509 \cdot 10^{-09}$

The data in Table 1 indicate that the combustible components burn completely inside the afterburner, while the concentration of nitrogen oxides at the outlet of the fuel burner approaches zero values.

4 Conclusions

Theoretical studies of the efficiency of the ejector-type afterburner of fuel cell exhaust gases have been conducted, which allowed us to draw the following conclusions.

1. A mathematical model of the afterburner for the prospective marine SOFC-GT system has been developed comprising stacks of solid oxide fuel cells alongside a contact gas turbine unit that operates with over-expansion.
2. Three-dimensional calculations of flow aerodynamics and combustion of combustible components in the ejector-type afterburner have been conducted.
3. It has been established that the proposed afterburner design ensures stable and complete combustion of combustible components in the exhaust gases under conditions of low calorific value and high water vapor content in the anode gases of solid oxide fuel cells.
4. The content of nitrogen oxides at the exit of the ejector-type afterburner approaches zero values.
5. Further theoretical and experimental studies are needed to optimize the performance characteristics of such types of afterburners and to eliminate the possibility of combustion of the combustible mixture in the ejector channel.

References

1. Faleh, S., Khir, T.: SOFC-Gas turbine hybrid power plant: Exergetic study. In: Li, Y. (ed.) Rotating Machines (2022)
2. Wang, Z., Chen, H., Xia, R., Han, F., Ji, Y., Cai, W.: Energy, exergy and economy (3E) investigation of a SOFC-GT-ORC waste heat recovery system for green power ships. Therm. Sci. Eng. Prog. **32**, 101342 (2022)

3. Zhang, B., et al.: Rapid load transition for integrated solid oxide fuel cell—gas turbine (SOFC-GT) Energy Systems: A demonstration of the potential for grid response. Energy Convers. Manag. **258**(15), 115544 (2022)
4. Serbin, S., Washchilenko, N., Cherednichenko, O., Burunsuz, K.: Energy system with solid oxide fuel cells and steam-injected gas turbine. In: 2023 IEEE 5th International Conference on Modern Electrical and Energy System (MEES) (2023)
5. Chen, D., Serbin, S., Washchilenko, N., Burunsuz, K.: Parameter evaluation for a hybrid marine system combining solid oxide fuel cells and overexpanded steam-injected gas turbine. Int. J. Thermofluids. **21**, 100565 (2024)
6. Li, M., et al.: A Comprehensive Review of Thermal Management in Solid Oxide Fuel Cells: Focus on Burners, Heat Exchangers, and Strategies. Energies. **17**, 1005 (2024)
7. Pianko-Oprych, P., Jaworski, Z.: Numerical investigation of a novel burner to combust anode exhaust gases of SOFC stacks. Pol. J. Chem. Technol. **19**(9), 20–26 (2017)
8. Lingstädt, T., Grimm, F., Kutne, P., Aigner, M.: Design and setup of a low calorific SOFC off-gas combustion chamber for a pressurized MGT hybrid power plant test rig. In: E3S Web of Conferences, vol. 113, p. 02016 (2019)
9. Yanyi, Y., Weijuan, Y., Jingyi, C., Xiaoyu, Z., Junhu, Z.: Numerical investigation on the exhaust gas combustion of an SOFC in a catalytic multichannel burner. Front. Energy Res. **12**, 1322956 (2024)
10. Lingstädt, T., Grimm, F., Krummrein, T., Bücheler, S., Aigner, M.: Experimental investigation of an SOFC off-gas combustor for hybrid power plant usage with low heating values realized by natural gas addition. In: Proceedings of GPPS Forum 18 Global Power and Propulsion Society Montreal (2018)
11. Kashmiri, S.A., Tahir, M.W., Afzal, U.: Combustion Modeling and Simulation of Recycled Anode-off-Gas from Solid Oxide Fuel Cell Energies, vol. 13, p. 5186 (2020)
12. Launder, B.E., Spalding, D.B.: Lectures in Mathematical Models of Turbulence. Academic, London (1972)
13. Serbin, S.I., Kozlovskyi, A.V., Burunsuz, K.S.: Investigations of non-stationary processes in low emissive gas turbine combustor with plasma assistance. IEEE Trans. Plasma Sci. **44**(12), 2960–2964 (2016)
14. Matveev, I.B., Serbin, S.I., Vilkul, V.V., Goncharova, N.A.: Synthesis Gas Afterburner Based on an Injector Type Plasma-Assisted Combustion System. IEEE Trans. Plasma Sci. **43**(12), 3974–3978 (2015)
15. ANSYS, Inc. ANSYS Fluent Theory Guide. (2013)
16. Bilger, R.W., Starner, S.H., Kee, R.J.: On Reduced Mechanisms for. Methane-Air Combustion in Nonpremixed Flames. Combust. Flame. **80**, 135–149 (1990)
17. Magnussen, B.F., Hjertager, B.H.: On mathematical models of turbulent combustion with special emphasis on soot formation and combustion. 16-th Symp. (Int.) on Combustion: the Combustion Institute 16(1), 719–729 (1977)
18. Matveev, I., Serbin, S., Butcher, T., Tutu, N.K.: Flow structure investigations in a Tornado combustor. In: 4th International Energy Conversion Engineering Conference (2006)
19. Serbin, S.I., Matveev, I.B., Mostipanenko, G.B.: Plasma-assisted reforming of natural gas for GTL: Part II—Modeling of the methane-oxygen reformer. IEEE Trans. Plasma Sci. **43**(12), 3964–3968 (2015)
20. Matveev, I.B., Washchilenko, N.V., Serbin, S.I.: Plasma-assisted reforming of natural gas for GTL: Part III—Gas turbine integrated GTL. IEEE Trans. Plasma Sci. **43**(12), 3969–3973 (2015)
21. Serbin, S., Burunsuz, K., Chen, D., Kowalski, J.: Investigation of the characteristics of a low-emission gas turbine combustion chamber operating on a mixture of natural gas and hydrogen. Pol. Marit. Res. **29**(2), 64–76 (2022)

Open Access This chapter is licensed under the terms of the Creative Commons Attribution-NonCommercial-NoDerivatives 4.0 International License (http://creativecommons.org/licenses/by-nc-nd/4.0/), which permits any noncommercial use, sharing, distribution and reproduction in any medium or format, as long as you give appropriate credit to the original author(s) and the source, provide a link to the Creative Commons license and indicate if you modified the licensed material. You do not have permission under this license to share adapted material derived from this chapter or parts of it.

The images or other third party material in this chapter are included in the chapter's Creative Commons license, unless indicated otherwise in a credit line to the material. If material is not included in the chapter's Creative Commons license and your intended use is not permitted by statutory regulation or exceeds the permitted use, you will need to obtain permission directly from the copyright holder.

Coastal Protection with Tensioned Viscoelastic Wave Barriers

Jianjun Chen[1,2], Chunxiao Li[3], Xin Liu[1,2], Jianjian Zhao[4,5], Cheng Bi[1,2(✉)], Yong Zhang[3], Maquan Wang[3], Mingjun Nan[3], Hao Hu[4,5], Hewen Hu[1,2], Ruichao Liu[1,2], and Yu Lei[1,2]

[1] China Huaneng Clean Energy Research Institute, Beijing, China
bicheng630@163.com
[2] National Energy R&D Center of Offshore Wind Power Engineering and Operation, Beijing, China
[3] Yantai Power Plant of Huaneng Shandong Power Generation Co., Ltd, Yantai, China
[4] Huaneng Jiangsu Clean Energy Branch, Nanjing, China
[5] Shengdong Rudong Offshore Wind Power Generation Co., Ltd, Nantong, China

Abstract. A vertical tensioned flexible sheet can function as a low-cost removable wave barrier, providing essential sheltering for coastal waters and floating facilities located behind it. In this study, we investigate the complex interactions between surface waves and a thin viscoelastic barrier that is only partially submerged, extending from the water surface into the water column. Three distinct materials: nonwoven geotextile, PA 6, and HDPE, are considered for barrier construction. Through comprehensive analysis, we evaluate the effects of these materials on wave attenuation, aiming to determine the optimal dimensions, tension, and material type for various coastal environments. Using the eigenfunction expansion method, the study reveals that wave transmission significantly decreases with increased rigidity of the barrier. Additionally, the viscous effects become more pronounced for barriers demonstrating greater flexibility. Notably, upon comparing the performance of the three materials, it is found that the HDPE barrier, designed with a draft constituting 40% of the water depth and a thickness of 15 cm, provides an excellent balance between effective wave attenuation and cost efficiency. These findings underscore the potential of using innovative materials in coastal engineering to enhance the resilience of marine infrastructure against wave action.

Keywords: Viscoelastic Barrier · Wave Attenuation · Eigenfunction Expansion Method

1 Introduction

Over the past few decades, there has been growing interest in utilizing submerged vertical flexible barriers as an economical substitute for the conventional fixed rigid breakwaters. These barriers can safeguard coastal facilities and shorelines against wave impacts with the advantage of cost-effectiveness and adaptability.

The performance of the vertical flexible barrier for wave protection has been investigated extensively. Thin barriers with a certain flexural rigidity are commonly modelled using the elastic Euler-Bernoulli beam equation. Mahmood et al. [1] investigated the wave interactions with submerged elastic plates that have semi-infinite length, finite length or circular shape. Behera and Sahoo [2] presented a study on the wave protection by a submerged flexible porous plate-like structure. Chakraborty et al. [3] adopted the hypersingular integral equation approach to investigate a submerged vertical plate in deep water. Same method was also used by Kundu et al. [4] to study the wave attenuation by an inclined plate. In addition, tensioned flexible barriers with membrane-like dynamic behaviors have also been proposed for the wave protection. Lo [5] employed the eigenfunction expansion method to investigate the wave scattering by a single membrane. The same approach was used by Lee and Lo [6] to investigated multiple surface piercing membranes. Karmakar et al. [7] considered multiple moored membranes as a protection system. Recently, Koley and Sahoo [8] studied the wave interactions with a vertical submerged porous membrane using the eigenfunction expansion and boundary element method.

Barrier material is an important consideration for the real application. Geotextile fabrics are potential materials for barriers since they have been widely used in fabricating silt curtains which uses a composite of woven and nonwoven geotextile with multiple layers to withstand strong loads in coastal environments [9]. Geotextile fabrics are soft solids with extremely small flexural rigidity and therefore they are usually tensioned with a chain weight. Another type of material to make barriers is engineering plastics which have been widely used in marine applications for many years due to the advantages of light weight, easy fabrication, corrosion resistance and durability [10]. They usually possess high specific strength and good mechanical properties, making them good candidates to replace the traditional concrete- or metal structures applied in sea water.

In the previous studies, the dynamic behavior of the wave barrier was assumed as either plate-like or membrane-like a priori. However, in real application, the wave barrier dynamics may vary between plate-like and membrane-like depending on the material characteristics and tension magnitude. Thus, the analysis should consider the relative dominance of tension and flexural rigidity of the barrier. In addition, materials that used to fabricate wave barriers are typically viscoelastic [11]. Therefore, a viscoelastic analysis should be adopted. Moreover, the appropriate barrier material and optimal dimensions should be suggested for the real application.

In the current study, we perform an analytical analysis on the surface wave interactions with a vertical viscoelastic thin barrier under tension. The appropriate barrier material and the corresponding optimal dimensions are further determined based on the analytical solution as well as cost analysis.

2 Mathematical Formulation

The BVP associated with the wave interaction with a viscoelastic barrier is analysed in the two-dimensional Cartesian x-y coordinate, as sketched in Fig. 1. The viscoelastic barrier submerged in water of finite depth h is assumed to be thin, incompressible and have small displacement, and it extends from a depth d (d < h) to the mean water level (y = 0) with the hinged and tensioned ends. The water is assumed to be inviscid, incompressible, irrotational and have time-harmonic motion.

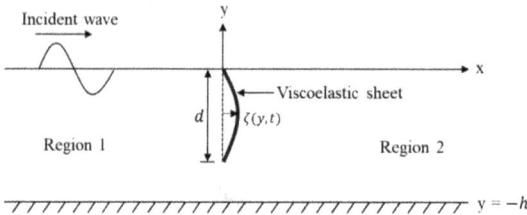

Fig. 1. Schematic diagram of a vertical submerged viscoelastic barrier.

The analysis is nondimensionalized with the water depth h being the length scale and $\sqrt{g/h}$ being the time scale. The nondimensional variables denoted by overbars can be defined as

$$\bar{x} = \frac{x}{h}, \bar{y} = \frac{y}{h}, L = \frac{d}{h}, \bar{\zeta} = \frac{\zeta}{h}, \bar{t} = \sqrt{\frac{g}{h}}t, \bar{\omega} = \sqrt{\frac{h}{g}}\omega, \bar{k} = kh, \text{ and } \bar{\Phi} = \frac{\Phi}{h\sqrt{hg}} \quad (1)$$

where ω is angular frequency and k is angular wavenumber. Due to the time-harmonic assumption, the velocity potential $\overline{\Phi}_l$ ($l = 1, 2$) and the barrier displacement $\bar{\zeta}$ are expressed in the form of $\overline{\Phi}_l(\bar{x}, \bar{y}, \bar{t}) = Re\{\overline{\phi}_l(\bar{x}, \bar{y})e^{-i\bar{\omega}\bar{t}}\}$ and $\bar{\zeta}(\bar{y}, \bar{t}) = Re\{\bar{\xi}(\bar{y})e^{-i\bar{\omega}\bar{t}}\}$, where $\overline{\phi}_l$ is the spatial velocity potential and $\bar{\xi}(\bar{y})$ is the complex displacement amplitude of the barrier. In the following analysis, all the variables are nondimensionalized with the overbar dropped.

Based on the Linear Wave Theory, the velocity potentials in the fluid regions R_1 and R_2 satisfy the Laplace equation and the following linearized free surface, bottom and far field conditions:

$$\nabla^2 \phi_l = 0, l = 1, 2 \quad (2)$$

$$\frac{\partial \phi_l}{\partial y} - \omega^2 \phi = 0, \text{ at } y = 0 \quad (3)$$

$$\frac{\partial \phi_l}{\partial y} = 0, \text{ at } y = -1 \quad (4)$$

$$\phi(x, y) = \begin{cases} \phi^I(x, y) + R_r \phi^I(-x, y), & x \to -\infty \\ T_r \phi^I(x, y), & x \to +\infty \end{cases} \quad (5)$$

In addition, the kinematic and dynamic boundary conditions at the barrier are:

$$\frac{\partial \phi_1}{\partial x} = \frac{\partial \phi_2}{\partial x} = -i\omega\xi, \text{ at } x = 0, -L < y \leq 0 \quad (6)$$

$$D_1 \frac{\partial^4 \xi}{\partial y^4} - T_p \frac{\partial^2 \xi}{\partial y^2} - m_1 \xi = i\omega(\phi_1 - \phi_2), \text{ at } x = 0, -L < y < 0 \quad (7)$$

where $D_1 = \frac{E_1 h_p^3}{6h^4 \rho g} - i\bar{\omega} \frac{\eta_1 h_p^3}{6h^4 \rho g}\left(\frac{g}{h}\right)^{\frac{1}{2}} = Q_1 - i\omega Q_2$ is the nondimensional complex bending stiffness for the viscoelastic barrier, $T_p = \frac{T}{\rho g h^2}$ is the nondimensional tension and $m_1 = \omega^2 m$ is nondimensional mass coefficient with $m = \frac{\rho_p h_p}{\rho h}$ being barrier mass per unit area. Q_1 represents the elastic component of the barrier material resisting the wave attack by combating the bending deformation while Q_2 denotes the viscous component dissipating mechanical energy into heat [12].

The hinged boundary conditions of the barrier gives:

$$\xi = 0, \frac{d^2\xi}{dy^2} = 0 \text{ at } (0, -L) \text{ and } (0, 0) \quad (8)$$

It is noted that at barrier edges, the gradient of the velocity potential has a square-root singularity [13] governed by the relation of $\nabla \phi = O(r^{-1/2})$ as $r \to 0$, where r denotes the distance of an arbitrary point from the edge of the barrier. At the gap below the barrier, the continuity of water pressure and velocity gives:

$$\frac{\partial \phi_1}{\partial x} = \frac{\partial \phi_2}{\partial x}, \phi_1 = \phi_2 \text{ at } x = 0, -1 \leq y < -L \quad (9)$$

Finally, the velocity potential of the incident wave $\phi^I(x, y)$ with wavenumber k_0, wave frequency ω and wave height H can be expressed as $\phi^I(x, y) = a \cosh(k_0(1+y))e^{ik_0 x}$.

3 Analytical Solutions

Using the eigenfunction expansion method, the velocity potentials satisfying Eqs. (2), (3), (4) and (5) in each of the two regions are given by

$$\phi_1 = aY_0(y)e^{ik_0 x} + R_0 Y_0(y)e^{-ik_0 x} + \sum_{n=1}^{\infty} R_n Y_n(y) e^{k_n x} \quad (10)$$

$$\phi_2 = T_0 Y_0(y)e^{ik_0 x} + \sum_{n=1}^{\infty} T_n Y_n(y) e^{-k_n x} \quad (11)$$

where R_n, T_n ($n = 0, 1, 2...$) are unknowns to be determined. The vertical eigenfunctions are given by:

$$Y_n(y) = \begin{cases} \cosh(k_0(1+y)), n = 0 \\ \cos(k_n(1+y)), n \geq 1 \end{cases} \quad (12)$$

where k_0 and k_n satisfy the dispersion relation $\omega^2 = k_0 \tanh(k_0)$ and $\omega^2 = -k_n \tan(k_n)$, respectively. Substituting Eqs. (10) and (11) into the continuity of horizontal velocity across the barrier boundary as well as the gap yields

$$T_0 + R_0 = a, T_n = -R_n \ (n = 1, 2, \ldots) \quad (13)$$

The barrier displacement is expanded with eigenfunctions as

$$\xi(y) = \sum_{m=1}^{\infty} A_m \sin(\lambda_m (L+y)) \quad (14)$$

where $\lambda_m = m\pi/L$. Substituting Eqs. (10), (11) and (14) into Eq. (7) yields

$$A_m = \frac{i4\omega}{P_m L} \sum_{n=0}^{\infty} R_n I_{nm} \qquad (15)$$

where $P_m = D_1 \lambda_m^4 + T_p \lambda^2 - m_1$. The integral I_{nm} is given by

$$I_{nm} = \begin{cases} \int_{-L}^{0} \cosh(k_0(1+y)) \sin(\lambda_m(L+y)) dy, & n = 0 \\ \int_{-L}^{0} \cos(k_n(1+y)) \sin(\lambda_m(L+y)) dy, & n = 1, 2, \ldots \end{cases} \qquad (16)$$

Substituting Eqs. (15), (14), (11) and (10) into Eqs. (6) and (9) yields the mixed boundary conditions:

$$G(y) = 0 = \begin{cases} (a - R_0) i k_0 Y_0 + \sum_{n=1}^{\infty} R_n k_n Y_n - \frac{4\omega^2}{L} \sum_{n=0}^{N} R_n U_n, & -L < y < 0 \\ \sum_{n=0}^{\infty} R_n Y_n, & -1 < y < -L \end{cases} \qquad (17)$$

where $U_n = \sum_{m=1}^{\infty} \frac{I_{nm}}{P_m} \sin(\lambda_m(L+y))$. In addition, the singularity of the water velocity exists at $(0, -L)$. We adopt the least square approximation method which suggests to minimize $\int_{-1}^{0} |G(y)|^2 dy$ with respect to the R_i ($i = 0, 1, 2, \ldots$) and it yields

$$\int_{-1}^{0} G^*(y) \frac{\partial G(y)}{\partial R_i} dy = 0, \ (i = 0, 1, 2, \ldots) \qquad (18)$$

where * denotes the complex conjugate. Substituting Eqs. (17) into (18) gives:

$$\begin{aligned} i = 0, 0 = & \sum_{n=0}^{\infty} R_n^* Z_{0n} - k_0^2 (a^* - R_0^*) J_{00} - i k_0 \sum_{n=1}^{\infty} R_n^* k_n J_{0n} \\ & + \frac{4i\omega^2 k_0}{L} (a^* - R_0^*) \Lambda_{00} + \\ & \frac{4i\omega^2 k_0}{L} \sum_{n=0}^{\infty} R_n^* \Lambda_{n0}^* - \frac{4\omega^2}{L} \sum_{n=1}^{\infty} R_n^* k_n \Lambda_{n0} \\ & + \frac{8\omega^4}{L} \sum_{n=1}^{\infty} R_n^* C_{n0} \end{aligned} \qquad (19)$$

$$\begin{aligned} i = 1, 2 \ldots, 0 = & \sum_{n=0}^{\infty} R_n^* Z_{in} - i k_0 k_i (a^* - R_0^*) J_{i0} \\ & k_i \sum_{n=1}^{\infty} R_n^* k_n J_{in} - \frac{4\omega^2 k_i}{L} \sum_{n=0}^{\infty} R_n^* \Lambda_{ni}^* + \\ & \frac{4i\omega^2 k_0}{L} (a^* - R_0^*) \Lambda_{0i} - \frac{4\omega^2}{L} \sum_{n=1}^{\infty} R_n^* k_n \Lambda_{ni} \\ & + \frac{8\omega^4}{L} \sum_{n=1}^{\infty} R_n^* C_{ni} \end{aligned} \qquad (20)$$

with $Z_{in} = \int_{-1}^{-L} Y_i(y) Y_n(y) dy$, $J_{in} = \int_{-L}^{0} Y_i(y) Y_n(y) dy$, $\Lambda_{ni} = \sum_{m=1}^{\infty} \frac{I_{nm} I_{im}}{P_m}$, and $C_{ni} = \sum_{m=1}^{\infty} \frac{I_{nm} I_{im}}{|P_m|^2}$. By keeping a finite number of M terms for the barrier modes and N terms for the wave evanescent modes, Eqs. (19) and (20) yield $(N + 1)$ equations with the same number of unknowns R_i^*. Sufficiently accurate results can be obtained by using the truncation number $N = 100$ and $M = 30$. The wave reflection and transmission coefficients are defined as $R_r = \left|\frac{R_0^*}{a^*}\right|$; $T_r = \left|\frac{T_0^*}{a^*}\right|$.

4 Barrier Material Consideration

Three materials, 500 GSM nonwoven geotextile, PA 6 and HDPE, are selected as candidates for the fabrication of barriers in real application. Since they possess viscoelastic properties, the elastic and viscous effects are quantified by measuring the storage modulus G' and loss modulus G'' through dynamic mechanical analyser (DMA) test. The material constants E_1 and η_1 in Eq. (7) are related to G' and G''through the relationship [14]: $G' = E_1$ and $G'' = \omega\eta_1$. The shear mode of DMA was carried out for the 500 GSM geotextile and PA 6 while the tension mode was adopted for HDPE. Frequency sweep was used for both modes with a constant strain amplitude of 0.1% to ensure the linear viscoelastic range. The DMA test results are shown in Fig. 2. In general, the nonwoven geotextile is a soft solid with the smallest rigidity among the three materials, while the HDPE is the most rigid solid, and the rigidity of PA 6 sits in between.

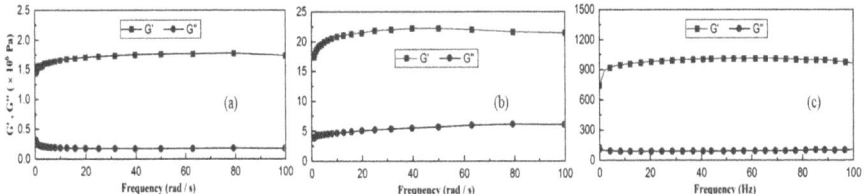

Fig. 2. Storage and loss modulus of (a) nonwoven geotextile, (b) PA 6, and (c) HDPE.

5 Results and Discussion

In this section, the performance of a single barrier on wave attenuation is investigated for different materials, flexural rigidity and tension. The measured storage and loss modulus are used to quantify the elasticity and viscosity of the three materials. The barrier length is fixed at 40% of water depth to balance the water circulation and wave attenuation. The plotted range of nondimensional wave number k0h covers the wave periods from 1.65–3 s for the water depth of 10 m. The barrier thickness ranges from 5 cm to 15 cm. The tension per unit length applied on barriers is selected as 5, 10, and 15 kN/m which can be achieved by hanging cylindrical recyclable concrete at the bottom edge of barriers.

5.1 Nonwoven Geotextile

Figure 3 shows the transmission coefficient Tr for nonwoven geotextile barrier against wavenumber for different barrier thickness. Both the viscoelastic case and the purely elastic case are examined. It can be observed that lesser wave energy can transmit through the barrier for thicker nonwoven geotextile barriers due to the increase in flexural rigidity. However, the effects of barrier thickness are not as significant as the tensioning effect due to the soft nature of the nonwoven geotextile. In other words, to achieve satisfactory wave attenuation, it is more effective to increase the tension exerted on nonwoven geotextile barriers rather than thicken them. Moreover, to mitigate the long waves, heavier clump weights need to be adopted to increase tension, which may lead to high cost due to more concrete consumption. In addition, it shows that the viscous effect of the nonwoven geotextile barrier is highly dependent on the barrier thickness as well as exerted tension. As the barrier becomes thicker, lesser wave energy transmission can be observed, especially for the cases of small tension of 5 kN/m, due to more energy dissipation. However, the viscous effect becomes less significant as tension increases, and the difference between the viscoelastic and elastic case is almost negligible when the tension is 15 kN/m. This could be because the barrier becomes more rigid as tension increases and the resulted small barrier motion leads to less internal energy dissipation.

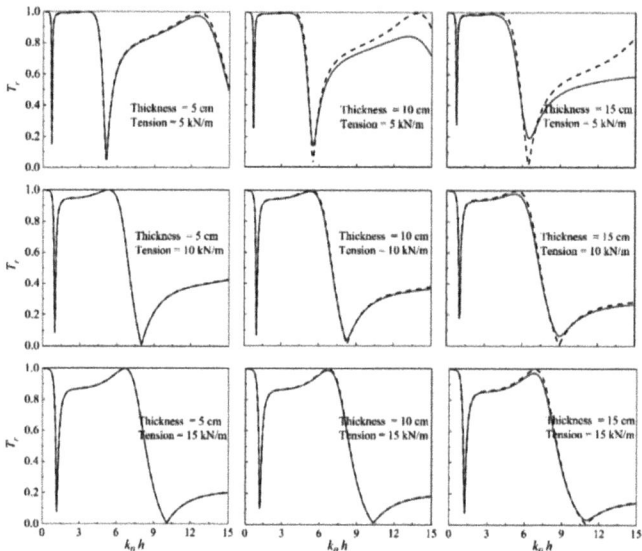

Fig. 3. Tr of the nonwoven geotextile barrier for different thickness and applied tension with a length of 40% of water depth. Viscoelastic case (solid line) and the purely elastic case (dashed line) are presented.

5.2 PA 6

Figure 4 shows the results of transmission coefficient Tr for PA 6 (nylon) barrier. It can be observed that when the nylon barrier has a relatively small thickness, e.g. 5 cm, Tr decreases significantly as tension increases. However, Tr is less sensitive to tension as the barrier thickness increases, exhibiting different patterns from nonwoven geotextile barrier. This is because the PA 6 has a more solid nature compared to nonwoven geotextile, and thus the tension is no longer the only mechanism to control the wave attenuation. In other words, the tension effect dominates the waver resistance when the thin barrier has relatively small flexural rigidity, while the dominance shifts from tension to flexural rigidity as the barrier thickness increases. Moreover, more reduction in wave transmission can be accomplished by increasing the barrier thickness from 5 to 15 cm compared to increasing tension from 5 to 15 kN/m, which is contrary to the trend for the geotextile case. Furthermore, comparing the viscoelastic with elastic cases, it is observed that more Tr reduction resulted from material viscosity occurs for thicker barrier, showing the similar pattern as observed for geotextile barriers. However, the overall viscous effects on Tr for PA 6 barrier is more obvious than that for geotextile due to its larger loss modulus. For the incident waves with longer periods, the viscous effect on dissipating incident energy is more significant than short periods, especially when the barrier is as thick as 15 cm.

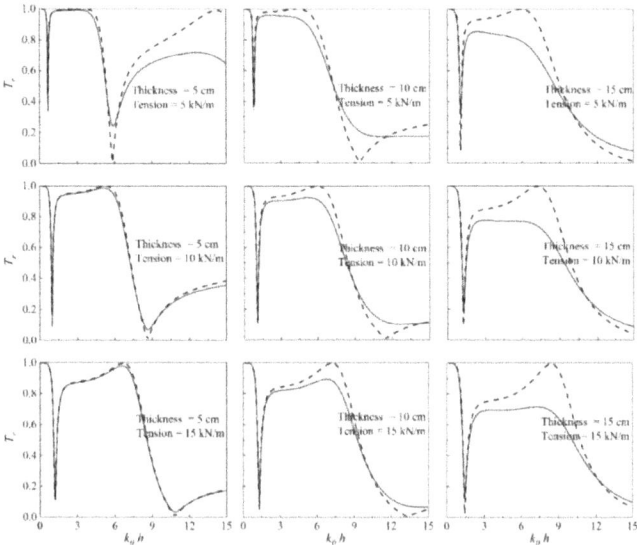

Fig. 4. Tr of the PA 6 barrier for different thickness and applied tension with a length of 40% of water depth. Viscoelastic case (solid line) and the purely elastic case (dashed line) are presented.

5.3 HDPE

Figure 5 shows the results of transmission coefficient Tr for HDPE barrier. Since the HDPE possesses the largest rigidity compared to the other two materials, increasing the barrier thickness from 5 cm to 15 cm can reduce wave transmission in the most effective way. For example, for the cases of 15 cm thickness, less than 20% of the incident wave energy with the wave period ranging from short ($k_0 h = 15$) to long ($k_0 h = 2$) can be transmitted through the HDPE barrier. Moreover, for the thick HDPE barrier, the wave resistance is dominated by the flexural rigidity and thus the tensioning effect on the Tr reduction is almost negligible. However, as the thickness becomes smaller, the tensioning effect starts to play a role in decreasing Tr, especially for the longer waves. Nevertheless, the overall wave attenuation is achieved through the flexural rigidity of HDPE barrier, which shows a different wave resistance mechanism from nonwoven geotextile barrier. Therefore, thickening the HDPE barrier is more effective than increasing tension in terms of wave protection. In addition, when the HDPE barrier is less rigid with a small thickness, the viscous effect on reducing Tr is more obvious than the other two cases due to its larger loss modulus. As the thickness of HDPE barrier increases from 5 to 10 cm, the effectiveness of material viscosity on Tr reduction increases significantly, especially at small wavenumbers where Tr peak dampening occurs. However, when the HDPE barrier is thick (15 cm), the viscous effect is almost negligible on wave attenuation when comparing with the elastic cases due to the increasing rigidity of the barrier, although the wave transmission is already reasonably small.

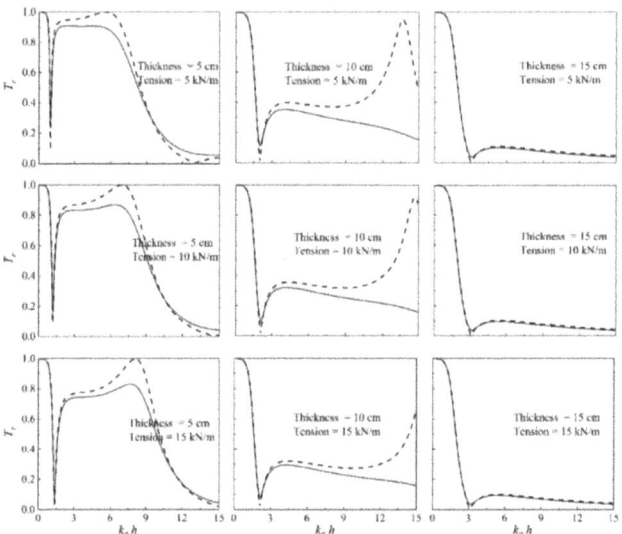

Fig. 5. Tr of the HDPE barrier for different thickness and applied tension with a length of 40% of water depth. Viscoelastic case (solid line) and the purely elastic case (dashed line) are presented.

5.4 Cost Analysis

An illustrative case study delves into the cost analysis of a comprehensive wave protection system, encompassing barriers, clump weights, modular pontoons, mooring lines, and anchors. Employing recyclable concrete, the clump weights serve to apply tension to the barrier, while modular pontoons provides buoyancy for the system. The assessment includes barriers constructed from the three previously mentioned materials, combined with the incorporation of other essential components, all with the goal of identifying the most economically feasible solution. A diverse range of pricing sources has been gathered for the analysis. The wave condition includes a water depth of 10 m, wave period of 3 s, and a wave height of 0.7 m ($k_0h = \sim 4.5$). The performance of all the barriers is kept the same by fixing the transmission coefficient at 0.5. Table 1 outlines the cost of several components used for the barrier system. It should be noted that this is a rough estimation of component cost proposed by different vendors, and the price might be even lower when ordering large amounts of each component.

For the nonwoven geotextile barrier, the clump weight used to apply tension needs to be 3500 kg/m to achieve Tr = 0.5 when the barrier has a draft of 4 m. The area per unit length of 1000 GSM nonwoven geotextile is 4 m and the cost is as low as $ 7/m. ~ 45 floating modular pontoons are needed to provide enough buoyancy, resulting in a cost

Table 1. Summary of costs of barriers made of different materials.

Barrier material	Estimated material cost (US $ / m)
1000 GSM nonwoven geotextile	957
PA 6 (Nylon)	2770
HDPE	1148

of $ 675/m. Moreover, assuming the concrete density of 2400 kg/m3, the volume of the clump weight per unit length is ~ 1.5 m2 with a cost of $ 90/m. Assuming a mooring rope of 14.7 m and an anchor of 1.5 ton per unit length, the costs of them reaches $ 185/m. Therefore, the total cost of the current design of barriers made of nonwoven geotextile is US $ 957/m. In addition, for the PA 6 barrier with a draft of 4 m, the clump weight needs to be 2500 kg/m. The PA 6 sheet with a 4 m area per unit length as well as a 10 cm thickness has a cost of $ 2000/m. ~ 35 modular pontoons are needed to provide enough buoyancy with a cost of $ 525/m. The clump weight has a volume of ~ 1 m2 per unit length with a cost of $ 60/m. Assuming that the same mooring lines and anchors as the geotextile case are used with the cost of $ 185/m, the total cost of the design of the barrier made of PA 6 is US $ 2770/m. Finally, for the HDPE barrier, only 500 kg/m clump weight is needed to keep the barrier in position. 10 cm thick HDPE sheet with an area of per unit length of 4 m costs ~ $ 800/m. The small clump weight costs $ 13/m, and ~ 10 modular pontoons with a cost of $ 150/m are needed to provide buoyancy. With the same cost of mooring lines and anchors as the case of geotextile barriers, the total cost of HDPE barriers is US $ 1148/m. It should be noted that these cost estimations are based on material costs. Other related costs such as transportation, installation and maintenance are not considered here. The costs of the barrier made of different materials are summarized in Table 1.

6 Conclusion

In the present study, the surface wave interactions with a thin vertical viscoelastic barrier are examined analytically using the eigenfunction expansion method. Numerical results are obtained for transmission coefficients to investigate the effects of tension, flexural rigidity and viscoelasticity on wave attenuation and determine the optimal material for barriers. The results show that the tension effect dominates the wave transmission for the nonwoven geotextile, whereas for PA 6 and HDPE barriers, the barrier thickness (or flexural rigidity) takes precedence. Additionally, the impact of the barrier material viscosity on wave energy dissipation increases with increasing thickness, while it diminishes with heightened barrier rigidity. Comparatively, the cost analysis unveils that a thick HDPE barrier bears a cost akin to that of a nonwoven geotextile, however, it necessitates a lesser amount of concrete for clump weights. On the contrary, the PA 6 barrier proves to be the most expensive due to the elevated material costs. Overall, the HDPE barrier with a draft of 40% of water depth and a 15 cm thickness emerges as the optimal performer in wave attenuation when compared with the other two materials. The choice of a 500 kg/m clump weight effectively maintains the vertical position of the barrier, given that tension does not dominate in wave protection. This configuration gains significance when stringent wave protection criteria are imperative, with an allowable wave transmission of less than 20%.

Acknowledgement. This research was funded by the Science and Technology Program of China Huaneng Group Co., Ltd. (HNKJ23-H21), China Huaneng Group Science and Technology Project (No. HNKJ20-H54) and Science and Technology Program of China Huaneng Clean Energy Research Institute (QNYJJ22-12).

References

1. Mahmood-Ul-Hassan, M., Meylan, M.H., Peter, M.A.: Water-wave scattering by submerged elastic plates. Q. J. Mech. Appl. Math. **62**, 321–344 (2009)
2. Behera, H., Sahoo, T.: Hydroelastic analysis of gravity wave interaction with submerged horizontal flexible porous plate. J. Fluids Struct. **54**, 643–660 (2015)
3. Chakraborty, R., Mondal, A., Gayen, R.: Interaction of surface water waves with a vertical elastic plate: A hypersingular integral equation approach. Z. angew. Math. Phys. **67**, 115 (2016)
4. Kundu, S., Gayen, R., Datta, R.: Scattering of water waves by an inclined elastic plate in deep water. Ocean Eng. **167**, 221–228 (2018)
5. Lo, E.Y.M.: Performance of a flexible membrane wave barrier of a finite vertical extent. Coast. Eng. J. **42**, 237–251 (2000)
6. Lee, W.K., Lo, E.Y.M.: Surface-penetrating flexible membrane wave barriers of finite draft. Ocean Eng. **29**, 1781–1804 (2002)
7. Karmakar, D., Bhattacharjee, J., Guedes Soares, C.: Scattering of gravity waves by multiple surface-piercing floating membrane. Appl. Ocean Res. **39**, 40–52 (2013)
8. Koley, S., Sahoo, T.: Scattering of oblique waves by permeable vertical flexible membrane wave barriers. Appl. Ocean Res. **62**, 156–168 (2017)
9. Dembicki, E., Niespodzińska, L.: Geotextiles in coastal engineering practice. Geotext. Geomembr. **10**, 147–159 (1991)
10. de Leon, A.C.C., da Silva, Í.G.M., Pangilinan, K.D., Chen, Q., Caldona, E.B., Advincula, R.C.: High performance polymers for oil and gas applications. React. Funct. Polym. **162**, 104878 (2021)
11. Heibaum, M.: Geosynthetics for waterways and flood protection structures – Controlling the interaction of water and soil. Geotext. Geomembr. **42**, 374–393 (2014)
12. Bi, C., Wu, M.S., Law, A.W.-K.: Surface wave interaction with a vertical viscoelastic barrier. Appl. Ocean Res. **120**, 103073 (2022)
13. Deng, Z., Huang, Z., Law, A.W.K.: Wave power extraction from a bottom-mounted oscillating water column converter with a V-shaped channel. Proc. R. Soc. A Math. Phys. Eng. Sci. **470**, 20140074–20140074 (2014)
14. Sree, D.K.K., Law, A.W.K., Shen, H.H.: An experimental study on the interactions between surface waves and floating viscoelastic covers. Wave Motion. **70**, 195–208 (2017)

Open Access This chapter is licensed under the terms of the Creative Commons Attribution-NonCommercial-NoDerivatives 4.0 International License (http://creativecommons.org/licenses/by-nc-nd/4.0/), which permits any noncommercial use, sharing, distribution and reproduction in any medium or format, as long as you give appropriate credit to the original author(s) and the source, provide a link to the Creative Commons license and indicate if you modified the licensed material. You do not have permission under this license to share adapted material derived from this chapter or parts of it.

The images or other third party material in this chapter are included in the chapter's Creative Commons license, unless indicated otherwise in a credit line to the material. If material is not included in the chapter's Creative Commons license and your intended use is not permitted by statutory regulation or exceeds the permitted use, you will need to obtain permission directly from the copyright holder.

Design and Optimization for the Base of Gantry Used in Wreck Salvage

Shude Chen, Wei Deng, Yuantao Bi, Wei Zhang$^{(\boxtimes)}$, Xinxin Wang, and Yijian Han

Technology Center of Yantai Salvage, Yantai 264012, China
donsh328@163.com

Abstract. The engineering of salvaging a wreck requires a variety of auxiliary equipment. It often affects the overall construction efficiency because of the cumbersome installation and disassembly of auxiliary fittings and placement problems. In order to solve the problems, based on the size and working condition of a 120 t gantry lifting device used for wreck salvage, this paper designs and transforms the base of gantry. In this paper, Abaqus was used to check the strength of the redesigned base. And we conducted dock trials. The result shows that the base can meet the strength requirements of 120 t gantry lifting device. In addition, it can be quickly installed and disassembled. This device can be matched with other auxiliary equipment used for salvage. It provides reference for the quick installation and disassembly of other equipment used for salvage. In order to improve the construction efficiency.

Keywords: Abaqus · The Base of Gantry · Design and Optimization

1 Introduction

Salvaging wreck [1, 2] is a huge and complicated project. All kinds of auxiliary equipment are used in the project. The storage and use of auxiliary equipment depends on the deck of working ship. Sometimes equipment needs to be secured to the deck of a transport vessel in order to be fully effective. However, the fixed equipment will not only damage the deck of the working ship, but also delay the development of other processes because the equipment is improperly fixed on the deck. And it will even affect the construction efficiency of the whole salvage project.

The gantry is an important part of the salvage engineering to improve the salvage efficiency. It can quickly lower the underwater boring [3, 4] from the deck of the work ship to the water. However, in the previous construction, the gantry was often welded directly to the deck. In this paper, according to the parameters of 120 t door lifting device (hereafter referred to gantry), a base of gantry is cleverly designed and optimized. The base can be fixed to the deck and the gantry can be quickly bolted to the base. After the construction is finished, the gantry can be removed by removing the bolts, and the base can be left on the working ship. In the next salvage operation, the gantry can be quickly placed at the base, which not only protects the deck of working ship, but also improves the salvage efficiency [5]. In addition, the base can be matched with other auxiliary fittings and can be used for rapid installation of other auxiliary fittings.

2 Engineering Background

2.1 The Operating Conditions of Gantry

During the working of the gantry lifting device, when the boom works to 35° inside the ship and 35° outside the ship, it is two limit working conditions. The loading condition of the base under these two working conditions and the installation diagram of the gantry are shown in Figs. 1 and 2.

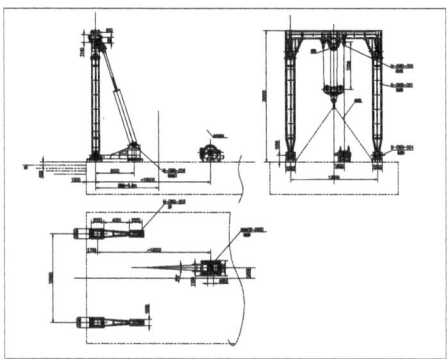

Fig. 1. Installation diagram of 120 T gantry.

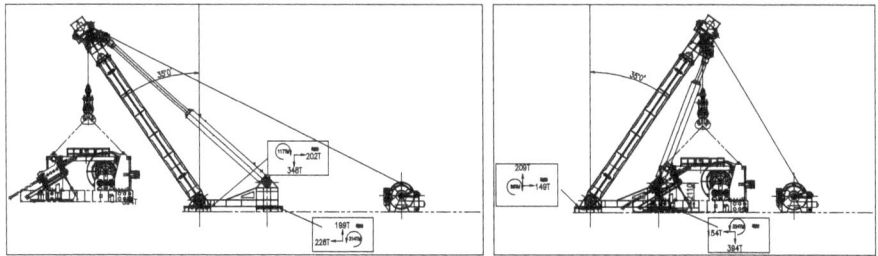

Fig. 2. The 120 T gantry works at 35° outside & 35 inside the ship.

2.2 The Base of Gantry

In this paper, based on the existing base, it is redesigned and reformed to meet the above 120 T gantry operating conditions. The existing base material is Q235 and its structure diagram is shown in Fig. 3.

In the figure, both transverse and longitudinal bases are H-shaped steel. Among which the transverse H-shaped steel specifications are 585*300*12*18 mm, and the longitudinal H-shaped steel specifications are 350*350*12*18 mm.

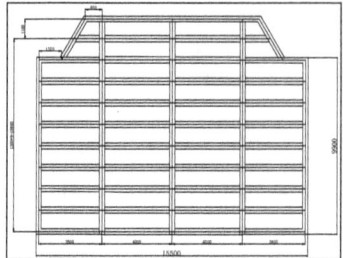

Fig. 3. Structure drawing of existing base.

3 Design Optimization of the Base and Strength Check

3.1 Design Optimization of the Base

Since the size of the existing base cannot meet the span of the 120 t gantry, it is necessary to cut the existing base along the transverse centre position. The separated bases are placed under the pile legs on both sides of the 120 t gantry to meet the span requirements of the 120 t gantry. At the same time, replace the panel at the contact position between the existing base and the 120 t frame with a panel made of Q355 with a thickness of 40 mm. The strength of the base is strengthened by welding the elbow plate of Q355 material at the bottom of the added panel and increasing the thickness of the base web plate. The panel is provided with a different number of bolt holes, and the 120 t door frame is fixed on the base through bolts. Weld 25*600*200 pads under each base. In the actual salvage operation, the base and the deck of the working ship can be fixed by the pad. The structure diagram of the optimized base is shown in Fig. 4.

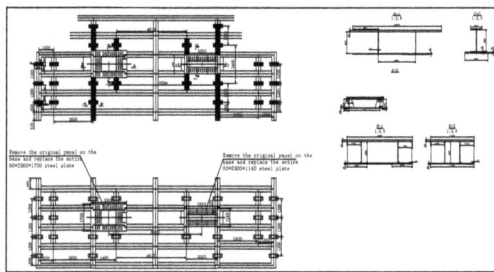

Fig. 4. The structure diagram of the optimized base.

3.2 The Model of the Base and Strength Check

Safety Factor and Allowable Stress. According to the CCS Classification Code for Offshore Mobile Platforms (2023), the **yield strength** of plates is specified as follows:

Sheet press down for yield check:

$$\sigma_{eq} \leq [\sigma] \tag{1}$$

$$[\sigma] = \frac{\sigma_s}{S} \tag{2}$$

Else, σ_{eq} represents the equivalent stress, $\sigma_{eq} = \sqrt{\sigma_x^2 + \sigma_y^2 - \sigma_x\sigma_y + 3\tau_{xy}^2}$, N/mm^2, the midplane stress value (film stress) at the center of the plate element is taken into account;.
σ_x is the stress in the unit x direction, N/mm^2;.
σ_y is the stress in the unit y direction, N/mm^2;.
τ_{xy} is the shear stress of the unit xy plane, N/mm^2;.
σ_s is the yield strength of the material, N/mm^2;.
S is the safety factor, and its values are shown in Table 1.

Table 1. Equivalent stress safety factor.

Static load condition	Combined condition
1.43	1.11

The optimized base material in this paper includes Q235 and Q355. And the yield strength of the two materials is 235 MPa and 355 MPa. The calculation in this paper belongs to the static load condition, so the allowable stress of Q235 steel is $[\sigma_1] = 235/1.43 = 164.3 MPa$. And the allowable stress of Q355 steel is $[\sigma_2] = 355/1.43 = 248.3 MPa$.

Model of The Base. Since the structure and force of the four positions under the gantry are basically the same. We selected one side of the gantry leg for analysis when used Abaqus [6–8] modeling analysis. And model the left and right sides of the base separately. When calculating, the strength of the left and right bases under two working conditions can be calculated respectively. Taking the left base as an example, its Abaqus model is shown in Fig. 5.

In the figure, "RP-1" is the coupling point of the bolt hole used, where the vertical force is applied. "RP-2" is the coupling point between the base and the contact surface of the frame, where the lateral force is applied. "RP-3" is the base counterweight coupling point where the counterweight weight is applied. Lay a backing plate on the base to support the crane equipment. And the corresponding position of the lower part of the base is also paved with plates to strengthen the strength of the base.

The model of the right base is shown in Fig. 6.

Plate Thickness and Reinforcement of The Base. The material and thickness of each part of the reformed base are shown in Figs. 7 and 8. and Tables 2 and 3. And the strengthening of the base is shown in Fig. 9.

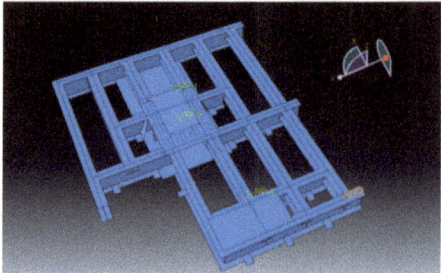

Fig. 5. Model drawing of the base on the left.

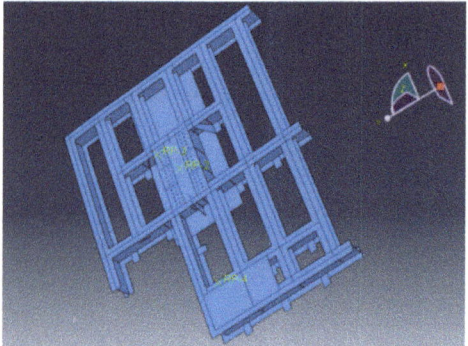

Fig. 6. Model drawing of the base on the right.

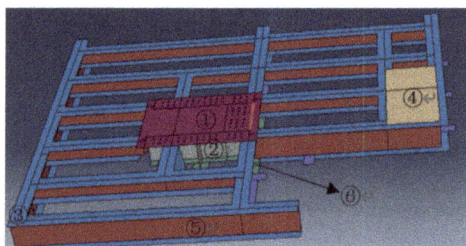

Fig. 7. Plate thickness diagram of the left base.

Stress Condition. Combined with the working condition of 120 t gantry, the magnification factor of 1.2 times is taken. The force of the base under different working conditions is shown in Table 4.

Boundary Conditions. Since the base is calculated separately, symmetrical constraints are applied to the center line of the base, and fully fixed constraints are applied to the base plate. As shown in Fig. 10.

Calculation Result. In this paper, Abaqus is used to check the local strength of the base. During the lifting operation, the gantry is located at 35° outside the ship and 35°

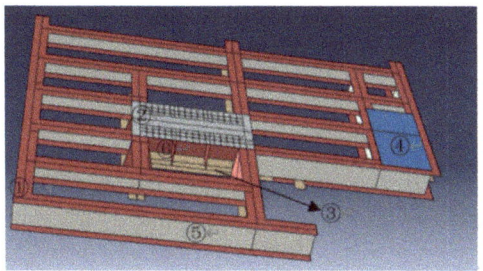

Fig. 8. Plate thickness diagram of the right base.

Table 2. Plate thickness and material of the left base.

Number	Color	Material	Thickness /mm
1	wine red	Q355	40
2	silver	Q355	18
3	blue	Q235	18
4	yellow	Q235	20
5	red	Q235	12
6	green	Q355	43

Table 3. Plate thickness and material of the right base.

Number	Color	Material	Thickness /mm
1	Red	Q235	18
2	Silver	Q355	40
3	Orange	Q355	43
4	Dark blue	Q235	20
5	Gray	Q235	12
6	Wine red	Q355	18

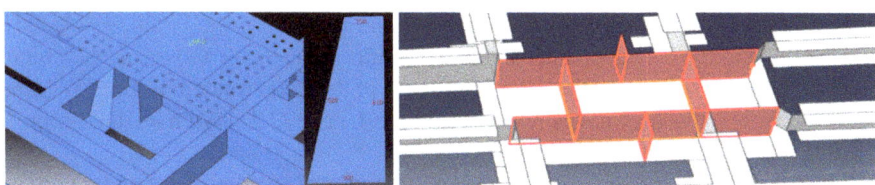

Fig. 9. Reinforced drawing of the base.

Table 4. Stress condition.

Location		The direction of the force (Right and top are positive/N)
35° outside the ship	left	Horizontal force 2.02×10^6 Vertical force -3.48×10^6 Counter weight $- 5 \times 10^5$
	right	Horizontal force -2.28×10^6 Vertical force 1.99×10^6 Counter weight $- 5 \times 10^5$
35° inside the ship	left	Horizontal force 1.49×10^6 Vertical force 2.09×10^6 Counter weight $- 5 \times 10^5$
	right	Horizontal force -1.54×10^6 Vertical force -3.94×10^6 Counter weight $- 5 \times 10^5$

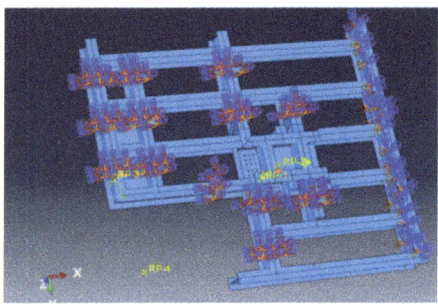

Fig. 10. Boundary conditions of the base.

inside the ship for two limit working conditions. Therefore, only the strength of the base under these two working conditions is checked.

Case 1:35° Outside The Ship

(1) Left base

Under this condition, the base is subjected to a horizontal force of 202 t and a vertical force of 348 t. The calculation results are shown in Fig. 11.

According to the finite element calculation results in the figure above, the maximum deformation is 0.968 mm and the maximum stress of the base structure is 122.271 MPa, which is less than the allowable stress of Q235 steel, i.e. 122.271 MPa < 164.3 Mpa.

(2) Right base

Under this condition, the base is subjected to a horizontal force of 228 t and a vertical force of 199 t. The calculation results are shown in Fig. 12.

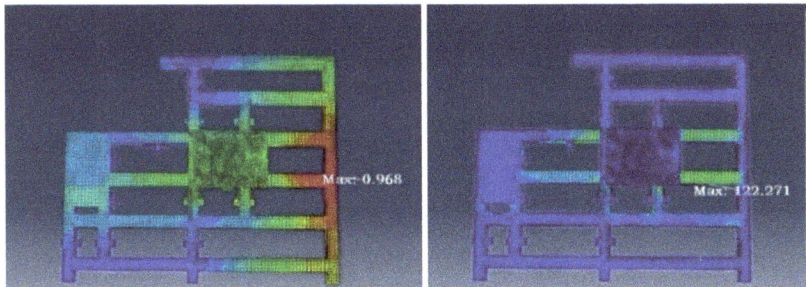

Fig. 11. Deformation(left) and stress(right) results of the left base.

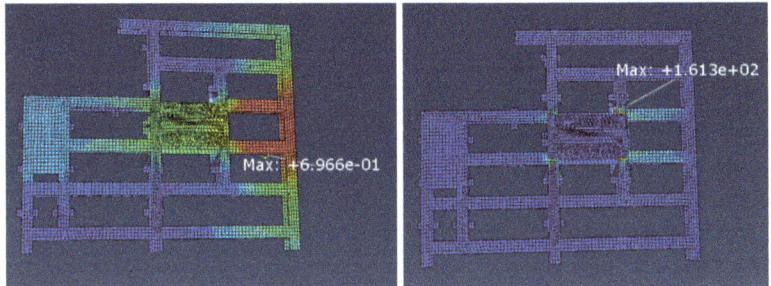

Fig. 12. Deformation(left) and stress(right) results of the right base.

According to the finite element calculation results in the figure above, the maximum deformation is 0.697 mm and the maximum stress of the base structure is 116.3 MPa, which is less than the allowable stress of Q235 steel, i.e. 161.3 MPa < 164.3 Mpa.

Case 2: 35° Inside The Ship.
(1) Left base
Under this condition, the base is subjected to a horizontal force of 149 t and a vertical force of 209 t. The calculation results are shown in Fig. 13.

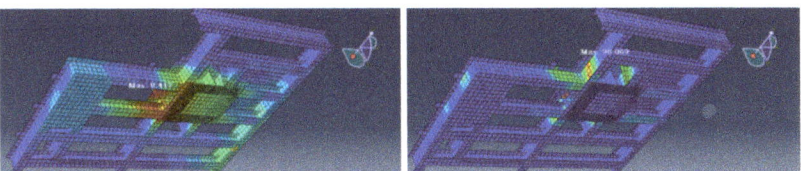

Fig. 13. Deformation(top) and stress(below) results of the left base.

According to the finite element calculation results in the figure above, the maximum deformation is 0.417 mm and the maximum stress of the base structure is 98.01 MPa, which is less than the allowable stress of Q235 steel, i.e. 98.01 MPa < 164.3 Mpa.
(2) Right base

Under this condition, the base is subjected to a horizontal force of 154 t and a vertical force of 394 t. The calculation results are shown in Fig. 14.

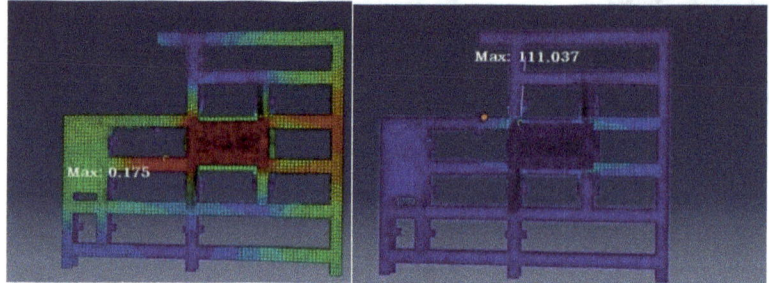

Fig. 14. Deformation(left) and stress(right) results of the right base.

According to the finite element calculation results in the figure above, the maximum deformation is 0.175 mm and the maximum stress of the base structure is 111.03 MPa, which is less than the allowable stress of Q235 steel, i.e. 111.03 MPa < 164.3 Mpa.

In summary, the calculation results of the gantry base under various working conditions are shown in Table 5.

Table 5. The calculation result of the gantry base.

		Case 1	Case 2
Left base	Max stress	122.27 MPa	98.01 MPa
	Max displacement	0.968 mm	0.417 mm
Right base	Max stress	161.3 MPa	111.03 MPa
	Max displacement	0.686 mm	0.175 mm
Allowable stress of Q355 steel		248.3 MPa	243.8 MPa
Allowable stress of Q235 steel		164.3 MPa	164.3 MPa
Meet the requirements or not		Y	Y

Practical Application. After the optimization and transformation of the gantry base according to the above scheme is completed, the reliability test of the base is carried out at the wharf. The field test situation is as shown in Fig. 15.

Through the test, it is found that the optimized design of the base can meet the requirements of the 120 t gantry lifting device in various working conditions. The reference is provided for the next gantry base used in Marine salvage operation.

Fig. 15. The trial at the wharf.

4 Conclusion

In this paper, the original base is redesigned and optimized to realize reuse and saves unnecessary waste. Through finite element software analysis and test, it is found that the strength of the reformed base meets the standard requirements. It meets the requirements of 120 t gantry lifting device in various working conditions, and also has the conditions for Marine salvage. In the actual salvage operation, the optimized design base can be used not only for the 120 t gantry lifting device described in this paper, but also for the installation of other similar equipment on the ship. It facilitates the installation and removal of auxiliary equipment, optimizes deck layout, and improves the efficiency of salvage operations.

References

1. ROGOWSKA, J., WOLSKA, L., NAMIESNIK, J.: Impacts of pollution derived from ship wrecks on the marine environment on the basis of s/s "Stuttgart" (Polish Coast, Europe). Sci. Total Environ. **408**(23), 5775–5783 (2010)
2. Jingjing, L., Baotao, T., Wei, Z.: Hydraulic lifting equipment to assist barge unilateral lifting floating salvage wreck design. Journal. **24**(07), 1–3 (2024)
3. Fei, W.A.N.G., Guo-liang, H.U.A.N.G., De-heng, D.E.N.G.: The design and steady-state simulation of underwater towed system. Journal. **42**(2), 69–684 (2008)
4. Pengfei, X.: Effect of the bending and straightening on mechanical behaviors of coiled tubing in working of underwater mud-penetrator. Sh. Eng. **43**(12), 77–81 (2021)
5. RIPPA, M., et al.: Gridded emissions of air pollutants for the period 1970-2012 within EDGAR v4.3.2. Earth Syst, Sci, Data. **10**(4), 1987–2013 (2018)
6. Congyang, H., Xufeng, L.: Static and Dynamic Analysis of Bearing Pedestal Based on ABAQUS. Journal. **58**(11), 125–128 (2020)
7. Liu, K., Wang, Z., Tang, W., et al.: Experimental and numerical analysis of laterally impacted stiffened plates considering the effect of strain rate. Ocean Eng. **99**, 44–54 (2015)
8. u Weigang, Yang Chun, Cheng Wenming.: The FEM analysis of structure static mechanics of C-type gantry crane. Journal. **43**(02), 5–7 (2016)

Open Access This chapter is licensed under the terms of the Creative Commons Attribution-NonCommercial-NoDerivatives 4.0 International License (http://creativecommons.org/licenses/by-nc-nd/4.0/), which permits any noncommercial use, sharing, distribution and reproduction in any medium or format, as long as you give appropriate credit to the original author(s) and the source, provide a link to the Creative Commons license and indicate if you modified the licensed material. You do not have permission under this license to share adapted material derived from this chapter or parts of it.

The images or other third party material in this chapter are included in the chapter's Creative Commons license, unless indicated otherwise in a credit line to the material. If material is not included in the chapter's Creative Commons license and your intended use is not permitted by statutory regulation or exceeds the permitted use, you will need to obtain permission directly from the copyright holder.

Research on Evaluation Methods for Lighting Layout of Manned Submersibles

Weizhe Xu[1,2](✉), Jun Cao[1,2], Wenxin Qu[1,2], Dan Xing[1,2], and Hongtao Ji[1,2]

[1] State Key Laboratory of Deep-sea Manned Vehicles, Wuxi, China
xu.weizhe@163.com
[2] China Ship Scientific Research Center, Wuxi, China

Abstract. The lighting system is a crucial component of manned submersibles, as it directly influences the operational efficiency and safety of the vehicle in complex underwater environments. This paper presents an innovative evaluation method for the lighting layout of manned submersibles, aiming to enhance their performance and reliability. Firstly, the study reviews current standards and practical application requirements for manned submersibles to establish clear objectives for the lighting design. Subsequently, a self-developed lamp is utilized to collect measurement data, which is then used to match an optical model and design a preliminary lighting layout scheme. To further validate and refine the lighting layout, DIALux software is employed to simulate the illumination of the target plane. Additionally, Lambert-Beer's law and deep-sea water sample measurement data are incorporated to complete the equivalent conversion of illumination in deep-sea environments, ensuring the accuracy and reliability of the evaluation results. Based on the comprehensive analysis of the simulation and measurement data, an optimization strategy for the lighting layout is proposed. This strategy not only improves the lighting efficiency and uniformity but also enhances the safety and comfort of the manned submersible. Overall, this paper provides new insights and methodologies for the design and optimization of lighting systems in underwater equipment, contributing to the advancement of manned submersible technology.

Keywords: Manned Submersibles · Lighting Layout · Evaluation Methods

1 Introduction

With the increasing demand for ocean resource development and scientific exploration by humans, the importance of manned submersibles as deep-sea exploration tools has become increasingly prominent. Manned submersibles can not only carry scientists and engineers deep into the seabed for direct observation and operations, but also obtain rich seabed data through various advanced instruments and equipment, providing important support for research in multiple fields such as marine science, geology, and biology [1]. In deep-sea environments, natural light rapidly decays with increasing water depth due to the absorption and scattering of light by seawater. When it exceeds a certain depth, the seabed will become extremely dim. Therefore, the importance of lighting systems for manned submersibles is self-evident.

Firstly, the lighting system is the foundation for manned submersibles to carry out underwater operations. Only through the light provided by the lighting system can scientists and engineers clearly see the seabed terrain, landforms, organisms, etc., and make detailed observations and records. Secondly, the lighting system is also extremely important for ensuring the safety of submarines. There are many unknown risks in deep sea environments, and lighting systems can help divers detect and avoid danger in a timely manner. In addition, the camera and photography equipment carried by manned submersibles also require lighting systems to provide sufficient light.

In recent years, light environment design has received widespread attention and research in multiple industries. Yongling Gai verified the rationality of the lighting layout in the aerospace testing plant by accurately calculating the illuminance [2]. Meng Li designed a municipal road lighting scheme that meets the standards by comparing and analyzing the illumination results of different schemes [3]. Yue Li established a combined illuminance simulation model to simulate the apron floodlighting environment and optimized the parking spot lighting scheme [4]. Jinchi Fu proposed a lighting simulation method for aircraft cabin lighting design, which quantitatively evaluates factors such as average illuminance, glare, and illuminance uniformity, and designs lighting schemes that meet the evaluation criteria [5]. However, there are currently few specialized literature on illuminance calculation and lighting layout in the field of manned submersibles.

This article uses DIALux software to simulate and calculate a self-developed underwater lamp, and combines the theory of equivalent conversion of illuminance to evaluate the deep-sea lighting effect of a manned submersible, providing reference for the layout of submersible lighting.

2 Evaluation Process for Lighting Layout of Manned Submersibles

The evaluation method for the lighting layout of manned submersibles mainly consists of six parts: determining design goals, simulating lamps, designing lighting layouts, simulating calculations, water sample analysis, and equivalent conversion of illuminance and result analysis. The process is shown in Fig. 1.

1) Determine design goals: Refer to existing lighting evaluation standards to understand the basic requirements of lighting design. Determine the calculation plane and target illuminance for light environment assessment based on the operational requirements of the submersible.
2) Lamp simulation: Collect optical simulation models of the tested lamps for DIALux simulation calculations. For lamps that cannot obtain IES files, test their illuminance distribution and match it with the database lighting model.
3) Lighting layout design: Build a manned submersible model and complete the lighting layout design according to the design objectives.
4) Simulation calculation: Based on the calculation module of DIALux software, obtain the illuminance of each calculation plane.
5) Water sample analysis: Based on the collected water samples from the sea area, the transmittance of the water samples is measured in the laboratory.

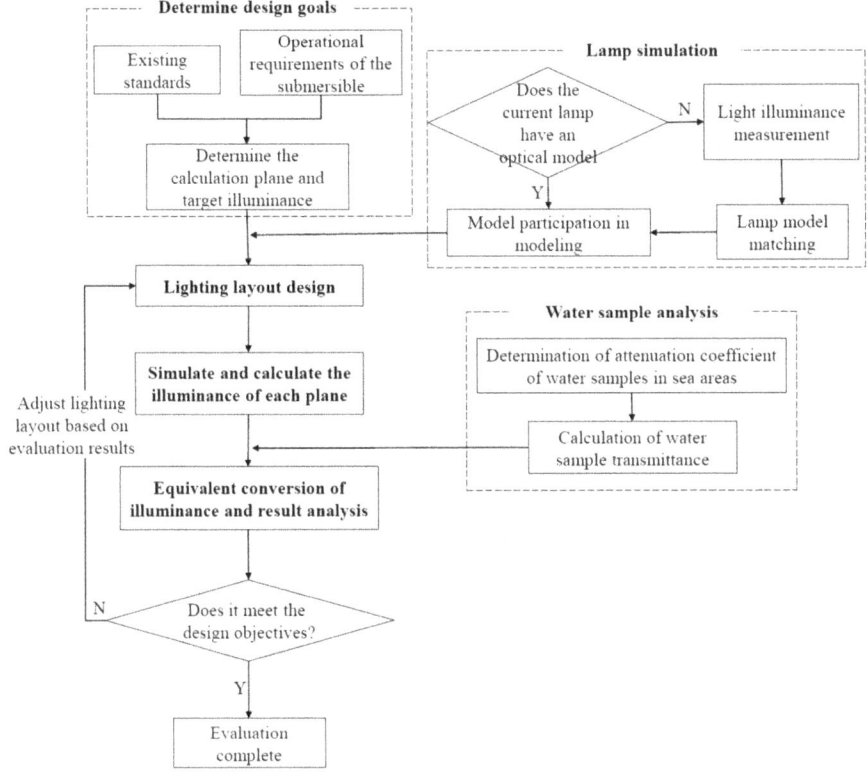

Fig. 1. Evaluation flowchart.

6) Equivalent conversion of illuminance and result analysis: Multiply the calculated illuminance by the corresponding transmittance to obtain the final underwater equivalent illuminance, compare the illuminance with the design objective, and propose optimization suggestions.

3 Establish an Evaluation Model

3.1 Determine Design Goals

According to literature review, there is currently no corresponding standard or specification for underwater lighting of manned submersibles for reference and implementation. According to the relevant provisions of the similar standard GJB 4000–2000 "General Specification for Ships", during the lighting control period, it is still necessary to adhere to the low light illumination of outdoor deck surfaces, semi outdoor deck areas, berth duty areas, and similar places, which should generally be 5-10 lx. According to the relevant provisions of GB 50034–2013 "Design Standard for Building Lighting", the illuminance of horizontal evacuation passages should not be less than 1 lx, and the illuminance of vertical evacuation areas should not be less than 5 lx.

According to feedback on the actual use of manned submersibles, in order to ensure the safety of navigation and observation needs, sufficient illumination should be ensured within a range of 7 m in front, 3 m around, and in front of the seating surface.

Therefore, the goal of lighting environment design is determined to be: the minimum equivalent illuminance of the following calculation planes shall not be less than 5 lx (as shown in Fig. 2):

1) The vertical field of view (7 × 4 m) at a distance of 3 m in front of the submersible;
2) The core vertical field of view (5 × 4 m) at a distance of 7 m in front of the submersible;
3) The horizontal field of view (9.2 × 5 m) at the submersible's bottom plane;
4) The core vertical field of view (5 × 4 m) at a distance of 3 m behind the submersible;
5) The core vertical field of view (5 × 4 m) at a distance of 3 m from the side of the submersible.

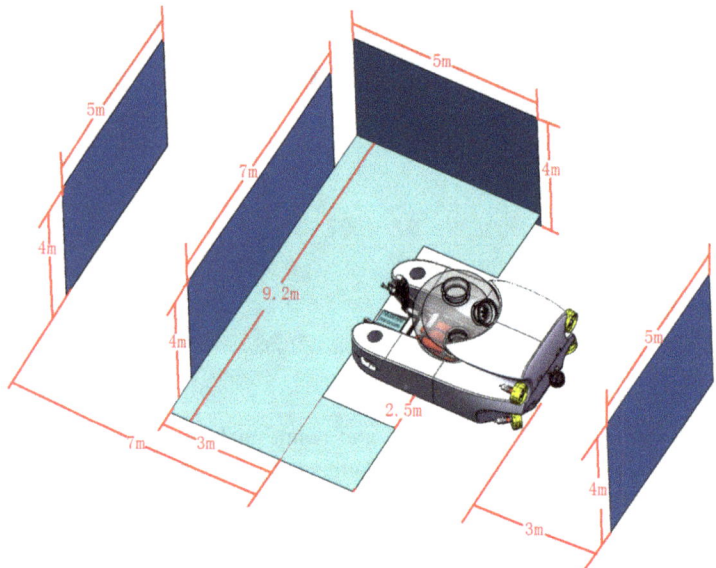

Fig. 2. Calculation planes.

3.2 Lamp Simulation

The underwater lighting developed by the project team is designed for specific underwater environments. In order to obtain its optical parameters, the project team sets up measurement points every 1 m on open land (as shown in Fig. 3) and measures the illumination distribution of the light through an illuminometer (as shown in Table 1).

Build calculation planes in DIALux software that is at the same distance as the measured point, and match a optical model from the lamp database that is basically consistent with the illumination of the self-developed underwater lighting lamp (as shown in Figs. 4 and 5).

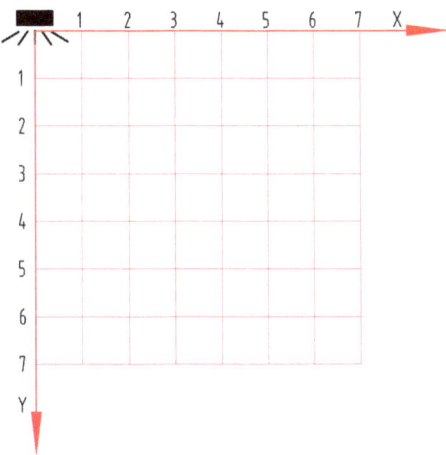

Fig. 3. Schematic diagram of illuminance measurement coordinates.

Table 1. Illumination of each measuring point of the lamp.

Illumination statistics for each measuring point (lx)								
	x(m)							
y(m)	0	1	2	3	4	5	6	7
0		220	21	8	4	3	2	1.6
1	11,900	2950	475	117	39	16	9	5.5
2	2720	1900	710	319	116	57	27	15
3	1346	1143	690	328	164	85	50	28
4	743	678	526	312	167	99	64	44
5	440	425	361	269	164	111	74	54
6	327	311	278	210	153	110	76	55
7	247	234	207	170	125	97	80	60

3.3 Lighting Layout Design

Firstly, the manned submersible model is subjected to lightweight processing, retaining key structural elements that affect the evaluation results, removing unnecessary complex details as much as possible, and improving computational simulation efficiency. After importing the model into DIALux software, a certain number of lamps are arranged in various directions of the submersible according to the illumination goals of each plane, completing the preliminary lighting layout plan (as shown in Fig. 6).

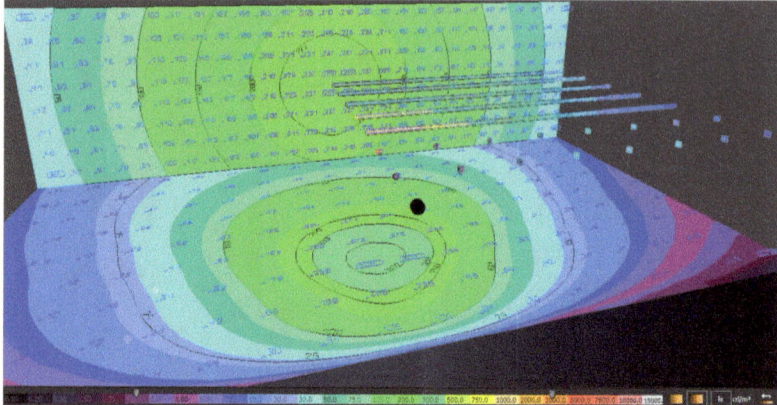

Fig. 4. Lamp matching diagram.

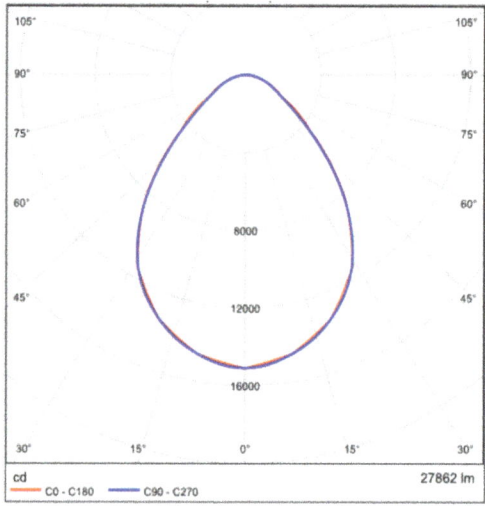

Fig. 5. Light distribution curve of the matched optical model.

4 Simulation Calculation and Analysis

4.1 Simulation Calculation Based on DIALux

Set calculation planes at 3 m in front, 7 m in front, bottom plane, 3 m behind, and 3 m on the side of the submersible. After simulation calculation, pseudo color diagrams of each calculation plane were obtained, which visually displayed the distribution of illumination in each area (as shown in Fig. 7). At the same time, the software will also provide specific illuminance data such as maximum and minimum illuminance.

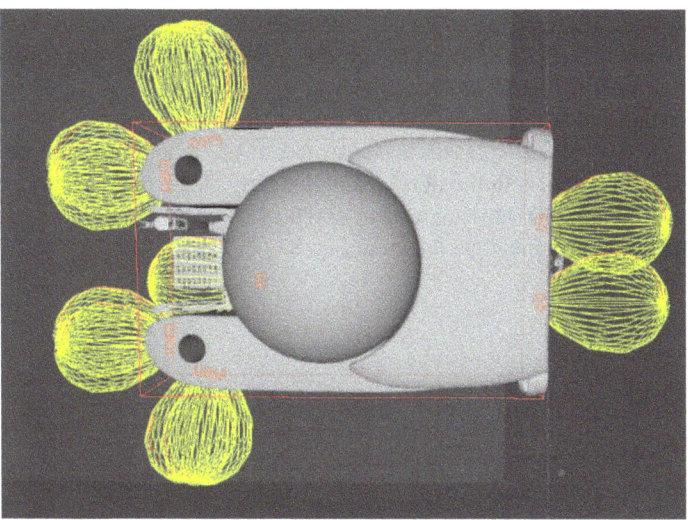

Fig. 6. Preliminary lighting layout plan.

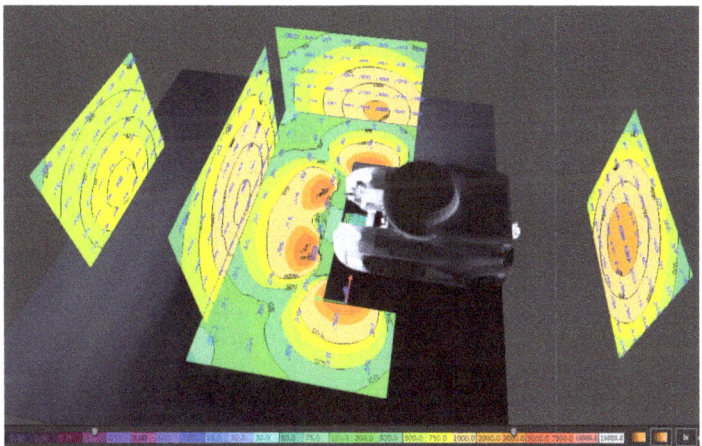

Fig. 7. Pseudo color diagrams.

4.2 Water Sample Analysis

According to Lambert Beer's law, the light transmittance of a medium is related to its attenuation coefficient and the actual distance of light passing through the medium:

$$T = \frac{I_{out}(\lambda)}{I_{in}(\lambda)} = \cdot e^{c(\lambda) \cdot d} \tag{1}$$

In the formula, T is the transmittance of the medium, I_{out} and I_{in} are the outgoing and incoming light intensities, $c(\lambda)$ is the attenuation coefficient of the medium, and d is the actual distance of light passing through the medium.

Therefore, the illuminance in the air calculated by DIALux is converted into the equivalent illuminance in seawater through the formula, which is the actual deep-sea illuminance. The d in the formula is the distance from the illumination lamp to the calculation plane and then reflected back to the human eye, which can be directly measured from the 3D model. $c(\lambda)$ can be measured in the laboratory. The project team obtained seawater samples in the deep sea in the early stage and measured the visible near-infrared (350 nm–800 nm) attenuation coefficient spectrum of the seawater sample in the laboratory environment, as shown in Fig. 8.

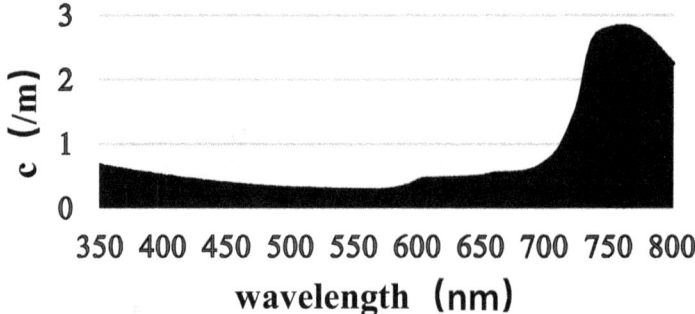

Fig. 8. Attenuation coefficient spectral lines of water samples in the visible near-infrared band.

The attenuation coefficient $c(\lambda)$ is a spectral line related to wavelength, and the illuminance in illumination evaluation is the total intensity of the spectral line. Therefore, in the calculation, it is necessary to integrate the light intensity to obtain the transmittance of the water sample. The calculation formula is as follows:

$$T = 100\% \times \frac{\int_{350}^{800} I_{out}(\lambda)d\lambda}{\int_{350}^{800} I_{in}(\lambda)d\lambda} \tag{2}$$

4.3 Equivalent Conversion of Illuminance

The equivalent illuminance of the lamp after passing through seawater is equal to the product of the calculated illuminance and transmittance of each plane (as shown in Table 2). By calculation, we can more accurately evaluate the lighting effect of lamps in the actual working environment of submarines.

4.4 Analysis of Calculation Results and Design Iteration

According to the simulation results, it can be concluded that:

1) The maximum illumination on a vertical plane 7 m ahead is 3 lx. To meet the illumination requirement of 5 lx at a distance of 7 m, it is necessary to increase the number of lighting fixtures in front;

Table 2. Equivalent illuminance of each plane.

Calculation planes		Calculated illuminance	Light path length at minimum/maximum illumination (m)	Seawater transmittance (%)	Equivalent illuminance (lx)
3 m in front	Minimum illumination	317	11.2	1.7	5
	Maximum illumination	1546	7.9	4.9	76
7 m in front	Minimum illumination	449	17	0.3	1
	Maximum illumination	819	15.8	0.4	3
bottom plane	Minimum illumination	71	11.5	1.6	1
	Maximum illumination	5347	4	18.8	1005
3 m behind	Minimum illumination	143	12.2	1.3	2
	Maximum illumination	2103	7.6	5.5	116
3 m on the side	Minimum illumination	282	9.2	3.2	9
	Maximum illumination	2445	6	9.3	227

2) The maximum illumination on the bottom plane is 1005 lx. When observing nearby targets, it is necessary to manually reduce the brightness of the front lighting to prevent local highlights;
3) The minimum illumination on the bottom plane is 1 lx, and the minimum illumination on the 3-meter side plane is 2 lx, both located in the left and right front of the submersible. Suggest deflecting the installation angle of the two front lights outward to increase the illumination of the left and right front of the submersible, ensuring the continuity of diver observation of the deep sea environment;
4) A vertical plane 3 m behind, with a maximum illumination of 227 lx and a minimum illumination of 9 lx, meeting observation requirements.

Based on the above analysis results, carry out the optimization design of the lighting layout plan, and after multiple rounds of iteration, until the equivalent illuminance of each plane meets the requirements of the design objectives.

5 Conclusion

The evaluation method for the lighting layout of manned submersibles proposed in this article not only comprehensively considers the current standards and operational requirements of submersibles, but also conducts an in-depth evaluation of the lighting layout scheme through measured data and simulation analysis. During the evaluation process, special attention was paid to the propagation characteristics and equivalent conversion issues of light in seawater, further improving the accuracy of the evaluation results. Finally, optimization suggestions were proposed based on the evaluation results, providing practical and feasible guidance for the lighting design of manned submersibles. This study can also provide reference for practical applications in underwater related fields.

Fund Project: Supported by the National Key Research and Development Program of China (2021YFC2800600)

References

1. Ye, C., et al.: Application of ergonomics to layout design of manned submersible. J. Northwest. Polytech. Univ. **39**(02), 233–240 (2021) (in Chinese)
2. Gai, Y., et al.: Research on construction reliability of lighting system in aerospace test plant. Environ. Adapt. Reliab. **42**(01), 19–24 (2024) (in Chinese)
3. Li, M.: Design of dakar municipal road lighting based on DIALux software. Technol. Innov. Appl. **13**(31), 18–21 (2023) (in Chinese)
4. Li, Y., et al.: Simulation of apron floodlight environment and analysis of combined aircraft stand lighting. J. Civ. Aviat. Univ. China. **42**(02), 51–57 (2024) (in Chinese)
5. Jinchi, F., et al.: Aircraft cabin lighting design method based on DIALux. J. Northwest. Polytech. Univ. **42**(01), 53–61 (2024) (in Chinese)

Open Access This chapter is licensed under the terms of the Creative Commons Attribution-NonCommercial-NoDerivatives 4.0 International License (http://creativecommons.org/licenses/by-nc-nd/4.0/), which permits any noncommercial use, sharing, distribution and reproduction in any medium or format, as long as you give appropriate credit to the original author(s) and the source, provide a link to the Creative Commons license and indicate if you modified the licensed material. You do not have permission under this license to share adapted material derived from this chapter or parts of it.

The images or other third party material in this chapter are included in the chapter's Creative Commons license, unless indicated otherwise in a credit line to the material. If material is not included in the chapter's Creative Commons license and your intended use is not permitted by statutory regulation or exceeds the permitted use, you will need to obtain permission directly from the copyright holder.

An Analytical Study of the Multi-body Motion of the Installation Operation between A Crane Ship and Gravity-based Wind Turbine Foundation in Waves

Pengfei Chen[1], Yu Lin[1,2(✉)], Bibo Li[1], Xiao Zhang[1], Jun Huang[1], and Yan Nie[1]

[1] Shanghai Investigation, Design & Research Institute Company Limited, Shanghai 200335, China
lin_yu1@ctg.com.cn

[2] Sino-Portuguese Centre for New Energy Technologies (Shanghai) Company Limited, Shanghai 200335, China

Abstract. With the rapid increase of the wind power capacity of installation in China, gravity-based foundation (GBF) of offshore wind become an important form of offshore wind turbine foundation in Chinese southeastern sea area especially. The GBF relies on its own and internal ballast weight to resist wind turbine loads and marine environmental loads. The GBF is suitable for shallow overburden and rock geological conditions. Fujian sea area is one of the main construction areas of offshore wind industry in China as well as one of the areas with the most abundant wind resources. However, there are large areas of shallow overburden and seabed with rock foundation in Fujian sea area. The gravity foundation have superior advantages than monopile or jacket type of foundations due to that the GBF can avoid offshore piling resulting in shorter construction time and lower cost. In this paper, the GBF designed on the basis of the environmental conditions in Fujian sea area is taken as the study case. Based on the numerical simulation method, the overall performance of GBF installation process under various combinations of wind and wave is simulated. Meanwhile, the operation window for the installation are studied which can provide reference for the offshore wind GBF application.

Keywords: Offshore Wind Turbine Foundation · Gravity-based Foundation · Installation Operation

1 Introduction

1.1 A Subsection Sample

By 2023, the total capacity of operating offshore wind farm reached 75.2 GW which is incredible and 10.8 GW of offshore wind was installed showing a strong uptrend [1]. The gravity-based foundation (GBF) plays an important role in the offshore wind industry in China which is suitable for shallow cover and rocky geological conditions which is typical in the costal area of Fujian Province, China. Through its self-weight and internal ballast weight, the GBF is designed to resist the heeling moment and slipping force caused by wind turbine load and the environmental loads. The sea area in Fujian region owns the richest wind resource in China while there exists a wide range of shallow cover as well as rocky foundation seabed. Due to the particularity of the seabed, other types of wind turbine foundation namely the monopile and jacket were facing difficulties in the installation operation phase. In this case, the GBF shows comparative advantage in the construction phase avoiding the piling operation in the shallow cover, rocky areas. The gravity foundation can be directly cast at onshore prefabrication yard and transported to the site by using the semi-submersible barge which is a good choice when it comes to the cost and time duration of installation of the foundations.

Subsequent paragraphs, however, are indented. The GBF has been used in Europe in a number of offshore wind farms, and in terms of the construction convenience and cost control, it shows greater superiority [2]. Regarding the manufacture of the GBF, prestressed concrete is a mainstream choice. Commonly, after the foundation is prefabricated at onshore shipyard, it will be transported to the site by large barge ship or crane ship. If the foundation is self-floating, it can also be wet towed to the site [3]. The fabrication and transportation of gravity foundation is technically mature, but the main difficulty lies in the lifting and sinking phase at sea. The gravity foundation is not only affected by wind, wave and current in the operation, but also affected by the motions of the crane ship. In Zhiyu Jiang's [4] work, the comprehensive review about many aspects of offshore wind turbine installation has been presented.

Referring to gravity foundation installation experience, Esteban et al. [5] suggest that the lifting and sinking operation of gravity foundation with significant wave height Hs shall not exceed 1–1.5 m, and the operation window duration should be strictly calculated. Dudson et al. [6] have developed a kind of transportation and installation barge (TIB), which can transport the gravity foundation under the condition of Hs equals to 3 m, and lifting and sinking the gravity foundation should be conducted under the condition of H_s equals to 1.25 m, which allows worse environmental conditions in the operation phase.

The 'Design and Construction of Gravity Wharves' (JTS 167-2-2009) [7] provides detailed instructions for caisson installation indicating the wave height should not be greater than 0.5 m within the installation process. When floating docks or semi-submersible barges are submerged, the wave height of the submerged area should not be greater than 1 m, the wind speed should not be greater than grade 6, and the current speed should be less than 1.0 m/s, etc. Li Fowen et al. [8] conducted a research on the construction process of 6000 t super-large caisson slipping and installation, focusing on the slipping process of caisson water injection, lifting ship lifting, changes in environmental conditions, and other construction operation precautions were explained in detail.

Current researches on gravity foundation installation shows that the overall bias are towards the construction instructions and construction precautions, the analysis of motion analysis of the crane ship, GBF lifting and sinking process is rather absent. Based on the engineering projects, this paper focuses the motion response of the crane ship and GBF in the installation process providing references to analyse the safety and application of gravity foundations in Fujian region offshore wind power development in the area.

2 Methodology

To study the characteristic motion of the installation process of GBF, the coupling system of crane ship and GBF is simulated. Regarding the crane ship, the wave load, additional mass and damping coefficient of the ship in the frequency domain can be calculated based on the potential flow theory [9]; then, using the Cummings impulse theory, the wave load to which the crane ship is subjected in the frequency domain is converted to the time domain, and the equation of motion of the crane ship monohull in the time domain can be expressed as [10, 11]:

$$[M + A(\infty)]\ddot{x}_1 + \int_0^t h(t-\tau)\dot{x}_1(\tau)d\tau + D_1 x_1 + k x_1 = F_1 \qquad (1)$$

In the above equation, M is the mass matrix, $A(\infty)$ is the additional mass matrix at frequency infinity; $\int_0^t h(t-\tau)\dot{x}_1(\tau)d\tau$ is the delay function, which can be obtained by converting the additional mass and damping coefficient in the frequency domain range; D1 is the linear damping matrix, k is the hydrostatic force response matrix; F1 is the external loads on the crane ship, including wave loads, wind loads, and mooring constraints, etc., in which the wind loads are calculated according to empirical formula [12].

For GBF, the wave current loads and inertial forces are calculated based on Morrison's formula, which is given below [13]:

$$F = \frac{1}{2}\rho C_D D (U - \dot{X})|U - \dot{X}| + \rho C_M \frac{\pi D^2}{4} \frac{\partial U}{\partial t} - \rho C_m \frac{\pi D^2}{4} \ddot{X} \qquad (2)$$

In the above equation, U denotes the current speed, X denotes the wave amplitude, CD denotes the drag force coefficient, CM denotes the inertia coefficient, and Cm denotes the additional mass coefficient, where $C_M = 1 + C_m$.

The equation of motion in the time domain of a gravity based monolith can be expressed as [14]:

$$m\ddot{x}_2 + D_2 \dot{x}_2 = (\rho_w V + C_m)a - m\ddot{x}_2 + C_D |u - \dot{x}_2|(u - \dot{x}_2) \qquad (3)$$

In the above equation, m is the mass matrix, D2 is the linear damping matrix; w is the seawater density, V is the gravity-based drainage volume, a is the acceleration of the water mass point motion; C_D, C_m denote the drag force coefficients and the additional mass coefficients; u is the current velocity.

In the multi-body motion analysis, the environmental loads, inertia force, damping force and connecting cable force on the crane ship and gravity foundation need to be considered separately, etc. Combined with Eq. (1) and Eq. (3), the time-domain equations of motion for the gravity foundation lifting and sinking process can be expressed as follows [15]:

$$(M + A_\infty)\ddot{x}_1 + \int_0^t h(t-\tau)\dot{x}_1(\tau)d\tau + D_1 \dot{x}_1 + k x_1 =$$

$$q_{WA}^{(1)} + q_{WA}^{(2)} + q_{current}^1 + q_{wind}^1 + q_{posi} + q_{lift} \qquad (4)$$

$$m\ddot{x}_2 + D_2\dot{x}_2 = q_{lift}(t, x_1, \dot{x}_1, x_2, \dot{x}_2) \tag{5}$$

Where x1 and x2 are the six-degree-of-freedom motion vectors of the crane ship and GBF, respectively; M and m are the mass matrices of the crane ship and GBF; A_∞ is the additional mass matrix at infinity frequency; $h(t - \tau)$ is the matrix of the time-delay function; $q(t, x_1, \dot{x}_1, x_2, \dot{x}_2)$ is the vector of the combined external force of the crane ship; D_1 and D_2 are the linear damping matrices of the crane ship and the GBF; k is the array of the hydro static restoring moments of the crane ship; $q_{WA}^{(1)}$ is the first-order wave force; $q_{WA}^{(2)}$ is the second-order mean wave drift force; q_{posi} is the mooring constraint force of the crane ship; q_{lift} is the reaction force on the rope between the hull and the weight; q_{wind}^1 and $q_{current}^1$ are the wind force of the crane ship and the current force of the crane ship.

3 Numerical Simulation

3.1 Case Study

The case presented in this paper is referred to an engineering project. The crane ship which is used for installing the foundations is equipped with 4 main hooks, 2 auxiliary hooks, the maximum lifting capacity of 3600 t. The main parameters of the crane ship are shown in Table 1.

Table 1. Parameters of the crane ship.

Parameter	Value	Unit
Length	118.9	m
Width	48	m
Depth	8.8	m
Design Draft	4.6	m
Displacement	26,900	t
Center of Gravity	20	m

The crane ship is equipped with eight positioning anchors, with four anchor points on both sides of the bow and stern, and each anchor point is connected to two anchor cables, with an angle of 30° between the two anchor cables, in a symmetrical arrangement. The anchor chain parameters are shown in Table 2, and the mooring arrangement diagram is shown in Fig. 1.

Regarding the GBF used in this paper, the height is 47 m, the diameter of the bottom is 32 m, the diameter of the column is 6.7 m, and the diameter of the reinforced ring beam at the top of the foundation is 7.8 m. The thickness of the foundation base plate is 1.0–1.5 m, and the thickness of the wall of the cylinder above the base plate is 60 cm, and the weight of the concrete gravity foundation is about 5733 t (the weight of the

Table 2. Parameters of the mooring system.

Parameter	Value	Unit
Diameter	56	mm
Length	1000	m
Weight in Air	676.614	N/m
Wet - Dry Weight Ratio	0.87	/
Anchor Weight	12	t
Modulus of elasticity	6.40E + 10	Pa
Minimum breaking strength	2.43E + 06	N

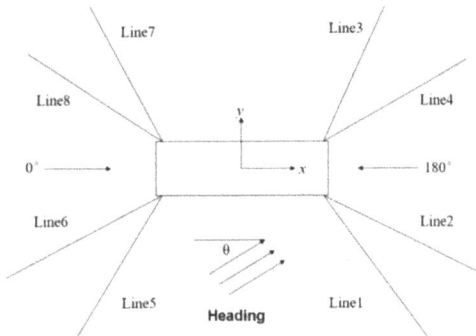

Fig. 1. Arrangement of Mooring System.

foundation on the land itself, excluding filler material), and the upright floating position gives a draft of about 15.5 m, and the foundation is designed to be filled with sand and gravel material. The diagram of the GBF is shown in Fig. 2(a).

The gravity foundation was transported by semi-submersible barge to the designated machine position. When the semi-submersible barge sinks to 23 m, the gravity foundation is released from the semi-submersible barge, and in order to ensure that the GBF has enough weight for sinking, a certain amount of ballast water will be injected into the gravity foundation at the initial time, and the GBF will be lifted and sunk by the crane ship. The in-situ arrangement is shown in Fig. 2(b).

3.2 Numerical Model

Numerical modeling and analysis is carried out by applying the software Sesam developed by DNV, mainly using the modules including GeniE, HydroD and Simo. The GeniE is used to establish the finite element model for hydrodynamic analysis of the crane ship. And HydroD is implemented to simulate the hydrodynamic coefficient namely additional mass, damping coefficients, etc. The nonlinear time-domain coupled analysis of the system is carried out by software Simo. The time-domain model is shown in Fig. 3(a).

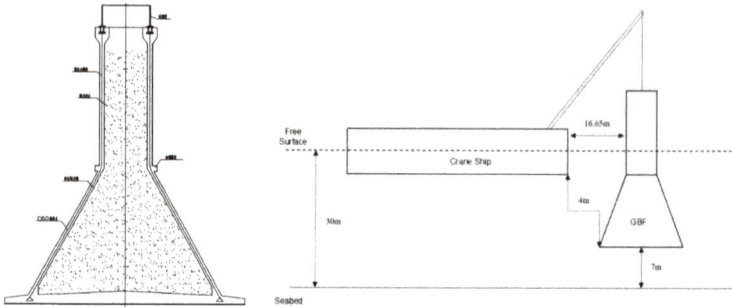

(a)The side view of the GBF.　　(b)Arrangement of gravity foundation in place.

Fig. 2. Section and layout of the GBF.

Subsequently, the GBF is simulated by Morrison unit and the variable cross-section part is modeled by segmental modeling, so as to establish the GBF single-body time-domain model as shown in Fig. 3(b). The spreader cables, spreader and balancing beams are established, coupling the crane ship and GBF. The multi-body numerical model of the installation process is shown in Fig. 3(c).

(a)Time-domain model of crane ship.　(b)Time-domain model of gravity foundation.

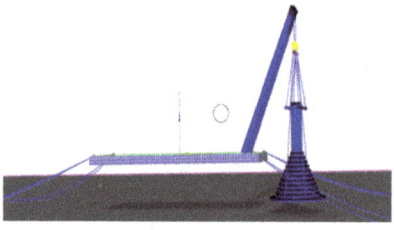

(c)Time-domain model of installation

Fig. 3. Numerical Model in Time-domain.

3.3 Environmental Parameters

The water depth of the operational sea area is 30 m, the surface current speed is 0.5 m/s. The JONSWAP spectrum is selected to describe the wave condition with a significant wave height of 1.25 m and peak period of 7 s. The wind spectrum is NPD spectrum, with an hourly average wind speed of 8.0 m/s. The environmental conditions are listed in Table 3. It is worthy to mention that wind, wave, and current are assumed to be at the same heading.

Table 3. Environmental Conditions.

Parameter		Value	Unit
Wind	Spectrum	NPD	mm
	Wind Speed	8	m
Wave	Spectrum	JONSWAP	N/m
	Hs	1.25	/
	Tp	7	t
	steepness parameter	3.3	Pa
Current	Current Speed	0.5	N

4 Analysis and Results

4.1 Response Motion of the System

When the GBF is relieved from the semi-submersible barge, the displacement of gravity foundation is 6441 t while the structure weighs 5733 t. Therefore, the buoyancy is 707 t greater than gravity and the ballast is needed to make the GBF sink successfully.

GBF could have different initial ballast weights and different static equilibrium positions. When the GBF leaves the semi-submersible barge, it will cause a downward bending moment to the crane ship, resulting in a certain pitch angle of the crane ship, meanwhile, the foundation will keep sinking to reach equilibrium. In order to define the appropriate initial ballast, the initial ballast is set at 707 t, 1000 t, 1500 t, 2000 t, 2500 t and 3000 t for analysis, and the headings of wind and wave current is taken as 180° during the calculation.

The results of different equilibrium positions with various ballast weights are shown in Table 4, and positions remain roughly invariant in the three DOFs namely, Y, Rx and Rz. According to Table 4, with the increase of initial ballast, the equilibrium position of the crane ship and BGF converges at lower positions. The initial ballast has the most significant effect on the pitch angle of the crane ship and the vertical position of the GBF. When the initial ballast is increased to 3000 t, the steady pitch angle of the crane ship is 2.9°, and the center of gravity (COG) of GBF stays at 6.4 m affecting the safety of operation.

Table 4. Equilibrium position with different ballast weights.

Initial Ballast (t)	Difference Between Gravity and Buoyancy at Draft of 23 m (t)	Difference Between Gravity and Buoyancy at Draft of 30 m (t)	Value			Unit		
			X(m)	Z(m)	Ry(°)	X(m)	Z(m)	Ry(°)
707	0	−203.32	0	0	0	0	0	0
1000	292.5	89.7	−0.05	−0.08	0.41	0.02	−1	0.42
1500	792.5	589.7	−0.28	−0.19	1	0.39	−2.4	0.43
2000	1292.5	1089.7	−0.52	−0.3	1.6	0.85	−3.7	0.44
2500	1792.5	1589.7	−0.76	−0.41	2.3	1.4	−5	0.43
3000	2292.5	2089.7	−1	−0.53	2.9	1.9	−6.4	0.4

The time series of steady vertical positions of COG of GBF with different initial ballast weights are shown in 5. The steady position of vertical position reflects whether the gravity foundation can sink or not. According to Fig. 4, when the initial ballast is 707 t the GBF can not sink, and finally floats on the water surface at an inclined position under wind and wave, as shown in Fig. 5(a). When the initial ballast is 1500 t, the gravity is greater than buoyancy, the gravity foundation can sink, the maximum value of sinking is −5.76 m, but can't sink to the seabed surface of -7 m; this is because with the increase of sinking depth, the buoyancy of the gravity foundation increases, and at the same time, the waves have a buoyant force on the gravity foundation, as shown in Fig. 5(b). When the initial ballast exceeds 2000 t, the larger difference between gravity and buoyancy can make the gravity foundation sink successfully, and the final equilibrium depth exceeds -7 m, as shown in Fig. 5(c).

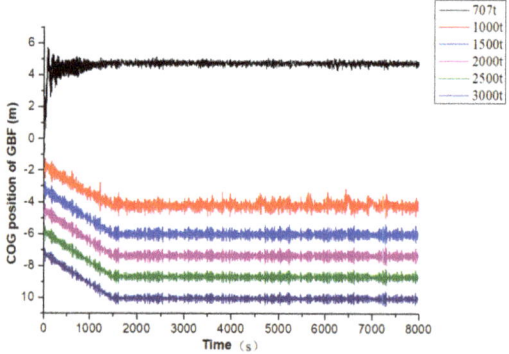

Fig. 4. COG position of GBF with different ballast weights.

An Analytical Study of the Multi-body Motion 103

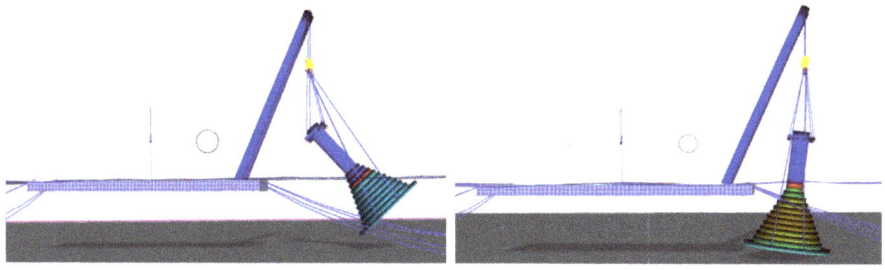

(a) Equilibrium position of ballast at 707t.   (b) Equilibrium position of ballast at 1500t.

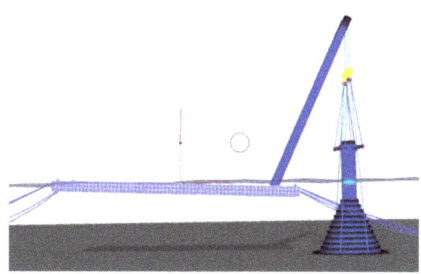

(c) Equilibrium position of ballast at 2000t.

Fig. 5. Equilibrium positions of ballast weights.

As shown in Fig. 6, it can be seen that the initial ballast has a great influence on the pitch motion, and the statistical values of the pitch motion with different initial ballasts are shown in Table 5. The larger the initial ballast is, the smaller the steady pitch position is reached and the whole operation is smoother.

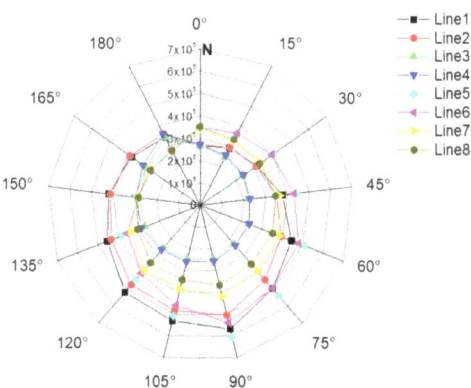

Fig. 6. Mooring line tension rose diagram.

Table 5. Equilibrium position with different ballast weights.

Initial Ballast (t)	Pitch motion(°)		
	Maximum	Minimum	Averaging value
707	0.1	−44.5	−41.3
1000	14.72	−0.52	7.2
1500	3.7	−2.1	0.77
2000	2.27	−1.38	0.39
2500	1.7	−1.17	0.26
3000	1.4	−1	0.2

Combining the calculation results of Table 4 and Table 5, the initial ballast of 1500 t is selected for following analysis. In this sense, the vertical descent distance between the GBF and seabed is 2.4 m which is safe. Moreover, the pitch motion is relatively small. According to Fig. 5, it noted that the maximum depth of sinking position is 5.76 m when the initial ballast is 1500 t, so in order to make it sink to the seabed surface, ballast water shall be increased slowly into the gravity foundation during the sinking process to sink to the seabed.

4.2 Sensitivity Study of Wave Headings

To determine the safe wave heading of the installation operation, the sensitivity study is implemented. The initial ballast of 1500 t is defined for the GBF and the environmental input data consisting of wave height, wind speed, current speed and other parameters are kept constant as shown in Table.3. The simulations are carried out at various wave headings with a step of 15° from 0° to 180° to study the influence of wave headings on motion response of the crane ship - GBF system. The maximum tension of each mooring line under different wave headings is shown in Fig. 6, and the motion response in 6 DOFs of the system is shown in Fig. 7.

According to norms, API specification [16] requires that under the combined excitation force of wind and wave, the horizontal motion of the floating structure shall not exceed 10% of the water depth. In this paper, horizontal displacement shall not exceed 3 m. By using the dynamic method of mooring analysis, the safety factor of break strength of the chain is not less than 1.67.

According to Fig. 7, it can be seen that the tension characteristics of mooring lines under different wave headings shows various results. The maximum value of chain tension is 6.07E + 05 N occurring in Line5, under wave heading 90°. The breaking force of the chain is 2.43E + 06 N and the safety factor is 4. Therefore, the strength of the anchor chain in each wave direction meets the requirements of API specification.

It can be seen from the results shown in Fig. 8 that the motion response of GBF is larger than that of the crane ship. The wave direction has little effect on the heave motion of the GBF and has obvious effect on the yaw motion which does not affect the installation process. Since the GBF is lifted at the bow of the crane ship, the crane ship and

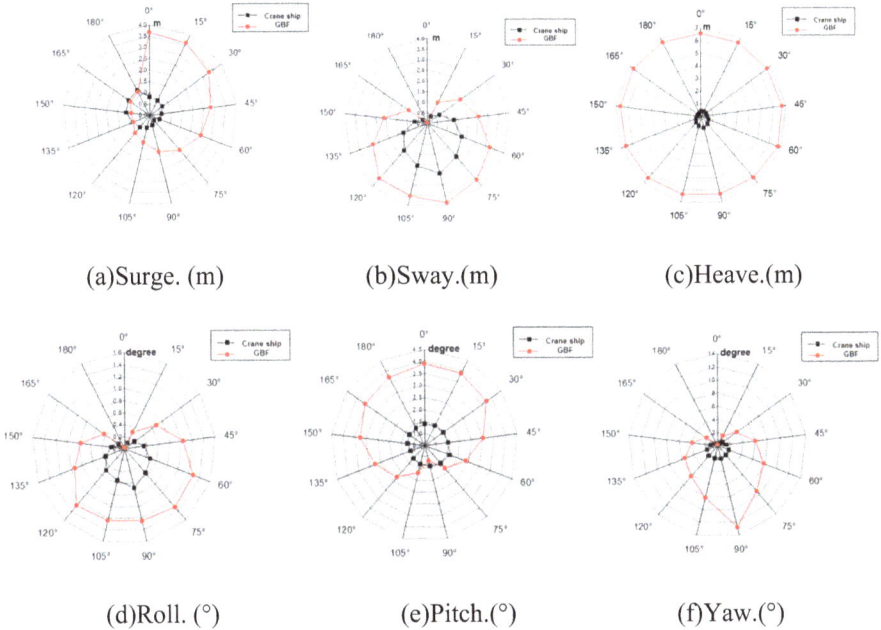

(a)Surge. (m) (b)Sway.(m) (c)Heave.(m)

(d)Roll. (°) (e)Pitch.(°) (f)Yaw.(°)

Fig. 7. Motion response of the system at 6 DOFs.

the gravity foundation have an initial pitch angle at equilibrium position. That explains the pitch motion amplitude has larger value compared to the roll motion amplitude. The sway and surge motion amplitude of GBF are more than 3 m under 0° ~ 30° and 60° ~ 120° respectively, which do not meet the requirements of API specification. Wave headings has less influence on the pitch motion of the crane ship. The maximum pitch angle is 1.63° and the minimum longitudinal rocking angle is 1.31°.

Oblique waves and beam waves will lead to roll and yaw motion of the crane ship. Under oblique waves and beam waves condition, the maximum pitch motion of the crane ship is 0.68° and the maximum yaw motion of the crane ship is 2.05°, which has lighter influence on the installation process. The maximum surge and sway motion amplitude of the crane ship under all wave headings are 1.24 m and 2.43 m respectively meeting the requirements of API specification. The heave motion of the crane ship plays an important role to secure the safety of the installation process. According to the relevant literature recommendations [17], the heave motion of the crane ship should not exceed 0.5 m during the installation phase while the heave motion amplitude is more than 0.5 m under 60° ~ 120° wave headings, which is not conducive to the safety of the operation.

4.3 Sensitivity Study of Wave Heights

Based on the analysis of 3.1 and 3.2, the initial ballast of 1500 t is selected and the wave heading is 0°. The environmental parameters like wind speed, current speed and wave period are kept unchanged. In order to study the safe wave height limit of the installation process, the significant wave heights of 1.6 m, 1.8 m, 2.0 m and 2.5 m were taken for simulation and the results are shown in Table 6.

Table 6. Response amplitude of the system under various wave heights.

Significant wave height (m)	Maximum of surge motion (m)		Maximum of pitch motion (°)	
	Crane ship	GBF	Crane ship	GBF
$H_S = 1.6$	1.25	0.86	1.386	2.926
$H_S = 1.8$	1.483	1.03	1.433	3.271
$H_S = 2.0$	2.098	1.61	1.49	3.779
$H_S = 2.5$	/	/	/	/

It can be seen from Table.6 that the surge and pitch motion amplitude increase non-linearly with the increase of wave height. When the Hs is 2.5 m, the time-domain simulation does not converge and the installation can not be implemented safely under this wave condition.

5 Conclusion and Discussion

In this paper, the GBF for wind turbine designed for Fujian sea area is taken as the study case and the main object is to find the critical environmental boundary conditions in the installation process of the BGF. The multi-body system which consist of crane ship, mooring system and the GBF is studies. Moreover, the mooring line tension and motion response amplitude of the system during its installation are analyzed by using the time-domain coupled analysis method. Under the sea state conditions specified in Sect. 3.3, the GBF installation operation is safe and the crane ship motion as well as the mooring line tension meet the specification requirements. It is recommended that significant wave height shall not exceed 2 m during the installation process. In this case, the motion of crane ship and GBF are within the controllable range.

The future work will mainly focus on more complex combination of the environmental conditions, since the wind, wave and current acting in the same direction might not be the most disadvantageous situation.

Acknowledgments. This research is sponsored by the research funding of Shanghai Investigation, Design and Research Institute Co., Ltd. (Project Number: 2019FD(8)-010, 2023FD(83)-001) and by project number 22dz1206003.

References

1. GLOBAL WIND REPORT 2024, GWEC

2. Wei, Y.A.N.G., Yifeng, L.I.N., Quan, Z.H.A.N.G.: Review of the development of offshore wind gravity foundation [J]. Wind Energy. **9**, 38–42 (2018) (in Chinese)
3. Xiaoqian, S.U.N., Hongbiao, G.A.O.: The key technologies of gravity base foundations for offshore wind farms [J]. Wind Power. **6**, 20–30 (2014) (in Chinese)
4. Jiang, Z.: Installation of offshore wind turbines: A technical review. Renew. Sustain. Energy Rev. **139**, 110576 (2021) ISSN 1364-0321
5. Esteban, M.D., et al.: Gravity based support structures for offshore wind turbine generators: Review of the installation process [J]. Ocean Eng. **110**, 281–291 (2015)
6. Ed Dudson, Mark Willbourn, William Brook-Hart. The transport and installation of concrete foundations [C]. In: European offshore wind conference & exhibition 2009:EOW 2009, vol. 3 of 3, Curran Associates, Inc. Stockholm, Sweden, 14–16 September 2009
7. Ministry of Communications of the People's Republic of China. Design and Construction Code for Gravity Quay [S]. China communication press London, 2009. (in Chinese)
8. Fuwen, L.I., Peng, D.A.I., Zhijie, R.E.N.: Launching and placing of ultra-large caissons [J]. Port Waterw. Eng. **11**, 198–202 (2012) (in Chinese)
9. Tiebing, S.H.A.N., Fanghao, P.A.N., Wen, Z.O.U., et al.: Study on characteristics of side-by-side mooring between the offshore floating structures and supply ship [J]. Ocean Eng. **35**(5), 1–11 (2017) (in Chinese)
10. Hongkang, Z.H.O.U., Xin, L.I., Jianmin, Y.A.N.G., et al.: Experimental and numerical researches for a dock mooring FSRU [J]. Ocean Eng. **35**(1), 12–20 (2017) (in Chinese)
11. Faltinsen, O.M.: Sea Loads on Ships and Offshore Structures. Cambridge University Press (1991)
12. OCIMF: Prediction of Wind and Current Loads on VLCCs, 2nd edn. Oil Companies International Marine Forum, London (1994)
13. Yuanlin, L.I.: Offshore Structural Hydrodynamic [M]. South China University of Technology, China (1999) (in Chinese)
14. Kvittem, M.I., Bachynski, E.E., Moan, T.: Effects of hydrodynamic modelling in fully coupled simulations of a semi-submersible wind turbine [J]. Energy Procedia. **24**, 351–362 (2012)
15. Xin, X.U., Xin, L.I., Jian-min, Y.A.N.G.: Numerical and experimental analysis for lifting of a semi-submersible crane vssel [J]. J. Sh. Mech. **18**(07), 799–808 (2014) (in Chinese)
16. API: Recommended Practice for Design and Analysis of Station Keeping Systems for Floating Structures. API RP. 2SK (2005)
17. Hongbin, W.: Multi-body Structure Dynamic Analysis During Spar Block Sitting [D]. Harbin Engineering University (2012) (in Chinese)

Open Access This chapter is licensed under the terms of the Creative Commons Attribution-NonCommercial-NoDerivatives 4.0 International License (http://creativecommons.org/licenses/by-nc-nd/4.0/), which permits any noncommercial use, sharing, distribution and reproduction in any medium or format, as long as you give appropriate credit to the original author(s) and the source, provide a link to the Creative Commons license and indicate if you modified the licensed material. You do not have permission under this license to share adapted material derived from this chapter or parts of it.

The images or other third party material in this chapter are included in the chapter's Creative Commons license, unless indicated otherwise in a credit line to the material. If material is not included in the chapter's Creative Commons license and your intended use is not permitted by statutory regulation or exceeds the permitted use, you will need to obtain permission directly from the copyright holder.

Research on the Influencing Factors of Water Hammer Caused by Large Spacing Water Transfer of Ship Based on Transient Flow Model

Chao Fang, Bin Li, Weibo Pan[✉], and Yu Guo

Wuhan Second Ship Design and Research Institute, Wuhan, China
ship-designer@qq.com

Abstract. The phenomenon of water hammer impact caused by the opening and closing of valves in large distance water transfer system of ships is prominent, which seriously threatens the operation safety and overall stealth performance of the system. In order to effectively guide the treatment of water hammer impact in water transfer system and explore the main influencing factors of water hammer pressure rise in large-distance water transfer system, a one-dimensional system simulation model is established based on the transient flow model and the design of large-distance water transfer system on a real ship. The relationship between the influencing factors such as valve closing time, water transfer pressure, pipe diameter and water hammer pressure rise is analyzed by numerical calculation. Based on the numerical test principle of single variable, the calculation results show that the pressure rise of water hammer can be effectively reduced by extending the valve closing time, reducing the water transfer pressure and increasing the pipe diameter. The research results point out the direction for the treatment of water hammer impact in the large spacing water transfer system. In the future, combined measures such as balancing the water transfer pressure of the water tank, increasing the pipe diameter, adjusting the valve controller to appropriately extend the valve closing time can be taken to suppress the destructive positive pressure water hammer and negative pressure water hammer at the moment of valve closing in the system pipeline.

Keywords: Large Spacing · Water Hammer Impact · System Simulation · Single Variable

CCS Concepts: Insert CCS text here

1 Introduction

When a ship experiences longitudinal tilt due to daily fuel consumption and other load changes, the static load weight distribution of the ship can be adjusted by moving water between the bow and the stern balance tanks to achieve longitudinal tilt adjustment. To maximize the longitudinal adjustment torque obtained from a single quantitative water transfer, the balance tank is usually arranged at the bow and stern ends of the ship. Traditional water transfer systems use pneumatic methods to transfer water, where compressed air drives seawater from one end of the tank to the other through long-distance pipelines [1]. The water transfer distance between the balance tank is long and the flow rate is high. At the end of the water transfer, it is necessary to quickly close the electro-hydraulic ball valve to cut off the water flow, causing drastic changes in seawater flow rate, causing water column separation and bridging, and inducing serious water hammer impact problems, which have a significant impact on the system safety and overall concealment of the ship [2–7].

The article focuses on the serious water hammer problem caused by the large spacing water transfer systems of ships. Starting from the mechanism of water hammer generation, the main influencing factors related to the pressure rise of the water hammer are identified. Using system simulation methods to compare and calculate single variables affecting factors such as valve closing time, water transfer pressure, and pipeline diameter, analyze the correlation between variable parameters and water hammer pressure increase, and propose feasible measures for water hammer control in large spacing water transfer systems.

2 Principle of Large Distance Water Transfer System

As shown in Fig. 1, the water transfer system mainly consists of a control system, a supply and exhaust system, and a seawater system. Ships are usually equipped with two independent water transfer systems, each responsible for the task of moving water from bow to stern or from stern to bow. The balance water tank at one end of the water transfer system is filled with compressed air, while the balance water tank at the other end is connected to the atmospheric environment inside the ship. When water needs to be moved from bow to stern, the bow balance water tank is filled to the specified pressure through the control console, and the exhaust valve of the stern balance water tank is opened to communicate with the atmospheric environment inside the tank. Open the seawater system control valve remotely from the control console. The seawater in the bow balance water tank is pushed by air pressure and transferred to the stern balance water tank through pipelines. When the instructed water transfer volume is reached, close the valve to complete a single water transfer action.

To ensure the maneuverability of ship maneuvering control, there are generally high requirements for the flow rate and control accuracy of the system's water transfer. Design parameters such as balancing the water transfer pressure inside the tank, opening time of control valves, pipeline diameter, and pipeline length are closely related to the flow rate and accuracy of water transfer. Due to the influence of the overall layout, the length of the pipeline is an inherent parameter of the system that cannot be changed. Therefore, this article selects three design parameters: water transfer pressure, valve opening time, and pipeline diameter to explore their relationship with the pressure increase of the water hammer in the water transfer system, in order to support the optimization and treatment of water hammer impact in the water transfer system.

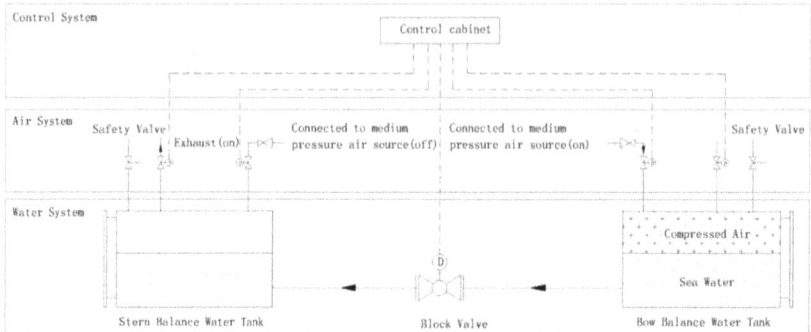

Fig. 1. Schematic diagram of a certain type of ship centralized cooling system.

3 Mathematical and Simulation Models

3.1 Basic Equation of Water Hammer

By organizing and simplifying the equations of motion and continuity of unsteady flow, the basic differential equations of water hammer can be derived, including the equations of motion and continuity. The equation of motion is based on the momentum conservation equation of the fluid, while the continuity equation is based on the mass conservation equation of the fluid. All water hammer calculation software follows the basic differential equation of water hammer and belongs to one-dimensional wave equation [8].

Equation of water hammer motion:

$$\frac{\partial H}{\partial x} + \frac{V}{g}\frac{\partial V}{\partial x} + \frac{1}{g}\frac{\partial V}{\partial t} + \frac{fV}{2g}|V| = 0 \quad (1)$$

Continuity equation of water hammer:

$$\frac{\partial H}{\partial t} + V\frac{\partial H}{\partial x} + \frac{a^2}{g}\frac{\partial V}{\partial x} = 0 \quad (2)$$

In the formula: f is the resistance coefficient along the way; H is the pressure head of the fluid in the pipeline, m; V is the flow velocity of the fluid in the pipeline; D is the diameter of the pipeline, m; G is the gravitational acceleration, m/s^2; a is the water hammer wave velocity, m/s.

3.2 Physical Simulation Model and Parameter Settings

According to the actual design parameters of the ship's water transfer system, a one-dimensional simulation model of the ship's water transfer system pipeline network was established using Flowmaster software, as shown in Fig. 2. The pipeline network includes pressure water tanks, several pipelines, elbows, ball valves, and switch valve controllers. The pressure water tank is used to simulate the balance water tank at the bow and stern, and the ball valves and switch valve controllers simulate the electro-hydraulic ball valves for cut-off control. The pipelines, variable diameter pipe fittings, and elbows are configured according to the actual ship design.

The initial condition of the pressure water tank is set to a volume of 20 m³, outer diameter 1800 mm, effective height 6733.5 mm, water level 2700 mm, roughness 0.1 mm. The characteristic of the elbow are set to a bending radius to diameter ratio of 1.5 and a diameter of 100 mm. The valve controller sets the closing time according to the working conditions. The valve closing law is a 90° linear valve closing, with different valve closing speeds set for different valve closing duration.

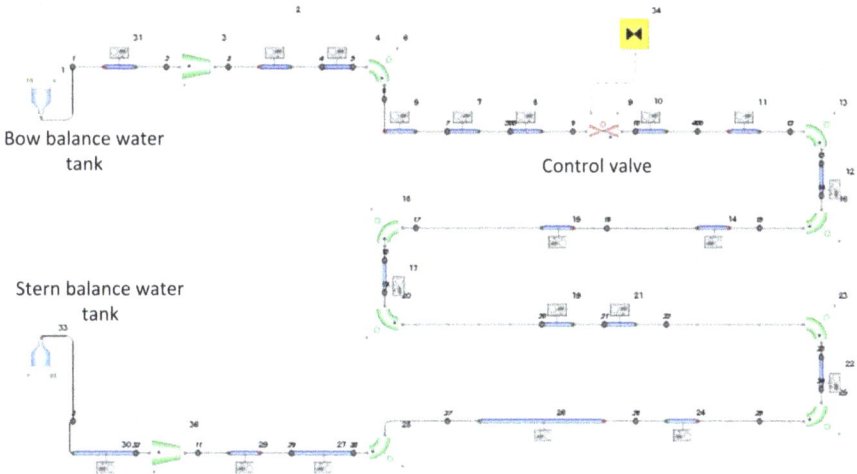

Fig. 2. Pipeline model of water transfer system.

4 Analysis of Factors Affecting Water Hammer

4.1 Valve Closing Time

The rapid cut-off of the control valve during the water transfer process leads to liquid column separation and bridging in the pipeline, and the speed of valve closure theoretically affects the degree of liquid column separation. This section takes the working conditions of the water transfer direction from bow to stern and the water transfer pressure of 0.4 MPa as an example. Without changing other parameters, the valve closing time is changed to simulate the transient characteristics of fluid pressure in the system pipeline after the control valve is opened, and to explore the influence of valve closing time on water hammer impact in the system. The calculation conditions for different valve closing time are shown in Table 1.

Table 1. Calculation of working conditions for different valve closing times.

Condition	Water transfer direction	Water transfer pressure (MPa)	Pipeline diameter(mm)	Valve closing time(s)
1	Bow to stern	0.4	100	1
2	Bow to stern	0.4	100	2
3	Bow to stern	0.4	100	3
4	Bow to stern	0.4	100	4
5	Bow to stern	0.4	100	5

The pressure curves before and after the valve corresponding to a closing time of 1 s are shown in Fig. 3. Due to space limitations, the pressure curves under other closing times will not be repeated. From Fig. 3, it can be seen that after closing the valve at t = 10 s, due to the blocking effect during the valve closing process, the pressure at the measuring point in front of the valve rapidly increases until the valve is completely closed. As the liquid column in front of the valve begins to separate and close, the pressure at the measuring point shows a periodic oscillation pattern. The pressure at the measuring point behind the valve decreases continuously after the valve starts to close, until it reaches the lowest saturated steam pressure after the valve is completely closed. The liquid column behind the valve experiences empty separation, and then under the separation and bridging effects of the liquid column, periodic oscillation patterns begin to appear. Observing the pressure oscillation phenomenon at the measuring points in front and behind the valve, it can be observed that the time interval of peak pressure is continuously decreasing. The main reason is that the filling point of the liquid column in front of the valve is constantly moving forward, and the filling point of the liquid behind the valve is constantly moving backward. The distance of pressure wave propagation between the two is decreasing, which is reflected in the trend of decreasing peak pressure period at the measuring point. Therefore, the frequency of peak pressure at the measuring point is continuously increasing.

The peak pressure at the measuring points before and after each valve closing time is shown in Table 2. From the simulation calculation results, it can be seen that, with other parameters remaining unchanged, as the valve closing time increases, the peak impact pressure before and after the control valve decreases.

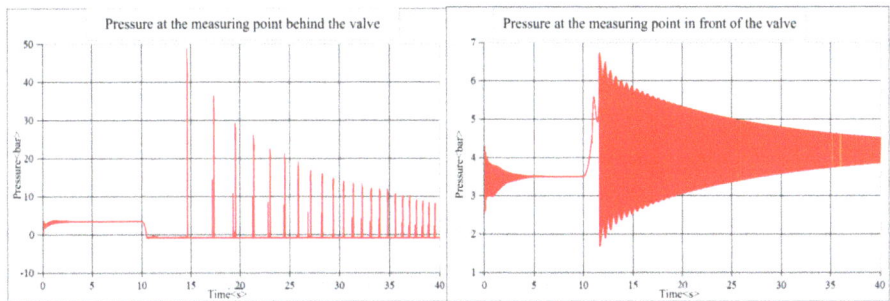

Fig. 3. Pressure at measuring points before and after the valve for 1s valve closing time.

Table 2. Peak pressure at measuring points before and after the valve at different closing times.

Condition	Valve closing time (s)	Peak pressure after valve (MPa)	Peak pressure before valve (MPa)
1	1	4.87	0.66
2	2	3.87	0.63
3	3	2.63	0.57
4	4	0.59	0.54
5	5	0.52	0.45

During the water transfer process, the rapid closure of the valve sends a pressure drop wave towards the back of the valve, causing liquid column separation near the valve, and reducing the pressure to near steam pressure in most areas of the pipeline. Due to the effects of friction and reverse pressure gradient, the liquid continues to move downstream under deceleration conditions. However, the pressure behind the valve is very low, and the liquid will eventually flow back in the direction of the valve under the pressure. At this time, the steam cavity is destroyed, causing destructive high pressure in the system. While extending the valve closing time appropriately, such as from 1 s to 5 s, the resistance of the valve gradually increases during the valve closing process, causing a decrease of flow rate and a decrease of the flow rate change. At this time, the water hammer pressure generated by the liquid reflux to the valve is smaller. For water transfer systems, the valve closing time can be appropriately extended to suppress excessive water hammer pressure increase.

4.2 Water Transfer Pressure

The pressure of water transfer in the balanced water tank is the driving force for the system's water transfer, which has a significant impact on the flow rate of seawater medium in the pipeline during the water transfer process. In this section, under the working condition of moving water from bow to stern and closing the valve for 1 s, different water transfer pressure are set while other parameters remain unchanged to explore the influence of water transfer pressure on water hammer impact. The calculation conditions are shown in Table 3.

Table 3. Calculation conditions for different water transfer pressures.

Condition	Water transfer direction	Watertransfer pressure(MPa)	Pipeline diameter (mm)	Valve closing time(s)
1	Bow to stern	0.8	100	1
2	Bow to stern	0.6	100	1
3	Bow to stern	0.4	100	1

Figure 4 shows the pressure curves before and after the valve corresponding to a water transfer pressure of 0.8 MPa. The pressure oscillation pattern presented by the water hammer impact in the figure is basically consistent with that in Fig. 3, both of which satisfy the law of decreasing cycle and increasing frequency of the impact pressure. The reason for this phenomenon is consistent and will not be repeated here. The peak pressure values at the measuring points before and after the valve under water transfer pressures are shown in Table 4. From the simulation calculation results, it can be seen that, with other parameters remaining constant, as the water transfer pressure increases, the peak impact pressure at the front and near measuring points of the control valve also increases. The main reason is that the greater the water transfer pressure, the faster the fluid velocity inside the pipeline. After the control valve is cut off, the liquid column separation caused by the inertial force of the fluid becomes more severe, and the impact pressure generated after the closure is greater. For water transfer systems, while ensuring a constant amount of water transfer, it is possible to consider appropriately reducing the air pressure inside the equilibrium water tank before water transfer.

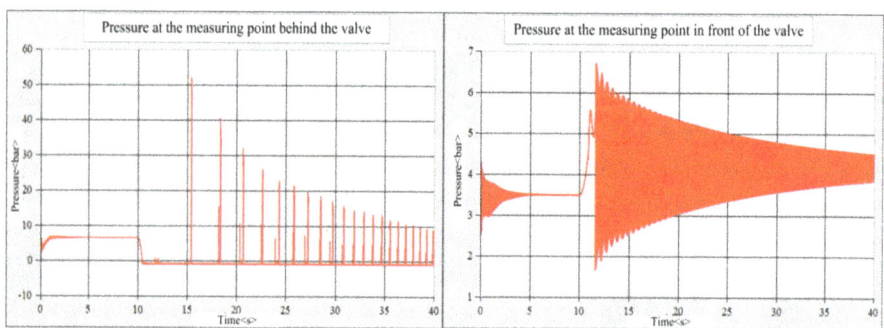

Fig. 4. Pressure at measuring points before and after the valve for water transfer pressure of 0.8MPa.

Table 4. Peak pressure at measuring points before and after the valve under different water transfer pressures.

Condition	Water transfer pressure(MPa)	Peak pressure after valve (MPa)	Peak pressure before valve(MPa)
1	0.8	5.38	1.03
2	0.6	5.19	0.86
3	0.4	4.87	0.66

4.3 Pipeline Diameter

Taking the working conditions of water transfer direction from bow to stern, water transfer pressure of 0.4 MPa, and valve closing time of 1 s as an example, the influence of pipe diameter on water hammer pressure increase is explored by changing the pipe diameter without changing other parameters. The calculation conditions are shown in Table 5.

Table 5. Calculation conditions for different pipeline diameters.

Condition	Water transfer direction	Water transfer pressure(MPa)	Pipeline diameter (mm)	Valve closing time (s)
1	Bow to stern	0.4	80	1
2	Bow to stern	0.4	100	1
3	Bow to stern	0.4	150	1

Figure 5 shows the pressure curves before and after the valve corresponding to a pipe diameter of 100 mm. The pressure oscillation pattern presented by the water hammer impact in the figure is basically consistent with that in Fig. 3, both of which satisfy the law of decreasing cycle and will not be repeated here. The peak pressure values at the measuring points before and after the valve for different pipe diameters are shown in Table 6. According to the calculation results, while keeping other parameters constant, as the pipe diameter increases, the water hammer pressure at the measuring points before and after the control valve decreases. The main reason is that the water transfer pressure remains unchanged as the driving force of the fluid inside the pipe. Increasing the pipe diameter will cause the fluid inside the pipe to slow down, and the inertia of the water flow will correspondingly decrease, thereby reducing the pressure increase and decrease of the water flow. Therefore, in order to suppress the water hammer impact pressure of the system, the diameter of the water transfer system pipeline can be appropriately increased under the allowable conditions of the overall layout.

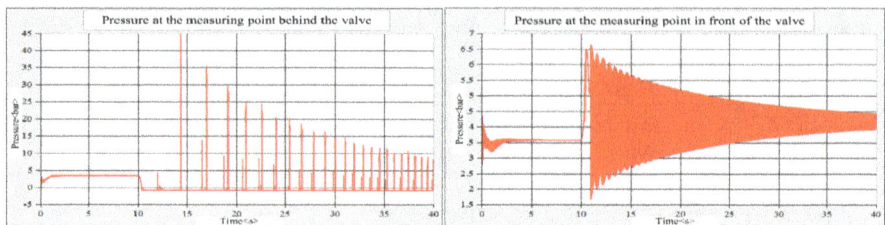

Fig. 5. Pressure at measuring points before and after the valve for pipeline diameter of 100mm.

Table 6. Pressure at measuring points before and after the valve under different pipe diameters.

Condition	Pipeline diameter(mm)	Peak pressure after valve (MPa)	Peak pressure before valve(MPa)
1	80	0.66	4.87
2	100	0.65	4.48
3	150	0.61	3.24

5 Conclusion

This article focuses on the water hammer impact problem generated during the water transfer process of a large spacing water transfer system. Based on the principle of single variable testing, a system simulation method is adopted to analyzed the relationship between factors such as valve closing time, water transfer pressure, pipeline diameter, and water hammer pressure increase. The following conclusions are drawn:

(1) Prolonging the valve closing time, reducing the water transfer pressure, and increasing the pipeline diameter can suppress the pressure increase of water hammer before and after the valve, while keeping other parameters unchanged.
(2) In response to the water hammer impact problem in the actual ship's water transfer system, measures can be taken to appropriately reduce the water transfer pressure in the balance tank and synchronously increase the pipeline diameter without changing the water transfer flow rate and accuracy indicators, while adjusting the valve controller to appropriately extend the valve closing time.

References

1. Cai, B.: The water hammer characteristic simulation and test study in warship system pipeline. Ship Sci. Technol. **33**(9), 52–55 (2011)
2. Fu, Y., Zhao, K., Qi, X., et al.: Modeling and reducing of water hammer of proportional direction valve controlled ship steering system based on AMESim. J. Beijing Univ. Aeron. Astron. **36**(6), 640–644 (2010)
3. Li, B., Cai, B., Yu, J., et al.: Research on method of water hammer suppression for marine moving water system by pump and valves. Ship Sci. Technol. **39**(1), 96–103 (2017)

4. Wang, B., Duan, Y., Li, Z.: Research of application of mechanical trim balance system in large unmanned underwater vehicle. Ship Sci. Technol. **44**(24), 45–49 (2022)
5. Shen, W., Liu, G.: Simulation on water hammer suppression on the piping of a trim balance system. J. Harbin Eng. Univ. **41**(3), 325–331 (2020)
6. Qian, Y., Xu, R., Hu, H.: Simulation and optimization of underwater complex equalization system based on AMESim. Ship Sci. Technol. **44**(19), 38–41 (2022)
7. Zhang, Y., Wu, M., Zhao, D., et al.: Research on transient flow characteristics of ship pressure pipeline. Mach. Tool Hydraul. **51**(10), 42–48 (2023)
8. Jiang, Z., Li, S., Wang, Y., et al.: Water hammer risk analysis and comprehensive protection for long distance wastewater pressurized pipeline based on transient fluid model. Water Purificat. Technol. **42**(s1), 234–239 (2023)

Open Access This chapter is licensed under the terms of the Creative Commons Attribution-NonCommercial-NoDerivatives 4.0 International License (http://creativecommons.org/licenses/by-nc-nd/4.0/), which permits any noncommercial use, sharing, distribution and reproduction in any medium or format, as long as you give appropriate credit to the original author(s) and the source, provide a link to the Creative Commons license and indicate if you modified the licensed material. You do not have permission under this license to share adapted material derived from this chapter or parts of it.

The images or other third party material in this chapter are included in the chapter's Creative Commons license, unless indicated otherwise in a credit line to the material. If material is not included in the chapter's Creative Commons license and your intended use is not permitted by statutory regulation or exceeds the permitted use, you will need to obtain permission directly from the copyright holder.

Development of a Pivoting Wave Energy Converter Concept

Gianmaria Giannini[1,2(✉)], Victor Ramos[1,2], Paulo Rosa-Santos[1,2], Esmaeil Zavvar[1,2], Tomás Calheiros-Cabral[1,2], and Francisco Taveira-Pinto[1,2]

[1] Department of Civil Engineering, Faculty of Engineering of the University of Porto (FEUP), Porto, Portugal
ggiannini@ciimar.up.pt

[2] Interdisciplinary Centre of Marine and Environmental Research CIIMAR) of the University of Porto, Porto, Portugal

Abstract. There is a need for innovative solutions to significantly increase reliability and commercial-viability of wave energy technology to take advantage of the vast wave energy resources available worldwide. Effective designs must demonstrate resilience to harsh conditions while efficiently capturing energy in normal sea states. Initial commercial efforts are likely to focus on near-shore technologies because of their cost advantages, including easier power transmission and maintenance. In this study, an innovative pivoting WEC concept suitable for near-shore uses, which has a survivability mode of operation to mitigate wave-induced structural loads during extreme sea conditions, is proposed. Recent analyses have highlighted the concept's favourable resonance properties, which enhance power production efficiency during relevant sea states. Numerical and experimental methods were employed to assess the behaviour and wave loads acting on the WEC's initial design. A potential flow method (time-domain based on diffraction radiation theory) was used to estimate motions and hydrodynamic loading, evaluating second-order wave loads, allowing for comprehensive evaluations across various sea states (of +20 minutes length) at reasonable computational costs. To assess numerical uncertainties, physical testing was conducted using an assembled 1:20 scale physical model. Initial performance assessments, validated by experimental measurements, indicate the capture width ratios of up to 40%, underscoring the concept's potential. Furthermore, recent survivability tests demonstrated up to 65% decrease of loads on the main moving floater when employing an advanced survivability mode across different critical sea states.

Keywords: Marine renewable energy · WEC Concept Analysis · WEC Survival Mode · Numerical and Physical Modelling · Hinged Device · Near-Shore Rotation WEC

1 Introduction

Wave energy converters can generate substantial energy, especially for coastal communities with increasing energy demand. Converting wave energy from the ocean can well complement the other, more mature, existing renewable energies such as solar and wind. Despite the worldwide massive wave energy resource available [1, 2], to present, the use of waves to produce energy for human activities was negligible. Currently, wave energy remains underutilized due to low readiness level and low reliability associated with existing technologies, causing available wave energy converter (WEC) technologies to provide a unfeasible levelized cost of energy impeding widespread commercialization. Key requirements of WECs relate to survivability, power performance, cost-effective maintenance and reliability aspects. In addition, it is important to notice that in the industry no merging to a specific concept is occurring yet, a shortcoming preventing teaming towards the development of a single concept. Unlike wind or solar power, where a few standard technologies dominate, wave energy has the potential for multiple types of technologies to reach commercial success. This is because different WEC systems can be tailored to specific water depths—whether shallow or deep—and to varying wave energy conditions at different locations, each with their own typical wave climate.

In the near future, close-to-land WECs are more viable compared to WECs to be installed far from the coast that instead would require large-scale deployments of many devices for ensuring economic viable projects, for which transmission and maintenance higher costs are justifiable. For near-shore WECs, their proximity facilitates accessibility for maintenance and reduces the cost of electrical connections due to shorter cable lengths [3, 4]. Typically, these devices are attached to the seabed via rigid structural components. These systems can be hinged devices that are fully submerged [5, 6] or semi submersed [7, 8]. For all type of systems the primary challenges are to survive extreme wave-induced loads during storms and maximizing power capture operational sea conditions [9].

For efficient WEC development, numerical and physical methodologies can be used to evaluate hydrodynamic loads, dynamic responses, as well as power absorption efficiency. Potential flow methods (PFM), advantageous for their computational efficiency, are widely used for hydrodynamic assessments and annual energy production (AEP) calculations. PFM are implemented in existing well-known numerical tools and software like WAMIT and ANSYS AQWA [10, 11]. PFM can be combined with a time-domain solution approach for including transient effects, such as those related to PTO components or mooring systems. When drag forces are significant, more advanced methods like computational fluid dynamics (CFD) based Navier-Stokes equations can be adopted, though they are computationally intensive and less suitable for early development stages [12]. Experimental validation follows numerical assessments, with physical models providing measurements of wave loads and WEC motions, especially in near-linear wave conditions. Various studies have effectively used experimental methods to analyze different WECs [13].

Despite over a century of WECs development [14], reliable designs remain elusive. To address these challenges, a pivoting Wave Energy Converter (PWEC) was conceptualized based on previous work [15–17] and initially developed at the University of Porto, Portugal. This paper details to latest the development of PWEC. The innovative contribution focuses a quantitative assessment of wave loads acting on PWEC for different mode of operation. The paper provides insights into the technology and explains two potential implementations. It also covers a preliminary numerical and physical modelling assessment of a new survival mode for extreme sea conditions, a key design feature.

2 PWEC Concept and Implementation Variants

The PWEC solo concept, Fig. 1 (a), was preliminarily developed and verified at FEUP. This device addresses the limitations of end-stops found in other systems. The PWEC benefits from a favourable oscillatory direction and eliminates the need for end-stops. PWEC uses a single floating block element (FBE), rotating around a central point atop a fixed supporting structure (FSS). This rotation drives a generator located at the top of the FSS. PWEC not only offers structural benefits due to its simplicity but also allows for greater oscillations due to the absence of end-stops, potentially leading to increased power absorption compared to the original system.

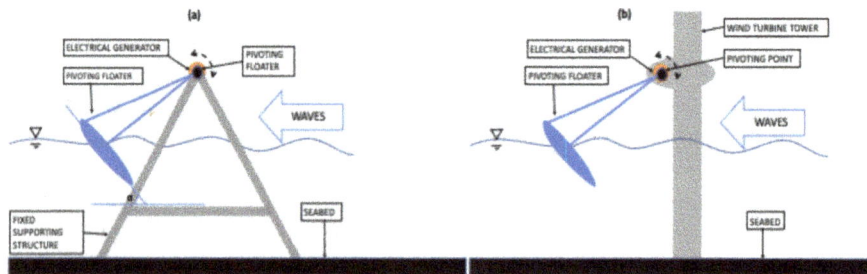

Fig. 1. (a) Pivoting WEC solo concept; (b) Pivoting WEC (two modules) integrated with fixed offshore wind turbine (side view)

Alternatively, the PWEC concept can be integrated with offshore fixed turbines, this integration offers high potential due to cost-sharing of the supporting structure for both the wind turbine and wave energy converter, Fig. 1 (b). During extreme weather, the two side-by-side floaters can be positioned either vertically at the highest rotation or submerged at the lowest position.

PWEC in these two variants is under open development for encouraging wide collaboration.

3 Previous Results, Case Study and Resource Assessment

An initial configuration of PWEC was developed and preliminarily assessed numerically by a combination of techno-economic indexes (e.g. power output, annual energy and capital cost of the device). In sea states starting from 3.5 meters Hs, the PWEC generates above about 200 KW, Fig. 2. However, these power estimates were based on linear Power Take-Off (PTO) systems, and further research is needed to explore PTO control for maximizing power absorption. Additionally, an initial economic analysis suggests that PWEC could be more cost-effective than its predecessor, with potential capital cost savings of 28% and a payback period reduced by approximately 2.5 years. For these Fig. [9], as for the novel results presented in this paper, are relative to the case study site located nearby the Port of Leixões in Matosinhos, Portugal, (41.176237 N, 8.712030 W). As for the mentioned previous studies, a potential flow numerical model is suitable for a computational time efficient hydrodynamic analysis of the device.

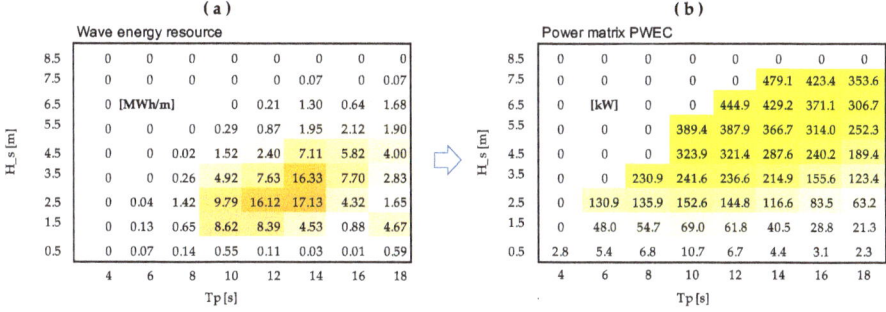

Fig. 2. (a) Wave energy resource of the case study site (Matosinhos, Portugal), and (b) Power matrix of PWEC

4 Numerical Methods

A numerical model is used to estimate wave loads relative to relevant sea states to evaluate the effectiveness of a new survival mode for the PWEC. The model is based on radiation-diffraction theory, using coupled frequency and time domain calculations, as described by Eq. 1. In this equation, m represents the FBE's mass (including added mass), while $\ddot{\xi}$ is its acceleration. The forces considered include wave radiation (f_{rad}), PTO force (f_{pto}), hydrostatic restoring force (f_{hs}), and wave forces (f_{exc}).

$$m\ddot{\xi} = -f_{rad} - f_{pto} + f_{hs} + f_{exc} \tag{1}$$

The model applies Cummins' theory [18], utilizing causal and non-causal convolution functions to calculate the wave radiation and wave excitation forces. The FBE is treated as a rigid body (no elastic deformation and only rotating along the hinge point), so flexibility is neglected in the load and motion calculations. The software ANSYS AQWA implementing radiation/diffraction theory and Wheeler stretching, for second order wave loads, is used for numerical simulations. This approach can be used to provide a preliminary evaluation of PWEC's survival mode.

5 Physical Modelling

Wave load results are compared to experimental data from hydraulic lab tests at the wave tank of the Hydraulic Lab of the University of Porto. For these tests, the tank was filled up to 0.6 m of water and the physical model was tested using 4 paddles (~3.08 m width). Three scenarios are considered, with cases for which the FBE is blocked or operational cases under typical sea conditions, Fig. 3.

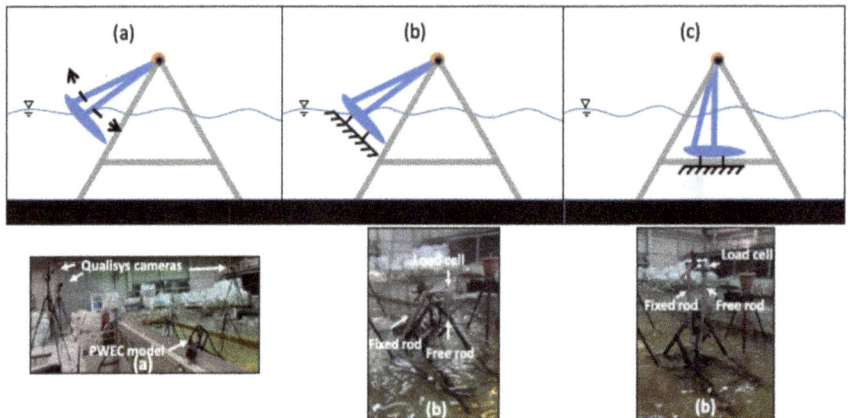

Fig. 3. Configurations assessed e relative experimental set-ups: (**a**) device in operation; (**b**) standard locked in place; and (**c**) survival locked in place

A 1:20 scale model of the PWEC allowed assessing hydrodynamic behaviour, total hydrodynamic loads and potential power production. Main system parameters and physical model dimensions are provided in Table 1. Froude scaling laws were applied, as they are typically suitable for WEC testing, especially for simple geometries like the PWEC's FBE, where viscous forces are minimal compared to inertial forces. The scaled FBE was connected to a cycle hub within a fork, allowing precise, low-friction rotation, with a hydraulic disk brake simulating the rotational PTO force. The brake system was designed to provide finely adjustable mechanical damping. This damping dissipates mechanical energy, simulating the Power Take-Off (PTO) load in the system. The brake was fine-tuned using a handle fixed by a screw rod (10 cm M4) and nut, which were regulated using caliper for ensuring repeatability of tests. The fork was securely bolted and welded to the FSS to align the rotational axis correctly. A PTO damping value was used, calibrated through dry oscillation tests. Wave probes monitored surface elevation, while the Qualisys system using infrared cameras tracked the FBE motion. A 10 kg force transducer measured wave-induced loads, with the transducer positioned out of the water to facilitate good quality measurements. Sensors' data from wave probes and loads cells was collected using the HR Wallingford data acquisition (hardware and software) system (HR DAQ). A high sampling frequency (above 100 Hz) for ensuring data high time resolution was adopted at all times. All instruments were calibrated according to recommended practices [19, 20].

Table 1. Main parameters of the study

	Model scale	Unit	Real scale	Unit		Model scale	Unit	Real scale	Unit
Overall width	88	cm	17.6	m	Floater height	64	cm	12.8	m

(*continued*)

Table 1. (*continued*)

	Model scale	Unit	Real scale	Unit		Model scale	Unit	Real scale	Unit
Total height	115	cm	23	m	Length from pivot	46	cm	9.2	m
Overall length	177	cm	38.9	m	Floater mass	13.4	kg	107	tons
Floater width	45	cm	9	m	Depth	60	cm	12	m

6 Results

To evaluate the survival mode of the PWEC, both numerical and experimental tests were conducted, focusing on horizontal wave loads. Five regular wave and five irregular wave cases were selected (based on wave tank capability) and tested.

It was found that the survival mode significantly reduced wave loads, with total hydrodynamic load on the FBE in survival locked position is about 40% reduced compared to the standard position, Fig. 4. While regular wave tests provided useful initial insights, irregular wave tests offered a more accurate representation of ocean conditions, further validating the PWEC's performance under real sea states.

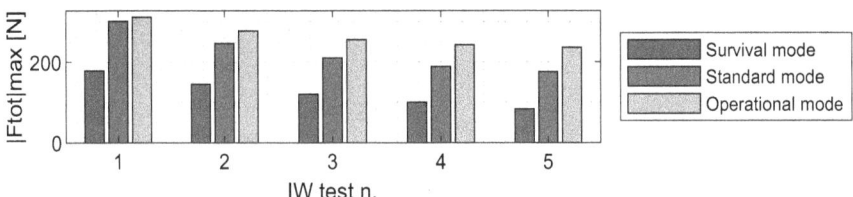

Fig. 4. Wave load (total resultant) for irregular sea states (numerical model) tests for survival (dark grey), standard (grey), and operational (light grey) modes

The comparison of experimental and numerical results was qualitatively good and quantitatively matching with exception of peak values, as illustrated in Fig. 5. Error metrics like RMSE, MAE, and R-squared were calculated, with average values of 13.21, 8.82 N, and 0.64, respectively, Table 2.

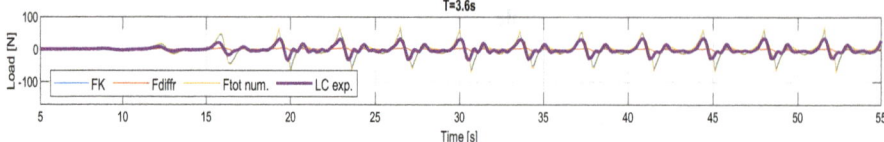

Fig. 5. Illustration of numerical calculated and experimental measured horizontal hydrodynamic loads F_x on PWEC (survival position) for a regular wave test (H = 0.2 m, T = 3.6 s)

Table 2. Irregular waves tests results

Period (s)	Froude-Krylov (N)	Diffraction (N)	Total (N)	Load cell (N)	RMSE (N)	MAE (N)	R2
1.79	46.49	5.48	49.34	33.68	13.58	9.75	0.70
2.24	45.50	5.24	49.26	34.48	12.92	9.02	0.68
2.68	41.16	4.36	44.89	29.86	10.72	7.49	0.67
3.13	72.37	7.62	75.51	32.82	14.78	9.06	0.59
3.58	70.07	8.17	73.90	36.40	14.02	8.78	0.56

7 Conclusions

This paper covers an initial assessment of the PWEC concept with a focus on its survival mode, aimed at reducing wave-induced loads. Using both numerical and physical modelling, the study demonstrated that the survival mode can significantly mitigate wave loads on the device's floater. Load reductions of up to 40% were observed compared to the standard mode, and up to 65% compared to the operational mode, under both regular and irregular wave conditions.

The use of the developed potential flow model for the scope of the study and cases considered was supported by physical tests using a 1:20 scale physical model. Despite limitations, such as the exclusion of viscous forces, the experimental results validated the numerical predictions with an average root mean square error (RMSE) of 13.21 N and a mean absolute error (MAE) of 8.82 N for wave load estimations. Despite their validity, these results underscore the importance of incorporating viscous forces in future analyses, particularly through CFD simulations and additional testing using larger waves or a smaller model.

Further work is required to validate the survival mode under more extreme and realistic sea states, which were not fully addressed due to the limitations of the current experimental setup. Additionally, the study suggests that implementing the survival mode could be beneficial during pre-commercial trials, potentially activating it at wave heights below extreme conditions to ensure device safety.

For commercial viability, the study recommends conducting a comprehensive cost-benefit analysis, factoring in manufacturing, installation, and maintenance costs. This analysis, alongside environmental impact assessments, is crucial to determine the economic feasibility and environmental safety of the PWEC technology before scaling up to full-size deployments.

References

1. Gunn, K., Stock-Williams, C.: Quantifying the global wave power resource. Renew. Energy. **44**, 296–304 (2012)
2. Mørk, G., et al.: Assessing the global wave energy potential. In: Proceedings of the ASME 2010 29th International Conference on Ocean, Offshore and Arctic Engineering, vol. 3, pp. 447–454 (2010)
3. Cheng, Y., et al.: Wave energy extraction for an array of dual-oscillating wave surge converter with different layouts. Appl. Energy. **292**, 116899 (2021)
4. Wilkinson, L., et al.: The power-capture of a nearshore, modular, flap-type wave energy converter in regular waves. Ocean Eng. **137**, 394–403 (2017)
5. Energy, C.C. CETO. 2020 15/06/2020]; Available from: https://www.carnegiece.com/technology/
6. Henry, A., et al., Advances in the Design of the Oyster Wave Energy Converter. 2010
7. Eco-Wave-Power. [15/06/2023]; Available from: https://www.ecowavepower.com/. 2023
8. Coiro, D.P., et al.: Wave energy conversion through a point pivoted absorber: Numerical and experimental tests on a scaled model. Renew. Energy. **87**, 317–325 (2016)
9. Giannini, G., et al.: Wave energy converters design combining hydrodynamic performance and structural assessment. Energy. (2022)
10. WAMIT. WAMIT®—The state of art in wave interaction analysis. [15/04/2024]; Available from: https://www.wamit.com/. 2024
11. ANSYS: Aqwa Theory Manual, Canonsburg (2015)
12. Davidson, J., Costello, R.: Efficient nonlinear hydrodynamic models for wave energy converter design—A scoping study. J. Mar. Sci. Eng. **8**(1) (2020)
13. Pecher, A., Kofoed, J.P.: Handbook of Ocean Wave Energy. Springer, Cham (2017)
14. Falnes, J.: A review of wave-energy extraction. Mar. Struct. (2007)
15. Giannini, G., et al.: Geometry assessment of a sloped type wave energy converter. Renew. Energy. **171**, 672–686 (2021)
16. López, M., Taveira-Pinto, F., Rosa-Santos, P.: Numerical modelling of the CECO wave energy converter. Ren. Energy. **113**, 202–210 (2017)
17. Marinhero, J., et al.: Feasibility study of the CECO wave energy converter, in Maritime Technology and Engineering, pp. 1259–1267. C.-T.a.F. Group. Editor (2014)
18. Cummins, W.E.: The impulse response function and ship motions. Schiffstechnic, 101–109 (1962)
19. Giannini, G., et al.: Wave energy converter power take-off system scaling and physical modelling. J. Mar. Sci. Eng. **8**(9), 632 (2020)
20. Payne, G.: Guidance for the experimental tank testing of wave energy converters. In: SUPERGEN Marine Project. University of Edinburgh (2008)

Open Access This chapter is licensed under the terms of the Creative Commons Attribution-NonCommercial-NoDerivatives 4.0 International License (http://creativecommons.org/licenses/by-nc-nd/4.0/), which permits any noncommercial use, sharing, distribution and reproduction in any medium or format, as long as you give appropriate credit to the original author(s) and the source, provide a link to the Creative Commons license and indicate if you modified the licensed material. You do not have permission under this license to share adapted material derived from this chapter or parts of it.

The images or other third party material in this chapter are included in the chapter's Creative Commons license, unless indicated otherwise in a credit line to the material. If material is not included in the chapter's Creative Commons license and your intended use is not permitted by statutory regulation or exceeds the permitted use, you will need to obtain permission directly from the copyright holder.

Research on Hydrodynamic Characteristics of A SWATH Based on CFD

Yilin Yang[1,2(✉)], Hao Hao[1,2], and Minmin Zheng[1,2]

[1] Shanghai Ship and Shipping Research Institute Co., Ltd, Shanghai, China
yang.yilin@coscoshipping.com
[2] State Key Laboratory of Maritime Technology and Safety, Wuhan, China

Abstract. The research object of this article is a SWATH, and numerical calculations were conducted on a SWATH model under various operating conditions using viscous CFD software STAR-CCM+. Explored the hydrodynamic performance of SWATH in still water and waves, as well as their longitudinal motion stability at different speeds. By observing and analyzing the time history curves of longitudinal motion at different speeds, it was determined whether the ship had lost its longitudinal motion stability under various speed conditions, and the motion response of SWATH on regular waves was studied.

Keywords: SWATH · Resistance Calculation · Longitudinal Motion · Regular Wave

1 Introduction

Compared to conventional monohull ships, SWATH have better wave resistance due to their small waterline area and low wave disturbance force; Having a longer natural period, less prone to resonance, and smaller motion amplitude. At the same time, the ship type is a catamaran structure, which effectively improves the lateral motion and provides a spacious deck area. Because SWATH have a very small waterplane area, they can only provide very small longitudinal restoring moment.

Latore R [1]. conducted longitudinal motion tests on a SWATH under different wave direction conditions, and compared the experimental results with theoretical calculation results. The conclusion was drawn that viscous damping cannot be ignored in predicting the longitudinal motion amplitude of this ship type. Brizzolara and Stefano [2] conducted in-depth research on the design direction of SWATH in terms of ship type concepts, such as changing hull dimensions and optimizing hydrodynamic performance, using CFD methods. Based on the slicing theory, Mao Xiaofei [3] from Wuhan University of Technology applied a complete set of hydrodynamic calculation formulas for SWATH to expand the wave resistance prediction program for conventional catamarans, and wrote the WAVMO-SWATH program for calculating and predicting the wave resistance of small waterline area catamarans. Xuyu Ouyang [4] verified the optimization effect of SWATH is verified by high precision numerical simulation method.

2 Research on SWATH Resistance Characteristics

The resistance performance of a ship is one of the basic properties of a ship. There are many classification methods for resistance based on different needs. In the study of the resistance performance of SWATH, based on the small waterline surface area and slender hull, which generates small waves, the total resistance is divided into frictional resistance and residual resistance in this article to study the resistance characteristics of this ship type. Frictional resistance is the combined force of tangential forces on the surface of the ship due to the viscous effect of water; The residua resistance refers to the wave making resistance generated by the rising waves during the motion of the ship, and the viscous effect of water causes the pressure in the front and rear parts of the ship to be unequal, resulting in viscous pressure resistance.

This article adopts the validated calculation scheme in "CFD and Model Test Study on the Effect of Stabilized Fins and Speed on SWATH" [5] to study the resistance performance of SWATH and analyze the resistance characteristics of this ship type. This article selects a SWATH model as a calculation example, and its main dimensions are shown in Table 1:

Table 1. Main dimensions of SWATH model

Displacement (kg)	Total length of hull (m)	Design draft (m)	Maximum pillar thickness (m)	Pillar length (m)	Submersible radius (m)	Waterline surface area (m2)
164	2.95	0.263	0.815	0.095	2.5119	0.0985

The schematic diagram of the model is shown in Fig. 1:

Fig. 1. The schematic diagram of the model

This article calculated 8 velocity points, and the resistance calculation results are shown in Fig. 2. The displacement provided by the SWATH for buoyancy is concentrated on the submerged body far from the water surface. The narrow waterline surface effectively reduces the wave making resistance, especially when the Froude number is high, which has a more significant effect on the reduction of wave making resistance. However, due to the twin hull structure of the small waterline area catamaran and the increased wet surface area to ensure deep submergence, it is much larger than conventional single hull ships with the same displacement. So, at the same speed, the frictional resistance of a SWATH is significantly greater than that of a conventional monohull.

However, when the ship is sailing at medium to high speeds, as shown in the following figure, due to its special small waterline surface hull type, the proportion of wave making resistance to total resistance begins to increase compared to frictional resistance. This gradually reflects the performance advantage of SWATH in terms of resistance.

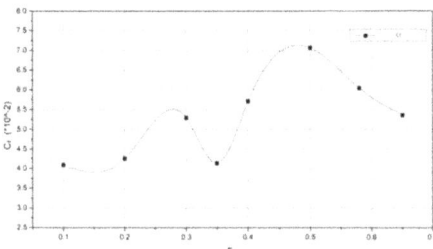

Fig. 2. Total resistance coefficient curve.

From Fig. 3, it can be seen that when the Fr value is around 0.3 and 0.5, there are two peaks in the residual resistance coefficient. Within this range, there is a clear valley, indicating that at this speed, favorable wave added interference is generated after the first peak, effectively reducing the wave added resistance; After Fr > 0.5, the pattern changes and the residual resistance coefficient begins to decrease. This is beneficial for the special ship type of SWATH, which effectively controls the wave making resistance at high speeds.

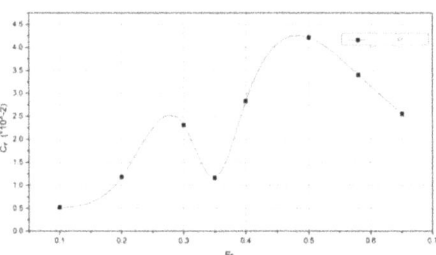

Fig. 3. residual resistance coefficient curve

This article uses the VOF method to capture the free liquid surface and simulate the changes in the flow field at different speeds. The following are the free liquid surface waveforms with Fourier numbers of 0.3 and 0.5, respectively:

3 Study on Longitudinal Motion Stability of SWATH

The definition of motion stability problem is the ability of a ship to return to its initial equilibrium position after being subjected to external interference and leaving its initial position. If it can return to the initial position, it is judged as stable longitudinal motion,

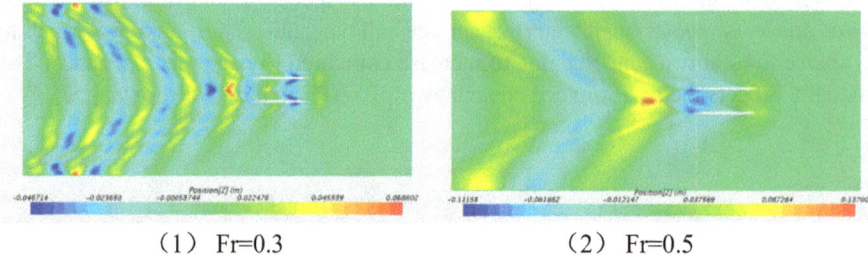

(1) Fr=0.3　　　　　　　　(2) Fr=0.5

Fig. 4. Free liquid surface waveform diagram

and vice versa, it is considered unstable. When a SWATH is in motion, its stability in the longitudinal plane is much more complex than that of conventional ship types. This article explores the dynamic longitudinal stability of this ship type, which considers the stability problem of the hydrodynamic longitudinal moment generated by the ship's movement on the hull. Due to its unique ship type characteristics, SWATH have a smaller water surface area, resulting in a decrease in the longitudinal recovery moment. When the speed of SWATH reaches medium to high speeds, the MUNK moment acting on the underwater hull will have an effect, making it easy for the ship to lose longitudinal stability. Therefore, when the speed of this ship type reaches a certain value, the stability of its longitudinal motion becomes a must consider issue (Fig. 4).

From the time history of longitudinal motion at different speeds in Fig. 5, it can be visually judged whether the ship has lost stability by the time history of longitudinal motion of the SWATH. If the time history curve of longitudinal motion of the SWATH can reach equilibrium at a certain speed or oscillate and gradually converge towards the equilibrium position, then the ship has longitudinal stability, and vice versa, longitudinal stability is lost.

The SWATH is stable in motion when sailing at a lower speed (Fr = 0.1 ~ 0.3). At the beginning of the sailing motion, there is some amplitude of oscillation in the longitudinal motion, and it gradually returns to the equilibrium position in the form of amplitude attenuation; When the Froude number approaches 0.4, it can be seen from the time history chart that although its longitudinal tilt value oscillates and decays, when the oscillation amplitude decreases to a certain value, it still fails to return to the equilibrium position and is basically in a state of losing longitudinal stability; When the Froude number reached 0.5, the SWATH was already sailing at a relatively high speed and could not return to its equilibrium position due to disturbance. The longitudinal and heave time charts clearly showed that the ship had a significant reciprocating motion, indicating that the SWATH had lost its longitudinal stability at this time.

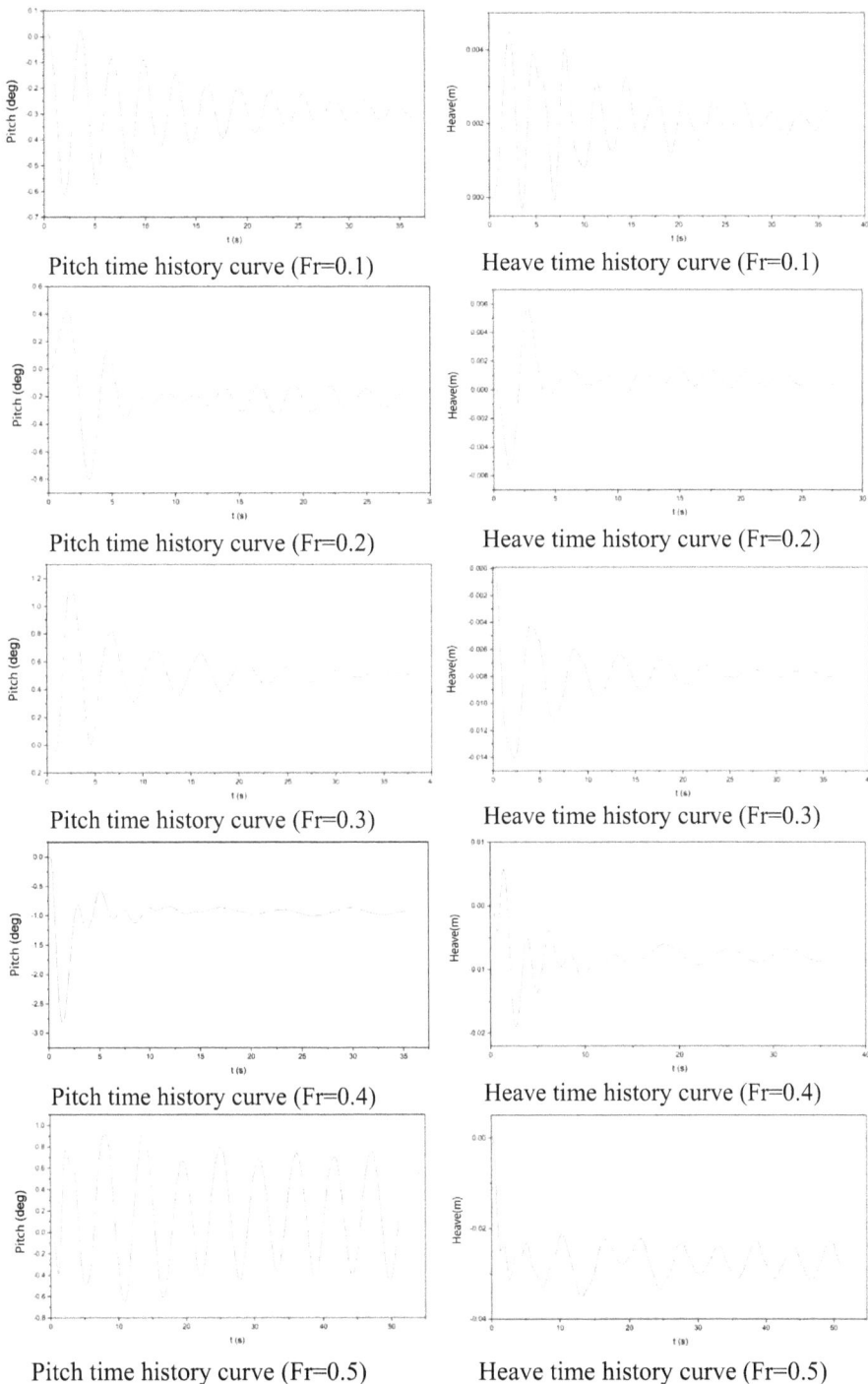

Fig. 5. Longitudinal motion time history.

In the research process of Lin Zheng [6] on the longitudinal motion stability of SWATH based on potential flow theory, it was found that the ship type began to be in an unstable state according to the judgment of characteristic roots when the Froude number reached 0.4. In this study, viscous CFD technology was used to analyze the longitudinal motion time history curve of the bare hull of a SWATH. It was also found that the ship type began to lose stability around the Froude number of 0.4 and exhibited significant reciprocating motion when the Froude number reached 0.5. The comparative statistics are shown in Table 2.

Table 2. Longitudinal motion stability of SWATH at different speeds.

Fr	0.1	0.2	0.3	0.4	0.5
CFD	stable	stable	stable	unstable	unstable
Potential flow theory	stable	stable	stable	unstable	unstable

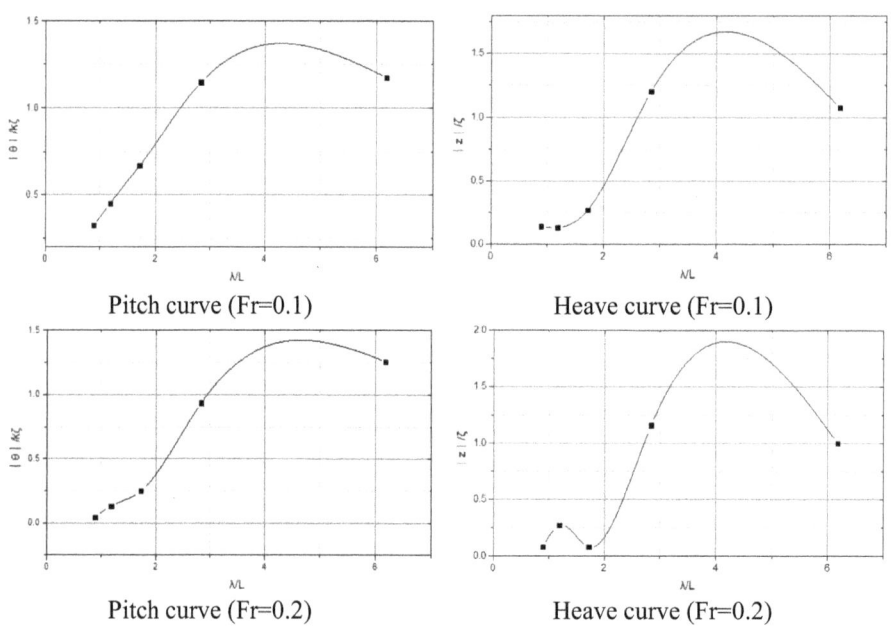

Pitch curve (Fr=0.1) Heave curve (Fr=0.1)
Pitch curve (Fr=0.2) Heave curve (Fr=0.2)

Fig. 6. Longitudinal motion transfer function.

4 Study on Motion Response of SWATH

In order to facilitate the comparison and analysis of calculation results, a dimensionless motion response transfer function is defined, where the transfer function of the heave response is $|Z_a|/\zeta_a$, the transfer function of the pitch response is $|\theta_a|/(k\zeta_a)$, where ζ_a is the amplitude of the incident wave and k is the number of waves of the incident wave. Fig. 6 shows the calculated longitudinal motion of the ship under different incident wavelengths of Fr $= 0.1$ and 0.2, and plots the transfer functions of the ship's heave and pitch motion as a function of the dimensionless encounter frequency.

5 Conclusion

(1) The static water resistance performance of a SWATH is significantly different from that of a conventional monohull, with two peaks in its residual resistance coefficient. As the speed increases, favorable wave making interference is generated due to the special ship type, which significantly reduces the wave making resistance of the ship type; Moreover, due to the slender and elongated body at the water surface, as the speed reaches medium to high speeds, the residual resistance does not account for the majority of the total resistance compared to conventional monohull ships, but rather decreases relative to frictional resistance. Therefore, in terms of resistance performance, SWATH are more suitable for medium to high speed navigation.

(2) By analyzing the longitudinal motion history of a SWATH at different speeds, it can be concluded that when Fr $= 0.4$, the ship type has basically lost its longitudinal motion stability, and the longitudinal motion fluctuates slightly in the equilibrium position. When Fr $= 0.5$, the longitudinal motion has begun to fluctuate violently, which is sufficient to determine that the longitudinal motion stability has been lost at this speed.

(3) In the numerical calculation of the motion of a SWATH in waves, this article simplifies the upper hull part in the numerical calculation. In future research work, precise modeling of the hull of a SWATH can also be carried out to study the motion characteristics of the ship on waves at higher speeds.

References

1. Latorre, R., Vasconsellos, J.: Study of hull angle influence on SWATH heave and pitch motions. Nav. Eng. J. **113**(1), 63–70 (2001)
2. Brizzolara, S., Chryssostomidis, C.: The second generation of unmanned surface vehicles: design features and performance predictions by numerical simulations. In: ASNE Day, vol. 2013, pp. 1–10 (2013)
3. Xiaofei, M.: Prediction of motion response of SWATH in waves. Mar. Eng. **34**(4), 13–15 (2005)
4. Ouyang, X., Hao, Z., & Shi, S. (2024, June). Optimization of resistance performance of semi-small waterplane area Twin-Hull. In ISOPE International Ocean and Polar Engineering Conference (pp. ISOPE-I). ISOPE
5. Yang, Y., Zheng, M., Wu, Z. and Li, J., (2020, October). Numerical study of the effect of stabilized fins on A SWATH in head waves. In ISOPE International Ocean and Polar Engineering Conference (pp. ISOPE-I). ISOPE
6. Lin, Z.: A Study on the Motion and Stability of SWATH in Waves. (Master's thesis,. Wuhan University of Technology (2009)

Open Access This chapter is licensed under the terms of the Creative Commons Attribution-NonCommercial-NoDerivatives 4.0 International License (http://creativecommons.org/licenses/by-nc-nd/4.0/), which permits any noncommercial use, sharing, distribution and reproduction in any medium or format, as long as you give appropriate credit to the original author(s) and the source, provide a link to the Creative Commons license and indicate if you modified the licensed material. You do not have permission under this license to share adapted material derived from this chapter or parts of it.

The images or other third party material in this chapter are included in the chapter's Creative Commons license, unless indicated otherwise in a credit line to the material. If material is not included in the chapter's Creative Commons license and your intended use is not permitted by statutory regulation or exceeds the permitted use, you will need to obtain permission directly from the copyright holder.

Influence of Stable Fins on SWATH Motion

Yilin Yang[1,2(✉)], Minmin Zheng[1,2], and Hao Hao[1,2]

[1] Shanghai Ship and Shipping Research Institute Co., Ltd, Shanghai, China
yang.yilin@coscoshipping.com
[2] State Key Laboratory of Maritime Technology and Safety, Wuhan, China

Abstract. For a SWATH equipped with stable fins, numerical calculations were conducted at a speed of 0.5 with the Froude number. The results showed that the SWATH's heave and pitch motion stability was guaranteed after the installation of stable fins, proving the advantage of stable fins for heave and pitch motion stability. By studying the influence of the longitudinal installation position, area, and combination method of the stable fins on the heave and pitch motion stability, it was analyzed that the tail fin area of the ship needs to reach 6% of the submerged body projection area to ensure heave and pitch motion stability. Taking 10% to 20% of the tail fin area for the head fin can achieve good heave and pitch motion performance and have a significant beneficial effect on the wave resistance performance.

Keywords: SWATH · Stable Fin · Heave and Pitch Motion · Regular Wave

1 Introduction

Due to the slender pillars at the waterline surface of the SWATH, the small waterline surface area results in insufficient longitudinal recovery torque that can be provided. When a ship is sailing at medium to high speeds, the hydrodynamic longitudinal moment acting on the hull will increase to a point where it cannot be balanced by the restoring moment. Therefore, whether the heave and pitch motion stability of a SWATH meets the requirements becomes a necessary consideration.

Dong Zushun and Dong Wencai [1] from Naval Engineering University developed a theoretical prediction program for the seakeeping performance of SWATH, and analyzed the effects of ship type, stabilizer fin size, longitudinal installation position, and combination mode changes on the seakeeping performance of SWATH. Mao Xiaofei [2] from Wuhan University of Technology used a 390 t oilfield small water surface traffic ship as the calculation object to analyze the heave and pitch of the ship under various sea conditions in regular and irregular waves. The influence of setting the head and tail fins on wave resistance was compared, and the calculation showed that the setting of stable fins greatly reduced the peak frequency response of heave and pitch motion. Fang, M. -C [3] applied real-time simulation technology to control stable fins to reduce the pitch motion of SWATH in long peak waves, and compared the results with the pitch motion results of SWATH equipped with fixed stable fins. It was found that self adjusting stable fins can make the control of ships more effective at medium and high speeds. Wang, J [4]. specifically addresses the heave and pitch motion stability of a variable-structure Small Waterplane Area Twin-hull Ship (SWATH) equipped with twin hydrofoils, proposed strategies to mitigate bow-diving during high-speed tests.

2 Analysis of the Impact of Stable Fins

This article selects a SWATH model as a calculation example, with the main dimensions shown in Table 1. The NACA symmetric airfoil series is used because of its good aerodynamic performance. The projection profile of the fins is trapezoidal, and the aspect ratio ranges from 1.0 to 1.8. This article selects the stable fins of NACA symmetric airfoils with an average aspect ratio of 1.2 as the research object, and the selected navigation state is the ship sailing in still water at a speed of $Fr = 0.5$.

Table 1. Main dimensions of SWATH model.

Displacement (kg)	Total length of hull (m)	Design draft (m)	Maximum pillar thickness (m)	Pillar length (m)	Submersible radius (m)	Waterline surface area (m2)
164	2.95	0.263	0.815	0.095	2.5119	0.0985

The schematic diagram of the model is shown in Fig. 1:

Fig. 1. The schematic diagram of the model.

The initial installation position of the tail fin is set at a distance of 0.35 L from the middle of the ship behind the ship, and the head fin is installed at a distance of 0.23 L from the middle of the ship in front of the ship. The projection area of the head fin is 20% of the projection area of the tail fin, and the projection area of the tail fin is 8% of the projection area of the submerged body. This article uses the validated calculation scheme in the "CFD and Model Test Study on the Effect of Stabilized Fins and Speed on SWATH" article [5] to numerically simulate and calculate the amplitude and pitch angle of the heave motion of the SWATH with a head and tail fin, and compares it with the heave and pitch motion time history curve of the bare hull. As shown in Fig. 2.

According to Fig. 2, it can be seen that when the stable fins are installed on the SWATH, under the condition of a Froude number of 0.5, its pitch and heave movements quickly return to the equilibrium position without reciprocating oscillation. Compared with the naked ship, the stable sailing state in still water at a speed of 0.5 has been greatly

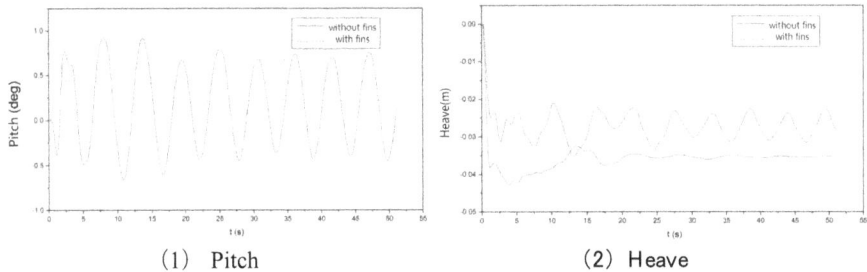

(1) Pitch (2) Heave

Fig. 2. Time history curve of SWATH with and without stable fins (Fr = 0.5).

improved, effectively improving the longitudinal stability of the SWATH at medium and high speeds, and providing sufficient recovery torque.

3 Design of Stable Fin Scheme

The geometric factors that have a decisive impact on the hydrodynamic lift of fins include the geometric shape, size, installation position, and angle of attack of the fins. This article selects three parameters that have a significant impact, such as the installation position, combination form, and size of the fins, to analyze the heave and pitch stability of a SWATH with stable fins.

3.1 Only Changing the Combination of Fins

The initial installation position of the tail fin is set at 0.35 L aft of the ship, and the installation position of the head fin is 0.23 L forward of the ship. The projection area of the tail fin accounts for 8% of the projection area of the submerged body. The combination form is that the area of the head fin is 5%, 10%, and 20% of the area of the tail fin. The calculation results are shown in Fig. 3.

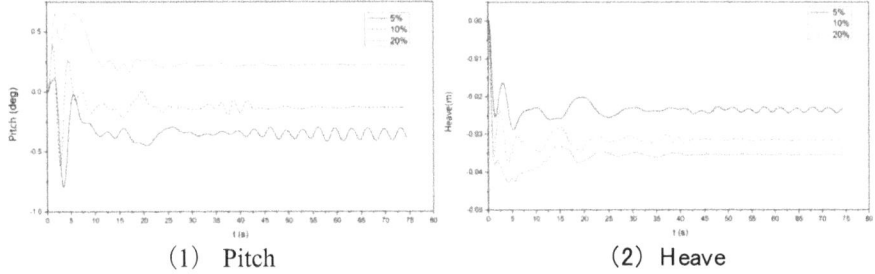

(1) Pitch (2) Heave

Fig. 3. Time history curves of different combinations of fins.

From the chart, it can be seen that when the head fin area/tail fin area is 5%, the pitch and heave curves cannot fully recover to the equilibrium position, and there are still small oscillations that cannot converge. However, when the head fin area/tail fin area is 10% and 20%, the final pitch and heave movements can be observed on the chart to converge, and the curve converges faster when the head fin area/tail fin area is 20%. Therefore, as the area of the head fin increases, the convergence of the oscillation response is also accelerating, which is beneficial for improving the seakeeping performance of the ship type. In Lin Zheng [6]'s research results, as the area of the head fin increases, the damping of the heave and pitch motion of the ship also increases, resulting in an increase in the natural pitch period and a decrease in the half life period. The CFD calculation results in this article are compared with the research results of Lin Zheng. There is a deviation in the results when the head fin area/tail fin area is 5%, the comparison results are shown in Table 2. The small head fin area cannot provide sufficient motion stability for the ship type.

Table 2. Longitudinal stability after changing the combination of fins.

Head fin area/ tail fin area	5%	10%	20%
CFD	unstable	stable	stable
Potential flow theory	stable	stable	stable

3.2 Only Change the Longitudinal Installation Position of the Fins

At this point, the projection area of the tail fin is set to 8% of the projection area of the submerged body, and the area of the head fin is set to 20% of the tail fin area and remains unchanged. Now, only the longitudinal installation position of the fin is changed. The schemes are to make changes to the longitudinal installation position of the tail fin, and the longitudinal installation position of the head fin remains unchanged at distances of 0.25 L, 0.3 L, and 0.35 L from the ship. The calculation results are shown in Fig. 4.

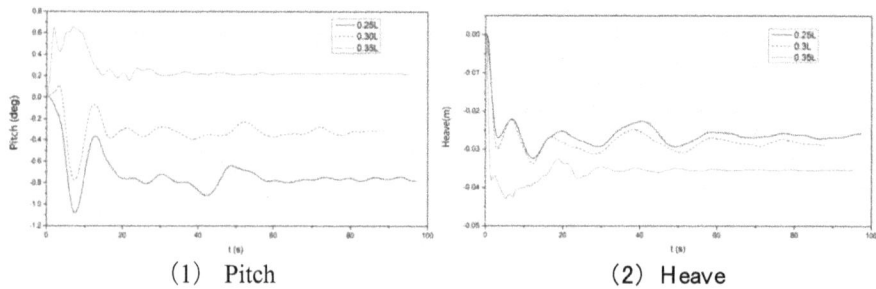

(1) Pitch (2) Heave

Fig. 4. Time history curves of different tail fin installation positions.

From the chart, it can be seen that when the front and tail stable fins are installed on a SWATH, the longitudinal installation position of the fins has little effect on the heave and pitch motion stability of the ship. When the installation position of the tail fin is 0.25 L away from the center of the ship, the pitch time history chart of the SWATH under this scheme is observed. The pitch value is the smallest among the three schemes and cannot fully return to the stable position, resulting in small amplitude oscillation. On the other hand, the installation position of the tail fin is 0.35 L away from the center of the ship. Under this scheme, the SWATH can quickly and effectively return to the equilibrium state, which is more reasonable than other schemes. In theory, as the installation position of the tail fin moves backward, the greater the torque generated, the better the longitudinal stability of the ship. The comparison results are shown in Table 3.

Table 3. Longitudinal stability after changing the longitudinal installation position of the tail fin.

Longitudinal position between the tail fin and the middle of the ship	0.25 L	0.3 L	0.35 L
CFD	unstable	stable	stable
Potential flow theory	unstable	stable	A

3.3 Only Change the Area of the Fins

The longitudinal installation position of the tail fin is set to be 0.35 L from the middle of the ship behind the ship, and the longitudinal installation position of the head fin is 0.23 L from the middle of the ship in front of the ship. The area of the head fin is selected as 20% of the tail fin, and it changes with the change of the tail fin area. The variables selected in this summary are the change of the tail fin area, and the formulas are tail fin area/submarine projection area = 4%, 6%, 8%, and 15%. The calculation results are shown in Fig. 5.

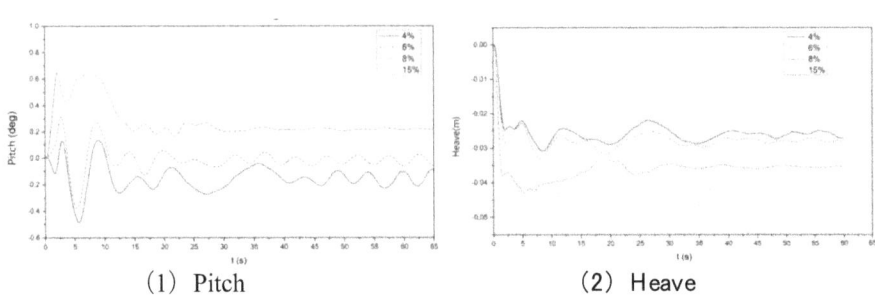

(1) Pitch (2) Heave

Fig. 5. Time history curve of changing fin area.

From the chart, it can be seen that when the area of the stable fins is too small, the ship still lacks stability in heave and pitch motion. As the fin area increases, the amplitude of the longitudinal reciprocating oscillation motion of the ship decreases significantly. When the area of the tail fin/projected area of the submersible is 6%, the ship's motion reaches a stable state. When the tail fin area/submerged body projection area is 8%, the stable fin completely overcomes the instability of the SWATH at this speed. When the area of the stable fins increases to a certain value, the ship loses longitudinal stability again and begins to oscillate repeatedly, indicating that excessive fin area also has adverse effects. The excess hydrodynamic lift provided by fins can also cause the ship to lose heave and pitch motion stability. Generally speaking, the higher the speed, the greater the requirement for the tail fin area. The comparison results are shown in Table 4.

Table 4. Longitudinal stability after changing fin area.

Tail fin projection area/submersible projection area	4%	6%	8%	15%
CFD	unstable	unstable	stable	unstable
Potential flow theory	unstable	stable	stable	stable

4 The Influence of Stable Fins on the Motion of Ships in Waves

This section compares and analyzes a SWATH equipped with stable fins with a bare hull. In this section of the calculation, five different wave conditions with a wavelength to length ratio of 3 to 7 were selected under the condition of heading waves. The calculated speed is Fr = 0.4, as shown in Fig. 6.

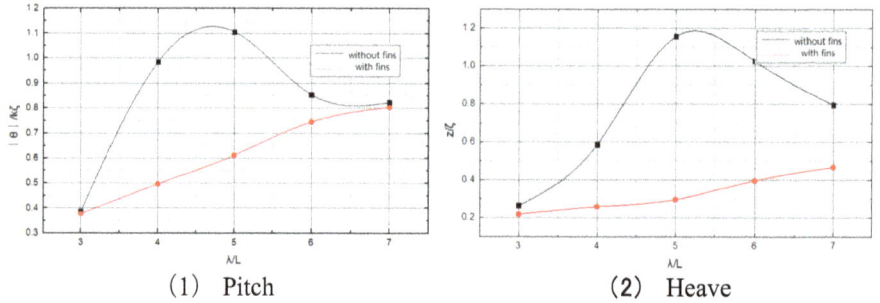

(1) Pitch (2) Heave

Fig. 6. Heave and pitch motion amplitude with and without stable fins in waves.

From this section, it can be seen that the installation of stable fins has a beneficial effect on the motion of SWATH in waves, effectively reducing the amplitude of the ship's motion in waves. Moreover, after installing stable fins, the SWATH can maintain a balanced floating state when moving in waves. Therefore, stable fins have become an integral part of SWATH in actual navigation, ensuring not only the stability of heave and pitch motion but also providing better seakeeping performance for this ship type.

5 Conclusion

This article evaluates the impact of installing stable fins on SWATH, mainly considering the influence of the installation position and combination form of fins on the stability of the ship type at a certain speed. The heave and pitch motion time history curves of

SWATH under different fin design schemes are drawn, and the numerical simulation results and curves are analyzed. Some rules for selecting stable fins are summarized:

(1) After installing stable fins, the heave and pitch motion stability of the SWATH during medium to high speed navigation has been effectively improved. After receiving disturbances, it can quickly oscillate and converge to the equilibrium position, and the motion of the ship in waves has been significantly improved.
(2) At a given ship type and speed, in order to ensure the heave and pitch motion stability of a SWATH, the area of the head fin needs to reach 10% of the area of the tail fin, while keeping the area of the tail fin constant. Taking into account the ship's motion performance, taking 20% of the area of the head fin is more conducive to improving the heave and pitch motion stability.
(3) When the head fin does not move, only the longitudinal position of the tail fin is changed, and within a certain range, as the tail fin moves towards the stern direction, the heave and pitch motion stability of this ship type is better. In theory, as the tail fin moves towards the stern direction, the longer its force arm, the greater the applied torque, which can improve the heave and pitch motion stability of the ship.
(4) The fin area needs to be moderate, and fins that are too large or too small cannot maintain stability in heave and pitch motion.

References

1. Zushun, D., Wencai, D.: Simplified Criteria and Analysis of Heave and Pitch Motion Stability for Small Waterline Area Catamarans. Shipbuilding of China (1994)
2. Xiaofei, M.: Prediction of motion response of SWATH in waves. Mar. Eng. **34**(4), 13–15 (2005)
3. Chiou, S.C., Fang, M.C.: SWATH ship motion simulation based on a self-tuning fuzzy control. J. Ship Res. **44**(02), 108–119 (2000)
4. Wang, J., Zhuang, J., Jin, M., Xu, F., Su, Y.: Experimental and numerical study on the adjustment of twin hydrofoils of a variable-structure SWATH. Ocean Eng. **303**, 117712 (2024)
5. Yang, Y., Zheng, M., Wu, Z., Li, J.: Numerical study of the effect of stabilized fins on A SWATH in head waves. In: ISOPE International Ocean and Polar Engineering Conference (pp. ISOPE-I). ISOPE (2020)
6. Lin, Z.: A Study on the Motion and Stability of SWATH in Waves (Master's thesis, Wuhan University of Technology) (2009)

Open Access This chapter is licensed under the terms of the Creative Commons Attribution-NonCommercial-NoDerivatives 4.0 International License (http://creativecommons.org/licenses/by-nc-nd/4.0/), which permits any noncommercial use, sharing, distribution and reproduction in any medium or format, as long as you give appropriate credit to the original author(s) and the source, provide a link to the Creative Commons license and indicate if you modified the licensed material. You do not have permission under this license to share adapted material derived from this chapter or parts of it.

The images or other third party material in this chapter are included in the chapter's Creative Commons license, unless indicated otherwise in a credit line to the material. If material is not included in the chapter's Creative Commons license and your intended use is not permitted by statutory regulation or exceeds the permitted use, you will need to obtain permission directly from the copyright holder.

Discussion on Active Fire Protection Systems for Offshore Substation of Wind Power Station

Song Li([✉]), Min Dai, Zhuoping Yuan, and Wei Jin

POWERCHINA Kunming Engineering Co. Ltd., Kunming, China
2015047_kmy@powerchina.cn

Abstract. The offshore wind development activity is accelerating worldwide in recent years and the fire protection of the offshore substation gets more and more important accordingly. In this article, the commonly adopted active fire protection systems including portable extinguisher, water mist system, deluge system, pressure water-spraying system, sprinkler system, gaseous system, foam system recommended in three representative specifications or standards namely *Offshore Standard* (DNV-OS-J201) *Offshore Substations for Wind Farms* compiled by DET NORSKE VERITAS, *Code for 110 kV ~ 220 kV Offshore Substation Design of Wind Power Projects* (NB/T 31115–2017) compiled by China electricity council, *Standard for design of offshore wind farm* (GB/T 51308–2019) compiled by China renewable energy engineering institute, are listed, compared and analyzed. The preferred system for each areas in offshore substation of wind power station is proposed. Considering the fire-fighting performance, environmentally-friendly characteristics, maintenance ease and cost of use, the water mist system is recommended for main transformer room, diesel engine room, reactor room, switch gear room, GIS room and temporary refuge or shelter area. The fixed pressure water- spraying system is recommended for machinery spaces. The deluge system is recommended for rooms containing cylinders with compressed gas. The sprinkler system is recommended for accommodations. The water monitors are recommended adopted on helideck. This article could provide some references for similar project engineering designs.

Keywords: Offshore Substation · Offshore Wind Power Station · Active Fire Protection System · Engineering Design

1 Introduction

The offshore wind energy has been developed for a relatively long history in Europe. The total offshore wind installations in Europe passed 30 GW by the end of 2022, making up 47% of the global total offshore [1]. In the US, the total installed offshore wind power capacity could reach up to 29 GW before 2030 [2]. For China, the offshore wind power capacity is expected up to be 1500 GW in 2050 [3]. Global Wind Energy Council (GWEC) market intelligence expects that over 380 GW of new offshore wind capacity will be added over the next decade (2023–2032), bringing the global total offshore wind capacity to 447 GW by the end of 2032 [1]. Predictably, with the depletion of finite energy resource such as oil and coal, as well as the land resources, renewable, clean and land-saving energy sources such as offshore wind resource development is expected to be more and more important [4]. As the scale of wind farms continues to increase, the fire safety is the key factor that cannot be ignored during their design, construction, operation and maintenance. For example, the installed capacity of Rudong offshore wind farm reaches 4380 MW, while the water depth reaches 60 m in Ledong offshore wind farm [5]. For the fire safety requirements of offshore substation of wind power stations, *Offshore Standard (DNV-OS-J201) Offshore Substations for Wind Farms* compiled by DET NORSKE VERITAS provides relatively complete and comprehensive technical provisions and acceptance criteria. Subsequently, other countries like China have also prepared their own specifications or standards for offshore substation fire safety based on DNV-OS-J201, latest research achievements and construction experiences. In this paper, the active fire protection systems recommended in three representative specifications or standards are listed, compared and analyzed. The three specifications or standards are i. *Offshore Standard* (DNV-OS-J201) *Offshore Substations for Wind Farms* compiled by DET NORSKE VERITAS, ii. *Code for 110 kV ~ 220 kV Offshore Substation Design of Wind Power Projects* (NB/T 31115–2017) compiled by China electricity council and iii. *Standard for Design of Offshore Wind Farm* (GB/T 51308–2019) compiled by China renewable energy engineering institute. Through comparison and analysis, the preferred active fire protection system for each areas in offshore substation of wind power station has been proposed, in order to provide reference for the design of similar projects.

2 Configuration Requirements of Active Fire Protection Systems

2.1 Portable Extinguisher

There are detailed description for the configuration of portable extinguisher in all three specifications or standards. The capacity and location of the portable extinguishers have been specified in the three codes with subtle differences. In particular, fluid extinguisher can be adopted according to *DNV-OS-J201*, while this kind of extinguisher has not been mentioned in the codes *NB/T 31115–2017* and *GB/T 51308–2019*.

2.2 Water Mist System

The code *NB/T 31115–2017* has the most detailed provisions for water mist systems. It specifies the minimum operating pressure, minimum duration, maximum system response time, maximum filling time of water distribution pipes. And it also makes clear provisions on the location of the protective nozzles and the water mist nozzles.

In the *DNV-OS-J201*, the recommended minimum capacity of mist systems is also specified, which is $10 \text{ L min}^{-1} \text{ m}^{-2}$ to provide exposure protection to equipment within that area.

As for the code *GB/T 51308–2019*, it mainly specifies the minimum operating pressure of the water mist system and the location of the water mist nozzles.

2.3 Deluge System

There is no description of deluge system in the codes *NB/T 31115–2017* and *GB/T 51308–2019*.

According to the *DNV-OS-J201*, deluge system could be adopted in storage areas for compressed gas. The system should be designed to achieve local control and remote control. The deluge firewater distribution pipe-works and deluge valves setting are also specified in this code.

2.4 Pressure Water-Spraying System

The application areas of pressure water- spraying system are not listed in the codes *NB/T 31115–2017* and *GB/T 51308–2019*, thus the setting requirements of pressure water-spraying system specification are also not be mentioned in these two codes.

DNV-OS-J201 specifies that the minimum effective average distribution of water should be 5 L min^{-1} m^{-2} in the spaces such as machinery spaces. In order to improve the effectiveness and timely response of the system, the system pressures in standby and startup condition are also specified. The requirements of main power supply and backup power supply used to drive the pressure water-spraying system pumps have been emphasized. Furthermore, the requirements for prevention the blockage of the nozzle by impurities in the water have been put forward in *DNV-OS-J201*.

2.5 Sprinkler System

Similar to the pressure water-spraying system, there is no description for sprinkler system in the codes *NB/T 31115–2017* and *GB/T 51308–2019*.

In the *DNV-OS-J201*, the application of the system is limited to such areas such as accommodation, where slow fire growth ($\alpha = 0.003$ kW s^{-2}) is expected.

2.6 Gaseous Systems

The requirements of gaseous system in *DNV-OS-J201* are specific and detailed. The setting requirements of fire extinguishing medium conveying pipeline, openings requirements for gaseous system rooms, the alarm device setting and personnel evacuation, are specified in detail. The design of the pressure level of the container shall take into account the pressure code requirements as well as the highest possible ambient temperature. The fire extinguishing medium should be stored in a safe, accessible and well-ventilated room. The entrance of the room should be connected with the open deck and independent from the protected area. The enclosure structure of the room, including the doors and windows should be air-tight, and the door should be open outwards.

The requirements of *NB/T 31115–2017* and *GB/T 51308–2019* for gaseous systems are relatively briefer. They only stipulate the selection principles of fire extinguishing mediums. For example, The fire extinguishing mediums that can release toxic gases and those can cause secondary damage to equipment are forbidden.

2.7 Foam System

DNV-OS-J201 specifies the foam discharge rate, the quantity of foam-forming liquid and the maximum expansion ratio of the foam. The codes *NB/T 31115–2017* and *GB/T 51308–2019* however only specify the suitable application areas of this system such as the diesel engine room, as well as the the suitable type of the foam system namely the high-expansion foam extinguishing system.

2.8 Fire Water Pump System

For the configuration requirements of fire pumps including marine growth prevention in fire water system, the codes *NB/T 31115–2017* and *GB/T 51308–2019* are described relatively in principle, while the concrete measures such as installation of the inlet strainers of fire pumps have been put forward in the *DNV-OS-J201*.

3 Methodology

To scientifically and rationally select the appropriate fire extinguishing system for each functional room of an offshore substation in a wind power station, a methodological approach is employed. Firstly, candidate fire extinguishing systems are screened based on their compliance with relevant standard requirements. Secondly, a comprehensive evaluation of the screened systems is conducted with respect to multiple criteria, including effectiveness, environmental impact, and maintenance feasibility and cost. This multi-criteria analysis aims to identify the most suitable fire extinguishing system that balances performance, environmental sustainability, and economic feasibility. The specific selection technology pathway is illustrated in Fig. 1 below.

4 Analysis and Discussion

4.1 Active Fire Control Systems Recommendation in Three Representative Specifications or Standards

The recommended active fire protection systems of three relevant specifications or standards are shown in Table 1.

4.2 Fire Suppression Effectiveness

Water mist system, gaseous system and foam system can all be used to extinguish fires [6]. However, when considering the factor of preventing the fire from reigniting, the gaseous system should be excluded [7].

4.3 Environmental Impact

Water mist system is a new type of environmentally-friendly fixed extinguishing system which contains the effect of heat absorption and oxygen insulation [8]. The foam concentrate however contains surface active agent, which may be harmful to the marine environment after fire extinguishing use [9, 10].

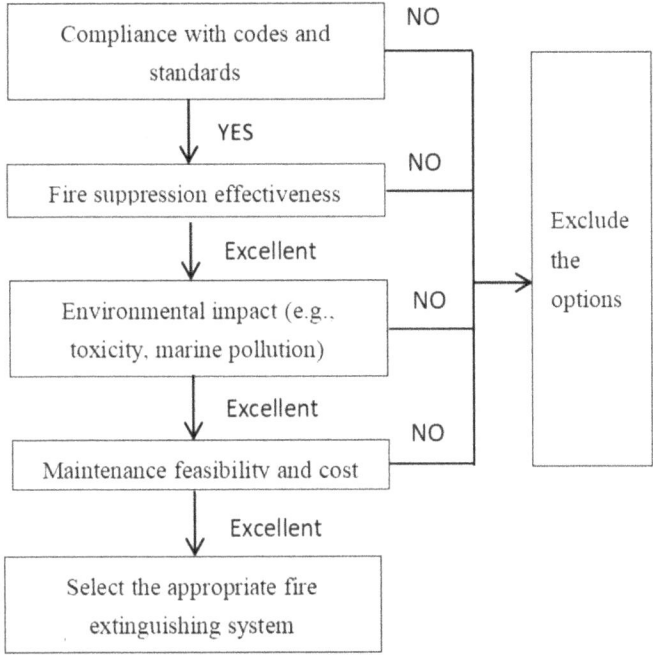

Fig. 1. Overview of method for selection of appropriate fire extinguishing system for each functional room of an offshore substation in a wind power station

4.4 Maintenance Feasibility and Cost

The foam concentrate used in foam fire extinguishing systems must be periodic replacement to ensure its fire-fighting performance. However, the offshore substations are generally far from land, so that the replacement and maintenance work of the foam concentrate are inconvenience. Moreover, considering the fire water source is usually seawater, seawater foam concentrate, which is more expensive compared to the conventional concentrate, has to been adopted.

4.5 Fire Extinguishing System Selection for Each Functional Room

4.5.1 Main Transformer Room

Water mist system, gaseous system and foam system can all be adopted for main transformer room fire control. For offshore substations with adequate area to set fire water tanks and fire pumps, gaseous system which is unable to prevent the fire from reigniting is not recommended. Although foam system is also effective for fire extinguishing, water mist system is more suitable for offshore substation of wind power projects considering the environmental impact.

Table 1. The recommended active fire protection systems

Areas	DNV-OS-J201	NB/T 31115–2017	GB/T 51308–2019
①All area	Portable fire extinguishers	Portable fire extinguishers	Portable fire extinguishers
②Main transformer room	Water mist system or gaseous system	Water mist system or foam system	Water mist system or foam system
③Diesel engine room, reactor room	Water mist system or gaseous system	Water mist system or foam system	Water mist system or foam system
④Switch gear room, GIS room	Water mist system or gaseous system	Water mist system or gaseous system	Water mist system or gaseous system
⑤Communication relay room, accumulator room, low voltage switchboard room	Water mist system or gaseous system	Water mist system or gaseous system	Water mist system or gaseous system
⑥Temporary refuge or shelter area	/	Water mist system	/
⑦Machinery spaces	Pressure water-spraying system or foam system	/	/
⑧Accommodation	Sprinkler system	/	/
⑨Rooms containing cylinders with compressed gas	Deluge system	/	/
⑩Helideck	Water monitors or foam system and dual-purpose nozzle and hoses	Water monitors or foam system	Foam and water dual water gun hydrant system

Note: / represents no description in the corresponding specifications or standards

4.5.2 Diesel Engine Room, Reactor Room, Switch Gear Room and GIS Room

The points of the fire extinguishing system selection for diesel engine room, reactor room, switch gear room and GIS room are similar to that of main transformer room. Therefore, the water mist system is the preferred extinguishing system for these areas.

4.5.3 Temporary Refuge or Shelter Area

Water mist system is recommended for temporary refuge or shelter area fire extinguishing and it is also the recommended fire extinguishing system in accommodation, public spaces and service areas on passenger ships by International Maritime Organization (IMO).

4.5.4 Machinery Spaces

Based on the DNV-OS-J201 and previous study [11], fixed pressure water-spraying system is suitable for extinguishing fires in machinery spaces, and it is also the recommended fire extinguishing system in Regulation 10 of Safety of Life at Sea (SOLAS). Considering the defects of foam system mentioned in part 3.2, fixed pressure water-spraying system is the preferred extinguishing system for machinery spaces.

4.5.5 Accommodation

Sprinkler system is widely used in cruise ship accommodation [12], transient accommodation buildings [13], student's accommodation [14] and residences [15]. This active fire protection system has a long record of property and life protection. In terms of reliability, safety and economy, sprinkler system is very suitable for fire protection for accommodations of offshore substations.

4.5.6 Rooms Containing Cylinders with Compressed Gas

Rooms containing cylinders with compressed gas has a relatively high risk of fire and explosion. Meanwhile, deluge system has been proved to been effective for fire protection for such places with high fire and explosion risk such as explosive processing building [16], LPG tanks [17] and even LPG vessels [18]. Hence, for rooms containing cylinders with compressed gas, deluge system is recommended for fire protection.

4.5.7 Helideck

The specified active fire protection systems for helicopter decks are basically consistent in the three specifications or standards. The relevant provisions are based on the requirements of the *Guide to helicopter/ship operations* published by International Chamber of Shipping (ICS).

According to the specifications and standards, water monitor system and foam system can both been adopted for fire fighting on helideck. For fire water monitor systems, they are widely installed on fire trucks or large space buildings as the fire fighting equipment and it's effectiveness for fire protection has been proved [19]. According to the previous study [7], fire monitor system is the most effective fire extinguishing method comparing to the foam fire extinguishing system (column type), water spray fire extinguishing system, foam fire extinguishing system (pipe network type) and water mist fire extinguishing system. Therefore, fire monitors are recommended for helideck fire protection. As for dual-purpose nozzle and hoses, they are used for the fire control by manual operation on the helideck.

5 Conclusions and Recommendations

In this paper, the selection of the appropriate fire extinguishing system for each functional room of an offshore substation in a wind power station was conducted through a hierarchical screening process, prioritized from the most critical to the least critical criteria. These criteria encompassed compliance with codes and standards, fire effectiveness, environmental impact, and maintenance feasibility and cost. By systematically evaluating various fire extinguishing systems against these criteria, the optimal fire extinguishing system was identified and selected for each functional room of an offshore substation. The selection results are listed in Table 2 and each active fire protection systems should meet the requirements of corresponding specifications or standards.

Table 2. The optimal active fire protection systems for each functional room of an offshore substation

Areas	①	② ③ ④ ⑤ ⑥	⑦	⑧	⑨	⑩
Optimal active fire protection systems	Portable fire extinguishers	Water mist system	Fixed pressure water-spraying system	Sprinkler system	Deluge system	**Fire monitors**

Note: The areas numbered ① to ⑩ referred to Table 1

Considering the huge development prospect of offshore wind power station in the future, the optimal design of the active fire protection systems for the offshore substations is worthy of further research and discussion.

References

1. Williams, R., Zhao, F.: Global Offshore Wind Report 2023. Global Wind Energy Council, Brussels, Belgium (2023)
2. Musial, W., Constant, C., Cooperman, A., et al.: Offshore Wind Electrical Safety Standards Harmonization: Workshop Proceedings. National Renewable Energy Laboratory (2020)
3. Guo, X., Chen, X., Chen, X., et al.: Grid integration feasibility and investment planning of offshore wind power under carbon-neutral transition in China. Nat. Commun. **14**(1), 2447 (2023)
4. Finn, J., Sandeberg, P.: AC Offshore Substations Associated with Wind Power Plants. In: Krieg, T., Finn, J. (eds.) Substations, pp. 591–728. Springer International Publishing, Cham (2019)
5. Zhang, J., Wang, H.: Development of offshore wind power and foundation technology for offshore wind turbines in China. Ocean Eng. **266**, 113256 (2022)
6. Liu, Y., Li, B., Wu, C., et al.: Effectiveness test and evaluation of transformer fire extinguishing system. Fire. Technol. **58**(5), 3167–3190 (2022)
7. Yong-Shik, H., Byung-Il, C., Myung-Bae, K.: Application of water mist system for a power transformer room—Fire extinguishment (Part 1). Fire Sci. Eng. **19** (2005)
8. Huo, Y., Chen, M., Li, T., et al.: Experimental study on fire suppression performance of the high pressure water mist in the engine room of an offshore platform. J. Loss Prev. Process Ind. **83**, 105052 (2023)
9. Zhang, G., Jiao, J., Wu, J., et al.: Environmentally friendly fluorine-free fire extinguishing agent based on the synergistic effect of silicone, hydrocarbon surfactants and foam stabilizers. Colloids Surf. A Physicochem. Eng. Asp. **694**, 134216 (2024)
10. Back, G.: An evaluation of the firefighting effectiveness of fluorine-free foams. Fire. Technol. **59** (2020)

11. Hansen, R.L., Back, G.G.: Water Spray Protection of Machinery Spaces. Fire. Technol. **37**(4), 317–326 (2001)
12. Aarnio, M.: Hotel Systems. In: Aarnio, M. (ed.) Cruise Ship Handbook, pp. 157–182. Springer International Publishing, Cham (2023)
13. Thomas G, Harding D. Fire Safety in New Zealand Transient Accommodation Buildings (2014)
14. Ismail, I., Taib, M., Othuman Mydin, M.A.: Appraisal of passive and active fire protection systems in student's accommodation. In: MATEC Web of Conferences, vol. 10, (2014)
15. Runefors, M., Andersson, R., Delin, M., et al.: Residential Fire Safety: An Interdisciplinary Approach, pp. 177–196. Springer International Publishing, Cham (2023)
16. Vishnu Teja, G., Meikandan, M.: Design of Automatic Deluge Fire Protection System for Explosive Processing Building. Springer Singapore (2021)
17. Roberts, T.A.: Effectiveness of an enhanced deluge system to protect LPG tanks and sensitivity to blocked nozzles and delayed deluge initiation. J. Loss Prev. Process Ind. **17**(2), 151–158 (2004)
18. Davies, G.F., Nolan, P.F.: Characterisation of two industrial deluge systems designed for the protection of large horizontal, cylindrical LPG vessels. J. Loss Prev. Process Ind. **17**(2), 141–150 (2004)
19. Li, W., Wang, G., Yang, Z., et al.: Prediction method of jet trajectory of fire water monitor. In: CSAA/IET International Conference on Aircraft Utility Systems (AUS) (2020)

Open Access This chapter is licensed under the terms of the Creative Commons Attribution-NonCommercial-NoDerivatives 4.0 International License (http://creativecommons.org/licenses/by-nc-nd/4.0/), which permits any noncommercial use, sharing, distribution and reproduction in any medium or format, as long as you give appropriate credit to the original author(s) and the source, provide a link to the Creative Commons license and indicate if you modified the licensed material. You do not have permission under this license to share adapted material derived from this chapter or parts of it.

The images or other third party material in this chapter are included in the chapter's Creative Commons license, unless indicated otherwise in a credit line to the material. If material is not included in the chapter's Creative Commons license and your intended use is not permitted by statutory regulation or exceeds the permitted use, you will need to obtain permission directly from the copyright holder.

Risk Assessment of Ship Pilotage Based on WBS-RBS and Cloud Model: A Case Study of Qinzhou Port

Peng Liu[1], Yongtao Xi[1,1(✉)], Yi Su[2], and Haibo Huang[2]

[1] Shanghai Maritime University, Merchant Marine College, Shanghai, China
xiyt@shmtu.edu.cn
[2] Beibu Gulf Port, Qinzhou Pilotage Association, Qinzhou, China

Abstract. An essential component of waterborne cargo movement is ship pilotage. This study offers a risk assessment model based on the Work Breakdown Structure-Risk Breakdown Structure (WBS-RBS) and cloud model to evaluate and control the risks during the full process of ships entering and leaving port. To thoroughly identify the risk elements at every stage of ship pilotage, utilize the WBS-RBS. The primary risk factors are identified using the Technique for Order Preference by Similarity to an Ideal Solution (TOPSIS). The comprehensive weights of the risk factors are obtained using the Analytic Hierarchy Process (AHP) and The Criteria Importance through the Inter-Criteria Correlation (CRITIC) approach. The risk assessment is then conducted using the cloud model. A case study of Qinzhou Port is conducted to assess the risks throughout the entire pilotage process. The results show that the risk level of pilotage in Qinzhou Port is medium. For ships entering Qinzhou Port, the preparation stage and the stage of completion and reporting are low-risk, while the coordination & communication and pilotage operation stages are medium-risk. This paper provides a reference for ship pilotage risk assessment research and a theoretical basis for safety management departments to develop risk control measures.

Keywords: Safety engineering · Ship pilotage · Risk assessment · WBS-RBS

1 Introduction

A crucial component of port safety management is ship pilotage safety. The volume of international trade imports and exports has increased recently, and with it, so has the domestic and international shipping industry. As a result, China's port pilotage stations have seen an increase in the number of ships requiring pilotage, and the conditions and work environment for pilots have also grown more complex. Simultaneously, as large-scale ships have evolved, so too have the challenges and demands of piloting, as well as the associated safety dangers, of this profession. Ensuring the safety of ships and ports is a practical importance of the risk assessment of port pilotage.

Through the efforts of numerous experts and scholars, the risk of ship pilotage has achieved significant study progress in recent years as part of the research on water traffic concerns. Chen et al. [1] and Guo et al. [2] conducted comprehensive research on the dangers associated with pilotage in port waters using the combined quantitative and qualitative techniques and the unknown measurement mathematical model, respectively. For the risk of the pilot's operation transfer process, Tuncel et al. [3] used the fuzzy fault tree analysis (FFTA) to conduct a comprehensive risk analysis of the pilot's operation transfer process at sea. For the construction of the pilotage risk assessment model, Jiang et al. [4] proposed a pilotage risk assessment model based on the multivariate correlation coefficient and set pair analysis theory, combining the entropy weight method and analytic hierarchy process, which takes into account both the objectivity of the data and the subjective value of the decision-makers. In terms of establishing the key risk indicator system of the pilotage process, Guo et al. [5] used the HHM method to establish the key risk indicator system of the ship pilotage process, and combined with the scenario analysis of the pilotage process of a certain container ship, conducted dynamic simulation research. Specifically for the different stages of pilotage operations, Pan et al. [6] divided the pilotage operations into 3 different stages, analyzed the division of pilotage responsibilities in different stages, analyzed the division of pilotage responsibilities in several special cases, and proposed safety recommendations during the pilotage responsibility period. In terms of establishing a comprehensive evaluation model, Chu et al. [7] used the Delphi method to determine the membership degree of each indicator, and combined it with the geometric characteristics of triangular fuzzy numbers to calculate the fuzzy weights, and established a comprehensive evaluation model of pilotage risk in Xiamen Port, providing an effective tool for the comprehensive assessment of port pilotage risks.

Many academics have started to concentrate on the effects of human variables, such as pilots, on pilotage safety as a result of the growing significance of human factors in ship pilotage hazards. Fatemeh et al. [8] used a causal-comparative design of post-human error data to explore the cognitive correlates of the history of human errors by maritime pilots. Zhang et al. [9] integrated human factors analysis and classification system (HFACS) into the framework of comprehensive safety assessment (FSA) to identify and assess the risks of pilotage operations and established a Bayesian network (BN) --based ship pilotage risk HOF coupling correlation model for a specific port.

The study of ship pilotage risk has advanced recently, and a variety of qualitative and quantitative techniques, including fault trees, event trees, the comprehensive evaluation method, and fuzzy comprehensive judgment, are employed to estimate the risk. They all have certain restrictions, though. As an illustration, the fault tree analysis that is frequently employed in risk assessment targets specific accidents with a given level of localization as opposed to specific processes. Ship pilotage in port waters is a relatively complex procedure, and the fault tree that is constructed is frequently big. This can make it difficult to do further analysis; if the complete operation system is not sufficiently familiar, there is a chance of omission. Therefore, the purpose of this paper is to identify the risks associated with the pilotage process by using the following methods: using the Work Breakdown Structure-Risk Breakdown Structure (WBS-RBS) coupling matrix method to prevent risk omission; screening risk factors and building a risk indicator system; utilizing the improved TOPSIS method in conjunction with the AHP-CRITIC method to determine the weight of each indicator at different pilotage stages; and, finally, utilizing the uncertain artificial intelligence cloud model for risk level visualization to obtain the assessment results.

2 Description of Problem

2.1 Risk of Ship Piloting Process

Risk is the result of combining the likelihood of an accident happening with the potential repercussions of that accident. The probability of risk and the ensuing repercussions show a constantly evolving process over time because of the instability of external environmental elements and the uncertainty of the prospective evolution of internal components. Equation (1) illustrates the functional relationship.

$$R = f(F_1, F_2, F_3, \cdots, F_n) \qquad (1)$$

Note that the value at risk at the moment, is the performance of risk indicators over time.

The pilotage process of a ship can be divided into several distinct stages, from the commencement of pilotage to the completion of berthing. These stages include the preparation for port entry, coordination and communication, pilotage operation, and subsequent tasks. The events that transpire during these phases display notable temporal and spatial disparities and uncertainties, which collectively comprise the ship's pilotage risk.

2.2 Identification of Risk Factors During Ship Pilotage

The Construction of the Work Breakdown Structure. The work breakdown structure (WBS) can be used to break down the pilotage work (W) into four main stages based on the actual operational workflow of ship pilotage: port entry preparation (W1), coordination and communication (W2), pilotage operation (W3), and pilotage completion and reporting (W4). Each stage's primary WBS is then further refined, producing a secondary WBS that is more detailed and is displayed in Table 1.

Table 1. Work breakdown of ship pilotage operations

Goal level	Guideline level	Indicator level
Pilotage work W	Port entry preparation W_1	Obtain ship and port information W_{11}
		Analyse waterway and hydrological conditions W_{12}
		Determine the pilotage plan W_{13}

(*continued*)

Table 1. (*continued*)

Goal level	Guideline level	Indicator level
	Coordination and communication W_2	Assign the pilot W_{21}
		Pre-pilotage ship handover W_{22}
		Notify the vessel to be piloted W_{23}
		Pilot boarding W_{24}
	Pilotage operation W_3	Weigh anchor W_{31}
		Navigate the waterway W_{32}
		Berth or depart the port W_{33}
	Pilotage completion and reported W_4	Post-pilotage ship handover W_{41}
		Pilot disembarkation W_{42}
		Submit reports and records W_{43}

The Construction of the Risk Breakdown Structure. The ship pilotage risk factors have been identified and categorized along four primary dimensions: man, machine, environment, and management, using the principles of safety system engineering. After further refining the core indicators, 16 risk variables that had a major influence or serious repercussions were chosen to serve as secondary indicators. These indicators reflect the smallest risk units within the risk breakdown structure (RBS) and are shown in Table 2.

Table 2. Risk decomposition structure for ship pilotage operations

Goal level	Guideline level	Indicator level
Pilotage risk R	Personnel risk R_1	Pilot R_{11}
		The crew of the piloted vessel R_{12}
		Mooring personnel R_{13}
		Crew of other vessels R_{14}
		Tug boat crew R_{15}
	Equipment risk R_2	Piloted vessel R_{21}
		Other vessels R_{22}
		Tug boat R_{23}
	Environmental risk R_3	Wind and waves R_{31}
		Tides R_{32}
		Visibility R_{33}
		Waterway conditions R_{34}

(*continued*)

Table 2. (*continued*)

Goal level	Guideline level	Indicator level
		Berth conditions R_{35}
		Traffic volume R_{36}
	Managing risk R4	VTS conditions R_{41}
		Pilotage management services R_{42}

The Construction of the Coupling Matrix of WBS-RBS. Based on accident investigation reports and expert survey data, the links between the sub-processes (i = 1,2,3,4; j = 1,2,3) and risk levels (m = 1,2,3,4; n = 1,2,3,…,6) were investigated to develop the coupling matrix between the work breakdown structure (WBS) and risk breakdown structure (RBS). The appropriate cell in the matrix was given a value of 1 if a risk factor was found; on the other hand, 0 was indicated if there was no risk factor or if the risk was minimal. Table 3 displays the entire list of risk factors while Table 4 displays the resulting coupling matrix.

Table 3. Inventory of risk factors in the ship pilotage process

Risk factor	Risk description	Risk factor	Risk description	Risk factor	Risk description
$W_{11}R_{12}$	Failure to apply for pilot services at the port or untimely application	$W_{31}R_{14}$	Collision caused by mishandling of the other vessel	$W_{33}R_{11}$	Pilot error or improper command
$W_{11}R_{42}$	Untimely handling of pilot requests	$W_{31}R_{21}$	Malfunction of the piloted vessel's navigational system	$W_{33}R_{12}$	Crew error or improper operation on the piloted vessel
$W_{12}R_{41}$	Failure to integrate port information resources promptly	$W_{32}R_{11}$	Pilot error in navigation instructions or improper command	$W_{33}R_{13}$	Non-standard operations by dock workers
$W_{12}R_{42}$	Inaccurate analysis of navigation channel and hydrological conditions	$W_{32}R_{12}$	Crew error or improper operation on a piloted vessel	$W_{33}R_{15}$	Tug crew error or improper operation
$W_{13}R_{42}$	Unreasonable pilotage plan	$W_{32}R_{14}$	Violation of rules by the other vessel	$W_{33}R_{21}$	Malfunction of own vessel's steering system
$W_{21}R_{11}$	Untimely response from the pilot	$W_{32}R_{15}$	Tug crew error or improper operation	$W_{33}R_{23}$	Tug equipment malfunction
$W_{21}R_{42}$	Untimely dispatch from the pilot station	$W_{32}R_{21}$	Malfunction of piloted vessel's steering or navigation system	$W_{33}R_{31}$	Inability to dock or undock normally due to severe weather conditions
$W_{22}R_{11}$	Incomplete or inaccurate handover information	$W_{32}R_{22}$	Malfunction of the other vessel's steering or navigation system	$W_{33}R_{32}$	Failure to consider tidal cycles or improper consideration

(*continued*)

Table 3. (*continued*)

Risk factor	Risk description	Risk factor	Risk description	Risk factor	Risk description
$W_{22}R_{12}$	Incomplete or inaccurate handover information	$W_{32}R_{23}$	Tug equipment malfunction	$W_{33}R_{33}$	Poor visibility preventing normal docking or undocking
$W_{23}R_{12}$	Untimely response from the guided vessel	$W_{32}R_{31}$	Inability to navigate normally due to severe weather conditions	$W_{33}R_{35}$	Insufficient depth at the waterfront of the dock
$W_{23}R_{42}$	Failure to notify the guided vessel or untimely notification	$W_{32}R_{32}$	Failure to consider tidal cycles or improper consideration	$W_{33}R_{36}$	Heavy traffic within the harbour
$W_{24}R_{12}$	Failure to place boarding equipment or improper placement	$W_{32}R_{33}$	Poor visibility resulting in the inability of own vessel to navigate normally	$W_{41}R_{11}$	Incomplete or inaccurate exchange of information
$W_{24}R_{21}$	Aging or malfunction of pilot boarding equipment	$W_{32}R_{34}$	Inadequate depth or width of navigational channel	$W_{41}R_{12}$	Incomplete or inaccurate exchange of information
$W_{24}R_{31}$	Risk of pilot leaving the vessel due to rough sea conditions	$W_{32}R_{36}$	Heavy traffic within the navigational channel	$W_{42}R_{12}$	Improper placement or absence of gangway
$W_{31}R_{12}$	Incorrect or improper operation by crew members of the vessel	$W_{32}R_{41}$	VTS coordination failure or delayed coordination	$W_{42}R_{21}$	Aging or malfunction of pilot boarding equipment
$W_{42}R_{31}$	Risk of pilot leaving the vessel due to rough sea conditions	$W_{43}R_{11}$	Untimely or incorrect reporting	$W_{43}R_{42}$	Reports not collated or untimely

Table 4. The coupling matrix of WBS-RBS

RBS decomposition system			WBS decomposition system												
			W												
			W_1			W_2				W_3			W_4		
			W_{11}	W_{12}	W_{13}	W_{21}	W_{22}	W_{23}	W_{24}	W_{31}	W_{32}	W_{33}	W_{41}	W_{42}	W_{43}
R	R_1	R_{11}	0	0	0	1	1	0	0	0	1	1	1	0	1
		R_{12}	1	0	0	0	1	1	1	1	1	1	1	1	0
		R_{13}	0	0	0	0	0	0	0	0	0	1	0	0	0
		R_{14}	0	0	0	0	0	0	0	1	1	0	0	0	0
		R_{15}	0	0	0	0	0	0	0	0	1	1	0	0	0
	R_2	R_{21}	0	0	0	0	0	0	1	1	1	1	0	1	0
		R_{22}	0	0	0	0	0	0	0	0	1	0	0	0	0

(*continued*)

Table 4. (continued)

RBS decomposition system		WBS decomposition system W												
		W_1			W_2				W_3			W_4		
		W_{11}	W_{12}	W_{13}	W_{21}	W_{22}	W_{23}	W_{24}	W_{31}	W_{32}	W_{33}	W_{41}	W_{42}	W_{43}
R_3	R_{23}	0	0	0	0	0	0	0	0	1	1	0	0	0
	R_{31}	0	0	0	0	0	0	1	0	1	1	0	1	0
	R_{32}	0	0	0	0	0	0	0	0	1	1	0	0	0
	R_{33}	0	0	0	0	0	0	0	0	1	1	0	0	0
	R_{34}	0	0	0	0	0	0	0	0	1	0	0	0	0
	R_{35}	0	0	0	0	0	0	0	0	0	1	0	0	0
	R_{36}	0	0	0	0	0	0	0	0	1	1	0	0	0
R_4	R_{41}	0	1	0	0	0	0	0	0	1	0	0	0	0
	R_{42}	1	1	1	1	0	1	0	0	0	0	0	0	1

2.3 TOPSIS Method for Selecting and Determining Key Risk Factors

Quantification of Risk Factors. Taking into account the uncertainties and restricted data availability, we used TOPSIS to quantitatively examine the identified risk variables from two angles: the likelihood of occurrence and the extent of loss. This allowed us to appropriately simplify hazards. High-importance threats are identified using this method.

In this study, we utilized the expert scoring approach to numerically evaluate risks at various levels according to the likelihood of occurrence and the extent of loss. These two characteristics eventually define the risk level. Table 5 provides specific risk-level grading criteria.

Table 5. Rating assessment and scoring criteria

Risk severity level	I	II	III	IV	V
Risk occurrence probability	seldom occur	Rarely occur	occasionally occur	relatively frequent occurrence	Frequent occurrence
Risk loss magnitude	negligible	considered	moderate	serious	very serious
Score value	1	2	3	4	5

Basis for Ranking and Selection of Risk Factors. TOPSIS uses a mathematical approximation method to rank objects according to how close they are to a perfect target. Enhancements to TOPSIS include removing the impact of various indicator dimensions, making complete use of raw data, and improving the distance formula between positive and negative ideal solutions to more accurately represent an object's superiority or inferiority [10]. The following are the steps for enhancing TOPSIS in terms of risk factor selection and ranking.

Standardization of Risk Factors. Let the number of risk factor indicators be denoted as n, the number of evaluation attributes as p, and each evaluation attribute represented as h1, h2, ..., hp, Suppose there are m retrieved survey questionnaires. The survey data that was retrieved can be arranged and analyzed to create the following first matrix:

$$A = \begin{bmatrix} a_1^T \\ a_2^T \\ \cdots \\ a_4^T \end{bmatrix} = \begin{bmatrix} a_{11} & a_{12} & \cdots & a_{1p} \\ a_{21} & a_{22} & \cdots & a_{2p} \\ \vdots & \vdots & \vdots & \vdots \\ a_{n1} & a_{n2} & \cdots & a_{np} \end{bmatrix} = \begin{bmatrix} \frac{\sum a_{11}}{m} & \frac{\sum a_{12}}{m} & \cdots & \frac{\sum a_{1p}}{m} \\ \frac{\sum a_{21}}{m} & \frac{\sum a_{22}}{m} & \cdots & \frac{\sum a_{2p}}{m} \\ \vdots & \vdots & \vdots & \vdots \\ \frac{\sum a_{n1}}{m} & \frac{\sum a_{n2}}{m} & \cdots & \frac{\sum a_{np}}{m} \end{bmatrix}$$

The standardized matrix $E = (e_{ij})_{m \times n}$ is obtained by normalizing the original matrix and uniformly standardizing it into performance-oriented indicators [10]. It is described by the formula below:

$$e_{ij} = \frac{\max u_j - u_{ij}}{\max u_j - \min u_j} \quad (2)$$

Note that e_{ij} represents the normalized value of the j-th attribute indicator of the i-th risk factor. u_{ij} represents the value of the j-th attribute indicator of the i-th risk factor. max u_j represents the maximum value of the j-th attribute indicator.

Determination of Attribute Weights and Ranking of Risk Factors Based on their Importance. An optimization model based on goal programming is built under the standardized matrix to lessen the subjective influence of experts. The weights of attribute indicators in this model are determined objectively through the use of mathematical techniques. The following is a summary of the solution process: The weighted sum of squared distances to the positive and negative ideal solutions for each risk factor is calculated as, and the weight of risk attributes is indicated as:

$$f_i(w) = \sum_{j=1}^{n} w_j^2 (1 - e_{ij})^2 + \sum_{j=1}^{n} w_j^2 e_{ij}^2 \quad (3)$$

Note that $W_j \geq 0$, j = 1,2,3, Meeting conditions $\sum_{j=1}^{n} w_j = 1$; a smaller value of $f_i(w)$ indicates that the risk factor is closer to the ideal point, resulting in better outcomes. Given $f_i(w) \geq 0$ and i = 1,2, ..., n, the following single-objective optimization model is established:

$$\min f(w) | f(w) = \sum_{i=1}^{m} f_i(w) \quad (4)$$

Consequently, we construct the Lagrangian function:

$$F(w, \lambda) = \sum_{i=1}^{m}\sum_{j=1}^{n} w_j^2((1-e_{ij})^2 + e_{ij}^2) - \lambda\left(1 - \sum_{j=1}^{n} w_j\right) \qquad (5)$$

It can be obtained that $\omega_j = \dfrac{\alpha_j}{\sum_{j=1}^{n}\alpha_j}$, among them, $\alpha_j = \dfrac{1}{\sum_{i=1}^{m}((1-e_{ij})^2 + e_{ij}^2)}$, $j = 1,2,\cdots,p.$

the risk factors are further ranked according to $f_i(w)$ derived from Eq. (3) [10].

Risk screening is based on the statistical analysis of risk data that has been rated by professionals to ascertain the relative importance of various risks. To attain consistent dimensions, the data matrix is quantified first. Normalization is not required because this research assesses both the likelihood that a risk will occur and the possible seriousness of the consequences, and these two evaluation criteria are consistent. Standardization is achieved by immediately applying the TOPSIS formula. Cost-type indicators are those that take into account both the likelihood that a risk will materialize and the magnitude of possible damages. The value of $f_i(w)$ is obtained through calculations using information gathered from questionnaire surveys, as indicated in Table 6.

Table 6. The fi(w)-values of each risk factor indicator

Risk Factors	$f_i(w)$	Risk Factors	$f_i(w)$	Risk Factors	$f_i(w)$
$W_{11}R_{12}$	0.4161	$W_{31}R_{21}$	0.587	$W_{33}R_{13}$	0.4708
$W_{11}R_{42}$	0.3029	$W_{32}R_{11}$	0.4161	$W_{33}R_{15}$	0.4161
$W_{12}R_{41}$	0.4708	$W_{32}R_{12}$	0.6971	$W_{33}R_{21}$	0.587
$W_{12}R_{42}$	0.1635	$W_{32}R_{14}$	0.7051	$W_{33}R_{23}$	0.5292
$W_{13}R_{42}$	0.4161	$W_{32}R_{15}$	0.2949	$W_{33}R_{31}$	0.1635
$W_{21}R_{11}$	0.413	$W_{32}R_{21}$	0.4708	$W_{33}R_{32}$	0.7051
$W_{21}R_{42}$	0.4161	$W_{32}R_{22}$	0.4161	$W_{33}R_{33}$	0.5292
$W_{22}R_{11}$	0.587	$W_{32}R_{23}$	0.5292	$W_{33}R_{35}$	0.4161
$W_{22}R_{12}$	0.1635	$W_{32}R_{31}$	0.4161	$W_{33}R_{36}$	0.4708
$W_{23}R_{12}$	0.4708	$W_{32}R_{32}$	0.2949	$W_{41}R_{11}$	0.4161
$W_{23}R_{42}$	0.5839	$W_{32}R_{33}$	0.7051	$W_{41}R_{12}$	0.587
$W_{24}R_{12}$	0.5292	$W_{32}R_{34}$	0.395	$W_{42}R_{12}$	0.7051
$W_{24}R_{21}$	0.1635	$W_{32}R_{36}$	0.6971	$W_{42}R_{21}$	0.4911
$W_{24}R_{31}$	0.7051	$W_{32}R_{41}$	0.605	$W_{42}R_{31}$	0.7051
$W_{31}R_{12}$	0.395	$W_{33}R_{11}$	0.4708	$W_{43}R_{11}$	0.5292
$W_{31}R_{14}$	0.4161	$W_{33}R_{12}$	0.413	$W_{43}R_{42}$	0.7051

A final risk screening for ship navigation can be accomplished using the $f_i(w)$-values screened from the data in the preceding table. The four stages of ship navigation: port entry preparation (W_1), coordination and communication (W_2), navigation operations (W_3), and navigation completion and reporting (W_4)—are identified by the screening risk indicators. The risk factor indicator system that results from this categorization is displayed in Table 7.

Table 7. Table of ship navigation risk factor indicators

Target Layer	Primary Indicators	Secondary Indicators
Vessel Pilotage Risk U	Port entry preparation W_1	Failure to apply for pilot services at the port or untimely application $W_{11}R_{12}$
		Failure to integrate port information resources in a timely manne $W_{11}R_{42}$
		Inaccurate analysis of navigation channel and hydrological conditions $W_{12}R_{42}$
		Failure to integrate port information resources promptly $W_{12}R_{41}$
		Unreasonable pilotage plan $W_{13}R_{42}$
	Coordination and communication W_2	Untimely dispatch from the pilot station $W_{21}R_{42}$
		Incomplete or inaccurate handover information $W_{22}R_{12}$
		Untimely response from the guided vessel $W_{23}R_{12}$
		Failure to notify the guided vessel or untimely notification $W_{23}R_{42}$
		Aging or malfunction of pilot boarding equipment $W_{24}R_{21}$
		Risk of pilot leaving the vessel due to rough sea conditions $W_{24}R_{31}$
	Pilotage operation W_3	Incorrect or improper operation by crew members of the vessel $W_{31}R_{12}$
		Collision caused by mishandling of the other vessel $W_{31}R_{14}$
		Malfunction of piloted vessel's navigational system $W_{31}R_{21}$

(*continued*)

Table 7. (*continued*)

Target Layer	Primary Indicators	Secondary Indicators
		Pilot error in navigation instructions or improper command $W_{32}R_{11}$
		Tug crew error or improper operation $W_{32}R_{15}$
		Malfunction of piloted vessel's steering or navigation system $W_{32}R_{21}$
		Inability to navigate normally due to severe weather conditions $W_{32}R_{31}$
		Inadequate depth or width of navigational channel $W_{32}R_{34}$
		VTS coordination failure or delayed coordination $W_{32}R_{41}$
		Heavy traffic within the harbor $W_{33}R_{36}$
		Insufficient depth at the waterfront of the dock $W_{33}R_{35}$
	Pilotage completion and reported W_4	Incomplete or inaccurate exchange of information $W_{41}R_{11}$
		Improper placement or absence of gangway $W_{42}R_{12}$
		Aging or malfunction of pilot boarding equipment $W_{42}R_{21}$
		Reports not collated or untimely $W_{43}R_{42}$

3 Risk Simulation Model of Ship Pilotage Based on Cloud Model

3.1 Cloud Model

The cloud model effectively manages randomness and fuzziness in evaluations, facilitating the conversion of uncertainty between quantitative and qualitative assessments. The numerical characteristics of a cloud include three variables: expected *Ex*, entropy *En*, and super entropy *He*. Here, *Ex* represents the centroid position of the cloud, *En* measures the fuzziness of the concept, and *He* quantifies the uncertainty of entropy.

Forward and backward cloud generators make up cloud generators. The forward cloud generator creates a large number of cloud droplets by converting qualitative ideas into quantitative representations. Drop(x_i, μ_i) according to the cloud's three numerical properties (*Ex*, *En*, *He*). The reverse cloud generator uses a known quantity of exact numerical values to compute the cloud's three numerical characteristics *Ex*, *En*, and *He*. In order to translate quantitative representations back into qualitative conceptions. By merging two or more linguistically comparable values into a more generalized linguistic value, the integrated cloud model advances notions by synthesizing many similar sub-clouds into a higher-level parent cloud [11].

3.2 Cloud Model

Four levels of danger are suggested based on comments and professional judgments. This paper uses a ten-point rating system under ship navigation safety regulations. The following are the assessment levels and the score ranges that correspond to them: There are four levels of risk: very low (0, 2.5), low (2.5, 5), medium (5.5, 7.5), and high (7.5, 10). There is a positive correlation between scores and risk. Equations (6) to (8) are used to calculate the standard cloud model parameters, and Table 8 displays the results.

$$Ex_i = \frac{x_i^{min} + x_i^{max}}{2} \quad (6)$$

$$En_i = \frac{x_i^{max} - x_i^{min}}{6} \quad (7)$$

$$He = k \quad (8)$$

Note that k is a constant, whose value is determined by the degree of comment ambiguity.

Table 8. Pilot risk assessment cloud model parameters

Risk level	Risk range	Cloud model parameters	Description of risk levels
I	[0, 2.5]	(1.25, 0.42, 0.1)	Very low risk
II	(2.5, 5]	(3.25, 0.42, 0.1)	Low risk
III	(5, 7.5]	(6.25, 0.42, 0.1)	Medium risk
IV	(7.5, 10]	(8.25, 0.42, 0.1)	High risk

The Analytic Hierarchy Process (AHP) and The Criteria Importance through the Inter-Criteria Correlation (CRITIC) technique are applied individually to handle the identified key risk factors to fully identify the weights of each level of indicators based on their subjective and objective qualities. To determine indicator weights for intricate multi-objective issues, the subjective AHP integrates both qualitative and quantitative elements. It determines the relative weights of the following four main indicators: management, equipment, people, and the environment. Comparison matrices are first established by pairwise comparisons between indicators, and then these matrices are consistently checked.

$$CI = \frac{\lambda_{max} - n}{n - 1} \quad (9)$$

Note that $\lambda_{\max}$ represents the matrix's largest eigenvalue, CI describes the degree of inconsistency in the matrix and n denotes the order of the matrix. Based on the matrix's order, the corresponding average random consistency index CR is determined, and the random consistency ratio is denoted as CR. When CR < 1 the comparison matrix exhibits good consistency. Utilizing the summation method for normalizing the comparison matrix allows the determination of the weights for each primary indicator.

$$CR = \frac{CI}{RI} \qquad (10)$$

The CRITIC method synthesizes objective indicator weights based on the conflict and intensity of comparisons between indicators. The conflict between indicators is expressed by their correlations, while the contrast intensity between indicators is quantified in terms of standard deviation. The conflict of the j-th indicator with other indicators can be expressed as

$$f_j = \sum_{i=1}^{m} (1 - r_{ij}) \quad j = 1, 2, \cdots, m \qquad (11)$$

Note that r_{ij} represents the correlation coefficient between the i-th and j-th indicators.

Set up σ_j to stand for indication j's standard deviation and let C_j be the information content of the j-th indicator, as given by Eq. (14). Greater information content in the indicator, and hence its greater importance, is indicated by larger values of C_j. Weight coefficients ω_j can be computed for each secondary indicator by normalizing C_j. The thorough AHP-CRITIC approach is used to determine the final weights of the secondary indicators ω_{ij}.

$$C_j = \sigma_j \sum_{i=1}^{m}(1 - r_{ij}) \quad j = 1, 2, \cdots, m \qquad (12)$$

$$\omega_{ij} = \omega_i \omega_j \qquad (13)$$

Evaluate each higher-level indication one at a time using the complete cloud model calculation approach. This will provide the upper-level indicator's evaluation cloud, which will lead to the final evaluation cloud. The following equation yields the entire evaluation cloud result A.

$$A = \sum_{i=1}^{n} W_i M_i / \sum_{i=1}^{n} W_i \qquad (14)$$

Note M_i represents the evaluation cloud corresponding to the next level's i-th indicator, and W_i represents the weight of the i-th indicator.

3.3 Process of Ship Navigation Risk Assessment

With an improved TOPSIS technique, the primary risk factors are chosen based on the thorough identification of ship navigation risk sources using WBS-RBS. The chosen risk variables are then combined using the AHP-CRITIC approach to establish indicator weights. These weights are then used to run the cloud model simulation, in which different elements' cloud model parameters are converted into cloud droplets via a cloud generator, making it easier to translate between numerical numbers and conceptual ideas. Different risk variables' fuzziness and uncertainty are reflected through entropy and hyper-entropy. Refer to Fig. 1 for the process flow.

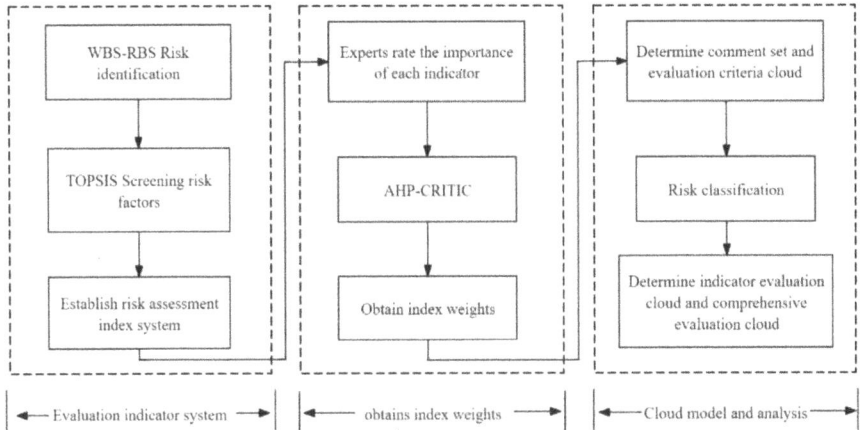

Fig. 1. Flow chart of risk simulation during ship pilotage

4 Case Study

4.1 Analysis of Ship Pilotage Task Scenario

For application and simulation, a bulk carrier that was piloted from the western channel into the Qinzhou Port and berthed was used as an example. The procedure is broken down into four phases based on the scenario analysis of ship piloting: preparation for port entry, coordination and communication, piloting activities, and completion and reporting of the pilotage. These phases enable an examination of the dangers associated with ship piloting at every level.

4.2 Determination of Index Weight

To account for the uncertainty of both subjective and objective data during the indicator system measurement process, this research uses the combined AHP-CRITIC technique to quantitatively calculate indicators at various levels. Primary and senior pilots, frequent vessel captains berthing at the port, and station management staff were the intended audience for the expert surveys and observations that produced the data. Nine of the ten responses to the ten surveys that were sent were considered valid.

Ten experts were asked to assess each indicator using the ship piloting risk factor index approach presented in Table 8. The AHP was used to compute subjective weights, which minimized errors caused by individual variances. The final weights were then determined by using the CRITIC approach to the objective weights. Tables 9 and 10 include the specific weights for the primary and secondary indicators, respectively.

Table 9. Weights of primary indicators for ship piloting risks

Primary indicators	Port entry preparation	Coordination and communication	Pilotage operation	Pilotage completion and reported
Weights	0.15	0.31	0.43	0.11

Table 10. Weights of secondary Indicators for ship piloting risks

Primary indicators	Weights	Secondary Indicators	Weights
Port entry preparationW1	0.15	Failure to apply for pilot services at the port or untimely application W11R12	0.226
		Failure to integrate port information resources in a timely manneW11R42	0.159
		Inaccurate analysis of navigation channel and hydrological conditions W12R42	0.102
		Failure to integrate port information resources promptly W12R41	0.043
		Unreasonable pilotage plan W13R42	0.047
Coordination and communicationW2	0.31	Untimely dispatch from the pilot station W21R42	0.062
		Incomplete or inaccurate handover informationW22R12	0.263
		Untimely response from the guided vessel W23R12	0.131
		Failure to notify the guided vessel or untimely notification of W23R42	0.038
		Aging or malfunction of pilot boarding equipment W24R21	0.112
		Risk of pilot leaving the vessel due to rough sea conditions W24R31	0.048

(continued)

Table 10. (*continued*)

Primary indicators	Weights	Secondary Indicators	Weights
Pilotage operation W3	0.43	Incorrect or improper operation by crew members of the vessel W31R12	0.077
Pilotage operation W3		Collision caused by mishandling of the other vesselW31R14	0.177
		Malfunction of the piloted vessel's navigational systemW31R21	0.151
		Pilot error in navigation instructions or improper command W32R11	0.091
		Tug crew error or improper operation W32R15	0.088
		Malfunction of piloted vessel's steering or navigation system W32R21	0.024
		Inability to navigate normally due to severe weather conditions W32R31	0.060
		Inadequate depth or width of navigational channelW32R34	0.033
		VTS coordination failure or delayed coordinationW32R41	0.073
		Heavy traffic within the harborW33R36	0.061
		Insufficient depth at the waterfront of the dockW33R35	0.054
Pilotage completion and reportedW4	0.11	Incomplete or inaccurate exchange of informationW41R11	0.055
		Improper placement or absence of gangway W42R12	0.036
		Aging or malfunction of pilot boarding equipmentW42R21	0.030
		Reports not collated or untimely W43R42	0.265

4.3 Analysis of Evaluation Results

The computation procedure, as presented in Table 11, establishes the distinctive parameters of each key indicator using Eqs. (6) to (8).

Table 11 makes it clear that the predicted value for the end of piloting and reporting is 3.744, which is relatively low and suggests that there are not many piloting hazards involved in ending piloting and reporting. Comparably, the anticipated value for port entrance preparation is 4.281, indicating that there is a minimal amount of piloting risk associated with this process. Coordination and communication are projected to be at a moderate risk level, with a rating of 5.549, and piloting operations, at 6.097. MATLAB

Table 11. Characteristic parameters of indicator cloud numerical features

Indicators	Ex	En	He
Port entry preparation W_1	4.281	0.643	0.334
Coordination and communication W_2	5.549	0.561	0.150
Pilotage operation W_3	6.097	0.573	0.161
Pilotage completion and reported W_4	3.744	0.491	0.266

software was utilized to calculate and plot evaluation indication cloud diagrams, as depicted in Fig. 2.

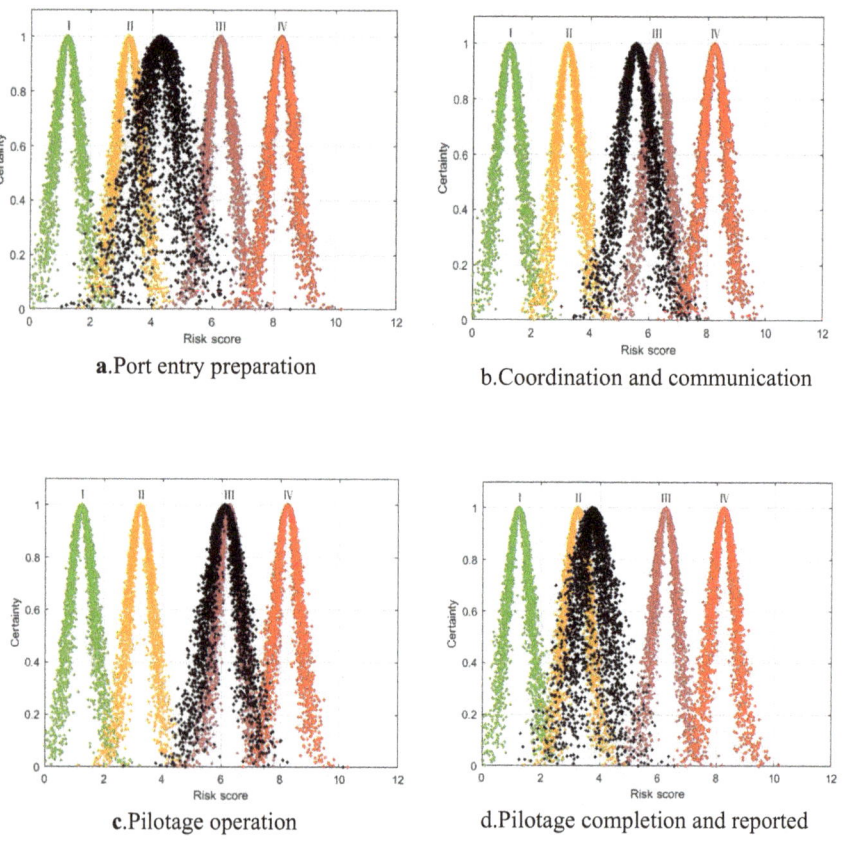

a. Port entry preparation

b. Coordination and communication

c. Pilotage operation

d. Pilotage completion and reported

Fig. 2. Cloud chart of primary indicator evaluation (a–d)

Piloting operations are associated with a moderate level of risk, as demonstrated by Fig. 2, which shows that they fall within the range of Level III indications. Indicators are positioned similarly, leaning more toward Level III than Level II, indicating that coordination and communication risks are likewise at a moderate level. In terms of the end of piloting and reporting, the indicator leans more toward Level II at this point, suggesting a lower level of risk. It is situated between Level II and Level III signs.

Equation (14) is used to produce a full evaluation cloud diagram, as seen in Fig. 3, by substituting the numerical characteristic parameters of the cloud indicators based on the estimated parameters of the primary indicator cloud model and their weights.

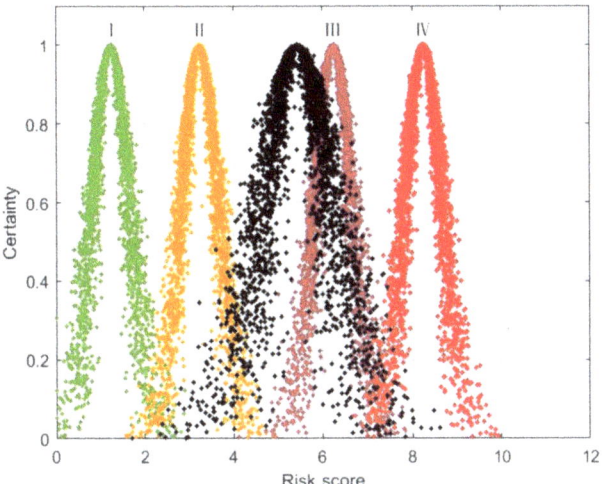

Fig. 3. Cloud chart of comprehensive assessment

As illustrated in Fig. 3, the complete risk assessment cloud diagram closely resembles the moderate risk level diagram and is situated between low and moderate risk levels. As a result, moderate danger has been identified for ship piloting in Qinzhou Port. The comprehensive risk assessment's strong credibility and dependability are indicated by the cloud diagram's thick cloud layers and narrow droplet span, which makes the evaluation results applicable.

5 Conclusions

In this paper, the risk detection procedure during ship piloting is examined using the WBS-RBS model. 48 risk indicators were found based on the piloting method; these included late or nonexistent requests for piloting to the port. These 48 risk indicators were then statistically analyzed using the TOPSIS model to determine the severity of the loss and the likelihood that they would occur. This process produced 26 risk indicators that were found.

The weights of the indicators were determined using the AHP-CRITIC approach. Furthermore, by utilizing the benefits of cloud diagrams, the relative strengths and weaknesses of the indicators were reflected more intuitively, improving the model's objectivity and accuracy.

The evaluation and analysis of ship piloting hazards at Qinzhou Port offer important pointers for improving the port's future lean management and piloting safety control. Nevertheless, the developed risk assessment indicator system is not adequately polished, and the analysis of risk variables in ship piloting is still not thorough enough. To make the system better, more survey respondents, an improved questionnaire, and additional scientific analysis of risk variables are required.

References

1. Chen, H., Hu, S., Wang, F., Xuan, S.: Assessment of risk situation for piloting a ship through the Pearl River Estuary involving unascertained measurement. Navig. China **46**(04), 38–45 (2023). (in chinese)
2. Guo, J., Wang, X., Bai, X., Xiao, Y.: Risks associated with piloting at Yangshan port in heavy weather and the expectation of economic benefits. Navig. China **46**(01), 9–15 (2023). (in chinese)
3. Tunçel, A.L., Akyuz, E., Arslan, O.: Quantitative risk analysis for operational transfer processes of maritime pilots. Marit. Policy Manag. **50**(3), 375–389 (2022)
4. Jiang, F., Zhou, C., Ma, Q.: Risk assessment for the inland river pilotage based on the set pair analysis model via integrated bestow. J. Saf. Environ. **21**(03), 990–996 (2021). (in chinese)
5. Guo, Y., Zhang, T., Hu, S.: Risk simulation model of ship piloting process in multi-period windows at tidal ports. J. Saf. Environ. **21**(01), 49–55 (2021). (in chinese)
6. Pan, J., Shao, Z., Huang, W., Fang, Q.: Pilotage responsibility period of marine pilot. Navig. China **43**(02), 46–49 (2020). (in chinese)
7. Chu, L., Shao, D.: Pilotage risk evaluation of Xiamen Port based on triangular fuzzy number AHP. J. Shanghai Maritime Univ. **40**(04), 61–65 (2019). (in chinese)
8. Seyfzadehdarabad, F., Sadeghi-Firoozabadi, V., Shokri, O.: Cognitive correlates of maritime pilots' human errors. Saf. Sci. **165**, 106196 (2023)
9. Zhang, X., Hu, S., Chen, Y., Wang, M.: Impact analysis of human-organizational factors on pilotage risks at the harbor. China Saf. Sci. J. **29**(12), 78–84 (2010). (in chinese)
10. Shi, Z., Li, J., Wang, Y., Song, W.: Study on environmental risk screening of petrochemical enterprises based on TOPSIS-AHP. J. Nankai Univ. (Nat. Sci.) **53**(1), 17–24 (2020). (in chinese)
11. Li, X.: TOPSIS model with entropy weight for eco-geological environmental carrying capacity assessment. Microprocess. Microsyst. **82**, 10380 (2020)

Open Access This chapter is licensed under the terms of the Creative Commons Attribution-NonCommercial-NoDerivatives 4.0 International License (http://creativecommons.org/licenses/by-nc-nd/4.0/), which permits any noncommercial use, sharing, distribution and reproduction in any medium or format, as long as you give appropriate credit to the original author(s) and the source, provide a link to the Creative Commons license and indicate if you modified the licensed material. You do not have permission under this license to share adapted material derived from this chapter or parts of it.

The images or other third party material in this chapter are included in the chapter's Creative Commons license, unless indicated otherwise in a credit line to the material. If material is not included in the chapter's Creative Commons license and your intended use is not permitted by statutory regulation or exceeds the permitted use, you will need to obtain permission directly from the copyright holder.

Fluid-Solid Coupling Analysis of Marine Stern Bearing Considering Viscosity-Temperature and Cavitation Effect

Xianjun Wu[1], Qianwen Huang[1,2(✉)], and Minghui Sheng[1]

[1] Hubei Key Laboratory of Mechanical Transmission and Manufacturing Engineering, School of Machinery and Automation, Wuhan University of Science and Technology, Wuhan 430081, China
qwhuang@wust.edu.cn

[2] Key Laboratory of Metallurgical Equipment and Control Technology, Ministry of Education, School of Machinery and Automation, Wuhan University of Science and Technology, Wuhan 430081, China

Abstract. The oil film lubrication performance of marine stern bearing is impacted by raised temperature and decrease local pressure as it operates with low speeds and heavy loads. The oil film's level of lubrication has a direct impact on the bearing's operating condition and lifespan. When operating at low speeds and heavy loads, the lubrication performance (including load capacity and oil film pressure) is greatly affected by the decrease in lubricating oil viscosity brought on by temperature increases and the cavitation of the oil film generated by low local pressure. This study examines the lubrication performance of low-speed plain bearings from the perspectives of cavitation and viscosity-temperature effects. Based on the flow field calculation results, a fluid-structure coupling system of oil-film bearings is built by constructing the viscosity-temperature equation and UDF program. The lubricating oil film and bearing lubrication qualities under different operating situations are systematically computed while taking into account the cavitation effect of the mixed multiphase flow model. The results show that the cavitation region's maximum oil film pressure, maximum temperature, maximum bearing capacity, and maximum cavitation volume fraction are all lower than those at a particular viscosity. As eccentricity and axial velocity increase, so does the maximum cavitation volume portion.

Keywords: Fluid-Solid Coupling · Marine Stern Bearing · Viscosity-Temperature Lubrication · Cavitation Effect

1 Introduction

Low-speed oil-lubricated marine stern bearing can minimize wear due to their excellent lubrication efficiency and stability. However, the viscosity of the lubricating oil plays a crucial role in the lubrication performance [1]. The cavitation induced by local pressure significantly impacts the lubrication state. It is more practical to evaluate the oil film lubrication state under gas-liquid two-phase flow conditions while accounting for cavitation.

The Reynolds equation has laid the foundation for major advancements in fluid lubrication for marine stern bearing [2]. It effectively describes the distribution of oil film pressure and contributes to hydrodynamic lubrication research [3].

Furthermore, researchers have focused on investigating gas-liquid two-phase flow. HiroshiYabe [4] scrutinized the cavitation issue and found that cavitation phenomena significantly impact the lubrication performance of oil-lubricated bearings. The Computational Fluid Dynamics (CFD) calculation method is a successful outcomes in fluid dynamics and rapid advancements in computer technology. Zenglin [5] originally suggested a CFD-based model to solve the requisite bearing characteristics.

In light of this, a study on viscosity-temperature and cavitation effect in marine stern is conducted with CFD method. It considers the viscosity-temperature effect, pressure, cavitation distribution, and bearing capacity of the oil film under various speeds and eccentricities. Additionally, it addresses issues such as bearing stress, strain, deformation while comparing results under constant viscosity lubricant.

2 Fundamental Equation

The form of the mass conservation equation for gas-liquid two-phase flow is:

$$\frac{\partial}{\partial t}(\rho_m) + \nabla \cdot (\rho_m \vec{v}_m) = 0 \tag{1}$$

where $\vec{v}_m = \sum_{k=1}^{2} \alpha_k \rho_k \vec{v}_k / \rho_m$, $\rho_m = \sum_{k=1}^{2} \alpha_k \rho_k$, ρ_m, is the mixing density of gas-liquid two-phase, $\vec{v}_m$ is the velocity vector, α_1, α_2 are the volume fraction of two phases respectively.

The momentum conservation equation can be obtained:

$$\begin{aligned}\frac{\partial}{\partial t}(\rho_m \vec{v}_m) + \nabla \cdot (\rho_m \vec{v}_m \vec{v}_m) &= -\nabla p \\ &+ \nabla \cdot \left[\mu_m (\nabla \vec{v}_m + \nabla \vec{v}_m^T)\right] \\ &+ \rho_m \vec{g} + \vec{F} + \nabla \cdot \left(\sum_{k=1}^{n} \alpha_k \rho_k \vec{v}_{dr,k} \vec{v}_{dr,k}\right)\end{aligned} \tag{2}$$

where $\mu_m = \sum_{k=1}^{n} \alpha_k \mu_k$, $\vec{v}_{dr,k} = \vec{v}_k - \vec{v}_m$, n is the number of phases, $\vec{F}$ is the external body force, μ_m is the viscosity, $\vec{v}_{dr,k}$ is the slip velocity of K phase, p, $\rho_m \vec{g}$ is the fluid pressure and gravity.

The component mass conservation equation can be given as:

$$\begin{aligned}\frac{\partial}{\partial t}(\alpha_v \rho) + \nabla \cdot (\alpha_v \rho \vec{v}_m) &= -\nabla \cdot (\alpha_v \vec{v}_{dr,v}) \\ &+ \sum_{l=1}^{n} (\dot{m}_{lv} - \dot{m}_{vl})\end{aligned} \tag{3}$$

where l and v represent liquids and gases, α_v and $\vec{v}_{dr,v}$ are volume fraction and slip velocity of the gas phase, $\dot{m}_{lv}$ is the mass transfer rate of gas-liquid. The fluid energy equation is:

$$\rho \frac{Du}{Dt} = \nabla(k\nabla T) + q \tag{4}$$

where, ρ, Du/Dt, k, T, $\nabla(k \nabla T)$ and q denotes fluid density, change rate of fluid internal energy, thermal conductivity, fluid temperature, heat conduction term and internal energy term.

The concentric flow can be assumed inside the bearing with Re_c is:

$$\text{Re}_c = 41.1\sqrt{\frac{r}{c}} \tag{5}$$

where r and c are journal radius and radius gap, respectively.

The Newton's second law provides the governing equation of the solid part:

$$\rho_s \ddot{d}_s = \nabla \cdot \sigma_s + f_s \tag{6}$$

where ρ_s is solid density, $\ddot{d}_s$ is the local acceleration vector of the solid domain, σ_s is Cauchy stress tensor, f_s is volume force vector. The thermal deformation term of the solid part can be caused:

$$f_T = \alpha_T \cdot \nabla T \tag{7}$$

where f_T, α_T is the amount of expansion, expansion coefficient related to temperature.

At the fluid-solid coupling interface, variables should satisfy conservation equations or equality:

$$\tau_f \cdot n_f = \tau_s \cdot n_s, d_f = d_s, q_f = q_s, T_f = T_s \tag{8}$$

where τ, d, q and T are the stress, displacement, heat flux and temperature.

3 Model Establishment

Figure 1(a) shows the marine stern bearing model diagram. o_1 is the bearing center, o_2 is the journal center, e is the eccentricity, the ratio of eccentricity e to radial clearance c is the eccentricity ε.

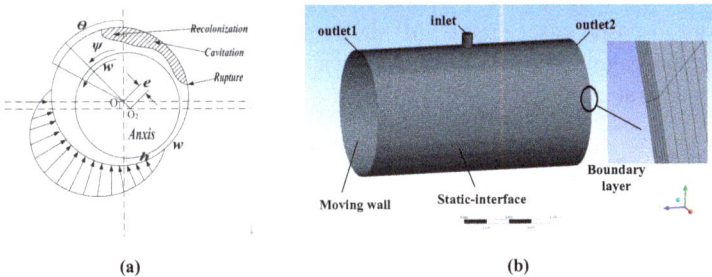

Fig. 1. Oil film cavitation of marine stern bearing and Oil film grid diagram with boundary layer.

The grid coupling is established with grid overlap surface set as an interface. The interaction between fluid and solid surfaces is described by boundary layers, so four layers of grids are placed in the oil outlet, resulting in a total of 186,642 grids as shown in Fig. 1(b). The grid was confirmed to be reliable when compared to the literature [6]. The maximum pressure and bearing capacity of the oil film at v t circumstances, with the number of grids in reference 502,425, are 0.015 MPa and 2390 N, compared with 0.101 MPa and 2375 N in this work, with errors of just 3.9% and 0.63% between the two. The parameters are shown in Table. 1.

Table 1. Basic parameters of sliding bearing.

Parameters	Values	Parameters	Values
Shaft neck radius R	70 mm	Thermal conductivity λ	0.135 W/(m °C)
Bearing length L	250 mm	Pressure inlet	0.1 MPa
Radial clearance c	0.1 mm	Inlet temperature	295 K
Oil film density ρ	890 kg/m^3	Bearing thickness	30 mm
Rotor rotary speed ω	200 rpm	Oil inlet diameter d	20 mm
Eccentricity ε	0.2	Heat capacity c	2307 J/(kg °C)
Dynamic viscosity v	12.5 mPa·s	Poisson's ratio	0.3

4 Analysis of Flow Result

4.1 Lubrication at Different Rotational Speeds

The Fig. 2(a) shows the maximum pressure gradually rises as the rotational speed increases, leading to negative pressure area (point A). Figure 2(b) shows that when rotational speed increases, the temperature rise in the central part of the bearing accelerates quicker than with constant viscosity, whereas the temperature rise at the two extremities of the bearing climbs more slowly. Figure 2(c) displays the circumferential temperature rise of the inlet port under viscosity-temperature effect is greater than the constant viscosity case. Figure 2(d) shows that when speed increases, the cavitation maximum volume fraction in the constant viscosity increases, in contrast to the viscosity-temperature effect.

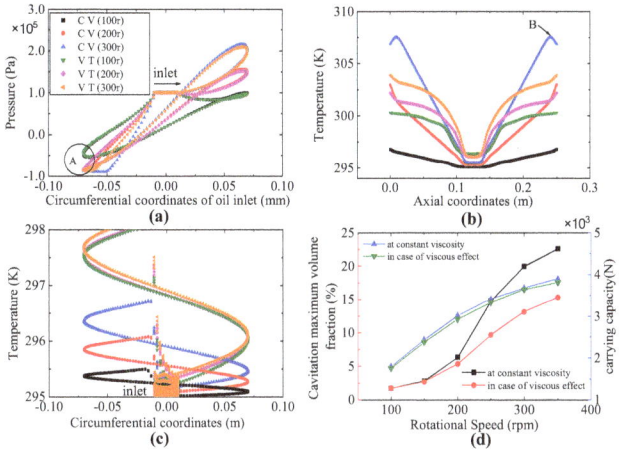

Fig. 2. Lubrication characteristics of oil film at different rotational speeds.

4.2 Lubrication with Different Eccentricity

Figure 3(a) shows how the pressure progressively rises with increasing eccentricity. The results in Fig. 3(b) displays the temperature along the axial direction, which is the area of pressure convergence (point E). It is observed that the maximum temperature is located in the convergence area of the end part with constant viscosity, slightly larger than viscosity-temperature effect. Figure 3(c) illustrates that there is a gradual increase in temperature distribution along with axial direction from center towards end parts. Figure 3(d) shows that the maximum cavitation volume fraction difference between the two viscosity conditions is: rather small as the eccentricity increases.

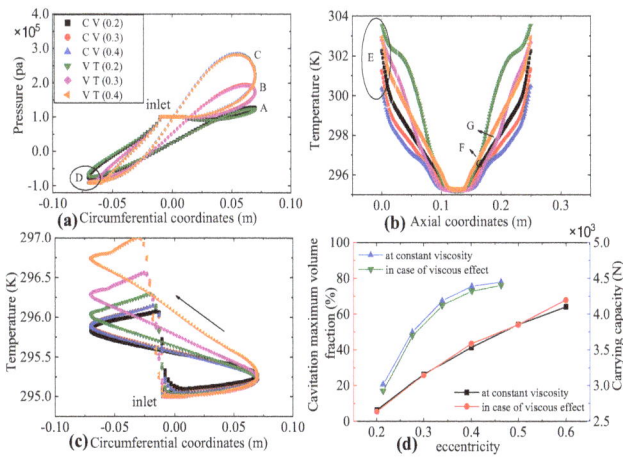

Fig. 3. Lubrication characteristics of oil film with different eccentricity.

4.3 Lubrication Under Different Inlet Pressure

Figure 4(a) illustrates that as the inlet pressure increases, the pressure gradually shifts from the maximum pressure point of convergence area to the inlet location. It is evident from Fig. 4(b) that the temperature in axial direction gradually increases from the inlet to the end part. The convergence area of oil inlet in circumferential direction is point B. In terms of circumferential direction in Fig. 4(c), there is a gradient in temperature related to rotational direction, but this gradient is not significantly affected by changes in inlet pressure. Figure 4(d) shows the maximum volume fraction of cavitation decreases from 0.064 to 0.02 with the inlet pressure increases from 0.1 MPa to 0.3 MPa. The cavitation is more concentrated while taking into account the viscosity-temperature effect.

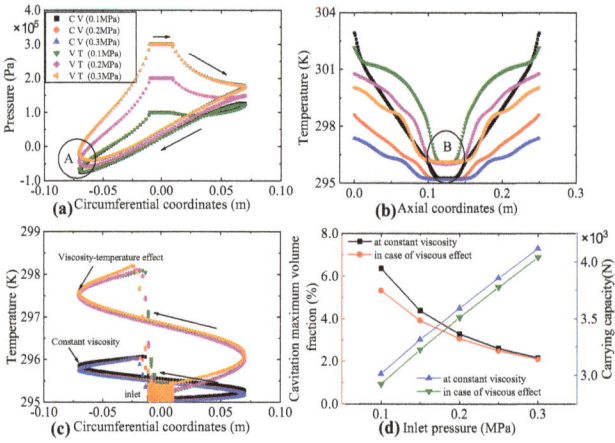

Fig. 4. Lubrication characteristics of oil film under different inlet pressure.

5 Discuss on Fluid-Solid Coupling

The bearing's inner wall serves as the fluid-structure coupling interface, while fixed limitations are added to the bearing's outer wall. The coefficient of thermal expansion is defined as 1.2×10^{-5} °C^{-1} with initial structural parameters shown in Table 1. It is evident from Fig. 5(a) that the maximum total deformation under constant viscosity increases from 3.12 μm to 5.96 μm with rotational speed ranges from 150 rpm to 350 rpm. The maximum radial deformation changes from 0.96 μm to 4.33 μm. Considering the viscous temperature effect, both the results range from 3.59 μm to 5.23 μm for total deformation and from 1.91 μm to 3.76 μm for radial deformation. Figure 5(b) shows that the maximum total deformation under constant viscosity exhibits an increasing trend from 5.69 μm to 6.45 μm as the eccentricity increased from 0.2 to 0.6. Additionally, the results changes from 3.84 μm to 4.15 μm with viscous temperature effect.

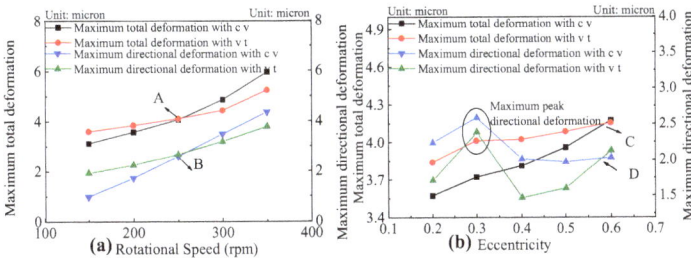

Fig. 5. Maximum deformation of bearings under different speeds and eccentricity.

5.1 Equivalent Stress and Thermal Strain

It is noticed from Fig. 6(a) that the maximum equivalent stress with constant viscosity increases from 74.739 MPa to 136.84 MPa as the rotational speed ranges from 150 rpm to 350 rpm. While that of the viscous temperature effect changes from 85.123 MPa to 118.53 MPa. It can be found in Fig. 6(b) that the maximum equivalent stress does not exhibit a unidirectional increasing trend as eccentricity increases. There is a pattern that the stress increases to maximal, then decreases to minimal, and subsequently increases again. The largest maximum equivalent force occurs at an eccentricity of 0.3, which is approximately at 85.326 MPa for constant viscosity and 92.341 MPa for viscosity-temperature effects, representing an increase by 8%.

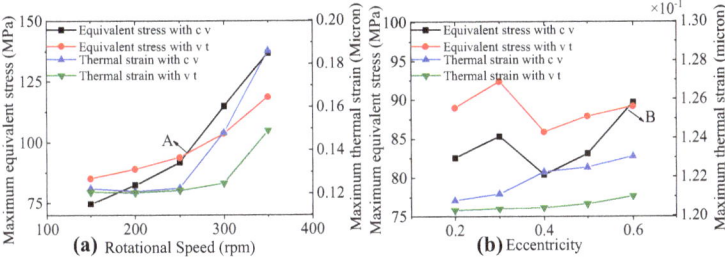

Fig. 6. Maximum stress and strain of bearings under different speeds and eccentricity.

6 Conclusion

This study investigates the effects of hole effect and viscosity-temperature effect on the lubrication performance of marine stern bearings. The research conclusions are as follows:

1. Increased rotating speed, eccentricity, and inlet pressure result in varying degrees of higher maximum pressure, bearing capacity, and oil layer temperature.
2. The axial deformation, stress, and strain rise progressively from the inlet towards both ends of the bearing, peaking at the end face. Radial deformation at the convergence and divergence zones of the oil film ends exhibits a symmetrical elliptical pattern.

3. The viscosity-temperature effect causes smaller maximum deformation, elastic stress, and elastic strain compared to constant viscosity conditions.

Acknowledgments. This research was funded by the National Natural Science Foundation of China (No. 52272377).

References

1. Lahmar, M., Bou-Saïd, B.: Nonlinear dynamic response of an unbalanced flexible rotor supported by elastic bearings lubricated with piezo-viscous polar fluids. Lubricants. **3**(2), 281–310 (2015)
2. Liming, Z., et al.: Nonlinear vibration induced by the water-film whirl and whip in a sliding bearing rotor system. Chin. J. Mech. Eng. **29**(02), 260–270 (2016)
3. Toshiyuki, D.: Generalized Reynolds equation for microscale lubrication between eccentric circular cylinders based on kinetic theory. J. Fluid Mech. **974**(413), A13–A13 (2023)
4. Yabe, H., Sakurai, T., Hirayama, T.: A theoretical analysis considering cavitation occurrence in oil-lubricated spiral-grooved journal bearings with experimental verification. J. Tribol. **126**(3), 490–498 (2004)
5. Zenglin, G., Toshio, H., Gordon, K.R.: application of cfd analysis for rotating machinery—part I: Hydrodynamic, hydrostatic bearings and squeeze film damper. J. Eng. Gas Turbines Power. **127**(2), 445–451 (2005)
6. Wang, L.: Performance analysis of hydrodynamic journal bearing considering viscosity-temperature effect of lubricating oil. lubr. seal. **45**(01), 54–58 (2020) (in Chinese)

Open Access This chapter is licensed under the terms of the Creative Commons Attribution-NonCommercial-NoDerivatives 4.0 International License (http://creativecommons.org/licenses/by-nc-nd/4.0/), which permits any noncommercial use, sharing, distribution and reproduction in any medium or format, as long as you give appropriate credit to the original author(s) and the source, provide a link to the Creative Commons license and indicate if you modified the licensed material. You do not have permission under this license to share adapted material derived from this chapter or parts of it.

The images or other third party material in this chapter are included in the chapter's Creative Commons license, unless indicated otherwise in a credit line to the material. If material is not included in the chapter's Creative Commons license and your intended use is not permitted by statutory regulation or exceeds the permitted use, you will need to obtain permission directly from the copyright holder.

Analysis and Experimental Study on Water Yield of Seawater Desalination Device by Light Distillation

Lucai Wang, Tianhua Xie, and Zhilin Zheng(✉)

Department of Navigation Dalian Naval Academy, Dalian, China
navy_zzl@163.com

Abstract. In order to solve the problem of fresh water demand for ocean survival on ships, a light distillation seawater desalination device for life rafts for ocean survival is proposed. The water production of the device under different degrees, seasons and weather conditions is analyzed by combining theory with experiment. It can be concluded that the total daily solar radiation and the total efficiency of the device can be used to estimate the water production of the light distillation desalination device. By employing the high light absorbing material "Dark Metallic Nano-materials", the total efficiency of the device can reach 47.58%, which is 58.60% higher than that of ordinary disk solar distillers. The proposal of the light distillation seawater desalination device, as well as the calculation, analysis and actual measurement of its water production, can provide a reference for the equipment of life raft desalination device on oceangoing ships.

Keywords: Seawater Desalination · Light Distillation · Ocean Rescue · Water Production · Life Raft

1 Introduction

Oceangoing ships generally sail far from land. Compared to the near-shore ships, once a oceangoing ship meet with a mishap, it is much more difficult to rescue, and the crew may not be rescued a few days or dozens of days, and can only drift in a life raft waiting for rescue, relying on the life support material inside the raft to survive. Fresh water is the key to maintain the crew's vital signs after a shipwreck, and is more important than food to those who survive at sea. Studies have shown that with fresh water but no food, survivors can still live for 30–50 days, but without fresh water and only food, they can only sustain life for a few days.[1] Although life rafts are equipped with bagged water, the amount is limited. Currently, life rafts are equipped with bagged water at a rate of 1.5 L per person per rated raft occupant[2, 3] and no fresh water is consumed for the first 24 h after entering the raft, unless injured or sick, and then no more than 500 mL per person per 24 h; when fresh water is about to run out, no more than 100 mL per person per 24 h can be afford. If people can not be rescued within 7 days after a shipwreck, fresh water will be the key to their survival.

The current life raft fresh water guarantee cycle is based on near-shore conditions, combined with the near-shore rescue capability to set, but in the ocean environment, especially far from the shore, the crew search and rescue will be much more difficult. If the crew can not be rescued for a long time, the current life raft fresh water equipment standard is difficult to meet the requirements to maintain their vital signs. If we want to improve the supply of fresh water on life rafts, on one hand we can increase the amount of water carried in bags, and on the other hand we can equip with the highly-effective desalination device. As life rafts have strict requirements on their own size and weight, it is difficult to increase the amount of fresh water they can carry. Therefore, there is an urgent need to study small seawater desalination devices applicable to life rafts to solve the demand for fresh water for long periods of time when the crew is not rescued away from shore conditions, and greatly extend the rescue time for those who fall overboard.

In this paper, for the problem of limited fresh water carrying capacity of life raft and difficulty in adapting to the life-saving demand of ocean-going ships, a light distillation seawater desalination device applicable to life raft is proposed, and its water production volume is calculated and analyzed and experimentally studied, in order to provide reference for the equipment of desalination device for life raft of ocean-going ships.

2 Device Structure

The basic structure of the light distillation desalination device is shown in Figs. 1 and 2, including a floating tire, a conical transparent air bladder, a freshwater groove, a freshwater collection bag, a seawater chamber, a water-absorbing strip, light-absorbing material and a seawater intake bag. The device is made of flexible material and can be folded and put away, as shown in Fig. 3. When used, the device is inflated and shaped (Fig. 2), and seawater is filled into the seawater chamber from the seawater inlet bag, and the suction strip will draw seawater into the surface of the light- absorbing material in a continuous stream and wet the light-absorbing material. Under the irradiation of sunlight, the light-absorbing material continuously absorbs sunlight and carries out photothermal conversion, heating the seawater absorbed by the absorbent strip. Then the seawater evaporates to produce water vapor and condenses on the inner surface of the conical transparent air bag, and the condensed water enters the freshwater groove and flows into the freshwater collection bag.

The water production performance of the light distillation desalination device is related to the spectral absorbance and photothermal conversion efficiency of the light-absorbing material. The higher spectral absorbance and photothermal conversion efficiency can lead to better water production performance of the device. In order to improve the water production of this device, the light-absorbing material was selected from the "Dark Metallic Nano-materials" material developed by the team of Professor Jia Zhu from Nanjing University [4–8], as shown in Fig. 4, which is a two-dimension nanomaterial with a porous structure. Its full-spectrum light absorption efficiency was above 99% and the photothermal conversion efficiency was above 90%. Additionally, this material can be produced in large quantities.

Analysis and experimental study on water yield of seawater 181

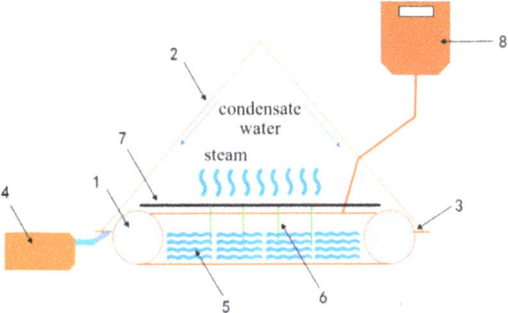

Fig. 1. Structure of light distillation seawater desalination device. (1- floating tire; 2- conical transparent air bladder; 3- fresh water groove; 4- fresh water collection bag; 5- seawater chamber; 6- water absorbing strip; 7- light absorbing material; 8- seawater intake bag.)

Fig. 2. Physical object of light distillation seawater desalination device. (inflation service status).

Fig. 3. Physical object of light distillation seawater desalination device. (folding storage status).

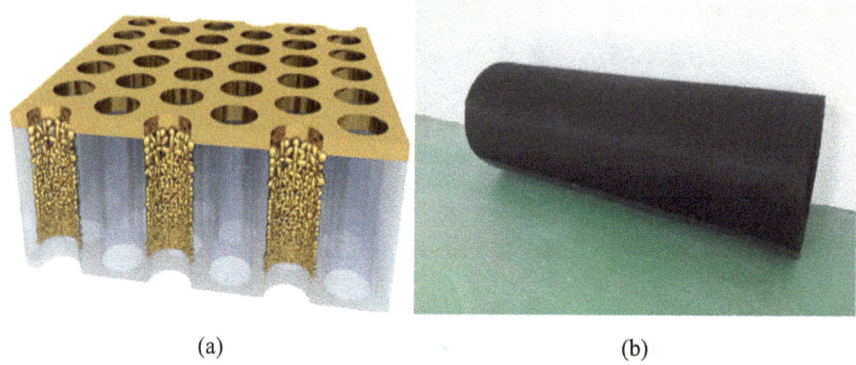

Fig. 4. Nano black gold material. (Fig. 4(a)– Microstructure, Fig. 4(b)- Physical Picture).

3 Method of Estimating Water Production of the Device

There are many factors affecting the water production of a light distillation seawater desalination device, including meteorological parameters such as solar radiation, seawater temperature, ambient temperature, wind speed, etc. And design parameters such as the light transmission rate of the conical transparent gas capsule, the light absorption rate of the light-absorbing material, the photothermal conversion efficiency, condensation efficiency, and heat loss of the unit also play key roles. In this paper, Eq. (1) is used to estimate the water production of the light distillation seawater desalination device:

$$Q = \frac{EGA}{\lambda} \qquad (1)$$

Where: Q is the daily water production of the device, L; E is the total efficiency of the device, dimensionless; G is the total daily solar radiation, MJ / m2; A is the area of the absorbing material; The total daily solar radiation for cloudy and overcast weather was obtained by a statistical of the device, m2; λ is the latent heat of vaporization of water, taken as 2.35 MJ/L.

The total daily solar radiation at ground level is mainly influenced by factors such as solar altitude angle, atmospheric transparency and weather conditions. Solar altitude angle can be calculated from Eq. (2).

$$\sin h = \sin\varphi \sin\delta + \cos\varphi \cos\delta \cos t \qquad (2)$$

Where: h is the solar altitude angle; ϕ is the geographical latitude; δ is the solar declination; and t is the hour angle (local time).

On a clear day, the total daily solar radiation is:

$$G = \sum_t \beta S . \sin h \qquad (3)$$

Where: β is the ratio of the energy radiated to the ground to the solar constant, taken as 73%; S. is the value of the solar constant, taken as 1368 W/m^2.

The total daily solar radiation for cloudy and overcast weather was obtained by a statistical method. Table 1 lists the average daily total solar radiation for sunny, cloudy and overcast days in 2016 at the Nanjing Observatory. The average total daily solar radiation for sunny days is used as the benchmark to calculate the ratio of the average total daily solar radiation in cloudy and overcast conditions to the average total daily solar radiation in sunny days for different months. Then the average value of these ratios by month are used as the standard ratio of the average total daily solar radiation in cloudy and overcast conditions to that in sunny days.

Table 1. Meteorological data of Nanjing Observatory (2016).

Month	Total daily solar radiation(MJ/m^2)			Cloudy/Clear skies	Overcast/Clear skies
	Clear skies	Cloudy	Overcast		
1	13.78	9.17	3.29	0.67	0.24
2	16.53	14.88	8.98	0.90	0.54
3	19.34	14.84	10.74	0.77	0.56
4	24.17	18.90	14.25	0.78	0.59
5	27.22	22.58	12.84	0.83	0.47
6	27.52	21.01	12.62	0.76	0.46
7	25.08	19.90	13.78	0.79	0.55
8	23.90	19.80	12.33	0.83	0.52
9	20.27	18.23	12.15	0.90	0.60
10	15.08	13.20	7.00	0.88	0.46
11	14.56	12.69	7.82	0.87	0.54
12	11.04	10.12	6.31	0.92	0.57
Average				0.82	0.51

From Table 1, the ratio of average daily total solar radiation under cloudy and overcast conditions to total daily solar radiation in sunny days is 0.82 and 0.51 for cloudy and cloudy weather, respectively.

If the total daily solar radiation at a location and the total efficiency of the desalination device could be obtained, the daily water production of the device can also be estimated. The total daily solar radiation at a given location can be obtained from meteorological office statistics, so it is crucial to obtain the total efficiency of the device if the daily water production of the device is to be estimated.

4 Measurement of the Total Efficiency of the Device

The daily water production test of the light distillation seawater desalination device was carried out on the Xiaoyang River in Sheyang County, Yancheng City, Jiangsu Province, and the device was placed on the water surface, and the light intensity data of the day was collected, and the collected fresh water was measured. Figure 5 shows the test scene in the Xiaoyang River, with fresh water in the river and the device filled with 3.5% NaCl solution (used to simulate seawater).

Fig. 5. Xiaoyang River test scenario.

The test results for different weather conditions are presented in Table 2 and the total efficiency of the unit is calculated from Eq. (1), the average value of the total efficiency of the unit is 47.58%.

Table 2. Results of water production of the unit.

Weather	Total daily solar radiation(MJ/m^2)	Water Production(L)	Total Unit Efficiency(%)
Sunny(August)	24.32	5.05	48.80
Sunny(August)	23.84	4.9	48.30
Sunny(August)	21.8	4.75	51.20
Cloudy(August)	20.21	4.05	47.09
Cloudy(August)	18.92	3.8	47.20
Sunny(September)	18.24	3.5	45.09
Cloudy(August)	17.59	3.55	47.43
Cloudy(September)	15.6	3.2	48.21
Overcast(September)	12.3	2.35	44.90
Average			47.58

At present, the total efficiency of the ordinary disc solar still is 30%,[9] and the device adopts a strong absorbent material "Nano Black Gold" on the basis of the ordinary disc solar still, which increases the total efficiency of the device by 58.60%.

5 Calculation and Analysis of the Water Production of the Device

The water production of the light distillation seawater desalination device is mainly related to the total solar radiation of the day, and Table 3 lists the water production of the device under the total solar radiation of different days.

Table 3. Water production (Typical Solar total radiation per day).

Total daily solar radiation (MJ/m^2)	Daily Water Production(L)	Total daily solar radiation (MJ/m^2)
20	4.05	20
17	3.44	17
12	2.43	12

The water production of light distillation seawater desalination devices in different latitudes was calculated and analyzed, and the three cities of Sanya (18° north latitude), Yancheng (33° north latitude) and Heihe (49° north latitude) were selected as the calculation basis within three latitudes: 0° to 23° north latitude (tropic of Cancer), 23° to 43° north latitude, and 43° to 66° north latitude (Arctic Circle).

Table 4 lists the total solar radiation on sunny days on the 15th of each month in Sanya, Yancheng and Heihe, averages the total solar radiation of sunny days in three months of each quarter as the total solar radiation of sunny days in that quarter, and calculates the average total solar radiation of cloudy and overcast days, as shown in Table 5. Based on the data in Table 5, the daily water output of the installation is calculated, as shown in Table 6.

Table 4. Solar total radiation on sunny days(clear-sky).

Season	Date	Solar Total Radiation Per Day(MJ/m^2)		
		Sanya	Yancheng	Heihe
Winter	12.15	18.61	12.66	5.68
	1.15	19.76	13.61	6.76
	2.15	22.61	17.62	10.91
Spring	3.15	24.63	21.37	16.26
	4.15	27.68	26.23	22.70
	5.15	27.37	27.29	26.61
Summer	6.15	28.73	29.58	28.29
	7.15	28.70	27.85	27.65
	8.15	27.07	27.64	23.42

Table 5. Solar total radiation per day(different weather).

City	Weather	Solar Total Radiation Per Day (MJ/m^2)		
		Winter	Spring	Summer
Sanya	Sunny	20.33	26.56	28.17
	Cloudy	16.67	21.78	23.10
	Overcast	10.37	13.55	14.37
Yancheng	Sunny	14.63	24.96	28.36
	Cloudy	12.00	20.47	23.26
	Overcast	7.46	12.73	14.46
Heihe	Sunny	7.78	21.86	26.45
	Cloudy	6.38	17.93	21.69
	Overcast	3.97	11.15	13.49

Table 6. Water production.

Latitude	City	Weather	Daily Water Production(L)		
			Winter	Spring	Summer
0° ~ 23°	Sanya	Sunny	4.12	5.38	5.70
		Cloudy	3.38	4.41	4.68
		Overcast	2.10	2.74	2.91
23° ~ 43°	Yancheng	Sunny	2.96	5.05	5.74
		Cloudy	2.43	4.14	4.71
		Overcast	1.51	2.58	2.93
43° ~ 66°	Heihe	Sunny	1.58	4.43	5.36
		Cloudy	1.29	3.63	4.39
		Overcast	0.80	2.26	2.73

In order to further test the water production of the device, the water production capacity of the installation was tested in Dalian City, Liaoning Province, the specific location is the Free River (38.9° north latitude) in Zhongshan district of Dalian City, the Free River is connected to the Yellow Sea, with seawater in the river, and the device is placed on the water surface and the collected fresh water is measured daily. Figure 6 shows the test scenario in Dalian. In July, the measurement result of the summer sunny day was 5.50 L/m^2, the cloudy day was 2.89 L/m^2, and the calculation error of Table 6 was 4.18% and 1.37%, and the analysis was that the latitude of Dalian was higher than that of Yancheng, and the total solar radiation per day was slightly lower, therefore, the water production is also slightly lower, but the error is not large. It can be shown that the calculation results of Table 6 can basically reflect the water production of the device in different regions and different seasons, and further verify the estimation method of the water production of the light distillation seawater desalination device in this paper.

Fig. 6. Free River test scenario.

6 Conclusion

In this paper, a light distillation seawater desalination device suitable for the life raft is proposed to solve the problem that the freshwater carrying capacity of the life raft is limited, and it is difficult to meet the lifesaving needs of ocean-going ships,. The total water production efficiency of the device was measured, and the water production capacity of the device was calculated and experimentally studied in different latitudes, different seasons and different weather conditions.

1) The estimation method to the water production of the light distillation seawater desalination device in this paper is feasible;
2) After the addition of the high light absorbing material "Dark Metallic Nanomaterials", the total water production efficiency of the device reached 47.58%, which was 58.60% higher than that of the ordinary disc solar still.

The proposal of the light distillation seawater desalination device, as well as the calculation analysis and actual measurement of its water production, can provide a reference for the equipment of the seawater desalination device for the life raft of oceangoing ships.

References

1. Jing, H.: Dangers and countermeasures in survival at sea. Sci. Technol. Inf. (2012) 483+485
2. China Maritime Service Center: Operation and Management of Lifeboats, Rafts and Rescue Boats. Dalian Maritime University Press, Dalian (2014) Beijing: China Communications Press
3. Shen, F.: Elements and precautions for survival at sea. J. Zhejiang Transp. Coll. **7**, 35–36 (2006) Author, F.: Contribution title. In: 9th International Proceedings on Proceedings, pp. 1–2. Publisher, Location (2010)

4. Zhou, L., et al.: Self-assembly of highly efficient, broadband plasmonic absorbers for solar steam generation. Am. Assoc. Adv. Sci. **2**(4), e1501227 (2016)
5. Li, X.Q., et al.: Three-dimensional artificial transpiration for efficient solar waste-water treatment. Natl. Sci. Rev. **5**(1), 70–77 (2018)
6. Xu, N., et al.: A water lily-inspired hierarchical design for stable and efficient solar evaporation of high-salinity brine. Am. Assoc. Adv. Sci. **5**(7), eaaw7013 (2019)
7. Liu, H.Z., et al.: Plasmonic nanostructures for advanced interfacial solar vapor generation. Sci. Sin.-Phys. Mech. Astron. **49**(12), 124203 (2019)
8. Zhu, J.: Solar Photothermal Conversion Based on Mocro Mano Structure. In: Abstracts of the 4th China Energy Materials Chemistry Symposium. MDPI, Xi'an, China (2019)
9. Gao, C.J., Ruan, G.L.: Seawater Desalination Technology and Engineering. Chemical Industry Press, Beijing (2016)

Open Access This chapter is licensed under the terms of the Creative Commons Attribution-NonCommercial-NoDerivatives 4.0 International License (http://creativecommons.org/licenses/by-nc-nd/4.0/), which permits any noncommercial use, sharing, distribution and reproduction in any medium or format, as long as you give appropriate credit to the original author(s) and the source, provide a link to the Creative Commons license and indicate if you modified the licensed material. You do not have permission under this license to share adapted material derived from this chapter or parts of it.

The images or other third party material in this chapter are included in the chapter's Creative Commons license, unless indicated otherwise in a credit line to the material. If material is not included in the chapter's Creative Commons license and your intended use is not permitted by statutory regulation or exceeds the permitted use, you will need to obtain permission directly from the copyright holder.

Analysis on the Application of Unmanned Equipment in Lifesaving Missions at Sea

Guanghui Yu(✉), Lei Su, and Wei Li

Department of Navigation, Dalian Naval Academy, Dalian, China
muxin1230@163.com

Abstract. After accidents and disasters occur on the vast sea, the process of maritime rescue is extremely complicated and difficult, there is bad sea environment, instantaneous disaster, insufficient target search and detection ability and various uncertainties, which seriously restrict the timeliness, safety and accuracy of Marine lifesaving tasks. With the rapid development of information technology and intelligent technology, unmanned technology has been quite mature, unmanned equipment has boarded the stage of history and has received general attention in the military field. With the rapid development of unmanned equipment, it has been applied to many fields such as reconnaissance and early warning, enemy attack, target guidance and so on. Unmanned surface craft, unmanned aerial vehicle, unmanned underwater vehicle and other unmanned equipment can carry out diversified tasks such as material delivery, target search and identification, and search and rescue in a number of lifesaving missions such as lifesaving, maritime rescue and joint search and rescue. Focusing on the future, unmanned equipment, with its advantages of intelligence, clustering, modularization, large-scale and multi-functional, has broad application prospects in lifesaving missions. Strengthening the application research of unmanned equipment is of great significance to improve the combat capability of the defense and rescue forces in lifesaving missions at sea and saving people's lives and property safety.

Keyword: Unmanned Equipment · At Sea Lifesaving · Application

1 With the Rapid Development of Unmanned Technology and Intelligence

1.1 Wide Range of Application

Unmanned equipment has greatly affected modern warfare and human life. At present, unmanned equipment such as drones, unmanned surface boats, as shown in Fig. 1, and underwater robots have played a huge role in the fields of reconnaissance and early warning, electronic countermeasures, fire strike and anti-mine warfare by their superior characteristics of unmanned autonomy, strong mobility and good stealth. These unmanned equipment can also be used in flood relief, lifesaving, maritime transportation, film and television shooting and other fields [1, 2].

Fig. 1. Surface unmanned vehicle.

1.2 Strong Adaptability to the Environment and Strong Survivability

Because unmanned equipment is not limited by human factors, it can better complete diverse tasks under extremely harsh environmental conditions. Similarly, unmanned equipment does not need to consider the existence of the operator, so it has a very large flexibility in structural design, and has a rich variety of structures on the basis of meeting the dynamics, while its small gas and small load capacity, but also indirectly enhance its environmental survivability [3].

1.3 Low Cost, High Efficiency Cost of, Flexible Configuration, Diversified Tasks, Higher Activity Concealment

Unmanned equipment reduces the carrying of a variety of life support systems, making the manufacturing cost plummeting, in addition, the use of unmanned equipment reduces the pilot, flight personnel training of major investment, and the training of control operators can be skilled in only a few hours. Unmanned equipment due to the comprehensive advantages in volume, use and maintenance, so that its configuration in the task is quite flexible, miniature unmanned equipment and even a full set of systems can be configured by carrying the way. At the same time, through the platform and modular design of unmanned equipment, reasonable optimization of its configuration, can perform a variety of tasks. Unmanned equipment is usually small in size, does not need to consider people's living environment and working conditions, and has greater freedom and space in the shape and size design, so in order to meet the needs of the battlefield, reduce the probability of being discovered by the enemy, and ensure the smooth implementation of the task, most unmanned equipment has the characteristics of high concealment.

2 Technical Application of Unmanned Equipment in Lifesaving Missions

2.1 Target Search and Identification

In the task of organizing the rescue of drowning personnel, the search area is usually large, limited to the performance of the platform and the nature of the target, it is often difficult to complete the task indicators, according to the statistics of the rescue results of drowning personnel in recent years, because no drowning personnel are found, the task cut-off probability is large. And even if they did, they missed the best time to treat them. Similarly, in lifesaving and salvage missions, the search and discovery of the target is crucial [4]. At present, a large number of UAV and underwater unmanned robots have played a significant role in target search and discovery instead of manned platforms. For example, a large number of unmanned aerial vehicles, as shown in Fig. 2, can be used for searching when organizing joint search and rescue for people falling into the water, and underwater robots can be used to replace divers for submarine operations when organizing search for crashed aircraft. Target identification in Marine lifesaving missions not only involves the sea surface and under the seabed, but also may be in the thick smoke and fog or even in the cabin of a fire, which brings higher requirements to the equipment for performing the task, the characteristics and performance of unmanned equipment meet these requirements, unmanned equipment can not only achieve mobile and flexible target discovery in time, It can also compare the size and physical characteristics of the target with the data of the database, and give an instant identification conclusion on the found target, which provides the judgment and decision basis for the command center. Therefore, unmanned equipment is incomparable to other platform equipment in target recognition and has an irreplaceable position. For example, the recognition function of unmanned underwater robots when searching for targets on the seabed, and the proximity recognition of drones when searching for and confirming targets in dense fog [5].

Fig. 2. Unmanned aerial vehicle.

2.2 Materials Transportation and in the Data Transmission

In lifesaving tasks at sea materials transportation is often particularly important, but affected by many reasons such as complex environment, poor personnel transportation time and dangerous site conditions, lifesaving materials and equipment are often difficult to be transported to the scene at the first time, unmanned equipment can replace the manned platform to carry out materials transportation tasks. When rescuing people in the water, the drone can carry rescue materials such as life jackets, hand-held radio, rescue ropes, medicine and food to the hands of the drowning person at the first time, and strive for more favorable time for subsequent lifesaving missions. In the Marine firefighting task, when the large platform can not get close to the damaged ship, unmanned aerial vehicle (UAV) and unmanned boat can be used to transport the damage control materials to the damaged ship. In the lifesaving task, when the rescue task cannot be completed in a short time, the underwater unmanned boat can be used to transport some materials and equipment [6]. When towing at sea is organized under complex environmental conditions, a drone can be used to transmit one end of the tether to the towed ship, reducing safety risks. In lifesaving task, it is necessary to grasp the specific situation of the rescue site to provide a basis for the implementation of the next rescue mission, however, the rescue site is usually more complex, the risk factor is relatively large, manned platforms or equipment is difficult to access the rescue site, especially in some emergency and dangerous rescue sites, unmanned equipment makes up for the gap in this respect. In the rescue organization, unmanned aerial vehicles and other unmanned equipment can carry cameras (electronic eyes), voice recorders and other reconnaissance and video equipment, the implementation of the rescue scene and the target of the implementation of photography, video recording and recording, and has a wide coverage, can be flexibly collected from multiple angles, timely transmission of data back to the command post, providing a favorable basis for the implementation of the rescue mission. The technology application of unmanned equipment in lifesaving tasks is far more than these, but also can complete fire monitoring, underwater lifesaving, remote delivery, hydrometeorological monitoring, danger monitoring and other tasks, these technologies can be applied at the same time, can also be reused, and there are overlapping parts between them.

3 The Demand Direction of Unmanned Equipment for Lifesaving Mission

3.1 The Ability to Operate in Harsh Environments

Most of the lifesaving tasks occur in harsh environments, or the complex meteorological environment of strong winds, waves and thick fog, or the hydrology environment of the deep sea, or the on-site environment of heavy smoke and fire, which puts higher requirements on the ability of unmanned equipment to operate in harsh environments. In particular, the enhancement of the ability to withstand wind and waves in the air, pressure resistance in the seabed, fire resistance and heat resistance has brought great convenience to the organization of lifesaving in harsh meteorological environment, the organization under complex terrain conditions in the deep sea, and the loss prevention management at sea under the condition of fire and large damage, greatly reducing the safety risks and costs of lifesaving tasks, and providing the possibility for lifesaving tasks to complete the task under various conditions in the whole sea. For example, the current unmanned fire boat, as shown in Fig. 3, carries foam close to the ship that is in a fire to extinguish it, and the intelligent life buoy, as shown in Fig. 4, rescues people who fall into the water in a large wind and wave environment.

Fig. 3. Remote control rescue device.

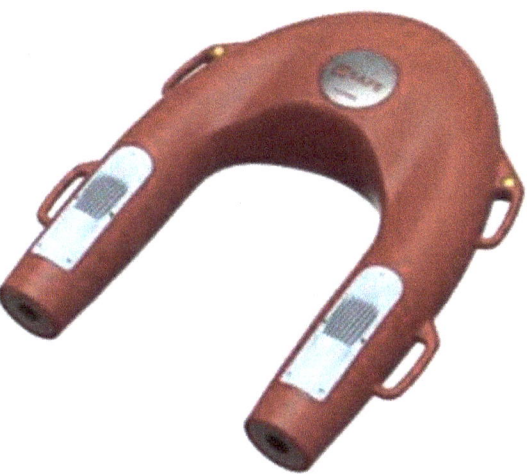

Fig. 4. Remote control life buoy.

3.2 Good Load Performance and Strong Endurance

To some extent, load performance and endurance are two contradictory aspects, the application of unmanned equipment is to deal with the relationship between the two, and the second is to strive to improve the performance of both. The load performance of unmanned equipment determines the type and quantity of rescue equipment carried by it. The current load capacity of unmanned equipment is relatively small, which is far from meeting the current needs of life saving. The foam carried by the unmanned fire boat mentioned above only has the ability to put out small fires. In addition, the lack of endurance of unmanned equipment is also a restricting factor that restricts its use in lifesaving missions. In harsh environments, continuous operation is often required for a long time, and the lack of endurance seriously limits its role in lifesaving missions.

3.3 Rapid Response, Rapid Rescue (Timeliness is Good)

The key to maritime rescue is "fast", the so-called "a small difference, a thousand miles" to describe the rescue mission is not too much, rapid response, rapid discovery of drowning personnel, wrecked submarine and rescue; The fire or damaged water vessels at sea need to quickly extinguish the fire and quickly plug the leak [7]. The use of unmanned equipment in the organization of sea rescue, must meet the first time to arrive at the scene, quickly start the rescue mission, the task of search and rescue personnel in the water is to be able to quickly find the drowning person, found the drowning person can quickly provide medicine, fresh water and food according to their physiological needs and rescue, in the sea fire task can quickly approach the fire site as soon as possible to extinguish the initial fire.

3.4 Various Detection Means, High Accuracy of Finding Targets

Through the sea rescue mission, most of the search and detection means are single, difficult to find the target, positioning accuracy is not high, resulting in the rescue mission efficiency is not high or difficult to complete. Unmanned equipment can carry a variety of detection equipment to carry infrared detection, thermal detection, electromagnetic detection, image detection and other detection means to search and find the target; In order to improve the target discovery probability and positioning accuracy, the aerial, surface and underwater three-dimensional transcendental search and distributed search can be established by using the advantages of unmanned equipment applied to the whole sea area. In addition, the target's characteristic database such as appearance characteristics and electromagnetic performance is input to the unmanned equipment performing the task, and the target's attribute and location are determined by detailed comparison.

4 The Application Prospect of Unmanned Equipment (Technology) in Life-Saving Tasks

4.1 Intelligence

Intelligence is the ultimate goal of the development of unmanned equipment, and its goal is to pass autonomous task planning, situation integration perception, deep autonomous learning and task autonomous decision-making. Unmanned equipment can independently complete specific lifesaving tasks such as target discovery and identification, maritime search and detection, material delivery and data transmission, and build a future system of maritime lifesaving forces. At present, the application of unmanned equipment technology is only limited to remote control or semi-autonomous remote control, and the intelligence of unmanned equipment cannot be realized. In the future, it is necessary to further improve the comprehensive perception of rescue scene situation, relying on large databases and deep autonomous learning. Continuously improve the intelligent level of unmanned equipment in life-saving missions, such as task planning ability, autonomous navigation operation ability and autonomous decision-making ability.

4.2 Clustered

Unmanned equipment system is a disruptive technology that will affect future life-saving missions at sea, and clustering is an important development direction of unmanned equipment system. With the development of artificial intelligence, wireless AD hoc networking and navigation technology, many unmanned equipment can be aggregated to form a cluster that can complete lifesaving tasks together, including unmanned aerial vehicle cluster, surface unmanned boat cluster, unmanned underwater robot cluster, which can complete a number of tasks such as sea surface (seabed) search, hydrometeorological monitoring and data transmission, and material delivery. And can greatly improve the search and discovery range and target discovery probability, improve the response speed and delivery capacity, can make up for the shortcomings of a single platform with a number of advantages, with a large number, small size, low-cost distributed unmanned platform to replace the traditional expensive large manned platform, can play a huge role in lifesaving tasks, is of great significance for future Marine lifesaving missions.

4.3 Large-Scale

Unmanned equipment can solve the current load capacity and endurance of the disadvantage of insufficient, in order to fully enhance the role of unmanned equipment in lifesaving tasks at sea, the use of large and medium-sized unmanned equipment will be the future focus of development [8]. Marine lifesaving missions often adopt the "non-contact" mode to reduce the loss of rescue forces, especially in the transport and delivery of materials, maritime target search tasks respectively put forward higher requirements for the payload and endurance of unmanned equipment, large-scale unmanned equipment can not only improve its endurance by installing batteries, but also carry a variety of equipment to perform a variety of tasks. Realize multi-function of unmanned equipment. In particular, unmanned boats for Marine fire fighting can carry enough foam after large-scale development, which is suitable for the rescue of oil tanker fire, dangerous goods ship fire and chemical ship fire.

4.4 Modular

Unmanned equipment ontology, load and interface will be modular in the future [2]. Marine lifesaving tasks are diversified, covering lifesaving, joint search and rescue, marine firefighting, off-ship damage management, rescue of drowning personnel, underwater salvage, sea towing and other tasks, and each task is composed of multiple subjects, so on the basis of common unmanned platforms, according to the different tasks and needs, modular payloads and excuses are adopted. Reduce the coupling between the platform and various loads, improve the adaptability of unmanned equipment, and form unmanned equipment systems to improve the ability to meet tasks through rapid iterative design, thereby shortening the development cycle and costs, and reducing risks. In addition, the development and upgrading of various sensors, electronic equipment, software and other systems are faster than the update speed of the unmanned equipment platform, and the use of modular design undoubtedly greatly improves the service life of unmanned equipment.

5 Conclusion

The navy is actively promoting the overall transformation of construction. Lifesaving at sea is an important task, mast also adapt to the requirements of the overall transformation, accelerate the pace of modernization. At present unmanned equipment has been applied in many fields in many fields such as anti-mine intelligence and reconnaissance and maritime attack, and showed a rapid development trend. Lifesaving troops should invest more in the application of unmanned technology in rescue operations, construction of lifesaving unmanned equipment system suitable for the needs of the navy, to achieve fast, safe and efficient lifesaving capabilities.

References

1. Da Qing, L.: Research on combat capability constitution of naval unmanned combat force. Command Control Simul. **42**(6), 9–13 (2020) (in Chinese)
2. Wen, D.Q.: Research on the new trend and influence of unmanned equipment developmeng. J. Equip. Acad. **27**(1), 76–79 (2016) (in chinese)
3. Jin Ke Fan: Propect of key technologies and intelligent evolution of marine unmanned equipment. Chinese. J. Ship Res. **13**(6), 1-7(in (2018) chinese)
4. Hong, G.S.: Foreign military field artificial intelligence developmemt planning and ship intelligent technology application. Sh. Sci. Technol. **42**(7), 175–177 (2020)
5. Shi, W.: Research on development status and typical combat application of unmanned craft abroad. Fire Control&Command Control. **44**(2), 11–14 (2019) (in chinese)
6. Liang, L.J.: Development and application of surface unmanned boat. Fire Control&Command Control. **37**(6), 203–206 (2012) (in chinese)
7. Wei, L.: Development and prospect of surface unmanned boat technology. Sh. Electron. Eng. **322**(4), 1–3 (2021) (in chinese)
8. Rong, Z.H.: Research status of surface unmanned boat. Shipbuild. China. **61**(1), 228–232 (2020) (in chinese)

Open Access This chapter is licensed under the terms of the Creative Commons Attribution-NonCommercial-NoDerivatives 4.0 International License (http://creativecommons.org/licenses/by-nc-nd/4.0/), which permits any noncommercial use, sharing, distribution and reproduction in any medium or format, as long as you give appropriate credit to the original author(s) and the source, provide a link to the Creative Commons license and indicate if you modified the licensed material. You do not have permission under this license to share adapted material derived from this chapter or parts of it.

The images or other third party material in this chapter are included in the chapter's Creative Commons license, unless indicated otherwise in a credit line to the material. If material is not included in the chapter's Creative Commons license and your intended use is not permitted by statutory regulation or exceeds the permitted use, you will need to obtain permission directly from the copyright holder.

An Improved Method of Constructing Mercator Chart

Yong Zhang, Junting Xiong(✉), and Qi Yang

Dalian Naval Academy, Dalian, China
fj120125571@163.com

Abstract. This paper introduces two main drawing methods of Mercator chart, and their advantages and disadvantages, and then analyzes the causes of errors and the existing limitations of the traditional simple drawing method of Mercator chart. Aiming at the problem of low accuracy when the simple drawing method is used in the case of a large latitude difference span, an improved drawing method of Mercator chart is proposed. By using the middle latitude as a parameter to calculate the average latitude magnification between two latitudes during drawing, which can effectively solve the error caused by using the average latitude calculation. Through the comparison and analysis with the method of Meridian parts, the effectiveness of this simple drawing method is verified.

Keywords: Mercator Chart · Middle Latitude · Average Latitude · Improved Drawing Method

1 Introduction

The Mercator chart was first created by the Dutch cartographer Gerards Mercator in 1569 AD. Due to its two characteristics which they represent rhumb line is a straight line on the chart and the angle measured on the chart is equal to the corresponding angle on the ground, it greatly facilitates the navigation operations of navigators. Therefore, it has become the most commonly used type of chart in navigation. Currently, about more than 95% of the charts are Mercator charts [1]. Sometimes, due to different operational requirements of navigators, the existing printed charts may be cannot meet the special requirements of users. Navigators can only draw charts by hand on blank paper to deal with urgent needs. Currently, there are mainly two methods for drawing the grid of Mercator charts: namely, the method of Meridian parts [2] and the Simple method. The method of Meridian parts is accurate in calculation and is often used in the production of chart publishing, but the steps are cumbersome and the efficiency is low when drawing by hand. The simple method is easy to operate and is currently a commonly used method for hand-drawing Mercator charts. Due to the phenomenon of increasing latitude in Mercator charts, the magnification ratio between two latitudes after the chart projection is variable, that is means, the magnification at each point is inconsistent [3]. For the convenience of drawing, when using the simple method to draw the chart, the magnification of the average latitude between two latitudes is used as the average magnification along the meridian between adjacent latitudes. When drawing the chart, if the latitude difference between two latitudes is large, the drawing accuracy will decrease significantly and cannot meet the needs of precise navigation support and military navigation support. Therefore, researching and improving the method of constructing Mercator charts has certain significance in navigation applications.

2 Construct Mercator Charts by Using the Method of Meridian Parts

2.1 Basic Principle of Drawing Mercator Charts

The Mercator chart adopts the conformal cylindrical projection [4], as shown in Fig. 1. After the projection, the meridians on the chart are parallel to each other, and the parallels are perpendicular to the meridians [5, 6]. Since there is deformation outside the equator, and the farther away from the equator, the greater the deformation, the intervals of parallels with the same latitude difference on the chart are unequal. Therefore, when drawing parallels, the increasing latitude difference of the parallel intervals must be calculated by mathematical analysis methods, and then the parallels are drawn one by one.

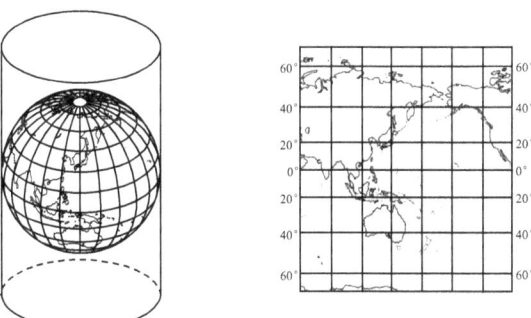

Fig. 1. Cylindrical projection.

2.2 Procedure of Constructing Mercator Charts

In order to simplify the rendering steps, When drawing Mercator charts., the length of $1'$ of longitude is usually taken as the calculation unit of the nautical chart, called the nautical chart unit (e). The length of $1'$ of latitude is usually expressed as a multiple of the nautical chart unit.

Calculating the Nautical Chart Unit (e).

$$e = S \times C_B \tag{1}$$

Where: S—The arc length of 1′ of longitude on the reference latitude circle of the Earth ellipsoid; C_B—The reference scale of the nautical chart.

And the arc length S of 1′ longitude on the reference latitude circle can be obtained by the radius r of the latitude circle using the following formula:

$$S = r \cdot \text{arc} 1' = \frac{a\cos\phi}{\sqrt{1-e^2\sin^2\phi}} \cdot \frac{1}{3437'.74677} \quad (2)$$

Where: a—The major radius of the Earth ellipsoid; ϕ—The latitude of the reference latitude circle.

Construct the Meridians Grid. Determine the longitude range for drawing the nautical chart according to the requirements of the task, calculate the transverse extent of the nautical chart. Then, determine the longitude interval Δ for drawing, and the longitude grid can be drawn.

$$L_h = D\lambda \times e \quad (3)$$

Where: $D\lambda$—The longitude range for drawing the nautical chart.

Construct the Parallels Grid. Due to the phenomenon of Meridian parts in Mercator charts, when drawing the latitude lines, first calculate the Meridian parts of the latitude, and then draw the latitude lines successively according to the calculated Meridian parts.

$$D = 7915'.70447 \lg\left[\left(\frac{1-e\sin\phi}{1+e\sin\phi}\right)^{\frac{e}{2}} \text{tg}\left(45° + \frac{\phi}{2}\right)\right] \quad (4)$$

Where: D—The difference of Meridian parts; ϕ—Latitude.

This method requires determining the latitude interval first when drawing the latitude grid, then calculating the Meridian parts of each latitude line, and then starting from the reference latitude to draw the latitude grid successively. Obviously, this method is rather cumbersome in practical operation.

3 Improved Method for Constructing Mercator Charts

3.1 Basic Principle of Simple Constructing Mercator Charts

When using the simple method to draw Mercator charts, the Earth is regarded as a sphere first [7]. Thus, the relationship between the radius r of the latitude circle and the radius R of the Earth is:

$$r = R\cos\phi \quad (5)$$

Then the relationship between the arc length W enclosed by two longitudes on the latitude circle and the arc length λ enclosed on the equator is:

$$\Delta W = \Delta\lambda \cos\phi \quad (6)$$

On the Mercator chart, the longitudes are parallel straight lines, that is to say, when drawing, △W is first enlarged by secϕ times and then drawn on the chart [8]. According to the Mercator projection principle, the local scale in the longitude direction at any point on the chart is equal to the local scale in the latitude direction. Therefore, the local scale of the latitude on the Mercator chart also needs to be enlarged by secϕ times [9]. However, due to the phenomenon of increasing latitude, the latitude between two latitudes is changing. Therefore, for the convenience of drawing in the simple drawing method, the magnification factor secϕm of the average latitude between two latitudes is used as the average magnification factor on the longitude between adjacent latitudes [10]. In this way, the drawing steps can be simplified. Take the latitude difference interval of 1° as an example, as shown in Fig. 2. But this method will produce a large drawing error when the latitude difference is large in drawing.

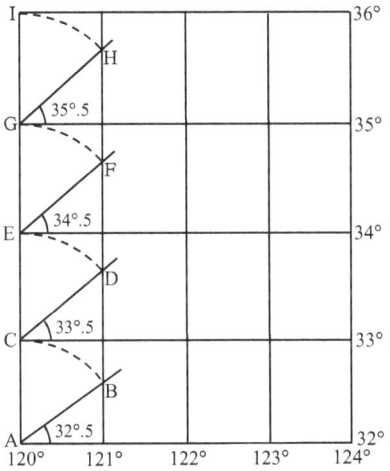

Fig. 2. Simple Method for Mercator chart.

3.2 Analysis of the Improved Method

Through the above analysis, it can be concluded that the main accuracy error of the traditional simple drawing method of Mercator charts is that when drawing the latitude lines, to simplify the method, the magnification factor of the average latitude between two latitudes is used as the average magnification factor within the latitude interval. Therefore, the larger the grid of the latitude lines drawn, the larger the error. In fact, there is such a latitude ϕn between two latitudes. The magnification factor of the latitude at ϕn is equal to the median of the gradual magnification factors of these two latitude intervals. In navigation, we call ϕn the middle latitude. When drawing the latitude lines, using the middle latitude instead of the average latitude can also simplify the drawing steps and improve the accuracy of drawing the latitude lines.

3.3 Calculation Method of Middle Latitude

On the nautical chart, for a certain line segment. When measured in different units, the ratio of difference in longitude(Dλ) and the departure(Dep) is equal to the inverse ratio of the lengths of the units used, that means, Dλ × length 1' of longitude on the chart is equal to the Dep × length 1' of latitude at ϕn on the chart. The length 1' of latitude at ϕn on the chart is equal to R × arc1' × m, where 'R' is the radius of the Earth and 'm' is the local scale in the longitude direction of the Mercator chart. The length 1' of longitude on the chart is equal to r × arc1' × n, where 'r' is the radius of the latitude circle and 'n' is the local scale in the latitude direction of the Mercator chart. Due to the equal-angle characteristic of the Mercator chart, m = n [11].

Thereby,

$$\sec\phi_n = \frac{D\lambda}{Dep} = \frac{\Delta D}{D_\phi} \quad (7)$$

Where: ΔD—Meridional Parts, Dϕ—Differece of Latitude, Dep—Departure.

$$\phi_n = \arcsec\frac{\Delta D}{D_\phi} = \arcsec\frac{7915.70447\left\{\lg[(\text{tg}(45° + \frac{\phi_2}{2}))] - \lg[[(\text{tg}(45° + \frac{\phi_1}{2}))]]\right\}}{\phi_2 - \phi_1} \quad (8)$$

4 Precision Analysis of Improved Method

In this paper, the accuracy of the improved method is studied, which means the error between the Meridian parts method and the improved chart drawing method.

4.1 The Value Calculated by the Meridian Parts Method

In the calculation, the basic latitude of 0° (equator) is taken, and the distance from any parallels line to the equator can be obtained by the formula described above. The latitude interval is 1° when drawing, and the calculation results of the Meridian parts are shown in Table 1.

Table 1. Meridian parts of the MP method.

φ	D	φ	D	φ	D	φ	D	φ	D	φ	D
0	0	15	904.5	30	1876.9	45	3013.6	60	4507.4	75	6948.1
1	59.6	16	966.4	31	1946.2	46	3099	61	4629.1	76	7187.7
2	119.2	17	1028.6	32	2016.2	47	3185.9	62	4754.6	77	7444.7
3	178.9	18	1091.1	33	2087	48	3274.4	63	4884.4	78	7722

(continued)

Table 1. (*continued*)

φ	D	φ	D	φ	D	φ	D	φ	D	φ	D
4	238.6	19	1154	34	2158.6	49	3364.7	64	5018.7	79	8023.1
5	298.4	20	1217.3	35	2231.1	50	3456.9	65	5157.9	80	8352.5
6	358.3	21	1280.9	36	2304.5	51	3550.9	66	5302.5	81	8716.3
7	418.2	22	1345.1	37	2378.8	52	3647	67	5452.8	82	9122.6
8	478.4	23	1409.6	38	2454.1	53	3745.4	68	5609.4	83	9582.9
9	538.6	24	1474.7	39	2530.4	54	3846	69	5773	84	10,114
10	599.1	25	1540.3	40	2607.9	55	3949.1	70	5944.3	85	10741.7
11	659.7	26	1606.4	41	2686.5	56	4054.8	71	6123.9	86	11509.5
12	720.5	27	1673.21	42	2766.3	57	4163.3	72	6312.9	87	12499.1
13	781.6	28	1740.4	43	2847.4	58	4274.8	73	6512.4	88	13893.4
14	842.9	29	1808.3	44	2929.8	59	4389.4	74	6723.6	89	16276.5

4.2 The Value Calculated by the Improved Method

When drawing the interval of parallels line, the latitude interval is also 1°, then the gradual latitude difference at any latitude can be obtained by the following formula:

$$d_\phi = 60 \times \sec\phi_n \tag{9}$$

Then the Meridian parts at any latitude is:

$$D_j = d_{\phi-1} + d_\phi \tag{10}$$

In $1 \leqq \phi \leqq 90$, the calculation results are shown in Table 2.

Table 2. Meridian parts of the Improved method.

φ	Dj	φ	Dj	φ	Dj	φ	Dj	φ	Dj	φ	Dj
0	0.0	15	910.5	30	1888.4	45	3029.9	60	4527.4	75	6970.3
1	60.0	16	972.7	31	1958.0	46	3115.5	61	4649.2	76	7210.1
2	120.0	17	1035.3	32	2028.4	47	3202.7	62	4775.0	77	7467.2
3	180.1	18	1098.2	33	2099.5	48	3291.5	63	4904.9	78	7744.6
4	240.2	19	1161.5	34	2171.5	49	3382.1	64	5039.4	79	8045.7
5	300.4	20	1225.1	35	2244.3	50	3474.5	65	5178.8	80	8375.2
6	360.7	21	1289.2	36	2318.0	51	3568.8	66	5323.5	81	8739.1
7	421.0	22	1353.7	37	2392.6	52	3665.2	67	5474.0	82	9145.5

(*continued*)

Table 2. (*continued*)

φ	Dj	φ	Dj	φ	Dj	φ	Dj	φ	Dj	φ	Dj
8	481.6	23	1418.6	38	2468.3	53	3763.8	68	5630.8	83	9605.8
9	542.2	24	1484.1	39	2544.9	54	3864.6	69	5794.6	84	10136.9
10	603.1	25	1550.0	40	2622.7	55	3968.0	70	5965.9	85	10764.6
11	664.1	26	1616.5	41	2701.6	56	4073.9	71	6145.7	86	11532.5
12	725.3	27	1683.5	42	2781.7	57	4182.6	72	6334.8	87	12522.1
13	786.8	28	1751.2	43	2863.1	58	4294.3	73	6534.4	88	13916.4
14	848.5	29	1819.4	44	2945.8	59	4409.1	74	6745.7	89	16299.6

4.3 Error of Improved Method

Error refers to the difference between the observed or calculated value of a quantity and the true value [12]. When verifying the improved method, the calculated value of Meridian parts is taken as the true value, and the accuracy error of the improved method can be expressed as follows:

$$\Delta = D_j - D \tag{11}$$

The error of the calculated value of the improved method is shown in Table 3, and the variation of the error is shown in Fig. 3.

Table 3. Error of Improved method.

φ	Δ	φ	Δ	φ	Δ	φ	Δ	φ	Δ	φ	Δ
0	0.0	15	6.0	30	11.5	45	16.3	60	20.0	75	22.2
1	0.4	16	6.3	31	11.8	46	16.5	61	20.1	76	22.4
2	0.8	17	6.7	32	12.2	47	16.8	62	20.4	77	22.5
3	1.2	18	7.1	33	12.5	48	17.1	63	20.5	78	22.6
4	1.6	19	7.5	34	12.9	49	17.4	64	20.7	79	22.6
5	2.0	20	7.8	35	13.2	50	17.6	65	20.9	80	22.7
6	2.4	21	8.3	36	13.5	51	17.9	66	21.0	81	22.8
7	2.8	22	8.6	37	13.8	52	18.2	67	21.2	82	22.9
8	3.2	23	9.0	38	14.2	53	18.4	68	21.4	83	22.9
9	3.6	24	9.4	39	14.5	54	18.6	69	21.6	84	22.9
10	4.0	25	9.7	40	14.8	55	18.9	70	21.6	85	22.9
11	4.4	26	10.1	41	15.1	56	19.1	71	21.8	86	23.0

(*continued*)

Table 3. (*continued*)

φ	Δ	φ	Δ	φ	Δ	φ	Δ	φ	Δ	φ	Δ
12	4.8	27	10.3	42	15.4	57	19.3	72	21.9	87	23.0
13	5.2	28	10.8	43	15.7	58	19.5	73	22.0	88	23.0
14	5.6	29	11.1	44	16.0	59	19.7	74	22.1	89	23.1

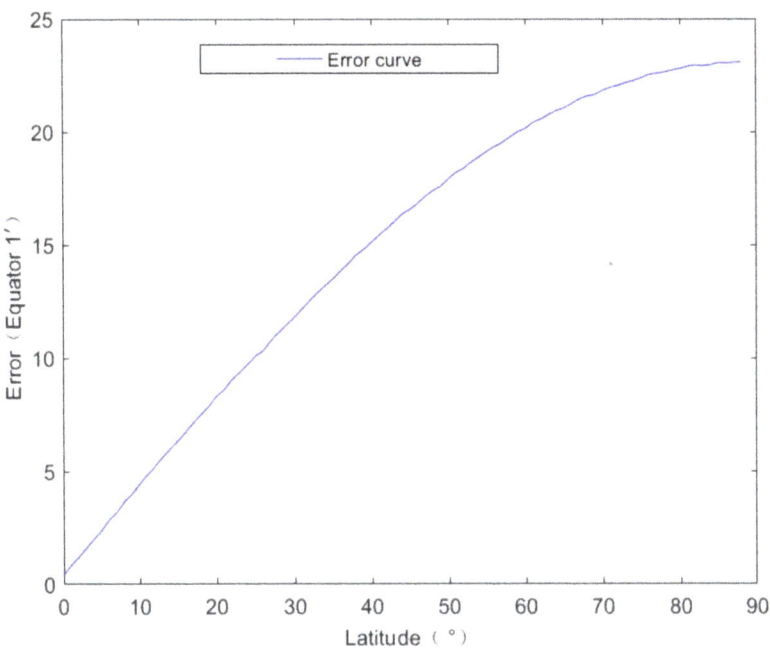

Fig. 3. Error curve of the Improved method.

Through the analysis of the calculation results, the improved method of constructing Mercator Charts has the following characteristics:

(1) The drawing accuracy of the improved method decreases as the latitude increases;
(2) In the low latitude area (0° ~ 30°), the error rate of the improved method is less than 0.61%;
(3) Taking the example of drawing a map of the 30° parallel circle with a scale of 1: 1 and a reference latitude of 30°, the length of the arc with an accuracy of 1' can be calculated to be 1608.132 m. Through calculation, it can be found that in the low latitude area (0° ~ 30°), the drawing error of the improved simplified drawing method for latitude 1' is less than 10 m, and in the mid-latitude area (30° ~ 60°) it is less than 17 m.

5 Conclusion

Improve the method of constructing Mercator chart by using Middle Latitude (ϕn), the procedures of constructing Mercator chart can be simplified, and also the drawing efficiency can be improved. Through calculation and verification, the drawing accuracy of the method in the mid-low latitude area can meet the needs of navigation operations and has certain application value for navigators to deal with emergency tasks at sea.

References

1. Xuanbin, X.: Basic Navigation. Dalian Naval Academy, China (2020) (in chinese)
2. Zang, J., Chen, S.: Marine Navigation. Dalian Maritime University, China (2013) (in chinese)
3. Liu, M., Jin, l.: Analysis of some mathematical properties of complex Gaussian projection [J]. Hydrogr. surv. charting. **43**(6), 78–81 (2023) (in chinese)
4. Jiao, J., Zeng, Q.: Cartography, vol. 1, p. 94. Peking University Press, Beijing, China (2009) (in chinese)
5. Liu, W., et al.: An improved method of polar ellipsoid transverse Mercator projection based on double projection [J]. Geomat. Inf. Sci. Wuhan Univ. **44**(8), 1138–1142 (2019) (in chinese)
6. Li, X., et al.: Direct expansions for common curvature radii in terms of different latitude variables [J]. J. Geom. **49**(2), 45–48 (2024) (in chinese)
7. Zou, D., et al.: Efficient mapping of common projected geodetic lines and Mathematica implementation [J]. Mod. Navig., 422–426 (2023) (in chinese)
8. Pan, Q.: Arc length of equal latitude circle with longitude difference of 1° on Mercator Chart. J. Shanghai Marit. Inst. **22**(2), 9–10 (2001) (in chinese)
9. Sun, J., Cao, X., Qiu, L.: Application analysis of Mercator chart projection deformation. China Water Transp. **16**(12), 73–74 (2016) (in chinese)
10. Zhao, R.: Navigation, pp. 52–53, People's Communications Press, Beijing (2009) (in chinese)
11. Zhang, Q., Liu, X.: Characteristic analysis of Mercator chart. J. Qingdao Ocean Shipp. Mar. Coll. **32**(4), 16–17 (2011) (in chinese)
12. Yan, Z., Wei, Z.: Statistical Analysis of Experimental Data, pp. 266–267, National Defense Industry Press, Beijing (2001) (in chinese)

Open Access This chapter is licensed under the terms of the Creative Commons Attribution-NonCommercial-NoDerivatives 4.0 International License (http://creativecommons.org/licenses/by-nc-nd/4.0/), which permits any noncommercial use, sharing, distribution and reproduction in any medium or format, as long as you give appropriate credit to the original author(s) and the source, provide a link to the Creative Commons license and indicate if you modified the licensed material. You do not have permission under this license to share adapted material derived from this chapter or parts of it.

The images or other third party material in this chapter are included in the chapter's Creative Commons license, unless indicated otherwise in a credit line to the material. If material is not included in the chapter's Creative Commons license and your intended use is not permitted by statutory regulation or exceeds the permitted use, you will need to obtain permission directly from the copyright holder.

Typhoon Prediction Analysis of Pangu Weather Model

Peng Yuehua[1(✉)], Yao Libo[2], Song Jun[3], and Yuchen Zhang[4]

[1] Dalian Naval Academy, Dalian 116018, China
`Pengyuehua303@163.com`
[2] Naval Aviation University, Yantai 264000, China
[3] Dalian Ocean University, Dalian 116023, China
[4] Tsinghua University, Beijing 100084, China

Abstract. Pangu weather model is the first of the "Top 10 Scientific Advances in China" in 2023 released by the National Natural Science Foundation of China. In this paper, the model is tested and analyzed for typhoon prediction. The research results show that: firstly, Pangu weather model is suitable for typhoon prediction; secondly, the typhoon prediction error of the model is not higher than 1% after error analysis. With such a high accuracy, the Pangu weather model could be used to do a lot of things not confined to prediction. Due to its global and 3D transformation, it can be used for artificial modification of typhoons. we can add the artificial intervention of typhoon and introduce the PID (proportional-integral-differential) controller in engineering cybernetics; and thirdly, through the modeling of the typhoon, we can build a simulation platform for the artificial modification of typhoon based on engineering cybernetics.

Keywords: Pangu Weather Model · Typhoon Prediction · Artificial Intelligence · Error Analysis

1 Introduction

Weather prediction is one of the most important application scenarios in scientific computing [1], by providing the ability to predict future weather, especially the occurrence of extreme weather events. In the past ten years, with the rapid development of high performance scientific computing technology, numerical weather prediction in extreme disaster warning, climate change prediction has achieved great success [2–4], but with the growth of slow, and prediction accuracy requirements gradually improve, physical model gradually complicated, the bottleneck of existing numerical prediction technology increasingly prominent. So based on the deep learning method of weather prediction paradigm has become the new direction, but in the numerical method of the most widely used long-term forecast, the prediction accuracy of traditional deep learning method is still significantly lower than the traditional numerical prediction method, and inadequate interpretability, extreme weather is not allowed. The main reason is that the previous weather prediction models are mostly based on two-dimensional information, which has an essential gap with the highly three-dimensional phenomenon of meteorology.

Therefore, based on the large model method, the 3D neural network method using high-resolution global meteorology is allowed [5]. The Pangu weather model was first proposed by researchers from Huawei Cloud, and was published in the official journal of Nature [6] on July 6, 2023. National natural science fund committee released on February 29,2024, "the 2023 annual China ten scientific progress", the first is the artificial intelligence model for accurate weather prediction [7–9], with the pangu weather model, review for the work to build independent controllable weather prediction system, in social production, people's life, disaster prevention and mitigation is of great significance.

The model is the first accuracy more than the traditional numerical prediction method of AI method [10–14], 1 h-7 days prediction accuracy is higher than the traditional numerical method, namely the European meteorological center operational IFS, at the same time in the local reasoning time cost of about 1.4 s over 10,000 times faster than the traditional numerical method, which is the engineering theory can be applied to the typhoon control method. Because the engineering cybernetics method needs the iterative optimization of closed-loop, if the traditional numerical prediction method is used, considering the time cost, the closed-loop method can not achieve high-precision convergence, which may also be the fundamental reason why the world does not combine the idea of engineering cybernetics with typhoon regulation.

The architecture used by the model, which is called the 3D Earth-specific transformer. The meteorological data of the 13th layer and the surface information variables are input into a deep network and combined into a 3D cube based on various downsampling methods. The data predicts the future with an encoding-decoder. The structure, derived from Swin transformer, is a variant form of Vision transformer that restores the original resolution by upsampling interpolation. Each 3D deep network contains about 64 million parameters.

2 The Results of the Typhoon Forecasts by Pangu Weather Model

Based on this method, we selected 88 tropical cyclones that emerged in 2018 for comparison of numerical accuracy and for the direct positional error at the eye position. As shown in Fig. 1, the prediction advantage of the Pangu model over the traditional methods. For the most important path forecast in this paper, the results of the two strongest typhoons in the western Pacific Ocean, Connie and the Jade Rabbit, are also compared. It is obvious that the model has very high accuracy for the typhoon path forecast in the western Pacific Ocean, which is very high fit with this paper.

The Pangu weather model is not so much a prediction model of extreme weather, as the Pangu weather model is a three-dimensional prediction model of the whole earth's climate environment, which is also the biggest advantage of the Pangu model compared with the traditional mesoscale meteorological model prediction method.

Another point that must be mentioned about the advantage of the Pangu weather model is its highly three-dimensional characteristics. For well-known reasons, the simulation of the geophysical system is very large. Chairman MAO has a poem: "Sitting on the ground and travel 80,000 miles a day, patrol the sky and see a thousand rivers away." The earth can be simplified to a sphere with a large circular circumference of 40,000 km. Using the paste grid to discretize the earth's atmosphere, the physical field

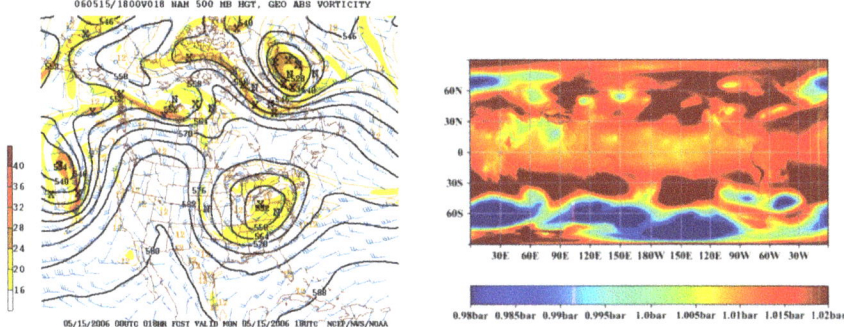

Fig. 1. Iso-potential lines calculated based on numerical weather prediction patterns on 500 mbar level. (left).

of the earth's weather can be regarded as a rectangle of 20,000 km wide and 40,000 km long, which is a huge physical field much larger than the calculation range of general computational fluid mechanics. Therefore, discretion needs to be careful enough. Once the size of the grid is slightly smaller, the calculation amount increases by hundreds of times. As a result, many weather simulations are limited by the computing power of the hardware and can only calculate two-dimensional information.

With the development of numerical computing ability, three-dimensional numerical weather prediction gradually become feasible, but due to the three-dimensional dispersion multiplied grid number, even now the weather simulation is mostly confined to local physical information field, unable to do the global weather simulation at the same time, it also weakened the reliability of the numerical weather prediction model.

The Pangu model, using the computing power of CUDA compared to the CPU rolling, can achieve almost real-time simulation of global meteorology. It can be seen in the global surface pressure prediction chart as shown in Fig. 2. However, the advantage of the Pangu model is not only there. Based on the network structure, the Pangu model divides the vertical direction into 13 grids. The vertical direction does not use the uniform physical elevation as the variable, but takes the pressure level with more physical information as the variable. Considering the combined effect of the number of parameters to be trained and the existing computing force, the 13 typical air pressure levels in the vertical direction were 50 hPa, 100 hPa, 150 hPa, 200 hPa, 250 hPa, 300 hPa, 400 hPa, 500 hPa, 600 hPa, 600 hPa, 700 hPa, 850 hPa, 925 hPa and 1000 hPa.

The output of the Pangu weather model is 69 different data from the input, including 5 high-altitude variables (13 pressure levels (50 hPa, 100 hPa, 150 hPa, 200 hPa, 250 hPa, 500 hPa, 500 hPa, 500 hPa, 600 hPa, 7500 hPa, 850 hPa, 9250 hPa, and 1000 hPa), and 4 surface data (pressure, wind speed), air temperature). As shown in Fig. 2 is the distribution map of earth pressure in the surface data, whereas shown in Fig. 3 is the distribution map of 1000 hPa, or air pressure under conventional atmospheric pressure, which is clearly the result of the highly three-dimensional prediction of meteorological results. Figure 4 shows the elevation distribution of the isopressure surface at 200 hPa. Considering that the meteorological dynamics is more related to the spatial elevation

210 P. Yuehua et al.

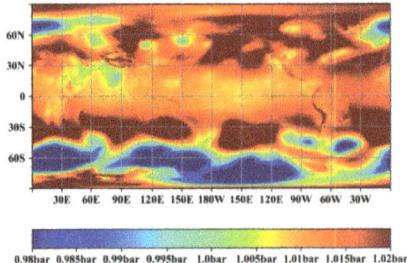

Fig. 2. Distribution of global surface pressure (bar) predicted by the Pangu model. (right).

distribution than to the pressure gradient, the isopressure surface information used by the Pangu model is more convincing than the traditional elevation information.

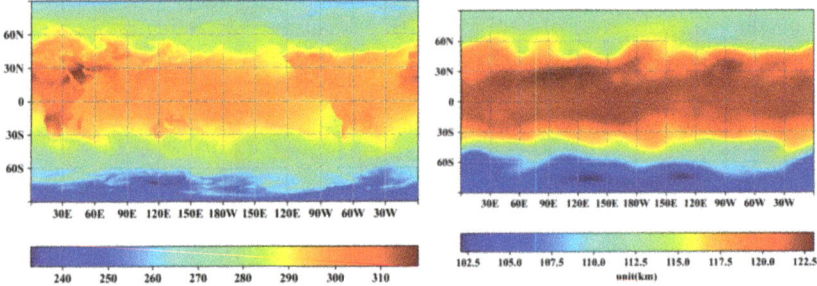

Fig. 3. Distribution of air temperature (K) on 1000 hPa level. (left).

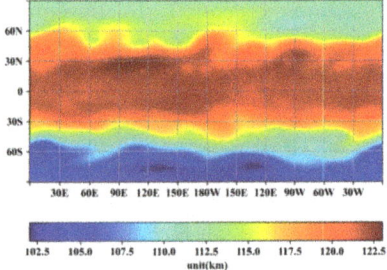

Fig. 4. Distribution of geopotential height (km) on 200 hPa level. (right).

3   Error Analysis of the Typhoon Prediction Results

We perform an error analysis on the data-driven large models of meteorology. Because the use of the system is the global interpolated correction data of the European Medium-term weather prediction Center, and it is the most common in the weather prediction.nc document. Therefore, the error analysis of the prediction results of typhoons can be directly compared based on the raw data of the European Meteorological Center as well as the prediction data of the large meteorological models. Because the typhoon prediction model is trained based on the global meteorological data, as well as the predicted. Therefore, considering that the meteorological prediction results far away from typhoons have less correlation with this paper, the error analysis of the prediction results of typhoons focuses on the error analysis of the surrounding area of the typhoon.

According to the provisions of typhoon intensity forecast error verification index in GB / T38308–2019 weather prediction, there are three error indicators that can be used, namely absolute error, root mean square error and trend consensus rate. However, none of these three error forms stipulates the relative error, so the definition of the relative error can only be put forward on the basis of these three provisions. That is, the basic law of error analysis is to compare the error between the predicted value and the true value, by calculating the percentage of the difference value and the true value, to obtain the specific percentage of the error. The first is the error value of the measurement and forecast of the most basic typical physical quantities. For the highly transient physical process of typhoon, the most typical physical quantities are air pressure, velocity and air temperature. Therefore, we conduct the error analysis of these three physical quantities respectively. GB / T38308- -2019 weather prediction also conceptually defines the national standard for typhoon intensity, that is, to represent the maximum average wind speed value of the bottom layer (near the ground or near the sea surface) near the typhoon center or the pressure value at the lowest sea level in the center. According to this regulation, we also corrected the error analysis calculation of speed, i.e

$$\text{Error}_{\text{rate}} = \frac{\sum I_{\text{predict}} - \sum I_{\text{measure}}}{\sum I_{\text{measure}}} \quad (1)$$

Next, an error analysis of the prediction model will be performed. The first is the pressure. Since the modeling of the typhoon mentioned in the following category is highly based on the pressure information, the error analysis of the pressure is very necessary.

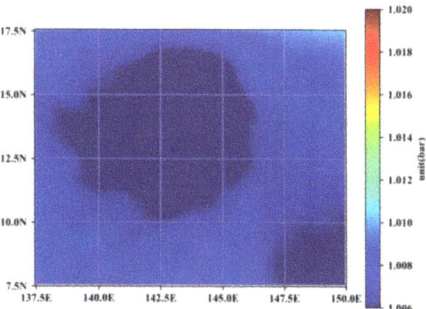

Fig. 5. Local predicted pressure field in the Pacific Ocean six hours after 0:00 on September 30,2018.

Figure 5 shows the weather data inferred after 6 h based on the meteorological data at time zero on September 30, 2018. While similar figure shows the actual air pressure field collected from the same location as the European Meteorological Center, we can see that the changing trend is completely consistent for both. The specific error expression should be calculated using the error analysis formula mentioned above. Since the calculated surface pressure field is a two-dimensional distribution map, the error distribution obtained by the calculation will also be a two-dimensional distribution map.

As shown in Fig. 6 is the distribution map of air pressure error. It can be seen that the error is much lower than 0.1% for most areas. In some specific areas around the typhoon eye, the calculation error is slightly larger, but the estimate of air pressure value is also of the order of magnitude lower than 0.3%.

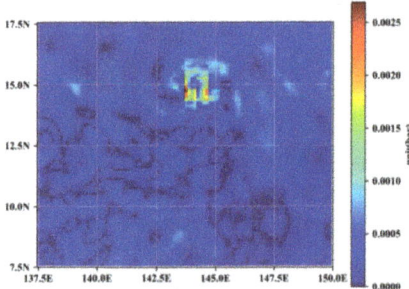

Fig. 6. The calculated error distribution map around the typhoon.

Next analysis speed, because the speed itself is a vector, more complex, so we directly analyze the size of the speed of a scalar error, due to the local speed size is usually more important than speed direction this information, and considering the effect of the typhoon cyclone, the direction of the speed change is more intense, not convenient as air pressure to the full data, so the speed space averaging, through the error analysis of the average speed, is intuitive and meaningful contrast.

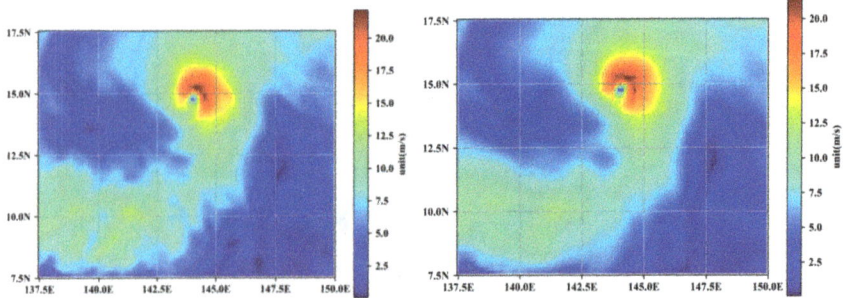

Fig. 7. The Pacific local forecast velocity field from the six hours after 00:00 on September 30,2018. (left).

By comparison, with the velocity field of Figs. 7 and 8, we can find that the overall trend of the velocity distribution is consistent, but due to the velocity change of the extremely high gradient around the typhoon, the direct numerical error analysis using the full physical field like air pressure does not explain the accuracy of the system, so the spatial average is taken. Through the calculation, the error value of the speed is about 0.3%, which also far ensures that the error is not higher than 1%.

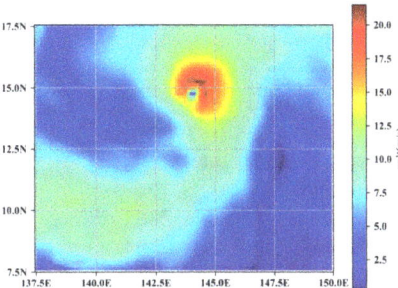

Fig. 8. Same-location Pacific Speed Field at time 06:00 on September 30,2018. (right).

4 Conclusions

The research results of this paper show that: firstly, Pangu weather model is suitable for typhoon prediction; secondly, the typhoon prediction error of the model is not higher than 1% after error analysis. With such a high accuracy, the Pangu weather model could be used to do a lot of things not confined to prediction. Due to its global and 3D transformation, it can be used for artificial modification of typhoons.

References

1. Alam, S.R., Gila, M., Klein, M., Martinasso, M., Schulthess, T.C.: Versatile software-defined hpc and cloud clusters on alps supercomputer for diverse workflows. Int. J. High Perform. Comput. Appl. **37**(3–4), 288–305 (2023)
2. Bauer, P., Thorpe, A., Brunet, G.: The quiet revolution of numerical weather prediction. Nature. **525**(7567), 47–55 (2015)
3. Bauer, P., Dueben, P.D., Hoefler, T., Quintino, T., Schulthess, T.C., Wedi, N.P.: The digital revolution of earth-system science. Nat. Comput. Sci. **1**(2), 104–113 (2021)
4. Bauer, P., et al.: Deep learning and a changing economy in weather and climate prediction. Nat. Rev. Earth Environ. **4**(8), 507–509 (2023)
5. Kaifeng Bi, Lingxi Xie, Hengheng Zhang, Xin Chen, Xiaotao Gu, and Qi Tian. Pangu-weather: A 3d high-resolution model for fast and accurate global weather forecast. arXiv preprint arXiv: 2211.02556, 2022
6. Bi, K., et al.: Accurate medium-range global weather forecasting with 3D neural networks. Nature. **619**, 533–538 (2023). https://doi.org/10.1038/s41586-023-06185-3
7. Allen, M. R., Kettleborough, J. A., Stainforth, D. A.: Model error in weather and climate forecasting. In: ECMWF Predictability of Weather and Climate Seminar, pp. 279–304. European Centre for Medium Range Weather Forecasts, Reading, UK 2022. http://www.ecmwf.int/publications/library/do/references/list/209

8. Palmer, T.N., et al.: Representing model uncertainty in weather and climate prediction. Annu. Rev. Earth Planet. Sci. **33**, 163–193 (2005)
9. Cristian Bodnar, et al.: Aurora: A foundation model of the atmosphere. arXiv preprint arXiv: 2405.13063, 2024
10. LeCun, Y., Bengio, Y., Hinton, G.: Deep learning. Nature. **521**, 436–444 (2015)
11. Bonavita, M., Laloyaux, P.: Machine learning for model error inference and correction. J. Adv. Model. Earth Syst. **12**(12), e2020MS002232 (2020)
12. Bouallègue, Z.B., Weyn, J.A., Clare, M.C.A., Dramsch, J., Dueben, P., Chantry, M.: Improving medium-range ensemble weather forecasts with hierarchical ensemble transformers. Artif. Intell. Earth Syst. **3**(1), e230027 (2024)
13. Bouallègue, Z.B., et al.: The rise of data-driven weather forecasting: A first statistical assessment of machine learning-based weather forecasts in an operational-like context. Bull. Am. Meteorol. Soc. **105**(6), E864–E883 (2024)
14. Cheon, M., Mun, C.: The climate of innovation: AI's growing influence in weather prediction patents and its future prospects. Sustainability. **15**(24), 16681 (2023)

Open Access This chapter is licensed under the terms of the Creative Commons Attribution-NonCommercial-NoDerivatives 4.0 International License (http://creativecommons.org/licenses/by-nc-nd/4.0/), which permits any noncommercial use, sharing, distribution and reproduction in any medium or format, as long as you give appropriate credit to the original author(s) and the source, provide a link to the Creative Commons license and indicate if you modified the licensed material. You do not have permission under this license to share adapted material derived from this chapter or parts of it.

The images or other third party material in this chapter are included in the chapter's Creative Commons license, unless indicated otherwise in a credit line to the material. If material is not included in the chapter's Creative Commons license and your intended use is not permitted by statutory regulation or exceeds the permitted use, you will need to obtain permission directly from the copyright holder.

Research on Intelligent Optimization of Deployment for High-Precision Positioning of Submerged Buoys

Leilei Han, Hairui Ma[✉], and Minghai Li

Dalian Naval Academy, Dalian, China
mahairui@hotmail.com

Abstract. In the long-baseline positioning of submerged buoys, the geometric structure of deployment significantly impacts positioning accuracy. Meanwhile, overly dense deployment of measurement stations has limited improvement on positioning accuracy, leading to redundant measurement station resources. To enhance positioning accuracy performance under limited measurement station resources while ensuring effective coverage of core areas, this paper proposes the use of an improved Particle Swarm Optimization (PSO) algorithm for iterative optimization of deployment positions. First, to avoid the optimization results from falling into local optimal solutions, the inertia weight was improved. Second, to ensure the coverage capability of the measurement area, the proportion of unmeasured areas was used as a penalty function to guide the array to cover as large a measurement area as possible. The simulation calculation results show that, the measurement accuracy of the underwater beacon array for pulse sound signals has been improved after optimization, with the coverage rate of three beacons increasing from 74.7% to 89.2%, and the coverage rate of four beacons increasing from 41.1% to 70.2%, while maintaining stable measurement accuracy. Meanwhile, this method has the potential for further application in different scenarios of long baseline positioning for submerged buoys.

Keywords: Particle Swarm Optimization · Long-Baseline Positioning · Optimized Deployment

1 Introduction

In the field of underwater acoustic sensor node deployment, research and design of positioning structures primarily focus on simple geometric configurations such as equilateral triangles, squares, and regular polygons. Deploying sensors on the water surface can be considered as a two-dimensional space, while integrating surface and underwater sensors into measurement nodes can be regarded as a three-dimensional space. Literature [1] proposes a D-optimal positioning structure design scheme, demonstrating that regular polygons can improve the accuracy of underwater positioning systems. Literature [2] summarizes underwater positioning algorithms, systematically analyzes underwater differential GPS positioning systems, and provides functions describing the impact of surface baseline networks on positioning accuracy. Finally, it analyzes the relative relationship between DOP changes and positions for radial and spiral distribution networks.

Excessively dense deployment of stations within the detection network coverage area can result in fewer sensors covering certain regions, reducing the network's robustness against sensor failures and generating redundant information, thereby wasting measurement resources. Reasonably utilizing multiple types of information related to target positions can improve positioning accuracy and robustness. To this end, numerous researchers have engaged in the study of positioning methods that combine multiple measurement modalities. Literature [3] utilizes TOA and AOA measurement information to establish positioning equations, which are then solved using weighted least squares. When sensor node distribution is suboptimal, this approach reduces the estimation error of target positions. Literature [4] establishes a positioning model using TDOA and AOA measurement information from targets, providing corresponding closed-form solutions. This algorithm utilizes the orthogonal relationship under the TDOA-AOA positioning model to construct positioning equations, achieving CRLB performance when measurement errors are small. In addition to the aforementioned positioning methods, hybrid positioning algorithms can also leverage RSSI information from targets, such as RSS-AOA [5–7] and RSS-TOA [8, 9] based target positioning methods.

Optimization algorithms that mimic the collective behavior of biological groups are an important branch of evolutionary algorithm research, also known as swarm intelligence algorithms [10]. In 1995, scholars Eberhart and Kennedy proposed the PSO algorithm [11], which is a swarm intelligence algorithm derived from simulating bird foraging behavior. After continuous development, PSO has become a representative algorithm in the field of evolutionary algorithms due to its simple design and efficient optimization ability, widely applied in scientific research and engineering practice [12].

2 Model Establishment

2.1 GDOP Model for Submersible Buoy Positioning

Assuming the position of the acoustic signal is $X = [x,y,0]^T$, and the position of the measurement stations is $X_i = [x_i,y_i,z_i]^T$, where X_0 is the main station position and the others are the positions of auxiliary stations. The sonar measurement equation is given by:

$$DR_i = R_i - R_0, i = 1, \cdots, N \tag{1}$$

The R_i and R_0 are the distances from the measurement stations to the target. The estimation error of the target position is denoted as $\Delta X = [\Delta x, \Delta y, \Delta z]^T$, and the measurement error of the ith station is denoted as $\Delta X_i = [\Delta x_i, \Delta y_i, \Delta z_i]^T$.

The resulting observation error vector is denoted as $[\Delta DR_1, \cdots, \Delta DR_N]$, and the error propagation equation is as follows:

$$\begin{bmatrix} \Delta DR_1 \\ \Delta DR_2 \\ \vdots \\ \Delta DR_N \end{bmatrix} = J_{DR} \Delta X + \text{diag}\left(A \begin{bmatrix} \Delta x_1 & \Delta x_2 & \cdots & \Delta x_N \\ \Delta y_1 & \Delta y_2 & \cdots & \Delta y_N \\ \Delta z_1 & \Delta z_2 & \cdots & \Delta z_N \end{bmatrix} \right) - B \Delta X_0 \tag{2}$$

From this, the state estimation error vector can be solved as:

$$\Delta X = \left(J_{DR}^T J_{DR}\right)^{-1} J_{DR}^T \left(\begin{bmatrix} \Delta DR_1 \\ \Delta DR_2 \\ \vdots \\ \Delta DR_N \end{bmatrix} + B\Delta X_0 - \text{diag}\left(A\begin{bmatrix} \Delta x_1 & \Delta x_2 & \cdots & \Delta x_N \\ \Delta y_1 & \Delta y_2 & \cdots & \Delta y_N \\ \Delta z_1 & \Delta z_2 & \cdots & \Delta z_N \end{bmatrix}\right)\right) \quad (3)$$

The above equation represents the theoretical error in the state estimation of the submerged buoy array. If the observation error covariance matrix is Λ_{DR}, and the station errors are independently and identically distributed with a mean of 0 and a variance of σ^2 following a Gaussian distribution, then the positioning error covariance matrix is given by:

$$P_{DR} = \left(J_{DR}^T J_{DR}\right)^{-1} J_{DR}^T \left(\Lambda_{DR} + \sigma_S^2 B I_3 B^T + \sigma_S^2 A I_N A^T\right) J_{DR} \left(J_{DR}^T J_{DR}\right)^{-T} \quad (4)$$

Here I_N is an n × n covariance matrix. The GDOP for the sonar array measurement subsystem is calculated as:

$$GDOP = \sqrt{trace(P_{DR})} \quad (5)$$

2.2 PSO Model for Station Deployment

The PSO algorithm facilitates collaboration through the exchange of information among individuals within the swarm, enabling each individual to be inspired by the experiences of its neighbors, guiding its search direction and velocity, and ultimately directing the swarm towards the global optimum. In PSO, each particle is in a state of motion, dynamically adjusting its velocity. The update formulas for the velocity and position of each particle in the d-dimensional search space at the gth iteration are given by Eq. (6):

$$v_{i,g+1} = v_{i,g} + c_1 \times \text{rand}_{i,1} \times \left(x_{i,pbest} - x_{i,g}\right) + c_2 \times \text{rand}_{i,2} \times \left(x_{i,gbest} - x_{i,g}\right)$$

$$x_{i,g+1} = x_{i,g} + v_{i,g+1} \quad (6)$$

In Eq. (6), $x_{i,g} = (x_1, x_2, \cdots, x_d)$ represents the position of the individual i in the d-dimensional search space at the iteration g of PSO, $v_{i,g} = (v_1, v_2, \cdots, v_d)$ represents the velocity of the individual i in the d-dimensional search space, $x_{i,gbest}$ denotes the best position found by the individual i throughout its search history in the d-dimensional space, and $x_{i,pbest}$ represents the best position found by any individual in the swarm during the optimization process in the d-dimensional space. And c_1 and c_2 are the learning rates for the individual's best and the global best, while $\text{rand}_{i,1}$ and $\text{rand}_{i,2}$ are random numbers within [0, 1].

To endow the PSO algorithm with better global search ability in the early stages of optimization and superior local search capability in the later stages, Shi and other scholars introduced the concept of inertia weight ω [13]. The inertia weight ω is determined by Eq. (7):

$$\omega = \omega_{start} + \frac{g}{g_{max}} \times (\omega_{start} - \omega_{end}) \qquad (7)$$

In experiments, the initial ω_{start} and final inertia weights ω_{end} are typically set to 0.9 and 0.4. Here, g is the current iteration number, and g_{max} is the maximum number of iterations. According to Eq. (7), the inertia weight ω decreases linearly with the increasing number of iterations. Consequently, the velocity update formula in PSO becomes Eq. (8):

$$v_{i,g+1} = \omega \times v_{i,g} + c_1 \times \text{rand}_{i,1} \times (x_{i,pbest} - x_{i,g}) + c_2 \times \text{rand}_{i,2} \times (x_{i,gbest} - x_{i,g}) \qquad (8)$$

This variant of PSO, incorporating the inertia weight, is commonly referred to as the standard PSO due to its widespread adoption [14].

To further avert the risk of convergence to local optima and enhance the curiosity during the optimization process, thereby expanding the exploration capabilities of the swarm, further refinements are made to the calculation method of the inertia weight ω in the PSO algorithm.

$$\omega_{id} = sigmoid\left(\alpha \frac{\Delta h}{v_{id}}\right) \qquad (9)$$

Here, the sigmoid function, an activation function commonly utilized in deep neural networks, is utilized. Its expression is given by:

$$sigmoid(x) = \frac{1}{1 + e^{-x}} \qquad (10)$$

During the optimization of the fitness function, the inertia weight become large when significant changes occur. Due to the nonlinear characteristics of the activation function, as the fitness function gradually changes less with each iteration, the inertia weight also decreases rapidly, gradually fixing the position of the individual and allowing it to explore local details.

Simultaneously, a greedy algorithm is employed to ensure that when some individuals become trapped in local optima, they can escape and start a new iteration round. For non-optimal individuals that have not shown significant improvement in the fitness function after multiple iterations, a certain probability is assigned to allow these individuals to undergo a larger positional change Δx during the current iteration.

3 Optimization Iteration Analysis

3.1 Mization of Station Deployment Model Based on Multi-objective PSO

When the deployment of measurement stations is unreasonable, significant errors may occur in areas with sparse stations, resulting in poor data quality and limited effectiveness of the entire measurement system. In contrast, in areas with concentrated stations, redundancy occurs, contributing little to improving measurement accuracy and leading to a waste of measurement resources. The fitness function is set as follows:

$$\min Fitness(\boldsymbol{P}) = f(GDOP, punish) \qquad (11)$$

Here Fitness(P) is the fitness function, P represents the population position, GDOP stands for positioning accuracy, and *punish* is a penalty function, represents the ratio of the area covered by a specified number of measurement stations to the total measurement area.

The fitness function f is defined as:

$$f(x, y) = x - 10y \tag{12}$$

The penalty function *punish* is used to handle constraints in the optimization problem. By assigning a penalty value to solutions that violate constraints, it guides the search process towards satisfying the constraints. To ensure high coverage, the ratio of grid points that are not covered by sufficient measurement stations to the total number of grid points is used as the penalty function *punish*:

$$punish = \frac{P}{N} \tag{13}$$

In Eq. (13), N is the total number of grid points in the calculation area, and P is the number of grid points covered by sonar stations that meet the basic station requirement.

3.2 Optimization of Station Deployment Model Based on Multi-objective PSO

Assuming that a subsurface buoy at a depth of 4000 m has a detection range of 12 km for a 200 dB @ 1KHz acoustic signal. As shown in Fig. 1, the preset measurement area is 24 km × 24 km, with an "electronic fence" area of 20 km × 20 km, which constrains the solutions during the optimization process within the corresponding area. Meanwhile, in the PSO algorithm, the number of particles is set to 500, with the learning rates c_1 for individual best experience and c_2 for global best experience both set to 1. The maximum velocity of particles is constrained to 2. The initialization of solutions significantly affects the convergence speed and results of the optimization process, so to improve computational efficiency and accelerate convergence, it is crucial to set reasonable initial solutions. The initial positions of subsurface buoys are shown in Table 1.

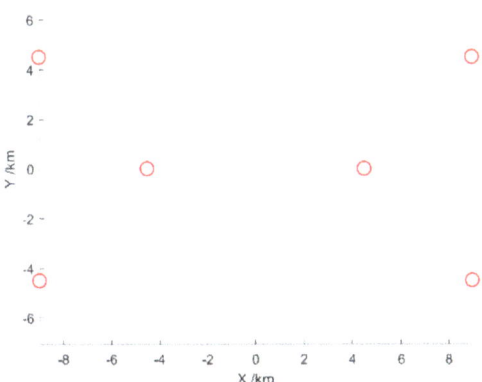

Fig. 1. Initial Deployment of Measurement Stations before Optimization

Table 1. Initial Positions for Optimization of Deployment in a Typical Deep-Sea Scenario

Platform type	Serial number	Coordinates/km
Submerged buoy	1	(4.5, 0, 4)
	2	(9, 4.5, −4)
	3	(−4.5, 0, −4)
	4	(−9, 4.5, −4)
	5	(9, 4.5, −4)
	6	(−9, -4.5, −4)

3.3 Results of Iterative Optimization

Based on the optimization model described above, the positions of the shipborne sonar array and subsurface buoys were optimized. Figure 2 shows the optimized positions of each station obtained through the improved PSO algorithm, with specific coordinates listed in Table 2.

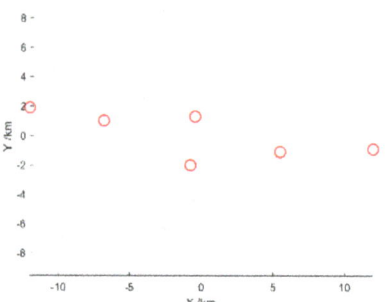

Fig. 2. Optimized Deployment of Measurement Stations

Table 2. Optimized Positions for Deployment in a Typical Deep-Sea Scenario

Platform type	Serial number	Coordinates/km
Submerged buoy	1	(−0.7409, −2.017, −4)
	2	(5.4894, −1.0944, −4)
	3	(−0.4187, 1.2814, −4)
	4	(−12, 1.8725, −4)
	5	(12, −0.9142, −4)
	6	(−6.8511, 0.9977, −4)

As shown in Fig. 3, the measurement accuracy of the long baseline array of the submerged buoy represented by the GDOP remains stable. Throughout multiple iterations, the precision is maintained at around 340. Guided by the penalty function, the measurement capability coverage rate of the designated area is significantly improved. The coverage rate of three buoys has been increased from 74.7% to 89.2%, and that of four buoys has been increased from 41.1% to 70.2%. During the iterative optimization process, the fitness function value continues to decrease with the increase in the number of iterations.

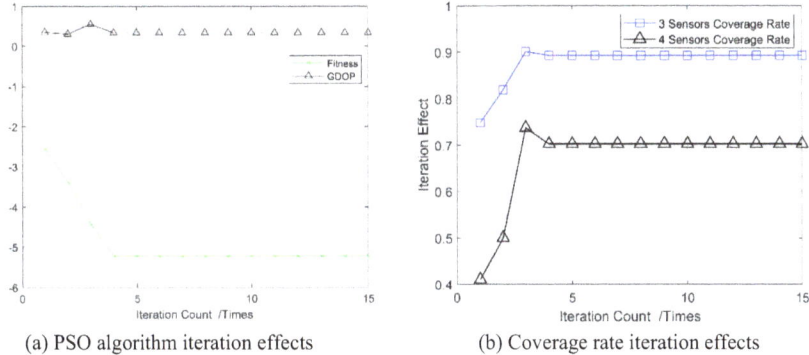

(a) PSO algorithm iteration effects (b) Coverage rate iteration effects

Fig. 3. Iteration effects

An excellent geometric structure of the station layout helps to achieve an ideal balance between two important indicators: coverage range and positioning accuracy. Figure 4(a) and (b) show the distribution maps of the coverage range before and after optimization by the PSO algorithm. Figure 4(c) and (d) illustrate the GDOP distribution of each subsystem before and after the optimization of the station layout. Although the differences may not appear significant at first glance, the positioning accuracy at the edge of the area was extremely poor before the optimization, to the extent that the position of the sound source was unmeasurable. After optimization, the effective measurable area has been significantly increased.

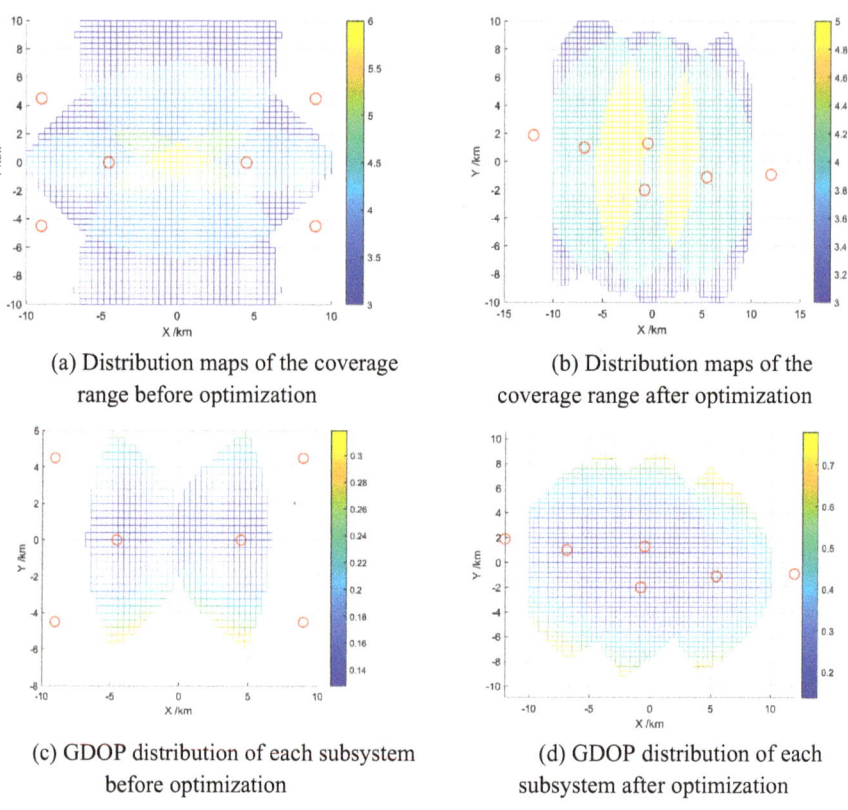

(a) Distribution maps of the coverage range before optimization

(b) Distribution maps of the coverage range after optimization

(c) GDOP distribution of each subsystem before optimization

(d) GDOP distribution of each subsystem after optimization

Fig. 4. Visualization maps of station deployment effects

4 Conclusion

The article discusses the optimization of the station layout for the long baseline array of underwater acoustic beacons in the measurement scenario of pulse sound signals. In order to balance both the effective measurement coverage and positioning accuracy while avoiding redundancy in measurement station resources, an improved PSO algorithm was used to optimize and iterate the positions of the stations. In the application of the PSO algorithm, some improvements were made based on the station layout optimization problem. In the paper, the inertia weight was improved to effectively avoid the optimization results from falling into local optima. Additionally, the proportion of unmeasured areas was used as a penalty function to ensure the coverage capability of the measurement area. The results show that the optimized submerged buoy array improves the measurement accuracy of pulsed acoustic signals, with coverage rates increasing from 74.7% to 89.2% for three buoys and from 41.1% to 70.2% for four buoys under stable measurement accuracy conditions.

The paper only discusses the problem-solving capabilities of convex optimization algorithms, and in subsequent work, there is potential to explore the space for improvement that more efficient algorithms could bring to this case.

References

1. Xue, S., et al.: Dynamic positioning configuration and its first-order optimization. J. Geod. **88**(2), 127–143 (2014)
2. Yanhui, C.: Research on Key Technologies for Integration of Differential GPS Underwater Positioning System [D]. Fuxin: Liaoning University of Technology, 2007. (in Chinese)
3. Sun, Y., et al.: 3D hybrid TOA-AOA source localization using an active and a passive station. In: Proceedings of IEEE International Conference on Signal Processing, 2016, Piscataway, NJ: IEEE, pp. 257–260 (2016)
4. Jia, T., et al.: Target localization based on structured total least squares with hybrid TDOA-AOA measurements. Signal Process. **14**(3), 211–221 (2018)
5. Tomic, S., Beko, M., Dinis, R.: Distributed RSS-AOA based localization with unknown transmit powers. IEEE Wirel. Commun. Lett. **5**(4), 392–395 (2016)
6. Tomic, S., et al.: A closed-form solution for RSS/AOA target localization by spherical coordinates conversion. IEEE Wirel. Commun. Lett. **5**(6), 680–683 (2016)
7. Omic, S., Beko, M., Dinis, R.: 3-D target localization in wireless sensor networks using RSS and AOA measurements. IEEE Trans. Veh. Technol. **66**(4), 3197–3210 (2017)
8. Xiong, H.L.: A novel hybrid RSS and TOA positioning algorithm for multi-objective cooperative wireless sensor networks. IEEE Sensors J. **18**(22), 9343–9351 (2018)
9. Tomic, S., Beko, M.: A robust NLOS bias mitigation technique for RSS-TOA-based target localization. IEEE Signal Process. Lett. **26**(1), 64–68 (2019)
10. Yang, X.S.: Swarm intelligence based algorithms: A critical analysis. Evol. Intel. **7**(1), 17–28 (2014)
11. Kennedy, J., Eberhart, R.C.: Particle swarm optimization. In: Proceedings of the IEEE International Conference on Neural Networks, pp. 1942–1948. IEEE, Piscataway, NJ (1995)
12. Zhang, Y., Wang, S., Ji, G.: A comprehensive survey on particle swarm optimization algorithm and its applications. Math. Probl. Eng. **2015**, 1–38 (2015)
13. Hua, L.: An inertia weight adaptive particle swarm optimization algorithm. J. Electron. Sci. Technol. **30**(3), 30–36 (2017)
14. Xiaoli, Z., Qinfei, W., Wenli, J.: An improved adaptive inertia weight particle swarm algorithm. Microelectron. Comput. **36**(3), 66–70 (2019)

Open Access This chapter is licensed under the terms of the Creative Commons Attribution-NonCommercial-NoDerivatives 4.0 International License (http://creativecommons.org/licenses/by-nc-nd/4.0/), which permits any noncommercial use, sharing, distribution and reproduction in any medium or format, as long as you give appropriate credit to the original author(s) and the source, provide a link to the Creative Commons license and indicate if you modified the licensed material. You do not have permission under this license to share adapted material derived from this chapter or parts of it.

The images or other third party material in this chapter are included in the chapter's Creative Commons license, unless indicated otherwise in a credit line to the material. If material is not included in the chapter's Creative Commons license and your intended use is not permitted by statutory regulation or exceeds the permitted use, you will need to obtain permission directly from the copyright holder.

Analyses on Consistency Between Level 1 and 2 Vulnerability Criterion for Dead Ship Stability

Kaibo Yu, Feng Cai(✉), and Jianjun Hou

Dalian Naval Academy, Jiefang Road 667, Dalian, China
820437895@qq.com

Abstract. The work on the second generation intact stability criterion is on trial now, and the vulnerability criterion assessment method of dead ship stability has already been confirmed. The dead ship stability is formulated to improve the safety of ship navigation at sea, compared with other failure modes in the second generation intact stability, there are some problems in the principle consistency of the dead ship stability vulnerability criteria. In order to make the criteria more scientific and early comprehensive application, the calculation principle and evaluation method of the dead ship stability vulnerability criteria are studied, the principle similarities and differences between the first and second level are analyzed, and the reasons affecting the consistency are explored through the calculation results of the sample ship. Firstly, the theory of dead ship stability assessment method is summarized, and the similarities and differences between level 1 and level 2 assessment method are analyzed, then sample ships have been calculated, including sea guard ships, patrol ships, fishery administration vessels and oil tankers. According to calculation results, the consistency between level 1 and level 2 is studied and the corresponding conclusion is obtained, which provides test data support to the completion of dead ship stability vulnerability criterion.

Keywords: Second Generation Intact Stability Criterion · Dead Ship Stability Failure Mode · Vulnerability Criterion · Sample Calculation · Analyses on Consistency

1 Introduction

The second generation intact stability criterion is on trial phase now [1], and the vulnerability criterion for dead ship stability is divided into two levels. The level 1 follows the original standard [2], and the level 2 uses probabilistic assessment method proposed by Italy [3], which adopt to calculate the rolling motion characteristics of dead ship under irregular cross wind and wave during a specific exposure time and obtain the long-term failure index, and the index is taken as the assessment index of the level 2 vulnerability criterion. Compared with the other four failure modes, the consistency of dead ship stability vulnerability criterion still has some problems, because the level 1 and 2 adopt different assessment methods.

2 Vulnerability Criterion Analysis of Dead Ship Stability

2.1 Overview of the Level 1 Calculation Theory

Please note that the first paragraph of a section or subsection is not indented. The first paragraphs that follows a table, figure, equation etc. does not have an indent, either.

Subsequent paragraphs, however, are indented.

The level 1 vulnerability criterion for dead ship stability are based on the existing meteorological criterion in the International Intact Stability Code 2008 (Part 2.3 of Rule) [4], supplemented by the wave slope table in MSC.1/Circ.1200 [5]. The content of the first level of criterion is as followed.

$$1.\ \varphi_0 \leq \min \begin{Bmatrix} 16° \\ 0.8 \cdot \varphi_{jb} \end{Bmatrix}$$

$$2.\ b \geq a \qquad (1)$$

where, φ_0 is the heeling angle under the action of steady wind, φ_{jb} is the flooding angle at the edge of the deck, b and a are the areas of different parts surrounded by the GZ curve of the unsteady wind action arm and the righting lever, as shown in the Fig. 1.

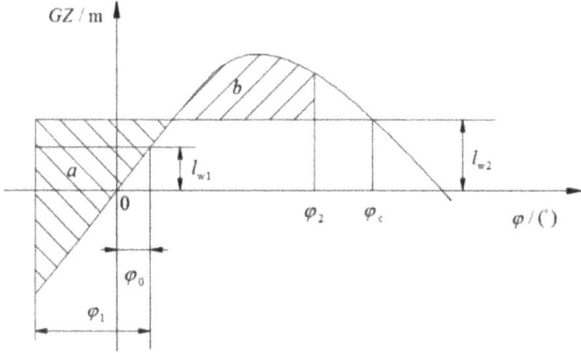

Fig. 1. Diagram of area b and area a in the level 1 vulnerability criterion.

2.2 Overview of the Level 2 Calculation Theory

The level 2 vulnerability criterion for dead ship stability is based on the rolling motion characteristics of dead ship, and then combined with the probability model, the short-term failure index CS under various environmental conditions is calculated, and the long-term failure index C is calculated according to the weighted wave distribution map of the given sea area. The long-term failure index C is used to evaluate the level 2 vulnerability of dead ship under certain loading conditions [3].

It is assumed that the ship is at zero speed and subjected to the combined action of random cross wind and waves. The motion equation is described by single-degree-of-freedom nonlinear roll motion equation, and the main features of roll motion are the variance of roll angle and the variance of roll angular velocity. The capsizing event of the dead ship complies with the Poisson distribution probability model [6], and the capsizing probability under each working condition is the short-term failure index CS [7], and the formula for calculating CS under given environmental working condition is as followed [8].

$$\begin{cases} C_s = 1 - \exp(-\lambda_{EA} \cdot T_{\exp}) \\ \lambda_{EA} = \frac{1}{T_{z,C_s}} \cdot \left[\exp\left(-\frac{1}{2RI_{EA+}^2}\right) + \exp\left(-\frac{1}{2RI_{EA-}^2}\right) \right] \\ RI_{EA+} = \frac{\sigma_{C_s}}{\Delta\phi_{res,EA+}}; \Delta\phi_{res,EA+} = \phi_{cap,EA+} - \phi_s \\ RI_{EA-} = \frac{\sigma_{C_s}}{\Delta\phi_{res,EA-}}; \Delta\phi_{res,EA-} = \phi_s - \phi_{cap,EA-} \\ T_{z,C_s} = 2\pi \frac{\sigma_{C_s}}{\sigma_{\dot{C}_s}} \end{cases} \quad (2)$$

where, the exposure time $T_{\exp}$ is taken as 3600 s, σ_{C_s} is the standard deviation of roll angle and $\sigma_{\dot{C}_s}$ is the standard deviation of roll angular velocity, and other parameters are shown in Fig. 2.

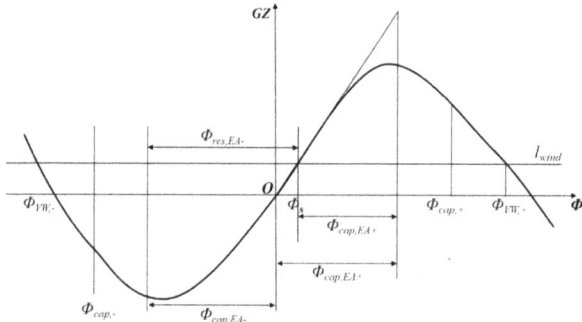

Fig. 2. Diagram of calculation parameters for the level 2 vulnerability criterion

2.3 Analysis of the Similarities and Differences Between the Two Levels Assessment Methods

Based on the level 1 and 2 assessment methods for the stability of dead ships, the similarities and differences of the two are analyzed and summarized as follows.

The Similarities of Level 1 and Level 2 Assessment Methods.

(1) Methods suppose the calculated ship loses power and is at zero speed state.
(2) Considering the ship's ability to resist the capsizing moment under the conditions of transverse wind and waves.
(3) The GZ curve of ships is based on the righting lever curve in still water, and the influence of free surface is considered.

(4) The definition and value of the flooding angle of the ship.
(5) The definition of ship capsizing angle is the same, which is the minimum value of flooding angle, 50° and residual stability disappearance angle.
(6) The calculation methods of wind tilt lever and static balance angle of ship under steady wind.
(7) The calculation methods of ship windward area and the distance from the center of the area to the waterline.

Differences Between Level 1 and Level 2 Assessment Methods. There is an essential difference between the vulnerability criteria of dead ship and the other four failure modes in the second generation intact stability criterion. The first level criterion of the other four failure modes are simplified from the second level, while the first level of dead ship directly follows the meteorological criterion in International Intact Stability Code, and some modifications are made on the basis of them. This results in the inconsistencies in the mathematical models and principles of the first and second floor weakness criterion for the stability of the dead ship, which may have certain influence on the consistency between two levels.

3 Calculation Results of Sample Ship

The 24 sample vessels with 3 loading conditions were selected, including sea area guard vessels, patrol vessels, fisheries administration vessels, oil tankers, etc., which could ensure the diversity and universality of the sample vessels. The calculation results of each loading condition were summarized in Table 1, the loading conditions numbered from 1 to 72, and the horizontal coordinate in the subsequent figure is the loading conditions number. According to the judgment of the heeling angle under the action of steady wind in the first level of criterion, all loading conditions meet the criterion requirements, so the data are not included in the table. The gray part of data in table is the load condition that does not meet the criterion requirements, where the criterion value of the second level is 0.06.

Table 1. Dead ship vulnerability criterion calculation results of ships.

Condition number	Ship name	L_{pp}/m	Loading condition	Average draft/m	b/a	long-term failure index
1	Fishing vessel	24	Departure of full load	3.956	1.462	2.29E-01
2			Arrival of full load	3.765	2.077	8.61E-02
3			In fishing	3.111	6.952	6.58E-04
4	Passenger ferry	30	Departure of full load	1.267	4.938	5.88E-01

(*continued*)

Table 1. (*continued*)

Condition number	Ship name	L_{pp}/m	Loading condition	Average draft/m	b/a	long-term failure index
5			Arrival of full load	1.247	4.531	6.14E-01
6			Arrival in light	1.173	4.432	6.30E-01
7	Patrol boat	31.6	Departure with icing	2.727	2.051	2.20E-01
8			Departure of full load	2.712	2.1	2.11E-01
9			Arrival of full load	2.482	2.909	6.20E-02
10	Trawler	32.3	Departure for fishing	3.414	3.415	1.43E-03
11			Arrival of full load	3.072	4.083	3.55E-05
12			Arrival in light	2.514	3.931	3.70E-05
13	Fishery administration vessel 1	35	Departure of full load	2.366	2.045	6.67E-02
14			At sailing	2.19	2.298	5.46E-02
15			Arrival in light	2.051	1.545	4.77E-02
16	Fireboat 1	39.06	Departure of full load	2.533	2.235	3.08E-01
17			Departure	2.515	2.256	2.96E-01
18			Arrival of full load	2.275	2.365	2.05E-01
19	Guard-boat	40	Departure of full load	2.486	3.689	9.45E-02
20			At sailing	2.344	3.746	5.65E-02
21			Arrival in light	2.231	3.536	3.98E-02
22	Fishery administration vessel 2	40	Departure of full load	2.634	1.788	3.44E-02
23			At sailing	2.445	2.03	1.96E-02
24			Arrival of full load	2.429	1.938	2.16E-02
25	Light Seine fishing boat	40.5	Return of full load	3.439	0.227	5.64E-01

(*continued*)

Table 1. (*continued*)

Condition number	Ship name	L_{pp}/m	Loading condition	Average draft/m	b/a	long-term failure index
26			In fishing	3.027	0.483	4.70E-01
27			Shipwreck condition	2.869	0.025	3.53E-01
28	Fishery administration vessel	42.4	Departure of full load	2.604	3.095	4.25E-02
29			At sailing	2.417	3.183	2.00E-02
30			Arrival in light	2.255	2.719	2.56E-02
31	Fishery administration ship 3	47	Departure of full load	2.398	4.574	2.38E-02
32			At sailing	2.224	4.078	1.81E-02
33			Arrival in light	2.106	3.493	1.96E-02
34	Fireboat 2	47	Departure of full load	2.97	1.627	2.78E-01
35			Arrival of full load	2.598	2.226	9.26E-02
36			Arrival in light	2.473	2.374	6.34E-02
37	Fishery administration vessel 4	48.6	Departure of full load	2.85	4.078	9.02E-02
38			At sailing	2.613	4.763	5.22E-02
39			Arrival in light	2.428	5.228	3.80E-02
40	Fishery administration vessel 5	49.2	Departure of full load	3.187	1.951	1.08E-02
41			Arrival of ballast	3.05	0.886	2.23E-02
42			Sailing of Half-load	3.021	1.306	1.44E-02
43	Oil tanker 1	72	Departure of full load	2.331	0.78	7.97E-01
44			Departure of half load	1.517	1.262	8.42E-01
45			Departure in light	0.82	0.792	7.11E-01
46	Oil tanker 2	78	Overloading departure	3.2	0.885	7.06E-01

(*continued*)

Table 1. (*continued*)

Condition number	Ship name	L_{pp}/m	Loading condition	Average draft/m	b/a	long-term failure index
47			Departure of full load	2.9	1.003	7.75E-01
48			Departure in light	1.2395	1.082	2.04E-01
49	Oil tanker 3	83	Departure of full load	3.2	0.455	7.83E-01
50			Departure of half load	2.13	0.788	8.47E-01
51			Departure in light	1.306	1.069	5.01E-01
52	passenger roll ship 1	83.3	Departure of full load	3.956	5.357	7.75E-05
53			Arrival of full load	3.884	4.896	8.64E-05
54			Departure in light	3.51	4.111	2.60E-04
55	passenger roll ship 2	83.8	Departure of full load 1	4	5.897	6.46E-05
56			Departure of full load 2	3.956	5.8	6.85E-05
57			Arrival of full load	3.41	3.667	2.90E-04
58	Large fishing trawlers	94.8	Return of full load	7.154	9.457	6.92E-07
59			Departure of full load	6.338	7.318	5.20E-04
60			Arrival of small loading	5.32	7.635	4.47E-06
61	LNG vessel	112	Departure of full load	5.703	4.133	1.68E-03
62			Departure of partial loading	5.248	4.5	7.88E-04
63			Arrival of full load	5.208	4	1.44E-03
64	Oil tanker 4	134	Departure of full load 1	8.8	3.459	1.30E-04

(*continued*)

Table 1. (*continued*)

Condition number	Ship name	L_{pp}/m	Loading condition	Average draft/m	b/a	long-term failure index
65			Departure of full load 2	8.2	4.505	1.38E-06
66			Departure of partial loading	7.25	5.258	8.46E-07
67	Oil tanker 5	164	Departure of full load 1	10.1	4.155	1.06E-05
68			Departure of full load 2	9.5	5.142	3.80E-07
69			Departure of ballast	6.5	4.437	3.74E-06
70	Oil tanker 6	186	Arrival of full load	12.371	3.463	3.80E-05
71			Departure of partial loading	10.858	4.831	2.94E-06
72			Arrival of partial loading	8.893	5.99	5.10E-05

As can be seen from Table 1, the area ratio b/a of the level 1, including the three loading conditions of light Seine fishing vessels, the five loading conditions of fishery administration vessel 5 and three oil vessel 1–3, did not meet the requirements, and the other 63 conditions can pass the first level. In the level 2, there are more loading conditions that do not meet the long-term failure index requirements, and there are 29 loading conditions that do not meet the requirements, and only 43 loading conditions meet the requirements.

According to the requirements of consistency, the loading conditions which can pass the assessment of level 1 definitely pass the assessment of level 2. There are still 21 kinds of loading conditions in the calculation results that do not meet the consistency requirements.

4 Consistency Analysis Between Level 1 and Level 2

The shape of the *GZ* curve, the capsizing angle and the roll angle of the ship under the action of external torque are the keys to determine the calculation results of the two levels criteria of the dead ship. The definition of *GZ* curve in the level 1 and 2 is consistent. This paper analyzes the consistency between level 1 and 2 of dead ship from the roll angle, capsizing angle and level 2 criterion value combined with the calculation results of the sample ship.

For the convenience of comparison and analysis with the level 1, the following data about the level 2 of the dead ship are all taken from the environmental conditions when significant wave height is 8.5 m, and it will not be repeated in the following paragraphs.

4.1 Influence Analysis of the Roll Angle

The roll angle is calculated according to the given formula in level 1, and roll angles of various ships under various loading conditions is given in Fig. 3.

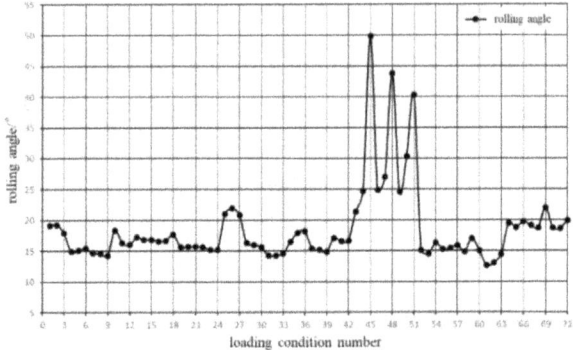

Fig. 3. Rolling angle in level 1 vulnerability assessment of dead ship.

It can be seen from Fig. 3 that the roll angles of most of the sample ships loading conditions is between 15° to 20°, and the roll angle of the oil tankers 1–3 is relatively large. The factors causing the excessive roll angle are analyzed as follows. The formula for calculating the roll angle of in the level 1 is shown in formula (3).

$$\phi_1 = 109 * k * X_1 * X_2 * \sqrt{r * s} \tag{3}$$

It can be seen that the roll angle value of the level 1 is determined by five parameters. In order to analyze the influence of each parameter on the roll angle, the corresponding five parameters under various loading conditions are given in Fig. 4.

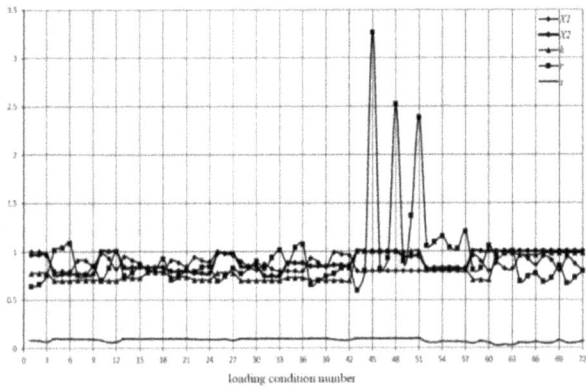

Fig. 4. The various parameters in level 1 vulnerability assessment of dead ship.

It can be seen from figure that the curve trend of parameter r is close to that of roll angle curve. Especially for several loading conditions of the three oil tankers, parameter r plays a decisive role.

Through in-depth study, it can be found that the other four parameters have corresponding value ranges, except for parameter r, which can be reflected in Fig. 4. Parameter r is determined by the ratio of the center of gravity height to draft, the ratio of the three inland oil tankers is too large, so these loading conditions are not suitable for the assessment of the level 1. In the level 2, the roll angle is not calculated according to the fixed formula, but is expressed by the variance of the roll angle and roll angular velocity. The variance of roll angle and roll angular velocity for each load condition of the sample ship are given in Fig. 5.

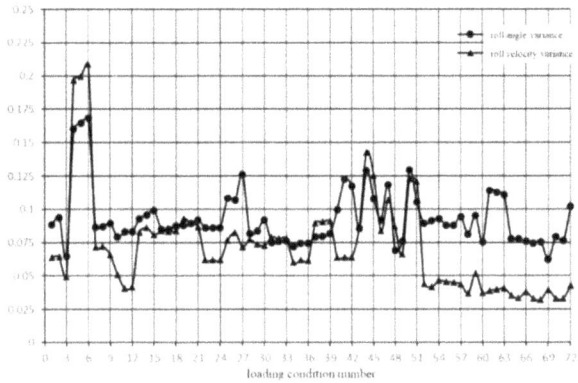

Fig. 5. Roll angle and roll velocity variance in level 2 vulnerability assessment of dead ship.

As can be seen from Fig. 5, the variance of roll angle and roll angular velocity of the passenger ferry and the three inland oil vessels are very large in several loading conditions. It can also be found that the variance of the angular velocity of large ships is generally smaller than that of small boats.

The calculation of the variance of the roll angle and roll angular velocity in the level 2 depends on the equivalent linear roll damping coefficient, which is given in Fig. 6 under different loading conditions of the sample ship.

As can be seen from Fig. 6, the equivalent linear roll damping coefficients of the passenger ferry and the three inland oil tankers are all small, so the corresponding roll angle variance is large. The passenger ferry is a hard chine type ship, and the damping calculation of this type of ship is not considered in Ikeda method, which results in small damping assessment. The three inland oil tankers have no bilge keels and a relatively large ratio of width and draft, and the damping assessment of the Ikeda method for this type of ship is still small, so these ships are not suitable for the assessment method of the level 2 vulnerability assessment of dead ship.

It can be seen from the above analysis that the loading condition is the most significant factor affecting the roll angle in the level 1, and the equivalent linear roll damping coefficient is the most significant factor in the level 2.

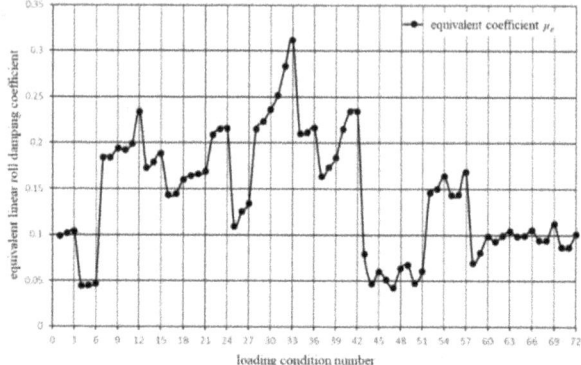

Fig. 6. Equivalent linear roll damping coefficient in level 2 vulnerability assessment.

4.2 Influence Analysis of the Capsizing Angle

The definitions of capsizing angle in the first and second tiers of dead ships are consistent, but the definitions of wind heeling moment arm are different, so the stability loss angle of the two is different. The capsizing angle of the first and second tiers under different loading conditions of the sample ships are given in Fig. 7.

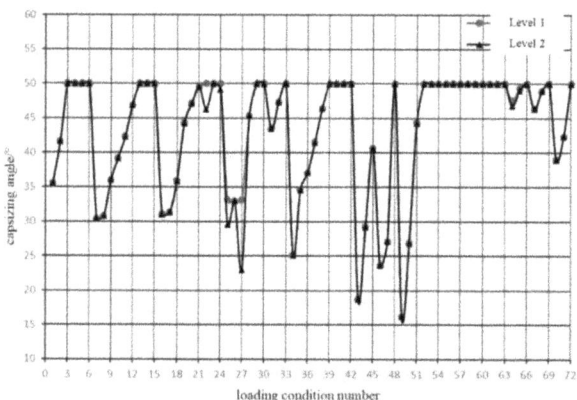

Fig. 7. Capsizing angle Comparison between level 1 and level 2 vulnerability assessment.

It can be seen from Fig. 7 that the capsizing angles of the level 1 and 2 of most sample ships are the same, and the capsizing angles of the level 1 are smaller than those of the level 2 for a few inconsistent loading conditions.

Combined with Table 1, it is believed that although the capsizing angle of the level 1 and 2 is consistent, the calculation results of the sample vessels show that the capsizing angle corresponding to various loading conditions, including patrol vessel, two fireboats, light Seine fishing vessel and three inland oil tankers, is relatively small, and the capsizing angle is around 30°, or even lower than 30°. Most of these loading conditions can meet the assessment requirements of the level 1, but they cannot meet the assessment requirements of the level 2. In the level 1, the original capsizing angle is directly used for assessment and calculation, but in the level 2, the capsizing angle is equivalent and the equivalent capsizing angle is used to calculate the capsizing probability, and the equivalent capsizing angle is basically less than the original value of the capsizing angle, so the assessment results of the level 2 are more conservative, and the actual angles used for capsizing judgment of the level 1 and 2 are inconsistent.

4.3 Influence Analysis of the Level 2 Criterion Value

This study considers that the criterion value can be determined according to the length of ship. As can be seen from the calculation results of the sample ships, the capsizing probability of each large ship is small, and the capsizing probability of each small boat is big. In order to ensure the consistency between the level 1 and 2, this paper suggests that the criterion value should be formulated according to the captain. The criterion value of the first 24 m fishing vessel should be 0.25, the criterion value of the last 186 m oil tanker should be 0.06, and the criterion value of the middle length can be obtained through linear interpolation, and the criterion value R_d is calculated as follows.

$$R_d = 0.06 + \frac{0.25 - 0.06}{186 - 24} * (186 - L_{PP}) \tag{4}$$

According to the formula, except for several ships that are not applicable to the assessment method and navigation area, it can be seen that according to the proposed criterion value, the loading conditions of other sample ships can guarantee the consistency between the level 1 and 2 calculation results.

5 Conclusion

Based on the principle of vulnerability criterion assessment method of dead ship stability, this paper analyzes the differences and similarities between the level 1 and 2 assessment methods, and collates and analyzes the calculation results of a total of 72 loading conditions of 24 sample ships, and draws the following conclusions.

(1) For most of the sample ships, the load condition of the largest b/a in the level 1 is the lowest capsizing probability in the level 2. The load condition with the smallest b/a has the largest capsizing probability in the level 2, which meets consistency requirement.
(2) Regarding the influence of roll angle on the consistency between the level 1 and 2, a fixed formula is used to calculate the roll angle in the level 1, and the damping coefficient of hard chine type is considered. In the level 2, the variance of roll angle and the variance of roll angular velocity are used to represent the roll amplitude, but the damping calculation of Ikeda method does not take hard chine type into special consideration, so the calculation results for the level 1 and 2 of this ship type is inconsistent.

(3) Regarding the influence of capsizing angle on the consistency between the level 1 and 2, the definition of capsizing angle of the level 1 and 2 is consistent, but the equivalent capsizing angle is used to calculate the capsizing probability in the level 2, and the equivalent capsizing angle is generally smaller than the original capsizing angle, so the actual angle used for capsizing judgment of the level 1 and 2 is inconsistent.

(4) The level 2 criterion value should be formulated according to the length of the ship, and the calculation formula of the criterion value is given, so that the calculation results of the level 1 and 2 of the sample ships can meet the consistency requirements. In the process of setting the level 2 of criterion value, countries generally do not consider sample ships with small length, so it is recommended that IMO add sample data for small boats.

References

1. IMO. Interim guidelines on the Second Generation Intact Stability Criteria. MSC.1/Cric.1627 (2020)
2. IMO. Draft amendments to part B of the IS CODE with regard to vulnerability criterion of levels 1 and 2 for the dead ship condition failure mode. SDC 3/WP.5.Annex1 (2016)
3. IMO. Draft explanatory notes on the vulnerability of ships to the dead ship stability failure mode (Japan). SDC 4/5/1-Add.4 (2016)
4. IMO. Adoption of the International Code on Intact Stability 2008. MSC.267(85).(2008)
5. IMO. Interim guidelines for alternative assessment of the weather criterion. MSC.1/Circ.1200 (2006)
6. IMO. Vulnerability assessment for dead ship stability failure mode (Italy and Japan). SDC 1-INF.6 (2013)
7. Lifen, H., et al.: Research on capsizing probability of stability under dead ship condition in beam wind and wave. China J. Hydrodyn. A. **33**(6), 794–800 (2018). (in Chinese)
8. Chen, R.: Influence and analysis of hull lines on dead-ship condition vulnerability criterion. Dalian University of Technology (2022). (in Chinese)

Open Access This chapter is licensed under the terms of the Creative Commons Attribution-NonCommercial-NoDerivatives 4.0 International License (http://creativecommons.org/licenses/by-nc-nd/4.0/), which permits any noncommercial use, sharing, distribution and reproduction in any medium or format, as long as you give appropriate credit to the original author(s) and the source, provide a link to the Creative Commons license and indicate if you modified the licensed material. You do not have permission under this license to share adapted material derived from this chapter or parts of it.

The images or other third party material in this chapter are included in the chapter's Creative Commons license, unless indicated otherwise in a credit line to the material. If material is not included in the chapter's Creative Commons license and your intended use is not permitted by statutory regulation or exceeds the permitted use, you will need to obtain permission directly from the copyright holder.

Research on Evaluation of Personnel Evacuation Paths in Ship Fire

Xiaogang Jiang(✉), Zhijang Yuan, and Mingdong Lv

Department of Navigation Dalian Naval Academy, Dalian, China
forgething@163.com

Abstract. The law of smoke spread in ship fires has a significant impact on the safe evacuation of personnel. This article takes a certain type of passenger ship as a modeling prototype and uses the fire simulation software Pyrosim to analyze the smoke spread law after a fire in the restaurant and engine room. Using the personnel safety evacuation simulation software Pathfinder, the evacuation situation of personnel located on the second and third floors was evaluated for safety. The results indicated that the height of fire smoke was a key factor affecting personnel evacuation, and the research conclusions could provide important references for optimizing personnel evacuation routes.

Keywords: Ship · Flue Gas Flow · Escape

1 Introduction

With the development of global trade, the passenger and freight transportation business of ships is becoming increasingly prosperous, and the resulting ship fire accidents are also on the rise [1]. As a means of water transportation, ships are densely populated with personnel and property, with a large number and density of flammable and explosive materials. Especially during sea navigation, due to their distance from land, once a fire occurs, rescue forces are difficult to arrive in a timely manner and can only rely on the existing personnel and firefighting equipment on board for rescue [2]. The sea waves are strong, and the wind can also promote the spread of the fire, making firefighting relatively difficult [3]. Therefore, the evaluation of personnel evacuation and escape plans is crucial.

2 Numerical Simulation of Ship Cabin Fire

The location of a ship fire had a direct impact on the evacuation path of personnel. Therefore, this article take one location where people were prone to gather, namely the restaurant, as examples to conduct numerical analysis of the spread of fire smoke [4].

2.1 Establishment of Ship Cabin Fire Model

Select a certain type of passenger ship with a captain of 120 m, a width of 23 m, a draft of 4.6 m, and a displacement of 4000 tons. Based on the overall cabin structure of the ship, the second and third floors where the restaurant and engine room were located are modeled as a whole, as shown in Fig. 1. The model involves many cabin doors. In the model studied in this article, the cabin doors of compartments with frequent personnel access were set to open according to the principle of the least favorable fire situation, and the cabin doors of compartments with less personnel access were set to close.

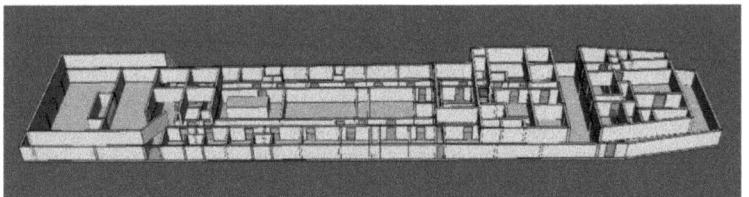

Fig. 1. The Second and Third Floor Model of Ship.

2.2 Determination of Fire Hazard Time

According to the analysis and statistics of the causes of death in fires, smoke causes the greatest harm to people, and the proportion of deaths caused by smoking was relatively high up to 78.9%. The direction of smoke flow determines the direction of personnel evacuation. So, the study of fire smoke was the focus of fire research. In the process of a ship fire, when the smoke layer drops to the point of contact with human eyes, it will have a serious impact on personnel evacuation and cause smoke injuries. In addition, due to the irregular movement of smoke during the fire process, turbulent flow of smoke may occur. Therefore, the criteria for determining the harm caused by smoke to human bodies should be appropriately relaxed, and the criteria for determining danger should not be based on the contact between the smoke layer and human eyes, but on formulas [5]

$$H_s > H_c = H_p + 0.1H_b \tag{1}$$

In the formula: H_s is the height of the smoke layer; H_p Height of human eye features; H_b The height of the building ceiling; H_c To determine the critical height that poses a danger to people, based on the actual situation of the studied population, the height of the human eye is set to 1.6 m. In this physical model, the height of the cabin roof is 2.1 m. Therefore, according to the above formula, the critical height at which the smoke layer poses a danger to people in the event of a fire was 1.81 m. Therefore, when the height of the smoke layer decreases to the critical height of 1.81 m that poses a danger to the human body, the height of the smoke layer reaches the simulated critical danger value.

In addition to the impact of smoke, another important cause of casualties in fires was the heat generated by combustion. When the height of the smoke layer drops to direct contact with people, the harm of smoke to people was direct burns. According to the data, to cause second degree burns on the skin, it only takes 60 s for the skin to be continuously exposed to smoke at 71 °C, 30 s at 82 °C, and 15 s at 1000 °C.

In summary, the smoke from ship fires has a greater impact on personnel evacuation and is the main factor causing casualties. It was also the first factor causing harm to human health in fires. In addition, high-temperature smoke could also cause damage to human health and is another factor that cannot be ignored in causing casualties in fires. Therefore, this article mainly judges the fire hazard situation by combining the spread of fire smoke and the distribution of temperature fields, and then obtains the fire hazard value.

2.3 Numerical Calculation of Ship Cabin Fire

Due to the fact that the model can be approximated as a rectangular prism, a rectangular grid was taken that fully includes the model. The size of each grid calculation unit is $0.2562 \times 0.2812 \times 0.2762$ m, and the total number of grids is 435456. The initial conditions for fire simulation were:

(1) Initial temperature of the environment: 20 °C;
(2) Initial relative humidity of the environment: 50%;
(3) Environmental atmospheric pressure: 101325Pa (1 atm);
(4) Wind speed (Velocity): 0 m/s;
(5) Fire simulation running time (Time): 1200 s;
(6) Type of fire growth: rapid fire;
(7) Structural material (Floor): set to steel;
(8) Fire conditions: equipped with smoke detectors, no sprinkler or mechanical smoke exhaust;
(9) Combustion Type: Fuel and composite materials;
(10) Building Height: 4.35 m.

Under the above simulation conditions, a restaurant fire was simulated with the ignition point set at the right rear corner of the restaurant. After the fire, smoke continuously gathered indoors and the temperature rapidly increased. As the smoke increased, it spread through the front door of the first restaurant to the left and right passages, and then through the staircase at the entrance to the third classroom. The smoke gathered on the ceiling of the passage and then spread to the cabins and living quarters on both sides of the passage. Due to the research needs of this article, only the distribution of flue gas and temperature field will be studied.

The initial and final stages of smoke spread during a fire were shown in Fig. 2.

The temperature distribution after the fire was shown in Fig. 3.

According to the simulation results, the following conclusion can be drawn: after a fire occurs, the smoke accumulates on the roof of the restaurant. With the continuous accumulation of smoke and the intensification of the roof jet, the thickness of the smoke layer continues to increase. When the height of the smoke layer drops to the height of the room door, the smoke will spread through the restaurant to the third floor passage

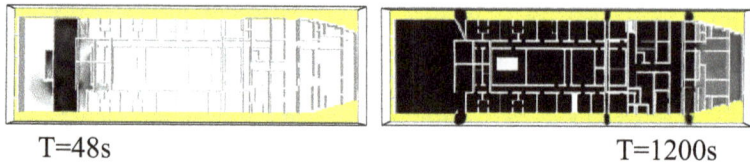

T=48s   T=1200s

Fig. 2. Smoke accumulates at the top of the restaurant and spreads outward.

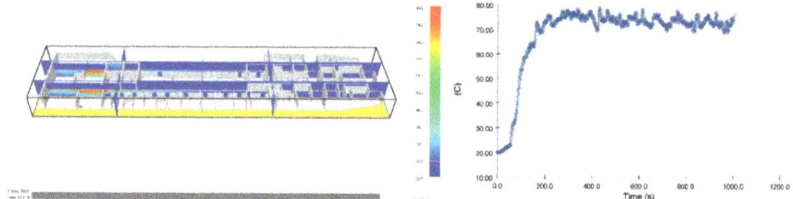

Fig. 3. Temperature Field Distribution (T = 1200s).

and through the stairs to the rear of the second floor. After entering the channel, the smoke continuously spreads along the channel towards the front, and at the same time, it also spreads to the cabin connected to the channel and opens the door. With the increase of time, smoke continues to accumulate and the height of smoke begins to decrease. According to the smoke height detector installed at the entrance of the restaurant, the height of the smoke layer drops to the dangerous value of 1.81m at around 43s, and the smoke spreads to the rear through the stairs at around 130s. According to the simulation results, it can be seen that during the entire fire period, except for the temperature in the restaurant being above 60 °C, the temperature in other parts was below 60 °C, so temperature is not the main reason affecting personnel evacuation. Due to the fire occurring in the restaurant, personnel inside the restaurant are the most dangerous and have the shortest response time. It can be considered that there is no response time. Simulation results show that personnel inside the restaurant can leave before the smoke completely fills the restaurant. Therefore, the fire hazard value in the fire scenario is the time it takes for smoke to spread to the exits of the two evacuation routes, which is 324 s.

3 Numerical Simulation of Ship Fire Evacuation

3.1 Establishment of Personnel Evacuation Model

Based on the above analysis of the fire spread law, the smoke spread mainly involves the second and third decks. The personnel evacuation model of the second and third floors of the passenger ship was established by using the personnel evacuation simulation software Pathfinder. The 3D evacuation effect of the second and third floors of the passenger ship and the designed escape route were shown in Fig. 4.

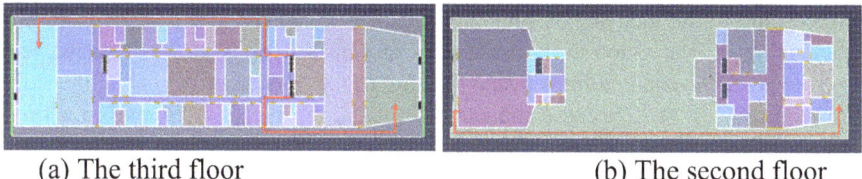

(a) The third floor (b) The second floor

Fig. 4. Simulation diagram of personnel evacuation.

3.2 Numerical Calculation of Personnel Evacuation

In case of fire, the way to notify people to evacuate is different, the function of the building and the indoor environment are different, and the time for people to get the news of the fire and prepare for evacuation is also different. Based on the least favourable principle, the response time was hereby determined to be 60 s.

According to the carrying capacity of the passenger ship, the number of staff in the restaurant where the fire occurred is 40, the accommodation is set at 6, and the remaining places where the fire did not occur are set at 100. The default speed of personnel in Simulex evacuation model is divided into four types: men, women, children and the elderly, and the speed ranges from 0.8 m/s to 1.35 m/s. Considering that the passenger ship personnel are relatively complex and will have an impact on each other, a relatively slow speed is taken here, assuming that the evacuation speed of personnel is 1 m/s.

Considering the uncertainty in the analysis of the timing of danger and evacuation actions, it was necessary to add a safety margin. In general, it was recommended to set the safety margin at 0 to 1 times the evacuation action time. For the ship studied in this article, the personnel density was high and the safety evacuation signs are obvious, but the internal structure was relatively complex and the space was narrow. Based on the above conditions, the safety margin value should not be too small or too large, so a safety margin of 0.2 was taken.

The remaining parameters were set to:

(1) Sport Mode: SFPE mode and steering mode;
(2) Evacuation preparation Time (Time): 60 s
(3) Number of personnel: 146;
(4) Safety Margin: 0.2.

According to the original escape road map of the passenger ship, simulation results are shown in Fig. 5.

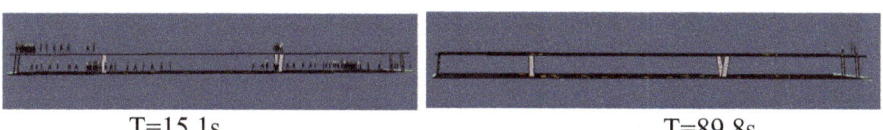

T=15.1s T=89.8s

Fig. 5. Evacuation status of personnel at different times.

According to the analysis of simulation results, it can be seen that the time from the beginning of evacuation to the front deck and back deck is 95.3 s, because when the personnel are evacuated from inside the ship to the outside of the cabin, the safety evacuation time is 34 s, because the safety evacuation time of personnel in a fire is composed of evacuation preparation time and evacuation action time. Through analysis, it has been determined that the safe evacuation preparation time of the personnel evacuation model is 60 s, so the safe evacuation time of all personnel in the model is 94 s. Since the safety margin of personnel evacuation is set at 0.2, the final safe evacuation time is 110.8 s.

This article quantitatively analyses and determines the fire hazard time under different fire scenarios using the fire simulation software Pyrosim, and obtains the fire hazard time T_a, which is the allowed evacuation time; Then, the personnel evacuation simulation software Pathfinder was used to quantitatively simulate and analyse the evacuation of personnel in different fire scenarios, and the time required for all personnel to evacuate safely, that is, the necessary evacuation time T_n, was obtained. By comparing and analysing with T_a and T_n, the safety status of personnel evacuation was determined.

The same numerical method was used to simulate and evaluate the fire in the ship's engine room again. The safety assessment of ships in fire scenarios was shown in Table1:

Table 1. Judgment of ship fire safety.

Fire scene	$T_a(s)$	$T_n(s)$	Safe or not
Restaurant	324	110.8	yes
Engine room	103	110.8	no

4 Conclusion

Combined with the specific conditions of the second and third floors of the passenger ship studied, for the second and third floors of the passenger ship studied in this paper, the results of fire simulation and safety evacuation simulation under different scenarios are compared and analysed, and the following conclusions can be drawn:

(1) For the restaurant fire, through comparative analysis of the fire spread process and the safe evacuation process set in the fire scenario, it can be seen that all personnel can be evacuated before the fire danger time. It can be seen from the simulation results that the smoke will quickly spread to the back exit of the left and right channels after the fire, so people should be evacuated from other exits as far as possible.
(2) For the engine room fire, through comparative analysis of fire spread process and personnel evacuation process, it can be seen that all personnel could not be evacuated before the fire danger time. As the accommodation is close to the internal stairs, once the fire starts, the smoke will quickly spread to the stairway and the exit of the left and right channels, so the personnel located on the second floor should evacuate to the assembly point through the external straight stairs on the second and third floors as far as possible.

From the above simulation and comparative analysis, it can be seen that the evacuation plan of the second and third floors of the passenger ship was not the best choice, and the evacuation risk was greater after the fire.

References

1. Xiao, M., Zhou, X., Pan, X., et al.: Simulation of emergency evacuation from construction site of prefabricated buildings. Sci. Rep. **12**(1), 2732 (2022)
2. Nguyen, D.-T., Shen, Z., Honda, K., et al.: A GIS-based model for integrating risk estimations of residential building damage and shelter capacity in the case of earthquakes **21**(2) (2020)
3. Hlatka, M., Kampf, R., Krile, S., Kubasakova, I.: Streamlining the logistics evacuation process using the specific simulation software. Trans. Res. Procedia **44** (2020)
4. Shanshan, F., Song, Q., Zhuang, H., Zhang, C.: Risk analysis of ship fire accident based on tripod-beta model. J. Saf. Environ. **20**(01), 9–19 (2020)
5. Zhong, M.-H., Li, P.-D., Liu, T.-M., et al.: Study on simulation experiment of smoke movement in multi-storey multi-compartment building indoor fire. Chin. Sci. (Part E: Mater. Sci. Eng. Sci.) **35**(5), 490–502 (2005)

Open Access This chapter is licensed under the terms of the Creative Commons Attribution-NonCommercial-NoDerivatives 4.0 International License (http://creativecommons.org/licenses/by-nc-nd/4.0/), which permits any noncommercial use, sharing, distribution and reproduction in any medium or format, as long as you give appropriate credit to the original author(s) and the source, provide a link to the Creative Commons license and indicate if you modified the licensed material. You do not have permission under this license to share adapted material derived from this chapter or parts of it.

The images or other third party material in this chapter are included in the chapter's Creative Commons license, unless indicated otherwise in a credit line to the material. If material is not included in the chapter's Creative Commons license and your intended use is not permitted by statutory regulation or exceeds the permitted use, you will need to obtain permission directly from the copyright holder.

Crashworthiness Analysis of New Double-Layer Side Structure

Ming-dong Lv[1], Xi-wei Wang[2], Zhi-jiang Yuan[1(✉)], Xiao-gang Jiang[1], Zhong-chang An[3], and Peng-yao Yu[4]

[1] Dalian Naval Academy, Dalian, China
1213730780@qq.com
[2] Tianjin International Marine Engineering Co., Ltd., Tianjin, China
[3] Naval Research Institute, Beijing, China
[4] Dalian Maritime University, Dalian, China

Abstract. In order to improve the crashworthiness of the side structure of a ship's engine room, a new crashworthiness structure of double-layer side plates is proposed in this paper. The nonlinear explicit finite element software LS-DYNA is used to simulate the collision of two ships under different schemes, and the stress situation, energy absorption and deformation of the new double-layer side plate structure are discussed and analyzed. The calculation results show that when the striking ship strikes the double-layer side plates at a speed of 5 m/s and at different impact angles of 45°, 60° and 90°, no rupture occurred in the inner plates of the double-layer side plates. The double-layered side plate crashworthiness structure can effectively prevent seawater from flooding into the engine room after a ship collision, and provide a reference for improving the side crashworthiness of the engine room during ship design and construction.

Keywords: Ship Collision · Double-layer side Plate · Numerical Simulation

1 Introduction

Ship collision is a common traffic accident at sea, threatening the safety of people and property on board. Especially the breach caused by the collision may cause a large amount of seawater to flood into the cabin, which will seriously damage the stability and strength of the ship and is an important reason of the sinking of the ship. The engine room is the power core of the ship. Core components such as the engine, gearbox, and transmission shaft are installed in the engine room. A large amount of water is filled in the engine room, which will cause the ship to lose power for navigation, causing greater risks [1]. Nonlinear numerical simulation technology is an effective method for analyzing ship collision damage [2, 3]. SUN [4] derived finite element simulation and analytical methods to analyze the structural response of the wedge bow impacting the broadside outer plate. HU [5] derived nonlinear numerical simulation technology to analyze the crashworthiness of the side structure of the Y-type floating production and storage ship. ZHANG [6] derived experimental and numerical simulation methods to simulate the tearing characteristics of ship shell plates after impact rupture. Ringsberg [7] derived a nonlinear finite element method to simulate the collision experiment between the indenter and the ship structure.

In this paper, a new type of double-layer side plates is proposed, which improves the crashworthiness of the side by arranging the inner shell. Using the nonlinear explicit finite element software LS-DYNA, collision simulations of ship-ship collision accidents under different schemes are carried out. The research shows that the double-layer side plates can effectively improve the crashworthiness of the cabin side structure, prevent the inner shell from rupturing and causing seawater to flood into the cabin, and improve the ship's anti-sinking ability.

2 Finite Element Collision Model

2.1 Finite Element Collision Model

As an explicit nonlinear finite element simulation software, LS-DYNA can better handle large structural deformations, dynamic responses and elastic-plastic problems of materials. It is widely used in ship collision simulation. The structural dynamics equations discretized by this software are:

$$M\ddot{x}(t) = P(t) - F(t) + H(t) - C\dot{x}(t) \tag{1}$$

where M is a concentrated quality matrix, $P(t)$ is the external load vector, $F(t)$ is the internal force vector, $H(t)$ is the hourglass resistance vector, $\ddot{x}(t)$ is the node acceleration vector, C is the damping matrix, $\dot{x}(t)$ is the node velocity vector. The internal force vector and external load vector are calculated by the following two equations:

$$F(t) = \sum_e \int_{V_e} B^T \sigma dV \tag{2}$$

$$P(t) = \sum_e \left(\int_{V_e} N^T f dV + \int_{\partial b_{2e}} N^T \overline{T} dS \right) \tag{3}$$

where ∂b_{2e} is the boundary stress condition, f is the voxel load vector, $\overline{T}$ is the surface force vector. LS-DYNA uses the explicit central difference method to solve the above discretized structural dynamics equations and obtains the following formula:

$$\ddot{x}(t_n) = M^{-1}\left[P(t_n) - F(t_n) + H(t_n) - C\dot{x}(t_{n-1/2})\right] \tag{4}$$

$$\dot{x}(t_{n+1/2}) = \dot{x}(t_{n-1/2}) + \ddot{x}(t_n)(\Delta t_{n-1} + \Delta t_n)/2 \tag{5}$$

$$x(t_{n+1}) = x(t_n) + \dot{x}(t_{n+1/2})\Delta t_n \tag{6}$$

where $\ddot{x}(t_n)$ is the node acceleration vector at time t_n, $\dot{x}(t_{n-1/2})$ is the node velocity vector at time $t_{n-1/2}$, $x(t_{n+1})$ is the displacement vector at time t_{n+1}, in the expression $t_{n-1/2} = (t_n + t_{n-1})/2, t_{n+1/2} = (t_{n+1} + t_n)/2, \Delta t_{n-1} = t_n - t_{n-1}, \Delta t_{n-1} = t_{n+1} - t_n$.

For explicit time integration, the following equation must be satisfied in order to ensure the convergence time step:

$$\Delta t \leq \Delta t_{cr} = \frac{2}{\omega_{max}} \quad (7)$$

where ω_{max} is the highest natural frequency of the collision system.

In ship-ship collision damage accidents, because the bulbous bow is located below the waterline, it is more likely to cause the shell of the other ship to rupture, causing seawater to flood into the cabin. Therefore, in a ship-ship collision accident, bulbous bow will cause more serious damage to the hull of the struck ship than other bows. In this paper, a 65000-t container cargo ship with a bulbous bow is selected as the striking ship, and a ship with a double-layer side plates is selected as the struck ship. The collision response of the double-layer side plates under the impact of a bulbous bow is analyzed. Where, the size of the 65000-t container ship main dimensions is as shown in Table 1.

Table 1. The main dimensions of 65000 t container(in m).

Main dimension name	Size	Main dimension name	Size
Length O.A./L	299.95	Depth moulded/D	24.60
Breadth moulded/B	48.20	Scantling Draft/ T_m	14.80

In order to ensure the strength of the bulbous bow structure of the striking ship, the distribution and plate thickness of each structure at the main impact positions of the bulbous bow of the striking ship should be further clarified. Where, the bulbous bow structure on the bow of the 65000-t container ship mainly includes structures such as planking, floor, stiffener, bracket and bulkhead. The material is mild steel. A partial enlarged view of the bulbous bow of the 65000-t container ship is shown in Fig. 1, and the plate thickness of the bulbous bow structure of the 65000-t container ship is shown in Table 2.

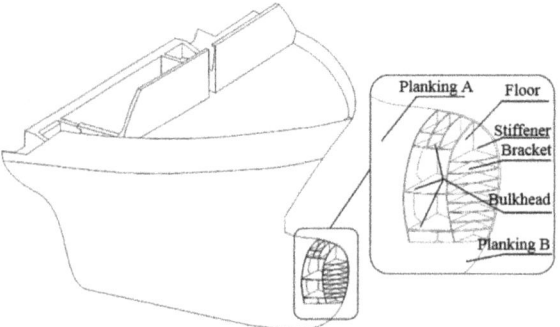

Fig. 1. Partial enlarged view of bow and bulbous bow structure of 65000 t container ship.

Table 2. Plate thickness of bulbous bow structure of 65000 t container ship(in mm).

Bulbous bow structure part	Thickness	Bulbous bow structure part	Thickness
Planking A	20	Stiffener	14
Planking B	23	Bracket	14
Floor	12	Bulkhead	12

Because the collision damage is local damage, only the local cabin section of the struck ship needs to be modeled. The cross-section of the double side structure is shown in Fig. 2, and the thickness of the main structural plates of the struck ship is shown in Table 3. The inner shell of the double side structure is the final protective structure of the engine room, and its collision response in a ship-ship collision accident is an important indicator to measure the crashworthiness of the engine room structure.

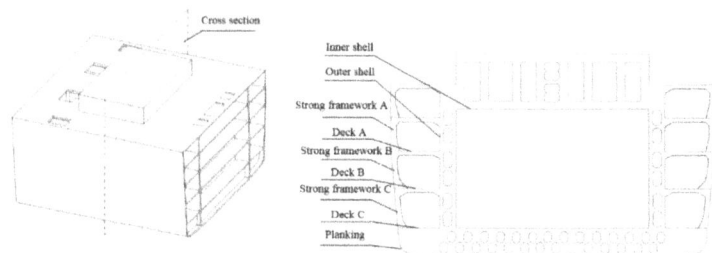

Fig. 2. Schematic diagram of the location of a cross section of the struck ship.

Table 3. Plate thickness of bulbous bow structure of 65000 t container ship(in mm).

Main structural parts of struck ship	Thickness	Main structural parts of struck ship	Thickness
Inner shell	26	Deck B	10
Outer shell	26	Strong framework C	14
Strong framework A	16	Deck C	16
Deck A	10	Planking	31
Strong framework B	15		

In order to ensure the accurate mass and center of gravity position of the striking ship and the struck ship, the 65000-t container cargo ship adopts a full-ship modeling. In order to save computing resources, the bow part of the collision ship retains its rib and beam structure, and only the other non-collision areas retain the plate and shell structure, and adjust the mass of the entire ship in stages to ensure that the center of gravity position of the entire collision ship remains unchanged. The finite element model of the 65000-t container cargo ship is shown in Fig. 3; In order to improve calculation efficiency, the cabin section of the double-layer side plates section parallel to the midbottom part of the struck ship is modeled, and all the ribs and beam structures in the double-layer side plates are retained. In order to ensure that the double-layer side plates are consistent with the center of gravity of the original ship, the bow and stern of the collided ship are extended in the bow and stern direction to the length of the entire ship in proportion to the profile lines of both ends, and the quality of the extended part is adjusted to ensure that the center of gravity of the model is consistent with the entire ship model. The collision finite element model of the double-layer side plates cabin section is shown in Fig. 4.

Fig. 3. Finite element model of 65000 t container ship.

Fig. 4. Finite element model of struck ship and double-layer side plate.

2.2 Finite Element Collision Model

Elastoplastic materials were selected to simulate the steel materials of the double-layer side plates cabin sections of the struck ship and the bulbous bow collision area structure of the striking ship. The PLASTIC-KINEMATIC material model was selected in LS-DYNA. The material used for the internal structure of the struck ship is mild steel, and the material used for the outer panels and open decks of the ship is high-strength steel AH36. See LS-DYNA Material Card Parameter Setting Table 4 for the common mechanical properties and parameters of the materials.

Because the interaction forces on the surfaces of ships are very complex during the collision process of ships, it is very important to choose an appropriate contact algorithm to simulate the collision process of ships. The contact algorithm selected in this paper is CONTACT-AUTOMATIC-SURFACE-TO-SURFACE, which uses primary and secondary contact. The collision between the striking ship and the struck ship is defined as the contact from the contact point to the main contact surface of the ship.

Table 4. LS-DYNA material parameter setting

Parameter	Mild Steel	High strength steel
Material density/□□ (g/mm³)	7.83×10^3	7.83×10^3
Elastic modulus/E (Pa)	2.07×10^{11}	2.15×10^{11}
Yield strength /□□ (Pa)	3.15×10^8	3.55×10^8
Hardening parameter□□	1885	
Cowper-Symonds	$D = 40.4s^{-1}$, q = 5	
Failure coefficient/F_s	0.35	

The contact algorithm will automatically search for contact boundaries during collision numerical simulation to ensure relevant calculation accuracy. In the ship collision model studied in this paper, all component surfaces are contactable areas after collision. The dynamic friction coefficient between the contact surfaces is 0.3, and the static friction coefficient is 0.1.

3 Numerical Simulation Analysis

3.1 Design of Collision Scheme

This paper combines some scholars' research on failure mode analysis of broadside structures using striking parameters [8] and determines that the collision speed of the striking ship is 5 m/s. When the collision angles of the striking ship relative to the struck ship are different, the damage to the hull of the struck ship is also different. In order to obtain the dynamic impact response of the double side structure, the collision angles of the impacting ship are selected in this paper as 90°, 60°, and 45°. Combined with the different angles of the striking ship and the different striking positions of the struck ship, the specific collision scheme is shown in Table 5, and the schematic diagram of the collision scheme is shown in Fig. 5.

Table 5. Crash condition scheme.

Scheme	Collision angle(°)	Collision velocity(m/s)
1	90	5
4	45	5
5	60	5

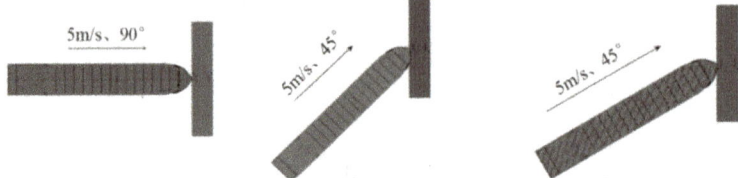

Fig. 5. Schematic diagram of collision scheme 1–3.

3.2 Design of Collision Scheme

The numerical simulation results of dynamic collision responses of double-layer side plates under different parameters (different striking positions and different angles) are analyzed and summarized. The outer shell stress, inner shell stress, deformation and deformation energy of the double-layer side plates under different schemes are shown in Figs. 6, 7 and 8.

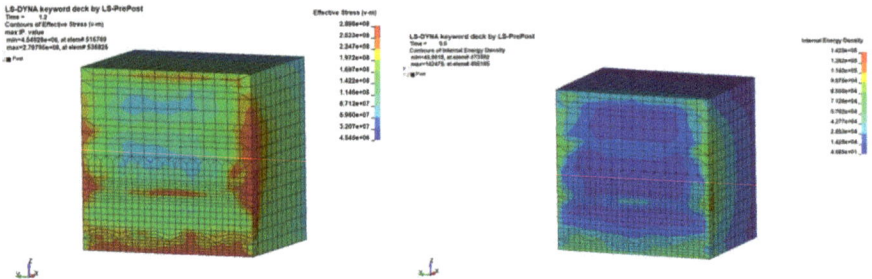

(c)deformation of inner shell in scheme 1 (d)deformation energy of inner shell in scheme 1

Fig. 6. Von-Mises stress of outer plate and Von-Mises stress, deformation, deformation energy of outer plate and inner shell in scheme 1.

It can be seen from Figs. 6, 7, 8 and Table 6 that when the striking ship strikes the double side structure at 5 m/s and at 45°, 60°, and 90°, the stress in the inner shell does not exceed the material yield limit, and the inner shell will not rupture, indicating that the double-layer side plates can effectively resist the collision of the striking ship's bulbous bow, effectively prevent the bulbous bow from crashing the inner shell, causing a large amount of seawater to flood into the cabin. When the bulbous bow of the striking ship strikes the double-layer side plates at a speed of 5 m/s and 90°, the inner shell is subjected to the greatest stress and produces significant deformation, indicating that the outer plate strong frame will transfer the collision force to the inner shell, which is the most likely to cause the inner shell to rupture.

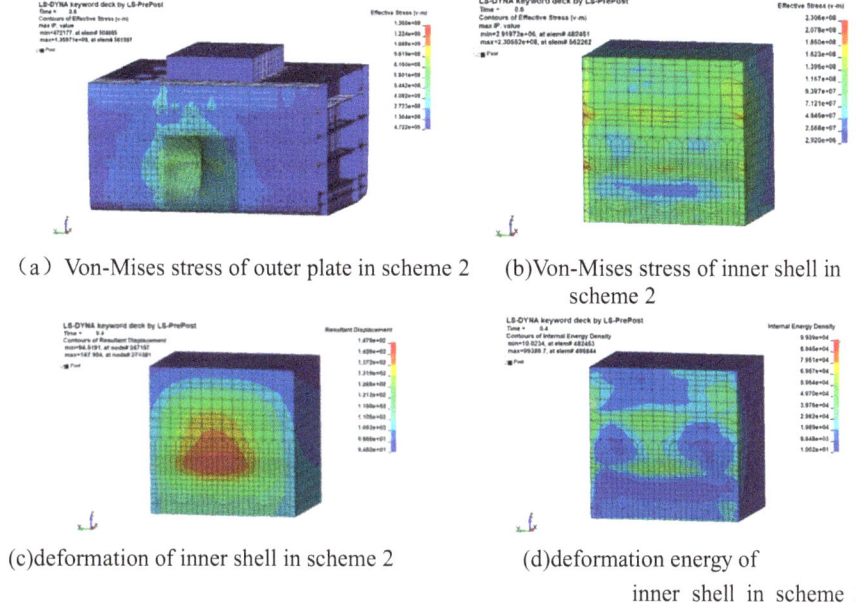

(a) Von-Mises stress of outer plate in scheme 2   (b) Von-Mises stress of inner shell in scheme 2

(c) deformation of inner shell in scheme 2   (d) deformation energy of inner shell in scheme 2

Fig. 7. Von-Mises stress of outer plate and Von-Mises stress, deformation, deformation energy of outer plate and inner shell in scheme 2.

4 Conclusions

In this paper, the finite element model of the double ships are modeled according to the hull structures and relevant parameters of the striking ship and the struck ship, and the parameter settings of various LS-DYNA cards in the collision simulation calculation are determined. According to the striking angle of the striking ship and the striking position of the struck ship, the schemes are used to verify the crashworthiness of the double side structure. According to the simulation results, the following conclusions are obtained:

(1) The double-layer side plates can effectively improve the crashworthiness of the side structure, especially the inner shell is not easy to rupture, which can prevent large amounts of seawater from flooding into the cabin;
(2) When the striking ship strikes the double side structure at 45° and 60°, the outer plate of the double side has a small area of damage, the internal structure is basically complete, the shape of the inner shell becomes elastic deformation, and the peak energy absorption value of the inner shell is much lower than that at the same speed. The response results of the collision simulation at an angle of 90° finally conclude that the damage caused by the striking ship striking the side structure at 45° and 60° at the same speed is much smaller than the damage caused by the striking ship striking the side structure at 90°.

When the striking ship strikes the strong frame of the side structure at 90° and 5 m/s, the inner shell has the most severe deformation and the highest energy absorption. The strong frame structure should be further optimized.

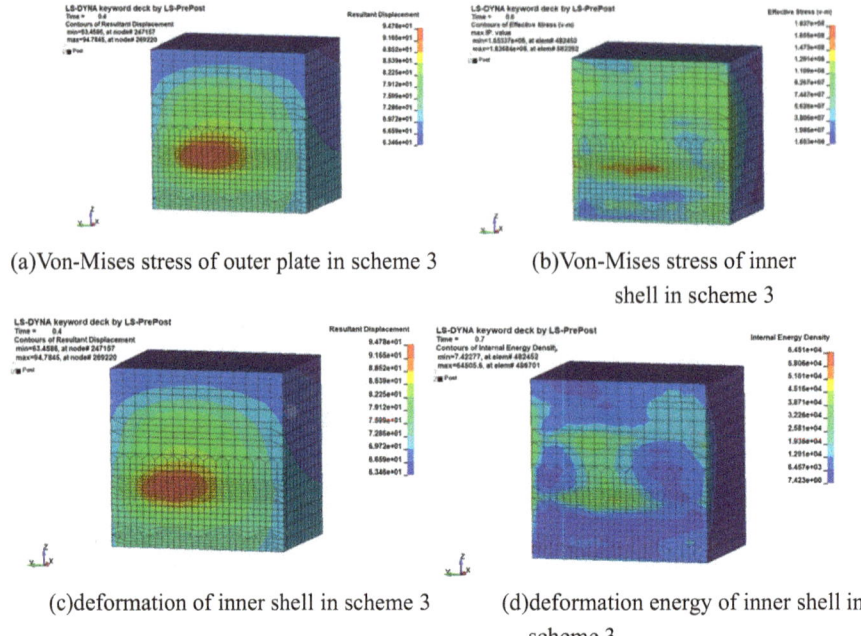

(a) Von-Mises stress of outer plate in scheme 3

(b) Von-Mises stress of inner shell in scheme 3

(c) deformation of inner shell in scheme 3

(d) deformation energy of inner shell in scheme 3

Fig. 8. Von-Mises stress of outer plate and Von-Mises stress, deformation, deformation energy of outer plate and inner shell in scheme 5.

Table 6. Inner shell maximum deformation and deformation energy in different schemes.

Scheme	Maximum deformation(mm)	Maximum deformation energy(kJ)
1	37	4.5×10^5
4	22	4.0×10^5
5	17.5	3.8×10^5

References

1. Luo, K., Yu, L.: Discussion on emergency response after ships lose control in port. Marine Technol., 10–13 (2023). (in Chinese)
2. Zhang, S., Pedersen, P.T., Villavicencio, R.: Probability and Mechanics of Ship Collision and Grounding, pp. 1–4. Butterworth-Heinemann, Oxford (2019)
3. Liu, B., Pedersen, P.T., Zhu, L.: Review of experiments and calculation procedures for ship collision and grounding damage. Mar. Struct. **59**, 105–121 (2018)
4. Sun, B., Hu, Z., Wang, J.: Structural response analysis for a ship side plating impacted by raked bow. J. Vibr. Shock **35**(23), 46–50 (2016). (in Chinese)
5. Hu, Z., Gao, Z., Gu, Y.: Research on the crashworthiness of a Y-shape side structure design for FPSO. J. Shanghai Jiaotong Univ. **39**(5), 706–710 (2005). (in Chinese)

6. Zhang, M., Zhang, X., Zhao, Y., et al.: Experimental and numerical studies of tearing characteristics of ship plates after fracture subjected to impact loads. J. Ship Mech. (008), 026 (2022)
7. Ringsberg, J., Amdahl, J., Chen, B.Q., et al.: MARSTRUCT benchmark study on non-linear FE simulation of an experiment of an indenter impact with a ship side-shell structure. Mar. Struct. **59**, 142–157 (2018)
8. Zhang, J., et al.: Research on the failure mode of the impact parameter to the side structure. Ship Sci. Technol. **42**(1), 5 (2020)

Open Access This chapter is licensed under the terms of the Creative Commons Attribution-NonCommercial-NoDerivatives 4.0 International License (http://creativecommons.org/licenses/by-nc-nd/4.0/), which permits any noncommercial use, sharing, distribution and reproduction in any medium or format, as long as you give appropriate credit to the original author(s) and the source, provide a link to the Creative Commons license and indicate if you modified the licensed material. You do not have permission under this license to share adapted material derived from this chapter or parts of it.

The images or other third party material in this chapter are included in the chapter's Creative Commons license, unless indicated otherwise in a credit line to the material. If material is not included in the chapter's Creative Commons license and your intended use is not permitted by statutory regulation or exceeds the permitted use, you will need to obtain permission directly from the copyright holder.

Ship Navigation Risk Assessment in The Arctic Northeast Passage Via Stochastic Petri Nets Model

Cuiwen Fang[1], Xiaoyu Du[2], Xinxin Zhang[3], and Shenping Hu[3]($\boxtimes$)

[1] College of Ocean Science and Engineering, Shanghai Maritime University, Shanghai, People's Republic of China
[2] Patent Examination Cooperation (Henan) Center of the Patent Office, Zhengzhou, Henan, People's Republic of China
[3] Merchant Marine College, Shanghai Maritime University, Shanghai, People's Republic of China
sphu@shmtu.edu.cn

Abstract. Navigational risk for ships has always been a key focus in maritime traffic safety research, with a variety of risk assessment methods available, each possessing its own advantages. In the era of big data, with the diversification of information and the deep application of system safety, this study addresses the navigational risks of ships in Arctic ice regions from a complex systems perspective. Taking the Arctic Northeast Passage as an example, a quantitative assessment model for navigational risk in Arctic waters is constructed using stochastic Petri nets (SPN). This model analyzes the mechanism of the navigational process and provides risk control schemes specifically targeting high-risk factors. The research findings indicate that sea ice density is a pivotal factor in the navigational risk of ships traversing the Arctic Northeast Passage. The risk for ships navigating the East Siberian Sea, Laptev Sea, and Kara Sea (especially the Vilkitsky Strait) is heightened due to persistent sea ice even during the summer season. Icebreaker assistance is deemed necessary when sailing in spring and autumn.

Keywords: Risk Assessment · SPN · Arctic · Ship Navigation

1 Introduction

With the rise in global temperatures, there is a significant decreasing trend in Arctic sea ice, particularly evident during the summer months [1]. As a result, countries worldwide have shown great interest in the Arctic region, seeking to promote further development in its resources, shipping, and tourism. As a major near-Arctic country, China actively participates in research related to Arctic shipping and has proposed the "Polar Silk Road" to promote sustainable development in the Arctic region. The development of Arctic routes shortens navigation distances and reduces logistics costs to some extent. At the same time, the safety of navigation in ice-covered areas has garnered continuous attention from scholars. China's polar research ship, "*Xuelong*", has ventured into the Arctic region multiple times, accumulating valuable navigation experience for Chinese ships and marking a milestone in China's maritime history [2].

The development and utilization of Arctic shipping routes have become focal points in maritime transport, with an increasing number of ships navigating the Northeast and Northwest Passages annually [3]. Due to the unique natural environment of Arctic waters, ships encounter various risks during navigation, making the safety of Arctic maritime navigation a worthy research topic. Sea ice is one of the critical factors affecting the safe navigation of ships [4]. Given the limited sea ice monitoring and prediction technology in the early stages, ship-ice collision accidents occurred frequently. Thus, initial risk assessments of Arctic navigation focused primarily on accident studies [3, 5, 6]. Mawuli Afenyo et al. utilized Bayesian networks to analyze the interrelationships among factors in ship-ice collision incidents [5]. Bushra Khan et al. [7] proposed and applied an object-oriented Bayesian network for the dynamic analysis of ship-ice collision risks in Arctic waters, validating the model's uniqueness with a case study of an oil tanker. Zhuang Li et al. developed a standard paradigm for the dynamic association (DA) effect of risk factors and used Bayesian networks to establish a risk assessment model for ship-ice collision accidents [8]. Bushra Khan et al. [9] created a dynamic Bayesian network model to assess the risk of ship-ice collisions and quantify environmental risks. Abayomi Obisesan et al. [10] proposed a conceptual framework for the performance characterization and quantitative risk assessment of iceberg-ship collisions, validated using a fault tree model for its applicability to ship-ice collision accidents. Mingyang Zhang et al. employed the Human Factor Analysis and Classification System (HFACS) model and a fault tree to qualitatively analyze the risk of ship-icebreaker collision incidents and formulated risk control options (RCOs) [11]. Shanshan Fu et al. combined fuzzy event trees and quantitative analysis results to address the uncertainty of event probabilities, using Monte Carlo simulation and fuzzy set theory to calculate event correlations, and validated the quantitative model with a case study of ship-ice entrapment accidents [6].

The aforementioned studies primarily utilized event trees, fault trees, and Bayesian networks as their main research models. The essence of these qualitative or quantitative analysis models references accident causation theories. However, their underlying logic is based on the unidirectional linear reasoning of accident causation. The overly simplified causal relationships and accident processes in accident causation models exclude many system factors and the indirect or non-linear interactions between events [12]. To move away from focusing solely on accidents and instead focus on the system itself, system models have become the mainstream paradigm in accident modeling [13] To adopt complex system safety as a qualitative analysis method, quantitative research tools based on system thinking are required. Petri nets are widely used in fields such as computer engineering and control systems, due to their unique features of concurrency, conflict management, synchronization, and resource sharing, which are of significant importance for system engineering modeling and analysis. Petri nets have also demonstrated good modeling capabilities in safety and risk analysis [14]. As a tool for risk and reliability analysis, Petri nets have gained widespread attention and application in recent years. For instance, Nývlt et al. applied stochastic Petri nets to model complex accident scenarios [15]. Their work not only accomplished the modeling and analysis of dynamic accident scenarios but also optimized the model for computational efficiency, thereby demonstrating the applicability of Stochastic Petri nets (SPN) in risk analysis. Similarly, Chao

Wu et al. utilized stochastic Petri nets to model and analyze train rear-end collision accidents [16], further validating the rationality and effectiveness of employing Petri nets for accident modeling. SPN, as an extension in the time series domain, have proven their applicability in risk assessment studies [17–19].

Considering the research challenges of complex systems and the navigational risks for ships in the Arctic, this paper conducts a risk assessment of the complex system of ship navigation in the Arctic under multi-factor coupling. It addresses the issue of risk control for polar navigation. This paper qualitatively models the Arctic ship navigation process and combines it with stochastic Petri nets to perform quantitative risk assessment and risk control. This approach provides a new systematic perspective for the study of navigation risks in Arctic waters.

2 Method and Data

2.1 SPN

The Stochastic Petri Net (SPN) is a category of Petri Net (PN) and an extension of PN in the temporal domain. It associates a random exponentially distributed time delay with each transition between its enabling and firing, thereby enhancing the model's capability for evolutionary analysis [20]. SPN can accurately describe the dynamic and probabilistic values of transitions by associating timed firing functions [21]. The SPN framework provides a robust set of components for describing the state transition mechanisms and event scheduling mechanisms of discrete event stochastic systems.

The Arctic ship navigation risk assessment system not only explores the risk components of the maritime traffic system from a systems theory perspective but also considers ship navigation risk as a process risk issue. Therefore, it also conforms to the Markov stochastic process. The ship navigation risk system moves forward along the time dimension, meaning the system's state varies at different times, and the state at a later time is influenced by the state at the previous time. This reflects the temporal dependency of the system's state. As an extended form of Petri Nets, SPN allows the introduction of randomness during state transitions and have asynchronous isomorphic properties. This means that state transitions in the network are asynchronous—transitions occurring in each state are not influenced by other states—but transitions between states are isomorphic, following the same rules and conditions. Based on SPN, a risk assessment model can be constructed, as shown in Fig. 1.

In Fig. 1, the threshold range of environmental data for the Northeast Passage from June to November is quite large, necessitating a focus on its impact on navigational risks for ships. Combined with the analytical evolution capabilities of the stochastic Petri net, a quantitative risk assessment model for ship navigation in the Arctic Northeast Passage can be developed. Figure 1 illustrates the control feedback process of this model, with the meanings of each node represented in Table 1.

In the constructed stochastic Petri net model for the polar Northeast Passage, the processed transition rate values are input. The environment-related transition rate values are derived by normalizing the environmental data and using the median as the input value. To avoid model deadlock situations when the transition rate value is zero and to minimize the impact on overall navigational risk, transition rate values less than or equal

Ship Navigation Risk Assessment 257

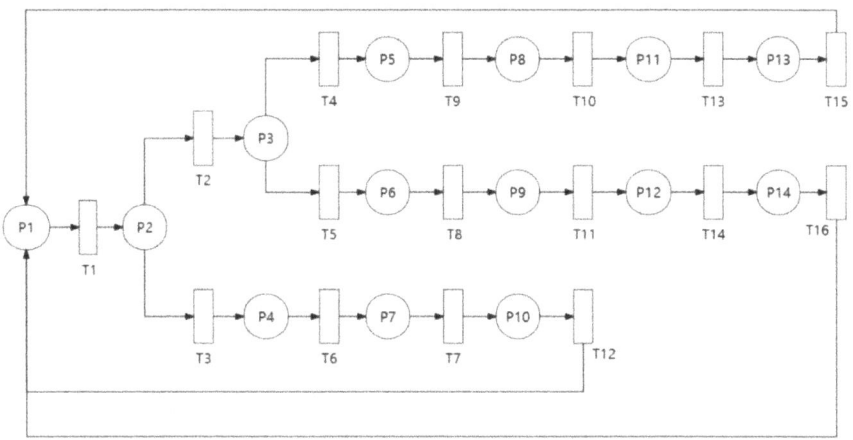

Fig. 1. SPN model for ship navigation risk in Arctic

Table 1. Meanings of places and transitions

Place	Meaning	Transition	Meaning
P_1	Risk of ship navigation	T_1	Risk information feedback
P_2	Captain	T_2	Give instructions to the mate
P_3	Mate	T_3	Give instructions to the engine department
P_4	Engine Department	T_4	Monitoring temperature
P_5	Temperature	T_5	Give command to helmsman change course
P_6	Helmsman	T_6	Give command to engineer to change speed
P_7	Engineer	T_7	Control marine propulsion system
P_8	Ice concentration	T_8	Monitor compass
P_9	Compass	T_9	Monitor ice concentration
P_{10}	Marine propulsion system	T_{10}	Monitor wind speed
P_{11}	Wind speed	T_{11}	Operate rudder
P_{12}	Helm	T_{12}	Control ship speed
P_{13}	Wave height	T_{13}	Monitor wave height
P_{14}	Ship route	T_{14}	Design course
		T_{15}	Environmental impact
		T_{16}	Control course

to zero are set to 1. Other node transition rate values are collected through expert questionnaires, gathering subjective evaluation data based on the expert evaluation method to obtain node parameters. It is assumed that when a ship is navigating a certain segment, it is only affected by changes in external environmental disturbances in different months, while other factors remain constant with minor variations. Therefore, the transition rates of environmental nodes change over time.

When the firing times of transitions followed an exponential distribution and markings were countable, the model could be isomorphic to a continuous-time Markov chain. Each state marking of the SPN could be mapped to the Markov state, indicating that the reachability graph of the SPN was isomorphic to the state space of the Markov chain. Therefore, if directed arcs represented the dynamic transitions between different markings or states, an equivalent isomorphic Markov chain could be obtained from the SPN mode. The evolution process of the system state was analysed based on the relevant theorems of the steady-state distribution of the Markov chain, as shown in Fig. 2.

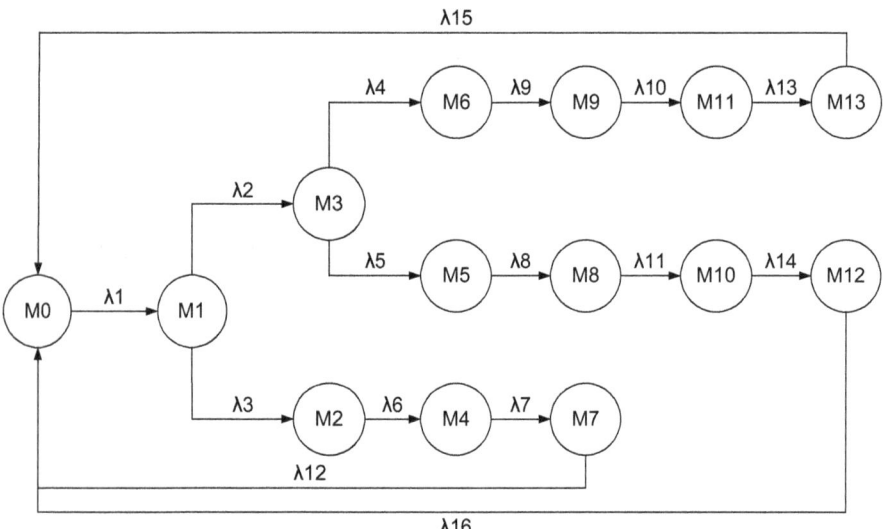

Fig. 2. The Markov chain of the SPN model

By defining the firing rate for each arc transition in the directed graph of the stochastic Petri net model, the following transition firing rate matrix can be obtained. In continuous-time SPN, there is a certain delay time required for a transition to be enabled and then actually fired. This delay time is considered a continuous random variable that follows an exponential distribution. $M0, M1, \ldots M13$ is the 14 marks or states in the SPN model of Arctic navigation risk evolution.

$M_0 = (1,0,0,0,0,0,0,0,0,0,0,0,0,0)$; $M_7 = (0,1,0,0,0,0,0,0,0,0,0,0,0,0)$;
$M_1 = (0,0,0,0,0,1,0,0,0,0,0,0,0,0)$; $M_8 = (0,0,0,0,0,0,0,0,0,0,0,0,0,1)$;
$M_2 = (0,0,0,0,0,0,0,1,0,0,0,0,0,0)$; $M_9 = (0,0,0,0,0,0,0,0,0,0,0,0,1,0)$;
$M_3 = (0,0,0,0,0,0,0,1,0,0,0,0,0,0)$; $M_{10} = (0,0,0,1,0,0,0,0,0,0,0,0,0,0)$;
$M_4 = (0,0,0,0,0,0,0,0,0,0,1,0,0)$; $M_{11} = (0,0,1,0,0,0,0,0,0,0,0,0,0,0)$;

$M_5 = (0,0,0,0,0,0,0,0,0,1,0,0,0)$; $M_{12} = (0,0,0,0,0,1,0,0,0,0,0,0,0)$;
$M_6 = (0,0,0,0,0,0,0,0,0,1,0,0,0)$; $M_{13} = (0,0,0,0,1,0,0,0,0,0,0,0,0)$;

To measure the dynamic risk during ship navigation in Arctic waters, this paper, conducts dynamic simulation analysis of ship navigation in Arctic waters based on environmental data, using Petri net theory and relevant theorems of the steady-state distribution of Markov chains. The steady-state probabilities of various key nodes throughout the entire process of ship navigation in the Northeast and Northwest Passages of the Arctic are obtained. The simulation process represents that the initial state of the place, $M0$, triggers a series of transitions, causing tokens in the place to transfer to the next place according to the direction of the arcs, until the final place is reached. The state changes in the system are denoted by M. Starting from the current state marking $M0$, a reachable set of markings for different trigger transitions can be obtained. Let $P(Mi)$ be the probability of each risk node Mi that may be encountered during the ship navigation process. The system input-output function satisfies the math expression (1) and (2):

$$\begin{cases} \lambda_{12}P(M_7)+\lambda_{15}P(M_{13})+\lambda_{16}P(M_{12})=\lambda_1 P(M_0) \\ \lambda_1 P(M_0)=\lambda_2 P(M_1)+\lambda_3 P(M_1) \\ \lambda_2 P(M_1)=\lambda_5 P(M_3)+\lambda_4 P(M_3) \\ \lambda_5 P(M_3)=\lambda_8 P(M_5) \\ \lambda_8 P(M_5)=\lambda_1 P(M_8) \\ \lambda_{11}P(M_8)=\lambda_{14}P(M_{10}) \\ \lambda_{14}P(M_{10})=\lambda_{16}P(M_{12}) \end{cases} \quad (1)$$

$$\begin{cases} \lambda_4 P(M_3)=\lambda_9 P(M_6) \\ \lambda_9 P(M_6)=\lambda_{10}P(M_9) \\ \lambda_{10}P(M_9)=\lambda_{13}P(M_{11}) \\ \lambda_{13}P(M_{11})=\lambda_{15}P(M_{13}) \\ \lambda_3 P(M_1)=\lambda_6 P(M_2) \\ \lambda_6 P(M_2)=\lambda_7 P(M_4) \\ \lambda_7 P(M_4)=\lambda_{12}P(M_7) \end{cases} \quad (2)$$

Based on the distribution of ship navigation environment node data in the Northeast Passage, the median method is used to extract the raw data of environmental risk indicators for different months. These are then converted into transition rate values corresponding to the indicator impacts in the stochastic Petri net model. By having multiple experts evaluate and score the impact of ship facilities and crew on the navigational risks in the Northeast Passage, the transition rates for the remaining nodes can be obtained. By changing the transition rates of the environmental nodes for different months, the comprehensive navigational risk values for the Northeast Passage from June to November can be determined.

2.2 Data Processing

Before pre-processing the data on wind, waves, temperature, and sea ice concentration, a preliminary understanding of the central tendency and distribution range of the objective environmental data was conducted. The data sources and accuracy are detailed in Table 2.

Table 2. Arctic environment data source

Item	Casual event	Data type	Means of evidence
Environment	Ice concentration	Objective Data	Climate Data Store:*https://cds.climate.copernicus.eu/cdsapp#!/search?type = dataset*
	Wind speed	Objective Data	
	Temperature	Objective Data	
	Wave height	Objective Data	

The obtained NC data for the four variables were interpolated and normalized using some defined formulas, and the polar coordinate projection visualization was completed. The visualizations of 2 m temperature above sea level, sea ice concentration, 10 m wind speed above sea level, and significant wave height are presented according to the real scenarios.

When pre-processing objective environmental data, it is necessary to consider inconsistencies in data accuracy, data types, and data units. Therefore, normalization algorithms and interpolation algorithms need to be applied to ensure that the data are complete and have uniform dimensions.

3 Case Study

3.1 Scenario Description

The Northeast Passage refers to a set of sea routes starting from the North Cape of Norway in northwestern Europe, passing along the northern coast of the Eurasian continent and Siberia, and then through the Bering Strait to the Pacific Ocean.

The Northeast Passage, located near the Arctic Circle, has a high latitude with extremely low temperatures during the year, making the analysis of ship navigation safety crucial. This passage is a major ship route in Arctic waters, which can be divided into the five segments: the Barents Sea, the Kara Sea, the Laptev Sea, the East Siberian Sea, and the Chukchi Sea. Figure 3 shows the study area of this work, which focuses on the waters of the five segments of the Northeast Passage.

As shown in Fig. 4, the summer temperatures in the Arctic Northeast Passage region are relatively stable, while the autumn temperatures begin to gradually decrease. According to meteorological data, the average temperature in this region during June to August is approximately 0 °C to 5 °C. Entering autumn, temperatures experience a steep drop,

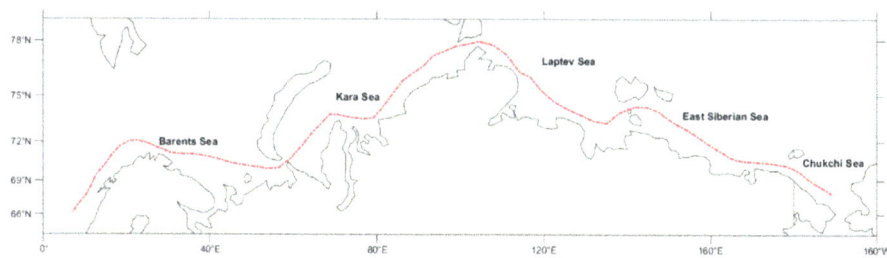

Fig. 3. Route of the Arctic northeast passage

with November being the coldest period, where temperatures fall to around -15 °C. The sea ice concentration reaches its annual low point of about 0% between July and September, then begins to gradually increase. In the central region of the Northeast Passage, influenced by geographical location and atmospheric circulation, these changes exhibit obvious regional and seasonal characteristics. The average wind speed in summer reaches about 6 m/s, and in autumn, it approaches 7 m/s. Due to the presence of considerable floating ice in the Arctic waters during this period, the wave height is relatively low. In the Northeast Passage, both summer and autumn have relatively small and stable wave heights, ranging between approximately 0.5 to 1 m.

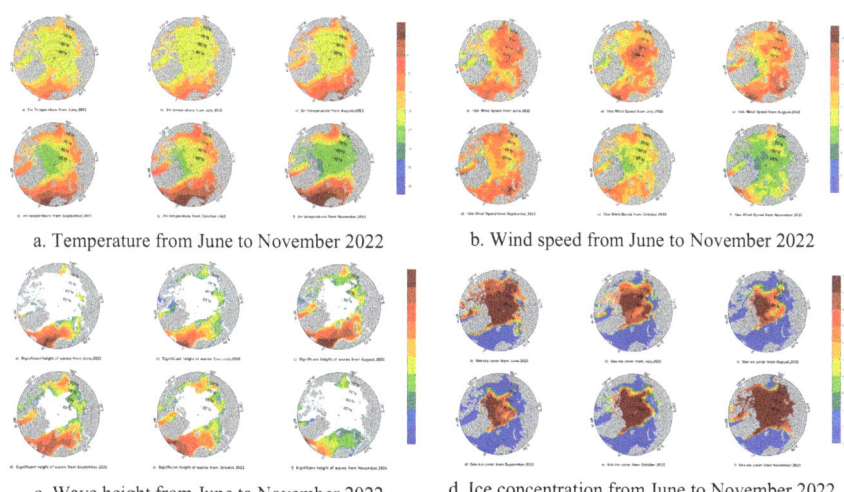

a. Temperature from June to November 2022   b. Wind speed from June to November 2022

c. Wave height from June to November 2022   d. Ice concentration from June to November 2022

Fig. 4. Visualization of Arctic environmental data

3.2 Risk Assessment of Ship Navigation in Arctic Waters

The simulation results, as shown in Fig. 5, indicate that the steady-state probability distribution trends of the nodes across the five navigation segments are generally similar, with the sea ice factor node having a relatively greater impact on navigational risk compared to other factors.

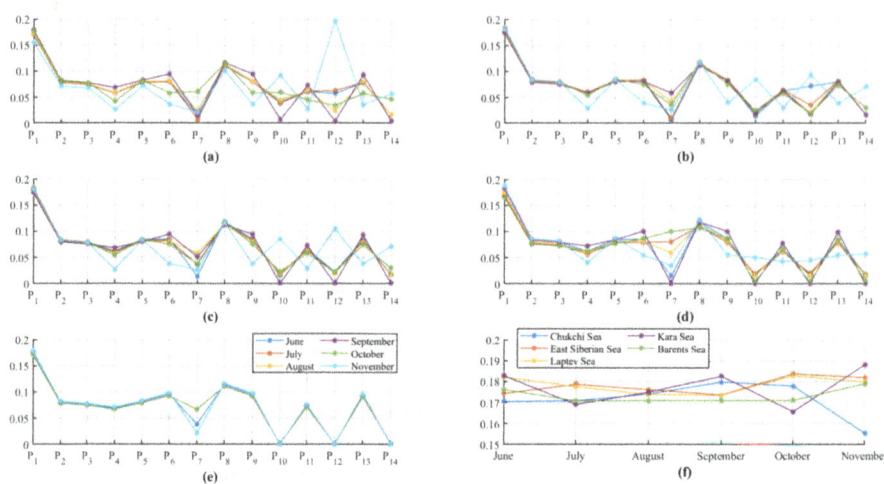

Fig. 5. SPN steady-state probability variation chart

Based on the risk assessment results, risk control standards can be established: when the comprehensive risk value is below 0.17, the navigational risk level is considered low; when the comprehensive risk value is between 0.17 and 0.18, the navigational risk level is considered moderate; and when the comprehensive risk value is above 0.18, the navigational risk level is considered high. Therefore, when ships navigate the Arctic Northeast Passage, the navigational risk level in August and September is low, with low sea ice coverage, allowing for free navigation. In July and October, the navigational risk level is moderate, and there is sea ice coverage in some navigation areas, so icebreaker assistance can be chosen as needed. In June and November, the navigational risk is high, with most waters covered by sea ice, and it is recommended to use icebreaker assistance for navigation.

When ships navigate the Northeast Passage, the comprehensive risk values for the five segments from June to November are shown in Table 3. The overall trend of comprehensive risk shows that the risk is lower in summer and relatively higher in spring and autumn, with the highest risk observed in June and November. This indicates that the navigational risk in the polar Northeast Passage exhibits seasonal variation characteristics, significantly influenced by changes in sea ice. The steady-state probability of crew-related nodes may change with variations in their environment and other additional physical and mental factors, also impacting navigational risk to some extent.

Table 3. Multi-segment combined risk result from June to November

Sea\Month	6	7	8	9	10	11
Chukchi Sea	0.17047	0.17087	0.17404	0.17976	0.17789	0.15531
East Siberian Sea	0.17434	0.17873	0.1762	0.1737	0.18377	0.18191
Laptev Sea	0.18217	0.1776	0.17386	0.17352	0.18283	0.17984
Kara Sea	0.18291	0.16916	0.17496	0.1826	0.16572	0.1881
Barents Sea	0.17617	0.17085	0.17099	0.17103	0.17106	0.17898

4 Conclusion

This paper considers the risk assessment of ship navigation from the perspective of complex systems, establishing a quantitative SPN assessment model. It evaluates the dynamic risks of ships navigating the Northeast Passage in polar waters from June to November and provides corresponding risk control measures. Firstly, the ship navigation system in polar waters is analyzed, considering risks from the perspectives of environment, human factors, ships, and management. Environmental big data is visualized to intuitively display regional characteristics. By integrating objective environmental data and expert opinions, it is found that sea ice concentration significantly affects navigation across the five segments of the Northeast Passage.

Risk studies on simple systems, such as human-machine-environment interactions, are relatively common. However, the actual relationships between maritime traffic risk factors are complex and variable, with varying degrees of mutual influence between factors. Analyzing these with a simple system approach can easily overlook the interdependencies of factors at different levels. Compared to other waters, polar regions present even more complex and variable environments. Traditional simple system safety theories and models struggle to accurately represent the risk landscape of ship navigation in such waters and tend to overlook the dynamic characteristics of the system. Therefore, this paper employs system safety theory to elucidate the interrelationships and state change trends of internal and external factors within the navigation system of ships. This approach helps to clarify the functions of various stakeholders within the polar navigation system and delineate their roles during risk events. In order to explore new navigation risk control options in Arctic waters, in-depth research on navigation modes in Arctic waters will be conducted in the future to assess navigation risks under different modes, so as to obtain risk control options for different navigation modes, which will help improve the safety of shipping navigation in Arctic waters under different navigation modes.

References

1. Changqing, K., et al. (2020), Comparison of Antractic and Arctic sea ice variations and their impact factors. Chin. J. Polar Res. 32(1):1–12. (in Chinese)

2. Bing, Z., Y. Dongfang, and X. Yinyue (2022), Practice voyage and exploration of the Northwest Passage by the R/V Xuelong in Arctic. Chin. J. Polar Res. 34(1):101–113. (in Chinese)
3. Kum, S., Sahin, B.: A root cause analysis for Arctic Marine accidents from 1993 to 2011. Saf. Sci. **74**, 206–220 (2015)
4. An, L., et al.: Research on navigation risk of the Arctic Northeast Passage based on POLARIS. J. Navig. **75**(2), 455–475 (2022)
5. Afenyo, M., et al.: Arctic shipping accident scenario analysis using Bayesian Network approach. Ocean Eng. **133**, 224–230 (2017)
6. Fu, S., et al.: A quantitative approach for risk assessment of a ship stuck in ice in Arctic waters. Saf. Sci. **107**, 145–154 (2018)
7. Khan, B., et al.: An operational risk analysis tool to analyze marine transportation in Arctic waters. Reliab. Eng. Syst. Saf. **169**, 485–502 (2018)
8. Li, Z., et al.: Risk reasoning from factor correlation of maritime traffic under arctic sea ice status association with a Bayesian belief network. Sustainability. **13**(1), 147 (2020)
9. Khan, B., Khan, F., Veitch, B.: A Dynamic Bayesian Network model for ship-ice collision risk in the Arctic waters. Saf. Sci. **130**, 104858 (2020)
10. Obisesan, A., Sriramula, S.: Efficient response modelling for performance characterisation and risk assessment of ship-iceberg collisions. Appl. Ocean Res. **74**, 127–141 (2018)
11. Zhang, M., et al.: Use of HFACS and fault tree model for collision risk factors analysis of icebreaker assistance in ice-covered waters. Saf. Sci. **111**, 128–143 (2019)
12. Leveson, N.G.: Applying systems thinking to analyze and learn from events. Saf. Sci. **49**(1), 55–64 (2011)
13. Zhang, Y., et al.: Systems theoretic accident model and process (STAMP): A literature review. Saf. Sci. **152**, 105596 (2022)
14. Taleb-Berrouane, M., Khan, F., Amyotte, P.: Bayesian Stochastic Petri Nets (BSPN)—A new modelling tool for dynamic safety and reliability analysis. Reliab. Eng. Syst. Saf. **193**, 106587 (2020)
15. Nývlt, O., Haugen, S., Ferkl, L.: Complex accident scenarios modelled and analysed by Stochastic Petri Nets. Reliab. Eng. Syst. Saf. **142**, 539–555 (2015)
16. Wu, C., et al.: Modeling and analysis of train rear-end collision accidents based on Stochastic Petri Nets. Math. Probl. Eng. **2015** (2015)
17. Liu, Z.Y., Jia, L.C., Dong, S.H.: Refined oil loading and unloading process risk assessment using Stochastic Colored Petri Nets integrated with risk factors. Teh. Vjesn. **31**(1), 70–78 (2024)
18. Kamil, M.Z., et al.: Dynamic domino effect risk assessment using Petri-nets. Process Saf. Environ. Prot. **124**, 308–316 (2019)
19. Ghazel, M.: Using Stochastic Petri Nets for level-crossing collision risk assessment. IEEE Trans. Intell. Transp. Syst. **10**(4), 668–677 (2009)
20. David, R., Alla, H.: Petri nets for modeling of dynamic systems: A survey. Automatica. **30**(2), 175–202 (1994)
21. Haas, P.: Stochastic Petri Net Models: Modeling and Simulation. Springer, New York (2002)

Open Access This chapter is licensed under the terms of the Creative Commons Attribution-NonCommercial-NoDerivatives 4.0 International License (http://creativecommons.org/licenses/by-nc-nd/4.0/), which permits any noncommercial use, sharing, distribution and reproduction in any medium or format, as long as you give appropriate credit to the original author(s) and the source, provide a link to the Creative Commons license and indicate if you modified the licensed material. You do not have permission under this license to share adapted material derived from this chapter or parts of it.

The images or other third party material in this chapter are included in the chapter's Creative Commons license, unless indicated otherwise in a credit line to the material. If material is not included in the chapter's Creative Commons license and your intended use is not permitted by statutory regulation or exceeds the permitted use, you will need to obtain permission directly from the copyright holder.

Preparation of the Super-sophobic Oil Surface on Steel Substrate by Composite Electrodeposition

Hong Cheng(✉), Dongliang Gu, and Yong Zhang

Dalian Naval Academy, Dalian 116018, China
847376891@qq.com

Abstract. The goal is to propose a new method for preparing super-sophobic oil Cu-Ni-nSiO$_2$ composite coating on a Q235 steel surface and to avoid damaging the surface of the base steel physical properties at the same time. Through composite electroplating and liquid deposition method, the appropriate binary micro-nano rough structures on the substrate are constructed on the surface of steel base. The preparation process parameters of the super-sophobic oil surface on steel substrate are obtained by changing the composite electroplating parameters. Finally, When the current density is 13 A/dm^2, the speed of magnetic fluid mixing is 200 r/min, the electroplating time is 35 min, the temperature is 65 °C, the super-sophobic oil surface can be prepared with glycerin, diethylene glycol, Peanut oil contact angle of 150°, 140°, 135°.

Keywords: Super-sophobic Oil · Composite Electrodeposition · Preparation Process Parameter

1 Introduction

The "lotted leaf effect" in the nature is the most common superhydrophobic surface. The self-cleaning property of superhydrophobic, anti-pollution and water friction resistance are widely excellent. The concept of super-sophobic oil surface is proposed on the basis of superhydrophobic surface. The contact angle between oily liquid and solid surface is greater than 150° [1]. It can solve the problems of ship pollution, micro flow control and coexistence of oil and water. The super-sophobic oil surface can be applied to the tank wall of ships and other marine equipment or the surface of energy transportation pipeline, which can reduce the corrosion of oil pollutants and the adhesion of microorganisms. Meanwhile, it can be used on energy waste and self-cleaning and environmental pollution; In addition, the anti-viscous and self-cleaning properties of ultra-phobic oil on oil-water separation have a bright prospect in the cleaning of offshore oil leakage [2].

The wettability of solid surfaces is usually expressed by static contact angles. The contact angle refers to the angle from solid-liquid interface to gas-liquid interface at the junction of solid, liquid and gas. The wettability of solid surfaces depends mainly on the microstructure and surface free energy of the surface (surface free energy is related to the chemical composition of the surface). So, the key to the construction of Super-sophobic oil surface: one is to construct micro-nano-rough structure on solid surface, and the other is to modify micro-nano-rough structure with low surface energy for super-sophobic oil surface [3]. The micro-nano-rough structure is more demanded on the ultra-hydrophobic surface, and the low surface energy substance is more superior [4]. At present, the super-sophobic oil surface can be successfully prepared by introducing fluorine-containing low-energy materials and constructing the microstructure of the surface. The preparation methods of super-sophobic oil have electrochemical method, template method, plasma etching method and spraying method [5].

In this paper, the super-sophobic oil surface is prepared by composite electrodeposition. The copper-nickel ions and SiO_2 nanoparticles are deposited together to construct rough structures on the surface of the Q235 steel, then these structures are modified by perfluorooctanoic silane to reduce the surface energy. This method of composite electrodeposition is easy to operate and can construct large area of super-sophobic oil surface, but its disadvantages are poor economy, easily damaged and short life.

2 Mechanism of Preparation of Super-sophobic Oil Surface on Steel Substrate

2.1 Composite Electrodeposition Method for Constructing Micro-Nano-Rough Structure

The hydrophobicity of super-sophobic oil surface is closely related to the microstructure of the material surface. Micro-nanometer structure on steel-based surface is the key to ultra-oleophobic. A uniform and stable metal surface with microcosmic double rough geometry is constructed on the surface of steel substrate by composite electrodeposition technology. Copper-nickel metal layer is plated on the surface of steel and SiO_2 nanoparticles are added simultaneously to form binary micro-nanometer double rough metal surface by composite electrodeposition technology, which can meet the requirement of ultra-oil-phobic. Copper-nickel ions and silica particles are combined and deposited on the surface of steel base to meet the requirements of oleophobic formation. With the excellent performance of SiO_2 nanoparticles, it can meet the advantages of abrasion resistance and corrosion resistance of super-sophobic oil surface. The composition of plating solution, temperature of plating solution, electroplating current density, electroplating time and particle size and content of nanoparticles SiO_2 during composite electrodeposition have significant influence on the preparation of super-sophobic oil surface on steel surface. The relevant factors described above are the keys to the construction of micro-nano rough structures on steel based surfaces.

2.2 Modification of Low Surface Energy Substances

The realization of super-sophobic oil surface on steel substrate is not only related to the microstructure of nanometer structure on steel surface, but also with the chemical composition of micro-nanostructure surface. The surface energy of the steel base surface is reduced and the oleophobic property of the steel base surface is increased by modification of the corresponding chemical substances. Hydrophobicity of nanometer rough structure is with modification of low surface energy substances. In this paper, the low surface energy material of modified super-sophobic oil surface is perfluorooctanoic silane. Perfluorooctane is modified with steel based surface material by corresponding chemical ways to improve its oleophobic properties. The carbon chain is used as the framework and the carbon chain is encapsulated by fluorine atoms, which makes the chemical property stable. Meanwhile, its fluorane group has lower surface energy, the fluoroalkyl group is bound to the coating material on the steel based surface to improve the oleophobic properties of the micro and nano structures on the steel base surface. The perfluorooctanoic acid was modified by liquid phase deposition method onto the nanometer coarse substrate on the surface of steel substrate to reduce its surface energy, The liquid phase deposition method can better bind the perfluorooctanoic acid to the steel substrate surface to obtain a more uniform surface, thus achieving super oil-phobic effect.

3 Mechanism of Preparation of Steel-based Super-Oil-phobic Surface

3.1 The Design Ideas of Experiments

In this paper, the preparation process of Cu-Ni-nSiO$_2$ super-sophobic oil surface on steel substrate by composite electrodeposition was studied. Because of the complexity of composite electrodeposition technology, the advantages and disadvantages of electroplating surface properties are related to the composition of plating solution, temperature of plating solution, electroplating time, electroplating current density and electroplating stirring speed. Considering the limited experimental conditions and the errors that may be caused by the experimental operation process, this paper keeps the composition of the plating solution and the stirring speed of the plating solution and the particle size of the nanoparticle SiO$_2$ unchanged, and changes the electroplating temperature, plating time and current density. The optimum technological parameters for preparing super-sophobic oil surface on steel substrate are obtained through experiments.

3.2 Preparation of Experimental Solution

The particle size of nanoparticles SiO$_2$ in composite electrodeposition is 30 nm and the stirring speed of electroplating solution is 200 r/min. This deposition of nanoparticles SiO$_2$ and copper-nickel ions is favorable to ensure the uniform stability of the coating, and at the same time exerting the unique properties of nanoparticles SiO$_2$. The chemical drugs used during the experiment are analytically pure. The following is the formulation of the test step solution as shown in Table 1.

The anode selected in this experiment is pure nickel electrode, the size is 60 × 80 × 5 mm, the cathode is Q235 steel, the size is 40 × 30 × 5 mm. The modified solution of low surface energy substance is perfluorooctanoic acid silane ethanol solution, whose solution is 25 g/L perfluorooctanoic acid silane ethanol solution.

Table 1. The formula of solution required each step process

Name	Including	Content(g/L)
Electric clean liquid	NaOH	15
	Na_2SiO_3	15
	Na_2CO_3	30
	Na_3PO_4	75
Activated liquid	HCl(12 mol/L)	340
Composite plating solution	$NiSO_4$	60
	$CuSO_4$	10
	$NiCl_2$	40
	H_3BO_3	35
	$C_{12}H_{25}SO_4Na$	0.5
	SiO_2	8
Low Surface Energy Liquid	C_2H_5OH	1000
	$C_8H_7F_{13}Si$	25

3.3 Procedure of Experiment

The experimental procedure for preparing Cu-Ni-nSiO$_2$ super-sophobic oil surface on steel substrate by composite electrodeposition includes three steps: pretreatment of steel-based surface, composite electrodeposition process and modification of low surface energy substance. Steel base surface pretreatment is divided into sand paper mechanical polishing, acid derusting, electric clean oil removal and activation treatment. Clean steel surface can be obtained by pretreatment. Subsequently, composite electrodeposition was carried out to construct micro-nano-rough structure, and finally modified with low surface energy substance perfluorooctanoic acid to obtain the super-sophobic oil surface. The modification of low surface energy substances is divided into two steps: one is heated in ethanol solution of perfluorooctanoic acid for four hours at a constant temperature of 65 °C; Secondly, the modified steel base surface is heated and cured for four hours at a constant temperature of 150 °C. Experimental procedures: mechanical polishing→acid rust removal→electric clean oil removal→activation treatment→electroplating working layer→low surface energy material modification.

3.4 The Data of Experiment

In the experiment, keep the composition of electroplating solution and the stirring speed unchanged, change the electroplating current density, electroplating temperature and electroplating time, then measure the average contact angle between the surface of steel base and oil by using the quantity angle method. The selected oil is glycerol, diethylene glycol and peanut oil. The below is the data of experiment in Table 2.

Table 2. Experiment data.

Numner	Stirring Speed (r/min)	Current Density (A/dm^2)	Plating Temperatur (°C)	Plating Time (min)	Contact angle of Glycerol (°)	Contact angle of Diethylene Glycol (°)	Contact angle of Peanut Oil (°)
1	200	11	55	35	120	109	100
2	200	11	55	55	118	105	98
3	200	11	65	35	131	113	105
4	200	11	65	55	128	110	101
5	200	11	75	35	122	105	100
6	200	11	75	55	117	100	96
7	200	13	55	35	128	120	115
8	200	13	55	55	125	118	110
9	200	13	65	35	150	140	135
10	200	13	65	55	140	132	125
11	200	13	75	35	126	120	115
12	200	13	75	55	123	118	110
13	200	14.5	55	35	122	115	108
14	200	14.5	55	55	117	110	105
15	200	14.5	65	35	135	127	118
16	200	14.5	65	55	130	122	116
17	200	14.5	75	35	125	116	110
18	200	14.5	75	55	120	112	106

3.5 The Treatment and Optimization of Experimental Results

For further analysis of the experimental data obtained, it can be concluded that the electroplating time has great influence on the super-sophobic oil surface. Compared with the electroplating time of 55 min, the electroplating time of 35 min is better for the super-sophobic oil surface. The electroplating time will affect the micro-nano-rough structure of the coating, the longer the electroplating time will reduce surface roughness. At the same time, electroplating current density is also an important factor in the construction of micro-nano-rough structure surface. Suitable current density is beneficial to the construction of micro-nano-rough structure. The low current density or too large current density not only affects the processing efficiency, but also the microstructure of nanometer structure is poor, which cannot meet the requirement of super-sophobic oil surface. Finally, the electroplating temperature will also affect the effect of super-sophobic oil surface Suitable electroplating temperature is beneficial to the deposition of nanoparticles SiO2 and the deposition of copper-nickel metal ions. The electroplating temperature is too low or too high, which will cause the unevenness of the plating layer and the phenomenon of dendrite and peeling of the plating layer, which is unfavorable for the formation of ultra-oleophobic nanometer coarse structure.

The optimum parameters for preparing Cu-Ni-nSiO$_2$ super-sophobic oil surface can be obtained from Table 2 for this paper. Electroplating time is 35 min, current density is 13 A/dm^2 and electroplating temperature is 65 °C. The contact angle with glycerin, diethylene glycol and peanut oil is 150°, 140° and 135°. As the wettability of the super-sophobic oil surface under these parameters is better in this paper.

4 The Analysis of Micro Surface Structure

Under the optimum technological conditions, the prepared super-sophobic oil surface is uniform and the oil-phobic property is good. The contact angle measured by the measuring angle method is shown in Fig. 1, Fig. 2, Fig. 3. The microstructure of micro-nano-roughness is analyzed by SEM under the optimum process parameters.

Fig. 1. The contact angle with glycerin (150°).

Fig. 2. The contact angle with diethylene glycol (140°).

From the microstructure, as shown in Fig. 4, when observed under SEM, the surface of the coating is very rough and uneven, forming many sizes of heteroplasts.

These processes formed under the composite electrodeposition condition satisfy the requirements of micro-nano-rough structure of super-sophobic oil surface. The reasons for the formation of these microstructures are as follows: (1) Metallic copper-nickel ions

Fig. 3. The contact angle with Peanut oil (135°).

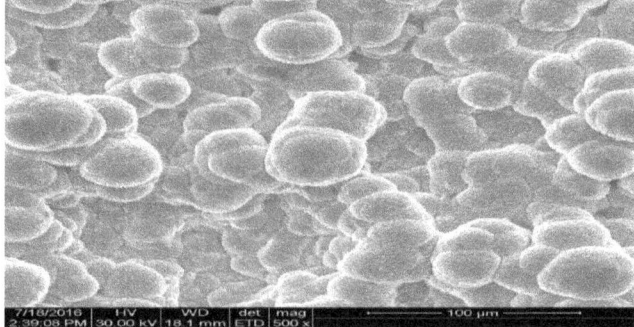

Fig. 4. The SEM images of Cu- Ni-SiO2 composite plating on surface

and nano-SiO_2 particles are deposited on the steel-based surface under the composite electrodeposition conditions. However, due to the nano-SiO_2 particles, the crystal grains of nickel atoms in the deposition process are more refined, reaching the micro-nanometer level. The papillary structure similar to the "lotus leaf effect" is formed; (2) In the composite electrodeposition process, due to the change of copper-nickel ion concentration in the electroplating solution and cathodic polarization, the nucleation velocity of the crystal on the surface of steel substrate is larger than the growth rate of crystal nuclei, forming different sizes of papillary structures. These papilla structures are also nanoscale (3) Although gaps exist between various papillary structures, however, the plating inside is relatively tight and not easy to fall off, enhanced corrosion resistance. In conclusion, the complex nano-rough structure formed by metal nickel ions and nano-SiO_2 particles is necessary to meet the requirements of super-sophobic oil surface.

5 Conclusion

In this paper, the Cu-Ni-$nSiO_2$ super-sophobic oil surface is constructed without destroying the steel-based structure, and the optimum parameters for preparing super-sophobic oil surface on steel substrate are investigated. The composite electrodeposited super-sophobic oil surface of steel base is not only economical and stable, but also can be prepared in large area by electroplating technology. Specific conclusions are as follows:

(1) In the process of composite electrodeposition to prepare super-sophobic oil surface, the micro-nano-rough structure satisfying super-sophobic oil surface can be constructed by controlling electroplating current density, electroplating time and electroplating temperature.
(2) The selection of low surface energy substances and the method of low surface energy modification are also the keys to preparing the super-sophobic oil surface on steel substrate.
(3) During the experiment, the composition of electroplating solution is stable and the magnetic stirring speed remains unchanged. Meanwhile, pay attention to the distance of the anodes and try best to reduce the influence of other irrelevant factors on the construction of super-sophobic oil surface.
(4) It is necessary to further improve the hydrophobic properties of steel-based super-sophobic oil, and further satisfy the effect of super-sophobic oil on various oils.

References

1. Cao, I., Gao, D.: Design and fabrication for microtextures for inducing a superhydrophobic behavior hydrophilic materials. Langmuir. **23**(8), 4310 (2007)
2. Savoy, E.S., Escobedo, F.A.: Molecular simulations of wetting of a rough surface by an oily fluid: effect of topology, chemistry, and droplet size on wetting transition rates. Langmuir. **28**(7), 3412–3419 (2012)
3. Zhang, G., et al.: Robust superamphiphobic coatings based on silica particles bearing bifunctional random copolymers. ACS Appl. Mater. Interfaces. **5**(24), 13466–13477 (2013)
4. Buist I., et al.: Harding surfactants to contract and thicken oil spills in pack ice for in situ burning. Cold Reg. Sci. Technol., 2011, 67 (1/2): 3–23
5. Tian, Y., Liu, H., Deng, Z.: Electrochemical growth of gold pyramidal nanostructures: Toward super-amphiphobic surfaces. Chem. Mater. **18**, 5820–5822 (2006)

Open Access This chapter is licensed under the terms of the Creative Commons Attribution-NonCommercial-NoDerivatives 4.0 International License (http://creativecommons.org/licenses/by-nc-nd/4.0/), which permits any noncommercial use, sharing, distribution and reproduction in any medium or format, as long as you give appropriate credit to the original author(s) and the source, provide a link to the Creative Commons license and indicate if you modified the licensed material. You do not have permission under this license to share adapted material derived from this chapter or parts of it.

The images or other third party material in this chapter are included in the chapter's Creative Commons license, unless indicated otherwise in a credit line to the material. If material is not included in the chapter's Creative Commons license and your intended use is not permitted by statutory regulation or exceeds the permitted use, you will need to obtain permission directly from the copyright holder.

Assessment of Dynamic Maritime Navigation Risks During Cold Surges in Bohai Bay and Northern Yellow Sea: An Integration of Atmospheric and Wave Numerical Models

Haoyu Chen and Chen Chen(✉)

School of Navigation, Wuhan University of Technology, Wuhan, China
cc198895@whut.edu.cn

Abstract. Cold surges can significantly impact maritime transportation safety due to strong winds, high waves, significant temperature drops, and dense fog. Therefore, it is crucial to assess the risk to ships during cold surges. This study integrates meteorological and wave numerical models with risk assessment methods to dynamically assess the navigation risks of ships in the Bohai Bay and the northern Yellow Sea during two consecutive cold surges. First, a sensitivity analysis of six sets of Planetary Boundary Layer (PBL) parameterization schemes was conducted using the Weather Research and Forecasting (WRF) atmospheric numerical model to obtain high-resolution and accurate meteorological variables. Subsequently, the wave field during the cold surge was simulated using the WAVEWATCH III wave model. The wind speed, visibility, temperature, wave height, and wave steepness output by the numerical models were then used as assessment factors and categorized into risk levels. The weights of the assessment factors were calculated using the Analytic Hierarchy Process (AHP) method. Finally, a fuzzy risk assessment matrix was established by evaluating the membership of the assessment factors at different risk levels, and a risk distribution map for navigation risk assessment was generated. The risk assessment model that utilizes numerical model results as input can effectively demonstrate the spatiotemporal dynamic evolution of risk during cold surges.

Keywords: Maritime Transportation Safety · Atmospheric and Wave Numerical Models · Risk Assessment · PBL Parameterization Schemes · AHP

1 Introduction

Maritime navigation safety has always been highly valued worldwide. However, due to human, maritime environment, traffic, and other factors, ship navigation safety has consistently faced significant uncertainty. Especially in rough sea conditions, where wind, waves, currents, and fog are the main environmental factors that affect ship navigation, and thus are also the main risk factors selected for assessing navigation risks. Wang et al. [1] conducted a safety assessment of shipping routes in the South China Sea in different seasons based on the AHP method. The results show that the risks in the South China Sea in spring and winter were higher than those in summer and autumn. Based on the meteorological and oceanographic data obtained from NOAA (National Oceanic and Atmospheric Administration) and satellites, Du et al. [2] determined the weights of different risk factors using the AHP-sequential weighted average operator method and utilized GIS (geographic information system) technology to perform navigation risk assessment. The dynamic risk assessment results can intuitively display the evolution process of navigation risks.

Therefore, integrating ship environmental risk assessment with atmospheric and ocean numerical models is a wise decision because they can capture additional variables related to wind and waves by model downscaling. Liu [3] integrated the risk assessment model with the operational wind and wave numerical prediction system to investigate the ship risk assessment system under severe wind and wave conditions in the Yellow Sea and Bohai Sea. Chen et al. [4] conducted a statistical analysis of the impact of waves on ship navigation using high-resolution numerical wave simulation and ship measurement data. Liu et al. [5] combined numerical models and fuzzy risk matrices to develop a ship risk warning system in the South China Sea.

In summary, previous studies indicate that meteorological and oceanographic data sources lack sufficient details while the introduction of numerical models can address this shortcoming. Moreover, few studies have examined the parameterization schemes performance of such numerical models in cold surge events.

In this study, firstly, a cold surge occurred in 2022 was simulated by the WRF model, and sensitivity analysis of six sets of PBL parameterization schemes was conducted. The best result was selected as surface driven forces of the WW3 model, generating wave data. Then, based on the simulated meteorological and ocean data, the AHP method was utilized to determine the weights of various assessment factors. Finally, these weights were combined with the fuzzy risk assessment matrix, to assess the navigational risk during the focused cold surge event.

2 Case Study and Methods

2.1 Model Configurations of WRF and WW3

WRF version 4.4 was used in this study to simulate the cold surge. The initial boundary conditions were derived from the ERA-5 dataset with a resolution of 0.25°. A three-layer nested grid and the time steps was set to 18 km, 6 km, 2 km and 60 s, 30 s, 10 s, respectively. The simulation time spanned from 0000 UTC on November 11, 2022, to 0000 UTC on December 1, 2022, encompassing the entire process of two cold surges and advection fog.

To assess the uncertainty of the boundary layer parameterization scheme, we designed 6 sets of sensitivity experiments, as shown in Table 1.

Table 1. Six groups of experiments with different parameterization scheme combinations.

Group number	Micro Physics	PBL Physics	Surface Layer	Land Surface	Radiation
1	WSM5	YSU	Revised MM5	Noah	New Goddard
2		MYJ	Revised MM5		
3		MYNN2.5	MYNN Scheme		
4		ACM2	Revised MM5		
5		QNSE	QNSE Scheme		
6		SHINHONG	Revised MM5		

The spatial resolution uses a fine grid of 0.08°. 38 frequencies were set from 0.0345 Hz to 1.1731 Hz with an increment factor of 1.1. Thirty-six directions were set with 10° intervals. The parameterization schemes of WW3 are shown in Table 2. Both WRF and WW3 models are driven by the ERA-5 dataset.

Table 2. WW3 parameterization scheme configurations.

Linear input term	LN1
Nonlinear interactions	NL1
Wave–bottom interaction	BT1
Wave breaking	DB1
Wind input–dissipation term	ST2

2.2 Algorithms of the Visibility and Wave Steepness

The visibility algorithm refers to the FSL algorithm based on relative humidity and dew point temperature difference proposed by NOAA [6]. The formula of the algorithm is as follows:

$$\text{vis (n mile)} = 6000 \times \frac{T - T_d}{RH^{1.75}} \times 1.151 \tag{1}$$

where T and T_d represent the temperature and dew point temperature at a height of 2 m, respectively, and RH represents the relative humidity at a height of 2 m. When the relative humidity is 100%, the visibility is 0 n mile. The above meteorological variables can all be extracted from the simulation results of the WRF model.

The calculation formula for wave steepness is based on the research of Toffoli et al. [7], and the formula is as follows:

$$WS = \frac{2\pi H_S}{gT_{m02}^2} \quad (2)$$

where H_s and T_{m02} are the significant wave height and the zero-crossing period respectively, and g represents the gravity acceleration. The variables in the formula can be obtained from the simulation results of the WW3 model.

2.3 Evaluation Methods for Numerical Model Results

To evaluate the simulation effects of different PBL parameterization schemes on cold surges, Bias and root-mean-squared-error (RMSE) were utilized as quantitative criteria of simulation performance according to Eq. (1) and Eq. (2):

$$Bias = \frac{\sum(W_i - O_i)}{T} \quad (3)$$

$$RMSE = \sqrt{\frac{\sum(W_i - O_i)^2}{T}} \quad (4)$$

where W and O represent the model calculations and observations of a certain time T, respectively. Bias indicates systematic errors, while the RMSE provides the total errors.

2.4 Navigation Risk Assessment Methodology

The environmental factors used in this study mainly include wind speed, wave height, wave steepness, visibility, and temperature, which were generated by the WRF and WW3 models.

The important relationship between different factors is collected through questionnaires. The survey subjects were several personnel with shipboard work experience and researchers in ship weather routing. The weight of each factor is then determined by the AHP method. Finally, the normalized eigenvector is tested for consistency and the consistency ratio is calculated. When the consistency ratio (CR) is less than 0.1, the result is considered to have reliable consistency and pass the consistency test. The calculation formula of the CR is as follows:

$$CR = \frac{CI}{RI} \quad (5)$$

where the RI and CI are the random consistency and consistency index. CI can be calculated by the following formula:

$$CI = \frac{\lambda_{max} - n}{n - 1} \quad (6)$$

where n is the order of the matrix and λ_{max} is the largest eigenvector.

After the above operations, the weights of the assessment factors in this study can be obtained. Then, the thresholds of different environmental variables under different risk levels are divided. The ship navigation risks are divided into 4 levels in this study:

$$U = \begin{bmatrix} -1 & 0 & 1 & 2 \end{bmatrix} \tag{7}$$

Corresponding to "Low", "Medium", "High" and "Extreme".

Subsequently, through the expert questionnaire survey, the probability of different types of ships encountering different risk levels in different environments can be obtained, and the risk matrix affecting ship navigation is obtained as follows:

$$R(t) = \begin{matrix} m_1 \\ m_2 \\ m_3 \\ m_4 \end{matrix} \begin{bmatrix} r_{11}(t) & r_{12}(t) & r_{13}(t) & r_{14}(t) \\ r_{21}(t) & r_{22}(t) & r_{23}(t) & r_{24}(t) \\ r_{31}(t) & r_{32}(t) & r_{33}(t) & r_{34}(t) \\ r_{41}(t) & r_{42}(t) & r_{43}(t) & r_{44}(t) \end{bmatrix} \tag{8}$$

where $r_{ij}(t)$ represents the probability of occurrence of level j when different meteorological and oceanic states m_i occur under certain navigation conditions t.

After inputting the meteorological and oceanographic data obtained by the numerical models, the membership degree of the risk levels corresponding to the five different environmental factors is obtained from R(t) to form a new matrix E. The risk value can be calculated using the following formula:

$$RL = B \circ E \circ U^T \tag{9}$$

where B is the weight matrix of different environmental factors.

3 Results

3.1 Validation of Numerical Simulations

Observational data from two coastal stations (Dalian and Chengshantou) along Bohai Bay, provided by the CMA (China Meteorological Administration), were utilized to validate the numerical simulations generated by the WRF model. The time interval of observation data from the two stations is 1 h, which is utilized to validate the simulation results of WRF (Fig. 1). As can be seen from Fig. 1, WRF performs well at peak wind speeds but mostly overestimates at lower wind speeds. In addition, the temperature and dew point temperature values of the WRF model align well with the observed values, whereas the relative humidity shows the opposite trend.

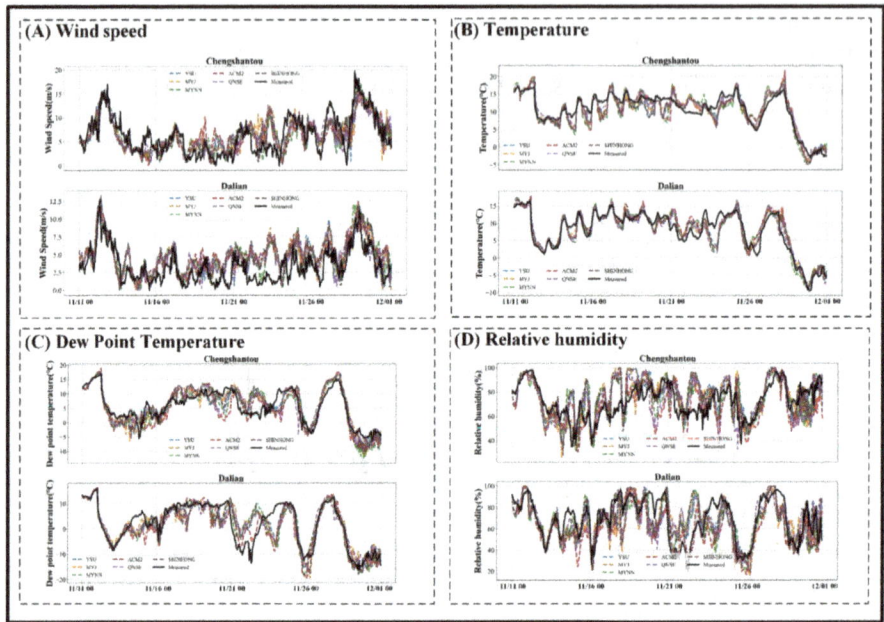

Fig. 1. Comparisons of WRF simulated wind speed and observations for two metrological stations. Black lines represent observations, and others represent results using different parameterization scheme groups in the WRF model.

The Bias and RMSE of four variables are shown in Fig. 2. Overall, the WRF model tends to overestimate weak wind, leading to an overestimation in wind speed simulation.

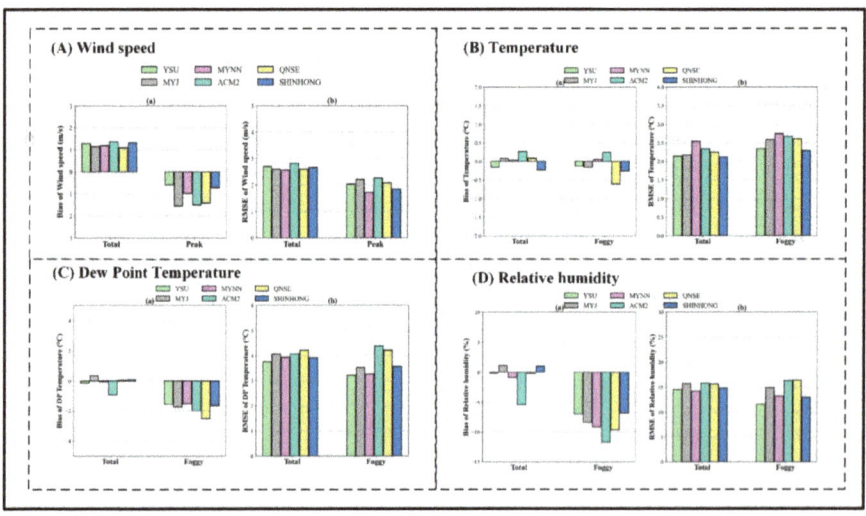

Fig. 2. Average Bias and RMSE values of four variables in the stations.

And a slight underestimation can be found in the "Peak" group. Moreover, the relative humidity and dew point temperature are significantly underestimated in the "Foggy" group.

The MYNN scheme presents the lowest RMSE value in wind speed simulation, particularly for strong wind. The SHINHONG is more advantageous in temperature simulation compared to all other schemes. The YSU scheme performs better in simulating dew point temperature and relative humidity. And the ACM2 performs poorly in all four variables.

3.2 Risk Assessment in Cold Surge Events

The Yellow and Bohai Sea waters focus on bulk carriers of 4000T-5000T [3]. Therefore, this study takes this type of ship as the research object, and the risk thresholds and membership of wind speed and wave height also refer to Liu's study [3]. The risk threshold of visibility refers to the research of Xing et al. [8]. Previous risk assessment studies that consider temperature factors mainly focused on the Arctic route, but the emergence of cold waves in the Bohai Bay area will also bring a significant drop in temperature (the lowest temperature reaches -9.93 °C), so we refer to the temperature risk threshold on the Arctic route by Hu et al. [9]. There was no risk assessment research related to wave steepness in the Bohai Bay waters as a reference, currently. According to the research of Chen et al. [10], most ship accidents occur when the wave steepness is higher than 0.035. In the study of the numerical response of ship motion under bow wave breaking and strong nonlinear waves, Sun [11] pointed out that when the wave steepness exceeds 0.06, the hull will have a violent impact on the wave, and the water will surge onto the deck. Therefore, the risk threshold of wave steepness was also finally determined.

By summarizing previous studies and consulting experts in ocean engineering, the thresholds of risk factors for each meteorological element have been categorized as presented in Table 3. Table 4 displays the weight values of each factor determined using the AHP method.

Table 3. Risk levels and thresholds.

	Low (−1)	Moderate (0)	High (1)	Extreme (2)
Wind speed(m/s)	[0, 8]	(8, 13]	(13, 17]	(17, +∞)
Visibility (n mile)	[5, +∞)	[2, 5)	[1, 2)	[0, 1)
Temperature (°C)	[5, +∞)	[0, 5)	[−5, 0)	(−∞, −5)
Wave height (m)	[0, 2]	(2, 3]	(3, 4]	(4, +∞)
Wave steepness	[0, 0.016]	(0.016, 0.035]	(0.035, 0.06]	(0.06, +∞)

The fuzzy relation matrices of wind speed and wave height are based on the study of Liu [3]. However, no relevant references exist for factors such as wave steepness, temperature, and visibility. Therefore, a questionnaire was administered to 6 senior captains, 1 s officer, 1 third officer, 1 professor, and 6 postgraduates majoring in ship navigation. Subsequently, the fuzzy relation matrices for wave steepness, temperature, and visibility were established.

Table 4. Risk factor weights.

Environmental factors	Weights
Wind speed	0.30906977
Visibility	0.25906976
Temperature	0.05976744
Wave height	0.24906977
Wave steepness	0.12302326

Based on the ship navigation risk assessment method in Sect. 2.4, the meteorological field from the MYNN and YSU parameterization schemes, and the wave field from WW3 are used as environmental inputs. Finally, risk distribution maps with a time interval of

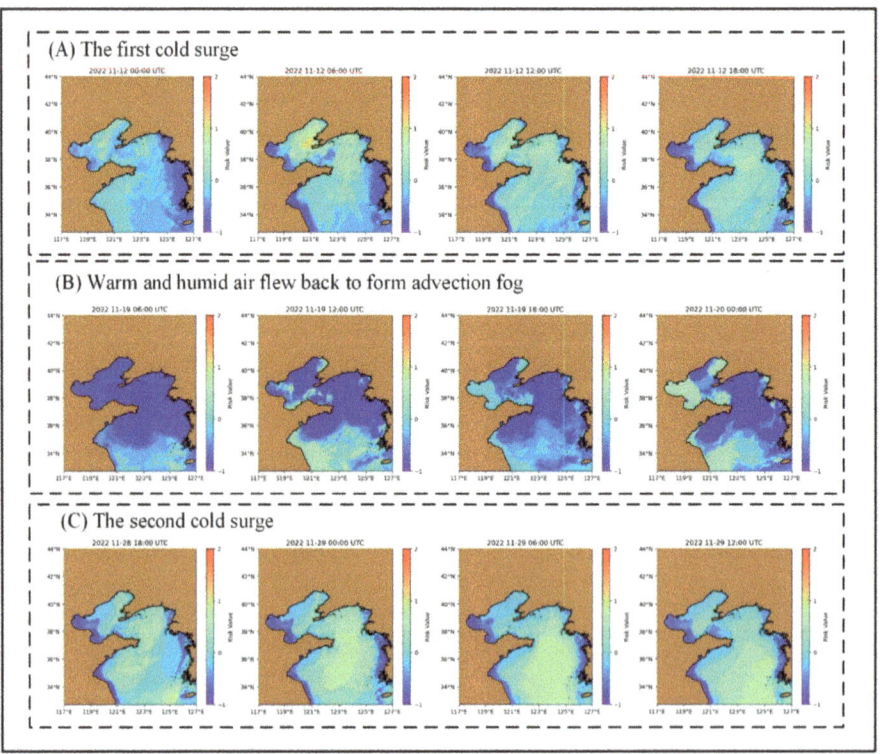

Fig. 3. Spatial distribution map of risks during cold surges and fog.

1 h from 0000 UTC on November 11, 2022, to 0000 UTC on December 1, 2022, have been obtained.

Figure 3 illustrates the risk distribution during the two cold surges and advection fog. Starting at 0000 UTC on Nov.12th, the first cold surge brought about strong winds and waves in the northwest of Bohai Bay. The wind and waves gradually weakened upon entering the northern Yellow Sea. About seven days later, the warm and humid airflow shifted northward (Fig. 3-A). Starting at 0600 UTC on Nov. 19th, advection fog appeared in the southern Yellow Sea and the warm and humid airflow continued to move northward and finally reached Bohai Bay. Visibility decreased steadily until 0000 UTC on Nov. 20th, when sea fog covered a large area of Bohai Bay (Fig. 3-B). After 1800 UTC on Nov. 28th, the second, stronger cold surge arrived in Bohai Bay. After passing the Shandong Peninsula, it brought strong winds, waves, and cooling, revealing a large high-risk area in the Yellow Sea (Fig. 3-C).

4 Conclusions

This study advances maritime risk assessment during cold surges through three principal innovations. First, the investigation uniquely addresses the dual impacts of cold surge-induced gales and subsequent advection fog in Bohai Bay—a critical yet underexplored hazard integrating meteorological and navigational risks. Second, systematic evaluation of six planetary boundary layer (PBL) parameterization schemes within the WRF model, rigorously validated against observational data, achieves high simulation accuracy for weather variables during a cold surge event. Third, coupling high-resolution WRF outputs with the WW3 model generates dynamically consistent wind-wave fields, significantly enhancing hazard simulation fidelity. Integration of these refined meteorological and wave outputs into the AHP-based risk framework enables spatiotemporally risk warnings. According to the experimental results, we draw the following conclusions:

(1) For the overall performance of wind simulation, WRF model provide more accurate results for strong wind ("Peak" group) than for weak wind during these two cold surge events.
(2) For the performance of different parameterization schemes on wind simulation, the MYNN is an appropriate parameterization scheme for wind speed during the cold surge period in the region because it can best simulate strong wind speed, which is vital to the ship safety in cold surges. Meanwhile, YSU and SHINHONG are the best PBL schemes for dew point temperature difference and relative humidity, respectively.
(3) In addition to wind speed and wave height, visibility and wave steepness were also calculated using the results of the WRF and WW3 models and utilized as important risk factors, which were integrated with the AHP method to assess the navigational risk. Subsequently, a spatiotemporal risk map of Bohai Bay and the northern Yellow Sea in the cold surges was developed. The additional factors contribute to a more comprehensive assessment of the risks in the navigation area.

References

1. Wang, J., Li, M., Liu, Y., Zhang, H., Zou, W., Cheng, L.: Safety assessment of shipping routes in the South China Sea based on the fuzzy analytic hierarchy process. Saf. Sci. **62**, 46–57 (2014)
2. Du, P., et al.: Navigation risk assessment method based on dynamic marine environmental factors. Mar. Sci. **45**(5), 121–129 (2021). (in chinese)
3. Liu, Z.: Research on Risk Evaluation System for Vessels in Yellow Sea and Bohai Sea during Severe Weather (2016). (in chinese)
4. Chen, C., Sasa, K., Prpić-Oršić, J., Mizojiri, T.: Statistical analysis of waves' effects on ship navigation using high-resolution numerical wave simulation and shipboard measurements. Ocean Eng. **229**, 108757 (2021)
5. Liu, D., Feng, A., Wang, K., Zhao, W., Zhang, Z., Zhang, Y.: Navigational risk pre-evaluation system for bulk ships carrying cargo subject to fluidization in heavy sea. Navig. China **45**(01), 18–23 (2022). (in chinese)
6. Doran, J.A., et al.: The MM5 at the Air Force Weather Agency-New products to support military operations. In: The 8th conference on aviation, range, and aerospace meteorology, Dallas, Texas, vol. 10 (1999)
7. Toffoli, A., Lefevre, J.M., Bitner-Gregersen, E., Monbaliu, J.: Towards the identification of warning criteria: analysis of a ship accident database. Appl. Ocean Res. **27**(6), 281–291 (2005)
8. Xing, H., Zhou, Z.S., Lv, A.Q., Duan, S.L., Yu, H.L.: Risk early warning for shipwreck based on fuzzy comprehensive evaluation. China Saf. Sci. J. **21**(4), 115–120 (2011). (in chinese)
9. Hu, S.P., Xuan, Y.S., Liu, Y., Fu, S.S., Xi, Y.T.: Dynamic simulation of process risk on ship navigation at the arctic northeast route. Chin. J. Polar Res. **31**(1), 84–93 (2019). (in chinese)
10. Chen, C., Sasa, K., Ohsawa, T., Kashiwagi, M., Prpić-Oršić, J., Mizojiri, T.: Comparative assessment of NCEP and ECMWF global datasets and numerical approaches on rough sea ship navigation based on numerical simulation and shipboard measurements. Appl. Ocean Res. **101**, 102219 (2020)
11. Sun, B.K.: Numerical study on breaking bow waves of a medium and high-speed ship and ship motion response under strong nonlinear waves(2016). (in chinese)

Open Access This chapter is licensed under the terms of the Creative Commons Attribution-NonCommercial-NoDerivatives 4.0 International License (http://creativecommons.org/licenses/by-nc-nd/4.0/), which permits any noncommercial use, sharing, distribution and reproduction in any medium or format, as long as you give appropriate credit to the original author(s) and the source, provide a link to the Creative Commons license and indicate if you modified the licensed material. You do not have permission under this license to share adapted material derived from this chapter or parts of it.

The images or other third party material in this chapter are included in the chapter's Creative Commons license, unless indicated otherwise in a credit line to the material. If material is not included in the chapter's Creative Commons license and your intended use is not permitted by statutory regulation or exceeds the permitted use, you will need to obtain permission directly from the copyright holder.

Research on Significant Wave Height Forecast Based on Extreme Learning Machine

Bo Yang, Xu Wang(✉), and Xiao Wang

Department of Navigation, Dalian Naval Academy, Dalian, China
nudtwangxu@126.com

Abstract. Wave information forecast is of great significance to ship navigation safety. Aiming at the problem of significant wave height forecast, a forecast method based on extreme learning machine is proposed to improve the capability of wave information forecast. Based on the measurement data of the real ship, the optimal hidden layers of the extreme learning machine forecast model are first determined, and then the forecast results of the extreme learning machine algorithm and other three classical algorithms are compared to verify the forecast advantages of the model. The forecast accuracy of the proposed models is tested under the condition of small sample. The generalization ability of the models is verified and analyzed by using the sample data of different time periods. The results show that the significant wave height forecast model based on extreme learning machine has obvious advantages over the other three algorithms, and can provide support and reference for the navigation safety of ships.

Keywords: Extreme Learning Machine · Significant Wave Height · Forecast · Small Sample · Generalization

1 Introduction

Wave forecasting refers to the estimation and prediction of ocean wave characteristics (such as height, period, direction, etc.) over future periods through scientific methods and technological means. Its core objective is to reveal the laws of wave motion and provide decision support for human activities. The forecasting process typically integrates physical models (such as numerical simulations based on fluid mechanics), statistical methods (like regression analysis), and modern monitoring technologies (including satellite remote sensing, buoy observations, and radar detection). By combining historical data with real-time observations, it enables dynamic deduction of wave generation, propagation, and attenuation. Wave forecasting holds irreplaceable application value in various fields such as maritime safety (route planning, disaster avoidance warnings), coastal engineering (port design, breakwater construction), marine resource development (offshore wind power, oil and gas exploration), and environmental research (ecological assessment, climate change monitoring). Especially in the context of frequent extreme weather, the accuracy of wave forecasting directly affects the safety of lives and property, as well as socio-economic benefits [1].

With the continuous development of machine learning, the neural network method in machine learning has shown obvious advantages in the field of ocean wave information forecast [2]. Zhihui Zhu et al. applied the Error Back Propagation (BP) algorithm to the numerical forecast of ocean waves [3]. Salcedo Sanz et al. used the regression method in Support Vector Machine (SVM) to predict the meaningful wave height of ocean waves [4]. James et al. used SVM to predict the characteristic period of ocean waves, and Multi-Layer Perception (MLP) algorithm to predict the meaningful wave height [5]. Xinyu Zhang et al. used Echo State Networks (ESN) to forecast ocean wave spectrum in real time [6]. Xiaochen Zhang et al. proposed a meaningful wave height inversion method for SAR ocean waves based on ELM model [7]. Zhou et al. proposed an Empirical Mode Decomposition (EMD) method mixed with long and short term memory for wave height forecast to improve the accuracy of long-term forecast [8].

The Extreme Learning Machine (ELM) algorithm, proposed by Professor Huang Guangbin et al. [9], uses random construction of the hidden layer of the network and linear regression method to train the output weights. It has the characteristics of simple structure, high forecast accuracy, fast training speed, etc., and can effectively overcome the defects of traditional neural networks. It shows good generalization performance in both classification and forecast fields. Therefore, this paper proposes a method of wave height forecast based on extreme learning machine. Based on wave data collected by a real ship, a series of simulation tests are carried out to compare the forecast results of four algorithm models and verify the forecast ability of the extreme learning machine model.

2 Algorithm and Forecast Model of ELM

2.1 Algorithm of ELM

The ELM algorithm is a single-hidden layer feedforward neural network. It adopts non-iterative learning method without modifying the synaptic weights and bias values between the hidden layer and the output layer, and uses Moore-Penrose generalized inverse to solve the synaptic weights between the hidden layer and the output layer. Generally, ELM network consists of input layer, hidden layer and output layer, and its algorithm structure is shown in Fig. 1.

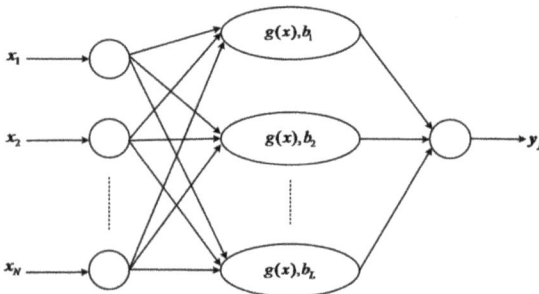

Fig. 1. Structure diagram of ELM algorithm

The equation of ELM is shown in Eq. (1).

$$\sum_{j=1}^{L} \beta j g(\omega j, bj, xj) = \sum_{j=1}^{L} \beta j g(\omega j \cdot xj, bj) = \quad (1)$$

$$oi, i = 1, 2, \ldots, N$$

where L represents the number of hidden layer nodes and oi represents the output. $\omega j = [\omega j1, \omega j2, \ldots, \omega jn]^T$ is the weight vector corresponding to the j hidden layer node and input layer node, and $\beta j = [\beta j1, \beta j2, \ldots, \beta jm]^T$ is the weight vector corresponding to the j hidden layer node and output layer node. $\omega j \cdot xj$ represents the inner product of ωj and xj, and bj represents the bias value of the hidden neuron of the j layer. Input weight ωj and bias value bj are random at the beginning and remain fixed during learning process. If the single output node network model is required, set $m = 1$. In this article, only the multi-output nodes model is considered.

Suppose that the output of the ELM with L hidden layer nodes and activation function $g(\cdot)$ is close to yi with zero error, that is $\sum_{i=1}^{L} \|oi - yi\| = 0$, then $\omega j, \beta j, bj$ satisfy the formula as followed.

$$\sum_{j=1}^{L} \beta_j g(\omega_j \cdot x_j + b_j) = y_j \quad (2)$$

$$H = \begin{pmatrix} g(\omega 1, b1, x1) \cdots g(\omega L, bL, x1) \\ \vdots \\ g(\omega 1, b1, xN) \cdots g(\omega L, bL, xN) \end{pmatrix} \quad (3)$$

where, $\beta = [\beta 1^T, \beta 2^T, \ldots, \beta L^T]_{L \times m}^T$ represents the output weight, and $Y = [y1^T, y2^T, \ldots, y_N^T]_{L \times m}^T$ represents the output of ELM, the activation function $g(\cdot)$ is generally selected Sigmoid function, H is defined as the output matrix of the hidden layer, and the corresponding solution can be expressed as Eq. (4).

$$\hat{\beta} = H^+ Y \quad (4)$$

where H^+ represents the pseudo inverse of the hidden layer output matrix[10].

2.2 Forecast Model

The steps of ELM algorithm can be summarized as follows:

Step 1: The initial N sample sequence $x1, x2, \ldots xN$ is converted into training sets $(x_1, y_1), (x_2, y_2), \ldots, (x_N, y_N)$, and $xi = [xi, xi + 1, \ldots, xi + m - 1]$ are selected as inputs, where $k = N - m$.

Step 2: The number of hidden layer neurons L is determined, and the activation function is $f(x)$, and input weights and bias values are randomly generated.

Step 3: The neuron output matrix H is calculated
Step 4: The output weights are calculated.

According to the steps of ELM algorithm, the forecast model of ELM algorithm is shown in Fig. 2.

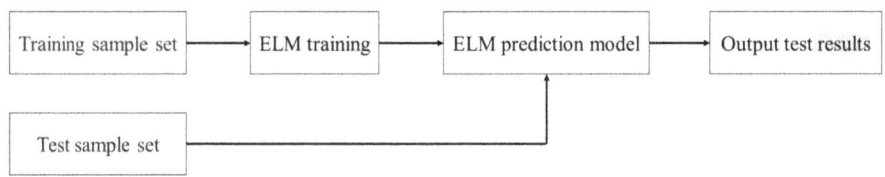

Fig. 2. Forecast model of ELM algorithm

The forecast steps of ELM model can be summarized as follows.

1. The training sample set and the test sample set is selected, and the forecast model needs enough samples.
2. The ELM model is created and trained, and the appropriate number of hidden layer nodes according to the trial-and-error method plays an important role in forecasting performance;
3. The test value is obtained by simulation test.
4. According to the results, the accuracy and robustness of the forecast model were evaluated.

3 Significant Wave Height Forecast

The experimental data originates from measured wave data from a certain type of naval vessel, encompassing observation records from two independent voyages: Data 1 spans from September 19th to 22nd of a given year, while Data 2 covers March 25th to 27th of the following year. Synchronous measurements were conducted using a shipborne X-band wave radar and a pressure-type wave height gauge, and data fusion was employed to generate 64-dimensional spectral density matrices of ocean waves every 2 min (with a frequency range of 0.04–0.5 Hz).

Data preprocessing includes: Removing 3.2% (for Data 1) and 2.8% (for Data 2) of abnormal samples using the 3σ rule. Aligning multi-sensor data to the UTC time base with an error of $\leq \pm 1$ s. Extracting 64-dimensional spectral density as input features and performing Z-score standardization. Normalizing the significant wave height (Hs) labels using Min-Max normalization.

Sample division follows a time series segmentation strategy: Data 1 is divided into a training set of 2160 samples (September 19th to 21st), a validation set of 180 samples (00:00–06:00 on September 22nd), and a test set of 180 samples (06:00–12:00 on September 22nd).Data 2 is divided into a training set of 1440 samples (March 25th to 26th), a validation set of 180 samples (00:00–06:00 on March 27th), and a test set of 180 samples (06:00–12:00 on March 27th).

A small sample test set is constructed by randomly selecting 10% (216 samples) from the training set of Data 1. Data quality control involves verifying temporal consistency by comparing with ECMWF reanalysis data (with an Hs deviation of ≤ ±0.1 m) and removing samples with missing spectral density peaks or abnormal frequency resolution to ensure coverage of typical sea conditions, including wind waves of grades 4 to 6 (Hs = 1.5–4.0 m) and mixed waves (with swell proportions ranging from 30% to 70%) (Figs. 3 and 4).

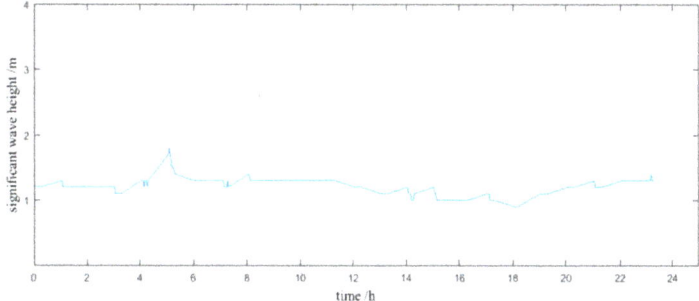

Fig. 3. Partial wave information of Data 1

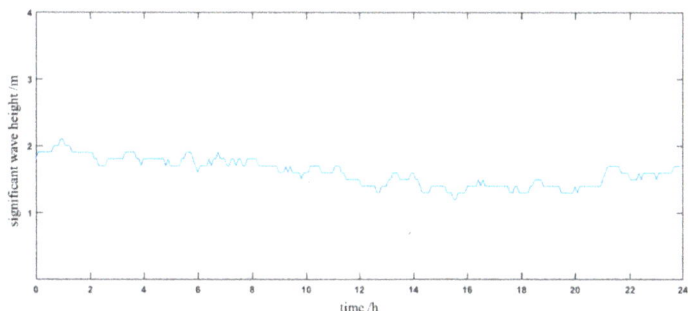

Fig. 4. Partial wave information of Data 2

3.1 Setting the Optimal Number of ELM Hidden Layers

In order to determine the influence of the number of hidden layers on the forecasting performance of ELM algorithm, part of the data in data 1 is used to conduct simulation tests. The training data has 212 sample data on September 21, and the test data is 45 sample data from 6:00 to 10:00 on September 22. The activation function of ELM selects Sigmoid function, and uses cross-validation method to determine the number of hidden layers. The number of hidden layers ranges from 10 to 100, and increases by 10 each time. A total of 20 groups were conducted in the experiment, and the average value of the results was taken as the final result. The root mean square error (RMSE) and forecast error rate (PER) were used as evaluation indicators.

Table 1. RMSE and PER for different hidden layers

Number of hidden layers	RMSE	PER
10	0.0878	0.0583
20	0.0937	0.0637
30	0.0867	0.0575
40	**0.0759**	**0.0491**
50	0.0828	0.0532
60	0.0900	0.0564
70	0.1073	0.0691
80	0.1231	0.0776
90	0.1494	0.0968
100	0.1770	0.1136

As can be seen from Table 1, when the number of hidden layers is 40, the RMSE and PER are 0.0759 and 0.0491, which are the lowest values. Therefore, when the number of hidden layers is 40, the simulation effect of ELM algorithm is the best.

3.2 Comparison of Each Model Forecast

Model Settings. On the basis of the simulation test in Sect. 2.1, the ELM algorithm, SVM algorithm, BP algorithm and ESN algorithm are simulated and compared. The activation function in ELM algorithm uses Sigmoid function, and the number of hidden layers is 40. In SVM algorithm, the kernel function uses RBF, and the regularization parameter C can be set to 16 and the kernel function parameter γ to 1 by using cross-validation method. According to the literature research, in the BP algorithm, the number of neurons in the optimal hidden layer is set to 5. According to literature research, in the ESN algorithm, the spectral radius is set to 0.8, and the output regularization factor λ is 1, and the input scale factor ωin is $0.1 \times [1, 1, \ldots, 1]$[6].

Model Simulation Analysis. The training data used 643 sample data from September 19 to 21, and the test data used from 2 different time periods, the first time period was 45 sample data from 6:00 to 10:00 on September 22, and the second time period is 18 sample data from 10:00 to 12:00 on September 22. The four algorithms were simulated respectively, and 20 groups were tested. The average value was taken as the final result, and the forecast results of the first time period are shown in Table 2, and the forecast results of the second time period are shown in Table 3.

From Tables 2 and 3, it can be seen that the root mean square error and forecast error rate obtained by ELM algorithm are the lowest, and the forecast effect of ELM algorithm

Table 2. Forecast results of the first time period

Algorithm	RMSE	PER
ELM	**0.0575**	**0.0356**
SVM	0.0593	0.0395
BP	0.0633	0.0396
ESN	0.1619	0.0823

Table 3. The forecast result of second time period

Algorithm	RMSE	PER
ELM	**0.0549**	0.0379
SVM	0.1011	0.0660
BP	0.0565	**0.0372**
ESN	0.2951	0.1477

on the selected data is the best. Comparing the forecast results of the two tables, it can be seen that the root mean square error and forecast error rate of the ELM algorithm for the data with a longer time interval are larger, and the forecast effect is worse. However, the forecast effect of BP algorithm on the data of the second time period is slightly better than that of the first time period. It can be seen that the BP algorithm model has better adaptability to the data of the second time period.

According to the above data, the significant wave height is forecasted within 10 h. The forecast results curve and the real wave data curve are shown in Fig. 5. The relative error curve is shown in Fig. 6. It can be seen from the figure that the relative error of the forecast near 10 h is within 10%.

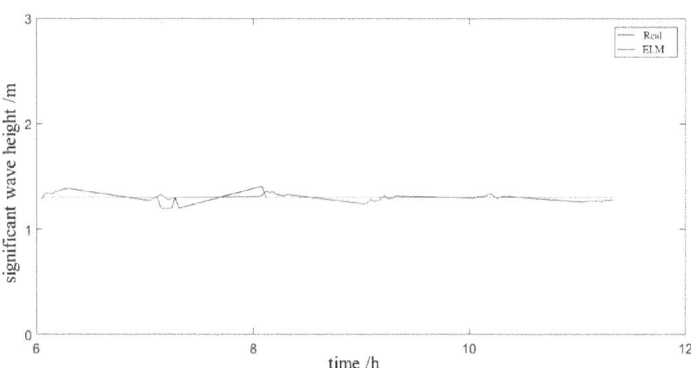

Fig. 5. 10-h forecast result and real wave data curve

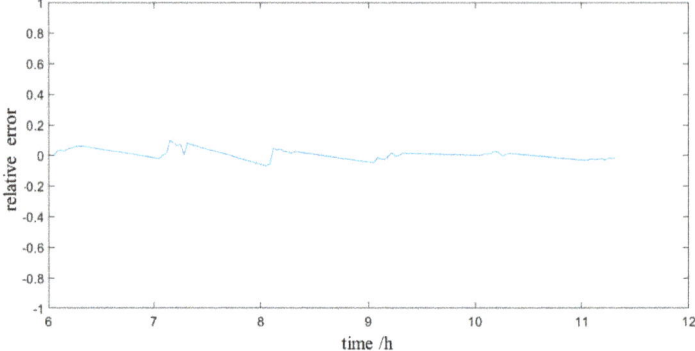

Fig. 6. Relative error of 10-h forecast

3.3 Forecast Analysis Under Small Sample Conditions

In order to test the forecast performance of ELM algorithm with small sample, the training data in Sect. 3.1 is used for the training data, and the test data is the same as that in Sect. 3.2. The ELM, SVM, BP and ESN algorithms were respectively used to conduct comparative tests, and 20 groups of tests were conducted, and the average value was taken as the final result. The forecast results of the first time period are shown in Table 4, and the forecast results of the second time period are shown in Table 5.

Table 4. Forecast results of the first time period

Algorithm	RMSE	PER
ELM	**0.0759**	**0.0491**
SVM	0.0942	0.0650
BP	0.0810	0.0507
ESN	0.1369	0.0797

From Tables 4 and 5, it can be seen that the root mean square error and forecast error rate obtained by ELM algorithm are the lowest, and the forecast effect of ELM algorithm on the selected data is the best. Comparing the forecast results of the two tables, it can be seen that the trend of the forecast results of the two time periods presented by the ELM algorithm is consistent with that of the last simulation test. The root mean square error and forecast error rate of the data forecasted with a longer time interval are larger, and the forecast effect is worse. Compared with the test results in Sect. 3.2, it can be seen that the less training data, the worse the forecast effect of the algo-rithm, and with the reduction of training data, the greater impact on the forecast effect of data with large time intervals.

Table 5. Forecast results of the second time period

Algorithm	RMSE	PER
ELM	**0.0760**	**0.0506**
SVM	0.0987	0.0709
BP	0.0845	0.0539
ESN	0.0845	0.0539

3.4 Generalization Ability of Forecast Model

Setting the Optimal Number of Hidden Layers. In order to test the generalization performance of ELM algorithm, data 2 is used for simulation experiments. First of all, the optimal number of hidden layer nodes in the ELM network should be confirmed. The test method is the same as in Sect. 2.1. The training data adopts 456 samples from March 25 to 26 in data 2, and the test data is 48 samples from 6:00 to 10:00 on March 27. The optimal number of hidden layer nodes in ELM is finally determined as 20.

Simulation Analysis. The training data used 456 samples from March 25 to 26 in Data 2, and two different time periods were used as test data, the first time period was 48 sample data from 6:00 to 10:00 on March 27, and the second time period was 24 sample data from 10:00 to 12:00 on March 27. A comparative test was conducted on the four algorithms respectively. The number of nodes in the hidden layer of ELM was 20, and the parameters used by the other three algorithms were the same as in the Sect. 2.2. The forecast results of the first time period are shown in Table 6, and the forecast results of the second time period are shown in Table 7.

Table 6. Forecast results of the first time period

Algorithm	RMSE	PER
ELM	0.2014	**0.1247**
SVM	0.3599	0.2234
BP	**0.1946**	0.1287
ESN	0.5166	0.2144

From Tables 6 and 7, it can be seen that, in the first time period, the BP algorithm obtains the lowest root mean square error, while the ELM algorithm obtains the lowest forecast error rate. The root mean square error of ELM is slightly higher than that of BP algorithm, with little difference. In the second time period, ELM algorithm obtains the lowest root mean square error and forecast error rate. It can be seen that the ELM algorithm has the best forecast effect for the first and second time periods. Combining the above two tables, it can be seen that the forecast error of the next 10 h is within 20%, and the algorithm used has achieved a good effect in forecasting the next 10 h of the

Table 7. Forecast results of the second time period

Algorithm	RMSE	PER
ELM	**0.2216**	**0.1859**
SVM	0.6736	0.6070
BP	0.4372	0.4260
ESN	1.3558	0.5990

wave data. The comparison between the forecast data of ELM algorithm and the real data is shown in Fig. 7, and the relative error of 10-h forecast is shown in Fig. 8.

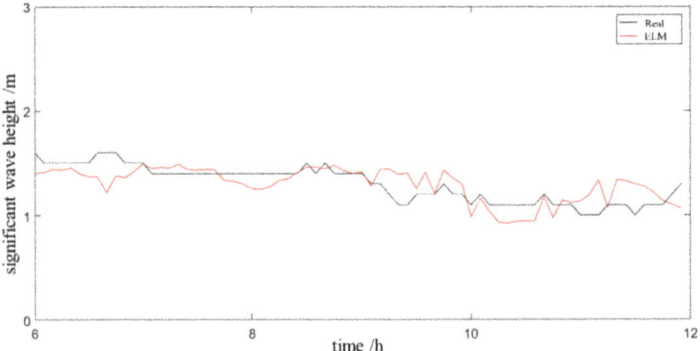

Fig. 7. 10-h forecast results

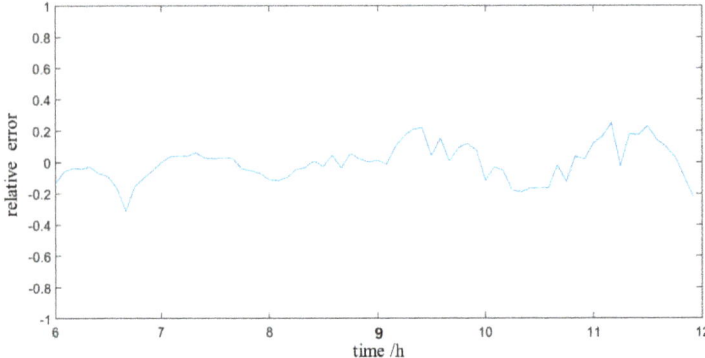

Fig. 8. Relative error of 10-h forecast

4 Conclusion

In this paper, significant wave height forecast model based on ELM is proposed. By comparing the forecast results with those of BP, SVM and ESN, the advantages of ELM method are verified. Based on several sets of ocean wave measured data and a series of simulation tests, the optimal hidden layer of ELM algorithm is determined, the model forecast accuracy under the condition of small samples is tested, and the generalization ability of the model is verified. The results show that the 10-h forecast error of the meaningful wave height forecast model based on ELM model is less than 20%, which has a good forecast effect and can provide support and reference for the navigation safety of ships.

References

1. Li, L.: Design and Implementation of a Fast Forecast Model of Wave Height based on Convolution Neural Network. Shandong University (2022). (in Chinese)
2. Lou, R.: Research on Wave Height Forecast for Different Time Spans Based on Machine Learning. Qingdao University (2022). (in Chinese)
3. Zhu, Z., et al.: Application of neural networks to wave forecast in coastal areas of Shanghai. Marine Forecasts **35**(5), 25–33 (2018). (in Chinese)
4. Wang, Y., et al.: Application of support vector regression in significant wave height forecasting. Marine Forecasts **37**(2), 29–34 (2020). (in Chinese)
5. James, S.C., et al.: A machine learning framework to forecast wave conditions. Coastal Eng. **137**, 1–10 (2018)
6. Zhang, X., et al.: Real-time forecast method of wave spectrum based on echo state network. Marine Forecasts **35**(5), 34–40 (2018). (in Chinese)
7. Wang, X., et al.: Research on SAR Sea significant wave height inversion method based on ELM model. Chin. J. Lasers **47**(7), 1–9 (2020). (in Chinese)
8. Zhou, S., et al.: Improving significant wave height forecasts using a joint empirical mode decomposition–long short-term memory network. J. Mar. Sci. Eng. **9**(7), 744 (2021)
9. Huang, G.B., et al.: Extreme learning machine: theory and applications. Neurocomputing **70**(1–3), 489–501 (2006)
10. Wang, T.: Significant Wave Height Forecast Based on Extreme Learning Machines and Beyonds. China University of Petroleum (2019). (in Chinese)

Open Access This chapter is licensed under the terms of the Creative Commons Attribution-NonCommercial-NoDerivatives 4.0 International License (http://creativecommons.org/licenses/by-nc-nd/4.0/), which permits any noncommercial use, sharing, distribution and reproduction in any medium or format, as long as you give appropriate credit to the original author(s) and the source, provide a link to the Creative Commons license and indicate if you modified the licensed material. You do not have permission under this license to share adapted material derived from this chapter or parts of it.

The images or other third party material in this chapter are included in the chapter's Creative Commons license, unless indicated otherwise in a credit line to the material. If material is not included in the chapter's Creative Commons license and your intended use is not permitted by statutory regulation or exceeds the permitted use, you will need to obtain permission directly from the copyright holder.

Research on the Recognition Method of Unsafe Action in Mooring Operations Based on Improved YOLOv7

Zhen Fang[1,2], Wenjun Zhang[1,2(✉)], Xiangkun Meng[1,2], Xue Yang[1,2], and Xiangyu Zhou[1,2]

[1] Navigation College, Dalian Maritime University, No. 1, Linghai Road, Dalian, China
wenjunzhang@dlmu.edu.cn
[2] Dalian Key Laboratory of Safety & Security Technology for Autonomous Shipping, No. 1, Linghai Road, Dalian, China

Abstract. Mooring operations are vital for ship safety, where unsafe actions can result in severe accidents. This paper enhances the YOLO v7 network for identifying unsafe actions through video monitoring. The approach includes optimizing anchor points with the k-means++ algorithm, improving the match with unsafe actions, and integrating the Squeeze-and-Excitation (SE) attention mechanism to enhance focus on critical features, boosting recognition performance. Experimental results demonstrate that the improved model outperforms the original YOLO v7, achieving a 0.78% increase in mean Average Precision (mAP) while maintaining the same computational load. This model effectively aids in monitoring unsafe mooring actions. This improved model can serve as a reference for recognizing unsafe action in mooring operations.

Keywords: YOLOv7 · Unsafe Action · SE

1 Introduction

With global trade and logistics evolving, ships are vital for maritime transportation. Human factors notably contribute to maritime accidents, with 80–85% of incidents resulting from human errors, as per the 2023 European Maritime Safety Agency (EMSA) statistics [1]. Mooring operations, essential for berthing and unberthing, significantly impact ship and port safety and personnel protection. European Harbour Masters' Committee statistics reveal that 95% of personal injuries involve ropes, with most occurring during mooring [2].

Despite the importance of safety in mooring operations, practical supervision of unsafe actions is lacking. Advances in deep learning offer solutions to these challenges. For example, Yu et al. [3] introduced a real-time method for detecting unsafe actions in construction, while Lei et al. [4] analyzed driver behavior using a bimodal neural network. Gao et al. [5] implemented intelligent monitoring to enhance safety in construction through real-time warnings, and Liu et al. [6] focused on dangerous driving recognition. Li et al. [7] identified violations in substations using Mask R-CNN, and Hägele et al. [8] improved safety management with drones.

The YOLO model by Redmon et al. [9], known for its speed and accuracy, has been applied in various safety-related fields. For instance, Cuma et al. [10] used YOLOv5 for driver action recognition, and Nguyen et al. [11] leveraged it for real-time human detection. Zhao et al. [12] enhanced maritime safety with YOLOv4 by identifying unsafe mooring actions. However, improvements are needed for better detection in complex mooring scenarios [13]. Integrating attention mechanisms has significantly enhanced target detection performance.

Based on this, this paper improves the YOLOv7 algorithm by utilizing the k-means++ algorithm to obtain better-matched anchor box sizes and incorporating the SE (Squeeze-and-Excitation) attention mechanism. These enhancements aim to improve the detection accuracy of unsafe action in mooring operations, providing technical support for preventing operational accidents.

2 Methodology

The YOLOv7 model is an end-to-end deep learning framework that optimizes architecture and training, featuring three main components: Backbone, Neck, and YOLO Head. The Backbone extracts features from images, the Neck enhances these features, and the YOLO Head performs object detection using anchor boxes at feature points. YOLOv7 builds on the YOLO series architecture with a denser skip connection structure that improves feature transmission and extraction capabilities.

2.1 Selection of Anchor Boxes

In the YOLOv7 model, selecting anchor boxes is crucial for improving object detection performance across different sizes and aspect ratios. Bounding box predictions rely on these anchors, and choosing the right ones reduces prediction errors and enhances precision and recall. YOLOv7 uses the k-means clustering algorithm to determine anchor sizes and aspect ratios directly from the data, aiming to minimize the sum of squared distances between points and cluster centers. To avoid issues with local optima caused by extreme values, the k-means++ algorithm is applied to refine the clustering results.

After improving the clustering method, a comparison between k-means and k-means++ (as shown in Fig. 1) shows that k-means converges to local optima, while k-means++ provides a more balanced selection of cluster centers. This leads to more accurate anchor boxes, enhancing YOLOv7's object detection performance.

2.2 SE Attention Mechanism

The SE attention mechanism improves convolutional neural networks (CNNs) performance by modeling channel interdependencies. It consists of two operations: Squeeze and Excitation [14]. The Squeeze operation uses Global Average Pooling to reduce the spatial dimensions of the feature map to a global descriptor for each channel. The Excitation operation learns the importance of each channel using a fully connected network, applying ReLU and Sigmoid functions to scale the importance weights. These weights are used to rescale the original feature map.

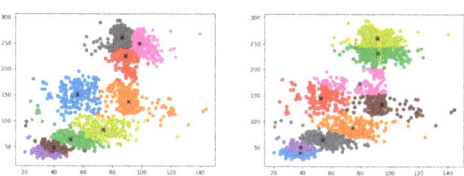

Fig. 1. Comparison of Methods for Selecting Cluster Centers.

2.3 Improved Network Structure

In this study, the YOLOv7 model structure is enhanced by integrating the SE attention mechanism into the P4 feature layer, improving detection of key unsafe actions in mooring operations. This adjustment increases the model's sensitivity to relevant features and reduces the influence of irrelevant data, allowing for better differentiation of specific unsafe actions and postures. Consequently, the enhanced model achieves higher accuracy in identifying critical unsafe actions, enhancing the reliability of safety monitoring in mooring operations.

3 Experiments

3.1 Preparation

The Safety Management System (SMS) documents define unsafe actions in mooring operations. Five typical actions identified are: sitting on the bollard, stepping on the cable, stepping on the bollard, crossing the cable, and crossing the taut cable. Seafarers often rest on bollards or step on tightly stretched cables, and crew members frequently cross cables to move quickly. These actions are prohibited in the SMS as they can lead to hazardous consequences, including falls and injuries, particularly in unstable conditions or adverse weather.

This research focuses solely on these five actions from the SMS. While other unsafe actions may exist, they are not included to maintain the study's focus. The dataset consists of images from crew mooring operation videos from the Yukun ship of Dalian Maritime University and the Changjiang No. 1 ship, yielding 1818 images after screening. Figure 2 shows some sample images.

Fig. 2. Dataset Images.

LabelImg software was used to manually annotate the images, and the dataset was split in a 9:1 ratio for training and testing to ensure adequate data for model learning and assessment.

3.2 Training Process

The experimental setup for this study included a Windows 11 operating system with an NVIDIA GeForce RTX 3070 GPU, utilizing the PyTorch deep learning framework. The network input image size was configured to (640, 640, 3). The model underwent training for 100 epochs, with a batch size of 5. Parameter optimization was conducted using the stochastic gradient descent (SGD) algorithm, featuring a weight decay coefficient of 0.0005 and a maximum learning rate of 0.002. Model weights were saved at intervals of every 10 epochs.

3.3 Experimental Results and Comparative Analysis

Figure 3 displays the improved YOLOv7 model's detection results for unsafe actions in mooring operations, demonstrating high accuracy. The study evaluates model performance using mean Average Precision (mAP), parameters, and Giga Floating-point Operations (GFLOPs).

Fig. 3. Detection Results of Unsafe Action in Mooring Operations Across Different Scenarios.

According to Table 1, the YOLOv7 with SE attention has a higher mAP compared to other models, and its performance in terms of parameters and GFLOPs is competitive, indicating superior recognition of unsafe actions in mooring operations.

Table 1. Performance comparison between different detection methods.

Method	mAP	Parameters/M	GFLOPs/G
YOLO v7	97.42%	37.620	106.472
SE-YOLO v7	98.20%	37.628	106.473
Fast RCNN	97.41%	137.099	370.210
SSD	96.71%	26.285	62.747

4 Conclusion

This paper proposes a method for recognizing unsafe actions in mooring operations using an improved YOLOv7 model, supported by a custom annotated dataset of real mooring operations. By enhancing the clustering center selection algorithm for better anchor sizes and incorporating the SE attention mechanism, the model's recognition capabilities are improved. The improved YOLOv7 model is evaluated for accuracy and computational cost, outperforming three other algorithms in accuracy while maintaining comparable computational efficiency, meeting the needs of mooring operation safety recognition.

References

1. EMSA: EMSA Annual Overview of Marine Casualties and Incidents [R]. EMSA, Lisbon (2023)
2. DNV. Taking a new look at mooring safety—Industry insights. In Maritime Impact; DNV: Bærum, Norway, 2020
3. Yu, Y., Guo, H., Ding, Q., Li, H., Skitmore, M.: An experimental study of real-time identification of construction workers' unsafe behaviors. Autom. Constr. **82**, 193–206 (2017)
4. Lei, Z., Zengcai, W., Xiaojin, W., Yazhou, Q., Qing, L., Guoxin, Z.: Human fatigue expression recognition through image-based dynamic multi-information and bimodal deep learning. J. Electron. Imaging. **25**(5), 053024 (2016)
5. Gao, J., Liu, J., Han, J.: A study for real-time identification of unsafe behavior of taking off safety helmet based on VSM model. In: Proceedings of the 11th International Conference on Computer Modeling and Simulation, pp. 84–90. Association for Computing Machinery, North Rockhampton, QLD, Australia (2019). https://doi.org/10.1145/3307363.3307400
6. Liu, W., Li, H., Zhang, H.: Dangerous driving behavior recognition based on hand trajectory. Sustainability. **14**, 12355 (2022)
7. Li, J., Ren, Q., Wu, J.: Intelligent video recognition of violation behavior in substation site based on mask R-CNN. In: International Conference on Image Processing, Computer Vision and Machine Learning (ICICML), vol. 2022, pp. 183–186. IEEE, Xi'an, China (2022). https://doi.org/10.1109/ICICML57342.2022.10009754
8. Hägele, G., Sarkheyli-Hägele, A.: Situational risk assessment within safety-driven behavior management in the context of UAS. In: 2020 International Conference on Unmanned Aircraft Systems (ICUAS), pp. 1407–1415. IEEE, Athens, Greece (2020). https://doi.org/10.1109/ICUAS48674.2020.9214072
9. Redmon, J., Divvala, S., Girshick, R., Farhadi, A.: You Only Look Once: Unified, Real-Time Object Detection. In: 2016 IEEE Conference on Computer Vision and Pattern Recognition (CVPR), pp. 779–788. IEEE, Las Vegas, NV, USA (2016). https://doi.org/10.1109/CVPR.2016.91.S
10. Cuma, M.U., Dükünlü, Ç., Yirik, E.: Smart Driver Behavior Recognition and 360-Degree Surround-View Camera for Electric Buses. Electronics. **12**, 2979 (2023). https://doi.org/10.3390/electronics12132979
11. Nguyen, H.H., Ta, T.N., Nguyen, N.C., Bui, V.T., Pham, H.M., Nguyen, D.M.: YOLO based real-time human detection for smart video surveillance at the edge. In: 2020 IEEE Eighth International Conference on Communications and Electronics (ICCE), pp. 439–444. IEEE, Phu Quoc Island, Vietnam (2021). https://doi.org/10.1109/ICCE48956.2021.9352144
12. Zhao, C., Zhang, W., Chen, C., Yang, X., Yue, J., Han, B.: Recognition of unsafe onboard mooring and unmooring operation behavior based on improved YOLO-v4 algorithm. J. Mar. Sci. Eng. **11**, 291 (2023). https://doi.org/10.3390/jmse11020291
13. Deng Y.P., Li Y.J.: Review of YOLO algorithm and its applications to object detection in autonomous driving scenes. J. Comput. Appl., 1–12. http://kns.cnki.net/kcms/detail/51.1307.TP.20230904.1321.006.html. (in Chinese)
14. Hu, J., Shen, L., Sun, G.: Squeeze-and-Excitation Networks. In: 2018 IEEE/CVF Conference on Computer Vision and Pattern Recognition, pp. 7132–7141. IEEE, Salt Lake City, UT, USA (2018). https://doi.org/10.1109/CVPR.2018.00745

Open Access This chapter is licensed under the terms of the Creative Commons Attribution-NonCommercial-NoDerivatives 4.0 International License (http://creativecommons.org/licenses/by-nc-nd/4.0/), which permits any noncommercial use, sharing, distribution and reproduction in any medium or format, as long as you give appropriate credit to the original author(s) and the source, provide a link to the Creative Commons license and indicate if you modified the licensed material. You do not have permission under this license to share adapted material derived from this chapter or parts of it.

The images or other third party material in this chapter are included in the chapter's Creative Commons license, unless indicated otherwise in a credit line to the material. If material is not included in the chapter's Creative Commons license and your intended use is not permitted by statutory regulation or exceeds the permitted use, you will need to obtain permission directly from the copyright holder.

A Framework of Predictive Maintenance for Maritime Autonomous Surface Ships Considering Component Degradation

Hongqiang Li[1,2], Xiangkun Meng[1,2(✉)], Jing Liu[3], Wenjun Zhang[1,2], Xiang-Yu Zhou[1,2], and Xue Yang[1,2]

[1] Navigation College, Dalian Maritime University, No. 1, Linghai Road, Dalian, China
mxk0117@dlmu.edu.cn
[2] Dalian Key Laboratory of Safety & Security Technology for Autonomous Shipping, No. 1, Linghai Road, Dalian, China
[3] Shengli Xinda New Material Co., Ltd., Dongying, China

Abstract. With the emergence and development of Maritime Autonomous Surface Ships (MASS), reliable operation over extended periods for safety is crucial. Regular maintenance is vital for maintaining reliability, and predictive maintenance is particularly efficient in minimizing maintenance time and reducing resource waste. This paper proposes a condition prediction framework that considers component degradation to aid in developing effective predictive maintenance strategies for MASS. Five machine learning models were utilized to analyze a collected dataset. This framework evaluated the models using mean squared error (MSE), mean absolute error (MAE), and determination coefficient (R^2) to identify the most effective one for condition prediction. To enhance the accuracy, the framework also considers component degradation. The framework's applicability was demonstrated using a gas turbine system as a case study. The results show that the multilayer perceptron is the most fitting one for the condition prediction of the gas turbine system. The framework can support predictive maintenance in decision-making to improve MASS safety.

Keywords: Condition Prediction · Machine Learning · Maritime Autonomous Surface Ships · Predictive Maintenance

1 Introduction

With developments in the shipping industry, Maritime Autonomous Surface Ships (MASS) have advanced rapidly, thanks to the maturation of intelligent technologies [1]. However, the numerous subsystems within ships demand specialized skills and expertise that are difficult to substitute, particularly in unforeseen situations [2]. Therefore, MASS presents new reliability challenges with reduced crews or even no one on board [3].

In this context, a systematic approach is required to furnish adequate insights into the reliability of mechanical equipment, enabling MASS to perform missions autonomously for extended periods [4]. indicate that the autonomous missions may last over a month.

Safe operation in maritime transportation is typically ensured through reactive maintenance (RM), and preventive maintenance (PM) [5]. MASS will ultimately operate without maintenance personnel to perform RM, and traditional PM relies on fixed maintenance intervals based on experience [6]. Therefore, [7] proposed condition-based maintenance (CBM), which focuses on system maintenance by assessing component degradation conditions. Although this method is cost-effective, the accumulated degradation of the system may reach failure thresholds, increasing the risk of system failure [8]. By utilizing real-time monitoring and data analysis, Predictive maintenance (PdM) predicts equipment failures [9], thereby ascertaining the optimal timing for maintenance activities. PdM optimizes maintenance strategies, and serves as a foundation for further research and application, offering precise predictions of equipment conditions.

With the development of big data, traditional methods encounter significant challenges in managing large amounts of data. Machine learning (ML) technologies have markedly improved the ability to process complex data sets and graphical information [10, 11]. utilized machine learning techniques for failure prediction and maintenance prioritization to enhance the resilience of ship propulsion systems. However, the PdM for MASS is challenged by the operational data. To address the data scarcity issue, [4] suggested a reliability analysis framework grounded in hierarchical Bayesian inference utilizing data from conventional ships. However, the impact of component degradation in real operation is not considered.

This paper proposes a framework that introduces a predictive maintenance approach for MASS that considers component degradation, supporting maintenance strategy in decision-making to improve MASS safety. Additionally, various ML models are used to analyze the condition data. By systematically comparing the mean squared error (MSE), mean absolute error (MAE), and coefficient of determination (R^2) of different models, the most fitting model in forecasting the future condition of components is chosen for enhancing predictive accuracy. This evaluation highlights the predictive capabilities of various models and lays the foundation for future research. Finally, to further verify the necessity of considering component degradation in predictive maintenance, this paper conducts a comparative analysis of the metrics for components with and without consideration of degradation.

2 Framework

The methodology employed consists of four steps:

(1) Data collection and processing: The process began with data collection and processing, including describing the data sources and the specific processing methods used in this study. To aid further analysis, the processed dataset was normalized.
(2) Model building and training: This step detailed the algorithm construction process. Considering component degradation, this paper used multilayer perceptron (MLP), convolutional neural network (CNN), random forest (RF), support vector machine (SVM), and K-nearest neighbor (KNN) to train the feature variables.
(3) Model selection: In the algorithm selection phase, errors of different algorithms were compared using the MSE, MAE, and R^2. This analysis informed the selection of the algorithm best suited to the dataset.

(4) Results analysis: In this step, the selected algorithm was applied to forecast the condition. Before making predictions, the data was denormalized to provide more intuitive results, offering clearer insights into the prediction outcomes.

3 Case Study

3.1 Dataset and Processing

The dataset used in this study was generated by a sophisticated simulator for a GT installed on a naval frigate equipped with a Combined Diesel Electric and Gas (COD-LAG) propulsion system, sourced from the UCI [12]. The data represents the degradation of turbines and compressors through decay state coefficients [7]. The GT component features are used as prediction targets for state prediction (see Table 1).

Table 1. Feature description.

ID	Features	Units
1	Lever position (lp)	
2	Ship speed (v)	knots
3	Gas Turbine shaft torque (GTT)	kNm
4	Gas Turbine rate of revolutions (GTn)	rpm
5	Gas Generator rate of revolutions (GGn)	rpm
6	Starboard Propeller Torque (Ts)	kN
7	Port Propeller Torque (Tp)	kN
8	High-Pressure (HP) Turbine exit temperature (T48)	C
9	GT Compressor outlet air temperature (T2)	C
10	HP Turbine exit pressure (P48)	Bar
11	GT Compressor outlet air pressure (P2)	bar
12	Turbine Injection Control (TIC)	%
13	Fuel flow (mf)	kg/s
14	GT Compressor decay state coefficient (CDSC)	
15	GT Turbine decay state coefficient (TDSC)	

The lp contains nine distinct ship speeds: 3, 6, 9, 12, 15, 18, 21, 24, and 27 knots. The speed and weight used in this paper are 12: 20%, 15: 60%, and 18: 20%, respectively.

3.2 Model Comparison

Given the degradation conditions of turbines and compressors, state predictions for various GT components were performed using five ML models. The performance of

Table 2. Mean values of metrics.

Model	MSE	MAE	R^2
MLP	1.36E-03	1.93E-02	9.69E-01
CNN	2.90E-01	4.77E-01	−4.82E+00
SVM	3.03E-03	4.25E-02	9.38E-01
RF	1.64E-03	1.91E-02	9.62E-01
KNN	1.55E-03	1.92E-02	9.65E-01

these algorithms was evaluated using three metrics. The mean values of these evaluation metrics are presented in Table 2.

As shown in Table 2, the MLP model outperformed the other models, delivering superior results across all three metrics.

4 Results Analysis

This section focuses on the prediction results for a ship speed of 15 knots. The data were denormalized before prediction to provide a more intuitive representation of the outcomes. This paper presents the GTT and Ts as examples in Fig. 1.

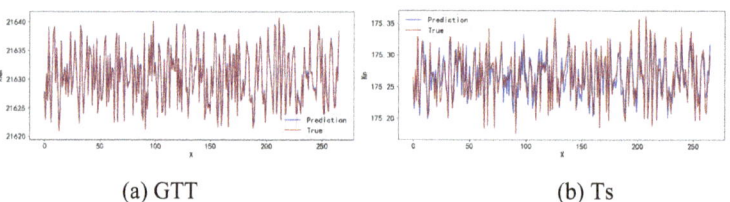

(a) GTT (b) Ts

Fig. 1. Prediction results of MLP.

As shown in Fig. 1, a comparison between the MLP model's predictions and the actual values is presented. When considering compressor and turbine degradation, the prediction errors for features related to these components are minimal, where the predicted values deviate significantly from the actual values. These results underscore the importance of considering component degradation in enhancing prediction accuracy and the significant impact of degradation on prediction.

5 Conclusion

The results show that prediction errors related to the propeller are significantly higher than the other two, highlighting the importance of considering component degradation. The approach effectively predicts the future condition of MASS mechanical equipment and offers valuable insights for maintenance scheduling. However, some ML models exhibited high errors, due to the limited dataset. A more diverse and reliable dataset could improve the framework's accuracy, and provide more reliable and cost-effective maintenance strategies for practitioners.

Acknowledgments. This work is supported by the National Key R&D Program of China (Grant No. 2023YFB4302300), the Natural Science Foundation of Liaoning Province (under Grant No. 2023-MS-123), the National Natural Science Foundation of China (under Grant No. 52004142), and the Fundamental Research Funds for the Central Universities of China (under Grant No. 3132024136, 3132023130).

References

1. Yang, X., Zhou, T., Zhou, X.Y., Zhang, W.J., Mu, C.R., Xu, S.: A framework to identify failure scenarios in the control mode transition process for autonomous ships with dynamic autonomy. Ocean Coast. Manag. **249**, 107003 (2024)
2. Chaal, M., Ren, X., Bahoo Toroody, A., Basnet, S., Bolbot, V., Banda, O.A.V., Gelder, P.V.: Research on risk, safety, and reliability of autonomous ships: A bibliometric review. Saf. Sci. **167**, 106256 (2023)
3. Tao, J.C., et al.: Hazard identification and risk analysis of maritime autonomous surface ships: A systematic review and future directions. Ocean Eng. **307**, 118174 (2024)
4. Abaei, M.M., Hekkenberg, R., Bahoo Toroody, A.: A multinomial process tree for reliability assessment of machinery in autonomous ships. Reliab. Eng. Syst. Saf. **210** (2021)
5. Kong, X.F., Yang, J., Li, L.: Remaining useful life prediction for degrading systems with random shocks considering measurement uncertainty. J. Manuf Syst. **61**, 782–798 (2021)
6. BahooToroody, A., Abaei, M.M., Banda, O.V., Kujala, P., Carlo, F.D., Abbassi, R.: Prognostic health management of repairable ship systems through different autonomy degree: From current condition to fully autonomous ship. Reliab. Eng. Syst. Saf. **221**, 108355 (2022)
7. Cipollini, F., Oneto, L., Coraddu, A., Murphy, A.J., Anguita, D.: Condition-based maintenance of naval propulsion systems: Data analysis with minimal feedback. Reliab. Eng. Syst. Saf. **117**, 12–23 (2018)
8. Zhang, T.P., et al.: Joint multi-objective optimization method for emergency maintenance and condition-based maintenance: Subsea control system as a case study. Reliab. Eng. Syst. Saf. **250**, 110307 (2024)
9. Ellefsen, A.L., Æsøy, V., Ushakov, S., Zhang, H.X.: A comprehensive survey of prognostics and health management based on deep learning for autonomous ships. IEEE Trans. Reliab. **68**(2), 1–21 (2019)
10. Park, J., Oh, J.: Anomalistic symptom judgment algorithm for predictive maintenance of ship propulsion engine using machine learning. Appl. Sci. **13**(21), 11818 (2023)
11. Elmdoost-gashti, M., Shafiee, M., Bozorgi-Amiri, A.: Enhancing resilience in marine propulsion systems by adopting machine learning technology for predicting failures and prioritizing maintenance activities. J. Mar. Eng. Technol. **13**(1), 18–32 (2023)
12. Coraddu A, Oneto L, Ghio A, Savio S, Anguita D, Figari M.: Machine learning approaches for improving condition-based maintenance of naval propulsion plants. In: Proceedings of the Institution of Mechanical Engineers, Part M: Journal of Engineering for the Maritime Environment. 230(1):136–153 (2016)

Open Access This chapter is licensed under the terms of the Creative Commons Attribution-NonCommercial-NoDerivatives 4.0 International License (http://creativecommons.org/licenses/by-nc-nd/4.0/), which permits any noncommercial use, sharing, distribution and reproduction in any medium or format, as long as you give appropriate credit to the original author(s) and the source, provide a link to the Creative Commons license and indicate if you modified the licensed material. You do not have permission under this license to share adapted material derived from this chapter or parts of it.

The images or other third party material in this chapter are included in the chapter's Creative Commons license, unless indicated otherwise in a credit line to the material. If material is not included in the chapter's Creative Commons license and your intended use is not permitted by statutory regulation or exceeds the permitted use, you will need to obtain permission directly from the copyright holder.

Experimental Study on the Rudder Angle Optimization for a Twin Rudder Ship Resistance and Self-propulsion Performance

Lei Xing[1(✉)], Li Zhang[1,2], Weimin Chen[1,2], and Yongyue Li[3]

[1] Shanghai Ship and Shipping Research Institute Co., Ltd., Shanghai, China
xingsanshi@126.com
[2] College of Shipbuilding Engineering, Harbin Engineering University, Harbin, China
[3] Shanghai Merchant Ship Design and Research Institute, Shanghai, China

Abstract. Ship-propeller-rudder interaction affects the overall performance of the ship. For twin-propeller ships which equipped with twin-rudder, the variation of rudder angle will aggravate the interaction of rudder & propeller & ship and have a significant influence on the overall ship performance. Based on the experimental method, both the resistance and self-propulsion performance of a twin-propeller ship which equipped with twin-rudder at different symmetrical rudder angles near the design speed are studied regarding the ITTC recommended guidelines for resistance tests and self-propulsion tests, and the optimal rudder angle at the inflection point is considered. For the resistance test, the total ship resistance at the rudder angle of 0, 2, 4, 6 and 8 degrees is studied, and the optimal rudder angle is 6 degree. For the self-propulsion test, the ship's self-propulsion performance at the rudder angle of 0, 2, 4 and 6 degrees is studied, and the best rudder angle is 4 degree. And the optimal rudder angle is similar at different speeds. Therefore, the symmetrical rudder angle has a significant effect on the resistance and self-propulsion performance of a twin-propeller ship which equipped with twin-rudder, and the optimal rudder angle can effectively reduce the energy consumption of the twin rudder type of ship.

Keywords: Experiment · Rudder Angle · Twin Rudder Ship · Resistance and Self-Propulsion

1 Introduction

Rudder is one of the important structures which affect ship design and ship performance. In addition to the influence on ship maneuvering performance, rudder also has an important influence on ship resistance and self-propulsion performance. Through the resistance model test and numerical simulation, regarding the impact of the rudder on the resistance performance of single-rudder ships, there is research indicating that the influence of the rudder is approximately between 1% and 2%. For twin-rudder ships, the contribution of conventional streamlined rudders to the hull resistance is approximately 2% to 4% [1]. Compared with single-rudder ships, the rudder angle arrangement of twin-stern ships is also one of the key factors affecting resistance and self-propulsion performance.

The research of Molland and Turnock shows that the lift coefficient of the open water rudder and the rudder behind the propeller is roughly linear with the angle of attack, while the drag coefficient is roughly parabolic with the angle of attack. At the same time, the propeller has an effect on the pressure center position of the rudder [2]. Charles et al. used the CFD method to conduct numerical simulations on a single propeller and single rudder with different inflow angles and different angles between the propeller and rudder. The results indicated that changes in the rudder angle significantly alter the lift and drag coefficients of the rudder, as well as the thrust coefficient of the propulsor [3]. Jialun and Robert's research indicates that the angle of attack of the flow over the rudder surface changes due to the rotation of the propeller [4]. Therefore, the optimal arrangement angle between rudder and propeller can be found to minimize the ship resistance coefficient [5]. Roberto et al. studied on the rudder-propeller interaction of the two-axis bracket ship by CFD method. The results show that there is a significant influence on the rudder force when the twin rudders with the same angle and the same direction [6]. Yue Jiao numerically simulated the rudder system which designed/developed by herself. The results show that the wake behind propeller can generate torque on the rudder and affect the rudder lift performance [7]. Wu Sichuan studied on the self-propulsion performance of a twin-podded propulor luxury cruise ship influencing by different rudder angle. The results show that an optimal rudder that minimizes the received power can be found [8].

In the previous work, the numerical simulation of the twin-tailed ship under different rudder angles was studied. In order to further verify the performance of different rudder angles at different speeds, resistance test and self-propulsion test were carried out to study and analyze the optimal rudder angle.

2 Model Tests Method

Model test is one of the important methods to study the performance of ships, and it is also a benchmark reference method among many different research methods. The resistance test is to study the resistance of the ship at the corresponding speed in the corresponding state, and the self-propulsion test is to study the ship propulsion performance when the propeller is installed. The resistance test is based on the ITTC recommended procedure 'Resistance tests [9] and the self-propulsion test is based on the ITTC recommended procedure 'Propulsion/Bollard Pull Test [10], and the test results of the ship in the corresponding state are analyzed accordingly.

3 Model Tests of the Twin Rudder Ship

3.1 Ship Parameters

The main parameters of the ship are shown in Table 1, and the arrangement of the aft ship is shown as Fig. 1.

Table 1. Main parameters of the ship.

Designation	Symbol	Full scale	Model scale
Length between perpendicular	L_{PP}(m)	290.00	6.7845
Breadth moulded on WL	B(m)	45.80	1.0715
Draft moulded on FP	T_F(m)	11.50	0.2690
Draft moulded on AP	T_A(m)	11.50	0.2690
Block coefficient	Cb	0.753	

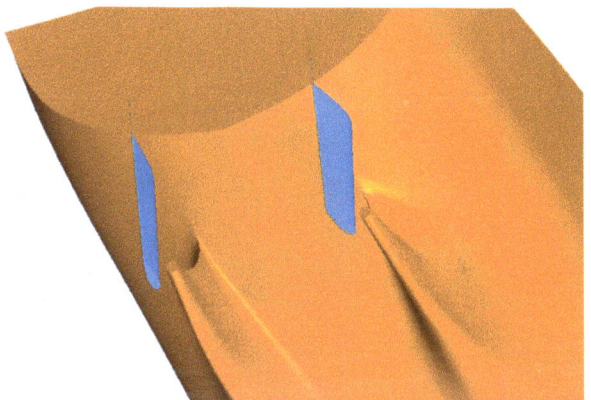

Fig. 1. Arrangement of the aft ship.

3.2 Model Tests Matrix

In the previous work, the numerical simulation of resistance and self-propulsion at a speed of 19.5 kn was carried out. The results show that the ship total resistance is the smallest with rudder angle of 6°. After installing the propeller, the total power consumption of the ship is the smallest with rudder angle of 4°. Based on the results of numerical simulation, in order to verify the resistance and self-propulsion performance of different rudder angles at different speeds, the model test of resistance and self-propulsion was carried out. Resistance tests at 3 speeds × 5 rudder angel and self-propulsion tests at 3 speeds × 4 rudder angles had been carried out. The test matrix of the model tests is shown in Table 2, and Fig. 2 gives the photo of the tested ship model.

3.3 Resistance Tests

For the resistance test, the total resistance R_{Tm} is measured, and the parameters such as the total resistance coefficient C_{Tm}, the residual resistance coefficient C_r, and the effective power E_m at the model scale are analyzed in the design draft state. The calculation formula is shown below, and the rudder angle 0° is used as a reference to compare the

Table 2. Tests matrix of the model tests.

Item	Speed(kn)	Model speed(m/s)	Fr	Rudder angle
Resistance tests	17.5, 18.5, 19.5	1.377, 1.456, 1.534	0.167, 0.177, 0.187	0, 2, 4, 6, 8
Self-propulsion tests	17.5, 18.5, 19.5	.377, 1.456, 1.534	0.167, 0.177, 0.187	0, 2, 4, 6

Fig. 2. Model tests of the ship.

resistance results at different rudder angles.

$$C_{Tm} = \frac{R_{Tm}}{0.5 \, \rho S_m V_m^2} \quad (1)$$

ρ is the water density, S_m is the wetted surface area of ship model, V_m is model speed.

Residuary resistance coefficient:

$$C_r = C_{Tm} - C_{fm} \quad (2)$$

where, Frictional resistance coefficient $C_f = 0.075/(lgR_n - 2)^2$, Reynolds number $R_n = VL_{WL}/v$

$$E_m = V_m(R_{Tm} - F_D) \quad (3)$$

$$F_D = 0.5\rho_m V_m^2 S_m (C_{fm} - C_{fs} - \Delta C_f) \quad (4)$$

ΔC_f is the roughness allowance coefficient, C_{fs} is the friction coefficient of ship, F_D is full-scale ship self-propulsion points.

3.4 Self-Propulsion Tests

For the self-propulsion tests, the rotation speed n_m, thrust T_m and torque Q_m of the propeller are measured at different rudder angles and different speeds under the designed draught state. According to the measurement results, the self-propulsion factor is analyzed and the self-propulsion performance is compared.

$$t = 1 - (R_{Tm} - F_D)/T_m \quad (5)$$

$$P_m = 2\pi n_m Q_m \quad (6)$$

$$\eta_D = V_m(R_{Tm} - F_D)/(2\pi n_m Q_m) \quad (7)$$

t is the thrust deduction, P_m is received power of propeller model, η_D is propulsion efficiency.

4 Results

4.1 Resistance Test Results

Taking the symmetrical rudder angle as the x-axis, the total resistance R_{Tm} and the effective power E_m as the y-axis, the resistance test results are as Figs. 3 and 4.

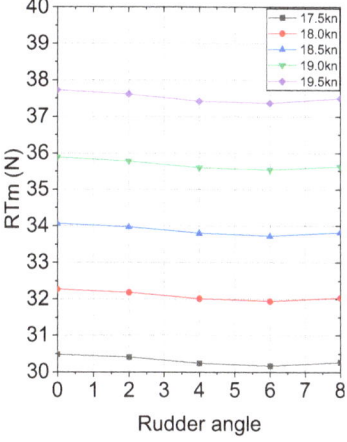 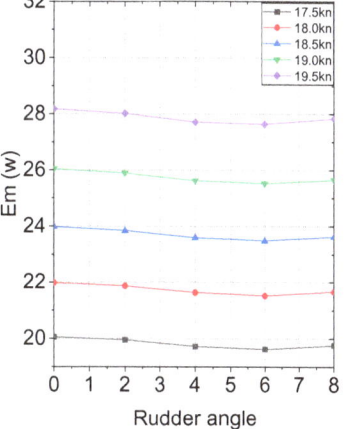

Fig. 3. R_{Tm} at different rudder angle.   **Fig. 4.** E_m at different rudder angle.

Taking the total resistance of the 0° rudder angle as the reference benchmark, the change of resistance when rudder angle changes are shown in the following Tables 3 and 4, where the negative value indicates that the total resistance decreases.

Taking the effective power E_m of 0 degree rudder angle as the reference, the effective power changes at different rudder angles are as follows.

Through the comparison of the above resistance test results, as the rudder angle gradually increases from 0° to 6°, both the total resistance R_{Tm} and the effective power E_m show a downward trend. The resistance R_{Tm} and effective power E_m of different speeds are the lowest with the rudder angle 6°, which is the best rudder angle of resistance. When the rudder angle increases to 8°, the total resistance and effective power begin to increase.

Table 3. R_{Tm} at different rudder angle and different speed.

Vs	Rudder angle				
kn	0	2	4	6	8
17.5	0.00%	−0.25%	−0.79%	−1.02%	−0.70%
18.0	0.00%	−0.27%	−0.78%	−1.01%	−0.70%
18.5	0.00%	−0.28%	−0.77%	−1.00%	−0.71%
19.0	0.00%	−0.28%	−0.78%	−0.97%	−0.72%
19.5	0.00%	−0.28%	−0.81%	−0.94%	−0.59%

Table 4. Em at different rudder angle and different speed.

Vs	Rudder angle				
kn	0	2	4	6	8
17.5	0.00%	−0.52%	−1.66%	−2.14%	−1.46%
18.0	0.00%	−0.55%	−1.62%	−2.10%	−1.46%
18.5	0.00%	−0.58%	−1.60%	−2.06%	−1.47%
19.0	0.00%	−0.59%	−1.61%	−2.00%	−1.48%
19.5	0.00%	−0.58%	−1.66%	−1.92%	−1.21%

4.2 Self-propulsion Test Results

The thrust reduction t, propulsion power P_m and propulsion efficiency η_D of the self-propulsion test results are analyzed. The results are shown in Figs. 5–7 and in Tables 5–7, respectively.

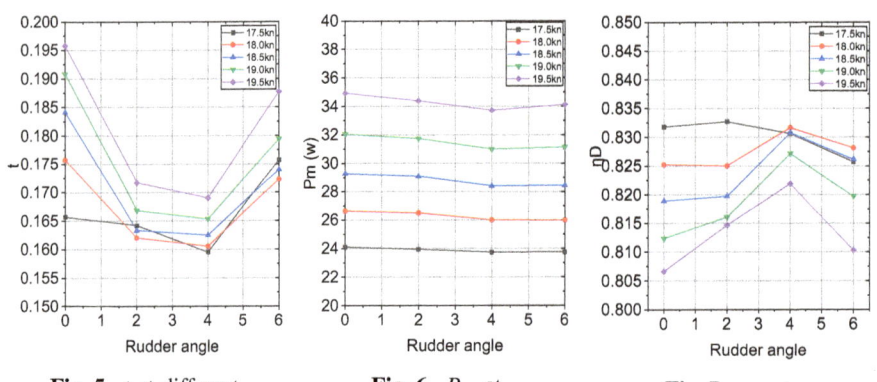

Fig. 5. t at different rudder angle.

Fig. 6. P_m at different rudder angle.

Fig. 7. η_D at different rudder angle.

Table 5. t at different rudder angle.

Vs	Rudder angle			
kn	0	2	4	6
17.5	0.00%	−0.91%	−3.74%	6.04%
18.0	0.00%	−7.80%	−8.65%	−1.94%
18.5	0.00%	−11.30%	−11.73%	−5.49%
19.0	0.00%	−12.58%	−13.36%	−5.97%
19.5	0.00%	−12.31%	−13.69%	−4.14%

Table 6. Pm at different rudder angle.

Vs	Rudder angle			
kn	0	2	4	6
17.5	0.00%	−0.63%	−1.52%	−1.41%
18.0	0.00%	−0.53%	−2.37%	−2.44%
18.5	0.00%	−0.68%	−2.99%	−2.92%
19.0	0.00%	−1.03%	−3.36%	−2.88%
19.5	0.00%	−1.56%	−3.49%	−2.38%

Table 7. η_D at different rudder angle.

Vs	Rudder angle			
kn	0	2	4	6
17.5	0.00%	0.11%	−0.15%	−0.74%
18.0	0.00%	−0.03%	0.78%	0.35%
18.5	0.00%	0.10%	1.44%	0.89%
19.0	0.00%	0.45%	1.82%	0.91%
19.5	0.00%	0.99%	1.90%	0.46%

According to the analysis of test results, as shown in the table. As the rudder angle varieties from 0° to 4°, the thrust reduction t gradually decreases, and the propulsion power P_m decreases gradually. Therefore, the propulsion performance will be improved when rudder angle changes, and the propulsion power is reduced by about 1.52% -3.36%. When the rudder angle is 6°, the propulsion power P_m decreases slightly at 18kn, but increases slightly at other speeds. For the propulsion efficiency η_D, when the speed is 17.5 kn, the improvement of the propulsion efficiency η_D is not obvious, while at other speeds, the increase of the rudder angle effectively improves the efficiency η_D. Therefore, on the whole, the rudder angle of 4° can be used as the best rudder angle.

5 Conclusions

To study the resistance performance and propulsion performance of the rudder of a twin rudder ship at various rudder angles and different speeds, the resistance tests were carried out, the total resistance R_{Tm} and effective power E_m were analyzed. The self-propulsion tests were carried out, the thrust reduction t, propulsion power P_m and propulsion efficiency η_D were analyzed. Based on the test results of the ship model described in this paper, the conclusions were obtained:

1) For the resistance performance, the resistance R_{Tm} and the effective power E_m are the smallest with the rudder angle 6°. At this time, it is the best rudder angle for the resistance performance. The total resistance R_{Tm} is reduced by about 1% and the effective power E_m is reduced by about 2%.
2) For the self-propulsion performance, when the speed is 18.5–19.5 kn and the rudder angle is 4°, the thrust reduction t is the smallest, the propulsion power P_m is the smallest, and the propulsion efficiency η_D is the most improved. At 18.5 kn, the propulsion power P_m is the smallest at 6°, which is similar to the result at 4°. Therefore, on the whole, the rudder angle of 4° can be considered as the best rudder angle.
3) For the twin-stern ship, the optimal arrangement of rudder angle can have a certain energy-saving effect without additional costs.

References

1. Franceschi, A., Piaggio, B., Villa, D., Viviani, M.: Development and assessment of CFD methods to calculate propeller and hull impact on the rudder inflow for a twin-screw ship. Appl. Ocean Res. **125**, 103227 (2007)
2. Molland, A.F., Turnock, S.R.: Marine rudders and control surfaces: Principles, data, design and applications. Elsevier (2007)
3. Badoe, C.E., Phillips, A.B., Turnock, S.R.: Influence of drift angle on the computation of hull–propeller–rudder interaction. Ocean Eng. **103**, 64–77 (2015)
4. Villa D., Franceschi A., Viviani M.. Numerical analysis of the rudder–propeller interaction. J. Mar. Sci. Eng., 8 12 990 2020
5. Liu, J., Hekkenberg, R.: Sixty years of research on ship rudders: effects of design choices on rudder performance. Ships Offshore Struct. **12**(4), 495–512 (2017)
6. Muscari, R., Dubbioso, G., Viviani, M., Mascio, A.D.: Analysis of the asymmetric behavior of propeller–rudder system of twin screw ships by CFD. Ocean Eng. **143**, 269–281 (2017)
7. Jiao, Y.: The development of ship rudder design system and numerical simulation of the hydrodynamic performance. Shanghai Jiaotong University (2020)
8. Sichaun, W.: Experimental research on propulsive characteristics of a mid-sized cruise ship with podded propulsion under static azimuthing conditions. Ship Ocean Eng. **51**(04), 11–15 (2022)
9. ITTC. Resistance Tests: ITTC: 2021
10. ITTC. Propulsion/Bollard Pull TEST: 2021

Open Access This chapter is licensed under the terms of the Creative Commons Attribution-NonCommercial-NoDerivatives 4.0 International License (http://creativecommons.org/licenses/by-nc-nd/4.0/), which permits any noncommercial use, sharing, distribution and reproduction in any medium or format, as long as you give appropriate credit to the original author(s) and the source, provide a link to the Creative Commons license and indicate if you modified the licensed material. You do not have permission under this license to share adapted material derived from this chapter or parts of it.

The images or other third party material in this chapter are included in the chapter's Creative Commons license, unless indicated otherwise in a credit line to the material. If material is not included in the chapter's Creative Commons license and your intended use is not permitted by statutory regulation or exceeds the permitted use, you will need to obtain permission directly from the copyright holder.

Hull Form Optimization of a Twin-Screw Full Ship Based on RANSE Method

Li Zhang[1,2(✉)], Haikui Ren[2], and Jianting Chen[2]

[1] College of Shipbuilding Engineering, Harbin Engineering University, Harbin, China
vzhangli@hrbeu.edu.cn
[2] Shanghai Ship and Shipping Research Institute Co., Ltd., Shanghai, China

Abstract. To optimize a hull form of a twin-screw vessel with a block coefficient of about 0.9 and velocity Fr number of about 0.1, to reduce its resistance. Firstly, numerical simulations of the parent hull form were carried out at design draft and design speed by using CFD methods to analyze the dynamic pressure distribution, wave pattern, streamlines around the hull, etc. and determine the optimization direction. Then, under displacement and geometrical constraints, modifications were made to the bow entrance angle and bow shape to reduce wave resistance and make pressure distribution more uniform. Then the transom length at the aft was increased and the run angle in longitudinal sections was reduced to avoid possible flow separation. Around the stern fins, fullness was increased and run angle decreased to refine and fair the tail-fin shape, to avoid shoulder interference. After several optimization iterations, while satisfying all constraints, the total resistance was reduced by over 5%, free surface wave making was significantly reduced, bow pressure distribution was more uniform, and no obvious flow separation occurs near the stern. The hull form optimization achieved the goal of reducing resistance and saving energy.

Keywords: Hull Form Optimization · Full Ship · Twin-Screw · Resistance · Numerical Simulation

1 Introduction

Twin-screw ships are a common and widely used type of ships. Compared with traditional single-screw propeller ships, twin-screw ships have many advantages in ship maneuverability, propulsion efficiency and safety [1]. Under similar block coefficient, the delivered power of twin-screw ships decreased significantly compared to single-screw ships [1], and wake fraction is smaller. For full ships, due to the shielding effect at the stern, the wake fraction around the propeller disk increases, the non-uniformity rises, and in order to improve the wake at the stern, full ships often adopt the twin-screw type.

The form features of large block coefficient twin-screw ships are mainly: 1) long length of dead flat body; 2) tunnel-type twin-screw; 3) the spacing between the two propeller shafts is generally about 0.45–0.65 times of the ship breadth; 4) due to the fullness of the ship, they generally trial at low speed.

In order to further improve ship performance, modern ship design process often adopts multi-round optimization design to enhance hull performance. Hull form optimization is an important means to improve ship hydrodynamic performance. By optimizing hull form to improve pressure distribution on hull surface, reducing total resistance, and finally achieving optimization of hull performance [2].

In the hull form optimization process, computational fluid dynamics (CFD) is a commonly used method and serves as an important evaluation tool. These works provide good reference for hull form optimization. Zhang Baoji [3] and Chen Hongmei [4] carried out optimization on medium block coefficient ships based on wave-making theory and potential flow software respectively, believing that different tools and optimization methods have guiding and reference significance for reducing resistance. For the optimization design of large ships, Chen Kang et al. optimized and verified through model tests on a second generation 400,000 DWT ultra large ore carrier, achieving good optimization results [5]. Liu Yingliang et al. combined potential flow software, comprehensive optimization tools and viscous flow software to carry out optimization design on a LNG carrier with medium block coefficient, and put forward a recommended design procedure [6]. Dong Yuansheng optimized a large block coefficient twin screw ship and proposed an optimization method based on ship form characteristics [7]. Wang Mingchen used CFD methods to optimize the fins of a river-going twin-screw ship [8]. Pan Lufeng et al. comprehensively optimized the twin skeg hull form of a Ro-Ro ship and obtained ideal optimization results [9]. The development of these works has good reference value for hull form optimization of present work.

The research background of the present work is to design and construct a twin-screw ship with large block coefficient and trial at low speed, for the transportation purpose of general cargo in the coastal region of China. Since the parent ship is an earlier design solution that no longer meets the low carbon requirements, it is planned to optimize the ship based on the mother ship to reduce the total resistance. Firstly, analyze the flow field information of the original ship, formulate optimization strategies combined with its characteristics, compare and analyze optimization schemes, and finally achieve the expected resistance reduction.

2 CFD Method

As a kind of CFD method, Reynolds Averaged Navier-Stokes (RANS) has been widely used in engineering, and its computational method is as follows:

$$\frac{\partial \rho}{\partial t} + \nabla \cdot (\rho \bar{V}) = 0 \qquad (1)$$

$$\frac{\partial}{\partial t}(\rho \bar{V}) + \nabla \cdot (\rho \bar{V} \otimes \bar{V}) = -\nabla \cdot \overline{p_{\text{mod}}} I + \nabla \cdot (\bar{T} + T_{RANS}) + f_b \qquad (2)$$

Where ρ is density, ∇ is Hamiltonian operator, $\bar{V}$ is average velocity, p_{mod} is modified pressure, I is unit tensor, $\bar{T}$ is average viscous tensor, T_{RANS} is additional stress term, fb is object force (such as gravity etc.). Among the various forms of RANS equations, the SST k-ω turbulence model has better advantages in both near-field and far-field adaptability [10], the numerical simulations were performed with CFD software Star-CCM+.

3 Hull Form Optimization

Hull form optimization is the optimization of the target function under a series of constraint conditions by improving the geometric shape. In general, the optimization constraints and target functions are mainly: 1) geometric constraints: hard points at bow and stern, block coefficient, engine room at stern; 2) minimum total resistance.

In order to optimize a full twin-screw ship, optimization analysis under model scale was carried out. The main dimensions of the target ship model are listed in Table 1, with typical characteristics of large block coefficient and medium/low speed.

Table 1. Main Parameters of Target Ship.

L_{pp}(m)	F_r	L_{cb}	B(m)	C_b	Hard points
6.500	About 0.1	0%–2%	1.250	≥0.9	Bow and stern

3.1 Parent Ship Analysis

CFD analysis was performed on the parent ship to obtain mechanical and flow field information. According to ITTC recommended procedures, namely 75–03–02-03 (Practical Guidelines for Ship CFD Applications) [11], For the boundary conditions, the inflow, domain side and domain bottom were the inlet velocity conditions, the pressure outflow was the outlet condition, the ship body surface was the wall condition, the middle plane along the ship was the symmetry plane condition. For ship CFD simulations, the lift force is negligible and the ship surface is streamlined, then the inlet length of upstream can be about one time of the ship length. The outflow surface depends on the free surface and lift, is recommended at least one length downstream of the ship. For the boundary conditions, flow velocity was set as the inlet condition, pressure outflow was the outlet condition, slip wall and the symmetry conditions were used in the present case. The present case mainly focus on the resistance force along the upstream flow. With the benchmark validation data and the authors' experience, the upstream domain is about 1.4 × Lpp, side domain about 1.5 × Lpp, downstream domain about 2.5 × Lpp. The setting of computational domain and mesh generation method are shown in Figs. 1 and 2.

After grid convergence verification with different grid density, results of wave pattern around the hull, hull surface pressure of the fore body and aft body, streamlines of the stern about 0# -3# station and velocity contour at the 2#station are shown in Figs. 3, 4, 5 and 6, respectively.

According to the above results, in the calculation for the parent ship, large bow waves and troughs behind the transom are observed from wave pattern visualizations. In terms of pressure distribution, large areas encountered obvious negative pressure at the bow, peaks of pressure at the shoulders change sharply, mainly due to the local curvature change of bow geometry.

Fig. 1. Computational Domain.

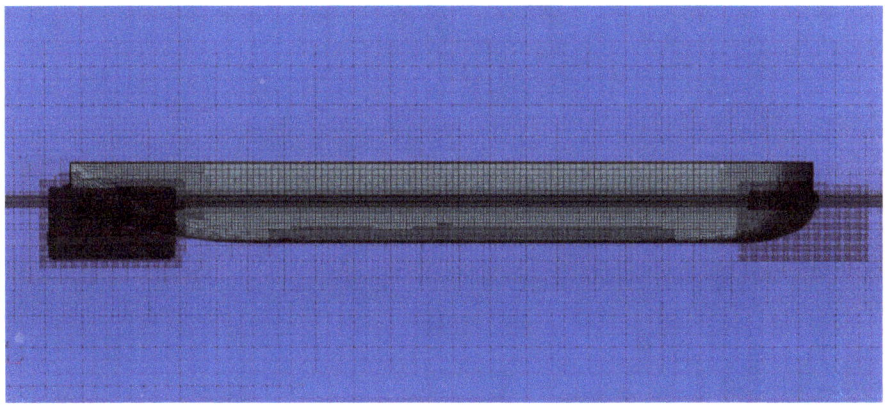

Fig. 2. Mesh Generation around the Ship.

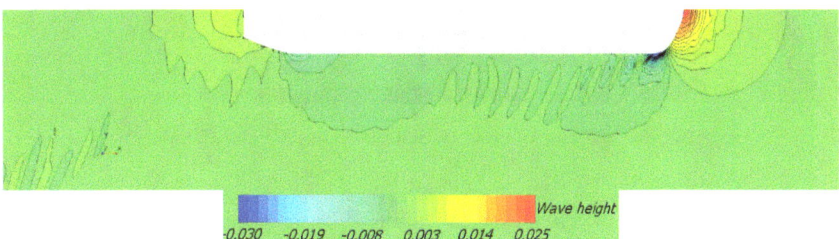

Fig. 3. Wave Pattern around the Hull Surface of the Parent Ship.

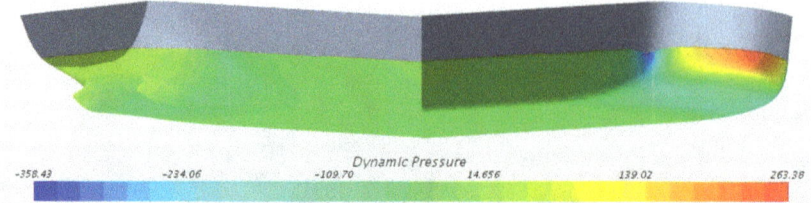

Fig. 4. Pressure Distribution of the Parent Ship.

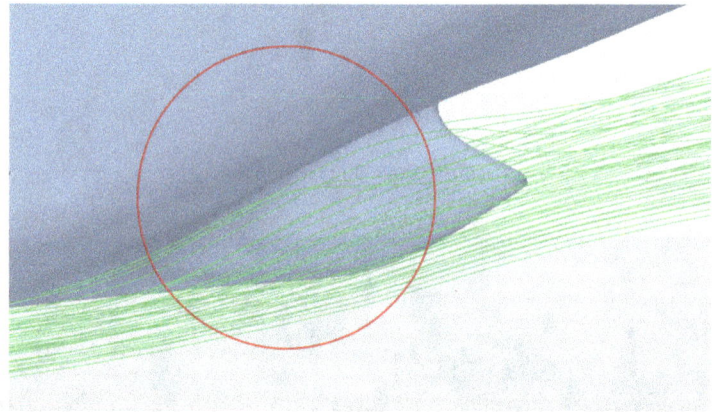

Fig. 5. Streamline of the Stern of Parent Ship.

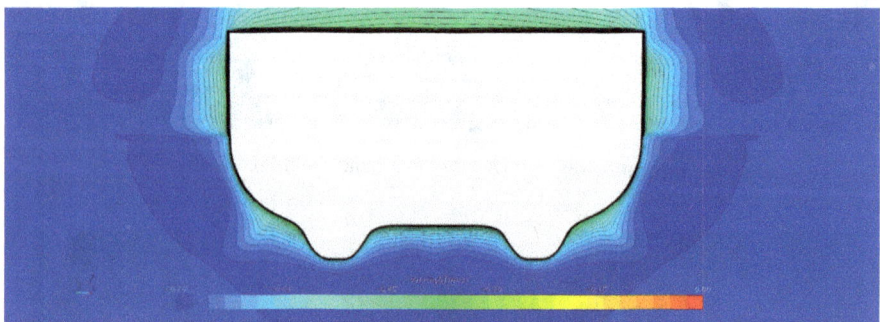

Fig. 6. Velocity Contour of Station 2# of the Parent Ship.

The pressure gradient distribution near station 2# by the transom decreases gradually. According to the streamlines, fluid near bottom of station 1# starts to separate from hull, and trajectory of streamlines fluctuates greatly, indicating obvious stern flow separation.

3.2 Optimization Steps

The analysis of the original ship focuses on reducing resistance by improving the pressure distribution and decreasing the wave making at bow, and reducing flow separation at stern. The block coefficient of hull is about 0.9, main optimization choices are fullness at bow, height of lifting transom, slope at stern and fullness of fins. Main optimization strategies are listed as follows:

1) Both automated optimization tools and manual experience are good choices for the optimization work, but due to the constraints of the hardpoints, and limited amount of space that can be transformed, therefore the manual cut-and-try approach allows for a more efficient selection of optimization directions in a limited number of plans. The hull surface transformation proceeded with in-house geometric tool.
2) For full ships, since the working speed is often relatively low, wave making resistance accounts for a small proportion, so more focus should be placed on friction and viscous pressure resistance optimization.
3) Under the geometric constraints, sharp changes in pressure distribution often lead to significant increase of total resistance, so hull surface pressure distribution should be as uniform as possible. Bow shape approximates streamlined as much as possible, inflow angle does not change abruptly to reduce bow waves caused by high pressure zone.
4) Under the premise of ensuring displacement, lengthen the separation flow region and reduce runbody angle.
5) Reduce probability of flow separation, minimize slope of centerplane section at stern.
6) The purpose of the optimization is to reduce the resistance at the design speed of Fr about 0.1, since the speed is low, the optimization results will not change much at around speeds, therefore the target speed focuses on the design speed only.

4 Results and Discussions

The first step was to reduce the wave resistance, Scheme A was selected among 6 plans. The second step was to reduce the stern resistance and make the flow field uniform, Scheme B was selected among 4 plans. Based on the above analysis and optimization steps, geometric transformations and numerical calculations, Scheme A and Scheme B were obtained and chosen as typical results of the 10 optimization plans. Compared with the parent ship, forebody and aft body was transformed in scheme A, and scheme B changed the aftbody based on scheme A. The line plan comparisons are presented in Figs. 7–9 and Table 2, where the red lines are the parent ship, the green lines are scheme A, and the black lines are scheme B. In Figs. 7 and 8, the green line of scheme A and black line of scheme B were in the same position.

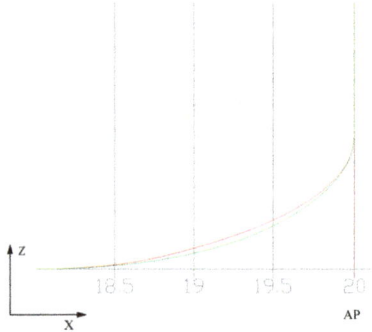

Fig. 7. Mid-Section of station 18#-20#.

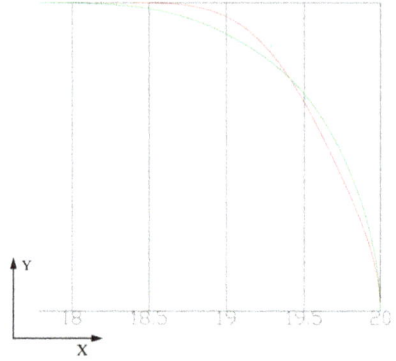

Fig. 8. Waterline of station 18#-20#.

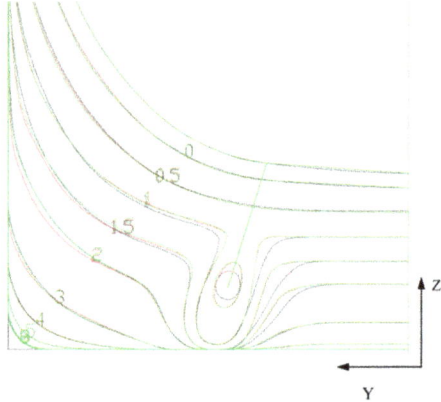

Fig. 9. Frameworks of station 0#-9#.

Table 2. Summary of the Results

Parameters	Parent ship	Scheme A	Scheme B
C_b	0.901	0.903	0.902
L_{cb}	1.363	1.405	1.365
$S(m^2)$	11.947	11.953	11.929
$R_{tm}(\%)$	100%	95.1%	94.7%
$C_r(\times 10^{-3})$	0.735	0.554	0.552

4.1 Summary of Schemes

Parameters and results of schemes were summarized, as shown in Table 2. Original model was referred with parameters as baseline values. Compared with the parent ship, block coefficient of Scheme A and Scheme B increased slightly for both schemes under limited space, met requirement of longitudinal position of buoyancy center. The total resistance and the residual resistance coefficient decreased obviously. Optimization objective of hull form was achieved.

4.2 Scheme A

Considering the hardpoints at bow, it was unrealistic to significantly increase forebody length. In Scheme A, circular fairing transition was implemented at bow to make the angle between shoulder and incoming flow direction smaller, so as to achieve the goal of reducing wave height. And local area of the stern also has geometric changes. The final optimization results are shown in Figs. 10, 11, 12 and 13.

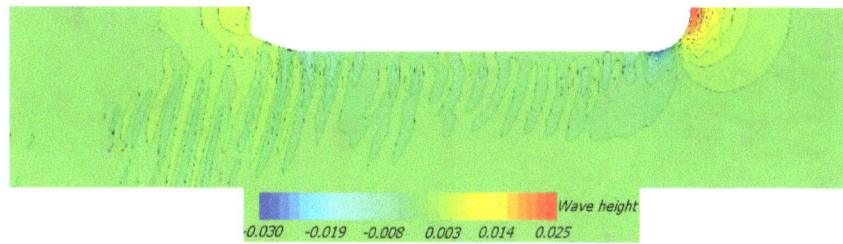

Fig. 10. Wave Pattern around the Hull Surface of Scheme A.

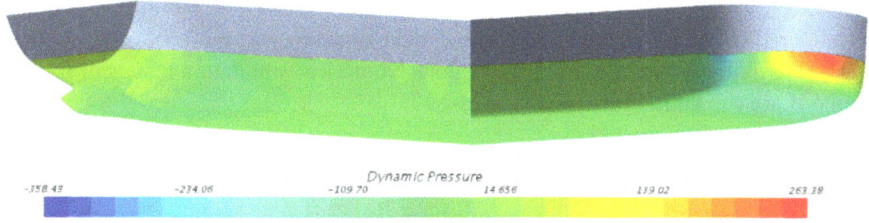

Fig. 11. Pressure Distribution of Scheme A.

According to the graphical results above, wave height at bow was significantly reduced in Scheme A. Wave height variation alongside hull became a series of smaller wave heights, and wave height behind transom was markedly decreased. High negative pressure zone at bow shoulders was reduced and pressure distribution became more uniform. Flow speed around hull was slightly reduced compared to the original model. Observing streamlines at stern, they adhere better to bottom for Scheme A.

4.3 Scheme B

In order to further reduce total resistance, the flow separation region and angle at stern were optimized. The angle beween the flow and the hull surface in longitudinal direction

Fig. 12. Streamline of the Stern of Scheme A.

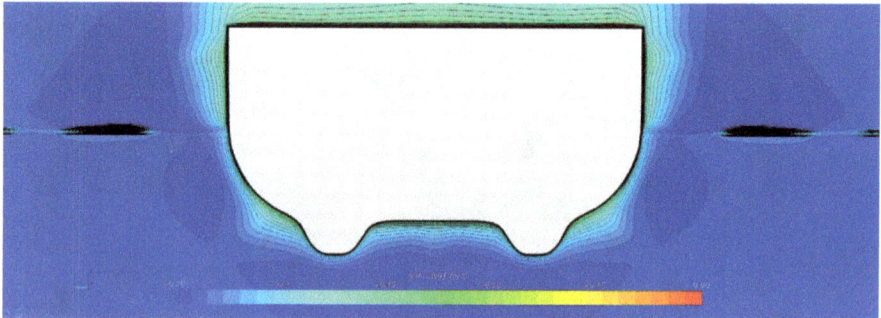

Fig. 13. Velocity Contour of Station 2# of Scheme A.

was reduced, curvature change of hull lines near transom was more smooth, so as to mitigate stern flow separation from water flow. Numerical simulation results of formed Scheme B are shown in Figs. 14, 15, 16 and 17.

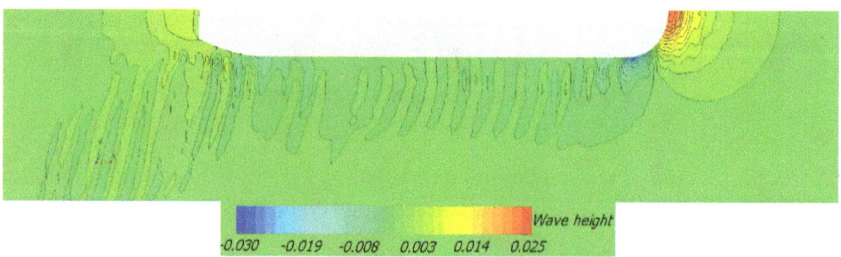

Fig. 14. Wave Pattern around the Hull Surface of Scheme B.

Observing results of Scheme B, wave pattern, hull surface pressure distribution and flow speed around hull had no significant changes compared to Scheme A. But from streamline charts, separation point of some streamlines from stern was postponed, meaning reduced flow separation.

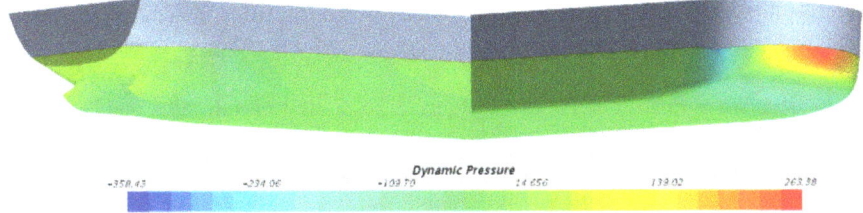

Fig. 15. Pressure Distribution of Scheme B.

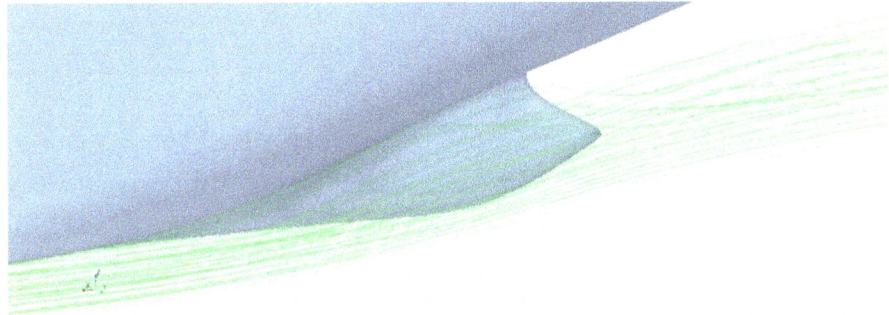

Fig. 16. Streamline of the Stern of Scheme B.

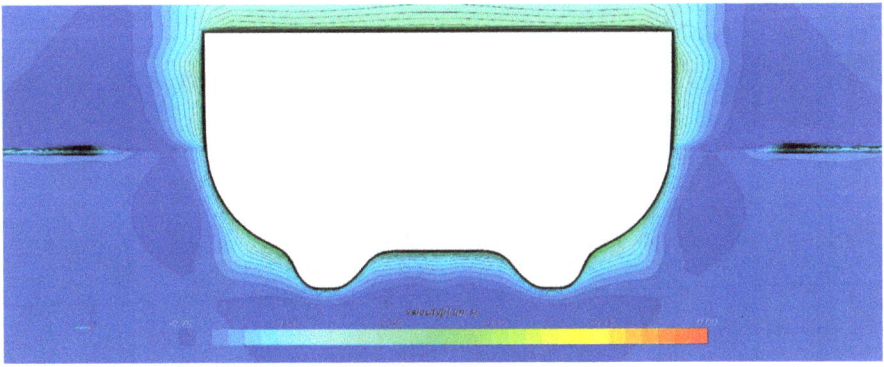

Fig. 17. Velocity Contour of Station 2# of Scheme B.

5 Conclusions

In order to realize hull form optimization of a twin-screw ship with block coefficient up to 0.9 and Fr about 0.1, target parameters and numerical simulation results of original model were analyzed to formulate optimization strategies. Scheme A and Scheme B were obtained with corresponding numerical simulation results. Total resistance of both schemes were reduced by about 5%, achieving optimization goal of hull form. Meanwhile following conclusions were obtained:

1) Due to bow hardpoint constraints, circular fairing transition was implemented at forebody, effectively reducing wave making and avoiding large pressure changes at bow.
2) Excessive wave making leads to flow acceleration around hull, adversely affecting hull form.
3) Longitudinal runbody angle at stern should not be too large, curvature change of separation flow region should be as smooth as possible, otherwise flow separation will be caused and total resistance will be adversely affected.

List of Abbreviations

B(m)	Breadth
C_b	Block Coefficient, $C_b = DisV/(Lpp \times B \times D)$
D(m)	Draft of the ship
DisV(m3)	Displacement Volume of the Ship
Fr	Froud Number, $Fr = V/\sqrt{gL_{pp}}$
g(m/s2)	Acceleration of Gravity
L_{cb}	Buocyancy Center of Longtitude Direction, Positive at Forward Midship
L_{pp}(m)	Length Between Perpendiculars
R_{tm}	(N) Total Resistance
S(m^2)	Wetted Surface
V(m/s)	Velocity of the Ship
C_r	Residual Resistance Coefficient

References

1. Jingzhou, Z.: Catamaran-stern hull's lines character and local transformation method research. Wuhan University of Technology (2007)
2. Huang, F., Chi, Y.: Hull form optimization of a cargo ship for reduced drag. J. Hydrodyn., Ser. B. **28**(2), 173–183 (2016)
3. Baoji, Z.: Research on Optimization Design of Hull Lines and Minimum Re-sistance Hull Form. Dalian University of Technology (2009)
4. Hongmei, C., Rongquan, C.: Study on Applicability of Potential Flow Theory on Optimization Lines. Ship. Eng. **34**, 9-11+19 (2012)
5. Chen Kang, W., Xiaoping, Gang, C.: Optimization of Hydrodynamic Per-formance of New Generation of 400k VLOC. Shipbuilding of China. **57**, 73–79 (2016)
6. Ying, L., liang, Jiang Wu jie, Liu Song.: Hull Form Optimization for Very Large Gas Carriers Based on CFD. Ship & Ocean. Engineering. **47**, 1–5 (2018)
7. Dong Yuansheng. Technical development status and ship type characteristics of large square coefficient twin-stern ship. Transportation Science& Technology, 119-120 (2005)
8. Mingchen, W.: Research and Optimization on Tail of Twin-Skeg in Shal-low-Water Based on CFD Theory. Harbin Engineering University (2018)
9. Lufeng, P., Huawei, Z., Liwei, M., Sujunl, Y.: Comprehensive Optimization of Twin-Skeg Hull Lines of a Ro-Ro Ship. Ship & Boat. **32**, 25–32 (2021). https://doi.org/10.19423/j.cnki.31-1561/u.2021.05.025

10. Menter, F.R.: Two-equation eddy-viscosity turbulence models for engineering applications. AIAA J. **32**, 1598–1605 (1994)
11. ITTC. Practical Guidelines for Ship CFD Applications Practical Guidelines for Ship CFD Applications. 7.5–03–02-03 (2021)

Open Access This chapter is licensed under the terms of the Creative Commons Attribution-NonCommercial-NoDerivatives 4.0 International License (http://creativecommons.org/licenses/by-nc-nd/4.0/), which permits any noncommercial use, sharing, distribution and reproduction in any medium or format, as long as you give appropriate credit to the original author(s) and the source, provide a link to the Creative Commons license and indicate if you modified the licensed material. You do not have permission under this license to share adapted material derived from this chapter or parts of it.

The images or other third party material in this chapter are included in the chapter's Creative Commons license, unless indicated otherwise in a credit line to the material. If material is not included in the chapter's Creative Commons license and your intended use is not permitted by statutory regulation or exceeds the permitted use, you will need to obtain permission directly from the copyright holder.

Effects of Mass Ratio on the Flow-Induced Vibration Characteristics of Rigid Cylindrical Oscillator Supported by Maglev

Guoqiang Lei, Xu Bai(✉), Wen Zhang, Zhenbang Yang, Renwei Ji, and Jianjie Niu

Jiangsu University of Science and Technology, Zhenjiang, China
baixu@just.edu.cn

Abstract. Flow-induced vibration power generation devices are suitable for low-velocity ocean currents and are a current focus of research. Compared to metal springs, oscillator systems supported by permanent magnetic levitation offer greater advantages in capturing ocean current energy. To investigate the influence of mass ratio on the vortex-induced vibration characteristics of a rigid cylindrical oscillator, this study developed a magnetically levitated oscillator model. The magnetic force-displacement curve of the permanent magnetic spring was obtained using COMSOL software, and the magnetic force-displacement function was fitted by the least squares method. Simulations of a single cylindrical oscillator were conducted in STAR-CCM+.The results indicate that when the mass ratio increases, both the vibration amplitude ratio and vibration frequency of the oscillator decrease. At a mass ratio of $m^* = 1.095$ and flow velocity of $U = 0.6$ m/s, the amplitude ratio A^* reached a maximum value of 0.93. At $m^* = 1.095$ and flow velocity $U = 1.0$ m/s, the oscillator frequency f reached a maximum of 1.98 Hz. Under the five different mass ratios, the vortex shedding modes of the single cylindrical oscillator were primarily 2S and 2P, showing significant similarities.

Keywords: Flow-Induced Vibration · Permanent Magnetic Levitation · Mass Ratio · Vibration characteristics

1 Introduction

At present, the research and utilization of ocean energy have become a focal point. The phenomenon in which alternating fluid forces are applied to a solid surface, causing the solid to undergo oscillatory motion, is known as flow-induced vibration [1]. Bernitsas et al. [2, 3] varied the damping ratio at different flow velocities to obtain the maximum amplitude ratio and energy conversion efficiency. Tan Junzhe and Wang Shujie [4] analyzed the effect of turbulence intensity on vortex-induced vibration devices. Bai Xu [5] demonstrated through their research on E-c section oscillators that they exhibit better energy harvesting characteristics under low flow velocity and high damping conditions.

Most current tidal energy harvesting devices focus on cylindrical oscillators supported by metal springs; however, metal springs are unable to optimize the oscillator's energy capture efficiency across different water flow velocities. To address this issue, experts and scholars have proposed using magnetic levitation technology for vibration energy harvesting. Compared to metal springs, magnetic levitation systems allow for easier stiffness adjustment, enabling the oscillator to achieve optimal vibration states at varying flow velocities, thereby enhancing the device's energy capture efficiency.

There have been significant achievements in the research of energy capture by magnetic levitation system. Peter et al. [6] proposed a broadband responsive and adjustable magnetic levitation system that achieves vibration through three longitudinally arranged annular magnets, with a prototype average power density of 600 W/m^3. Masoumi [7] studied a magnetic levitation repulsive wave energy harvesting device, analyzing the system's vibration response using the Duffing equation and estimating the output voltage and frequency using Faraday's law. Bai Xu et al. [8] proposed a vortex-induced vibration device supported by magnetic levitation, and studies indicate that this system outperforms metal springs in terms of vibration performance.

The mass ratio is a key factor influencing the flow-induced vibrations of the magnetic levitated oscillator. This paper integrates the magnetic levitation system with the flow-induced vibration energy harvesting device, utilizing COMSOL software to compute the restoring force-displacement curve of the magnetic spring, and fitting a function for magnetic force and displacement using the least squares method. The fitted function is then incorporated into STAR-CCM+ to calculate the oscillator's amplitude ratio and vibration frequency at five mass ratios, analyzing the effect of mass ratio on the vortex-induced vibration characteristics.

2 Construction and Analysis of the Magnetic Force-Oscillator-Hydraulic System Model

Figure 1 shows the model of the rigid cylindrical oscillator with permanent magnetic levitation support for flow-induced vibration. The oscillator undergoes vortex-induced vibration under the action of water flow force, deviating from its original equilibrium position. The permanent magnets at both ends of the guide rails produce a repulsive force with the permanent magnets on the oscillator, causing it to move back toward the equilibrium position. The oscillator undergoes reciprocating motion under the influence of the water flow force F_{fluid}, the system damping force F_{damp}, and the elastic restoring force F_{spr}, as shown in the force model in Fig. 2.

According to the principle of single-degree-of-freedom systems in advanced structural dynamics, the equation of motion for the vibration of the magnetically levitated oscillator is:

$$my'' + cy' + ky = F_{fluid} + F_{spr} \tag{1}$$

The restoring force experienced by the oscillator depends only on its displacement. Since the magnetic force changes only with the displacement of the oscillator, the restoring force acting on the oscillator is provided by the magnetic spring. Therefore, the equation of motion is:

$$m_{osc}y'' + (c_{structure} + c_{fluid})y' + F_{mag} = F_{fluid} \tag{2}$$

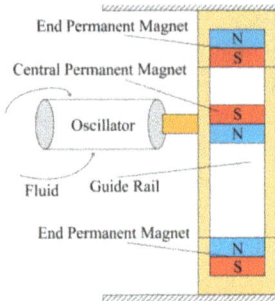 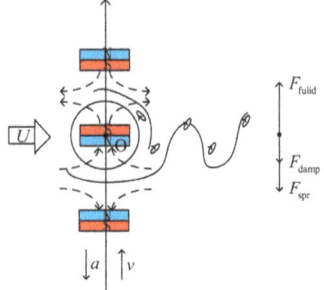

Fig. 1. Physical model of the flow-induced vibration device

Fig. 2. Force model of flow-induced vibration

The fluid dynamic pressure F_{fluid} includes the fluid's viscous force and inertial force [9], which can be expressed as:

$$(m_{osc} + m_a)y'' + (c_{structure} + c_{fluid})y' + F_{mag} = \frac{1}{2}c_y(t)\rho U^2 DL \tag{3}$$

In the equation, m_{osc} is the mass of the vibrating system, m_a is the oscillator's added mass, $c_{structure}$ is the structural damping, c_{fluid} is the fluid damping, $c_y(t)$ is the instantaneous lift coefficient, U is the water flow velocity, D is the diameter of the vibrator, and L is the length of the vibrator.

3 Numerical Simulation Methods and Validation

3.1 Numerical Simulation and Validation of the Magnetic Levitation System

The finite element software COMSOL can be used to simulate and solve for magnetic forces. Figure 3 shows the established geometric model of the magnetic levitation system. The arrows on the fixed permanent magnets at the top and bottom indicate the magnetization direction of the permanent magnets.

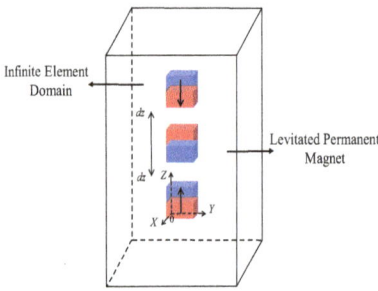 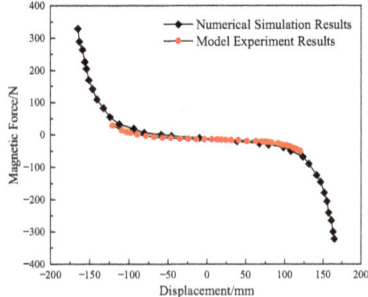

Fig. 3. Geometric model of the magnetic suspension system

Fig. 4. Comparison of magnetic force calculation results

All domain materials are set to Water, Liquid, with a relative permeability of 1.0. The permanent magnets are made of N35 neodymium iron boron, with a residual magnetic flux density set to 1.23 T. The computed results are compared with experimental results for permanent magnets with a side length L = 60 mm and an end-to-end permanent magnet spacing MS = 3.8 D, as shown in Fig. 4. The total trend of the magnetic force is akin to the experimental results, with an average relative error of 5.09%.

3.2 Numerical Simulation and Validation of Flow-Induced Vibration

Establishment of the Fluid-Structure Interaction Model. The fluid-structure interaction calculation model for the oscillator established in this paper is shown in Fig. 5. The entrance of the computing domain is set to the velocity entrance and the exit is set to the pressure exit. The front and back borders are symmetric planes, and the upper and lower borders are set as sliding walls. The surface of the cylindrical oscillator is set to a non-slip wall. Table 1 lists the main parameters of the model.

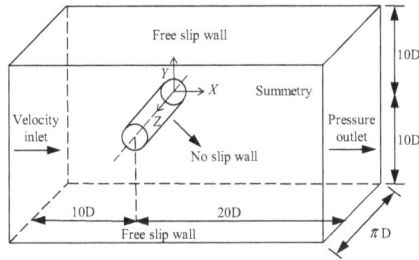

Fig. 5. Diagram of the computation domain and boundary conditions

Table 1. Parameters of the computational model

Oscillator Diameter D/m	Oscillator Length L/m	Vibrating Mass m/kg	Dynamic Viscosity Coefficient η/Pa·s	Flow Velocity U/m·s^{-1}
0.0889	0.2793	2.71	1.074×10^{-3}	0.4 ~ 1.1

Validation of the Fluid-Structure Coupling Model. To verify the correctness of fluid-structure coupling model, this study compared the results with the experimental findings of JH. Lee [3] under conditions of high damping ratio and high Reynolds number. The numerical model is established with the same parameters as the experiment, with specific parameters listed in Table 1.The comparison between simulation results and experiments is shown in Figs. 6 and 7. Figure 6 indicates that the trends of the amplitude ratio curves are similar, with a maximum error of 5.04%, which is within a reasonable range. Figure 7 shows that the overall error is less than 5%, confirming the accuracy of the numerical model. To verify the rationality of the grid, this study set up four different grid models with varying densities, calculating the fluid-induced vibration responses under the parameters listed in Table 1 under the same conditions. Comparing Grid 3 and Grid 4, although the total number of grids increased by 100.08%, the amplitude changed only by 1.10%.Considering the accuracy and efficiency of the calculation, this paper selectes grid three for simulation. This study selected the time step sizes based on the Courant number principle, choosing three time steps: $\Delta t = 0.001$ s, 0.002 s, 0.005 s for the simulations. The results obtained using time steps 0.001 s and 0.002 s were very close, with an error rate of only 0.86%.To ensure accuracy and reduce computation time, this study used a time step of $\Delta t = 0.002$ s for the simulation calculations.

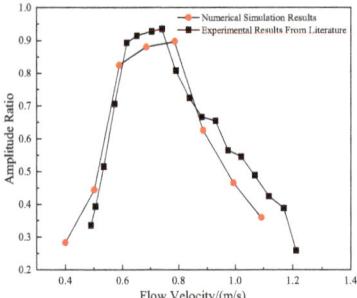
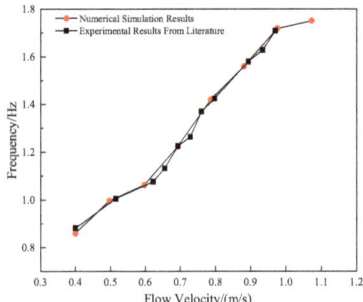

Fig. 6. Comparison of amplitude ratio results

Fig. 7. Comparison of frequency results

3.3 Magnetic-Fluid-Structure Interaction Calculation Method

The magnetic field is simulated using COMSOL Multiphysics software, where a magnetic levitation oscillator model is established to compute the magnetic force, and the magnetic force and displacement curve is fitted as polynomial functions. Next, a fluid-structure interaction simulation is performed in STAR-CCM+ using DFBI six-degree-of-freedom settings. The polynomial function is applied as an external force to the oscillator, and a displacement report is generated. The instantaneous vibration displacement is then fed into the polynomial function to achieve real-time coupling and iterative calculation of the magnetic-fluid-structure interaction.

4 Effects of Mass Ratios on the Flow-Induced Vibration Characteristics of a Magnetically Levitated Cylindrical Oscillator

4.1 Calculation Parameter Settings

To investigate the effects of mass ratio on the flow-induced vibration of a single cylindrical oscillator with magnetic levitation support, this paper refers to the experimental data from J.H. Lee [3]. In the experiments, the mass ratio was set to $m^* = 1.564$. This paper adjusts this value by ±15% twice, resulting in a total of five different mass ratios. According to previous research, a maximum compression of 1.2D for the permanent magnet spring and a magnet edge length of 0.06 m are more suitable [10]. Table 2 lists the model's main parameters.

Table 2. Calculation parameters of the vibrator for different mass ratios

Parameters	Symbols/Units	Numerical Values
Oscillator Diameter	D/m	0.0889
Oscillator Length	L/m	0.2793
Mass Ratio	m^*	1.095, 1.329, 1.564, 1.799, 2.033
Flow Velocity	U/m·s^{-1}	0.4, 0.5, 0.6, 0.7, 0.8, 0.9, 1.0
Damping	c/N·s·m^{-1}	1.344
Dynamic Viscosity Coefficient	η/Pa·s	1.074×10^{-3}

4.2 Amplitude Ratio Analysis

Simulation analysis of the oscillator amplitude under five small mass ratio conditions was performed, and Fig. 8 shows the results. Figure 8 indicates that the overall vibration trend of the oscillator is similar under different mass ratio conditions. The amplitude ratio increases with the increase of flow velocity in the range of 0.4 to 0.6 m/s. The amplitude ratio decreases with the increase of flow velocity in the range of 0.6 ~ 1.0 m/s, and reaches the maximum value in the 0.5 to 0.6 m/s range. Longitudinal comparison indicates that the oscillator's amplitude ratio decreases with the increase of mass ratio. When the mass ratio is 1.095 and the flow velocity is 0.6 m/s, the oscillator's amplitude ratio A^* is at its maximum, with a value of 0.93. Under high mass ratio conditions, oscillator's amplitude variation is more pronounced and sensitive to changes in flow velocity.

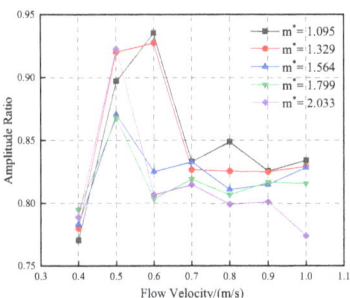

Fig. 8. Amplitude ratio curves for different mass ratios

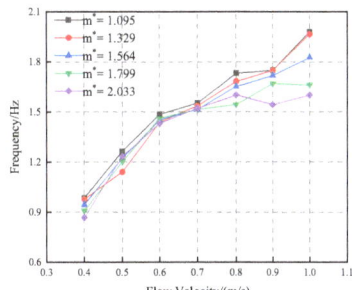

Fig. 9. Frequency curves for different mass ratios

4.3 Vibration Frequency Analysis

As shown in Fig. 9, the oscillator's vibration frequency shows an increasing trend when the velocity increases, reaching its maximum at a flow velocity of 1.0 m/s. Longitudinal comparison shows that the oscillator's vibration frequency decreases with increasing mass ratio. When the mass ratio *is* 1.095 and flow velocity *is* 1.0 m/s, the oscillator's frequency f is at its maximum, with a value of 1.98 Hz. In the range of 0.4 to 0.7 m/s, the frequency variation of oscillators under five mass ratios is nearly the same. However, when flow velocity exceeds 0.7 m/s, the frequency variation of low mass ratio oscillators is more pronounced. To increase the oscillator's vibration frequency, the mass ratio can be reduced.

4.4 Vortex Shedding Pattern Analysis

During vortex-induced vibration, vortices are alternately shed from both sides of the oscillator, and this shedding affects the oscillator's amplitude and frequency. In this study, mass ratios of 1.095, 1.564 and 2.033 were selected to analyze the wake vortex patterns after the oscillation stabilized, under flow velocities of 0.5 m/s, 0.7 m/s and 0.9 m/s. The results is shown in Fig. 10. At flow velocity of 0.5 m/s, two vortices of equal strength and opposite rotation are shed per cycle behind the oscillator, following the 2S pattern [11]. At flow velocities of $U = 0.7$ m/s and 0.9 m/s, two unequal-strength vortex pairs are shed each cycle, following the 2P pattern [11]. As the mass ratio increases, the wake width shows little change at low flow velocities but gradually narrows at higher flow velocities. At the same flow velocity, larger mass ratios result in narrower wakes and lower vortex shedding frequencies, with a corresponding reduction in amplitude and frequency.

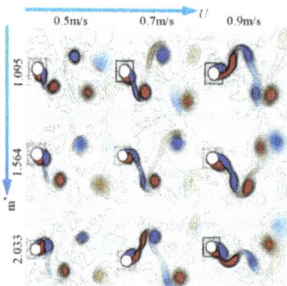

Fig. 10. Comparison of vortex wake diagrams

5 Conclusion

Based on the magnetically levitated cylindrical oscillator used in vortex-induced vibration devices, this study calculated the amplitude ratio, frequency, and vortex shedding patterns of the device at flow velocities ranging from 0.4 to 1.0 m/s by varying the oscillator's mass ratio. The following conclusions were drawn:

1) When the flow velocity increases, the cylindrical oscillator's amplitude ratio first increases and then decreases. The maximum amplitude ratio decreases when the mass ratio increases, with the max amplitude ratio occurring in the range of 0.5 to 0.6 m/s. The oscillator achieves the maximum amplitude ratio of 0.93 when the mass ratio *is* 1.095 and flow velocity *is* 0.6 m/s.
2) The vibrator's frequency increases when the velocity increases and decreases as mass ratio increases. The oscillator frequency f reaches its maximum value of 1.98 Hz as the mass ratio is 1.095 and flow velocity is 1.0 m/s.
3) The tail vortex shedding patterns of the cylindrical oscillator are similar, primarily featuring 2S and 2P modes. Under different mass ratios, the wake shedding frequency and wake width will change, leading to variations in the oscillator's amplitude and frequency.

Acknowledgments. The authors would like to thank the National Natural Science Foundation of China (Grant No.42276225) and the Natural Science Foundation of Jiangsu Province (Grant No. BK20211342) for their financial supports. The authors would like to sincerely thank the reviewers for their valuable and constructive feedback.

References

1. Lian, J., Yan, X., Liu, F.: Development and prospect of study on the energy harness of flow-induced motion. South-to-North Water Transfers and Water Sci. Technol. **16**(1), 176–188 (2018)
2. Bernitsas, M.M., Ben-Simon, Y., Raghavan, K., Garcia, E.M.H.: The VIVACE converter: model tests at high damping and reynolds number around 10^5. J. Offshore Mech. Arc. Eng. **131**(1), 011102 (2009)
3. Lee, J.H., Bernitsas, M.M.: High-damping, high-Reynolds VIV tests for energy harnessing using the VIVACE converter. Ocean Eng. **38**(16), 1697–1712 (2011)
4. Tan, J., Wang, B., Yuan, P., Wang, S., Chen, C., Zheng, Z.: Study on turbulence intensity influence on cylindrical oscillator response of viv tidal energy conversion device. Acta Energiae Solaris Sinica. **41**(10), 20–26 (2020)
5. Bai, X., Sun, M., Zhang, W., Wang, J.: A novel elli-circ oscillator applied in VIVACE converter and its vibration characteristics and energy harvesting efficiency. Energy. **296**, 131143 (2024)
6. Constantinou, P., Mellor, P.H., Wilcox, P.D.: A magnetically sprung generator for energy harvesting applications. IEEE/ASME Trans. Mechatron. **17**(3), 415–424 (2012)
7. Masoumi, M., Wang, Y.: Repulsive magnetic levitation-based ocean wave energy harvester with variable resonance: Modeling, simulation and experiment. J. Sound and Vib. **381**, 192–205 (2016)
8. Bai, X., Shao, L., Niu, J., Han, C.: Analysis of flow-induced vibration amplitude-frequency characteristics of rigid cylindrical oscillator supported by permanent magnet suspension. Acta Energiae Solaris Sinica. **44**(02), 210–217 (2023)
9. Long, T., Tong, S.: Fluid Mechanics, 2nd edn. Chongqing University Press, Chongqing (2018)

10. Shao, L., Bai, X., Niu, J., Sun, M.: Analysis of flow induced vibration response of single cylinder vibrator supported maglev under different flow. Ship Sci. Technol. **45**(04), 14–17 (2023)
11. Song, L., Ji, C., Zhang, X.: Vortex-induced vibration and wake tracing mechanism of harbor seal whisker: A direct numerical simulation. Chin. J. Theor. Appl. Mech. **53**(02), 395–412 (2021)

Open Access This chapter is licensed under the terms of the Creative Commons Attribution-NonCommercial-NoDerivatives 4.0 International License (http://creativecommons.org/licenses/by-nc-nd/4.0/), which permits any noncommercial use, sharing, distribution and reproduction in any medium or format, as long as you give appropriate credit to the original author(s) and the source, provide a link to the Creative Commons license and indicate if you modified the licensed material. You do not have permission under this license to share adapted material derived from this chapter or parts of it.

The images or other third party material in this chapter are included in the chapter's Creative Commons license, unless indicated otherwise in a credit line to the material. If material is not included in the chapter's Creative Commons license and your intended use is not permitted by statutory regulation or exceeds the permitted use, you will need to obtain permission directly from the copyright holder.

Dynamic Risk Analysis for Escort Operations in Arctic Waters Considering the Time Lag Feature of Risk Factors

Jiaxuan Luo[1], Xiaofang Luo[1(✉)], Yingfei Zan[2], and Xu Bai[1]

[1] Jiangsu University of Science and Technology, Zhenjiang, China
luoxiaofang@just.edu.cn
[2] Harbin Engineering University, Harbin, China

Abstract. The escort operation serves as an effective operational mode in Arctic waters. However, this mode carries inherent risks, including potential collisions between icebreakers and escorted ships or instances of escorted ships becoming trapped in ice. To mitigate such risks, this paper conducts a dynamic risk analysis specifically for escort operations in Arctic waters. A dynamic Bayesian network (DBN) model is developed to address the temporal lag associated with risk factors, employing conditional probability distribution functions. This model enables real-time computation of risk probabilities associated with escort operations, thereby offering critical early warning support for informed operational decision-making. Additionally, an analysis of the importance of various risk factors reveals that ice thickness (X5), ice concentration (X6), speed (X7), and distance (X8) are critical factors to the risk associated with escort operations. Based on these key risk factors, ship operators can implement targeted measures to reduce the overall risk of such operations risk analysis.

Keywords: Escort Operation · Dynamic Risk Analysis · Time Lag Feature

1 Introduction

The escort operation is a common mode for icebreakers in Arctic waters, where an icebreaker creates an ice channel while the escorted ship navigates behind it at a recommended distance and speed. From the point of view of safety and operability, the distance between the stern of the icebreaker and the bow of the ship is very important in the escort operation. Because the reduced distance heightens the risk of collisions, while excessive distance may leave the following ship vulnerable to becoming trapped in ice. Thus, the challenge of escort operation is how to prevent accidents. Although accidents cannot be eliminated, their risks can be mitigated through risk analysis to enhance operational safety.

In recent years, numerous scholars have explored the safety of escort operations from various perspectives. Goerlandt et al. [1] quantitatively examined safe navigation conditions for escort operations, focusing on sea ice conditions, ship speed, and the distance between ships in Arctic waters. Khan et al. [2] investigated collisions between the icebreaker and the assisted ship, predicting the maximum traffic flow density and collision probabilities. Xu et al. [3] developed a Bayesian Network model to assess the risk of ice-trapped vessels, considering human, ship, and environmental dimensions along the North Sea Route (NSR). Zhang et al. [4] introduced the 'risk field' concept, highlighting the potential for collisions when the distance between the icebreaker and assisted ship is either too close or too far. They quantitatively analyzed the interactions between the ship risk field and the sea ice risk field, providing a quantitative assessment of safe distances. Xu et al. [5] proposed a hybrid causal logic model to estimate icebreaker-ship collision risk throughout the entire voyage, incorporating human factors specific to escort operations in the Northeast Passage, along with risk management strategies. While these studies offer valuable insights into various aspects of escort operations, dynamic risk analysis remains in its nascent stages. Future research should emphasize the time lag feature of risk factors, particularly in Arctic waters, where the risk at the current time may be affected by the dynamic risk factors at the previous time or earlier.

DBN is particularly effective in addressing the time lag feature of risk factors, making them valuable for reliability analysis, dynamic risk assessment, and elastic evaluation in various fields, including deep water drilling, emergency response, and construction operations. Qian et al. [6] developed a DBN model for ship navigation scenarios, focusing on index selection and data processing to evaluate dynamic environment risks at key nodes in the Arctic Northwest Passage. Their findings indicate that DBNs are highly applicable for inferring risk evolution in maritime operations. Therefore, this paper employs DBNs for dynamic risk analysis in escort operations, considering the time lag feature of risk factors.

The remaining of this paper is as follows: Sect. 2 identifies risk factors through literature review. Section 3 utilizes the constructed DBN model to dynamically calculate risk probabilities and analyze key risk factors. Finally, conclusion is in Sect. 4.

2 Identification of Risk Factors for Escort Operations

2.1 Analysis of Escort Operations

To accurately identify the factors in escort operations, the feature analysis of escort operation is carried out in this paper. In fact, the arrangement of icebreakers and escorted ships is like the one-way road in traffic transportation. According to the car-following theory, ship driving in the following state also has three basic features:

Restriction. The speed of the escorted ship and the distance between two ships are restricted by the speed of the icebreaker.

Delay. After the icebreaker changes its operational state, the escorted ship will change its operational state accordingly. This change does not occur simultaneously.

Transitivity. Changes in the operational state of the icebreaker affect the motion state of the escorted ships in the fleet, propagating backward step by step.

2.2 Risk Factors for Escort Operations

In the literature review, due to few research on the risk factors of escort operations, this paper also sorts out the risk factors of accidents in Arctic waters, and combines expert knowledge and escort operation features analysis, and finally selects 12 risk factors for escort operations [3, 5, 7–9]. These risk factors can be divided into environment factors (Y1) and ship factors (Y2). Environment factors (Y1) can be further divided into meteorology factors (Y11) and hydrology factors (Y12), while ship factors (Y2) can be further divided into ship navigation information (Y21) and ship system (Y22). Meteorology factors include wind (X1), visibility (X2), and wave (X3). Hydrology factors include current (X4), ice thickness (X5), and ice concentration (X6). Speed (X7) and distance (X8) are considered in the Ship navigation information (Y21). Steering system failure (X9), engine failure (X10), navigation system failure factors (X11), and communication equipment failure (X12) are considered in the ship system factors (Y22). It is noted that although personnel factors may also have an impact on the risks in the escort operation, due to the difficulty in obtaining relevant risk data and the strong subjectivity, this aspect is not covered in the selected risk factors.

3 Dynamic Risk Analysis for Escort Operations

3.1 DBN Model Considering Time Lag Feature of Risk Factors

In DBN model, time lag feature refers that the risk at the current time may be affected by the dynamic risk factors at the previous time or earlier. To more clearly explain the time lag feature of risk factors, Fig. 1 presents an example of the DBN model. Among them, Y represents the total risk during the escort operations. X1, X2, X3 and X4 represent dynamic risk factors, and their risk probabilities change over time. X2 has a time-step risk impact lag, while X3 has two time-step of risk impact lag. The probabilities of X1, X2, X3, X4 and Y change over time.

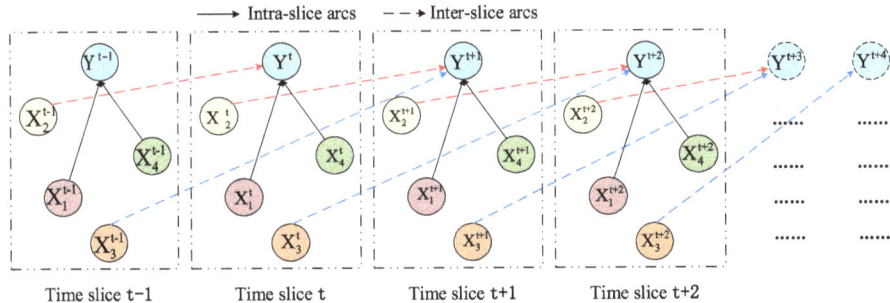

Fig. 1. An example of DBN model considering time lag feature of risk factors.

A DBN can be regarded as a network structure composed of multiple SBN model. The transition network between two time slices of SBN models can be represented as follows [10]:

$$y(X(t)|X(t-1)) = \prod_{i=1}^{n} P(x_i(t)|s_i(t)) \tag{1}$$

where $X(t) = \{x_1(t), \cdots, x_n(t)\}$ is a set of n random variables that construct a SBN within the time slice t. $x_i(t)$ represents the nodes at time t. $s_i(t)$ is the parent node of $x_i(t)$.

It is noteworthy that the parent notes of $x_i(t)$ may be present not only within the current time slice, but also in the preceding time slice.

The joint probability distribution function of nodes in the DBN model is expressed as follows [10]:

$$y(X_{1:T}) = \prod_{t=1}^{T} \prod_{i=1}^{n} P(x_i(t)|s_i(t), G) \qquad (2)$$

where T is the number of the time slice in the DBN model. X represents a set of n random variables that collectively form the DBN model.

According to the topological structure of the time axis extended Bayesian network, the change of factors with time is described to complete the establishment of the DBN model structure of the risk evolution of the escort operation, as shown in Fig. 2, in which node T represents the overall risk of escort operations. The arc with time in the figure expresses the lag of risk impact.

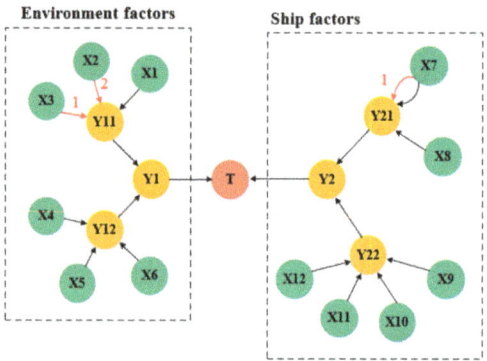

Fig. 2. Risk evolution model of escort operations in Arctic waters.

3.2 Prior Probability of Root Node

Dynamic risk analysis is carried on an actual escort operation in this paper. In 2018, the cargo ship TIANHUI was escorted by the icebreaker VAYGACH and sailed eastward along the Northeast Passage. The prior probability data of the risk factors are derived from the meteorological data of the escort operating environment and Ref. [5]. The state classification criteria of risk factors are shown in Ref. [5]. The root node prior probability of the risk of the escort operation is shown in Table 1.

3.3 Conditional Probability Table of Intermediate Node

Regarding the conditional probability distribution within the risk analysis model, the limited availability of data samples for escort operations necessitates the exclusion of certain established logical relationships. As a result, this study employs the Best-Worst Method (BWM) to effectively allocate the conditional probabilities.

Table 1. The prior probability of root node.

Node	State	Prior probability	Node	State	Prior probability
X1	Fast	0	X7	High	0.715
	Moderate	0.775		Moderate	0.271
	Slow	0.225		Low	0.014
X2	Good	0.625	X8	Short	0
	Moderate	0.125		Moderate	0.125
	Low	0.250		Long	0.875
X3	Good	0	X9	Functioning	0.999
	Moderate	0.625		Failed	0.001
	Harsh	0.375	X10	Functioning	0.999
X4	High	0.233		Failed	0.001
	Low	0.767	X11	Functioning	0.999
X5	Thick	0.500		Failed	0.001
	Moderate	0.125	X12	Functioning	0.944
	Light	0.375		Failed	0.056
X6	High	0.375			
	Medium	0.375			
	Low	0.250			

3.4 Risk Probability Calculation for Escort Operations

In this paper, the DBN model is established by GeNIe software. First, the prior probability of each root node and the conditional probability table of each intermediate node are input. Then, the node state of risk factor is taken as the dynamic evidence input to realize the dynamic calculation of the risk value of the escort operation.

The state of node Y11 is set to be {Normal, Normal, Normal, Abnormal, Abnormal, Normal}, while the state of node Y12 is set to be {Normal, Abnormal, Normal, Normal, Abnormal, Abnormal}. The risk probability of Y1 in each time slice is 0.238, 0.617, 0.238, 0.768, 0.811, and 0.617, respectively. The state of node Y21 is set to be {Normal, Normal, Normal, Abnormal, Normal, Abnormal}, while the state of node Y22 is set to be {Normal, Abnormal, Abnormal, Abnormal, Normal, Normal}. The risk probability of Y2 in each time slice is 0.197, 0.570, 0.570, 0.805, 0.197, and 0.662, respectively. Based on the variations in both ship risk and environment risk states, the changes in the overall risk value of escort operations are illustrated in Fig. 3.

3.5 Importance Analysis of Escort Operations Risk Factors

Using the diagnostic reasoning ability of BN, the posterior probability of root nodes in a certain state can be obtained. The absolute difference between the prior probability and

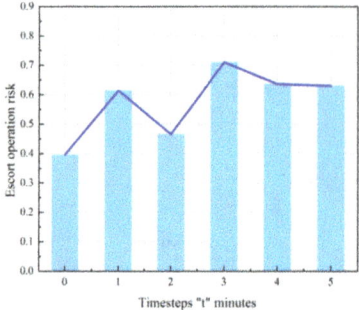

Fig. 3. Risk probabilities of the escort operations (min-to-min).

the posterior probability is calculated, and the absolute difference of the risk factors in different states is accumulated and summed to obtain the importance of each risk factor to the occurrence of a certain state of the target node.

In this paper, the risk node T is the target, with states {Low, Low, High, High, Low, Low} across six time slices. Focusing on the third moment, the absolute difference between prior and posterior probabilities of each root node is shown in Fig. 4, indicating that in {High} state, main risk factors are ice thickness (X5), ice concentration (X6), speed (X7), and distance (X8). Existing research on icebreaker-assisted navigation generally identifies important factors related to environmental (e.g., harsh ice, low visibility) and technical ship factors (e.g., speed, relative distance), which aligns with the factors identified in this paper [8].

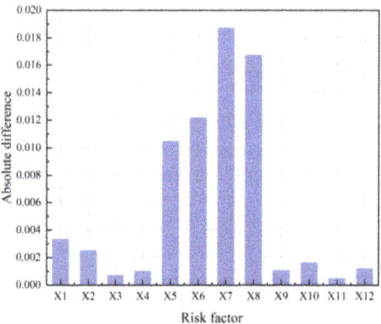

Fig. 4. Absolute difference between prior probability and posterior probability.

During the escort operations in Arctic waters, ship operators need to closely monitor environmental changes, especially the impact of complex ice conditions on icebreaker capacity, speed and hull safety. Crews must strictly implement lookout rules and maintain safe speed and spacing to effectively prevent collision risks.

3.6 Discussion

The uncertainties of risk factor selection, model construction and expect judgement for conditional probability that may influence the results are considered in this paper.

The model and its risk factor correlations, validated by experts experienced in Arctic navigation and risk analysis, have low uncertainty in risk factor selection and model construction. These experts also estimate conditional probabilities within the Bayesian network. However, differing perceptions of risk factors result in variability in their evaluation criteria. Thus, while their insights are valuable, inherent cognitive differences and subjective judgments introduce some uncertainty into the evaluation results.

4 Conclusion

Despite improvements in the Arctic navigation environment, its harsh and variable conditions still challenge safety. Current research on dynamic risk analysis for escort operations is insufficient, prompting this study to develop a dynamic risk analysis model using the DBN method, which incorporates the time lag of risk factors. The model employs a conditional probability distribution function to address the time lag feature of risk factors over time, enabling real-time calculations of overall risk probabilities and providing early warning support for decision-making and accident prevention.

The proposed model serves as a valuable tool for dynamic risk analysis, offering risk control measures to enhance safety and prevent accidents. The limited existing literature on this topic underscores the significance of this research for ship operators and the academic community.

Acknowledgments. The authors would like to thank the National Key R&D Program of China (No. 2024YFC2816400), National Natural Science Foundation of China (No. 52371274) and the Postgraduate Research & Practice Innovation Program of Jiangsu Province (No. KYCX24_4055) for their financial supports. The authors would like to sincerely thank the reviewers for their valuable and constructive feedback.

References

1. Goerlandt, F., Montewka, J., Zhang, W., Kujala, P.: An analysis of ship escort and convoy operations in ice conditions. Saf. Sci. **95**, 198–209 (2017)
2. Khan, B., Khan, F., Veitch, B.: A cellular automation model for convoy traffic in Arctic waters. Cold Reg. Sci. Technol. **164**, 102783 (2019)
3. Xu, S., Kim, E., Haugen, S., Zhang, M.: A Bayesian network risk model for predicting ship besetting in ice during convoy operations along the Northern Sea Route. Reliab. Eng. Syst. Saf. **223**, 108475 (2022)
4. Zhang, D., Han, J., Wu, D., Mao, W.: The model of ship navigation risk field for risk assessment of icebreaker convoy operations. IEEE Trans. Intell. Transp. Syst. **25**(1), 682–696 (2023)
5. Xu, S., Kim, E.: Hybrid causal logic model for estimating the probability of an icebreaker–ship collision in an ice channel during an escort operation along the Northeast Passage. Ocean Eng. **284**, 115264 (2023)

6. Qian, H., Zhang, R., Zhang, Y.: Dynamic risk assessment of natural environment based on Dynamic Bayesian Network for key nodes of the arctic Northwest Passage. Ocean Eng. **203**, 107205 (2020)
7. Zhu, X., Hu, S., Li, Z., Wu, J., Yang, X., Fu, S., Han, B.: Risk performance analysis approach for convoy operations via a hybrid model of STPA and DBN: A case from ice-covered waters. Ocean Eng. **302**, 117570 (2024)
8. Zhang, M., Zhang, D., Goerlandt, F., Yan, X., Kujala, P.: Use of HFACS and fault tree model for collision risk factors analysis of icebreaker assistance in ice-covered waters. Saf. Sci. **111**, 128–143 (2019)
9. Fu, S., Zhang, Y., Zhang, M., Han, B., Wu, Z.: An object-oriented Bayesian network model for the quantitative risk assessment of navigational accidents in ice-covered Arctic waters. Reliab. Eng. Syst. Saf. **238**, 109459 (2023)
10. Chen, J., Zhong, P., An, R., Zhu, F., Xu, B.: Risk analysis for real-time flood control operation of a multi-reservoir system using a dynamic Bayesian network. Environ. Model. Softw. **111**, 409–420 (2019)

Open Access This chapter is licensed under the terms of the Creative Commons Attribution-NonCommercial-NoDerivatives 4.0 International License (http://creativecommons.org/licenses/by-nc-nd/4.0/), which permits any noncommercial use, sharing, distribution and reproduction in any medium or format, as long as you give appropriate credit to the original author(s) and the source, provide a link to the Creative Commons license and indicate if you modified the licensed material. You do not have permission under this license to share adapted material derived from this chapter or parts of it.

The images or other third party material in this chapter are included in the chapter's Creative Commons license, unless indicated otherwise in a credit line to the material. If material is not included in the chapter's Creative Commons license and your intended use is not permitted by statutory regulation or exceeds the permitted use, you will need to obtain permission directly from the copyright holder.

Risk Analysis of Container Ships Navigating in Deep-Water Channels of Yangtze River Estuary in Severe Restricted Visibility

Tingrong Qin(✉), Pingping Luo, Idwimoh Daniel Ogooluwa, and Xiaojing Zhang

Shanghai Maritime University, Shanghai 1550 Haigang Avenue, Pudong New Area, 201306, China
trqin@shmtu.edu.cn

Abstract. Shanghai Port has been the world's largest one in terms of container throughput for many years. Meanwhile, its capacity hits a bottleneck currently due to complex waterways, congested traffic flows, variable meteorological conditions, etc. In order to further increase the container throughput capacity, the Yangtze River Estuary Deep-water Channel, which serves as a vital route leading to the Shanghai Port, is focused on. Especially, risk analysis of container ships navigating in the channel in severe restricted visibility is conducted. The FMEA (Failure Mode and Effects Analysis) method is adopted to evaluate the risk. Based on the method, potential hazardous events that may occur in different sections of the waterway are obtained, and frequency index levels and severity index values are determined to construct a risk matrix. Subsequently, quantitative evaluations were carried out based on this risk matrix for dangerous events that could arise during container ship navigation within the Yangtze River Estuary Deep-water Channel in severe restricted visibility. Following the evaluation process, corresponding risk values and their respective grades in the risk matrix under different visibility conditions are obtained. Then, dangerous events occurring within the ALARP (As Low as Reasonably Practicable) area within 0–200 meters visibility range are analyzed specially, and measures aimed at reducing its frequency and mitigating its consequence are provided correspondingly. After a reevaluation, the risk events in the ALARP area are successfully lowered into the low-risk zone. This study has certain guiding significance for improving the container throughput of Shanghai Port and promoting the high-quality development of Shanghai International Shipping Center.

Keywords: Navigation in Fog · Failure Mode and Effects Analysis · Risk analysis · ALARP

1 Introduction

Amidst the backdrop of the "The Belt and Road" initiative and the expedited execution and promotion of national strategies, the shipping industry stands at a pivotal and historic juncture. The foggy weather that has been prevalent in coastal cities of the East China Sea, has significantly hindered the safe navigation of vessels. It is worth noting in particular fog is the primary factor causing severe restricted visibility, and collision accidents during fog navigation account for more than 70% of all maritime collision accidents, thereby seriously affecting the allocation efficiency of shipping resources within ports. The close correlation between the Yangtze River Estuary Deepwater Channel and Shanghai Port, coupled with the intricate and complex navigation environment of the channel, undoubtedly significantly increases the difficulty and potential risks associated with container ship navigation. Based on the above conditions, the research on the deepwater channel of the Yangtze River Estuary is carried out. In order to better analyze the risks in the entry and exit of container ships within the Yangtze River Estuary Deepwater Channel in severe restricted visibility, this paper has undertaken a risk analysis utilizing the FMEA methodology. FMEA stands as both an accurate and classic means of identifying and evaluating risks.

The following is the current research status of using FMEA method on ship navigation. Zhang Hongqi, Shao Xiaodong, Su Chun and Liu Chun (2017) proposed the ultimate objective of FMEA has been to enable cost-effectiveness of the product, a strategy that identifies problems and avoids or minimizes the potential for failure, thereby enabling the identification of potential failure modes and the analysis of their consequential impacts [1]. Veysi Bashab, Hakan Demirel and Muhammet Gul (2020) jointly studied the risks faced by ship navigation under FMEA conditions, moreover, 23 kinds of basic risks such as extreme weather conditions, ship maneuverability failure and ship personnel injury are taken into account, and some preventive action measures and management methods are put forward [2]. Dong Xiang (2020) introduced the basic methods and analysis steps of FMEA by taking ships as an example, and used this method to conduct an in-depth analysis of the navigation safety of ships, and verified the navigation safety of ships IBS [3].

In summary, based on the current research status, FMEA is indeed a highly effective method. It has achieved certain results in the application of ship navigation safety. By identifying potential failure modes, analyzing risks, and proposing preventive measures, FMEA contributes to enhancing the safety and reliability of ship navigation. The main purpose is to ensure the safety of container ships entering and leaving the deep-water channel of the Yangtze River Estuary in severe restricted visibility. In this paper, the term "severe restricted visibility" pertains specifically to the circumstances where container ships encounter dense fog during navigation. This limitation is subsequently categorized into three distinct levels based on the visibility of the fog: below 200 meters, 200 meters to 500 meters, and 500 meters to 1000 meters.

2 Risk Measure Criteria

The study of ship navigation in severe restricted visibility has always been a subject of great interest, with numerous scholars and research institutions conducting in-depth investigations in this field. These studies encompass various aspects, including the formation mechanism of fog, its impact on navigation safety, and measures to ensure safe navigation of ships under foggy conditions etc. The main research contents of this study include (1) Risk factor identification: systematically identifying various potential risk factors faced by container ships navigating the deep-water channel of Yangtze River Estuary in severe restricted visibility. (2) Risk quantification model construction: based on the identification of key risk factors, selecting a combination of FMEA method and ALARP principle to construct an appropriate model structure, in order to achieve precise quantification of these risks. (3) Risk management and control measures: proposing a series of effective risk management and control measures based on the quantified risk results [4].

2.1 Risk Definition and Expression

According to the risk assessment principles, risk has been literally defined as a set of formulas that encompass both the frequency of occurrence and the severity of its potential consequences, and the expression formula is below.

$$Risk = frequency \times severity \tag{1}$$

In practice, widely used is the multiplicative representation of risk can be transformed using logarithms, as demonstrated below.

$$\log(risk) = \log(frequency) + \log(severity) \tag{2}$$

Thus, the expression formula for the risk index can be derived.

$$Risk\ index\ (RI) = frequency\ index\ (FI) + severity\ index\ (SI) \tag{3}$$

By obviously defining the scores and their values for both the frequency index and the severity index, the risk matrix is derived by a straightforward summation.

2.2 Classification and Value of Ship Accidents

A unified five-level definition and valuation of the frequency of potential accidents has been established during fog navigation, and the levels can be defined and assigned values as frequently, regularly, moderately, seldom, and rarely. Specific details are outlined in Table 1.

In this paper, the severity index for the potential consequences that arise during navigation in foggy conditions is expanded into five levels: catastrophic, particularly serious, serious, minor, and slight. Specific details are outlined in Table 2.

2.3 Risk Matrix

The ALARP principle (As Low as Reasonably Practicable) has been commonly recognized as an approach that can reduce costs to bring profits and make goals more reasonable. The United Kingdom (1974) applied this standard to the law (the HSW Ac) and required the risk management process to comply with this regulation [5]. Curtis C. Thravis and Holly A. Hmatter-Frey (1988) proposed applying the ALARP principle of acceptable risk to the American environment [6]. Gao Yongyi and Dou Yong (2024)

Table 1. Definition and value of accident frequency index.

FI	Definition	Explanation	F(Per ship year)
5	Frequently	An event may occur once a year for every 10 container ships	10–1
4	Regularly	An event may occur once a year for every 100 container ships	10–2
3	Moderately	An event may occur once a year for every 1000 container ships	10–3
2	Seldom	An event may occur once a year for every 10,000 container ships	10–4
1	Rarely	An event may occur once a year for every 100,000 container ships	10–5

Table 2. Definition and value of consequence severity index.

SI	Definition	Explanation(Let's say 200,000 per TEU)	Property damage(RMB)
5	Catastrophic	Multiple fatalities, or total loss of the vessel, or the loss of more than 1000 TEU containers or cargo.	More than 200 million
4	Particularly Serious	A single fatality, multiple severe injuries, or severe damage to the ship, or loss of 100–1000 TEU containers or cargo.	Between 20 million to 200 million
3	Serious	A single severe injury, multiple minor injuries, or non-critical damage to the ship, or loss of 10–100 TEU containers or cargo.	Between two million and 20 million
2	Minor	A single minor injury, partial damage to the ship, or loss of 1–10 TEU containers or cargo.	Between 200,000 to two million
1	Slight	No personnel injuries, minor damage to a part of the ship or equipment, or loss of less than 1 TEU container or cargo.	Less than 200,000

thought its essential goal was to maintain a relatively stable balance between minimizing risk and ensuring cost-effectiveness [7]. Zhang Fengli (2011) proposed hierarchical risk management, and risks were categorized into three different zones: the unacceptable zone, the ALARP zone, and the acceptable zone [8]. The principle of ALARP was divided into three zones, and the ALARP zone means the risk is considered acceptable if the benefit of the action taken to reduce the risk is greater than the cost of the action. As shown in Fig. 1.

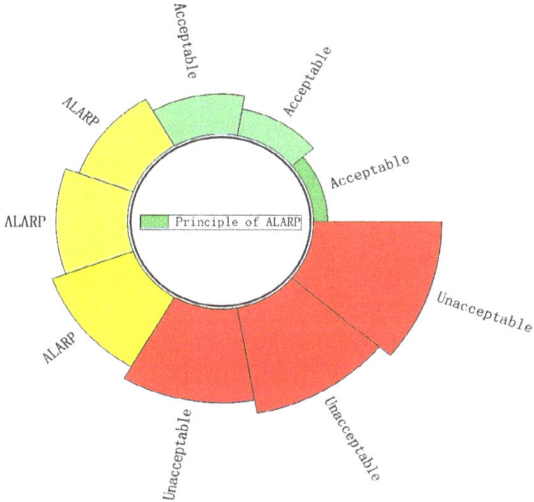

Fig. 1. Principle of ALARP.

By meticulously defining the levels and values of both the frequency index and the severity index and subsequently computing them in strict adherence to the prescribed expression for the risk index, a risk calculation table has been systematically derived, as shown in Table 3.

Table 3. Calculation of shipping safety risk of container ships in severe restricted visibility.

SI	FI	1 Rare	2 Seldom	3 General	4 Usual	5 Frequent
5	Catastrophic	6	7	8	9	10
4	Particularly Serious	5	6	7	8	9
3	Serious	4	5	6	7	8
2	Minor	3	4	5	6	7
1	Slight	2	3	4	5	6

Matrix the risk calculation table and classify the hazard levels. As shown in Fig. 2.

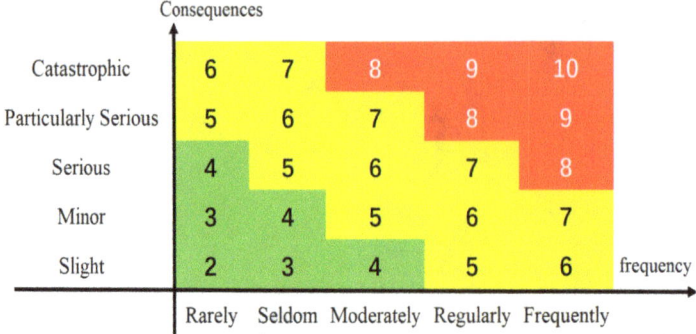

Fig. 2. Schematic diagram of risk matrix partition.

3 Risk Analysis of Navigation in the Yangtze River Estuary Deep-water Channel

3.1 The Location of the Yangtze River Estuary Deep-water Channel

The Yangtze River Estuary Deep-water Channel, which spans a total length of 43 nautical miles, encompasses the area situated between the western boundary line of the A warning zone and the eastern boundary line of the Yuanyuansha warning zone. Commencing from the western boundary of the A warning zone to buoy D12, the channel bottom width measures a substantial 400 meters, thereafter narrowing to 350 meters as it extends from buoy D12 to the eastern boundary of the Yuanyuansha warning zone. The location of the Yangtze River Estuary Deep-water Channel has been depicted in Fig. 3.

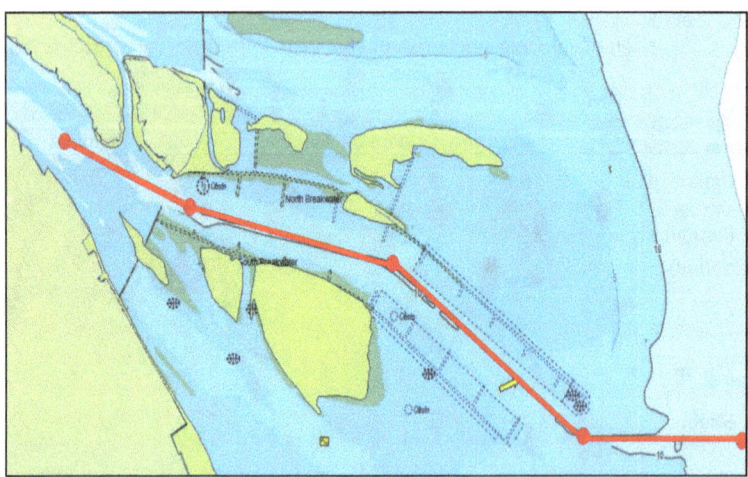

Fig. 3. Schematic diagram of the Yangtze River Estuary Deep-water Channel.

3.2 Influencing Factors

The Yangtze River Estuary Deep-water Channel has been designed to accommodate the navigation requirements of container ships. This research explores in depth the navigation safety issues of container ships in critical shipping lanes where visibility is severely limited. The risk factors of container ships passing through the Yangtze River Estuary Deep-water Channel in severely restricted visibility have been analyzed in four aspects: human factors, machine factors, environment factors, and management factors [9]. Summarise as follows: 1. Human factors: In foggy conditions, the crew's field of view is severely limited, leading to problems such as misjudgment, communication barriers, and operational errors due to severe restricted visibility; 2. Machine factors: The humidity in the air and reduced environmental visibility, due to the severe restricted visibility, which can potentially impact the functionality of the ship's navigation and communication equipment [10]; 3. Environmental factors: The visibility of ships in the surrounding waters is reduced, and obstacles such as other ships, buoys and reefs pose potential hazards to ships; 4. Management factors: The Port Authority is responsible for overseeing departments to implement their own works and port control to ensure that vessels strict compliance with vessel navigation procedures [11].

According to the detailed analysis already carried out above, the various dangerous situations that may occur to container ships entering and leaving the Yangtze River Estuary Deep-water Channel in severe restricted visibility are listed in detail, in Table 4. In general, it can be divided into the following items: 1. Rubbing with navigational aids after yaw; 2. Collision with small-size vessel; 3. Collision with large-scale ship; 4. Grounding after yaw; 5. Striking a reef after yaw; 6. Collision in North Channel anchorage; 7. Collision in Hengsha cross-channel; 8. Collision at South Trough intersection.

Table 4. Risk analysis of container ships in severe restricted visibility.

boundary	Identification of risk		Risk analysis	
	Serial number	Hazardous events	The reason	Consequences
Yangtze River Estuary light boat -D12 buoy	1	Rubbing with buoys and other navigational AIDS after yaw	Severe restricted visibility	rubbing
	2	Collided with a passing small size vessel	Severe restricted visibility	collision
	3	Collided with an opposing large-scale ship	Severe restricted visibility	collision

(*continued*)

Table 4. (*continued*)

boundary	Identification of risk		Risk analysis	
	Serial number	Hazardous events	The reason	Consequences
D12 buoy -D25 buoy	1	Rubbing with buoys and other navigational AIDS after yaw	Severe restricted visibility	rubbing
	2	The aground occurred after the yaw	Severe restricted visibility	aground
	3	strike a reef after yaw	Severe restricted visibility	strike a reef
	4	Collided with a passing small size vessel	Severe restricted visibility	collision
	5	Collided with an opposing large-scale ship	Severe restricted visibility	collision
D25 buoy – Yuanyuansha Light boat	1	Rubbing with buoys and other navigational AIDS after yaw	Severe restricted visibility	rubbing
	2	The aground occurred after the yaw	Severe restricted visibility	aground
	3	Collision occurred with ships in the North Channel anchorage after yaw	Severe restricted visibility	collision
	4	Collision with a vessel in the cross channel(hengsha channel)	Severe restricted visibility	collision
	5	Collided with a passing small size vessel	Severe restricted visibility	collision
	6	Collided with an opposing large-scale ship	Severe restricted visibility	collision
	7	Ship collision occurred at the intersection of the South Trough channel	Severe restricted visibility	collision

3.3 Conduct FMEA Risk Assessment

FMEA mainly includes planning and preparation, structure analysis, function analysis, failure analysis, risk analysis, optimization, result documentation. These processes aimed at improving the quality and reliability of products and processes by identifying potential failure modes and taking appropriate actions to reduce risks. A quantitative assessment of the risks related to ship movements into and out of the Yangtze River Estuary Deep-water Channel in severely restricted visibility has been carried out after adopting the FMEA evaluation approach and developing the frequency and severity indices. The purpose of this study is to assess the risk index for three different vision ranges: below 200 meters, 200–500 meters, and 500–1000 meters. Equation (3) has been used to calculate risk levels, and Table 5 presents the findings of this thorough risk assessment for navigation through the deep-water channel.

Table 5. Results of navigation risk assessment in different visibility.

Boundary	Hazardous events	Visibility of 500-1000 m Risk index	Zone	Visibility of 200-500 m Risk index	Zone	Visibility of 0-200 m Risk index	Zone
Yangtze River Estuary light boat -D12 buoy	Rubbing with buoys and other navigational AIDS after yaw	3	Low-risk	5	ALARP	5	ALARP
	Collided with a passing small size vessel	6	ALARP	6	ALARP	7	ALARP
	Collided with an opposing large scale ship	5	ALARP	6	ALARP	6	ALARP
D12 buoy - D25 buoy	Rubbing with buoys and other navigational AIDS after yaw	4	Low-risk	5	ALARP	5	ALARP
	The aground occurred after the yaw	3	Low-risk	4	Low-risk	4	Low-risk
	strike a reef after yaw	4	Low-risk	5	ALARP	5	ALARP
	Collided with a passing small size vessel	6	ALARP	6	ALARP	7	ALARP
	Collided with an opposing large scale ship	5	ALARP	6	ALARP	6	ALARP
D25 buoy - Yuanyuansha Light boat	Rubbing with buoys and other navigational AIDS after yaw	4	Low-risk	5	ALARP	5	ALARP
	The aground occurred after the yaw	3	Low-risk	4	Low-risk	4	Low-risk
	Collision occurred with ships in the North Channel anchorage after yaw	4	Low-risk	5	ALARP	5	ALARP
	Collision with a vessel in the cross channel (hengsha channel)	6	ALARP	6	ALARP	7	ALARP
	Collided with a passing small size vessel	6	ALARP	6	ALARP	7	ALARP
	Collided with an opposing large scale ship	5	ALARP	6	ALARP	6	ALARP
	Ship collision occurred at the intersection of the South Trough channel	7	ALARP	7	ALARP	7	ALARP

3.4 Risk Quantification and Risk Control Measures

For minimizing risk while maintaining cost-effectiveness, this study has meticulously combed through and analyzed the 13 hazardous events within the ALARP zone, pertaining to visibility ranging from 0 to 200 meters, as outlined in Table 6. The measures are summarized as follows: 1. Maritime Regulation Strengthened: Focus on monitoring small vessels near large container ships during fog. Law enforcement vessels actively patrol; 2. Tugboat at Niupi Reef: Monitors and assists small vessels, executes emergency

Table 6. Risk analysis and control measures of ALARP zone.

Border	Hazardous events	Visibility of 0–200 m		Control measures	Risk index
		Risk index	Zone		
Yangtze River Estuary light boat -D12 buoy	Rubbing with buoys and other navigational AIDS after yaw	5	ALARP	Maritime authorities need to supervise the illegal navigation of small vessels, and the pilot should carefully check the movements of passing vessels, especially the activities of small vessels in the channel, and put their vessels in the right position as the first priority.	3
	Collided with a passing small size vessel	7	ALARP		3
	Collided with an opposing large scale ship	6	ALARP		3
D12 buoy -D25 buoy	Rubbing with buoys and other navigational AIDS after yaw	5	ALARP	Maritime authorities strictly monitor illegal small vessels and have a tugboat stationed near the Niupi Reef (buoy D16) and a tugboat stationed near buoy D25. The pilot should pay close attention to the movement of vessels, especially small vessels in the channel. Buoys and other surrounding objects can be used for multi-target positioning to ensure the accurate position of the hull.	3
	strike a reef after yaw	5	ALARP		3
	Collided with a passing small size vessel	7	ALARP		3
	Collided with an opposing large scale ship	6	ALARP		3

(*continued*)

Table 6. (*continued*)

Border	Hazardous events	Visibility of 0–200 m		Control measures	Risk index
		Risk index	Zone		
D25 buoy – Yuanyuansha Light boat	Rubbing with buoys and other navigational AIDS after yaw	5	ALARP	Maritime authorities supervise illegal sailing vessels, a tugboat stationed near Yuanyuansha. And the pilot should pay close attention to the ships passing through the main channel and the Hengsha channel.	3
	Collision occurred with ships in the North Channel anchorage after yawing	5	ALARP	Maritime authorities strictly monitored small vessels sailing illegally, and arranged for one tugboat stationed near the D25 buoy and another tugboat stationed near Yuanyuansha. The pilot should closely observe and pay attention to the movement of passing ships in the channel, especially the small ships in the channel. In addition, the pilot should also pay close attention to the ships crossing and entering the main channel and the Nancao channel.	3
	Collision with a vessel in the cross channel(hengsha channel)	7	ALARP		3
	Collided with a passing small size vessel	7	ALARP		3
	Collided with an opposing large scale ship	6	ALARP		3
	Ship collision occurred at the intersection of the South Trough channel	7	ALARP		3

towing, offers technical assistance; 3. Tugboat at D25 Buoy: Deployed due to limited visibility, performs emergency towing, warns small vessels; 4. Tugboat near Yuanyuansha: Equipped with infrared telescopes and speakers, monitors and warns small vessels, disperses illegal vessels in tight traffic area.

3.5 Innovations and Advantages

Through the systematic application of a combined approach of FMEA and ALARP principles, a comprehensive assessment of the identified hazardous events was conducted. The results revealed that the identified hazardous events were predominantly distributed within the low-risk and ALARP areas of the risk assessment matrix. Following the implementation of control measures, the risk indices within the ALARP area were significantly reduced to the low-risk area. The aforementioned experimental results have validated the effectiveness of the combined approach utilizing both FMEA and ALARP principles and the advantages can be summarized in two aspects [12].

1) A comprehensive and systematic identification of various hazardous events and their key indicators faced by navigation in severe restricted visibility has been conducted.
2) By employing a comprehensive approach that integrates FMEA methodology and ALARP principles, an in-depth analysis of the identified risks was conducted, involving both qualitative and quantitative assessments. This approach enabled a comprehensive quantification of risks, providing a solid foundation for the formulation of scientific and effective control measures.

4 Concluding Remarks

The primary research problem focus is on how the integrated application of the FMEA and ALARP principle can be leveraged to mitigate the navigation risks faced by container ships in the deep-water channel of the Yangtze River Estuary, particularly in severe restricted visibility. The FMEA approach is first introduced in this study, after which risk is defined and risk measure criteria are established. Subsequently, utilizing the FMEA approach in conjunction with the risk measure criteria that have been established, a risk assessment is carried out for container ships navigating in the Yangtze River Estuary Deep-water Channel in severe restricted visibility. Eventually, the risk assessment results and cost-effectiveness are analyzed, adopting the ALARP principle to find the most beneficial zone to determine improvement initiatives meant to lower the risk indexes [13].

References

1. Hongqi, Z., Xiaodon, S., Chun, S., Chun, L.: Integrated fault diagnosis method based on improved FMEA and failure propagation model. Mech. Sci. Technol. Aerosp. Eng. **1**, 23–28 (2017) (in chinese)
2. Bahan, V., Demirel, H., Gul, M.: An fmea-based topsis approach under single valued neutrosophic sets for maritime risk evaluation: the case of ship navigation safety. Soft. Comput. **24**(24), 184749–118764 (2020)

3. Xiang, D.: Navigation safety analysis of integrated bridge system based on FMEA method. Nav. Archit. Ocean Eng. **02**, 72–77 (2020) (in chinese)
4. Rong, Y., Yaru, W., Yanping, H., Qi, T.: Application of FMEA in Port Equipment Risk Analysis. Port Sci. Technol. **11**, 42–45 (2016) (in chinese)
5. Li, B.: Study on criteria of acceptable risk (MA thesis, Jiangsu University). (in Chinese)
6. Travis, C.C., Hattemer-Frey, H.A.: Determining an acceptable level of risk. Environ. Sci. Technol. **22**(8), 873–876 (1988)
7. Yongyi, G., Yong, D.: Application of cost-benefit analysis to ALARP principles. PRO. **01**, 36–38 (2024) (in chinese)
8. Zhang, F.: Study on quantitative risk assessment criteria of tankers oil spill based on ALARP principle (MA thesis, Dalian Maritime University) (2011). (in Chinese)
9. Weizhong, Q., Yong, M.: Discussion on the implementation of ship voyage plan management for the deep water fairway of the Yangtze river estuary. China Marit. Saf. **05**, 50–52 (2015) (in chinese)
10. Liu, J., Fei, L.: Assessment of water diversion ratio and channel stability in the Yangtze estuary deep-water channel. Port Water. Eng. **02**, 53–57 (2021) (in chinese)
11. Tang, Y.: Movement law of floating mud and fine maintenance dredging in deep water channel of Yangtze River Estuary. Port Water. Eng. 07 89–95 2024. (in Chinese)
12. Hong, C.: The investigation into the safety measures for ships collision accidents in fog. Tianjin of Navigation. **04**, 16–17 (2018) (in chinese)
13. Jianhua, K.: Study on the measures of ships sailing in fog. science & technology. Information. **22**, 112–115 (2022) (in chinese)

Open Access This chapter is licensed under the terms of the Creative Commons Attribution-NonCommercial-NoDerivatives 4.0 International License (http://creativecommons.org/licenses/by-nc-nd/4.0/), which permits any noncommercial use, sharing, distribution and reproduction in any medium or format, as long as you give appropriate credit to the original author(s) and the source, provide a link to the Creative Commons license and indicate if you modified the licensed material. You do not have permission under this license to share adapted material derived from this chapter or parts of it.

The images or other third party material in this chapter are included in the chapter's Creative Commons license, unless indicated otherwise in a credit line to the material. If material is not included in the chapter's Creative Commons license and your intended use is not permitted by statutory regulation or exceeds the permitted use, you will need to obtain permission directly from the copyright holder.

Enhancing Maritime Safety via a Hybrid TTT-LSTM Model for Vessel Trajectory Prediction

Xiangxing Zhou, Yuhao Li, Zhengchuan Qin, and Qing Yu(✉)

Maritime Risk &: Behavioral Science Lab, School of Navigation, Jimei University, Xiamen, China
qing.yu@jmu.edu.cn

Abstract. Real-time prediction of ship trajectory is important to prevent collision and eliminate potential navigational conflicts, and has significant significance for improving navigational safety. However, the declining accuracy of the long-term prediction and the inadequacy of the model to adapt to the complex situation have seriously affected the reliability of ship trajectory prediction. To overcome this problem, this study proposes an advanced prediction model that combines a Test-Time Training (TTT) layer with a long short-term memory network (LSTM). The proposed TTT-LSTM model utilizes LSTM layers to capture the dependency of time series data in ship movement. Moreover, it achieves real-time parameter adjustments through the integrated TTT layer, enabling the model to maintain both real-time responsiveness and predictive efficiency as new data is received. This paper collects the data of the Automatic Identification System (AIS) in the sea area around Xiamen Port and uses it as an empirical case for model validation. Experimental results show that compared with the baseline model, the proposed TTT-LSTM model has a significant improvement in both prediction accuracy and response time. Therefore, it can provide more reliable navigation decision support, reduce navigation risk and optimize route planning.

Keywords: AIS Data · Vessel Trajectory Prediction · Test-time Training · Long Short-term Memory Network

1 Introduction

As global maritime traffic volume grows, the risk of channel congestion and ship collisions has significantly increased [1]. Advanced communication technologies and automated systems allow for real-time analysis and optimization of maritime navigation, improving safety and efficiency. Additionally, using Automatic Identification System (AIS) data enables effective vessel trajectory prediction, crucial for managing complex maritime environments [2].

Although progress in vessel trajectory prediction, challenges persist with data quality and integrity. Dynamic ocean conditions and complex ship-to-ship interactions also complicate real-time predictions. To address these issues, this paper introduces a novel method combining Test-Time Training (TTT) with a Long Short-Term Memory (LSTM) network. This approach incorporates filtered AIS historical data into the TTT-LSTM model, which adjusts and optimizes in real-time during testing. This enhancement significantly improves prediction accuracy in complex scenarios, promising better performance in practical applications.

This paper introduces the main progress at present:(1) A method combining Test-Time Training (TTT) with LSTM is introduced, which enhances the real-time adaptability of ship motion and the understanding of time dynamics, thus improving the prediction accuracy. (2) The dynamic adjustment strategy allows LSTM parameters to be modified in real time according to current conditions during the test process, overcoming the rigidity of the traditional fixed parameter model.

2 Related work

2.1 Methods Based Machine Learning

Machine learning methods like SVM, Decision Trees, RF, and GMM are commonly used in vessel trajectory prediction to analyze patterns from historical data. These methods are effective for simpler, smaller datasets and can capture key trajectory features. However, they struggle with high-dimensional data and long-term dependencies.

Rong et al. modeled vessel movement patterns as Gaussian processes, cleverly addressing the uncertainty on the ship's future position [3]. Hexeberg et al. proposed a single-point neighbor search method based on AIS data for predicting vessel trajectories, predicting the positions within the next 30 minutes by analyzing the course and speed of nearby vessels [4]. Liu et al. proposed a model for trajectory prediction based on SVM, optimized using an adaptive chaotic differential evolution algorithm [5]. However, SVMs have limited generalization ability and can get stuck in local optima. Zhang et al. combined various clustering and regression techniques to assess the dynamics of ship movement and operating conditions for ship trajectory prediction [6]. Traditional vessel trajectory prediction methods like SVM and GMM are limited by strict assumptions, such as Gaussian distributions, which may not apply in dynamic maritime environments, reducing accuracy.

Traditional vessel trajectory prediction methods, such as SVM and GMM, are often limited by strict mathematical assumptions. Although RF can handle large datasets, it falls short in capturing dynamic interactions between ships. While these methods are effective in specific settings, they may struggle to provide reliable predictions in complex and variable maritime environments.

2.2 Methods Based Deep Learning

In recent years, the field of vessel trajectory prediction has seen the emergence of various improved methods based on deep learning, significantly enhancing prediction accuracy and stability. Liu et al. proposed an improved LSTM network method (QSD-LSTM), which incorporates Quaternion Ship Domain (QSD) and trajectory clustering to better consider the dynamic interactions among neighboring vessels, thereby improving prediction accuracy [7]. In the realm of sequence-to-sequence deep learning models, Capobianco et al. proposed a model based on an encoder-decoder RNN for vessel trajectory prediction. Experiments show that it has significant advantages over traditional linear regression and multilayer perceptron methods when processing AIS data [8].

To address this issue, Wang et al. proposed a method based on a Sparse Multi-Graph Convolutional Hybrid Network (SMCHN model), which combines spatial and temporal sparse graphs to capture adaptive interactions and movement trends between vessels [9]. While current methods enhance vessel trajectory prediction, they still struggle to capture dynamic interactions in complex maritime environments, needing further optimization.

3 Methodology

3.1 Model Framework

This paper proposed a vessel trajectory prediction model that combines Test-Time Training (TTT) with Long Short-Term Memory (LSTM) networks to enhance prediction accuracy in dynamic maritime environments. The model utilizes the TTT mechanism to dynamically adjust LSTM weights during the testing phase, improving adaptability to variable conditions. The LSTM component is crucial for capturing long-term dependencies and nonlinear traits from historical trajectory data, which bolsters the model's robustness against noise and outliers. By updating model parameters in real-time based on new inputs, the TTT-LSTM model adapts to changing environments and varying navigation conditions, significantly enhancing its predictive performance. The structural framework of the model, illustrating the integration of TTT and LSTM, is shown in Fig. 1.

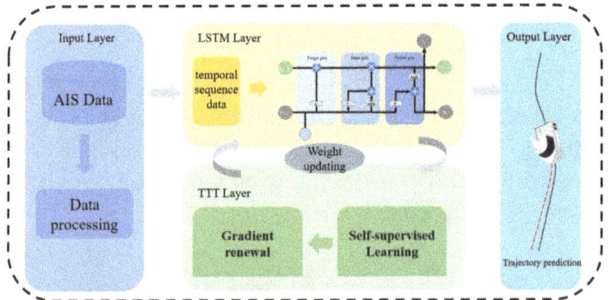

Fig. 1. TTT-LSTM model framework diagram.

Long Short-Term Memory. Vessel trajectory prediction requires forecasting future vessel positions based on historical data, a task challenged by environmental factors such as wind and ocean currents. This paper utilizes Long Short-Term Memory (LSTM) networks, renowned for their proficiency in managing long-term dependencies in time series data through advanced memory cells and gating mechanisms [10]. LSTMs incorporate three gates: the forget gate, which determines what information to discard; the input gate, which selects data to be updated; and the output gate, which releases the final state based on the current cell state. These features enable LSTMs to overcome the gradient issues typically seen in standard RNNs, enhancing their capability to handle complex sequence data. By updating its internal states at each timestep using specific activation functions and operations, the LSTM model ensures robust and accurate predictions in variable maritime conditions, with further operational details provided in Fig. 2.

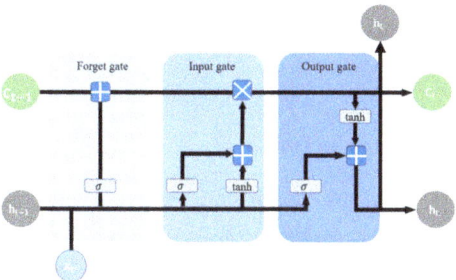

Fig. 2. The unit structure of LSTM

Test-Time Training Layer. The Test-Time Training (TTT) model significantly enhances vessel trajectory prediction by dynamically adjusting its weights during the testing phase, utilizing a self-supervised learning mechanism [11]. This innovative approach enables the model to continuously adapt to changing inputs and environmental conditions, thereby improving prediction accuracy and robustness. Unlike traditional models that maintain fixed parameters after training, the TTT model actively optimizes these parameters in real-time based on a self-supervised loss function, represented by

$$w_t = w_{t-1} - \eta \nabla \ell(w_{t-1}; x_t) \qquad (1)$$

where w_t represents the model parameters at time step; η denotes the learning rate; $\nabla \ell(w_{t-1}; x_t)$ is the gradient of the loss function with respect to the parameters w_{t-1}.

This dynamic adjustment is crucial for handling the variability and unpredictability typical in maritime environments. By applying updates derived from the self-supervised loss at each testing timestep, the TTT model ensures high adaptability and accuracy, even under fluctuating conditions. This continual refinement of parameters helps the model stay aligned with the latest data trends and environmental changes, significantly reducing prediction errors in long-term trajectory forecasting. Overall, the TTT model's ability to adjust its weights in real time during the testing phases makes it particularly effective at managing the complexities of maritime trajectory prediction, ensuring consistent performance amidst diverse and evolving maritime conditions. This process is known as Test-Time Training (TTT) and is programmed into our TTT layer, as shown in Fig. 3.

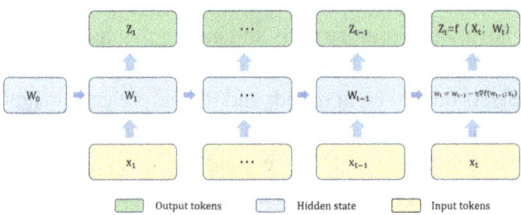

Fig. 3. The unit structure of Test-Time Training (TTT)

4 Case Study

4.1 Data Collection and Preprocessing

To validate our vessel trajectory prediction model, we conducted training and evaluations using AIS data from the Port of Xiamen waters collected in August 2020. This dataset included essential features like latitude, longitude, speed over ground, and course over ground, which are crucial for precise trajectory predictions. We prioritized these features for model input and implemented a rigorous data preprocessing regimen. This included cleaning to eliminate noise and outliers, interpolation to address data gaps, standardization to normalize feature scales, and sliding window sampling to capture the dynamic nature of vessel movements. This preprocessing protocol, crucial for maintaining the accuracy and stability of our predictions, is detailed in Fig. 4.

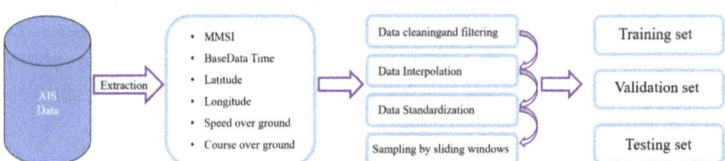

Fig. 4. AIS data preprocessing process.

4.2 Experimental Results and Analysis

This paper evaluates the TTT-LSTM model for vessel trajectory prediction, using 20 frames of historical data to predict the next 10 frames with features including longitude, latitude, speed, and course. Figure 5 shows the loss functions for the TTT-LSTM, Seq2Seq, CNN-LSTM and LSTM models during training, demonstrating a consistent reduction in loss across all models as training progresses.

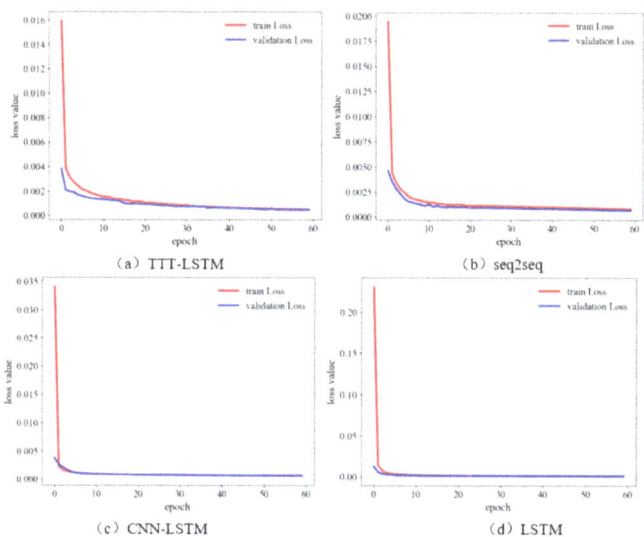

Fig. 5. Loss function comparison of vessel trajectory prediction models. (a) TTT-LSTM model. (b) Seq2seq mode. (c) CNN-LSTM model. (d) LSTM model.

For real-world applicability, comparative experiments were conducted against baseline models in four navigation scenarios: straight-line, initial straight then turn, turn followed by straight, and significant trajectory fluctuations. These tests assessed the model's adaptability and prediction accuracy in both simple and complex navigational tasks, confirming the TTT-LSTM model's effectiveness and its potential for real-time applications. Figure 6 visualizes the comparative results, underscoring the model's robust performance across diverse vessel operations.

In the evaluation of the TTT-LSTM model for vessel trajectory prediction, the results across various scenarios demonstrate its superior performance. In Fig. 6(a), the model shows high precision and adaptability, particularly in scenarios involving complex maneuvers such as turning after a straight path, where it tracks the actual trajectory smoothly and accurately. Figure 6(b) highlights the TTT-LSTM's accuracy in straight-line trajectories, aligning closely with the observed path, outperforming other models which show deviations. In scenarios with a turn followed by a straight path, Fig. 6(c), and in conditions with significant trajectory fluctuations, Fig. 6(d), the TTT-LSTM consistently captures the complex dynamics and sharp changes accurately, maintaining stability and reducing prediction errors.

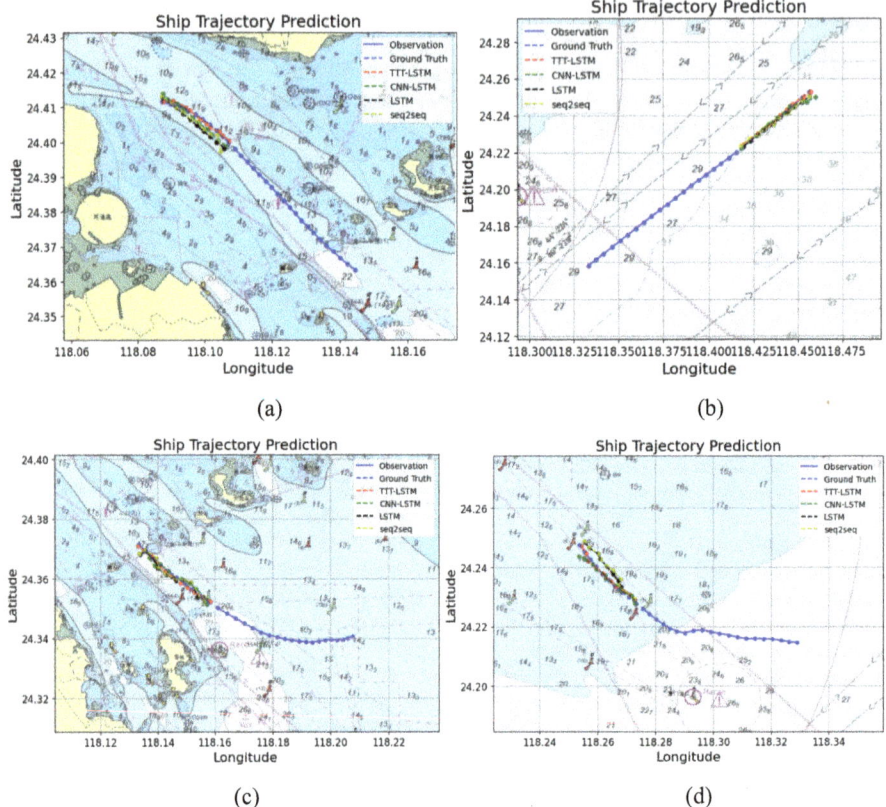

Fig. 6. Prediction results of ship trajectories of different models in different scenarios.

Compared to baseline models like CNN-LSTM, Seq2Seq, and LSTM, the TTT-LSTM model exhibits higher accuracy and robustness across all tested scenarios, including straight-line navigation and complex turning maneuvers. Its strong adaptability and precision in scenarios with fluctuating trajectories further confirm its effectiveness, making it exceptionally suitable for complex navigation conditions. These advantages clearly showcase the TTT-LSTM model's capability to outperform traditional models in accurately predicting vessel trajectories under varied navigational challenges.

Furthermore, a quantitative evaluation of the TTT-LSTM model and baseline models (LSTM, Seq2Seq, and CNN-LSTM) conducts using metrics such as MAE, RMSE, ADE, and FDE to assess their predictive performance. The results, to be detailed in Table 1 and illustrated in Fig. 7, demonstrate that TTT-LSTM significantly outperforms the other models in all metrics. Specifically, TTT-LSTM achieved the lowest error rates, with MAE, RMSE, ADE, and FDE values of 1.198×10^{-3}, 1.637×10^{-3}, 1.987×10^{-3}, and 2.454×10^{-3}, respectively. These findings indicate TTT-LSTM's superior precision and stability in vessel trajectory prediction, confirming its effectiveness and reliability in practical applications.

Table 1. Quantitative Analysis of Prediction Accuracy for Vessel Trajectory Models.

Metric	Models			
	LSTM	Seq2Seq	CNN-LSTM	TTT-LSTM
MAE	2.278×10^{-3}	2.123×10^{-3}	2.199×10^{-3}	1.198×10^{-3}
RMSE	3.352×10^{-3}	2.928×10^{-3}	3.285×10^{-3}	1.637×10^{-3}
ADE	4.237×10^{-3}	3.395×10^{-3}	3.697×10^{-3}	1.987×10^{-3}
FDE	6.143×10^{-3}	6.288×10^{-3}	6.098×10^{-3}	2.454×10^{-3}

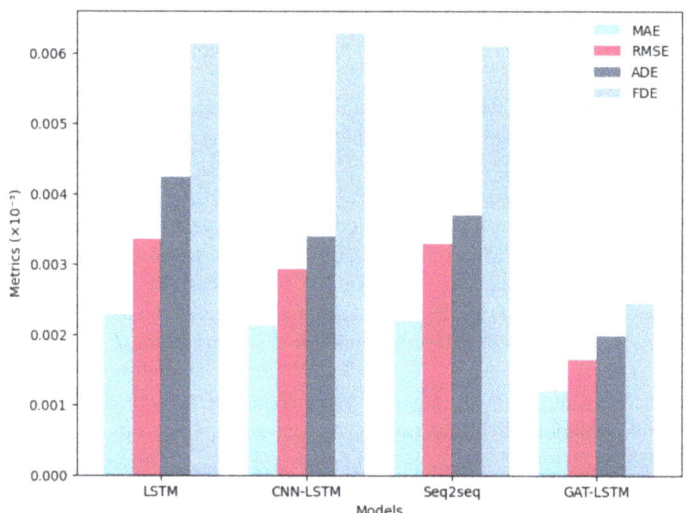

Fig. 7. Performance Comparison of Trajectory Prediction Models.

5 Conclusions

Vessel trajectory prediction plays a pivotal role in enhancing maritime navigation safety and efficiency. It aids in optimizing route planning, reducing collision risks, and improving traffic management and resource allocation, thereby ensuring the safety of waterborne traffic. This paper introduces the TTT-LSTM model, which integrates Test-Time Training (TTT) with Long Short-Term Memory (LSTM) networks to enhance the accuracy of vessel trajectory prediction. By incorporating the TTT mechanism into the LSTM framework, the model constructs a hidden state with linear complexity and enhanced expressive power, achieving superior prediction accuracy and stability in long sequence contexts. A comprehensive evaluation of the TTT-LSTM model across various navigation scenarios indicates that, compared to traditional baseline models, the TTT-LSTM model significantly reduces MAE, RMSE, ADE, and FDE, outperforming LSTM, CNN-LSTM, and Seq2Seq models, thus demonstrating its superiority in vessel trajectory prediction. Although the inclusion of additional features may further improve prediction accuracy, it can also increase model complexity, extend training time, and slow down convergence.

References

1. Lind, M., Hägg, M., Siwe, U., et al.: Sea traffic management–beneficial for all maritime stakeholders. Transp. Res. Procedia. **14**, 183–192 (2016)
2. Veitch, E., Alsos, O.A.: A systematic review of human-AI interaction in autonomous ship systems. Saf. Sci. **152**, 105778 (2022)
3. Rong, H., Teixeira, A.P., Soares, C.G.: Ship trajectory uncertainty prediction based on a Gaussian Process model. Ocean Eng. **182**, 499–511 (2019)
4. Hexeberg, S., Flåten, A.L., Brekke, E.F.: AIS-based vessel trajectory prediction. In: 2017 20th International Conference on Information Fusion (Fusion), pp. 1–8. IEEE, Xi'an, China (2017)
5. Liu, J., Shi, G., Zhu, K.: Vessel trajectory prediction model based on AIS sensor data and adaptive chaos differential evolution support vector regression (ACDE-SVR). Appl. Sci. **9**(15), 2983 (2019)
6. Zhang, M., Kujala, P., Musharraf, M., et al.: A machine learning method for the prediction of ship motion trajectories in real operational conditions. Ocean Eng. **283**, 114905 (2023)
7. Liu, R.W., Hu, K., Liang, M., et al.: QSD-LSTM: Vessel trajectory prediction using long short-term memory with quaternion ship domain. Appl. Ocean Res. **136**, 103592 (2023)
8. Capobianco, S., Millefiori, L.M., Forti, N., et al.: Deep learning methods for vessel trajectory prediction based on recurrent neural networks. IEEE Trans. Aerosp. Electron. Syst. **57**(6), 4329–4346 (2021)
9. Wang, S., Li, Y., Zhang, Z., et al.: Big data driven vessel trajectory prediction based on sparse multi-graph convolutional hybrid network with spatio-temporal awareness. Ocean Eng. **287**, 115695 (2023)
10. Zhao, J., Yan, Z., Zhou, Z.Z., et al.: A ship trajectory prediction method based on GAT and LSTM. Ocean Eng. **289**, 116159 (2023)
11. Sun Y, Li X, Dalal K, et al. Learning to (learn at test time): Rnns with expressive hidden states. arXiv preprint arXiv:2407.04620, 2024

Open Access This chapter is licensed under the terms of the Creative Commons Attribution-NonCommercial-NoDerivatives 4.0 International License (http://creativecommons.org/licenses/by-nc-nd/4.0/), which permits any noncommercial use, sharing, distribution and reproduction in any medium or format, as long as you give appropriate credit to the original author(s) and the source, provide a link to the Creative Commons license and indicate if you modified the licensed material. You do not have permission under this license to share adapted material derived from this chapter or parts of it.

The images or other third party material in this chapter are included in the chapter's Creative Commons license, unless indicated otherwise in a credit line to the material. If material is not included in the chapter's Creative Commons license and your intended use is not permitted by statutory regulation or exceeds the permitted use, you will need to obtain permission directly from the copyright holder.

Study on Ice Resistance of Real Ship under Different Navigation Environments and Ice Thickness

Yujia Zhang[1,2], Bin Mei[1,2], Jiayun Ye[1,2], Congcong Zhao[1,2], Weifeng Li[1,2], and Guoyou Shi[1,2(✉)]

[1] Navigation College, Dalian Maritime University, Dalian 116026, China
sgydmu@dlmu.edu.cn
[2] Key Laboratory of Navigation Safety Guarantee of Liaoning Province, Dalian Maritime University, Dalian 116026, China

Abstract. A field investigation was conducted to analyze the interaction between a tugboat and ice floes during ice-breaking operations in actual ice conditions and the resulting ice channel contraction. The Hertz-Mindlin contact model, implemented through the CFD-DEM numerical analysis software, was used to simulate the fluid-solid coupling between the ice floes and the vessel. A full-scale two-car tugboat was selected as the study object, and the forces acting on the hull under different speeds and ice thicknesses were calculated. The interaction forces between the ship and the ice were obtained and compared with empirical formulas for validation. In the post-processing phase, the simulation results were analyzed to determine the impact of the propeller wake on the vessel's resistance performance, as well as the variation of ice resistance with different experimental factors such as ship speed and ice thickness. This research provides valuable data and practical insights for studies on winter navigation conditions.

Keywords: Navigation technique · CFD-DEM · Ship-ice impact · Ice resistance

1 Introduction

Research on winter navigation conditions focuses on the mechanical properties of ice. Ice resistance is the time-averaged value of all longitudinal forces acting on a ship due to ice. Estimation of a ship's resistance in ice-covered waters is an important factor in the preliminary design of the ship and decision-making during navigation, because it is closely related to the ship's propulsion force, determines the engine power of the ship, and has attracted the attention of many experts in the shipping industry. Li Zhou et al. [1] conducted ice-layer experiments using the icebreaker tanker Uikku model. The effect of ice load on broken ice formation and accumulation under different towing speeds and hull heading angles was analyzed by measuring the broken ice in the broken ice channel. Zhong Wan et al. [2] conducted a series of model tests in an ice tank to study the ice loads of Arctic LNG carriers under different ice conditions. Jakub Montewka et al. [3] and others have completed a two-probability event-oriented model for predicting the speed of a ship on ice and the possible two probability events of ice entrapment.

Through field photography of ice-area ports, we found that in the broken ice channel formed after the icebreaker opens up the shipping channel, due to the long distance between the ships, the broken ice will agglomerate again during the movement of the ship, which will lead to the contraction of the shipping channel. By summarizing previous on-site measurements of actual ships, Xu Ying et al. [4] reviewed the methods of ship-ice interaction and concluded that the experimental method is one of the best methods for studying the characteristics of sea ice and ship-ice interaction, analyzing the resistance characteristics under different navigation environments and ice thicknesses.

2 Actual Ship Experiment Process

Ice load is the main environmental load encountered during polar navigation, which can affect the structural safety of the ship, resulting in serious structural damage or fatigue damage. Wang Jianwei et al. [5] divided the on-site monitoring of ice loads on ship structures into two categories: local ice load on-site monitoring and overall ice load on-site monitoring. Daisuke Ishibashi and others [6] studied ship collisions with floating ice by conducting small-scale model tests with synthetic ice and numerical analysis. Li et al. [7] conducted ship model experiments under broken ice conditions using non-freezing model ice in the room temperature towing tank of Harbin Engineering University. According to the ice tank experiment process recorded in the summary report of the ice resistance sub-conference of the 28th International Towing Tank Conference (ITTC) [8], the ice characteristics parameters to be recorded during the process are ice thickness, elastic modulus, flexural strength, friction coefficient, and ice/water density. The towing facilities and measurement equipment are arranged in the tank experiment to obtain the ship speed and resistance results.

The test speed was selected as 9 kn, and the ice thickness was 0.1088 m. The drone video taken of the brash ice channel under experimental conditions shows the phenomenon of the influence of the propeller wake on the contraction of the brash ice. The brash ice channel behind the wake contracts in a V-shape until the channel closes (see Fig. 1).

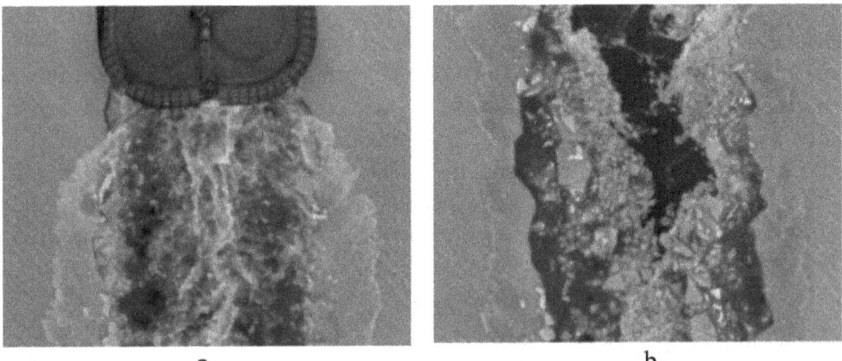

Fig. 1. Diagram of changes in the contraction of the ice channel

3 Modeling Process of the Broken Ice Channel

3.1 Full-scale Ship Model Parameters and Establishment Process

In the development of ice pool technology, the characteristics of ice will be defined as a real scenario that reflects the conditions of floating structures, many scholars use the CFD-DEM model-based STAR-CCM+ modeling software to refine the modeling of the experimental properties of discrete ice. Zhang et al. [10] used the discrete element coupling method and Euler's multiphase flow model to perform CFD-DEM coupling, the collision, accumulation, extrusion, rolling, and sliding characteristics of the ship-ice interaction process are in good agreement with the observations of the ice tank test [11], indicating that the proposed full-scale CFD-DEM method for direct assessment of ship-ice interaction shows reasonable results and practical value.

The experimental modeling started with a full-scale model of the Jacob tugboat with the same square coefficient, which facilitated later comparison with the experimental footage and the results of other tugboat ice load experiments [12]. By loading the geometric surface mesh model, the mesh was refined mainly at the bow and stern (see Fig. 2).

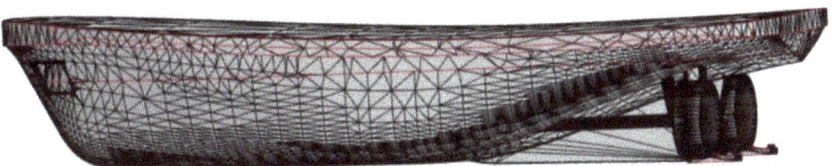

Fig. 2. Tugboat hull mesh refinement

Then, within the created towing tank, the flow field around the hull is refined based on the characteristics of the experiment to generate a continuous and enclosed virtual

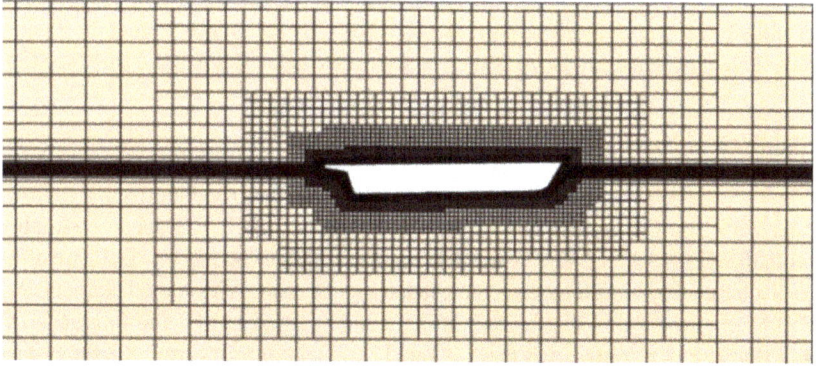

Fig. 3. Fluid mesh refinement

towing tank that includes the ship [13]. The hydrostatic air layer and the contact layer between the hull and water are then refined. The visualization of the fluid grid in the towing tank (see Fig. 3).

3.2 CFD-DEM Coupling Process

The shape of the discrete phase ice particles in the numerical fairway is set as a typical particle based on the shape and thickness of the salvaged ice from real ship tests [14]. The shape is mapped to an irregular polyhedral particle, which is generated by combining multiple basic spherical particles through geometry. The established numerical discrete phase ice particles (see Fig. 4.a). Then, in the discrete particle During the spraying process, the Rosin-Rammler method [15] is used to describe the particle size distribution, and the flat ice is established (see Fig. 4.b), on the premise of meeting the requirements of the experimental ice thickness and channel width, so that the contact force between discrete units can be calculated through the Hertz-Mindlin contact model [16], and the interaction between the simulated ship and broken ice in the channel can be established (see Fig. 4.c).

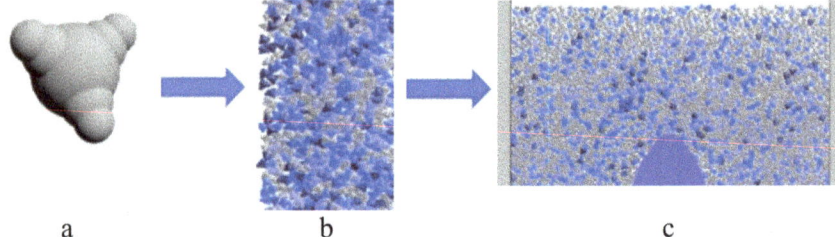

Fig. 4. Discrete element numerical model of the created crushed ice

When modeling the overall ship-ice-water interaction, the mechanical properties of the ice, such as its flexural strength, were set using Bohai Bay winter ice data. To leave enough layers of ice to interact with the ship, an area for generating and ejecting enough ice particles in front of the ship in the channel of the basin was set up. The calculation and establishment process of the layer ice channel is based on the typical ice basin test, and the relevant dimensions are scaled up based on the basic scale of the ship (see Fig. 5).

3.3 Analysis of Numerical Simulation Results

It can be seen from the ship-ice interaction diagram that the propeller discharge flow has a greater effect on the contraction of the broken ice channel. The specific situation is similar to the process of broken ice movement in real ship experiments. Both are affected by the discharge flow on the broken ice particles to generate a Y-shaped channel. As the simulation time increases, the broken ice channel is squeezed by the influence of the inter-ice and ship-ice forces, and the Y-shaped channel contracts from the far end to the near end of the stern until the channel tends to close (see Fig. 6).

Fig. 5. Towboat sailing in the layer of ice channel

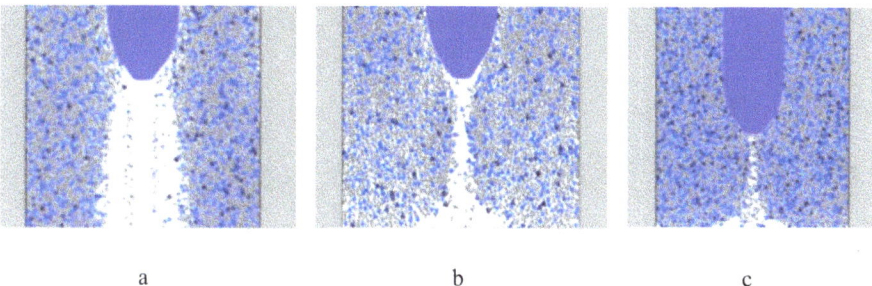

a   b   c

Fig. 6. Simulation diagram of the channel contraction process

By detecting the interaction process between the hull and the ice, the pressure waveform between the hull and the ice can be obtained. It can be seen that the impact of ice particles on the ship is mainly distributed near the bow and stern, and the extrusion effect on the hull is small. The waveform distribution near the stern exhaust flow is more complex (see Fig. 7).

To verify the validity of this tug modeling, typical ice thickness and vessel speed were selected and substituted into the Jacob tug's numerical modeling process to output the ice resistance results. After numerical simulation for several sets of conditions, the mean, standard deviation, and maximum values of the ice load in the ship ice action model results for the full-size tugboat are in good agreement in terms of order of magnitude and dispersion compared with the results of the MV ReneeG tug [17] tested for ice resistance in the Mississippi River and Keinon's formula. Therefore, the results of the real ship ice resistance study at different speeds and ice thicknesses are output (see Fig. 8).

The figure contains three curves corresponding to the results of ice resistance for working conditions at ship speeds of 3, 5, 7, and 9 knots for ice thicknesses of 0.1009 m, 0.129 m, and 0.1375 m, respectively. The overall trend of ice resistance shows a linear increase with increasing ship speed.

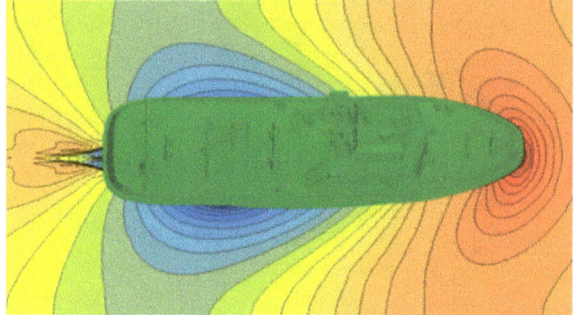

Fig. 7. The waveform of ship-ice interaction in the broken ice channel

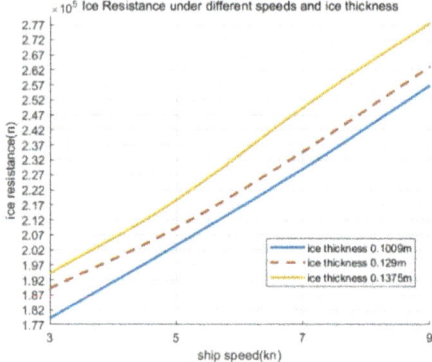

Fig. 8. Results of ice resistance under different speeds and ice thicknesses

The difference in ice resistance between different ice thicknesses gradually widens as the ship speed increases, showing a linear increase in ice resistance follows the principle of hydrodynamics that drag is proportional to speed, i.e. The increase in ice thickness leads to an increase in ice resistance, and thicker ice increases the surface area of the vessel in contact with the ice, which increases the effect of shear and viscous forces, resulting in greater drag for the vessel to overcome.

4 Conclusion and Outlook

By analyzing the research status of winter navigation at home and abroad, it is found that tugboats are more suitable for the working conditions of winter harbors due to their dynamic characteristics, so field tests are carried out on real ship, it is found that the ice-broken channel is contracted in a Y-shape at the stern of the ship. Therefore, the influence of the mechanical properties of ice and the working conditions such as the speed on the ship's ice-broken and ice-broken channel is analyzed and researched. The experimental results of ice resistance under different navigational environments and ice thicknesses are output by establishing a towed ice pool test on a full-size tugboat ship model. And compare it with the ice resistance experiment of MV ReneeG tugboat in the Mississippi River, select the conditions that meet the application conditions of Keinonen's empirical formula for comparison, verify the validity of the experimental ship model, and then carry out multiple sets of resistance experiments under navigational conditions and ice conditions in line with the conditions of winter navigation, and validate the principle that hydrodynamic resistance is directly proportional to the speed for the tugboat's ice resistance experiments.

Experience and experimental comparison results verify the importance of hull design and propulsion system optimization when navigating in icy areas, especially in high-speed and thick ice conditions, and that ships should give full consideration to sea ice conditions when navigating in icy areas. Determining reasonable speeds and routes effectively reduces ice resistance, and these factors need to be taken into account when designing and operating ships in ice areas to optimize the energy consumption and safety of the ship. At high speeds or in thick ice conditions, stronger power systems or special designs may be required to overcome the increased drag to effectively meet the challenges of winter navigation.

References

1. Zhou, L., Riska, K., Von Bock Und Polach, R., Moan, T., Su, B.: Experiments on level ice loading on an icebreaking tanker with different ice drift angles. Cold Reg. Sci. Technol. **85**, 79–93 (2013)
2. Wan, Z., Yuan, Y., Tang, W.: Experimental investigation on ice resistance of an Arctic LNG carrier under multiple ice breaking conditions. Ocean Eng. **267**, 113264 (2023)
3. Montewka, J., Goerlandt, F., Lensu, M., Kuuliala, L., Guinness, R.: Toward a hybrid model of ship performance in ice suitable for route planning purposes. J. Risk Reliab. **233**(1), 18–34 (2019)
4. Xu, Y., Hu, C., Chen, G., Xu, Y.J.: A review of ship-ice interaction research methods. Ship Mech. **23**(1), 110–124 (2019) (in Chinese)
5. Jianwei, W., Qinglin, D., Shunying, J.: Advances in in situ measurement and inversion of ice loads on ship structures during navigation in icy regions. Adv. Mech. **50**, 93–123 (2020) (in Chinese)
6. Ishibashi D.; Shigihara T.; Konno A. Experimental and numerical investigation of model-scale collision of ship with ice floe 2014
7. Xiayan, L.: Research on the resistance performance of ships navigating in ice areas. Harbin Engineering University, M.S. (2018) (in Chinese)
8. 28th ITTC meeting report 2017 28Th ITTC. 2017. https://www.ittc.info/media/7825/19-sc-ice.pdf (accessed 2024-06-21)
9. Hu, J., Zhou, L.: Experimental and numerical study on ice resistance for icebreaking vessels. Int. J. Nav. Archit. Ocean Eng. **7**(3), 626–639 (2015)
10. Wang, C., Feng, Z., Li, X., Li, P.: An analysis of ice resistance and ice response of ships sailing in broken ice areas. China Ship Res. **13**(1), 73–78 (2018) (in Chinese)
11. Zilin, L., Yu, L., Shanshan, S., Yunliang, L., Shunying, J.: Discrete element model and ice load analysis for ship navigation in broken ice area. Chin. J. Theor. Appl. Mech. **45**(6), 868–877 (2013) (in Chinese)
12. Luo, W., et al.: Numerical simulation of an ice-strengthened bulk carrier in brash ice channel. Ocean Eng. **196**, 106830 (2020)

13. Tuhkuri, J., Polojärvi, A.: A review of discrete element simulation of ice-structure interaction. Phil. Trans. R. Soc. A. **376**(2129), 20170335 (2018)
14. Cundall, P.A., Strack, O.D.L.: A discrete numerical model for granular assemblies. Géotechnique. **29**(1), 47–65 (1979)
15. Jin, Q.: Numerical prediction of ice resistance of polar floating/broken ice ships based on CFD-DEM. Dalian Maritime University, Master (2021) (in Chinese)
16. Zhang, J., Zhang, Y., Shang, Y., Jin, Q., Zhang, L.: CFD-DEM based full-scale ship-ice interaction research under FSICR ice condition in restricted brash ice channel. Cold Reg. Sci. Technol. **194**, 103454 (2022)
17. Ashton G. D.; DenHartog S. L.; Hanamoto B. Icebreaking by Tow on the Mississippi River 1973

Open Access This chapter is licensed under the terms of the Creative Commons Attribution-NonCommercial-NoDerivatives 4.0 International License (http://creativecommons.org/licenses/by-nc-nd/4.0/), which permits any noncommercial use, sharing, distribution and reproduction in any medium or format, as long as you give appropriate credit to the original author(s) and the source, provide a link to the Creative Commons license and indicate if you modified the licensed material. You do not have permission under this license to share adapted material derived from this chapter or parts of it.

The images or other third party material in this chapter are included in the chapter's Creative Commons license, unless indicated otherwise in a credit line to the material. If material is not included in the chapter's Creative Commons license and your intended use is not permitted by statutory regulation or exceeds the permitted use, you will need to obtain permission directly from the copyright holder.

Observations on Size Distribution Characteristics of Ice Floes in Icebreaking Channel

Congcong Zhao, Bin Mei, Yujia Zhang, Jiayun Ye, Weifeng Li, and Guoyou Shi[✉]

Navigation College, Dalian Maritime University, Dalian 116026, China
yejiayun1999@dlmu.edu.cn

Abstract. This study used a drone to conduct on-site aerial observations of the status of the ice floes on the frozen channel after the tugboat broke the ice, and collected relevant video image data. These data are used to analyse the size distribution (FSD) characteristics and geometric shape characteristics of floating ice in the channel after icebreaking. Research results show that floating ice has a power-law number density, and its power-law index threshold is like that of natural floating ice. Moreover, the area-weighted FSD of the floating ice shows a modal distribution, indicating that after the channel ice is broken, a group of floating ice with a dominant size is formed. The probability density distribution of ice floe size is fitted by Weibull function. In addition, through statistical analysis of the area, circumference and mean caliper diameter (MCD) of the ice floe, it was found that the size of the ice floe has little impact on its shape parameters (ice floe roundness, ratio of maximum and minimum caliper diameters). This research data set provides an important basis for exploring the geometric characteristics and distribution characteristics of floating ice after ships break ice, and provides data support for analysing the physical characteristics of the interaction between ships and natural sea ice.

Keywords: Floe Size · Floe Shape · Broken Ice · Floe Drift · Power Law

1 Introduction

At present, the calculation of ice resistance and the simulation of floating ice movement after the ship breaks ice have become research hotspots. Key parameters for establishing a numerical model of sea ice include ice concentration, ice thickness and ice floe geometry. Studies have explored the impact of these parameters on ship ice resistance. For example, Yang et al. [1] found that the shape of ice floe significantly affects ice load; Tang et al. [2] analyzed ice resistance, ship speed, ice concentration and other factors. Relationship. The digital ice pools constructed by these studies are usually based on the characteristics of naturally cracked ice floes, and their data are derived from macro-geographical statistics, such as Denton et al. [3], Qin Zhang and Nick Hughes et al. [4], and Li et al. [5] respectively. The size distribution of ice floe was studied using aerial photography, satellite images, etc.

However, the above-mentioned research mainly focuses on floating ice in its natural state, and there is a lack of research on ice breaking after ship icebreaking. This article aims to explore the size, shape data and geometric distribution characteristics of ice floes after icebreaking, and provide more realistic data support for polar shipping safety assessment, icebreaking operation strategy optimization and marine environmental impact assessment. Future research will be dedicated to improving the numerical model and improving the understanding of the changing state of floating ice after icebreaking through on-site observations and experimental research. An in-depth understanding of the distribution and movement patterns of ice floes after icebreaking is of great significance for optimizing icebreaking navigation strategies, determining the optimal following distance and time [6], and building a more realistic digital ice pool.

2 Measurements

To obtain the distribution data of broken ice after the tug breaks the ice, this paper established the following experimental platform. The experimental site selected a large and flat floating ice area in a frozen port to ensure that the floating ice was intact and not cracked. The icebreaking operation is undertaken by a tugboat with a length of 34 m, a width of 10 m, and a draft of 3.5 m. Aerial video monitoring uses the DJI Inspire 2 drone, which has the advantages of autonomy, low temperature resistance, hovering stability and high-resolution camera, and the camera has been calibrated by DJI factory. The air pressure sensor of DJI Inspiration 2 can provide an altitude accuracy of ± 0.5 m, and the GPS and GLONASS positioning can provide an overall positioning accuracy of ± 1 m.

During the experiment, the weather conditions were good (clear, cloudless, wind speed about level 3). The experimental speed is 4.3 kn. The tugboat was pushed ahead of time before approaching the ice-breaking position to stabilize the speed at the predetermined value. Affected by ocean currents, the speed fluctuates by ± 0.1 kn, but it can be regarded as breaking ice at a constant speed. Since there is a deviation between the heading angle of the manually operated drone and the tugboat, angle correction processing is required before data analysis.

On the scheduled route, the drone hovered at an altitude of 27 m for aerial video monitoring. The UAV adjusts the camera depression angle to 90° to record the icebreaking status of the ship perpendicular to the sea level, and keep the horizontal shooting angle consistent with the route heading, as shown in Fig. 1. The video sampling frame rate is 30 frames/second, the frame image size is 1280 × 720, and the resolution is 2.2 cm/px after actual scaling. The experiment mainly records the changes in floating ice after the tugboat breaks ice at different icebreaking speeds, including information such as the size, shape, distribution and drift speed of the broken ice, to provide data support for subsequent analysis.

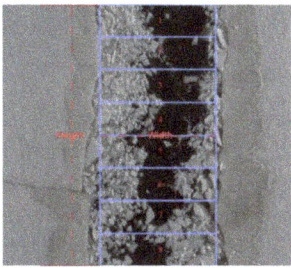

Fig. 1. Experimental diagram.

3 Experimental Images Processing

Identification of ice floe edges has always been a challenge. Traditional threshold segmentation methods, snake models (snake) and watershed methods often have poor results when processing images with complex local gray threshold gradients, and it is difficult to form closed contours that are consistent with reality. To this end, this article adopts a deep learning image segmentation model (Segment Anything Model, SAM) based on feature sample points [7]. This model uses deep learning technology to achieve image segmentation. Its structure is like the U-Net encoding-decoding network and can automatically identify and segment different objects in the image. SAM uses a multi-layer convolutional neural network to extract image features and focuses on areas related to the segmentation task through an attention mechanism, thereby effectively distinguishing foreground objects and background, and using context information to assist target recognition and segmentation.

Nonetheless, SAM may still suffer from over- or under-segmentation problems in some cases. To solve these problems, the image was preprocessed in the experiment, including removing water pixels, smoothing the image, and enhancing edges. The floating ice segmentation method based on characteristic sample points performs well when dealing with larger areas of floating ice, but the recognition effect is still unsatisfactory for smaller areas of broken ice and flocs. Therefore, this paper adopts the method of manual outline drawing for correction to ensure that each ice floe outline is basically independent and matches the visually observed ice floe pattern as much as possible. The image contour effect based on feature sample point learning is shown in Fig. 2(a), and the correction effect is shown in Fig. 2(b).

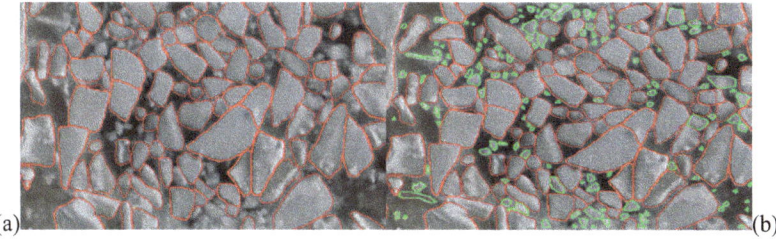

Fig. 2. (a) Image contour recognition renderings. (b) Manually corrected edges. The red ones are the contours obtained by the computer, and the green ones are the manually corrected contours.

4 Results

4.1 Area Size and Shape Distribution

To quantify the geometric shape characteristics of the ice floe, this paper follows the previous evaluation method and uses the radius $d_s = (S/\pi)^{1/2}$ of a circle that is the same as the area of the ice floe as an indicator of the size of the ice floe. According to the research of Horvat et al. [8], the extracted broken ice area data is converted into a radius data set N(d), and its probability density distribution can be represented by a power law function, such as Formula 1. Figure 3 shows the logarithmic plot of the cumulative number distribution obtained. The analysis found that when the ice floe size is less than 0.20 m, the power law exponent α is about 0.69; When the ice floe size is larger than 0.20 m, the power law index α rapidly increases to more than 3.2, and gradually decreases after about 0.35 m.

$$\frac{N(>L)}{N_0} = C_0 L^{-\alpha} \tag{1}$$

In addition to cumulative numbers, we also use number density to more clearly reflect the size distribution characteristics of ice floe. To establish the probability density distribution histogram of floating ice, we follow the standardization method proposed by Stern et al. [9] and set the floating ice size category di (i = 1...M), where the number of boxes M = N^0.5, and N is the identified the total number of ice floes), the box width Δd is the ratio of the size threshold range to the number of boxes. The probability density function of floating ice size distribution calculated from the number density is called NFSD (Formula 2). The obtained NFSD histogram is shown in Fig. 4. Its mode shape mean and standard deviation are (0.21 m, 0.12 m), which is like results obtained by Dumas-Lefebvre et al. [10] were similar and consistent with the power-law distribution characteristics of natural systems. Use the least squares method to fit the Weibull function [11] to NFSD. The fitting probability density formula is as follows, where k is the shape parameter (exponent) and λ is the scale parameter. The obtained fitting parameters are (0.21, 1.54) as shown in the red line in Fig. 4 shown.

$$p_N(d_i) = \frac{n_i}{N\delta d}, \sum_{i=1}^{M} p_N(d_i)\delta d = 1 \tag{2}$$

$$p_A(d_i) = \frac{1}{\Delta d} \sum_{j=1}^{n_i} \frac{a_j}{A}, \sum_{i=1}^{M} p_A(d_i)\Delta d = 1 \tag{3}$$

$$A = \sum_{k=1}^{N} a_k \tag{4}$$

However, there are limitations to relying solely on NFSD to analyze ice floe size distribution. As shown in Fig. 4, most of the ice-breaking area is covered by large ice floes of similar size, and small ice fragments are filled between the cracks. Although NFSD shows that the number density of small-sized broken ice is greater than that of large-sized broken ice, it fails to reflect the differences in the contribution of different sizes of floating ice to the ice-breaking area. To overcome the limitations of NFSD, this paper introduces the area weighting method, uses the floating ice area as a weighting factor, re-evaluates the probability density function, and obtains an AFSD that more accurately reflects the contribution of the floating ice area.

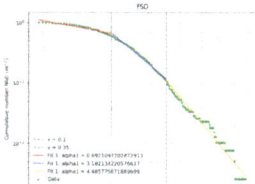 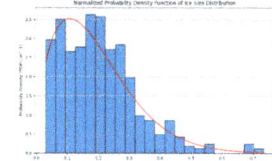 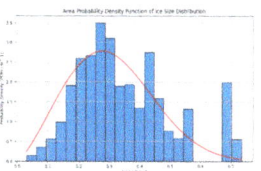

Fig. 3. Cumulative number chart. (Corresponding to four groups of experiments).

Fig. 4. NFSD floe ice probability distribution map.

Fig. 5. AFSD floe ice probability distribution map.

The area weighted number density distribution AFSD formula is as formula 3, where A is the sum of the areas of floating ice fragments. After binning according to the same procedure as NFSD, the resulting AFSD histogram is shown in Fig. 5, and its mode shape mean and standard deviation are (0.35 m, 0.15 m). Fitting the Weibull distribution to the AFSD yields parameters of (0.35, 2.37). Comparing the results of AFSD and NFSD, it can be found that AFSD can more accurately reflect the size distribution characteristics of actual floating ice, and the modal ice floating size is slightly shifted to a larger value compared with NFSD, which is consistent with the phenomenon observed in orthographic broken ice images.

4.2 Ice Floe Shape Distribution

When ships navigate in ice areas, the shape of the ice floe is one of the key factors affecting ice resistance. Crushed ice under natural conditions has different shapes, but its shape information such as perimeter, area, and characteristic diameter has certain statistical characteristics. In this experiment we used two indicators to evaluate the shape of the ice floe - roundness R and the maximum/minimum caliper diameter ratio (d_{max}/d_{min}). Roundness R is defined as d_p/d_s, $d_s = (S/\pi)^{1/2}$, $d_p = P/2\pi$, where P is the circumference of the ice floe. The closer its value is to 1, the greater the circumference of the ice floe. The closer the shape is to a circle; conversely, the larger.

the value, the more the shape of the floating ice deviates from the circle and the higher the degree of distortion. Therefore, roundness R is considered an important parameter to measure the degree of distortion of a geometric shape.

Figure 6 is the data scatter plot of d_p/d_s. After linear fitting these data points, at the 95% significance level, the obtained ratio is estimated to be 1.31 ± 0.05, which is slightly higher than Toyota and Kohout [12] Reported 1.29. It can be seen from the scatter plot and linear fitting results that they show a strong linear correlation.

To further verify the correlation between ice floe shape and size, we assume that the ice cube is elliptical. When the d_p/d_s ratio is 1.33, the corresponding ratio of the major and minor semi-axes of the ellipse is 3.6. The relationship between the maximum and minimum caliper diameters d_{max} and d_{min} of the ice cube is shown in Fig. 7. As can be seen from the figure, although the correlation between d_{max} and d_{min} is not as strong as the correlation between dp and ds, they also show a certain correlation. At the 95%

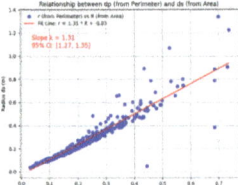 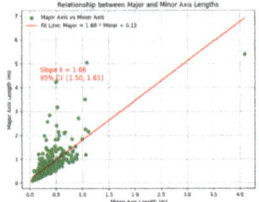

Fig. 6. Relationship between dp and ds.   **Fig. 7.** Relationship between maximum and minimum caliper diameters.

significance level, the resulting ratio is estimated to be 1.60 ± 0.14, which is consistent with the range of 1.2 to 2.2 obtained by Toyota and Enomoto [13] for multi-year ice.

Taken together, the experimental results indicate that ice floes have an average aspect ratio of about 1.6 and generally exhibit some degree of distortion rather than being completely round. And as the size of the ice floe changes, the values of d_p/d_s and d_{max}/d_{min} are almost the same, indicating that these shape characteristics are hardly affected by the size of the ice cube.

5 Conclusion

This study used aerial photography equipment to obtain video images of ice breaking in the tug channel, and used image processing technology to extract geometric parameter data of the ice breaking. Through statistical analysis, we obtained the geometric shape distribution characteristics of floating ice in the broken ice channel and drew the following main conclusions:

1. The size distribution of floating ice shows the characteristics of power law distribution. When performing a cumulative number analysis on ice floe size, we observed an obvious bend in the cumulative number curve under logarithmic coordinates, which may be related to the truncation effect of data sampling and the weight distribution of the ice floe size distribution. Data sets containing such biases can be better fitted using the Weibull function, whose scale coefficient λ is intrinsically linked to the average ice floe size.
2. The size distribution of broken ice produced by tugboat icebreaking shows a modal distribution pattern. The NFSD and AFSD histograms clearly demonstrate this feature, indicating that ship icebreaking produces fragmented ice of preferential sizes.
3. The size distribution of broken ice produced by tugboat icebreaking has a certain similarity with the size distribution of naturally split ice floes. Although the power-law distribution index obtained in this study is different from the previous research results on naturally split ice floes, its modal distribution is like the previous research results, and both have power-law distribution characteristics, which implies that two ice-breaking processes may exist Some commonalities.

References

1. Yang, B., Sun, Z., Zhang, G., Wang, Q., Zong, Z., Li, Z.: Numerical estimation of ship resistance in broken ice and investigation on the effect of floe geometry. Mar. Struct. **75**, 102867 (2021)
2. Tang, X., Zou, M., Zou, Z., Li, Z., Zou, L.: A parametric study on the ice resistance of a ship sailing in pack ice based on CFD-DEM method. Ocean Eng. **265**, 112563 (2022)
3. Denton, A.A., Timmermans, M.L.: Characterizing the sea-ice floe size distribution in the Canada Basin from high-resolution optical satellite imagery. Cryosphere. **16**(5), 1563–1578 (2022)
4. Zhang, Q., Hughes, N.: Ice floe segmentation and floe size distribution in airborne and high-resolution optical satellite images: towards an automated labelling deep learning approach. Cryosphere. **17**(12), 5519–5537 (2023)
5. Li, Z., Lu, P., Zhou, J., Zhang, H., Huo, P., Yu, M., et al.: Evolution of the Floe Size Distribution in Arctic Summer Based on High-Resolution Satellite Imagery. Remote Sens. **16**(14), 2545 (2024)
6. Liu, Y., Zhou, Y., Zhong, R.Y.: A multi-objective optimisation strategy for ice navigation under ship safety-following scenarios. Ocean Coast. Manag. **243**, 106727 (2023)
7. Kirillov, A., et al.: Segment anything. In: Proceedings of the IEEE/CVF International Conference on Computer Vision, pp. 4015–4026. IEEE, Paris, France (2023)
8. Horvat, C., Tziperman, E.: The evolution of scaling laws in the sea ice floe size distribution. J. Geophys. Res. Oceans. **122**(9), 7630–7650 (2017)
9. Stern, H.L., Schweiger, A.J., Zhang, J., Steele, M.: On reconciling disparate studies of the sea-ice floe size distribution. Elem. Sci. Anth. **6**, 49 (2018)
10. Dumas-Lefebvre, E., Dumont, D.: Aerial observations of sea ice breakup by ship waves. Cryosphere. **17**(2), 827–842 (2023)
11. Bolat, F., Ercanlı, İ.: Modeling diameter distributions by using Weibull function in forests located Kestel-Bursa. Kastamonu Univ. J. For. Faculty. **17**(1), 107–115 (2017)
12. Toyota, T., Kohout, A., Fraser, A.D.: Formation processes of sea ice floe size distribution in the interior pack and its relationship to the marginal ice zone off East Antarctica. Deep-Sea Res. II Top. Stud. Oceanogr. **131**, 28–40 (2016)
13. Toyota, T., Enomoto, H.: Analysis of sea ice floes in the Sea of Okhotsk using ADEOS/AVNIR images. In: Paper Presented at 16th International Symposium on Ice. Int. Assoc. for Hydraul Res. , Dunedin, New Zealand (2002)

Open Access This chapter is licensed under the terms of the Creative Commons Attribution-NonCommercial-NoDerivatives 4.0 International License (http://creativecommons.org/licenses/by-nc-nd/4.0/), which permits any noncommercial use, sharing, distribution and reproduction in any medium or format, as long as you give appropriate credit to the original author(s) and the source, provide a link to the Creative Commons license and indicate if you modified the licensed material. You do not have permission under this license to share adapted material derived from this chapter or parts of it.

The images or other third party material in this chapter are included in the chapter's Creative Commons license, unless indicated otherwise in a credit line to the material. If material is not included in the chapter's Creative Commons license and your intended use is not permitted by statutory regulation or exceeds the permitted use, you will need to obtain permission directly from the copyright holder.

Brash Ice Contraction Phenomenon in Ice-Covered Channels

Jiayun Ye, Bin Mei, Yujia Zhang, Congcong Zhao, Weifeng Li, and Guoyou Shi[✉]

Navigation College, Dalian Maritime University, Dalian 1160206, China
yejiayun1999@dlmu.edu.cn

Abstract. This study aims to investigate the phenomenon of brash ice re-gathering within ice-covered channels after an icebreaker passes through at different wind speeds. The research utilizes combined Computational Fluid Dynamics (CFD) and Discrete Element Method (DEM) numerical simulation techniques to model the behavior of brash ice when a vessel navigates through an ice-covered channel. Using a single-coupling strategy, the study calculates the forces exerted by the flow field on the brash ice and analyzes the formation process of the brash ice contraction. The numerical simulation results indicate that changes in the flow field around the vessel significantly affect the movement trajectories of the brash ice. Specifically, in the stern region, due to the recirculation effect, the brash ice is attracted to the center of the channel, forming a V-shaped contraction. The study demonstrates that the distance at which the brash ice begins to contract increases as the wind speed increases, moving farther away from the icebreaker. By optimizing navigation strategies, particularly by considering the characteristics of the flow field, the impact of brash ice contraction on channel navigability can be effectively reduced, thereby enhancing the safety and efficiency of navigation in ice-covered areas.

Keywords: Brash Ice Contraction · Discrete Element Method (DEM) · Computational Fluid Dynamics (CFD)

1 Introduction

Navigation in polar regions is a critical component of maritime operations [1]. As icebreakers carve out channels in ice-covered waters, the phenomenon of brash ice re-gathering within these channels becomes significant. This phenomenon refers to the re-concentration of ice fragments back into the cleared channel, potentially obstructing vessels and increasing operational risks. Understanding the dynamics of brash ice contraction is crucial for the navigation of vessels following icebreakers, optimizing navigation strategies, and ensuring the safety of operations in polar waters.

Researchers recognize that the contraction of brash ice within channels after ice-breaking poses a major challenge to maintaining clear channels. This realization has led to a focus on the fluid dynamics and environmental factors affecting brash ice movement. The introduction of advanced modeling techniques and experimental methods has made in-depth studies of brash ice contraction possible. For example, Spencer et al. (2001) [2]

conducted extensive experiments to measure the forces acting on icebreakers and their resulting brash ice behavior. These studies emphasize the importance of fluid dynamics and external forces (such as ocean currents and wind) in driving brash ice movement. In recent years, the combination of remote sensing technology, Computational Fluid Dynamics (CFD), and field observations has significantly advanced the study of brash ice contraction. Researchers now use high-resolution satellite imagery [3] and advanced simulation tools to analyze brash ice movement with unprecedented precision. A notable example is the work by Luo et al. (2019) [4], who used CFD to simulate brash ice contraction patterns in icebreaker channels. Their findings provided valuable insights into the impact of hull design and operating speed on brash ice behavior.

Future research may further refine these models and explore the impact of climate change on ice dynamics. With global temperatures continuing to rise, the extent and thickness of polar ice are expected to decrease, which could alter the patterns of brash ice contraction.

2 Brash Ice Contraction Phenomenon and Numerical Simulation

Brash ice contraction refers to the phenomenon where ice fragments, after being broken by an icebreaker or other vessels passing through ice-covered channels, re-gather in the channel due to the interaction of wind and the ice itself. This phenomenon presents challenges for navigation, especially for vessels traversing the same channel multiple times, as it increases resistance and can lead to safety issues.

Computational Fluid Dynamics (CFD) combined with the Discrete Element Method (DEM) was initially used to study particle flow and fluid phenomena, primarily focusing on the interactions between water flow and small particles. As research progressed, to address complex situations involving free surfaces, researchers introduced Volume of Fluid (VOF) techniques into the existing coupling methods and adopted virtual mesh technology to optimize grid handling, particularly for larger particle sizes. In recent years, this CFD-DEM coupling technique has been extended to the study of ship-ice-water interactions.

2.1 Brash Ice Contraction in Actual Ice-Covered Channels

Brash ice contraction typically manifests in the following forms (as shown in Fig. 1):

1. Re-aggregation of Ice Blocks: After the icebreaker passes, the brash ice gathers behind the vessel, forming new ice areas that cover the previously cleared channel.
2. Drift of Ice Blocks: Under the influence of water flow and wind, ice blocks drift along the channel, gradually filling it.
3. Accumulation of Ice Blocks: In some cases, ice blocks pile up within the channel, forming thicker ice layers, which increase the icebreaking resistance for vessels.

Brash ice contraction is a common problem encountered by icebreakers and other vessels navigating in ice-covered channels. By understanding the primary influencing factors and their impact on vessel navigation, we can implement measures to mitigate the adverse effects of this phenomenon. Numerical simulation, optimization of channel design, improvements in icebreaker design, and the application of intelligent icebreaking technologies can all effectively address brash ice contraction. This paper uses numerical simulation to model brash ice contraction in ice-covered channels.

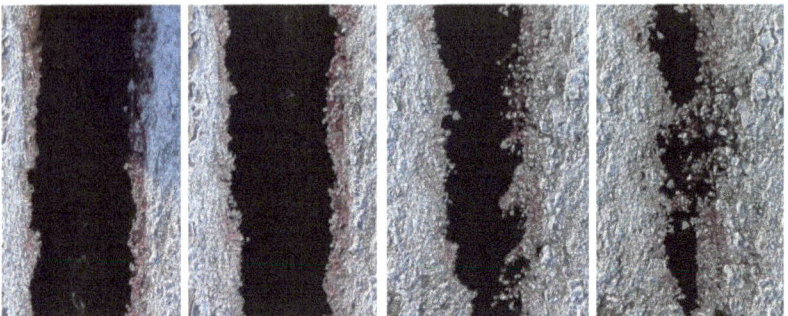

Fig. 1. Actual Brash Ice Contraction Phenomenon in Ice-Covered Channels.

2.2 Discrete Element Model (DEM) of Sea Ice

For polar floating/brash ice regions, sea ice exhibits strong discrete distribution characteristics, and the physical properties and size distribution of the sea ice are random [5]. During the coupled interaction between ships, ice, and water, brash ice is composed of several basic DEM particles. The navigation process of vessels in brash ice channels is numerically simulated through the interactions among DEM particles, between DEM particles and walls, and between DEM particles and the vessel hull. To generate brash ice models of varying sizes in floating/brash ice domains, based on the sizes of floating/brash ice in ice tank tests, the Rosin-Rammler distribution method is used to randomly distribute the model ice sizes within the range of [0.35, 0.5] m [6]. Different-sized spherical particles are used to fill the physical model generated in Fig. 2. The more particles used for filling, the closer the model resembles the created physical entity; however, an increase in the number of particles leads to greater memory usage and computational time [7]. Therefore, an appropriate number of particles should be selected for computation. In this study, the number of particles used is 50 for a pentagonal pyramid. The generated floating/brash ice DEM model is shown in Table 1 with the physical parameters of the brash ice.

Fig. 2. The physical entity model and discrete element numerical model of the generated brash ice.

Table 1. The physical parameters of brash ice.

Physical Quantity	Parameter Value
Young's Modulus (Pa)	7.4×10^9
Density (kg/m^3)	916.72
Poisson's Ratio	0.361

2.3 Ship Model

The hull model selected for this study is the DTMB 5415. The propeller has been removed to establish a three-dimensional geometric ship model. The scale ratio is 1:24.824. The main parameters of the actual ship and the model ship are shown in Table 2.

Table 2. DTMB 5415 Main parameters.

Parameter	Model	Actual Ship
Scale Ratio	24.824	1
Length Between Perpendiculars (m)	5.72	142
Breadth (m)	0.76	18.9
Draft (m)	0.248	6.16
Displacement (tons)	0.55	36,084
Displacement Volume (m^3)	0.55	25.6

Using geometric modeling software, the 3D model built is as follows (as shown in Fig. 3):

Fig. 3. DTMB 5415 Model.

2.4 Establishment of Calculation Domain and Mesh Generation

For the numerical simulation of the brash ice area, a rectangular calculation domain is selected with dimensions of 35 m in length, 16 m in width, and 10.5 m in height. The calculation domain contains water, air, and brash ice phases. The Volume of Fluid (VOF) method is used to mark the free surface [8], with air above the free surface and water below, both being fluids of constant density. The depth from the liquid surface to the bottom of the domain is 5.5 m, with brash ice particles floating on the free surface.

For the mesh generation within the calculation domain, the base mesh size is set to 0.1 m. Five prism layers are established with a 1.0 extension factor, giving a total thickness of 0.02 m. For the ship's hull, local refinement is applied to the bow and stern areas, which are the primary contact regions with the ice. After mesh processing, the total number of generated mesh cells is 5,026,603 (as shown in Fig. 4).

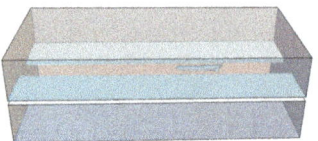
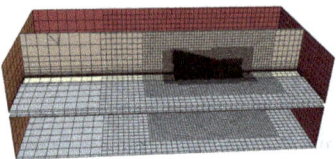

Fig. 4. Establishing the Computational Domain and Mesh Refinement Results.

2.5 Theoretical Analysis of Brash Ice Contraction

Factors of Ship Wake. After an icebreaker passes through an ice-covered channel, the turbulence and negative pressure zone at the stern significantly affect the movement of the brash ice within the channel. First, we can describe the impact of the stern vortices on the movement of brash ice using the fundamental equations of fluid dynamics.

Assuming the fluid velocity at a certain point is v, the fluid motion can be described by the Navier-Stokes equation:

$$\rho\left(\frac{\partial v}{\partial t} + (v \cdot \nabla)v\right) = -\nabla p + \mu \nabla^2 v + f \tag{1}$$

Here:
- ρ is the density of the fluid,
- μ is the velocity vector field,
- p is the pressure,
- f represents any external forces acting on the fluid.

In the negative pressure zone at the stern, the pressure gradient ($-\nabla p$) creates an attractive effect on the brash ice, causing it to converge toward the centreline of the channel, forming a V-shaped contraction.

Wind Factors. During the experiment, we also specifically studied the effect of wind speed on the brash ice contraction phenomenon. This can be quantitatively analyzed using equations of motion and force balance

$$\theta = \arctan\left(\frac{F_{wind}}{F_{fluid}}\right) \tag{2}$$

Here:
- θ is the opening angle of the V-shaped contraction of brash ice,
- F_{wind} is the force exerted by the wind on the brash ice,
- F_{fluid} is the force exerted by the fluid on the brash ice.

As the wind speed increases, F_{wind} (the force exerted by the wind on the brash ice) increases, and the opening angle θ of the V-shaped contraction also increases, manifesting as a spreading effect of the brash ice.

3 Analysis of Experimental Results

The following is an analysis of the experimental results: After the DTMB 5415 vessel passes through the brash ice area, the high-pressure zone in the bow region significantly affects the movement of the brash ice within the channel. The experimental results show that the vortex structures formed by the interaction between the vessel and the water flow create a noticeable suction effect near the DTMB 5415 vessel's track. This effect causes the brash ice to gradually converge toward the centerline of the channel (as shown in Fig. 5, depicting the pressure distribution on the sea surface near the DTMB 5415 vessel). Detailed analysis of the flow field reveals that the vortex structures on either side of the channel typically exhibit a certain symmetry. This symmetry directly leads to the convergence of brash ice from both sides toward the center, forming a V-shaped contraction. The trajectories of the brash ice observed in the experiment match the fluid velocity fields in the numerical simulation, further validating this viewpoint. The low-pressure zone at the stern significantly influences the movement path of the brash ice. In this region, the brash ice, driven by the vortex, quickly converges inward due to the attraction of the water flow, ultimately forming a noticeable V-shaped accumulation at the stern (as shown in Fig. 6).

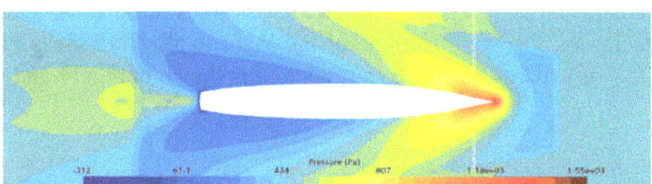

Fig. 5. DTMB 5415 Vessel's Sea Surface Pressure Distribution.

3.1 Brash Ice Contraction Phenomenon in Ice-Covered Channels

In this experiment, we conducted a detailed analysis of the movement trajectories of brash ice in the channel after the passage of the DTMB 5415 vessel through experimental observations and numerical simulations. The formation process of the V-shaped contraction phenomenon was clarified, as shown in Fig. 6, Numerical Simulation of the DTMB 5415 Ship and Variation Curve of the Number of Particles in the Waterway (Ship Speed: 5 m/s, Wind Speed: 3.2 m/s). Under these conditions, the particles contracted and filled the channel at a distance of 0.7 times the ship's length. The numerical simulation results show images where the particle density in the contracted channel occupies 85%–100% of the open water area. Additionally, both the particle injection and wind direction are oriented from the bow towards the stern.

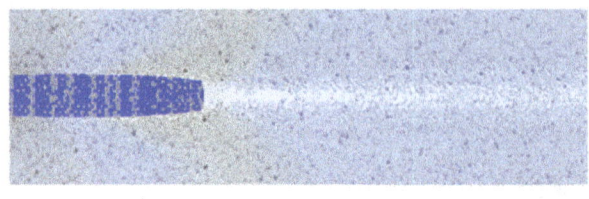

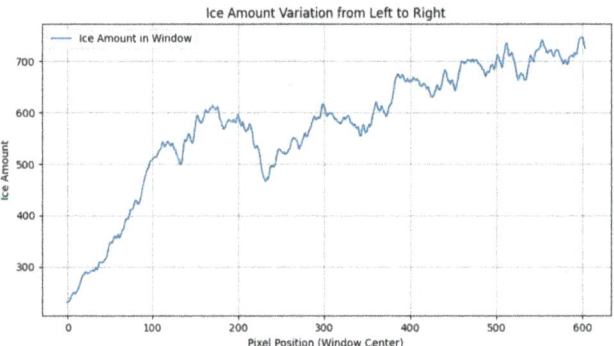

Fig. 6. Numerical Simulation of the DTMB 5415 Ship and Variation Curve of the Number of Particles in the Waterway. (Speed 5 m/s, Wind Speed 3.5 m/s).

3.2 The Modulating Effect of External Environmental Conditions

Wind speed plays a significant regulatory role in the V-shaped contraction of brash ice. The experiment observed that at a wind speed of 4 m/s, the particles contracted and filled the channel at a distance of 1.3 times the ship's length, while at a wind speed of 3.2 m/s, the particles contracted and filled the channel at 0.4 times the ship's length. When the wind speed increases, the opening angle of the V-shape also increases, indicating the dispersing effect of wind on brash ice contraction (as shown in Figs. 7 and 8).

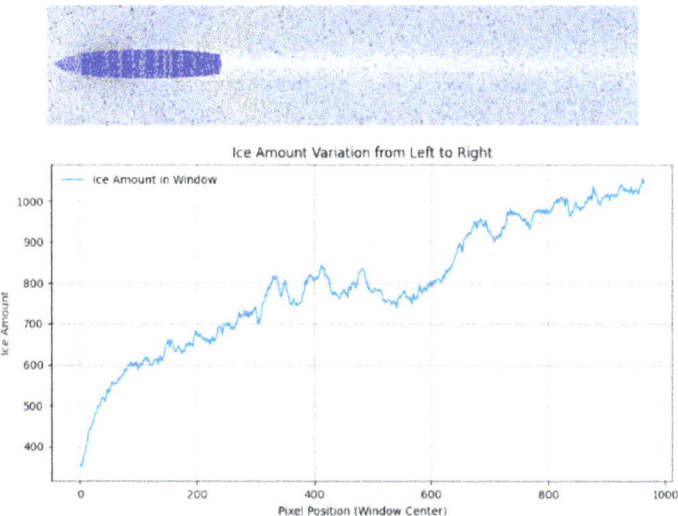

Fig. 7. Numerical Simulation of the DTMB 5415 Ship and Variation Curve of the Number of Particles in the Waterway. (Speed 5 m/s, Wind Speed 4 m/s).

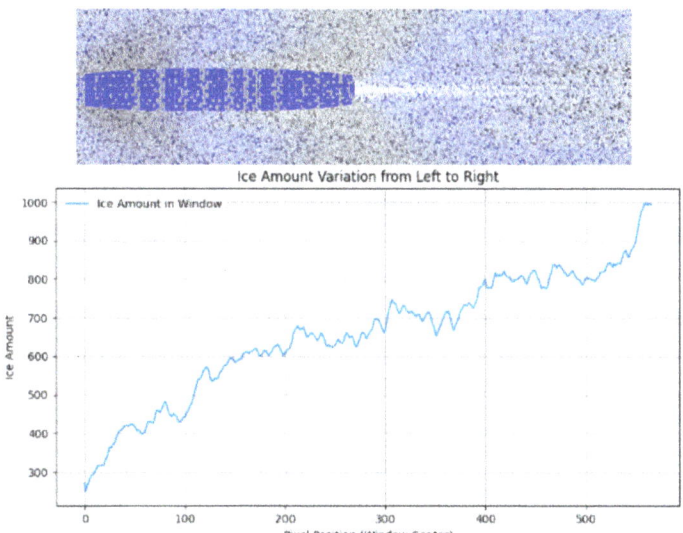

Fig. 8. Numerical Simulation of the DTMB 5415 Ship and Variation Curve of the Number of Particles in the Waterway. (Speed 5 m/s, Wind Speed 3.2 m/s).

4 Conclusions

This study, through the combination of Computational Fluid Dynamics (CFD) and the Discrete Element Method (DEM), thoroughly investigates the formation mechanism and influencing factors of brash ice contraction in ice-covered channels after icebreaker passage. The main conclusions are summarized as follows:

1. Flow Field Influence: After the icebreaker passes, the flow field around the vessel significantly affects the movement trajectories of the brash ice. Particularly, the recirculation effect in the stern region causes the brash ice to gather towards the center of the channel.
2. V-Shaped Contraction: The brash ice contraction phenomenon is characterized by a distinct V-shaped structure. This phenomenon is caused by the interaction between the vessel and the fluid, reflecting the complexity of brash ice movement within the channel.
3. Wind Speed Effect: The study finds that increased wind speed causes the starting point of brash ice contraction to move farther away from the vessel, indicating that external environmental conditions play a significant role in regulating brash ice movement.
4. Future Research Directions: Further research is recommended to study the impact of climate change on ice dynamics to better understand and address potential future brash ice contraction issues.

Although this study revealed the formation mechanism of brash ice contraction in ice channels influenced by ship wake and wind using the CFD-DEM coupling method, the following limitations still exist:

1. Incomplete Incorporation of Environmental Factors: The study did not account for dynamic environmental factors such as ocean currents, temperature variations, and changes in ice thickness. These factors could significantly affect the drifting and aggregation behavior of brash ice.
2. Limitations in Ship Type and Scale Ratio Applicability: The study was based on the DTMB 5415 model ship (scale ratio of 1:24.824). Its hydrodynamic characteristics may differ from those of full-scale ships or other ship types (e.g., wide-body icebreakers). The generalizability of the conclusions needs further validation.
3. Omission of Propeller Effects: The simplified treatment of removing the propeller may overlook the additional impact of the propeller wake on the movement of brash ice.

References

1. Song, Y., Zhang, A.: The Economy Analysis of Sailing in the Arctic Northeast Passage. Appl. Mech. Mater. **409-410**, 1253–1257 (2013)
2. Spencer, D., Jones, S.J.: Model-scale/full-scale correlation in open water and ice for Canadian Coast Guard 'R-Class' icebreakers. J. Ship Res. **45**(4), 249–261 (2001)
3. Hyun, C.-U., Kim, H.: A Feasibility Study of Sea Ice Motion and Deformation Measurements Using Multi-Sensor High-Resolution Optical Satellite Images. Remote Sens. **9**(9), 930 (2017)

4. Luo, W. Z. (2019). Study on the Resistance and Wake Field Characteristics of Ship-ice water Coupling in Floe Areas [Doctoral dissertation, Harbin Engineering University]. Doctorate. (in chinese)
5. Marquart, R., Bogaers, A., Skatulla, S., Alberello, A., Toffoli, A., Schwarz, C., Vichi, M.: A Computational Fluid Dynamics Model for the Small-Scale Dynamics of Wave. Ice Floe and Interstitial Grease Ice Interaction. Fluids. **6**, 176 (2021)
6. Jin, Qiang. (2020). The numerical forecast of ice resistance for the ships navigating in Polar floating/broken ice area based on the method of CFD-DEM [Master's thesis]. Dalian Maritime University. (in chinese)
7. Horvat, C.: Floes, the marginal ice zone and coupled wave-sea-ice feedbacks. Philos. Trans. R. Soc. A Math. Phys. Eng. Sci. **380**, 20210252 (2022)
8. Herman, A.: Wave-Induced Surge Motion and Collisions of Sea Ice Floes: Finite-Floe-Size Effects. J. Geophys. Res. Oceans. **123**, 7472–7494 (2018)

Open Access This chapter is licensed under the terms of the Creative Commons Attribution-NonCommercial-NoDerivatives 4.0 International License (http://creativecommons.org/licenses/by-nc-nd/4.0/), which permits any noncommercial use, sharing, distribution and reproduction in any medium or format, as long as you give appropriate credit to the original author(s) and the source, provide a link to the Creative Commons license and indicate if you modified the licensed material. You do not have permission under this license to share adapted material derived from this chapter or parts of it.

The images or other third party material in this chapter are included in the chapter's Creative Commons license, unless indicated otherwise in a credit line to the material. If material is not included in the chapter's Creative Commons license and your intended use is not permitted by statutory regulation or exceeds the permitted use, you will need to obtain permission directly from the copyright holder.

Establishment of Multi-ship Joint Navigation Without Satellite Navigation System Supporting

Yundong Han, Fanjun Meng[✉], Qian Huang, Guohui Lan, and Hairui Ma

Department of Navigation, Dalian Naval Academy, Dalian 116018, Liaoning, China
fanjunmengg@163.com

Abstract. In order to solve the problem of navigation support capability degradation in the absence of satellite navigation in ship formation, the method of multi-ship joint navigation was proposed. The multi-ship joint navigation system was established by taking the ship with inertial navigation system as the mother ship and the ship only equipped with platform compass as the sub-ship. The mother ship obtained relevant information of sub-ships through measurement and calculation, and sent it to the sub-ships, so as to improve the reliability and effectiveness of navigation information of sub-ships, and thus to ensure the accuracy and continuity of navigation information of the ship formation. The results show that the positioning error of a single ship has a significant decrease after adopting the joint navigation method, which provides a theoretical reference for ensuring the navigation of the ship formation.

Keywords: Joint Navigation · Navigation Information · Inertial Navigation System · Platform Compass

1 Introduction

With the development of navigation equipment, ship navigation has developed from traditional navigation to multi-system integrated navigation, in which the combination of satellite navigation with inertial navigation system or platform compass has become the main navigation mode of many large ships. However, under complex electromagnetic environment, the satellite navigation system may be seriously interfered, or even can't be used normally. How to obtain relatively accurate navigation information for the ship formation without satellite navigation has become the focus of research.

With the development of technology, integrated navigation has become the main navigation mode that most surface ships rely on [1]. Although the satellite navigation system can provide users with all-weather, uninterrupted, high-precision, real-time three-dimensional position, its weak anti-interference makes it vulnerable to electromagnetic interference and unable to work normally, which may lead to the integrated navigation becomes the mode of INERTIAL navigation or platform compass working alone when the ship is performing special tasks in complex electromagnetic environment [2–4]. Both inertial navigation and platform compass are autonomous navigation system, but the platform compass only have a short-time inertial navigation function, without the support of satellite navigation system, the navigation information of platform compass, such as position, speed, heading, will increase and eventually lead to a large error over a period of time. Compared with the platform compass, the amplitude of inertial navigation error increases is much smaller [5–8].

Therefore, when multiple ships are performing tasks in formation and the satellite navigation system cannot be used, inertial navigation or platform compass can only be used to provide necessary navigation information for the ship. As not all ships are equipped with inertial navigation system, after a period of time, ships only equipped with platform compass will lose the support ability of navigation information due to the increase of positioning error of platform compass [9, 10].

In order to solve this problem, this paper will adopt the method of formation joint navigation, that is, to build a joint navigation system in the formation, which can realize real-time sharing of key navigation information, and ensure that the platform compass of each ship in the formation can provide more reliable and effective navigation information.

2 Background of Joint Navigation

2.1 The Concept of Joint Navigation

Joint navigation is aimed at ensuring the real-time and relative accuracy of navigation information between ships and improving the reliability of navigation system in formation. Therefore, the idea of joint navigation between ships in formation is put forward, and its concept is described as follows.

Joint navigation: When a ship is performing a task in formation, a relative coordinate system is established on the ship with good and high precision navigation equipment as the origin of the relative reference system, and the ship is taken as the mother ship. The error of the mother ship can be measured in this coordinate system. The mother ship determines the relative bearing, position and speed information of the other sub-ships through certain observation means, and then the relative navigation information of the sub-ships is obtained through a series of calculations. The navigation information is provided to the sub-ships by existing means, so that the equipment with large errors on the sub-ships can be recalibrated to meet the requirements of navigation information.

2.2 The Meaning of Joint Navigation

The significance of joint navigation is to solve the following two problems:

1. When the onboard satellite navigation system is disrupted or destroyed and cannot be trusted, the integrated navigation system is downgraded to an inertial navigation mode or platform compass mode alone. After a period of time, the platform compass is difficult to meet requirements of ensuring the completion of special tasks, even it is difficult to guarantee the navigation safety.
2. When a ship in the formation suffers serious damage and is lack of necessary navigation information, the navigation safety of the ship will not be guaranteed, the special equipment system which needs navigation information will not work effectively.

For the above two cases, the joint navigation of ship formation can solve the above problems. By establishing a formation joint navigation system, each ship in the formation can be regarded as a part of the system. In this formation joint navigation system, the ship which can provide complete navigation information in high precision is chosen as a mother ship, all ships establish the formation coordinate system with the mother ship as the center. Through observing sub-ship's relative navigation information, mother ship can send the navigation information in higher precision to the sub-ship after a series of calculation.AS the mother ship navigation information accuracy is much higher than the accuracy of the sub-ships, the information obtained through measuring and calculating will be more reliable and accurate, and can obviously increase the accuracy of ship navigation information, also can provide ship all-weather navigation support to perform a task with continuous navigation information, and thus improve the navigation accuracy of whole formation.

3 Establishment of Joint Navigation System

In the research process of establishing the joint navigation system of a ship formation, a ship equipped with inertial navigation and a ship equipped with platform compass form a ship formation, as shown in Fig. 1.

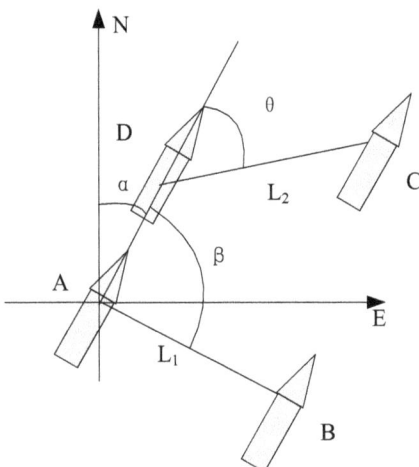

Fig. 1. Joint navigation coordinate system of ship formation

Assuming that the sea is in good condition, A and B in the coordinate system are respectively taken as the starting points of the two ships in the formation. Ship A is equipped with inertial navigation as the mother ship, while ship B is only equipped with platform compass as the sub-ship. In the coordinate axis, due east and due north are taken as the positive side of the horizontal axis and the vertical axis respectively. In the measurement process, it is assumed that both the mother ship and the sub-ship sailing at a fixed speed and course. At the beginning, the course of ship A is α, the speed is

$\vec{V}$, the azimuth Angle of ship B relative to ship A is β, the distance observed by ship A through shipborne navigation radar is L_1, and the observation error of the radar is ΔL. If the coordinates of A is $P_A(X_1, Y_1)$, then:

$$D_\phi = (L + \Delta L) \cos(\alpha + \beta) \tag{1}$$

$$P_{BY} = D_\phi + Y_1 \tag{2}$$

$$Dep = (L_1 + \Delta L) \sin(\alpha + \beta) \tag{3}$$

$$D_i = Dep \sec\varphi_m \tag{4}$$

$$\varphi_m = Y_1 + \frac{P_{BY}}{2} \tag{5}$$

$$P_{EX} = D_i + X_1 \tag{6}$$

where, D_ϕ is the latitude difference, Dep is the departure, D_λ is the longitude difference, and φ_m is the average latitude. When the latitude difference is not large, the average latitude is used to replace the middle latitude, $0 \leq \alpha \leq 2\pi$, $0 \leq \beta \leq 2\pi$, and after conversion, $P_B(P_{BX}, P_{BY})$ is obtained. Due to the working nature of the navigation radar, it is assumed that the mother ship arrives at position D after time T and is located at position C when the second scan reaches the sub-ship. The scanning distance of the navigation radar is L_2, and the azimuth Angle of mother ship A and sub-ship B is θ. Through Eqs. (1)–(6), the position of the mother ship at Point D and the sub-ship at point C can be obtained:

$$P_{DR} = vT \cos\alpha + Y_1 \tag{7}$$

$$P_{DX} = vT \sin\alpha \sec\frac{Y_1 + P_{DX}}{2} + X_1 \tag{8}$$

$$P_{CY} = (L_2 + \Delta L) \cos(\alpha + \theta) + P_{DY} \tag{9}$$

$$P_{CX} = (L_2 + \Delta L) \sin(\alpha + \theta) \sec\frac{P_{DY} + P_{CY}}{2} + P_{DX} \tag{10}$$

$P_C (P_{CX}, P_{CY})$ is the arrival point C of sub-ship, $P_D (P_{DX}, P_{DY})$ is the arrival point D of mother ship, the position P_C and speed $\vec{V} = \frac{P_C - P_B}{T}$ of the sub-ship can be measured by the mother ship and transmitted to the sub-ship through a data link, so that the sub-ship can reset the platform compass positioning error and meet the requirements by the navigation information transmitted from mother ship after the shipborne platform compass positioning error becomes out of the valid interval. Therefore, the sub-ship can re-enter the effective working state without satellite system calibration even if the platform compass exceeds the specified working time.

In addition, the instantaneous course and attitude information provided by the shipborne navigation equipment such as platform compass are not affected by the satellite, can still be used normally. The sub-ship after reset has more accurate position, speed, heading, attitude and other information, which can effectively meet the requirements of other special task systems.

4 The Experimental Analysis

A two-type vessel is formed into a formation to perform a task. The formation of the vessel sails in a single column with starting point $(30°00'.0\,N, 125°00'.0\,E)$. Assuming that the course of the formation is 090°.0, after the formation accelerates uniformly from the anchorage state for 600 s with an acceleration of 0.017 m/s^2, the formation maintains a speed of 20 nm/h (10 m/s), and the sailing time is 8 h.

Suppose that the satellite navigation system is in the state of being interfered and unable to work during the navigation of the formation. The ship equipped with inertial navigation system is set up as the mother ship while the ship only equipped with platform compass as the sub-ship. Position A and B are respectively taken as the starting points of the two ships in the formation while position C and D are respectively the arrival points after 8 h. Since the uncertain quantities in the whole simulation process are the radar observation distance L and its measurement error ΔL respectively, these two variables can be determined by the navigation radar equipment and its user manual. The distance between the sub-ship and the mother ship observed by navigation radar is L_1 and L_2 respectively. According to Eqs. (1)–(6), the position coordinates of the sub-ship at point B and point D are as follows:

$$P_{BY} = L_1 + P_{AY} \quad (11)$$

$$P_{BY} = P_{AX} \quad (12)$$

$$P_{DY} = L_2 + P_{CY} \quad (13)$$

$$P_{DX} = P_{CX} \quad (14)$$

The positions of mother ship and sub-ship can be obtained by deck reckoning, and the rough location information of sub-ship position C also can be measured and calculated by mother ship at position D. Since the formation keeps a fixed speed when sailing, the speed information could be directly sent to the sub-ship through a data link with the calculated sub-ship's position information, so that the sub-ship with platform compass can compare the navigation information from mother ship with its own compass navigation information.

Through simulation, the positioning error of the mother ship's inertial navigation system and the sub-ship's platform compass within 8 h could be obtained, as shown in Fig. 2 and Fig. 3. In the simulation figure, the horizontal axis is time and the unit is hour (h), and the vertical axis is positioning precision and the unit is meter (m).

Since there is only latitude change in the position information, and the change of longitude is ignored, the curves in the two figures respectively represent the change of latitude positioning error of the mother ship's inertial navigation system and the sub-ship's platform compass within 8 h. The four vertical lines represent four time points of 2 h, 4 h, 6 h and 8 h, for subsection comparison. It can be seen from Fig. 2 and Fig. 3 that the positioning error of the latitude of the inertial navigation system changes steadily with the growth of time. In contrast, the positioning error of the latitude of the

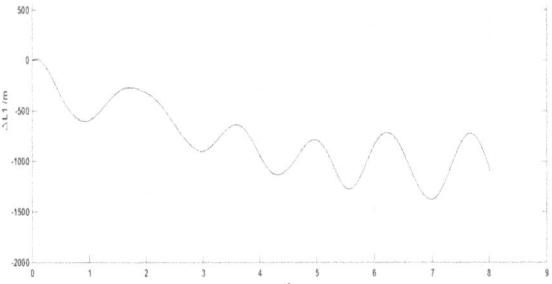

Fig. 2. Variation of inertial navigation positioning error

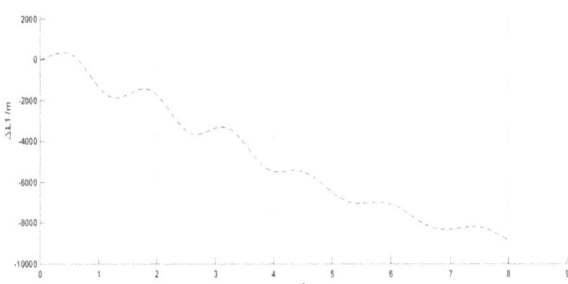

Fig. 3. Variation of positioning error of platform compass

platform compass gradually increases with the growth of time, and the latitude error finally increases to 9000 m.

Therefore, according to the above analysis, the mother ship readjusts and corrects the position information of the sub-ship every 2 h, only the error of the latitude positioning information of the sub-ship's platform compass is changed while the rest information remains unchanged. The sub-ship keeps sailing in its original state, and the simulation figure is shown in Fig. 4.

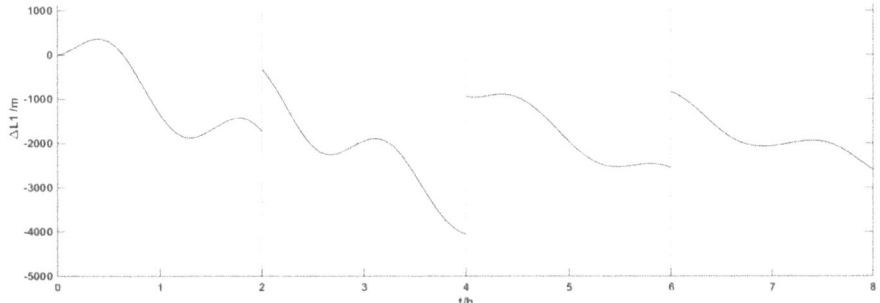

Fig.4. Comparison of corrected information of sub-ship

The curve in Fig. 6 shows that the positioning error of the platform compass decreases from 2000 m to 500 m due to the position information correction at the time point 2 h with the other platform compass parameters remaining unchanged. When it works for 4 h, the navigation information will be corrected again, and the latitude positioning error is reduced from 6000 m to 500 m. When it worked for 6 h, the navigation information will be corrected the third times, and the latitude positioning error was reduced from 7000 m to 2500 m. And finally the latitude positioning error of sub-ship platform compass increased to 3000 m when it worked after 8 h. This is a significant decrease from the previous 9000 m.

The simulation results show that without the support of satellites navigation system, the positioning accuracy of navigation information of sub-ships can be significantly improved by establishing the formation joint navigation system, and the improvement effect becomes more obvious with the accumulation of time.

5 Conclusion

Joint navigation can effectively increase the ship formation navigation information accuracy without satellite navigation support, and thus improve the whole navigation support capability of formation. The joint navigation method not only can improve the safety and reliability of ship formation, but also can effectively increase the navigation precision of the whole ship formation. This paper only puts forward the concept of joint navigation, and designs a simple experiment for analyzing the construction of multi-ship joint navigation system. Before the joint navigation can be applied in practice, there are still many problems that need to be studied and explored.

References

1. Shi, W., Zhang, Y., Gao, J.: Study on the integrated navigation system of medium-sized and pint-sized warships. Chin. J. Ship Res. **1**(4), 56–57 (2006)
2. Cao, N., Huang, Z.: Study on relative navigation technology of air fleet formation. Modern Navig. (2), 87–89 (2012)
3. Zhou, H., Zhong, Y., Wei, L.I.: Comparison of relative navigation methods for large vessel formation. J. Traffic Transport. Eng. **16**(1), 149–157 (2016)
4. Fan, J., Lei, C.: Research on relative navigation technology of aircraft formation. Mod. Navig. **3**, 161–165 (2016)
5. Cui, N., Wang, X., Guo, J.: Study of relative navigation system based on INS/GPS for cooperative intelligent missile. Tactical Missile Technol. **2**, 1–5 (2010)
6. Chu, R.: Relative navigation for spacecraft formation flying using UKF. Electron. Sci. Technol. **22**(7), 5–8 (2009)
7. Liu, M., Feng, X., Shi, J., et al.: Optimal positioning algorithm for integrated relative navigation of ships. Navig. China **37**(4), 6–10 (2014)
8. Zhang, J., Li, Y., Ge, D.: Study on relative navigation for underwater vehicle and buoy. Command Control Simul. **29**(5), 83–85 (2007)

9. Zhang, G., Cheng, Y., Cheng, C., et al.: A Joint correcting method of multi-platform INS error based on relative navigation. Acta Aeronautica Et Astronautica Sinica. **32**(2), 271–280 (2011)
10. Kong, D., Liu, Z., Cui-Can, et al.: Correction of platform compass system of naval ship. Navig. China **32**(1), 10–13 (2009)

Open Access This chapter is licensed under the terms of the Creative Commons Attribution-NonCommercial-NoDerivatives 4.0 International License (http://creativecommons.org/licenses/by-nc-nd/4.0/), which permits any noncommercial use, sharing, distribution and reproduction in any medium or format, as long as you give appropriate credit to the original author(s) and the source, provide a link to the Creative Commons license and indicate if you modified the licensed material. You do not have permission under this license to share adapted material derived from this chapter or parts of it.

The images or other third party material in this chapter are included in the chapter's Creative Commons license, unless indicated otherwise in a credit line to the material. If material is not included in the chapter's Creative Commons license and your intended use is not permitted by statutory regulation or exceeds the permitted use, you will need to obtain permission directly from the copyright holder.

Risk Analysis of LNG Ship Collision Accident in Arctic Waters Based on Bayesian Network

Zhuang Li[1], Haoran Jiang[1], Xiaoming Zhu[2(✉)], Hongbo Wang[1], and Kaixian Gao[1]

[1] Guangdong Ocean University, Zhanjiang 524000, China
[2] Shanghai Maritime University, Shanghai 201306, China
xiaoming_zhu2021@163.com

Abstract. The opening of the Northeast Arctic route allows the rich oil and gas resources in the Arctic region to be easily transported to Asian and European countries, and the navigation safety of LNG carriers needs to be guaranteed. In order to effectively reduce the navigation risk of LNG carriers in Arctic waters, a Bayesian network model of risk evolution based on accident scenario analysis is proposed. By combining subjective and objective data sources, the collision risks with LNG carriers in the five seas of the Arctic Chukchi Sea, Eastern Siberia Sea, Laptev Sea, Kara Sea and Barents Sea is assessed. Based on Bayesian evidence reasoning, the main risk causation paths of LNG carrier traffic accidents in Arctic waters are analyzed. Results show that the collision risk of LNG carriers in the East Siberian Sea is the highest among the five waters, with a quantitative value of 0.062. In addition, the collision risk of LNG carriers on the Northeast Arctic route is most likely to come from human and the environment. There are three main risk causation paths, and the corresponding risk management strategy can be carried by cutting off these risk transmission paths.

Keywords: LNG Carriers · Northeast Arctic Route · Risk Analysis · Bayesian Network

1 Introduction

In recent years, the number of ship navigation in Arctic waters has shown a rising trend year by year [1]. Due to the remote location of the Arctic waters, the harsh natural environmental characteristics such as low temperature, high latitude, and many ices lead to ships navigating in the Arctic waters are more prone to collision accidents [2]. Therefore, it is necessary to carry out risk management and control research on the risk of collision accidents for LNG carriers navigating in Arctic waters to ensure the safety of ship navigation.

At present, scholars focus on risk identification and risk assessment for ships navigating in Arctic waters. The main purpose of risk identification is to find out the key risk factors, so as to take targeted control measures for the main risk sources. Vegard B. Marken et al. [3] tried to perform risk analysis qualitatively using fuzzy set theory to model uncertainty in the data. Kum Serdar et al. [4] established a fault tree model for collision accidents by analyzing the accident database and found that carelessness was the main cause of accidents. Mawuli Afenyo et al. [5] established a Bayesian evaluation model for collision accidents of ships in Arctic waters, and found that equipment failures and human failures were the most important risk factors through network sensitivity analysis. Mingyang Zhang et al. [6] analyzed the main causative factors of collision accidents of ships navigating in formation in Arctic waters by improving the method of classifying and analyzing human failures. The purpose of risk assessment is to provide a decision basis for risk management by quantifying the risk of ship navigation in Arctic waters. For example, Al-Amin Baksh et al. [7] established a Bayesian assessment model for the risk of collision of ships navigating in Arctic waters, assessed the risk of ship navigation in the Arctic Northeast Passage, and derived the main risk factors of collision accidents in the Chukchi Sea through the sensitivity analysis of the model, which provided a basis for the risk management and control of the sea. Bushra Khan et al. [8] established a dynamic Bayesian risk assessment model for the risk of collision accidents of oil tankers navigating in Arctic waters, so as to provide support for risk analysis and risk decision-making. Shanshan Fu et al. [9] presented a Frank copula-based fuzzy event tree analysis approach to assess the risks of major ship accidents in Arctic waters. In recent years, scenario analysis based on the analysis of accident scenarios has been gradually applied to risk control, risk management and control by analyzing the accident causation pathways has achieved good results [10, 11].

This study establishes a Bayesian risk assessment network model based on the analysis of collision accident scenarios of LNG carriers navigating in Arctic waters, and identifies the high-risk waters for LNG carriers navigating in Arctic waters based on the quantitative assessment of the navigation risk. On this basis, the main paths leading to collision risks are quantitatively analyzed through Bayesian evidence-based reasoning, so as to formulate targeted risk management and control measures and improve the emergency management capability of LNG carriers navigating in Arctic waters.

2 Problem Description

2.1 Definition of Accident Risk

Ship collision accidents are formed by a series of risk factors through a complex correlation. According to the analysis of accident scenarios, in ship collision accidents, the initial risk factor induces a certain risk event with which it has a correlation, and through the transferring effect, a series of risk events eventually lead to the occurrence of the collision accident. Since there are multiple types of initial risk factors, in addition, even for the same risk factor, there are multiple possibilities for its path leading to the accident. Therefore, by representing the evolution of collision risk as nodes and directed edges, risk control can be targeted by identifying the shortest path in the evolutionary network. This process can be represented by Eq. (1):

$$E = I \otimes M \otimes A \tag{1}$$

Where: I represents the initial risk factor that triggers the risk of ship collision accident, M represents the intermediate event, A represents the risk event ship collision accident. $\otimes$ represents the logical relationship and quantitative value between risk factors. The final purpose of this arithmetic process is to find the shortest risk causal path by constructing the LNG carrier collision accident evolution network.

2.2 Bayesian Network

Bayesian network is a graphical model that describes the probabilistic relationships between variables [12]. In Bayesian networks, directed edges between variables indicate logical relationships between them and are qualitative in nature. Probabilistic transfer between variables that have a logical relationship is achieved through Bayes' rule. Based on the observed values of the variables, the network can be updated using Bayes' theorem. The variables represented by the nodes in a Bayesian network usually contain three types:

  i. Root nodes: such nodes have no parent.
  ii. Intermediate nodes: such nodes have both parent and child nodes.
  iii. Leaf nodes: such nodes have no children.

Assume that the variable set D = {d1, d2, ..., dn} satisfies conditional dependency in a Bayesian network. There is a directed edge from di to dj, indicating that di is the parent node of dj and dj is the child node of di. The dependency between child nodes and parent nodes is quantified by a conditional probability table. For nodes without parent nodes, a specified probability is given. Then, for this Bayesian network with n variables, the process of Bayesian reasoning by the chain rule can be expressed by Eq. 2:

$$P(d1, d2, \ldots, dn) = P(d1| d2, \ldots, dn) * P(d2| d3, \ldots, dn) * \cdots * P(dn-1| dn) * P(dn) \quad (2)$$

3 Risk Analysis Model for Collision Accidents of LNG Carriers in Arctic Waters

3.1 Collision Scenario Risk Factor Identification

Collisions during navigation of LNG carriers in Arctic waters can arise from a variety of sources. For example, operational errors can be caused by factors such as unregulated crew operations and insufficient crew skill levels, increasing the risk of operational failure of the ship. Another example is the vulnerability of LNG storage equipment on LNG carriers, which may lead to LNG leakage under the influence of harsh environmental conditions such as low temperatures and high winds in the Arctic, and these leaks will bring about a rapid drop in local temperatures in the process of rapid evaporation, resulting in an increased risk of equipment failure on the ship. In addition, the ship machinery and equipment itself also has a certain possibility of failure. When ship equipment failure leads to personnel operation failure, or ship manoeuvring error due to personnel error, or ship power is insufficient, if there are obstacles in the fairway that threaten ship navigation, then LNG may have the risk of collision accidents. According to the concept of system engineering, the risk of collision accidents for ships navigating in Arctic waters can be summarized in the three risks of personnel (C1), ship machinery and equipment (C2) and the navigational environment (C3) in which they are located, and the specific risk factors are analyzed as shown in Table 1.

Table 1. Risk factors for collisions in Arctic waters for LNG carriers.

Form	Risk factor
C1	Insufficient training (X1), low skill level (X2), irregular operation (X3)
C2	Engine shutdown (X4), propeller failure (X5), damage to LNG transfer components (X6), tank damage (X7)
C3	Low temperatures (X8), high winds (X9), large waves (X10), other ship obstructions (X11), obstacles such as reefs (X12), large ice floes (X13)

3.2 Modelling the Evolution of Collision Risk for LNG Carriers Navigation in Arctic Waters

During the normal voyage of LNG carriers in Arctic waters, the rise in the probability of random occurrence of the initial risk event will lead to an increase in the probability of occurrence of the risk factors in the intermediate nodes associated with it, and the risk will be transferred between the nodes with correlation, which will eventually manifest itself in the rise in the risk of collision accidents. According to the risk factors in Table 1, and in accordance with the logical relationship between the risks, the possible paths of the accident are sorted out, and a collision accident risk evolution model for LNG carriers sailing in Arctic waters is constructed, as shown in Fig. 1. The model contains thirteen risk factors of three types (X), six intermediate events (M) and one target event (R).

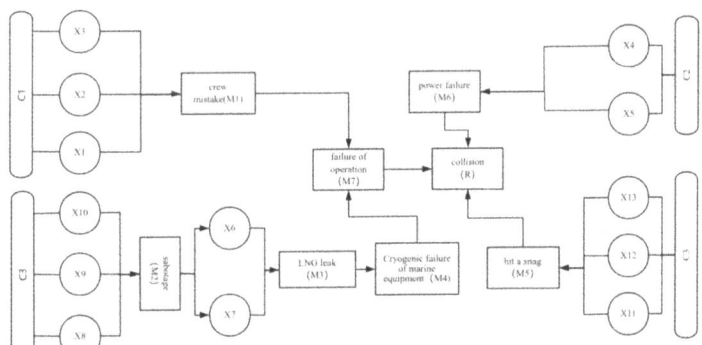

Fig. 1. Evolutionary model of collision risk for LNG carriers in Arctic waters.

3.3 Bayesian Approach to Risk Analysis and Management

A Bayesian network structure for collision risk analysis is constructed by combining the LNG carrier collision risk evolution model in Arctic waters shown in Fig. 1. For each node in the network, its risk status is set to "yes" and "no" according to whether the event occurs or not. Depending on the type of node in the Bayesian network, the root node needs to input the prior probability distribution, which is mainly obtained through accident data statistics or expert knowledge. The intermediate nodes and leaf nodes need to input conditional probability tables, which are often difficult to obtain directly and need to rely on expert judgement to obtain.

After model input, Bayesian network inference is performed through eq. 2 to obtain the risk of collision accidents of LNG carriers during navigation. In addition, the leaf nodes are set up with evidence information so that the probability of their state being "yes" is 100%, and the most sensitive risk factors in each category can be obtained through Bayesian evidence inference. Taking the sensitive risk factors as the starting point, the risk causation path of the collision with the highest probability of occurrence is analyzed, so that risk control measures can be implemented in a targeted manner.

4 Case Study

The Northeast Arctic Route mainly includes five sea areas, namely, Chukchi Sea, East Siberian Sea, Laptev Sea, Kara Sea and Barents Sea, and LNG carriers need to pass through these five sea areas or some of them when they travel to Europe and eastern Asia through the Northeast Arctic Route, as shown in Fig. 2. The month of August 2019 is selected to conduct risk control analysis for each sea area based on the risk assessment of collision accidents of LNG carriers navigating in these five sea areas respectively.

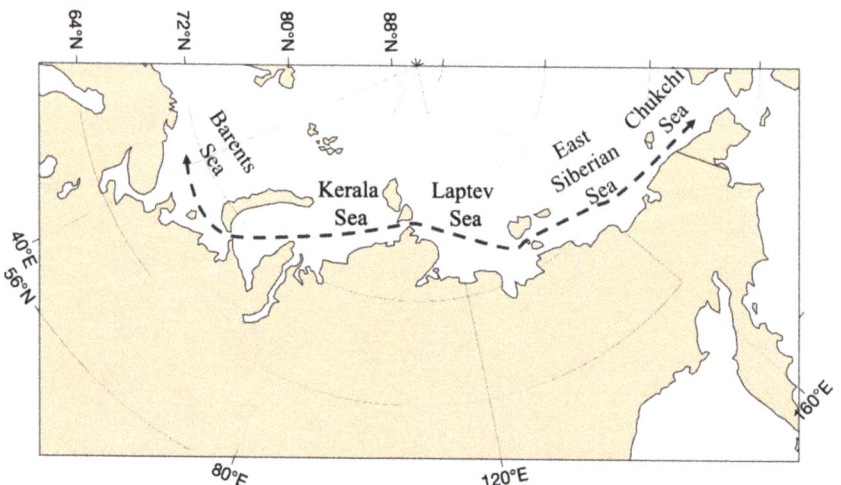

Fig. 2. Characteristics of the Northeast Arctic Route.

4.1 Accident Scenario Probability Calculation

In the Bayesian network model, the probability distributions of the environmental nodes have an objective data source as a reference basis. The information on these data is taken from the ERA5 reanalysis data published by the European Centre for Medium-Range Weather Forecasts (ECMWF), as shown in Fig. 3. The probability distributions of the other nodes in the network need to be obtained by relying on expert knowledge for judgement. They are mainly academics in relevant research organizations and captions involved in Arctic navigation operations.

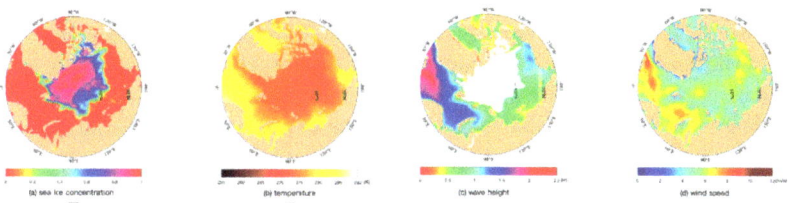

Fig. 3. Environmental information for Arctic waters for August 2019.

Based on eq. 2, Bayesian network inference is performed to obtain the probability of occurrence of collision accident scenarios for ships in the Barents Sea, as shown in Fig. 4. Similarly, in the other waters of the Northeast Arctic route, model inputs are performed based on node occurrence probabilities to obtain the probability of occurrence of collision accident scenarios in each water.

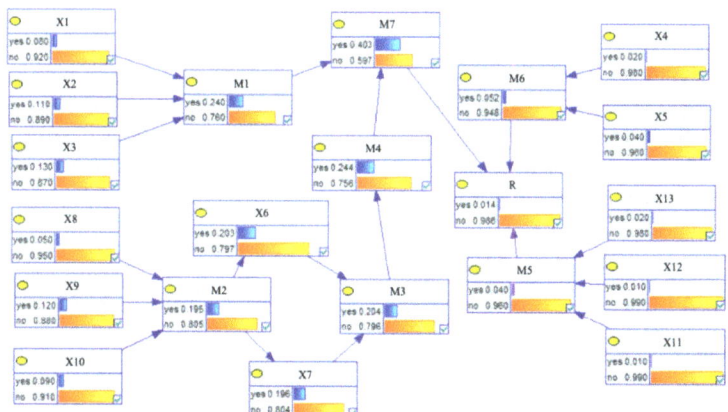

Fig. 4. Bayesian network inference for Barents Sea collision scenarios.

4.2 Risk Analysis and Management of LNG Carrier Collision Accident

As can be seen from the results of the risk analysis in Fig. 5(a–e), the main risk factors affecting LNG carrier collisions in the Chukchi Sea, the East Siberian Sea, the Laptev Sea, the Kara Sea, and the Barents Sea show a similar pattern while also having some differences. In these five waters, large ice floes (X13) are the most important risk factor affecting collisions. High winds (X9) and irregular operation (X3) are the key risk factors for collision accidents, but their effects varied in each water. Risk factor "large waves (X10)" affects collisions mainly in the Chukchi and Laptev Seas, and risk factor "low temperatures (X8)" affects collisions mainly in the East Siberian Sea. In the Kara and Barents Seas, the problem of insufficient skills of ship operators may lead to the risk of ship collisions. According to Fig. 5(f), the risk of collision accidents for LNG carriers in each water area on the Northeast Arctic route can be obtained. The highest risk of ship collision accidents is found in the East Siberian Sea with a quantitatively assessed value of 0.062. The second highest risk of collision accidents is found in the Laptev Sea with a quantitatively assessed value of 0.027. The next highest risk is found in the Chukchi Sea, the Kara Sea, and the Barents Sea, which have an assessed risk value of 0.02, 0.017, and 0.014, respectively.

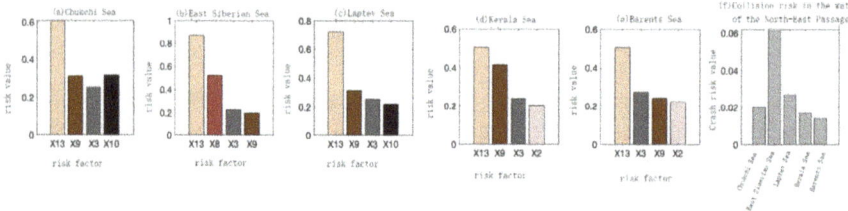

Fig. 5. Analysis of the main risk factors in each of the waters of the North-East Arctic shipping lanes (a–e) and the risk of collision accident (f).

It can be seen that the main factors affecting the collision risk of LNG carriers on the Northeast Arctic route include low skill level (X2), irregular operation (X3), low temperatures (X8), high winds (X9), and large ice floes (X13). Combining the classification of risk factors in Table 1 and the Bayesian network inference for quantitative analysis of collision risk in Fig. 4, the main risk causal paths of these risk factors can be obtained, as shown in Table 2.

Table 2. Accident causation pathways for major risk factors.

Form	Serial number	Crash Risk Causation Pathways
C1	①	C1 → X2 → M1 → M7 → R
	②	C1 → X3 → M1 → M7 → R
C3	③	C3 → X8 → M2 → X6 → M3 → M4 → M7 → R
	④	C3 → X9 → M2 → X7 → M3 → M4 → M7 → R
	⑤	C3 → X13 → M5 → R

When conducting risk management for LNG carriers sailing in Arctic waters, emphasis should be placed on the two aspects of personnel (C1) and the navigational environment (C3). The risk control of ship navigation on the whole Northeast Arctic route can be started from the management of the risk causation path ②, ④ and ⑤. In addition, in the Kara Sea and the Barents Sea, attention should be paid to the control of risk cause path ①, and in the East Siberian Sea, attention should be paid to the control of risk cause path ③. The specific risk control analysis is as follows:

(1) Risk causation paths ① and ② show that non-standard operation or low skill level will directly lead to crew errors, which will be manifested as operational failure. Therefore, crew members participating in the commercial voyage practice of LNG carriers in Arctic waters should be strictly selected, their operational level should be comprehensively assessed, and regular skills assessment should be carried out.
(2) The risk causation path ③ indicates that low temperature in Arctic waters may cause damage to LNG transmission components and lead to LNG leakage, which in turn leads to low-temperature damage to ship's equipment near the leakage source, resulting in ship's operational failure. An effective way to cut off the transmission of this risk could be to consider covering the LNG transmission parts with insulation materials.
(3) Path ④ shows that high wind speed mainly damages LNG storage tanks and causes LNG leakage and low-temperature failure of ship's equipment, which leads to operational failure and increases the risk of collision. When LNG carriers experience high winds, inspections should be carried out promptly and fastening equipment should be reinforced promptly to prevent the LNG tank from shaking and causing damage.
(4) The threat of risk causation path ⑤ is highest in every water. Although most LNG carriers in Arctic waters have been strengthened with special ice level, in order to effectively cut off the risk transfer, ships should apply for ice-breaking escort operation when passing through the waters with poor ice condition.

5 Conclusion

This study analyzed accident scenarios and constructs a risk evolution model for LNG carrier collision accidents in Arctic waters. Based on the analysis of the risk and risk causation paths of LNG carrier collision accidents in five main waters of the Arc-tic Northeast Passage, the risk management and control measures for ships sailing in Arctic waters are analyzed. The risk control method established in this study enables quantitative analysis of ship collision accidents, identifies the main risk factors af-fecting LNG carrier collision accidents in multiple Arctic waters and the main risk causation paths, and facilitates the relevant decision-making departments to carry out risk control in a more targeted manner.

Acknowledgments. This work is financially supported by National Natural Science Foundation of China (NSFC) (Grant No. 52402422). This work is also financially supported the Zhanjiang Science and Technology Plan Project (Project No. 2024B01001). The authors wish to express their many thanks to the reviewers for their useful comments.

References

1. A Gregory K. Silber, et al.: Vessel operations in the arctic, 2015–2017. Frontiers in Marine Science 6:573 (2019)
2. FarréA, A.B., et al.: Commercial arctic shipping through the northeast passage: Routes, resources, governance, technology, and infrastructure. Polar Geogr. **37**(4), 298–324 (2014)
3. Marken, V.B., et al.: Delay risk analysis of ship sailing the northern sea route. Sh. Technol. Res. **62**(1), 26–35 (2015)
4. Serdar, K., et al.: A root cause analysis for Arctic Marine accidents from 1993 to 2011. Saf. Sci. **74**, 206–220 (2015)
5. Afenyo, M., et al.: Arctic shipping accident scenario analysis using Bayesian Network approach. Ocean Eng. **133**, 224–230 (2017)
6. Zhang, M., et al.: Use of HFACS and fault tree model for collision risk factors analysis of icebreaker assistance in ice-covered waters. Saf. Sci. **111**, 128–143 (2018)
7. Baksh, A.-A., et al.: Marine transportation risk assessment using Bayesian Network: Application to Arctic waters. Ocean Eng. **159**, 422–436 (2018)
8. Khan, B., et al.: An operational risk analysis tool to analyze marine transportation in Arctic waters. Reliab. Eng. Syst. Saf. **169**, 485–502 (2018)
9. Shanshan, F., et al.: A quantitative approach for risk assessment of a ship stuck in ice in Arctic waters. Saf. Sci. **107**, 145–154 (2018)
10. wu, C., et al.: A new accident causation model based on information flow and its application in Tianjin Port fire and explosion accident. Reliab. Eng. Syst. Saf. **182**, 73–85 (2019)
11. Li, X., et al.: Modelling and assessment of accidental oil release from damaged subsea pipelines. Mar. Pollut. Bull. **123**(1–2), 133–141 (2019)
12. Li, Z., et al.: Using DBN and evidence-based reasoning to develop a risk performance model to interfere ship navigation process safety in Arctic waters. Process Saf. Environ. Prot. **162**, 357–372 (2022)

Open Access This chapter is licensed under the terms of the Creative Commons Attribution-NonCommercial-NoDerivatives 4.0 International License (http://creativecommons.org/licenses/by-nc-nd/4.0/), which permits any noncommercial use, sharing, distribution and reproduction in any medium or format, as long as you give appropriate credit to the original author(s) and the source, provide a link to the Creative Commons license and indicate if you modified the licensed material. You do not have permission under this license to share adapted material derived from this chapter or parts of it.

The images or other third party material in this chapter are included in the chapter's Creative Commons license, unless indicated otherwise in a credit line to the material. If material is not included in the chapter's Creative Commons license and your intended use is not permitted by statutory regulation or exceeds the permitted use, you will need to obtain permission directly from the copyright holder.

Research Status of Resilience Governance of Transportation Safety on the Maritime Silk Road

Liu Zhu, Jianjun Wu[✉], Shenping Hu, and Xiangqian Meng

Merchant Marine College, Shanghai Maritime University, Shanghai 201306, China
jjwu@shmtu.edu.cn

Abstract. Maritime transportation serves as a vital foundation for promoting economic globalization, ensuring the smooth flow of trade and goods. Enhancing the resilience of transportation safety is therefore a critical measure for achieving sustainable, reliable, and efficient international transportation systems in the face of evolving global challenges. Currently, ports and waterways along the Maritime Silk Road (MSR) face a range of complex and uncertain factors, such as frequent natural disasters, intricate non-traditional safety issues, the development of large-scale and low-carbon shipping equipment and so on, which put forward urgent requests for the construction of the resilient shipping system. By examining maritime transportation safety research from a resilience perspective, this paper took the Maritime Silk Road as the research target. It reviewed significant achievements in this area, focusing on the current state of academic research, the measurement of transportation resilience, and the application of safety resilience in the context of the Maritime Silk Road. Additionally, it proposed the directions and development suggestions for deepening the research on the resilience of maritime transportation safety from the perspectives of characteristics of resilience curve, theoretical research framework and objectivity. Future work should consider the resilience spiral curve, Environmental-Social-Governance (ESG) framework, and resilience governance research for new safety situations along the Maritime Silk Road.

Keywords: The Maritime Silk Road · Transportation Safety · Resilience Governance · Spatial Differentiation

1 The Research Framework of this Paper

1.1 Purpose

China is currently in a critical phase of developing a robust transportation and shipping infrastructure. Enhancing the coordination of risk prevention and control within transportation infrastructure is essential for improving the resilience of transportation system. In alignment with China's strategies to become a leading maritime, shipping, and transportation nation, and within the context of the 21st Century MSR initiate, this paper focused on MSR as the research target. The objective is to summarize the current status domestic and international research on waterway transportation safety resilience. The paper examined the research status, summarized key findings, identified existing challenges, and proposed academic outlooks for future studies.

1.2 Research Methodology

To clarify the core concepts and research scope of this field, and to organize the main contents and methods of existing studies, this paper utilizes CNKI, SCI and SSCI as the citation databases for literature retrieval within the database. The search string was confirmed and the search period was defined from "2000 to 2023". The region was set as the MSR, and the disciplines were filtered as "road and waterway transportation" and "safety science and disaster prevention", and on this basis, the retrieved literature was analyzed subsequently by content analysis method and method of induction.

2 Current Status of Academic Research

2.1 Research Theme

According to the knowledge graph method, 52 English papers were retrieved, and after manual screening, 27 English papers and 16 Chinese papers related to resilience in maritime transportation safety were obtained, with fewer Chinese papers, reflecting that domestic research is still in beginning stage. 43 papers with high relevance were statistically analyzed by year, and resilience as a research topic received significant attention in 2015 and more attention after 2019 (see Fig. 1).

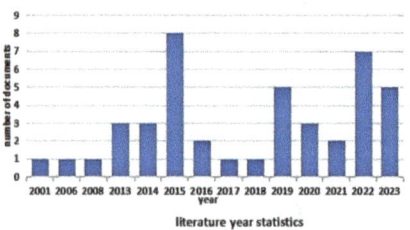

Fig. 1. Results of literature statistics for the period 2001–2023.

Using VOSviewer for bibliometric visualization, 49 high-frequency keywords were identified and divided into five clusters, primarily focusing on maritime transportation safety, systems, and models (see Fig. 2). Since the early 21st century, interest in maritime transportation safety resilience has grown, with Bayesian networks and management emerging as key topics.

Recently, due to rising demands for resilient systems against natural disasters and non-traditional safety threats, research has delved into areas like collision risk and safety supervision. Scholars have recognized that maritime accidents are the result of multiple

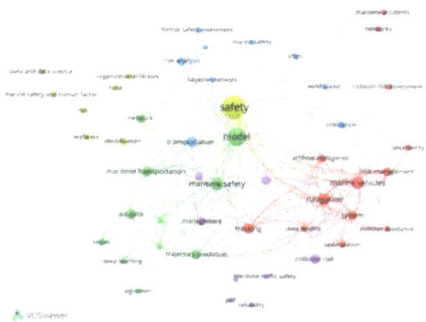

Fig. 2. The keywords co-occurrence network in the field of maritime traffic safety resilience.

factors, and have conducted coupled analysis research. And with advancements in computer technology, deep learning and algorithms have become critical tools, alongside complex network analysis. Additionally, human factors are increasingly recognized as pivotal in maritime accidents.

2.2 Research Perspective

Maritime transportation safety resilience undergoes an "absorption-recovery-adaptation" process. Based on the related literature [1], the main research perspectives on maritime traffic safety resilience are engineering resilience, social-ecological resilience and resilience to long-term uncertainty.

Initially, maritime traffic safety resilience focused on engineering resilience, emphasizing quick recovery to a single equilibrium after disturbances. However, this approach overlooked environmental interactions. Ecological resilience expanded the scope to multiple equilibrium states, highlighting system flexibility and resistance to disruptions. As researchers recognized the interdependence between human society and ecosystems, the social-ecological resilience perspective emerged, focusing on dynamic, adaptive, and learning capacities (see Fig. 3). In this rapidly evolving world, social-ecological systems are increasingly confronted with long-term uncertainties such as climate change, economic crises, and geopolitical shifts. Addressing these challenges requires a broader perspective and comprehensive coping strategies. It has led to a resilience perspective focused on coping with long-term uncertainty. The holistic, dynamic and future-adaptive natures of the system are emphasized. Its diagram under long-term uncertainty forms a spiral, achieving dynamic equilibrium through multi-party game theory, with periodicity and steadily increasing resilience levels.

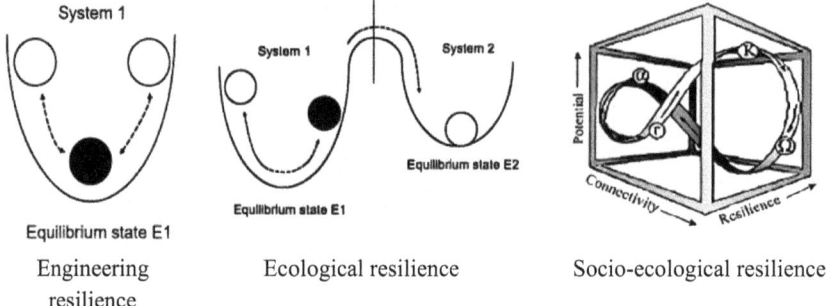

Fig. 3. Evolution of resilience perspectives.

3 Measurement and Application of Maritime Transportation Safety Resilience

3.1 Measurement of Transportation Resilience

Transportation safety resilience measurement methods are divided into qualitative and quantitative categories. Qualitative approaches, focusing on frameworks like reliability, redundancy, robustness, and recoverability [2]. Over the past decade, the resilience definitions and frameworks proposed in the literature are often only applicable to certain specific domains, so Rioshar et al. [3] proposed a resilience assessment indicator that includes absorptive, recovery, and adaptive capacities, and also considered the factors of speed and performance loss, providing a new perspective.

Due to the limitations of qualitative methods in quantifying resilience levels, scholars have shifted towards quantitative research. Many have sought resilience performance from a temporal perspective based on the resilience curve. Ortiz et al. [4] and Baroud et al. [5] used recovery time and disruption handling efficiency to measure resilience. However, these papers overlook dynamic changes in routes and cargo transfers. Additionally, relying on a single scalar often fails to comprehensively represent system resilience, particularly in complex systems. A comprehensive evaluation of system performance in different aspects is needed for accurate resilience assessment.

As resilience to long-term uncertainty becomes a research focus, methodologies have evolved towards network structures and multi-indicator approaches. Hu et al. [6] evaluated the resilience of Ro-Ro ship navigation safety in the Qiongzhou Strait using performance-based and logically associated topological indicators. Liu et al. [7] assessed resilience of European port network with a graph-theoretical topological indicator-based method. Centrality measures have analyzed the connectivity of the port shipping network along the MSR, considering its small-world and scale-free characteristics [8]. However, Kuang et al. [9] noted that relying solely on geographical distance or direct connections fails to comprehensively assess node importance. They proposed a more holistic approach by incorporating connection number and weight into node centrality. Research could develop directed and weighted shipping networks using actual traffic volume data in future, aligning analysis more closely with reality, as current studies focus primarily on directed unweighted graphs.

Although network methods have a mathematical foundation, they may not fully reflect system complexities. So, exploring comprehensive and easily applicable indicator-based methods has been promising. However, scholars mostly rely on expert consultations, data analysis to estimate parameters, with few using real first-hand ship data. This may undermine assessment accuracy, making the development of more realistic resilience measurement methods an important future challenge.

3.2 Increased Resilience in Transportation Safety

From the engineering resilience perspective, the primary focus is on enhancing transportation safety resilience through technological means. However, introducing new technologies may increase system's complexity and unpredictability, making it difficult to accurately assess and manage resilience. So, optimization under it typically yields limited improvement in resilience enhancement.

3.3 Spatial Characteristics of Transportation Resilience on the MSR

Direction of Maritime Safety. Take the areas of pirate activities as one of the spatial differentiation indicators. Piracy was particularly rampant at key nodes such as the Strait of Malacca, the Gulf of Aden, and the Gulf of Guinea along the MSR. Pirate attacks in the Strait of Malacca area are mainly robberies, primarily targeting ships with poor anti-piracy awareness, slow speed, moored or anchored, usually occurring at night, so more dedicated piracy patrol personnel and increased safety patrol frequency are needed in this area. 80% of incidents in the Gulf of Aden occurring during the day, so crews can develop navigation plans based on pirate activity patterns, and enforcement mechanisms need be improved. The Gulf of Guinea faces weak coastal protection, so ships should promptly gather piracy intelligence, strengthen watch protocols, and cooperate with shore forces on reporting systems.

Ship Emission Control. Emission Control Area (ECA) serve as a spatial differentiation indicator in maritime emission management. Different seas and ports enforce varying emission control measures, with some regions requiring the use of fuels with lower sulfur content. For instance, the North American ECA implementing a 0.1% sulfur content in 2015. Starting from January 1, 2012, in any Turkish port or inland waterway, if the ship stays for more than 2 h, it must use low-sulfur oil with a sulfur content of no more than 0.1%. Starting from January 2022, all ships entering the Hainan Emission Control Area in China must use fuel with a sulfur content <0.1% m/m ultra-low sulfur fuel oil [10, 11]. The establishment of ECAs has significantly reduced air pollutant emissions, improved resource utilization, and promoted sustainable development in the shipping industry and environment.

Risk of Maritime War. Using the occurrence of war as a spatial differentiation indicator, the Red Sea crisis provides a pertinent case study. Scholars generally agree that rising ship insurance premiums effectively address the increased risks in the region. Notteboom et al. [12] noted that certain strategies can enhance regional resilience, such as shipping companies imposing additional fees like war risk surcharges and emergency adjustment fees. Shipping companies like Maersk, MSC confirmed suspending Red Sea routes or rerouting around the Cape of Good Hope. However, Houthi rebels in Yemen later threatened to block ships linked to Israel from using the Cape route, rendering it ineffective. Therefore, further exploration of ways to enhance resilience will be needed as the situation evolves.

Maritime Natural Disaster. Taking the predicted arrival region of the typhoon as an indicator of the spatial differentiation, typhoon warnings define control zones based on arrival time and intensity to facilitate appropriate defense measures. Within the 48 h control zone, personnel should monitor the typhoon's dynamics and enhance patrols, while managers track path and intensity changes and initiate emergency plans. Within the 24 h control zone, measures become more urgent, including navigation restrictions, and safety reminders for ships and crews. The 12 h control zone is the most critical stage, requiring intensified rescue efforts and strict compliance from ships to moor safely and avoid navigation. This delineation allows for tailored, timely implementation of necessary precautions and emergency actions.

3.4 Dynamic Assessment of Transportation Resilience on the MSR

The resilience process can be divided into three stages: disturbance avoidance, disturbance control, and safety recovery. The disturbance avoidance stage emphasizes risk perception to monitor disruptions, with scholars assessing risks in MSR areas, including ports in Guangzhou, Fujian, and the Malacca and Qiongzhou Straits, using methods like FTA, multiple regression, random forest, and STAMP [13–16]. The disturbance control stage includes emergency response and rescue capabilities, system monitoring and forecasting, using an indicator system to evaluate maritime rescue efficiency based on information perception, analysis, and decision-making. The safety recovery stage emphasizes optimization and restoration, improving system adaptability and recovery mechanisms, such as unmanned boat rescue technology, enhancing emergency plans, management, and resource support.

From the perspective of resilience in the face of long-term uncertainties, traditional safety theories and models are inadequate for addressing increasingly complex issues. Enhancing system resilience has become critical. Meng et al. [17] proposed a ship stability safety analysis method considering the navigation environment and climate factors. Traditional methods often assess individual ports or components independently. Jiang et al. [18] explored the impact of mitigation and adaptation strategies on ports like Shenzhen and Xiamen, extending to the entire network. In recent years, more and more scholars have focused climate issues in the resilience of maritime transportation. Wang, T. found that the existing literature related to climate change adaptation planning is still in its infancy, therefore calling for a global perspective to integrate climate risk assessments of different transportation modes in different regions, promoting the overall adaptation of multimodal transportation systems to climate change [19]. Yang et al. [20] proposed adaptation strategies such as monitoring tides and waves to enhance port resilience against extreme climate events. These multidimensional approaches aim to strengthen the resilience of complex maritime systems in the face of prolonged uncertainty.

4 Governance Perspectives on Transportation Safety Resilience on the MSR

4.1 Dynamic Balance and Steadily Increasing Resilience Enhancement

In resilience research, the evolution of various elements and subsystems and their interactions form the foundation of system resilience. Current studies primarily focus on dynamic changes within these subsystems and their complex interactions, which are crucial for enhancing overall resilience (see Fig. 4). The spiral resilience curve in Fig. 4(b) reflects a dynamic balance. This cyclical change indicates that each loop not only advances the system but also fosters steady improvement under various influences. Subsystems enhance overall resilience through mutual influences, support, constraints, or competition. With each completed cycle, the system level rises, highlighting the synergistic effects of internal and external factors. Therefore, in order to better promote sustainable transportation safety resilience, research should explore measures to enhance resilience and the mechanisms of this spiral ascent in future.

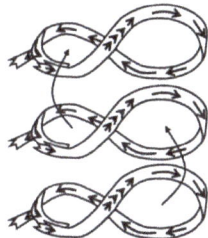
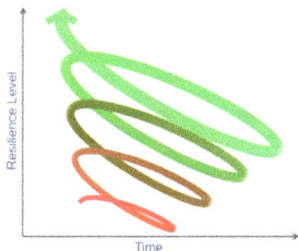

(a) Resilience dynamic interaction spiral diagram between subsystems.
(b) Spiral upward overall resilience curve.

Fig. 4. Diagram of resilience enhancement.

4.2 Research Methodologies Applying the "Environmental-Social-Governance" Theory

The transportation industry is currently transforming towards sustainable development, with ESG (Environmental - Social - Governance) providing a framework for analyzing maritime transportation safety resilience. ESG helps identify key environmental, social, and governance factors, develop targeted response strategies, and enhance system stability and sustainability. Dos Santos [21] addressed the lack of ESG application in the port industry by proposing a method to quantify ESG performance for international ports. Implementing ESG principles can improve safety, sustainability, and social responsibility, establishing robust resilience to future challenges. Research should integrate theoretical approaches and perspectives on ESG to explore their complementarity and differences in future.

4.3 Resilience Governance of the MSR Transportation Facing the New Safety Situation

The Ancient Silk Road Endpoints-the New Situation (Turkey and Nearby Waters). Taking the Turkish Straits and surrounding waters as an example, domestic research often appears within the broader context of the "Silk Road" network. Foreign studies, using simulations and accident data, identify the Bosporus Strait as a high-risk area due to its geography and heavy traffic. Accident-prone zones include the eastern and northern Mediterranean, southern and eastern Greece, and the Bosporus Strait coastlines. Additionally, in the current global context, geopolitical tensions are continuously intensifying, particularly in critical waterways such as the Black Sea situation and the Red Sea crisis, hence safety resilience can be investigated from the perspectives of warfare, emergency response and risk management, as well as policies and regulations.

The 21st Century Century MSR Endpoints-the New Situation (Gulf of Guinea and Adjacent Waters). Piracy activities pose a significant threat to the shipping industry and international maritime safety. Piracy in the Gulf of Guinea, near the endpoint of the 21st century century MSR, poses a major threat due to its geographic location, political complexity, and resource richness. Escalating piracy, combined with other safety issues, has brought global attention to maritime security in this region. To address MSR transportation safety resilience, new frameworks drawing on cross-disciplinary approaches, such as psychological and sociological theories, could help tackle both global and region-specific challenges.

5 Conclusion

Through the previous discussion, we have gained a comprehensive understanding. The paper systematically reviews various perspectives on maritime transportation safety resilience, elaborating on methods for measuring resilience, including qualitative analysis, quantitative measurement, and exploratory studies. Secondly, this paper has also discussed the applications of safety resilience in the MSR transportation, such as the enhancement of transportation safety resilience, the spatial characteristics of MSR transportation resilience, and the dynamic assessment of MSR transportation resilience. Overall, the enhancement of transportation safety resilience is still in an exploratory phase, presenting several challenges. In the future, based on the spiral upward trend of the resilience curve, ESG framework can be applied to explore the content in depth and strengthen industry practice. It is believed that through positive interaction between academic research and industry practice, we can advance the construction of maritime transportation safety resilience.

Acknowledgments. This research is funded by the Humanities and Social Sciences Foundation of Ministry of Education of the People's Republic of China [Grant No. 23YJAZH157].

References

1. Li, T.: New progress in study on resilient cities. Urban Planning International. **32**(05), 15–25 (2017) (in chinese)
2. Wan, C., Yang, Z., Zhang, D., et al.: Resilience in transportation systems: a systematic review and future directions. Transp. Rev. **38**(4), 479–498 (2018)
3. Rioshar, Y., Chuan, G., Faisal, K.: A simple yet robust resilience assessment metrics. Reliab. Eng. Syst. Saf. **197**, 106810 (2020)
4. Ortiz, D.S., Ecola, L., Willis, H.H.: Adding Resilience to the Freight System in Statewide and Metropolitan Transportation Plans: Developing a Conceptual Approach (Project No. 8–36, Task 73). Transportation Research Board of the National Academies, NCHRP, Washington, D.C. (2009)
5. Hiba, B., Kash, B., Ramirez-Marquez Jose, E., et al.: Importance measures for inland waterway network resilience. Transp. Res. E-logist. Transp. Rev. **62**, 55–67 (2014)
6. Shenping, H., Liu, W., Han, B., et al.: Resilience dynamics simulation of passenger-roller ship channel navigation safety considering factor mutation. J. Saf. Environ. **23**(10), 3408–3418 (2023) (in chinese)
7. Liu, Q., Yang, Y., Ng, A.K.Y., Jiang, C.: An analysis on the resilience of the European port network. Transp. Res. A Policy Pract. **175**, 103778 (2023)
8. Jing, W., Zhang, D., Wan, C., Zhang, J., Zhang, M.: Novel approach for comprehensive centrality assessment of ports along the maritime silk road. Transp. Res. Rec. **2673**(9), 461–470 (2019)
9. Kuang, Z., Liu, C., Jinran, W., Wang, Y.-G.: An effective distance-based centrality approach for exploring the centrality of maritime shipping network. Heliyon. **8**(11), 11474 (2022)
10. Weng, J., Han, T., Shi, K., et al.: Impact analysis of ECA policies on ship trajectories and emissions. Mar. Pollut. Bull. **179**, 113687 (2022)
11. Tan, Z., Liu, H., Shao, S., et al.: Efficiency of Chinese ECA policy on the coastal emission with evasion behavior of ships. Ocean Coast. Manag. **208**, 5691 (2021)
12. Notteboom, T., Haralambides, H., Cullinane, K.: The Red Sea Crisis: Ramifications for vessel operations, shipping networks, and maritime supply chains. Marit. Econ. Logist. **26**(1), 1–20 (2024)
13. Ugurlu, H., Cicek, I.: Analysis and assessment of ship collision accidents using fault tree and multiple correspondence analysis. Ocean Eng. **245**, 110514 (2022)
14. Shenping, H., Li, W., Xi, Y., et al.: Evolution pathway of process risk of marine traffic with the STAMP model and a genetic algorithm: A simulation of LNG-fueled vessel in-and-out harbor. Ocean Eng. **253**, 111133 (2022)
15. Uyanik, T., Karatu, C., Arslanoglu, Y.: Machine learning based visibility estimation to ensure safer navigation in strait of Istanbul. Appl. Ocean Res. **112**, 102693 (2021)
16. Afenyo, M., Khan, F., Veitch, B., Yang, M.: Arctic shipping accident scenario analysis using Bayesian network approach. Ocean Eng. **133**, 224–230 (2017)
17. Meng, X., Jianjun, W., Zhu, Q., et al.: Analysis method for ship stability and safety under the environmental impact of the maritime silk road. J. Tsinghua Univ. (Sci. Technol.). **64**(06), 1060–1069 (2024) (in chinese)
18. Jiang, C., Zheng, S., Ng, A.K.Y., Ge, Y., Fu, X.: The climate change strategies of seaports: Mitigation vs. adaptation. Transp Res Part D: Transp Environ. **89**, 102603 (2020)
19. Wang, T., Poo, M.C.-P., Ng, A.K.Y., Yang, Z.: Adapting to the impacts posed by climate change: Applying the climate change risk indicator (ccri) framework in a multi-modal transport system. Sustainability. **15**(10), 8190 (2023)
20. Yang, Y., Ge, Y.: Adaptation strategies for port infrastructure and facilities under climate change at the Kaohsiung port. Transp. Policy. **97**, 232–244 (2020)

21. Dos Santos, M., Pereira, F.: ESG performance scoring method to support responsible investments in port operations. Case Studies on. Transp. Policy. **10**(1), 664–673 (2022)

Open Access This chapter is licensed under the terms of the Creative Commons Attribution-NonCommercial-NoDerivatives 4.0 International License (http://creativecommons.org/licenses/by-nc-nd/4.0/), which permits any noncommercial use, sharing, distribution and reproduction in any medium or format, as long as you give appropriate credit to the original author(s) and the source, provide a link to the Creative Commons license and indicate if you modified the licensed material. You do not have permission under this license to share adapted material derived from this chapter or parts of it.

The images or other third party material in this chapter are included in the chapter's Creative Commons license, unless indicated otherwise in a credit line to the material. If material is not included in the chapter's Creative Commons license and your intended use is not permitted by statutory regulation or exceeds the permitted use, you will need to obtain permission directly from the copyright holder.

Warship Formation Time Unification Based on High Precision Celestial Observation

Zhiyou Zhang, Fu Yu, and Yongxin Jiang(✉)

PLA Dalian Naval Academy, Dalian 116000, China
1017552756@qq.com

Abstract. On the stage of information warfare, time communication plays a cornerstone role for joint operations In this paper, the method of time acquisition of war formation based on high precision cellular observation is proposed Through mathematical derivation, the mathematical model of time, the position of the observer and the height of the cellular body can be observed, and the iterative algorithm is used to solve the model Based on this algorithm, the universal time can be observed by observing the height of any cellular body and combining with the existing position information With this method, we are like to have an additional means to obtain accurate time when the timing system is disrupted on the sodium battlefield in the future, which is also the significance of this study, The results show that: The method is effective and independent in the observation time Even for the determination of different cellular bodies, the source of time is unique It can be applied to most combo units of the navy With the advancement of industrialization, the high accuracy of cellular observation will be improved, which will be more conductive to the accurate acquisition of time in this method.

Keywords: Time Acquisition · Celestial Observation · Astrological Positioning · Celestial Navigation

1 Introduction

Information warfare requires coordination and consistency among multiple branches of the military to achieve precise command, control, and collaboration. However, without a unified military standard time, it is impossible to achieve true joint operations. For the navy, collaborative operations between different branches require coordination of various warships, submarines, aircraft, and various command and control agencies on land [1]. Therefore, time unity can be regarded as the cornerstone of joint operations. This article proposes a method for obtaining ship time based on high-precision celestial observation. By observing the height of a certain navigation celestial body and combining it with existing position information, it is expected to directly solve the equation to obtain accurate world time, thereby better improving the current time unification system.

2 The Proposal of a Method for Obtaining Ship Time Based on Celestial Observation

Through investigation and analysis of the time unification system and its application status in naval forces, it is found that there is currently no well-established time unification system and system, and the contradictions and problems in the application capabilities of the troops are still prominent [2, 3]. Propose a ship time acquisition method based on celestial observation to address the above issues.

2.1 Current Ship Time Acquisition Methods

Modern technology typically uses satellite timing, longwave timing, and shortwave timing systems to achieve long-range remote replication and transmission of standard time by naval forces. Under undisturbed background conditions, naval forces can choose any timing method to obtain time [4]. However, considering convenience and timing accuracy, satellite timing has been widely used in the military.

At present, electronic interference technology has been able to interfere with satellite navigation and positioning systems. The reliability and effectiveness of wartime satellite positioning and navigation systems will be greatly reduced, and even completely ineffective. Relying solely on satellites to provide time guarantee for maritime combat units has become extremely fragile [5, 6].

Especially with the increasingly fierce maritime struggle between China and the United States, multiple types and types of satellite navigation equipment have been subject to varying degrees of interference during the tracking, monitoring, and evacuation missions of naval forces, resulting in abnormal operation and even antenna burnout, which has a significant impact on ship time acquisition and military operations [7].

Even though there are currently new equipment to support the acquisition of time for ships, there are still various technical, institutional, and construction issues in the existing naval forces. In the current situation, there is an urgent need to find a unified and efficient time tracking equipment of a unified model to standardize time traceability and improve system time uniformity [8].

2.2 Main Principles of Ship Time Acquisition Technology Based on Celestial Observation

The principle of astronomical positioning: When a ship observes the height of a certain celestial body, it calculates the geographical position of the celestial body's sub star point at the observation time, as well as the true height and top distance of the celestial body. The spherical circle made with the geographical position of the sub star point of the celestial body as the center and the top distance as the spherical radius is the astronomical position circle at the observation time, and the ship at the observation time is on the astronomical position circle at the observation time. By simultaneously observing two celestial bodies or using line shifting positioning, the observation vessel position can be obtained. In summary, the principle of astronomical positioning is to obtain the Green's time angle and declination of the observed celestial body through accurate world time and the name of the observed celestial body, and then combine it with the height of the celestial body to obtain the astronomical position circle to achieve positioning.

After determining the observation of celestial bodies, the longitude and latitude coordinates of the ship can be obtained by accurately measuring the world time and the height of the celestial body.

Using the principle of astronomical positioning, accurate world time can be calculated by reverse engineering. The calculation principle is as follows: The height of a celestial body is calculated by solving an astronomical triangle based on the declination of the celestial body during observation, the local semicircle time angle, and the estimated latitude of the ship. If we can accurately obtain the height of celestial bodies (i.e. the calculated height of celestial bodies) and know the longitude and latitude of ships, we can use astronomical triangles to calculate the declination and local time angle of celestial bodies. Based on the determined names of celestial bodies, we can obtain the accurate world time. Based on the longitude and time zone, the ship can further obtain its local time and zone time.

3 Feasibility Analysis of 2 Methods

The method of obtaining time by observing the height of celestial bodies is applicable to various navigation celestial bodies, and the theoretical feasibility of this method is verified through the derivation of mathematical models.

3.1 Mathematical Model Derivation

T_0 After converting the inaccurate ship time on board to world time, the declination δ and local semicircle time of the celestial body at this time can be obtained by checking the astronomical calendar and appendix t_{half}. φ Based on the premise of having accurate latitude.

Calculate the height of celestial bodies using the formula h_c:

$$h_c = arcsin(sin\,\varphi\,sin\,\delta + cos\,\varphi\,cos\,\delta\,cos\,t_{half}) \tag{1}$$

Calculate the T_0 corresponding height by inputting accurate longitude, latitude, and declination values h_c.

By observing with a sextant (or other star measurement methods) and supplementing and correcting observation values, the true observation height at this time can be obtained h_0.

For taking the derivative of (1):

$$\frac{dh}{dt} = \frac{-cos\,\varphi\,cos\,\delta\,sin\,t}{\sqrt{1 - (sin\,\varphi\,sin\,\delta + cos\,\varphi\,cos\,\delta\,cos\,t)^2}} \tag{2}$$

Derived from the sine formula of a spherical triangle (A is the semi circular orientation of the celestial body):

$$sinA = sin\,t\,cos\,\delta\,sech \tag{3}$$

Combining Eq. (1), it can be obtained that:

$$sech = \frac{1}{\sqrt{1-(\sin\varphi\sin\delta+\cos\varphi\cos\delta\cos t)^2}} \quad (4)$$

Substituting Eqs. (3) and (4) into Eq. (1) yields the following conclusion:

$$\frac{dh}{dt} = -\cos\varphi\sin A \quad (5)$$

The change in height dh can also be expressed as:

$$dh = h_0 - h_c = h_0 - arcsin(\sin\varphi\sin\delta+\cos\varphi\cos\delta\cos t_{half}) \quad (6)$$

According to Eq. (4), the variation of the local semicircle angle can be obtained dt:

$$dt = -\frac{dh}{\cos\varphi\sin A} \quad (7)$$

Combining Eqs. (3), (5), (6), and (7), it can be found that:

$$dt = -\frac{h_0 - arcsin(\sin\varphi\sin\delta+\cos\varphi\cos\delta\cos t_{half})}{\cos\varphi\cos\delta\, sec\, h_0 \sin t_{half}} \quad (8)$$

It should be noted that in this equation, it dt represents the change in the local semicircle angle, not directly the change in time, and needs to be transformed Change units. If using to dT represent the change in time, first dt convert the radian value into an angle value, and then use: $dT = 4 \times dt$ to convert it into a change in time (1 rad = 4 min).

3.2 Using Iteration and Recurrence to Accurately Calculate Real Time T_s

$$T_1 = T_0 + dT_0 \quad (9)$$

$$T_n = T_{n-1} + dT_{n-1} \quad (10)$$

The above equation dT depends on the initial T_n time, and as it T_n approaches T_s, the dT value of also decreases. The accuracy can be artificially designed, such as using the time calculated when it is dT less than 0.1 s as the exact time. So we can solve the above-mentioned problem based on this algorithm.

4 Research on the Impact of Time Acquisition Accuracy on 3 Times

The impact of latitude, observation time, and other conditions on the accuracy of time acquisition is discussed. Due to the numerous types of celestial bodies, the study of the impact on the accuracy of the Sun is more typical and commonly used. Therefore, taking the Sun as an example, the impact of changes in various factors on the accuracy of time acquisition is discussed below.

4.1 Time Acquisition Error

Further analysis DT of Eq. (5), considering that h_0 there is a certain degree Dh of error due to inaccurate height during observation, the errors in time and time angle Dt exist $DT = 4 \times Dt$, and can be obtained as follows:

$$DT = \frac{-4Dh}{\cos \varphi \sin A} \quad (11)$$

Due to the fact that the real time is unknown in real situations, when analyzing the error of this algorithm, it is assumed that the real time is T_S^{real} known. Based T_S^{real} on traditional table lookup, the actual observation value should be obtained h_0^{real}, and the accuracy of the unknown obtained on the ship T_0 can be calculated h_c, The possible T_S^{real} errors that may occur when human control changes the observation Dh will h_0 result in the actual time calculated according to this algorithm, which is called a certain time error DT.

If the control Dh is fixed at 1 point, then according to the error formula (11), a three-dimensional graph of the error variation with latitude and orientation can be obtained, as shown in Fig. 1.

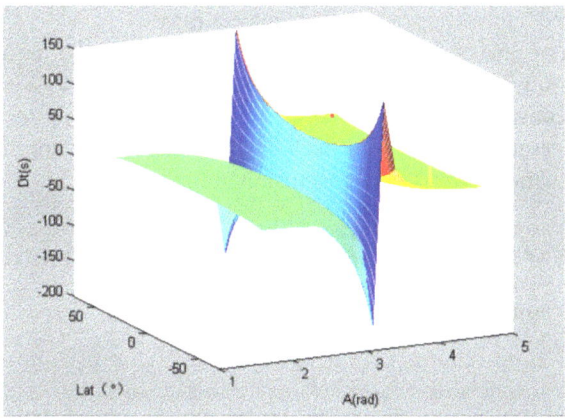

Fig. 1. The error variables with latitude and orientation.

4.2 The Impact of Time Changes on Time Acquisition Errors

Considering only the time variation factor, Fig. 2 is obtained:

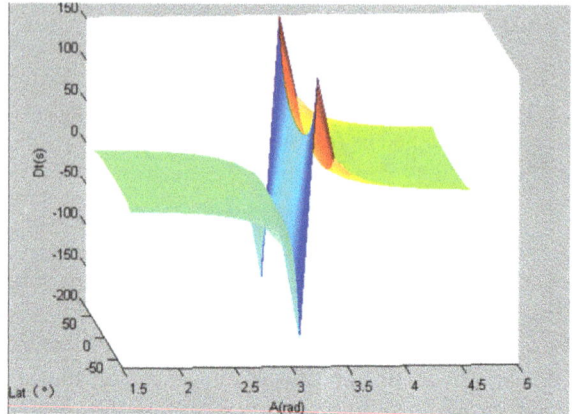

Fig. 2. Variation of the error with the sun's position.

According to Fig. 2 analysis, it can be concluded that when the latitude is constant, the time error changes with the sun's orientation, which is also known as time variation. When the sun's orientation approaches around 3.14 (radians) in the figure, the maximum time error is approximately at noon. Except for this time period, the time error is all within 10 s, which can meet the needs of practical applications.

The variation of time on the sun is reflected in the variation of the sun's orientation A. For Eq. (11), since the variation of A is roughly from 90° to 270° $sinA \in [0,1]$ during the day, considering only the variation of time, it can be seen that the error generated by observation around noon each day has a significant impact on the error of time acquisition, which confirms the conclusion in Fig. 2.

4.3 The Impact of Latitude Changes on Time Acquisition Errors

Considering only latitude variation factors, Fig. 3 is obtained:

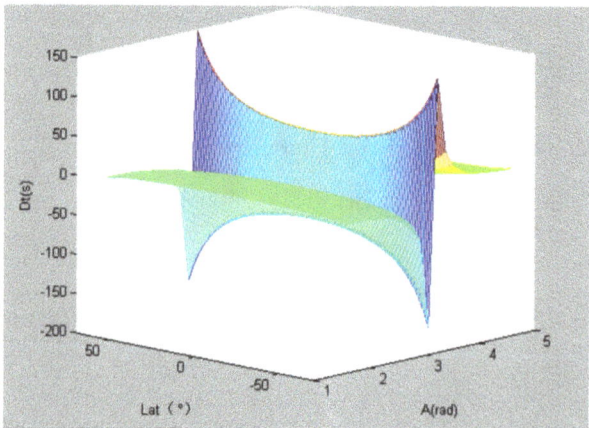

Fig. 3. The error variables with latitude.

From the analysis, it can be concluded that under a fixed observation time, the time error gradually moves from the two poles towards the equator, and the error value shows a decreasing trend. That is, under the same conditions, the closer to the equator, the smaller the error.

The change in the latitude $cos\varphi$ of the measurer affects the magnitude of Eq. (8), thereby affecting the error in time acquisition. According to the variation law of the cosine function, the impact of latitude change on time acquisition is relatively small in the middle and low latitudes, while the higher the latitude of the measurer, the greater the impact.

4.4 Example Analysis

Assuming the date is July 23, 2020, the following situations may occur:

① Our ship is located at ($\varphi 11°$ N, $\lambda 121°$ E) (low latitude), and the true altitude of the sun is calculated using real time. At the same time, the error of the true altitude is manually adjusted to 1 min and incorporated into this algorithm for calculation. It is found that the time calculation error exceeds 10 s between 11 h30 min and 12 h 30 min at ship time.

② When our ship is at ($\varphi 39°$ N, $\lambda 121°$ E) (mid latitude), using the same method as ①, the error between 11 h15 min and 12 h45 min at ship time exceeds 10 s.

③ When our ship is at ($\varphi 59°$ N, $\lambda 121°$ E) (high latitude), using the same method as ①, it is found that the error between 09 h45 min and 14 h30 min on board exceeds 10 s.

In summary, this algorithm is not suitable for the sun due to the large error before and after noon; The second is that the lower the latitude of the measurer, the smaller the time error. Therefore, it is recommended to avoid using this technology in high latitude areas around noon. The precautions for observing and calculating any celestial body used in navigation are the same as above.

5 Application of 4 Technologies

The time-series method for naval formation based on celestial observations is applicable to various locations in the navy:

① Ship formation: A formation composed of various types of ships, with limited equipment due to differences in model and age, and there are inevitably differences in equipment. Relying solely on Beidou and GPS timing is very easy to be interfered with. When conducting joint operations, if ships can have accurate longitude and latitude, combined with celestial body altitude, accurate world time can be calculated. Even without accurate latitude and longitude, the distance between ships within a naval formation is generally maintained within 1–2 nautical miles. With accurate celestial measurement altitude and suitable conditions, a time system with a maximum error of 10 s can be obtained, which is of great significance for joint operations of naval formations.

② Remote naval units, islands, etc.: These units have limited means of obtaining time, and time information is obtained through traditional methods [8]. When conducting important combat operations, it often delays the aircraft. Remote naval units or islands have precise geographic latitude and longitude, and precise time information can be obtained through celestial observations.

6 Conclusion

This article proposes a method for obtaining ship time based on high-precision celestial observation. The feasibility of this method is verified by solving the astronomical triangle in traditional astronomical positioning, using mathematical reverse derivation and iterative calculation methods. According to the verification of examples, when the observation position and time are appropriate, there is a 1 "error in observation accuracy, and the timing accuracy can be maintained within 5". The mathematical model is solved to obtain a trend chart of time error. Further analysis of the factors affecting the error shows that this technology has a wide range of applications, and only needs to avoid using it in high latitude areas before and after noon to achieve high time accuracy.

This method is of great significance for the formation of a sound time unified system and system, as well as for naval collaborative operations. However, there are still issues of low observation accuracy and inaccurate acquisition of location information at this stage, and it is hoped that this can be addressed in future research. This is likely to be widely applied in naval formation movements and various naval combat units such as remote islands in the future, providing strong support for our military operations.

References

1. Yao, K.: Study on the BVR cooperative air combat based on BP neural network. J. Phys. Conf. Ser. **1744**(4) (2021)
2. He, W., Wand, D.: Development of navigation systems and technologies for foreign naval ships. Ship Sci. Technol. **42**(11), 185–189 (2020)
3. Shen, Y.: Research on time service technology of naval command and control system. Ship Electron. Eng. (02), 42–45 (2005)
4. Song, Y., Liu, M., Wu, H., Lin, J., Yang, G.: Implementation of NTP network time service for marine time measurement equipment. Navig. Position. Time Serv. **8**(01), 163–167 (2021)
5. Liu, F., Shu, Z., Xie, W.: Current status and development trends of satellite navigation adversarial capability. J. Navig. Position. **8**(06), 1–5+13 (2020)
6. Kim, H., Lee, J., Oh, S.H., So, H., Hwang, D.H.: Multi radio integrated navigation system M&S software design for GNSS backup under navigation warfare. Electronics **8**(2) (2019)
7. Cheng, K., Chen, X.: Preliminary analysis of time system application in naval warfare systems. Ship Electron. Eng. (03), 39–42 (2004)
8. Li, D., Yang, W., Shen, Y.: Research on the development status of GPS independent PNT technology in the United States. Aviation Missile (12), 93–98 (2020)

Open Access This chapter is licensed under the terms of the Creative Commons Attribution-NonCommercial-NoDerivatives 4.0 International License (http://creativecommons.org/licenses/by-nc-nd/4.0/), which permits any noncommercial use, sharing, distribution and reproduction in any medium or format, as long as you give appropriate credit to the original author(s) and the source, provide a link to the Creative Commons license and indicate if you modified the licensed material. You do not have permission under this license to share adapted material derived from this chapter or parts of it.

The images or other third party material in this chapter are included in the chapter's Creative Commons license, unless indicated otherwise in a credit line to the material. If material is not included in the chapter's Creative Commons license and your intended use is not permitted by statutory regulation or exceeds the permitted use, you will need to obtain permission directly from the copyright holder.

The Use of PNT System in the War and Enlightenment to Our Army's Future Operations

Zhenyu Zheng, Zhiyou Zhang, and Yulong Kong(✉)

PLA Dalian Naval Academy, Dalian 116000, China
19824377927@163.com

Abstract. Review of the construction and development of positioning, navigation, and timing (PNT) system of different countries, analysis of current technical means and coping methods of each PNT system which is satellite navigation as the center and multiple systems as the auxiliary. Explore the application cases of PNT technology and its interference technology under different war backgrounds and its impact on modern war. On this basis, provide enlightenment for our military operations: including promoting the upgrading of China's PNT system, accelerating the construction of new independent navigation technology and independent timing system, etc.

Keywords: PNT System · Satellite Navigation System · Positioning Navigation and Timing

1 Introduction

The Positioning, Navigation and Timing (PNT) system is an abbreviation for a project that involves the integration of multiple technical means of land, sea, air and space, including positioning, navigation and timing. The system is generally centered on the satellite navigation system, and also includes radars of each platform and band, auxiliary radio systems, underwater acoustic baseline positioning systems and other positioning systems.

Due to the diversification of applications, users now have higher requirements for the accuracy and reliability of spatio-temporal information in complex environments. Even advanced and accurate satellite systems face issues such as signal interference, obscuration, and deception [1, 2]. Furthermore, satellite navigation services are limited in their ability to function indoors and in culverts. Additionally, any abnormality in the space load or failure at the ground control end can result in service interruptions. The limitations of relying on a single system are becoming increasingly apparent. To meet usage requirements, individuals naturally combine various positioning, navigation, and timing technologies to create a safe, reliable, high-performance, globally available, and consistently unified spatio-temporal service system, known as the PNT system. (This study strictly complies with academic publishing ethics and national security review requirements. All analyzed technical parameters are derived from publicly available sources including academic journals, official technical reports, and open-access defense white papers.)

2 The Development of the PNT System and Its Impact on Warfare

2.1 Foreign PNT System

Currently, the development of the PNT system is mainly undertaken by Russia, the United States, and the European Union, with other countries basing their development on it.

The mathematical principles underlying satellite navigation systems primarily rely on measuring the signal propagation time between satellites and user receivers to calculate the position of the user receiver. Here, we outline the key mathematical formulas involved in this process.

(a)Satellite Navigation Principle. *Pseudorange Measurement.* The first step is to measure the pseudorange, which is the apparent distance between a satellite and the user receiver, taking into account the clock errors:

$$\rho = c\,(\tau\_received - \tau\_transmitted) \quad (1)$$

where ρ is the pseudorange, c is the speed of light in vacuum, $\tau_{received}$ is the time the signal is received by the user receiver, and $\tau_{transmitted}$ is the time the signal was transmitted by the satellite.

Spherical Intersection Equations. To determine the three-dimensional position (latitude, longitude, and altitude) of the user receiver, measurements from at least four satellites are typically required. Each satellite provides an equation representing a sphere centered at the satellite's position with a radius equal to the pseudorange.

For four satellites, we have the following set of equations:

$$\rho_1 = \sqrt{(x - x_1)^2 + (y - y_1)^2 + (z - z_1)^2} + c \cdot \delta t \quad (2)$$

$$\rho_2 = \sqrt{(x - x_2)^2 + (y - y_2)^2 + (z - z_2)^2} + c \cdot \delta t \quad (3)$$

$$\rho_3 = \sqrt{(x - x_3)^2 + (y - y_3)^2 + (z - z_3)^2} + c \cdot \delta t \quad (4)$$

$$\rho_4 = \sqrt{(x - x_4)^2 + (y - y_4)^2 + (z - z_4)^2} + c \cdot \delta t \quad (5)$$

where (x,y,z) is the position of the user receiver, (x_i, y_i, z_i) is the position of the i-th satellite, ρ_i is the pseudorange measured from the i-th satellite, and $c \cdot \delta t$ represents the clock bias error, which is common to all equations.

Solving for Position. The above equations are nonlinear and are typically solved iteratively using methods such as the least squares method. In practice, additional corrections are applied to account for factors like atmospheric delays, relativistic effects, and other systematic errors.

(b) Satellite Navigation Types. Russia's PNT system is primarily based on enhancing its own global navigation satellite system, GLONASS.

GLONASS can meet the requirements for timing, land, sea, and air positioning and navigation, geodesy, and cartography. It is also used by Russia for battlefield support and operations [3]. Currently, Russia has deployed a new generation of GLONASS-K2 satellites, accelerated the expansion of the supporting ground facilities of the GLONASS system, developed a new satellite clock, established an interference detection system and an elastic navigation receiver, and further improved the robustness, anti-jamming ability, and navigation and positioning accuracy of the system. Additionally, Russia has been developing radio-aided navigation systems. Russia is actively promoting the development of the 'Consul' positioning system and the 'Octopus'-N1 (Sprut) radio high-precision navigation system, in addition to the original LORAN C system 'Chayka' and the ground-based ultra-long-range navigation system 'Alpha'. Currently, Russia has installed a new radio navigation system called 'Octopus'-N1 (Sprut) for the Navy. This system provides more accurate location, altitude, and speed information than the GLONASS system. The Octopus-N1 signal is resistant to interference and has high signal safety. It cannot be suppressed [4]. Regarding autonomous navigation, Russia's inertial navigation technology is in the second tier and has significant independent research and development capabilities. Additionally, Russia has developed autonomous navigation technologies such as terrain matching and astronomical navigation. These navigation aids work together to establish Russia's independent PNT system based on the GLONASS system. This system can effectively prevent the failure of a single PNT method during a war, which could have a significant impact on operations.

The primary objective of constructing the PNT system in the United States is to establish a functional framework centred around GPS (Global Positioning System) and various non-GPS systems to enhance PNT capabilities in complex and challenging environments. To achieve this objective, the United States has implemented two main strategies: strengthening the GPS system and expanding the use of non-GPS systems. The United States is committed to integrating technologies and systems to bring together all PNT resources and capabilities into a cohesive and synergistic whole to address challenges.

The US military has invested significant resources in upgrading the GPS system, including the space, ground, and user segments, increasing the number of satellites, signal power, and navigation and warfare capabilities. These enhancements include high-speed inter-satellite satellite-to-ground links, point beam augmentation, and passive laser ranging [5, 6]. On the other hand, the United States is deploying a new generation of GPSIII.F satellites while also continuously developing various new anti-jamming technologies at the user level.

To expand beyond GPS, the focus should be on developing technical approaches such as positioning, navigation and timing microtechnology (Micro-PNT) including inertial devices and quantum clocks. Autonomous navigation systems, such as precision inertial navigation systems, should also be developed. Strive to incorporate a range of principles, diverse information, and unified benchmarks while maintaining high integration.

The European Union's PNT is primarily constructed using the Galileo Satellite Navigation System (GALILEO), the world's first civilian global navigation and positioning satellite system. The European Union's PNT is primarily constructed using the Galileo Satellite Navigation System (GALILEO), the world's first civilian global navigation and positioning satellite system. The European Union's PNT is primarily constructed using the Galileo Satellite Navigation System (GALILEO), the world's first civilian global navigation and positioning satellite system. GALILEO enables non-military control and operation, providing non-confrontational global positioning and navigation functions. However, its navigation capabilities are relatively weak in confrontational environments. To address this issue, the system is equipped with a special situation handling system that ensures continuous service in exceptional cases and notifies users within seconds if the system fails.

Additionally, the system is capable of achieving mutual compatibility with both the GPS of the United States and the GLONASS system of Russia. This allows users to collect data from each system using a multi-system receiver or rely on a combination of data from both systems for positioning and navigation. The system's operation allows users worldwide to receive more signals from navigation and positioning satellites using multi-standard receivers, significantly enhancing navigation and positioning accuracy.

2.2 Domestic PNT System

Currently, China's positioning, navigation, and timing (PNT) system primarily relies on the Beidou global satellite navigation system as its core. It also collaborates with auxiliary systems, such as inertial navigation, and auxiliary and backup systems, such as land-based radio navigation, to establish a high-precision spatio-temporal information service system with multiple means, information sources, and platforms.

The Beidou navigation system is currently the world's leading positioning system. It offers global coverage, high precision, multi-mode fusion, and a wide range of applications. In addition to providing positioning services, the Beidou satellite navigation system also offers functions such as Beidou clock, Beidou altimetry, Beidou ephemeris, and short message two-way communication. The system utilises a multi-mode fusion strategy to integrate signals from GPS, GLONASS, Galileo, Beidou, and other systems to enhance positioning accuracy and reliability.

In the presence of interference, the reception of Beidou satellite signals may be affected. However, the Beidou system utilises a combination of active and passive signal transmission and is equipped with three-channel transmission. When one channel experiences interference, data can still be transmitted smoothly using the other two channels without compatibility issues. This reduces the impact of interference on signal reception to some extent.

Additionally, the Beidou system has anti-interference capabilities and can self-repair quickly to ensure continuous and stable operation of the Beidou navigation system. However, the type and degree of interference can affect the normal operation of the Beidou system. The specific situation depends on the actual circumstances.

In the future, the space-based PNT system will remain the main focus of our development, but we will also explore other directions.

2.3 Application of the PNT System in Warfare

In recent decades, global warfare has undergone significant changes. Modern warfare is now characterized by drone attacks, precision-guided strikes, and electronic warfare, all of which are closely related to the PNT system. The precise PNT system has had a profound impact on the pattern of warfare, bringing about a subversive change.

The precise PNT system can effectively locate enemy targets, cooperate with other systems to collect information, and infer their strategic intentions. If necessary, it can also collaborate with weapon systems to carry out strategic strikes.

The United States confirmed bin Laden's hiding place during the Neptune Spear operation using intelligence and high-altitude thermal imaging technology. They achieved precise positioning and global monitoring through the use of drones, satellites, airborne and ground-deployed electronic warfare equipment. The mission was successful in carrying out a typical assassination.

During the Middle East conflict, the US military utilised the PNT system's technology to precisely target Qassem Soleimani. Subsequently, three missiles were launched to eliminate him.

The Precision Navigation and Timing (PNT) system enhances the strategic position of unmanned equipment. With an accurate PNT system, unmanned equipment can perform precise position changes and accurately locate enemy targets, thereby improving traditional reconnaissance methods. In modern warfare, UAVs are frequently utilized by all sides to strike strategic targets. The cost-effective and high-yield benefits of this technology were evident during the war.

Currently, most Western countries have achieved precise positioning and navigation through the mature PNT system in Europe and the United States. This allows their artillery units to utilize various types of UAVs for multiple precision artillery attacks on enemy positions, maximizing the use of artillery ammunition reserves. The use of UAVs to guide traditional artillery has revolutionised the artillery strike mode, reducing costs and improving weapon efficiency. The integration of unmanned equipment with the PNT system has injected new vitality into traditional weapons and equipment, allowing for greater benefits from inexpensive weapons.

The impact of unmanned aerial vehicles (UAVs) on modern warfare is significant, particularly those that rely on precise positioning and navigation. In UAV warfare, an effective means of defence and attack is to interfere with the UAV's positioning, navigation, and timing (PNT) system. For instance, during a war, the enemy's PNT system can be effectively suppressed by electronic jamming, which would result in the loss of the enemy's advantage in precision strikes. This electronic countermeasure will improve our survivability and help us develop more effective follow-up strategies.

3 PNT System Interference Technology and Countermeasures

3.1 Existing Techniques for Interfering With PNT

China's current positioning, navigation and timing (PNT) system revolves around a satellite navigation system. It includes land-based radio systems, various platform inertial navigation equipment, multi-band radar systems, and underwater acoustic baseline systems. The jamming means of each system can be divided into two categories: soft kill and hard kill. This section will cover the signal transceiving, propagation, and system equipment of each subsystem. Additionally, we will provide a summary of the various interference methods that can affect the PNT system.

(1) Soft Kill. Its main principle is to interfere with, trick, or even suppress the enemy's detection equipment by emitting a specific signal. This reduces its positioning accuracy or disables it. Spectrum confrontation is at the heart of this tactic. High-power, wideband signals are typically used in complex interference environments. Frequency sweeping, broadband blocking, and specific frequency repetition are employed to submerge the signal in a background full of clutter, weakening the effectiveness of the enemy's equipment and suppressing it. Furthermore, it can analyse and simulate the signal structure of the opponent, send false signals, or delay the navigation signal to induce the enemy equipment to produce positioning errors, thereby reducing its positioning accuracy.

There are two main types of soft-kill techniques: suppressive jamming and spoofed jamming.

Suppressive jamming submerges the signal in a signal background full of clutter by means of broadband blocking, frequency sweeping, and specific frequency repetition. This is achieved by transmitting high-power and wideband interference signals, resulting in the inability of the receiving equipment to receive the actual signal normally. Due to the weak nature of satellite signals, high-power suppressive jammers can be used to transmit broadband noise signals that cover all frequency bands and block the enemy's signal reception. Additionally, the signal can be transmitted in a certain frequency range using step scanning to achieve the purpose of frequency sweeping, resulting in the satellite signal receiving equipment being unable to receive the actual satellite signal normally. To achieve accurate targeting scanning jamming, the radar's frequency range should be aimed at the frequency band where the enemy's communication equipment is located, which typically falls within the 3–20 MHz range. It is important to ensure a frequency repetition of at least 75%. Similar noise signals can be used to cover or drown the target echo signal. Other systems can apply a similar principle to achieve interference.

The principle of deceptive jamming involves analyzing and simulating the signal structure of the opposing party. Deception jamming equipment can then transmit a forged signal that resembles the system parameters. This can cause the receiver to receive false or low-quality PNT information. Alternatively, the equipment can delay and relay the navigation signal from the transmission path to mislead the enemy and cause errors.

(2) Hard Kill. A hard-kill threat is the use of weapons of destruction to destroy the basic unit or entire structure of an enemy system, with the goal of rendering it inoperable.

In the case of satellite navigation systems, space anti-satellite weapons are typically employed to directly destroy the satellites in the system. The United States, Russia, and other military powers have conducted research and development in the field of space weapons, including advanced technologies such as anti-satellite missiles and laser weapons. Other systems can also cause damage through the use of specific weapons.

The use of weapons of hard destruction can lead to international disputes in peacetime, so their use is relatively limited.

3.2 The Use of PNT System Jamming Technology in Modern Warfare

Electronic Interference. The accuracy of enemy strikes in modern warfare is increasing due to the continuous development of various positioning technologies. Electronic jamming has become an ideal countermeasure to such strikes.

The Battle of the Bekaa Valley is considered a model of electronic warfare victory. During the campaign, the Israeli army utilised electronic warfare planes to conduct reconnaissance and monitor the movements of Syrian troops. They also employed electronic communication jamming and radar jamming to disrupt the Syrian fighters' contact with the ground. Throughout the war, the Israeli army coordinated active and passive jamming techniques, resulting in significant achievements.

During the Russian-Georgian war, the Russian military employed the MI-8SMV-PG helicopter, which was equipped with the Daqing electronic warfare system. This system was used to suppress electromagnetic signals, resulting in a significant reduction of the Georgian army's missile guidance radar detection range to 40%, leaving only 10 kilometers. Furthermore, the Russian army effectively disrupted the radio communication lines of the Georgian army and jammed their drone signals, resulting in the loss of a significant number of drones. This provided a strategic advantage for the Russian army.

During the operation in Syria, the Russian military extensively utilised electronic warfare, deploying systems such as 'Krasukha-4' and 'Shibine'. "Krasuha-4" is a multifunctional jamming system deployed at air bases in northern Syria. Its main function is to jam air-based radars and cooperate with other air defense systems. The "Hibine" electronic countermeasures pod is installed on Su-30SM, Su-34, Su-35S and other fighters, and is deployed at the Latakia base. Its main function is to suppress noise or deceive jamming signals, and interfere with the enemy's electronic systems and pulsed signal radar. The 'Learjet-3' electronic warfare system is carried on 2–3 'Sea Hawk-10' UAVs. A vehicle-mounted ground control station is used to carry out mobile phone network jamming up to 100 kilometers away. It can even take over the virtual base station, send false information, identify and destroy the enemy's communication system, and weaken the communication and coordination ability of the Syrian troops. During the operation in Syria, electronic warfare systems played a crucial role in air defense strikes, providing significant support for the ground offensive of the Russian army.

Similarly, in the Gulf War, the United States conducted electronic intelligence reconnaissance months before the war, detecting and studying Iraq's radar defense capabilities. Five hours before the start of the war, the United States interfered with Iran's radar system and dismantled Iran's air defense system, resulting in Iraq's loss of air supremacy.

The role of electronic jamming in modern warfare is becoming increasingly important. Electronic jamming can significantly impact the enemy's critical equipment at a low cost, causing unpredictable consequences on the battlefield.

4 Enlightenment for the Development of the PNT System of Our Army

4.1 PNT System Upgrades

To comprehensively improve the safety, reliability, and accuracy of the PNT system, attention must be paid to satellite systems, ground-based radio systems, and radar systems of various platforms and bands, as well as information integration technology of various platforms.

Satellite System. The development of a PNT system is closely linked to the advancement of satellite technology. By improving the satellite system, the accuracy, reliability, security, and anti-interference capabilities of the PNT system can be enhanced.

One way to achieve this is by improving atomic clock technology, which can lead to more precise time measurement and distance calculation, resulting in more accurate navigation positioning. The atomic clock is a crucial component of navigation satellites, as it is responsible for time accuracy. It is considered a core secret of satellite technology in many countries.

Additionally, satellite accuracy is dependent on both the number and quality of satellites. In general, the number of satellites determines the number of equations required for position solving. The more satellites a user receives simultaneously, the higher the accuracy of the solution and the positioning.

It is also important to enhance the signal broadcast power of Beidou satellites, reduce the possibility of signal interference, and improve the security of satellite systems.

To address the issue of weak signal penetration in satellite navigation systems, the associated signal vectors can be intensified. The μ mesons, which travel at the same speed regardless of the material they pass through, can be used to develop a new type of positioning system called the Micromeasurement Wireless Navigation System (MuWNS). This system can work underground, indoors, and underwater. China should develop a micro-measurement wireless navigation system (MuWNS) to enhance the multi-domain service capabilities of satellite systems. To achieve this, research and development of chip-level atomic clocks and other means are necessary. It is important to improve China's underwater, indoor, and underground PNT service capabilities to keep pace with the times.

LORAN C System. Currently, the coverage of LORAN C signals in China is limited to the central and eastern regions, as well as offshore waters. However, the western region, particularly remote areas in the northwest, lacks comprehensive coverage of long-wave timing signals. To address this issue, we must prioritize promoting the comprehensive development of the radio navigation system and expanding the construction of the LORAN C system:

1. Improve the propagation effect of LORAN C signal in mountainous areas: In the western mountainous areas, the terrain and landform are complex, and the signal propagation is restricted to a certain extent. To this end, we need to research and apply advanced signal propagation technology to improve the propagation effect of LORAN C signal in mountainous areas.

2. Optimize the layout of LORAN C system platforms: In the western region, especially in the northwest remote areas, increase the construction of LORAN C system platforms, optimize the site layout, and improve the coverage and reliability of the system.
3. Strengthen the construction of navigation stations: In the western region, gradually promote the construction of navigation stations, expand their coverage to the west, and provide accurate and reliable full-time positioning, navigation and timing support for the whole territory.

The westward expansion and construction of the navigation station will improve the coverage rate of LORAN C in the western region. This will further expand the coverage of the positioning, navigation and timing system of the land LORAN C system, filling the PNT gap in China's western border territory. It will provide accurate and reliable full-time positioning, navigation and timing support for China's entire territory.

Platform Information Integration System. The integration of platform information is a current trend, and it is necessary to establish a comprehensive PNT information integration system. This system should integrate satellite navigation, auxiliary radio navigation, network information, and related platforms of various services and arms. The information from each platform should be used to integrate with AI large model algorithms to form a complete set of information links. The goal is to achieve a PNT platform system with improved accuracy, stability, reliability, and independent operation. This results in a complete and reliable PNT system, that can achieve all-weather, all-scenario positioning needs.

Furthermore, the integration of the network information system allows the PNT platform to acquire real-time data, thereby enhancing the accuracy of positioning. By utilizing AI large model algorithms, we can intelligently process diverse data to achieve precise positioning. Additionally, AI technology can be utilized for fault diagnosis and system optimization, ensuring the stable operation of the PNT platform in complex environments.

The system has a wide range of potential applications. It is important to increase R&D efforts, make full use of platform advantages, and combine AI technology to create a high-precision and high-stability PNT platform. This will help enhance China's overall strength in the fields of positioning, navigation and timing, and meet the needs of national development and national defense construction.

Atomic Radio Technology. Atomic radio technology can accurately perceive hard-to-reach space electromagnetic field information by detecting state changes of atoms in the field. It has many advantages, including high accuracy, sensitivity, and ultra-wideband capabilities. Additionally, it is not affected by other electromagnetic interference signals, requires no calibration, and has strong anti-interference and anti-damage capabilities. Even in complex electromagnetic environments, atomic radio technology can still provide a strong guarantee for the development of China's communication field.

Moving forward, it is essential to increase research efforts on atomic radio technology, with a focus on maintaining its stability and reliability in extreme environments, extracting useful information from interfering signals, and improving communication rates. Simultaneously, the advancement of technologies such as quantum computing means that atomic radio technology may face security threats. Therefore, it is crucial to enhance the protection capabilities of atomic radio technology to improve its application in various fields.

4.2 Explore New Autonomous Navigation Technologies

Geomagnetic Navigation Technology. Geomagnetic navigation technology utilises magnetometer sensors to determine location by measuring the difference in magnetic field magnitude as the sensor passes through and comparing it with known readings on the magnetic field map. Geomagnetic navigation offers high stability and no interference compared to satellite positioning, and can provide 24-hour passive navigation for domestic warships. Easy to use, simple device requirements, low dependence, even without other positioning means to assist verification, can achieve accurate positioning. Geomagnetic navigation has the advantages of all-weather, all-day, all-regional, medium and high precision, high concealment and strong anti-interference, and is a safe and reliable new positioning technology.

New Inertial Navigation Technology. The development of application requirements is increasing, which is leading to higher demands for inertial technology. For instance, inertial systems used in high-precision long-endurance applications must meet higher standards of reliability, accuracy, and time retention. On the other hand, low-precision and large-dynamic applications require systems that can withstand harsh environmental conditions and have greater adaptability. In the field of weapons and equipment, there are stringent requirements for bandwidth, measurement range, fast start-up time, adaptability to complex environments, and long-term calibration-free operation.

Gravity Gradient Navigation Technology. The navigation system that matches the gravity field measures the Earth's gravity field using a gravity gradiometer for positioning. It does not require the transmission and reception of radio signals, making it less susceptible to external interference. The Earth's gravity field has the advantage of strong regularity and high coverage as a navigation signal source, which can meet the requirements of high precision, long endurance, autonomy, and passivity. During the flight of long-range missiles and aircraft, they inevitably pass through areas with drastic changes in gravity gradients, such as high mountains. This improves the accuracy of gravity gradient-assisted navigation, which is significant for long-range missile navigation. Therefore, the gravity field matching navigation system has become an important research direction in the field of autonomous navigation.

Astronomical Navigation Technology. Currently, the field of starlight navigation is anticipating the arrival of miniaturised high-precision vertical gyroscopes. Radio and infrared navigation are crucial in overcoming the effects of adverse weather conditions and ensuring uninterrupted and continuous operation of navigation systems day and night. Therefore, there is an urgent need for technological innovation and development. The astronomical navigation system is expected to develop towards miniaturization, high precision, global coverage, full automation, and all-weather functionality in the future.

In the future, astronomical navigation systems are expected to become more compact, precise, and capable of operating globally, day and night, automatically, and in all weather conditions. To achieve this goal, China must increase research on astronomical navigation technology and make breakthroughs in key core technologies. Technological innovation will improve the performance and reliability of the astronomical navigation system, enabling it to play a greater role in various fields.China should actively invest in the field of astronomical navigation, increase research and development efforts, and promote China's astronomical navigation technology to become a world leader.

4.3 Accelerate the Construction of a New Type of Autonomous Timing System

The timing system is a significant national infrastructure. China should develop its national timing system architecture for space-ground integration, accelerate the improvement and upgrading of the existing timing system, and focus on the research and development of advanced multi-means timing systems. This will enable high-precision continuous autonomous measurement and service of world time integration across multiple technologies, extending timing services to deep space, deep sea, and other fields. Strive to develop a distinctive national timing system by integrating celestial and terrestrial elements in a three-dimensional intersection within the world.

LORAN C Autonomous Timing System. The high-precision ground-based timing system is a strategically oriented national major scientific and technological infrastructure. Its purpose is to improve the safety, reliability, and timing accuracy of China's timing system, and to meet the needs of high-precision time frequency for basic scientific research and major engineering applications.

Currently, China's national timing system uses the LORAN C timing system for BPL long-wave timing. The LORAN C signal is stable and reliable, with good antiinterference performance. However, due to its inherent technical characteristics, there are also limitations, such as the provision of only single timing information and the inability to schedule independently. To increase the information content, an additional modulation channel can be established in the LORAN C system. This will enable the LORAN C self-timing and real-time broadcast time signal correction, thereby improving the system timing function and performance. Additionally, the coverage area of the timing signal should be increased.

New Type of Atomic Clock. Atomic clocks are a crucial component for space applications, including Beidou satellites, space satellites, global starlink Internet, and satellite television. The ability of a country to independently develop atomic clocks with independent intellectual property rights is not only a reflection of its scientific research strength in related fields, but also an important indicator of its national security capabilities. Therefore, it is essential to conduct research and apply independent new atomic clock technology.

Based on the existing atomic clock technology, we should increase research and development efforts to achieve breakthroughs in miniaturized and high-performance atomic clock products to meet the application needs of naval navigation, Beidou navigation, 5G communication, and Arctic scientific research, among other fields.

Autonomous Astronomical Timing. Astronomical timing primarily utilises star sensors to collect information. The sensors project stars onto an image sensor to match and identify star maps. This process allows for the retrieval of information on the corresponding stars in the navigation star library, which is then used to determine the ship's position. The timing is then completed based on this information.

Currently, Zhang Zhiyou and colleagues from the Dalian Naval Academy have proposed a 'time-integrated support method for ship formations based on high-precision celestial body observation'. This method can serve as an effective means of time support in future naval battlefields by observing celestial bodies and combining the observations with existing position information.

5 Conclusion

Regarding future construction, the Chinese military should consider the advantages and disadvantages of the PNT system in modern warfare, as demonstrated in conflicts such as those in Russia and Ukraine. It would be beneficial to learn from the experiences of all parties involved in the construction of their positioning, navigation, and timing systems. The military should also conduct self-examinations to identify any deficiencies and work to fill any gaps. The focus should remain on satellite navigation, supplemented by other means of navigation, timing, and positioning, such as LORAN C, in order to build a reliable PNT system. It is also important to conduct research on interference and anti-interference measures for existing PNT systems to ensure their defensive and offensive capabilities.

Additionally, China should integrate advanced technology with traditional means and develop new high-precision PNT-related technologies. We aim to develop independent and advanced technologies, striving to enhance the construction of China's PNT system to new heights.

References

1. Sanjin, V., Antonio, Š., Lovro, M., et al.: GMDSS equipment usage: Seafarers' Experience. J. Mar. Sci. Eng. **9**(5), 476–476 (2021)
2. Jie, Z., Yingjun, Z.: Modelling of Pilotage Service Portfolio for e-navigation and its application in approaching port. J. Phys. Conf. Ser. **1827**(1), 012033 (2021)
3. Adam, W., Jacek, P., Wiesław, P., et al.: e-Navigating in highly-constrained waters: A case study of the Vistula Lagoon. J. Navig. **74**(3), 505–514 (2021)
4. Osés, D.M.X.F., Juncadella, U.À.: Global maritime surveillance and oceanic vessel traffic services: Towards the e-navigation. WMU J. Marit. Aff. **20**(1), 1–14 (2021)
5. Luo, J., Wan, X., Duan, J.: A new model of environment-aware geographic information services in E-navigation. J. Navig. **73**(2), 471–484 (2020)
6. Rivkin, S.B.: e-Navigation: Five years later. Gyroscopy Navig. **11**(2), 176–187 (2020)

Open Access This chapter is licensed under the terms of the Creative Commons Attribution-NonCommercial-NoDerivatives 4.0 International License (http://creativecommons.org/licenses/by-nc-nd/4.0/), which permits any noncommercial use, sharing, distribution and reproduction in any medium or format, as long as you give appropriate credit to the original author(s) and the source, provide a link to the Creative Commons license and indicate if you modified the licensed material. You do not have permission under this license to share adapted material derived from this chapter or parts of it.

The images or other third party material in this chapter are included in the chapter's Creative Commons license, unless indicated otherwise in a credit line to the material. If material is not included in the chapter's Creative Commons license and your intended use is not permitted by statutory regulation or exceeds the permitted use, you will need to obtain permission directly from the copyright holder.

Internal Wave Wake Characteristics of Submerged Vehicle in Different Fr Numbers

Hang Sun[1], Yongjie Zhang[3], Haoyu Yang[2], and Dawei Li[1(✉)]

[1] Navigation Department, Dalian Naval Academy, Dalian, China
lidawei01150115@126.com
[2] The Second Team, Dalian Naval Academy, Dalian, China
[3] The Third Group, Dalian Naval Academy, Dalian, China

Abstract. Internal wave wake characteristics of submerged vehicle in a stratified ocean are research hotspot at home and abroad. Horizontal Digital schlieren measurement system was constructed to measure the wave height field in pycnocline. Measured wave height and angle were used to verify the reliability of the numerical simulation, and the conclusion of regularity was obtained. With the increase of the velocity, the angles of the internal wave wake were decreased. The maximum wave heights were increased and then decreased with the increased the velocity, and the maximum wave height presents a maximum value when the Fri = 1 in all experimental states.

Keywords: SUBOFF · Simulation · Digital Schlieren Measurement System · Internal Wave Wake

1 Introduction

The density of seawater increases with the increase of depth, and there is a depth interval of rapid density change - pycnocline. Ocean internal waves are that occurs in the pycnocline, its maximum amplitude appears in the ocean interior, and the wave frequency is between inertial frequency and floating frequency. The disturbed seawater particles in such a background field will oscillate up and down near the equilibrium position and eventually evolve into ocean internal waves [1–4].

There are many mechanisms of internal wave generation, one of them is source-induced internal wave, that is internal wave formed by movement of submerged vehicle in stratified ocean. The motions of submerged vehicle in stratified ocean includes: horizontal, vertical, oscillation and so on. From the perspective of internal wave generation mechanism, internal waves generated by submerged vehicle in density stratified fluid can be divided into two types: The first is the internal wave caused by the interaction between the flow separation of the submerged vehicle surface and the background density stratified fluid, which is also called the volume effect internal wave. The second type is wake internal wave, which is produced by the interaction between turbulence in the wake of the submerged vehicle and the background density stratified fluid, or by the gravitational collapse of turbulent mixing region in the background density stratified fluid. The former is also called Kelvin type internal wake, and the latter is called non-Kelvin type internal wake [5–7].

The research on the mechanism of source-induced internal waves can not only provide theoretical basis for the exploration of submarine detection methods, but also provide necessary data for the research on drag reduction, seaworthiness and seakeeping of underwater vehicles. Therefore, it is of great practical significance to carry out research on internal waves generated by underwater vehicles.

In this paper, drag experiments based on three-dimensional internal wave tank and numerical calculations based on SUBOFF scale model were carried out to research the internal wave wake. The horizontal digital schlieren system that independently developed by project team is used to collect and analyze the data of the wave height field of the internal interface. The error is less than 10% compared with the simulation results. The relationship between internal Froude number, stratification, drag position and internal wake characteristics of the scale model are given by numerical simulation and pool experiment.

2 Digital Schlieren System and Pool Experiment Designs

2.1 Digital Schlieren Measurement System Designs

The horizontal Digital schlieren measurement system is composed of industrial camera (CCD), large size schlieren screen, internal wave three-dimensional experiment tank and schlieren analysis module. The CCD pixel resolution is 5 million, the maximum frame rate is 50 frames, the CCD is placed above the experiment area, the height is 3 m. The schlieren screen is 1.5 m × 0.8 m and is placed under the internal wave tank to provide the schlieren background. The experimental schematic diagram is shown in Fig. 1. In these experiments, the tank size is 5.5 m × 1.5 m × 0.5 m. The diameter D of scale model SUBOFF is 8 cm, the length L is 66 cm, the length ratio λ is about 8, and the critical Fr_c is about 3.7.

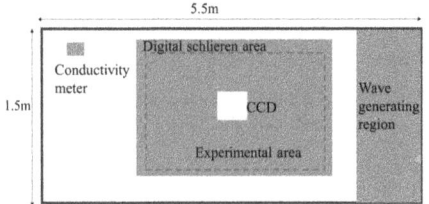

Fig. 1. Experimental Setup

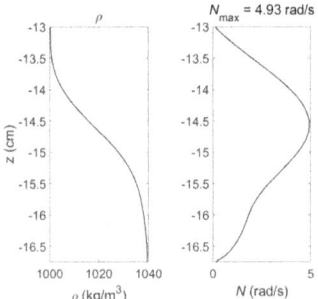

Fig. 2. Density and N_{max} profile

2.2 Stratification Condition Design

The total water depth of the experiments was 40 cm, and the stratification thickness was 15 cm in the upper layer and 25 cm in the lower layer. In order to study the influence of stratification conditions on wake characteristics, two kinds of stratification were set in the experiment, strong stratification (upper layer fluid density is 1000 kg/m^3, lower layer fluid density is 1040 kg/m^3) and weak stratification (upper layer fluid density is 1020 kg/m^3, lower layer fluid density is 1040 kg/m^3).

In order to ensure the stratification conditions, the experiment set a single point conductivity to collect stratification data, the collection water depth range of 13 cm–17 cm, vertical resolution of 0.25 cm. The experimental profile of strong stratification, lower towing position and towing speed (v) of 22 cm/s can be found in Fig. 2. The maximum floating frequency of the data is $N_{max} = 4.93$ rad/s [8.9]. In order to ensure the accuracy of experiments, the density profile was re-measured before each towing experiment, and the maximum floating frequency error of each density profile was less than 10%.

The towing speed is shown in Table 1. For each towing speed, 5 repeated experiments are performed. In order to investigate the effect of the towing position, the towing position was upper layer (10 cm) and lower layer (22 cm).

Table 1. Towing speed and Fri

Towing speed cm/s		10	15	20	25	30	35	40	45	50
Strong stratification upper layer	Fri	0.31	0.46	0.62	0.77	0.93	1.08	1.24	1.39	1.55
Strong stratification lower layer	Fri	0.31	0.47	0.63	0.78	0.94	1.09	1.25	1.41	1.56
Weak stratification upper layer	Fri	0.46	0.69	0.93	1.16	1.39	1.63	1.85	2.08	2.32
Weak stratification lower layer	Fri	0.40	0.61	0.81	1.01	1.21	1.41	1.61	1.82	2.02

3 Numerical Simulation Method Design

The numerical model SUBOFF has a length of 87 m, a radius of 10 m, and a simulated underwater displacement of 5600 tons. The upstream and downstream length of the computing domain are 200 m and 1000 m respectively, and the width is 400 m. The mesh refinement was carried out at the pycnocline and the free liquid surface, as well as at the bow and rudder surface of SUBOFF. The mesh models located in the upper fluid were shown in Fig. 3.

The k-Epsilon (k-ε) turbulence model is a two-equation model that solves transport equations for the turbulent kinetic energy (k) and the turbulent dissipation rate ε in order to determine the turbulent eddy viscosity.

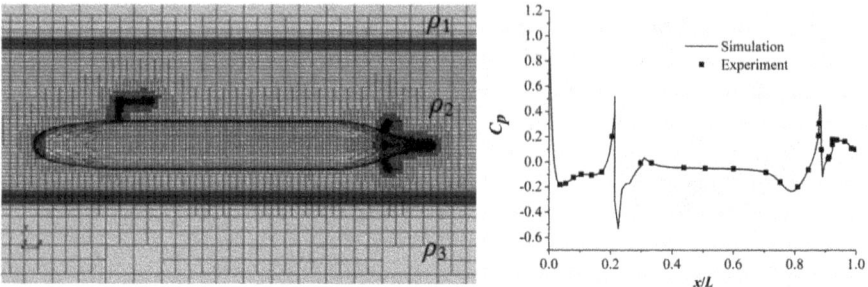

Fig. 3. SUBOFF mesh model   Fig. 4. Results verification

The transport equations for the kinetic energy (k) and the turbulent dissipation rate (ε) are:

$$\frac{\partial}{\partial t}(\rho k) + \nabla \cdot (\rho k \overline{V}) = \nabla \cdot \left[\left(\mu + \frac{\mu_t}{\sigma_k}\right)\nabla k\right] + P_k - \rho(\varepsilon - \varepsilon_0) + S_k \quad (1)$$

$$\frac{\partial}{\partial t}(\rho \varepsilon) + \nabla \cdot (\rho \varepsilon \overline{V}) = \nabla \cdot \left[\left(\mu + \frac{\mu_t}{\sigma_\varepsilon}\right)\nabla \varepsilon\right] + \frac{1}{T_e}C_{\varepsilon 1}P_\varepsilon - C_{\varepsilon 2}f_2\rho\left(\frac{\varepsilon}{T_e} - \frac{\varepsilon_0}{T_0}\right) + S_\varepsilon \quad (2)$$

Where: $\overline{V}$ is the mean velocity; μ is the dynamic viscosity; σ_k, σ_ε, $C_{\varepsilon 1}$ and $C_{\varepsilon 2}$ are Model Coefficients. P_k and P_ε are Production Terms; f_2 is a Damping Function; S_k and S_ε are the user-specified source terms. ε_0 is the ambient turbulence value in the source terms that counteracts turbulence decay. The possibility to impose an ambient source term also leads to the definition of a specific time-scale T_0 that is defined as:

$$T_0 = \text{Max}\left(\frac{k_0}{\varepsilon_0}, C_t\sqrt{\frac{v}{\varepsilon_0}}\right)$$

where: C_t is a Model Coefficient.

When the SUBOFF model was located on the pycnocline and the inflow velocity is 3 m/s, the calculated surface pressure coefficient (C_p) at the symmetric section of SUBOFF is compared with the experimental value in Fig. 4 [10]. It can be found that the numerical results are accurately.

4 Internal Wave Wake Features Analysis

The influence of the towing position on the evolution characteristics of internal wake is shown in Fig. 5. Figure (a) shows the evolution law of the internal wave wake with time when the towing position is at the upper level, and Figure (b) shows the evolution law of the internal wave wake with time when the towing position is at the lower level.

It can be found from the Fig. 5, the head wave amplitude of the towed underwater vehicle is positive when the towed position is in the upper layer, while the head wave amplitude of the towed underwater vehicle is negative when the towed position is in the

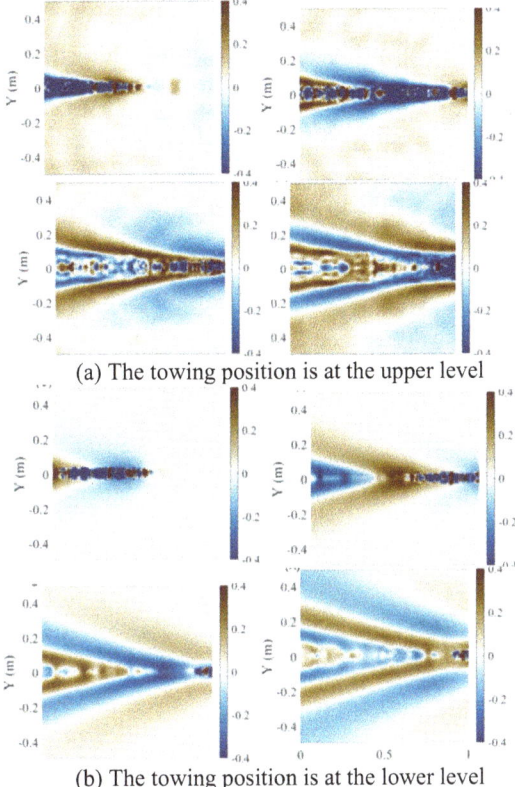

(a) The towing position is at the upper level

(b) The towing position is at the lower level

Fig. 5. The influence of the towing position (strong stratification, V = 22 cm/s).

lower layer. As shown in Fig. (b), for a wave front with positive wave amplitude, the water quality points at the wave interface tend to move upward, and the corresponding upper laminar flow will diverge to meet the mass conservation:

After systematic experiments, the curve of maximum wave height () of internal wave wake with Fri is shown in Fig. 6.

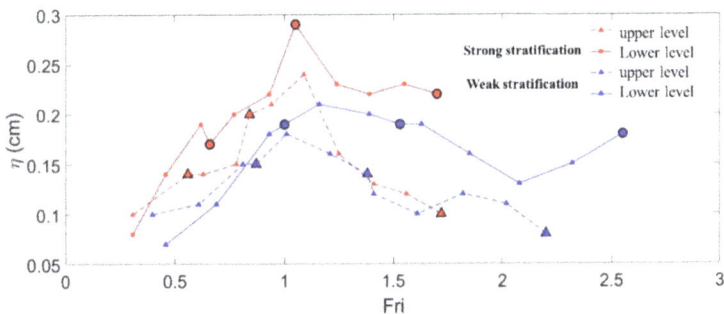

Fig. 6. Curve of maximum wave height in different Fri

It needs to point that: The color indicates the stratification condition, red indicates strong stratification, blue indicates weak stratification; The scatter shape indicates the towing position, the round point indicates the upper towing depth, and the triangle indicates the lower towing water depth. As can be seen from the figure, with the increase of the drag velocity (or Fri), the amplitude of the internal wave wake wave reaches its maximum value around Fri = 1, and presents a unimodal structure with Fri.

When the simulation model is located under the pycnocline and the drag speed is 33 cm/s, the numerical simulation results of the time evolution of the internal wake wave height field are shown in Fig. 7:

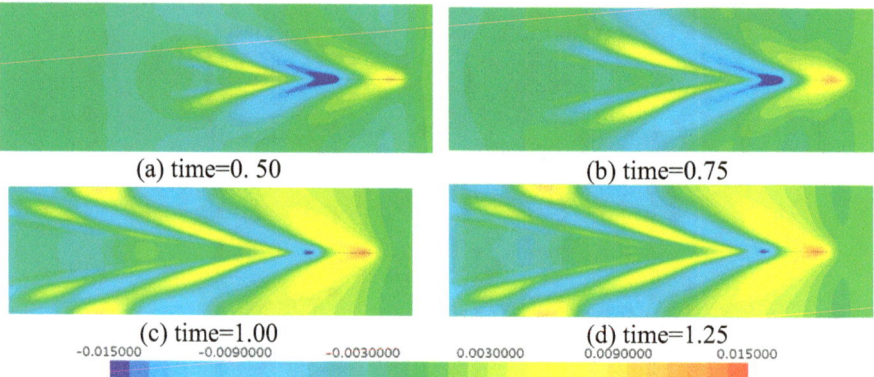

Fig. 7. Internal wake wave height field (lower layer, U = 33 cm/s).

The open angle of internal wave wake and the selection point of the maximum wave height value are shown in Fig. 8.

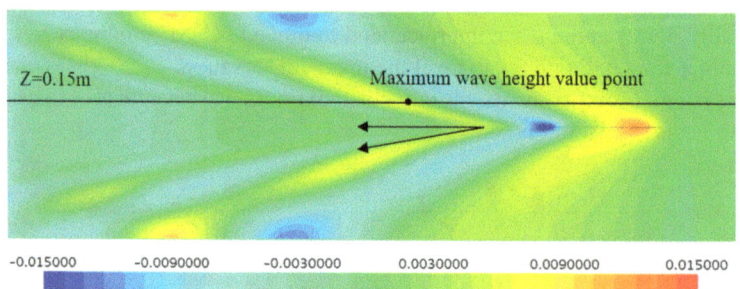

Fig. 8. Open angle and the maximum wave height value point

When the scale model is located under the pycnocline and the drag velocity is V = 33 cm/s, the comparison between the experimental results of the maximum wave height of the internal wave wake and the opening Angle of the wake with the numerical simulation results is shown in Table 2. It can be seen that the simulation results are very close to the experimental results.

Table 2. Comparison of internal wave wake features (lower layer, V = 33 cm/s)

Fri	Maximum wave height			Opening Angle of the wake		
0.84	Experiment	Simulation	error	Experiment	Simulation	Error
	0.0029	0.0032	9.3%	14.10°	13.1°	7.1%

Figure 9 shows the comparison between the simulation results of internal wave wake characteristics at different sailing speeds and the experimental results of the pool when the model is located under the cline. Similarly, both the variation law and the absolute value of typical speed are in good agreement with the pool experiment:

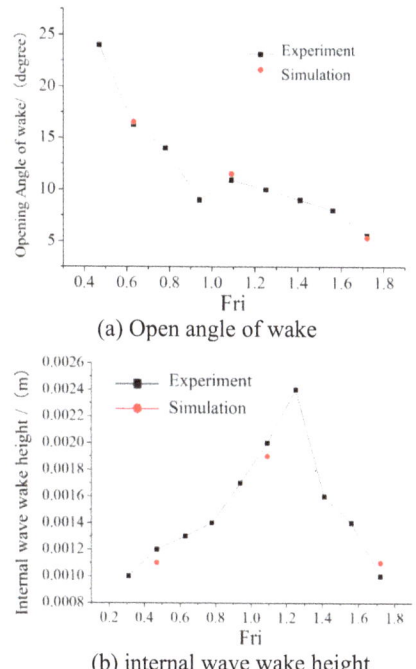

(a) Open angle of wake

(b) internal wave wake height

Fig. 9. Internal wave wake characteristics

5 Conclusion

The digital Schlieren system was used in this paper to capture the evolution of internal wave wake. The open angle of the internal wave wake amplitude field decreases with the increase of towing velocity. These results are consistent with the numerical simulation results. The amplitude field of internal wave wake also has obvious wave height characteristics at low drag velocity. However, compared with the high drag velocity, the internal wave wake travels closer at low velocity. With the increase of drag velocity, the wake propagation distance was increased gradually, and the wake opening angle was decreased gradually. The maximum wave height increases first and then decreases, and presents a maximum value when the Fri = 1.

Due to the limitations of the model sizes, and the absence of a wave-absorbing device in the water tank and other experimental conditions, the far-field characteristics of the internal wave wake have not been systematically studied. In the next step, the author will improve the experimental conditions and conduct research on more far-field hydrodynamic physical field characteristics.

Acknowledgement. The article is supported by Dalian NAVAL Academy research and development fund.

References

1. Newman, J.N.: Marine hydrodynamics. The MIT Press, United States (2018)
2. D Q, and Chen T T.: Surface and interfacial gravity waves induced by an impulsive disturbance in a two-Layer inviscid fluid. J. Hydrodyn. **21**(1), 26–33 (2009)
3. Xu Z, Xu C, Izolda V, et al. I (2006). Surface characters of internal waves generated by Rankine ovoid. Acta Mech. Sinica, 22(5):417–423
4. Brandt, A., Rottier, J.R.: The internal wave field generated by a towed sphere at low Froude number. J. Fluid Mech. **769**, 103–129 (2015)
5. Nematollahi, A., Dadvand, A., Dawoodian, M.: An axisymmetric underwater vehicle-free surface interaction: A numerical study. Ocean Eng. **96**, 205–214 (2015)
6. Shariati, S.K., Mousavizadegan, S.H.: Identification of underwater vehicles using surface wave pattern. Appl. Ocean Res. **78**, 281–289 (2018)
7. Shariati, S.K., Mousavizadegan, S.H.: The effect of appendages on the hydrodynamic characteristics of an underwater vehicle near the free surface. Appl. Ocean Res. **67**, 31–43 (2017)
8. Mackinnon, J.: Mountain waves in the deep ocean. Nature. **501**(7467), 321 (2013)
9. Wang, J., Chen, X., Wang, W., et al.: Laboratory experiments on the resonance of internal waves on a finite height subcritical topography. Ocean Dyn. **65**(9–10), 1269–1274 (2015)
10. Moisy, F., Rabaud, M.: A synthetic Schlieren method for the measurement of the topography of a liquid interface. Exp. Fluids. **46**(6), 1021–1036 (2009)

Open Access This chapter is licensed under the terms of the Creative Commons Attribution-NonCommercial-NoDerivatives 4.0 International License (http://creativecommons.org/licenses/by-nc-nd/4.0/), which permits any noncommercial use, sharing, distribution and reproduction in any medium or format, as long as you give appropriate credit to the original author(s) and the source, provide a link to the Creative Commons license and indicate if you modified the licensed material. You do not have permission under this license to share adapted material derived from this chapter or parts of it.

The images or other third party material in this chapter are included in the chapter's Creative Commons license, unless indicated otherwise in a credit line to the material. If material is not included in the chapter's Creative Commons license and your intended use is not permitted by statutory regulation or exceeds the permitted use, you will need to obtain permission directly from the copyright holder.

Simulation Research on Ship Drag Reduction Technology Based on Air Blowing

Xiao Wang[1], Haoyu Yan[2], Yongjie Zhang[3], and Dawei Li[1(✉)]

[1] Navigation Department, Dalian Naval Academy, Dalian, China
lidawei01150115@126.com
[2] The Second Team, Dalian Naval Academy, Dalian, China
[3] The Third Group, Dalian Naval Academy, Dalian, China

Abstract. Boundary layer flow control technology which is a significantly reduce the resistance method of surface/underwater transport equipment has been concerned by researchers at home and abroad. In this paper, a numerical simulation method of two-phase flow was used to study the drag reduction effect of ship controlled by blowing air jet flow. The researches show that the air blowing control on the bow underwater hull can effectively change the resistance characteristics of the ship. With the increase of blowing velocity, there were an obvious air isolation layers on both sides of the ship model, and the air separation flows phenomenon were appeared in the downstream of the flow field, which is conducive to reducing the friction resistance of the ship. Two mutually perpendicular velocities are applied to the blowing port to form a jet perpendicular to the ship surface at a 45°, thereby to affect the boundary layer of the ship surface. The blowing control velocity is not the greater the better, but there is an optimal value that have the best the drag reduction effect. Therefore, when the ship is sailing, the blowing control speed should be adjusted according to the sailing speed to achieve the optimal drag reduction effect.

Keywords: Blowing Control · Drag Characteristics · Pressure Distribution · Two-Phase Flow

1 Introduction

Reducing drag and increasing speed has always been a research hotspot in the field of ship building both domestically and internationally. In order to reduce ship resistance, researchers have proposed many drag reduction techniques, such as using a bulbous bow or optimizing the hull line to reduce wave making resistance; In order to reduce frictional resistance, super hydrophobic coatings, microbubbles, etc. are installed on the surface of the ship. Although active drag reduction techniques such as super hydrophobic surfaces, grooved surfaces, and Bionic surfaces technology [1–3]; Passive drag reduction techniques such as polymer injection and wall deformation [4–6] also have a certain degree of friction drag reduction effect, but these methods still need further exploration in terms of drag reduction sustainability and robustness against external interference.

The article is supported by Dalian NAVAL Academy research and development fund.

Sayyaadi et al. [7] conducted experimental researches on a catamaran and found that the air jet volume has a significant impact on the drag reduction effect, and there exists a critical air jet velocity; Wang Jiamei et al. [8] used porous silicon plates to generate microbubbles and conducted experimental research on the drag reduction of microbubbles on ship models with different drafts. They found that the drag reduction of ships was better when the drafts were relatively small. In 2013, Aljallis et al. [9] studied the drag reduction effect of large-area hydrophobic nanoparticle spray coatings and found that a drag reduction effect of about 30% could be achieved at high Reynolds numbers. In 2015, Cohen's research group [10] studied the drag reduction ability of super hydrophobic rotor surfaces in Taylor Couette flow and proposed that the effective slip length is related to the square root of the Reynolds number in high Reynolds numbers.

The boundary layer flow control method of ventilated partial cavity (VPC) is one of the most promising turbulent boundary layer control techniques [11]. The control principle of VPC is to create step surfaces or discontinuous grooves on the surface of a moving object immersed in water, so that flow separation is formed at the surface discontinuity, and artificial ventilation is carried out in the low-pressure area of the flow separation. Finally, the surface of the ship will be covered by a relatively stable air cavity. When the thickness of the air cavity can be compared with the thickness of the turbulence boundary layer, the shear stress on the coverage surface of air cavity will be significantly reduced, and the flow characteristics of the boundary layer will be significantly modulated. Therefore, if this boundary layer flow control method is appropriately applied to surface vessels and underwater submarines, the control methods can significantly reduce their frictional resistance and change the flow state of the boundary layer. Wu Hao and Ou Yongpeng [12] simplified the bottom of the ship to a flat plate and used the Mixture model to study jet drag reduction, effectively simulating gas-liquid mixed flow. It can be found that the current simulation of ship bubble drag reduction has simplified the numerical model to a certain extent, and the geometric model is simple, without fully considering the interaction between gas-liquid and bubble.

Blowing control drag reduction technology include the bubble drag reduction flow and gas layer drag reduction. Essentially, bubble drag reduction flow is to discrete bubble flow, while gas layer drags reduction and cavity (or gas chamber) drag reduction flow belong to gas-liquid stratified flow. The definition of bubble drag reduction originates from that bubbles exist in an independent or discontinuous state in the boundary layer. Gas is injected into the liquid boundary layer through methods such as strips, porous media, and porous plates. The gas is then torn apart by turbulent transport into individual free bubbles. Although these independent bubbles, which are larger in scale than Hinze scale and smaller in thickness than the boundary layer, tend to coalesce and disintegrate. Therefore, in addition to the free state, there are also flow states such as aggregation and clustering. However, overall, the gas-liquid interface is not continuous.

In order to further study the drag reduction effect of blowing control in the actual model, this paper uses numerical simulation method to study the drag reduction method of KCS standard ship model based on blowing control. The blowing drag reduction works by injecting air into the hull surface below the waterline to create bubbles or layers of air, creating a buffer between the hull and the water. The effect of blowing speed on drag reduction was considered. With the increase of blowing velocity, there is an obvious air isolation layer on both sides of the ship model, and the air separation phenomenon appears in the downstream of the flow field, which is conducive to reducing the friction resistance of the ship.

2 Simulation Model

2.1 Control Principle

Ship air blowing drag reduction, also known as air lubrication drag reduction, uses artificial injection of gas in the boundary layer to change the original single-phase (liquid phase) boundary layer state, and changes the wall shear stress by changing the momentum transfer, eddy current, flow state, density, viscosity and other fluid flow states of the boundary layer, thus achieving the goal of reducing friction resistance. The schematic diagrams can be found in Fig. 1.

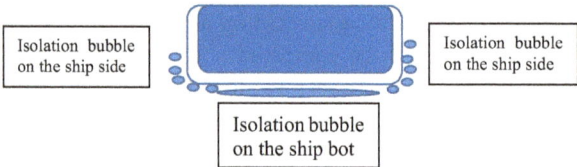

Fig. 1. Schematic diagrams of blowing drag resistance.

2.2 Grid Model

In this paper, surface reconstruction, polyhedral mesh generator and automatic surface repair techniques are used to construct polyhedral mesh.

In order to improve the overall quality of an existing surface and optimize it for a volume grid model, a surface reconstruction can be used to quadrilateralized the surface. The reconstruction is mainly based on the provided target side length, and can also include feature encryption based on curvature and surface proximity values. Local encryption based on part surfaces or borders can also be included.

Polyhedral mesh provides smooth solution for complex grid generation problems. Their construction process is relatively easy and efficient, requiring no more surface preparation than the equivalent tetrahedral mesh. In addition, for a given starting surface, a polyhedral grid contains about five times fewer grid cells than a tetrahedral grid. Multi-region meshes with conformal mesh interfaces are allowed. The polyhedral mesh model uses arbitrary polyhedral mesh cell shapes to construct the core mesh.

The Automatic Surface Repair tool provides an automated process for correcting a range of geometric type problems that may exist in the reconstructed surface after the surface reconstruction process has been completed. When performing surface refactoring on a cladding surface, the automatic repair option is usually used. Up to three different measures can be used to determine whether automatic repair is needed, including: perforated surfaces (surfaces intersecting including heterotopic intersecting), surface proximity value and surface quality.

The geometric and grid models of KCS standard ship models that based on the above method were found in the Fig. 2. The waterline of a ship.

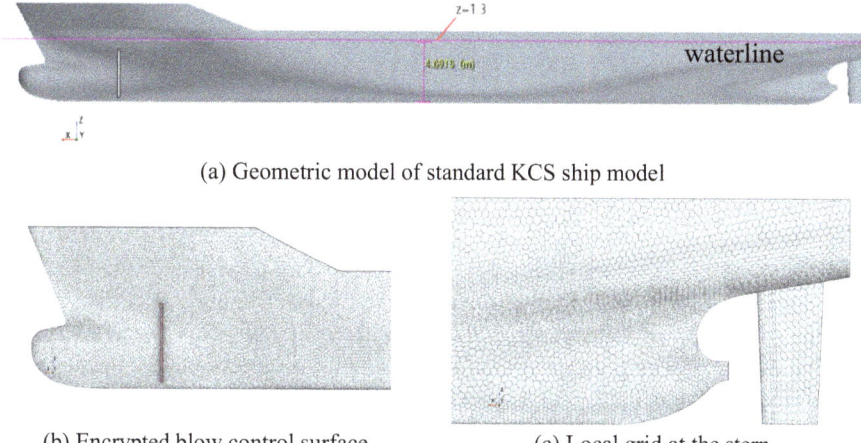

(a) Geometric model of standard KCS ship model

(b) Encrypted blow control surface   (c) Local grid at the stern

Fig. 2. Geometric and grid models of KCS.

2.3 Turbulence Model

The numerical model KCS has a length of 78 m, wide of 10, and has a draft of 4.69 m. To solve the Reynolds mean Navier-Stokes equation, it is necessary to decompose each solving variable ϕ in the instantaneous Navier-Stokes equation into its mean $\overline{\phi}$ and its fluctuation component ϕ':

$$\phi = \overline{\phi} + \phi' \tag{1}$$

Where: the ϕ represents the velocity component, pressure, energy, or component concentration.

The average can be considered as the time average of the steady-state case and the overall average of the repeatable transient case. The mean quantity equation is essentially the same as the original equation, except that there is now an extra term in the momentum transfer equation.

This additional term is the stress tensor and is defined as follows:

$$T_{RANS} = -\rho \begin{pmatrix} \overline{u'u'} & \overline{u'v'} & \overline{u'w'} \\ \overline{u'v'} & \overline{v'v'} & \overline{v'w'} \\ \overline{u'w'} & \overline{v'w'} & \overline{w'w'} \end{pmatrix} + \frac{2}{3}\rho k \tag{2}$$

where: ρ is density, u, v, w are velocity components, k is Turbulent kinetic energy, I is unit tensor.

It is difficult to model TRANS based on the average flow to close the governing equation. Two basic methods are used in the actual calculation: eddy current viscosity model and Reynolds stress transfer model.

The k-Epsilon (k-ε) turbulence model is a two-equation model that solves transport equations for the turbulent kinetic energy (k) and the turbulent dissipation rate ε in order to determine the turbulent eddy viscosity. The transport equations for the kinetic energy (k) and the turbulent dissipation rate (ε) are:

$$\frac{\partial}{\partial t}(\rho k) + \nabla \cdot [\rho k \bar{V}] = \nabla \cdot \left[\left(\mu + \frac{\mu_t}{\sigma_k}\right)\nabla k\right] + P_k - \rho(\varepsilon - \varepsilon_0) + S_k \quad (3)$$

$$\frac{\partial}{\partial t}(\rho \varepsilon) + \nabla \cdot [\rho \varepsilon \bar{V}] = \nabla \cdot \left[\left(\mu + \frac{\mu_t}{\sigma_\varepsilon}\right)\nabla \varepsilon\right] + \frac{1}{T_\varepsilon}C_{\varepsilon 1}P_\varepsilon - C_{\varepsilon 2}f_2 \rho \left(\frac{\varepsilon}{T_\varepsilon} - \frac{\varepsilon_0}{T_0}\right) + S_\varepsilon \quad (4)$$

Where: $\bar{V}$ is the mean velocity; μ is the dynamic viscosity; σ_k, σ_ε, Cε1 and Cε2 are Model Coefficients. Pk and Pε are Production Terms; f2 is a Damping Function; Sk and Sε are the user-specified source terms. ε0 is the ambient turbulence value in the source terms that counteracts turbulence decay. The possibility to impose an ambient source term also leads to the definition of a specific time-scale T0 that is defined as:

$$T_0 = Max\left(\frac{k_0}{\varepsilon_0}, c_t\sqrt{\frac{\nu}{\varepsilon_0}}\right) \quad (5)$$

where: Ct is a Model Coefficient.

3 Result Analysis

The blowing control surfaces were located on both sides of the bow underwater, with a length of 3.6 m*0.2 m. The blowing control gas is an ideal gas, and the flow rate can be control by controlling the blowing speed. The position of control surface and flow line from the blowing control surface are shown in Fig. 3. The blowing velocity includes two equal components, transverse and longitudinal, and the values are 1 m/s, 5 m/s and 10 m/s respectively. Therefore, the combined velocities are 1.414 m/s, 7.07 m/s and 14.14 m/s respectively.

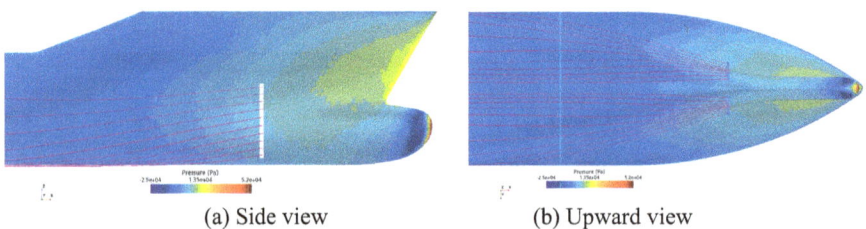

(a) Side view   (b) Upward view

Fig. 3. Control surfaces and flow lines.

When the control surfaces have different blowing control velocities, the air percentage distribution on the side of the ship model is shown in Fig. 4. It can be seen that with the increase of blowing speed, there is an obvious air isolation layer on both sides of the ship model, and the air separation phenomenon occurs in the downstream of the flow field. The air isolation layer is conducive to reducing the friction resistance of the ship.

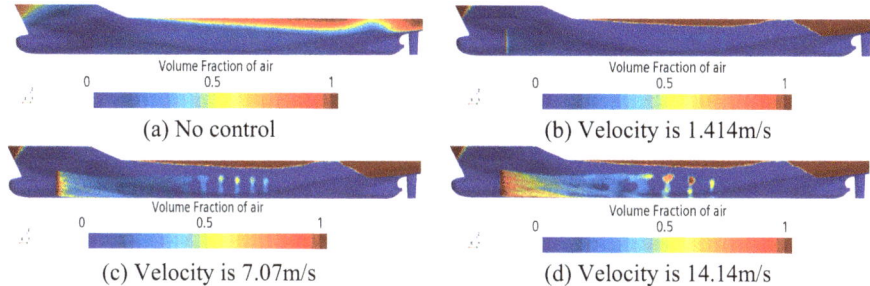

Fig. 4. Air percentage distribution at different control speeds.

The ship surface pressure distribution is shown in Fig. 5. It can be found that: when there is no control, the bow positive pressure area automatically transitions to the control surface when the control is removed, and there is no obvious negative pressure area on the downstream hull surface. With the increase of the blowing control speed, the pressure that close the control surface changes intermittent, which leads to the increase of the ship hull downstream surface pressure, and the most obvious pressure change occurs when the control speed is 7.07.

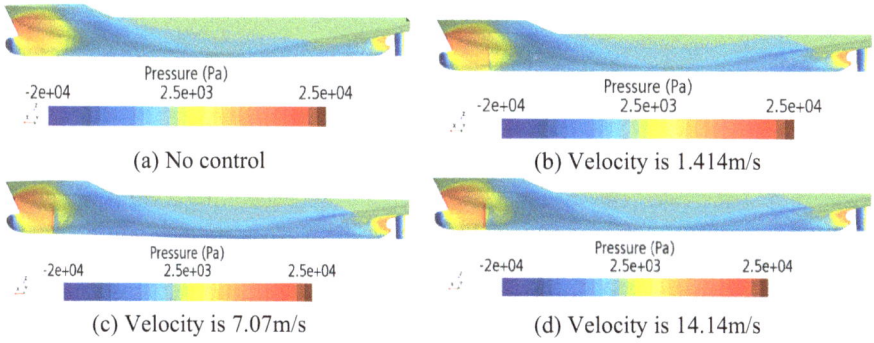

Fig. 5. Ship surface pressure distribution.

In order to further explain the pressure changes on both sides surfaces of the ship, the ship's surface pressure distribution at 0.3 m, 1.3 m, 2.3 m and 3.3 m below the waterline was intercepted respectively, as shown in Fig. 6. What needs illustration is that ship's surface pressure displays range from 0 to 73 m, and the rest ranges in the Fig. 6 were the upstream and downstream flow field pressure distribution of ship model. It can be seen that, due to the blowing control, there is an obvious change in pressure distribution in the middle of the ship (30 < X < 50), and the pressure is the highest when there is no control, and the pressure is the minimum when the blowing speed is 7.07.

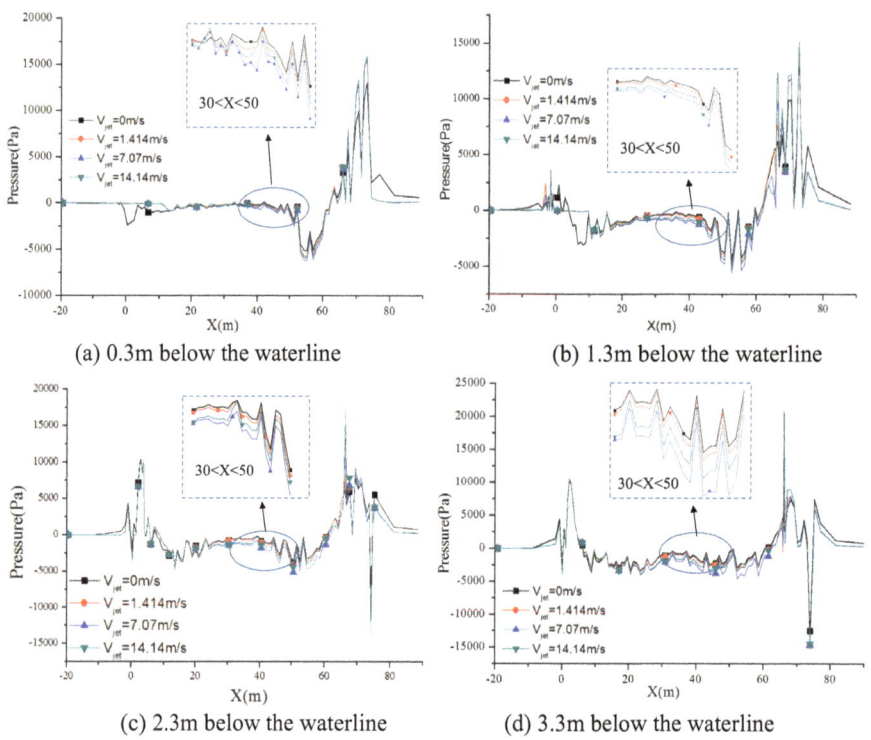

Fig. 6. Pressure changes on both sides surfaces of the ship.

Table 1 shows the change of the total resistance. Similar to the above results, the total resistance first decreases and then increases with the increase of the control velocities. When the control speed is 7.07, the total resistance is the least, so that control speed is not always better.

Table 1. The change of the total resistance.

Blowing velocity (m/s)	0	1.414	7.07	14.14
The total resistance (N)	215647.13	189561.887	155827.67	193857.101

4 Conclusion

The drag reduction effect of blowing control is studied by numerical simulation in the paper. The results show that with the increase of blowing velocity, there is an obvious air isolation layer on both sides of the ship model, and the air separation phenomenon occurs in the downstream of the flow field, which is conducive to reducing the friction resistance of the ship. However, the blowing control speed is not the greater the better, but there is an optimal value to acquire the best control effect, so the ship should adjust the blowing control speed according to the sailing speed.

References

1. Xiaojie, Z., Zhi, Z., Wang, J.: Numerical study on microbubble drags reduction of low-speed ships. J. Sh. Mech. **28**(3) (2024) (in Chinese)
2. Nagy, P.T., Paal, G.: The effect of spanwise and streamwise flexible coating on the boundary layer transition. J. Fluids Struct. **110**, 103521 (2018)
3. Xi, L.: Turbulent drag reduction by polymer additives: Fundamentals and recent advances. Phys. Fluids. **31**(12), 121302 (2019)
4. Perlin, M., Dowling, D.R., Ceccio, S.L.: Freeman scholar review: Passive and active skin-friction drag reduction in turbulent boundary Layers. J. Fluids Eng. Trans. ASME. **138**(9), 091104 (2016)
5. Mingjie, L., Wu, Q., Hao, Y.: Progress and challenges of bionic drag reduction surfaces. J. Beijing Univ. Aeronaut. Astronaut. **48**(9), 1782–1790 (2022) (in Chinese)
6. Tian, G., Fan, D., Feng, X., et al.: Thriving artificial underwater drag-reduction materials inspired from aquatic animals: Progresses and challenges. RSC Adv. **11**(6), 3399–3428 (2021)
7. Sayyaadi, H., Nematollahi, M.: Determination of optimum injection flow rate to achieve maximum micro bubble drag reduction in ships: An experimental approach. ScientiaIranica. **20**(3), 535–541 (2013)
8. Jiamei, W., Xiaowei, Z., Mansong, J.: Experimental study of the effects of ship draft on microbubble drag reduction. Sh. Eng. **26**(6), 9–12 (2004)
9. Aljallis, E., Sarshar, M.A., Datla, R., et al.: Experimental study of skin friction drag reduction on super hydrophobic flat plates in high Reynolds number boundary layer flow. Phys. Fluids. **25**(2), 351–412 (2013)
10. S, S., A, K.J., J B, G., et al.: Sustainable drag reduction in turbulent Taylor Couette flows by depositing sprayable super hydrophobic surfaces. Phys. Rev. Lett. **114**(1), 014501 (2015)
11. Wang, L.: Investigation on ventilated partial cavity two-phase flow and its effect on drag reduction and stability enhancement. ZheJiang Univ. (2021) (in Chinese)
12. Wu Hao, Ou Yongpeng (2016). Numerical study of method of flat plate viscous flow field with bubble. Sh. Sci. Technol., 38(15): 47–51. (in Chinese)

Open Access This chapter is licensed under the terms of the Creative Commons Attribution-NonCommercial-NoDerivatives 4.0 International License (http://creativecommons.org/licenses/by-nc-nd/4.0/), which permits any noncommercial use, sharing, distribution and reproduction in any medium or format, as long as you give appropriate credit to the original author(s) and the source, provide a link to the Creative Commons license and indicate if you modified the licensed material. You do not have permission under this license to share adapted material derived from this chapter or parts of it.

The images or other third party material in this chapter are included in the chapter's Creative Commons license, unless indicated otherwise in a credit line to the material. If material is not included in the chapter's Creative Commons license and your intended use is not permitted by statutory regulation or exceeds the permitted use, you will need to obtain permission directly from the copyright holder.

Structural Study on High-Speed Catamaran in Regular Head Wave

Yikang Chen[1,2], Zhipeng Deng[3(✉)], and Jinglei Yang[3]

[1] Haikou Branch, Guangzhou Bureau, Ultra-High Voltage Transmission Company, China
Southern Power Grid Co., LTD, Hainan, China
[2] Joint Laboratory of HVDC Equipment and Submarine Cable Safety Operation, Guangzhou, China
[3] College of Marine Engineering Jimei University, Xiamen, China
dzp01@outlook.com

Abstract. This study aimed to investigate the forces of a high-speed catamaran under the action of regular head waves at speed of 20 kn, and provided data support for the structural design of high-speed catamarans. This paper employs Computational Fluid Dynamics (CFD) and Finite Element Analysis (FEA) methods to construct a unidirectional fluid-structure interaction (FSI) numerical model for a high-speed hull with an aluminum alloy shell material, based on the Reynolds-averaged Navier–Stokes (RANS) equations. Structural dynamic analysis is performed using the numerical results, focusing on the study of the hull's structural forces and deformation characteristics under head wave regular wave conditions. By adjusting the local structure of the hull frame, a high-speed vessel design and optimization were achieved in accordance with regulatory requirements. The results show that the maximum equivalent stress and maximum shear stress both were occurred when the wavelength-to-ship-length ratio was approximately 1.25. The maximum stress was occurred at the wavelength from 14 to 16 m. The maximum deformation was 3 mm at the bow of longitudinal girder under the wave load. The maximum equivalent stress and shear stress were typically at the bow of ship bottom. Through appropriate structural reinforcement, the equivalent stress could be significantly reduced, thereby improving the structural strength.

Keywords: High-speed catamaran · Finite element analysis · Regular head wave · CFD · Fluid-structure interaction

1 Introduction

Catamarans offer enhanced stability and seakeeping over monohulls, yet their wide demi-hull separation complicates wave-induced loads. The asymmetric motion of demihulls in waves imposes significant bending and torsional stresses on the cross-deck, necessitating robust structural design, as shown in Fig. 1. Analyzing the structural stress of high-speed catamarans is crucial for optimizing design and ensuring vessel safety.

Currently, the fluid-structure interaction (FSI) research method of structure is prevalent both domestically and internationally. Tu Jianjun et al. [1] developed a one-way FSI

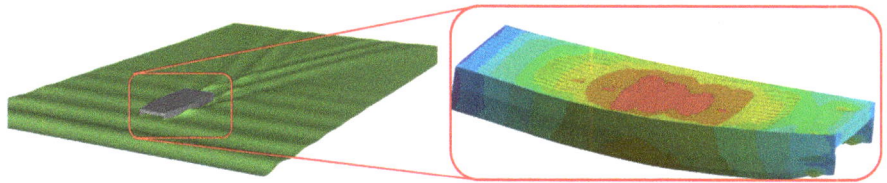

Fig. 1. Structural Stress and deformation distribution of a catamaran in head wave.

method in ANSYS Workbench incorporating three-dimensional wave loads to verify planing boat structural strength. Their analysis process provided more accurate loading characteristics and structural assessments compared to standard formula results. Tian Xiaojie et al. [2] utilized ANSYS software to analyze the FSI characteristics of marine risers in a gas-liquid two-phase flow, examined changes in vibration modes with and without FSI and the effects of fluid boundary conditions on the natural frequencies and configurations of the riser. The findings provided important theoretical insights for optimizing the design and ensuring the operational reliability of marine risers. Liu Jihai et al. [3] utilized the ACP module in Workbench to model a composite wing, analyzing its flow field at high angles of attack and pre-stress modal. They employed CFX-Static and Transient Structure modules for one-way steady-state and transient FSI analyses, and a two-way weak coupling method to assess dynamic response and aeroelastic stability. They found steady-state results more conservative than transient analysis. Tang Ying et al. [4] proposed a fully nonlinear numerical method for hydroelasticity to address FSI problems in flexible ships by using a fully nonlinear coupled solver. Sug W. C. et al. [5] studied the rudder stability issue considering the effect of FSI phenomenon, analyzed the vibrations caused by FSI in rudders and other common underwater appendages, and proposed a method for estimating the structural response and stability of rudder. WU Y et al. [6] investigated wave-induced motions and structural deformations, including wave loads, rigid and flexible body responses, and elastic and fatigue behaviors. In order to accurately predict the structural strength of ships and further enhance the quality of ship design, Yun Yajie [7] created a full-ship model and employed a one-way FSI method to compute ship resistance, flow fields, and surface forces under hogging and sagging conditions. Utilizing ANSYS's data transfer, surface loads were analyzed in the static structural module to assess overall strength, yielding stress and deformation data for ship components. Similar methods were used by Zhou Zhilu et al. [8–12] for numerical studies on wind turbine blades, sail wings, impellers, propellers, and other components.

The aforementioned research findings demonstrate the feasibility and reliability of using fluid-structure interaction (FSI) methods to address fluid-related structural issues. In this study, a finite element model (FEM) of a 12 m high-speed aluminum catamaran was developed. Computational Fluid Dynamics (CFD) and Finite Element Analysis (FEA) methods were utilized, with the CFD software Fluent employed to calculate the static and dynamic pressures acting on the hull. The resultant forces were then transferred to the Static Structure module in ANSYS, where the one-way FSI method was used to analyze the stress and deformation of the hull under the applied pressures. Based on the 2022 edition of the "Rules for Classification and Construction of High-Speed Craft at

Sea" (hereinafter referred to as the "Rules") [13], preliminary optimization of the frame structure was conducted.

The main contributions are as follows:

(1) A simple and effective topology treatment of the frame was carried out, ensuring an efficient connection between the frame and the hull.
(2) The stress conditions of the entire hull under two different wave heights and nine different wavelengths were calculated and analyzed, and assessed the forces acting on the hull in regular head water conditions.
(3) Preliminary optimization of the hull frame was conducted while ensuring that the stress remained within the requirements specified by the Rules.

2 Numerical Calculation Theory and Method

2.1 Numerical Calculation Theory

In this paper, the fluid dynamics model utilizes the k-epsilon turbulence model, which is widely used in engineering applications. This model is suitable for predicting the mean flow field and turbulence quantities [14]. The k-epsilon model is a semi-empirical model based on the Reynolds-averaged Navier–Stokes (RANS) equations, which are expressed as follows:

$$\frac{\partial \rho}{\partial t} + \frac{\partial}{\partial x_i}(\rho u_i) = 0 \tag{1}$$

$$\frac{\partial (\rho u_i)}{\partial t} + \frac{\partial (\rho u_i u_i)}{\partial x_j} = -\frac{\partial p}{\partial x} + \frac{\partial}{\partial x_j}\left(\mu \frac{\partial u_i}{\partial x_j} - \rho \overline{u'_i u'_j}\right) + F_i \tag{2}$$

$$\tau_{ij} = \rho \overline{u'_i u'_j} \tag{3}$$

where u_i represents the velocity field in the i-direction; ρ denotes the fluid density; μ is the dynamic viscosity coefficient of the fluid; u_i' and u_j' represent the fluctuating components of the velocity; p stands for pressure; F represents the body force; τ is the viscous stress.

This model employs two equations to describe turbulence kinetic energy (k) and its dissipation rate (epsilon). Here, k represents the turbulence kinetic energy, which is the energy per unit mass of fluid due to turbulent motion. Epsilon denotes the dissipation rate of turbulence energy, which is the rate at which turbulence kinetic energy is converted into internal energy via viscous action per unit time per unit mass of fluid. The transport equations are given as follows [15]:

$$\frac{\partial (\rho K)}{\partial t} + \frac{\partial (\rho \overline{u_j} K)}{\partial x_j} = \frac{\partial}{\partial x_j}\left[\left(\mu + \frac{\mu_t}{Pr_K}\right)\frac{\partial K}{\partial x_j}\right] + P_K + G_b - \rho \varepsilon - Y_M + S_K \tag{4}$$

$$\frac{\partial \rho \varepsilon}{\partial t} + \frac{\partial (\rho \overline{u_j} \varepsilon)}{\partial x_j} = \frac{\partial}{\partial x_j}\left[\left(\mu + \frac{\mu_t}{Pr_\varepsilon}\right)\frac{\partial \varepsilon}{\partial x_j}\right] + C_{\varepsilon 1}\frac{\varepsilon}{K}(P_K + C_{\varepsilon 3} G_b) - C_{\varepsilon 2}\rho \frac{\varepsilon^2}{K} + S_\varepsilon \tag{5}$$

where P_K is the production term for turbulence kinetic energy K due to mean velocity gradients; G_b represents the production term for turbulence kinetic energy K due to buoyancy; for incompressible fluids, $G_b = 0$; $-\rho\varepsilon$ is the dissipation term; Y_M is the Compressibility correction term, which accounts for the contribution of fluctuating expansion in compressible turbulence; $C_{\varepsilon 1}\frac{\varepsilon}{K}P_K$ is the production term for ε; $C_{\varepsilon 1}\frac{\varepsilon}{K}C_{\varepsilon 3}G_b$ is the buoyancy correction term; $-C_{\varepsilon 2}\rho\frac{\varepsilon^2}{K}$ is the dissipation term; S_K and S_ε represent the source terms for the K equation and ε equation, respectively.

The Coupled algorithm used in this paper is a pressure-based coupled solver. Unlike segregated solvers such as SIMPLE, SIMPLEC, and PISO, the Coupled algorithm simultaneously solves the continuity, momentum, and energy equations to achieve pressure-velocity coupling.

The wave-making equation is as follows [16–17]:

$$v_h = \zeta_a \omega \frac{chk(z+d_w)}{shkd_w} \cos k(x-ct) \tag{6}$$

The vertical velocity equation is given by:

$$v_v = \zeta_a \omega \frac{shk(z+d_w)}{shkd_w} \sin k(x-ct) \tag{7}$$

Where ζ_a represents the wave amplitude; k denotes the wave number; ω is the angular frequency; d_w is the water depth; c stands for speed; t stands for time.

The wave period is defined as follows:

$$T = \frac{2\pi}{\omega} \tag{8}$$

The relationship between wave period and wavelength under finite water depth conditions is expressed as:

$$T = \left[\frac{g}{2\pi\lambda}\tanh\left(\frac{2\pi d}{\lambda}\right)\right]^{-1/2} \tag{9}$$

In Eq. (9), λ denotes the wavelength; g is the gravitational.

2.2 Computational Model and Fluid Domain

The computational model in this paper is based on a 12 m aluminum catamaran. The principal dimensions are shown in Fig. 2 and Table 1.

The external fluid domain was established using a direct stretching method, with dimensions as follows: 75 m in the lengthwise direction, with 12 m from the bow to the inlet, roughly equivalent to one ship length; 40 m in the widthwise direction; and 24 m in depth, with 12 m below the baseline. Grid encryption is carried out within the range of the ship's perimeter and at the free surface. The final fluid domain setup is shown in Fig. 3.

Based on the ship's principal dimensions and referencing wave load conditions in the direct calculation of catamaran, the wave length was chosen to be close to the ship length to simulate the extreme hogging and sagging conditions during navigation. Therefore, the wave height ζ set to 0.5 m and the wavelength λ set to 8 m, 9 m, 10 m, 11 m, 12 m, 13 m, 14 m, 15 m, and 16 m. The speed set to 20 knots.

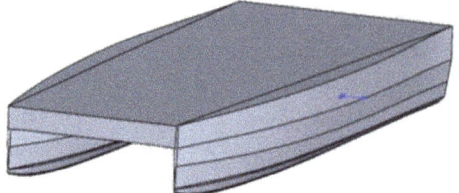

Fig. 2. Catamaran model.

Table 1. Principal dimensions of the catamaran model.

Parameter	Value
Overall length	12.37 m
Breadth	4.76 m
Depth	1.50 m
Draft	0.56 m
Designed displacement	7.98 t
Speed	20 kn

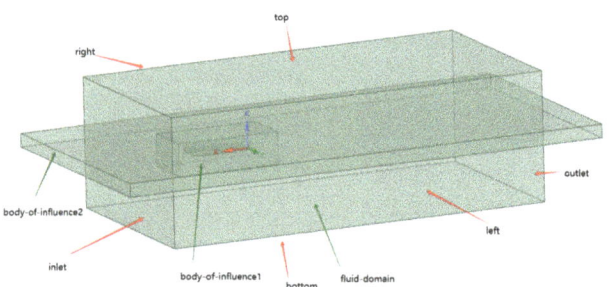

Fig. 3. Fluid domain.

2.3 Mesh Generation

In this study, the poly-hexcore volumetric meshing method was employed, resulting in most of the fluid domain being filled with hexahedral elements. Near the ship's surface and wall boundaries, polyhedral elements were used to achieve node-sharing connections between the hexahedral and polyhedral meshes. The mesh resolution was designed to ensure that there are no fewer than 10 elements within a single wave height range. The total number of volume mesh elements reached 1.8864 million. The mesh generation is shown in Fig. 4.

The frame mesh was generated using adaptive meshing, based on the smallest feature size of the frame, ensuring that most features were covered by at least two layers of elements. The maximum element size was set to 20 mm, with a resolution level of 2. In

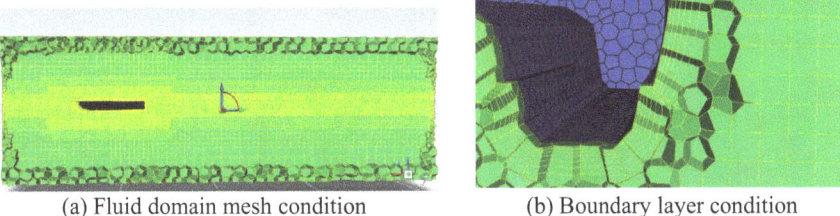

Fig. 4. Mesh generation.

the overall mesh quality control, the number of low-quality elements was kept below 5% to ensure computational accuracy and efficiency. The frame mesh had 654,000 nodes and 629,000 elements. The frame mesh and the hull surface mesh are depicted in Fig. 5.

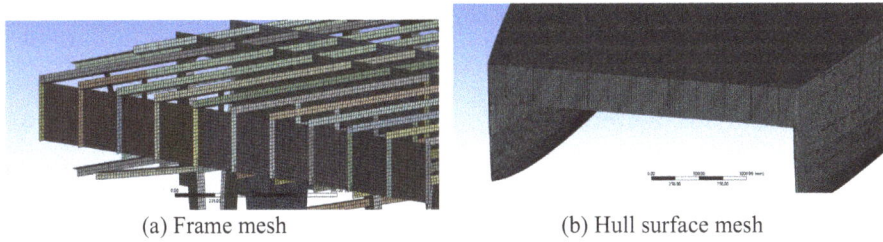

Fig. 5. Frame and hull surface mesh.

2.4 Structural Model

This study utilized the modeling functions of Ansys Workbench module to establish the structural model. The entire ship's framework is set to shell elements for calculation which is composed entirely of welded angle aluminum and flat aluminum, the framework is shown in Fig. 6. The material property parameters are shown in Table 2.

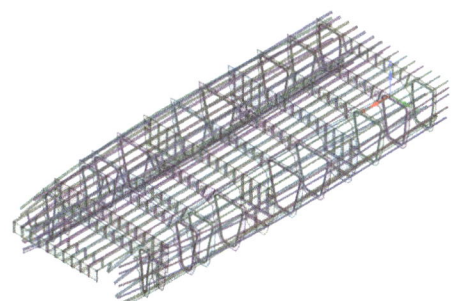

Fig. 6. Ship framework.

Table 2. Material property.

Property	Value
Density	2770 kg/m^3
Young's modulus	7.10E+10 pa
Poisson's ratio	0.33
Bulk modulus	6.96E+10 pa
Shear modulus	2.67E+10 pa
Tensile yield strength	2.80E+08 pa
Compressive yield strength	2.80E+08 pa
Tensile ultimate strength	3.10E+08 pa

3 Results and Analysis

3.1 Constraints and Solver Settings

Considering that the ship is subjected to wave forces, gravity, and buoyancy during navigation, these loads were simplified into the ship's total surface pressure and self-weight. The total pressure was divided into static pressure and dynamic pressure, as distinguished in the Fluent calculation report. Under wave action, the ship remains in dynamic equilibrium, experiencing periodic accelerations under ideal wave conditions. To prevent excessive rigid body displacement during calculations, a "321" constraint method was employed. Displacement constraints were applied at two points along the longitudinal girders at the bow and stern on the center longitudinal plane. The bow point was constrained in the x, y, and z directions, while the stern point was constrained in the y and z directions. A third constraint point was located at the aft longitudinal girder, restricting displacement in the y direction. Detailed constraints are shown in Fig. 7.

Fig. 7. Constraint settings.

3.2 Calculation Results

Figs. 8 and 9 show the residual time history curve and the free surface condition of the ship sailing in head seas under $\lambda = 8$ m wavelength, respectively. It can be observed that the overall numerical convergence is satisfactory. The waves generated by the ship on the free surface are relatively smooth, demonstrating good seakeeping performance.

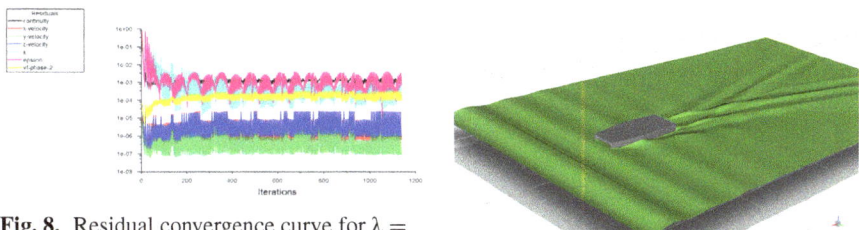

Fig. 8. Residual convergence curve for $\lambda = 8$ m.

Fig. 9. Free surface condition at $\lambda = 8$ m.

Figs. 10–13 display stress and strain diagrams at the pressure peak and trough for different wavelengths. Fig. 10 shows that the curve of maximum deformation at the pressure peak decreases as the wavelength increases, while the curve of maximum deformation at the pressure trough first increases and then decreases, reaching a maximum near $\lambda/L = 1.25$. This pattern mirrors the trends in maximum equivalent stress and maximum shear stress seen in Figs. 12 and 13, with the locations of peak forces being close, indicating that the forces are directly reflected in the areas of maximum hull deformation. Fig. 11 illustrates the average deformation under different pressures and wavelengths, showing a clear difference from the force representation, suggesting that when deformation factors are considered, maximum deformation better reflects the characteristics of the forces acting on the hull.

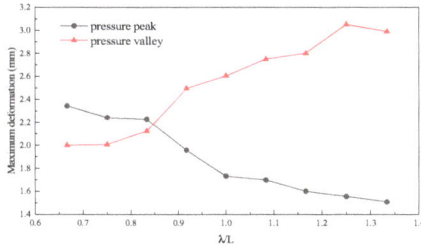
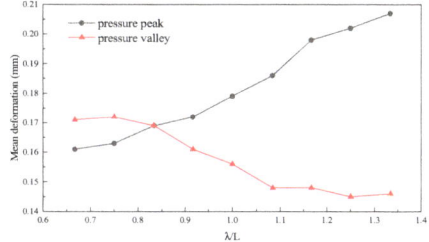

Fig. 10. Maximum deformation at different wave lengths and pressures.

Fig. 11. Average deformation at different wave lengths and pressure.

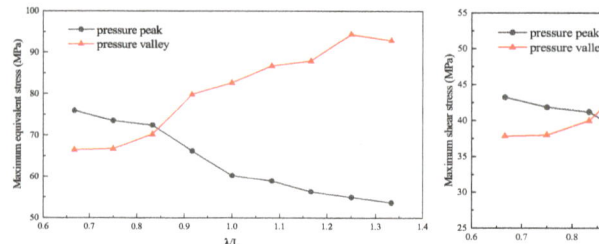 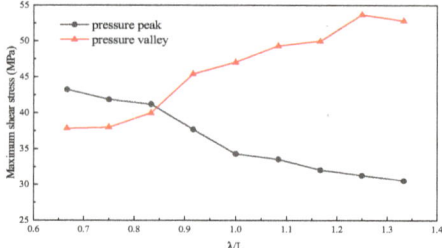

Fig. 12. Maximum equivalent stress at different wave lengths and pressures.

Fig. 13. Maximum shear stress at different wave lengths and pressures.

According to section. 4.4 of Appendix 2 of the "Rules", the allowable equivalent stress for aluminum alloy plate elements should not exceed 0.75 σ_{sw}, or 210 MPa. The "Rules" also stipulate that the allowable shear force should not exceed 0.41 σ_{sw}, or 114.8 MPa. From the maximum deformation data in Fig. 10 and the maximum equivalent stress in Fig. 12, it can be seen that the maximum deformation value is around 3 mm, and the maximum equivalent stress is approximately 94 MPa. Although the data complies with the regulatory requirements, the margin is insufficient, indicating that the hull structure requires reinforcement.

Fig. 14 shows the stress and deformation contours under different wavelengths. The left part displays the deformation, while the right part shows the stress distribution. It is evident that the maximum deformation is concentrated at the bow, with smaller deformations in other parts of the hull. This is primarily due to the weakened strength caused by the extension of the framework at the bow, leading to significant deformation under head water. Fig. 15 depicts the deformation of the internal framework of the catamaran, revealing that under high deformation magnitudes, the two frameworks at the bow begin to intersect. Combined with the deformation of the hull surface, this deformation can be classified as inward concavity.

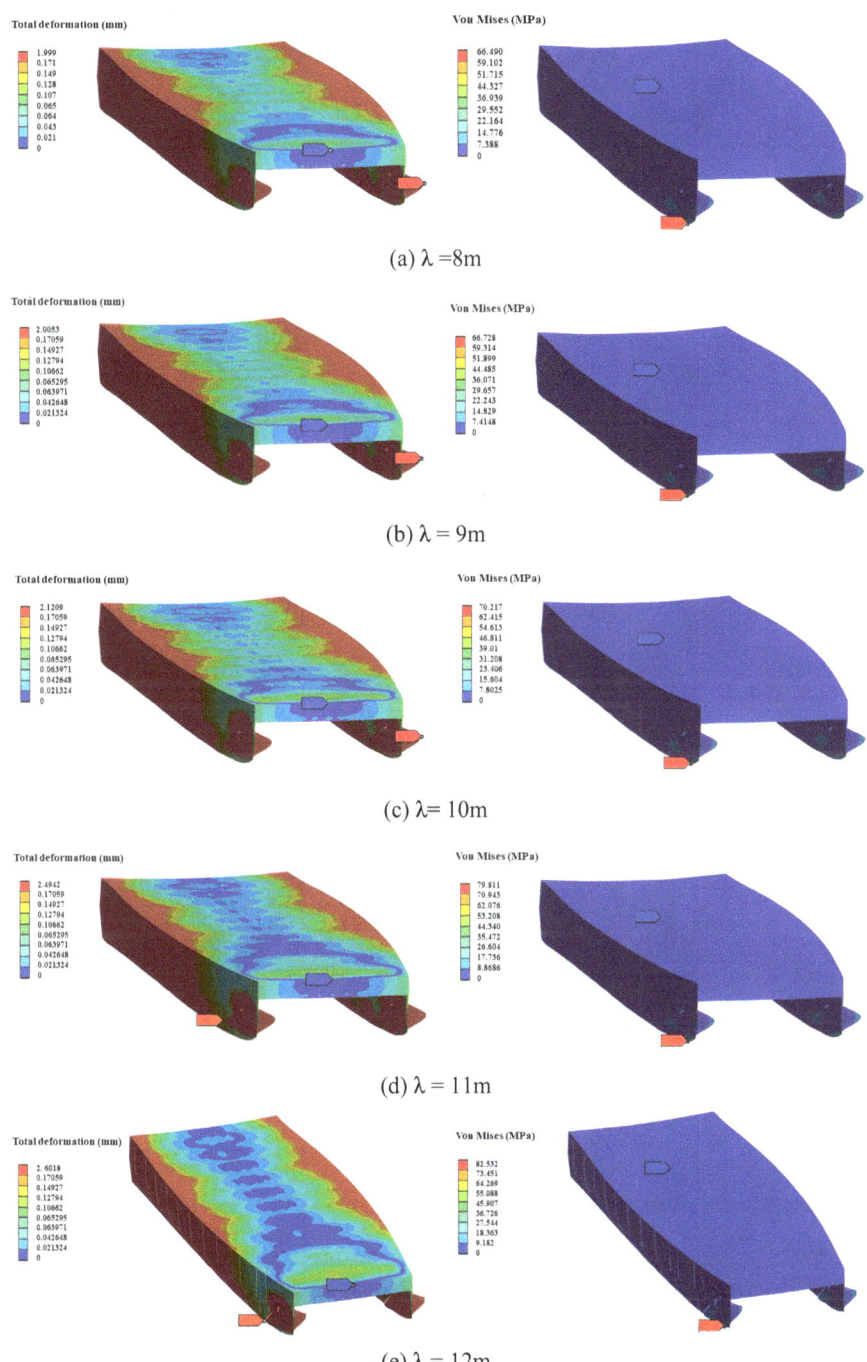

(a) λ = 8m

(b) λ = 9m

(c) λ = 10m

(d) λ = 11m

(e) λ = 12m

Fig. 14. Deformation and stress diagrams for head water.

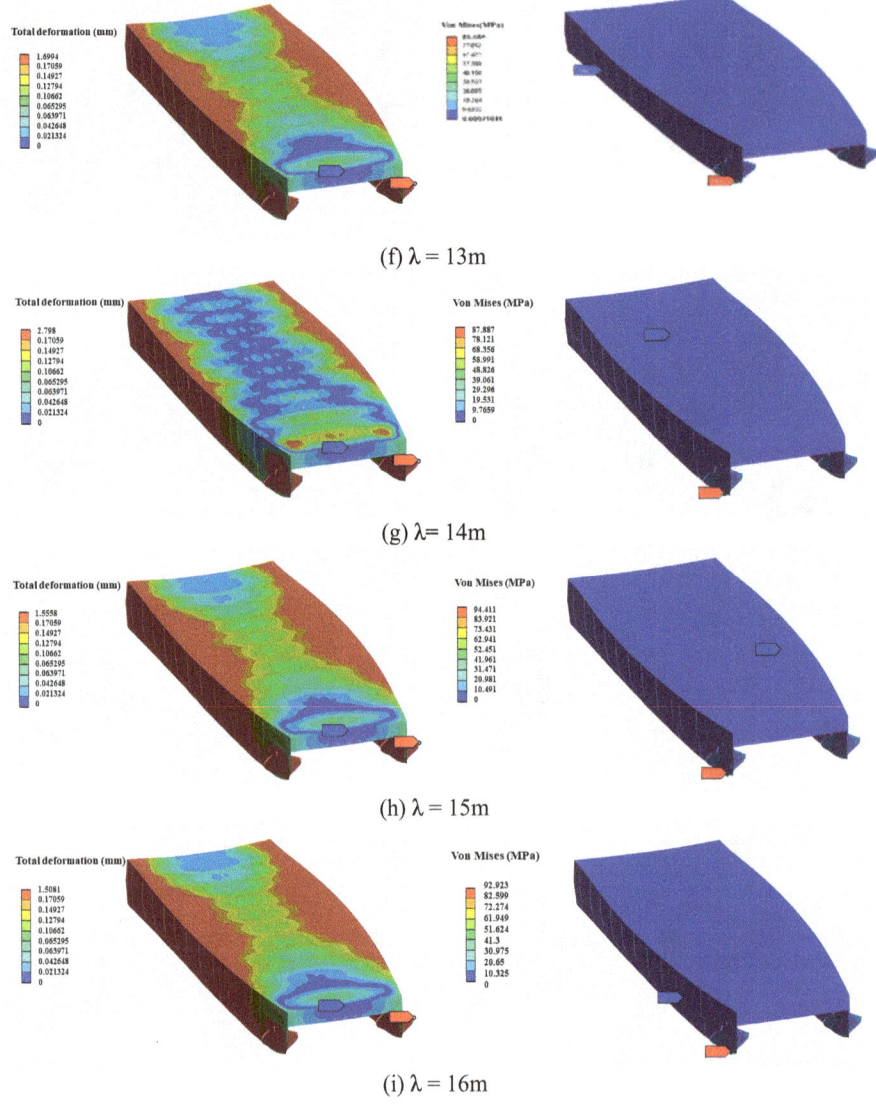

(f) λ = 13m

(g) λ = 14m

(h) λ = 15m

(i) λ = 16m

Fig. 14. (*continued*)

3.3 Structural Optimization and Analysis

To address the issues of insufficient allowable shear force and excessive deformation at the bow identified in the previous calculations, this study has optimized the hull framework by adding transverse reinforcements at the bow and extending the longitudinal stiffeners. The aim is to reduce deformation and stress. The optimized framework is shown in Fig. 16.

As illustrated in Figs. 17–20, the stress levels in the optimized framework have significantly decreased under all conditions, particularly shear stress, which has been

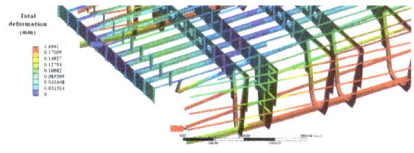
Fig. 15. Structure frame deformation.

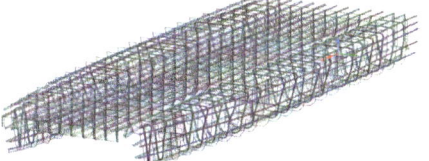
Fig. 16. Optimized frame.

reduced to less than half of the allowable value, thereby greatly improving the safety factor. Comparing the deformation and stress cloud plots in Fig. 21 with those of the unoptimized framework, it is evident that while the location of major deformation regions remains relatively unchanged, most of these regions have turned blue. This indicates a substantial reduction in deformation values compared to the red deformation regions, demonstrating the effectiveness of the optimization.

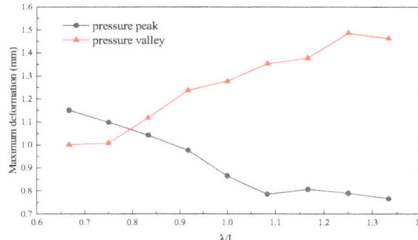
Fig. 17. Maximum deformation at different wave lengths and pressures.

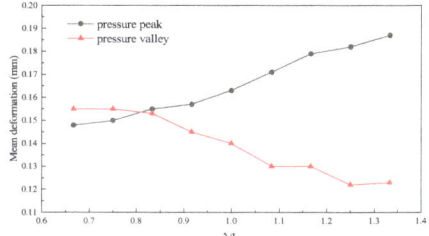
Fig. 18. Average deformation at different wave lengths and pressures.

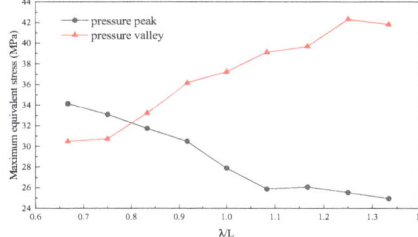
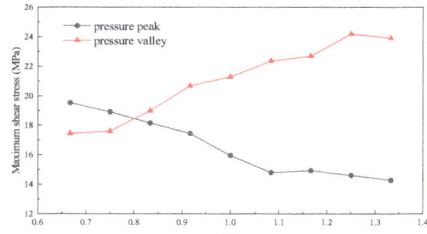

Fig. 19. Maximum equivalent stress at different wave lengths and pressures.

Fig. 20. Maximum shear stress at different wave lengths and pressures.

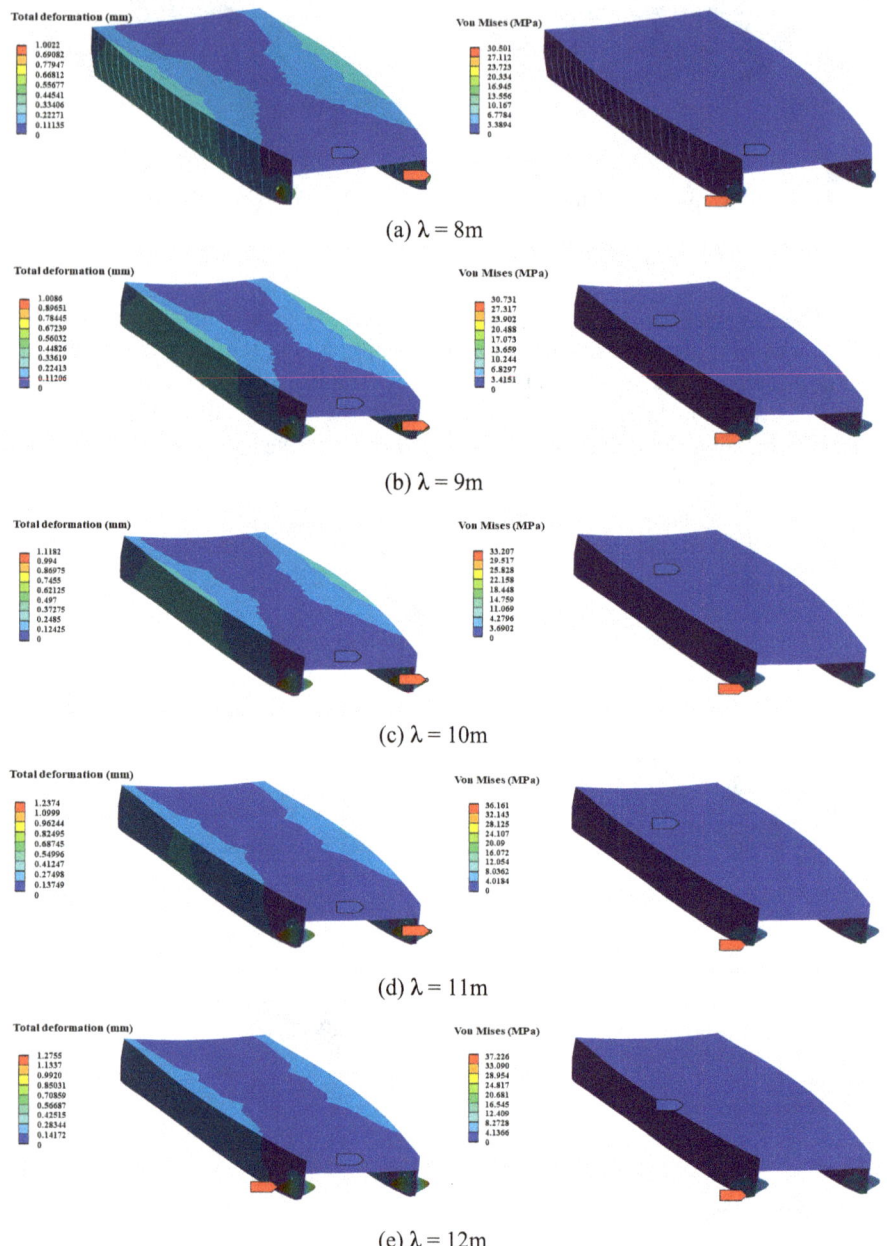

Fig. 21. Deformation and stress diagrams for head water.

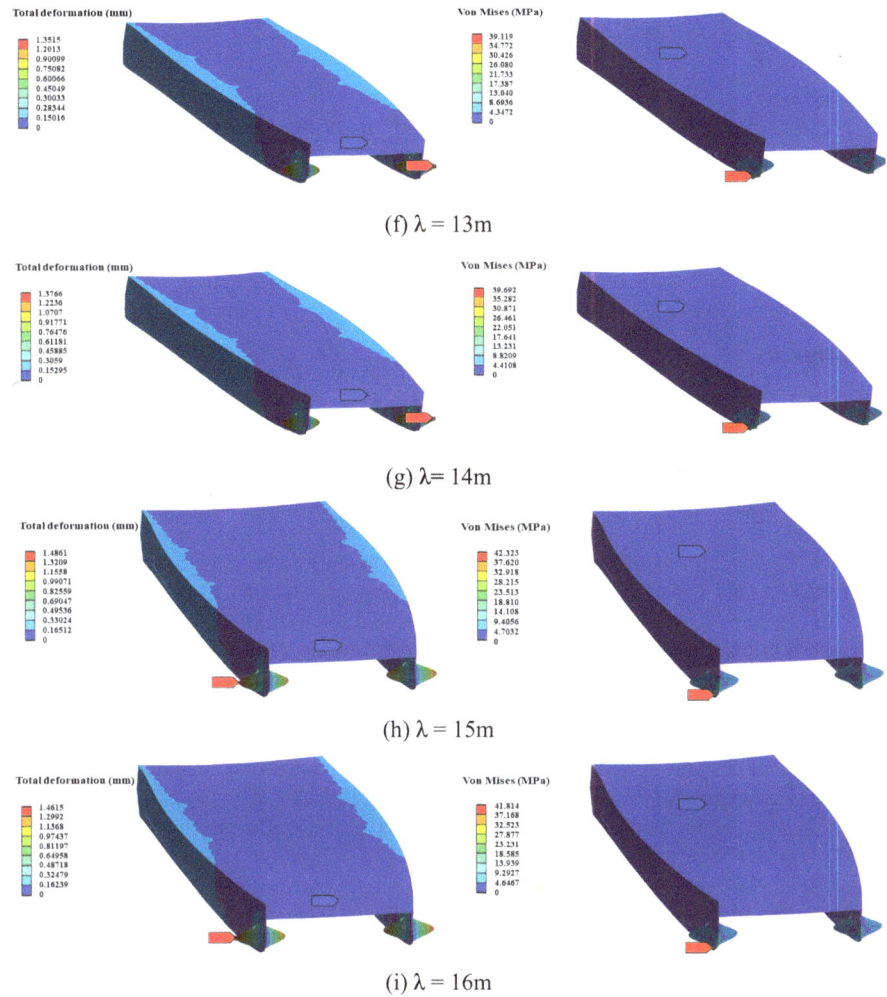

(f) λ = 13m

(g) λ = 14m

(h) λ = 15m

(i) λ = 16m

Fig. 21. (*continued*)

4 Conclusion

This study conducted structural analysis and optimization of a 12 m aluminum catamaran under head water loads. Using CFD and FEA methods with Ansys software for fluid and finite element analysis, the following conclusions were drawn:

(1) The framework is predominantly composed of T-shaped and X-shaped structures. It is best to use the shared topology operations in Space Claim within Ansys to ensure node sharing during mesh generation.
(2) Constraints should be applied with consideration of inertial release, and it is preferable for all three constraints to be located on the longitudinal section of the ship.

(3) For head water conditions, based on hull pressure and deformation data, the maximum equivalent stress and maximum shear stress occur at the place of maximum deformation and also reaches a maximum near $\lambda/L = 1.25$ at different wavelength ranges.
(4) The maximum deformation of the framework under head water loads occurs at the bow longitudinal girder which is 3 mm. The maximum equivalent and shear stresses are also located at the bow's bottom. Therefore, strengthening the bow area is crucial. Post-optimization, there was a significant improvement in deformation and stress conditions.

References

1. TU Jianjun, et al (2023). Local strength analysis of planning boat based on fluid-structure interaction. Shipbuild. China, **64**(5):86–97. (in Chinese)
2. Xiaojie, T.I.A.N., et al.: Analysis of fluid-structure interaction characteristics of gas-liquid two-phase flow marine riser based on ANSYS. J. Vib. Shock. **40**(7), 260–267 (2021) (in Chinese)
3. Jihai, L.I.U., et al.: Application analysis of one-way and two-way fluid-structure coupling method in UAV wing. Appl. Sci. Technol. **49**(4), 70–91 (2022) (in Chinese)
4. Ying, T., et al.: A fully nonlinear BEM-beam coupled solver for fluid–structure interactions of flexible ships in waves. J. Fluids Struct. **121** (2023)
5. Sug, W.C., et al.: Methods for assessing the ship rudder stability under lock-in phenomena considering fluid-structure interactions. Proc. Inst. Mech. Eng. Part m: j. Eng. Marit. Environ. **236**(1), 113–124 (2022)
6. Y, W.U., et al.: Theory and applications of coupled fluid-structure interactions of ships in waves and ocean acoustic environment. J. Hydrodyn. B. **28**(6), 923–936 (2016)
7. Yajie, Y.: Strength analysis of the whole ship based on fluid-structure coupling. Dalian Maritime University (2024) (in Chinese)
8. Zhilu, Z.: Research on the Fluid-Structure Coupling Characteristics of Large-scale Flexible Blade Wind Turbine. Harbin Engineering University (2022) (in Chinese)
9. xiaoshan, L.E.I., et al.: Numerical simulationon aerodynamic performance of a sail wing at different wind speeds based on unidirectional fluid – Structure interaction. J. Wuhan Inst. Phys. Educ. **53**(11), 95–100 (2019) (in Chinese)
10. Chaoping, M., et al.: Research on calculation method of dynamic mode of wind turbine based on one-way FSI. Renew. Energy Resour. **34**(6), 884–888 (2016) (in Chinese)
11. Zixue, D., et al.: Strength and vibration of one-way fluid-solid coupling turbocharger impeller. J. Chongqing Jiaotong Univ. (Nat. Sci.). **33**(2), 142–145 (2014) (in Chinese)
12. Bang, A.N., et al.: Numerical comparison analysis of propeller performance based on fluid solid coupling. Mach. Tool & Hydraul. **46**(17), 155–160 (2018) (in Chinese)
13. China Classification Society: Specification for Classification and Construction of Offshore High Speed Vessels. China Classification Society (2022) (in Chinese)
14. Ping, C.H.E.N., et al.: Numerical simulation of propeller loads on ships navigating in broken ice. Sh. Sci. Technol. **45**(8), 15–19 (2023) (in Chinese)

15. Yihan, T.: Research On Ship Wave Force Modeling In Regular Waves Based On CFD, Dalian Maritime University (2014) (in Chinese)
16. Wang Y-Fan (2022). Development of Ship Structure Program Based on Inertial Release Principle. South China University of Technology. (in Chinese)
17. Pei, Z.Y., et al.: Strength and Structural Design of Ship Hull. Beijing Science &Technology Press (2017) (in Chinese)

Open Access This chapter is licensed under the terms of the Creative Commons Attribution-NonCommercial-NoDerivatives 4.0 International License (http://creativecommons.org/licenses/by-nc-nd/4.0/), which permits any noncommercial use, sharing, distribution and reproduction in any medium or format, as long as you give appropriate credit to the original author(s) and the source, provide a link to the Creative Commons license and indicate if you modified the licensed material. You do not have permission under this license to share adapted material derived from this chapter or parts of it.

The images or other third party material in this chapter are included in the chapter's Creative Commons license, unless indicated otherwise in a credit line to the material. If material is not included in the chapter's Creative Commons license and your intended use is not permitted by statutory regulation or exceeds the permitted use, you will need to obtain permission directly from the copyright holder.

Research on Distortion Control Indexes of Large-Scale Temporary Hatch Opening on Deck

Zhiyuan Wei, Jianjun Hou(✉), and Xiao Wang

Department of Marine Navigation, Dalian Naval Academy, Dalian, China
jianjungbd@qq.com

Abstract. In recent years, the issue of structural safety has arisen due to the increasing ship repairing and refitting, particularly with regards to large-scale temporary hatch openings applied on deck or bulkhead. These openings lead to a loss of structural continuity, resulting in significant local deformations under higher loads. Therefore, it becomes necessary to implement real-time monitoring of opening deformations during repairs. This study focuses on investigating deformation control for two typical hatch openings on deck by analyzing their behavior under wave bending and torsional moments using finite element analysis. Based on the observed deformation behaviors, indexes such as hatch flatness and variation in diagonal length are proposed for evaluating deformation control. Furthermore, the effects of different hatch sizes on these indexes are examined, leading to an optimization design for opening sizes and locations aimed at achieving effective deformation control.

Keywords: Temporary Hatch Opening · Index of Deformation Control · Opening Flatness · Opening Optimization

1 Introduction

In recent years, transport ships have been continuously undergoing repair and modification. Consequently, corresponding issues regarding the design of modifications and structural safety have also emerged. During these repairs and modifications, it is often necessary to create temporary openings on the main deck or hull structure for easy access by equipment and personnel (which will be sealed after construction is completed). Some of these openings can span over ten meters in length and several meters in width, extending from the main deck all the way through to the hull structure. Such extensive modifications significantly disrupt the structural continuity [1]. This situation is analogous to container ships, bulk carriers, sectional barges, and other vessels with large openings on their decks. The common characteristic of these types of hulls is that their open structure has low resistance to bending and torsion [2], but the structural strength of temporary openings is much lower than that of vessels with larger openings. Especially when construction work cannot be immediately completed or needs to continue at another location, if a ship needs to remain anchored or navigate in water for an extended period under significant wind and wave loads, substantial deformation can occur at opening locations along with severe stress concentration phenomena [3]. Once deformations and stresses at opening locations exceed allowable values, they become hidden risks for ship structural safety [4]. There have been multiple cases both domestically and internationally where modified ships experienced uneven force distribution in their structures leading ultimately to local deformations and damage in hulls [5].

To ensure the structural safety throughout the construction process of large opening ships, it is imperative to implement real-time monitoring of deformations at the opening area. Monitoring deformations requires tracking measurements for specific deformation indexes, making it essential to propose a set of temporary control indexes for assessing opening area deformations effectively. These control indexes should accurately describe the deformation characteristics while quantifying the extent of deformation. This paper initially examines the forces and corresponding deformation responses of opening structures in waves using a finite element numerical calculation model specifically designed for large opening hull sections. Subsequently, two distinct control indexes are proposed based on their distinctive deformation characteristics to effectively monitor temporary opening area deformations. Finally, an analysis is conducted to investigate the relationship between opening size, position, and deformation control indexes.

2 Numerical Modeling of Structural Deformation in Large Opening Sections

Based on the principles of elastic-plastic mechanics and finite element analysis, this study presents a segment model featuring two representative openings for investigating structural deformations. The elastoplastic static analysis of this segment primarily considers external loads such as static water bending moment, wave-induced additional bending moment, and wave torsion moment acting on the hull in waves, along with boundary conditions that simulate the free state of the hull.

2.1 Finite Element Model of Large Opening Sections

In order to derive more generalized conclusions and minimize the influence of distant structural features, this study disregards the specific shape of the ship's outer panel and instead employs a standardized square section model [6], as depicted in Fig. 1(a). The dimensions of this compartment structure are 35 m in length, 30 m in width, and 18 m in height. It comprises three decks, two transverse bulkheads, two longitudinal bulkheads, as well as supporting elements such as transverse beams, longitudinal frames, and longitudinal ribs on the deck.

Typically, there are two types of openings on the deck: those arranged longitudinally along the ship's length and those arranged laterally across its width. To represent these scenarios accurately within this main deck section model, we have introduced two openings at its central position: Opening A measures 9.4 m × 3.4 m along the length direction (Fig. 1(b)), while Opening B measures 4 m × 14.6 m along the width direction (Fig. 1(c)). Both openings feature fillets with a size of 0.1 m; their respective local grids at fillet locations are illustrated in Figure (d).

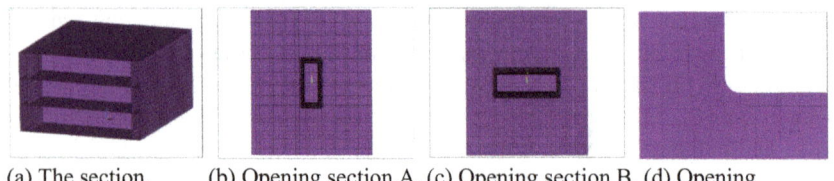

(a) The section. (b) Opening section A. (c) Opening section B. (d) Opening chamfering.

Fig. 1. Finite element model of large opening sections.

2.2 Wave Loads on the Hull

According to the theory of elastic-plastic mechanics, when analyzing the deformation of a thin plate with an opening, primary consideration should be given to the horizontal tensile (or compressive) stress and horizontal shear stress exerted on the plate [7]. Therefore, for a ship in water with a temporary opening, when examining the deformation at the opening location, it is necessary to calculate various types of loads including static water bending moment, wave-induced bending (Mw) moment, and wave torsion (MT) [8]:

$$M_w(+) = +190 MCL^2 BC_b \times 10^{-3} \; kN \cdot m \cdot \text{(hogging)} \tag{1}$$

$$M_w(-) = -110 MCL^2 B(C_b + 0.7) \times 10^{-3} kN \cdot m \cdot \text{(sagging)} \tag{2}$$

$$M_{T(x)} = 9.81 e^{-0.00295L} \frac{LB^3 C_T}{20000} \left(1.75 + 1.5 \frac{\varepsilon}{D}\right)\left(1 - \cos\frac{2\pi}{L}x\right) kN \cdot m \tag{3}$$

3 Deformation Index for Temporary Opening

A finite element calculation model is employed to evaluate the deformation response of the large open hatch section under three distinct conditions, namely structural wave bending moment (static water bending moment + wave-induced bending moment), wave torsion moment, and combined action of bending and torsion moments, through elastic-plastic static analysis.

3.1 Deformation Characteristics and Index of Opening Under Bending Moment

The deformation characteristics and control indexes under the influence of wave bending moment at the opening position are discussed based on the calculation results of bending deformation.

Calculation of Opening Deformation Under Bending Moment The deformation of the section with opening A under bending moment is illustrated in Fig. 2.

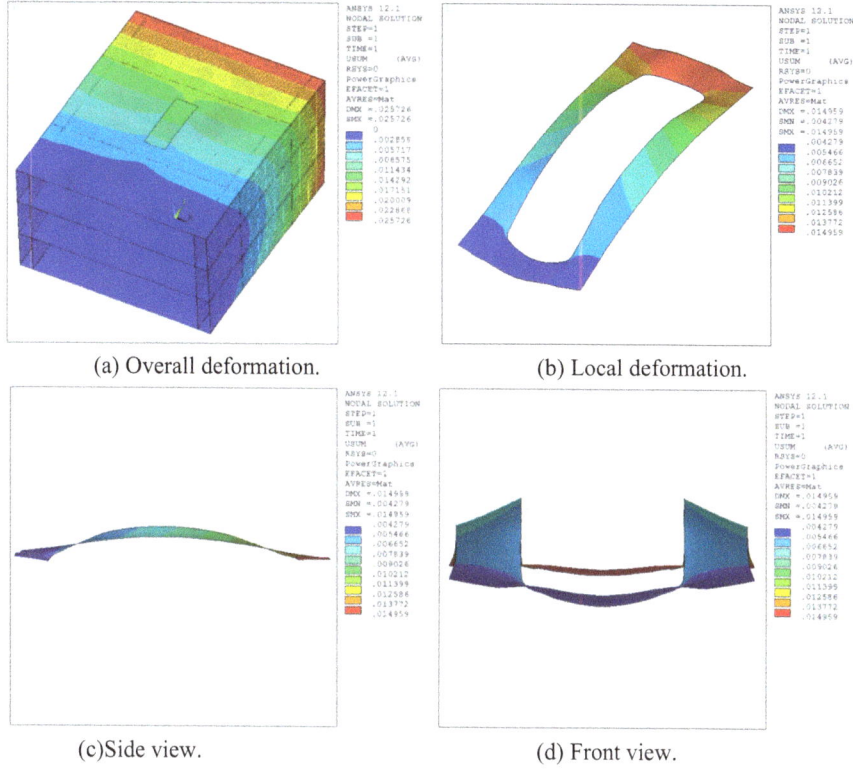

(a) Overall deformation. (b) Local deformation.

(c) Side view. (d) Front view.

Fig. 2. Deformation calculation of opening A (bending).

The deformation of the section with opening B under bending moment is illustrated in Fig. 3.

After analyzing Figs. 2 and 3, we can preliminarily infer the following: when subjected to a bending moment, the opening's shape deviates from its original plane. The longitudinal edge parallel to the ship's length (referred to as "longitudinal edge") exhibits an upward protrusion, while the transverse edge parallel to the ship's width (referred to as "transverse edge") experiences a downward sinking effect, with maximum deformation occurring at each midpoint of these edges [9]. Upon comparing both sets of figures, it becomes evident that deformation is more pronounced on the longer side of the opening in comparison to its shorter side.

Opening Flatness Under Bending Moment According to the deformation characteristics of the opening under bending moment, "opening flatness" can be used as an index for evaluating opening deformation.

Definition of Opening Flatness. According to the standards GB/T 11337–2004 [10], the flatness error is defined as the deviation between the actual plane and its ideal plane, where the position of the ideal plane must meet minimum requirements [11].

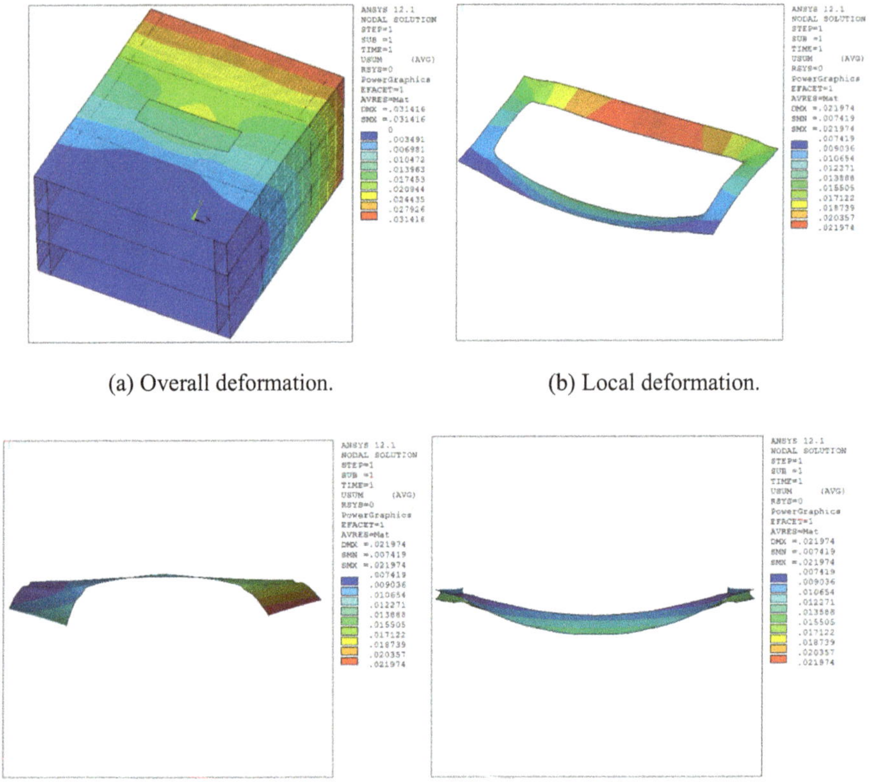

480 Z. Wei et al.

(a) Overall deformation.

(b) Local deformation.

(c) Side view.

(d) Front view.

Fig. 3. Deformation calculation of opening B (bending).

The magnitude of flatness error is quantified by a numerical value using the width f of the minimum encompassing area for flatness, as illustrated in Fig. 4.

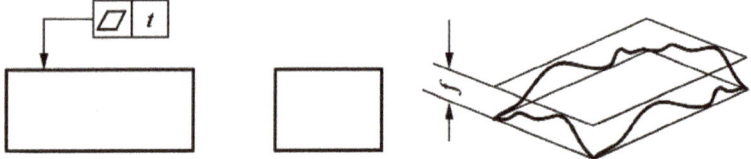

Fig. 4. Definition of opening flatness.

The methods used to assess flatness error encompass the minimum encompassing area method, least squares method, diagonal plane method, and three-point plane method. Among these techniques, the minimum zone method yields evaluation results that are either smaller or equal to those obtained from the other three methods. By employing the standard equation for a plane, it is possible to derive a calculation formula using the minimum zone method. The general equation for a plane can be expressed as follows:

$$Ax + By + Cz + D = 0 \tag{4}$$

The distances from each point to the plane can be determined by measuring the coordinates xi, yi, zi (i = 1,......,n) of n points:

$$d_i = \frac{Ax_i + By_i + Cz_i + D}{\sqrt{A^2 + B^2 + C^2}} \tag{5}$$

The equation presented below represents the function that should be satisfied by the minimum encompassing area, which is constructed based on the distances between each point:

$$\min F(A, B, C, D) = [d_i]i_{minmax} \tag{6}$$

The variables $[d_i]_{max}$ and $[d_i]_{min}$ respectively denote the upper and lower bounds of di in the equation.

Calculation of Opening Flatness. According to the definition and calculation method of flatness, the flatness values fh for opening A and B were separately determined and are presented in Table 1.

Table 1. Opening flatness under bending moment.

Items	dmax /mm	dmin /mm	f_h /mm
Opening A	−3.515	−7.071	3.556
Opening B	1.417	−6.034	7.451

The flatness value f_h of opening B is observed to be 2.1 times higher than that of opening A, as evident from Table 1. This finding suggests a greater deformation in opening B.

Variation in the Diagonal Length Under Bending Moment. The length of the diagonal has always been a significant index for large openings. Therefore, besides proposing flatness as an index for the opening, it is also essential to consider the change in diagonal length (△L) as an index for monitoring opening deformation [12]. The variations in the diagonal length under bending moment are calculated and presented in Table 2.

Table 2. Variation in the diagonal length under bending moment..

Items	L_0 /mm	L_1 /mm	△L /mm
Opening A	9921	9925	4.397
Opening B	15,066	15,063	−2.693

The results presented in Table 2 demonstrate that, when subjected to a bending moment, opening A experienced an elongation of 4.397 mm along its diagonal length, whereas opening B underwent a reduction of 2.693 mm in its diagonal length.

3.2 Deformation Characteristics and Index of Opening Under Torsion Moment

Based on the calculation of torsional deformation, this study examines the deformation characteristics and index associated with the opening position under wave torsion moment.

Calculation of Opening Deformation Under Torsion Moment The deformation of the section with opening A under torsion moment is illustrated in Fig. 5.

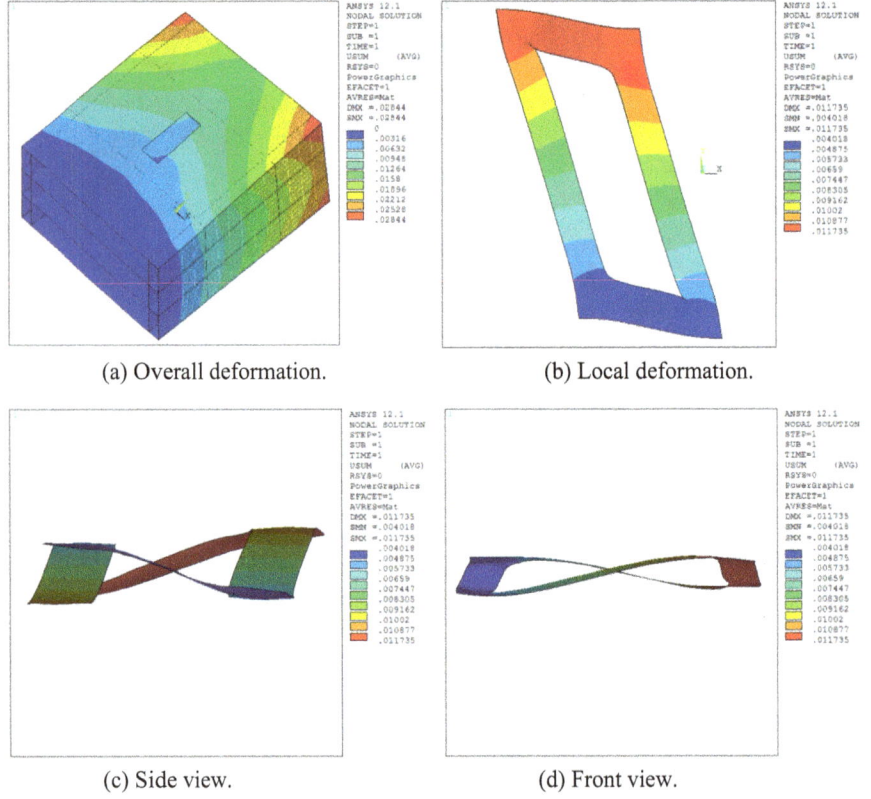

(a) Overall deformation. (b) Local deformation.

(c) Side view. (d) Front view.

Fig. 5. Deformation calculation of opening A (torsion).

The deformation of the section with opening B under torsion moment is illustrated in Fig. 6.

The torsional stiffness of the large open ship significantly decreases, as observed from Figs. 5 and 6. The opening area undergoes bending deformation with maximum

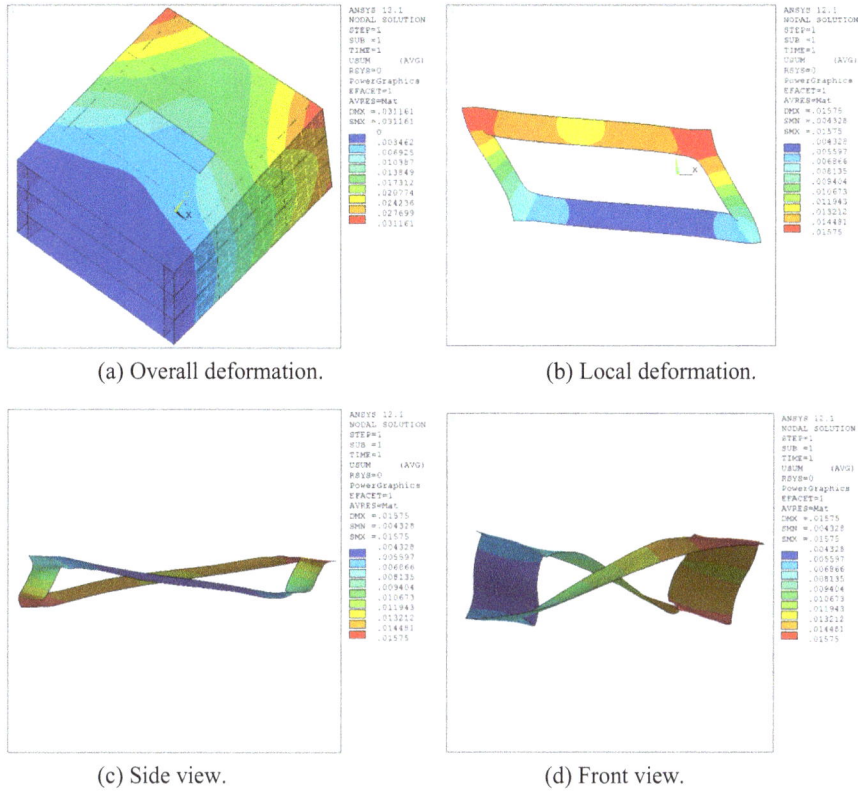

Fig. 6. Deformation calculation of opening B (torsion).

deformation occurring at the chamfered corners, where a significant change in diagonal length is also evident. Hence, we can describe the opening deformation under torsion using two indexes: ΔL to represent changes in diagonal length and fh to represent changes in flatness of the opening [13].

Deformation Index Under Torsion Moment. The calculation of fh and ΔL should be performed separately for opening A and B, as indicated in Table 3 and 4.

Table 3. Opening flatness under torsion moment.

Items	dmax /mm	dmin /mm	f_h /mm
Opening A	2.815	−2.815	5.63
Opening B	8.273	−8.271	16.544

Table 4. Variation in the diagonal length under torsion moment.

Items	L_0 /mm	L_1 /mm	ΔL /mm
Opening A	9921	9925	4.271
Opening B	15,066	15,073	6.712

4 Relationship Between Opening Deformation Index and Opening Size

By quantifying the structural deformation of various opening sizes, a strong correlation between f_h or ΔL values and the opening size was observed.

4.1 Variation of Opening Deformation Index with Opening Size

The variations of f_h and ΔL with the length-to-width ratio L/B are illustrated in Figs. 7–9, depicting the opening length along the ship's longitudinal direction and the opening width along its transverse direction.

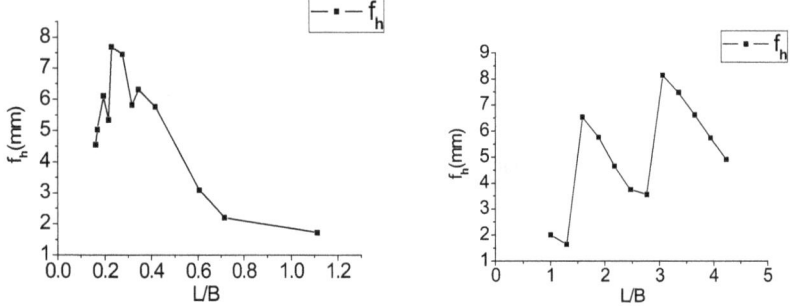

Fig. 7. Relationship between L/B and f_h under bending moment.

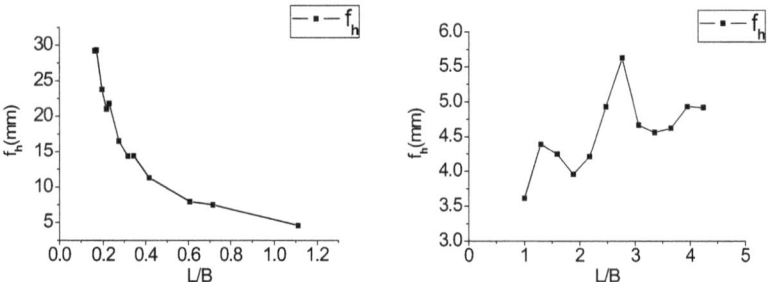

Fig. 8. Relationship between L/B and f_h under torsion moment.

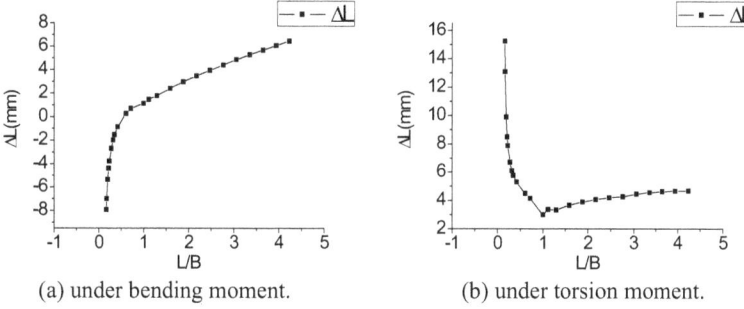

Fig. 9. Relationship between *L/B* and Δ*L*.

As depicted in Fig. 7, under the bending moment, as the *L/B* increases: for *L/B* values less than 1, the f_h value exhibits an alternating pattern of increase and decrease. This behavior arises due to a sudden drop in f_h value caused by each truncation of a vertical strut with increasing opening width (B), followed by a gradual rise until the next vertical strut is severed; for *L/B* values greater than 1, the f_h value also displays an alternating trend of increase and decrease. This phenomenon occurs because each truncation of a horizontal beam resulting from an increment in opening length (L) leads to a sudden surge in f_h value, which then gradually diminishes until the subsequent horizontal beam is removed.

As depicted in Fig. 8, the influence of wave torsion is observed with varying *L/B*: for *L/B* values less than 1, f_h exhibits a decreasing trend; for *L/B* values greater than 1, f_h demonstrates overall fluctuations; when the opening length is positioned between adjacent beams, f_h displays a trough; and when the opening length is situated near the beam, f_h exhibits a peak.

As depicted in Fig. 9, the Δ*L* value exhibits an increasing trend with the rise of *L/B* under the bending moment. Specifically, when *L/B* is less than 1, this upward trend becomes more pronounced; whereas for *L/B* greater than 1, it becomes relatively smaller. Similarly, as illustrated in Fig. 9, under the impact of wave torsion and varying *L/B*: when *L/B* is less than 1, a decreasing trend is observed for Δ*L*; however, for *L/B* greater than 1, an upward trend can be observed.

4.2 Optimization of Opening Size Based on Deformation Control

Based on the relationship between deformation index and opening size, the opening length, width, and position can be optimized during the process of opening design.

(1) Setting the opening length: With an increase in the opening length, both bending deformation and torsional deformation exhibit an upward trend. To ensure compliance with usage requirements, it is recommended to position the transverse edge of the opening precisely at a location where it does not intersect any beam. At this specific point, the bending deformation of the opening remains relatively minimal.

(2) Setting the opening width: Increasing the opening width leads to an upward trend in both bending and torsional deformations of the opening. To ensure compliance

with usage requirements, it is crucial to position the longitudinal edge of the opening precisely where it intersects a specific longitudinal beam, thereby minimizing both bending and torsional deformations [14].

(3) The influence of the opening width on the deformation surpasses that of the opening length. As the opening width increases, all deformation indexes at the opening location exhibit a significant increase. Therefore, it is recommended to align the longer side of the opening with the ship's longitudinal direction for effective deformation control in this model [15]. From a deformation management perspective, opening A is more favorable than opening B.

5 Conclusion

The primary objective of this study is to investigate the deformation issues associated with two typical temporary large openings on the ship deck under wave-induced bending and torsional moments. Based on the distinctive characteristics of bending and torsion deformations observed in temporary openings, we propose two indexes for controlling deformation and discuss their relationship with opening size. Furthermore, we conduct an optimization design for opening size and position based on these findings. The conclusions derived from this study are as follows:

(1) The temporary large opening on the deck will experience significant deformation due to wave-induced bending and torsional moments, resulting in loss of planarity at the opening location and substantial changes in its shape: under bending moment, it is primarily characterized by upward warping and depression along the longer side of the opening; under torsional moment, it exhibits diagonal buckling deformation.
(2) The flatness f_h and the change in diagonal length ΔL can serve as the index for temporary opening deformation, based on its deformation characteristics.
(3) The values of f_h and ΔL are closely correlated with L/B, and when the value of L/B aligns precisely with the truncation (or non-truncation) point of a vertical strut or horizontal beam, there will be an abrupt change in f_h value. Leveraging this characteristic, the optimization of opening size and position can be achieved.
(4) The influence of the opening width on the deformation surpasses that of the opening length. With an increase in the opening width, all deformation indexes at the opening location exhibit a significant rise. Therefore, it is recommended to align the longer side of the opening parallel to the ship's longitudinal direction as much as possible.

References

1. Zhou, B., Liu, Y., Ji, Z.: Numerical and experimental investigations on temperature and stress distribution in oxygen cutting. J. Sh. Prod. **25**(1), 14–20 (2009)
2. Sun, H., Soares, C.G.: An experimental study of ultimate torsional strength of a ship-type hull girder with a large deck opening. Mar. Struct. **16**(1), 51–67 (2003)
3. Liu, J., Yongning, G.: A study on stress concentration at hatch corner for ship with large openings. Sh. Eng. **6**, 9–12 (2000) (in Chinese)
4. Huang, T.D., Dong, P., Decan, L.: Fabrication and engineering technology for lightweight ship structures, part 1: Distortion and residual stresses in panel fabrication. J. Sh. Prod. **20**(1), 43–59 (2004)

5. McPherson, N.A.: Thin plate distortion-the ongoing problem in shipbuilding. J. Sh. Prod. **23**(2), 94–117 (2007)
6. Zhou, B., Liu, Y., Wei, Z.: Study of reinforcing design of ship structures with cutting hatch openings. J. Dalian Univ. Technol. **51**(2), 215–220 (2011) (in Chinese)
7. Zhilun, X.: Elastic Mechanics. Higher Education Press, Beijing (2016) (in Chinese)
8. CCS: Classification Rules of Steel Sea-going Vessel. China Communications Press, Beijing (2018) (in Chinese)
9. Zhou, H., Kong, X., Yuan, T., et al.: Experimental study on the ultimate bearing capacity of deck grillage structure with multiple openings. Chin. J. Sh. Res. **14**(2), 45–50 (2019) (in Chinese)
10. China National Standardization Management Committee: GB/T 11337–2004, Flatness error detection. China National Standardization Management Committee, Beijing. (in Chinese)
11. Cui, Y., Yang, Y., Wei, M., et al.: Multi-objective topology optimization design of opening decks using improved fourfold combination weighting model of game theory. J. Mech. Eng. **59**(9), 263–273 (2023) (in Chinese)
12. Yuren, H., Chen, B.: Limit state of torsion of ship hulls with large hatch openings. J. Shanghai Jiaotong univ. **35**(4), 556–561 (2001) (in Chinese)
13. Yang, Y., Zhang, X., Li, Z., et al.: Failure mode and deformation control design of large open deck of cruise ship. Sh. Eng. **44**(5), 18–26 (2022) (in Chinese)
14. A, K.: Multi-objective topology and geometry optimization of statically determinate beams. Struct. Eng. Mech. **70**(3), 367–380 (2019)
15. M, T., M, A.: Multi-objective BESO topology optimization for stiffness and frequency of continuum structures. Struct. Eng. Mech. **72**(2), 181–190 (2019)

Open Access This chapter is licensed under the terms of the Creative Commons Attribution-NonCommercial-NoDerivatives 4.0 International License (http://creativecommons.org/licenses/by-nc-nd/4.0/), which permits any noncommercial use, sharing, distribution and reproduction in any medium or format, as long as you give appropriate credit to the original author(s) and the source, provide a link to the Creative Commons license and indicate if you modified the licensed material. You do not have permission under this license to share adapted material derived from this chapter or parts of it.

The images or other third party material in this chapter are included in the chapter's Creative Commons license, unless indicated otherwise in a credit line to the material. If material is not included in the chapter's Creative Commons license and your intended use is not permitted by statutory regulation or exceeds the permitted use, you will need to obtain permission directly from the copyright holder.

Development and Advantages of the BeiDou Navigation Satellite System Using in Maritime Search and Rescue

Yulong Kong, Zhenyu Zheng(✉), and Ming Wu

PLA Dalian Naval Academy, Dalian, China
19824377927@163.com

Abstract. With the completion of the three-step construction phase of the BeiDou Navigation Satellite System (BDS), achieving the overall goal of global service coverage, the system has significantly enhanced its role across a multitude of sectors. The BDS is not only pivotal in daily life and national defense but also extends its influence into emergency response, particularly maritime search and rescue (SAR) operations. As the utilization of BDS in SAR missions grows, so too does the sophistication of policies, regulations, and construction standards that govern its implementation.This article delves into the institutional improvements that have been made to integrate the BDS more effectively into maritime SAR services. It outlines the system's architecture, emphasizing the robust infrastructure that supports its functionality. Moreover, it examines the current status of equipment used in conjunction with the BDS, highlighting advancements in technology that have bolstered operational capabilities.The unique features of the BDS short message service and Radio Determination Satellite Service (RDSS) functions are critical to its effectiveness. These components offer distinct advantages, such as strong autonomy and controllability, which ensure reliable operation even under adverse conditions. The system's resistance to deception and interference enhances its reliability, while its two-way communication capabilities provide real-time data exchange between rescuers and those in distress. These advantages will provide a theoretical basis for the BDS to play a greater role in the development of maritime lifesaving systems in the future.

Keywords: Maritime Search and Rescue · BDS · RDSS · BDS Short Message Service

1 Introduction

The BDS began its journey with the launch of the first satellite in 2000 and has since completed three phases of construction, moving from serving domestic users, to serving the Asia-Pacific region, and finally to serving globally. In the first generation, only two Geostationary Earth Orbit (GEO) satellites were in operation, which limited the positioning function to cover only China and its surrounding areas. However, the active positioning mode led to the development of an additional feature the short message communication service. In subsequent updates and expansions, GEO orbit satellites have been continuously retained, along with the unique features they carry, such as Radio Determination Satellite Service (RDSS) and the short message service. These distinctive capabilities have become hallmarks of the BDS, contributing to its versatility and utility in various applications, including maritime SAR [1, 2].

During the 5·12 Wenchuan earthquake in 2008, the first-generation BDS ensured the effective transmission of rescue commands and disaster information when ground-based communication infrastructure failed, playing a significant role in the rescue efforts. In recent years, the BDS has rapidly developed in the field of maritime lifesaving in China, with extensive construction of lifesaving systems and the development of related equipment [3, 4]. This article reviews the development and current state of BDS equipment in the field of maritime lifesaving. It focuses on analyzing the advantages of BDS RDSS positioning and short message service functionalities in supporting maritime lifesaving efforts. The aim of this article is to provide a theoretical foundation for the BDS to play a greater role in maritime search and rescue applications in the future.

2 Overview of Development of BDS Using in Maritime SAR Services

China has long been considering the potential of using the BDS to enhance maritime lifesaving capabilities. By 2020, the third-generation BDS was fully operational, becoming the fourth major global satellite navigation system after GPS, GLONASS, and Galileo. In addition to expanding its communication capabilities, the BDS introduced global message services and utilized six MEO satellites to provide free distress alert services to maritime, aviation, and land users worldwide, in accordance with the standards set by the International Search and Rescue Satellite Organization (COSPAS-SARSAT) [5].

2.1 Development in Policy and Regulatory

From 2018 to 2022, Subsequent developments have further supported the integration of the BeiDou system into the Global Maritime Distress and Safety System (GMDSS). Here is a summary of the key milestones:

1. Submission to MSC: In May 2018, China submitted an application to the 99th session of the Marine Safety Committee (MSC) to recognize the BDS messaging service system as a service provider for GMDSS.
2. Data Exchange Protocol Standard: In December 2019, the Ministry of Transport released the "Specification for Shipborne Terminals of BeiDou Navigation Satellite System Part 2: Data Exchange Protocol" This document specified the formats for:

a. Distress Alarm Requests: The process of initiating a distress alarm.
b. Distress Alarm Acknowledgments: The confirmation of receipt of a distress alarm.
c. Distress Alarm Information Output: The format for transmitting distress alarm information.

3. IEC Standard Release: In March 2020, the International Electro technical Commission (IEC) formally published the first international standard for BDS shipborne receivers (IEC 61108-5). This standard provides technical specifications and testing procedures for BDS shipborne receivers, supporting the integration of the BDS into GMDSS.
4. COSPAS-SARSAT Approval: In November 2022, the 67th Public Council Meeting of COSPAS-SARSAT approved the revisions to the technical and procedural standards for global satellite search and rescue systems that included BDS-related content. This marked the formal recognition of China as a space segment provider for COSPAS-SARSAT [6].

2.2 Development in Practical Applications

Global Navigation Satellite Systems (GNSS) such as GPS, Galileo, GLONASS, and BDS have all integrated international SAR payloads, allowing them to participate in the global satellite search and rescue system. The fundamental mechanism behind this integration relies on the Radio Navigation Satellite Service (RNSS) positioning mode to obtain the location information of distress users, which is then transmitted via the 406 MHz global distress reporting channel to relay distress messages. The specific SAR system is shown in Fig. 1.

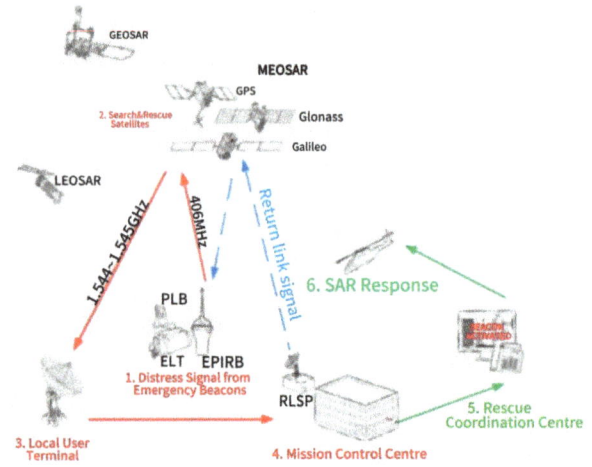

Fig. 1. Marine lifesaving alarm process based on navigation satellite.

In addition, the unique RDSS positioning function and regional/global short message service of the BDS can independently assist maritime search and rescue missions, without relying on the RNSS positioning mode or the international SAR satellite system. During BDS RDSS positioning, the distance measurement between the satellite and the user and the determination of the user's position are not handled independently at the client end but by the ground operation control center of the BDS. Its feature is that the user's location is reported to the ground center simultaneously through the user's response [7].

When the ship is in danger, or after the person wearing the BDS-based lifesaving equipment falls into the water, the distress information can be transmitted to the BDS main control station through the satellite. When the SOS is located abroad, the SOS message is first received by the overseas MEO satellite, then sent to the domestic satellite through the inter-satellite link, and sent to the main control station through the domestic satellite downlink. The main control station can push the distress information to the emergency search and rescue mission control center through the BDS satellite RDSS link or the BDS ground private network, and then forward it step by step to the nearby rescue department. Designated ships or helicopters are arranged to carry out search and rescue. Search-and-rescue helicopters and ships are equipped with 121.5 MHz homing signal receivers to guide the discovery of shipwrecks. After confirming the alarm information, the emergency search and rescue center can also send the receipt confirmation information to the Beidou emergency wireless position indicator through the Beidou inbound link. After feedback receipt confirmation information, the emergency search and rescue center controls the working mode of the indicator, increases the alarm interval of the indicator, prolongs the operation time of the equipment, facilitates the alarm confirmation, and prevents the chance of false alarm [8, 9]. The specific process is shown in Fig. 2.

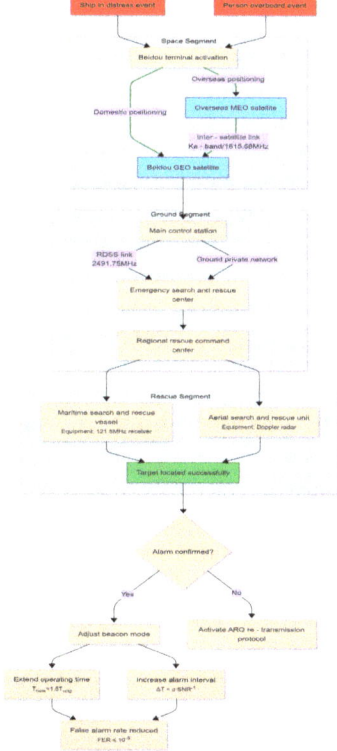

Fig. 2. Marine lifesaving alarm process based on BDS.

3 Current Situation of BDS-Based Lifesaving Equipment

The international common maritime distress positioning equipment includes two categories: one is the radio signal class. The most commonly used is the emergency position indicator. The disadvantages of this type of product are large size, large weight, inconvenient to carry, and short working distance. Usually this type of equipment is installed on ocean-going cargo ships and is not suitable for individuals in distress. The second is the use of flame signals, smoke signals, flares, floating lights, seawater dyeing, reflective bands and various combination signal classes. Its advantages are convenient to carry and simple to use; The disadvantage is that it cannot actively tell the search and rescue party its own location, and it is greatly affected by external factors when used, which is not conducive to the rapid and accurate identification and search and rescue of the ship and personnel in distress. The Marine lifesaving position beacon based on the BDS can not only realize real-time positioning, but also support long-distance communication, which can solve the above problems.

At present, the common BDS-based lifesaving position indicator can be divided into two categories: ship use and ship crew use. Among them, the shipborne search and rescue terminal is mainly installed on the ship, in the event of maritime danger, it can automatically alarm by touching the water, and send distress and location information to the command center. The personal distress terminal is mainly installed on the life jacket, when the wearer is in distress and falls into the water, the personal distress terminal can actively or passively send distress signals and location information to the search and rescue center, so that the search and rescue center can know the user is in the water in the shortest time, and timely rescue.

The ship-crew-use lifesaving position indicator is generally divided into two parts: surface and underwater. The surface part is mainly composed of BDS antenna, BDS module, LED indicator light and button. The underwater part is mainly composed of control module, power module, battery and water switch, and the two are connected by flexible watertight cables. The shell of the surface part and the underwater part of the BDS beacon are waterproof shells, which meet the waterproof requirements for reliable operation of the beacon when the personnel fall into the water [10].

4 RDSS Location Principle and Short Message Service Scope

In the process of BDS RDSS positioning, the distance measurement between the satellite and the user and the solution of the user's own position are not completed independently at the client end, but by the ground operation control center of the BDS. Its feature is that through the user's response, the user's location is reported to the ground center at the same time, and this process also realizes the deep integration of message communication [7].

The traditional RDSS positioning principle process is that the client sends a positioning application to the ground operations control center through the BDS GEO satellite, and then the ground operations control center sends a ranging signal to calculate the distance between the client and two GEO satellites according to the signal transmission delay. At the same time, the ground center also has a surface database, which can be used according to the principle of three-sphere convergence. The location of the user is calculated in the control center, and the coordinate information is sent to the client through the GEO satellite link.

At present, the operation basis of BDS-3 regional message is five GEO satellites (including two BDS-2 GEO satellites), whose fixed beam communication coverage is 75°E ~ 135°E and 10°N ~ 55°N, as shown in Fig. 3. In fact, the theoretical area is only to meet the minimum range of the traditional double GEO satellites positioning, if the generalized RDSS positioning principle is considered, the BDS RDSS positioning service range can be extended outward by 5°, and the boundary area can reach 70°E ~ 140°E and 5°N ~ 60°N. For the fact that only one GEO satellite is needed to establish a communication link, so the message coverage of the BDS region message service is consistent with the scope of the generalized RDSS positioning service.

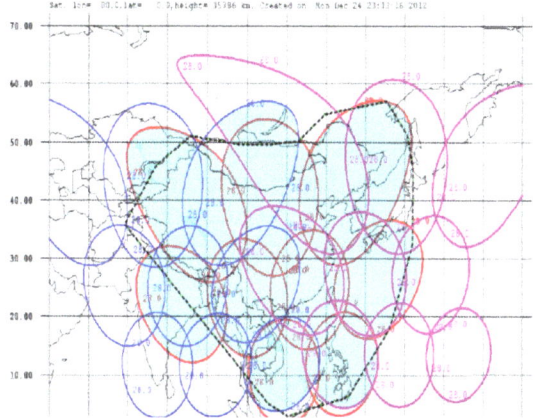

Fig. 3. BDS RDSS location and region message service coverage.

5 Application Advantages of the BDS in Maritime SAR Services

With the turbulence of the international situation and the introduction of the concept of navigation confrontation, the security problem of satellite navigation system application has become increasingly prominent, and in the field of maritime search and rescue, it also faces the risk of navigation interference and deception. The RDSS positioning mode of BDS in China is different from RNSS positioning mode of other satellite navigation systems, and has unique technical system advantages in the confrontation environment.

First, the degree of autonomy and controllability is high, and the usability is strong under the condition of confrontation. Considering that the COSPAS-SARAT service is an international open service, it is likely that the service will be shut down or the area of use will be restricted under the condition of confrontation. The self-reporting and short message services of the BDS in China are completely independent of international search and rescue services, and there is no problem of limited use in wartime. And the secret key encoding of BDS message communication is autonomous and controllable, which has stronger security and confidentiality compared with open maritime satellite communication services.

Second, the positioning principle is different from RNSS passive positioning, and the risk of navigation deception interference is small. Current research on navigation deception technology at home and abroad mainly focuses on RNSS positioning mode. It has been possible to carry out false simulation for BDS RNSS positioning signal, and realize serious deviation of self-solving position coordinates of distress users by means of recording, playback and forwarding, so as to achieve decoy and interference to our maritime search and rescue forces. According to the traditional RDSS positioning principle introduced in 3.1, RDSS positioning signal sending and location calculation rely on the ground operation control center, which cannot be simulated by strong enemies, and naturally has a strong ability to prevent deception interference. However, for general RDSS positioning and global location reporting services, there is also a risk of navigation spoofing as they involve the use of RNSS positioning signals. Therefore, BDS RDSS positioning can play a role in anti-deception interference, mainly limited to the coverage area of traditional GEO binary satellite positioning.

Third, the communication capability of BDS short message two-way feedback can enhance the confidence of those in distress, and indirectly extend the role of waiting for help. At the same time, due to the greatly increased capacity of the third-generation BDS message, in addition to transmitting basic time and coordinate information, it can also expand the reporting of personnel's physical conditions and injury threats, and lead the rescue team to prepare for medical treatment and diet in advance.

6 Summary

The construction of the BDS in maritime SAR has been increasingly mature and widely used in the civilian field. The unique RDSS positioning and message communication services of the system can ensure that the transmission of maritime lifesaving information is more complete, more reliable and more secure. In today's continuous disputes over maritime rights and interests, we not only need to have a sense of war at any time, establish a combat thinking transformation of peacetime and war, and be able to recognize the characteristics and functional advantages of the BDS, but also see the problems and challenges existing in the application of the BDS in maritime SAR under the condition of confrontation, so as to more comprehensively protect the safety of the lives and property of those in danger. It is of great significance to implement the national sustainable development strategy, promote the development of the national economy, safeguard the national maritime interests and national security, and promote the development of China's satellite navigation industry.

References

1. Yulong, K., et al.: Analysis of SPP result based on new satellites of BDS-3. Sci. Surv. Mapp. **44**(04), 152–157 (2019)
2. Jialin, L.: Design and Implementation of the Deep-Sea Profiling Float Monitoring System Based on BeiDou Navigation Satellite System. Tianjin University (2022)

3. Benyao, F., et al.: Application and development proposition of Beidou satellite navigation system in the rescue of Wenchuan earthquake. Spacecr. Eng. **04**, 6–13 (2008)
4. he, L., et al.: Design of drowning distress terminal based on Beidou System. Sci. Technol. Vis. **15**, 12–14 (2021)
5. Xianguo, C., et al.: The Discussion of BDS to Participate the Cospas-Sarsat system. In: Proceedings of the Eighth China Satellite Navigation Conference, pp. 146–149 (2017)
6. Xinjian, M., et al.: The application prospects of "BeiDou-3" in the search and rescue. Navigation. **03**, 3–5 (2023)
7. Xu, Y., et al.: Beidou RDSS data transmission terminal's Remote Firmware Upgrade. In: Proceedings of the Ninth China Satellite Navigation Conference, pp. 114–117 (2018)
8. qian, C., et al.: Application and standardization of Beidou satellite navigation system in ship emergency search and rescue. Top. New Gener. Inf. Technol. Stand. **06**, 18–22 (2020)
9. Xiang, S., et al.: Maritime emergency SAR system based on Beidou navigation satellite system. Command Control Simul. **40**(6) (2018)
10. Jiasheng, Q., et al.: A design and implementation of Beidou beacon technology for the rescue of the drowning personnel at sea. Electron. Meas. Technol. **42**(19), 63–68 (2019)

Open Access This chapter is licensed under the terms of the Creative Commons Attribution-NonCommercial-NoDerivatives 4.0 International License (http://creativecommons.org/licenses/by-nc-nd/4.0/), which permits any noncommercial use, sharing, distribution and reproduction in any medium or format, as long as you give appropriate credit to the original author(s) and the source, provide a link to the Creative Commons license and indicate if you modified the licensed material. You do not have permission under this license to share adapted material derived from this chapter or parts of it.

The images or other third party material in this chapter are included in the chapter's Creative Commons license, unless indicated otherwise in a credit line to the material. If material is not included in the chapter's Creative Commons license and your intended use is not permitted by statutory regulation or exceeds the permitted use, you will need to obtain permission directly from the copyright holder.

Study on the Effect of Free End on Flow-Induced Vibration and Energy Harvesting of PTC Cylinders

Ruitao Tang[1], Sihan Liu[1], Yichen Ma[1], Ruipu Zhao[1], Fenglai Huang[2], and Chunhui Ma[1(✉)]

[1] School of Mechanical Engineering, Nantong University, Nantong 226019, China
catch0226@163.com
[2] Wuhan Second Ship Design and Research Institute, Wuhan 430205, China

Abstract. This study investigates the influence of a free end on flow-induced vibration (FIV) and energy harvesting from Passive Turbulence Control (PTC) cylinders through three-dimensional numerical simulations. Specifically, the research focuses on PTC cylinders with a length-to-diameter ratio (L/D) of 3.14, comparing their vibration response and energy harvesting capabilities against infinitely long cylinders that lack a free end. The analysis demonstrates that the presence of a free end significantly modifies the wake dynamics and fluid-structure interactions involved, resulting in a higher flow speed threshold necessary to initiate FIV, as well as an increase in vibration amplitude. Although cylinders with free ends require greater flow speeds to begin energy collection, they ultimately exhibit a higher power output, albeit at a lower efficiency compared to their infinitely long counterparts. Furthermore, a comprehensive investigation is conducted on cylinders with varying length-to-diameter ratios (L/D = 1, 3.14, and 10), revealing that the influence of the free end diminishes as the length of the cylinder increases, with the shorter cylinder demonstrating the most pronounced effects. This research provides valuable insights into optimizing the design and application of PTC cylinders for energy harvesting in turbulent flow conditions.

Keywords: Flow-induced vibration · Energy harvesting · Free end effect · PTC cylinder

1 Introduction

Flow-induced vibration (FIV) is a complex phenomenon arising from the interaction between fluid and solid structures, primarily manifesting as vortex-induced vibration (VIV) and galloping. In the late 20th century, FIV was incorporated into marine energy harvesting studies, opening new avenues for renewable energy development [1]. This field intersects fluid dynamics [2] and solid mechanics [3], garnering attention in ocean engineering [4], wind energy [5], and bridge engineering [6]. Research shows that smooth cylindrical bodies induce VIV with low energy collection efficiency within specific Reynolds number ranges. In contrast, Passive Turbulence Control (PTC) cylinders, with their complex geometries, can generate various vibrations over a wider range of Reynolds numbers. Li et al. demonstrated that fluid around PTC cylinders with large symmetric control bands creates a recirculation zone, significantly disturbing the flow and enabling both VIV and galloping for sustained energy harvesting [7]. Zhang [8] and others have studied the energy harvesting capabilities of PTC cylinders at high Reynolds numbers (30,000 ≤ Re ≤ 110,000), providing valuable engineering insights. In practice, most cylinders have free ends, leading to complex vortex patterns that complicate experimental studies. The wake flow of finite-length cylinders [9] is inherently more complex than that of infinite-length ones, complicating parameter measurements and understanding of vortex interactions. Consequently, the impact of free end on vibration characteristics is gaining attention in FIV and energy harvesting research. While previous studies have addressed cylinder vibrations [10] and energy harvesting [11], the effects of free ends remain underexplored. This study aims to investigate these effects on FIV and energy harvesting of PTC cylinders, offering theoretical support and design recommendations for further optimization of energy harvesting device design.

2 Geometric Model and Numerical Methods

Figure 1 illustrates the single-degree-of-freedom oscillation model of an elastically supported cylinder in fluid, allowing oscillation only in the transverse direction perpendicular to the incoming flow. The PTC cylinder uses symmetric strips arranged on both sides of the cylinder, with the leading edge located at a distance of α, from the front stagnation point, covering an angle β, and a thickness of t. The cylinder diameter is D, the spring stiffness is K, the damping is C, and the flow speed is U_∞.

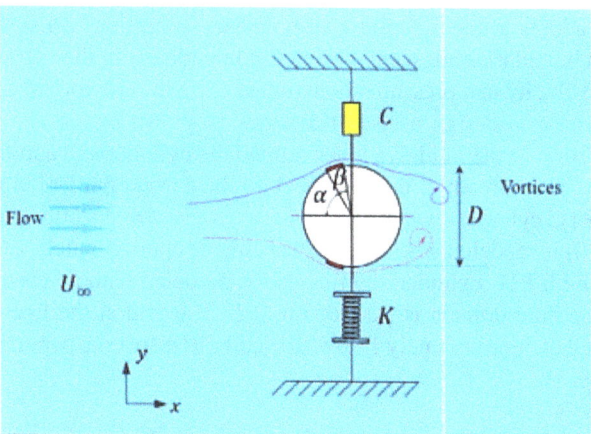

Fig. 1. Geometric model.

This study uses OpenFOAM and the Delayed Detached Eddy Simulation (SST-DDES) turbulence model, which combines the strengths of the SST model for improved accuracy and separated flow prediction. To enhance computational precision and minimize numerical dissipation, second-order accurate numerical methods are applied to the momentum and turbulence transport equations. The PIMPLE algorithm is employed for coupling velocity and pressure in solving the momentum equations.

3 Computational Details

As shown in Fig. 2, The field of computation is a cuboid of $30D \times 20D \times 2N$, Column length $L = \pi D$. It is elastically supported and oriented transversely perpendicular to the inflow (x-direction), The center of gravity of the cylinder serves as the coordinate origin, located downstream from the inlet boundary of $10D$. The y-direction is perpendicular to both the cylinder and the incoming flow, while the z-direction extends along the cylinder. Please note, this study considers the influence of free ends. Therefore, it separately simulates the Flow-Induced Vibration (FIV) characteristics for both an infinite-length cylinder of $N = 1/2L$ and a cylinder with a free end of $N = 1/2L$.

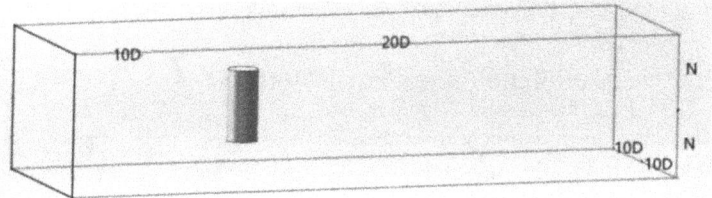

Fig. 2. Computational domain.

The boundary conditions are set as follows: at the inlet boundary U_∞, a uniform inflow velocity is applied and the normal gradient of pressure is set to zero. The surface of the cylinder adopts a no-slip solid wall boundary condition, while the front, back, top, and bottom boundaries are set as symmetric boundaries in the solver. At the outlet boundary, both velocity and pressure are assumed to have zero normal gradients.

The structured grid design used in this study is shown in Fig. 3. To enhance the numerical resolution of the fluid dynamics around the cylinder and additional grid cells have been added downstream of the flow field. In the current numerical simulation, the infinite-length PTC cylinder (with an angle of coverage $\beta = 20°$, thickness $t = 0.05D$, named P5–60-20) is modeled with a grid consisting of approximately 1.04×10^6 cells. For the finite-length PTC cylinder, the free end is modeled with an unstructured mesh, and the spanwise fluid domain is extended to $2L$. The grid at the free end is refined, resulting in a total of approximately 1.96×10^6 cells. The specific parameters are shown in the Table 1 below.

4 Results and Discussion

4.1 Vibration Response of Infinite-Length Cylinders

This section focuses on the vibration response analysis of an infinite-length cylinder. The analysis is used for numerical validation and serves as a comparison for the cylinder with a free end. Figure 4 shows amplitude ratio A^* and frequency ratio f^* for infinite-length cylinder, with the reduced velocity U^* and Reynolds number. The amplitude and

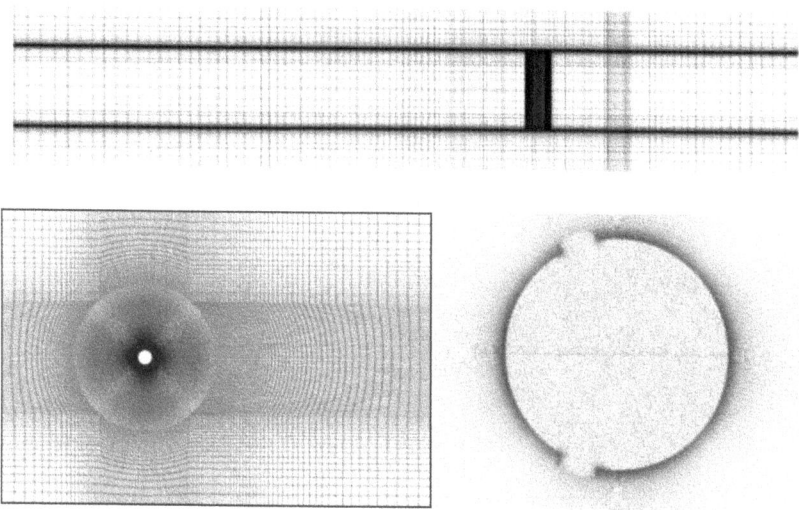

Fig. 3. Computational mesh.

Table. 1. specific parameters.

Parameter	Connotation	Unit
D	Cylinder diameter	$0.0381\ m$
L	Cylinder length	$0.1197\ m$
K	Spring stiffness	$7.56\ N/m$
C	Amping	$0.016\ N \bullet s/m$
U_∞	Flow speed	$0.0897 \sim 0.425\ m/s$
α	Angle	$60°$
β	Angle	$20°$
t	Thickness	$0.001905\ m$
m_{osc}	Total oscillating mass including 1/3rd of the spring mass	$0.315\ kg$
$f_{n.\,w}$	Natural frequency in still water	$0.651\ Hz$

frequency responses of the smooth cylinder demonstrate typical fluid-induced vibration characteristics, confirming the reliability of the solver and grid setup. The response of the PTC cylinder reveals the typical fluid-induced vibration phenomenon from vortex-induced vibration to galloping. The amplitude response before $U^* = 5.2$ is not significantly different from that of the smooth cylinder, and it remains around 0.9 within the range of $5.2 < U^* < 7.2$, which is divided into "passive upper branch". The PTC cylinder shows a small drop at $U^* > 7.2$, which is the transition region from VIV to galloping. When $U^* > 12$ is used, the amplitude of vibration is greater than that of the upper branch, and the subsequent vibration can be identified as pure galloping.

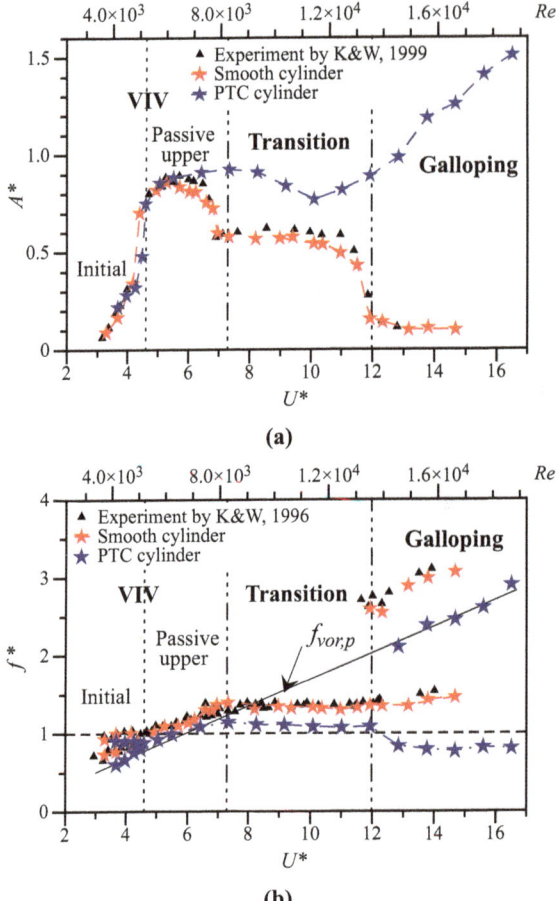

Fig. 4. Vibration response of an infinitely long cylinder: (a) amplitude ratio; (b) frequency ratio.

4.2 Vibration Response of PTC Cylinders with Free Ends

Figure 5 shows the vibration response of the PTC cylinder, the green star is the response result of the PTC cylinder with free end, and the infinite long cylinder is shown as a comparison. The figure shows that the end effect excites the initial branch nearby $U^* = 6.5$. The passive upper branch covers the reduced velocity range $8.2 < U^* < 11.5$, the transition branch is in the $11.5 < U^* < 13.8$ region, and the galloping occurs in the $U^* > 13.8$. Compared with the result of infinitely long PTC cylinder, the initial vibration equivalent flow rate increases and enters the initial branch later. In the passive branch, the amplitude curve of the wireless long cylinder reaches a peak value of 0.90 at $U^* = 7.2$ and $Re = 8055$, while in the numerical simulation under the influence of the end effect, the amplitude ratio curve roughly reaches a peak value of 1.09 at $U^* = 10.2$. For the transition branch, under the influence of the end effect, the reduced flow velocity into the transition branch increases to $U^* = 11.5$, aligning with the vibration response of an infinite cylinder. Figure 5(b) illustrates the relationship between normalized frequency response and reduced velocity derived from fast Fourier transform. As shown in the figure, under the influence of the end effect, the vibration frequency f* is not affected by the vortex shedding frequency $f_{vor,s}$, and vibrates with the natural frequency f_n during the initial branch and passive branch. When the vibration reaches the transition branch, it is consistent with frequency of the infinite cylinder.

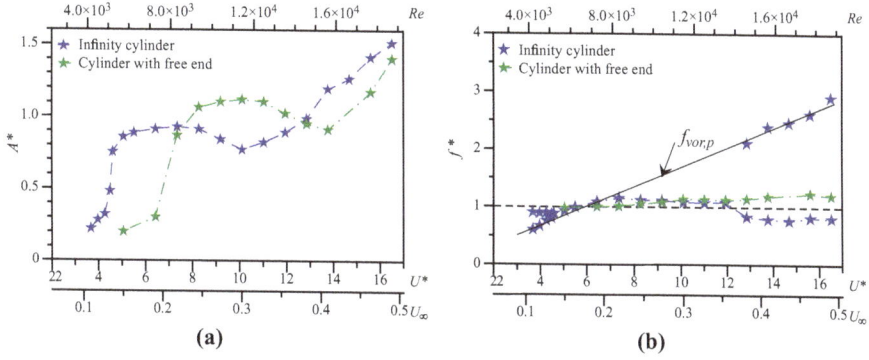

Fig. 5 Vibration response of PTC cylinder: (a) amplitude ratio (b) frequency ratio.

4.3 Flow Visualization of PTC Cylinder

Figure 6 presents a top view of the instantaneous wake vortex structures for both finite-length and infinite-length PTC cylinders. The left side shows the wake structure for VIV, while the right displays that for galloping. The figure reveals that the wake vortex structures around the free-end cylinder are more complex across all FIV branches, with recirculation on both sides extending into a wider diffusion zone downstream. In the galloping phase, distinct vortex contraction occurs behind the free-end cylinder, leading to end effects and arch vortex formation, which significantly increase the cylinder's vibration amplitude.

4.4 Energy Harvesting

This section discusses energy conversion from FIV in finite-length cylinders. Figure 7 shows that free-end cylinders need higher flow velocities for energy conversion compared to infinite-length cylinders. As flow velocity increases, power initially rises, then decreases before sharply increasing again. While free-end cylinders demonstrate stronger energy harvesting during galloping, their efficiency remains lower than that of infinite-length cylinders.

Building on this, the study investigates the FIV of finite-length cylinders with different length-to-diameter ratios L/D = 1, π, 10. Due to limited computational resources, only one reduced velocity $U^* = 12.5$ is discussed. The research indicates that as the

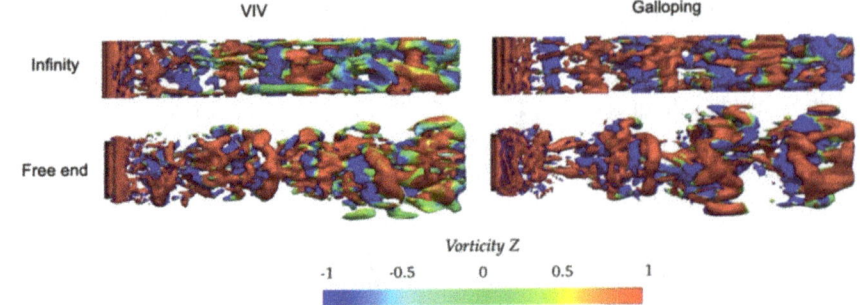

Fig. 6. Top-view of wake structures with isosurface of $\Omega = 0.52$. (Upper: Infinity cylinder; lower: Cylinder with free end).

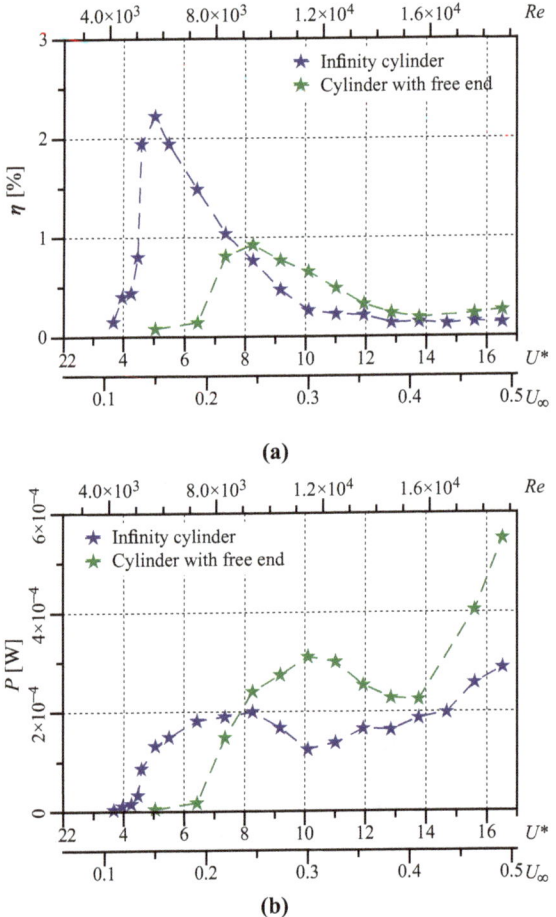

Fig. 7. Power and efficiency of energy harnessing: (a) power (b) efficiency.

length-to-diameter ratio increases, the vibration response and energy conversion characteristics of the finite-length cylinder approach those of the infinite-length cylinder. Interestingly, at a specific reduced velocity $U^* = 12.5$, L/D = 1 the vibration amplitude reaches a peak of $A^* = 1.32$. Determining the corresponding FIV branch for this vibration is challenging, emphasizing the need for further investigation into the FIV and energy harvesting of shorter cylinders, as their behavior and energy collection potential are complex.

5 Conclusion

This study analyzes the effects of free ends on the flow-induced vibration (FIV) and energy harvesting performance of PTC cylinders using three-dimensional numerical simulations. The results show that the free end significantly increases the excitation flow velocity and vibration amplitude, which are vital for optimizing energy harvesting systems. While cylinders with free ends can achieve higher power output in later stages, their overall efficiency is lower than that of infinite-length cylinders, highlighting the need to balance power and efficiency in design. Moreover, the impact of the free end decreases with increasing cylinder length, with the most pronounced effects seen in shorter cylinders. These findings provide valuable insights for designing more efficient energy harvesting devices.

References

1. Tong, Z., He, S., Tong, S., Chen, K.: Flow-induced vibration of turbo-expander impellers for industrial waste heat recovery: An analysis based on two-way fluid structure interaction. Sustain Energy Technol Assess. **70**, 103953 (2024)
2. Sun, Z., Wang, T., Qian, B., et al.: Influence of crosswind on the flow induced vibration profile of high-speed train's windshield. J. Fluids Struct. **120**, 103902 (2023)
3. Liu, W., Zeng, B., Zhou, Z., Ding, H., Lu, Y.: Wind-induced vibration and control of truss string structures subjected to thunderstorm downbursts. J. Constr. Steel Res. **222**, 108968 (2024)
4. Liu, Y., Jiang, Y., Zhao, H., Wang, S., Han, J.: Experimental investigation on vortex-induced vibration characteristics of a segmented free-hanging flexible riser. Ocean Eng. **281**, 115032 (2023)
5. Fan, X., Guo, K., Wang, Y.: Toward a high performance and strong resilience wind energy harvester assembly utilizing flow-induced vibration: Role of hysteresis. Energy. **251**, 123921 (2022)
6. Li, H., Xu, Y., Zhu, L., Zhang, G., Cheng, B.: Aerodynamic interference between road vehicles and bridge deck subjected to vortex-induced vibration. J. Wind Eng. Ind. Aerodyn. **252**, 105845 (2022)
7. Li, N., Park, H., Sun, H., Bernitsas, M.: Hydrokinetic energy conversion using flow induced oscillations of single-cylinder with large passive turbulence control. Appl. Energy. **308**, 118380 (2022)
8. Zhang, D., Wang, W., Sun, H., Bernitsas, M.: Influence of turbulence intensity on vortex pattern for a rigid cylinder with turbulence stimulation in flow induced oscillations. Ocean Eng. **237**, 109349 (2021)

9. Karthikeyan, S., Nallayarasu, S., Liu, Y.: Time-domain analysis of vortex-induced vibration of flexible cantilever cylinder using CFD simulation. Ocean Eng. **311**, 118875 (2024)
10. Yildizdag, M., Ardic, I., Ergin, A.: An isogeometric FE-BE method to investigate fluid–Structure interaction effects for an elastic cylindrical shell vibrating near a free surface. Ocean Eng. **251**, 111065 (2022)
11. Shen, Y., Wang, J., Alam, M.: FIV and energy harvesting using bluff body piezoelectric water energy harvester with different cross-sections. Ocean Eng. **288**, 116121 (2023)

Open Access This chapter is licensed under the terms of the Creative Commons Attribution-NonCommercial-NoDerivatives 4.0 International License (http://creativecommons.org/licenses/by-nc-nd/4.0/), which permits any noncommercial use, sharing, distribution and reproduction in any medium or format, as long as you give appropriate credit to the original author(s) and the source, provide a link to the Creative Commons license and indicate if you modified the licensed material. You do not have permission under this license to share adapted material derived from this chapter or parts of it.

The images or other third party material in this chapter are included in the chapter's Creative Commons license, unless indicated otherwise in a credit line to the material. If material is not included in the chapter's Creative Commons license and your intended use is not permitted by statutory regulation or exceeds the permitted use, you will need to obtain permission directly from the copyright holder.

Weather Processes of Strong Wind over China's Coastal Waters in 2023

Bin Zuo(✉), Shipeng Su, and Yuepeng Yang

Navigation Department, People's Liberation Army Dalian Naval Academy, 667 Jiefang Road, Dalian, China
howbigyouare@163.com

Abstract. Strong winds at sea not only pose a threat to ship navigation and coastal port facilities but also jeopardize the safety of life and property for residents in both urban and rural coastal areas. In order to reveal the patterns of strong winds at sea, continuous research is needed on its weather processes. In this article, we have compiled statistics on the strong wind weather over China's offshore waters in the year 2023, classified them based on the weather systems (such as cold anticyclone system and typhoon), and analyzed their weather processes of each category. It was found that the occurrence of strong wind weather in China's coastal waters is generally more common in spring and winter seasons, and the probability of strong wind weather in summer and autumn is significantly lower than that in spring and eastern seasons. This suggests that strong wind conditions in China's offshore waters are most closely associated with the southward movement of cold anticyclone systems originating from high latitudes. During the summer, typhoons are the primary cause of strong winds, predominantly affecting the Donghai Sea and posing a significant threat to maritime vessels. This research will enhance our understanding of the characteristics of strong wind weather along China's coastal waters, thereby improving navigation safety.

Keywords: Strong Wind · China's Coastal Waters · Cold High Pressure System · Tropical Cyclone · Frontal Cyclone

1 Introduction

Strong winds at sea not only threaten the safety of ships at sea, port facilities along the coast, but also the safety of people's lives and property in coastal areas. Research and analysis have shown that strong winds and sea fog are the main meteorological elements that cause major marine accidents, with strong winds affecting the most, accounting for about 62% [1]. Strong winds generate large waves, which can damage ships, coastal infrastructure, and buildings, and significantly disrupt navigation, offshore construction, and fishing operations. China has a 1.8-million-kilometer long continental coastline and 30 million square kilometers of maritime jurisdiction, with a total of about 580 million people in the 12 provinces with maritime interests. The marine environment is complex and dynamic, and the frequent occurrence of strong winds at sea not only impacts the livelihoods and daily activities of coastal and island residents, as well as tourists, but also hinders the development of China's marine economy [2].

Many studies have analyzed the distribution characteristics of the sea surface wind over China's coastal waters, including the annual average, seasonal average and extreme wind speed [3–6]. It has been observed that China's coastal waters exhibit a typical monsoon climate [3], and the sea surface wind field here has obvious seasonal and monthly variation characteristics: In summer, the prevailing southwest wind, the wind direction is changeable, the overall average wind speed is the lowest, but there is still a Nanhai sea gale area, the seasonal average wind speed is 7 m/s; Northeast winds prevail in winter, with high winter winds. The area of strong winds above force 6 is located in the Nanhai sea, the Taiwan Strait and the east sea of Japan Island, and there is a significant seasonal difference, with a higher probability of occurrence in winter than in the other three seasons [4]. Extreme wind speed also has similar distribution characteristics. The maximum value area of extreme wind speed in the northwest Pacific Ocean is mainly located in the east of Japan, and the secondary maximum value area is located in the east of Taiwan, the northern part of the Nanhai sea and the Bohai Sea, and the extreme wind speed in the near sea is obviously smaller than that in the ocean [5]. Also, the linear variation trend of the sea surface wind has been studied based on the ERA-40 sea surface wind data. The result shows that most of the sea areas outside the first island chain showed an increasing trend year by year, but there were different variations at different stages. During 1958–1974, the increasing trend of sea surface wind speed was more obvious. During 1975–2001, the linear trend was relatively smooth, especially during 1976–1983, the linear trend was not obvious [6].

In order to reveal the mechanism of strong winds at sea, continuous research is still needed. In this article, statistics on the strong wind weather over China's offshore waters in the year 2023 were compiled, classified based on the weather systems, and analyzed for each category's weather processes.

2 Data and Method

The wind data used in this article comes from the ERA5 reanalysis dataset [7], which is the fifth generation reanalysis dataset provided by the European Centre for Medium-Range Weather Forecasts (ECMWF). This dataset contains almost global meteorological data since 1979, covering the surface, atmosphere, and ocean surface. ERA5 provides a large number of meteorological variables, including temperature, pressure, humidity, wind speed, precipitation, as well as various statistical measures such as mean and standard deviation. ECMWF utilizes advanced meteorological models and algorithms to process and fuse historical meteorological observation data to ensure data quality and accuracy [8]. Therefore, ERA5 provides global meteorological data with high resolution, with a surface resolution of up to 4 kilometers. In addition, this article also uses the fax weather chart released by the Japan Meteorological Agency, which is broadcasted and can be received by ships sailing in the northwest Pacific using fax machines. These weather chart are the most commonly used maritime meteorological products for ships sailing in the China's coastal waters.

This article employs the "quantile method" to define and analyze strong wind weather conditions. Quantiles, also referred to as percentiles, are numerical points that partition the probability distribution of a random variable into equal intervals. Commonly used percentiles include the median (i.e., the 50th percentile), quartiles, and other specific percentiles. In this study, the 95th percentile is utilized as the threshold for identifying strong wind events. By calculating the 95th percentile of wind speed data for a specific region, days with an average daily wind speed exceeding this threshold are classified as strong wind days. This approach allows for the precise identification of dates experiencing exceptionally high wind speeds, which are critical for understanding and predicting strong wind weather patterns.

3 Results

3.1 The Climate Characteristics of Strong Wind Weather

In this paper, the daily average wind speed in four seas offshore of China's coastal waters was calculated (the Nanhai sea, the Donghai Sea, the Huanghai Sea and the Bohai Sea) from 1979 to 2023. After ranking the daily average wind speed in each offshore sea from small to large, days with wind speeds greater than the 95th percentile were identified as strong wind days (Fig. 1).

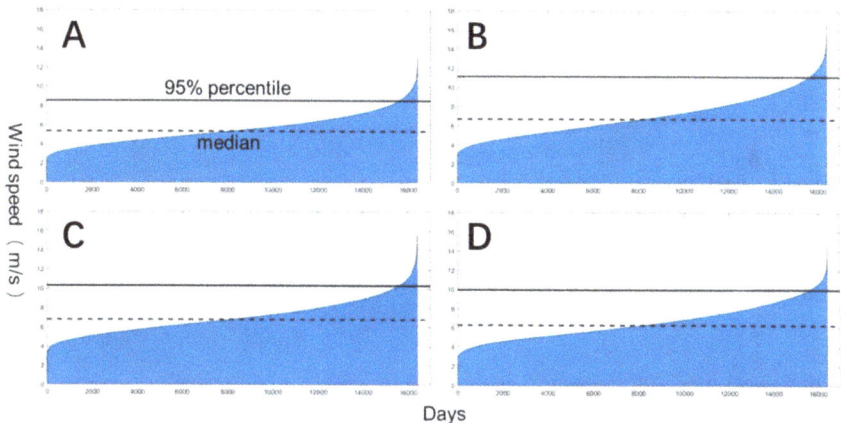

Fig. 1. Average daily wind speed in four seas offshore of the China's coastal waters. Figures A, B, C, and D represent the Bohai Sea, the Huanghai Sea, the Donghai Sea, and the Nanhai Sea, respectively.

The wind speeds in the four offshore seas, in descending order, are as follows: The Bohai Sea, the Nanhai Sea, the Donghai Sea, and the Huanghai Sea. The wind speed in the Bohai Sea is generally the lowest, which can be attributed to its landlocked nature and significant influence from topographic factors. The wind speed of the other three seas offshore is generally increasing from south to north, which may indicate that the strong wind weather in China's offshore waters is most closely related to the southbound cold anticyclone system in high latitudes, because the gale caused by the cold anticyclone system often presents the pattern of wind speed increasing from south to north.

The following figure (Fig. 2) shows the monthly distribution characteristics of strong wind weather in China's coastal areas. Strong winds in China's coastal waters often occur in winter, which further proves the close relationship between strong wind weather and the southbound cold anticyclone system. In summer, strong winds are relatively rare in China's coastal areas, but it should be noted that for the Donghai Sea, strong winds in summer are not uncommon because typhoons often occur in the Donghai Sea during the summer.

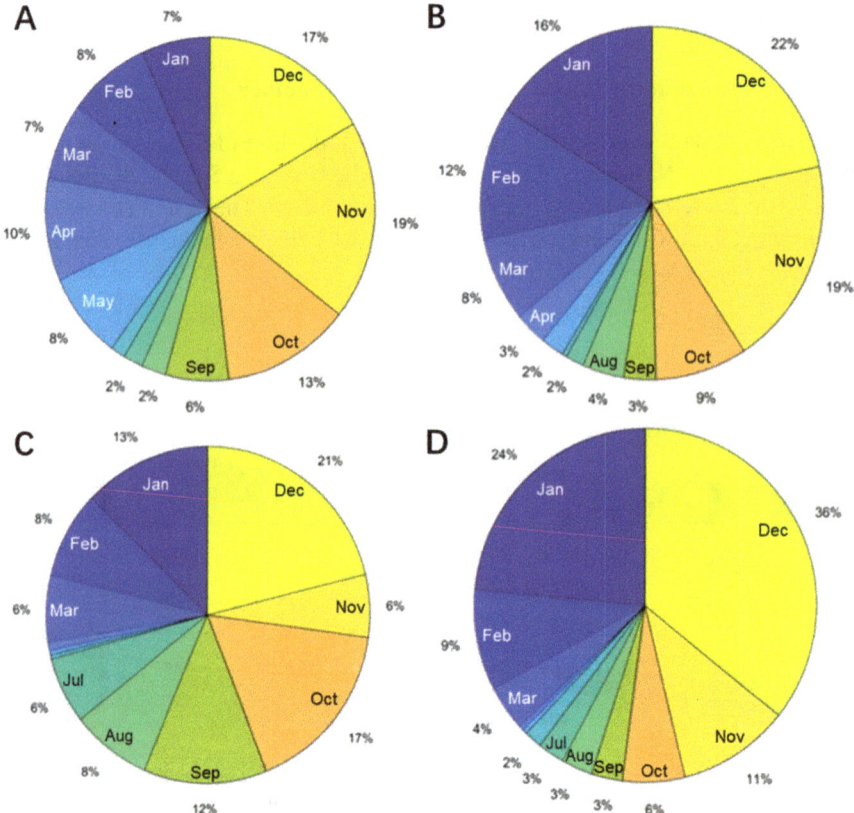

Fig. 2. Monthly distribution characteristics of strong wind weather in four seas offshore. Figures A, B, C, and D represent the Bohai Sea, the Huanghai Sea, the Donghai Sea, and the Nanhai Sea, respectively.

3.2 The Weather Processes of Strong Wind in 2023

Table 1 shows the date of strong wind weather in China's coastal waters in 2023. It can be seen from the table that the probability of strong winds occurring in winter is significantly higher than in the other three seasons. Through analysis of the fax weather charts corresponding to each strong wind event, it was found that the weather systems causing strong winds in China's coastal waters primarily include cold anticyclones, frontal cyclones, and typhoons. These are the three weather systems that have the greatest impact on China's coastal areas. From the perspective of the range of seas offshore affected by strong wind weather, some strong wind weather can cover the entire coastal waters of China, while others only affect individual sea offshore. Next, the weather processes affecting strong wind events are distinguished and discussed based on the varying impact ranges of strong wind weather.

Table 1. Date of strong wind weather in China's coastal waters in 2023.

Date	Sea offshore	Date	Sea offshore	Date	Sea offshore	Date	Sea offshore
1.3–1.6	Nan	2.20	Dong	5.5	Bo	11.9–11.10	Huang & Bo
1.14	entire	2.25–2.27	Nan	5.24	Bo	11.12	Dong & Huang
1.17	Nan	2.25	Nan & Dong	6.1	Dong	11.16–11.17	Entire
1.24	entire	3.11	Bo	7.16–7.17	Nan	11.23	Huang & Bo
1.27	entire	3.12	Huang	7.30	Bo	11.24	Dong
1.28–1.29	Nan	3.17	Nan & Dong	8.1–8.4	Dong	12.9–12.11	Huang & Bo
2.1	Huang	3.23	Huang	8.28	Huang & Bo	12.14–12.15	Huang & Bo
2.13–2.14	Huang	4.4	Huang & Bo	9.20	Huang	12.16–12.17	Nan & Dong
2.15–2.16	Nan	4.10	Huang	11.5–11.6	Huang & Bo	12.20–12.21	Nan & Dong

Figure 3 shows the ASAS (Surface Analysis of Asia) weather chart at 0:00 and 12:00 UTC on January 24, 2023. This is a date in which strong wind weather covers the entire coastal waters of China. It can be seen from the figure that there were two main weather systems in the western Pacific region on that day, a cold anticyclone system and a frontal cyclone. The frontal cyclone was far away from the coastal waters of China, and the main factor affecting the occurrence of strong wind weather in the coastal waters of China was the cold anticyclone system. The central pressure of the cold anticyclone system was 1040 hPa, moving southeast at a speed of 15 knots. The coastal waters of China were located in the front of the cold anticyclone system and on its central movement route. The front pressure gradient of the cold anticyclone system was large, forming strong wind weather along the coast of China, accompanied by severe cooling. The other three date in which strong wind weather covers the entire coastal waters of China has a similar weather process as January 24, that is the strong wind was caused by cold anticyclone system. In conclusion, cold anticyclones are typically the only weather systems capable of causing strong winds that cover the entire coastal waters of China, due to their large scale.

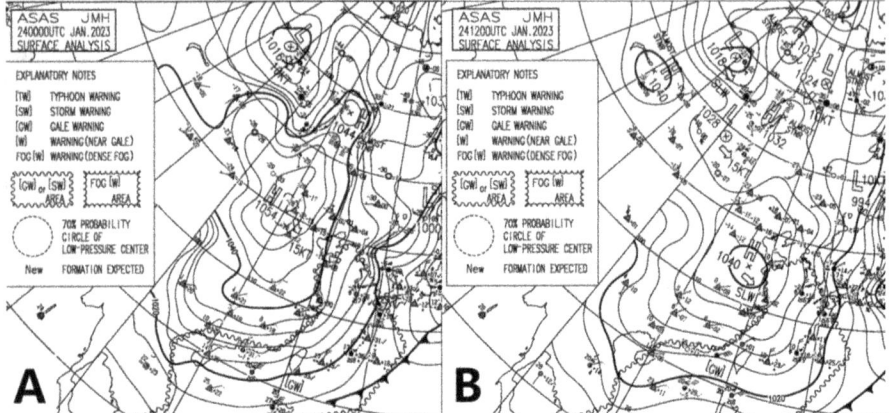

Fig. 3. ASAS (Surface Analysis of Asia) weather chart at 0:00 and 12:00 UTC on January 24, 2023

Due to the small area of the Bohai Sea and its close proximity to the Huanghai Sea, when the weather system passing through is relatively large, it often affects both areas simultaneously, resulting in strong winds in both areas. This often corresponds to the southward anticyclone process during winter. Taking the strong wind weather on November 23, 2023 as an example (Fig. 4), the Bohai Sea area is located in the northwest of the frontal cyclone cold front. This cold front was located in the northern part of the Huanghai Sea at 0:00 UTC, with a central pressure of 1000 hPa. It is located in the northeast of the Huanghai Sea and Bohai Sea area and is moving northeastward at a speed of 25 knots; The Huanghai Sea and Bohai Sea are both located on the moving path of a cold anticyclone system, with a central pressure of 1050 hPa and moving southeast at a speed of 10 knots. The isobars in the Bohai Sea are dense, while those in the Huanghai Sea are relatively sparse. Therefore, on that day, under the influence of both weather systems, the wind speeds in both sea areas were relatively high.

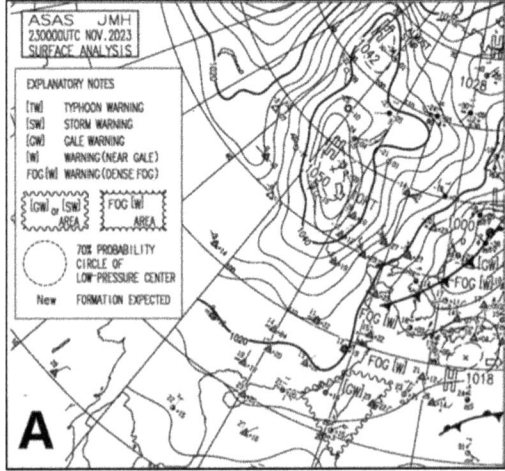

Fig. 4. ASAS weather chart at 0:00 UTC on November 23, 2023

Figure 5 shows the ASAS weather chart from August first to fourth at 12:00 UTC. During these four days, there was a prolonged strong wind weather process in the Donghai Sea caused by a typhoon. The center of the cyclone was located in the eastern part of the Donghai Sea and slowly moved westward, with a central pressure of about 930 hPa and high intensity. The central wind speed exceeded level 12, resulting in poor sea conditions, extremely poor visibility, unpredictable wind direction, and strong winds in the East China Sea during these four days. The strong winds in summer that only occur in the Donghai Sea are often caused by typhoons, as in this example.

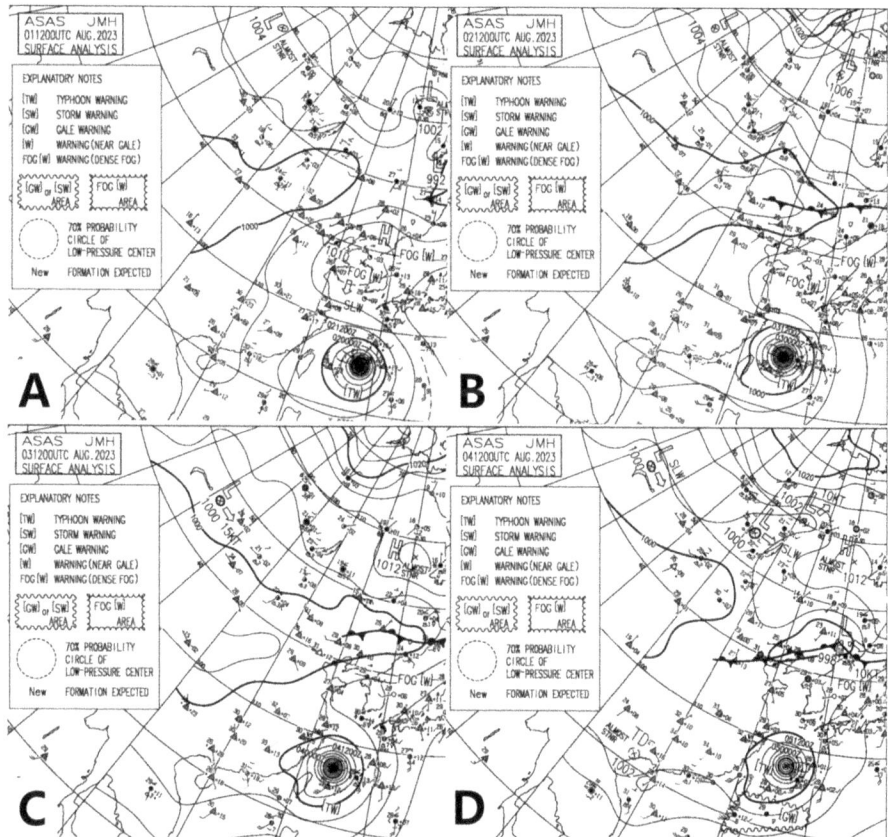

Fig. 5. ASAS weather chart at 12:00 UTC form August first to fourth, 2023

Figure 6 shows the weather chart for January third to sixth, corresponding to an example of strong winds only occurring in the Nanhai Sea. A prolonged period of strong winds occurred in the Nanhai Sea from January third to 6th. The main reasons for the strong winds during these four days were the cold anticyclone system moving southward from high latitude areas and the low-pressure center appearing in the southern part of the Nanhai Sea. On the third, the high-pressure center of the cold anticyclone system moved slowly towards the southeast on land, with the front of the high pressure entering the sea and a strong wind warning zone appearing in the southern Nanhai Sea. On the fourth, the high-pressure center of the cold anticyclone system approached the sea surface, and the strong wind warning zone in the Nanhai Sea expanded. A low-pressure center appeared in the southern part of the strong wind warning zone, slowly moving westward. On the fifth, the high-pressure center of the cold anticyclone system entered the sea, but the strong wind warning area in the Nanhai Sea shrank. The position of the low-pressure center in the southern part of the strong wind warning area did not change much, and the overall wind strength weakened. On the sixth, the high-pressure center of the cold anticyclone system entered the open sea, and the strong wind warning zone disappeared. However, another cold anticyclone system followed closely and moved southeastward. The position of the low-pressure center in the Nanhai Sea remained roughly unchanged, and the wind force greatly weakened. The reason for the extreme strong winds in the Nanhai Sea during these four days is closely related to the continuous generation and southward movement of cold anticyclone system in high latitude regions. The absence of strong winds in other sea areas during this period is due to their proximity to the center of the anticyclone.

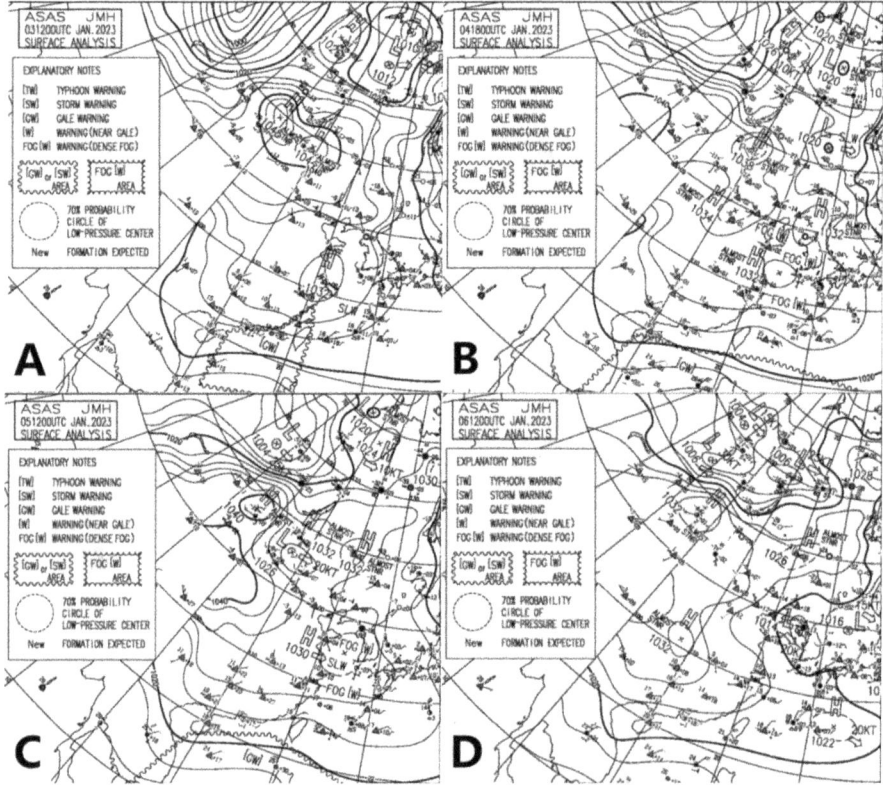

Fig. 6. ASAS weather chart form January third to sixth, 2023

4 Conclusion

Based on the preceding analysis, it can be concluded that the occurrence of strong wind weather in China's coastal waters is generally more common in spring and winter, and the probability of strong wind weather in summer and autumn is significantly lower than that in spring and eastern seasons. This is closely related to China's geographical location. In spring and winter, the Siberian region continuously generates a cold anticyclone system, which advances southeastward with a cycle of about seven days, and then moves towards the coastal areas of China. After entering the sea and moving away from the coastal waters, the next cold anticyclone system follows one after another, continuously affecting the weather in China's coastal waters. In the anticyclone system, the pressure gradient is large and the isobars are dense, which often brings about strong wind weather. This is a key factor contributing to the frequent occurrence of strong wind conditions in China's coastal waters during winter.

The impact of tropical cyclones on China's coastal areas is mainly from June to October each year, with the Donghai Sea being the most affected area. When a tropical cyclone passes through, it generates strong winds with enormous destructive power, posing a serious threat to the safety of ship navigation. Therefore, particular emphasis should be placed on typhoon prevention and mitigation measures.

Acknowledgments. This work was jointly supported by the National Natural Science Foundation of China (NSFC) Project (42305187) and the Shandong Natural Science Foundation Project (ZR2019ZD12).

References

1. Zhou, F.X.: The sea fog and its classification. Mar. Forecasts. **1**, 78–84 (1988) (in Chinese)
2. Huang, B., Mao, D.Y., Kang, Z.M., et al.: Synoptic and clinatic characteristics of Yellow Sea fog and causation analysis. J. Trop. Meteorol. **27**(6), 920–929 (2011) (in Chinese)
3. Ding, Y.H., Liu, Y.Y.: A study of the teleconnection in the Asian-Pacific monsoon region. Acta. Meteor. Sin. **66**, 670–682 (2008) (in Chinese)
4. Zhan, S.Y., Qi, L.L., Lu, W.: Analysis of sea surface wind in Northwest Pacific based on CCMP satellite data. Mar. Forecasts. **34**(2), 10–20 (2017) (in Chinese)
5. Zheng, C.W.: Sea surface wind field analysis in The China sea during the last 22 years with CCMP wind field. Meteorol. Disaster Reduct. Res. **34**(3), 41–46 (2011) (in Chinese)
6. Pan, J., Zheng, C.W.: Long-term trend analysis of sea surface wind speed in the Northwest Pacific Ocean. Mar. Forecasts. **29**(5), 48–52 (2012) (in Chinese)
7. Berrisford, P., Soci, C., Bell, B., et al.: The ERA5 global reanalysis: Preliminary extension to 1950. Q. J. R. Meteorol. Soc. **147**(741), 4186–4227 (2021)
8. Zhai, R., Huang, C., Yang, W., et al.: Applicability evaluation of ERA5 wind and wave reanalysis data in the South China Sea. J. Oceanol. Limnol. **41**(1), 495–517 (2023)

Open Access This chapter is licensed under the terms of the Creative Commons Attribution-NonCommercial-NoDerivatives 4.0 International License (http://creativecommons.org/licenses/by-nc-nd/4.0/), which permits any noncommercial use, sharing, distribution and reproduction in any medium or format, as long as you give appropriate credit to the original author(s) and the source, provide a link to the Creative Commons license and indicate if you modified the licensed material. You do not have permission under this license to share adapted material derived from this chapter or parts of it.

The images or other third party material in this chapter are included in the chapter's Creative Commons license, unless indicated otherwise in a credit line to the material. If material is not included in the chapter's Creative Commons license and your intended use is not permitted by statutory regulation or exceeds the permitted use, you will need to obtain permission directly from the copyright holder.

Prediction of Ship Motion Attitude Using an Improved Transformer Model

Lingyi Hou[1,2], Hang Sun[1(✉)], Jianjun Hou[1], and Dawei Li[1]

[1] Navigation Department, Dalian Naval Academy, Dalian, China
284521684@qq.com
[2] School of Mechanical Engineering, Dalian University of Technology, Dalian, China

Abstract. In maritime navigation, ships operating in open waters encounter complex dynamics that result in irregular six-degree-of-freedom motion responses. The inherent unpredictability and randomness of these movements pose substantial uncertainty and risks to various maritime operations, including missile launches and helicopter takeoffs and landings. Therefore, the ability to accurately predict a ship's motion attitudes within a short timeframe is of critical importance for enhancing safety in navigation and operational efficiency. This paper proposes an advanced methodology for predicting ship motion attitudes based on an enhanced Transformer model. The proposed framework optimizes the original Transformer architecture by integrating a convolutional layer into the encoder module, which facilitates more comprehensive feature extraction and enriches the representation process. Through the synergistic interaction between this convolutional layer and the multi-head attention mechanism, deeper correlations within time series data are effectively examined. Experimental results demonstrate that this improved Transformer-based approach for predicting ship motion attitudes achieves high accuracy levels, underscoring its practical relevance in engineering applications.

Keywords: Transformer · CNN · Ship Motion Attitude · Prediction

1 Introduction

During the sailing process, a vessel experiences sway due to environmental influences, with this motion becoming particularly pronounced in high sea states. The ship's movement can significantly impact critical operations conducted on board, such as ensuring the safe takeoff and landing of carrier aircraft during flight operations, navigating through wind and waves, and executing missile launches. These effects can have substantial repercussions, potentially jeopardizing the safety of life and property. Therefore, it is essential to accurately predict the ship motion attitude, thereby providing a solid foundation for decision-making [1].

Domestic and foreign scholars have proposed a variety of methods, which can be roughly divided into prediction methods based on hydrodynamics theory, time series model prediction method and data-driven machine learning model prediction method according to their prediction principles [2–4]. The time series prediction method establishes the corresponding time series model through regression analysis. For example, Auto-Regressive (AR) model [5] and Autoregressive Integrated Moving Average (ARIMA) model [6], etc. The prediction method does not require the motion equations or flow field information of the vessel, relying solely on its historical data to identify patterns for forecasting. This modeling approach is straightforward and entails a low computational burden. However, it is primarily applicable to linear problems and exhibits certain limitations when addressing the strong nonlinear challenges associated with predicting ship motion in wave conditions.

In recent years, the advancement of artificial intelligence and big data technologies has led researchers to apply machine learning methods in ship motion modeling and prediction due to their excellent nonlinear fitting capabilities. YIN et al. [7] integrated wavelet transform with a variable structure Radial Basis Function (RBF) neural network for real-time prediction of ship rolling motion. HUANG et al. [8] utilized wavelet neural networks to investigate real-time predictions of ship rolling motion. Additionally, HOU et al. [9] and CHEN et al. [10] employed SVMs for modeling and forecasting the rolling motions of vessels in both wave conditions and shallow waters. With advancements in GPU technology and machine learning algorithms, deep learning techniques represented by deep neural networks have emerged as a prominent research focus among scholars, facilitating nonlinear modeling and predictive studies on ship rocking motion. While these methods enable predictions of vessel motion attitudes, they do not incorporate the characteristics of ship rocking attitude data for adaptive design. The Transformer model, introduced in 2017, represents a significant advancement in addressing long sequence prediction challenges [11]. This model comprises an encoder and a decoder module that extract features, with both modules employing the self-attention mechanism for feature extraction [12]. The self-attention mechanism effectively emphasizes the long-term dependencies within the data, yielding impressive results in long sequence predictions [13].

Building upon the Transformer model, we introduce an enhanced version tailored for predicting ship motion attitudes. By facilitating interaction between the convolutional layer and the multi-head attention mechanism, the model deeply explores the correlations within time series data and effectively extracts relevant features. Training and testing were conducted using real ship motion attitude data collected during third, fourth, and fifth sea states. The results indicate that the suggested approach greatly improves the accuracy of predictions regarding ship motion attitudes.

2 Methodology

2.1 Model Structure

The motion attitude of a ship is influenced by various environmental factors, including wind, waves, and currents, resulting in non-stationary time series data. Capturing the dependencies within these time series presents significant challenges, leading to low predictive accuracy with conventional models. In this study, we propose an enhanced Transformer model that builds upon the original architecture. To address the specific characteristics of ship motion attitude data, we incorporate a convolutional module to augment the model's performance. The structure of the proposed model is illustrated in Fig. 1.

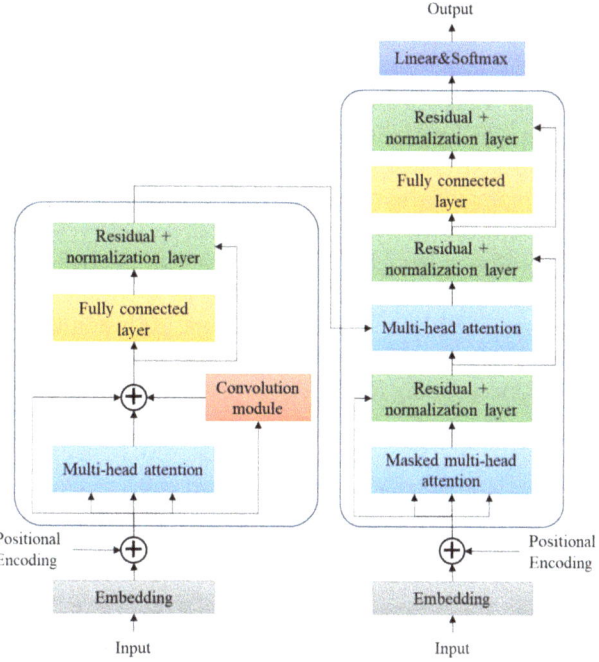

Fig. 1. Ship motion attitude prediction model.

2.2 Embedding

The Transformer model transforms the input sequence into a vector representation via embedding and incorporates positional encoding to capture the positional information of each element within the sequence. Positional encoding is represented as a vector that matches the dimensionality of the input embedding, corresponding to the position information for each element in the input sequence. The primary objective of positional encoding is to enable the model to differentiate between words at various positions, thereby facilitating its ability to manage order relationships within the sequence.

In this study, the values of sine and cosine functions are utilized to compute the numerical representation of positional encoding, which is then combined with the input embedding vector to derive the final input representation. The calculation of positional encoding is detailed as follows.

$$PE(p, 2i) = \sin\left(\frac{p}{10000^{\frac{2i}{d_{model}}}}\right) \quad (1)$$

$$PE(p, 2i + 1) = \cos\left(\frac{p}{10000^{\frac{2i}{d_{model}}}}\right) \quad (2)$$

where, p represents the position within the input sequence, i indicates the index of the dimension in the position encoding vector, and d_{model} refers to the dimensionality of the input embedding vector.

2.3 Encoder

Multi-head Attention Mechanism The attention mechanism utilizes three inputs: Query, Key, and Value, which are essential for computing the self-attention value matrix. Both the encoder and decoder incorporate a multi-head attention approach, with dimensions for Q, K, and V represented as $\boldsymbol{D_q}, \boldsymbol{D_K}$ and $\boldsymbol{D_V}$ respectively. To perform multi-head attention computation effectively, it is necessary to split the Query, Key, and Value into n separate vectors prior to executing self-attention. During the internal calculations involving Q and K, it is important to normalize by dividing by $\sqrt{D_k}$ after performing the dot product in order to mitigate issues related to gradient explosion that may arise post-calculation. Following this step with a Softmax function yields the attention weights denoted as $\hat{a}_{i,j}$. The corresponding formula is outlined below.

$$\hat{a}_{i,j} = \frac{exp(a_{i,j})}{\sum_m exp(a_{i,m})} = \frac{exp\left(\frac{q_i \cdot k_j}{\sqrt{D_k}}\right)}{\sum_m exp\left(\frac{q_i \cdot k_m}{\sqrt{D_k}}\right)} \tag{3}$$

The vector V is multiplied with the attention weights $\hat{a}_{i,j}$ to obtain v with attention weights.

$$attention(Q, K, v) = softmax\left(\frac{QK^T}{\sqrt{d_k}}\right)V \tag{4}$$

The revised Transformer neural network incorporates a multi-head attention mechanism to improve the multi-dimensional representation of self-attention. This mechanism divides a single attention process into n distinct groups, with each group containing its own set of independent Q, K, and V. Within each group, separate attention operations are conducted, and the results from all groups are then merged back to match the original size for the output of the multi-head attention layer. This approach enables the model to concentrate on information across various subspaces and capture more diverse features by synthesizing data from multiple perspectives, ultimately enhancing prediction accuracy.

The multi-head attention mechanism effectively maps the input into distinct subspaces, thereby enhancing information extraction. Let us denote the number of attention heads as h, with each head receiving an input dimension of $\frac{Dq}{h}$. Subsequently, each head independently performs self-attention calculations. The results from all heads are then aggregated into a matrix and subsequently forwarded to a fully connected layer for linear transformation and output generation.

Convolution Module. The model has been enhanced by incorporating a convolution module, tailored to the specific traits of ship motion attitude data, which is illustrated in

Fig. 2. This convolution module is effective in extracting feature information from the original dataset, aiding in the retention of both its characteristics and trends. By taking a weighted sum of the information derived from this convolution module alongside multi-head attention, we can not only capture the essential features of the data but also maintain its trend characteristics. Experimental findings indicate that this modified model significantly improves the alignment between predicted ship motion attitude values and actual measurements.

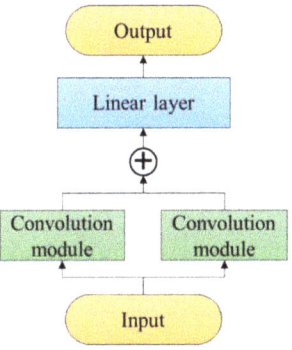

Fig. 2. Structure of convolution module.

2.4 Decoder

The primary distinction between the overall architecture of the decoder and that of the encoder lies in the inclusion of a masking mechanism within the attention process. When ship motion attitude data is fed into the decoder, it receives the entire sequence, which could inadvertently reveal information necessary for prediction. To mitigate this issue, a masking mechanism is integrated into the decoder to ensure that it does not leverage predictive information from subsequent elements in the sequence.

To implement the masking effect, a mask matrix can be constructed with zeros in the lower triangle and negative infinity in the upper triangle, as depicted in Fig. 3. The masked self-attention scores are then derived by performing element-wise addition between the self-attention matrix and this mask matrix.

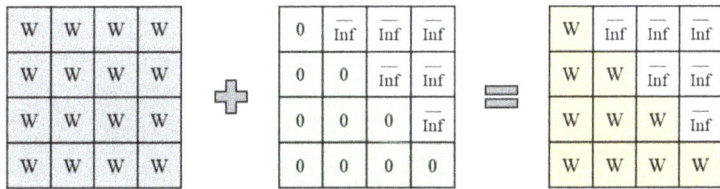

Fig. 3. Example of mask attention calculation.

3 Experiment

3.1 Data-preprocessing

In this study, the simulation utilizes measured data from a ship's motion under Level 3, 4, and 5 sea conditions. The sampling period is set at 0.2 s, resulting in a dataset length of 7500 s. This data encompasses real-time measurements of roll and pitch angles. In the simulation experiments presented in this paper, the roll and pitch angles are examined independently. Figure 4. presents an example of the roll and pitch data associated with Level 5 sea conditions.

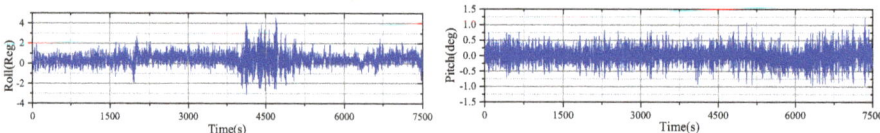

Fig. 4. Ship motion attitude sequence in Level 5 sea condition. (left: roll, right: pitch).

Given that the dataset utilized in this study is a time series dataset, random shuffling of samples is not feasible. As a result, the dataset is divided into training and test sets in a chronological order, following an 8:2 ratio. In this research, we employ the min-max normalization method for data standardization, as described by the following formula.

$$y = \frac{x - x_{min}}{x_{max} - x_{min}} \quad (5)$$

In the formula, x represents the original time series data, x_{max} denotes the maximum value within the time series, and x_{min} indicates the minimum value. The variable y corresponds to the normalized time series data. This approach allows us to constrain the range of the transformed data within [0, 1].

3.2 Evaluation Metrics

Indicators such as Mean Squared Erro, Mean Absolute Error, and Root Mean Square Error are commonly employed in regression tasks. A smaller value for these metrics indicates higher prediction accuracy of the model. Besides, coefficient of correlation (R_2) can reflect the mutual relationship between two variables and its correlation direction. The prediction accuracy increases as the value of R_2 approaches 1. Consequently, RMSE, MAE, MSE, and R^2 are chosen to assess the effectiveness of the predictive model. The corresponding equations can be found in Eq. (6) through Eq. (9).

$$MAE = \frac{1}{m} \sum_{i=1}^{m} |y_i - \hat{y}_i| \quad (6)$$

$$MSE = \frac{1}{m} \sum_{i=1}^{m} (y_i - \hat{y}_i)^2 \quad (7)$$

$$RMSE = \sqrt{\frac{1}{m}\sum_{i=1}^{m}(y_i - \hat{y}_i)^2} \qquad (8)$$

$$R^2 = 1 - \frac{\sum_i (y_i - \hat{y}_i)^2}{\sum_i (y_i - \bar{y}_i)^2} \qquad (9)$$

In the above formula, y_i refers to the actual value at time t, $\hat{y}_i$ indicates the forecasted value at that same time, and m represents the total count of actual values.

3.3 Result Analysis

The experimental environment for this study comprises Windows 10 (Intel i5-13600F, 3.50 GHz, 32 GB of RAM, and NVIDIA GeForce RTX 4070 Ti) along with the Python 3.10 programming language.

To evaluate the effectiveness of the proposed model, Transformer, LSTM, and AR models are selected for comparative analysis. The Transformer model serves as the baseline without any enhancements. The LSTM model is widely recognized as one of the most effective approaches in time series forecasting. The AR model represents a traditional linear regression approach and is among the two simplest models utilized in time series analysis.

All models are configured to run for 100 epochs, using a learning rate of 0.01 and the Adam optimizer, while applying a linear activation function in the fully connected layer.

Utilizing 7500 s (35,000 data points) historical ship motion attitude data points as training data, the model predicts conditions for the next 9.6 s (48 data points) in a single instance. The comparative outcomes of roll prediction for the four models under different sea conditions are shown in Fig. 5, along with accuracy evaluation metrics provided in Table 1. Furthermore, Fig. 6 illustrates the comparison results for pitch prediction among the same models, accompanied by relevant accuracy evaluation indices in Table 2.

The comparative results of the experimental predictions for roll and pitch motion across the four models, along with the evaluation metrics under sea conditions 3, 4, and 5, indicate that the improved Transformer model exhibits significantly lower values in MAE, MSE, and RMSE compared to the other three models. Additionally, its R^2 value is closer to 1 than those of the other models. This demonstrates that the predictive performance of the improved Transformer model surpasses that of its counterparts, followed by the LSTM model and RNN model; conversely, the AR model shows inferior performance.

The AR model is a classical statistical approach that, despite its simple structure, offers high computational efficiency and ease of understanding and implementation. However, it assumes a linear relationship in the data, meaning that future values are solely dependent on past values at a finite number of time points. In contrast, the motion attitude data of ships often exhibit complex nonlinear relationships influenced by factors such as waves, wind, and currents—relationships that cannot be accurately captured by linear models. Consequently, the AR model struggles to account for these nonlinear dynamics in ship motion data, resulting in lower prediction accuracy.

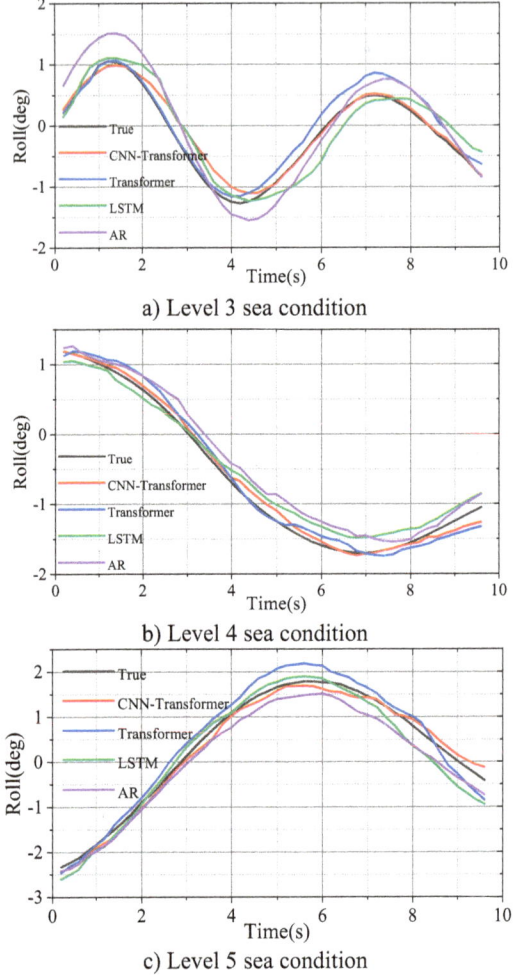

Fig. 5. Prediction result curves of roll under different sea conditions.

Table 1. Prediction accuracy of roll under different sea conditions.

Sea Condition	Metric	CNN-Transformer	Transformer	LSTM	AR
Level 3	MAE(°)	0.0986	0.1511	0.2225	0.2874
	MSE	0.0247	0.0380	0.0681	0.1067
	RMSE(°)	0.1573	0.1950	0.2609	0.3266
	R^2	0.9412	0.9259	0.8758	0.8829
Level 4	MAE(°)	0.0695	0.1129	0.1645	0.2107

(*continued*)

Table 1. (*continued*)

Sea Condition	Metric	CNN-Transformer	Transformer	LSTM	AR
	MSE	0.0078	0.0170	0.0323	0.0557
	RMSE(°)	0.0882	0.1304	0.1798	0.2360
	R^2	0.9927	0.9853	0.9586	0.9445
Level 5	MAE(°)	0.1179	0.2193	0.2033	0.2755
	MSE	0.0168	0.0615	0.0681	0.0867
	RMSE(°)	0.1297	0.2479	0.2610	0.2945
	R^2	0.9895	0.9688	0.9632	0.9421

The LSTM model is a variant of the RNN designed to address the gradient vanishing and exploding issues that standard RNNs encounter when processing long sequences of data. Its primary advantage lies in its ability to effectively manage long-term dependencies while maintaining good training stability. However, the LSTM model also exhibits high computational complexity and demands substantial computing resources, coupled with a large number of parameters, which complicates model interpretation and parameter adjustment.

The Transformer model demonstrates higher prediction accuracy compared to both the LSTM and AR models. Its self-attention mechanism enables it to concentrate on relationships between various positions in the sequence, applying different weights throughout the processing of sequence data. This capability enables the model to effectively capture long-term dependencies, such as periodic changes and trends in ship motion attitude data. However, the predictive performance of the model still falls short of our expectations, primarily because the Transformer is predominantly designed for natural language processing tasks and may not be well-suited for predicting ship motion attitudes. This paper enhances the Transformer by integrating a convolution module, which effectively captures nonlinear relationships in ship motion attitude data. This approach allows for a deeper extraction of data features. As a result, the improved Transformer model can more accurately predict future trends in ship attitude, thereby enhancing prediction accuracy. A comparison of prediction results under sea states three, four, and five reveals that the improved Transformer model also demonstrates superior generalization ability, achieving reliable predictions across different ships and navigation conditions.

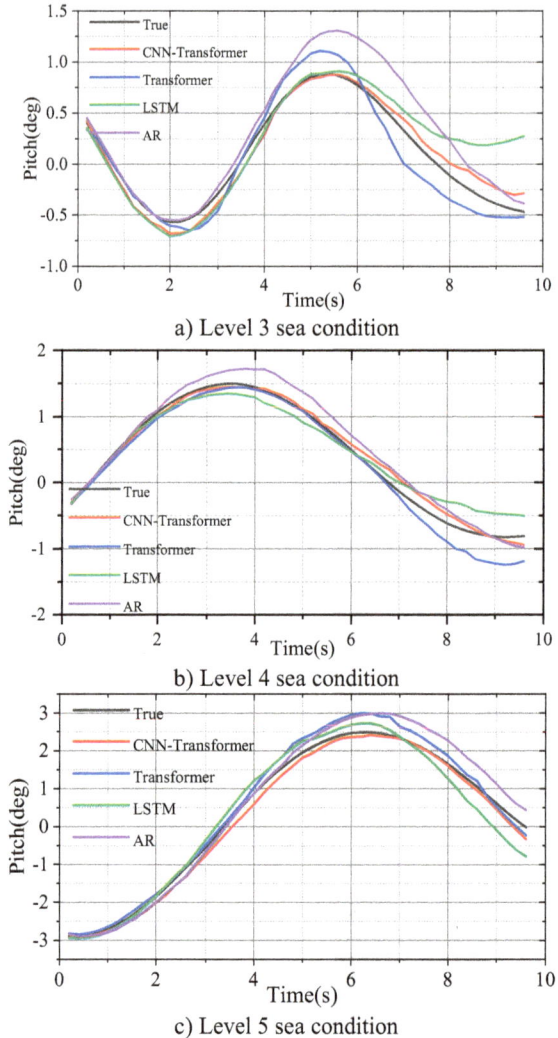

Fig. 6. Prediction result curves of pitch for different sea conditions.

Table 2. Prediction accuracy of pitch under different sea conditions.

Sea Condition	Metric	CNN-Transformer	Transformer	LSTM	AR
Level 3	MAE(°)	0.0857	0.1302	0.1893	0.2111
	MSE	0.0098	0.0255	0.0715	0.0744
	RMSE(°)	0.0988	0.1597	0.2673	0.2728
	R^2	0.9599	0.9233	0.7261	0.8247

(*continued*)

Table 2. (*continued*)

Sea Condition	Metric	CNN-Transformer	Transformer	LSTM	AR
Level 4	MAE(°)	0.0708	0.1250	0.1498	0.1638
	MSE	0.0080	0.0324	0.0352	0.0359
	RMSE(°)	0.0893	0.1800	0.1875	0.1895
	R^2	0.9872	0.9619	0.9152	0.9544
Level 5	MAE(°)	0.1343	0.2004	0.2493	0.2842
	MSE	0.0257	0.0617	0.1087	0.1282
	RMSE(°)	0.1605	0.2484	0.3298	0.3581
	R^2	0.9924	0.9841	0.9698	0.9701

4 Conclusion

This paper conducts an in-depth analysis of ship motion attitude prediction and proposes a model based on an improved Transformer. The experimental findings indicate that the proposed approach successfully attains a high level of accuracy in forecasting the ship motion attitude. The model not only accurately forecasts future motion attitudes over a specified period but also captures fluctuations, enabling predictions of potential peak and valley values, which is crucial for ensuring maritime safety. However, we recognize that there is still room for improvement. Future research should fully consider relevant factors that significantly impact ship sway attitude; incorporating these elements into the model could enhance its adaptability to real-world conditions and improve prediction accuracy. Additionally, optimizing the fluctuation peaks may further refine the model's predictive performance.

References

1. Sun, Q., Tang, Z., Gao, J., et al.: Short-term ship motion attitude prediction based on LSTM and GPR. Appl. Ocean Res. **118**, 102927 (2022)
2. Zhang, B., Wang, S., Deng, L., et al.: Ship motion attitude prediction model based on IWOA-TCN-Attention. Ocean Eng. **272**, 113911 (2023)
3. Gao, N., Hu, A., Hou, L., et al.: Real-time ship motion prediction based on adaptive wavelet transform and dynamic neural network. Ocean Eng. **280**, 114466 (2023)
4. Hou, L., Wang, X., Sun, H., et al.: A long sequence time-series forecasting model for ship motion attitude based on informer. Ocean Eng. **305**, 117861 (2024)
5. Khan, A., Bil, C., Marion, K., et al.: Real time prediction of ship motions and attitudes using advanced prediction techniques. In: Congress of the International Council of the Aeronautical Sciences (2004)
6. Suhermi, N., Prastyo, D.D., Ali, B.: Roll motion prediction using a hybrid deep learning and ARIMA model. Procedia Comput. Sci. **144**, 251–258 (2018)
7. Yin, J.C., Perakis, A.N., Wang, N.: A real-time ship roll motion prediction using wavelet transform and variable RBF network. Ocean Eng. **160**, 10–19 (2018)

8. Huang, B.G., Zou, Z.J., Ding, W.W.: Online prediction of ship roll motion based on a coarse and fine tuning fixed grid wavelet network. Ocean Eng. **160**, 425–437 (2018)
9. Hou, X.R., Zou, Z.J., Liu, C.: Nonparametric identification of nonlinear ship roll motion by using the motion response in irregular waves. Appl. Ocean Res. **73**, 88–99 (2018)
10. Chen, C., Ruiz, M.T., Delefortrie, G., et al.: Parameter estimation for a ship's roll response model in shallow water using an intelligent machine learning method. Ocean Eng. **191**, 106479 (2019)
11. Vaswani, A.: Attention is all you need. In: Advances in Neural Information Processing Systems (2017)
12. Zhang, M., Taimuri, G., Zhang, J., et al.: A deep learning method for the prediction of 6-DoF ship motions in real conditions. Proc. Inst. Mech. Eng. M: J. Eng. Marit. Environ. **237**(4), 887–905 (2023)
13. Huang, Z., Mo, X., Lv, C.: Multi-modal motion prediction with transformer-based neural network for autonomous driving. In: 2022 International Conference on Robotics and Automation (ICRA), pp. 2605–2611. IEEE (2022)

Open Access This chapter is licensed under the terms of the Creative Commons Attribution-NonCommercial-NoDerivatives 4.0 International License (http://creativecommons.org/licenses/by-nc-nd/4.0/), which permits any noncommercial use, sharing, distribution and reproduction in any medium or format, as long as you give appropriate credit to the original author(s) and the source, provide a link to the Creative Commons license and indicate if you modified the licensed material. You do not have permission under this license to share adapted material derived from this chapter or parts of it.

The images or other third party material in this chapter are included in the chapter's Creative Commons license, unless indicated otherwise in a credit line to the material. If material is not included in the chapter's Creative Commons license and your intended use is not permitted by statutory regulation or exceeds the permitted use, you will need to obtain permission directly from the copyright holder.

Analysis of Hydro-Meteorological Characteristics of the Yellow Sea and Bohai Sea Based on ERA-5

Yuepeng Yang, Hanjia Wang^(✉), and Peng Han

PLA Dalian Naval Academy, Dalian, China
hanaawang@foxmail.com

Abstract. Due to the special geographical location and climatic conditions of the Yellow Sea and Bohai Sea, the hydro-meteorology is very important to the safety of shipping routes. This statistical analysis was conducted on the hydrological and meteorological characteristics of the Yellow Sea and Bohai Sea from 1993 to 2023 based on ERA-5 (the fifth-generation European Centre for Medium-Range Weather Forecasts Re-Analysis). The main conclusions are as follows: (1) In winter, northerly strong winds prevail over the Yellow Sea and Bohai Sea, with the average of 5.23 m/s. Ice formation on the sea surface is the main factor affecting navigation in winter. (2) Affected by the monsoon, the waves are larger in autumn and winter, with seasonal average wave heights of 0.80 m and 0.94 m. The waves are weaker in spring and summer, with seasonal average wave heights of 0.69 m and 0.56 m. The duration of waves in the south or southeast direction is longer in summer. (3) The temperature distribution in the past 30 years has fully demonstrated the comprehensive effect of monsoon and terrain. The highest monthly temperature occurs in August with an average of 24.88 °C. The lowest monthly average occurs in January, and the average is −2.23 °C. The temperature change at sea is one month later than on shore.

Keywords: Yellow Sea and Bohai Sea · Temperature · Wind · Sea Wave · Seasonal Characteristics

1 Introduction

The Bohai Sea is an almost enclosed inland sea. The seabed is flat and mainly composed of silt and soft mud (Liu X et al., 2022). Due to the continuous erosion of water flow, the regional rivers are characterized by steepness and depth, and are often accompanied by strong winds and waves on both sides of the river. The North Yellow Sea is a semi enclosed sea area located at the junction of Shandong and the Korean Peninsula (east of Dalian). The different depths of the Yellow Sea and Bohai Sea, as well as the complex seabed topography, make wave prediction very difficult. Strong winds in the Yellow Sea and Bohai Sea regions are mainly of 8–9 and 6–7 levels, accounting for 63% of the total number of strong winds. Among them, the frequency of occurrence is highest in December and lowest in October(Darlington T M et al., 2018). Strong winds of 7–8 levels occur frequently during the late autumn and early winter months from October to November (Weiwei C et al., 2018). The area north of the Yellow Sea and Bohai Sea is the only frozen water body in Chinese marginal waters. Due to the fact that sea ice is an important member of the climate system and has a modulating effect on global climate, it is also significantly affected by climate change (Senftleben D et al., 2017). Therefore, the impact on the weather and climate characteristics of the Yellow Sea and Bohai Sea cannot be ignored.

The Sixth Assessment Report of the Intergovernmental Panel on Climate Change (IPCC) and related studies such as the China Meteorological Administration's Blue Book on Climate Change (2023) show that extreme heat events occur frequently in China, sea levels rise, extreme precipitation events increase, and the climate risk index shows an upward trend. The ecological impact caused by climate change is significant. With the increasingly severe extreme climate change in recent years, the distribution characteristics of hydrological and meteorological elements in various sea areas have also changed. Therefore, it is very important to reanalyze the distribution characteristics of hydrological and meteorological elements in various sea areas in recent years. This article uses the latest generation of atmospheric reanalysis data ERA-5 dataset released by the European Centre for Medium Range Weather Forecasts (ECMWF) to analyze and summarize the characteristics of hydro meteorological elements in Yellow Sea and Bohai Sea from 1993 to 2023.

2 Data and Methods

The ERA-5 (the fifth generation European Centre for Medium Range Weather Forecasts Re Analysis) reanalysis dataset is published by the European Centre for Medium Range Weather Forecasts (ECMWF), providing hourly and monthly data on various atmospheric, land surface, and oceanic state parameters, as well as estimates of uncertainty. This dataset integrates model data with global observation data. ERA-5 data can be accessed at the Climate Data Center and presented on a conventional latitude longitude grid at resolutions of 0.5 ° × 0.5 ° and 0.25 ° × 0.25 ° from 1940 to the present. In this study, monthly gridded ocean surface data and related meteorological field data from 1993 to 2023 were analyzed.

This article selects the monthly average data of near surface wind direction, wind speed, temperature, sea surface temperature, wave height, and wave direction in the Yellow and Bohai Sea area (117.5 ° -125.5 ° E, 36 ° -41 ° N) from 1993 to 2023 for statistical analysis, with a horizontal spatial resolution of 0.25 ° × 0.25 ° .

3 Analysis of Monsoon Characteristics in the Yellow and Bohai Seas

The Yellow and Bohai Sea areas are influenced by both the continental air masses in northern East Asia and the oceanic air masses in the Pacific, and have unique climatic characteristics. Controlled by the eastward migration of oceanic air masses, the winter in Bohai Sea is characterized by a continental climate, characterized by drought, low rainfall, strong and frequent winds, mainly northwest winds. In summer and autumn, under the influence of oceanic air masses in the western Pacific, southeast winds dominate, forming a southeast monsoon climate characterized by high temperature, high humidity, heavy precipitation, and gentle wind speed. Spring is the period of monsoon transition, and the monsoon in the Yellow Sea and Bohai Sea region changes from winter monsoon to summer monsoon. During this period, south and north winds alternate and prevail, with variable wind directions and suitable wind speeds.

The spring monsoon season starts in March and lasts for 2–3 months, depending on the climate conditions of the year. The seabed topography of the Yellow Sea gradually slopes towards the center and southeast from east, north, and west, with an average inclination angle of 121 °. This terrain characteristic may cause varying degrees of resistance or guidance during monsoon movement, thereby affecting its wind direction and speed. As shown in Fig. 1, this study selected U and V component data of 10 m winds in the Yellow Sea and Bohai Sea for 31 years to plot monthly average wind direction and speed. From the graph, it can be seen that the average wind speed in the Yellow Sea and Bohai Sea during winter is 2.41 m/s, with the highest average wind speed in December, reaching 2.76 m/s.

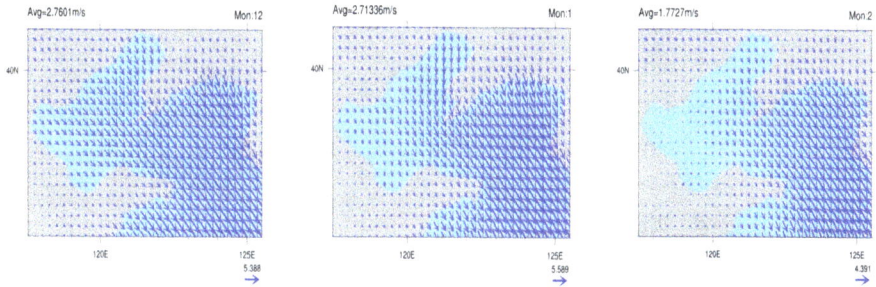

Fig. 1. Distribution of average wind in Yellow Sea and Bohai Sea during winter from 1993 to 2023.

The characteristics of autumn wind field in the Yellow Sea and Bohai Sea are basically consistent with those of winter wind field (figure omitted), with prevailing northerly winds but lower average wind speeds than in winter. The seasonal average wind speed in the Yellow Sea and Bohai Sea from September to November is 1.51 m/s, with the highest average wind speed in November at 1.82 m/s. In October, it decreased by 17.21% to 1.25 m/s. Affected by the turning of prevailing winds, the average wind speed in September is the lowest in autumn, with an average wind speed of 0.72 m/s and a northerly direction. The Yellow and Bohai Sea regions have winding coastlines and numerous islands, such as the Liaodong Peninsula and the Shandong Peninsula. These geographical features have a certain blocking or guiding effect on the monsoon, causing changes in direction or intensity during its movement.

The monthly and seasonal average wind direction and speed data of the sea surface wind field in the Yellow Sea and Bohai Sea from 1993 to 2023 were statistically analyzed and shown in Table 1. The wind field characteristics in spring are different from those in autumn and winter, mainly influenced by southerly winds. The overall wind speed is lower than that in autumn and winter, with a seasonal average wind speed of 1.33 m/s. In April and May, due to the influence of prevailing winds, the Yellow Sea and Bohai Sea area are southerly. The average wind speed in May is 1.64 m/s, and the average wind direction is southwesterly (as shown in Fig. 2). The characteristics of the wind field in each month of summer are shown in Fig. 2. The overall sea area is controlled by southerly winds, with an average wind speed of 1.47 m/s.

Table 1. Climate average wind in the Yellow Sea and Bohai Sea from 1993 to 2023

Month	Wind speed	Wind direction	Seasonal average
12	2.76 m/s	NW	
1	2.71 m/s	NW	2.41 m/s N
2	1.77 m/s	NW	
3	1.17 m/s	W	
4	1.19 m/s	SW	1.33 m/s S
5	1.64 m/s	SW	
6	1.87 m/s	S	
7	1.92 m/s	S	1.47 m/s S
8	0.62 m/s	S	
9	0.72 m/s	N	
10	1.25 m/s	NW	1.26 m/s N
11	1.82 m/s	NW	

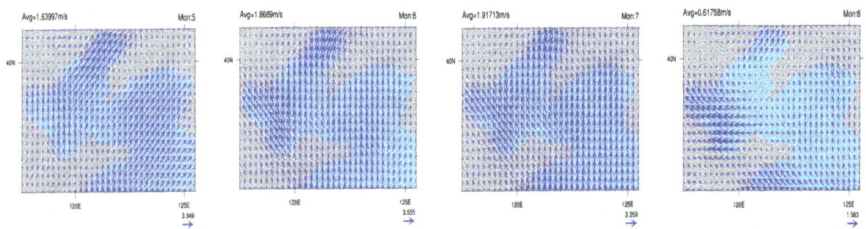

Fig. 2. Distribution of average wind direction and speed in the Yellow Sea and Bohai Sea from May to August from 1993 to 2023.

The winter monsoon has a long prevailing period, a high frequency of occurrence, and a wide range of impacts. The prevailing period of the southwest monsoon in summer is relatively short, and the wind direction in summer is influenced by alternating weather systems such as typhoons, East China Sea cyclones, and anticyclones, resulting in more complex changes in wind direction and a lower frequency of northerly winds.

As shown in Fig. 3, the annual and seasonal average wind speeds in the Yellow Sea and Bohai Sea from 1993 to 2023 vary year by year. The average wind speed in the

Yellow Sea and Bohai Sea from 1993 to 2023 was 1.95 m/s, with 2020 being the year with the highest annual average wind speed in the past 31 years, reaching 2.19 m/s, which is 10.96% higher than the average wind speed for many years. 1994 was the year with the lowest annual average wind speed, with an average wind speed of 1.74 m/s, which was 20.55% lower than the annual average wind speed. In the winter of 2001, the wind speed was significantly low, with an average wind speed of 2.05 m/s, which was 19.61% lower than the winter average wind speed (2.55 m/s). The wind speed in the autumn of 2017 and 2021 was abnormally high, at 1.84 m/s and 1.88 m/s respectively, which were 21.85% and 24.50% higher than the average wind speed (the average wind speed in autumn was 1.51 m/s). The wind speed in spring is generally low, with an average annual wind speed of 1.49 m/s.

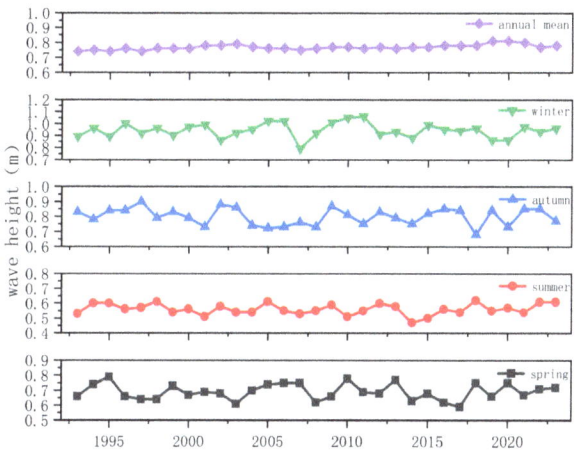

Fig. 3. Shows the annual and seasonal average wind speed changes in the Yellow and Bohai Sea from 1993 to 2023.

The wind field characteristics are greatly influenced by extra-tropical cyclones. The main characteristics of cyclones affecting the Yellow Sea and Bohai Sea are as follows: 70% of cyclones strengthen after entering the sea, 15% become explosive cyclones, 18% remain unchanged after entering the sea, and 12% show a decreasing trend in intensity after entering the sea. In terms of the degree of cyclone impact in the Yellow Sea and Bohai Sea areas, autumn is the most concentrated, followed by spring and winter, and summer rarely occurs(Zhu Nannan et al., 2021).

4 Analysis of Wave and Temperature Characteristics in the Yellow and Bohai Seas

The waves in the Yellow and Bohai Sea regions are mainly influenced by monsoons, with wind waves as the main source and swell waves as a supplement. The average period is 2.3–7.0 s, and the wave height shows a distribution pattern of high in autumn and winter and low in spring and summer. The wave direction is greatly affected by the monsoon, and in winter, the waves are mainly northward, and the swells are also northward. In spring, due to frequent changes in wind direction, both southerly and northerly winds and waves appear, with slightly more southerly winds and waves, but swells are still mainly in North direction. In summer, wind waves and swells are mainly in the south direction, while in autumn, wind waves are mainly in the west northwest direction, and swells are mainly in the north northwest direction. The average monthly wave height and direction in the Yellow Sea and Bohai Sea from 1993 to 2023 are shown in Fig. 4. Affected by the prevailing northwest wind in winter, cold air frequently affects, with an average wave height of 0.80 m and an average wave direction of 216.43 °. Among them, the maximum wave height is consistent with the maximum wind speed, which also occurs in December, with an average wave height of 1.04 m and a wave direction of 285.29 °. In June, the wave direction is variable, with the average wave height being the lowest of the year, only 0.53 m, and the average wave direction being 168.36 °. In winter, the Yellow Sea and Bohai Sea regions are mainly affected by the northwest monsoon, with waves mostly heading north. The period of wind and waves in the Yellow Sea is between 2.0–7.0 s, with the maximum period generally occurring outside the Haizhou Bay on the west side of the South Yellow Sea. In summary, the wave direction in the Yellow Sea and Bohai Sea is influenced by various factors such as season, wind direction, and terrain. The wind zone in Bohai is short, mainly consisting of wind and waves. Due to the influence of the circulation system and terrain, the Yellow Sea exhibits certain regional differences in wave direction and wave height.

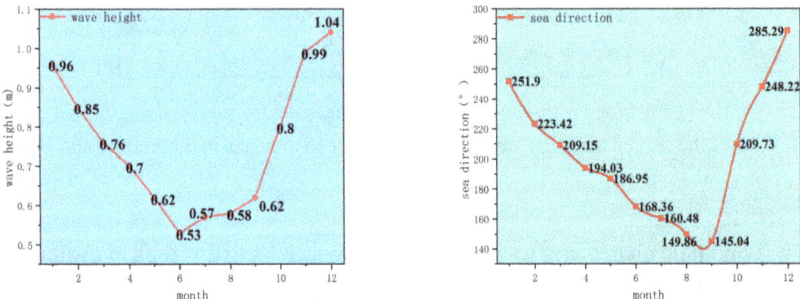

Fig. 4. Monthly average wave in the Yellow Sea and Bohai Sea from 1993 to 2023.

As shown in Fig. 5, the annual and seasonal average wave heights in the Yellow Sea and Bohai Sea from 1993 to 2023 vary year by year. The average wave height in the Yellow Sea and Bohai Sea from 1993 to 2023 was 0.80 m, with 2019 being the year with the highest annual average wave height in the past 31 years, reaching 0.81 m, which is 1.25% higher than the annual average wave height. The year 1995 was the smallest year in terms of average annual wave height, with an average annual wave height of 0.74 m, which was 7.5% below the multi-year average wave height. In the winter of 2007, the wave height was relatively low, with a seasonal average wave height of only 0.79 m, which was 15.96% lower than the winter average wave height (0.94 m). The wave direction in the Yellow and Bohai Sea is greatly influenced by seasons and weather systems. Based on the analysis of the graph, it can be concluded that the wave direction in the Yellow and Bohai Sea is counterclockwise in spring and summer, and clockwise in autumn and winter.

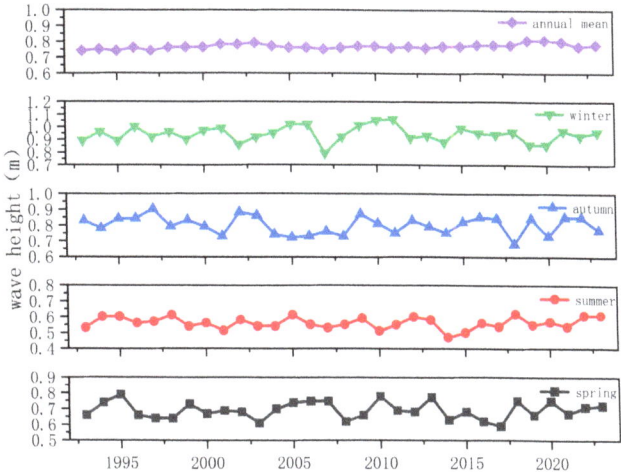

Fig. 5. Annual changes in seasonal and annual average wave height in the yellow and Bohai Seas from 1993 to 2023.

The Yellow Sea and Bohai Sea are significantly affected by monsoon climate, with northerly winds dominating in winter, with northwest winds being predominant. The wind is strong and the direction is stable. In summer, south winds are predominant, with southeast winds being the most common. The wind is weak and the direction is variable. From winter monsoon to summer monsoon, and from summer monsoon to winter monsoon, both go through an intermediate transition stage, with the former having a slightly longer transition period and the latter having a shorter transition period. The summer monsoon along the southeast coast of China begins to appear in April, but there are still cold air irregularly invading the south from April to May, so April is defined as the monsoon transition period. The summer monsoon mainly prevails from May to August, with the July August summer monsoon being the most prevalent. In September, the Siberian high gradually dominates, the summer monsoon gradually weakens, and the transition to the winter monsoon is faster.

Normally, from late November to early December, the coastline of the northern part of the Bohai Sea and Yellow Sea will gradually begin to freeze from north to south, and then the ice on the sea surface will gradually melt until the end of February or mid to early March of the following year. The entire freezing period lasts about 2 to 4 months. In general, the ice conditions are more severe from January to mid-February, known as the peak ice period. In years of extremely cold climate, according to historical records, the northern coastal areas of China have suffered four extremely severe sea ice events. The distribution of temperature fully reflects the climate characteristics of cold winters and warm summers in the Yellow Sea and Bohai Sea, with spring and autumn as transitional seasons. According to the monthly average temperature distribution map of the Yellow Sea and Bohai Sea from 1993 to 2023 (as shown in Fig. 6), the highest monthly average occurs in August, with an average temperature of 24.88 °C. The lowest value occurs in January, with an average temperature of −2.23 °C. In addition, through the analysis of temperature data over the past 30 years, the article found that the temperature distribution in the Yellow Sea and Bohai Sea fully demonstrates the comprehensive influence of monsoon and terrain on climate.

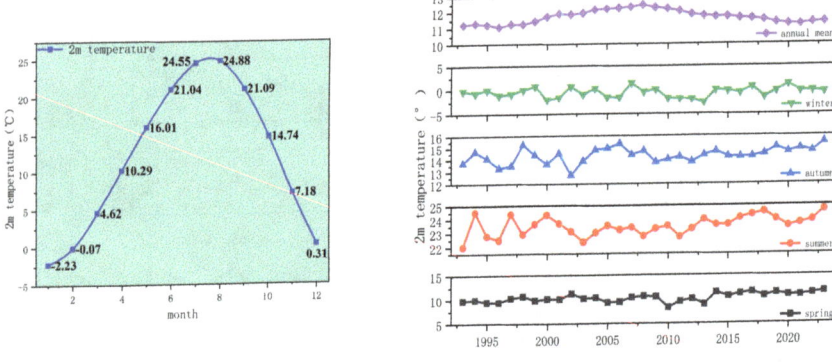

Fig. 6. Shows the annual variation of monthly average temperature (left) and seasonal average temperature in the Yellow Sea and Bohai Sea from 1993 to 2023 (right).

The annual average highest temperature appeared in 2008, reaching 12.40 °C, while the annual average lowest temperature appeared in 1996, at 11.08 °C. The change in monthly average temperature further confirms the influence of monsoons. August, as the month with the highest monthly average temperature, reached 24.88 °C, while January was the month with the lowest monthly average temperature, dropping to −2.23 °C, accompanied by the appearance of sea ice phenomenon, which fully reflects the feedback effect of the ocean on the atmosphere.

5 Summary

This study is based on ERA-5 reanalysis data and systematically analyzes the characteristics of hydro meteorological elements in the Yellow and Bohai Sea from 1993 to 2023. Northerly strong winds prevail over the Yellow Sea and Bohai Sea, with the average of 5.23 m/s. The severe phenomenon of sea ice formation has become a key factor affecting winter navigation safety, and further emphasizes the need to pay special attention to sea ice risk during winter navigation in this sea area. In summer, southeast winds dominate, with relatively low wind speeds, and the average wind speed drops to 2.45 m/s. Frontal cyclones are an important weather system that affects navigation during this season. Due to the significant influence of monsoon, the waves in the Yellow Sea and Bohai Sea exhibit distinct seasonal changes. The waves in autumn and winter are relatively turbulent, with seasonal average wave heights reaching 0.80 m and 0.94 m, posing a certain threat to navigation; In spring and summer, the waves are relatively gentle, with seasonal average wave heights of 0.69 m and 0.56 m. Moreover, the duration of waves in the south or southeast direction is longer in summer, providing favorable conditions for navigation. The temperature distribution in the past 30 years has fully demonstrated the comprehensive effect of monsoon and terrain. The highest monthly temperature occurs in August with an average of 24.88 °C. The lowest monthly average occurs in January, and the average is -2.23 °C. The temperature change at sea is one month later than on shore.

In terms of navigation safety, considering the significant tidal changes in the Bohai Bay, it is recommended that ships closely monitor tidal forecasts and arrange their navigation time reasonably; In winter, special precautions should be taken against sea ice and measures should be taken to avoid it.

In this study, only the ERA-5 dataset was used. The next step is to use satellite data and model simulation data to compare. In the future, efforts will be made to summarize the rules of extreme weather at sea to ensure the safety of navigation.

References

Liu, X., Chen, Y., Liu, C., et al.: Detailed seafloor geomorphology of the western region of the North Yellow Sea, China: The result of Holocene erosional and depositional processes sculpting the offshore continental shelf. Acta Oceanol. Sin. **41**(12), 38–47 (2022)

Nannan, Z., Qiufen, X., Tiantian, H., et al.: Statistic characteristics and strengthening analysis of cyclones over the Yellow Sea and the Bohai Sea in recent 10 years. Haiyang Xuebao **43**(10), 50–60 (2021)

Darlington, T.M., Brandon, C., Juliet, G., et al.: Localised human thermal discomfort assessment using high temporal resolution meteorological data: A case of University of Zimbabwe. Phys. Chem. Earth. **110**, 138–148 (2018)

Weiwei, C., Shichun, Z., Quansong, T., et al.: Regional characteristics and causes of haze events in Northeast China. Chin. Geogr. Sci. **28**(5), 836–850 (2018)

Senftleben, D., Eyring, V., Lauer, A., et al.: Temperature and sea ice hindcast skill of the MiKlip decadal prediction system in the Arctic. Meteorol. Z. **27**(3), 195–208 (2017)

Shannon, E., Eric, B.: Everglades restoration science and decision-making in the face of climate change: A management perspective. Environ. Manag. **55**(4), 876–883 (2015)

Liu, J., Shi, X., Ge, S., et al.: Identification of the thick-layer greigite in sediments of the South Yellow Sea and its geological significances. Chin. Sci. Bull. **59**(22), 2765775 (2014)

Open Access This chapter is licensed under the terms of the Creative Commons Attribution-NonCommercial-NoDerivatives 4.0 International License (http://creativecommons.org/licenses/by-nc-nd/4.0/), which permits any noncommercial use, sharing, distribution and reproduction in any medium or format, as long as you give appropriate credit to the original author(s) and the source, provide a link to the Creative Commons license and indicate if you modified the licensed material. You do not have permission under this license to share adapted material derived from this chapter or parts of it.

The images or other third party material in this chapter are included in the chapter's Creative Commons license, unless indicated otherwise in a credit line to the material. If material is not included in the chapter's Creative Commons license and your intended use is not permitted by statutory regulation or exceeds the permitted use, you will need to obtain permission directly from the copyright holder.

Calculation of the Existence Time of Bubbles in the Wave

Zhijiang Yuan, Xiaogang Jiang(✉), and Mingdong Lv

Navigation Department of Dalian Naval Academy, Dalian, China
forgething@163.com

Abstract. In order to calculate the existence time of bubbles in the wave, differential equations of heat and mass transfer, velocity and radius of bubbles are established respectively, and a coupled model of bubbles floating in high temperature in wave is constructed by combining two-dimensional deep-water wave model. The Runge-Kutta method with variable step size is used to solve the numerical problem. Three laws of the bubble existence time in different initial states are obtained by simulation. First, the bubble is pulled by waves and its upward trajectory is spiral. The smaller the bubble radius has the longer of bubble existence time. Second, the higher of the initial temperature, the longer of bubble existence time is affected by the radius change. Thirdly, in still water and wave, the difference of bubble radius and average velocity is small, so that the existence time is basically the same. The model and conclusion take into account the influence of wave on the existence time of different bubbles in real sea conditions, and can be used as a theoretical basis for the research on the bubbles floating existence phenomenon.

Keywords: Ship · Wave · Coupled Model · Existence Time

1 Introduction

The sea water density artificial jump technology, which was first proposed by the author, has the characteristics of selective gas generation, controllable gas generation rate and bubble formation radius. The generated sea water density artificial jump zone can change the sea water density and its acoustic characteristics, directly restrict the normal implementation of detection, identification, tracking and guidance of submarines and torpedo sonars, and has become an important military countermeasure technology research direction. The realization of "sea water density jump zone" is to load chemicals through special equipment and make them to react with sea water at the right time to produce gas quickly and form a large number of high temperature bubbles [1]. During the process of high temperature tail gas rising, non-isothermal heat transfer and non-equilibrium mass transfer reactions will occur rapidly with low temperature seawater liquids [2–7]. In the marine environment, wave is one of the main factor affecting the motion of bubbles. Therefore, considering the motion of bubbles in the wave, the process of bubbles floating in the marine environment can be simulated more accurately. In recent years, many scholars have studied the heat and mass transfer between bubbles and liquids, but the calculation of bubble mass transfer model neglects the effects of bubble density, approximate treatment of heat transfer model for isothermal conditions, and the fluctuation term of additional mass force in the process of bubbles buoyancy [8–10]. Because the process of bubbles floating is complex. Most studies fail to consider the factors affecting heat transfer, mass transfer and motion comprehensively [11, 12].In the existing studies, the bubbles floating motion is mainly studied in stationary fluid, and the effect of wave on bubbles motion is not fully considered.

Therefore, based on the characteristics of floating bubbles in still water, the motion law of bubbles in wave is studied in this paper. According to the mass transfer and heat transfer of bubbles and the forces acting on them, the non-isothermal and non-equilibrium mass transfer models of bubbles, the expressions of the transient acceleration of bubbles and the differential equations of bubble radius are constructed respectively. Based on the four sub-models, the two-dimensional deep-water wave model is used to describe the wave, and a coupled model is constructed to describe the whole process of high-temperature bubbles in waves. Finally, the variable step Runge-Kutta method is used to solve the model and simulate numerically. The motion characteristics of bubbles with different initial radius and initial temperature in waves and still water are compared and analyzed, and the effects of different wave elements on the motion process of bubbles are obtained.

2 Basic Model of Bubble Floating in Wave

In order to study the factors affecting the motion of bubbles in the wave, it is necessary to establish the non-isothermal heat transfer differential equation, the transient non-equilibrium mass transfer differential equation, the velocity differential equation and the radius differential equation of bubbles in the flow field. It is assumed that a stable spherical shape is maintained during the motion of bubbles, and the interaction between bubbles is neglected.

2.1 Differential Equation for Heat and Mass Transfer of Bubbles

From the physical nature of mass transfer and the derivation process [13], the instantaneous mass transfer model of bubbles in waves is similar to that of bubbles in still water. It is only need to replace the velocity v_b of bubbles with the relative velocity $|\vec{v_b} - \vec{v_L}|$ of bubbles to the fluid. $\vec{v_L}$ is the wave velocity, following expressions:

$$\frac{dm}{d\tau} = 3.1(D^2 R^4 |\vec{v_b} - \vec{v_L}|)^{1/3}(c_a - c_i) \tag{1}$$

m is the mass of bubbles. R is the radius of bubbles. D is the diffusion rate of gas molecules in liquids. c_i is the mass concentration of gas at the gas-liquid interface and c_a is the mass concentration of gas in liquids. The calculation is based on non-equilibrium mass transfer theory and satisfies $HP_g c_i = \exp[2(c_a - c_i)/(c_a + c_i)]$. $P_g = P + 2\sigma/R$ is the internal pressure of bubbles. P is the generalized pressure in wave. σ is the surface tension coefficient of liquid, and H is the solubility coefficient.

The average heat flux on the surface of bubbles at high temperature in still water is [13]:

$$q = 0.25\lambda \left(\frac{v_b}{\alpha R^2}\right)^{1/3}(t_l - t_g) \qquad (2)$$

α is the temperature conductivity coefficient. λ is the thermal conductivity coefficient. t_g is the temperature inside the bubble, and tl is the temperature of liquid. The average heat flux of bubbles in wave can be replaced by v_b in Eq. (2) by $|\vec{v_b} - \vec{v_L}|$, and the temperature t_g of bubbles in wave can be obtained as follows:

$$t_g = t_l - 4\left(\frac{|\vec{v_b} - \vec{v_L}|}{\alpha \lambda^3 R^2}\right)^{-1/3} q \qquad (3)$$

Because of $|\vec{v_b} - \vec{v_L}| \frac{d|\vec{v_b} - \vec{v_L}|}{d\tau} = (\vec{v_b} - \vec{v_L}) \cdot \frac{d(\vec{v_b} - \vec{v_L})}{d\tau}$, the differential of bubble temperature t_g to time is as follows

$$\frac{1}{t_g - t_l}\frac{dt_g}{d\tau} = \frac{1}{q}\frac{dq}{d\tau} + \frac{2}{3R}\frac{dR}{d\tau} - \frac{(\vec{v_b} - \vec{v_L})}{3|\vec{v_b} - \vec{v_L}|^2} \cdot \frac{d(\vec{v_b} - \vec{v_L})}{d\tau} \qquad (4)$$

Because of the energy relationship, it is that

$$c_g(m - \delta m)\delta t_g + c_g \delta m(t_g - t_l) = Sq\delta\tau \qquad (5)$$

c_g is the specific heat capacity of gas in bubbles, neglecting the second-order small quantity $\delta m \delta t_g$, and the differential treatment is taken to obtain:

$$\frac{dt_g}{d\tau} = \frac{4\pi R^2 q}{c_g m} - \frac{t_g - t_l}{m}\frac{dm}{d\tau} \qquad (6)$$

In combination with (4) and (5):

$$\frac{dq}{d\tau} = q\left[\frac{4\pi R^2 q}{c_g m(t_g - t_l)} - \frac{2}{3R}\frac{dR}{d\tau} + \frac{(\vec{v_b} - \vec{v_L})}{3|\vec{v_b} - \vec{v_L}|^2} \cdot \frac{d(\vec{v_b} - \vec{v_L}v_L)}{d\tau} - \frac{1}{m}\frac{dm}{d\tau}\right] \qquad (7)$$

Equations (1, 6) and (7) are the differential equations for heat and mass transfer of high-temperature bubbles.

2.2 Differential Equation of Bubble Velocity

Bubbles in waves are mainly affected by gravity, viscous drag, additional mass force, Basset force, pressure gradient force (buoyancy in still water), Magus lift force and Saffman lift force. Because the density difference between gas and liquid phases is three orders of magnitude, gravity is far less than the pressure gradient force on bubbles, and Magus lift and Saffman lift are generally in the same order of magnitude as the gravity of bubbles, gravity, Magus lift and Saffman lift can be neglected in examining the motion of bubbles in waves. The specific force expression is as follows:

Viscous resistance [14]:

$$\vec{F_D} = \frac{\pi}{2}C_d\rho_l R^2 |\vec{v_b} - \vec{v_L}| (\vec{v_b} - \vec{v_L}); C_d = \frac{12(1+0.15Re^{0.687})}{Re} Re = \frac{\rho_l R |\vec{v_b} - \vec{v_L}|}{\mu} < 1000\ [12] \tag{8}$$

Additional mass force [15, 16]:

$$\vec{F_M} = 2\pi\rho_l R^2 (\vec{v_b} - \vec{v_L})\frac{dR}{d\tau} + \frac{2}{3}\pi\rho_l R^3 \frac{d(\vec{v_b} - \vec{v_L})}{d\tau} \tag{9}$$

Basset force [17]:

$$\vec{F_B} = 6R^2(\pi\mu\rho_l)^{1/2} \int_0^\tau \frac{1}{\sqrt{\tau-t}} \frac{d(\vec{v_b} - \vec{v_L})}{dt} dt \tag{10}$$

Pressure gradient force:

$$\vec{F_B} = \frac{4}{3}\pi R^3 \nabla P \tag{11}$$

$\rho_g = 3m/(4\pi R^3)$ is bubble density and μ is dynamic viscosity coefficient. Newton's second law equation of bubbles in still water is:

$$\frac{4}{3}\pi R^3 \rho_g \frac{d\vec{v_b}}{d\tau} = -\vec{F_D} - \vec{F_M} - \vec{F_B} - \vec{F_P} \tag{12}$$

The expression of each force is substituted into Eq. (12) to obtain the differential equation of velocity

$$\frac{dv_b}{d\tau} = -\frac{3C_d}{4R}|\vec{v_b} - \vec{v_L}|(\vec{v_b} - \vec{v_L}) - \frac{3(\vec{v_b} - \vec{v_L})}{R}\frac{dR}{d\tau} - \frac{9}{R}\sqrt{\frac{\mu}{\pi\rho_l}}\int_0^\tau \frac{1}{\sqrt{\tau-t}}\frac{d(\vec{v_b} - \vec{v_L})}{dt}dt - \frac{2}{\rho_l}\nabla P + \frac{d\vec{v_L}}{d\tau} \tag{13}$$

The numerical calculation of the generalized integral term $\int_0^\tau \frac{1}{\sqrt{\tau-t}} \frac{d(\vec{v_b} - \vec{v_L})}{dt} dt$ in the above formula is shown in Eq. (14), in which the derivatives of $\vec{v_L}$ are all body-dependent derivatives.

2.3 Differential Equation of Bubble Velocity

The gas in the bubble can be regarded as ideal gas, which satisfies the Clapeyron equation.

$$P_g = N\rho_g t_g \tag{14}$$

N is the ratio of universal gas constant to gas molar mass. For bubbles

$$\frac{dm}{d\tau} = \frac{4}{3}\pi R^2 (R\frac{d\rho_g}{d\tau} + 3\rho_g \frac{dR}{d\tau}) \tag{15}$$

The differential equation of time for ρ_g in Eq. (14) is obtained, and the differential equation of bubble radius is obtained by paralleling Eq. (15):

$$(3\rho_g - \frac{2\sigma}{NRt_g})\frac{dR}{d\tau} = (\frac{3}{4\pi R^2}\frac{dm}{d\tau} + \frac{R\rho_g}{t_g}\frac{dt_g}{d\tau} - \frac{R}{Nt_g}\frac{dP}{d\tau}) \quad (16)$$

In formula (16), $dP/d\tau$ is the dependent derivative of pressure, which is solved as follows

$$\frac{dP}{d\tau} = \frac{\partial P}{\partial \tau} + (\vec{v_b} \cdot \nabla)P \quad (17)$$

2.4 Two-Dimensional Deep Water Wave Model

In order to accurately and completely describe the motion of bubbles in waves, the expressions of generalized pressure and wave velocity in differential equations of bubble motion and radius are need to be clarified. Therefore, a two-dimensional deep-water wave model is used to describe the wave. Let the wave occurs on the oxz plane, the x-axis is horizontal, the z-axis is vertical, and the origin is at the free surface at rest. The velocity potential φ can be expressed as:

$$\varphi = Ae^{kz}sink(x - c\tau) \quad (18)$$

According to the wave theory, the pressure distribution in the wave using the above velocity potential is as follows

$$P(x, z, \tau) = -g\rho_l z + A\rho_l kce^{kz}cosk(x - c\tau) + P_{atm} \quad (19)$$

The direction of wave propagation is on the x-axis.

$$v_{Lx} = Ake^{kz}cosk(x - c\tau), v_{Lx} = Ake^{kz}sink(x - c\tau) \quad (20)$$

The correlations among wave elements are as follows:

$$c^2 = g/k, H = \frac{2Akc}{g}, \lambda = \frac{2\pi}{k} \quad (21)$$

c is the wave velocity. k is the wave number. H is the wave height. λ is the wavelength. Each quantity is the national standard unit. The model mainly analyses the motion of wake bubbles in the main stream of the marine environment, so it can be considered as follows: 1. The fluid is incompressible, ρl is a constant; 2. The mass force is only gravity; 3. The flow of water is neither rotationalnor viscous; 4. The pressure at the surface of water is standard atmospheric pressure Patm. If the parameters of wave length and wave height are given, the other parameters of wave elements are calculated by Eq. (21). The accurate expressions of velocity and pressure distribution can be obtained by introducing Eqs. (19), (20).

3 Construction and Solution of Coupled Model

In order to more accurately characterize the interaction law of various influencing factors in high temperature bubbles floating motion, it is necessary to synthesize the above characteristic differential equations, establish a coupled model of bubbles motion, and solve it numerically according to the initial conditions.

3.1 Coupled Model

The coupled model of bubbles floating in waves is obtained by simultaneous mass transfer differential Eq. (1), bubble heat transfer differential Eq. (6) (7), velocity differential Eq. (13) and radius differential Eq. (16). If the differential equations of heat transfer are taken to zero, the isothermal bubbles upward motion model can be obtained.

The initial conditions of the model are the sea water temperature t_1, initial bubble temperature t_{g0}, release depth h, initial floating distance $z_0 = 0$, initial translation distance $x_0 = 0$, initial radius R_0, initial density $\rho_{g0} = (P_0 + 2\sigma/R_0)/(Nt_{g0})$, initial generalized pressure $P_0 = -g\rho_l h + A\rho_l kce^{kh} + P_{atm}$, initial mass $m_0 = 4\pi \rho_{g0} R_0^3 /3$, initial velocity $v_{bx0} = 0$, $v_{bz0} = 0$, initial heat flux $q_0 = 0.25\lambda a^{-1/3} R_0^{-2/3} (Ake^{kh})^{1/3}(t_1-t_{g0})$.

3.2 Solution Method of Model

Because the integral formula of Basset force contains singular endpoints, the generalized integral term of the velocity differential equation should be dealt with firstly. The bubble motion time τ is divided into discrete time points $\{\tau_0, \tau_1,...,\tau_n\}$, where $\tau_0 = 0$, $\tau_n = \tau$, $\Delta \tau_n = \tau_n - \tau_{n-1}$, the discrete expression of the integral term can be obtained from the complex trapezoid formula and the differential mean value theorem

$$\int_0^\tau \frac{1}{\sqrt{\tau-t}} \frac{d(\vec{v_b} - \vec{v_L})}{dt} dt = \frac{\Delta \tau_1}{2} \frac{\vec{a}(0)}{\sqrt{\tau_n}} + \sum_{i=1}^{n-2} \frac{\Delta \tau_i \vec{a}(\tau_i)}{\sqrt{\tau_n - \tau_i}} + \left(\frac{\Delta \tau_{n-1}}{2\sqrt{\tau_n - \tau_{n-1}}} + \frac{\Delta \tau_n}{2\sqrt{\Delta \tau_{n-1}}}\right) \vec{a}(\tau_{n-1}) + \frac{\sqrt{\Delta \tau_n}}{2} \vec{a}(\tau_n) \qquad \vec{a}(\tau_n) = \frac{d(\vec{v_b} - \vec{v_L})}{dt} \tag{22}$$

Note the $\vec{a}(\tau_n) = \vec{a}(\tau)$ in the above equation, so when we substitute the velocity differential equation, we need to reorganize the equation:

$$\begin{cases} \frac{d\vec{v_b}}{d\tau} = \frac{2(k_3-k_1k_2)}{2+k_1\sqrt{\Delta \tau_n}} \\ k_1 = \frac{9}{R}\sqrt{\frac{\mu \rho_l}{\pi}} \\ k_2 = \frac{\Delta \tau_1}{2} \frac{\vec{a}(0)}{\sqrt{\tau_n}} + \sum_{i=1}^{n-2} \frac{\Delta \tau_i \vec{a}(\tau_i)}{\sqrt{\tau_n-\tau_i}} + \left(\frac{\Delta \tau_{n-1}}{2\sqrt{\tau_n-\tau_{n-1}}} + \frac{\Delta \tau_n}{2\sqrt{\Delta \tau_{n-1}}}\right)\vec{a}(\tau_{n-1}) \\ k_3 = -\frac{3C_d}{4R}|\vec{v_b}-\vec{v_L}|(\vec{v_b}-\vec{v_L}) - \frac{3(\vec{v_b}-\vec{v_L})}{R}\frac{dR}{d\tau} - \frac{2}{\rho_l}\nabla P + \frac{d\vec{v_L}}{d\tau} \end{cases} \tag{23}$$

By using the variable step size Runger-Kutta method, the parameters of bubble radius, velocity, location and instantaneous mass transfer rate at any time t can be solved quickly. Using the bubbles motion model, the existence time of bubbles in liquid is calculated. The bubbles are released from 10 m underwater, and the bubbles are stored in water at different time points by variable step method. When the bubbles rise to 10 m, the existence time of bubbles in water is obtained at the end of the time.

4 Calculation of Bubble Existence Time in Wave

According to the bubbles coupled model in a wave, by setting wavelength, wave height, wave initial radius of bubble, bubble parameters such as initial temperature, bubbles in the static water yards can obtain different initial states (wavelength λ, wave height H are both 0), and bubbles movement of wave, through the bubble floating time step length calculation method, bubble existence time. The variation rules of bubble upwelling process were obtained by further simulation to explore the similarities and differences of bubble existence time. For the convenience of calculation, the analysis is carried out in a two-dimensional flow field. Suppose the flow occurs in the oxz plane, the x-axis is horizontal, the z-axis is vertical, and the origin is at the water surface.

4.1 The Existence Time of Bubbles with Different Initial Radius

Firstly, the existence time of bubbles at different scales in waves are calculated and analyzed. The five bubbles have initial radius R0 = 500, 800, 1000, 1200, 1500 μm, initial gas-liquid temperature difference, initial position (0, - 10) respectively, in still water, and in wave with wavelength λ = 100 m, wave height H = 2 m. Two kinds of bubbles with initial radius R0 = 500 and 1500 μm are selected. The trajectories of bubbles in static water and wave are shown in Fig. 1.

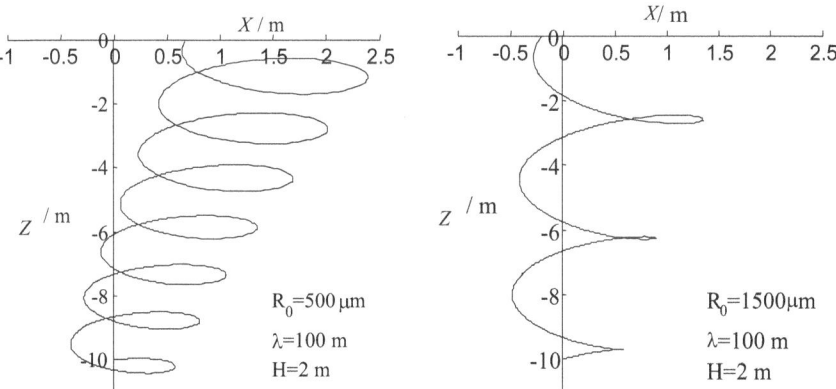

Fig. 1. The trajectories of bubbles at different scales in still water/waves.

From Fig. 1, it can be seen that the bubbles in the wave are spiral, which is different from that in still water. Under the wave-induced action, the horizontal displacement of bubbles in the direction of wave propagation is produced. The smaller the initial radius of bubbles is, the more significant the horizontal motion is relative to the vertical motion, and the more dense the overlap is. This is because, according to formula (18), on the one hand, in the x-axis direction, the viscous drag, additional mass force and Basset force acting on bubbles mainly act as the pulling force of waves. The smaller the radius of bubbles, the larger the pulling force is, that is, the larger the average acceleration in the x-direction, the larger the average velocity of bubbles; on the other hand, the value of bubbles following the wave motion and velocity will gradually tend to be larger. So the average velocity in z direction decreases with increasing. Therefore, for bubbles with different initial radius, the smaller the scale, the more obvious the trajectory bends to the x-axis. In fact, the motion of bubbles in the wave is the superposition of the induced motion of waves and the steady upward motion of bubbles. The smaller the radius of the bubble, the smaller the buoyancy speed of the bubble itself, the more significant the influence of the induced velocity of the wave on it. Therefore, the smaller the bubble radius, the stronger the wave-induced upward motion relative to itself, and the larger the horizontal displacement of the bubble. From this point of view, the macroscopic law of the bubble trajectory in Fig. 1 can also be explained.

On this basis, the existence time of five radius bubbles in still water and wave is calculated, respectively, as shown in Table 1.

Table 1. Existence time (T_{max}) of bubble in still water and waves ($\Delta t = 0\,°C$) unit: second.

Initial bubble radius radius of bubble	500 μm	800 μm	1000 μm	1200 μm	1500 μm
still water	52.5	35.7	30.0	26.1	22.1
waves ($\lambda = 100$ m, $H = 2.0$ m)	54.4	38.0	30.6	29.2	22.6

It can be seen from Table 1, with the increase of bubble radius, the bubble existence time is significantly shortened. According to the differential equation of bubble velocity, Eq. (18) shows that the average vertical acceleration of $\frac{dv_b}{d\tau} \sim -\frac{1}{R}$ increases with the increase of bubble radius, so the bubbles float faster, and the existence time of large bubbles is shortened under the same release conditions.

4.2 The Existence Time of Bubbles at Different Initial Temperatures

The existence time of bubbles in waves at different initial temperatures is calculated and analyzed. The initial gas-liquid temperature difference $-100, -50, 0, 200, 400, 600$, the initial bubble radius is 1000 μm, and the initial position is $(0, -10)$ of five bubbles, respectively, in still water, as well as in wave with wavelength $\lambda = 100$ m and wave height $H = 2$ m. Two kinds of bubbles with initial temperature difference of -100 and 600 are selected and their trajectories in hydrostatic and waves are shown in Fig. 2.

As we can see from Fig. 2, the higher the initial temperature of the bubble, the more significant the horizontal motion is relative to the vertical motion, and the more dense the overlap is. This law is just contrary to the law of motion of bubbles with different radius. On the basis of bubble trajectory, the existence time of five initial temperature bubbles in static water and wave is calculated, respectively, as shown in Table 2.

In order to research the motion characteristics of bubbles with different initial temperatures in waves and the difference of their existence time, a coupled model of bubble motion is used to simulate the radius variation of bubbles in static water and wave with

Calculation of the Existence Time of Bubbles in the Wave   545

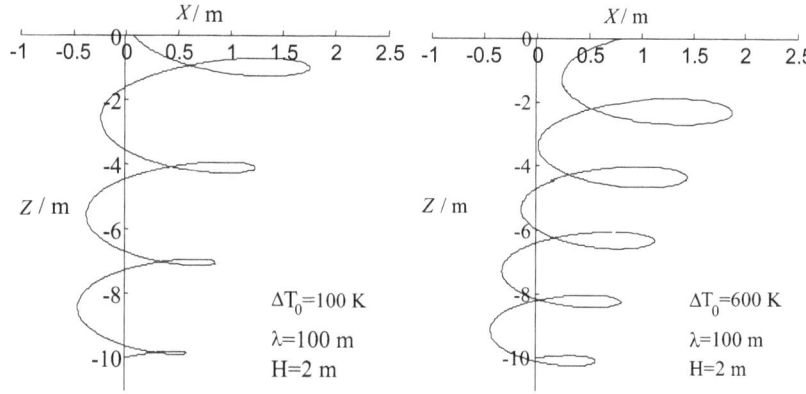

Fig. 2. Movement path of different bubbles in waves.

Table 2. The existence of bubble in still water and waves under different temperature unit: second.

Initial bubble temperature difference (radius of bubble)	−100 °C	0 °C	200 °C	400 °C	600 °C
still water	27.0	30.0	34.4	37.5	40.2
waves ($\lambda = 100$ m, $H = 2.0$ m)	29.5	30.6	37.6	38.5	39.5

initial radius of 1000 μm and different initial temperatures in the process of bubbles floating up in static water and wave. The results are shown in Fig. 3.

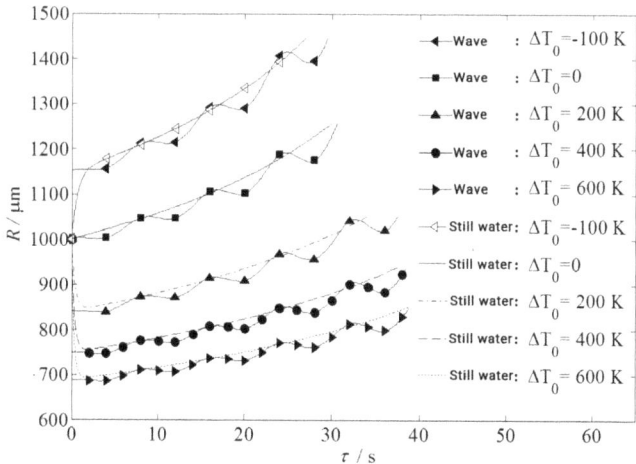

Fig. 3. Movement path of different bubbles in waves.

From Fig. 3, it can be seen that the radius of bubbles decreases rapidly at first and then rises with the increase of bubble temperature gently, whether in still water or in wave. The main reasons are as follows: in the initial stage of bubble motion, because of the short time of the rising stage, the bubble mass and pressure almost remain unchanged, which is roughly in line with $t_i \sim V \sim R^3$. The rapid decrease of temperature makes the bubble volume shrink sharply and radius decrease accordingly. When the bubble temperature drops to liquid temperature, the temperature no longer affects the radius of the bubble. The shrinkage process ends and the radius reaches a relatively stable value. After that, with the bubbles floating up, the volume of bubbles increases smoothly under the main effect of pressure.

From the analysis of Fig. 3 and above, it can be seen that the different trajectories of bubbles at different temperatures in waves are mainly caused by the radius change of bubbles in the process of buoyancy. Therefore, the bubbles with higher initial temperature show the same motion trajectory as those with smaller radius.

4.3 Bubble Existence Time in Different Waves

Based on the analysis of bubble existence time in still water and waves with different initial radius and temperature, the influence of different waves on bubble existence time is analyzed. The initial radius R0 = 1000 μm and the initial position is (0, −10). Four sets of wave elements are set as follows: (I) wavelength λ = 100 m, wave height H = 1.0 m; (II) wavelength λ = 100 m, wave height H = 2.0 m; (III) wavelength λ = 50 m, wave height H = 1.0 m; (IV) wavelength λ = 50 m, wave height H = 2.0 m. The trajectories of bubbles in two particular wave cases (II) and (III) are shown in Fig. 4.

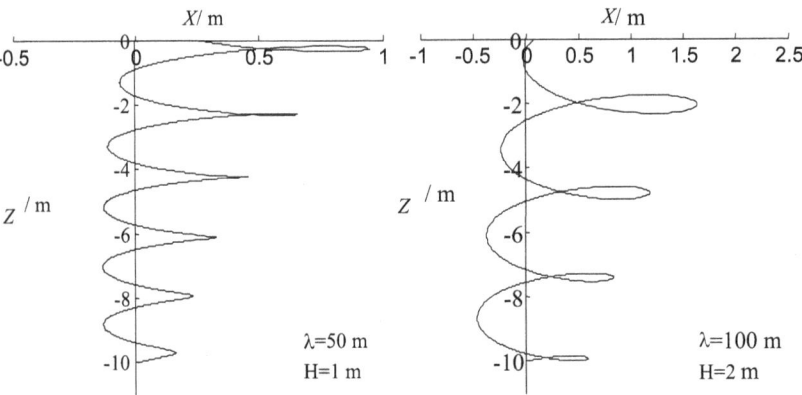

Fig. 4. Movement path of bubble (R_0 = 1000 μm) in different waves.

The existence time of bubbles in still water and four kinds of waves are calculated, respectively, as shown in Table 3.

From Table 3, it can be seen that there is no significant change in the initial wave wavelength λ and wave height H compared with the bubble existence time in still water.

Table 3. Existence time of bubble ($R_0 = 1000$ μm) in still water and waves unit: second.

environment	still water	$\lambda = 100$ m $H = 1.0$ m	$\lambda = 100$ m $H = 2.0$ m	$\lambda = 50$ m $H = 1.0$ m	$\lambda = 50$ m $H = 2.0$ m
existence time T_{max}	30.0	30.5	30.6	31.7	28.4

In order to explore the reasons, this paper studies the variation of bubble radius and velocity under different wave conditions.

1) Variation of Bubble Velocity.

The initial radius R0 = 1000 μm and the initial gas-liquid temperature difference $\Delta t = 0$ are chosen to simulate the horizontal and vertical velocity variation of bubbles in still water and wave. As shown in the Fig. 5.

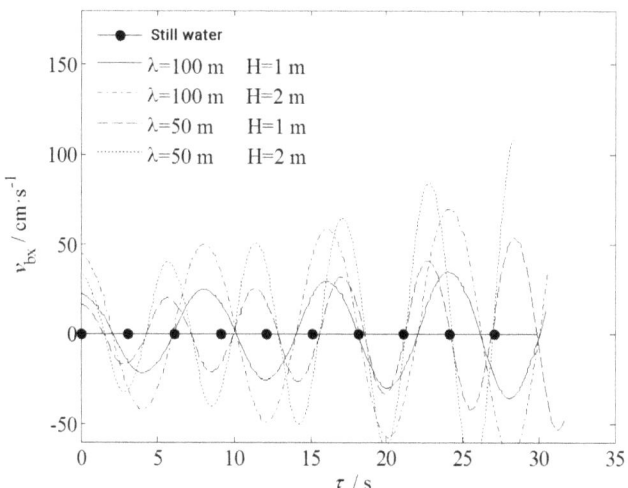

Fig. 5. The horizontal velocity of different bubbles in waves.

From Fig. 5 and 6, it is known that when bubbles float in still water, the horizontal velocity $v_{bx} = 0$ and the vertical velocity v_{bz} rise rapidly to about 28 $cm \cdot s^{-1}$ in the initial stage, and then increase slowly. In different waves, the horizontal/vertical velocity of bubbles fluctuates symmetrically and non-periodically around the horizontal/vertical velocity in still water. When the wave drags the water point downward, the bubbles move downward in half-cycle, which delays the bubbles rising, and then accelerates the bubbles rising in the other half-cycle. The effect of the two half-cycles almost cancels each other out. From the velocity differential Eq. (17), it is known that the bubbles fluctuate up and down at the symmetrical zero points of $f(v_{b0} - v_L)$ and v_L in the wave. On the whole, the sum of v_L velocities is 0. Therefore, the average velocity of bubbles moving in the wave is basically the same as that of free buoyancy in still water. Under the same release depth, the difference of bubble existence time was not obvious.

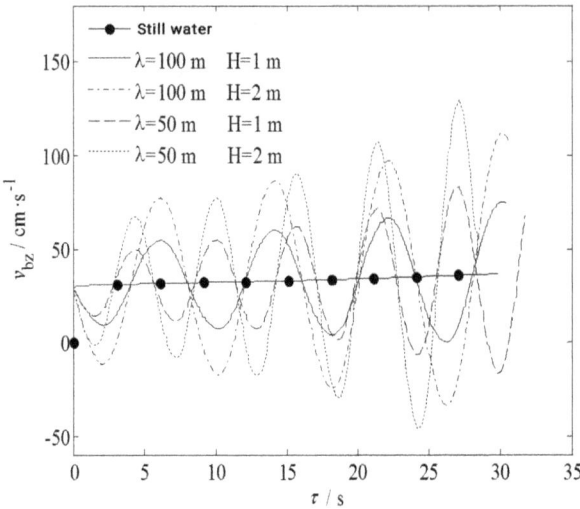

Fig. 6. The vertical velocity of different bubbles in waves.

2) Variation of bubble radius.
In 3.1.1, the radius variation of bubbles with different initial radius in still water and wave is simulated to explore the reasons for the influence of bubbles existence time.

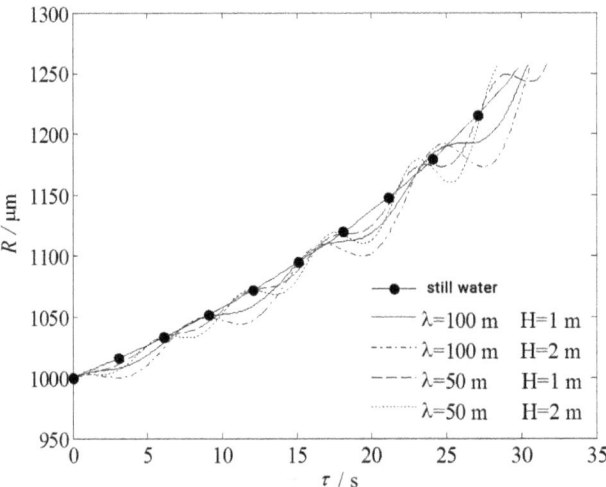

Fig. 7. Radius of different bubbles in waves.

From Fig. 7, it can be seen that the radius of bubbles increases slowly with time in still water and increases irregularly in different waves. On the whole, the radius variation of bubbles is not obvious in different flow fields, basically fluctuates up and down along the curve of radius variation in still water. The bubbles radius determines the buoyancy velocity and acceleration, thus affect the existence time. Therefore, it is shown from the graph that when $R \approx 1250\,\mu m$, or $\tau \approx 30\,s$, bubbles float out of the water surface and disappear.

5 Conclusion

By establishing differential equations of heat and mass transfer, velocity and radius of bubbles in waves, and combining with two-dimensional deep-water wave model, the coupled model of bubbles floating in high temperature in wave is constructed. The main conclusions are as follows:

1) Under the affection of waves, the bubbles floating trajectory is spiral and displacement is generated in the direction of wave propagation. When the initial radius of bubbles is larger, the absolute speed of bubbles floating is accelerated, and the remaining time of bubbles is shortened under the same release conditions.
2) The radius of bubbles with different initial temperatures decreases rapidly at first and then increases slowly. Therefore, the bubbles with higher initial temperature and smaller radius show the same motion trajectory, and the bubble duration is prolonged.
3) The average velocity of bubbles in still water field and wave tends to approximate, and the radius of bubbles have no obvious difference. Therefore, the existence time of bubbles in wave and still water is basically the same.

Reference:s

1. Jin, L., Zhang, Z., Yuan, Z., et al.: Problems in ship wake stealth technology and countermeasures. Shipbuilding China **58**(1), 177–185 (2017)
2. Zhang, B., Li, J.B., Wu, S.M.: Investigation on diesel engine modeling based on matlab and EU VI calibration method. Chinese Internal Combustion Engine Eng. **02**, 101–104 (2016)
3. Naval Command Naval High-tech Knowledge Textbook. Beijing: Naval Command, 121–123 (1997)
4. Hua, J.S, Lou, J.: Numerical simulation of bubble rising in viscous liquid. J. Comput. Phys. **222**(2), 769–795 (2007)
5. Wu, F., Wu, H.J., Cui G.Q.: . Visualization experiment for bubble generation of subcooled flow boiling. J. Eng. Thermophys. **37**(11), 2417–2423 (2016)
6. Yao, C.Y.Y., X L,: Dynamics of underwater explosion bubble near free surface. Shipbuilding of China **03**, 65–71 (2016)
7. Besagni, G., Brazzale, P., Fiocca, A.: Estimation of bubble size distributions and shapes in two-phase bubble column using image analysis and optical probes. Flow Meas. Instrum. **52**, 77–85 (2016)
8. Campos, F.B., Lage, P.L.C.: Simultaneous heat and mass transfer during the ascension of superheated bubbles. Inter. J. Heat Mass Trans. **43**(2), 179–189 (2000)
9. Bonazza, R., Anderson, M., Oakley, J., et al.: The bubble velocity research of rayleigh-taylor and richtmyer-meshkov instabilities at arbitrary atwood numbers. Acta Phys. Sinica **61**(7), 314–320 (2012)
10. Dai, G.C, Chen, M.H.: Chemical Engineering Fluid Mechanics, vol. 256, 276, 374. Chemical Industry Press, Beijing (1988)

11. Zhang, A., Ni, B.: Influences of different forces on the bubble entrainment into a stationary gaussian vortex. Sci. China Phys. Mech. Astronomy **56**(11), 2162–2169 (2013)
12. Carrica, P.M., Bonetto, F.J., Drew, D.A., et al.: The interaction of background ocean air bubbles with a surface ship. Inter. J. Numerical Methods Fluids **28**(4), 571–600 (1998)
13. Tian, H., Liangan, J., Wei, C., et al.: Impact of basset force on the movement of soluble bubble in fluid. Chin. J. Theoretical Appli. Mech. **4**, 680–687 (2011)
14. Yoshida, K., Fujikawa, T., Watanabe, Y.: Experimental investigation on reversal of secondary bjerknes force between two bubbles in ultrasonic standing wave. J. Acoustical Soc. America **130**(1), 135–144 (2011)
15. Tian, H., Jin,L., Ding, Z., et al.: Coupling model for bubble rise and mass transfer process in liquid. J. Chem. Indust. Eng. Soc. China **1**, 15–21 (2010)
16. Li, A., Zhang, L., Zang, Z.: Iterative and adjusting method for computing stream function and velocity potential in limited domains and its convergence analysis. Appli. Math. Mech. **6**, 651–662 (2012)
17. Brenn, G., Kolobaric, V., Durst, F.: Shape oscillations and path transition of bubbles rising in a model bubble column. Chem. Eng. Sci. **61**(12), 3795–3805 (2006)

Open Access This chapter is licensed under the terms of the Creative Commons Attribution-NonCommercial-NoDerivatives 4.0 International License (http://creativecommons.org/licenses/by-nc-nd/4.0/), which permits any noncommercial use, sharing, distribution and reproduction in any medium or format, as long as you give appropriate credit to the original author(s) and the source, provide a link to the Creative Commons license and indicate if you modified the licensed material. You do not have permission under this license to share adapted material derived from this chapter or parts of it.

The images or other third party material in this chapter are included in the chapter's Creative Commons license, unless indicated otherwise in a credit line to the material. If material is not included in the chapter's Creative Commons license and your intended use is not permitted by statutory regulation or exceeds the permitted use, you will need to obtain permission directly from the copyright holder.

Research on Vibration Isolation Performance of Active Vibration Isolation System Base Structure for Marine Power Equipment

Jianlong Hu[1], Lin Lin[2], and Zhaowang Xia[1(✉)]

[1] Jiangsu University of Science and Technology, Zhenjiang 212000, China
dlxzw@163.com
[2] Shanghai Marine Diesel Engine Research Institute, Shanghai 200000, China

Abstract. This study enhances the vibration reduction performance of marine equipment bases using Finite Element Analysis (FEA). It explores how structural parameters, such as elbow plate count and panel thickness, along with vibration mass positioning and damping layer application, affect performance. Key findings within the 10 Hz to 5000 Hz range include a 1.4 dB increase in vibration level difference with 11 elbow plates versus fewer, and a 1.9 dB increase with a 70 mm panel thickness. An 80mm hollow offset vibration mass boosts the difference by 2.5 dB in the 1000–5000 Hz band, while a 100mm mass improves it by 3.0 dB in the 10–1000 Hz band. These insights guide the design of more effective marine equipment base structures for vibration reduction.

Keywords: Base Equipment · Vibration Level Difference · Finite Element Analysis · Optimal design

1 Introduction

During navigation, vibrations generated by marine power equipment are transmitted to the ship's structure through the base, causing vibrations in different parts of the ship [1, 2]. These vibrations not only reduce the comfort of the crew, affecting their work efficiency and physical and mental health, but may also cause serious damage to the precision instruments on board, and even affect the integrity of the ship's structure under the action of high-frequency vibrations for a long time. Especially in high-stress areas of the ship, these vibrations may cause structural damage and pose a threat to the safety of the ship [3, 4]. The base, as a key link in the vibration transmission chain, directly affects the transmission of equipment vibration energy to the ship's structure [5]. Therefore, optimizing the base structure is crucial for improving the overall vibration reduction performance [6, 7]. By optimizing the structural parameters of the base (such as the number of elbow plates, panel thickness), arranging the vibration mass, and applying damping materials on the base surface, the vibration level difference of the base can be effectively improved [8]. This study takes the base of marine power equipment as the object, uses the Finite Element Analysis (FEA) method, and systematically optimizes the vibration reduction performance of the base structure [9]. In the study, the impact of elbow plate quantity, panel thickness, the size and arrangement of the vibration mass, and the thickness and position of the damping layer on the vibration reduction performance of the base were analyzed. Through a series of simulation verifications, the best plan to improve the base vibration level difference was determined [10].

2 Analysis Method of Base Mechanical Impedance and Vibration Transmission Characteristics

In the structural design of marine power equipment bases, understanding and calculating its mechanical impedance and vibration transmission characteristics are crucial. These parameters not only affect the vibration response of the base but also directly relate to the acoustic and vibration characteristics of the entire ship structure.

Mechanical impedance is one of the parameters describing the response of the structure to the exciting force and its transmission vibration characteristics, determining the size of the power received by the base. For a linear vibration system, mechanical impedance is defined as the complex ratio of the exciting force to the steady-state response of the structure it causes.

$$Z = \frac{F}{X} \tag{1}$$

where, Z is the mechanical impedance, F is the exciting force, and X is the structural response it causes.

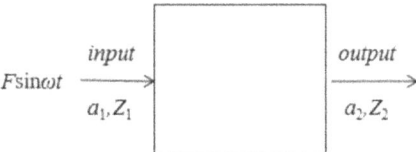

Fig. 1. Concept of Mechanical Impedance.

Figure 1 illustrate the concept of mechanical impedance in a coupling. For a forced vibration system, the input acceleration response at the position of the exciting force is L1L1, and the acceleration response at the output end of the coupling is L_2. Then the vibration level difference of the coupling is:

$$L_a = 20\lg(\frac{a_1}{a_2}) = 20\lg(\frac{a_1/a_{ref}}{a_2/a_{ref}}) = L_1 - L_2 \tag{2}$$

where, L_1 is the input acceleration vibration level at the position of the exciting force, L_2 is the output acceleration vibration level at the response end position, and L_a is the acceleration vibration reference value (taken as 1).

3 Comprehensive Analysis of Base Structure Optimization

This study selected the 10–5000 Hz frequency range for analysis, which covers the most critical vibration frequencies in marine applications. The low-frequency range (10–100 Hz) primarily affects crew comfort and may cause seasickness; the mid-frequency range (100–1000 Hz) influences structural fatigue and resonance in ship components; while the high-frequency range (1000–5000 Hz) impacts sensitive navigation equipment and precision instruments. This comprehensive frequency analysis enables evaluation of vibration suppression strategies across various operational conditions.

The very large and complex ship structure cannot be solved by pure analytical methods. Currently, the calculation of impedance characteristics of the base is often done using numerical analysis methods, but for large-scale structural finite element models, the calculation efficiency is often very low, which cannot meet the requirements for rapid calculation in the early design stage of the base. In actual calculations, due to the complexity of the compartment structure, it is basically unfeasible to model according to the actual ship size, and the base model occupies a small proportion in the compartment structure, so it is not necessary to use the complete compartment for calculation. In order to improve calculation efficiency, it is necessary to simplify the actual model. As shown in Fig. 2, it is the geometric model of the base and the compartment.

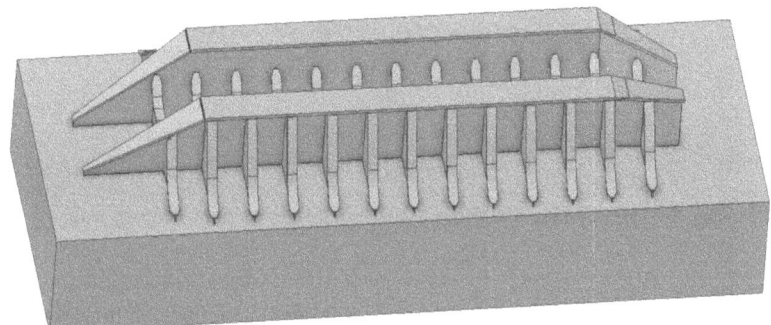

Fig. 2. Geometric model of the base and the compartment.

In the structural design of marine power equipment bases, it is crucial to comprehensively discuss the impact of different structural parameters on vibration reduction performance. This chapter will comprehensively discuss the impact of key parameters such as the number of elbow plates, panel thickness, and vibration mass on the vibration reduction performance of the base, and propose optimization suggestions through finite element simulation and experimental verification.

3.1 Optimization of Elbow Plate Quantity and Panel Thickness

Elbow plates and panels, as important components of the base structure, directly affect the stiffness and damping characteristics of the base, thereby affecting the transmission of vibration. Elbow plates serve as structural reinforcements that connect the base panels and provide additional stiffness to the overall structure. They create a more complex vibration path, which helps dissipate vibration energy through structural interfaces and increases the mechanical impedance of the system.

Impact of Elbow Plate Quantity. Simulation analysis was carried out by setting different numbers of elbow plates (6, 9, and 11). These specific numbers were selected based on standard shipbuilding practices and space limitations within typical engine. As shown in Fig. 3, compared with 6 elbow plates, the configuration of 11 elbow plates increased the base vibration level difference by 1.4 dB in the 10-1000Hz frequency band, effectively improving the vibration reduction performance. A higher vibration level difference indicates superior vibration isolation performance, as it represents a greater reduction in vibration energy transmission from the equipment to the ship structure.

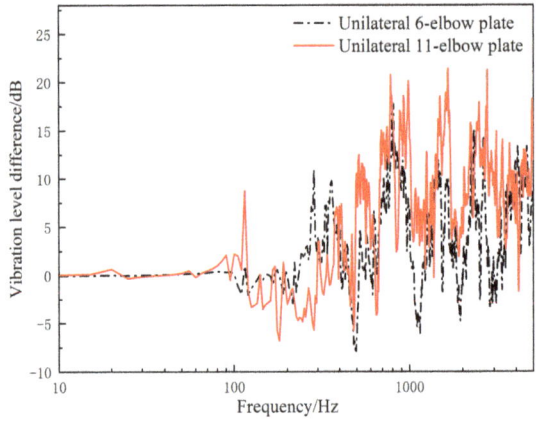

Fig. 3. Vibration level difference curve of 11 elbow plates on one side.

Impact of Panel Thickness. Changes in panel thickness also have an important impact on the vibration characteristics of the base. As shown in Fig. 4, compared with the 60 mm panel (standard thickness), the 70 mm panel increased the base vibration level difference by 1.9 dB in the 1000–5000 Hz frequency band, effectively suppressing high-frequency vibrations that typically affect sensitive electronic equipment.

While the study demonstrates significant improvements in vibration reduction performance, several practical limitations should be considered. Adding more elbow plates and increasing panel thickness adds considerable weight to the structure, which may affect the ship's overall weight distribution and fuel efficiency. More complex structures with additional components increase manufacturing and installation costs. Engine compartments in ships often have limited space, which may restrict the implementation of the optimal number of elbow plates or vibration masses. Based on these findings, future research could explore combined optimization strategies, advanced composite materials, active vibration control systems, and full-scale testing on operational vessels.

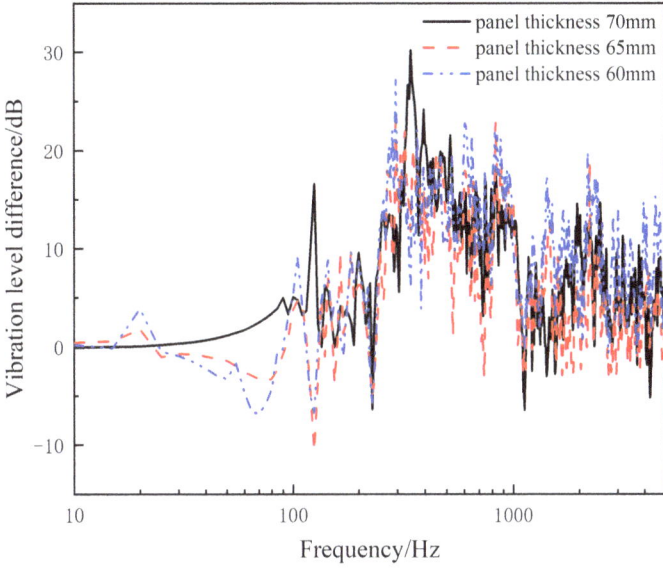

Fig. 4. Vibration level difference curve of different panel thicknesses.

3.2 Optimization of Vibration Mass

Vibration mass is an effective means to reduce vibration transmission. By setting the vibration mass at key positions of the base, the vibration level difference can be significantly improved.

Central Vibration Mass and Offset Vibration Mass. The analysis of central vibration mass and offset vibration mass was carried out. Central vibration mass is placed at the geometric center of the base, while offset vibration mass is placed at a non-central position. As shown in Fig. 5, compared with no vibration mass configuration, the hollow offset vibration mass increased the vibration level difference by 2.5 dB in the 1000–5000 Hz frequency band.

Size and Position of Vibration Mass. The size and position of the vibration mass have a significant impact on the vibration level difference. Simulations were performed on vibration masses of different sizes (80 mm × 80 mm, 100 mm × 100 mm) and positions. As shown in Fig. 6, the vibration mass of 100 mm × 100 mm size increased the vibration level difference by 3.0 dB in the 10-1000 Hz frequency band, compared to the 80 mm × 80 mm vibration mass, which increased the vibration level difference by 0.5 dB in the 10–1000 Hz frequency band compared to no vibration mass.

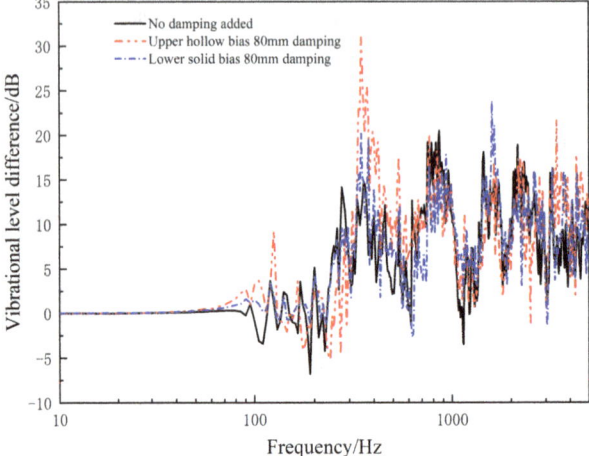

Fig. 5. Vibration level difference curve of 80 offset solid and hollow vibration masses.

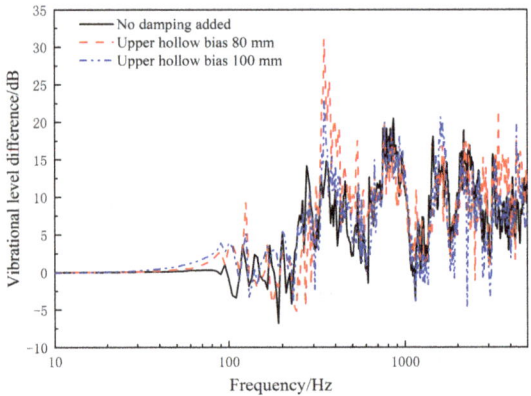

Fig. 6. Vibration level difference curve of 80 and 100 hollow vibration masses on offset.

4 Conclusion

The results of this study indicate that within specific frequency bands, increasing the number of elbow plates, panel thickness, and the rational arrangement of vibration masses can enhance the vibration level difference of the base. The main conclusions are as follows:

1. When the number of elbow plates is increased from 6 to 11, the vibration level difference of the base is increased by 1.4 dB in the 10–1000 Hz frequency band.
2. When the panel thickness is increased from 60 mm to 70 mm, the vibration level difference is increased by 1.9 dB in the 1000–5000 Hz frequency band.
3. Adding an 80 mm*80 mm hollow offset vibration mass compared to no vibration mass increases the vibration level difference by 2.5 dB in the 1000–5000 Hz frequency band.

4. Adding a 100 mm*100 mm size hollow vibration mass compared to the 80 mm*80 mm vibration mass increases the vibration level difference by 0.5 dB in the 10–1000 Hz frequency band.

However, limitations exist regarding weight increases, manufacturing costs, and space constraints. Future research should explore comprehensive optimization strategies, advanced composite materials, active vibration control systems, and validation through full-scale testing on operational vessels. The simulation of the vibration reduction performance of the base of marine equipment was carried out using the Finite Element Analysis method. The impact of the number of elbow plates, panel thickness, and arrangement of vibration mass on the vibration reduction provides practical reference for ship design engineers.

References

1. Du, M., Yang, Y., Yuan, Y., et al.: Vibration analysis and control technology of ship main engine base system. Ship Eng. **46**(03), 14–30 (2024)
2. Chidamparam, P., Leissa, A.W.: Vibrations of planar curved beams, rings, and arches. Appl. Mech. Rev. **46**(9), 467–483 (1993)
3. Charpie, J.P.: An analytic model for the free in-plane vibration of beams of variable curvature and depth. J. Acoust. Soc. Am. **94**(2), 866–879 (1993)
4. Markus, S., Nanasi, T.: Vibration of curved beams. Shock Vibr. Digest **13**(4), 3–14 (1981)
5. Kerwin, E.M.: Damping of flexural waves by a constrained viscoelastic layer. J. Acoust. Soc. Am. **31**(7), 952–962 (1959)
6. Rao, D.K.: Frequency and loss factors of sandwich beams under various boundary conditions. ARCHIVE J. Mech. Eng. Sci. 1959–1982 (vols 1–23) **20**(5), 271–282 (1978)
7. Johnson, C.D., Kienholz, D.A.: Finite element prediction of damping in structures with constrained viscoelastic layers. AIAA J. **20**(9), 1284–1290 (1982)
8. Ravi, S., Kundra, T.K., Nakra, B.C.: A response re-analysis of damped beams using eigenparameter perturbation. J. Sound Vib. **179**(3), 399–412 (1995)
9. Ramasamy, G.: Vibration and damping analysis of fluid filled orthotropic cylindrical shells with constrained viscoelastic damping. Comput. Struct. **70**(3), 363–376 (1999)
10. Araújo, A.L., Madeira, J.F.A., Soares, C.M.M., et al.: Optimal design for active damping in sandwich structures using the direct multisearch method. Compos. Struct.Struct. **105**(8), 29–34 (2013)

Open Access This chapter is licensed under the terms of the Creative Commons Attribution-NonCommercial-NoDerivatives 4.0 International License (http://creativecommons.org/licenses/by-nc-nd/4.0/), which permits any noncommercial use, sharing, distribution and reproduction in any medium or format, as long as you give appropriate credit to the original author(s) and the source, provide a link to the Creative Commons license and indicate if you modified the licensed material. You do not have permission under this license to share adapted material derived from this chapter or parts of it.

The images or other third party material in this chapter are included in the chapter's Creative Commons license, unless indicated otherwise in a credit line to the material. If material is not included in the chapter's Creative Commons license and your intended use is not permitted by statutory regulation or exceeds the permitted use, you will need to obtain permission directly from the copyright holder.

Structural Strength Analysis of Winch for Marine Platforms

Xin Ding[1], Lin Lin[2], and Zhaowang Xia[1](✉)

[1] Jiangsu University of Science and Technology, Zhenjiang, China
dlxzw@163.com
[2] Shanghai Marine Diesel Engine Research Institute, Shanghai, China

Abstract. This paper presents a finite element analysis (FEA) of a winch installed on a naval vessel, focusing on its static strength validation. Conducted in accordance with the China Classification Society (CCS) standards, the study evaluates the structural integrity of the winch under various static loading conditions. The results confirm that the winch meets the required safety thresholds, ensuring reliable operation in marine environments. The analysis also highlights the importance of material selection, loading conditions, and safety factors, which were chosen based on CCS guidelines to simulate real-world conditions and guarantee operational safety and durability of the winch design.

Keywords: Marine Winches · Static Strength Validation · Finite Element Analysis · Structural Integrity

1 Introduction

Winches are indispensable components on naval vessels, facilitating the deployment and retrieval of hoses used for various operations, including fluid transfer and underwater equipment servicing [1]. The structural integrity and strength characteristics of these mechanical systems are crucial for ensuring operational safety and efficiency [2, 3]. In recent years, significant advances have been made in understanding the relationship between structural properties and mechanical performance [4, 5], with particular emphasis on material strength and deformation characteristics [6]. Research has shown that structural steel properties, including retained austenite and bainite formation, play a crucial role in determining the overall strength and impact resistance of mechanical components [7]. Furthermore, the correlation between hardness and structural strength has been extensively studied [8], providing valuable insights for design optimization. This paper presents a finite element analysis (FEA) of a winch, focusing on its static strength validation and structural optimization [9]. The study was conducted in accordance with the China Classification Society (CCS) standards [10] to ensure the winch's compliance with the required safety thresholds for marine equipment. These standards provide essential guidelines for material selection, loading conditions, and safety factors, ensuring the winch design can withstand the harsh operating conditions of marine environments. The parameters for the FEA were chosen based on these standards, ensuring that the analysis reflects real-world scenarios and meets the necessary safety criteria.

2 Calculation Basis

The verification is based on the guidelines of the China Classification Society (CCS), specifically the "Regulations for the Classification of Mobile Offshore Units," 2021 edition. These standards provide the necessary framework for selecting design parameters, such as material properties, loading conditions, and safety factors. The use of these standards ensures that the winch design complies with marine safety regulations. Furthermore, citations and discussions on parameter design considerations, as highlighted by various studies in marine equipment design, have been referenced to justify the choice of parameters in this analysis. These standards play a critical role in ensuring that the winch's performance meets the required safety thresholds for marine operations.

3 Winch Model

Winch is primarily composed of several subcomponents, including the wall frame, Expansion frame, and clamp devices. The total mass of the original model is 17.7 tons, which remains unchanged in the simplified model used for analysis. Components like the pedestrian ladder and control box are represented by equivalent masses at appropriate positions, and distributed loads are applied to the simplified model as necessary. As shown in Fig. 1, it is Three-dimensional model of winch.

Fig. 1. Three-dimensional model of winch.

The three-dimensional (3D) model of the winch provides a foundation for finite element analysis and has been simplified to improve computational efficiency while preserving the structural integrity necessary for accurate analysis.

4 Material Parameters of the Winch

The primary materials used for the winch components are Q345B and Q235B steel, chosen for their mechanical properties. The material properties for the main components of the winch are summarized in Table 1.

Table 1. Material Properties of winch Components

Component	Material	Density (kg/m³)	Young's Modulus (GPa)	Poisson's Ratio	Yield Strength (MPa)	Tensile Strength (MPa)
Wall Frame	Q345B	7850	205	0.3	345	630
Expansion frame	Q345B	7850	205	0.3	345	630
Clamp	Q235B	7850	235	0.3	235	500

5 Cases and Loads

Three cases were considered for the static load strength analysis:
 Case One: The winch is fixed at the bottom, and the hose experiences a downward pull of 100 kN.
 Case Two: The winch is fixed at the bottom, and the hose experiences the pull of a submersible pump.
 Case Three: The telescopic frame is checked separately for strength under the action of two 100 kN forces.

6 Finite Element Modeling

The complex assembly of the winch was simplified for finite element analysis to ensure efficient meshing and calculation. The model was meshed with 734,837 elements and 1,457,242 nodes for Case One, 733,973 elements and 1,452,254 nodes for Case Two, and 32,670 elements and 65,844 nodes for Case Three.

7 Calculation Results

The finite element analysis provided results for the equivalent stress and deformation of key components under each load case. Table 2 summarizes the stress and deformation observed in major components, The stress deformation cloud diagram is shown in Fig. 2.

Table 2. Stress of Critical Components of winch.

Load Case	Component	Materia	Yield Strength (MPa)	Tensile Strength (MPa)	Stress (MPa)	Deformation (mm)
Case One	Wall Frame	Q345B	345	630	142.5	3.3
Case Two	Wall Frame	Q345B	345	630	87.1	2.2
Case Three	Expansion framel	Q345B	235	500	134.1	9.6

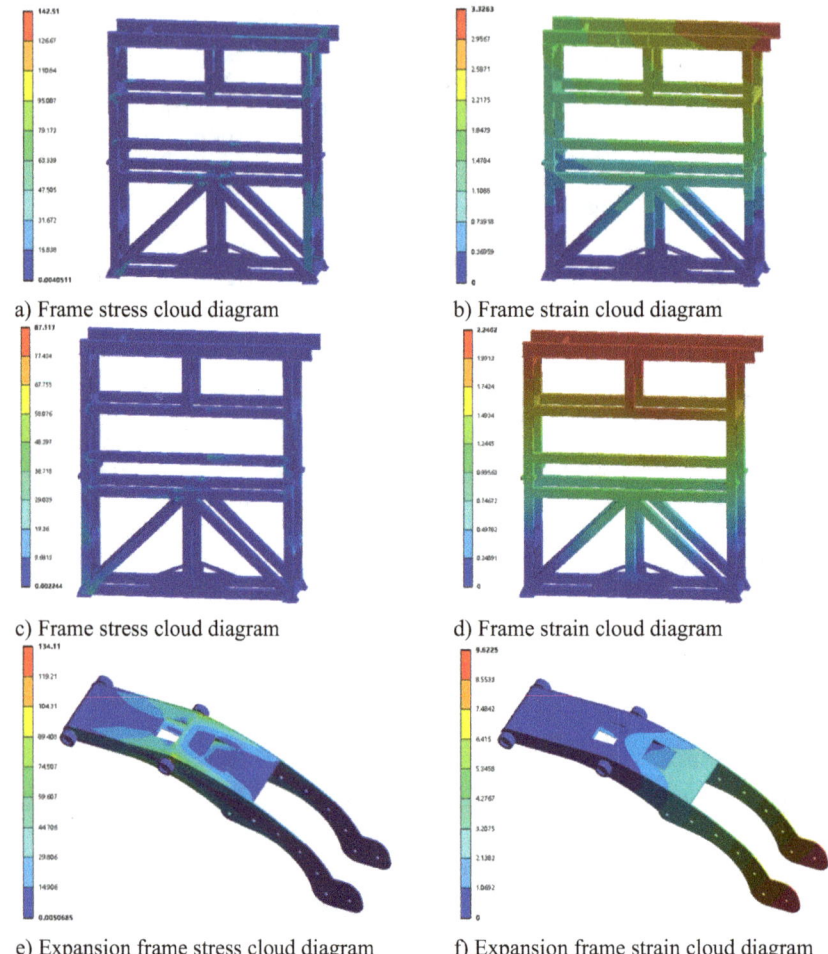

a) Frame stress cloud diagram
b) Frame strain cloud diagram
c) Frame stress cloud diagram
d) Frame strain cloud diagram
e) Expansion frame stress cloud diagram
f) Expansion frame strain cloud diagram

Fig. 2. Stress-strain cloud diagram.

8 Safety Factor Check

The allowable stress values for the structural components were determined using the safety factors provided by China Classification Society in Table 3. The allowable stress value of the structural member of the main frame of the platform for structural analysis can be determined $[\sigma]$:

$$[\sigma] = \sigma_s / S \tag{1}$$

In the formula: σ_s——Yield strength of materials, N/mm^2;
S——factor of safety, use according to Table 3.

Table 3. Safety factor.

Static load conditions	combined condition
1.67	1.25

Case 1 and 2 are combined conditions, and Case 3 is static load condition.
Case one:

$$[\sigma] = \sigma_s/S = 345/1.25 = 276.0 \tag{2}$$

The wall frame stress of working Case one is less than 276.0. Meet the design requirements.
Case two:

$$[\sigma] = \sigma_s/S = 345/1.25 = 276.0 \tag{3}$$

The stress of the wall frame in the second working Case is less than 276.0. Meet the design requirements.
Case three:

$$[\sigma] = \sigma_s/S = 345/1.67 = 206.6 \tag{4}$$

The wall frame stress of working Case three is less than 276.0. Meet the design requirements.

9 Conclusion

The strength verification of the winch under the given loading conditions confirms that the equivalent stress values for all components are below the material yield strength. However, it is important to note that the selection of these parameters, such as material properties and loading conditions, were based on the guidelines provided by the China Classification Society (CCS), which are crucial for ensuring the winch's compliance with safety thresholds. The rationale behind these parameters is to simulate real-world marine operating conditions, ensuring the winch's reliability under stress.

The simulation results highlight that the chosen parameters, including material properties and loading scenarios, are appropriate for evaluating the winch's structural integrity in real operational conditions. The results also demonstrate that the winch design meets the required safety margins, with the stress values well below the allowable limits. To further enhance the quality of this study, a more detailed interpretation of the simulation outcomes would provide a better understanding of how the parameters affect the final design. Additionally, incorporating these findings into the conclusions helps in justifying the relevance of the chosen standards and the success of the design in ensuring operational safety and structural integrity under expected conditions.

References

1. Shengxu, D., Qiang, L., Jing, L.: The design and mechanical study of the pipe joint structure in a high-pressure hydraulic system. J. Phys. Conf. Ser. **2658**(1), (2023)
2. Cheng, Y., Gao, H., Ma, J., et al.: Stiffness analysis and structural optimization design of an air spring for ships. Sci. Rep. **14**(1), 14650 (2024)
3. Han, B., Zhou, Y., Zhang, J., et al.: Kinematics and dynamics characteristics analysis of a double-ring truss deployable mechanism based on rectangular scissors unit. Eng. Struct. 307117900 (2024)
4. Wang, F.Y., Chen, H.Z., Chen, X.J., et al.: High strength and tackling structural relaxation by sub-grains synergistic deformation in W-Re alloy. Mater. Sci. Eng. A 914147160–147160 (2024)
5. Han, J., Wang, K., Wang, et al.: Tailoring the strength and low-temperature toughness of HSLA structural steel by adding trace Ce. Mater. Today Commun. 40109789–109789 (2024)
6. V.N.A., N.A.S., A.N.P., et al.: Influence of deformation on the structural–phase state of structural steel and on the formation of its yield strength and degree of hardening. Tech. Phys. **69**(1), 23–29 (2024)
7. Yu, A.K., V.Y.K., N.Y.S.: Retained austenite in the structure of carbide-free bainite and its effect on the impact strength of structural steels. Metal Sci. Heat Treat. **65**(11–12), 677–682 (2024)
8. Piotr, O., Bronisław, G., Michał, R.: Relationship between Brinell hardness and the strength of structural steels. Structures 59105701 (2024)
9. ' Marine Mobile Platform Classification Specification (2023) takes effect on January 1, 2023. Ship Standardization Eng. **56**(01), 3 (2023)
10. Weilong, H.: The specification for lifting equipment for ships and offshore facilities came into force on December 1, 1989. Shipbuild. Technol. **01**, 44 (1990)

Open Access This chapter is licensed under the terms of the Creative Commons Attribution-NonCommercial-NoDerivatives 4.0 International License (http://creativecommons.org/licenses/by-nc-nd/4.0/), which permits any noncommercial use, sharing, distribution and reproduction in any medium or format, as long as you give appropriate credit to the original author(s) and the source, provide a link to the Creative Commons license and indicate if you modified the licensed material. You do not have permission under this license to share adapted material derived from this chapter or parts of it.

The images or other third party material in this chapter are included in the chapter's Creative Commons license, unless indicated otherwise in a credit line to the material. If material is not included in the chapter's Creative Commons license and your intended use is not permitted by statutory regulation or exceeds the permitted use, you will need to obtain permission directly from the copyright holder.

Research on Intelligent Monitoring of Ship Piping System Based on Digital Twin

Tingyu Jiao[✉], Ji Zeng, Yu Zhang, Haoyun Gu, and Weiqi Ding

Shanghai Maritime University, Shanghai, China
jiaotingyu_1@163.com

Abstract. With the concept of Intelligent Manufacturing 2025, the shipbuilding industry, as one of the important manufacturing industries in China, needs to address the issue of intelligent and digital transformation. Ship piping system is distributed in various parts of the ship, the number is huge, the spacing is small, and contained in a variety of hull structure, not only is it not easy to find its abnormality, and the alarm cannot determine the location of the fault, which affects the operation and safety of the ship, so there is an urgent need for a visualisation of a new method for monitoring and management. This paper proposes an intelligent monitoring method of pipe system based on three-dimensional digital twin, studies the key technology and architecture of digital twin implementation in the field of ships, and proposes an effective monitoring method. By constructing an equal-scale digital pipe system model of the ship, real-time display of the physical information spatial data of the pipeline and the output data of the pipe line digital twin, and real-time monitoring and abnormal alarm of the ship's pipe line based on the digital twin, dynamically displaying the abnormal location, thus improving the safety of the ship.

Keywords: Ship Piping · Digital Twin · Visualization · Digital Piping System Modeling

1 Introduction

The ship piping system is primarily categorized into power piping and auxiliary piping, mainly connecting to the main and auxiliary engines as well as related equipment on board. It serves to supply a variety of fuels, lubricating oil, cooling water, compressed air, and other industrial materials for the ship's equipment to ensure their normal operation. Due to the year-round sailing of ships on water, most of the pipe systems are exposed to harsh and complex environments characterized by high temperature, high pressure, strong noise, vibration, and corrosion. Additionally, factors such as water exposure, aging, welding defects may also lead to system failure and subsequent leaks. Leakage in the piping system commonly results in damage and can even lead to equipment or structural failure or sinking accidents in severe cases. Currently onboard detection technology relies heavily on manual inspection which is low-cost but suffers from issues such as low efficiency speed and accuracy due to its reliance on crew experience for judgments. As ship systems become increasingly complex it becomes more challenging to rely solely on manual inspection for leak detection in pipelines. Early detection of leaks would enable prompt maintenance measures thus preventing accidents.

2 Current Status of Piping System Development

In recent years, with the occurrence of some subsea pipeline oil leakage incidents, Chinese researchers have realized the importance of pipeline leakage detection, and began to study, although the start is late, but also achieved certain research results. In 1997, Yan Dachon and Tang Xiujia et al. proposed for the first time to construct the characteristic parameters of leakage signals into a neural network input matrix to build a neural network model of pipeline operation conditions for detecting pipeline leakage faults, and verified its feasibility under experimental conditions [1]. Xia Habo et al. proposed in 2003 that GPS time pulse signal could be synchronized, and the velocity of negative pressure wave at pipeline leak point could be determined by converting sampling frequency and time label, and the location of pipeline leak could be determined by using the difference of time label of leakage characteristic signal detected at the leak point, which could effectively improve the accuracy of legal leakage point of negative pressure wave [2]. In 2005, according to the transient mathematical model of pipelines, Zhang Jibing et al. used the chart line method to monitor a gas pipeline with a wall thickness of 7 mm, a length of 23.5 km and an unequal temperature. The system successfully detected pipeline leakage and issued a pool leak alarm with an error of 0.9% [3]. In 2009, Zhang Yu et al. mainly studied the detection fields of negative pressure wave method and dynamic pressure signal, and proposed a new method of detecting pipeline leakage with additional dynamic micro-pressure excitation [4]. In 2021, Mina Fahimipirehgalin et al. proposed a liquid leakage detection and location method for chemical equipment based on infrared video data and machine vision technology [5].

All the above methods simulate leakage on specific pipelines or pipeline experimental devices. Due to the limitations of experimental conditions, they have certain limitations, and it is difficult to detect the change of force inside pipelines when leakage occurs [6]. Based on the intelligent monitoring method of pipeline system based on 3D digital twins, this paper modeled and simulated the pipeline structure of the ship, and checked the leaking pipeline through the comparison of the resulting data, and finally achieved the visual monitoring and management of the pipeline [7].

3 Simulation Experiment

The SolidWorks software is the world's first Windows-based 3D CAD system, which has been developed and widely utilized in various industries such as aerospace, marine, locomotive, machinery, defense, electronic communications and others. In this study, we will utilize SolidWorks to model the pipeline, followed by simulating and calculating the typical marine pipeline system using Ansys Workbench. Subsequently, we will analyze and visualize the results obtained with sensor data.

3.1 Model Construction and Meshing

The analysis of the static mechanical characteristics of the pipeline of the seawater cooling system belongs to the category of structural mechanics, so the article adopts the method of numerical simulation and analysis for the study. In order to establish the numerical model of seawater cooling system pipeline, combined with the actual engineering of the ship, the specific parameters of the selected pipeline are shown in Table 1.

Table 1. Pipe parameter.

parameters	retrieve a value
Pipe Material	A53GrandB
Corrosion margin(mm)	1.7
negative machining deviation (%)	12.5
fluid densitykg/(cu·cm)	0.001(water)

The seawater cooling pipeline model is a straight and elbow fit, the model dimensions are shown in Table 2 below. In Solidworks 1:1 to establish the 3D model of the seawater cooling system pipeline, the external material for the solid model, the internal filling part of the fluid model. The fluid part of the selection of water, the solid part of the selection of structural steel material, material specific parameters as shown in Table 3.

Table 2. Model dimensions.

parameters	Segment location	Outer diameter(D/mm)	inside diameter(d/mm)	wall thickness(δ/mm)
sizes	122S	42	36	3
	122P	114	106	4
	123	220	200	10
	M1	700	660	20

The parameters shown in Table 3 are set on the finite element model to get the 3D geometric model of seawater cooling system pipeline that is more in line with the actual project. Then mesh division, in ANSYS software provides a total of automatic division, tetrahedral division, hexagonal explicit division and other five mesh division methods. Considering the actual situation of the seawater cooling system pipeline, the tetrahedral meshing method is used in M1 segment of the article, and the hexagonal explicit meshing method is used in 122 and 123 segments, which can reduce the number of mesh cells on the basis of improving the convergence speed. The results are shown in Figs. 1–3. The piping part of ship M1 segment has 31,870 nodes and 5880 meshes. Ship 122 segmented

Table 3. Model material parameters.

parameters	densities/(kg/m^{-3})	modulus of elasticity (Gpa)	Poisson's ratio
structural steel	122P	206	0.3
	122S	206	0.3
	123	206	0.3
	M1	206	0.3

piping part has 56,032 nodes and 23,071 grids. Ship 123 segment piping part has 34,861 nodes and 17,401 grids.

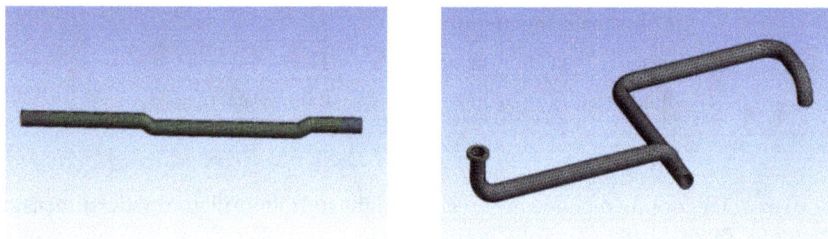

Fig. 1. M1subsection pipe mesh delineation. **Fig. 2.** 123subsection pipe mesh delineation.

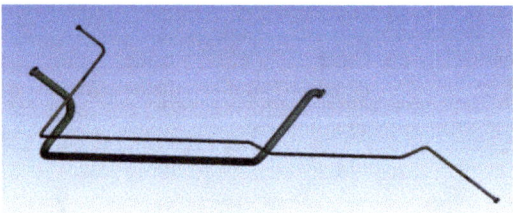

Fig. 3. 122 subsection pipe mesh delineation.

3.2 Parameter Settings

After completing the meshing of the finite element model of the seawater cooling system pipeline, it is necessary to define the contact type for the model, and the article adopts the bound contact type, that is, there is no relative sliding and separation between the contact surface and the contact edge in the finite element model. Then according to the actual operating characteristics of the seawater cooling system pipeline, the corresponding boundary conditions are applied to the finite element model. As the cooler, flange, pipe support belongs to rigid components, mainly showing its inertial nature, and the deformation at the connection with the pipeline is small, so the fixed constraint simulation is carried out at the cooler, flange and pipe support, and the corresponding pipeline unit simulation for the straight pipe section and elbow. According to the above content to complete the establishment of the finite element model of seawater cooling pipeline, that is, using static analysis.

3.3 Analysis of Results

After the establishment of pipeline model, cell division, input of pipeline parameters, pipeline constraints, and selection of analytical conditions, the equations of the whole pipeline system are calculated and solved.

In ANSYS finite element software for different segmental models to apply different external loads the results of the analysis are shown in Figs. 4–6. Comparing the results of the simulation analysis with the results of the data tested by the sensors, when the displacement offset $\geq 5\%D$, an alarm is needed to remind maintenance inspection to prevent fatigue damage. Then the different devices are numbered to collate the alarm file, when the displacement deflection exceeds the preset value, the data detected by the sensor is displayed as 1, and when the displacement deflection does not exceed the preset value, the sensor detects the normal operation is displayed as 0. Therefore, the displacement of the cooling water pipeline of 122 segments exceeds the permitted value for detection, and the table is displayed as 1.

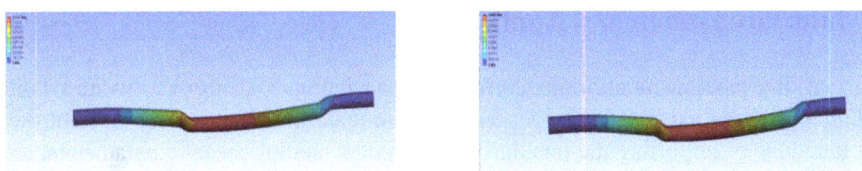

Fig. 4. Cloud diagram of cooling water pipe displacement and deformation in M1 subsection.

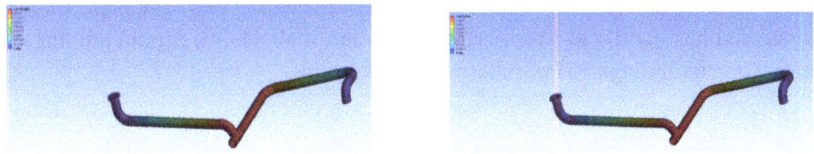

Fig. 5. Cloud diagram of cooling water pipe displacement and deformation in 123 subsection.

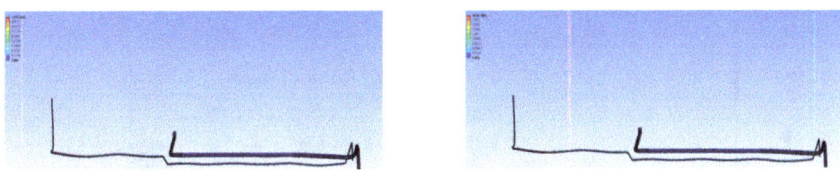

Fig. 6. Cloud diagram of cooling water pipe displacement and deformation in 122 subsection.

4 Sensor Selection

The reasonable layout of pressure sensors in the ship's pipe system is crucial to ensure the safe and stable operation of the ship. The layout types we choose are distributed and key node type. Distributed layout, the sensors are evenly dispersed along the pipe system, the layout of the spacing of this parameter is extremely critical, generally based on the pipe diameter and importance of the pipe system is set, such as the main fuel oil pipeline system, in order to accurately capture the pressure fluctuations, the spacing may be controlled at 1–2 m, in order to comprehensively monitor the pressure changes of the pipeline system; and key node layout, the sensors are concentrated in the pipeline system of the elbows, branches and other key areas prone to pressure abnormalities, and the layout angle should ensure that the pressure can be accurately perceived here. The layout angle should ensure that the pressure can be accurately sensed, for example, at 90 ° elbows, the sensor is installed at an angle of 45 ° to the pipeline axis, which is the best way to detect sudden pressure changes.

Sensor selection is also closely related to the layout. If the pipe system conveys corrosive media, corrosion-resistant sensors, such as molybdenum-containing stainless steel, should be used. At the same time, we should pay attention to the range of the sensor, generally selected 1.5–2 times the normal working pressure, in order to cope with possible pressure peaks, to ensure the measurement accuracy and service life.

5 Interface Design and Applications

The real-time pipeline monitoring interface is shown in Fig. 7. Using vs software to build a digital platform using C language, where in the main interface, we have established the functions of importing the pipeline system model, setting specific parameters, and manually stopping the monitoring at any time to achieve the function of querying the historical data. When the pipeline is abnormal and an alarm is issued, the pipeline displayed on the interface will turn red, and the researchers can visually observe the operation status of the system, which is convenient for debugging. The platform is easy to operate and has good real-time performance, and it also has a certain guiding effect on the leakage fault diagnosis of other pipeline systems.

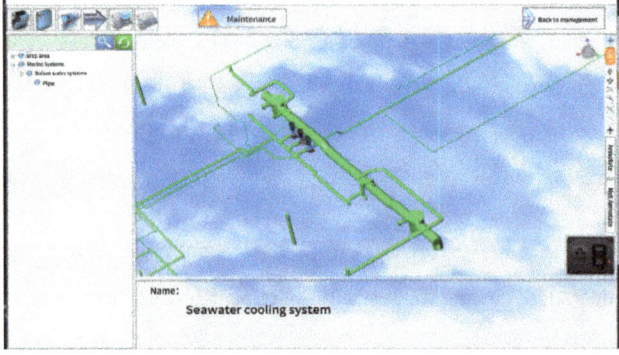

Fig. 7. Program main interface.

6 Conclusion

The article firstly establishes the three-dimensional model of the pipeline of the ship's seawater cooling system through solidworks software, and analyses the hydrostatic characteristics of the pipeline, and calculates the maximum displacement change of the pipeline of different segments of the ship under different external loads, and the simulation results are basically in line with the actual situation of the cooling water pipeline. In the actual application project of digital twin, once the data detected by the pipeline sensors in the actual operation of the ship exceeds the preset range, the main interface of the digital system will automatically issue an alarm, thus playing the role of visual detection and management and improving the safety of the ship's pipeline system.

References

1. Xujia, T., Dachun, Y.: Neural network based pipeline leakage detection method and instrument. J. Peking Univ. (Nat. Sci. Ed.). **03**, 49–57 (1997) (in Chinese)
2. Haibo, X., Laibin, Z., Thames, C.: Pipeline leakage localisation method based on GPS timestamp. Comput. Meas. Control. **0**(3), 161–162 (2003) (in Chinese)
3. Hongbing, Z., Changjun, L.: Research on leakage monitoring technology of long-distance gas pipeline. Oil Gas Chem. Ind. **02**, 146–147+80 (2005) (in Chinese)
4. Zhang, Y.: Research on New Methods and Key Technologies for Leakage Detection of Oil Pipelines: (Doctoral Dissertation). Tianjin University, Tianjin (2009) (in Chinese)
5. Fahimipirehgalin, M., Trunzer, E., Odenweller, M.: Birgit Vogel-Heuser. Automatic Visual Leakage Detection and Localization from Pipelines in Chemical Process Plants Using Machine Vision Techniques. Engineering. **7**(6), 758–776 (2021)
6. Yulong, W.: Fault Diagnosis and Analysis of Oil Slide Leakage in Marine Engine Room. Dalian Maritime University (2015) (in Chinese)
7. Fukushima, K., Maeshima, R., Kinoshita, A., et al.: Gas pipeline leak detection system using the online simulation method. Comput. Chem. Eng. **24**, 453–456 (2000)

Open Access This chapter is licensed under the terms of the Creative Commons Attribution-NonCommercial-NoDerivatives 4.0 International License (http://creativecommons.org/licenses/by-nc-nd/4.0/), which permits any noncommercial use, sharing, distribution and reproduction in any medium or format, as long as you give appropriate credit to the original author(s) and the source, provide a link to the Creative Commons license and indicate if you modified the licensed material. You do not have permission under this license to share adapted material derived from this chapter or parts of it.

The images or other third party material in this chapter are included in the chapter's Creative Commons license, unless indicated otherwise in a credit line to the material. If material is not included in the chapter's Creative Commons license and your intended use is not permitted by statutory regulation or exceeds the permitted use, you will need to obtain permission directly from the copyright holder.

Research on Inland Waterway Remote Driving Based on Scaled Ship Model

Tao Guo[1], Tao Li[1(✉)], Mao Zheng[2,3], Hao Wu[2,3], Yong Wu[4], and Chengguo Song[1]

[1] Changjiang Waterway Institute of Planning and Design, Wuhan, China
258359894@qq.com
[2] State Key Laboratory of Maritime Technology and Safety, Wuhan, China
[3] Wuhan University of Technology, Wuhan, China
[4] Jiangsu Maritime Institute Navigation College, Nanjing, China

Abstract. To study the feasibility of ship remote operating based on 4G/5G network. Considering the high risk of actual ship testing, a scaled ship model was used for remote driving testing, and driving methods such as heading keeping and path following were tested. Network transmission delay, packet dropout rate, and driver attention were tested. The results show that the scaled ship model can achieve accurate heading keeping and path following control under remote driving controller, the network transmission delay is about 1–1.7 s, and the data packet dropout rate is about 0.25%–0.32%. The driver focuses on the display of ship model navigation situation and rudder control more than other information. The remote driving of ships based on 4G/5G networks has certain feasibility, and the measurement methods of network latency, packet dropout rate, and driver attention can be used in actual ship experiments.

Keywords: Scaled Ship Model · Remotely Operated Vessels · Free Running Test · Path Following · Attention Value

1 Introduction

Inland waterway shipping faces fierce competition and meager profits all year round, and the traditional work mode of being on board for a long time has led to a severe talent crisis for inland waterway shipping. How to improve the working mode of crew members is the key to breaking through the scientific development of inland waterway navigation. In 2021, Academician Yan Xinping [1] proposed a new driving operation mode of "shore based driving as the main mode and ship end monitoring as the auxiliary". By constructing a new shipping system of green intelligent ships, intelligent shipping infrastructure, and shore based centralized control system, a safe, intelligent, green, and efficient new generation shipping system is established. At present, multiple foreign ship intelligent navigation research and development teams have regarded remote driving as an important breakthrough in the development of intelligent ships. In 2017, Wartsila conducted remote control ship testing on ships across 8000km through satellite communication. ABB, Rolls Royce, Det Norske Veritas, Samsung Heavy Industries [2] and other institutions have also completed remote ship driving tests. Domestic institutions such as Harbin Engineering University, Wuhan University of Technology, Dalian Maritime University, Smart Navigation, and Zhuhai Yunhang have also conducted research on remote ship driving [3]. Currently, China's inland vessels face the dilemma of low level of technological equipment informatization, and there is an urgent need for new technologies such as remote driving and autonomous navigation to improve transportation safety and efficiency, and to improve the working environment for crew members. At the same time, the inland waterway shipping system is unable to bear the cost of systematic upgrading and renovation such as satellite communication and wireless private networks. If relatively sound 4G/5G networks and cloud computing resources can be utilized, new driving technologies such as human-machine collaboration, ship shore collaboration, autonomous driving, and remote monitoring can be developed, and low-cost and high-efficiency intelligent systems can be built to achieve remote driving and autonomous navigation of inland vessels. This is currently a more feasible development path.

The economy in the Yangtze River Basin is relatively developed, and the existing 4G/5G wireless communication facilities have good coverage, which is a prerequisite for carrying out remote ship driving. However, automatic navigation control of ships, delay and packet loss of wireless signals, and ergonomics are important issues that plague remote driving. This article focuses on the above issues and conducts remote driving tests based on scaled model ships. Experimental research is conducted on ship model motion control, signal delay, driver attention, etc., verifying the feasibility of remote driving based on existing 4G/5G networks.

2 Test Platform

2.1 Analysis of Remote Driving Mode

Waterway transportation technology will develop in the following five directions. In terms of waterway infrastructure, the vessels will gradually require less man power, and inland river, off shore and deep sea vessels will form a spectrum. In terms of shipping infrastructure, the waterway facilities, energy supply, and information networks will be integrated. The ship-shore collaboration ability will be strengthened, the ship remote control can be realized, and the construction of shore-based ship control centre will become an important part of waterway transport infrastructure. In terms of ship power, increasingly stringent emission reduction targets will promote the development of ship clean energy, and the ship power system will be driven by multi-diversified energy and electricity. In terms of ship navigation, multi-ship cooperative transportation can improve transportation efficiency, and the inland river and offshore ship formation navigation becomes a new mode of transportation. In terms of maritime supervision and safety, the application of intelligent system makes ship human-caused accidents gradually reduce, and intelligent unmanned system rescue becomes a reality.

Remote driving of inland vessels is a new technology, which is not simply manual driving, but based on various automatic control algorithms, control targets are sent from the shore, and the ship's motion is precisely controlled by the ship's automatic control system. This greatly reduces the workload of drivers, improves driving efficiency, and makes it possible for one person to drive multiple ships. The remote driving of inland vessels mainly includes several driving modes such as remote manual driving, course keeping, and path following.

Remote manual driving refers to the real-time transmission of steering instructions from the shore driver to the ship through wireless communication, and direct control of the steering, similar to the direct operation of the ship by the driver on the shore. This driving mode is the simplest and most direct driving mode for manual remote intervention in emergency situations. In the event of automatic control program failure, manual driving can be used, which is the last barrier to ensure the safety of remote driving. In the actual remote driving process, it can be used for emergency operations such as collision avoidance and ship stopping of ships.

Heading hold mode refers to the wireless communication that sends the target heading set by the shore driver to the ship, and the ship's autopilot controls the servo to keep the ship sailing in the set heading. Heading hold mode is currently the most frequently used function of ship autopilots, which can effectively reduce the workload of the driver for remote driving, eliminating the need to constantly operate the ship's rudder to maintain the heading.

Path following mode refers to the remote driver setting a series of path nodes on a chart, sending them to the ship's end through wireless communication, and the ship's rudder is controlled by the ship's end program to keep the ship sailing on the set path, achieving precise tracking of the planned path. For remote driving, it can greatly reduce the workload of the remote driver, and only temporarily take over the ship in the event of obstacles.

Remote driving of inland river vessels is not simply a remote manual control, but is based on various automatic control algorithms. The shore sends control targets, and the automatic control system at the ship end precisely controls the ship's movement, greatly reducing the workload of the driver, improving the driving efficiency, and making it possible for "one person to control multiple ships". Remote driving of inland river vessels mainly includes several driving modes such as remote manual driving, course keeping, and path following.

Remote manual driving refers to the real-time transmission of the driver's steering instructions from the shore end to the ship end through wireless communication, and directly controls the steering. It is similar to the driver directly operating the ship from the shore end. This driving mode is the simplest and most direct remote manual intervention control in emergency situations. It can be used for manual driving when the automatic control program fails, and is the last barrier to ensure the safety of remote driving. In the actual remote driving process, it can be used for emergency operations such as emergency collision avoidance and stopping of ships.

The course keeping mode refers to the transmission of the target course set by the shore-based driver to the ship end through wireless communication, and the ship-end automatic rudder controls the steering gear to keep the ship sailing on the set course. The course keeping mode is currently the most frequently used function of marine automatic rudders. For remote driving, it can effectively reduce the workload of the driver, without having to constantly operate the rudder to maintain the course.

The path following mode refers to the process where the remote driver sets a series of path nodes on the nautical chart, sends the path nodes to the ship end through wireless communication, and the ship-end program controls the ship's rudder to keep the ship sailing on the set path, achieving precise tracking of the planned path by the ship. For remote driving, it can greatly reduce the workload of the remote driver, and only needs to temporarily take over the ship when obstacles appear.

2.2 Construction of Remote Control System for Scale Ship Model

Since conducting remote driving on actual ships has high risks, this article uses a 1:180 KVLCC2 ultra-large tanker self-propelled ship model as the research platform. The model is 1700 mm long and 284 mm wide, equipped with a self-developed scale ship model control system, which is composed of a shore-based remote driving test monitoring system and a ship-based autonomous navigation control system. It can achieve automatic control of the ship model under multiple navigation modes, and verify the remote driving performance by issuing control targets from the shore end, as shown in Fig. 1. The participating system equipment is divided into two parts: one is the test ship model, and the other is the shore-based remote driving test control system. The distance between the two is about 470 km. Since the ship model itself is unmanned, the automation level of this system belongs to the L2 level of remote supervision navigation without people on board [4].

2.3 Construction of Remote Driving Control System (RDCS) for Scaled Ship Models

Due to the high risk of remote driving on actual ships, this article uses a 1:180 KVLCC2 ultra large oil tanker self-propelled ship model as the research platform. The model is 1700 mm long and 284 mm wide and is equipped with a self-developed scaled ship model control system [5]. The system consists of a shore borne remote driving control system (RDCS) and a ship borne autonomous navigation control system (ANCS), which could control ship models in various navigation modes. By issuing control targets from the shore end, the remote driving performance is verified, as shown in Fig. 1. The test system is divided into two parts, one is the test ship model, and the other is the shore end remote driving test control system, with a distance of about 470 km between RDCS and ANCS. Due to the unmanned nature of the ship model, the automation level of this system belongs to the L2 level of remote supervision navigation for unmanned ships [4].

The experimental ship model is controlled by the onboard main controller. To ensure the reliable and stable operation of the ship model, a control process is designed as shown in Fig. 2. In the main thread, data packets from differential GPS, gyrocompass, and 4G/5G data transmission radio are received, analyzed, and motion control is carried out. The heading is maintained by an adaptive PID control algorithm, which adaptively adjusts the PID parameters based on the engine speed to calculate and obtain the optimal rudder angle. These kinds of data-driven controllers become feasible for ship course-keeping control with the development of data acquisition and processing technologies. Methods adopted are known as expert knowledge controllers, artificial neural networks, neuro-fuzzy systems, and multi-agent systems. These controllers are established based

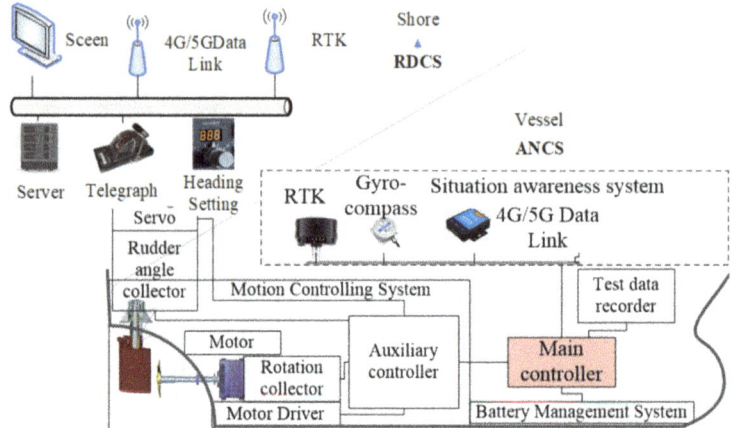

Fig. 1. Framework of remote driving control system (RDCS) for scaled ship models.

on empirical knowledge or navigation data. Researchers fused ship motion and operation data and processed it using various statistical and intelligent modeling methods to establish automatic ship course-keeping controllers. To improve the accuracy of course-keeping, these controllers have high requirements for the amount and quality of data in particular scenarios. As a result, such controllers are more effective in scenarios with a large amount of data on similar ships. Conversely, it is hard to realize accurate ship course-keeping in scenarios with less navigation data. Meanwhile, the selection of empirical knowledge or navigation data has a substantial impact on the effectiveness of ship course-keeping control, which further increases the uncertainty of the controller; Path following uses line of sight (LOS) method to obtain target heading, and adaptive PID control algorithm to obtain rudder angle. To ensure the safety of the test ship model, an automatic return function for low voltage or signal loss is added, and a watchdog program is added to the control chip to ensure that no crash faults occur.

The test ship model is controlled by the on-board main controller. To ensure the reliable and stable operation of the ship model, a control process as shown in Fig. 2 is designed. In the main thread, the data packets of differential GPS, gyrocompass, and 4G/5G data transmission radio are received and analyzed for motion control. Among them, the course keeping adopts an adaptive PID control algorithm to adaptively adjust PID parameters based on the main engine speed, thereby calculating and obtaining the optimal rudder angle. The path following adopts the line-of-sight (LOS) method to obtain the target course, and the rudder angle is obtained through the adaptive PID control algorithm. To ensure the safety of the test ship model, automatic return-to-base functions for low voltage or signal loss are added, and watchdog programs are added to the control chip to ensure that no crashes occur.

To support the completion of remote driving tests, 4G/5G cameras are installed on the ship model, and the shore end can view the real-time navigation status of the ship model, as shown in Fig. 3(a).

The shore-based control center includes a remote driving test monitoring system and a set of ship-end video return display (Fig. 3(b)), which realizes the functions of

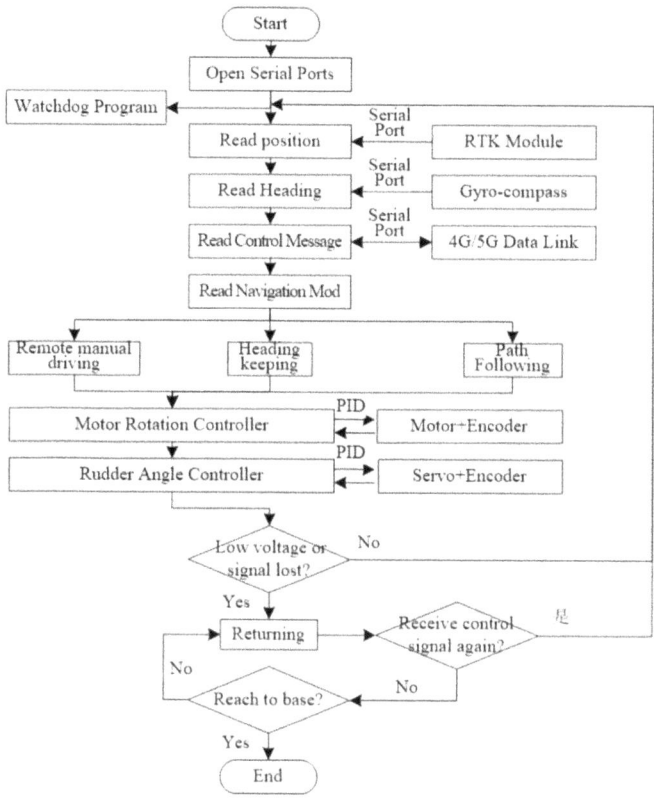

Fig. 2. Flow chart of scaled ship model contrlling system.

sending driving instructions to the test ship model, switching driving modes, returning operating status, and displaying video return from the ship-end camera. The shore-based equipment is connected to a 1000 M broadband network.

Test Indicators and Test Methods. Remote driving mainly focuses on the control precision of course keeping, communication delay and stability, and driver's attention. The test indicators include: control precision of course keeping, control precision of path following, communication delay, packet loss rate, and heat map of driver's eye position.

Measurement Scheme for Course Keeping Accuracy. To measure the control precision of course keeping, a gyrocompass is used to measure the heading angle of the ship model with a measurement accuracy of ±0.5°. The shipborne automatic control program uses a PID algorithm to achieve remote control of the ship model's course keeping, and the root mean square error (RMSE) is used to represent the precision of course keeping:

$$RMSE = \sqrt{\frac{1}{N}\sum_{i=1}^{N}[Y_i - f(x_i)]^2} \qquad (1)$$

where Y_i is heading measurements, $f(x_i)$ is target heading.

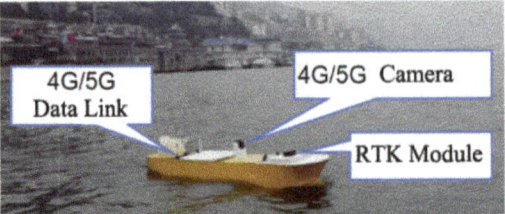

(a) Remote driving scaled ship model for test.

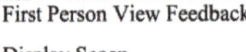

Remote Driving Controlling System

(b) Shore borne romote driving control system.

Fig. 3. Test system and equipments.

Path Following Accuracy Measurement Scheme. To conduct self-propelled tests on the scaled ship model, it is necessary to accurately perceive the position of the ship model. A differential GPS positioning system based on the ublox ZED-F9P differential positioning chip is used, which can achieve centimeter-level positioning of the ship model. The lateral deviation between the actual sailing trajectory of the ship model and the planned path is taken as the path following control error. This article uses the line-of-sight (LOS) method to conduct the path following control of the ship model (Fig. 4), set the previous path node to WP_{i-1}, $(i = 2, 3, 4, ..., N)$, the current path node is WP_i, The next path node is WP_{i+1}, P_H is the current position of the ship model, ψ_H is the current heading, take P_H as the center of the circle, R_{LOS} is the radius of the circle and the path $\overrightarrow{WP_{i-1}WP_i}$ Intersect at two points, and the end of the path WP_i the closer intersection point is the LOS point, The azimuth vector ψ_{LOS} from the current position P_H of the ship model to the LOS point is the LOS angle, and the radius R_{LOS} can be obtained from Eq. (2):

$$\begin{cases} R_{LOS} = \varepsilon + 3L_{PP} \\ \varepsilon = \sqrt{|\overrightarrow{WP_{i-1}P_H}|^2 - \left(\frac{|\overrightarrow{WP_{i-1}WP_i}|^2 - |\overrightarrow{WP_iP_H}|^2 + |\overrightarrow{WP_{i-1}P_H}|^2}{2|\overrightarrow{WP_{i-1}WP_i}|}\right)} \end{cases} \quad (2)$$

where ε is track error, this is the path following control error.

Network Latency Measurement Scheme. During the experiment, the ship-end uses GPS timing while the shore-end utilizes network timing to ensure the uniformity of their respective time, with the error within milliseconds. By adding timestamps to the control instructions and the ship model's operational data return messages, the network

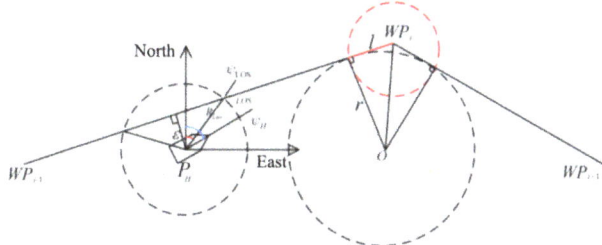

Fig. 4. Path-following control schematic.

delay between sending and receiving is calculated using the timer program [6]. Controls the downlink delay of the command T_{Di}. The uplink delay of the ship model's running data T_{Ui} and the total delay from the command to the receipt of the backhaul data T_{Ai}, which is shown in Fig. 5. The shore remote driving test monitoring system measures the uplink delay once per operation cycle T_{Ui} and total latency T_{Ai}, the ship-end autonomous navigation control system calculates the downlink delay once per control cycle T_{Di}.

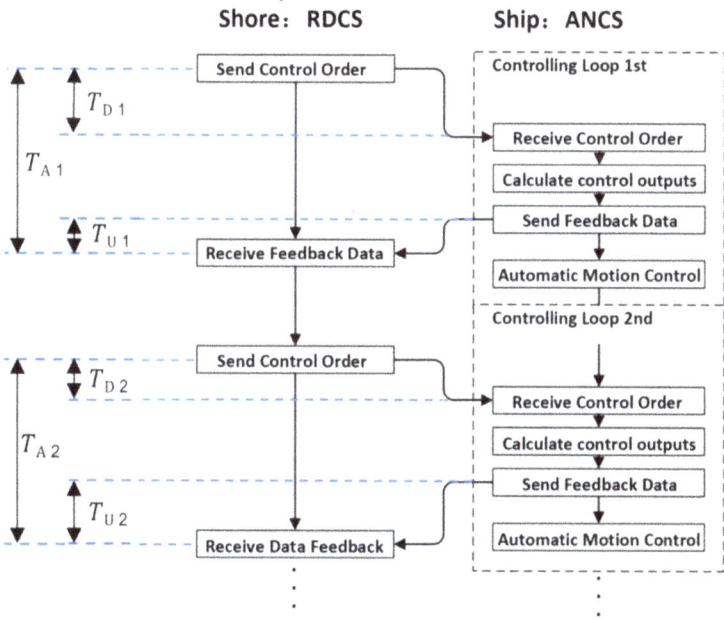

Fig. 5. Schematic diagram of network latency testing.

Data Packet Loss Rate Measurement Scheme. During the test, the remote driving test monitoring system on the shore-end and the autonomous navigation control system on the ship-end both sent messages at 1 Hz, and theoretically, the interval between the reception time of each message should be close to 1 s. Packet numbers were added to the downstream data packets of the control instructions and the upstream data packets of the navigation data. The receiving end automatically recorded the number sequence of the successfully received data packets. After the test, the number of unsuccessfully received data packets was counted and divided by the total number of sent and received packets to calculate the packet loss rate [7].

Remote Driving Attention Measurement Scheme. In order to measure the attention of the remote driver during the test and determine which information is more important for remote driving, an eye position meter was installed to collect the eye attention position of the remote driver in real time during the test, and the attention heat map was automatically drawn [8]. Analyze concerns to determine what information is more important for remote driving.

3 Experimental Validation

3.1 Test Waters

The test water area is located in Fengjie County of the Three Gorges Reservoir Area. The water area is calm with little wind and waves (the maximum wave height is about 0.2 m) and few ships. The 4G/5G signal coverage is good. The test water area includes the main channel of the Yangtze River and Zhuyi River (tributary), with a total navigable area of about 6000 m in length, as shown in Fig. 6. There are slightly more ships in the main channel of the Yangtze River, while few ships sail in the tributary of Zhuyi River, resulting in a simple navigation.

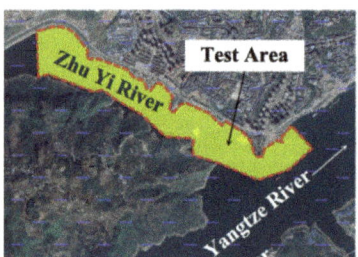

Fig. 6. Test area.

3.2 Test Data Analysis

Heading Hold Test. The heading maintenance test was conducted in the main channel of the Yangtze River. The target heading was issued by the shore-end, and the ship-end program automatically achieved heading maintenance control. The heading data of the ship model was recorded and plotted as shown in Fig. 7.

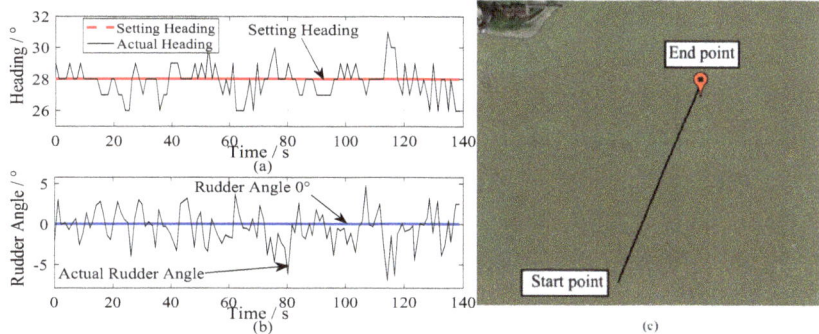

Fig. 7. Heading keeping accuracy test.

Figure 7(a) shows the heading curve. It can be seen that the target heading of the scaled ship model is set to 28°. Under the automatic control of the program, the actual heading error of the scaled ship model is 1.004°, and the control range of the automatic rudder is between –7° and 5° (Fig. 7(b)), and the sailing trajectory is approximately a straight line (Fig. 7(c)).

Path-Following Test. A series of waypoints (WP1 –WP9) are remotely set by the shore-end and sent to the ship-end. The ship-end control program uses the Line of Sight (LOS) method to control the navigation trajectory of the ship model. Among them, when approaching WP7, due to the presence of static obstacles (buoys), the mode is switched to remote manual driving mode to avoid obstacles, and after the clearance, it is switched back to the path following mode to automatically return to the original track.

Figure 8(a) shows the actual sailing trajectory of the ship model, which can accurately track the WP1–WP9 waypoints, with an actual sailing distance of 4984 m. Figure 8(b) is the curve of path following control error. It can be seen that between WP4 and WP9, the scaled ship model can accurately achieve path following control, with a path tracking error within 0.5 m. However, between WP2 and WP4, due to sailing in the main channel of the Yangtze River, where the wind and waves are relatively strong, a large tracking error is generated under the influence of wind and waves, with a maximum error exceeding 12 m. It can be seen that under the condition of weak environmental interference, the path following control has high accuracy, which can enable the ship model to sail accurately according to the planned path.

Network Latency Test. Combined with the path-following control test, the network delay test is carried out, and the program automatically records the downlink delay distribution of the control command, the uplink delay distribution of the ship model operation data, and the total delay distribution between the issuance of the control command from the shore end and the receipt of the return message from the ship end, as shown in Fig. 9.

As shown in Fig. 9(a), the downstream latency of control instructions is between 0.5 s and 1.5 s, with an average of about 1 s, and the latency is evenly distributed between 0.7 s and 1.2 s. In Fig. 9(b), the upstream latency of the ship model's operating data is between 0.5 s and 1.6 s, with an average of about 0.8 s. Relatively speaking, the upstream data transmission latency is lower and the distribution is more concentrated. In

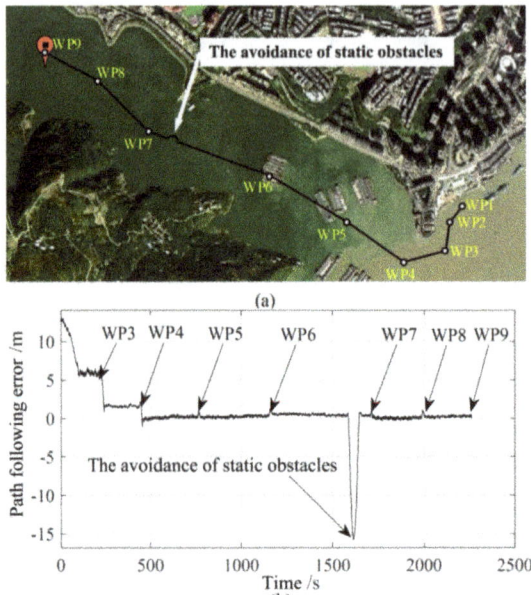

Fig. 8. Path-following control accuracy test.

Fig. 9(c), the total latency from issuing the instruction to receiving the return message is the shortest at 0.9 s, the longest at 2.2 s, mainly distributed between 1 s and 1.7 s, and the maximum probability latency is 1.3 s. It can be seen that the latency of the downstream control instructions and upstream ship model operating data are both in the order of 1 to 2 s. Since the path following control mode is adopted, the ship model is always under automatic control, and the latency has limited impact on remote driving of the ship.

Data Packet Loss Rate Test. In combination with the path following control test, a data packet loss rate test was conducted. As shown in Figs. 10(a) and (c), there was a small amount of packet loss in the control instruction downlink data packets, with a packet loss rate of 0.32%. The downlink data packet reception interval was approximately 1s, consistent with the transmission frequency of 1 Hz. The longest reception time interval was close to 18 s, occurring near the 2116th data packet. As shown in Figs. 10(b) and (d), similar to the downlink data pattern, there were also a few packet losses in the uplink data of navigation data, with a packet loss rate of 0.25%. The uplink data packet reception time interval was approximately 1 s, and the longest reception time interval was close to 18 s, occurring near the 2115th data packet, which was consistent with the time of the longest interval of the downlink data packet. The reason for this was that during the test, the ship model traveled from the main channel of the Yangtze River to tributary waters, crossing multiple signal base stations, and there was packet loss during the switching of base stations. However, due to the adoption of the path following control mode, the ship model was always under automatic control, and the shore-end driver did not perceive the data packet loss phenomenon significantly.

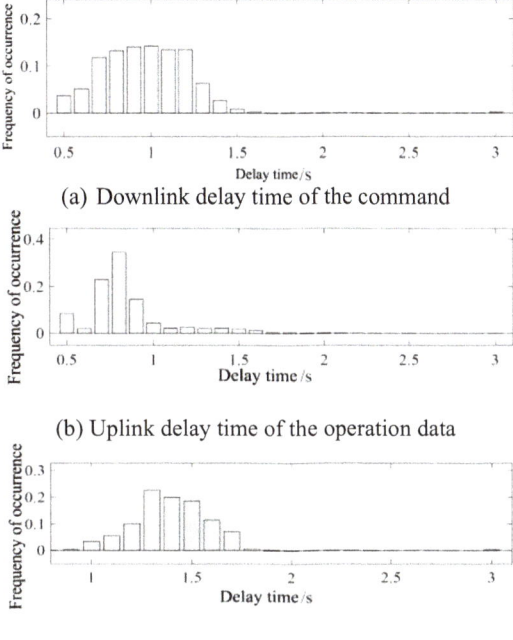

(a) Downlink delay time of the command

(b) Uplink delay time of the operation data

(c) Total delay time

Fig. 9. Network latency test.

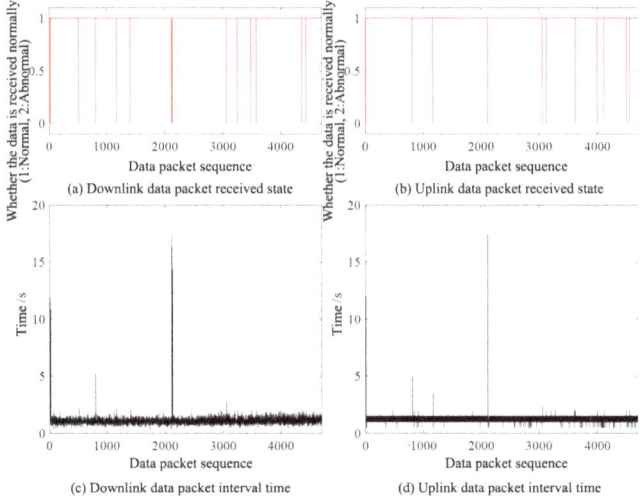

Fig. 10. Packet interval time.

Ocular Alignment Data Test. The attention of the driver is collected using an eye tracker at the shore end, and its heat map is shown in Fig. 11. The colored heat map in the figure represents the areas where the shore-end driver's eyes focus most frequently.

Fig. 11. Driver attention test.

During the process of remote driving, the driver's attention is focused on different factors in the following descending order: ship model's navigation situation > rudder angle control > heading display > propeller speed display > data feedback. It can be seen that for remote driving, the status display and control areas such as the ship's navigation situation and rudder angle control are more important. Meanwhile, the propeller speed, ship model's heading, and feedback data can reflect the running status of the ship model, which are also the key information that needs to be paid attention to during remote driving. For remote driving, the ship's navigation situation map can provide the most comprehensive and intuitive information, which is the most concerned by the driver. In the subsequent remote driving experiments and practical processes, it is necessary to focus on strengthening the feedback and display of key information, and optimize the display interface according to ergonomics.

4 Discussion

The data acquisition systems on the scaled ship model require further enhancement, particularly in capturing real-time hydrological parameters, visibility, and the height of wave. Additionally, the priority should be given to refining hydrodynamic parameters and introducing nonlinear response compensation techniques. This can be achieved through continuous optimization of control models based on multi-sensor feedback from the shipboard systems. Such advancements will directly address complex operational challenges, including environmental interference and sudden load variations during maneuvering. Furthermore, transmitting comprehensive shipboard sensor data to shore-based platforms will substantially improve the reliability of navigation decision-making. It is recommended to implement redundant sensor configurations under environmental disturbances, which ensures robust data acquisition while facilitating the integration of real-time environmental monitoring with remote-control intelligent decision-making frameworks.

5 Conclusion

To study the key issues of remote ship driving, a 4G-based remote driving test was conducted using a scaled ship model. The conclusions are as follows:

(1) Automatic control programs such as course keeping and path following can replace drivers to complete the ship's navigation control, thus reducing the work intensity of shore-based drivers;
(2) The total latency of 4G network transmission is about 1 to 2 s, which has limited impact on remote ship driving. The network packet loss rate is low, and overall, the impact on remote driving is effective;
(3) During the remote driving process, drivers pay more attention to information such as the ship's navigation situation, rudder angle control, propeller speed, heading, and ship operation feedback data.

References

1. Yan, X.P.: Reflection on the construction of the new generation inland navigation system. China Water Transp. **05**, 6–8 (2021)
2. Zhang, Y.T., Xu, L.X., Cao, L., et al.: The current status and prospects of technological development in intelligent ships. Marine Eng. **45**(S1), 185–192 (2023)
3. Wang, Y.Y., Liu, J.L., Ma, F., et al.: Review and prospect of remote control intelligent ships. Chin. J. Ship Res. **16**(1), 18–31 (2021)
4. Zhang, P.Z., Wu, B., Yan, X.P., et al.: Safety analysis for remote control system of inland ships. China Saf. Sci. J. **32**(8), 126–132 (2022)
5. Sun, N., Li, L., Zheng, M., et al.: Design and implementation of general control system for scaled ship model in free-running tests. Exp. Technol. Manag. **39**(9), 132–139 (2022)
6. Peng, G.J., Wang, J.J., Liu, M.T.: A test method of measuring time delay caused by data acquiring network for the flight test AFDX avionic system. Meas. Control Technol. **35**(4), 98–100 (2016)
7. Shi, C.T., Wang, W.F.: Research on measurement method of packet loss rate in AD hoc network performance. China Standard. **24**, 205–209 (2022)
8. He, M.S., Du, Z.G., Han, L., et al.: Research on the characteristics of driver's gaze behavior in the diversion area of urban tunnels. J. Wuhan Univ. Technol. (Transport. Sci. Eng.) **46**(2), 230–234, 241(2022)

Open Access This chapter is licensed under the terms of the Creative Commons Attribution-NonCommercial-NoDerivatives 4.0 International License (http://creativecommons.org/licenses/by-nc-nd/4.0/), which permits any noncommercial use, sharing, distribution and reproduction in any medium or format, as long as you give appropriate credit to the original author(s) and the source, provide a link to the Creative Commons license and indicate if you modified the licensed material. You do not have permission under this license to share adapted material derived from this chapter or parts of it.

The images or other third party material in this chapter are included in the chapter's Creative Commons license, unless indicated otherwise in a credit line to the material. If material is not included in the chapter's Creative Commons license and your intended use is not permitted by statutory regulation or exceeds the permitted use, you will need to obtain permission directly from the copyright holder.

Image Object Detection and Tracking of Ships in Complex Inland Water Areas

Fang Jin[1], Yong Wu[2,3](✉), Yongtao Wang[4], Xiao Liu[2,3], and Bing Han[5]

[1] Huazhong, Institute of Electro-Optics-Wuhan National Laboratory for Optoelectronics, Wuhan 430223, China
[2] Jiangsu Maritime Institute Navigation College, Nanjing 211170, Jiangsu, China
wy118116@163.com
[3] Digital Engineering Technology Research and Development Center for Maritime Safety and Security, Nanjing 211072, China
[4] Lianyungang Aids to Navigation Division of Eastern Navigation Service Center, Lianyungang 222042, Jiangsu, China
[5] National Engineering Research Center of Ship and Shipping Control System, Shanghai 200134, China

Abstract. As nowadays the targets detection and tracking of ships in complex inland water areas are mainly based on radars and AIS (Automatic Identification System), the accuracy and update rate of automatic targets detection and tracking could not support the auto-pilot navigation. In order to conduct the image object detection and tracking of ships in complex inland water areas, the image object detection based on YOLO v8 network and Byte Track algorithm were researched. The samples augmentation and image object detection network training method were putforward. Based on 2-step Byte track framework, the target tracking model was built. To prove the accuracy and practicality of this method, an experiment was conducted on 'Changhanghuoyun 001' vessel which operated on Yangtze River, the result shows that different kinds of targets including ships and beams were detected accurately. Compared with other sensors tracking trajectories, the 2-step Byte track framework could gain the trajectories with the same accuracy. The targets detection and tracking based on image sensors could used on ships in complex inland water areas.

Keywords: Inland Ships · Detection and Tracking · Complex Inland Water Areas · YOLO

1 Introduction

Situation awareness is an important field in human-machine ergonomics [1], the Endsley model is a widely used situational awareness cognitive model, which includes 3 levels: current environmental perception, feature analysis and association fusion, and prediction of future trend changes. ABB company achieved a ship environment real-time perception system in 2017. In the same year, Maersk group and Sea Machines Robotics improved their capabilities for maritime target recognition and tracking, as well as situational awareness, using AI technology. In 2018, Rolls Royce Shipbuilding launched an intelligent perception system for ship navigation situations (IA). The intelligent awareness system integrates ship 3D map and Light detection and ranging(LIDAR), linked to the global positioning system (GPS) data, create a 3D environments in virtual reality (VR) and augmented reality (AR). The augmented reality system provides navigation situation information to ship drivers, and has been tested by the Japanese merchant ship Mitsui's 165-m Sunflower ferry between Kobe and Oita.

In 2017, Wuhan University of Technology developed an intelligent assistance system for safe driving on board ships, which monitors the navigation situation of all ships in the navigation area in a three-dimensional, real-time, online, and dynamic manner, and provides early warning of traffic safety situation during sailing. In 2019, Mairun company developed a situation awareness and data fusion system based on optical perception, achieving real-time visualization of navigation situation 3D view reconstruction and real-time visualization connection between 'ship' and 'shore'. In the research of intelligent ship navigation environment perception and enhancement technology, Gaglione [2] proposed a navigation situation multi-target tracking method based on sum and integration. Hermann [3] estimated ship attitude and position information using Kalman filtering technology based on vision and radar. In response to the problems of unnatural regional transitions, block effects, and severe information loss in current enhancement technologies, Zhang [4] designed a ship visual image enhancement technology in complex environments. Regarding the issue of the utility of applying navigation situational awareness information provided by the current integrated bridge system, Vu [5] pointed out the phenomenon of alarm information overload and proposed an improvement strategy for human-machine interaction of comprehensive bridge system information. Jing [6] proposed a digital twin framework for ships based on virtual reality systems to enhance their situational awareness capabilities. Liu [7] proposed an AR-HUD based assisted lookout system architecture to help ship drivers quickly obtain collision avoidance information and other object information during navigation, using augmented reality head up display technology and enhanced information projection algorithms. In the research of understanding and predicting navigation situations, Du [8]used a nonlinear velocity obstacle algorithm to predict the intention of a passing ship, thereby improving the situational awareness ability of the vessel. Szlapczynski [9] proposed a non conventional collision threat parameter region (CTPA) based approach, The visualization method of ship collision avoidance information using Collision Threat Parameter Area technology enables navigators to quickly select effective collision avoidance strategies. Lei [7] used situational awareness theory and visual analysis technology to design a visual analysis based framework for ship navigation risk situational awareness in VTS regulated waters. Banda [10] used the number center to refer to the type of accident and combined it with a risk matrix to demonstrate the risks, probabilities, and consequences encountered by ships during winter navigation in Finnish waters.

Ship image object detection has important applications in areas such as oceans, inland rivers, ports, and ship locks. To accurately detect all kinds of targets, many researchers have conducted relevant research and practice. Currently, ship target detection methods mainly include machine learning methods and deep learning methods.

Traditional machine learning methods are divided into several stages, including preprocessing, region extraction, object segmentation, feature extraction, and object classification. Deep learning methods are mainly based on convolutional neural networks to extract target features, which require high hardware requirements [11].

Ship target detection based on deep learning methods has become an increasingly popular method in recent years. It has advantages such as high degree of automation, good adaptability to the environment, and good detection effect. The ship target detection methods based on deep learning methods are mainly divided into candidate region detection methods and regression detection methods [12]. As shown in Table 1.

Table 1. Comparative table of deep learning methods.

Method	Model	FPS	mAP	S/FRAME
Detection based on candidate region	Fast-R-CNN	–	61.1%	0.67
	Faster-R-CNN	–	84.0%	0.34
	RA-CNN	–	86.7%	1.50
Detection based on regression detection	YOLO v3	19	84.0%	0.05
	YOLO v5	25	87.8%	–
	Significant sensory awareness CNN	49	87.4%	–
	SSD	43	84.9%	–

From the above table, it can be seen that the candidate region based methods are mainly based on the Fast-R-CNN series, with high average detection accuracy (mAP), but each image takes a long time (S/Frame), making it difficult to adapt to real-time detection requirements. The YOLO series is the main method based on linear regression, MAP has basically reached over 80%. However, regardless of the type of deep learning method, there is a problem of low generalization ability.

With the improvement of computing technology, ship image target detection based on deep learning will be the future development direction. Based on this method, it can play a greater role in subsequent ship target tracking, behavior analysis, and understanding. However, the current deep learning video ship target detection methods still need to overcome the following problems: On one hand, they are easily affected by lighting and need to improve the detection accuracy of the algorithm in complex inland river environments by expanding and expanding the dataset. On the other hand, lightweight minimum combination detection technology should be researched to achieve fast and real-time ship target detection algorithms.

Due to the sufficient feature extraction ability of convolutional neural networks, it is possible to use deep learning algorithms for ship target detection and tracking. There are currently many algorithms for target recognition and tracking, including YOLO Faster R-CNN, SORT, and MOT, this article, firstly establishes a framework for target recognition and tracking, and constructs a process for detecting and tracking targets. Secondly, a video object detection method based on YOLO v8 was proposed, which trained and validated the detection algorithm by collecting image data. Finally, the Byte Track target tracking method was selected to achieve real-time detection and tracking of ship targets.

2 Image Object Detection for Inland River Ships

2.1 The Framework of Image Object Detection

The integrated algorithm structure for video object detection and tracking is shown in Fig. 1. The algorithm consists of three parts: video playing module, object detection module, and object tracking module [13]. The algorithm uses Byte Track as the target tracker, The YOLO v8 algorithm serves as the target detector. The framework of image target recognition and tracking is shown in Fig. 1. Firstly, the video playback module obtains a real-time video stream and divides the real-time video into video frames, which are then distributed to the target detector and tracker. Secondly, in the process of target recognition and tracking, the algorithm uses the target detector to detect the target boxes, and then sends these target boxes to the target tracker as the initial target boxes. When the target detector discovers an untracked target, the algorithm will add a new tracker accordingly and assign threads to it. This thread will add after the current tracker completes the update calculation, and perform the calculation after the next frame of image input. Finally, the trackers of each thread input camera video frames and output response quality, bounding boxes, and target tracking status. According to the target tracking status, when the tracker loses the target, the algorithm automatically clears the tracker and its threads.

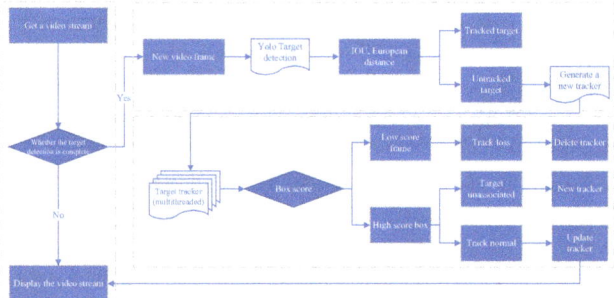

Fig. 1. Framework of image target recognition and tracking.

2.2 Image Object Recognition Based on YOLO V8

Image Target Recognition Algorithm Process. Ship image target recognition adopts a Darknet network model based on deep learning framework combined with YOLO algorithm. The Darknet deep learning framework is the open-source neural network framework YOLO (You Only Look Once) proposed by Joseph Redmon, which is the core object detection algorithm for this framework. In the YOLO algorithm, the object detection problem is treated as a regression problem, and a convolutional neural network structure can directly predict the bounding box and category probability from the input image. This method improves the basic classification network structure and target binary prediction method in traditional deep learning, achieving ship tracking and recognition of ship types [14].

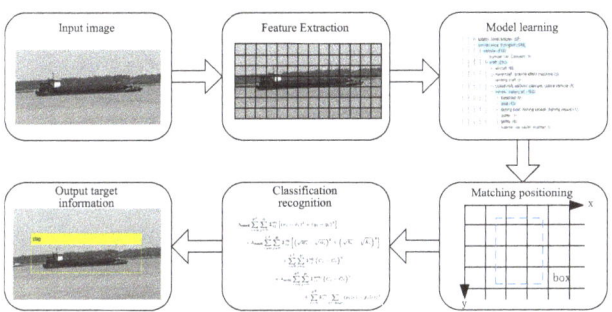

Fig. 2. Ship target recognition algorithm flowchart.

The real-time ship tracking and recognition algorithm based on the Darknet network and YOLO algorithm achieves a balance of speed and accuracy. The flowchart is shown in Figure 2 through ship image target recognition, information such as pixel position, region size, target credibility, and target classification of ship image targets can be obtained.

YOLO v8 Algorithm Principle. *Training data augmentation.* The data preprocessing of YOLO v8 follows the methods adopted by YOLO v5, including similar enhancement and mosaic enhancement, as shown in Fig. 3. In order to improve training effectiveness, the YOLO v8 algorithm has also adopts methods such as hybrid enhancement (Mixup) and color perturbation (HSV augment).

1) Mixed enhancement

Mixed enhancement [15] is a simple and effective data augmentation method that mixes different types of images using linear interpolation to form new training samples as supplements to the original training set, as shown in Fig. 4. This method can significantly improve training effectiveness in multiple fields such as image, text, speech, and adversarial sample defense. As a type of additive mixed class enhancement strategy, it has a significant impact on subsequent image data augmentation, thus giving rise to technologies such as cutMix FMix, co Mix, and other enhancement strategies that overlay images from different perspectives.

The core formula for hybrid enhancement is shown as:

$$\widetilde{x} = \lambda x_i + (1 - \lambda) x_j \tag{1}$$

$$\widetilde{y} = \lambda y_i + (1 - \lambda) y_i \tag{2}$$

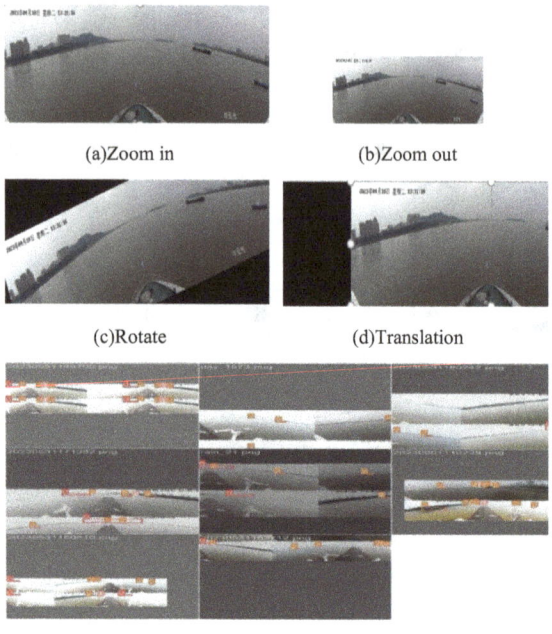

(a)Zoom in (b)Zoom out

(c)Rotate (d)Translation

(f)Mosaic enhancement

Fig. 3. Some preprocessing methods of YOLO v8.

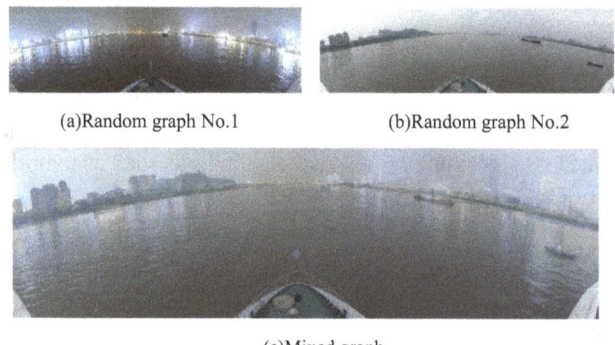

(a)Random graph No.1 (b)Random graph No.2

(c)Mixed graph

Fig. 4. Mixed augmentation.

where, (x_i, y_i) and (x_j, y_j) are two randomly selected samples from the training data, $\lambda \in [0,1]$.

2) Color perturbation

The way HSV expresses color images consists of three parts: Hue (color tone) Saturation and Value [16]. As shown in Fig. 5, Hue is measured by direction angle, with a value range of 0 ~ 360°, representing the color information of the image. The horizontal direction represents Saturation, with a range of 0–100%. The larger the value, the more saturated the color will be. The vertical direction represents Value, with a range of 0–100%. The larger the value, the brighter the color. By adjusting the size of color tone, saturation, and brightness, different styles of image transformations can be achieved. When these image transformations are converted into RGB format and input to the neural network, the values of each channel change, thereby enriching the dataset and improving the generalization performance of the model.

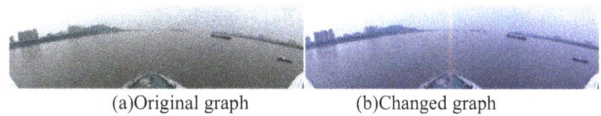

(a)Original graph   (b)Changed graph

Fig. 5. Color perturbation.

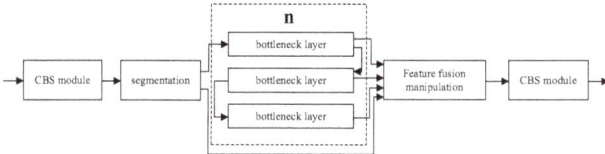

Fig. 6. C2f module network structure diagram.

Network Framework.

1) Core network structure

The backbone network structure of YOLO v8 is in line with the backbone network of YOLO v5. Its backbone network uses a 3x3 convolution with a step size of 2 for downsampling feature maps, followed by a C2f module to further enhance the features, The C2f module is to some extent inspired by YOLO v7's ELAN module, adding more branches to enrich the branches during gradient backpropagation. As shown in Fig. 6, the network structure diagram of the C2f module is shown.

2) PaFPN network structure

YOLO v8 still uses the PaFPN structure to construct the feature pyramid of YOLO, allowing for sufficient fusion of multi-scale information. In PaFPN, only one layer of 1x1 convolution was reduced during the upsampling process at the head, and the C2f module was used to replace the C3 module. The rest of the structure is basically consistent with YOLO v5's FPN-PAN.

3) Decoupled head network structure

From YOLO v3 to YOLO v5, the detection head has always been 'Coupled', using a layer of convolution to complete both classification and localization tasks simultaneously, until the advent of YOLO X, The YOLO series has only been equipped with a Decoupled Head for the first time, followed by the YOLO v6 YOLO v8 also adopts an understanding coupling structure, which is more in line with the design concept of advanced detection frameworks.

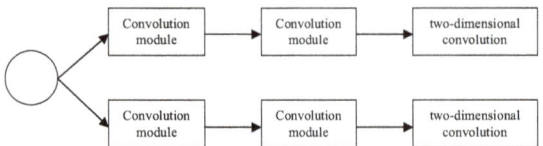

Fig. 7. Decoupled head network structure diagram.

Its network structure consists of two parallel branches that extract category features and position features respectively, and then use one layer of 1x1 convolution to complete the classification and localization tasks. The network structure is shown in Fig. 7.

Model Training. The training samples were collected from the navigation data of the 'Hangdao-1' and 'Changjianghuoyun001' ships in the Wuhan section of the Yangtze River. The target types are divided into boats, beams (navigation aids) and bridges. The number of labels is shown in Fig. 8. This paper selects 16944 training images, 4264 test images, and 2356 validation images. Use the image annotation tool MakeSense to annotate the dataset. Using a self built server for training, the graphics card is Gigabyte RTX4090-24G, the CUDA version is 10.1.

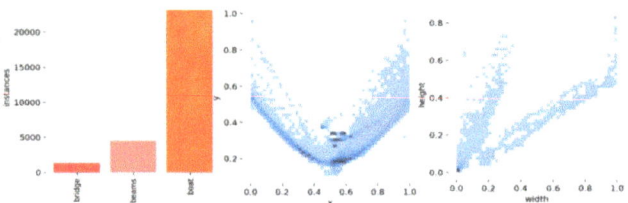

Fig. 8. Label quantity.

In this experimental analysis, a confusion matrix was used PR curve chart Analyze and explain the F1 distinguishing image and other indicators such as ship target detection accuracy, recall rate, and missed detection rate. Considering the network training time limit and final effect, as well as device performance, the training iteration of the algorithm in this paper is set to epoch = 300, batch_size = 16, The final training duration is 8 h. This article analyzes the overall performance of the object detection network and obtains the confusion matrix as shown in Fig. 9. The confusion matrix is a summary of the prediction results for classification problems, while subdividing the accuracy and recall of each category. It is one of the most intuitive performance representations for object detection classification. In this confusion matrix, the classification recall rate for ship targets reached 94%, the recall rate for navigational targets reached 68%, and the recall rate for bridge targets reached 100%.

Further analyze the relationship between F1 score and confidence (x-axis), as shown in Fig. 10. The F1 classification indicates that the larger the score, the stronger the algorithm's classification ability and better performance. From the graph, it can be seen that the confidence level of ship targets in the overall classification is between 0.8 and 1.0, demonstrating good discriminative classification ability. Observing the PR curve in

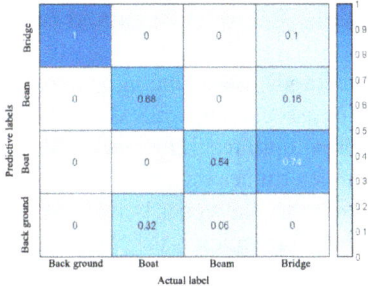

Fig. 9. Normalized confusion matrix.

the graph, after calculating the mean AP for each class, the mean mAP for all classes is 0.925, indicating that the performance of the trained model is relatively stable. As shown in Fig. 10. F1-curve Fig. 11, the training results of the detector show that the mean loss function (Box) shows a gradually decreasing gradient below 0.6, indicating that the box has high accuracy. After training, the mean loss of object detection also showed a range of values approaching 0.25, indicating that object detection is more accurate. The classification loss mean is also relatively good in terms of classification realization, with no significant fluctuations in accuracy and recall curves, indicating that the training is effective, the performance is shown in Fig. 12.

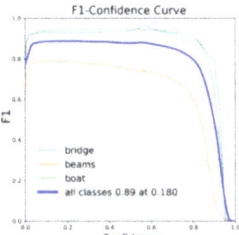
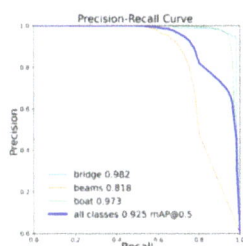

Fig. 10. F1-curve   **Fig. 11.** PR-curve

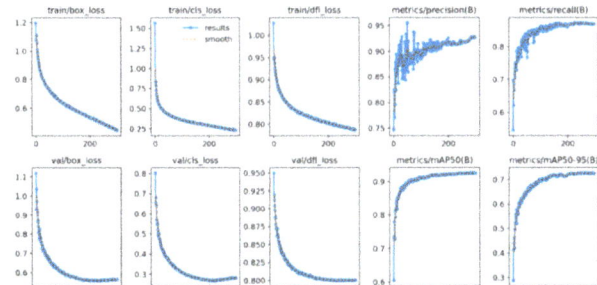

Fig. 12. Automatic evaluation results after training.

3 Multi Object Tracking Algorithm Based on ByteTrack

Most multi-target tracking methods use bounding boxes with detection scores higher than a certain set threshold for association, while discarding low score detection boxes caused by occlusion and motion blur, resulting in significant target loss and trajectory fragmentation. The detection determines the upper limit of the tracking task, while the correlation part determines whether the upper bound of the detection decision can be better achieved. Therefore, data correlation is the core of multi-target tracking tasks. To solve this problem, consider associating each detection box instead of just associating high score detection boxes, and use multiple matches and multi-step matches to improve tracking accuracy. After prioritizing the matching of high score detection boxes and trajectories, the low score detection boxes are matched with the remaining trajectories, and their similarity with previous trajectories is compared to restore the real target and filter the background[17].

3.1 Byte Track Framework

The core of the ByteTrack algorithm lies in the data association method Byte [18]. Its workflow is as follows: firstly, based on the score of the target detection box, it is divided into low score and high score detection boxes, and processed separately according to their classification. In the first matching, use the high score detection box to match the tracking trajectory before matching, and in the second matching, use the low score detection box to match the tracking trajectory that was not divided into the high score detection box before matching (such as a decrease in matching score due to target occlusion or high-speed motion blurring). Afterwards, a new tracking trajectory will be established for the target box with a high score but no matching tracking trajectory. Finally, for tracking trajectories that do not match the detection box, a certain amount of time is reserved for matching when the target reappears. Its working principle can be simply understood as occlusion often accompanied by a slow decrease in detection score from high to low. The occluded object is a visible object before being occluded, with a high detection score. A trajectory is established, and when an object is occluded, the occluded object can be extracted from the low bounding box by the degree of overlap between the detection box and the trajectory, maintaining the continuity of the trajectory.

3.2 Byte Association Method

The inputs for Byte include video clip V, detector Det, and Kalman filter KF, with three thresholds set simultaneously τ_{high}, τ_{low}, ε. The first two are detection score thresholds, and the last one is tracking score threshold. The output is the trajectory of the video, with each trajectory containing the detection box and identity of the target.

For each frame (picture) of input video, use the detector Det to predict its detection box and score. For detection boxes with scores higher than the upper threshold τ_{high}, place them in the high score detection box D_{high}. For detection boxes with scores above the lower threshold and below the upper threshold τ_{high}, put them in the low score detection box D_{low}. After separating the low and high detection boxes, a Kalman filter is used to predict the new position of each trajectory in the current frame for all trajectories in the trajectory set T. This process is discussed in detail as follows.

Firstly, the high score detection box is correlated with all trajectories T (including lost trajectories), and the similarity is calculated by IOU parameter between the detection box predicted by the detection network and the tracking box predicted by the Kalman filter. Then, the Hungarian algorithm is used for association matching. Reject matching detection boxes with similarity less than 0.2 and store unmatched detection boxes D_{remain}. Store the unmatched trajectories in the middle, and the first association process is roughly shown in Fig. 13.

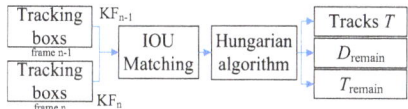

Fig. 13. Primary association.

Then, the second association is performed between the low score detection box and the remaining trajectories T_{remain}, using IOU as the similarity alone. As the low score detection box contains situations where objects do not appear or are blurred due to occlusion or high-speed motion during the detection process. The occurrence of these situations makes the appearance features detected at the current time point unreliable. Store the unmatched trajectories in the box $T_{re-remain}$ and delete the unmatched low score detection boxes. The second association process is roughly shown in Fig. 14.

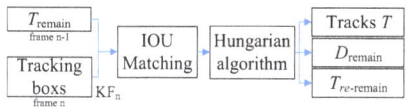

Fig. 14. Secondary association.

For the unmatched trajectories after the second association, it is considered that the target has been temporarily lost and placed in the box T_{lost}. For each trajectory in the box T_{lost}, it will only be deleted from trajectory T when it exceeds a certain time, otherwise, keep it in the trajectory.

Finally, for high score detection boxes that have never been matched after the first association, if the detection scores of these targets are high enough, and survive for more than two frames, they are treated as new trajectories. The output of each frame is the bounding box and identity of the trajectory simultaneously, however it does not output the bounding box and identities. In this research, the value τ_{high} was set as 0.4, τ_{low} was set as 0.1, ε was set as 0.5. The low score detection boxes gained by occlusion and motion blur issues were restored, matching the low score detection boxes with the trajectory, so that the algorithm can achieve better tracking results.

4 Experiments and Verification

To conduct this research, high defination cameras, navigation radars and AIS (Automatic Identification System) receivers were assembled on the vessel 'Changhanghuoyun 001' to collect video and camera image data for training and testing, as shown in Fig. 15. The collected sample data and scenarios are widely applicable, which can reflect the characteristics of inland ships.

Fig. 15. Test vessel

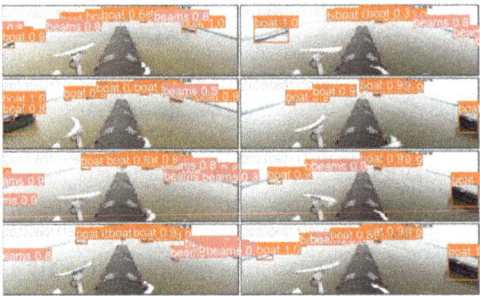

Fig. 16. Video target recognition and tracking effectiveness.

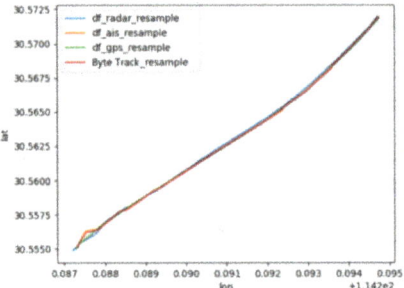

Fig. 17. Trajectory comparison.

Firstly, this paper adopts the RTSP (Real Time Streaming Protocol) to build an intelligent recognition platform by obtaining real-time camera images. Secondly, using YOLO deep learning algorithm to recognize ship targets within the image. Finally, the Byte Track algorithm was used to track the detected video targets, and the real-time output of target information was achieved. The accuracy of video target recognition was 87.2%, with a recognition accuracy of over 50% within the target recognition framework reaching 93.5%. This indicates that the video target recognition and tracking method proposed in this paper has good recognition accuracy and precision under complex navigation scene conditions, meeting the requirements of engineering use, the detection and tracking results were shown in Fig. 16. It can be seen that the targets including boats and beams were detected successfully with high matching scores. Figure 17 is the Byte Track method tracking performance compared with other sensors, such as radar, ais and GPS, which were assembled on the experiment vessel. It can be seen that the trajectory gained by Byte Track method based on image detection was almost the same as other sensors, shown the practicality of this method.

5 Conclusion

To conduct the ship image targets detection and tracking in complex traffic scenes and navigation environment in inland rivers, target recognition and tracking of perception data from AIS, radar, and cameras were carried out in the experimental scenarios of controlling river sections, bridge areas, and busy traffic areas, achieving estimation and tracking of ship trajectories in complex inland river scenes. The research results are as follows:

(1) A ship trajectory prediction method using Kalman filtering algorithm and long short-term neural network models has been proposed. The effectiveness of AIS trajectory prediction for ships in curved inland waterways was analyzed. The experimental results showed that, driven by a large amount of AIS data and combined with ship dynamic models, long and short-term neural network models had better accuracy in ship AIS trajectory prediction.
(2) An interactive multi-model trajectory filtering algorithm considering the encounter situation of inland ships and a radar target recognition and tracking algorithm for SORT multi-target tracking have been proposed. Proposed a radar target detection method based on connected domain detection method. A radar multi-target tracking algorithm based on SORT was proposed, and an interactive multi-model filtering method based on encounter situations was integrated to improve the radar target tracking performance in complex scenes, achieving an accuracy rate of 89.5% for radar target recognition and tracking.
(3) We have developed a video target recognition and tracking method based on YOLO v8 as the target detector and Byte Track as the target tracker, achieving an accuracy of 87.2% for identifying and tracking targets such as ships, navigational aids, and bridges in video images of inland ships.

References

1. Lei, J.Y., Chu, X.M., Jiang, Z.L., et. al.: Situation awareness system for vessel navigation based on visual analytics. Navigation of China **41**(03): 47–52 (2018)
2. Gaglione, D., Soldi, G., Meyer, F., et al.: Bayesian information fusion and multitarget tracking for maritime situational awareness. IET Radar Sonar Navig. **14**(12), 1845–1857 (2020)

3. Hermann, D., Galeazzi, R., Andersen, J.C., et al.: Smart sensor based obstacle detection for high-speed unmanned surface vehicle. IFAC-PapersOnLine **48**(16), 190–197 (2015)
4. Zhang, Y.: Vessel visual image enhancement technology in complex environment. Ship Science and Technology **41**(04), 178–180 (2019)
5. Vu, V.D., Lützhöft, M., Emad, G.R.: Frequency of use—the first step toward human-centred interfaces for marine navigation systems. J. Navig. **72**, 1089–1107 (2019)
6. Jing, Q.F., Shen, H.L., Ying, Y.: A ship digital twin framework based on virtual reality system. J. Beijing Jiaotong Univ. **44**(05), 117–124 (2020)
7. Liu, Y.W., Yang, S.H., Suo, Y.F., et. al.: Application of AR- HUD technology in ship lookout. J. Jimei Univ. (Nat. Sci.) **25**(01), 32–37 (2020)
8. Lei, D., Goerlandt, F., Banda, O.A.V., et al.: Improving stand-on ship's situational awareness by estimating the intention of the give-way ship. Ocean Eng. **201**, 107110 (2020)
9. Szlapczynski, R., Szlapczynska, J.: A method of determining and visualizing safe motion parameters of a ship navigating in restricted waters. Ocean Eng. **129**, 363–373 (2017)
10. Banda, V.A.O., Goerlandt, F,, Montewka, J., et al.: A risk analysis of winter navigation in Finnish sea areas. Accident Analy. Prevent. **79** (2015)
11. Bi, Z.B., Zhang, S.Y., Yang, H., et al.: Survey of ship detection in video surveillance based on shallow machine learning. J. Syst. Simulat. **33**(12), 2792–2807 (2021). (In Chinese)
12. Bousetouane, F., Morris, B.: Fast CNN surveillance pipeline for fine-grained vessel classification and detection in maritime scenarios. In: 2016 13th IEEE International Conference on Advanced Video and Signal Based Surveillance (AVSS), pp. 242–248. IEEE (2016)
13. Moon, T.K.: The expectation-maximization algorithm. IEEE Signal Process. Mag. **13**(6), 47–60 (1996)
14. Wang, Z., Zhang, C.Y., Yin, J.: Application realization of YOLO V3 model target recognition based on HXAI 100 of Hun Xin 5. China Integrated Circuit **32**(09), 15–19 (2023). (In Chinese)
15. Liang, D., Yang, F., Zhang, T., et al.: Understanding mixup training methods. IEEE Access **6**, 58774–58783 (2018)
16. Sural, S., Qian, G., Pramanik, S.: Segmentation and histogram generation using the HSV color space for image retrieval. In: Proceedings International Conference on Image Processing, vol. 2, pp. I-II (IEEE, 2002)
17. Zhang, Y., Sun, P., Jiang, Y., et al.: Bytetrack: multi-object tracking by associating every detection box. In: European Conference on Computer Vision., pp. 1–21. Springer Nature Switzerland, Cham (2022). https://doi.org/10.1007/978-3-031-20047-2_1
18. Gaglione, D., Braca, P., Soldi, G.: Belief propagation based AIS/radar data fusion for multi-target tracking. In: 2018 21st International Conference on Information Fusion (FUSION), pp. 2143–2150. IEEE (2018)

Open Access This chapter is licensed under the terms of the Creative Commons Attribution-NonCommercial-NoDerivatives 4.0 International License (http://creativecommons.org/licenses/by-nc-nd/4.0/), which permits any noncommercial use, sharing, distribution and reproduction in any medium or format, as long as you give appropriate credit to the original author(s) and the source, provide a link to the Creative Commons license and indicate if you modified the licensed material. You do not have permission under this license to share adapted material derived from this chapter or parts of it.

The images or other third party material in this chapter are included in the chapter's Creative Commons license, unless indicated otherwise in a credit line to the material. If material is not included in the chapter's Creative Commons license and your intended use is not permitted by statutory regulation or exceeds the permitted use, you will need to obtain permission directly from the copyright holder.

Human Error Probability Analysis on Pilot's Operation in Changshu Section of the Yangtze River

Haoxuan Yan[1]($\boxtimes$), Cunlong Fan[1], Linglong Yan[2], Min Wang[3], Yongtao Xi[4], Shenping Hu[4], and Bing Han[5]

[1] College of Transport & Communications, Shanghai Maritime University, Shanghai, China
engpex@163.com
[2] Qinhuangdao Maritime Safety Administration of the People's Republic of China, Qinhuangdao, China
[3] Shanghai Maritime Pilot's Association, Shanghai, China
[4] Merchant Marine College, Shanghai Maritime University, Shanghai, China
[5] Shanghai Ship and Shipping Research Institute Co., Ltd, Shanghai, China

Abstract. The critical role that pilots take part in the inland navigation has been widely recognized, emphasizing the great importance to conduct pilotage safety analysis. This paper analyzes Human Error Probability (HEP) of pilot's operation in the Changshu section of the Yangtze River. Firstly, a detailed hazard analysis is conducted to derive the Performance Shaping Factors (PSFs). Then, the HEP of the pilot's operation is calculated by using the Success Likelihood Index Method (SLIM). Four tasks of pilot's operation are examined, which are Navigation, Crossing the channel, Berthing operation, and Departure. In each task, the HEP of sub-tasks are quantified based on experts' judgement, and the highest and lowest HEPs of sub-tasks are identified. The results shed light on the management of pilotage safety in the downstream of the Yangtze River.

Keywords: Inland Navigation · Pilotage Safety · Human Factor · Success Likelihood Index Method

1 Introduction

Inland navigation has been steadily developed because of its large capacity and low freight price. However, inland navigation is an inherently risky activity [1]. It is mainly because the vessels are confronted with restricted conditions, e.g., confined depth. In such situations, even minor errors may lead to severe consequences, e.g., collision and grounding [2, 3]. In 2015, the Eastern Star capsized and resulted in more than 400 deaths [4]. To better prevent these hazard accidents, mandatory inland pilotage is often implemented in inland navigation, such as the compulsory pilotage in the downstream of the Yangtze River.

In the downstream of the Yangtze River, there are numerous ports, and high volume of vessel traffic including oceangoing vessels. These oceangoing vessels usually have larger dimensions, which pose more risk to pilotage safety [5–7]. Once accidents occur, the costal economy may be negatively impacted. Thus, the reliability of pilot's operation is important. More importantly, pilotage in confined waterways is faced with significant challenges. These challenges are complex hydrodynamic force and bank effect in navigation phrase [8]; passing the anchorage in crossing the channel [9, 10]; the length of berths [11] and layout of berths [12] in berthing operation. In almost every phrase of the pilotage, it is seen that there are some hazardous factors that impact pilotage safety significantly. The main responsibility for pilots is to guide vessels in and out of the port safely. In this process, the pilot needs to communication with tugboats and surrounding vessels. It is recognized that pilotage is essential to ensuring the maritime safety [13]. Due to the crucial role that pilots take in the inland navigation, their performance is often studied, e.g., pilots' safety behavior [14, 15]. Researchers also conduct Human Reliability Analysis (HRA) to pilotage safety [1, 16]. However, few study devotes to investigating HRA of pilot operation in the downstream of the Yangtze River. In this context, this paper conducts HRA of the pilot's operation in a berth in the Changshu section of the Yangtze River.

2 Methodology

Success Likelihood Index Method (SLIM) is developed by Embrey [17]. It is assumed that the occurrence of error is the result of various factors. Given that SLIM relies on PSFs to quantify the reliability of specific tasks, PSFs are required to be selected properly [18]. Generally, the process of SLIM is divided into the following four parts: PSF derivation, PSF rating, PSF weighing, SLI determination, and HEP value conversion [19]. These four parts are introduced in the following subsections.

2.1 PSF Derivation

PSFs are selected from environmental and navigational conditions, and related literature. In this process, experts' experience can assist in the PSFs determination.

2.2 PSF Weighing

The weight of a PSF indicates its relative importance, which varies from tasks. The relative importance is determined by experts' judgement, which is required to be normalized by Eq. (1):

$$Normalized\ Value(W_i) = \frac{Mean\ weight\ for\ i-th\ PSF}{\sum_{i=1}^{n} Mean\ Weight\ for\ i-th\ PSF} \quad (1)$$

where Wi denotes the weight of i-th PSF, Mean weight for i-th PSF is the sum of all experts' rating on a PSF, n denotes the numbers of PSFs.

2.3 Human Error Calculation

Each PSF is evaluated by experts. In this paper, a value from 10 to 100 in a linear scale is selected. Then, the consensus rating (R_i) is obtained according to Eq. (2):

$$R_i = \sum_{i=1}^{n} PSF_{ij} \times W_j \qquad (2)$$

The relative importance of experts is calculated based on the evaluation criteria shown in Table 1. In Table 1, the evaluation criteria include four aspects, i.e., the educational background, professional title, maximum length leading vessels, and tenure.

Table 1. Criteria for expert's relative importance evaluation.

Score	Educational Background	Professional title	Maximum length leading vessels (m)	Tenure (year)
25	Bachelor's degree and above	first-class	above 250	above 15
20	Secondary school	second-class	[200,250)	[10,15)
15	NULL	third-class	[150,200)	[5,10)

2.4 Human Error Calculation

After the derivation of Consensus rating (R_i) and Normalized value of PSFs (W_i), the Success Likelihood Index (SLI) is obtained by Eq. (3):

$$SLI = \sum_{i=1}^{n} W_i \times R_i \qquad (3)$$

Experts' judgement is integrated to determine the constant a and b. These two constants are assigned respectively for four tasks. Eventually, the SLI value is converted into HEP values by Eq. (4):

$$Log(HEP) = a \times SLI + b \qquad (4)$$

3 Case Study

3.1 Overall Situation

The waters in this case study are located in the lower reaches section of the Yangtze River, with the total south shoreline length of 37.5 km. It is under the jurisdiction of the Changshu Maritime Administration of the People's Republic of China. This Changshu section is divided into two parts, which are separated by the Sutong Yangtze River Bridge. As demonstrated by Fig. 1, The upstream part is called Shadong Waterway. Correspondingly, the downstream is called Baimaosha Waterway. The navigable mileage of the Changshu Section is approximately 82 km. Notably, with the completion of 12.5 m deep-water channel, the Changshu Section has witnessed a steadily increase of oceangoing vessels with large scale.

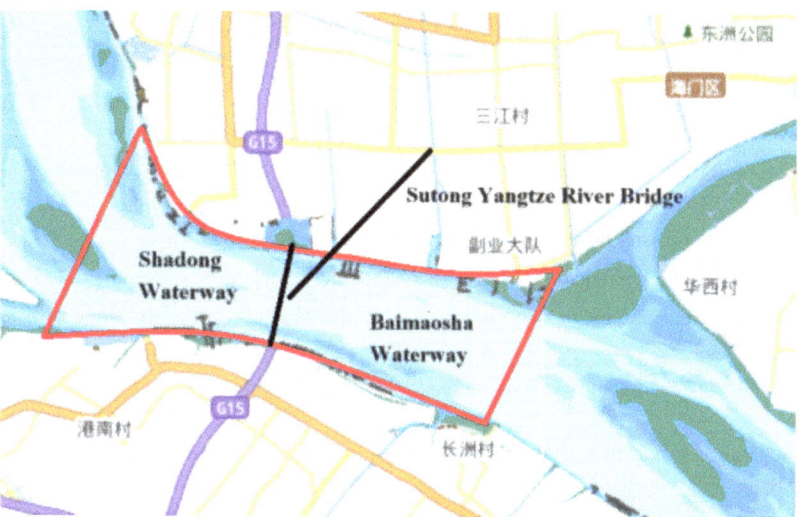

Fig. 1. The Changshu Section of the Yangtze River. (The map is from: https://www.cjhy.com.cn/home)

There are numerous berths in the Changshu section, and a wide variety of cargos are handled. Among these berths, approximately 25% of the berths are dedicated to hazardous cargos, including liquid gas, chemical, and petrochemical. Accidents involved with these hazardous cargos are usually more severe [20]. The release of these materials may severely threat life, property, and environment [21]. In addition, such dense distribution of berths also poses greater impacts on the pilotage of ultra-large vessels. It is because the hydrodynamic interaction between two ultra-large vessels is significant, requiring early action and effective communication [22].

The Sutong Yangtze River Bridge is located in the Changshu Section. The water flow under the bridge is turbulent. When the wind force reaches five degree or above, huge surging waves occur. In addition, vessels may pose great danger to the bridge. Vessel-bridge collisions are the most disastrous accidents in navigation [23]. Currently, more than 30 vessels of over 50,000 tons pass the Sutong Yangtze River Bridge every day, with the heaviest vessels reaching 400,000 tons. However, the anti-collision measure of Sutong Yangtze River Bridge is designed for vessels less than 50,000 tons. Therefore, once vessels lose control in the bridge area, it may pose catastrophic threat to the bridge.

Apart from these, there are three main anchorages, i.e., Changshu Anchorage, Baimaosha Anchorage, and Changshu Port lightering Anchorage, and two main ferry lines, i.e., the Tongchang Ferry Line and Haitai Ferry Line.

3.2 Result

Identification of Tasks. In the case study, a pilin's operation is generally decomposed into four tasks, i.e., Navigation, Crossing the channel, Berthing operation, and Departure operation. The four tasks are further broken down into several sub-tasks that are rated by pilots. All sub-tasks are presented in Table 2 and Table 3.

Table 2. Tasks 1 and 2 in the case study

Task 1	Task 2
Navigation	Crossing the channel
Exchange information with the captain	Communication (reporting vessel dynamics to VTS)
Enhance lookout during navigation	Control the vessel speed
Use safe speed during navigation	Position the vessel
Maintain communication with VTS and nearby vessels	Coordinate with other vessels to avoid collision
Avoid collision situations	Choose appropriate timing to cross a shipping lane
Get preparation to anchor during navigation	Adjust the vessel position timely
Take emergency anchoring measures	

Table 3. Task 3 and 4 in the case study.

Task 3	Task 4
Berthing operation	Departure
Control vessel speed	Work out the departure plan
Control the berthing angle	Exchange information with the pilot
Control the berthing transverse distance	Test the propulsion and steering system
Control the berthing normal velocity	Coordinate with tugboats
Turning maneuver	Release mooring lines
Coordinate with tugboats	Choose the appropriate departure time
Take emergency procedures	Choose appropriate time to enter the main channel of the waterway
Mooring operations	

PSF Derivation in This Study. SLIM relies on PSFs to calculate the human error probabilities. Hence, the PSFs are required to be properly selected. By detailed analysis of the Changshu Section and literature review, nine PSFs are selected and shown in Table 4.

PSF Weighting in this Study. Six experts are invited by questionnaire survey to assign ratings for each PSF of all sub-tasks between 10 and 100 on a linear scale that best expresses their judgements, with results shown in Table 5. Then, the normalized value of PSFs (W_i) is calculated by Eq. (1).

Table 4. Selected PSFs in this paper.

PSFs	Literature
Safety responsibility attitude	[16]
Competence	[8]
Psychological quality	[24, 25]
emergency response capability	[26, 27]
Ship seaworthiness	[28, 29]
Environmental conditions	[25, 30]
Navigational conditions	[31]
Communication	[1, 24]
Pilotage safety culture	[15, 32]

Table 5. PSF rating.

PSF	Ex1	Ex2	Ex3	Ex4	Ex5	EX6	Mean Weight	W_i
PSF1	70	85	85	90	70	90	490	0.15
PSF2	80	90	90	90	90	20	460	0.141
PSF3	20	50	80	85	50	10	295	0.09
PSF4	60	20	50	90	20	10	250	0.076
PSF5	70	20	70	90	10	80	340	0.104
PSF6	50	10	60	90	10	50	270	0.083
PSF7	10	10	20	90	10	50	190	0.058
PSF8	80	90	90	90	90	80	520	0.159
PSF9	70	80	50	85	85	85	455	0.139

PSF Rating in this Study. In this process, we first determine the relative importance of the j-th expert, i.e., W_j. Experts may have distinctive working experiences and educational backgrounds. Thus, they are likely to have various perceptions with the impact of each PSF. The information of involved experts is presented in Table 6. Based on Tables 1 and 5, the relative importance of experts is determined and shown in last column of Table 6. Then, experts give points for rating PSFs, and all ratings for each PSFs are shown in Table 7. Finally, based on Eq. (2), the Consensus Rating of each PSF is determined and also shown in Table 7.

SLI Derivation in this Study. After obtaining the Consensus Rating (R_i) and normalized value of PSFs (W_i), Success Likelihood Index (SLI) is generated by Eq. (3). For example, the SLI of sub-task 1.1 is generated and shown in Table 8.

Table 6. Information of experts.

Number	Educational Background	Professional title	Maximum length leading vessel (m)	Tenure (year)	W_j
1	Bachelor degree	first-class	255	18	0.18
2	Secondary school	first-class	202	11	0.17
3	Bachelor degree	second-class	174	10	0.17
4	Bachelor degree	first-class	295	14	0.17
5	Bachelor degree	third-class	300	5	0.15
6	Bachelor degree	first-class	254	12	0.15

Table 7. The calculation of Consensus Rating (R_i).

PSF	Ex1	Ex2	Ex3	Ex4	Ex5	Ex6	Consensus Rating (R_i)
W_j	0.18	0.17	0.17	0.17	0.15	0.15	
PSF1	80	90	90	90	90	80	85.8
PSF2	80	80	90	90	90	70	82.6
PSF3	70	80	90	90	90	60	79.3
PSF4	70	80	90	90	90	60	79.3
PSF5	60	70	70	90	90	50	70.9
PSF6	50	80	70	90	90	50	70.8
PSF7	50	70	70	90	90	40	67.6
PSF8	60	70	90	90	90	40	72.8
PSF9	60	50	50	80	70	30	56.4

Table 8. The calculation of SLI for sub-task 1.1.

Normalized Value of PSFs (W_i)	0.15	0.141	0.09	0.076	0.104	0.083	0.058	0.159	0.139
Consensus Rating (R_i)	85.8	82.6	79.3	79.3	70.9	70.8	67.6	72.8	56.4
SLI value	74.266								

Human Error Calculation in this Study. Human Error is calculated by Eq. (4). In this paper, absolute error probability judgement is completed by pilots. This method requires pilots to estimate human error probability of the best and worst scenario which is used to compute the constant a and b. Table 9 shows the constant a and b assigned for task 1 to 4.

Table 9. The constant a and b.

Task	Estimated HEP for the best scenario	Estimated HEP for the Worst scenario	constant a	constant b
Task1	1.00E-04	1.00E-02	4.32E-05	−4.36E-03
Task2	1.00E-03	1.00E-01	4.53E-04	−4.58E-02
Task3	1.00E-03	1.00E-01	4.53E-04	−4.58E-02
Task4	1.00E-05	1.00E-03	4.30E-06	−4.35E-04

After the constant a and b are calculated for all tasks, the HEP values of all sub-tasks are derived. HEP values of tasks are presented in Tables 10–13.

Table 10. HEP of sub-tasks in Task 1.

sub-tasks	HEP	Task 1 Navigation
1.1	2.66E-03	Exchange information with the captain
1.2	2.64E-03	Enhance lookout during navigation
1.3	2.65E-03	Use safe speed during navigation
1.4	2.64E-03	Maintain communication with VTS and surrounding vessels
1.5	2.61E-03	Avoid collision situations
1.6	2.60E-03	Get preparation to anchor during navigation
1.7	2.64E-03	Take emergency anchoring measures

Table 11. HEP of sub-tasks in Task 2.

sub-tasks	HEP	Task 2 Crossing the channel
2.1	2.70E-02	Communication (reporting vessel dynamics to VTS)
2.2	2.75E-02	Control the vessel speed
2.3	2.72E-02	Position the vessel
2.4	2.72E-02	Coordinate with other vessels to avoid collision
2.5	2.70E-02	Choose appropriate timing to cross a shipping lane
2.6	2.72E-02	Adjust the vessel position timely

Table 12. HEP of sub-tasks in Task 3.

sub-tasks	HEP	Task 3 Berthing operation
3.1	2.72E-02	Control vessel speed
3.2	2.72E-02	Control the berthing angle
3.3	2.72E-02	Control the berthing transverse distance
3.4	2.72E-02	Control the berthing normal velocity
3.5	2.73E-02	Turning maneuver
3.6	2.74E-02	Coordinate with tugboats
3.7	2.73E-02	Take emergency procedures
3.8	2.75E-02	Mooring operations

Table 13. HEP of sub-tasks in Task 4.

sub-tasks	HEP	Task 4 Departure
4.1	2.60E-04	Work out the departure plan
4.2	2.58E-04	Exchange information with the pilot
4.3	2.61E-04	Test the propulsion and steering system
4.4	2.64E-04	Coordinate with tugboats
4.5	2.63E-04	Release mooring lines
4.6	2.64E-04	Choose the appropriate departure time
4.7	2.63E-04	Choose appropriate time to enter the main channel of the waterway

4 Conclusion

This paper performed human reliability analysis of the inland pilotage in the Changshu section of the Yangtze River by utilizing the SLIM. The main findings of this research are as follows: In Task 1 (Navigation), the HEP of Exchange information with the captain is the highest, while the HEP of Get preparation to anchor during navigation is the lowest. In Task 2 (Crossing the channel), the HEP of Control the vessel speed is the highest, while the HEPs of Choose appropriate timing to cross a shipping lane and Communication (reporting vessel dynamics to VTS) are the lowest. In Task 3 (Berthing operation), the HEP of Mooring operations is the highest, while the HEPs of Control vessel speed, Control the berthing angle, Control the berthing transverse distance, and Control the berthing normal velocity are the lowest. In Task 4 (Departure), the HEPs of Coordinate with tugboats and Choose the appropriate departure time are the highest, while the HEP of Exchange information with the pilot is the lowest.

Acknowledgements. This work is supported by funding from the State Key Laboratory of Maritime Technology and Safety (Grant No. W24CG000042), National Natural Science Foundation of

China (Grant No. 52301419), Fund of National Engineering Research Center for Water Transport Safety (Grant No. A202404). The authors also greatly appreciate Mr. Lei Feng from Changshu pilot station, Yangtze River Pilotage Center for his assistance in questionnaire survey. The views expressed remain solely those of the authors.

References

1. Abreu, D.T.M.P., Maturana, M.C., Droguett, E.L., Martins, M.R.: Human reliability analysis of conventional maritime pilotage operations supported by a prospective model. Reliab. Eng. Syst. Saf. **228**, 108763 (2022)
2. Zhang, D., Yan, X.P., Yang, Z.L., Wall, A., Wang, J.: Incorporation of formal safety assessment and Bayesian network in navigational risk estimation of the Yangtze River. Reliab. Eng. Syst. Saf. **118**, 93–105 (2013)
3. Xi, Y.T., Yang, Z.L., Fang, Q.G., Chen, W.J., Wang, J.: A new hybrid approach to human error probability quantification–applications in maritime operations. Ocean Eng. **138**, 45–54 (2017)
4. Wang, Y., Zio, E., Wei, X., Zhang, D., Wu, B.: A resilience perspective on water transport systems: The case of Eastern Star. Int. J. Disaster Risk Reduct. **33**, 343–354 (2019)
5. Kevin Cullinane, M.K.: Economies of scale in large containerships optimal size and geographical implications. J. Transp. Geogr. **8**, 181–195 (2000)
6. Höffmann, M., Roy, S., Berger, A., Bergmann, W.: Wind affected maneuverability of tugboat-controlled ships. In: IFAC Conference Paper Archive, pp. 71–75 (2021)
7. Eliopoulou, E., Papanikolaou, A., Voulgarellis, M.: Statistical analysis of ship accidents and review of safety level. Saf. Sci. **85**, 282–292 (2016)
8. Fan, S., Yang, Z., Wang, J., Marsland, J.: Shipping accident analysis in restricted waters: Lesson from the Suez Canal blockage in 2021. Ocean Eng. **266**, 113119 (2022)
9. Liu, C.-P., Liang, G.-S., Su, Y., Chu, C.-W.: Navigation safety analysis in Taiwanese ports. J. Navig. **59**, 201–211 (2006)
10. Debnath, A.K., Chin, H.C.: Modelling collision potentials in Port Anchorages: Application of the Navigational Traffic Conflict Technique (NTCT). J. Navig. **69**, 183–196 (2015)
11. Das, S.N., Kulkarni, S., Kudale, M.D.: Design of safe mooring arrangement for large oil tankers. Procedia Eng. **116**, 528–534 (2015)
12. Khan, R.U., Yin, J., Mustafa, F.S., Liu, H.: Risk assessment and decision support for sustainable traffic safety in Hong Kong waters. IEEE Access. **8**, 72893–72909 (2020)
13. Wild, C.R.J.: The paradigm and the paradox of perfect pilotage. J. Navig. **64**, 183–191 (2010)
14. Xi, Y., Hu, S., Yang, Z., Fu, S., Qin, T.: The moderating effect of risk tolerance on the hazardous attitudes and safety behavior of maritime pilots: A Chinese case. In: The 5th International Conference on Transportation Information and Safety (2019)
15. Xi, Y., Zhang, Q., Hu, S., Yang, Z., Fu, S.: The effect of social cognition and risk tolerance on marine pilots' safety behaviour. Marit. Policy Manag. **48**, 1–18 (2020)
16. Aydin, M., Uğurlu, Ö., Boran, M.: Assessment of human error contribution to maritime pilot transfer operation under HFACS-PV and SLIM approach. Ocean Eng. **266**, 112830 (2022)
17. Embrey, D.E.: Incorporating management and organisational factors into probabflistic safety assessment. Reliab. Eng. Syst. Saf. **38**, 199–208 (1992)
18. Kim, J.W., Jung, W.: A taxonomy of performance influencing factors for human reliability analysis of emergency tasks. J. Loss Prev. Process Ind. **16**, 479–495 (2003)
19. Akyuz, E.: Quantitative human error assessment during abandon ship procedures in maritime transportation. Ocean Eng. **120**, 21–29 (2016)

20. Cao, X., Lam, J.S.L.: A fast reaction-based port vulnerability assessment: Case of Tianjin Port explosion. Transp. Res. A Policy Pract. **128**, 11–33 (2019)
21. Wu, C., Huang, L.: A new accident causation model based on information flow and its application in Tianjin Port fire and explosion accident. Reliab. Eng. Syst. Saf. **182**, 73–85 (2019)
22. Lee, C.-K., Moon, S.-B., Jeong, T.-G.: The investigation of ship maneuvering with hydrodynamic effects between ships in curved narrow channel. Int. J. Nav. Archit. Ocean Eng. **8**, 102–109 (2016)
23. Zhang, L., Chen, P., Li, M., Chen, L., Mou, J.: A data-driven approach for ship-bridge collision candidate detection in bridge waterway. Ocean Eng. **266**, 113137 (2022)

Open Access This chapter is licensed under the terms of the Creative Commons Attribution-NonCommercial-NoDerivatives 4.0 International License (http://creativecommons.org/licenses/by-nc-nd/4.0/), which permits any noncommercial use, sharing, distribution and reproduction in any medium or format, as long as you give appropriate credit to the original author(s) and the source, provide a link to the Creative Commons license and indicate if you modified the licensed material. You do not have permission under this license to share adapted material derived from this chapter or parts of it.

The images or other third party material in this chapter are included in the chapter's Creative Commons license, unless indicated otherwise in a credit line to the material. If material is not included in the chapter's Creative Commons license and your intended use is not permitted by statutory regulation or exceeds the permitted use, you will need to obtain permission directly from the copyright holder.

Research on Magnetic Field Distribution of Rotating Magnetic Target Under Water

Pengfei Lin[1], Ming Chang[2(✉)], Jun Li[1], Lei Xu[3], and Heda Zhao[4]

[1] Dalian Naval Academy, Dalian, China
[2] Naval University of Engineering, Wuhan, China
cm702165937@163.com
[3] Naval Research Institute, Tianjin, China
[4] Naval Research Institute, Beijing, China

Abstract. The magnetic field generated by the rotation of the magnetic source is a dynamic magnetic field with a certain frequency, which has strong ability to resist interference, and it can not only improve communication quality and system performance but also enhance information security and spectrum utilization efficiency, which hold significant application value in fields such as resource exploration and ocean exploration. In order to make use of the rotating magnetic for ocean exploration, the model of the rotating magnetic source (RMS) is modeled with the rotation axis as the z-axis in paper, and we analyze the spatial magnetic field distribution of the stationary magnetic source and the moving magnetic source, respectively. Besides, the variation of magnetic field amplitude with distance is analyzed, and the results show that the three-axis magnetic component is a sinusoidal signal with a certain frequency, and the triaxial magnetic components generated by the magnetic moment components mp and mf are symmetrically distributed about B = 0, and the static magnetic field generated by the vertical magnetic moment component ml is superimposed with the sinusoidal signals generated by the magnetic moment components mp and mf, which causes the amplitude to shift. The study is of great significance which provide a theoretical foundation for the ocean exploration.

Keywords: Ocean exploration · Rotating magnetic source · Magnetic moment · Magnetic field

1 Introduction

Magnetic positioning technology is a technique that uses magnetic field signals to locate targets. This technology is less affected by multi-path effects and meteorological and hydrological conditions, and possesses the benefits of strong robustness, high positioning accuracy, and low cost. Therefore, magnetic positioning has become an important part of ocean exploration and positioning [1, 2]. Among them, dynamic magnetic

---

This work was supported by the National Natural Science Foundation of China under Grant 42074074.

field positioning uses the changing magnetic field signal generated by the magnetic target for positioning, and the varying magnetic field generated by the rotating magnetic source is a dynamic magnetic field with a certain frequency, which possesses strong anti-interference capabilities and is less susceptible to disturbances from static magnetic field. Besides, its frequency is generally low, and can penetrate rocks, soil, buildings and other types of media. So, it is an effective way to achieve high-precision target positioning in ocean exploration [3, 4].

The rotating magnetic source (RMS) can be the target itself, for example, the propeller of a ship or a helicopter is a kind of RMS, whose rotation can generate a rotating magnetic field [5–7], or it can be a self-made rotating permanent magnet beacon, which is purposefully installed on the target and uses its radiated rotating magnetic field to locate the target. It was first used in the fields of cave surveying and coal mine safety. At present, it is also used in medical devices [8, 9], oil drilling, resource exploration, indoor positioning, robots and other fields. In addition, it can also be used for underwater frogman tracking, underwater enclosed space positioning, underwater emergency rescue, underwater construction and other applications [10–12].

Modelling and analysis of the spatial distribution of the Rotating Magnetic Field (RMF) radiated by the magnetic source is the theoretical basis of positioning research. This paper focuses on the modelling and analysis of the magnetic field of the RMS, and it analyzes the characteristics and distribution rules of the RMF distribution of the RMS under static and moving conditions, and the distribution characteristics of the RMF provide a theoretical basis for subsequent research on the target detection and reconstruction.

The rest of the paper is organized as followed. In Sect. 2, the magnetic field of the RMS is modelled. In Sect. 3, the spatial distribution of the RMF radiated by the magnetic source is simulated based on the model established in Sect. 2. Finally, the study is concluded in Sect. 4.

2 Magnetic Field Modeling of RMS

When the detection distance is much larger than the size of the target itself, the magnetic target can be equivalent to the magnetic dipole. Then the RMS can be equivalent to the rotating magnetic dipole, and the Cartesian coordinate system xyz is established with the rotation axis of the magnetic source as the z axis, and its equivalent diagram is shown in Fig. 1.

Assuming that the magnetic dipole rotating around z axis with a magnetic moment m, and rotation frequency f. The projection of the magnetic moment m on the xy plane and z axis are denoted as mxy and ml, respectively. The component of mxy on the x axis and y axis are mp and mf, respectively. The angular velocity of the magnetic moment on the xy plane is $\omega = 2\pi f$. The initial angle between the magnetic moment mxy and the x-axis is α0. The magnitudes of the magnetic moment components mp, mf and ml can be expressed as

$$\begin{cases} m_p = m_{xy}\cos(\omega t + \alpha_0) \\ m_f = m_{xy}\sin(\omega t + \alpha_0) \\ m_l = m_l \end{cases} \quad (1)$$

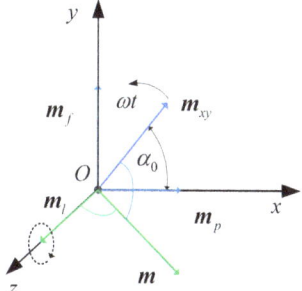

Fig. 1. The rotating magnetic dipole coordinate system

Where:

mp is the projection of the magnetic moment on the x-axis.
mf is the projection of the magnetic moment on the y-axis.
ml is the projection of the magnetic moment on the z-axis.
α0 is the initial angle between mxy and the x-axis.
ω is the angular velocity, calculated as 2πf.
t is the time variable, which accounts for the changing position of the magnetic moment as it rotates.

Based on the Biot-Savart law [13], the strength of the magnetic field can be determined:

$$\boldsymbol{B} = \frac{\mu_0}{4\pi r^3}\left[\frac{3(\boldsymbol{m}\cdot\boldsymbol{r})\times\boldsymbol{r}}{r^2} - \boldsymbol{m}\right] \quad (2)$$

Where:

B is the magnetic field vector.
μ0 is the vacuum permeability, calculated as 4π × 10-7 H/m.
r is the distance vector from the observation point to the magnetic source.
m is the magnetic moment of the RMS, m = (mp, mf, ml).

Assume that the coordinates of the observation point and the magnetic source are (x, y, z) and (x0, y0, z0), respectively. Then the distance vector r = (x-x0, y-y0, z-z0), and the tri-components of the magnetic field (Bx, By, Bz) at the observation point can be expressed as

$$\begin{bmatrix} B_x \\ B_y \\ B_z \end{bmatrix} = \frac{\mu_0}{4\pi r^5}\begin{bmatrix} 3(x-x_0)^2 - r^2 & 3(x-x_0)(y-y_0) & 3(x-x_0)(z-z_0) \\ 3(x-x_0)(y-y_0) & 3(y-y_0)^2 - r^2 & 3(y-y_0)(z-z_0) \\ 3(x-x_0)(z-z_0) & 3(y-y_0)(z-z_0) & 3(z-z_0)^2 - r^2 \end{bmatrix}\begin{bmatrix} m_{xy}\cos(\omega t + \alpha_0) \\ m_{xy}\sin(\omega t + \alpha_0) \\ m_l \end{bmatrix} \quad (3)$$

The RMF of the RMS target has uneven magnetic distribution, which results in that the magnetic moment of the RMS is not located in the center of the object, that is, the magnetic moment of the RMS does not rotate around the center of the object, but shakes within a certain range from the center of the object. The diagram illustrating the shaking RMS is depicted in Fig. 2.

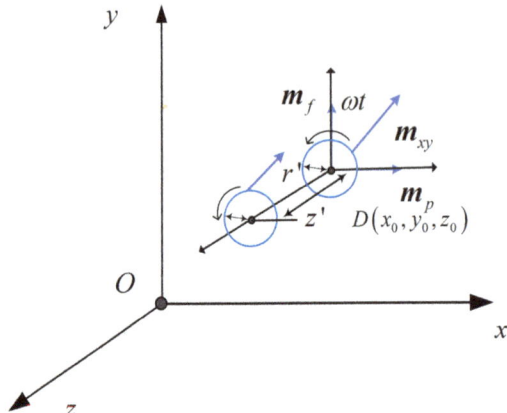

Fig. 2. Schematic diagram of magnetic moment motion caused by uneven magnetic distribution

As can be seen from Fig. 2, the magnetic moment m shakes around a circle with magnetic source D (x0, y0, z0) as the center point, and the r' is the radius in the xy plane. The shaking range in z axis is [z0-z', z0 + z'], and z' is the range of change. If the magnetic source is located at the coordinate origin D, the actual coordinate of the magnetic moment m is

$$\begin{cases} x'_0 = x_0 + r'\cos(\omega t + \alpha_0) \\ y'_0 = y_0 + r'\sin(\omega t + \alpha_0) \\ z'_0 = z_0 \pm [-z', z'] \end{cases} \quad (4)$$

x0', y0' and z0' are the actual triaxial coordinates of the RMS when there is uneven magnetic distribution and sloshing.

3 Simulation

3.1 Magnetic Field Distribution of Fixed Magnetic Source

Assume the magnetic moment of the RMS is about 1000 A . m², the magnetic moment $m = 1005$ A . m², the magnetic moment components $m_{xy} = 1000$ A . m², $m_l = 100$ A . m², the initial angle $\alpha_0 = -8/9\pi$, and the rotation frequency $f = 5$ Hz. The coordinates of the x and y axes of the observation point are 5 m and 4 m, respectively, and the range of z coordinates is [−80,80] m. And the spatial RMF distribution of the RMS is shown in Fig. 3. The RMF distribution of the RMS in the range z = [−80, 80] at a certain time is shown in Fig. 4. Taking the coordinates of the observation point as (5, 4, −80), and the time domain distribution of tri-components of the RMF at the point is shown in Fig. 5.

In Fig. 3, Bx, By and Bz are the RMF intensity of the RMS in the directions of the x-axis, y-axis, and z-axis, respectively. It is observed that in the time domain, the three-axis magnetic component manifests as a sinusoidal signal with a specific frequency, and in the space domain, the amplitude of the three-axis magnetic component initially increases and then decreases as the z-axis coordinate increase. This occurs because as

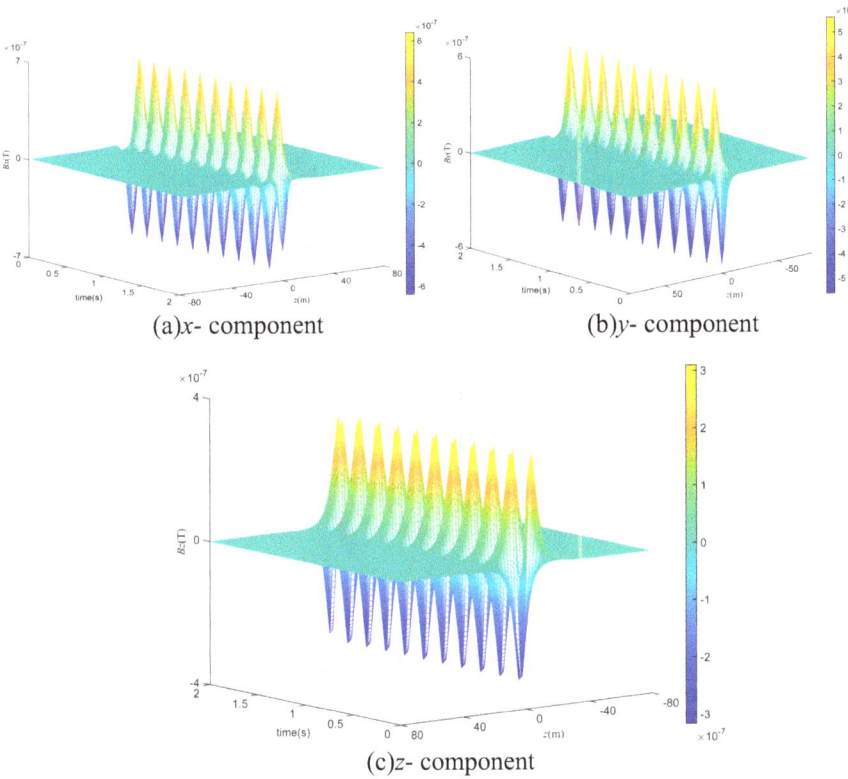

Fig. 3. Temporal and spatial distribution of the tri-components of the RMF along the z axis.

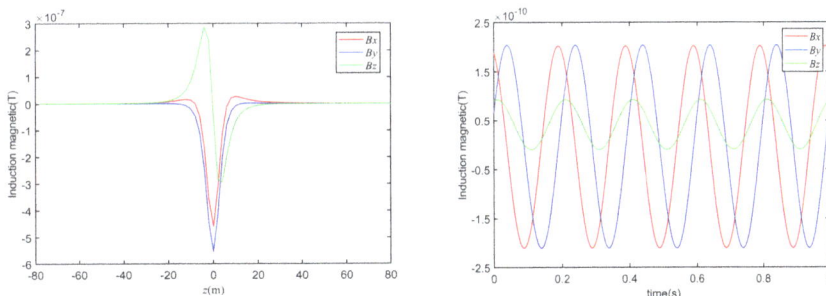

Fig. 4. Distribution of the tri-components of a specific RMF along the z axis

Fig. 5. Time-domain distribution of tri-components of RMF at a point

the observation point gradually approaches the magnetic source as the z-axis increase, and moves away from the magnetic source after passing through it.

In Figs. 4 and 5, the red curve denotes the RMF component Bx, the blue curve demotes the RMF component By, and the green curve denotes the RMF component Bz. It can be seen from Fig. 5, the triaxial magnetic component of the rotating magnetic dipole at a certain time meets the distribution characteristics of the static magnetic dipole, and the intensity of the magnetic component near the magnetic source reaches about $3 \times 10\text{-}7$ T. In Fig. 6, the intensity of the magnetic component generated by the magnetic source at this point is on the order of 10-10 nT, which is the lowest measurement threshold of the current vector magnetic sensor, when the coordinates of the observation point are (5, 4, -80) and the distance from the magnetic source is about 80.3 m. Moreover, the three-axis magnetic component at this point is a sine-cosine signal with a certain frequency. The frequency of the signal is consistent with the rotation frequency of the RMS at 5 Hz. The amplitude of the gyromagnetic three-component measured at the observation point is not symmetrical along $B = 0$. Equation (3) reveals that the static magnetic field generated by the vertical magnetic moment component ml is superimposed with the sinusoidal signals generated by the magnetic moment components mp and mf, which causes the amplitude to shift. By removing the static magnetic field generated by the magnetic moment component ml, the tri-components of the RMF whose amplitude is symmetric about the coordinate origin can be obtained, and their distribution is shown in Fig. 6.

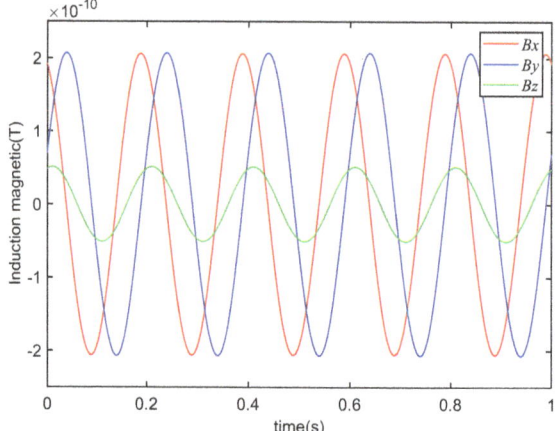

Fig. 6. Time-domain distribution of the tri-components of the RMF with static magnetic field removed

In Fig. 6, the red curve, the blue curve and the green curve represents the magnetic field component Bx, By and Bz, respectively. It can be seen that when the static magnetic field generated by the vertical magnetic moment component of ml is removed, the gyromagnetic tri-components generated by the magnetic moment components mp and mf are symmetrically distributed about $B = 0$. In actual signal processing, this part of sinusoidal signal is mainly processed.

3.2 Magnetic Field Distribution of Moving Magnetic Source

The simulation conditions are established on the basis of 3.1, the x and y axis coordinates of the observation point are set as x = 0 and y = 0, respectively. And the z axis coordinates are in the range [−80,80], the unit is m. The starting coordinate of the magnetic source is (−80, 4, 5), and the velocity v = 5 m/s passes through the observation point along the x-axis, then the temporal and spatial distribution of the tri-components of the RMF generated by the moving RMS at the observation point is shown in Fig. 7.

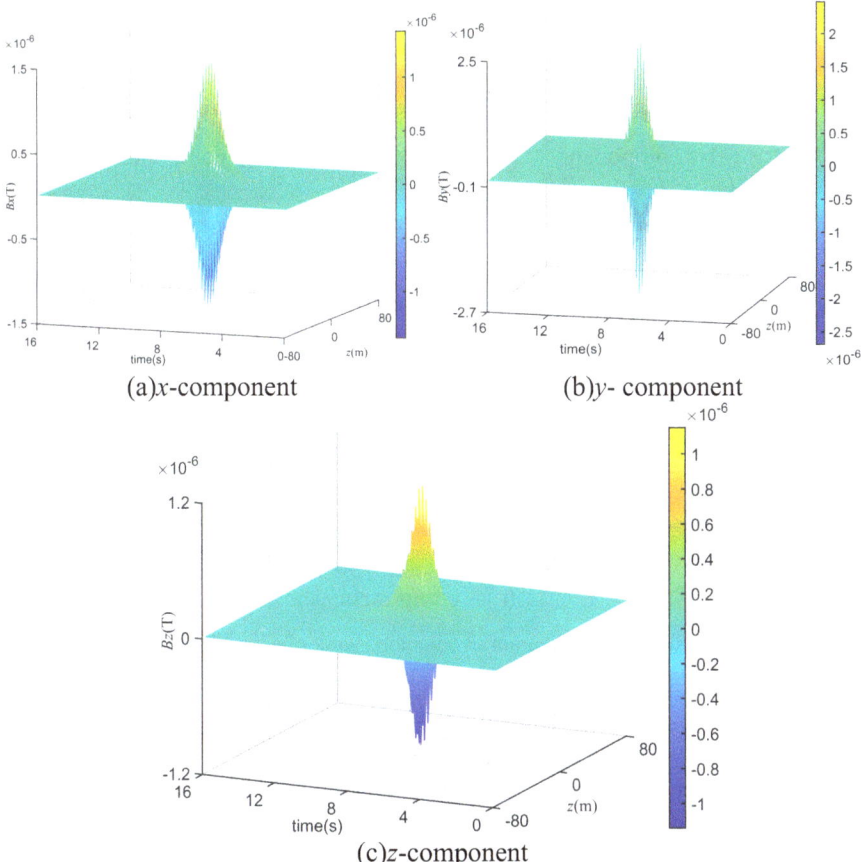

Fig. 7. Temporal and spatial distribution of the three magnetic components of a moving RMS along the z axis.

In Fig. 7, Bx, By and Bz are the RMF components of the moving magnetic source in the x, y and z axis directions, respectively. It can be observed that the RMF component of the moving magnetic source in the spatial triaxial direction is a harmonic signal with a certain frequency. Assume the coordinates of the observation point is (0, 0, 0), and the magnetic dipole move from the starting point (−80, 4, 5) along the positive direction of the x axis with the velocity v = 5 m/s, then the RMF at the observation point is shown in Fig. 8.

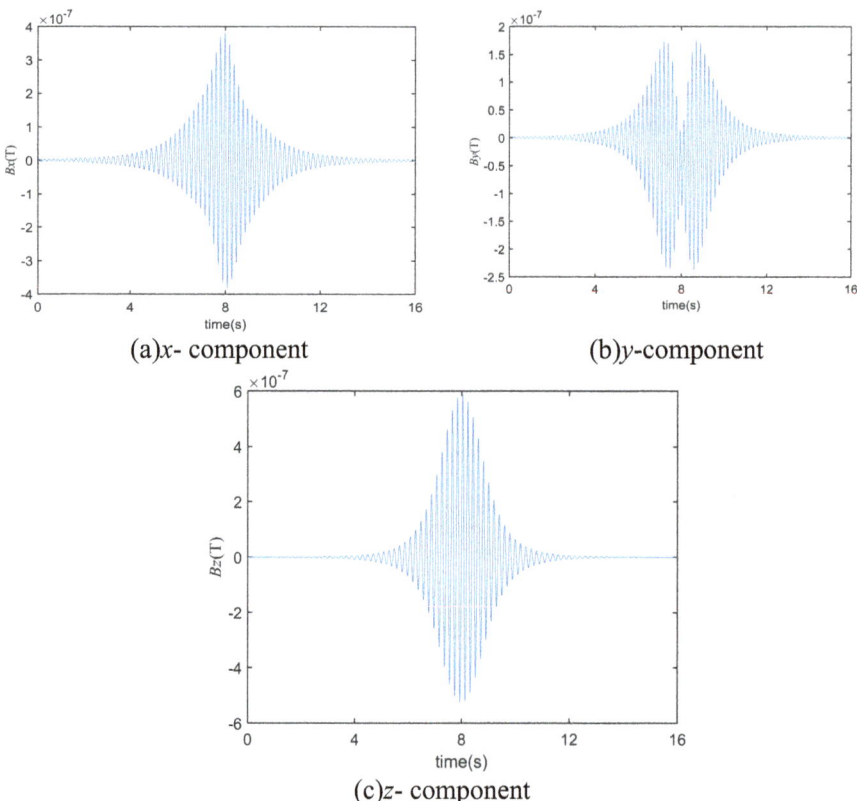

Fig. 8. Time domain distribution diagram of tri-components of a moving magnetic source at a point.

In Fig. 8, Bx, By and Bz are the RMF intensity of the moving magnetic source in the direction of xaxis, y axis and z axis, respectively. It can be seen that when t = 8 s, the RMS passes through the coordinate origin (0, 0, 0). And the components Bx, By and Bz are the signals with a certain frequency, whose frequency is the same as the dipole rotation frequency of 5 Hz, and the RMF strength has a certain offset, which is caused by the static magnetic field generated by the magnetic moment component ml. The amplitudes of Bx and Bz gradually increase and then decrease due to the gradual variation in the distance between the magnetic source and the observation point. The amplitude of by is symmetrically distributed along x = 0 and decays to both sides. When the static magnetic field generated by the magnetic moment ml is removed, the RMF measured at the observation point is shown in Fig. 9.

The Fig. 9 reveals that the RMF intensity of the RMS is symmetrically distributed after removing the static magnetic field generated by the magnetic moment component

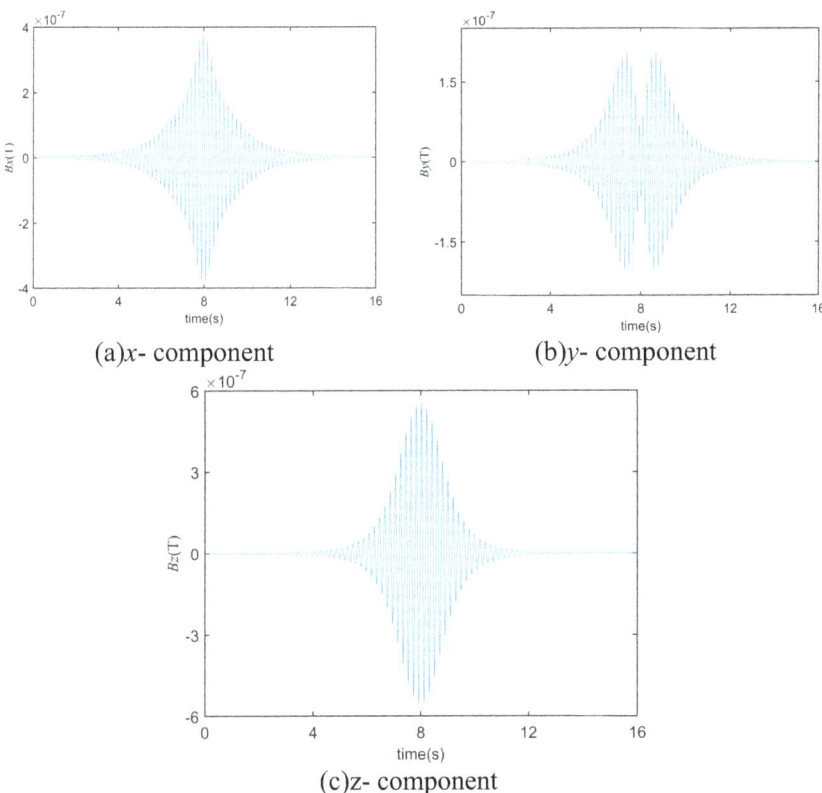

Fig. 9. Time domain distribution diagram of tri-components of moving magnetic source without vertical magnetic moment component.

ml. The actual signal processing is mainly to deal with the RMF after filtering out the static magnetic field signal in the detection study.

4 Conclusion

In this paper, the RMF distribution of the RMS is modelled, and the spatial RMF distribution of the fixed RMS and moving RMS is simulated and analyzed, respectively. The simulation results show that the frequency of the RMF is consistent with the rotational speed of the rotating magnetic dipole, and the amplitudes of the tri-components of the RMF generated by the magnetic moment components mp and mf are symmetric about the coordinate origin. The distribution of RMF studied in this paper provides a theoretical basis for the subsequent research of RMS for ocean exploration.

References

1. Chunsheng, L., Shenguang, G.: Ship Physics Field. Ordnance Industry Press (2007) (in Chinese)
2. Zili, Z.: Theory Research and Application of Ocean Electromagnetic Field. China University of Geosciences (Beijing), Beijing (2009) (in Chinese)
3. Lin, P., Zhang, N., Chang, M.: Research on the model and the location method of ship shaft-rate magnetic field based on rotating magnetic dipole. IEEE Access. **8**, 162999–163005 (2020)
4. Lin, P., Zhang, N., Lin, C.: Two-point magnetic field positioning algorithm based on rotating magnetic dipole. Measurement. **174**, 109059 (2021)
5. Yao, F., Zhiyu, W., Guangyou, F.: Research on axial frequency magnetic field signal detection method of unmanned helicopter. Electron. Meas. Technol. **40**(06), 180–183 (2017) (in Chinese)
6. Jianhua, W.: Research on Cathodic Protection and Electromagnetic Field Optimization Control of Underwater Hull Corrosion. Dalian University of Technology, Dalian (2012) (in Chinese)
7. Xiao, Y., Debnath, O.B., Chikaki, S.: Numerical and experimental evaluation of magnetic markers for localized tumor excision with a handheld magnetic probe. IEEE Trans. Magn. **57**(2), 1–5 (2021)
8. Yongshun, Z., Zhibo, W., Liu, X.: Posture detection method of a magnetically controlled dual hemisphere capsule robot using images. J. Huazhong Univ. Sci. Technol. (Nat. Sci. Ed.). **48**(10), 14–19 (2020) (in Chinese)
9. Kisseleff, S., Akyildiz, I.F., Gerstacker, W.H.: Survey on advances in magnetic induction-based wireless underground sensor networks. IEEE Internet Things J. **5**(6), 4843–4856 (2018)
10. Saeed, N., Alouini, M., Al-Naffouri, T.Y.: 3D localization for internet of underground things in oil and gas reservoirs. IEEE Access. **7**, 121769–121780 (2019)
11. Sheinker, A., Ginzburg, B., Salomonski, N.: Localization of a mobile platform equipped with a rotating magnetic dipole source. IEEE Trans. Instrum. Meas. **68**(1), 116–128 (2019)
12. Qi, H., Ye, J., Zhang, X.: Wireless tracking and locating system for in-pipe robot. Sensors Actuators A Phys. **159**(1), 117–125 (2010)
13. Rosser, W.G.V.: The Biot-Savart law. Phys. Educ. **9**(7), 493–493 (1974)

Open Access This chapter is licensed under the terms of the Creative Commons Attribution-NonCommercial-NoDerivatives 4.0 International License (http://creativecommons.org/licenses/by-nc-nd/4.0/), which permits any noncommercial use, sharing, distribution and reproduction in any medium or format, as long as you give appropriate credit to the original author(s) and the source, provide a link to the Creative Commons license and indicate if you modified the licensed material. You do not have permission under this license to share adapted material derived from this chapter or parts of it.

The images or other third party material in this chapter are included in the chapter's Creative Commons license, unless indicated otherwise in a credit line to the material. If material is not included in the chapter's Creative Commons license and your intended use is not permitted by statutory regulation or exceeds the permitted use, you will need to obtain permission directly from the copyright holder.

An Optimization Model for Water Search and Rescue Sites Considering Accident Black Spot

Tingrong Qin(✉), Xueli Chen, and Qinyou Hu

Shanghai Maritime University, Shanghai 201306, People's Republic of China
trqin@shmtu.edu.cn

Abstract. Water search and rescue (SAR) is the last line of security to protect the lives of people on the water, and the reasonable degree of the layout of the SAR site will directly affect the efficiency of SAR. For this reason, this paper proposes a SAR site layout and optimisation model based on accident black point combined with particle swarm algorithm. In the accident black spot identification stage, the equivalent number of accidents and the dynamic segmentation method are used to achieve the initial identification, followed by the introduction of the density space-based clustering algorithm (DBSCAN), which further achieves the accurate identification. Subsequently, the particle swarm algorithm (PSO) is used to find the optimal location of the SAR site with the target of the accident coverage in the water. Using the above model to validate the accident data of Shanghai Huangpu River, which is of guiding significance for the scientific layout and optimisation of the SAR sites in the Huangpu River waters.

Keywords: Accident black spot · Search and rescue · Site layout · Clustering algorithm · Particle swarm optimization

1 Introduction

SAR is related to the safety of people's lives, property loss and environmental pollution, while the scientific nature of the layout of search and rescue sites is one of the key factors affecting the effectiveness of SAR. Early research on site selection mostly focused on the distance minimization problem. Alfred Weber [1] first proposed to locate a warehouse in a two-dimensional plane to minimize the total distance between the warehouse and customers. Hakimi [2] summarized and proposed the "p-median" and "p-center "problems. With the enrichment of algorithms, the siting problem was gradually transformed from "points and lines" to "surfaces". Toregas [3] proposed the ensemble coverage problem to cover all the demand points with the fewest possible facility points. With the deepening of research, several papers proposed different types of multi-objective optimization models [4]. Quandang Ma [5] used K-means algorithm to find the accident black spots, and then determined emergency response sites and resource allocation through the long and short-term memory method and elite genetic algorithm. Jinfen Zhang [6] proposed a two-stage accident black spot identification model for identifying the frequent places of waterborne accidents. From the above literature analysis, taking ship traffic accidents as the research object, combining various clustering algorithms to obtain the accident black spot as the theoretical basis for site layout is a more scientific way of thinking about site construction.

2 Model Construction

2.1 Theoretical Basis

Equivalent Number of Accident Points In order to consider the severity and type of the accident, this paper refers to the literature [7] to propose the equivalent number of accident points, the calculation formula is shown in (1).

$$E_A = A + \alpha D + \beta S \tag{1}$$

Where: E_A is the total number of equivalent accident points; A is the number of accident points that originally occurred; D is the number of accidents that there are fatalities or disappearances; S is the number of shipwrecks; α and β indicate the weights of dead, missing and sunken ships, respectively. Drawing on domestic and international research results, the value of α is 2.0 and the value of β is 1.0.

Accident Black Spots. Accident black spot theory was first used for hazardous roadway identification for land traffic accidents [8]. The location may be a road section, or an area, it refers to certain locations where the number of accidents occurring over a longer period of time are significantly prominent compared with other locations. In recent years, the theory has been gradually introduced to the identification of risk areas on water [9].

2.2 Accident Black Spot Identification Model

There are various identification methods for accident black spots [10], and in this paper, we choose the accident black spot identification model proposed by Geng Chao et al. [11].

Initial Identification Phase. Dynamic segmentation method. This method is based on the introduction of a fixed segmentation of the mobile step, this paper selects 1 km as the fixed segmentation length of the watercourse unit, the 200-m moving step is chosen as the benchmark for dynamic segmentation.

Cumulative frequency method. The cumulative frequency is defined as the sum of the frequencies below or above a certain value. Using the number of equivalent accident points as the horizontal coordinate and the cumulative frequency as the vertical coordinate, a scatter plot of the cumulative frequency is plotted. The horizontal coordinate value corresponding to the abrupt point of the curve is the threshold value, which is used to define the accident-prone section.

Precise Identification Phase. Considering the economic issues as well as the pertinence and effectiveness of water safety management, it is necessary to further screen out the more detailed accident black spots. The DBSCAN is a density-based clustering method, which is able to deal with clusters of arbitrary shapes while not needing to pre-assign the number of clusters. The algorithm contains neighbourhood radius (*Eps*) and density threshold (*MinPts*), whose values have an important impact. According to the first stage and related literature, the value of *Eps* is 500 m, and the threshold value of *MinPts* is determined according to Eq. (2).

$$MinPts = 2Eps \cdot N/L \tag{2}$$

Where: *Eps* is the neighbourhood radius; *N* is the equivalent number of accidents in the black spot area of the waterway; *L* is the length of the black spot waterway.

2.3 Search and Rescue Site Determination

The identification model only counts the black spot channel segments. In order to further find the optimal position of the SAR site within each accidental black spot channel segment, the classical Particle Swarm Optimisation (PSO) algorithm proposed [12]. The algorithm can be formulated as follows:

$$V_i(t+1) = wV_i(t) + c_1 r_1 (P_{best} - X_i(t)) + c_2 r_2 (G_{best} - X_i(t)) \tag{3}$$

$$X_i(t+1) = X_i(t) + V_i(t+1) \tag{4}$$

Equation (3) is the formula for particle update velocity and eq. (4) is the formula for particle update position. Where the velocity update term of each particle contains three parts. Where w is called the inertia weight, the value of w is [0.8,1.2], and in this paper, w is taken to be 1. c_1 and c_2 is called the learning factor, r_1 and r_2 is a random number, the range of the random number is between 0 and 1. The particles in the population update their positions through Eq. (3, 4), and after many iterations, they gradually approach the global optimal solution, and finally obtain the optimal solution in the whole space.

2.4 Search and Rescue Site Optimization Model

In order to cover as many incident points as possible, it is necessary to build many SAR sites. However, in the actual construction of the sites, it is often constrained by the cost of the sites and other conditions, and it is necessary to consider the problem of maximum coverage of accident points under the limited number of sites. Combined with the accident black spot identification model, The model of SAR sites based on particle swarm algorithm is as follows:

$$maxZ = \sum_{i=1}^{n} q_i \tag{5}$$

$$\text{s.t.} \begin{cases} F(x_j, y_j) = 0, \forall j \\ \sum_{j=1}^{m} J_j = P \\ N_i = \{j | d_{ij} \leq r_j\}, j \in J \\ \sum_{j \in N_i} J_j \geq q_i, i \in I \\ J_j \in \{0, 1\}, j \in J \\ q_i \in \{0, 1\}, i \in I \end{cases} \quad (6)$$

Where $maxZ$ denotes the maximum number of equivalent incident points covered. n denotes the number of equivalent incident points, i denotes the ith equivalent incident point, q_i indicates whether the ith equivalent incident point is covered under the choice of P SAR sites, and if it is, it takes 1, otherwise it takes 0. F denotes the accidental black spot segment function, (x_j, y_j) denotes the coordinate point of a SAR site. m denotes the number of candidate sites, j denotes the jth SAR site, J_j indicating whether the jth SAR site is selected, and if it is, it takes 1, otherwise it takes 0, and P denotes the number of SAR sites selected. d_{ij} is the Euclidean distance from the ith equivalent incident point to the jth SAR site, r_j denotes the service coverage radius of the SAR site, and N_i denotes a non-empty set consisting of the subscripts of the SAR sites corresponding to the ith equivalent incident point. The constraints $\sum_{j \in N_i} J_j \geq q_i$ indicate that the ith equivalent incident point can be served only if the jth SAR site is enabled.

2.5 Algorithm Design

Comprehensively speaking, the design flow is shown in Fig. 1:

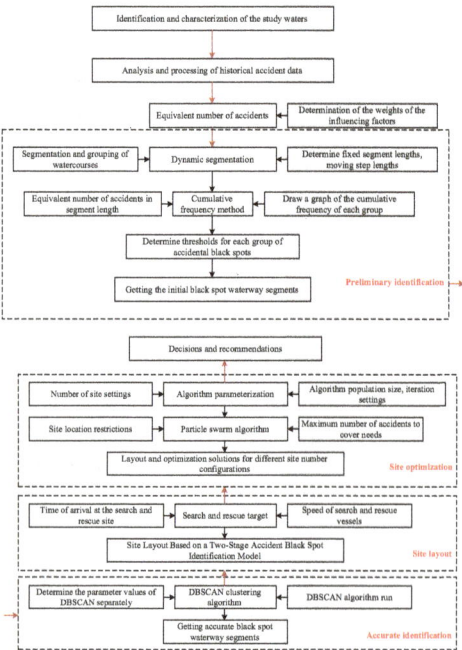

Fig. 1. Paper frame diagram

3 Case Study and Analysis

Take Huangpu River as an example to verify the feasibility and correctness of this model. The Huangpu River channel is narrow and long, which is highly adaptable to the model. According to the historical accident data provided by the Maritime Search and Rescue Centre (MSRC), a total of 537 accidents occurred during the period of 2018–2022, and 585 equivalent accident points were obtained by using Eq. (1).

3.1 Model Calculation Results

The dynamic segmentation method mentioned above is used to segment the Huangpu River, a total of 5 groups of datasets were obtained, each containing 84 segments, as shown in Table 1.

Table 1. Location of dynamic segmentation of Huangpu River channel

Subparagraph number	Group I		Group II		Group III		Group IV		Group V	
	initial point	endpoint	initial point	endpoint	initial point	endpoint	initial point	endpoint	initial point	endpoint
1	121.527E, 31.401 N	121.519E, 31.395 N	121.527E, 31.401 N	121.525E, 31.400 N	121.525E, 31.400 N	121.523E, 31.399 N	121.525E, 31.400 N	121.522E, 31.398 N	121.527E, 31.401 N	121.520E, 31.397 N
..	...	...	...	...	...	...	...	...	...	...
84	121.225E, 30.963 N	121.215E, 30.960 N	121.222E, 30.963 N	121.213E, 30.960 N	121.219E, 30.963 N	121.212E, 30.958 N	121.217E, 30.960 N	121.214E, 30.953 N	121.216E, 30.961 N	121.206E, 30.957 N

Calculating the cumulative frequencies for the five data sets in Table 1, a cumulative frequency plot can be obtained, as shown in Fig. 2.

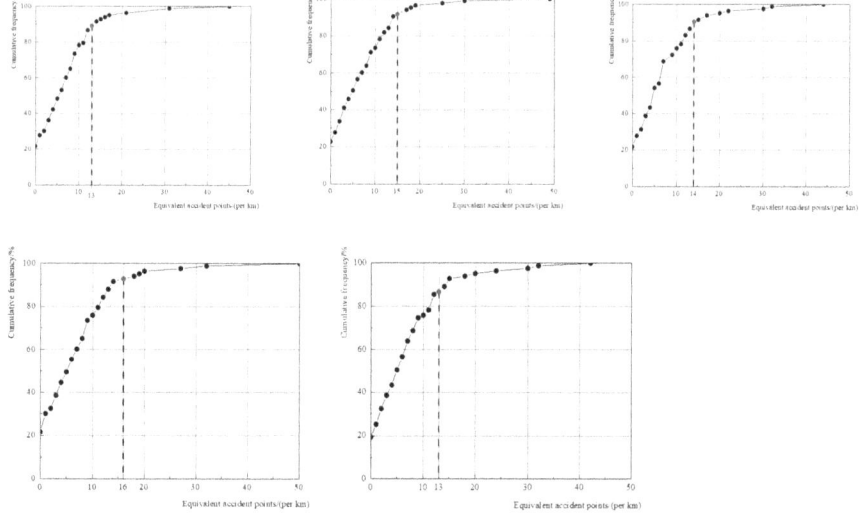

Fig. 2. Cumulative frequency graph

It can be determined from Fig. 2 that the accident thresholds of each data set are 13, 15, 14, 13 and 16, respectively. When the number of accidents in each unit segment in the 5 groups is greater than the accident threshold, it is identified as the accident black spot segment. Finally, the identification of 11 preliminary accident black spot segments is obtained by merging the 5 groups of black spot segments, its length accounts for 24% of the total length of the Huangpu River, and the number of equivalent accidents within the accident black spot segments accounts for 58.12% of the total, as shown in Fig. 3. This result indicates that more than half of the accidents in the Huangpu River waterway occur in only 24 percent of the waterway segments.

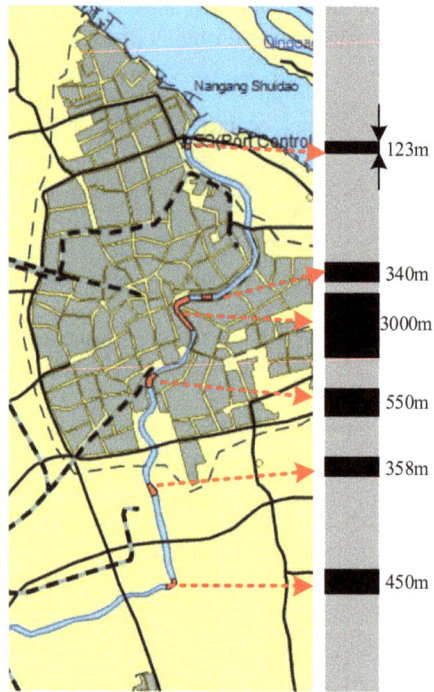

Fig. 3. Preliminary identification

The lengths of each of the 11 initially identified accident black spot waterways were substituted into Eq. (2) to determine the values of the two parameters of DBSCAN, and the DBSCAN code was run to identify six precise accident black spot segments, accounting for 5.81% of the total length and 32.48% of the total number of equivalent accidents. as shown in Fig. 4.

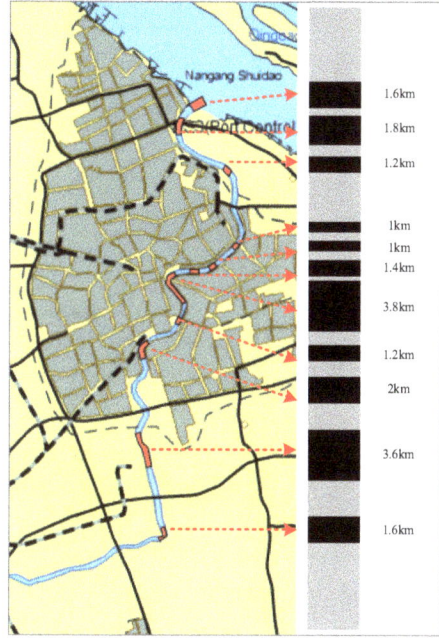

Fig. 4. Accurate identification

3.2 Site Layout and Optimization

This paper further refers to the usual practice of Shanghai Maritime Safety Administration's emergency rescue, for accidents occurring in the Huangpu River, the SAR force strives to reach the SAR in 15 min. This paper takes the average speed of SAR vessels as 15 knots, and the radius of coverage of SAR sites is about 3.75 nautical miles.

According to the six accident black spot segments calculated by the accident black spot identification model, the particle swarm algorithm is used to calculate the optimal location of the built-in SAR sites in each black spot segment using Eqs. (3, 4), the preliminary site setup is shown in Fig. 5.

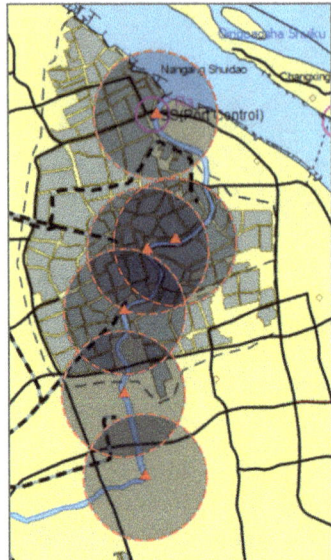

Fig. 5. Preliminary site layout

Based on Fig. 5, the overlap of the SAR coverage areas of several sites in the middle section is obvious, resulting in a waste of SAR resources. In the actual process of building sites, the number of sites will be limited by a variety of factors. Based on this, this paper proposes a maximum coverage model based on particle swarm algorithm to solve the overall layout optimisation problem under the limited sites, the points within the accident black spot segment are the SAR sites building area, and the particle swarm algorithm is used to perform the traversal search, and calculate the maximum equivalent number of accidents that can be covered when the SAR sites are limited to 3, 4 and 5 respectively.

The site layouts for the conditions of 3, 4 and 5 search and rescue sites are shown in Fig. 6:

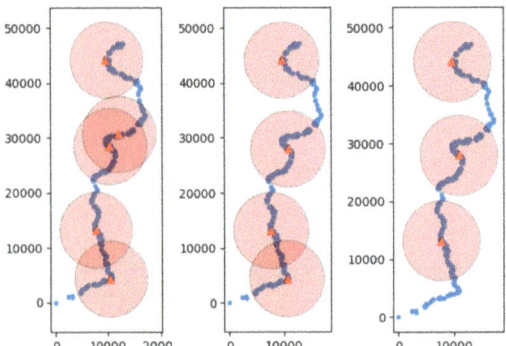

Fig. 6. The overall layout diagram of different number of search and rescue sites

Based on Table 2, it can be seen that when there are three SAR sites, the accident point coverage rate is 78.9 percent; when there are four SAR sites, the accident point coverage rate is 90.4 percent, which is an improvement of about 12 percent; and when there are five SAR sites, the accident point coverage rate is 94.1 percent, which is an improvement of about 4 percent. In terms of the degree of enhancement, the optimal layout plan for SAR sites produces a greater benefit from setting up four SAR sites. For other waters that cannot be covered, the optimised SAR sites can be used to cover the SAR needs of the Huangpu River by setting up smaller SAR duty sites.

Table 2. Longitude and latitude of search and rescue sites and its accident coverage

sites	3 sites			4 sites				5 sites				
Longitudes	31.37	31.23	31.10	31.37	31.10	31.02	31.23	31.02	31.37	31.10	31.25	31.23
Latitude	121.50	121.50	121.46	121.50	121.46	121.49	121.50	121.48	121.50	121.46	121.51	121.49
Accidents covered	462			529				551				
Percentage	78.90%			90.40%				94.10%				

4 Conclusions

(1) Aiming at the difficult problem of irrational layout of previous SAR sites, this paper proposes a model for the layout and optimisation of waterborne SAR sites. The accident black point theory solves the strategic forward movement of SAR sites to shorten the SAR reachable time. The particle swarm algorithm solves the optimal solution of the overall layout of SAR sites under the condition of the given number of sites.

(2) The model is applied to the Huangpu River waterway, Subsequently, under the condition that the SAR sites are limited to 3, 4 and 5 respectively, the particle swarm algorithm calculates that it is most appropriate to set up 4 SAR sites in Huangpu River waters, and for other waters that cannot be covered, they can be effectively supplemented by setting up smaller SAR duty sites, thus systematically solving the problem of the layout and optimisation.

(3) From the viewpoint of theoretical construction and its practical application effect, the water SAR site optimisation model considering the accident black spots helps to construct a more comprehensive and systematic, scientific.

(4) In this study, the theory of accident black spots is utilized to optimize the layout of SAR stations. Though this method can effectively decrease the response time to accidents, there may exist certain subjectivity in the selection and quantity of accident black spots, and it is confined to historical data. Hence, how to adjust the settings of accident black spots in accordance with real-time data, dynamic monitoring, and emergent situations will be the key to further optimizing the layout.

(5) Owing to the restriction on the quantity of SAR sites in the research, the model might not be applicable to some complex water environments, particularly in scenarios where multiple sites operate in a coordinated manner. In practical operations, water

conditions change rapidly and site configurations are various. Consequently, how to design a dynamic optimization model capable of adapting to the requirements of different scenarios is a direction meriting further discussion.

References

1. Chen, M.: Research on Location and Capacity Determination of Electric Vehicle Charging Sites Based on Two-stage Stochastic Programming Model. Southwest Jiaotong University (2021)
2. Duan, X., et al.: Subway network emergency sites location based on multi-objective distribution estimation algorithm. J. Saf. Environ. **19**(03), 923–930 (2019)
3. Jiang, Y., et al.: Optimization model of oil spill emergency material storage location based on Stackelberg game under failure scenario. J. Shanghai Marit. Univ. (2023)
4. Chen, Z., et al.: Multi-objective decision model for location selection of emergency rescue facilities in major emergencies. Manag. Sci. **19**(4), 45–46 (2006)
5. Ma, Q., et al.: Multi-objective emergency resources allocation optimization for maritime search and rescue considering accident black-spots. Ocean Eng. **261**, 112178 (2022)
6. Zhang, J., et al.: A two-stage black-spot identification model for inland waterway transportation. Reliab. Eng. Syst. Saf. **213**, 107677 (2021)
7. Liu, X.: Study on Safety Evaluation of Expressway Location Considering Accident Traffic Impact. Beijing Jiaotong University (2020)
8. Zeng, Q., et al.: Identification of highway accident black spots based on Bayesian space-time modeling. Traffic Inf. Saf. **38**(06), 87–94 (2020)
9. He, M., et al.: Water traffic accident black spot identification based on ISODATA algorithm. J. Saf. Environ. **17**(02), 413–417 (2017)
10. Zhang, C., et al.: Review on identification methods of highway accident-prone sections. J. Chang 'an Univ. (Nat. Sci. Ed.). **43**(05), 72–87 (2023)
11. Geng, C., et al.: Traffic accident blackspot identification method based on dynamic segmentation and DBSCAN algorithm. J. Chang 'an Univ. (Nat. Sci. Ed.). **38**(05), 131–138 (2018)
12. Gan, D.: A novel simulated annealing particle swarm optimization algorithm based on sinusoidal adaptive weights. Sci. Technol. Innov. Appl. **12**(05), 11–14 (2022)

Open Access This chapter is licensed under the terms of the Creative Commons Attribution-NonCommercial-NoDerivatives 4.0 International License (http://creativecommons.org/licenses/by-nc-nd/4.0/), which permits any noncommercial use, sharing, distribution and reproduction in any medium or format, as long as you give appropriate credit to the original author(s) and the source, provide a link to the Creative Commons license and indicate if you modified the licensed material. You do not have permission under this license to share adapted material derived from this chapter or parts of it.

The images or other third party material in this chapter are included in the chapter's Creative Commons license, unless indicated otherwise in a credit line to the material. If material is not included in the chapter's Creative Commons license and your intended use is not permitted by statutory regulation or exceeds the permitted use, you will need to obtain permission directly from the copyright holder.

Three Dimensional Stress Analysis Method of Ring-Stiffened Cylindrical Shell with New Design Features

Shagu Chen[1,2](✉), Yuan Gao[1,2], and Xiaozhong Xie[1,2]

[1] China Ship Scientific Research Center, Wuxi, China
`chenshagu@163.com`
[2] National Key Laboratory of Ship Structural Safety, Shanghai, China

Abstract. The traditional stress analysis method of ring-stiffened cylindrical shell under hydrostatic pressure is to simplify the cylindrical shell as a bending beam in bidirectional stress state. With the emergence of new design features such as large depth loads and large thickness in deep-sea pressure structure, the assumption of bidirectional stress state for cylindrical shell is clearly unreasonable. This paper proposed a new simplified mechanical model for ring-stiffened cylindrical shell, which transforming the stress problem of ring-stiffened cylindrical shell under hydrostatic pressure into the superposition of bending beam stress under axial force and plane annulus stress under radial force, and thereby a three dimensional stress analysis method of ring-stiffened cylindrical shell has been derived. The application example results showed that the calculation results of the three dimensional stress analysis method has been in good agreement with the finite element simulation results, and it has better calculation accuracy and safer compared with the traditional stress analysis method.

Keywords: Ring-stiffened Cylindrical Shell · Deep-sea Pressure Structure · New Design Features · Mechanical Model · Three Dimensional Stress

1 Introduction

Ring-stiffened cylindrical shell was widely used as pressure structure in underwater engineering products such as submarines and submersibles due to its excellent stress characteristics, space utilization, and processing performance [1–3]. Under hydrostatic pressure, the traditional stress analysis method of ring-stiffened cylindrical shell is to study the complex bending elastic foundation beam by dividing the cylindrical shell into two ends fixed on elastic supports [3–5]. The elastic supports are assumed to an equal deflection and stress on the cross section of the ring-stiffener. For some cylindrical shell with large ring-stiffener or especially large ring-stiffener, the assumption of elastic supports is obviously unreasonable. Reference [6] proposed a stress analysis method for ring-stiffened cylindrical shell considering the centroid of ring-stiffener, which disassembled the ring-stiffened cylindrical shell into three parts: ring-stiffener web, ring-stiffener flange, and cylindrical shell. This method can solve the stress of the cylindrical shell, as well as various parts of the inner-stiffener or outer-stiffener.

Both traditional stress analysis method and the method described in reference [6] essentially simplify the cylindrical shell as a bending beam in bidirectional stress state, ignoring the influence of normal stress on the curved surface. At the same time, ring-stiffener flange is also considered as unidirectional stress state, greatly simplifying its complex bending problem as a ring shell. At present, the design calculation for the strength of ring-stiffened cylindrical shell in current relevant standards of submarines and submersibles are based on these assumptions [7–11]. In the traditional structural characteristics and design parameter categories, the calculation errors caused by these assumptions are very small, which can meet the requirements of engineering applications.

However, the application depth of underwater engineering products is getting deeper and deeper with the development of deep-sea technology. The application depth of some deep-sea unmanned systems often reaches several thousand meters, and the hydrostatic pressure will be more than 10% compared to the strength of the structural materials [12, 13]. At the same time, to ensure the strength of the pressure structure, the thickness of the structural design will far exceed the scope of traditional thin shell structures [14, 15] (the t/R of traditional thin shell is generally much less than 5%, while the t/R of deep-sea pressure structures will reach or exceed 10%). In addition, sometimes due to limitations in overall layout and other reasons, the ring-stiffened cylindrical shell has to be reinforced with low and wide ring-stiffener in special situations. Obviously, the new design features of these ring-stiffened cylindrical shells do not conform to the traditional assumption of bidirectional stress state.

In order to analyze the stress of ring-stiffened cylindrical shell with new design features, this paper proposes a new simplified mechanical model. Using the analysis method of the axisymmetric plane problem and the bending differential equation based on triaxial stress state, a three dimensional stress calculation method of ring-stiffened cylindrical shell has been derived and analyzed.

2 Simplified Mechanical Model of Ring-Stiffened Cylindrical Shell

2.1 Structural Characteristic Symbols

The T-shaped ring-stiffener of the ring-stiffened cylindrical shell was generally arranged at equal intervals, as shown in Fig. 1. For the convenience of analysis, the definition of structural characteristic symbols are: inner radius is R_i, outer radius is R_o, middle radius is R, and cylindrical shell thickness is t; the height of the ring-stiffener web is H and its thickness is δ, as well as the radius corresponding to the height of the centroid is R_h; the width of the ring-stiffener flange is W, and its thickness is t_f, the inner radius of the flange shell is R_{fi}, and its middle radius is R_f, and its outer radius is R_{fo}; the distance between two adjacent ring-stiffener is l; the elastic modulus is E and Poisson's ratio is μ.

To consider the difference of the inner-stiffener and outer-stiffener, defining $F = 1$ to represent outer-stiffener, and $F = -1$ to represent inner-stiffener. Therefore, there are structural parameter geometric relationships such as: $R_i = R - 0.5t$, $R_o = R + 0.5t$, $R_h = R + 0.5F(t+H)$, $R_f = R + F(0.5t + H + 0.5t_f)$, $R_{fi} = R_f - 0.5Ft_f$, $R_{fo} = R_f + 0.5Ft_f$.

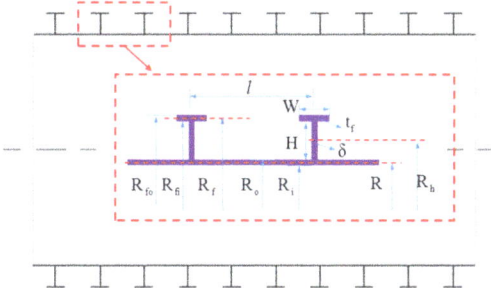

Fig. 1. Ring-stiffened cylindrical shell.

2.2 Mechanical Model

The ring-stiffened cylindrical shell consists of three parts: cylindrical shell, ring-stiffener web, and ring-stiffener flange. From the structural type, the cylindrical shell and ring-stiffener flange are ring shell, while the ring-stiffener web is a flat circular plate structure. Obviously, when assuming the interaction force between the cylindrical shell and the ring-stiffener web is f_1, and the interaction force between the ring-stiffener web and the ring-stiffener flange is f_2, the web can be regarded as a plane annulus with radial loads on its inner and outer surfaces. Therefore, the ring-stiffener web can be analyzed based on the method of the axisymmetric plane strain problem, as shown in Fig. 2.

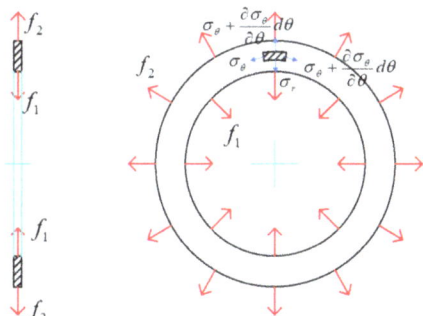

Fig. 2. Mechanical model of ring-stiffener web.

Owing to the load being axisymmetric, the deformation of the cylindrical shell and ring-stiffener flange as ring shell are also axisymmetric. To analyze the three dimensional stress, the ring shell under hydrostatic pressure is simplified as a combination of the beam under axial force and the plane annulus under radial load, where the cylindrical shell is a closed ring shell, as shown in Fig. 3, and the ring-stiffener flange is a non closed ring shell, as shown in Fig. 4.

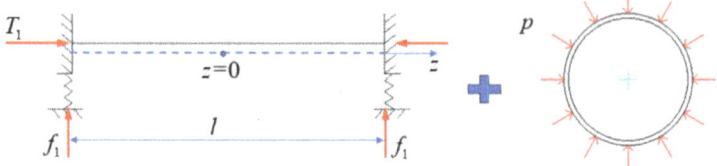

Fig. 3. Mechanical model of cylindrical shell.

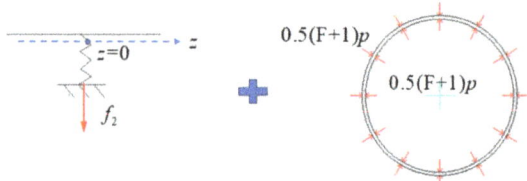

Fig. 4. Mechanical model of ring-stiffener flange.

3 Analysis of Deflection and Stress Based on Bending Differential Equations

3.1 Bending Differential Equation Considering Triaxial Stress

For the axisymmetric ring shell in the mechanical model of Fig. 3 and 4, a beam strip with a unit width (ds=Rdθ=1) can be cut from two diameter surfaces to study, as shown in Fig. 5.

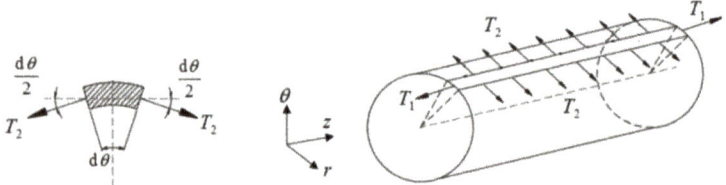

Fig. 5. Force of the axisymmetric ring shell.

For the ring shell beam with radius of R and thickness of t, the normal load can be expressed as

$$q = 2T_2 \sin\frac{d\theta}{2} \approx T_2 d\theta = \frac{T_2}{R} \qquad (1)$$

According to Hooke's Law, the circumferential strain of the beam can be expressed as

$$\varepsilon_2^0 = \frac{1}{E}\left(\frac{T_2}{t} - \mu\frac{T_1}{t} + \mu q\right) \qquad (2)$$

In addition, the circumferential strain of the beam is

$$\varepsilon_2^0 = \frac{2\pi(R-w) - 2\pi R}{2\pi R} = -\frac{w}{R} \tag{3}$$

where w is the deflection of the beam. Thus, the normal load of the beam can be obtained as

$$q = \frac{1}{1+\mu t/R}\left(\mu\frac{T_1}{R} - Et\frac{w}{R^2}\right) \tag{4}$$

And the bending differential equation of the ring shell can be obtained as

$$Dw'''' - T_1 w'' + \frac{1}{1+\mu t/R}Et\frac{w}{R^2} = \frac{\mu}{1+\mu t/R}\frac{T_1}{R} \tag{5}$$

where D is the bending stiffness of the ring shell.

3.2 Bending Deflection and Stress of Cylindrical Shell

According to mechanical model, the axial force of closed cylindrical shell is $T_1 = -0.5(1+0.5m)^2 pR$. The longitudinal stress of cylindrical shell is composed of two parts, one is caused by the axial force and remains constant along the thickness, the other is caused by the bending of the beam and change along thickness. Therefore, there are

$$\sigma_z(r,z) = -\frac{(1+0.5m)^2}{2m}p + \frac{Dw(z)''}{t^3/12}(r-R) \tag{6}$$

where r is the radial direction of the ring shell in cylindrical coordinate system, z is the longitudinal direction in cylindrical coordinate system, $w(z)$ is the bending deflection of the cylindrical shell under axial force, $D = Et^3/[12(1-\mu^2)]$ is the bending stiffness, and $m = t/R$ is the thickness characteristic parameters of the cylindrical shell. According to the method in Sect. 3.1, the bending differential equation of cylindrical shell is

$$Dw''''(z) + \frac{(1+0.5m)^2}{2}pRw''(z) + \frac{1}{1+\mu m}\frac{Et}{R^2}w(z) = -\frac{(1+0.5m)^2}{1+\mu m}\frac{\mu}{2}p \tag{7}$$

The general solution of the differential equation is

$$w(z) = -\frac{(1+0.5m)^2\mu R^2}{2Et}p + \begin{bmatrix} C_1\cosh\alpha_1 z\cos\alpha_2 z + C_2\cosh\alpha_1 z\sin\alpha_2 z \\ + C_3\sinh\alpha_1 z\cos\alpha_2 z + C_4\sinh\alpha_1 z\sin\alpha_2 z \end{bmatrix} \tag{8}$$

In the equation, C_1, C_2, C_3 and C_4 are undetermined constants. Where $\alpha = \frac{\sqrt[4]{3(1-\mu^2)}}{\sqrt{Rt}}$, $\gamma=(1+0.5m)^2\frac{\sqrt{3(1-\mu^2)}}{2}\frac{pR^2}{Et^2}$, $\alpha_1 = \alpha\sqrt{\frac{1}{\sqrt{1+\mu m}} - \gamma}$, $\alpha_2 = \alpha\sqrt{\frac{1}{\sqrt{1+\mu m}} + \gamma}$.

Due to being $z = 0$ set at the middle between two adjacent ring-stiffener, the deflection of the cylindrical shell is symmetrical about the coordinate axis, therefore $C_2 = C_3 = 0$. There are also boundary conditions:

1) The angle of rotation at the ring-stiffener is zero, that is $w'(z)\big|_{z=l/2} = 0$;
2) The shear force of the shell at the ring-stiffener is half of the force acting on the web, that is $2Dw'''(z)\big|_{z=0} = f_1$.

From this, it can be concluded that

$$\begin{cases} C_1 = Q_1 f_1 \\ C_4 = Q_2 f_1 \end{cases} \quad (9)$$

where $u_1 = \frac{\alpha_1 l}{2}$, $u_2 = \frac{\alpha_2 l}{2}$, $Q_1 = -\frac{R^2}{Etl}\frac{1}{\sqrt{1-(1+\mu m)\gamma^2}}\frac{u_2\sinh u_1 \cos u_2 + u_1 \cosh u_1 \sin u_2}{\sinh^2 u_1 + \sin^2 u_2}$, and $Q_2 = -\frac{R^2}{Etl}\frac{1}{\sqrt{1-(1+\mu m)\gamma^2}}\frac{u_2\cosh u_1 \sin u_2 - u_1 \sinh u_1 \cos u_2}{\sinh^2 u_1 + \sin^2 u_2}$.

Therefore, it can be concluded that the bending deflection of the cylindrical shell under axial force is

$$w(z) = -\frac{(1+0.5m)^2}{2m}\frac{\mu R}{E}p + f_1(Q_1\cosh\alpha_1 z \cos\alpha_2 z + Q_2\sinh\alpha_1 z \sin\alpha_2 z) \quad (10)$$

And the longitudinal stress of the cylindrical shell is

$$\sigma_z(r, z) = -\frac{(1+0.5m)^2}{2m}p + \frac{E(r-R)}{(1-\mu^2)}f_1(Q_3\cosh\alpha_1 z \cos\alpha_2 z + Q_4\sinh\alpha_1 z \sin\alpha_2 z) \quad (11)$$

where $Q_3 = Q_1\alpha_1^2 + 2Q_2\alpha_1\alpha_2 - Q_1\alpha_2^2$ and $Q_4 = Q_2\alpha_1^2 - 2Q_1\alpha_1\alpha_2 - Q_2\alpha_2^2$.

Based on the bending deflection $w(z)$ and longitudinal stress $\sigma_z(r, z)$, the radial stress $\sigma_{r1}(r, z)$ and circumferential stress $\sigma_{\theta1}(r, z)$ corresponding to the bending of cylindrical shell can be further obtained as

$$\begin{cases} \sigma_{r1}(r, z) = -q(z) = \frac{1}{1+\mu m}\left[\frac{(1+0.5m)^2}{2}\mu p + m\frac{E}{R}w(z)\right] \\ \sigma_{\theta1}(r, z) = \mu\sigma_z(r, z) + \frac{1}{1+\mu m}\left[\frac{(1+0.5m)^2}{2}\mu^2 p - \frac{E}{R}w(z)\right] \end{cases} \quad (12)$$

3.3 Bending Deflection and Stress of Ring-Stiffener Flange as Ring Shell

According to the mechanical model, the axial force of non closed ring-stiffener flange as ring shell is $T_1 = -0.5(F+1)pt_f$. The longitudinal stress of ring-stiffener flange is composed of two parts, one is hydrostatic pressure $-0.5(F+1)p$ ($-p$ for the outer-stiffener and zero for the inner-stiffener), the other is caused by the bending of the beam and change along thickness. Therefore, there are

$$\sigma_{zM}(r, z) = -0.5(F+1)p + \frac{D_f w_f''(z)}{t_f^3/12}(r - R_f) \quad (13)$$

where $w_f(z)$ is the deflection of ring-stiffener flange, and $D_f=\frac{Et_f^3}{12(1-\mu^2)}$ is the bending stiffness of ring-stiffener flange. According to the method in Sect. 3.1, the bending differential equation of the ring-stiffener flange as ring shell is

$$D_f w_f''''(z)+\frac{F+1}{2}pt_f w_f''(z)+\frac{1}{1+\mu k}\frac{Et_f}{R_f^2}w_f(z)=-\frac{\mu k}{1+\mu k}\frac{F+1}{2}p \quad (14)$$

where $k=t_f/R_f$ is the thickness characteristic parameter of ring-stiffener flange. The general solution of this differential equation is

$$w_f(z)=-\frac{F+1}{2}\frac{\mu R_f}{E}p+\begin{pmatrix}G_1\cosh\beta_1 z\cos\beta_2 z+G_2\cosh\beta_1 z\sin\beta_2 z\\ +G_3\sinh\beta_1 z\cos\beta_2 z+G_4\sinh\beta_1 z\sin\beta_2 z\end{pmatrix} \quad (15)$$

In the equation, G_1, G_2, G_3 and G_4 are undetermined constants. Where $\beta=\frac{\sqrt[4]{3(1-\mu^2)}}{\sqrt{R_f t_f}}$, $\kappa=\frac{\sqrt{3(1-\mu^2)}}{2}\frac{(F+1)p}{Ek}$, $\beta_1=\beta\sqrt{\frac{1}{\sqrt{1+\mu k}}-\kappa}$, $\beta_2=\beta\sqrt{\frac{1}{\sqrt{1+\mu k}}+\kappa}$. Obviously, when it comes to the inner-stiffener or neglecting the axial force of non closed ring shell, there are $\beta_1=\beta_2=\frac{\beta}{\sqrt[4]{1+\mu k}}$.

Due to being $z=0$ set at the ring-stiffener position, there are boundary conditions:

1) The angle of rotation at the ring-stiffener is zero, that is $w_f'(z)\big|_{z=0}=0$;
2) The shear force of the shell at the zero position is half of the force acting on the ring-stiffener web, that is $2D_f w_f'''(z)\big|_{z=0}=f_2$;
3) The edge of the flange is in a free state, both the bending moment and shear force is zero, that is $w_f''(z)\big|_{z=W/2}=0$ and $w_f'''(z)\big|_{z=W/2}=0$.

From this, it can be concluded that

$$\begin{cases}G_1=Q_5 f_2\\ G_2=Q_6 f_2\\ G_3=Q_7 f_2\\ G_4=Q_8 f_2\end{cases} \quad (16)$$

where $v_1=\frac{\beta_1 W}{2}$, $v_2=\frac{\beta_2 W}{2}$,

$Q_5=\frac{1}{4D_f}\frac{4\beta_1^2\beta_2^2+\beta_2^2(\beta_1^2+\beta_2^2)\sinh^2 v_1-\beta_1^2(\beta_1^2+\beta_2^2)\sin^2 v_2}{\beta_1\beta_2(\beta_1^2+\beta_2^2)^2(\beta_2\cosh v_1\sinh v_1+\beta_1\cos v_2\sin v_2)}$, $Q_6=\frac{1}{4D_f}\frac{1}{\beta_2(\beta_1^2+\beta_2^2)}$,

$Q_7=-\frac{1}{4D_f}\frac{1}{\beta_1(\beta_1^2+\beta_2^2)}$, $Q_8=\frac{1}{4D_f}\frac{2(\beta_2^2-\beta_1^2)-(\beta_1^2+\beta_2^2)\sinh^2 v_1-(\beta_1^2+\beta_2^2)\sin^2 v_2}{(\beta_1^2+\beta_2^2)^2(\beta_2\cosh v_1\sinh v_1+\beta_1\cos v_2\sin v_2)}$.

Therefore, it can be concluded that the bending deflection of the ring-stiffener flange is

$$w_f(z)=-\frac{F+1}{2}\frac{\mu r_f}{E}p+f_2\begin{pmatrix}Q_5\cosh\beta_1 z\cos\beta_2 z+Q_6\cosh\beta_1 z\sin\beta_2 z\\ +Q_7\sinh\beta_1 z\cos\beta_2 z+Q_8\sinh\beta_1 z\sin\beta_2 z\end{pmatrix} \quad (17)$$

And the corresponding longitudinal stress is

$$\sigma_{zM}(r,z) = -\frac{F+1}{2}p + \frac{E(r-R_f)}{(1-\mu^2)}f_2\left(\begin{array}{l}Q_9\cosh\beta_1 z\cos\beta_2 z + Q_{10}\cosh\beta_1 z\sin\beta_2 z \\ +Q_{11}\sinh\beta_1 z\cos\beta_2 z + Q_{12}\sinh\beta_1 z\sin\beta_2 z\end{array}\right) \quad (18)$$

where $Q_9 = Q_5\beta_1^2 + 2Q_8\beta_1\beta_2 - Q_5\beta_2^2$, $Q_{10} = Q_6\beta_1^2 - 2Q_7\beta_1\beta_2 - Q_6\beta_2^2$, $Q_{11} = Q_7\beta_1^2 + 2Q_6\beta_1\beta_2 - Q_7\beta_2^2$, $Q_{12} = Q_8\beta_1^2 - 2Q_5\beta_1\beta_2 - Q_8\beta_2^2$.

Based on the bending deflection $w_f(z)$ and longitudinal stress $\sigma_{zM}(r,z)$, the radial stress $\sigma_{rM1}(r,z)$ and circumferential stress $\sigma_{\theta M1}(r,z)$ corresponding to the bending of ring-stiffener flange can be further obtained as

$$\begin{cases}\sigma_{rM1}(r,z) = -q_f(z) = \dfrac{k}{1+\mu k}\left[\dfrac{F+1}{2}\mu p + \dfrac{E}{R_f}w_f(z)\right] \\ \sigma_{\theta M1}(r,z) = \mu\sigma_{zM}(r,z) + \dfrac{1}{1+\mu k}\left[\dfrac{F+1}{2}\mu^2 kp - \dfrac{E}{R_f}w_f(z)\right]\end{cases} \quad (19)$$

4 Analysis of Radial Deformation and Stress Based on Plane Annulus Problem

4.1 Analysis Method for Axisymmetric Plane Strain Problem

According to the analysis method of plane problems in elasticity, the radial deformation of an axisymmetric plane annulus can be expressed as

$$u_r(r) = \frac{1}{E}\left[-(1+\mu)\frac{A}{r} + 2(1-\mu)Cr\right] \quad (20)$$

where $u_r(r)$ is the radial deformation, A and C are undetermined constants related to loads. According to geometric equations and Hooke's law, the radial stress $\sigma_r(r)$ and circumferential stress $\sigma_\theta(r)$ of plane annulus can be obtained as

$$\begin{cases}\sigma_r(r) = \dfrac{A}{r^2} + 2C \\ \sigma_\theta(r) = -\dfrac{A}{r^2} + 2C\end{cases} \quad (21)$$

where r is the radial direction of the plane annulus in the cylindrical coordinate system, and θ is the circumferential direction of the plane annulus in the cylindrical coordinate system.

4.2 Radial Deformation and Stress of Cylindrical Shell as Plane Annulus

According to the method in Sect. 4.1, the radial stress of the cylindrical shell as plane annulus satisfies the boundary conditions: $\sigma_{r2}(R+0.5t) = -p$ and $\sigma_{r2}(R - 0.5t) = 0$. From which the radial deformation $u_{r2}(r)$, plane radial stress $\sigma_{r2}(r)$, and circumferential stress $\sigma_{\theta 2}(r)$ of the cylindrical shell can be calculated as

$$\begin{cases} u_{r2}(r) = Q_{13}\dfrac{p}{E}\left[-2mQ_{14}(1+\mu)\dfrac{R^2}{r} - (1-\mu)r \right] \\ \sigma_{r2}(r) = Q_{13}p\left(2mQ_{14}\dfrac{R^2}{r^2} - 1 \right) \\ \sigma_{\theta 2}(r) = Q_{13}p\left(-2mQ_{14}\dfrac{R^2}{r^2} - 1 \right) \end{cases} \quad (22)$$

where $Q_{13} = \dfrac{(1+0.5m)^2}{2m}$ and $Q_{14} = \dfrac{(1-0.5m)^2}{2m}$.

4.3 Radial Deformation and Stress of Ring-Stiffener Web

According to the method in Sect. 4.1, the radial stress of the ring-stiffener web satisfies the boundary conditions: $\sigma_{rF1}(R_h - 0.5FH) = -\dfrac{F+1}{2}p + F\dfrac{f_1}{\delta}$ and $\sigma_{rF1}(R_h + 0.5FH) = -\dfrac{F+1}{2}p + F\dfrac{f_2}{\delta}$. Therefore, the radial deformation $u_{rF1}(r)$, plane radial stress $\sigma_{rF1}(r)$, and circumferential stress $\sigma_{\theta F1}(r)$ of the ring-stiffener web can be calculated as

$$\begin{cases} u_{rF1}(r) = \dfrac{1}{E}\left[-(1+\mu)2nQ_{15}Q_{16}\left(\dfrac{f_1}{\delta} - \dfrac{f_2}{\delta}\right)\dfrac{R_h^2}{r} + (1-\mu)\left(Q_{15}\dfrac{f_2}{\delta} - Q_{16}\dfrac{f_1}{\delta} - \dfrac{F+1}{2}p \right)r \right] \\ \sigma_{rF1}(r) = 2nQ_{15}Q_{16}\left(\dfrac{f_1}{\delta} - \dfrac{f_2}{\delta}\right)\dfrac{R_h^2}{r^2} + Q_{15}\dfrac{f_2}{\delta} - Q_{16}\dfrac{f_1}{\delta} - \dfrac{F+1}{2}p \\ \sigma_{\theta F1}(r) = -2nQ_{15}Q_{16}\left(\dfrac{f_1}{\delta} - \dfrac{f_2}{\delta}\right)\dfrac{R_h^2}{r^2} + Q_{15}\dfrac{f_2}{\delta} - Q_{16}\dfrac{f_1}{\delta} - \dfrac{F+1}{2}p \end{cases} \quad (23)$$

where $Q_{15} = \dfrac{(1+0.5nF)^2}{2n}$ and $Q_{16} = \dfrac{(1-0.5nF)^2}{2n}$.

4.4 Radial Deformation and Stress of Ring-Stiffener Flange as Plane Annulus

According to the method in Sect. 4.1, the radial stress of the ring-stiffener flange as plane annulus satisfies the boundary condition: $\sigma_{rM2}(R_{fo}) = \sigma_{rM2}(R_{fi}) = -0.5(F + 1)p$. From which the radial deformation $u_{rM2}(r)$, plane radial stress $\sigma_{rM2}(r)$, and circumferential stress $\sigma_{\theta M2}(r)$ of the ring-stiffener flange can be calculated as

$$\begin{cases} u_{rM2}(r) = -(1-\mu)(F+1)\dfrac{p}{2E}r \\ \sigma_{rM2}(r) = \sigma_{\theta M2}(r) = -(F+1)\dfrac{p}{2} \end{cases} \quad (24)$$

5 Three Dimensional Stress of Ring-Stiffened Cylindrical Shell

5.1 Three Dimensional Stress Expression of Cylindrical Shell

By superimposing the derivation results of the cylindrical shell based on bending beam problem and plane annulus problem, according to Eqs. (10), (11), (12), and (22), it can be concluded that the combined radial deformation of the cylindrical shell $u_r(r, z)$ is

$$u_r(r,z) = u_{r2}(r) - w(z) = Q_{13}\frac{p}{E}\left[-2mQ_{14}(1+\mu)\frac{R^2}{r} - (1-\mu)r\right] - w(z) \quad (25)$$

And the longitudinal stress $\sigma_z(r, z)$, combined radial stress $\sigma_r(r, z)$, and combined circumferential stress $\sigma_\theta(r, z)$ of the cylindrical shell are

$$\begin{cases} \sigma_z(r,z) = -Q_{13}p + \dfrac{E(r-R)}{(1-\mu^2)}f_1(Q_3\cosh\alpha_1 z\cos\alpha_2 z + Q_4\sinh\alpha_1 z\sin\alpha_2 z) \\ \sigma_r(r,z) = \dfrac{m}{1+\mu m}\dfrac{E}{R}w(z) + Q_{13}p\left(\dfrac{\mu m}{1+\mu m} + 2mQ_{14}\dfrac{R^2}{r^2} - 1\right) \\ \sigma_\theta(r,z) = \mu\sigma_z(r,z) - \dfrac{1}{1+\mu m}\dfrac{E}{R}w(z) + Q_{13}p\left(\dfrac{\mu^2 m}{1+\mu m} - 2mQ_{14}\dfrac{R^2}{r^2} - 1\right) \end{cases} \quad (26)$$

5.2 Three Dimensional Stress Expression of Ring-Stiffener Web

Obviously, the longitudinal stress of the ring-stiffener web is $\sigma_{zF}(r) = -0.5(F+1)p$ (which is $-p$ for the outer-stiffener and zero for the inner-stiffener). Based on the results of 4.3, the radial deformation of the ring-stiffener web is

$$u_{rF}(r) = \frac{1}{E}\left[-(1+\mu)2nQ_{15}Q_{16}\left(\frac{f_1}{\delta} - \frac{f_2}{\delta}\right)\frac{R_h^2}{r} + (1-\mu)\left(Q_{15}\frac{f_2}{\delta} - Q_{16}\frac{f_1}{\delta} - \frac{F+1}{2}p\right)r\right] \quad (27)$$

Considering the lateral effect of longitudinal load on the ring-stiffener web plane annulus, by superimposing the longitudinal stress $\sigma_{zF}(r)$ with a coefficient of $\mu/(1-\mu)$ on the plane strain, the radial stress $\sigma_{rF}(r)$ and circumferential stress $\sigma_{\theta F}(r)$ of the ring-stiffener web can be obtained as

$$\begin{cases} \sigma_{rF}(r) = 2nQ_{15}Q_{16}\left(\dfrac{f_1}{\delta} - \dfrac{f_2}{\delta}\right)\dfrac{R_h^2}{r^2} + Q_{15}\dfrac{f_2}{\delta} - Q_{16}\dfrac{f_1}{\delta} - \dfrac{1}{1-\mu}\dfrac{F+1}{2}p \\ \sigma_{\theta F}(r) = -2nQ_{15}Q_{16}\left(\dfrac{f_1}{\delta} - \dfrac{f_2}{\delta}\right)\dfrac{R_h^2}{r^2} + Q_{15}\dfrac{f_2}{\delta} - Q_{16}\dfrac{f_1}{\delta} - \dfrac{1}{1-\mu}\dfrac{F+1}{2}p \end{cases} \quad (28)$$

5.3 Three Dimensional Stress Expression of Ring-Stiffener Flange

By superimposing the derivation results of the ring-stiffener flange based on bending beam problem and plane annulus problem, according to Eqs. (17), (18), (19), and (24), it can be concluded that the combined radial deformation $u_{rM}(r)$ of the ring-stiffener flange is

$$u_{rM}(r) = u_{rM2}(r) - w_f(z) = -(1-\mu)(F+1)\frac{p}{2E}r - w_f(z) \qquad (29)$$

And the longitudinal stress $\sigma_{zM}(r,z)$, combined radial stress $\sigma_{rM}(r,z)$, and combined circumferential stress $\sigma_{\theta M}(r,z)$ of the ring-stiffener flange are

$$\begin{cases} \sigma_{zM}(r,z) = -\dfrac{F+1}{2}p + \dfrac{E(r-R_f)}{(1-\mu^2)}f_2\begin{pmatrix} Q_9\cosh\beta_1 z\cos\beta_2 z + Q_{10}\cosh\beta_1 z\sin\beta_2 z \\ +Q_{11}\sinh\beta_1 z\cos\beta_2 z + Q_{12}\sinh\beta_1 z\sin\beta_2 z \end{pmatrix} \\ \sigma_{rM}(r,z) = -\dfrac{1}{1+\mu k}\left[\dfrac{F+1}{2}p - k\dfrac{E}{R_f}w_f(z)\right] \\ \sigma_{\theta M}(r,z) = \mu\sigma_{zM}(r,z) + \dfrac{1}{1+\mu k}\left[(\mu^2 k - \mu k - 1)\dfrac{F+1}{2}p - \dfrac{E}{R_f}w_f(z)\right] \end{cases} \qquad (30)$$

5.4 Simultaneous Solution of Force Parameters

In the three dimensional stress expression method for ring-stiffened cylindrical shell, there are two unknown parameters for the applied forces f_1 and f_2.

As a complete structure, the deformation at the connection between the cylindrical shell and the ring-stiffener web, as well as the connection between the ring-stiffener web and the ring-stiffener flange, is continuous and consistent, the deformation boundary conditions are

$$\begin{cases} u_r(R+0.5Ft, \dfrac{l}{2}) = u_{rF}(R_h - 0.5FH) \\ u_{rF}(R_h + 0.5FH) = u_{rM}(R_f - 0.5Ft_f, 0) \end{cases} \qquad (31)$$

Based on these two equations, two unknown parameters can be obtained as

$$\begin{cases} f_1 = \dfrac{Q_{19}Q_{22} - 2Q_{15}Q_{21}}{Q_{20}Q_{22} - 4Q_{15}Q_{16}}\delta p \\ f_2 = \dfrac{2Q_{16}Q_{19} - Q_{20}Q_{21}}{Q_{20}Q_{22} - 4Q_{15}Q_{16}}\delta p \end{cases} \qquad (32)$$

where $Q_{17} = \dfrac{1}{1+0.5mF}$, $Q_{18} = \dfrac{1}{1-0.5kF}$,

$Q_{19} = \dfrac{(1+0.5m)^2}{(1+0.5mF)^2}(1+\mu)Q_{14} + (1-\mu)Q_{13} - \mu Q_{13}Q_{17} - (1-\mu)\dfrac{F+1}{2}$,

$Q_{20} = (1+\mu)Q_{15} + (1-\mu)Q_{16} - Q_{17}(Q_1\cosh u_1\cos u_2 + Q_2\sinh u_1\sin u_2)\dfrac{E\delta}{R}$,

$Q_{21} = -\dfrac{F+1}{2}\mu Q_{18}$, $Q_{22} = (1+\mu)Q_{16} + (1-\mu)Q_{15} + Q_5 Q_{18}\dfrac{E\delta}{R_f}$.

644   S. Chen et al.

6 Stress Analysis of Ring-Stiffened Cylindrical Shell with New Design Features

Adopting the three dimensional stress analysis method to analysis ring-stiffened cylindrical shell with new design features such as large depth ($p = 90$ MPa), large thickness ($t/R > 0.05$), low and wide ring-stiffener ($W/H > 1.0$), and the results were compared with finite element simulation methods and traditional stress analysis methods.

The structural parameters are shown in Table 1, the elastic modulus of the material is 1.15×10^5 MPa, and the Poisson's ratio is 0.34.

Table 1. Parameter table of ring-stiffened cylindrical shell with new design features.

Parameter name	Symbol	Data(mm)
Inner radius of cylindrical shell	R_i	500
Thickness of cylindrical shell	t	50
Height of ring-stiffener web	H	90
Thickness of ring-stiffener web	δ	30
Width of ring-stiffener flange	W	140
Thickness of ring-stiffener flange	t_f	40
Distance between adjacent ring-stiffener	l	400
Outer-stiffener or inner-stiffener	F	1

Establishing an axisymmetric finite element model with plane solid element, the element mesh size is divided according to ESIZE = t/N (the element mesh size is N equal parts of the thickness of the cylindrical shell). Uniform line loads p are applied to the outer surface of the model to simulate hydrostatic pressure, and uniform line loads $p(R_i + t)^2/(2R_i t + t^2)$ are applied to one end of the model to simulate the axial force of closed cylindrical shell, and axial displacement constraints are applied to the other end of the model to simulate rigid body displacement constraints, as shown in Fig. 6.

The stress cloud map of the ring-stiffened cylindrical shell is shown in Fig. 7 (stress unit: MPa). Under hydrostatic pressure, the maximum longitudinal stress of cylindrical shell is located at the ring-stiffener, the maximum circumferential stress of cylindrical shell is located at the middle of adjacent ring-stiffener, and the maximum circumferential stress of ring-stiffener is located at the lower edge of the ring-stiffener web.

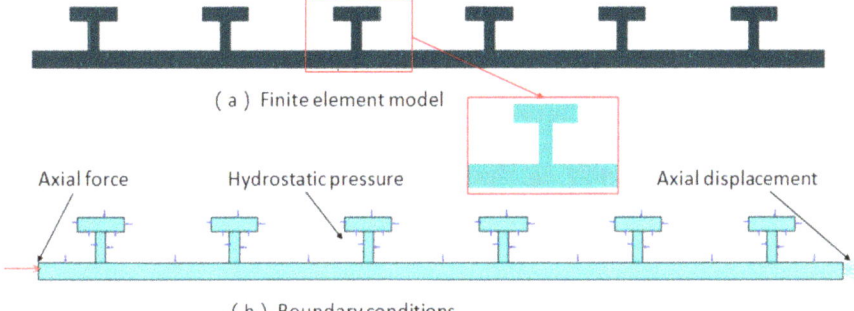

Fig. 6. Finite element model of ring-stiffened cylindrical shell.

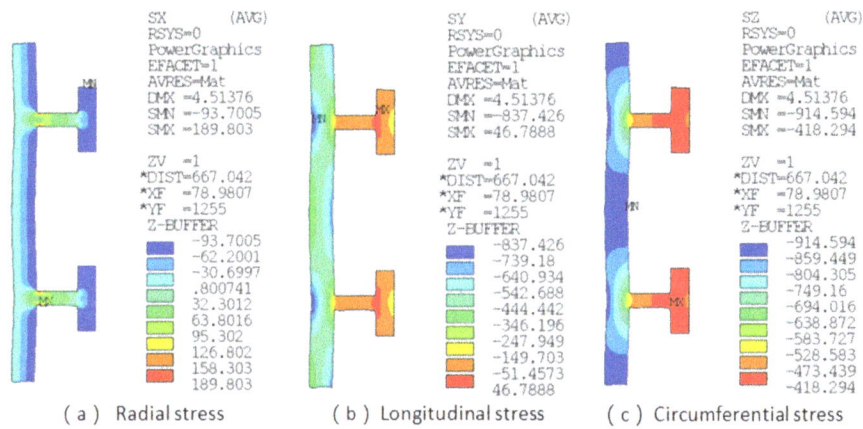

Fig. 7. Stress cloud map of ring-stiffened cylindrical shell.

To observe the influence of mesh size on the typical stress calculation results, a series of calculations were performed using different parameters N. Obviously, when N is set to 15, the stress calculation results of the ring-stiffened cylindrical shell tend to be stable, as shown in Fig. 8.

The three dimensional stress analysis method and finite element simulation results of typical stresses in ring-stiffened cylindrical shells are shown in Table 2. The results showed that: 1) Regarding the stress of cylindrical shell, whether it is at the middle of adjacent ring-stiffener or at the root of the ring-stiffener, the calculation results of the three dimensional stress analysis method are in good agreement with the finite element simulation results, and are relatively safe; 2) Regarding the stress on the ring-stiffener web, the calculated results of the two methods are also in good agreement; 3) Regarding the stress of ring-stiffener flange, the analytical calculation results of longitudinal and circumferential stresses are relatively high, while the analytical calculation results of radial stresses are relatively low. This indicates that the bending mechanics model for non closed ring-stiffener flange ring shell in the three dimensional stress analysis method is generally slightly conservative.

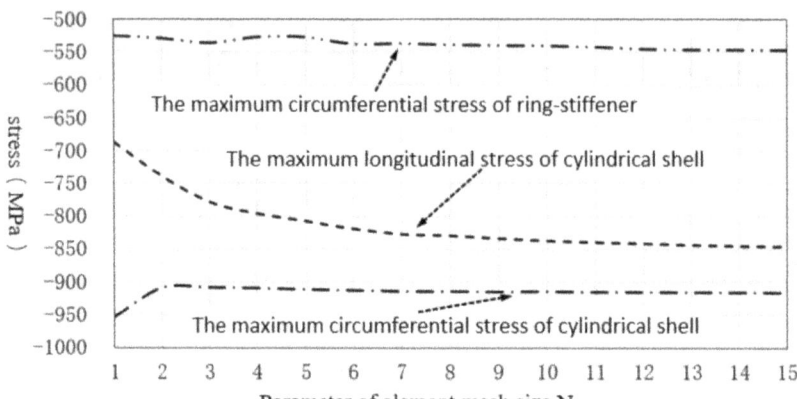

Fig. 8. The influence of element size on the stress simulation results.

The results of both methods show significant differences in the center and side stresses of the ring-stiffener flange, proving that assuming the ring-stiffener flange with one dimensional state is clearly unreasonable. In addition, it was found that the longitudinal stress (−915.7 MPa) on the inner surface of the ring-stiffener root shell was lower than its circumferential stress (−964.0MPa), and the circumferential stress (−550.1 MPa) of the ring-stiffener flange was greater than the circumferential stress (−477.2 MPa) on the upper edge of the ring-stiffener web, exhibiting significantly different stress characteristics from traditional thin ring-stiffened cylindrical shell.

Table 2. Stress results of three dimensional stress method and finite element simulation.

Typical stress name		Three dimensional stress method(MPa)	Finite element simulation(MPa)
cylindrical shell at the middle of adjacent ring-stiffener	Longitudinal stress of external surface	−697.8	−679.4
	Radial stress of external surface	−98.2	−85.5
	Circumferential stress of external surface	−921.6	−913.0
	Circumferential stress of internal surface	−889.8	−887.6
	Circumferential stress of middle surface	−905.7	−899.1
cylindrical shell at the ring-stiffener root	Longitudinal stress of internal surface	−915.7	−846.0

(*continued*)

Table 2. (*continued*)

Typical stress name		Three dimensional stress method(MPa)	Finite element simulation(MPa)
	Circumferential stress of external surface	−603.9	−584.5
	Circumferential stress of internal surface	−964.0	−906.3
	Circumferential stress of middle surface	−784.0	−745.4
Ring-stiffener web	Longitudinal stress of the centroid	−90.0	−89.9
	Radial stress of the centroid	57.2	55.7
	Circumferential stress of upper edge	−477.2	−436.4
	Circumferential stress of lower edge	−556.5	−546.6
External surface of ring-stiffener flange	Longitudinal stress of the center	−330.5	−212.9
	Longitudinal stress of the side	−90.0	−90.2
	Radial stress of the center	−66.5	−86.1
	Radial stress of the side	−67.5	−89.5
	Circumferential stress of the center	−550.1	−490.0
	Circumferential stress of side	−461.9	−431.4

Regarding the key stress (which is the strength verification stress term in the standard specifications) of the ring-stiffened cylindrical shell, the above calculation results are further compared with the results of traditional stress analysis methods such as reference [6], the results are shown in Table 3.

The results showed that the three dimensional stress analysis method is relatively more consistent with the finite element simulation results, and the calculated values of the shell circumferential stress at the middle of adjacent ring-stiffener and the circumferential stress of the ring-stiffener using the method in reference [6] are about 6.9% and 12.1% lower, while the shell longitudinal stress at the ring-stiffener root are about 10.7% higher. This indicates that traditional stress analysis methods used for the calculation of ring-stiffened cylindrical shell with design features such as large depth, large thickness, and low and wide ring-stiffener have dangerous results.

Table 3. Key stress results of ring-stiffened cylindrical shell.

Key stress name	Three dimensional stress method(MPa)	Method of reference [6] (MPa)	Finite element simulation (MPa)
Circumferential stress of middle surface of cylindrical shell at the middle of adjacent ring-stiffener	−905.7	−840.6	−899.1
Longitudinal stress of internal surface of cylindrical shell at the ring-stiffener root	−915.7	−947.2	−846.0
Maximum circumferential stress of ring-stiffener	−556.5	−487.6	−546.6

7 Conclusion

Focusing on research of three dimensional stress analysis method for ring-stiffened cylindrical shell, the conclusions are as follows:

(1) A new simplified mechanical model for the ring-stiffened cylindrical shell was proposed, which considers the cylindrical shell as closed ring shell and ring-stiffener flange as non closed ring shell, and the ring-stiffener web as a plane annulus. This model transforms the stress analysis of the ring-stiffened cylindrical shell under hydrostatic pressure into a combination of bending beam problem under axial force and axisymmetric plane annulus problem under radial force.
(2) A bending differential equation considering triaxial stress state is proposed, and combined with the analysis method of axisymmetric plane annulus problem, a three dimensional stress calculation formula which can be used to calculate the three dimensional stress at any position of ring-stiffened cylindrical shell is derived.
(3) Taking a ring-stiffened cylindrical shell with structural design features such as large depth, large thickness, and low and width ring-stiffener as an example. The results show that the three dimensional stress analysis method not only has good calculation accuracy, but also the calculation results are relatively safe.
(4) The calculation results show that not only are there significant differences in the center and side stresses of the ring-stiffener flange, but also for the ring-stiffened cylindrical shell with new design features, the stress characteristics of the ring-stiffener root shell and ring-stiffener stress are significantly different from those of traditional thin ring-stiffened cylindrical shell.

References

1. Lunchick, E., Overby, J.A.: Yield strength of machined ring-stiffened cylindrical shell under hydrostatic pressure. Exp. Mech. **1**(6), 178–185 (1961)
2. Amazigo, J.C., Fraser, W.B.: Buckling under external pressure of cylindrical shells with dimple shaped initial imperfections. Int. J. Solids Struct. **7**(8), 883–900 (1971)
3. Xu, B., Zhu, B.: Theory and Experiment of Structural Strength of Modern Submarine. National Defense Industry Press, Beijing (2007). (in chinese)
4. Wu, F., Zhu, X.: Structural Mechanics of Ships. National Defense Industry Press, Beijing (2010). (in chinese)
5. Ouyang, L., Ye, C.: Calculation Method for Compressive Strength of Deep-sea Submersible Structures. National Defense Industry Press, Beijing (2022). (in chinese)
6. Zhu, B., Wan, Z.: A method of stress analysis of ring-stiffened cylindrical shell. J. Ship Mech. **8**(4), 61–67 (2004). (in chinese)
7. GJB: Method for Design and Calculation of Submarine Structure. COSTIND, Beijing (2001). (in chinese)
8. CCS. Classification Specification for Diving Systems and Submersibles. People's Communications Press, Beijing (2018). (in chinese)
9. ABS. Rules for Building and Classing-Underwater Vehicles, Systems and Hyperbaric Facilities. The State of New York (2010)
10. RUS. Rules for The Classification and Construction of Manned Submersibles, Ship's Diving Systems and Passenger Submersibles No. 2-2-020201-005-E. Russian Federation (2004)
11. GL. Rules for Classification and Construction. Hamburg (2009)
12. Zhang, E., et al.: Research status and development trend of pressure resistant structure of deep submersibles. J. Ship Mech. **25**(10), 1427–1437 (2021)
13. Teguh, M., et al.: Optimisation of the design of a steel-welded pressure hull structure based on interactive nonlinear collapse strength analyses. Ships Offshore Struct. **17**(1/3), 76–91 (2022)
14. Wang, F., et al.: Preliminary evaluation of maraging steels on its application to full ocean depth manned cabin. J. Ship Mech. **20**(12), 1557–1572 (2016)
15. Yeh, M.K., Kao, C.M.: Finite element analysis of stress concentration at rounded crack tip with different physical parameters. In: Proceedings of 2013 2nd International Symposium on Quantum, Nano and Micro Technologies (ISQNM 2013) (2013)

Open Access This chapter is licensed under the terms of the Creative Commons Attribution-NonCommercial-NoDerivatives 4.0 International License (http://creativecommons.org/licenses/by-nc-nd/4.0/), which permits any noncommercial use, sharing, distribution and reproduction in any medium or format, as long as you give appropriate credit to the original author(s) and the source, provide a link to the Creative Commons license and indicate if you modified the licensed material. You do not have permission under this license to share adapted material derived from this chapter or parts of it.

The images or other third party material in this chapter are included in the chapter's Creative Commons license, unless indicated otherwise in a credit line to the material. If material is not included in the chapter's Creative Commons license and your intended use is not permitted by statutory regulation or exceeds the permitted use, you will need to obtain permission directly from the copyright holder.

Numerical Simulation on Planar Motion Mechanism of a Bulk Carrier with a Body-Force Propeller Model

Hao Hao[1(✉)], Weimin Chen[2], and Minmin Zheng[2]

[1] Shanghai Ship and Shipping Research Institute Co., Ltd, Shanghai, China
1103534933@qq.com
[2] State Key Laboratory of Maritime Technology and Safety, Shanghai, China

Abstract. To better study the maneuvering performance of a 210,000-ton bulk carrier, this paper utilizes commercial software STAR CCM+ to carry out numerical computations of three types of planar motion mechanism movements: The yaw motion, the yaw motion with drift angle, the yaw motion with steering angle. These simulations are based on the Reynolds-Averaged Navier-Stokes (RANS) equations, the Volume of Fluid (VOF) model, the body-force method, and six-degree-of-freedom model. The transverse forces and the yawing moments are compared, and their integral values are calculated. All computational results show good agreement with model experimental data, verifying the reliability of the numerical calculation method.

Keywords: Planar Motion Mechanism · Yaw Motion · Body-Force Method

1 Introduction

Maneuverability is an important navigational characteristic of ships, and it needs to be forecasted during the ship design phase. One of the most commonly used methods is to conduct constrained model tests using a planar motion mechanism (PMM). Although this method is highly accurate, it involves complex processes and high costs. Therefore, numerical calculations, which are efficient and can provide detailed information about the flow field, have also been applied to the constrained model motion.

Broglia [1, 2] conducted simulations of pure sway motion for the KVLCC2 ship, taking into account the effects of the free surface. He also simulated pure sway and pure yaw motion with propellers and rudders included. Stern [3] presented an overview of the progress in numerical studies on ship maneuverability at the SIMMAN2008 conference. He noted that fine mesh and Detached Eddy Simulation (DES) methods could enhance the accuracy of numerical calculations for maneuverability. Miller [4] conducted numerical simulations for both the bare hull and the full appendage of the DTMB 5415 model on PMM motion. The results indicated that the errors were smaller for the full appendage. Simonsen [5] used both Computational Fluid Dynamics(CFD) computed data and Experimental Fluid Dynamics (EFD) measured data of PMM to simulate the turning motion and zigzag motion of the KCS ship model in calm water based on a mathematical model of maneuverability. Turnock [6] used the CFD software CFX to numerically solve for the KVLCC2 ship model under conditions of straight ahead, oblique towing, and pure sway motion. He also analyzed shallow water conditions.

However, the above study did not consider the influence of ballast draft on yaw motion, nor did it take into account the characteristics of the yaw motion with steering angle. While simulating propeller rotation using an actual propeller model needs substantial computational workload and extended calculation periods, the body force method shows superior computational efficiency. In response to the above research status, this article uses the body-force method to numerically simulate three types of motion: The yaw motion, the yaw motion with drift angle, the yaw motion with steering angle.

2 The Mathematical Model of PMM Motion

In the Planar Motion Mechanism test, the pure yaw motion is achieved by mechanically forcing the ship model to move at a constant speed along the centerline of the basin in calm water. A horizontal displacement that varies sinusoidally in the vertical plane is superimposed on this motion. Additionally, a change in the heading angle is introduced, ensuring that the longitudinal axis of the ship model remains tangent to the trajectory of motion. The yaw motion with drift angle sets a fixed drift angle for the hull, while the yaw motion with steering angle sets a fixed steering angle for the rudder based on the yaw motion.

The mathematical expression for the yaw motion as follows:

$$\Psi = \Psi o sinwt \tag{1}$$

$$r = \Psi o \, w \, coswt \tag{2}$$

$$\dot{r} = -\Psi o \, w^2 sinwt \tag{3}$$

To ensure that the bow direction is always tangent to the motion trajectory, a certain coupling condition must be satisfied between the transverse speed and the longitudinal speed.

$$v = U \bullet \tan\Psi \tag{4}$$

In which: w represents the circular frequency of lateral vibration, and Ψ represents the amplitude of yaw.

The fundamental mathematical models for the second yaw motion and the third motion are the same as that of the yaw motion. The second motion includes an additional drift angle compared to the yaw, while the third motion includes an extra directional angle compared to the yaw motion. The model model of the yaw motion is shown in Fig. 1, the motion model of the second yaw motion is shown in Fig. 2, and the motion model of the third motion is the same as the yaw motion.

This article selects a 210,000-ton bulk carrier as a calculation example, and its main dimensions are shown in Table 1.

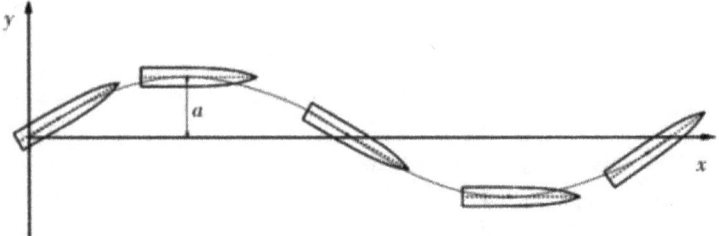

Fig. 1. The motion model of the yaw motion.

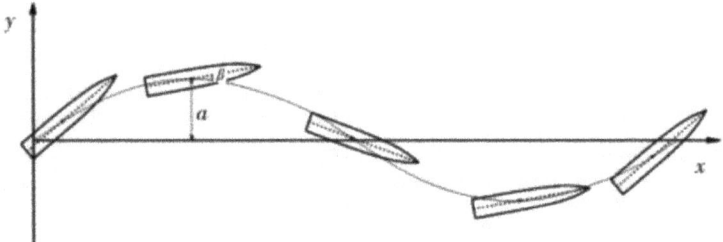

Fig. 2. The motion model of the yaw motion with drift angle.

Table 1. Main dimensions of ship model.

Displacement (kg)	Total length of hull (m)	Bow draft(m)	Stern draft (m)	Beam (m)	Square coefficient
508.98	4.998	0.12	0.192	0.83	0.798

The formulas for solving the integral values of the yaw motion are as follows:

$$Y_{in} = \frac{1}{4}\left(\int_0^{\frac{\pi}{2}} Y_E dwt - \int_{\frac{\pi}{2}}^{\frac{3\pi}{2}} Y_E dwt + \int_{\frac{3\pi}{2}}^{2\pi} Y_E dwt\right) \tag{5}$$

$$Y_{out} = \frac{1}{4}\left(\int_0^{\pi} Y_E dwt - \int_{\pi}^{2\pi} Y_E dwt\right) \tag{6}$$

$$N_{in} = \frac{1}{4}\left(\int_0^{\pi/4} N_E dwt - \int_{\pi/2}^{3\pi/2} N_E dwt + \int_{3\pi/2}^{2\pi} N_E dwt\right) \tag{7}$$

$$N_{out} = \frac{1}{4}\left(\int_0^{\pi} N_E dwt - \int_{\pi}^{2\pi} N_E dwt\right) \tag{8}$$

The formulas for solving the integral values of the combination of yaw and drift angle montion are as follows:

$$X_{out} = \frac{1}{4}\left(\int_0^{\pi} X_E dwt - \int_{\pi}^{2\pi} X_E dwt\right). \tag{9}$$

$$Y_C = \frac{1}{4}\left(\int_0^{\frac{\pi}{2}} Y_E dwt + \int_{\frac{\pi}{2}}^{\frac{3\pi}{2}} Y_E dwt + \int_{\frac{3\pi}{2}}^{2\pi} Y_E dwt\right) \tag{10}$$

$$N_{out} = \frac{1}{4}\left(\int_0^{\pi} N_E dwt - \int_{\pi}^{2\pi} N_E dwt\right). \tag{11}$$

The formulas for solving the integral values of the combination of yaw and steering angle montion are as follows:

$$Y_C = \frac{1}{4}\left(\int_0^{\frac{\pi}{2}} Y_E dwt + \int_{\frac{\pi}{2}}^{\frac{3\pi}{2}} Y_E dwt + \int_{\frac{3\pi}{2}}^{2\pi} Y_E dwt\right) \tag{12}$$

$$N_C = \frac{1}{4}\left(\int_0^{\frac{\pi}{2}} N_E dwt + \int_{\frac{\pi}{2}}^{\frac{3\pi}{2}} N_E dwt + \int_{\frac{3\pi}{2}}^{2\pi} N_E dwt\right) \tag{13}$$

In all calculations, the results are expressed using dimensionless coefficients normalized by the draught L, speed U, and density. The expressions are as follows:

$$X = \frac{X_E}{0.5\rho L^2 U^2} \tag{14}$$

$$Y = \frac{Y_E}{0.5\rho L^2 U^2} \tag{15}$$

$$N = \frac{N_E}{0.5\rho L^3 U^2} \tag{16}$$

3 Computational Domain and Boundary Conditions

The computational domain selected in this article is a rectangle. This domain moves at a constant speed U relative to the fixed coordinate system. According to the principle of relative motion, it can be considered that the computational domain is stationary, while the water flows into the domain at a velocity of -U. On one hand, the ship moves at a constant speed U along with the computational domain, and on the other hand, it performs various kinds of yaw motions within the domain. Table 1 is the main dimensions of the ship model and Fig. 3 is its schematic diagram. The numerical simulation uses a fully hexahedral unstructured grid, with refinement in the wave-making region generated by motion and near the free surface. The mesh division for the yaw motion is shown in Fig. 4 and Fig. 5, the mesh division around the stern for the combination of yaw and drift angle montion is shown in Fig. 6, and the mesh division for the combination of yaw and steering angle montion is shown in Fig. 7.

Fig. 3. The schematic diagram of the model.

Fig. 4. Mesh in the vertical direction for the yaw motion.

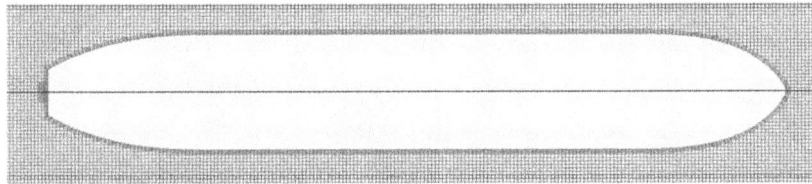

Fig. 5. Mesh in the horizontal direction for the yaw motion.

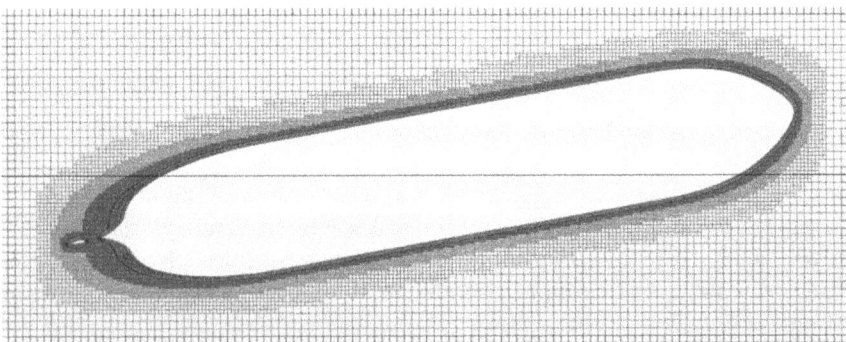

Fig. 6. Mesh in the horizontal direction for the yaw motion with drift angle.

The boundary conditions are set as follows:

1) Inflow boundary: It is set as velocity inlet and is used to specify the flow velocity.
2) Outflow boundary: Pressure outlet, the area above the free surface is set to standard atmospheric pressure.
3) Other interfaces: velocity inlet.

Figure 3 is the schematic diagram of the model:

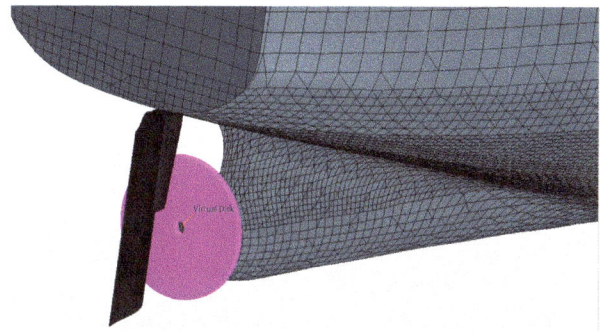

Fig. 7. Mesh of stern for the yaw motion with steering angle.

4 Numerical Result Analysis of Calculation Conditions

The calculation conditions for the three types of motion discussed in this article are in the table 2. The numerical simulation results within one period after the ship model stabilizes are analyzed. Figure 8 shows the computational results for the yaw motion, Fig. 9 displays the computational outcomes for the second yaw motion, and Fig. 10 presents the computational findings for the third yaw motion. The error between the computational and experimental results for the three types of motion are presented in following three tables respectively. The comparison between the calculation results and the experimental results are shown in Table 3, Table 4 and Table 5.

Table 2. The calculation conditions.

Velocity(m/s)	Frequency (hz)	Angular velocity (rad/s)	Drift angle(deg)	Steering angle (deg)	Amplitude(m)
0.863	0.0614	16.28	10	10	0.3

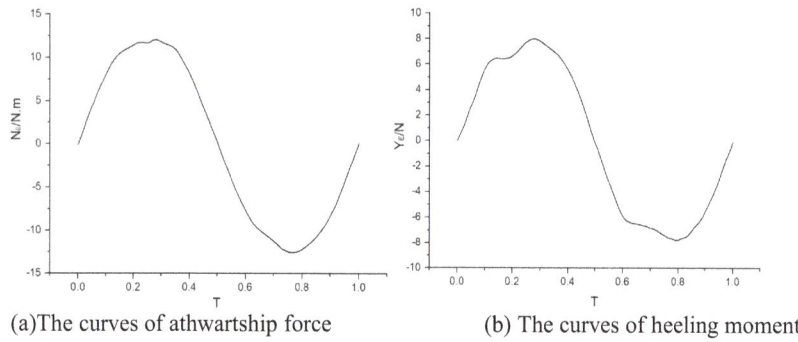

(a) The curves of athwartship force (b) The curves of heeling moment

Fig. 8. The curves of hydrodynamics of the yaw motion.

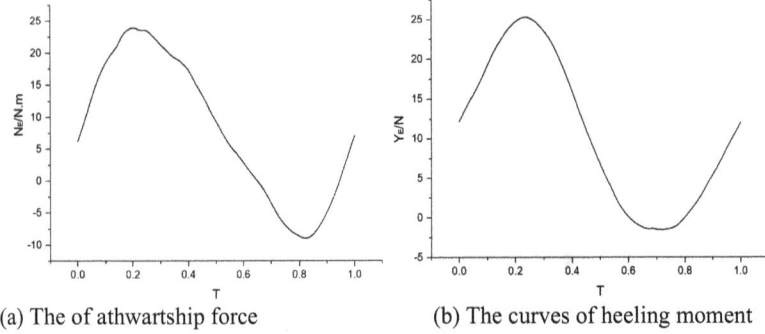

(a) The of athwartship force (b) The curves of heeling moment

Fig. 9. The curves of hydrodynamics of the yaw motion with drift angle.

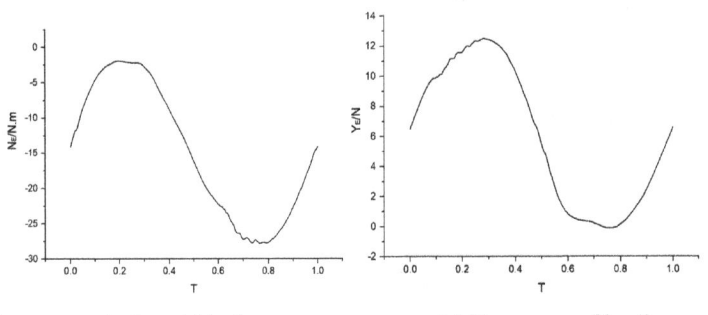

(a) The curves of athwartship force (b) The curves of heeling moment

Fig. 10. The curves of hydrodynamics of the yaw motion with steering angle. (a) The curves of athwartship force (b) The curves of heeling moment.

Table 3. The camprison of the yaw motion.

	$Y'_{out} \times 10^3$	$N'_{out} \times 10^3$	$Y'_{in} \times 10^3$	$N'_{in} \times 10^3$
experimental results	4.062	0.796	0.387	0.84
computational results	3.121	0.628	0.301	0.678
Error	−23.17%	−21.12%	−22.15%	−19.32%

Table 4. The camprison of the yaw motion with drift angle.

	$Y'_c \times 10^5$	$N'_{out} \times 10^3$	$X'_{out} \times 10^3$
experimental results	311.188	0.505	0.591
computational results	224.46	0.393	0.471
Error	−27.87%	−22.2%	−20.14%

Table 5. The camprison of the yaw motion with steering angle.

	$Y'_c \times 10^5$	$N'_c \times 10^5$
experimental results	125.602	−66.781
computational results	95.672	−50.04
Error	−23.48%	−25.06%

5 Conclusion

This study is based on the STAR-CCM+ software platform and focuses on a 210,000-ton bulk carrier. The body-force method is employed to simulate the actual propeller, and three types of the yaw motion are numerically simulated.

(1) The application of the body-force method to the treatment of the propeller significantly reduces the number of grids, thereby enhancing computational efficiency.

(2) The free motion of the ship was taken into account, which is more in line with the real situation of the experiment, and the calculation accuracy also meets the requirements of engineering applications.

(3) Although simulating propeller rotation with actual propellers requires a large amount of computation, it is more in line with the actual situation of motion. The next step will be to use real propellers for numerical calculations of the yaw motion.

References

1. Broglla, R., DiMascio, A., Amati, G.: A parallel unsteady RANS code for the numerical simulations of free surface flows. In: Proceedings of the 2nd International Conference on Marine Research and Transportation, Naples (2007)
2. Broglla, R., Muscari, R., DiMascio, A.: Numerical simulations of the pure sway and pure yaw motion of the KVLCC-1 and 2 tanker. In: Proceedings of SIM- MAN 2008 Workshop on Verification and Validation of Ship Maneuvering Simulation Methods, Lyngby, Demark (2008)
3. Stern, F., Agdrup, K., Kim, S.Y., et al.: Experience from SIMMAN2008–The first workshop on verification and validation of ship maneuvering simulation methods. J. Sh. Res. **55**(2), 135–147 (2011)
4. Miller, R.W.: PMM calculations for the bare and ap- pended DTMB 5415 using the RANS solver CFD- SHIP-IOWA. In: Proceedings of SIMMAN 2008 Workshop on Verification and Validation of Ship Maneuvering Simulation Methods. Lyngby, Demark (2008)
5. Simonsen, C.D., Otzen, J.F., Klimt, C., et al.: Maneuvering predictions in the early design phase using CFD generated PMM data. In: Proceedings of the 29th Symposium on Naval Hydrodynamics, pp. 26–31 (2012)
6. Turnock, S.R.: Urans simulations of static drift and dynamic maneuverers of the KVLCC2 tanker. In: Proceedings of SIMMAN 2008 workshop on verification and validation of ship Manoeuvring simulation methods. Lyngby, Denmark (2008)

Open Access This chapter is licensed under the terms of the Creative Commons Attribution-NonCommercial-NoDerivatives 4.0 International License (http://creativecommons.org/licenses/by-nc-nd/4.0/), which permits any noncommercial use, sharing, distribution and reproduction in any medium or format, as long as you give appropriate credit to the original author(s) and the source, provide a link to the Creative Commons license and indicate if you modified the licensed material. You do not have permission under this license to share adapted material derived from this chapter or parts of it.

The images or other third party material in this chapter are included in the chapter's Creative Commons license, unless indicated otherwise in a credit line to the material. If material is not included in the chapter's Creative Commons license and your intended use is not permitted by statutory regulation or exceeds the permitted use, you will need to obtain permission directly from the copyright holder.

Near-Field Low-Frequency Noise Prediction Analysis of Propeller Noise in Shallow Water Environment

He Yang[1], Chunyu Zhang[1(✉)], and Yun Liu[2]

[1] Jiangsu University of Science and Technology, Zhenjiang, China
hdzhangchunyu@163.com
[2] Shanghai Marine Diesel Engine Research Institute, Shanghai, China

Abstract. This paper focuses on the prediction and analysis of propeller-generated near-field low-frequency noise in shallow water environments. To address this challenge, we propose an innovative acoustic radiation calculation method for propellers, which comprehensively accounts for the acoustic reflection effects from the free water surface and the complex characteristics of the ocean. In developing the low-frequency propeller noise prediction model, the shallow water environment is accurately represented by simulating both the free surface and the seafloor impedance using specialized acoustic simulation software. Through detailed numerical simulations, we uncover the specific mechanisms by which the complex ocean environment influences the near-field noise radiation characteristics of propellers. The findings reveal that while the sound pressure level (SPL) radiated by the propeller follows a similar frequency-dependent trend, significant variations in amplitude are observed at different water depths (near the surface and closer to the seafloor). This insight deepens our understanding of noise propagation characteristics in shallow water regions.

Keywords: Propeller noise · Shallow water environment · Near-field radiation

1 Introduction

Propeller noise is the main noise source during high-speed navigation in underwater travel, featuring distinct line spectra and seriously affecting the underwater communication environment and concealment [1]. Current research on the impact of propeller noise mainly relies on infinite-depth modeling, leaving the influence of shallow water environments on the near-field radiated noise characteristics of propellers insufficiently understood [2]. However, in the real shallow sea environment, the boundary effects of the water surface and seabed have a significant impact on the reflection and propagation of noise in shallow water. Therefore, to accurately predict the near and far-field low-frequency noise generated by propellers in shallow water, the effects of the free water surface and seabed must be fully considered.

Zhu [3] employed the finite element calculation method, the influence law of the presence of internal waves on the propagation characteristics of low-frequency acoustic signals is simulated and analyzed with the acoustic energy flow as the research object. Donatas [4] introduced a modelling method for the noise footprint of a shallow-sea ship. Pan [5] suggested numerical prediction of propeller radiated noise in open water without bubbles, using DES method to obtain propeller winding information, and using FW-H equation in time domain to predict propeller radiated noise. Feng [6] employed A pilot study of accelerated rotating propeller noise testing was carried out in a large-scale circulating water tunnel of international standards. The test data contains the information of propeller speed change, which is helpful for further research on the characterization of radiated noise signals from accelerated rotating propellers and analysis methods. Eugeniusz [7] found that seabed sediments are an important determinant of underwater noise, and based on the numerical results of the actual data, the noise generated by ships was processed on selected characteristic spectral components.

In this paper, the near-field noise radiation of a propeller in a shallow sea environment is simulated and analyzed. The Finite Element Method (FEM) is employed to establish the hydrodynamic model, while the Boundary Element Method (BEM) is used for the acoustic calculations. The free water surface and seafloor boundaries are applied using appropriate boundary conditions to simulate the reflection and absorption of sound waves. Finally, the radiated noise of the propeller at various depths is presented and analyzed.

2 Modeling Geometry and Mathematical Model

In the numerical example, an INSEAN E1619 type 7-blade propeller with a diameter of $D = 3.12$ m is used, and the relevant parameters are presented in Table 1. The flow field domain is defined with the inlet surface positioned 5D upstream, the outlet surface positioned 10D downstream, and the radial boundary set at 5D, as shown in Fig. 1. A nonstructured grid is employed for the calculation, with a denser grid distribution around the propeller blade surface and the rotating region. The grid size is adjusted according to surface dimensions, with both areas having a grid size of 50 mm. In contrast, a sparser grid is used for the external stationary region. The total number of grid points in the computational domain is approximately 2.18 million.

Table 1. Seven-blade large side-slope paddle main parameters

No	Parametric	Value	Modelling
1	Diameter (D/m)	3.12 m	
2	Number of blades (Z)	7	
3	AE/AO	0.46	
4	0.7R Pitch ratio	1.15	
5	Hub-to-diameter ratio	0.226	
6	Direction of rotation	right-hand side	

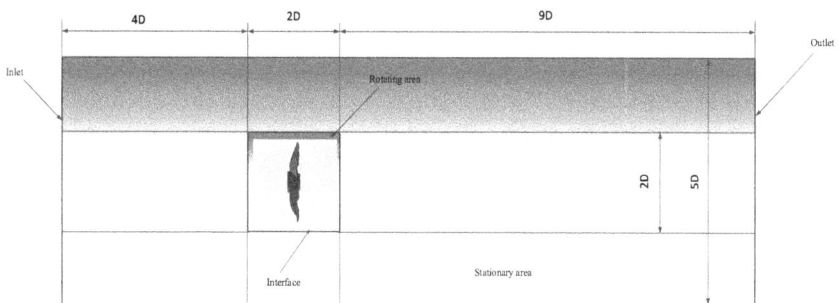

Fig. 1. Propeller computational domain distribution

For the calculation, the incoming velocity corresponds to submerged submarine speeds of 6 knots, with the propeller's rotational speed set at 70 r/min. This rotational speed is also applied to the rotating domain, which is modelled using the slip mesh (SM) method. The outlet is defined as a pressure outlet with a pressure value of zero. The interface between the rotating domain and the surrounding fluid domain is configured to allow interaction between the velocity fields of the dynamic and static regions. To simulate the effects of different water pressures on the submerged propeller, gravitational acceleration is enabled, and the operating density is set to zero. Pressure simulations at various water depths are performed by adjusting the operating pressures accordingly.

The Fluent constant calculation module is employed to compute the steady-state performance of the propeller under uniform incoming flow conditions. Once the steady-state calculation stabilizes, its results are used as the initial values for the Large Eddy Simulation (LES) calculation. Using a turbulence model and the slip grid method, the unsteady flow field of the propeller is calculated, yielding the periodic turbulent pressure pulsations. These pulsating pressure sources on the propeller surface are inputted into Fluent as a load. To simulate the shallow water environment, soft acoustic boundary conditions (p = 0) and impedance boundary conditions are set to represent the free water surface and the seafloor sediment, respectively. The acoustic calculation module of Fluent is then activated to obtain precise acoustic source data, followed by the use of the acoustic analogy method to calculate the radiated noise from the propeller.

The hydrostatic pressure at different depths is based on the following equation:

$$p = \rho g h \tag{1}$$

Where ρ is the density of water, 1000 kg/m^3. h is the acceleration of gravity, 9.8 m^2/s.

Lighthill analogized the complex phenomenon of fluid motion generating sound with the acoustic problem of an equivalent sound source in a static medium. He proposed that flow noise arises from flow instabilities caused by nonlinear interactions between velocity fluctuations, entropy fluctuations, and viscous stresses. Lighthill's equations are:

$$c_0^2 \nabla^2 \rho' - \frac{\partial^2 \rho'}{\partial^2 t^2} = -\frac{\partial Q}{\partial t} + \nabla \cdot \vec{F} - \frac{\partial^2 T_{ij}}{\partial x_i \partial y_j} \tag{2}$$

$$T_{ij} = \rho U_i U_j + \delta_{ij}\left(p - c_0^2 \rho'\right) - \tau_{ij}, T_{ij} = \rho_0 U_i U_j \tag{3}$$

where $\rho' = \rho - \rho_0$ is the undulation of the acoustic motion, $\frac{\partial Q}{\partial t}$ is the time rate of change of the fluid mass injection. $\nabla \cdot \vec{F}$ is the spatial variation of the acting body force, T_{ij} is the Lighthill stress tensor $\rho U_i U_j$ corresponds to the Reynolds stress in turbulent flow, τ_{ij} corresponds to the viscous stress.

3 Results and Discussion

The seven-bladed, large side-slope propeller is chosen as the research object. The Large Eddy Simulation (LES) method is used to calculate the unsteady flow field of the propeller; The Fluent acoustic calculation module is employed to compute the propeller noise. To analyse the radiated noise, two observation points are set up axially in the near-field of the propeller. The computed sound pressure levels are converted into the propeller's sound source level at 1 meter using the spherical wave attenuation equation. The coordinates of the observation points are as follows: P1(0,0,-1) and P2(0,-4,0.65). The calculations are based on a submerged speed of 6 knots, a propeller speed of 70 rpm, with the distance between the propeller, the water surface, the seafloor set to 9 m. The frequency distribution curves of the radiated sound pressure levels under different conditions are shown in Fig. 2 and Fig. 3.

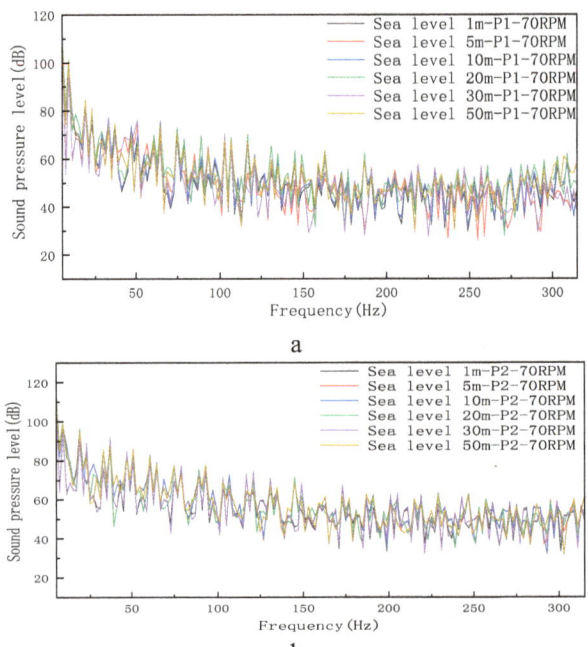

Fig. 2. Frequency distribution curves of radiated sound pressure levels at sea surface for different operating conditions

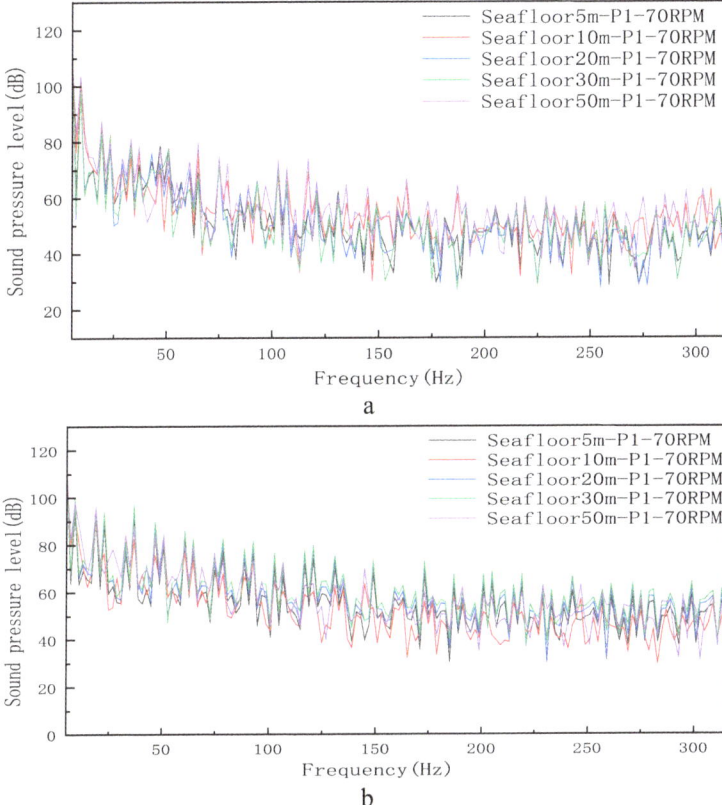

Fig. 3. Frequency distribution curves of radiated sound pressure levels on the seabed for different operating conditions

From Fig. 2 and Table 2, it is evident that while the trend of the sound pressure level with frequency remains generally consistent when the propeller is positioned at different distances from the water surface, the amplitude varies to some extent. For instance, when the propeller is 20 m, 30 m, and 50 m away from the water surface, the amplitude of the sound pressure level decreases in that order.

Table 2. Total sound level at monitoring points at different depths above the sea surface (dB)

Total sound pressure level	1 m	5 m	10 m	20 m	30 m	50 m
Measurement point 1	113.7	113.5	113.5	113.1	112.6	112.5
Measurement point 2	113.5	113.6	113.6	113.8	112.9	112.8

As the distance from the water surface increases, the low-frequency noise curve shifts towards lower frequencies, a phenomenon attributed to changes in the added mass of water. Furthermore, with greater distance from the surface, the overall hydrodynamic noise source level exhibits a decreasing trend. Specifically, as the depth increases from 1 m to 50 m, the average source level decreases by approximately 1 dB.

As shown in Fig. 3 and Table 3, when the propeller is positioned at varying distances from the seafloor, the trend of sound pressure level changes with frequency remains generally consistent, while the amplitude varies to some extent. Specifically, when the propeller is 5 m, 10 m, 20 m, 30 m, and 50 m from the seafloor, the amplitude of the sound pressure level decreases. This is primarily due to the impedance of the seafloor, which reflects acoustic waves at a high rate.

Table 3. Total sound level at monitoring points at different depths on the seabed (dB)

Total sound pressure level	5 m	10 m	20 m	30 m	50 m
Measurement point 1	113.2	112.7	112.7	112.6	112.5
Measurement point 2	113.3	112.8	112.6	112.4	112.3

As the distance from the seafloor increases, the low-frequency noise curve shifts toward higher frequencies, primarily due to changes in the properties of the surrounding water. Additionally, with increasing depth, the overall hydrodynamic noise source level exhibits a decreasing trend. Specifically, when the depth increases from 5 to 50 m, the average source level drops by approximately 0.9 dB.

4 Conclusion

The near-field noise prediction of propeller in shallow water environments is studied in this paper. Considering the acoustic reflection effect of free liquid surface and complex ocean features on propeller noise, this paper uses a calculation method of propeller acoustic radiation considering free liquid surface and complex ocean features, and establishes a low-frequency propeller noise prediction model in shallow water environment. Through numerical simulation, the influence of complex Marine environment on the noise characteristics of propeller near field radiation is revealed.

(1) When the propeller is positioned at different water depths, whether near the water surface or the seafloor, the radiated sound pressure level exhibits a similar trend across frequencies, although the amplitude varies. Specifically, the amplitude of the sound pressure level decreases as the distance between the propeller and the water surface or seafloor increases.

(2) Near the water surface, changes in the amplitude of the sound pressure level (SPL) are primarily attributed to reflections from the free liquid surface. In contrast, near the seafloor, the significant reduction in SPL amplitude is mainly due to the high impedance characteristics of the seafloor and its strong reflectivity to sound waves.

References

1. Baskaran, K., Jamaluddin, N.S., Celik, A.: Effects of number of blades on propeller noise. J. Sound Vib. **57**(2), 118–176 (2024)
2. Kuperman, W.A., Lynch, J.F.: Shallow-water acoustics. Phys. Today. **57**(10), 55–56 (2004)
3. Zhu, H.H., Xiao, R., Zhu, J.: Influence of isolated internal waves on the propagation of low-frequency acoustic energy flow in a three-dimensional shallow sea environment. J. Acoust. **46**(3), 365–374 (2021)
4. Brooker, A., Humphrey, V.: Measurement of radiated underwater noise from a small research vessel in shallow water. Ocean Eng. **120**, 182–189 (2016)
5. Pan, Y.C., Zhang, H.X., Zhou, Q.D.: Numerical prediction of propeller radiated noise based on the time-dominated algorithm of sound source. Ship Mechanics. **23**(09), 1139–1149 (2019)
6. Feng, Y., Tao, R., Zhou, Z.H.: Experimental study of accelerated rotating propeller noise testing in a water tunnel. J. Mil. Eng. **31**(12), 1611–1616 (2010)
7. Eugeniusz, K., Grelowska, G.: Propagation of ship-generated noise in shallow sea. Pol. Marit. Res. **25**(2), 37–46 (2018)

Open Access This chapter is licensed under the terms of the Creative Commons Attribution-NonCommercial-NoDerivatives 4.0 International License (http://creativecommons.org/licenses/by-nc-nd/4.0/), which permits any noncommercial use, sharing, distribution and reproduction in any medium or format, as long as you give appropriate credit to the original author(s) and the source, provide a link to the Creative Commons license and indicate if you modified the licensed material. You do not have permission under this license to share adapted material derived from this chapter or parts of it.

The images or other third party material in this chapter are included in the chapter's Creative Commons license, unless indicated otherwise in a credit line to the material. If material is not included in the chapter's Creative Commons license and your intended use is not permitted by statutory regulation or exceeds the permitted use, you will need to obtain permission directly from the copyright holder.

Research on Sea and Air Target Recognition Based on Otsu-Hough Sea Antenna Detection and YOLOv8 Algorithm

Wenjin Chen, Jingqiang Bi[✉], and Xizhen Qiao

Dalian Naval Academy, Dalian 116000, China
bijingqiangus@126.com

Abstract. To address the challenges of maritime and aerial target recognition under adverse conditions such as rain-fog interference, wave glare, and harsh weather, this paper proposes a detection framework integrating image enhancement and deep learning optimization. First, a dehazing algorithm based on guided filtering and atmospheric scattering models is constructed to resolve image degradation in low-visibility scenarios. Second, a sea-sky line detection module combining local Otsu segmentation and Hough transform is designed to achieve precise division of sea-sky regions through adaptive threshold segmentation, thereby narrowing the target search range. Finally, the ACMix hybrid convolution module is introduced into the YOLOv8 algorithm to enhance feature fusion capabilities, while a multi-scale detection head mechanism improves sensitivity to small and weak maritime targets. Experimental results demonstrate that the improved algorithm significantly enhances average precision (AP) and detection speed in complex environments such as rain-fog and salt fog crystallization. Compared to baseline models, it exhibits clear performance advantages, balancing recognition accuracy and real-time requirements, and provides a reliable technical solution for maritime security and emergency rescue.

Keywords: Maritime and aerial target recognition · complex environments · YOLOv8 algorithm · Otsu-Hough transform

1 Introduction

With the rapid development of the marine economy and the advancement of the "Smart Shipping" initiative, maritime and aerial traffic has experienced exponential growth, making accurate target recognition under complex sea conditions a critical challenge for navigation safety. Current shipborne sensors suffer from image degradation and contrast attenuation when encountering adverse environmental factors such as rain fog and wave glare, while conventional detection algorithms struggle to balance the trade-off between small-target omission and real-time processing in complex scenarios due to insufficient feature extraction capabilities.[1] To address these challenges, this paper proposes a multimodal collaborative intelligent detection framework. First, we construct a physics-prior-embedded Otsu-Hough sea-sky line detection model that achieves precise decoupling of marine and aerial regions through adaptive threshold segmentation and spatial constraints, overcoming the misjudgment limitations of traditional methods under dynamic wave interference. Subsequently, we develop an enhanced YOLOv8 network integrating ACMix hybrid convolution and multi-scale detection heads, combined with image enhancement preprocessing using guided filtering and atmospheric scattering modeling, thereby establishing an end-to-end solution encompassing "environmental perception - feature enhancement - cross-domain recognition". Experimental results demonstrate that our algorithm maintains an 85.7% mean detection accuracy under extreme conditions including fog and salt spray, with inference speed 23% faster than baseline models.[2] This work provides a novel technical paradigm for constructing high-reliability maritime intelligent perception systems, offering significant practical value for ship autonomous navigation and maritime search-and-rescue operations.

2 An Algorithm Based on Guided Filtering and YOLOv8 to Identify and Track Small and Small Targets Under Complex Sea Conditions

2.1 Algorithm Framework

Starting from the actual application environment conditions, the tracking and recognition of small and small targets on the sea surface under complex sea conditions need to solve two major problems:

(1) Frequent rain and fog at sea, water spray from ships affecting the photoelectric equipment, salt precipitation crystals in seawater attached to the photoelectric camera affecting target recognition;
(2) The problem of difficult target identification and poor real-time recognition algorithm.

To solve the above problems, two functional modules of image defogging enhancement and target recognition and tracking are designed. First of all, the historical sea surface target data is collected, and the image is enhanced by guided filtering. After labeling, the YOLOv8 model is trained. When the unmanned boat carries out patrol, reconnaissance and other tasks, the boat generally carries navigation radar, photoelectric camera, AIS system and other loads. After multi-sensor information fusion, suspicious ships can be warned and verified before independent planning. The real-time data obtained will be sent back to the mothership host, and after guided filtering for de-fogging enhancement, the de-fogging image will be sent to YOLOv8 for target identification and tracking. The algorithm implementation process is shown in Fig. 1.

2.2 Image De-Fog Module Based on Guided Filtering Algorithm

Guide Filtering Algorithm. In order to achieve accurate identification of maritime and air targets, it is also necessary to consider the impact of the weather in the actual mission area on the identification, such as the occurrence of rain and fog weather at sea, the

Fig. 1. Algorithm flow design diagram

splashing of waves during ship navigation, and the precipitation of salt from seawater evaporation attached to the photoelectric camera, which all lead to the blur or occlusion of the camera, thus affecting the imaging quality.

To solve the above problems, this algorithm introduces the image de-fog enhancement module, mainly using the guided filter algorithm. Compared with mean filtering and Gaussian filtering, it has better edge preserving effect. At the same time, as an edge-preserving filtering algorithm, it has the advantage of low time complexity compared with bilateral filtering, deep learning and other algorithms, which ensures that the real-time index of the proposed algorithm can meet the actual use needs.

The principle of the guided filtering algorithm adopted in this paper is that the input image itself or another different image is taken as the guided image, and the filtered output result is obtained by considering the content of the guided image. There is a linear relationship between the output image and the guided image, as shown in Fig. 2 below.

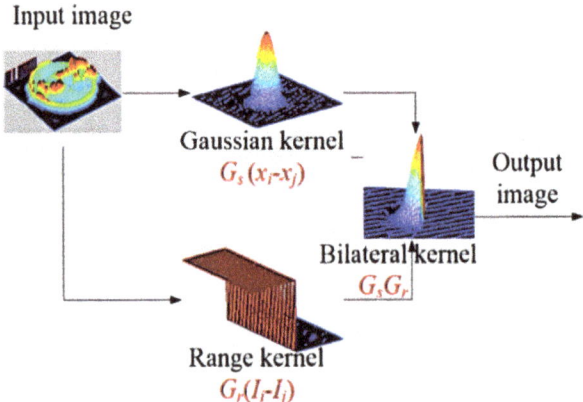

Fig. 2. Guided filter principle

Image Fog Removal Based on Atmospheric Scattering Model. On the basis before, fog imaging model commonly used in computer vision field is introduced:

$$I(x) = J(x)t(x) + A(1 - t(x)) \tag{1}$$

In formula (1), $I(x)$ is the original image with fog observed by the equipment; $J(x)$ is the fog free image after filtering processing; $t(x)$ is the transmission function of atmospheric environment to light; A is the background light intensity caused by atmospheric scattering.

According to Eq. (1), the fog removal image $J(x)$ can be obtained by estimating the transfer function $t(x)$ and the background light intensity A. According to the Dark Channel Prior theory proposed by He et al., [4] for a fog-free image $J(x)$, there are always some very low pixel values in at least one color channel in most areas excluding the sky.

Therefore, according to the dark channel prior theory, the specific steps of image de-fogging can be obtained as follows:

1. Estimate atmospheric background light intensity A: Estimate atmospheric light intensity by analyzing the dark channel of the image.
2. Estimate the transmission function $t(x)$: Use the dark channel prior to estimate the transmission rate of each pixel, that is, the proportion of light that is not scattered and absorbed during the process of reaching the imaging system from the scene.
3. Image restoration: According to the estimated transfer function and atmospheric light intensity, the fog-free image $J(x)$ is restored through the reverse model [5].

2.3 Tracking and Identification Module of Small and Small Targets on the Sea Surface

In order to achieve accurate and efficient tracking and recognition of small and small targets on the sea surface, two problems need to be solved: difficult recognition of small targets and poor real-time recognition algorithm [6]. The original network structure of YOLOv8 is shown in Fig. 3.

Four-Detection Head YOLOv8 Algorithm Identifies Dim Targets. In view of the problem that small targets are difficult to identify, it can be seen from the source code of YOLOv8 that YOLOv8 has 3 detection heads by default, which can detect targets at multiple scales. The size of the detection head is shown in Table 1.

Because the traditional YOLOv8 algorithm has only three detection heads, it has poor performance in the detection of very small and small targets. In this paper, based on the three detection heads of YOLOv8, a detection head of 160*160 dim targets is added, and a detection feature map is added to detect the above targets. The improved YOLOv8 network structure is shown in Fig. 4, and the size of the detection head is shown in Table 2.

Adding ACMix Hybrid Model Improves the Real-Time Performance of the Algorithm. Due to the limited computing power of the small unmanned boat, the real-time recognition and tracking time of the YOLOv8 with the detection head is about 0.125 s on average, and the frame number is 8 frames. To solve the problem of poor real-time

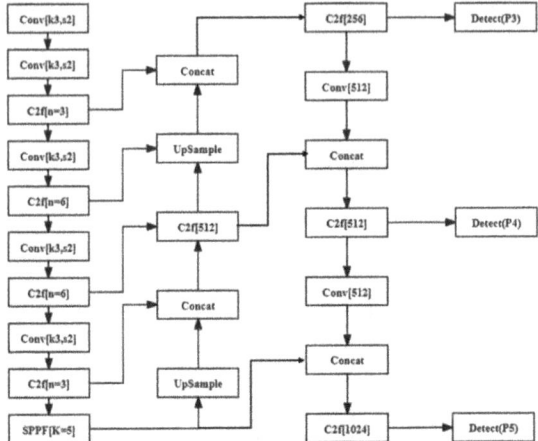

Fig. 3. YOLOv8 network structure diagram

Table 1. The size of the YOLOv8 detection head

Detection head	The size of Detection feature map	The size of Detection target
P3/8	80*80	8*8
P4/16	40*40	16*16
P5/32	20*20	32*32

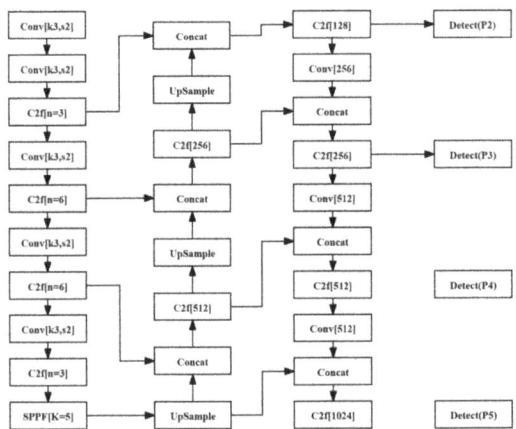

Fig. 4. YOLOv8 network structure diagram with dim target detection head added

performance, this product introduces the ACMix hybrid model, which integrates self-attention mechanism and convolutional neural network (CNN) to make full use of their respective advantages [7].

Table 2. The improved size of the YOLOv8 detection head

Heading level	Example	Font size and style
P3/8	80*80	8*8
P4/16	40*40	16*16
P5/32	20*20	32*32
P2/4	160*160	4*4

Specifically, the ACMix model first uses a self-attention mechanism to encode an input sequence to get a context-aware representation. This representation is then fed into a convolutional neural network to further extract local features. Finally, by integrating feature representations of self-attention and convolution, ACMix model can better capture semantic information and local structure in sequences [8]. On the basis of the network structure of the detection head of the dim target, the YOLOv8 model structure after adding the ACMix hybrid model is shown in Fig. 5.

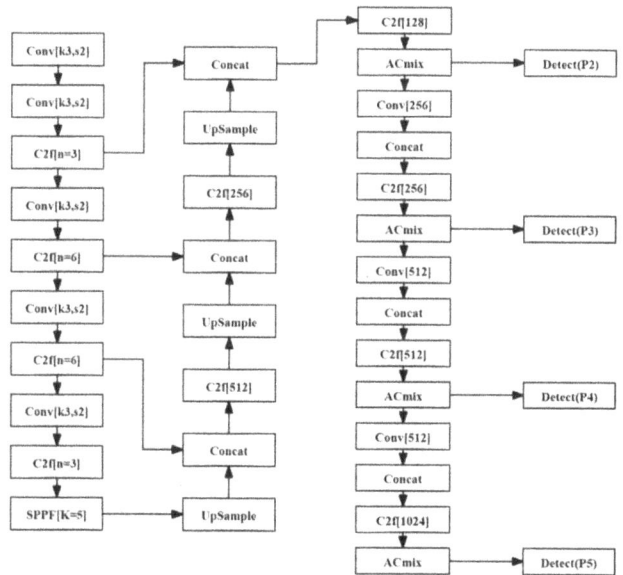

Fig. 5. Network structure of ACMix YOLOv8

3 Sea Antenna Detection Method Based on Local Otsu Segmentation and Hough Transform

Whether the sea antenna in the image can be accurately detected is very important for the accurate recognition of sea and air targets in this paper. At present, the detection methods of sea antenna can be summarized into the following two categories: the method based on edge point straight line detection and the method based on image segmentation.[9] In this paper, a sea antenna detection method based on local Otsu segmentation and Hough transform is adopted.

3.1 Image Preprocessing

Because sea antenna detection is mainly concerned with the longitudinal (vertical) features of the image, the longitudinal median filter is used to suppress the high frequency noise such as the spot in the image [10]. The specific method is to replace each pixel value in each column by calculating the median value of each column of the image, thereby reducing the impact of noise on the image.

The purpose of preprocessing fog removal images is to remove noise and interference and improve the accuracy of subsequent processing.

3.2 Local Otsu Segmentation

Local Otsu segmentation is a threshold segmentation method based on local image characteristics, which can adapt to the situation of uneven illumination in the image. The specific steps are as follows:

1. Image segmentation

The image is divided into several small blocks to compensate for the unevenness of illumination and limit the range of interference caused by ships on the sea surface to a small block.

In this paper, the specific size of the small block is determined according to the resolution of the fog removal image.

2. Image binarization by local Otsu threshold

In each small block, Otsu method is used to determine the optimal threshold by maximizing the inter-class variance, and then each small block is binarized according to the threshold to achieve image binarization.

3. Edge extraction

In the process of edge detection of binary image, Sobel operator is used to calculate the gradient of the image in the horizontal and vertical directions, so as to detect the edge.

4. Hough transform

Hough transform is used to detect the straight line feature in the image, and the line segment of the sea antenna is obtained by fitting the edge pixels. Finally, the position of the sea antenna is obtained [11].

4 Algorithm Effect Test

4.1 Image Defogging Enhancement Effect

The de-fog enhancement effects of guided filtering and bilateral filtering are compared, and the filtering effect diagram is shown in Fig. 6.

Fig. 6. Filter effect diagram

It can be seen that the color retention of bilateral filtering is better, but the processing time is much larger than that of guided filtering, and the guided filtering can reach 30 frames per second, basically meeting the real-time requirements.

4.2 Target Recognition Effect

This Paper Improves the Effect Comparison Between YOLOv8 Model and General YOLOv8 Model. A total of 470 target photos were selected, of which 310 were used for training model and the rest were used for verification. The effect comparison between the YOLOv8 model proposed in this paper, which combines the ACMix hybrid model and the multi-detection head mechanism [12], and the YOLOv8 model using the general network structure, is shown in Fig. 7.

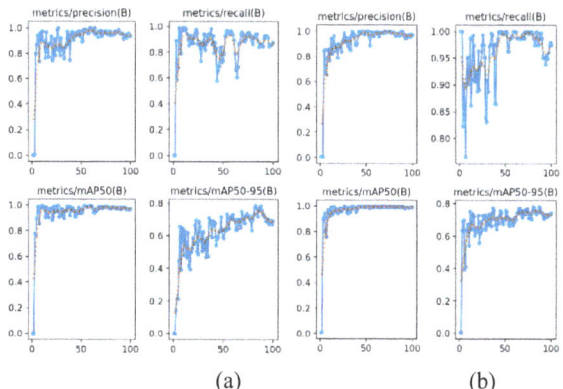

Fig. 7. Comprehension of effect of models

Figure 7(a) shows the general YOLOv8 model effect, and Fig. 7(b) shows the improved YOLOv8 model effect in this paper. Through comparison, it can be seen that the YOLOv8 model combining the ACMix hybrid model and the multiple detection head mechanism has an increase point of 3%–4% in the prediction accuracy, and has achieved better performance under various IOU thresholds.

This Paper Improves the Effect of YOLOv8 Model. In this study, a total of 5725 photos of Marine targets were collected, of which 4022 were used for training models and the rest for verification. Through the steps of labeling, model construction, type discrimination and target tracking, the Marine target can be identified and tracked. The decline curve of the loss function was selected to evaluate the training of the model, and the average accuracy (mAP) was selected as an indicator to measure the model detection performance. The model effect was shown in Fig. 8.

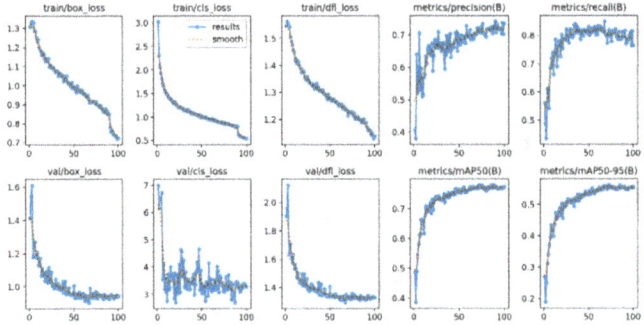

Fig. 8. The effect of improved YOLOv8 model

As you can see in Fig. 8, the model's performance on target positioning has steadily improved. The overall decrease of class loss and objective function loss indicates that the accuracy of classification and target detection is effectively improved. The mAP value of the model increases steadily and finally stabilizes at a high level, which indicates that the model has achieved good performance under various IOU thresholds.

5 Conclusion

Through the above series of methods and tests, this paper has achieved valuable research results in sea and air target recognition in complex sea conditions, and provides effective technical support for applications in related fields.

References

1. Shuai, Y., Liu, R., He, W.: Image Haze Removal of Wiener Filtering Based on Dark Channel Prior. IEEE (2013). https://doi.org/10.1109/CIS.2012.78

2. He, K., Sun, J., Tang, X.: Single image haze removal using dark channel prior. In: IEEE Conference on. Computer Vision and Pattern Recognition, 2009. CVPR 2009, pp. 1956–1963. IEEE (2009)
3. Zhenhua, D., Xiaolong, C., Wei, X., et al.: Multi-domain and multi-dimensional feature modeling and analysis of low, slow, and small targets via ubiquitous radar under sea and air background. J. Signal Process. **40**(5), 801–814 (2024). https://doi.org/10.16798/j.issn.1003-0530.2024.05.001
4. Xiang, W., Yumiao, W., Xingyu, C., et al.: Deep learning-based marine target detection method with multiple feature fusion. J. Radars. **13**(3), 554–564 (2024). https://doi.org/10.12000/JR23105
5. Shuwen, X., Xiaohui, B., Zixun, G., et al.: Status and prospects of feature-based detection methods for floating targets on the sea surface. J. Radars. **9**(4), 684–714 (2020). https://doi.org/10.12000/JR20084
6. Yaohui, H., Ke, Z., Chao, X.: Small and dim ship target detection based on sea-sky-line. J. Northwest. Polytech. Univ. **37**(1), 35–40 (2019)
7. Hao, X., Yibo, T., Wenzong, W., et al.: Image dehazing by incorporating Markov random field with dark channel prior[J]. J. Ocean Univ. China. **19**(03), 551–560 (2020)
8. Li, J., Zhou, Y.: Research on ADS-B spoof detection technology based on radio spectrum features. Appl. Math. Nonlinear Sci. **9**(1), 848 (2024)
9. Sun, P., Hou, M., Lyu, S., et al.: Virtual cleaning of sooty mural hyperspectral images using the LIME model and improved dark channel prior. Sci. Rep. **14**(1), 24807 (2024)
10. Dwivedi, P., Chakraborty, S.: TRMLGAN: Transmission multi loss generative adversarial network framework for image dehazing. J. Vis. Commun. Image Represent. **105**, 104324–104324 (2024)
11. Zhang, X., Wang, X., Yan, C., et al.: Polarization-based two-stage image dehazing in a low-light environment. Electronics. **13**(12), 2269–2269 (2024)

Open Access This chapter is licensed under the terms of the Creative Commons Attribution-NonCommercial-NoDerivatives 4.0 International License (http://creativecommons.org/licenses/by-nc-nd/4.0/), which permits any noncommercial use, sharing, distribution and reproduction in any medium or format, as long as you give appropriate credit to the original author(s) and the source, provide a link to the Creative Commons license and indicate if you modified the licensed material. You do not have permission under this license to share adapted material derived from this chapter or parts of it.

The images or other third party material in this chapter are included in the chapter's Creative Commons license, unless indicated otherwise in a credit line to the material. If material is not included in the chapter's Creative Commons license and your intended use is not permitted by statutory regulation or exceeds the permitted use, you will need to obtain permission directly from the copyright holder.

Design of Ship Air-Conditioning and Ventilation System in Low Latitude and High Altitude Area

Tianyu Tu[1,2], Liang Zhang[2], Dinghua Hu[1], and Lei Zhao[2(✉)]

[1] Nanjing University of Science and Technology, Nanjing 210094, Jiangsu, China
[2] Marine design and research institute of China, Shanghai 200011, China
zlei0112@163.com

Abstract. With the rapid development of inland river shipping, the total number of vessels has significantly increased, making the working and living environment for crew members more important than ever, especially in terms of creating a comfortable and healthy cabin environment. This paper takes a region in Kunming as an example, analyzing the climatic characteristics of the area, particularly the influence of factors such as solar radiation on the ship's air conditioning and ventilation system, in light of its unique geographical features of low latitude and high altitude. Considering the layout of the ship's cabins, where some areas are more concentrated and others are more scattered, the proposed design incorporates three different types of systems: Multi-split air conditioning, split-type air conditioning, and water-cooled air conditioning, complemented by an efficient auxiliary ventilation system. While ensuring the comfort of the crew, this design also fully considers the ship's environmental requirements, improving energy use efficiency, reducing energy consumption, and thereby enhancing the ship's economic performance. Furthermore, the scheme not only ensures crew comfort under the complex plateau climate conditions but also improves air quality, preventing discomfort caused by excessively high temperatures. This design approach provides valuable experience and theoretical guidance for future air conditioning and ventilation system development on vessels operating in high-altitude lake regions, playing a crucial role in optimizing the ship's environmental performance, energy utilization, and comfort. It offers strong support for the development of related technologies in the maritime field.

Keywords: Inland river · Ship · Solar radiation · Air conditioning · Ventilation

1 Introduction

In recent years, with the development of the economy and the advancement of urbanization, inland waterway (lake) shipping has played an increasingly significant role in regional economic development. The continuous growth of inland waterway (lake) vessels has consistently contributed to the local economic development. As the number of inland waterway (lake) vessels continues to increase, the shipping industry has not only played an important role in transportation but has also positively impacted the prosperity of the regional economy and the improvement of infrastructure. However, due to the differences in the navigational areas of inland waterway (lake) shipping, the design of air conditioning systems faces various challenges. How to design air conditioning systems based on the climate characteristics of different navigational areas to ensure the comfort of the crew has become an important issue.

This vessel operates in Kunming, located at longitude $102°10'$ to $103°40'$E and latitude $24°23'$ to $26°33'$N, within the subtropical humid monsoon climate zone. It is alternately controlled by tropical continental air masses and southwest monsoons, with an average temperature of 19.7 °C in the hottest month and 8 °C in the coldest month. The historical extreme temperatures have reached a high of 31.5 °C and a low of −7.8 °C. Additionally, the mountainous region exhibits significant vertical climatic variations and large diurnal temperature differences. The region's low latitude results in a large solar altitude angle, long sunshine hours, high atmospheric transparency, and substantial direct radiation reaching the ground, leading to strong environmental heat conduction and radiative heat transfer. Furthermore, with an average altitude of 1895 m, this area possesses the characteristics of low atmospheric pressure and hypoxia associated with high altitudes. Research indicates that atmospheric pressure decreases by approximately 10 kPa for every 1000 m of elevation, accompanied by a corresponding decrease in air density [1].

In such a complex external environment, the design of the air conditioning system is crucial. To ensure that the crew can work and live in a comfortable environment, the air conditioning and ventilation system design must fully consider the impact of factors such as solar radiation and air density on the cooling load. This paper aims to explore the specific impact of the climatic characteristics of the Kunming area on air conditioning system design and propose strategies for optimizing the system design, with the goal of providing theoretical support and technical guidance for the design of air conditioning systems in similar regions, thus promoting the sustainable development of the shipping industry.

2 Design Key Points

2.1 Impact of Solar Radiation Transmitted Through Glass

Kunming region enjoys an annual sunshine duration of 2000 to 2400 h, with a total solar radiation ranging between 5020 and 5440 MJ/m^2. Compared to the Yangtze River Basin and South China regions, Kunming experiences a higher solar radiation by 420 to 840 MJ/m^2. Prolonged direct sunlight exposure can lead to excessive indoor temperatures, particularly in the wheelhouse. Nowadays, the aesthetic design of a ship's superstructure has become a crucial consideration, balancing both aesthetics and functionality. To ensure a clear and unobstructed view for monitoring operational conditions, location, and movement, the wheelhouse is equipped with large glass windows on all four sides. Therefore, when calculating the air conditioning load, it is essential to take into account the impact of solar radiation transmitted through these glass windows.

The calculation of ship air conditioning loads is adapted from land-based methods, typically employing the steady-state heat transfer method. This approach selects the highest temperature within the entire navigation area as the external design value and calculates solar radiation heat based on the maximum sun-exposed area, neglecting heat storage and time-lag characteristics. This paper employs both conventional and modified calculation methods to estimate the load contributed by solar radiation transmitted through glass windows.

According to the "Practical Handbook of Ship Design - Marine Engineering Volume," the heat transmitted through glass windows is calculated as follows [2]:

$$Q = G_s A_s + h_g A_g (t_1 - t_2) \tag{1}$$

Where: A_s is the actual area of the glass window, m². A_g is the calculated area of the glass window, equal to the area of the glass window plus an additional 100 mm (0.1 m) width around its perimeter, m². h_g is the heat transfer coefficient per unit area of the glass window, W/m2 K. For single-pane glass, it is typically W/m² K, and for double-pane glass, it is 3.5 W/m² K. t_1 is the design temperature outside the cabin, °C. t_2 is the design temperature inside the cabin, °C. G_s is the solar radiation transmittance through the glass window, W/m², which can be obtained from Table 1.

Table 1. Solar radiation transmittance of glass windows

Type	Solar Radiation Transmittance(W/m²)
Single-layer ordinary glass window	350
Double-glazed ordinary glass window	310
Single-layer ordinary glass window with light-colored shading devices	240
Double-glazed ordinary glass window with light-colored shading devices	210

According to method 2, the solar radiation intensity I_t reaching the surface at some time is calculated as [3–5]:

$$I_t = I_D \cos \alpha + I_S \left(\frac{2 + \cos \beta}{3} \right) \tag{2}$$

Where: I_t represents the total solar radiation intensity received, W/m²; I_D denotes the direct solar radiation intensity, measured in W/m²; I_s signifies the diffuse solar radiation, W/m²; α is the incidence angle of direct solar radiation on the glass window, °; β represents the tilt angle of the cabin wall, which is assumed to be vertical, hence $\beta = 90°$ in the calculation process.

$$I_D = I_0 \cos \theta \tag{3}$$

$$\cos \theta = \sin \varphi \sin \delta + \cos \varphi \cos \delta \cos \omega \tag{4}$$

$$\delta = 23.45 \sin \left(360 \times \frac{284 + n}{365} \right) \tag{5}$$

$$I_0 = 1353(1 + 0.033\cos(360 \times n/365)) \tag{6}$$

$$\cos\alpha = \sin\varphi\cos\beta\sin\delta - \cos\varphi\sin\beta\cos\gamma\sin\delta + \cos\varphi\cos\beta\cos\delta\cos\omega \\ + \sin\varphi\sin\beta\cos\gamma\cos\delta\cos\omega + \cos\delta\sin\beta\sin\gamma\sin\omega \tag{7}$$

Where: I_0 is the solar constant, the corrected calculation method is used here [6]; n represents the day of the year ($n = 1$ for January first, $n = 365$ for December 31st); φ is the local latitude in degrees, °; δ is the solar declination angle in degrees, °, which is calculated using the Cooper formula; ω is the solar hour angle, where $\omega = 15(t - 12)$, with τ being the local time in hours, h; θ is the zenith angle; and γ is the azimuth angle, defined as the angle between the projection of the wall normal on the horizontal plane and the due south direction, which is mainly dependent on the ship's route and heading [6].

$$I_s = I_0 p^M \left(1 - p^M\right)/\left[2(1 - 1.4\ln p)\right] \tag{8}$$

$$M = \frac{1}{\cos\theta + 0.15(93.885 - \theta)^{-1.253}} \frac{P}{101325} \tag{9}$$

Where: P represents atmospheric pressure, Pa; p represents atmospheric transparency coefficient, with a value ranging from 0.67 to 0.7; M represents atmospheric optical mass.

The heat transmitted through the glass window is

$$Q = hgAg(t1 - t2) + AgGIt \tag{10}$$

Where: G represents the solar radiation heat transmittance.

Both calculation methods take single-layer ordinary glass as an example, with calculations performed for August 1st (solar declination at 18°N) at 3:00 PM. Since the ship's glass windows are usually distributed in multiple directions, considering the heat load of multiple glass surfaces simultaneously would significantly increase computational complexity. Therefore, the sun-facing side with the largest glass area, measuring 15.6 m², was selected for analysis. The calculation results are shown in Table 2.

Table 2. Radiation load calculation

Method	Heat Load Transmitted Through Single-Sided Glass Window (W)
1	6100
2	11,640

After considering the impact of altitude on solar radiation, method 2 shows an increase of approximately 90.8% in the heat load transmitted through the single-sided sun-facing side of the glass window, compared to method 1, for the same area of glass window.

2.2 Changes in Designed Airflow Volume

Ignoring the influence of longitude, latitude, and seasons, within an altitude of 1.1 km, the relationship between atmospheric pressure and altitude follows a concave exponential curve. After linear coupling, the calculation can be performed according to Eq. 11.

$$P = 101325\left(1 - \frac{Z}{44329}\right)^{5.255876} \tag{11}$$

Where: Z represents the altitude, m; P represents the atmospheric pressure, Pa.

$$\rho = 1.2 \times \left(1 - \frac{Z}{44300}\right)^{4.256} \tag{12}$$

The ventilation volume of a cabin is usually calculated based on either the number of air changes per hour or the heat generated within the cabin, and the larger value is taken to determine the final ventilation volume.

$$Q = nV \tag{13}$$

Where: n represents the number of air changes per hour, h^{-1}; V represents the volume of the cabin, m^3.

$$Q = \frac{q}{\rho C_\alpha \Delta T} \tag{14}$$

Where: q represents the heat generation, kW; C_α represents the specific heat capacity of air in kJ/kg·K; ρ represents the density of air in kg/m^3; ΔT represents the temperature difference between inside and outside of the cabin, K.

However, as altitude increases, the air gradually becomes thinner, with a decrease in the number of gas molecules per unit volume, resulting in a corresponding decrease in air density and oxygen content. Under other constant conditions, the same oxygen content would lead to an increase in gas volume. Therefore, the calculated ventilation rate must be corrected, as otherwise it would be underestimated, potentially leading to severe indoor pollution over extended periods of ventilation.

After adjusting the parameters of air density and specific heat capacity based on the altitude of Kunming, the ventilation rate of the cabin calculated based on heat generation is approximately 1.21 times that of a plain area.

2.3 Correction in Fan Selection

Due to the impact of atmospheric density changes caused by altitude, the equipment selection in high-altitude areas differs significantly from that in plain areas. Conventional equipment selection tables and resistance calculation charts are typically compiled under standard conditions of atmospheric pressure at 101,325 Pa, air density at 1.2 kg/m^3, air temperature at 20 °C, and relative humidity at 50%. If no conversion is performed, the equipment may fail to meet the design requirements. During fan selection, the fan type is generally determined based on the two parameters of air pressure and airflow in the air conditioning and ventilation system. The fan's air pressure is set to be 1.1 to 1.2 times the total resistance loss of the most unfavorable loop in the system, while the airflow is 1.1 times the total airflow. However, since the atmospheric pressure in high-altitude areas is lower than the standard atmospheric pressure, resulting in decreased air density, the volumetric flow rate at the fan inlet does not change with changes in atmospheric pressure and temperature at the fan inlet. But the total pressure generated by the fan decreases as the gas density at the fan inlet decreases [8]. Yan Tao et al. [9] conducted research based on the efficiency calculation method for tunnel fans at high altitudes and found that in the Balangshan Tunnel at an altitude of 3800 m, the fan efficiency was only 0.43 of that under plain conditions. Liu Xiang [10], on the other hand, derived through theoretical calculations that the fan efficiency in the Queershan Tunnel, which exceeds 4300 m in altitude, was 0.65 of that in plain areas. Therefore, during the fan selection process for this vessel, a conversion of performance parameters was performed. Under the conditions of ensuring equal fan speed and kinetic energy diameter, according to the similarity law, it can be obtained that:

$$\frac{H1}{H0} = \frac{\rho 1}{\rho 0} \tag{15}$$

$$Q1 = Q0 \tag{16}$$

$$\frac{N1}{N0} = \frac{\rho 1}{\rho 0} \tag{17}$$

$$\frac{\eta 1}{\eta 0} = \frac{\rho 1}{\rho 0} \tag{18}$$

$H1, Q1, N1, \eta 1$— These represent the fan's pressure, airflow, power, and efficiency, respectively, under actual operating conditions.

$H0, Q0, N0, \eta 0$— These represent the fan's pressure, airflow, power, and efficiency, respectively, under standard conditions.

As shown in Table 3, let's take the selection of a fan with a rated airflow of 800 m³/h as an example.

Table 3. Comparison of fan parameters between actual and standard conditions

Parameter	Actual Conditions	Standard Conditions
Air Density (kg/m³)	0.966	1.200
Atmospheric Pressure (Pa)	77,540	101,325
Airflow Rate (m³/h)	800	800
Fan Pressure (Pa)	317.98	395.00
Power (kW)	0.079	0.098
Fan Efficiency (%)	72.45	90.00

The comparison of fan performance curves between standard conditions and actual conditions is shown in Fig. 1.

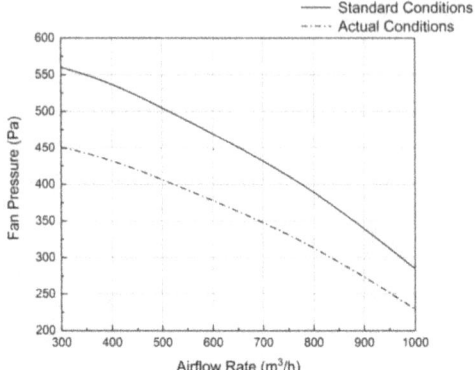

Fig. 1. Fan performance curve comparison

3 System Design Example

3.1 Air Conditioning System

The vessel type studied in this paper is a multipurpose ship navigating in a certain inland river (lake) area of Kunming. Due to draft limitations, it has a shallow hull depth, low deck height, and features locally concentrated cabins with some functional cabins scattered throughout. For the centralized air-conditioning systems commonly used in small and medium-sized ships, their processing units require dedicated cabins and large vertical space for arranging air-conditioning ducts, resulting in inefficient use of ship space. When a large number of air-conditioning pipes pass through non-air-conditioned cabins, it can affect the overall space layout, increase the resistance and energy consumption of the air-conditioning system, and compromise its cooling efficiency. When some cabins do not require air-conditioning, the entire system still needs to operate, leading to high energy consumption and a corresponding large number and weight of auxiliary accessories, which are not suitable for this ship. The air-conditioning cabins in the superstructure of this ship have similar operating hours, and to facilitate control and management on the same deck, a multi-split air-conditioning system is adopted. This system does not require a dedicated machine room, saving space, and does not need auxiliary equipment such as cooling towers, water pumps, and softened water systems, resulting in lower costs and shorter installation periods for shipyards [11].

To investigate the variation in indoor temperature under different solar radiation calculation methods, a simulation study was conducted using the air conditioning system loads obtained from the two calculation methods as different boundary conditions. A geometric model of the cockpit, as shown in Fig. 2, was established. The model is equipped with four ceiling-mounted indoor units, with dual-side air outlets. The air supply volume for a single outlet is 735 m^3/h, with a supply temperature of 18 °C and an air outlet angle of 60 °. The room contains two operators, and the indoor load is shown in Table 4. Various components inside the cockpit, such as trim panels, were simplified, and standard $k - \varepsilon$ model equations were used to solve the system.

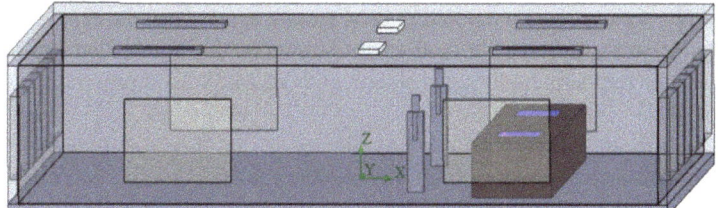

Fig. 2. Cab model

Table 4. Indoor load

Name	load /(W)
human body	120 x 2
toplight	60 x 2
control board	70 x 2

Given the selection of a cross-section at 1.5 m above the ground for temperature field analysis, as shown in Fig. 3.

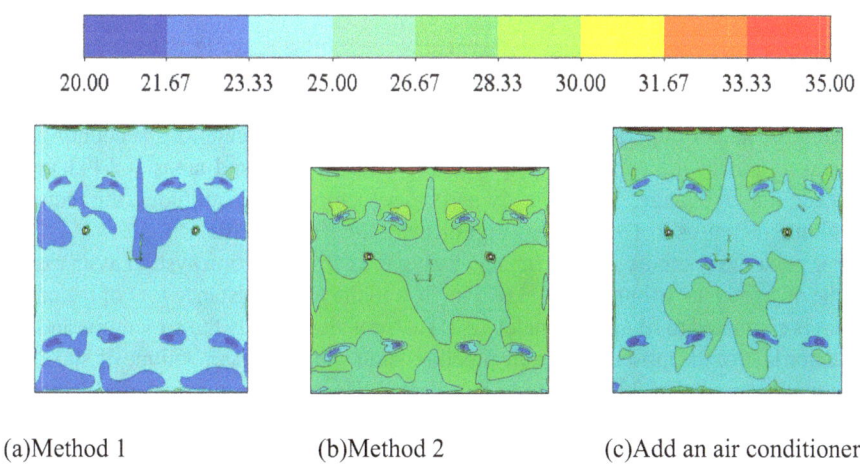

(a)Method 1　　　　　　　(b)Method 2　　　　　　(c)Add an air conditioner

Fig. 3. Temperature contour plot at a 1.5 m height section of the cab from the ground.

The glass windows facing the sun exhibit a pronounced temperature gradient due to intense solar radiation, with temperatures exceeding 30 °C in the vicinity. When calculating the cabin load using method 1, a relatively lower value is observed compared to method 2. Specifically, the average indoor temperature in the cross-section shown in Fig. 3 (a) is 23.76 °C, while in Fig. 3 (b), the average temperature rises to 26.7 °C, indicating an overall higher indoor temperature. Compared to the results obtained using method 1, the average cabin temperature increases by approximately 12.4%. To address this elevated temperature, an additional air conditioning unit was installed. Subsequently, the cross-section temperature dropped significantly to 24.88 °C, representing a decrease of about 6.8%.

Based on the simulation results, a variable refrigerant flow multi-split system with one outdoor unit and five indoor units was selected for the cab. In the design, hollow low-E glass was adopted, with a gas filling in the interlayer that significantly reduces the overall heat transfer coefficient. Additionally, the low-emissivity coating on its surface can reduce the total solar transmittance, leading to a substantial reduction in load.

3.2 Ventilation System

While air conditioning can effectively meet the indoor heating and cooling load demands, insufficient fresh air intake can negatively impact indoor air quality, leading to discomfort, particularly in public spaces such as conference rooms where there is a high turnover of people. When individuals congregate, it becomes essential to utilize a ventilation system to introduce an appropriate amount of outside fresh air, enhancing indoor air quality and mitigating odors and pollutants. The COVID-19 pandemic has further underscored the importance of ventilation systems for shipowners, ensuring that exhaust air or treated effluents do not recirculate back into the vessel. Below, we describe the ventilation system design for two different types of cabins as examples.

For functional cabins solely housing equipment, such as electronic equipment rooms, the substantial heat generated by the devices necessitates effective temperature control. Elevated ambient temperatures can lead to performance instability or reduced reliability of the equipment. Given the fully enclosed design of ships, mechanical ventilation is necessary to introduce outside cool air, effectively lowering the cabin temperature and ensuring the smooth operation of the equipment. Based on calculations accounting for the heat output of the cabin, the required air volume in flatland areas is 1493 m^3/h. However, due to changes in air density and other factors at high altitudes, the required air volume increases to 1812 m^3/h, a significant 21.4% increment.

For personnel activity compartments, taking the cab ventilation system as an example, the designed air volume is set at 675 m^3/h, and the total resistance loss of the entire system is estimated to be approximately 210 Pa. Considering only the air volume and the resistance loss of the most unfavorable pipeline, a fan with a rated air volume of 800 m^3/h and a wind pressure of 285 Pa should be selected. However, based on the influence of high-altitude wind pressure changes described earlier, after parameter correction, a fan with an air volume of 800 m^3/h and a total pressure of 395 Pa is chosen.

4 Conclusion

Currently, with the continuous development of shipping, the requirements for marine air conditioning systems are becoming increasingly demanding. To embody the people-oriented philosophy, this paper focuses on the regional characteristics of a certain inland river (lake) area in Kunming, conducts research on the key design points, and completes the system design based on these findings. The research indicates:

1. Due to Kunming's high altitude and intense solar radiation, for cabins with large glass areas such as the cab, it is not feasible to rely solely on simple selection coefficients for calculations. When considering parameters such as the angle of incidence, Method 2 results in an approximately 90.8% increase in heat load transmitted through the single-sided sunny side of the same-sized glass window compared to Method 1.
2. As altitude increases, air density and oxygen content decrease, necessitating adjustments to the required air volume and fan pressure. After modifying the parameters based on Kunming's conditions, the calculated ventilation volume for cabins based on heat output is approximately 1.01 times that of standard conditions, while the fan pressure required is 80.2% of the standard state.
3. Simulation studies were conducted under the same set of operating conditions using the loads obtained from two different calculation methods as boundary conditions. The results show that the average temperature inside the cabin using Method 2 is 12.4% higher than that using Method 1. Consequently, an additional indoor unit is required for the multi-split system in this cab. After installation, the temperature at the cross-section drops to 24.88 °C, a reduction of approximately 6.8%.

At the same time, existing research also has certain limitations, such as the calculation of solar radiation load being based on data from specific dates and assumed headings. Future studies could consider dynamic changes throughout the entire year to improve applicability.

References

1. Gao, R.: Study of Combustion Performance and Power Recovery for Heavy Duty Diesel Engine at High Altitudes. Beijing Jiaotong University (2012)
2. Huang, H.: Practical Handbook of Ship Design-Marine Engineering Volume. National Defense Industry Press (2023)
3. Tang, R., et al.: Optimal tilt-angles for solar collectors used in China. Appl. Energy. **79**(3), 239–248 (2004)
4. Notton, G., et al.: Performance evaluation of various hourly slope irradiation models using Mediterranean experimental data of Ajaccio. Energy Convers. Manag. **47**(2), 147–173 (2006)
5. Wei, S., et al.: Arrangement optimization of flat-plate solar collector arrays. Trans. Chinese Soc. Agri. Eng. **28**(14), 184–189 (2012)
6. Hou, S., et al.: Research and analysis on the solar radiation model of Huhehaote. Renew. Energy Resour. **02**, 79–82 (2008)
7. Shichuan, S., et al.: Numerical simulation on cabin airflow distribution of Ro-ro Passenger Ship. Sci. Technol. Eng. **20**(10), 4141–4148 (2020)
8. Cui, Y.: Some issues of HVAC design in high altitude areas. Heat. Vent. Air Cond. **06**, 39–42 (1999)
9. Yan, T., et al.: Review on key technologies of operational ventilation in high-altitude highway tunnels. Mod. Tun. Technol. **56**(S2), 88–95 (2019)
10. Liu, X.: Study on Key Technique of Construction Ventilation for Extra-Long Highway Tunnel at High Altitude on Queer Mountain. Southwest Jiaotong University (2016)
11. Nuo, H.: The Description of VRV Air-Condition and Heat Recover Screw Chiller are Designed in Hotel. South China University of Technology (2011)

Open Access This chapter is licensed under the terms of the Creative Commons Attribution-NonCommercial-NoDerivatives 4.0 International License (http://creativecommons.org/licenses/by-nc-nd/4.0/), which permits any noncommercial use, sharing, distribution and reproduction in any medium or format, as long as you give appropriate credit to the original author(s) and the source, provide a link to the Creative Commons license and indicate if you modified the licensed material. You do not have permission under this license to share adapted material derived from this chapter or parts of it.

The images or other third party material in this chapter are included in the chapter's Creative Commons license, unless indicated otherwise in a credit line to the material. If material is not included in the chapter's Creative Commons license and your intended use is not permitted by statutory regulation or exceeds the permitted use, you will need to obtain permission directly from the copyright holder.

Investigation of Surrounding Rock Failure Modes in Marine Soft Soil Strata for Shield Tunneling

Wenbin Xu, Yindong Sun, Heng Zhang, Wu Ke(✉), and Yajun Liu

School of Civil Engineering, Shandong University, Jinan, China
wuke@sdu.edu.cn

Abstract. This study investigates the failure modes of a shield tunnel passing through a layer of marine soft soil, employing the modified Cambridge model as the constitutive model for the soft soil. The objectives are to analyze the effects of varying groundwater levels and tunnel depths on the stress, strain, and to explore the characteristic failure modes during shield tunneling. A numerical computation model was established using the finite element software ABAQUS, simulating the processes of shield machine excavation, segment installation, and soil backfilling. The findings indicate that rising groundwater levels decrease the equivalent and shear stresses in the surrounding rocks, whereas increasing tunnel depth boosts both the equivalent and shear stresses in the surrounding rock. Settlement distribution around the tunnel is asymmetrical, with greater settlement at the top and lesser settlement at the bottom. The research provides insights for the design, construction, and safety assessment of shield tunnels within marine soft soil layers.

Keywords: Marine soft soil · Shield tunneling · Surrounding rock failure modes · Numerical computation

1 Introduction

The extensive presence of marine soft soil, characterized by high moisture content, high compressibility, low strength, and low permeability, significantly complicates engineering projects in coastal cities across China. The unique engineering properties of marine soft soil critically influence the stability and safety of subterranean constructions, such as tunnels. This study aims to investigate the failure mode characteristics of a shield tunnel passing through a layer of marine soft soil, considering the effects of varying groundwater levels and tunnel depths on the stress, strain, and segment stress of the tunnel's surrounding rock. Existing studies have explored various techniques and methodologies to assess and mitigate the hazards associated with marine soft soil, including simulation, numerical modeling, and field testing. Dongmei et al. [1] introduced a method for simulating the effect of grouting-induced soil volume strain on tunnel grouting. Wang et al. [2] conducted numerical simulations to study the excavation process of metro shield tunnels, investigating the law of surface settlement and the variation of settlement at different depths of the strata. Jing et al. [3] summarized settlement deformation characteristics in coastal soft soil areas. Several researchers, including He et al. [4], Yang et al. [5], Zhao [6], Wang et al. [7], and Yi et al. [8], analyzed various factors' effects on ground deformation during shield tunneling. Hou [9] explored the influence of hole diameter on large burial depth soft rock shield tunnels under creep effects. Li et al. [10] discussed pre-strengthening schemes for marine soft soil. However, despite these studies, there is a lack of comprehensive research investigating the characteristic failure modes of shield tunnels within marine soft soil layers, considering the combined effects of varying groundwater levels and tunnel depths. This study employs the modified Cambridge model as the constitutive model for marine soft soil, establishing a numerical computational model to investigate the failure modes of surrounding rock in coastal soft soil shield tunneling.

2 Project Overview

This study investigated a shield tunnel through a marine soft soil layer in a coastal city. The tunnel is an important part of the city's subway network. It is excavated using the shield method. Its dimensions are 6 m in inner diameter, 6.4 m in outer diameter, and 15 to 25 meters deep. The physical and mechanical parameters of the marine soft soil layer are summarized in Table 1.

Table 1. Physical-mechanical parameters of marine soft soil layers along the tunnel alignment

Soil Layer	Thickness(m)	Natural Density(g/cm^3)	Moisture Content(%)	Liquid Limit(%)	Plastic Limit(%)	Compression Coefficient(MPa^{-1})
Clay	2–50	1.76	44.9	1.43	17.0	0.88

Based on the actual conditions of the project, the effects of different working conditions were studied, as shown in Table 2.

Table 2. Different construction conditions

Condition	Groundwater Level (m)	Tunnel Depth (m)	Condition	Groundwater Level (m)	Tunnel Depth (m)
Condition 1	4	15	Condition 6	6	25
Condition 2	4	20	Condition 7	8	15
Condition 3	4	25	Condition 8	8	20
Condition 4	6	15	Condition 9	8	25
Condition 5	6	20			

3 Numerical Computation Model and Analytical Methods

3.1 Numerical Computation Model

To analyze the characteristics of surrounding rock failure modes in marine soft soil layers, a numerical computational model was established using the finite element software ABAQUS. The model simulated processes such as shield tunnel excavation and segment installation.

3.2 Constitutive Model and Mechanical Parameters of Marine Soft Soil

To capture the nonlinearity, non-elasticity, and non-uniformity of marine soft soil, this study employed the Modified Cambridge model as the constitutive model for marine soft soil. The Modified Cambridge model is an extension of the Cambridge model, incorporating secondary consolidation and shear-induced dilatancy to account for the time and history-dependent nature of marine soft soil. The modified Cambridge model parameters of marine soft soil used in the model are shown in Table 3.

Table 3. Modified Cambridge model parameters for marine soft soils

Soil Layer	Logarithm of Bulk Modulus	Poisson's Ratio	Stress Ratio	Logarithm of Plastic Volume Modulus	Intercept	Initial Void Ratio
Clay	**0.36**	0.35	0.45	0.0181	2.7	1.26

4 Analysis of Surrounding Rock Failure Modes

To analyze the characteristics of shield tunnel surrounding rock damage patterns in marine soft soil strata, this study investigates the variations and influencing factors of tunnel surrounding rock under different construction conditions from the perspectives of stress and settlement.

4.1 Stress Characteristics of Surrounding Rock

The stress characteristics of the surrounding rock reflect the stress state and changing trends during shield machine excavation, segment installation, and soil backfilling processes. The equivalent stress of the tunnel surrounding rock was selected as the primary stress component for analysis, as illustrated in Fig. 1.

Figure 1 depicts the stress distribution of tunnel surrounding rock, influenced by groundwater level and tunnel depth, as follows: (1) Rising groundwater levels reduce both equivalent and shear stresses in the surrounding rock due to buoyancy effects,

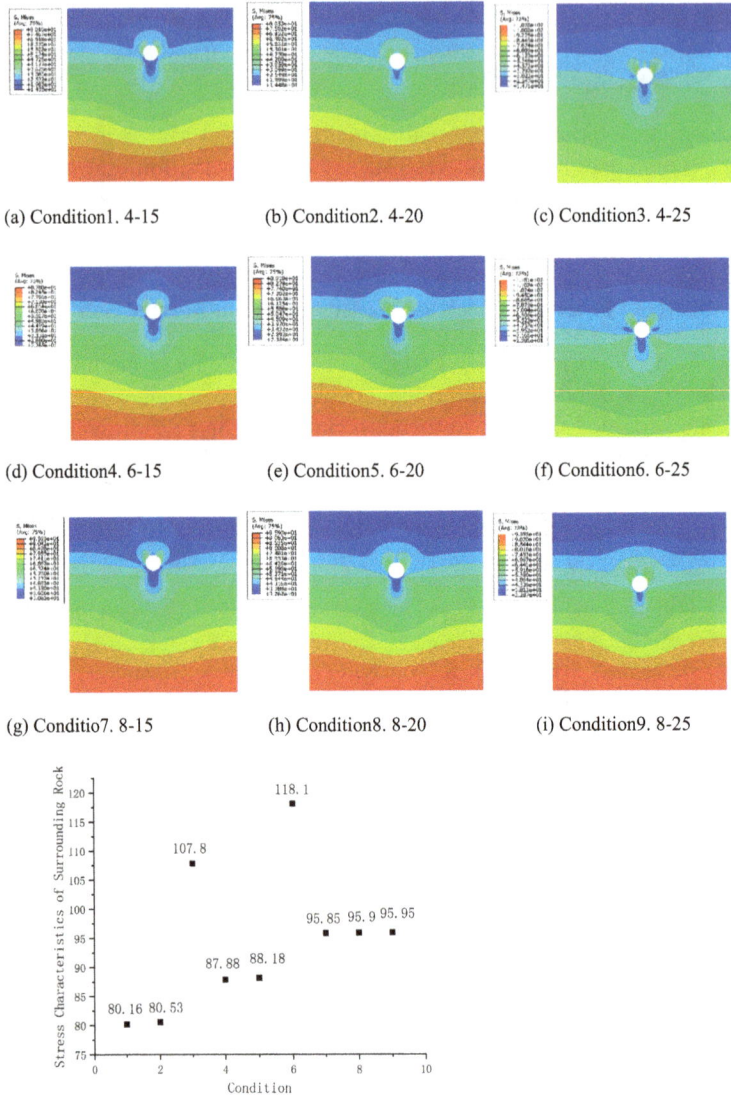

(a) Condition1. 4-15 (b) Condition2. 4-20 (c) Condition3. 4-25
(d) Condition4. 6-15 (e) Condition5. 6-20 (f) Condition6. 6-25
(g) Conditio7. 8-15 (h) Condition8. 8-20 (i) Condition9. 8-25

(j)Comparison chart of different operating conditions.

Fig. 1. Stress Distribution of Tunnel Surrounding Rock.

diminishing effective stress and soil strength. Higher water levels also exacerbate seepage, leading to soil consolidation settlement and further stress reduction. (2) Increasing tunnel depth elevates both equivalent and shear stresses due to self-weight of overlying soil, intensifying effective stress and soil strength. Greater depth also increases soil pressure, resulting in a more complex stress state. (3) Stress distribution around the tunnel is uneven, with higher stress at the top and bottom and lower stress on the sides due to stress redistribution from excavation.

4.2 Settlement Characteristics of Tunnel Surrounding Rock

The settlement characteristics of the tunnel surrounding rock reflect its deformation state and changing trends during shield machine excavation, segment installation, and soil backfilling. This study selected vertical settlement as the primary deformation components for analysis, as depicted in Fig. 2.

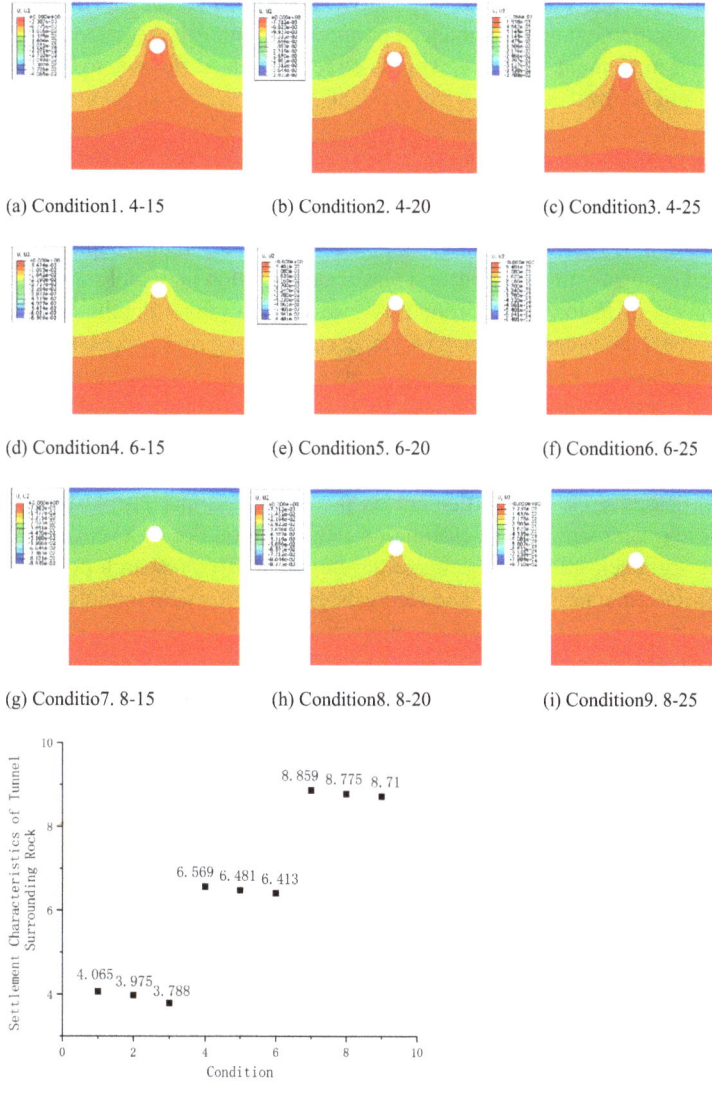

(a) Condition1. 4-15 (b) Condition2. 4-20 (c) Condition3. 4-25

(d) Condition4. 6-15 (e) Condition5. 6-20 (f) Condition6. 6-25

(g) Conditio7. 8-15 (h) Condition8. 8-20 (i) Condition9. 8-25

(j)Comparison chart of different operating conditions.

Fig. 2. Settlement Distribution of Tunnel Surrounding Rock.

Figure 2 illustrates that settlement distribution in the tunnel surrounding rock is influenced by groundwater level and tunnel depth, showing these characteristics: (1) Increasing tunnel depth reduces vertical settlement due to higher effective stress, enhancing soil strength and restraining deformation. (2) Settlement distribution around the tunnel is asymmetrical, with greater settlement at the top and lesser at the bottom. Sideways displacement is larger, while the center experiences less displacement, attributed to soil deformation redistribution post-excavation. Support from segments affects settlement at the tunnel's top and bottom. (3) Typically, Rising groundwater levels increase soil consolidation settlement due to enhanced permeability, intensifying seepage effects and horizontal soil movement. However, simulated results may differ due to the assumption of saturated soil below the groundwater level, affecting effective stress and deformation.

5 Conclusion

This study examines how varying groundwater levels and tunnel depths impact construction and analyzes stress, strain, and plastic deformation in tunnel surrounding rock. Conclusions are as follows: (1) Groundwater levels and tunnel depths inversely affect stress, settlement, and segment stress of tunnel surrounding rock, necessitating comprehensive consideration. Lower groundwater levels and tunnel depths generally favor the stability and safety of tunnel surrounding rock, but other factors such as soil properties, seismic risks, and environmental sensitivity should not be overlooked. (2) Failure mode characteristics primarily entail greater stress and settlement at the tunnel's top and bottom, lesser stress and settlement on the sides, and uneven plastic zone distribution. These reflect stress and deformation redistribution during excavation, segment installation, soil backfilling, and damage development. This research provides a reference for the design, construction, and safety assessment of shield tunnels in marine soft soil. However, it also carries certain limitations, such as the simplification of numerical computation models, applicability of constitutive models, and representativeness of conditions, which necessitate further enhancement and improvement in future research.

Acknowledgements. National Natural Science Foundation of China project (52179106).

References

1. Dong-mei, Z., Wei-biao, Z., Jing-ya, Y.: Effective control of large transverse deformation of shield tunnels using grouting in soft deposits. Chinese J. Geotech. Eng. **36**(12), 2203–2212 (2014)
2. Guo-cai, W., Da-jun, M., Yang, Y., et al.: 3-D finite element analysis of ground settlement caused by shield construction of metro tunnels in soft soils. Chinese J. Geotech. Eng. **33**(1), 266–270 (2011)
3. Fei, J., Feng, Z., Li, Z.: Monitoring analysis of settlement and deformation of soft marine clay in Jiangsu province. Chinese J. Geotech. Eng. **S2**, 554–557 (2010) (in Chinese)
4. He, S.: Study on three-dimensional deformation law of soft soil shield in Ningbo. Chinese Eng. Consult. **2**, 56–62 (2018) (in Chinese)

5. Bing-ming, Y.: Research on the law and prediction of long-term settlement of shield tunnel in soft soil. Strat. J. Railw. Eng. Soc. **32**(11), 87–92 (2015) (in Chinese)
6. Xuwei, Z.: The settlement law and construction control of railway hub traversed by a shield tunnel in soft ground. Hazard Control Tunn. Undergr. Eng. **4**(2) (2022) (in Chinese)
7. Huiwu, W., Jianning, G., Guodong, L., Yong, F.: Analysis of shield driving parameters for high rheological soft soil on Ningbo metro line 1. Railw. Eng. **7**, 49–54 (2016) (in Chinese)
8. Shun, Y., Jian, C., Wen-hui, K., Bin, C., Fu-sheng, L., Jue-hao, H.: Deformation of shield tunnels considering small strain in soft soil areas. Chinese J. Geotech. Eng. **42**(z2), 172–178 (2020)
9. Yongbing, H.: Study on the mechanical characteristics of large burial depth soft rock shield tunnels affected by hole diameter under creep effects. Railw. Constr. Technol. **09**, 108–113 (2023) (in Chinese)
10. Zunhao, L., Yuandong, P., Xiong, C.: Research on pre-reinforcement scheme of marine deep soft soil ground in Hengqin Mangzhou tunnel. Guangdong Archit. Civ. Eng. **29**(6) (2022) (in Chinese)

Open Access This chapter is licensed under the terms of the Creative Commons Attribution-NonCommercial-NoDerivatives 4.0 International License (http://creativecommons.org/licenses/by-nc-nd/4.0/), which permits any noncommercial use, sharing, distribution and reproduction in any medium or format, as long as you give appropriate credit to the original author(s) and the source, provide a link to the Creative Commons license and indicate if you modified the licensed material. You do not have permission under this license to share adapted material derived from this chapter or parts of it.

The images or other third party material in this chapter are included in the chapter's Creative Commons license, unless indicated otherwise in a credit line to the material. If material is not included in the chapter's Creative Commons license and your intended use is not permitted by statutory regulation or exceeds the permitted use, you will need to obtain permission directly from the copyright holder.

Hydrodynamic Analysis of Unilateral Salvage of Wreck by Semi-submersible

Wei Zhang[1,2(✉)], Shaoshi Dai[2], Zikai Xu[1], and Jingjing Liu[1]

[1] Technology Center, YanTai Salvage, Yantai, China
donsh328@163.com
[2] College of Shipbuilding, Harbin Engineering University, Harbin, China

Abstract. As a special engineering vessel, semi-submersible has high wave resistance and play a very important role in practical marine engineering, which can be applied to various operating conditions. This paper takes the unilateral salvage of wreck by semi-submersible as the background, based on the three-dimensional potential flow theory, and uses the hydrodynamic analysis software Moses to simulate the unilateral lifting and salvage of wreck by semi-submersible. For the salvage operation of a 2240-ton wreck in a 500 m deep water, the motion response of semi-submersible and wreck under different sea conditions is analyzed. Based on the calculation results, the maritime operation conditions of the unilateral salvage method for semi-submersible in this sea area are given, and the calculation results of the suspension force are analyzed, providing useful reference for the arrangement of salvage forces. The relevant research results have high reference value for the selection of operating sea conditions and the allocation of lifting cable forces in the design of unilateral salvage methods for semi-submersible.

Keywords: Lifting and salvage · 3D Potential flow theory · Kinematic response prediction

1 Introduction

Hydraulic lifting equipment assisted barge lifting and floating wreck salvage technology is a kind of efficient salvage technology, which can complete the wreck floating work in a short time after completing the preliminary preparation work. Combined with the characteristics of hydraulic lifting equipment, the process is designed according to the operating environment to adapt to different wreck salvage. For shallow water operations double barge lifting wreck operation has a greater advantage, can provide greater lifting force, wrecks can be lifted out of the water, such as "Seetsu" passenger ship salvage [1], but the deep-water environment, double barge lifting operations have a greater impact by the current state of the sea. In order to improve the efficiency of deep-water salvage operation, engineers designed single barge lifting wreck salvage operation, for example, "Kursk" submarine salvage [2], K-129 submarine salvage [3] and FC-35 fighter salvage operation [4], and so on. There are many factors to be considered when using a barge to lift wreck, among which the motion response of surface vessels and submerged wrecks in the operating current environment is one of the key factors. Tang Ji-Wei [5] and others simulated the complete process of semi-submersible vessel lifting and floating of "Seyue" through GHS special software, and checked the stability of semi-submersible vessel, Fengrui Zhang [6] and other scholars studied the motion analysis model of double barge salvage operation, and Hou Joyi [7] and other scholars investigated the motion response of single barge salvage ship. Liu Xu [8] and others studied the wave resistance of a 50,000-ton semi-submersible vessel using experimental methods and numerical simulation.

This paper takes the semi-submersible barge unilateral salvage wreck as the background, uses the semi-submersible barge unilateral salvage technology to salvage the wreck as a whole in 500 m water depth as a research case, based on the three-dimensional potential flow theory, with the help of the hydrodynamic analysis software Moses to simulate the working conditions located in different sea conditions, and analyses the motion response of semi-submersible barge and wreck based on the results of the calculations, to provide references to the sea conditions of the salvage operation in the actual project. The results are used to analyses the motion response of the semi-submersible barge and the wreck, and provide a reference for the selection of sea conditions for salvage operation in actual projects.

2 Figures and Tables Design of Single Barge Unilateral Salvage of Wrecks Under Mooring Conditions Semi-submersible Barge Unilateral Salvage of Wrecks

2.1 Demand Analysis Unilateral Lifting Salvage Design

The unilateral lifting and salvaging of wreck is arranging lifting equipment on one side of the barge deck to lift and salvage the wreck. In order to avoid significant lateral tilting caused by unilateral force on the barge, ballast needs to be carried out in the compartment

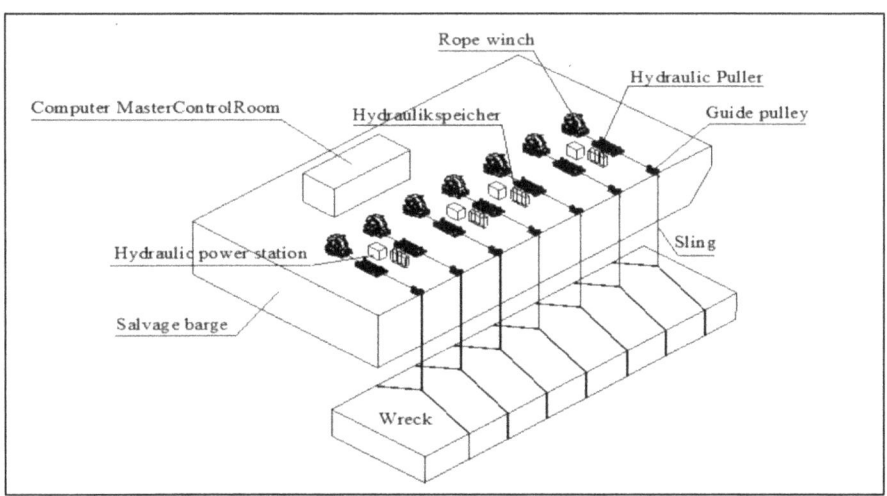

Fig. 1. Arrangement of single side lifting for salvage of wreck

on the other side of the barge. The design of unilateral lifting of the wreck is shown in Fig. 1. The wreck lifting equipment are hydraulic pullers, there are an equal number of guide wheels arranged on the side of the barge deck, slings which turning through the guide wheels to connected the wreck and the pullers. Multiple hydraulic pullers are controlled by a computer to jointly lift the wreck.

2.2 Semi-submersible Barge Parameters

Parameters of Semi-submersible Barge. Deep-water salvage operation requires the operating vessel to adapt to harsh sea conditions and strong wave resistance, especially for deep-sea long-wave surge adaptability. In order to achieve efficient salvage operations in deep water, it is a good choice to use a semi-submersible barge as an operational mother vessel. In the simulation of this paper, a 10wt semi-submersible barge is used as the salvage vessel, and the parameters are shown in Table 1.

Table 1. Parameter list of 10wt semi-submersible barge

Parameter	Value
Length	285 m
Breadth	76 m
Depth	15 m
Structure Draft	11 m
Centre of gravity	(150,-2.3, 12.6)
Inertial radius	$Rxx = 24.32$ m, $Ryy = 82.65$ m, $Rzz = 82.65$ m
Light Weight	67,000 t

Wreck Parameters. The wreck is 90 m long, 9 m wide, 5 m deep and weighs 2240 tones, the wreck has no detailed weight distribution parameters, in order to be able to accurately calculate the lifting cable force, the weight distribution of the wreck is analyzed and calculated. For the wreck along the direction of the ship's length of the weight division, follow the principle of static equivalence for the weight of the weight of the approximation and idealization of the distribution of processing [9]. The weight of a certain item distributed in 2 standard theoretical station distance with arbitrary law is P, and the distance of the center of gravity from 2 stations is α. According to the principle of distribution, the real weight distribution is replaced by the distributed loads in the 2 theoretical station distances of 1–2 and 2–3. Let the weights within the 2 theoretical station distances be P1 and P2 respectively, which can be obtained according to the principle of static equivalence:

$$P_1 + P_2 = P \qquad (1)$$

$$\frac{1}{2}P_1\Delta L - \frac{1}{2}P_2\Delta L = P\alpha \qquad (2)$$

Here, the wreck is divided into 6 equal theoretical station distances, i.e., $\Delta L = 15$ m. When the number of standard theoretical station distances within the theoretical station distance where the equipment is located is greater than or equal to 2, the weight of the equipment is equated to the standard theoretical station distances where the equipment is located in accordance with the principle of static equivalence, and the weight distribution within the standard theoretical station distances where the equipment is located is calculated as follows: Divide the number of theoretical station distances n in the equipment into 2 equivalent theoretical station distances. The specific calculation steps are as follows: The number of theoretical station distance n of the equipment is divided (n $\geq$ 2), equivalent into two equivalent theoretical station distance, calculate the equivalent weight distribution P1 and P2 of the two equivalent theoretical station distance, which can be calculated by the following formula:

$$P_1 = P \times \left(0.5 + \frac{\alpha}{\Delta L}\right) \qquad (3)$$

$$P_2 = P \times \left(0.5 - \frac{\alpha}{\Delta L}\right) \qquad (4)$$

where: P is the weight of the equipment, α is the distance between the point of action of the weight of this equipment and the geometrical center of the theoretical range of station distances in which it is located; ΔL is the length of the equivalent theoretical range of station distances after pair division. According to Table 2, the weight of each part of the wreck is estimated, and the calculation results are shown in Fig. 2.

Table 2. Proportion of weight and space in each part of the wreck

	Weight (%)	Space (%)	Weight (t)
Structure	43	–	731
Main engine and auxiliary engine	35	56	595
Living and working equipment	4	11	68
Stock	1	5	17
Fixed ballast	8	–	136
Other	9	28	153

3 Numerical Forecasting Programs

3.1 Mathematical Model

The velocity potential in the flow field is assumed to exist and to satisfy the Laplace equation and four types of boundary conditions: The free surface condition, the seafloor condition, the wet surface of the object condition, and the radiation condition. The

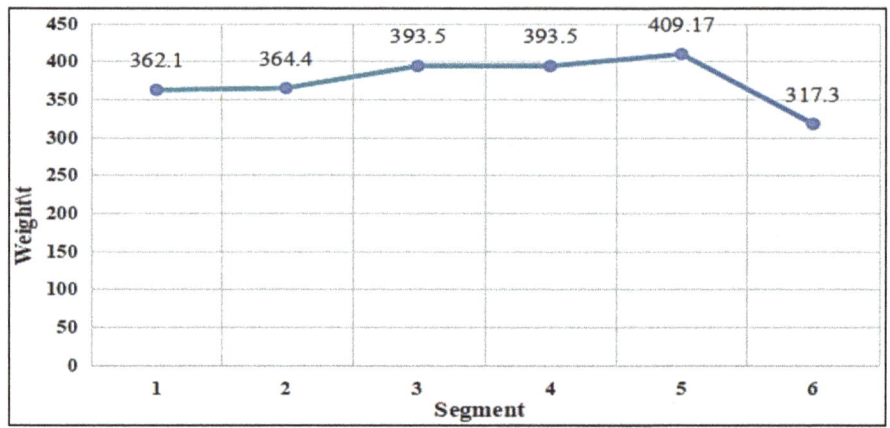

Fig. 2. Weight distribution map of wreck lifting

velocity potential can be uniquely determined from the Laplace equation and boundary conditions, and then the pressure on the wet surface of the object is calculated according to Bernoulli's formula. The free surface Green's function method is used to determine the velocity potential of the flow field by distributing the source and sink on the wet surface of the object.

The velocity potential in the fluid domain satisfies the governing equation Laplace equation:

$$\nabla^2 \Phi = 0 \tag{5}$$

Introducing the complex velocity potential, then the velocity potential in the fluid domain can be expressed as:

$$\Phi = \text{Re}\left(\phi e^{i\omega t}\right) \tag{6}$$

where: ω is the incident wave frequency and t is the time.

The linearised free surface boundary conditions are:

$$\phi z - k\phi = 0 \tag{7}$$

where: $k = \omega^2/g$

The incident wave velocity potential is defined as:

$$\phi_0 = \frac{igA}{\omega} \frac{\cosh(kz + H)}{\cosh kH} e^{-k(x\cos\beta + y\sin\beta)} \tag{8}$$

where: β is the direction of the incident wave. By the linearization assumption, the velocity potential φ can be decomposed into the diffraction potential φ_D and the radiating potential φ_R.

3.2 Calculation Model

The hydrodynamic analysis software Moses is used for numerical simulation. Combined with the ship parameters, the semi-submersible barge and wreck models were established in Moses based on the 3D potential flow theory, respectively, where the wreck model was equated with a square hull, as shown in Fig. 3 and Fig. 4.

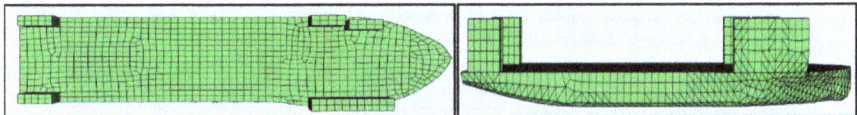

Fig. 3. Numerical modelling of semi-submersible barge

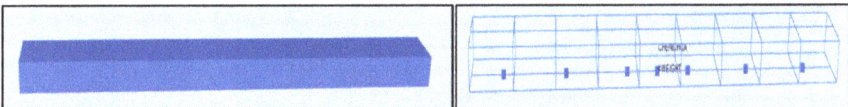

Fig. 4. Numerical modelling of wrecks

As the length of the wreck is 90 m and the width is 9 m, the distribution of the lanyard needs to be reasonably distributed. According to the actual size of the wreck, so take the distribution of ten lanyards, set up a lanyard on the starboard side of the semi-submersible barge to connect to the wreck, lanyard on the barge coordinates is shown in Table 3, the model of the wreck connecting to the barge is shown in Fig. 5.

Table 3. Table of coordinates of lifting cables on barge deck

No.	X	Y	Z
Sling1	114.9	38	15.5
Sling2	126.9	38	15.5
Sling3	129.05	38	15.5
Sling4	141.05	38	15.5
Sling5	143.2	38	15.5
Sling6	155.2	38	15.5
Sling7	157.35	38	15.5
Sling8	169.35	38	15.5
Sling9	171.5	38	15.5
Sling10	183.5	38	15.5

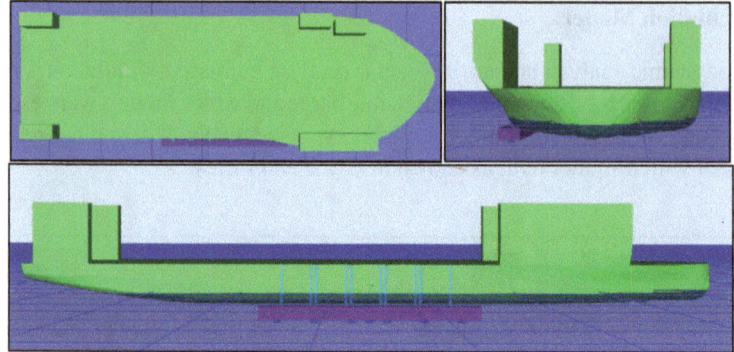

Fig. 5. Schematic diagram of a semi-submersible barge salvaging a wreck

3.3 Environmental Conditions

The simulated sea conditions were based on a depth of 500 m area. Monitoring and statistics were conducted on wave heights with a recurrence rate of over 50% for meaningful waves from March to September in the area. The results are shown in Fig. 6, and the probability of waves with meaningful wave heights below 1.2 m occurring from March to September is over 50%. Therefore, this paper uses the JONSWAP wave spectrum with a meaningful wave height of 1.2 m as the incident wave to simulate the working conditions. A ship motion response analysis was conducted for the condition of lifting a wreck from a 500 m seabed to a distance of 1 m from the bottom of the operating barge. The environmental conditions were wave periods of 6 s, 7 s, 8 s, and 9 s, with wave directions of 0 °, 45 °, 90 °, 135 °, and 180 °, respectively.

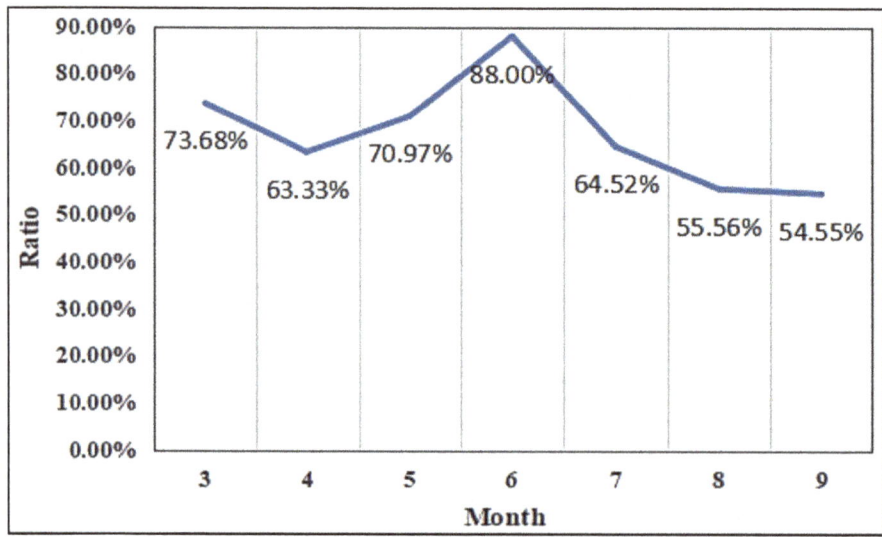

Fig. 6. Reproducibility of 1.2 m wave height at a point in the South China Sea

4 Motion Response Analysis

Based on the above conditions, simulations are carried out to give the motion response of the barge and the wreck when the wreck is located 1.0 m below the bottom of the semi-submersible barge and when it is located at a water depth of 500 m.

4.1 Movement Analysis

The RAO curves of six degrees of freedom of the semi-submersible barge and the wreck are calculated under the environmental conditions of wave height of 1.2 m, wave direction of 0 °, 45 °, 90 °, 135 °, 180 °, and wave period of 6 s, 7 s, 8 s, and 9 s, respectively. The calculation results are shown in Figs. 7-10, where Fig. 7 shows the RAO curve of the semi-submersible barge when the wreck is located in 500 m water depth condition, and Fig. 8 shows the RAO curve of the barge when the wreck is 1 m away from the bottom of the barge. Figure 9 shows the RAO curve of the wreck under the condition that the wreck is located in 500 m water depth, and Fig. 10 shows the RAO curve of the wreck under the condition that the wreck is located 1 m from the bottom of the barge.

4.2 Numerical Analyses

The Distance is 1 m from the Top of the Wreck to the Bottom of the Semi-submersible. According to the calculation results of the motion response of the semi-submersible barge, the extreme value of the surge amplitude of the semi-submersible barge under different sea states is about 0.04 m; the extreme value of the sway amplitude occurs when the wave direction is 90 °, and the maximum value is about 0.2 m. Similarly, the extreme value of the heave amplitude is 0.5 m when the wave direction is 90° and the wave period is 6 s. When the wave direction is 90 °, the extreme value of the roll angle amplitude reaches about 2° when the wave period is 9 s, and the extreme value of the pitch angle amplitude is about 1° under sea states with a wave period of 6 s and 7 s, and the extreme value of the roll angle amplitude is about 0.06°.

According to the calculation results of the motion response of the wreck, under different sea conditions, the extreme value of the surge amplitude of the wreck occurs at a wave direction of 135 ° and a wave period of 9 s, with an extreme value of about 0.05 m; the maximum amplitude value of the sway amplitude is about 0.4 m when the wave direction is 90 °; the maximum amplitude value of the heave amplitude is about 1.9 m when the wave direction is 90 °. The extreme value of the roll angle amplitude also occurs at a wave direction of 90 °, and the maximum amplitude angle reaches about 1.4° at a wave period of 9 s; the maximum amplitude angle of the pitch angle is about 0.3 °; and the maximum amplitude angle of the yaw angle is about 0.2°.

The Wreck is Located in a Water Depth of 500 m. According to the calculation results of the motion response of the semi-submersible barge, the maximum surge amplitude of the semi-submersible barge under different sea states is about 0.036 m; the extreme value of the lateral swing amplitude occurs when the wave direction is 90 °, and the maximum amplitude is about 0.3 m. Similarly, the extreme value of the heave amplitude occurs when the wave direction is 90 °, and the maximum amplitude is about 0.45 m; the roll

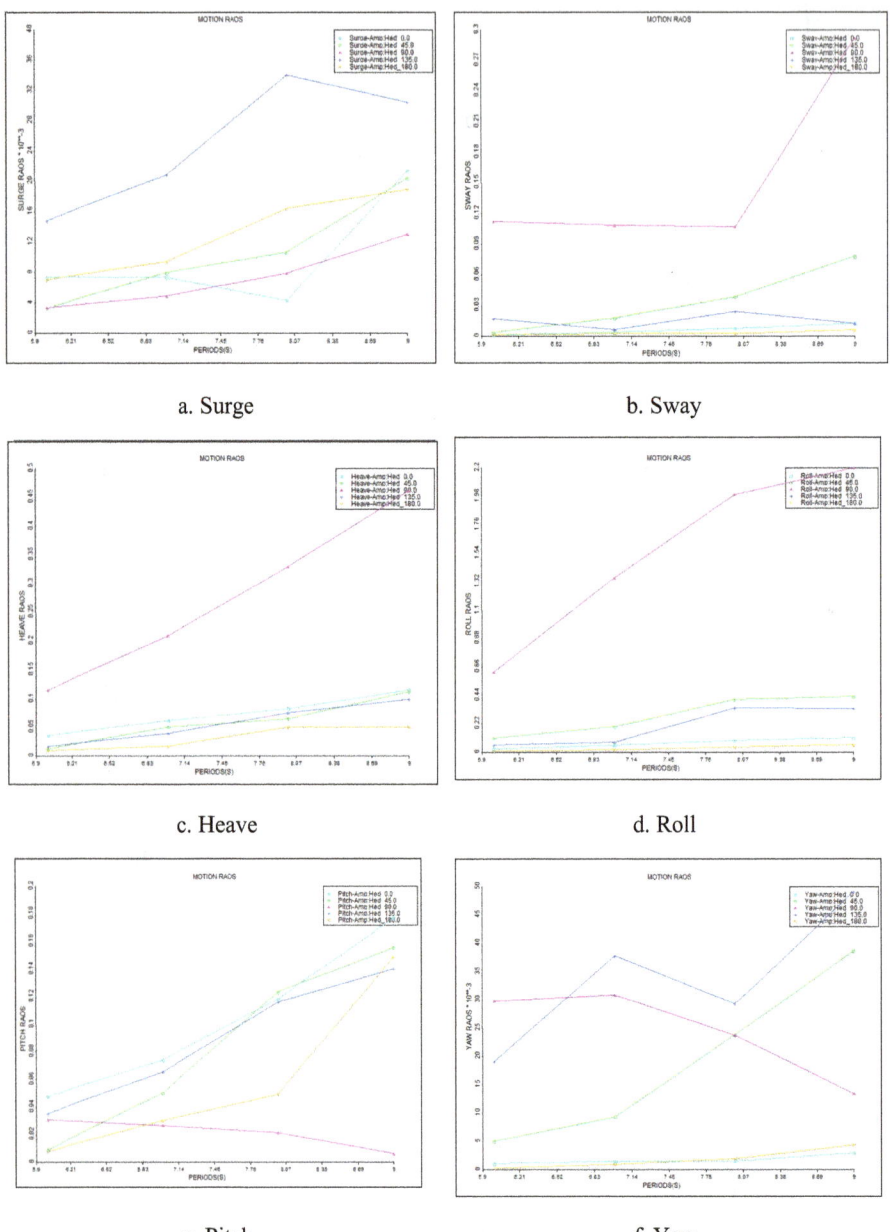

a. Surge b. Sway c. Heave d. Roll e. Pitch f. Yaw

Fig. 7. Semi-submersible barge RAO profile when the wreck is at 500 m water depth.

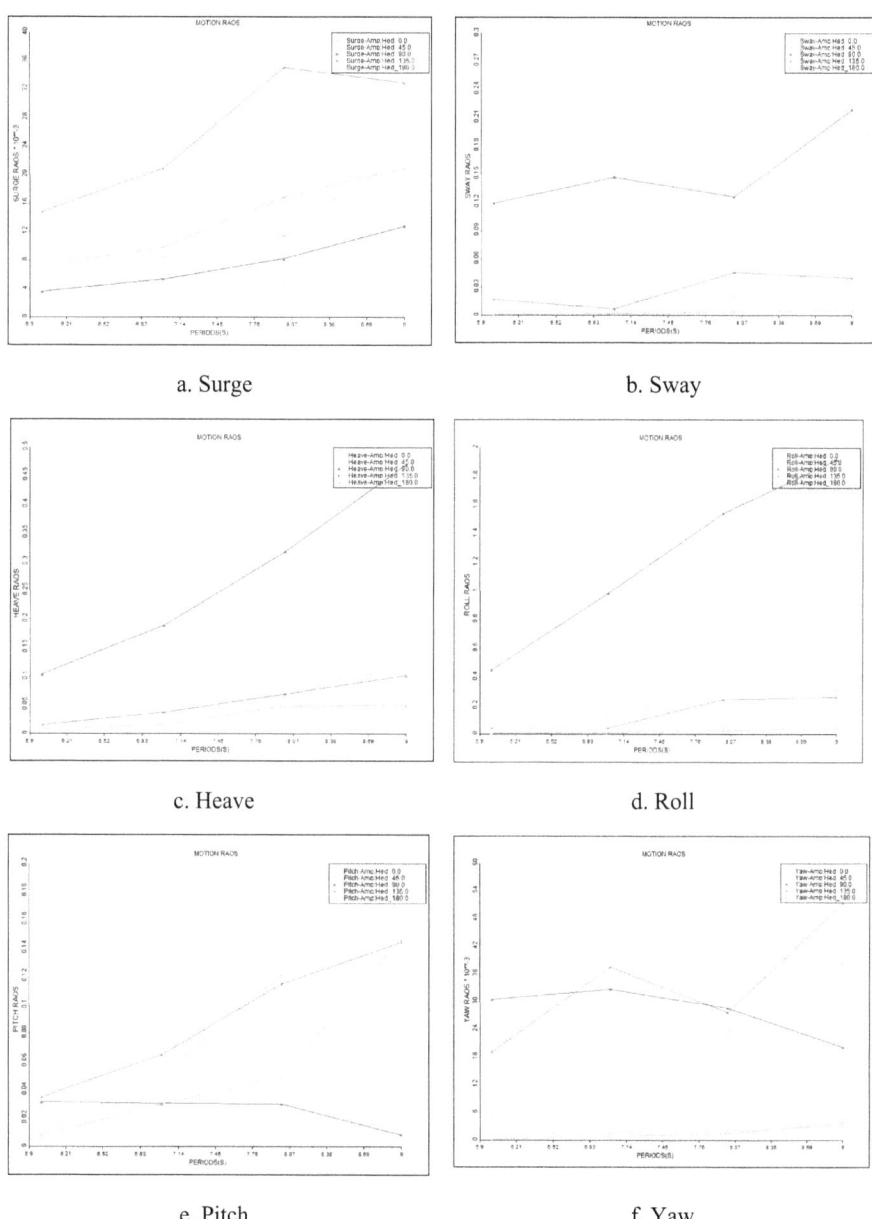

Fig. 8. Semi-submersible barge RAO profile when the wreck is 1 m below the barge bottom.

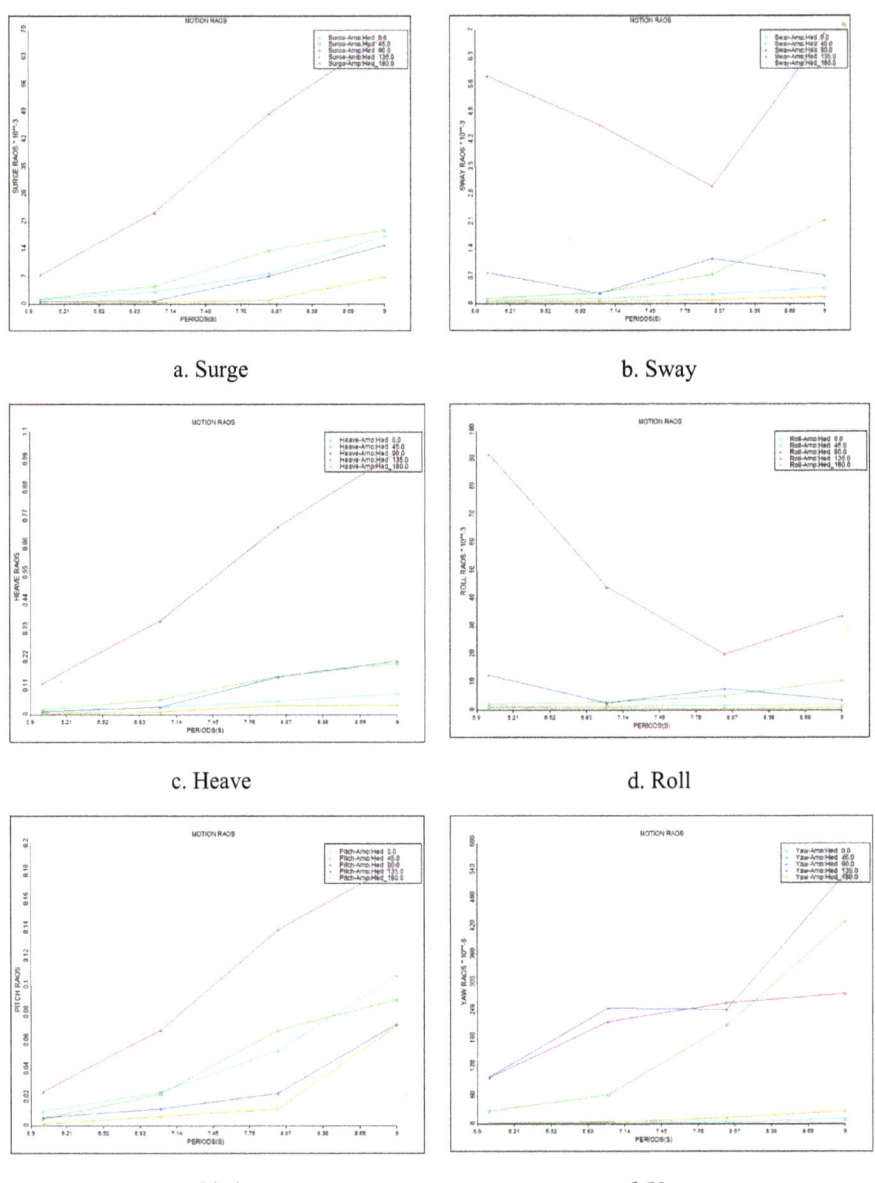

a. Surge　　　　　　　　　　b. Sway

c. Heave　　　　　　　　　　d. Roll

e. Pitch　　　　　　　　　　f. Yaw

Fig. 9. Wreck RAO profile when the wreck is at 500 m water depth.

Hydrodynamic Analysis of Unilateral Salvage of Wreck 705

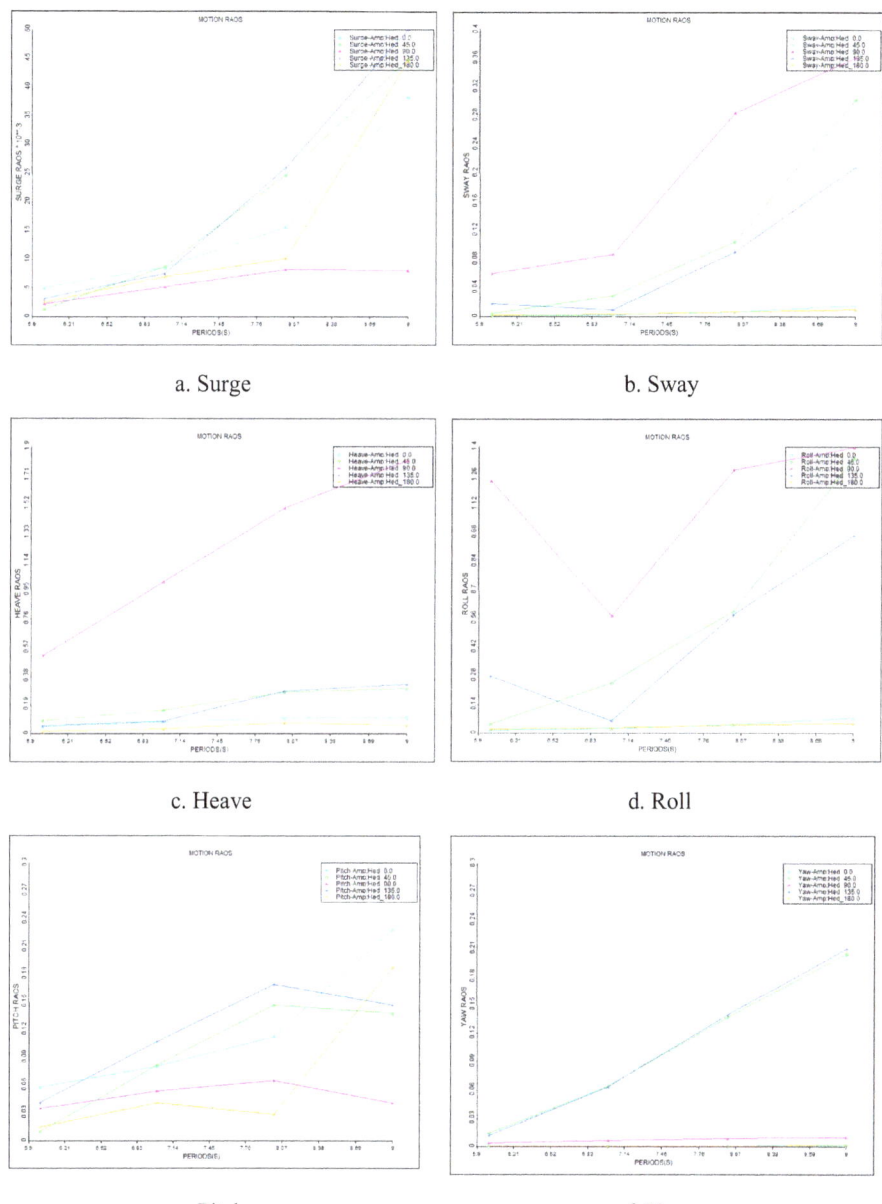

a. Surge

b. Sway

c. Heave

d. Roll

e. Pitch

f. Yaw

Fig. 10. Wreck RAO profile when the wreck is 1 m below the barge bottom.

angle amplitude also reaches its maximum value when the wave direction is 90 °, and the maximum amplitude reaches about 2.2 ° when the wave period is 9 s; the maximum pitch angle amplitude is about 1°, and under the sea conditions of wave periods of 6 s and 7 s, the pitch amplitude under all wave directions is less than 0.16°; the maximum roll angle amplitude is about 0.05°.

According to the results of the motion response of the wreck, under different sea conditions, the extreme value of the surge amplitude of the wreck appears when the wave direction is 135 °, and the maximum value appears when the wave period is 9 s, with an extreme value of about 0.07 m; the extreme value of the sway amplitude appears when the wave direction is 90 °, with a maximum amplitude of about 0.007 m; similarly, the extreme value of the heave amplitude appears when the wave direction is 90°, with a maximum amplitude of about 1.1 m; the roll angle amplitude also reaches its maximum value when the wave direction is 90°, with a maximum amplitude angle of about 0.9° when the wave period is 9 s; the maximum value of the pitch angle amplitude is about 0.2°; and the maximum value of the roll angle amplitude is about 0.006°.

In order to ensure the safety of the salvage barge and the wreck after the wreck is lifted near the surface, the design of the salvage barge and the wreck is to have a motion amplitude of less than 0.5 m. According to the calculation results and the above analysis, the amplitude of sway, surge, and heave is relatively large under the sea conditions with a wave direction of 90°, but the amplitude of sway, surge, and heave under other sea conditions is less than 0.5 m, which is an acceptable operating sea condition. Comprehensive evaluation and analysis show that the most suitable operating conditions are when the wave period is 6 s and the wave direction is 0° or 180°. Under this condition, both the surface ship and the wreck have six degrees of freedom motion response amplitude in an ideal state, meeting the operational requirements.

5 Force Analysis of Suspension Cable

From the analysis of unilateral salvage by semi-submersible barge, it can be concluded that the suitable sea conditions for operation are sea conditions with a wave period of 6 s, wave directions of 0 ° and 180 °. However, in practical engineering, it cannot be guaranteed that the on-site operation will always be in ideal sea conditions. Therefore, some intermediate sea conditions and ideal sea conditions are selected for the analysis of cable lifting force during the salvage process. After motion response analysis, the sea conditions with a wave period of 7 s, a wave direction of 45 °, and a wave period of 8 s and a wave direction of 135 ° will be analyzed. The total calculation time is 1500 s, with a time step of 1 s.

As shown in Figs. 11–14, the calculation results show that under sea conditions with a wave period of 6 s and wave directions of 0 °and 180 °, when the wreck is located at a depth of 500 m, the peak force of each suspension cable changes slightly, with an extreme fluctuation of no more than 5 tons. When the wreck floats to a distance of 1 meter from the bottom of the semi-submersible barge, the change in suspension cable force becomes significantly larger. Compared to under 180 ° waves, the suspension cable force under 0 ° waves is more severe, with an extreme fluctuation of about 10 tons; In the sea state with a wave period of 7 s and a wave direction of 45 °, the suspension force is in a relatively stable state around the first 600 s. However, in the subsequent simulation, the suspension force changes dramatically, and the maximum fluctuation of the suspension force for a single cable is about 50 tons. In this case, it is not the optimal operating sea state; In a sea state with a wave period of 8 s and a wave direction of 135 °, when the wreck is located at a depth of 500 m, the trend of changes in the force of each suspension cable is almost consistent. The extreme value of the force of a single suspension cable fluctuates about 15 tons. When the wreck floats to a distance of 1 m from the bottom of the semi-submersible barge, the change in the force of the suspension cable is more severe in the later simulation compared to the depth of 500 m, and the extreme value of the force of a single suspension cable also fluctuates about 15 tons.

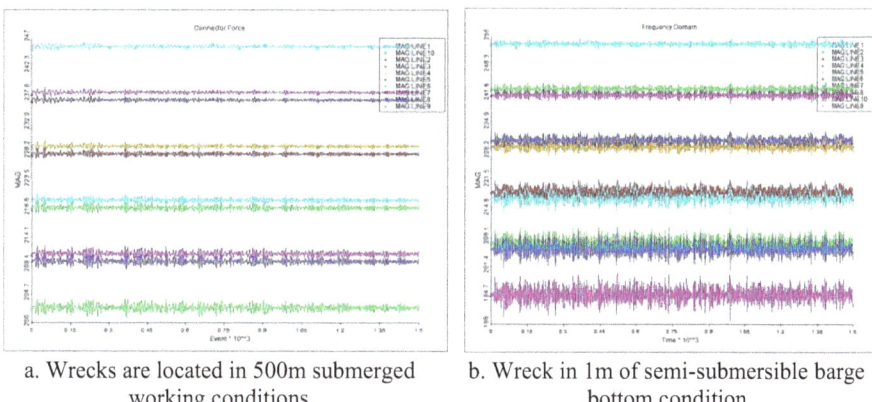

a. Wrecks are located in 500m submerged working conditions

b. Wreck in 1m of semi-submersible barge bottom condition

Fig. 11. Curve of cable suspension force over time under sea conditions with a wave period of 6 s and a wave direction of 0°.

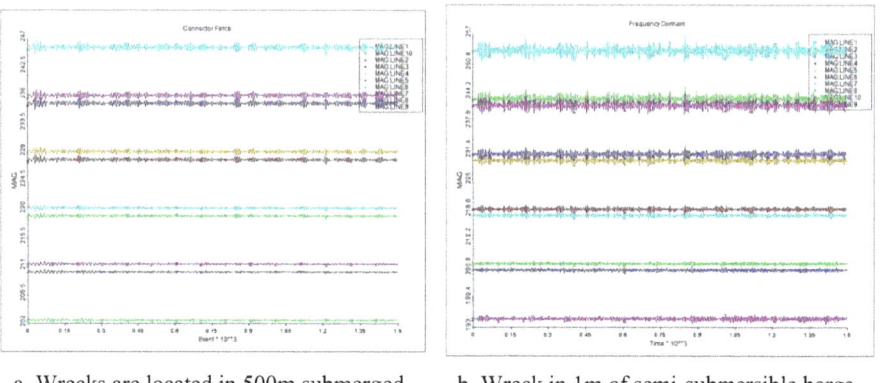

a. Wrecks are located in 500m submerged working conditions

b. Wreck in 1m of semi-submersible barge bottom condition

Fig. 12. Curve of cable suspension force over time under sea conditions with a wave period of 6 s and a wave direction of 180 °.

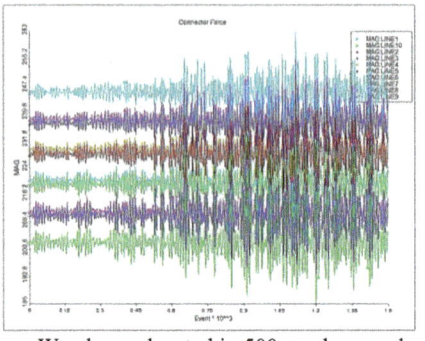 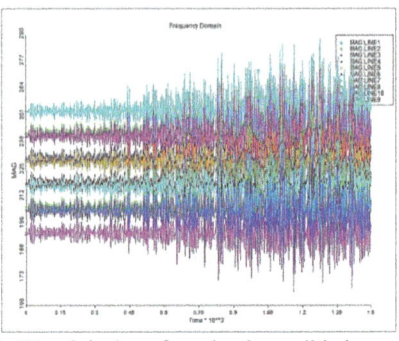

a. Wrecks are located in 500m submerged working conditions

b. Wreck in 1m of semi-submersible barge bottom condition

Fig. 13. Curve of cable suspension force over time under sea conditions with a wave period of 7 s and a wave direction of 45°.

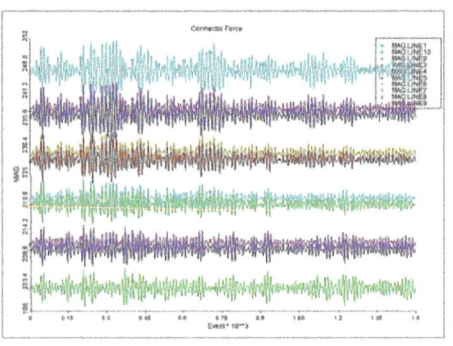 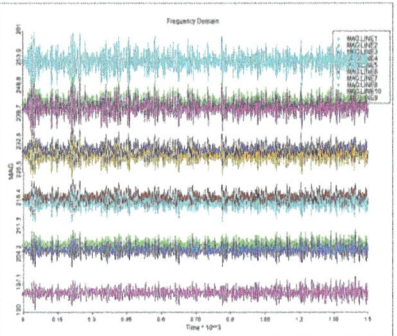

a. Wrecks are located in 500m submerged working conditions

b. Wreck in 1m of semi-submersible barge bottom condition

Fig. 14. Curve of cable suspension force over time under sea conditions with a wave period of 8 s and a wave direction of 135°

Based on the above analysis results, it can be seen that under sea conditions with a wave period of 6 s and wave directions of 0 and 180 °, the suspension force does not fluctuate violently and remains in a relatively stable state, which is a very good working sea condition for the salvage process of wreck; In the sea condition with a wave period of 8 s and a wave direction of 135 °, the extreme value of a single suspension cable force is relatively large compared to the above working conditions, but the overall suspension cable force time history curve has no significant fluctuations, which can still be considered as a working sea condition; However, under sea conditions with a wave period of 7 s and a wave direction of 45 °, the time history curve of the suspension cable force changes sharply, and the extreme value of a single suspension cable force is too large, which is not suitable for actual offshore operations.

6 Conclusion

This paper takes the unilateral salvage of wreck by semi-submersible barge as the background, based on the theory of three-dimensional potential flow, and uses the hydrodynamic analysis software Moses to simulate the motion response of semi-submersible barge and wreck under different sea conditions during the overall salvage operation of large tonnage wreck in deep water areas, as well as the results of suspension force. Based on the calculation results, the following conclusion can be drawn: Under all simulated sea conditions, the most suitable operating conditions for single barge unilateral salvage are sea conditions with a wave period of 6 s, wave directions of 0 ° and 180 °. Under these conditions, the six degree of freedom motion response amplitude of both surface ships and wreck is in a very small state, which is a very good operating sea condition. The analysis of the results of the lifting cable force further proves the correctness of the selection of operating sea conditions. This study provides valuable reference for the operation of single-sided salvage of sunken ships by semi-submersible barges, but there are still some limitations. In the simulation, only environmental factors such as wave height, wave period, and wave direction were considered, while in actual marine environments, there are other complex factors such as wind, current, tides, etc. These factors may have a significant impact on the motion response of semi-submersible barges and sunken ships, but this article does not provide a detailed analysis of them. Future research will consider introducing more environmental factors to further improve the accuracy and universality of the study.

References

1. Zong, Y., Wei-ping, W., Yan, J., et al.: Dynamic mooring force calculation of twin barges during 'Sewol' lifted above seabed. Chinese J. Hydrodyn. **33**(1), 17–27 (2018)
2. Just One Way To Salvage A Sub And Beat The Winter2001. https://www.mammoet.com/cases/Kursk, Accessed 21 June 2001
3. IAN H (2016). Project Azoria—recovery of the sunken Soviet submarine K-129. https://www.thevintagenews.com/2016/12/19/project-azoria-recovery-of-the-sunken-soviet-submarine-k-129/
4. USNI NEWS: UPDATED: Navy Recovers Crashed F-35C From Depths of South China Sea. https://news.usni.org/navy-recovers-crashed-f-35c-from-depths-of-south-china-sea. Accessed 03 Mar 2022
5. Jiwei, T., Shihai, C., Mingxin, D.: Stability checks of semi-submersible refitting and floating Sewol. World Shipp. **40**(09), 1–6 (2017)
6. Fengrui, Z., Jiaoyi, H., Dayong, N., et al.: Performance analysis of the passive heave compensator for hydraulic wreck lifting systems in twin-barge salvaging. Ocean Eng. **280**, 114469 (2023)
7. Fengrui, Z., Dayong, N., Jiaoyi, H., et al.: Semi-active heave compensation for a 600-meter hydraulic salvaging claw system with ship motion prediction via LSTM neural networks. J. Mar. Sci. Eng. **11**, 998 (2023)
8. Xu, L., Xiaoming, C., Liancai, C.: Seakeeping of 5000-ton semi-submersible vessels during navigation. China Navigation. **42**(04), 29–32+37 (2019)
9. Kangkang, Y., Da, S.: Research on the calculation method of load strength for ship weight distribution. Sh. Sci. Technol. **40**(19), 1–5 (2018)

Open Access This chapter is licensed under the terms of the Creative Commons Attribution-NonCommercial-NoDerivatives 4.0 International License (http://creativecommons.org/licenses/by-nc-nd/4.0/), which permits any noncommercial use, sharing, distribution and reproduction in any medium or format, as long as you give appropriate credit to the original author(s) and the source, provide a link to the Creative Commons license and indicate if you modified the licensed material. You do not have permission under this license to share adapted material derived from this chapter or parts of it.

The images or other third party material in this chapter are included in the chapter's Creative Commons license, unless indicated otherwise in a credit line to the material. If material is not included in the chapter's Creative Commons license and your intended use is not permitted by statutory regulation or exceeds the permitted use, you will need to obtain permission directly from the copyright holder.

Application and Discussion of Airbag in Salvage

Yingchun Qi(✉), Sihao Xing, Yingjie Li, and Yuantao Bi

China Yantai Salvage, Yantai Salvage Technology Center, Yantai, China
qiyc@ytsalvage.com

Abstract. The environment of wrecks and sunken objects in the sea is ever-changing, and the salvage methods are varied. Because of its unique performance and advantages, air bag has shown great potential and value in wrecks salvage and water transportation. Air bags have always been used as buoyancy AIDS in the field of salvage, usually tied to the two sides of the wreck or placed in the wreck cabin, and then combined with internal buoyancy or other large machinery to complete salvage. In some complex narrow waters, when large machinery cannot play a role, air bag buoyancy has become the best choice. This paper introduces the development and application scenarios of air bag salvage technology, summarizes the basic principle and technical characteristics of air bag salvage, and summarizes the key technologies of air bag salvage through the research of the construction technology of air bag salvage, and discusses the limitations and prospects of its application.

Keywords: Airbags · Salvage · Construction technology · Shipwreck

1 Introduction

China is a maritime power with a mainland coastline of over 18,000 km, as well as over 5000 rivers and over 500 lakes of various sizes. Its water transportation conditions are well-developed. With the continuous development of the economy, the maritime industry is also becoming increasingly prosperous, with a rapid increase in the number and tonnage of ships, an increase in the density of maritime transportation, and an increase in the range of navigation. Affected by meteorological changes, natural disasters pose a serious threat to the safety of people's lives and property, and major maritime accidents occur frequently. In addition to clearing obstacles in ports and waterways, the salvage of some special sunken targets also falls within the scope of emergency rescue and salvage, such as the salvage of the "5.7" air disaster black box [1], the maritime emergency support mission of spacecraft, and the clearance and salvage of hydropower stations. The main characteristics of such emergency rescue and salvage tasks are difficulty in underwater search or narrow construction waters, and relatively light salvage weight.

After more than half a century of continuous development, the equipment and techniques for salvage are also constantly being updated. The salvage methods can be divided according to the integrity of wrecks and objects into: Overall salvage, segmented salvage, salvage of broken ships, and dismantling salvage. According to the nature of buoyancy, it can be divided into internal buoyancy salvage method and external buoyancy salvage method [2]. The internal buoyancy salvage method can be divided into: Sealed pumping salvage, compressed air drainage salvage, sealed inflation drainage salvage, and filling material salvage; The external buoyancy salvage method can be divided into: Float bowl salvage method, floating crane salvage method, lifting and prying salvage method, etc. [3]. Airbag salvage can be used for both external and internal buoyancy salvage. The external buoyancy salvage of airbags is similar to the float bowl salvage method, in which the airbags are fixed to the wreck's sunken object using a sling or steel structure frame. Airbag internal buoyancy salvage is a part of the filling material salvage method, which involves inflating the airbag into the cabin to restore the buoyancy inside the ship.

The development process of airbag salvage technology is closely related to the continuous deepening of human exploration and utilization of marine resources. As early as ancient times, people had already used simple buoyancy principles to carry out salvage work using primitive tools such as animal skins filled with air. Entering modern times, with the progress of materials science and the development of industrial technology, rubber airbag salvage technology has undergone a transformation from primitive to modern. From Fig. 1, it can be seen that since the successful trial of rubber airbags for ship launching in 1981, the application of rubber airbags underwater [4] has gradually gained attention. In the 1980s, ship airbag launching technology was promoted in some inland shipyards in China, but due to structural and material strength limitations, the supported tonnage of ships was relatively small, usually with a self-weight of no more than 500 t. With the development of technology, the tonnage of ships that airbags can withstand has gradually increased, and by 2007, they were able to withstand the launch of ships with a self-weight of 9000 t underwater.

Fig. 1. Rubber bag launching

At present, rubber airbags are widely used in ship loading and launching construction due to their advantages of fewer limiting factors, no need for large unloading equipment, short construction period, and low cost. Two industry standards have been formed, CB/T-3795 "Rubber Airbags for Ship Loading and Unloading" and CB/T-3837 "Requirements for Rubber Airbags for Ship Loading and Unloading Process" [4]. In addition, rubber airbags have played an important role in fields such as construction [5], rescue, and pipeline sealing [6]. In addition to rubber airbags, there are currently materials such as polyurethane, PVC, and coated fabrics for airbags, which are widely used in fields such as automobiles and hovercraft [7].

Airbags mainly appear in the field of rescue in the form of buoyancy assistance, which is to cooperate with the distressed ship's own buoyancy or other main lifting equipment to complete the lifting and shallow work of the distressed ship. In the salvage industry, rubber airbags are often placed inside the cabin to restore buoyancy, and there are also some cases where external rubber airbags are used to provide auxiliary buoyancy, and there are few cases where external single rubber airbags are used to salvage wrecks and sunken objects. In conjunction with Fig. 2, in 2017, the Shanghai Salvage Bureau completed the salvage project of the "Sewol" ship and used 3.5-m diameter rubber airbags as buoyancy aids fixed to the ship's structure to assist in the lifting operation of the bow [8]. In conjunction with Fig. 3, in 2020, the China Yantai Salvage used a 4 m diameter external single rubber airbag for salvage operations in the "xxx hydropower station concrete foreign object" salvage project due to factors such as limited operating space and lack of large mechanical equipment. The concrete foreign object was successfully moved to the designated area.

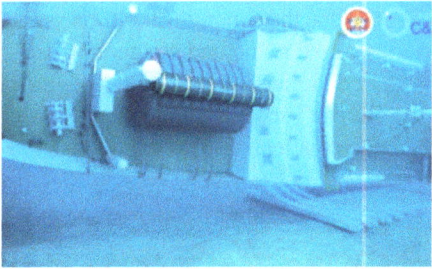

Fig. 2. The Sewol with external auxiliary rubber airbags [9].

Fig. 3. Single external rubber airbag salvaged foreign bodies in hydropower station

2 The Application of Airbags in Salvage Engineering

2.1 Application Case of Airbag External Buoyancy Salvage Method

In March 2020, during an underwater inspection of the sand discharge system at a certain hydropower station, divers discovered varying degrees of defects in the sand discharge tunnel and bottom hole through underwater exploration and inspection. Among them, there was a large concrete foreign object in front of the entrance inspection door of the 5 # sand discharge tunnel. Due to the limited strength of the dam crest of the hydropower station, it is unable to support large lifting equipment. Therefore, the foreign object can only be lifted and displaced by external buoyancy devices such as buoys and rubber airbags. Due to the width of the sand hole being only 4.5 m, the weight of the foreign object reaches nearly 120 t, and the length of the foreign object is only 7.7 m, making it impossible to place multiple rubber airbags or buoys. In order to provide more buoyancy in limited space and facilitate operation, an external single rubber airbag salvage solution was ultimately chosen. At the same time, when using rubber airbags for floating, the airbags are kept stable to ensure that the concrete blocks rise steadily horizontally during underwater floating, and to ensure that the underwater disposal of concrete blocks will not roll and collide with the edge piers on both sides of the gate or cause blockage. Finally, the Yantai Salvage Bureau successfully applied this technology and lifted and displaced foreign objects in the concrete. As shown in Figs. 4 and 5.

Fig. 4 Stabilizing air bag

2.2 Application Case of Buoyancy Salvage Method Inside Airbag

On April 16, 2014, the South Korean passenger ship Shiyue sank in the southwest waters of South Korea, resulting in 304 deaths and 9 missing. Due to the weight of the wreck on the seabed being over 7000 t, the construction unit considered factors such as the

Fig. 5 Rubber airbags help float concrete foreign bodies

structural strength of the hull, and fully utilized the internal and external buoyancy of the wreck during the lifting process of the bow. The internal buoyancy adopts the scheme of compressed air drainage and filling the cabin with airbags. As shown in Figs. 6.

Fig. 6 The wreck filled with air bags [9].

3 Key Technologies for Airbag Salvage

When using rubber airbags in salvage engineering, it is necessary to consider the size, placement, rooting point, strength, operability, and other factors of the rubber airbags.

3.1 Key Points of Airbag Structure Design

The structural design of airbags [10, 11] is the key to ensuring their optimal performance in underwater salvage operations, involving multiple aspects such as material selection, shape design, size calculation [12], pressure resistance [13], and buoyancy distribution. In terms of material selection, the mechanical properties and anti-aging capability required by the airbag must be taken into account. Generally, rubber and polyester fiber canvas with high strength, high abrasion resistance and high elasticity are used as the main materials. These materials not only need to have good elasticity and tear resistance, but also need to have sufficient resistance to ultraviolet radiation and seawater corrosion. The shape of the airbag used for salvage is usually cylindrical, which can increase the contact area and also be assembled in narrow spaces. The design of pressure resistance is the key to ensuring the safety of airbags. When designing, it is necessary to consider the water pressure effect of airbags at different depths and the impact of gas pressure on the expansion of airbag materials during inflation. At the same time, it is also necessary to consider the floating speed of wrecks to match suitable safety valves and quantities.

The general design process for airbags is:

1. Select the airbag body material according to the requirements, taking into account the material properties such as resistance to bursting strength, plasticity, and tensile strength;
2. Determine the geometric dimensions of the capsule body. The size of airbags, including diameter and length, needs to be determined based on factors such as the space in the wreck compartment and the buoyancy required for airbags.
3. Calculate the working pressure. Calculate the working pressure of the airbag based on factors such as water depth, and the inflation pressure of the airbag cannot exceed the allowable pressure.
4. To determine the number of safety valves, the following method can be used for calculation.

According to the gas pressure formula, assuming that the temperature inside the airbag is conserved during the lifting process, the volume change before and after lifting is:

$$P_1 V_1 = P_2 V_2 \tag{1}$$

Among them, P1 and P2 represent the pressure before and after airbag lifting; V1 and V2 are the volumes before and after airbag lifting, and the volume of exhaust before and after lifting can be calculated using this equation.

According to the Bernoulli equation, there are:

$$P_1 + \frac{\rho_1 v_1^2}{2} + \rho_1 g H_1 = P_2 + \frac{\rho_2 v_2^2}{2} + \rho_2 g H_2 \tag{2}$$

Among them: ρ1 and ρ2 is the density of gas inside the airbag; v1 and v2 are the exhaust velocities of the safety valve, with $v_1 = 0$ in the initial state; H1 and H2 are the water depths before and after lifting.

If the lifting speed of the ship is v, the lifting time is:

$$t = \frac{(H_2 - H_1)}{v} \tag{3}$$

Substitute the above results into the following gas flow rate formula:

$$v = v_2 St \tag{4}$$

The total exhaust hole area S can be obtained, and the number of safety valves can be calculated according to the following equation:

$$S = \frac{N\pi d^2}{4} \tag{5}$$

Among them, d is the aperture of the safety valve, and N is the number of safety valves.

3.2 Rubber Airbag Model Test

At present, the application of airbags in the field of salvage engineering is mainly based on rubber airbags. Therefore, this section takes a rubber airbag with a diameter of 1 m and an effective length of 6 m as an example for corresponding tests, with a working pressure of 0.5 MPa and a safety factor of 3 times.

Airbag Airtightness Test. The airtightness test mainly verifies that the airbag can maintain a certain pressure for a certain period of time, thereby continuously providing buoyancy and reducing the preparation work before the wreck floats. Inflate the airbag to the rated internal pressure, collect data every 12 h, use the pressure correction formula to correct the pressure to the initial temperature, and compare it with the initial pressure. Check whether the airbag has defects such as bubbling and delamination, and record them.

A) Pressure correction formula:

$$P' = \frac{P_2 \times (273.15 + T')}{273.15 + T_2} \tag{6}$$

Among them:

P' —— corrected pressure, in megapascals (MPa);

P_2 —— The actual pressure after supplementary inflation of the test airbag, in megapascals (MPa);

T' —— actual measured temperature, in degrees Celsius (°C);

T_2 —— Initial temperature of the test, in degrees Celsius (°C).

B) Calculate the decrease in airbag pressure K according to eq. (2):

$$K = \frac{P - P'}{P \times 100\%} \tag{7}$$

Among them:

K —— Airbag pressure drop rate, (%);

P —— The actual pressure after supplementary inflation of the test airbag, in megapascals (MPa);

P' —— The pressure corrected at a certain time interval, in megapascals (MPa).

From Fig. 7, it can be seen that, the pressure decreased from 0.594 MPa to 0.565 MPa in ten days, a decrease of 0.029 MPa, with a decrease rate of K of 4.88%, which meets the operational requirements.

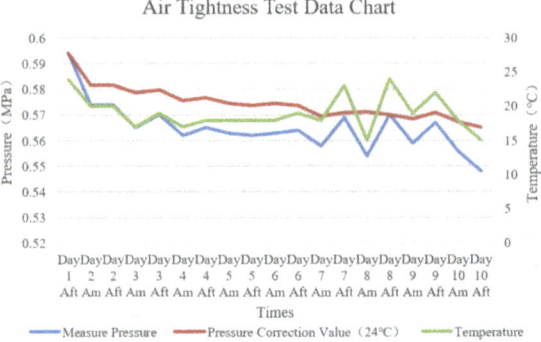

Fig. 7. Air tightness test data

Fig. 8. Air bag bursting test

Blasting Test. The blasting test mainly tests whether the strength of the airbag can reach the design safety **factor**. When the wreck rises rapidly, the external pressure continuously decreases, and the internal gas cannot be discharged in time, whether the airbag is safe. As shown in Figs. 8.

The working pressure of the airbag is 0.5 MPa, and the burst pressure value is 2.18 MPa, which exceeds the design requirement by three times the safety factor. Considering that when a wreck floats, the water pressure decreases and the pressure difference between the inside and outside of the airbag increases, according to the test results, the airbag will not rupture due to excessive pressure during the ascent operation with a water depth difference of no more than 100 m. However, for actual safety reasons, a pressure relief valve needs to be equipped for pressure relief [14].

Based on the experimentally measured burst pressure, the allowable working pressure can be obtained through conversion:

$$p_0 = \frac{D_1 p_e}{nD} \qquad (8)$$

Among them:

p_0—— Initial working pressure of airbag (MPa);
p_e—— measured blasting pressure (MPa);

D_1—— diameter of airbag used in blasting test (m);
D—— diameter of working airbag;
N—— Safety factor.

3.3 Safety Control of Airbags

Due to the softness of the airbag itself, when the pressure difference between the inside and outside reaches a certain threshold (the lower limit of the pressure difference between the inside and outside of the airbag), the airbag can provide maximum buoyancy. At this time, the hardness of the outer surface of the airbag is "enhanced" (similar to car tires). As the lifting depth increases, the external pressure will gradually decrease. When the pressure difference between the inside and outside reaches another threshold (the upper limit of the pressure difference between the inside and outside of the airbag), the airbag will tear. In actual salvage projects, it is necessary to choose an appropriate strength airbag based on the water depth. However, due to folding, friction, and other phenomena during transportation, deployment, and other operations, the strength of the airbag will decrease to varying degrees. Therefore, airbags should be equipped with pressure relief valves, and the number of pressure relief valves should be equipped according to the size of the airbags, lifting speed, etc.

3.4 Notes on Airbag Layout Plan

For the internal buoyancy salvage method, priority should be given to filling the airbags into the following types of compartments, including those that cannot restore internal buoyancy and those that are crucial for the buoyancy of wrecks. At the same time, attention should be paid to various factors such as the selection and quantity of airbags, the selection of sinking routes, the portability of inflation, the arrangement of inflation hoses, the selection of airbag fixing points, the verification of force points in the cabin, the anti-puncture treatment of airbags, and the recovery of airbags.

For the external buoyancy salvage method, the focus is on factors such as the placement of airbags, fixation methods, rooting points, stability of airbags, and the coordination between anti puncture treatment and the main lifting method.

3.5 Airbag Fixation and Sinking Control

Due to the irregular shape of rubber airbags before reaching the lower limit of internal and external pressure difference, it is difficult to control their shape during the sinking process, especially when filled in the cabin to provide internal buoyancy, making it difficult for divers to operate underwater.

When using the buoyancy salvage method inside the airbag for salvage operations, it is necessary to select a suitable position to lower the airbag [15], and multiple guides need to be welded in the cabin. Divers use the guides and ship equipment to tow the airbag to the designated position and fix it. Due to the fact that the airbag will float up and be located inside the cabin after inflation, it is necessary to clean the quick opening at the contact position between the airbag and the cabin in advance, and calculate the ship's strength at the contact position in advance. After sinking, connect the inflatable leather dragon in advance and mark it in order, extending it to the air distribution box.

For the external buoyancy salvage method, nets, steel cages [9], steel pipes, etc. are usually chosen to fix the airbags. As shown in Figs. 9. For small airbags that only require one rooting point, the airbag can be directly connected to the rooting point. External airbags are easier to sink compared to internal airbags. Use a crane to lift the airbag to the designated position, try to lay it flat, and the diver can fix it underwater. Similarly, it is necessary to place the inflatable leather dragon in advance.

3.6 Stability Control of Airbag Inflation

For external rubber airbags with multiple force points, the inflation process is the most unstable state. If not well controlled, the gas will cause one end of the airbag to be high and the other end to be low, resulting in excessive local force. In severe cases, it can cause the rubber airbag to fold and damage, leading to salvage failure. Therefore, before carrying out salvage operations, the external airbag should undergo corresponding inflation stability tests. If there is poor stability during the inflation process, one or more control methods can be combined to control the unstable rubber airbag during inflation, such as encrypted mesh bags, encrypted steel structure cages, and temporary rope fastening.

4 Opportunities and Challenges of Airbag Salvage Technology

The commonly used methods for the overall salvage of large wrecks and objects include float bowl salvage, floating crane salvage, etc. Currently, almost all pontoons used in the float bowl method are rigid pontoons made of hard materials. The rigid buoy has a large lifting capacity, but it is easily affected by the underwater environment when diving and being tied to wrecks. In addition, the buoy occupies a large space and has high storage and transportation costs. Large floating cranes are the main tool for salvaging wrecks and objects, but they are often limited by the lifting capacity, operating space, and high transportation costs of cranes. The produced airbag is flexible and versatile, and can be folded or rolled into a cylindrical shape for storage and transportation or diving, greatly improving the salvage ability. When in use, sink the rubber airbag into the seabed, fix it with the salvaged object, and then inflate it into the airbag to float it up, dragging the wreck and object out of the water. Airbags can be inserted into the flooded cabin or fixed to the deck of a wreck, exerting a small force on the unit area of the hull and providing strong support for ship safety. The impact of hydrological conditions on airbag diving is relatively small, and the efficiency of underwater operations is high. Airbags not only provide buoyancy for ship salvage, but also have great advantages in rescuing stranded ships. By means of excavation and other means, airbags can be inserted into the bottom of stranded ships. Inflation of airbags can lift the ship up, and under the towing effect of tugboats, the ship can smoothly resurface [16].

However, due to factors such as material strength and difficulty in lifting control, airbags cannot be scaled up, resulting in smaller drainage volume and lower lifting capacity. At the same time, in order to better control rubber airbags underwater, a method

(a) Steel structure cage fixing method. [9]

(b) Fixing method of net bag sling.

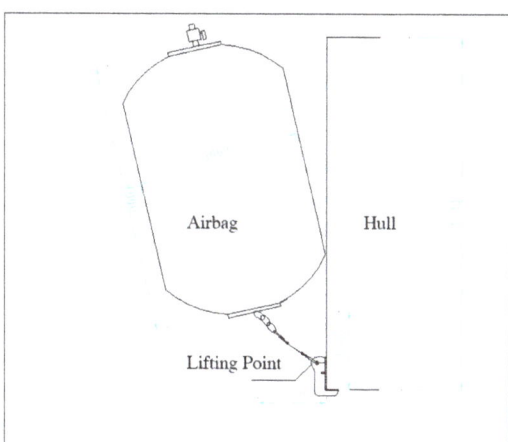
(c) Single root fixation method.

Fig. 9. Air Bag Fixation.

of "breaking them into smaller ones" is usually adopted to divide a large rubber airbag into multiple small rubber airbags, but it also further reduces the buoyancy of the airbag. Therefore, airbags are usually used as auxiliary buoyancy devices, and if they want to be applied on a larger scale in the field of salvage, further issues of material strength and operability need to be addressed.

5 Conclusion

The technology of using airbags to salvage wrecks and objects emerged earlier due to its low cost, low consumption, pollution-free, high efficiency, flexible mobility, and simple operation. However, due to material limitations, it cannot become the mainstream salvage method.

This article introduces the development history of airbag salvage technology for wrecks and sunken objects, and introduces the application scenarios of airbag salvage through two cases: Internal buoyancy salvage and external buoyancy salvage. This article outlines the basic principles and technical characteristics of airbag salvage, studies the construction process of airbag salvage, summarizes the key technologies of airbag salvage for wrecks and sunken objects, and points out the opportunities and challenges of airbag salvage in the future.

With the advancement of technology, airbag materials are constantly being updated and replaced. From natural rubber to butadiene styrene rubber, butyl rubber, chloroprene rubber [17], chlorosulfonated polyethylene, polyvinyl chloride, as well as thermoplastic polyurethane materials [18] and ultra-high molecular weight polyethylene [19], the strength, airtightness, and wear resistance have been greatly improved. Now it is possible to produce airbags with larger drainage volumes, and according to different manufacturing processes, airbags can be divided into multiple compartments to better control their stability. With the continuous improvement of performance, airbags will have a wider range of applications in the field of rescue and other fields in the future.

References

1. Yaohui, M., Yuantao, B.: Review of the search and salvage engineering of the 5.7 air crash aircraft [C]. In: The 11th China Coastal Engineering Academic Symposium and the 2003 Cross Strait Port and Coastal Development Seminar. China Ocean Engineering Society (2003)
2. Zuwen, W.: Current status and development of rescue and salvage equipment. J. Mech. Eng. **49**(20), 91–100 (2013)
3. Yongjun, G., Zengmeng, Z.: Salvage Engineering. Dalian Maritime University Press, Dalian (2012)
4. Peichao, Z.: Discussion on safety of airbag launching and launching in ships. Tianjin Navig. **4**, 52–55 (2023)
5. Li Xuguang, F., Sujuan, L.J.: Application research of high-pressure airbag handling technology in building translation. Build. Struct. **21**, 5 (2014)
6. Qiang, Z.: Application of pipeline sealing air bag in pipeline maintenance. Water Conserv. Constr. Manag. **9**, 3 (2015). https://doi.org/10.16616/j.cnki.11-4446/TV.2015.09.024
7. Guichun, L., Zhengjie, L.: Airbags and Air Cushions of Hovercraft. National Defense Industry Press (2011)

8. Jun, L., Juxiang, S., Wei, W.: The international standard boosts the launch technology of airbags towards the world. In: Proceedings of the 16th Annual Conference of the Chinese Association for Science and Technology (2014)
9. Peng, Y., Weiping, W., Yan, J., et al.: Design scheme of "Shiyue" fishing flexible airbag. Navigation. **4**, 35–37 (2017)
10. Jianhua, L.: Analysis on key points of longitudinal inclined berth design using air bag up-discharge and launching technology. Eng. Constr. **33**(02), 194–196 (2019)
11. Bing, C.: Research and application of self-regulating water sinking salvage device. Mech. Eng. Autom. **03**, 95–96 (2010)
12. Lin, L., Yihu, T.: Design of side inflatable airbag device. J. Chongqing Inst. Technol. (Nat. Sci. Ed.). **23**(07), 58–63 (2009)
13. Wenjun, W., Yong, L., Wei, Y.: Estimation of airship airbag pressure and skin tension. J. Astronaut. **05**, 1109–1112 (2007)
14. Dayong, S.: Research on the Damage Mechanism and Collision Prevention Performance of Ship Launching Airbags. Wuhan University of Technology (2021). https://doi.org/10.27381/dcnki.gwlgu.2021.000178
15. Mingyang, Z., Guojun, Y.: Sun Longquan research on the dynamic characteristics of water recovery of aircraft with airbag structure. J. Mech. **56**(04), 943–959 (2024)
16. Peng, Y., Youjun, W., Yan, J., et al.: The application of rubber lifting airbags in underwater engineering China water. Transport. **7**, 22–24 (2017)
17. Shiyang, G., Yuan Xiaocheng, A.: Manufacturing technology and application of large scale aging resistant offshore Buoyant airbags by mold pressing method. Util. Rubber Plast. Resour. **3**, 1–6 (2019)
18. Peng, L.: TPU adhesive tape and its application in inflatable capsule materials. Polyurethane Ind. **4**, 32–35 (2006)
19. Zhe, J., Ziyan, H.: Design and calculation of a new type of large tonnage wreck with built-in double-layer airbags for salvage. Navigation. **6**, 36–41 (2022)

Open Access This chapter is licensed under the terms of the Creative Commons Attribution-NonCommercial-NoDerivatives 4.0 International License (http://creativecommons.org/licenses/by-nc-nd/4.0/), which permits any noncommercial use, sharing, distribution and reproduction in any medium or format, as long as you give appropriate credit to the original author(s) and the source, provide a link to the Creative Commons license and indicate if you modified the licensed material. You do not have permission under this license to share adapted material derived from this chapter or parts of it.

The images or other third party material in this chapter are included in the chapter's Creative Commons license, unless indicated otherwise in a credit line to the material. If material is not included in the chapter's Creative Commons license and your intended use is not permitted by statutory regulation or exceeds the permitted use, you will need to obtain permission directly from the copyright holder.

Design and Optimization of Righting Hooks for Wreck Righting

Fei Wang, Shude Chen, Yijian Han, Jingjing Liu, and Wei Zhang(✉)

China Yantai Salvage Technology Center, Yantai 264012, China
donsh328@163.com

Abstract. In order to solve the problem of setting the hanging point (root point) of the wire rope during the process of righting sunken ship. The underwater non-welded rooting device (hook for righting) studied in this paper can be used to guide the operation of righting sunken ship in the process of salvage. This paper completed the modelling and structural analysis based on Abaqus. And this equipment has been used in the actual wreck salvage project. In order to improve the adaptation of this equipment to the sunken ship, the optimization design of the equipment is further studied in this paper. The results show that the optimized design of the hook for righting has stronger adaptability to the structure of the sunken ship. And it's easier to install and use underwater. The results of calculation and analysis show that the hook for righting meets the design expectation and the design requirement. The research results can provide some reference for the operation of righting capsized ship in the salvage.

Keywords: Salvage · Righting capsized ship · Hook for righting · Design and optimization

1 Introduction

A ship at sea or at anchor may sink due to a variety of causes. Wrecks with salvage value or obstructing navigation should be salvaged and removed in a timely manner. During the sinking process, the attitude of the ship will change, resulting in the state of turning over or reversing after sinking [1]. For wrecks that are not normally resting on the bottom of the sea, the key process in the salvage operation is to straighten the shipwreck first, which is called wreck righting [2]. The righting process mainly includes a single floating crane or a number of floating cranes jointly lifting the low side of the wreck to straighten the wreck, the pontoon connecting the bottom side of the wreck and the barge assisting to straighten the wreck [3]. The operation of wreck righting requires the use of wire rope to connect the shipwreck structure with strong strength, and then the use of floating crane and barge to provide force. If the hull of the ship is exposed above the water after the overturn, the lifting lugs can be welded at the exposed position as a wire rope connection point, and then the wreck can be righted using a floating crane [4]. Among them, the lifting lugs and other structures used to connect the steel wire are called "rooting points" in the salvage. When the wreck is completely submerged under water, the strength of the lifting lugs on the wreck cannot meet the requirements of the righting force due to the limitations of the underwater welding process and the external forces such as the soil adsorption force. For this reason, salvage engineers need to design special processes to straighten the wreck. For example, in the "Neftegaz 67" salvage project [5], steel wire ropes were used to wrap many circles around the wreck, and the floating crane adjusted the force of several hooks to straighten the wreck. This process was difficult to operate and the operation risk was high. This process is difficult to operate and has high operational risk. In addition, it is also a method of operation to open holes in the hull where the ribs cross the longitudinal, use a chain to cover the structure and then connect the wire rope. However, this process affects the self-floating of wrecks after they come out of the water. In order to solve the problem of underwater wreck righting, engineers at home and abroad have designed the "righting hook", which is used to get stuck on the edge of the wreck deck and connect the righting wire rope for wreck righting operation.

2 Engineering Background

2.1 Routine "Rooting Point"

In the process of wreck salvage, wreck righting is a routine operation. The existing technology is to weld "rooting points" on the high side of the wreck, and the commonly used rooting point structures include righting piles or lifting lugs (As shown in Fig. 1). The wire rope is connected to the "rooting point" from the other side of the wreck around the bottom of the ship, and the main hook of the floating crane is connected to the steel wire rope to straighten the wreck and restore the wreck to normal.

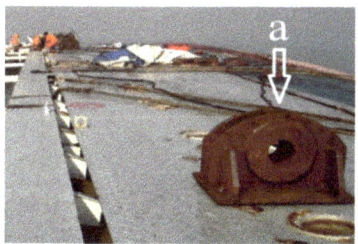
(a) Lifting lugs

(b) Righting piles

Fig. 1. Routine "Rooting Point"

2.2 Righting Hook

The rooting point welding process on underwater wrecks is immature and low efficiency, so salvage engineers designed a "righting hook" to be hung on the side of wrecks to replace the rooting point. This equipment is only suitable for the field of offshore construction. The design concept and shape of "righting hook" are similar to that of "sliding tail hook" used in offshore wind power construction [6]. The shape of the hook is L-shaped, with the short side hooked to the edge of the deck of the wreck, and the long side attached to the side of the ship and welded with piles or lifting lugs for hanging wire ropes. The righting hook can be used for the righting operation of wrecks that are completely submerged, and can also be used for wrecks whose sides are above the water. Righting hook and Sliding tail hook are shown in Fig. 2.

(a) Righting hook (b) Sliding tail hook

Fig. 2. Righting hook and Sliding tail hook in using

2.3 Application Engineering Background of Righting Hook

A salvage project of a ship in the northern waters of the Bohai Sea, which overturned and sank on the sea bed. Due to the weak structure of the ship, if the wreck is kept on its side and the floating crane is used to lift it out of the sea, it can be judged from previous salvage experience that the lifting rope will be squeezed into the hull, and even "cut" the wreck in serious cases. For this reason, it was decided to correct the wreck first and then carry out the work of lifting the wreck to ensure that the wreck could float safely. The process of sinking the wreck is shown in Fig. 3. The ship adopts the segmented salvage operation process, and the relevant parameters of the whole salvage part measured by divers under water are shown in Table 1, and the corresponding section parameters of the wreck after floating are shown in Fig. 4.

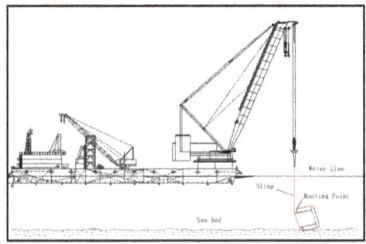 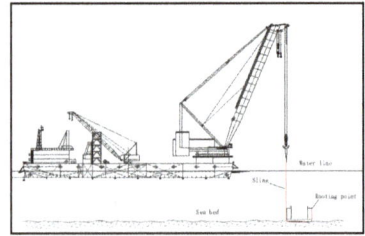

(a) Before the wreck was righted (b) After the wreck was righted

Fig. 3. Diagram of the righting process

Table 1. Wreck structure parameter

NO.	Item	Parameter
1	Depth	3000 mm
2	Double bottom height	300 mm
3	The distance from the bilge to the side cabin bracing	1200 mm
4	Side cabin width	1000 mm
5	Strong rib	B200mm × b80mm × d6 mm
6	Reinforcing material	B120mm × b60mm × d6 mm
7	Crossbeam	B50mm × b50mm × d6 mm
8	Deck strong beam	B120mm × b60mm × d6 mm
9	Outer plate thickness	6 mm
10	frame space	800 mm

Fig. 4. Ship section structure and plate thickness

3 Design and Application of Righting Hook

3.1 Design of Righting Hook

The righting hook is a rooting device that connects the righting wire rope when the wreck is rolling. Multiple rooting devices can be set according to the tilting state of the wreck and the pre-calculation situation, and then connect multiple groups of the wire rope, so as to achieve the effect of effectively righting the wreck.

The rooting device used for righting the wreck's roll state mainly includes righting pile, web plate, panel and elbow plate. The web and the panel are rectangular steel plates, and the panel is welded vertically on the bottom of the web and welded with elbow plates for strengthening and fixing. The righting pile is welded and fixed on the top surface of the web and welded with elbow plate for strengthening and fixing. The righting pile is welded with cross-shaped strengthening plate and the top of the pile is welded with sealing plate. The sealing plate is in the shape of water drop profile and the size is larger than the outer diameter of the straightener pile, which is used to prevent the wire rope from jumping out. When used, the rooting device panel is attached to the top deck of the sinking side, the web is attached to the top of the sinking side, one end of the straightening wire rope is hung on the main hook of the floating crane, and the other end is connected to the straightening pile around the wreck, the specific structure and application method are shown in the following figure. When used, the panel is attached to the top deck of the sinking side, the web is attached to the top of the sinking side, and one end of the steel wire rope is hung on the main hook of the floating crane and the other end is connected to the bending pile around the wreck, the specific structure and application method are shown in the Fig. 5.

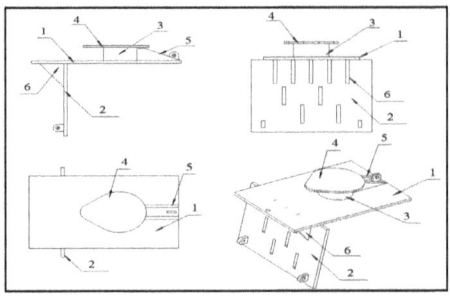

1-Web, 2-Panel, 3- Righting pile, 4- Seal plate, 5- The bracket of the righting pile, 6- The face of the righting pile

Fig. 5. Righting hook structure and use diagram Monitoring point

3.2 Structural Analysis

Calculation Model. This righting device is used for a salvage project of a ship in the northern waters of Bohai Sea. According to calculation, in addition to the fixed rooting point on board, 150 T extra straightening force is required for righting the ship, so two righting hooks are used in this design

The structural analysis calculation is based on Abaqus software, combined with the ship's structural size design and structural analysis. According to DNVGL-OS-H205 specification, the calculation needs to consider the design safety factor:

$$\gamma_{design} = \gamma_f \gamma_c \tag{1}$$

where: γ_{design} —design coefficient;

γ_f —— loading coefficient, value 1.3;

γ_c —— Coefficient of most importance, value 1.3.

γ_{design} value is 1.7 by calculation, the force of a single righting hook is 75 T (the total righting force is 150 T, and the force of two points is designed). After considering the safety factor, the load of the righting device is 75 T × 1.7 = 130 T.

Combined with the size of righting hook and the underwater measurement parameters of the wreck, the modeling of the righting hook and the hull section is carried out based on Abaqus software, as shown in Fig. 6.

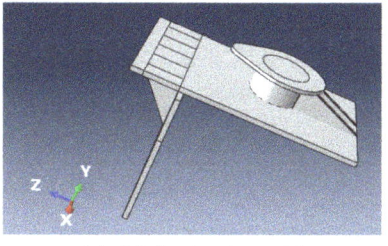 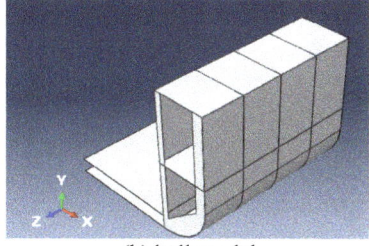

(a) righting hook model   (b) hull model

Fig. 6. The model of righting hook and hull

The righting hook is assembled with the hull segment. One righting hook is intended to be used in this hull segment, and the righting hook is arranged at the middle rib position of the hull segment. The upper plate of the hook and the upper deck surface of the hull segment are set as contact, the contact position between the side plate of the hook and the outer surface of the side plate of the hull segment are set as contact too. In practice, the side cylinder of the hook is subjected to the tension of the wire rope. For this reason, in the Abaqus model diagram, the side cylinder is coupled to its center, the tension of the wire rope is set to 130 T, and the two ends of the hull segment are constrained with full freedom.

Structural Analysis Calculation. According to API RP 2A-WSD, ASIC Steel structure design specifications and CCS offshore mobile platform classification specifications, the yield strength of the floating tank body and the reinforced area is checked, and the formula is as follows

$$\sigma_{eq} \leq [\sigma] \tag{2}$$

$$[\sigma] = \frac{\sigma_s}{S} \tag{3}$$

Where: σ ——permissible stress;
σ_{eq}——equivalent stress;
σ_s ——yield stress.
Where:

$$\sigma_{eq} = \sqrt{\sigma_x^2 + \sigma_y^2 - \sigma_x\sigma_y + 3\tau_{xy}^2} \tag{4}$$

σ_x is the stress in the x direction of the element, N/mm^2;
σ_y is the stress in the y direction of the element, N/mm^2;
τ_{xy} is the shear stress of the element in the xy plane, N/mm^2;
S is safety factor, and its value is shown in Table 2.

Table 2. Equivalent stress safety factor

Static load condition	Combined condition
1.43	1.11

The overall structure material of the righting hook is Q355B steel, and the structure material of the hull segment is Q235B steel. The allowable stress [σ] of Q235B and Q355B steel is 164.3 MPa and 248.3 MPa, respectively.

The structural analysis was carried out based on Abaqus software, and the stress nephogram of the righting hook and hull stress calculation is shown in Fig. 7. It can be seen that the maximum stress of the hull and the hook is 139 Mpa and 82 Mpa, both of which are less than the allowable stress, and meet the requirements of use.

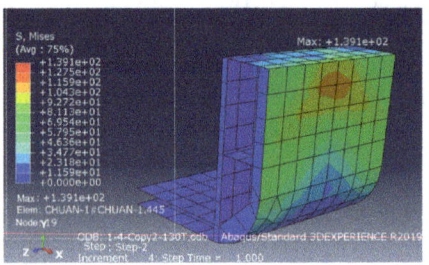
(a) Stress nephogram of hull

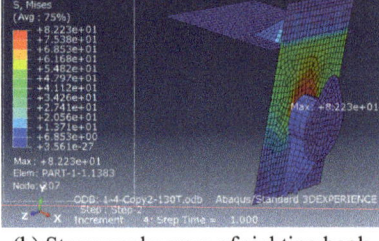
(b) Stress nephogram of righting hook

Fig. 7. Stress nephogram

3.3 Design of Righting Hook

The righting hook is machined according to the design and installed underwater by divers at the designed location of the wreck. To prevent falling off, the panels are slotted for installation and secured to the wreck deck using bent hook bolts, as shown in Fig. 8. After the completion of the righting operation, divers help to recover the leveling device, and operate the floating crane rope to float the wreck.

(a) The righting hook

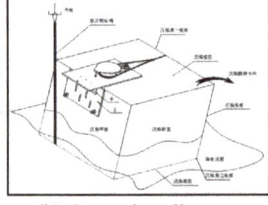
(b) Operation diagram

(c) The wreck righted

Fig. 8. Righting device and its application

4 Improvement and Optimization

4.1 Existing Problems

From the on-site usage situation, the main problems with this righting hook and rooting device are as follows: (1) Panel: The design of the righting device panel is L1000 mm × B800 mm. Due to the wreck being underwater for a long time, the deck is full of marine life, and there is a temporary welding structure near the design installation location. Before the installation of the righting and rooting device, divers carried out a lot of cleaning work, which affected the efficiency of the operation. (2) Web plate: In order to adapt to the welding of panels and panel brackets, the design of the web plate L1260mm × B800 mm (the 800 mm edge of the web plate is welded and fixed to the 1000 mm edge of the panel) also poses underwater installation problems. (3) Angle: There are no detailed drawings for the wreck, and the rib positions and ship plate thickness were measured underwater by divers, with significant differences in accuracy. The righting and rooting device is processed before the ship sets sail; the panels and belly plates are fixed and welded at a 90 ° angle. However, the angle between the ship deck and the hull is greater than 90 °, resulting in poor adhesion between the righting and rooting device and the ship deck, which is not conducive to structural stress.

4.2 Optimization Design

Taking into account the diversity of wreck, the design of the righting and rooting device is divided into two parts: The deck fitting seat and the righting hook. Design concept:

(1) The deck fitting seat is connected to the righting hook. The contact position of the righting hook is a semi-circular groove, and the contact position between the deck fitting base and the righting hook is a circular axis. (2) The righting hook is connected to the righting steel wire rope using a shackle. Reserve a pin hole at the bottom of the righting hook for hanging and disengaging the buckle, and the pin hole diameter matches the disengaging pin hole. (3) Force angle. The righting steel wire rope can be applied at an angle of 5 ° with the axis of the righting hook. Design a righting and rooting device based on the above ideas, as shown in Fig. 9.

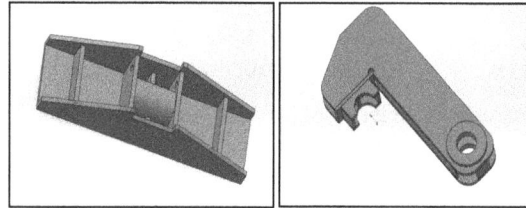
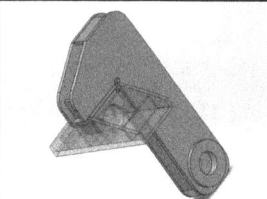

Fig. 9. Design diagram of base and righting hook

4.3 Structural Analysis

Based on the optimized dimensions of the righting hook, a new type of righting hook modelling was carried out using Abaqus software. The force verification of the righting hook was carried out using a standard bulk cargo vessel hull as an example.

Assemble the new type of righting hook with the bulk cargo vessel, arrange the new type of righting hook at the strong rib position of the bulk cargo vessel, and make contact with the upper surface of the deck of the bulk cargo vessel. Couple the lifting lugs on both sides of the new type of righting hook to the center point, and add steel wire rope tension at the coupling point. Referring to conventional salvage operations, taking a 100 T pulling force as an example, the pulling force is set to 100 T × γ design = 170 T. Perform structural analysis based on the above settings. The contact setting between the righting hook and the hull is shown in Fig. 10.

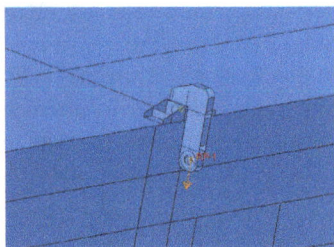 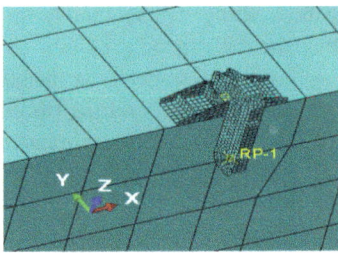

Fig. 10. Setting of contact between the righting hook and the hull

The overall structural material of the new type righting hook is planned to be Q355B, and the material of the segmented structure of the hull is Q235B. This calculation takes values based on static load conditions, and the stress calculation results of the new type righting hook and hull segmentation are shown in Fig. 11.

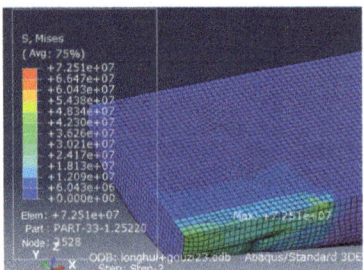 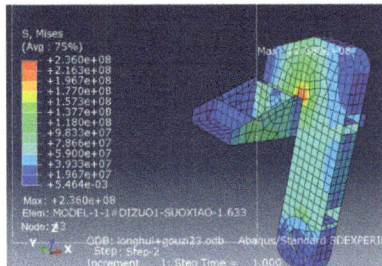

(a) Stress cloud map of vessel   (b) Stress cloud map of righting hook

Fig. 11. Stress cloud map for the analysis of the new type of righting hook structure

5 Conclusion

In summary, it was found through finite element software calculation that the designed righting hook meets the strength requirements under the set loading and constraints, and the segmented structure strength of the ship also meets the requirements. Moreover, through actual engineering, it was found that the designed righting hook had no abnormalities in the relevant structure during use. Therefore, this design is an effective one and provides important reference value for related fields such as wreck righting in the future. The application of finite element software during this period provides accurate and important reference value for practical engineering.

References

1. Pan, D., Lin, C., Sun, D.: Calculation and analysis of capsized ship straightening based on GHS software. J. Water Conserv. Transp. Eng. **6**, 78–83 (2014)
2. Pan, D., Lin, C., Zhou, Z.: Numerical simulation of righting process of damaged and capsized hull. J. Traffic Transp. Eng. **17**(5), 102–112 (2017)
3. Geng, J.Y.: Technical analysis of salvage engineering of costa concorde cruise ship. Technol Innov. **7**, 162–164+167 (2023)
4. Zhang, Y.: Numerical simulation and analysis of lifting and floating process of "Xia Chang" ship. Mar. Eng. **46**(6), 177–181 (2017)
5. Sun, Z.C., JianCheng, W., Zhang, Y.: Adopt the "Hua Tianlong" salvage "Neftegaz 67" vessel. In: The 5th China International Salvage Forum, pp. 166–170, Chengdu, Sichuan (2008)
6. Tan, L.: Structural improvement and strength analysis of tilting fixture for single pile foundation. Mar. Eng. **51**(6), 97–101 (2022)

Open Access This chapter is licensed under the terms of the Creative Commons Attribution-NonCommercial-NoDerivatives 4.0 International License (http://creativecommons.org/licenses/by-nc-nd/4.0/), which permits any noncommercial use, sharing, distribution and reproduction in any medium or format, as long as you give appropriate credit to the original author(s) and the source, provide a link to the Creative Commons license and indicate if you modified the licensed material. You do not have permission under this license to share adapted material derived from this chapter or parts of it.

The images or other third party material in this chapter are included in the chapter's Creative Commons license, unless indicated otherwise in a credit line to the material. If material is not included in the chapter's Creative Commons license and your intended use is not permitted by statutory regulation or exceeds the permitted use, you will need to obtain permission directly from the copyright holder.

Two-Stage Energy Management Strategy for Accommodation Area of Multi-Energy Ship

Lihao Wang, Jiahui Xue, and Can Cui(✉)

School of Engineering, Ocean University of China, Qingdao, China
cuican@ouc.edu.cn

Abstract. Maritime transportation industry is one of the major contributors to CO2 emissions. Energy conservation and emissions reduction in this industry are crucial for mitigating global warming. In recent years, the proportion of multi-energy ships in maritime transportation has been increasing, making energy management of multi-energy ship a hot research area. However, most scholars focus on energy management in propulsion systems of multi-energy ship, neglecting energy management in accommodation area. Therefore, we propose a two-stage energy management strategy for accommodation area of multi-energy ship. The proposed energy management strategy based on reinforcement learning and intelligent optimization algorithm to achieve rational scheduling of electrical appliances in ship accommodation area. To validate the effectiveness of the proposed method, we designed and conducted simulation experiments. The results indicate that the energy management strategy can effectively reduce energy consumption in multi-energy ship accommodation area, decrease pollutant emissions, and lower operating costs while meeting the needs of the crew.

Keywords: Two-Stage Energy Management Strategy · Reinforcement Learning · Intelligent Optimization Algorithm · Multi-Energy Ship · Ship Accommodation Area

1 Introduction

Maritime transportation as a crucial component of global trade, accounts for over 80% of the total freight volume worldwide [1]. However, fossil fuels remain the primary source of energy for most ships, and the ever-increasing demand for maritime transportation has made the industry became one of the major contributors to greenhouse gas emissions [2]. To address the environmental threats posed by the maritime transportation industry, countries are vigorously developing green maritime transportation [3]. The proportion of multi-energy ships in the maritime transportation sector are continue to increase, currently, ship energy management has emerged as a hot research area [4].

Nevertheless, most scholars placed their research emphasis on energy management strategy in the propulsion systems of multi-energy ships [5], overlooking energy management strategy in ship accommodation area. Ship accommodation area plays a significant role in the health, comfort and work efficiency of crew. Meanwhile, ship accommodation area is also a significant source of energy consumption for ship.

Therefore, we propose two-stage energy management strategy for accommodation area of multi-power ship (TS-EMS) to reduce energy consumption while ensuring the needs of crews are met.

2 Problem Formulation

The purpose of this study is to schedule the power of appliances for the next day based on the weather data for the following day. In this paper, the researched ship is powered by a combination of diesel generator, photovoltaic (PV) panels and energy storage battery (ESB). In addition, PV is the only source of the renewable system on the researched ship. The maximum power supplied by PV is 2 kW, if the power demand of total appliances exceeds 2 kW, the diesel generator will start supplying power. The ESB is utilized to store the electricity generated by PV, enabling power supply during night operations. The maximum battery capacity of the ESB is 25 kWh. The data of solar radiation intensity and temperature are from [6].

Due to the nonlinear and strongly coupled nature of heating ventilation and air conditioning (HVAC), establishing an accurate physical model for HVAC is indeed quite challenging [7]. On the other hand, the physical models of other appliances considered in this paper, such as washer and dryers, water heaters and so on, exhibit a high degree of linearity. Therefore, simple linear models can be used to describe them. As a result, we categorize appliances into two main groups: HVAC and non-HVAC. Figure 1 presents the framework the proposed energy management strategy for this paper.

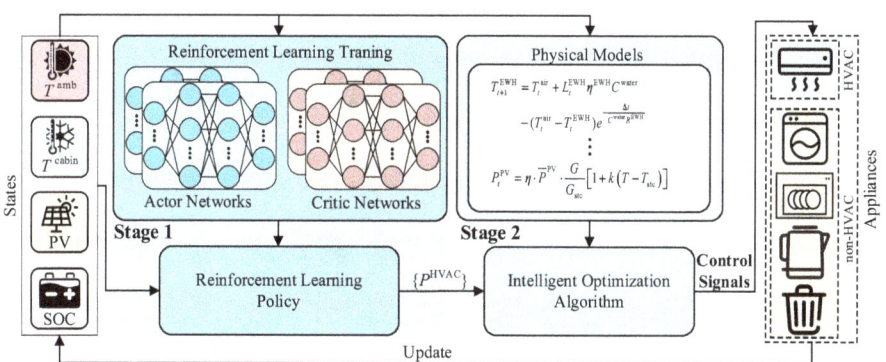

Fig. 1. Proposed TS-EMS framework.

2.1 Reinforcement Learning Stage for HVAC

According to [7], reinforcement learning (RL) is an effective method for addressing the challenges of HVAC modelling. RL offers advantages such as adaptability to complex environments, handling nonlinear relationships, dynamic decision-making, and performance optimization through learning. Hence, it can effectively address the challenges encountered in HVAC system modelling.

In this paper, we employ reinforcement learning method to address the challenges in HVAC modelling and train the agent using real-world data from [6]. This enables the agent to schedule HVAC power based on external environmental conditions.

The elements of reinforcement learning are defined as follows.

Design of State. At each time step t, the agent needs to perceive the external environment it is situated in, therefore the state s_t is designed as follows:

$$s_t = \left[T_t^{\text{amb}}, T_t^{\text{cabin}}, PV_t, S_t \right] \tag{1}$$

where T_t^{amb} is the ambient temperature at time step t; T_t^{cabin} is the cabin temperature at time step t; PV_t is the total photovoltaic power at time step t; S_t is the battery storage at time step t.

Design of Action. At each time step t, the agent needs to choose action to maintain the temperature within the comfort range in the cabin while minimizing energy consumption as much as possible, therefore the action a_t is designed as follows:

$$a_t = \left[P_t^{\text{HVAC}} \right] \tag{2}$$

where P_t^{HVAC} is the HVAC real time power at time step t. The value of P_t^{HVAC} is between [0, 2] kW.

Design of Reward. At each time step t, the HVAC need to ensure comfort while reducing energy consumption, therefore the reward r_t is composed of two parts:

$$r_t^1 = -P_t^{\text{HVAC}} \tag{3}$$

$$r_t^2 = \begin{cases} T^{\text{ref}} - T_t^{\text{cabin}}, & T_t^{\text{cabin}} > T^{\text{ref}} + \varepsilon \\ T_t^{\text{cabin}} - T^{\text{ref}}, & T_t^{\text{cabin}} < T^{\text{ref}} - \varepsilon \\ \sigma, & \text{else} \end{cases} \tag{4}$$

$$r_t = r_t^1 + \alpha \cdot r_t^2 \tag{5}$$

where r_t^1 is a negative number representing the penalty for energy consumption at time step t; r_t^2 is a reward for the degree of comfort at time step t; T^{ref} is the comfortable temperature; ε is a small positive number representing the range of comfortable temperature; σ is a positive number representing the reward for keeping comfortable temperature; r_t is the reward the agent gets at time step t, and α is the weight between r_t^1 and r_t^2.

2.2 Intelligent Method Stage for Non-HVAC Appliances

According to [7], the non-HVAC of this paper can be modelled as linear models. The models are as follows:

$$P_t^{\text{non-HVAC}} = \theta_t^{\text{non-HVAC}} \cdot P^{\text{non-HVAC}} \tag{6}$$

$$0 \leq P_t^{\text{non-HVAC}} \leq P_{max}^{\text{non-HVAC}} \qquad (7)$$

where $P^{\text{non-HVAC}}$ is power of the non-HVAC appliance, including washer and dryer, water heater, disinfection cabinet and wastewater treatment plant; $\theta_t^{\text{non-HVAC}}$ is a Boolean number, and the value is 1 when the appliance is running and 0 when it is turned off; $P_t^{\text{non-HVAC}}$ is power of the appliance at time step t; $P_{max}^{\text{non-HVAC}}$ is the maximum operating power of the appliance respectively. The data of reinforcement learning are shown in the Table 1. The comfort temperature range is determined by [8].

Table 1. The Data of Reinforcement Learning.

Parameter	Value	Parameter	Value
T^{ref}	22 °C	P_{max}^{DC} b	0.5 kW
ε	1 °C	P_{max}^{WT} c	1 kW
P_{max}^{WD} a	1 kW	P_{max}^{WH} d	1 kW

a P_{max}^{WD} is the power of washer and dryer

b P_{max}^{DC} is the power of disinfection cabinet

c P_{max}^{WT} is the power of wastewater treatment plant

d P_{max}^{WH} is the power of water heater

In this paper, we transform the non-HVAC appliances scheduling issue into an optimization problem and address it using intelligent optimization algorithm.

The objective of this optimization problem is to schedule the power of non-HVAC appliances in a manner that minimizes the peak-to-average ratio (PAR) of the renewable power system, while ensuring SOC of the battery remains above 30%. Additionally, it aims to ensure that the total power consumption of all appliances is less than or equal to the maximum power (P_{max}^{Grid}) that the renewable power (i.e., PV) system can supply. In cases where the total power consumption exceeds P_{max}^{Grid}, diesel generators would need to be utilized, resulting in increased CO_2 emissions.

Therefore, the objective function of this optimization problem is defined as follows:

$$\min F = f^1 + \beta_1 \cdot \sum_t^T f_t^2 + \beta_2 \cdot \sum_t^T f_t^3 \qquad (8)$$

$$f^1 = \max\left(P_t^{\text{all}}\right) / \text{mean}\left(P_t^{\text{all}}\right) \qquad (9)$$

$$f_t^2 = \begin{cases} 0.3 \cdot S_{\max} - S_t, & S_t < 0.3 \cdot S_{\max} \\ 0, & \text{else} \end{cases} \qquad (10)$$

$$f_t^3 = \begin{cases} P_t^{\text{all}}, & P_t^{\text{all}} > P_{max}^{\text{Grid}} \\ 0, & \text{else} \end{cases} \qquad (11)$$

where f^1 represents the value of PAR; $P_t^{\text{all}} = \sum_{i=1}^N P_t^i$ is the total electricity consumption of all appliances (including HVAC) at time step t, P_t^i is the power of i-th appliances at time step t, and N is the number of appliances; f_t^2 represents the extent to which SOC falls below 30% of its maximum capacity at time step t; S_{max} is the maximum capacity of the battery; S_t is the battery storage at time step t; f_t^3 represents the extent to which the total power consumption of all appliances exceeds the maximum power that the renewable power system can supply; P_{max}^{Grid} is the maximum power that the renewable power system can supply; β_1 and β_2 are the weights of the objective function; T is the maximum time step.

3 Two-Stage Energy Management Strategy for Accommodation Area of Multi-Energy Ship

We propose TS-EMS to accomplish the task of this paper; the method involves two stages. 1) Initially, a trained Deep Deterministic Policy Gradient (DDPG) reinforcement learning agent is employed to schedule HVAC power for the next 24 h based on weather forecasts, solar radiation intensity and other relevant information. The scheduling is conducted at 15-minute intervals, resulting in a total of 96 time points for HVAC power allocation. 2) Subsequently, utilizing the allocated HVAC power, a genetic algorithm (GA) is employed to schedule the power of remaining appliances while ensuring that SOC remains above 30%, the total power consumption of all appliances is less than or equal to P_{max}^{Grid}, and minimizing PAR as much as possible.

The workflow of TS-EMS is illustrated in Fig. 2.

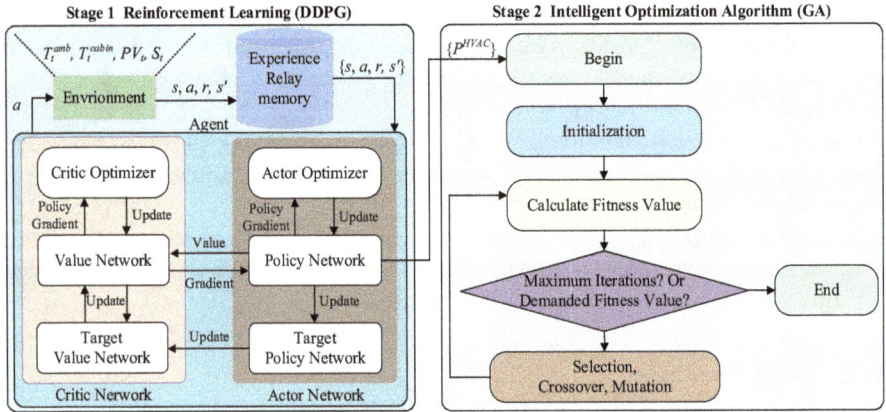

Fig. 2. The workflow of TS-EMS.

4 Simulation Experiment

4.1 Simulation Experiment Setup

To better train the RL agent, we extracted real-world temperature and solar radiation intensity data from [8], a region of the Yellow Sea, covering the period from March first, 2020, to July first, 2020. To make the simulation experiments more persuasive, we also introduced some noise into the test data to examine the robustness of the proposed method.

4.2 Simulation Experiment Results

The simulation experiment results are shown in Fig. 3 to Fig. 6. The test data is from [6], a region of the Yellow Sea, May 15th, 2021.

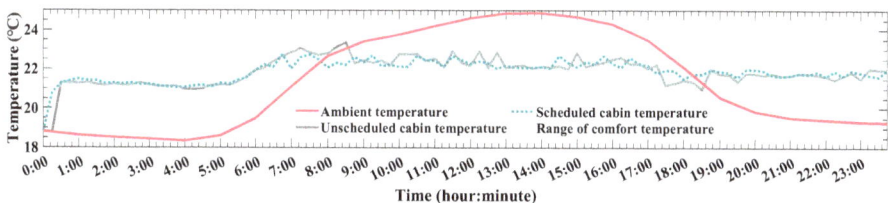

Fig. 3. Temperature curves of cabin and ambient.

From Fig. 3, it can be observed that the proposed method enables the cabin temperature to quickly reach the comfortable temperature range and maintains it within this range even under continuously changing external environmental temperatures. Figure 4 and Fig. 5 demonstrate that the proposed method can reasonably schedule the power of non-HVAC appliances to achieve peak shaving, thereby reducing the PAR of the renewable power system. Figure 5 also demonstrates the proposed method enables the total power consumption of all appliances is less than or equal to the maximum power (P_{max}^{Grid})

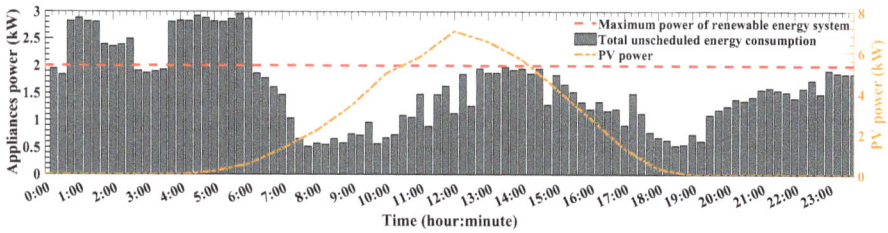

Fig. 4. Power of appliances in accommodation area of multi-energy ship without scheduling.

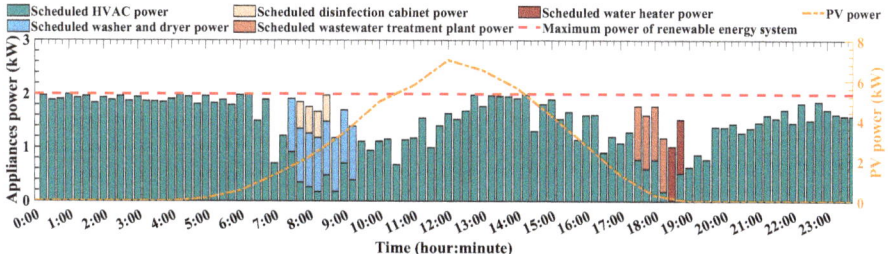

Fig. 5. Power of appliances in accommodation area of multi-energy ship with scheduling.

that the renewable power system can supply. And Fig. 6 shows that the proposed method can maintain the battery at over 30% of its maximum capacity for extended periods, thereby enhancing the ability of power system to handle unexpected issues, such as prolonged periods of rainy weather or sudden spikes in short-term electricity demand. Moreover, in this experiment, the proposed method's rational scheduling of electrical appliances ensures that all energy consumed in the ship accommodation area are derived from renewable sources, hence reducing CO_2 emissions.

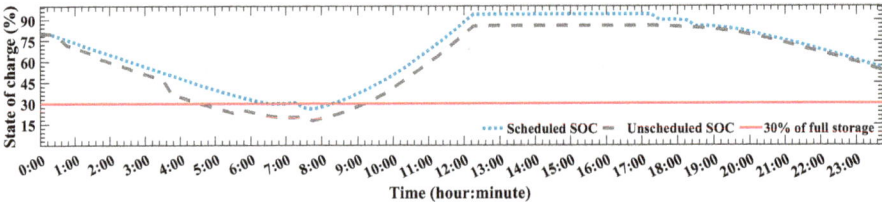

Fig. 6. Battery's state of charge.

The results of this simulation experiment are presented in Table 2.

Table 2. The Results of This Simulation Experiment.

Indices	Unscheduled	Scheduled
Energy consumption of all appliances	38.13 kWh	**37.68 kWh**
Energy supplied by diesel generator	12.39 kWh	**0**
PAR	1.86	**1.27**
Out of comfortable temperature range	2.25 h	**0.5 h**
Total duration of low SOC	4.5 h	**1.0 h**

5 Conclusion

The presented results show that TS-EMS can effectively schedule the usage time and power of various electrical appliances in ship accommodation area, ensuring crew comfort while maximizing the utilization of renewable energy sources and reducing the reliance on fossil fuels. Additionally, this method effectively reduces the PAR of renewable power system while ensuring that the battery maintains ample reserves over prolonged durations. This enhancement bolsters the power system's resilience in handling unforeseen contingencies.

Further studies will include energy management of multi-cabin ship accommodation areas; compartmental energy management considering emergency situations; integration of compartmental energy management with the propulsion system.

Acknowledgement. This work was supported by National Natural Science Foundation of China (Grant No. 52301348).

References

1. Wang, K., Yan, X., Yuan, Y., et al.: Dynamic optimization of ship energy efficiency considering time-varying environmental factors. Transp. Res. D: Transp. Env. **62**, 685–698 (2018). https://doi.org/10.1016/j.trd.2018.04.005
2. Xie, P., Tan, S., Guerrero, J.M., Vasquez, J.C.: MPC-informed ECMS based real-time power management strategy for hybrid electric ship. Energy Rep. **7**, 126–133 (2021). https://doi.org/10.1016/j.egyr.2021.02.013
3. Yuan, Y., Wang, J., Yan, X., Li, Q., Long, T.: A design and experimental investigation of a large-scale solar energy/diesel generator powered hybrid ship. Energy. **165**, 965–978 (2018). https://doi.org/10.1016/j.energy.2018.09.085
4. Fan, A., et al.: Development trend and hotspot analysis of ship energy management. J. Clean. Prod. **389**, 135899 (2023). https://doi.org/10.1016/j.jclepro.2023.135899
5. Yuan, Y., Wang, J., et al.: A review of multi-energy hybrid power system for ships. Renew. Sust. Energe Rev. **132**, 110081 (2020). https://doi.org/10.1016/j.rser.2020.110081
6. NASA "NASA Power Prediction of Worldwide Energy Resources". https://power.larc.nasa.gov/ (2024) Accessed 31 Mar.2024.
7. Gao, Y., Li, S., Xiao, Y., et al.: An iterative optimization and learning-based IoT system for energy management of connected buildings. IEEE Internet Things J. **9**(21), 21246–21259 (2022). https://doi.org/10.1109/JIOT.2022.3176306
8. Ventilation for Acceptable Indoor Air Quality: ASHRAE Standard 62–2010. In: American Society of Heating. Refrigerating and Air-Conditioning Engineers, Inc., ASHRAE, Atlanta, GA, USA (2010)

Open Access This chapter is licensed under the terms of the Creative Commons Attribution-NonCommercial-NoDerivatives 4.0 International License (http://creativecommons.org/licenses/by-nc-nd/4.0/), which permits any noncommercial use, sharing, distribution and reproduction in any medium or format, as long as you give appropriate credit to the original author(s) and the source, provide a link to the Creative Commons license and indicate if you modified the licensed material. You do not have permission under this license to share adapted material derived from this chapter or parts of it.

The images or other third party material in this chapter are included in the chapter's Creative Commons license, unless indicated otherwise in a credit line to the material. If material is not included in the chapter's Creative Commons license and your intended use is not permitted by statutory regulation or exceeds the permitted use, you will need to obtain permission directly from the copyright holder.

Cosine Wave Numerical Wave Tank Study and Its Force Analysis on Small-Scale Structures

ZiFeng Sun[1], MingKai Li[2], XinYu Zhang[2], YanJun Guan[1], YuXiang Niu[1], and JunSheng Zhang[2]($\boxtimes$)

[1] Ocean Department, Power China Guiyang Engineering Corporation Limited, Guiyang, China
[2] College of Ocean and Civil Engineering, Dalian Ocean University, Dalian, China
zhangjunsheng@dlou.edu.cn

Abstract. This investigation is committed to resolving the challenges of Fluent's accuracy in simulating cnoidal waves and the constraints of potential flow theory in real-world engineering contexts. By conducting secondary development on the Fluent platform, a numerical model grounded in viscous flow theory has been formulated. The study centers on the force analysis of small-scale structures subjected to varying wave parameters and, in alignment with practical engineering requirements, introduces a more refined numerical model and theoretical framework. This endeavor offers enhanced insights and solutions for the theoretical advancement and engineering application of hydrodynamic challenges. In the realm of nearshore and harbor engineering, wave characteristics undergo alterations due to the reduced water depth, leading to the emergence of shallow water waves. Nonlinear shallow water waves are conventionally characterized as cnoidal waves, which markedly diverge from Stokes waves. With the relentless progression of marine engineering, the demand for more precise nonlinear hydrodynamic analysis of marine structures has intensified. Accurately forecasting the wave forces exerted on small-scale members constitutes a pivotal concern in safety research. These issues are intricate, and the traditional potential flow theory falls short in delivering accurate analyses for such scenarios. Consequently, it is of paramount significance to investigate the wave action on small-scale members through the lens of viscous flow theory.

Keywords: Numerical wave tank · Small-scale structural members · Elliptical cosine waves · Nonlinear effects

1 Introduction

The prevailing paradigm in the investigation of hydrodynamic phenomena predominantly utilizes potential flow theory for wave simulation and dynamic computations, particularly when evaluating the impact of waves on large-scale rigid bodies. Nevertheless, with the escalating intricacy of marine engineering and the progressive refinement of fluid dynamics analysis, certain conundrums have extended beyond the purview of conventional computational techniques. Specifically, in instances where viscous flow interaction with objects is paramount, the exclusive reliance on potential flow theory and empirical formulae is inadequate to fulfill the exigencies of theoretical scrutiny and engineering application.

In the present investigation, it is acknowledged that the simulation of wave propagation through potential flow theory necessitates a comprehensive consideration of the intricacies of flow around objects, with a particular emphasis on the repercussions for diminutive-scale structures. Despite extant research elucidating the viscous flow forces on cylinders within the expansive confines of deep seas, potential flow theory remains the preeminent paradigm in the realm of shallow marine environments. Given the pronounced disparities in nonlinearity between cnoidal waves and Stokes waves, the empirical findings specifically attuned to Stokes waves cannot be indiscriminately extrapolated to their cnoidal counterparts. Scholars both domestically and internationally have engaged in analogous inquiries pertaining to cnoidal waves. R.L. Wiegel [1] advanced the cnoidal wave theory, transforming the computational outcomes into graphical representations, thereby facilitating the theory's extensive integration into engineering praxis. Zhu Yanrong [2] computed the horizontal velocity and acceleration of water particles in adherence to the cnoidal wave theory. Concurrently, Zhang Jie and Zheng Baoyou [3] devised a rapid computational algorithm for second-order cnoidal waves. Li Yanbao [4] employed the geometric mean method to compute cnoidal waves, thereby refining the computational technique and enhancing the methodology of the cnoidal wave theory. Subsequently, Gu Jialong and Shen Xianrong [5] introduced a diminutive parameter within the framework of the first-order cnoidal wave theory and extended this parameter to encompass other orders of cnoidal wave theories.

To delve into the nuanced nonlinear impact characteristics of cnoidal waves, this paper leverages the Fluent platform for secondary development to undertake numerical research, with the objective of elucidating the force effects of cnoidal waves on diminutive-scale rods. This endeavor not only proffers a novel vantage point on the extant labyrinthine hydrodynamic conundrums but also advocates for more meticulous deliberations in the selection of apt numerical models and theoretical constructs.

2 Construction of the Numerical Model

This chapter is predicated upon the boundary wave generation methodology, the damping wave absorption technique, and the Volume of Fluid (VOF) surface tracking technology to emulate cnoidal waves. To faithfully mirror the wave conditions extant in natural settings, the numerical simulation process incorporates the influences of gravity and fluid viscosity. This protocol substantiates the dependability of the wave generation and absorption techniques utilized, alongside the boundary conditions.

3 Governing Equations and Turbulence Models

In this investigation, a numerical model grounded in the Navier-Stokes equations is harnessed to scrutinize the flow of incompressible viscous fluids, with the intent of profoundly delving into the fluid dynamics within a wave flume. To more efficaciously emulate the flow field, the Renormalization Group (RNG) turbulence model is selected as the closure model for the flow field.

The momentum conservation equation constitutes a cornerstone methodology in fluid mechanics, employed to delineate the principle of momentum conservation within fluids. Accounting for viscosity, the differential form of the two-dimensional Navier-Stokes equations is articulated as follows:

$$\frac{\partial u}{\partial t} + u\frac{\partial u}{\partial x} + v\frac{\partial u}{\partial y} = -\frac{1}{\rho}\frac{\partial p}{\partial x} + g_x + v\left[\frac{\partial^2 u}{\partial^2 x} + \frac{\partial^2 u}{\partial^2 y}\right] \quad (1)$$

$$\frac{\partial u}{\partial t} + u\frac{\partial u}{\partial x} + v\frac{\partial u}{\partial y} = -\frac{1}{\rho}\frac{\partial p}{\partial y} + g_y + v\left[\frac{\partial^2 v}{\partial^2 x} + \frac{\partial^2 v}{\partial^2 y}\right] \quad (2)$$

The RNG k-ε turbulence model is utilized for turbulence closure, wherein the viscosity coefficient can be articulated as:

$$\mu_t = C_\mu \frac{k^2}{\varepsilon} \quad (3)$$

3.1 Free Surface Tracking Method

Fluent software utilizes the finite volume method for the computation of the flow field, and employs the Volume of Fluid (VOF) method to manage the water-air interface and the water surface within the pool. In this simulation, the water surface is considered as the boundary between air and water, and the wave flow is achieved by characterizing it as a two-phase flow involving air, water, and gas stratification. The volume fraction, referred to as the F function, is introduced to resolve the water-air interface, and the following equation must be adhered to at the free surface:

$$\frac{\partial a_i}{\partial t} + \vec{v}\nabla a_i = 0 \quad (4)$$

The computational conditions are as follows:

$$\sum_{i=1}^{n} a_i = 1 \quad (5)$$

In Fluent, when addressing a cell that encompasses the blending of two distinct substances, the following algorithm is employed to compute the density:

$$p = \sum_{i=1}^{n} a_i p_i \quad (6)$$

In Fluent, the generated velocity field is shared by all phases. The momentum equation is expressed as follows. The form of the momentum equation is influenced by the volume fractions of all phases for the properties (density) and (velocity):

$$\frac{\partial}{\partial t}(\rho\phi) + \nabla\frac{\partial}{\partial t}(\rho\phi u) = \nabla.(\Gamma\nabla\phi) + S_\phi \quad (7)$$

3.2 Wave Generation Method

The numerical simulation methodology employed in this study involves the construction of a numerical wave tank, which is adept at simulating both Stokes and cnoidal waves. This section is primarily dedicated to elucidating the wave generation technique for cnoidal waves.

The theory of cnoidal waves, initially posited by Korteweg and De Vries in 1895, postulates that periodic shallow water waves can be articulated through the utilization of the elliptic cosine function. The equation delineating the wave surface for cnoidal waves is as follows:

$$\eta = d(A_1 + A_2 c_1 + A_3 c_2) \tag{8}$$

The horizontal velocity u and the vertical velocity w of cnoidal waves are respectively expressed as:

$$u = \sqrt{gd}\left\{(B_1 + B_2 c_1 + B_3 c_2) - \frac{1}{2}\left(\frac{y}{d}\right)^2 (B_4 + B_5 c_1 + B_6 c_2)\right\} \tag{9}$$

$$w = \sqrt{gd}\frac{4K(k)d}{L}csd\left\{\left(\frac{y}{d}\right)(B_2 + 2B_3 c_1) - \frac{1}{6}\left(\frac{y}{d}\right)^3 (B_5 + 2B_6 c_1)\right\} \tag{10}$$

3.3 Wave Generation and Wave Absorption

This study successfully generates numerical waves by employing the User-Defined Function (UDF) module within the Fluent software. Concretely, by scripting a UDF file, a bespoke wave generation algorithm is integrated into the simulation model. Through the amalgamation of the wave velocity function and the wave height equation with the macros furnished by the Fluent software, a comprehensive UDF file is crafted. Subsequently, this file is compiled and implemented via the Fluent software to attain the wave generation effect through velocity boundary conditions. In scenarios where the simulated wave is a cnoidal wave, the velocity boundary is delineated in the form of the corresponding cnoidal wave velocity.

In numerical tank simulations, waves exhibit a propensity to reflect upon reaching the tank outlet, whereupon the reflected waves coalesce with the incident waves, potentially engendering wave resonance or amplification. To uphold the veracity of the simulation, countermeasures must be instituted to mitigate this detrimental phenomenon, commonly achieved through wave absorption treatment. This paper adopts the methodology of artificial viscosity, encapsulating the essence of emulating the wave absorption layer of an actual wave pool within a laboratory setting.

Within the damping absorption section, the momentum equation is formulated as follows:

$$\frac{\partial u}{\partial t} + u\frac{\partial u}{\partial x} + v\frac{\partial u}{\partial y} = -\frac{1}{p}\frac{\partial p}{\partial x} + g_x + v\left[\frac{\partial^2 u}{\partial x^2} + \frac{\partial^2 u}{\partial y^2}\right] - \mu u \tag{11}$$

$$\frac{\partial v}{\partial t} + u\frac{\partial v}{\partial x} + v\frac{\partial v}{\partial y} = -\frac{1}{p}\frac{\partial p}{\partial y} + g_y + v\left[\frac{\partial^2 v}{\partial x^2} + \frac{\partial^2 v}{\partial y^2}\right] - \mu v \tag{12}$$

$$\mu(x) = \alpha(x - x_1)(x - x_2) \tag{13}$$

In the aforementioned equation, μ, α, x_1, x_2 symbolize the wave absorption coefficient, empirical coefficient, the initial coordinate of the wave absorption region, and the final coordinate of the wave absorption region, respectively.

This approach meticulously modulates the momentum source term within the wave absorption region, progressively diminishing the wave energy, and thus adeptly eradicating reflected waves, engendering enhanced stability in numerical simulations. This is paramount for the faithful emulation of wave dynamics within the tank and for ensuring the precision of the simulation outcomes.

4 Numerical Wave Tank Model Validation

This chapter guarantees the veracity of wave simulation by corroborating the model's simulation outcomes of cnoidal waves with theoretical analytical results. Ultimately, the wave forces exerted upon a fully submerged horizontal cylinder are subjected to numerical simulation and juxtaposed with antecedent experimental findings, thereby substantiating the precision of the numerical model in computing fluid forces on structures.

4.1 Model Wave Generation Verification

In this segment, the efficacy of wave generation within the numerical tank shall be substantiated. For detailed wave generation parameters, kindly consult Table 1.

Table 1. Parameters for the test of wave generation

Waveform	Water Depth (m)	Wave Height (m)	Wave Length (m)	Ursell	H/d	H/gT^2	d/gT^2
Cnoidal Wave	3.5	1.6	40	59.7	4%	1.02×10^{-3}	1.02×10^{-2}

To juxtapose and scrutinize against theoretical wave patterns, wave height monitoring points were strategically positioned at intervals equivalent to each wavelength within the numerical tank, commencing from the initial locus (x = 0 m). Given that the wave-absorbing zone commences at x = 180 m, the terminal coordinates of the quartet of monitoring sections were delineated at x = 40 m, x = 80 m, x = 120 m, and x = 160 m. This meticulous configuration of monitoring sections facilitates an exhaustive comprehension of the wave evolution trajectory within the numerical tank and corroborates the veracity of the simulation outcomes through continuous surveillance. By benchmarking against theoretical wave patterns, we bolster our confidence in the precise prognostication of wave comportment by the numerical simulation, thereby augmenting the reliability of the numerical model.

Figures 1, 2, 3 and 4 delineate the temporal wave oscillation profiles at disparate horizontal coordinates within the numerical wave tank, specifically at the four sectional loci x = 40 m, x = 80 m, x = 120 m, and x = 160 m. The wave height profiles discerned in the numerical simulation manifest commendable stability and periodicity across time t, with only nominal discrepancies in relation to the theoretical undulations.

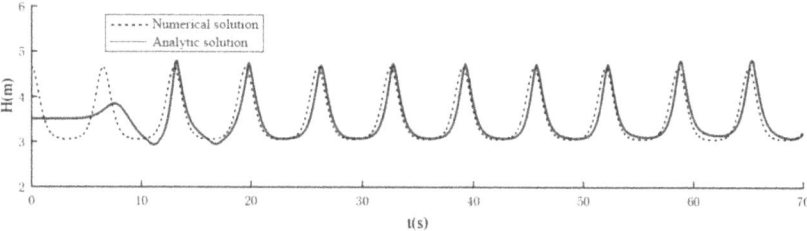

Fig. 1. Comparison of the numerical solution and theoretical solution of the wave surface time evolution for Coindal cnoidal waves at x/L = 1

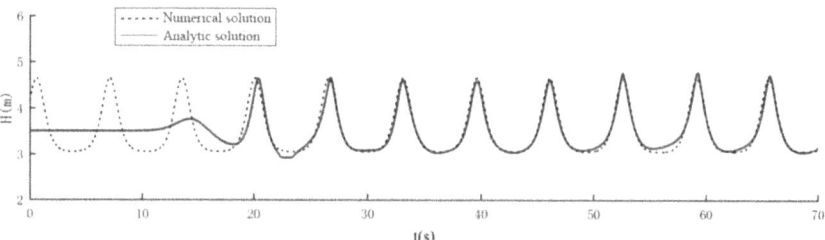

Fig. 2. Comparison of the numerical solution and theoretical solution of the wave surface time evolution for Coindal cnoidal waves at x/L = 2

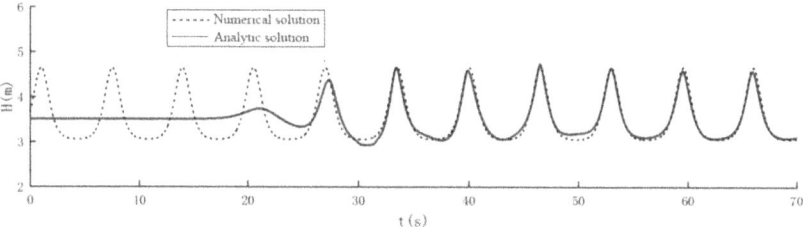

Fig. 3. Comparison of the numerical solution and theoretical solution of the wave surface time evolution for Coindal cnoidal waves at x/L = 3

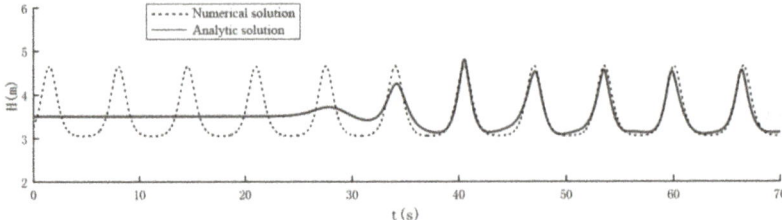

Fig. 4. Comparison of the numerical solution and theoretical solution of the wave surface time evolution for Coindal cnoidal waves at x/L = 4

4.2 Verification of Wave Forces on Submerged Horizontal Cylinders

In this segment, a numerical wave tank model predicated on viscous flow dynamics was employed to conduct numerical computations of wave forces exerted upon fully submerged horizontal cylinders. The parameters for numerical computation were judiciously chosen in accordance with the conditions posited in antecedent studies. The wave forces derived from disparate theoretical models were juxtaposed, and the precision of the numerical computations was deliberated.

In reference to the extant research, Fig. 5 encapsulates the schematic representation of the computational domain. The cylinder's radius is delineated as r = 0.051 m, and s signifies the vertical displacement from the cylinder's axis to the quiescent water surface, thereby denoting the cylinder's submergence depth. The cylinder's spatial orientation is situated 1.65 L adrift from the velocity inlet and 4 L in proximity to the wave-absorbing zone, with the latter extending over a span of 1.5 L. The parameters for the numerical calculations are delineated in the Table 2. The numerical examples are delineated in the Table 3.

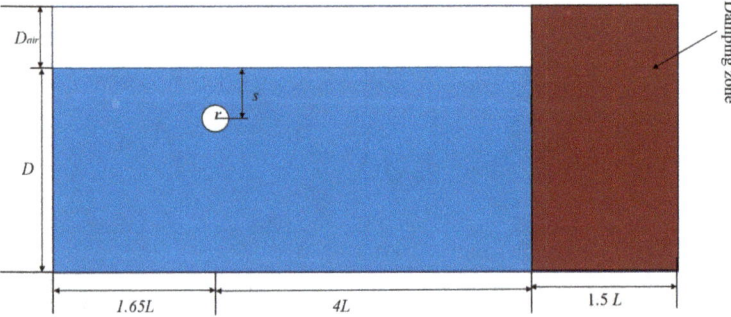

Fig. 5. Wave action on the computational domain with a fully submerged cylinder schematic diagram

Table 2. Computational parameters for the model case

T(s)	d(m)	s/r	kr	kd	β
1.0	0.85	2	0.206	3.43	9120

Table 3. Calculation parameters of Case

Case	A/r	Kc	H/L	Re
1	0.12	0.25	0.79%	2.28×10^3
2	0.24	0.50	1.57%	4.56×10^3
3	0.36	0.75	2.36%	6.84×10^3
4	0.47	1.00	3.15%	9.12×10^3
5	0.57	1.20	3.78%	1.09×10^4

Figures 6 and 7 elucidate a comparative analysis of the wave forces at the fundamental frequency as they fluctuate in accordance with the Kc number. The outcomes derived from the viscous flow numerical model in the current investigation harmonize convincingly with the findings of Mao Hongfei [6], as well as with the empirical data of Chaplin [7], and similarly resonate with the viscous flow numerical outcomes of Tavassoli and Kim [8]. It is discerned that both the viscous flow numerical outcomes and the experimental data are attenuated in comparison to those yielded by potential flow theory, and the divergence between the viscous flow and Ogilvie's [9] potential flow outcomes augments with the amplitude of the waves. The comparative outcomes signify that the numerical model in the present study exhibits commendable fidelity in computing the forces impinging upon the object.

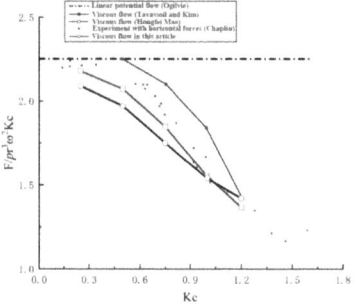
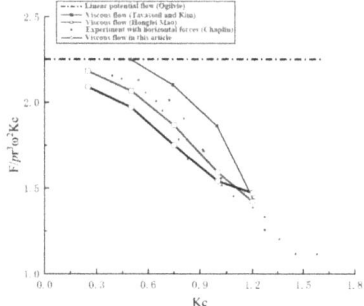

Fig. 6. Variation of one-frequency horizontal wave Force with Kc

Fig. 7. Variation of one-frequency vertical wave force with Kc

5 Force Analysis of Small-Scale Rods Under Different Wave Elements

The viscous flow model proffers a more precise methodology to delineate the ramifications of viscosity, thereby facilitating a more exhaustive contemplation of the nuances of wave-structure interaction. Conventionally, when kr is substantially less than 1, it may be construed as a diminutive-scale rod. It is an imperative undertaking to employ the viscous flow model to conduct numerical investigations into the wave forces exerted upon submerged diminutive-scale rods. Such numerical inquiry serves to deepen the comprehension of the genesis mechanism of wave forces on submerged diminutive-scale horizontal cylindrical rods, furnishing a basis for more meticulous engineering design and marine engineering applications.

5.1 The Influence of Incident Wave Amplitude on Wave Forces

In this segment of the investigation, meticulous numerical simulations were executed for two distinct sets of conditions across varying water depths, with a schematic depiction of the computational domain exhibited in Fig. 5. Within this figure, the radius of the cylinder is stipulated as r = 0.51 m, and s denotes the vertical separation from the axis to the quiescent water surface, that is, the submergence depth of the cylinder. The cylinder's placement in relation to the configuration of the flume is 2 L distant from the velocity inlet zone and 2.5 L removed from the wave absorption zone. The aggregate length of the numerical flume area is 6 L.

The numerical calculation parameters for the two sets of cases are shown in Table 4.

Table 4. Parameters for the two groups of cases

Case	T(s)	D(m)	s/r	kr	kd
A	5	20	5	0.078	3.14
B	6.5	3.5	5	0.078	0.55

In this division of the research, exhaustive numerical computations will be undertaken for two sets of conditions across disparate water depths, designated as Case A and Case B. Pertinent parameters, encompassing wave height, wave steepness, and the Ursell number, have been delineated in Table 5 and Table 6.

Table 5. Calculation parameters of Case A

CaseA	A(m)	H/L	Ursell
A1	0.2	1%	0.08
A2	0.4	2%	0.16
A3	0.6	3%	0.24
A4	0.8	4%	0.32

Table 6. Calculation parameters of Case B

CaseB	A(m)	H/L	Ursell
B1	0.2	1%	14.9
B2	0.4	2%	29.8
B3	0.6	3%	44.8
B4	0.8	4%	59.7

Force Analysis of Small-Scale Rods under Case A. Upon scrutinizing the wave force outcomes under Case A for the sake of analytical clarity, the dimensionless formulation of the wave force is $F/(\rho g \pi r^2)$, The Dimensionless Form of Time is t/T, The comparative analysis of the temporal evolution of both horizontal and vertical wave forces throughout a cycle is conducted.

Figures 8 and 9 provide a visual representation of the relationship between horizontal and vertical wave forces and wave amplitude. It has been observed that both horizontal and vertical wave forces demonstrate a marked escalation as the wave amplitude intensifies. This observation underscores a direct positive correlation between wave force and wave amplitude, implying that the impact of waves on structures becomes notably more pronounced with greater amplitudes. Initially, as the wave amplitude ascends from 0.2 to 0.4, the surge in wave force is most rapid, registering at approximately 50%; subsequently, from 0.4 to 0.6 and 0.6 to 0.8, the rates of increase are comparable, each averaging around 25%. In the context of numerical computations, the horizontal and vertical forces are discernibly similar yet not congruent, diverging from the predictions of potential flow theory.

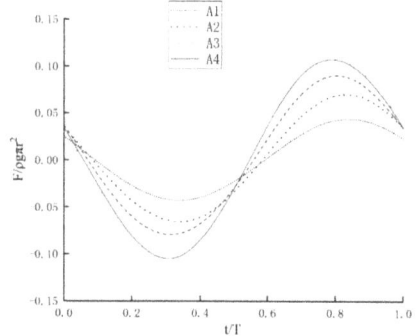
Fig. 8. Horizontal wave forces in a period (A)

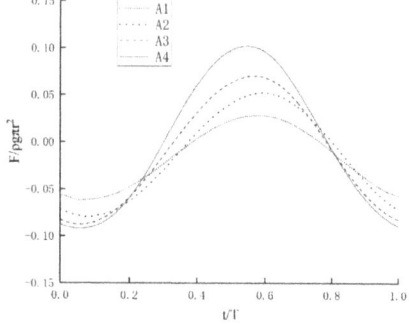
Fig. 9. Vertical wave forces in a period (B)

Force Analysis of Small-Scale Rods under Case B. Figures 10 and 11 offer a visual comparison of how horizontal and vertical wave forces change in relation to wave amplitude. As wave amplitude escalates, both horizontal and vertical wave forces exhibit a pronounced upward trend. This reinforces the established positive correlation between wave force and wave amplitude, highlighting that the force exerted by waves on structures becomes significantly more impactful with greater amplitudes. The transition from Stokes waves to cnoidal waves reveals the most pronounced increase in wave force, with the growth rate of horizontal force reaching 77%. This indicates that the rate of increase in wave force can vary substantially depending on the type of wave. For instance, when the wave amplitude is doubled, the force growth rate for cnoidal waves exceeds 35%. This suggests that cnoidal waves experience a more rapid increase in force for the same increment in wave amplitude, implying that they exert a more significant destructive force on small-scale structures under equivalent wave amplitude conditions. However, in contrast to deep water scenarios where horizontal and vertical wave forces are approximately equal, in finite water depths, horizontal wave forces are considerably larger than vertical wave forces. This demonstrates that different wave types exert varying forces on structures, with cnoidal waves being particularly responsive to changes in wave amplitude and possessing greater potential for destruction. These insights are invaluable for engineering design and the evaluation of a structure's wave resistance capabilities, offering practical guidance for mitigating the adverse effects of wave forces on coastal and offshore structures.

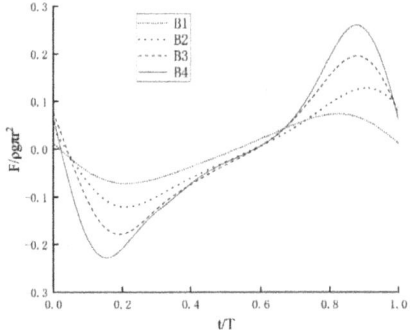

Fig. 10. Horizontal wave forces in a period (A)

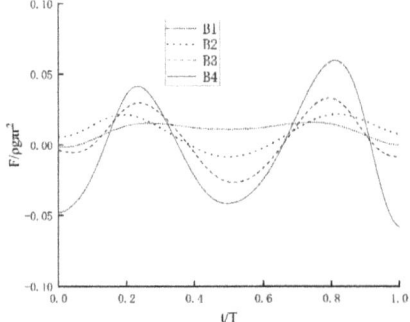

Fig. 11. Vertical wave forces in a period (B)

5.2 Research on Wave Forces at Different Wave Frequencies

In the preceding section of wave force research, the emphasis was placed on the dynamic assessment of small-scale rods subjected to varying wave amplitudes under a constant frequency scenario. This section shifts the analytical lens to a comprehensive investigation using a viscous flow numerical wave flume model, focusing on the same wave amplitude but with a spectrum of frequencies. The objective is to delve into the effects of these frequencies on the wave forces exerted upon submerged small-scale rods. The schematic representation of the computational domain for this study is depicted in Fig. 5; here, the incident wave period is denoted as T, the water depth is set at $d = 3.5$ m, the wave amplitude is $A = 0.8$ m, the small-scale rod is positioned 2 L units from the velocity inlet and 2.5 units away from the wave absorption zone, s represents the submergence depth, and the radius of the small-scale rod is $r = 0.5$ m. A detailed compilation of the various computational parameters is provided in Table 7. This setup allows for a meticulous examination of how frequency variations influence wave forces, offering critical insights into the dynamic interactions between waves and submerged structures across a range of frequencies.

Table 7. Parameters for the cases

Case	T(s)	L(m)	s/r	f(hz)
1	6.5	40	5	0.153
2	6.0	38	5	0.167
3	5.7	36	5	0.175
4	5.4	34	5	0.185
5	5.2	32	5	0.192

Upon observing Figs. 12 and 13, it can be noted that under the same wave height, the variations in horizontal wave forces are relatively minor and essentially equal. In contrast, the vertical wave forces gradually decrease as the wave frequency increases. Under the same wave height, as the wave frequency increases, the wave period decreases, and the distance between the wave crest and trough shortens. For horizontal wave forces, they are primarily influenced by the difference between the wave crest and trough, which remains relatively stable as the wave period shortens. Therefore, horizontal wave forces are relatively insensitive to changes in wave frequency. Conversely, vertical wave forces are more significantly affected by the distance between the wave crest and trough. When the wave frequency increases, the energy of the wave tends to be more concentrated within the brief period between the wave crest and trough. This means the transmission time of the wave force becomes shorter, thereby reducing the duration of its effect on the structure, leading to a decrease in the force exerted. Under the same water depth and wave height, the wave length decreases, the wave period decreases, the frequency increases, but at the same time, the wave speed decreases, due to factors such as the reduction in water depth, increased energy loss during wave propagation, and the viscous effects of the fluctuating fluid.

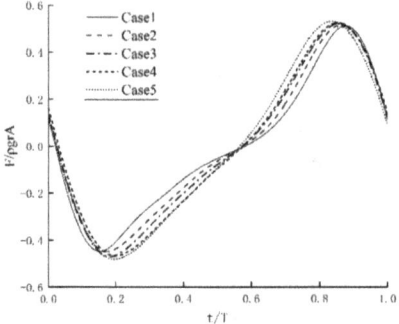
Fig. 12. Horizontal wave forces in a period

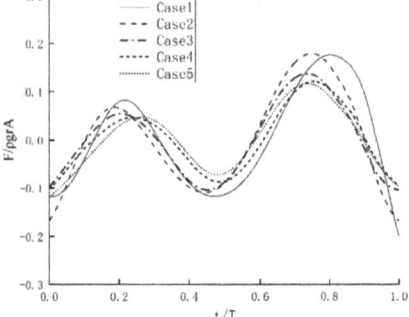
Fig. 13. Vertical wave forces in a period

6 Conclusion

The research, conducted within the virtual confines of a numerical wave flume simulated in Fluent, delves into the force dynamics experienced by small-scale rods under a diverse array of wave conditions. This investigation has led to a series of pivotal insights that serve as a robust foundation for practical engineering applications and the optimization of structural design. Initially, the study confirms the substantial effect of wave amplitude on small-scale rods, demonstrating that an escalation in wave amplitude correlates with a marked increase in the force imparted by waves onto the structure, with a particular emphasis on the horizontal and vertical wave forces. The comparative analysis of distinct wave types reveals that the shift from Stokes waves to cnoidal waves results in the most dramatic surge in horizontal wave forces, with cnoidal waves exhibiting a notably greater capacity for damage to small-scale structures. In scenarios involving finite water depths, the horizontal wave forces markedly exceed the vertical wave forces, underscoring the critical role of water depth in shaping the distribution and magnitude of wave forces. These findings underscore the necessity for tailored design considerations that account for the specific wave conditions and water depths encountered in various marine and coastal environments.

These findings not only furnish substantial empirical support for a nuanced comprehension of the dynamic responses of small-scale rods to wave action but also extend more holistic guidance for engineering practice. The implications of this research are of considerable significance for advancing future investigations into hydrodynamic phenomena and for the refinement of practical engineering design. By elucidating the intricate interplay between wave characteristics and structural responses, these observations pave the way for more informed and resilient design strategies, ultimately enhancing the safety and efficiency of coastal and offshore infrastructures. The insights gleaned from this study are poised to catalyze further innovation in the field, fostering a more sophisticated approach to the challenges posed by wave-structure interactions.

References

1. Wiegel, R.L.: Oceanographical Engineering (1964)
2. Yanrong, Z.: Computational Analysis of the Cnoidal Wave Theory. Port Eng. **03**, 23–29 (1982) (in Chinese)
3. Jie, Z., Baoyou, Z.: Fast Computational Method for Second-Order Cnoidal Waves [J]. Waterway and Harbor (1982) (in Chinese)
4. Yanbao, L.: Improvement of the Computational Method for Cnoidal Waves. Water Transp. Eng. **01**, 91002–94972 (1988) (in Chinese)
5. Jialong, G., Xianrong, S.: Numerical calculation and characteristic study of cnoidal waves. Ocean Eng. **01**, 30–40 (1989) (in Chinese)
6. Hongfei, M.: Numerical Examination On Wave Forces On Horizontal Circular Cylinders Based On Viscous Fluid Theory. Dalian University of Technology (2018) (in Chinese)
7. Chaplin, J.R.: Non-linear forces on a horizontal cylinder beneath waves. J. Fluid Mech. **147**, 449–464 (1984)
8. Tlavassoli, A., Kim, M.H.: Interactions of fully nonlinear waves with submerged bodies by a 2D viscous NWT. 11th Int. Offshore Polar Eng. Conf. **3**, 348–354 (2001)
9. Ogilvie, T.F.: First- and second-order forces on a cylinder submerged under a free surface. J. Fluid Mech. **16**, 451–472 (1963)

Open Access This chapter is licensed under the terms of the Creative Commons Attribution-NonCommercial-NoDerivatives 4.0 International License (http://creativecommons.org/licenses/by-nc-nd/4.0/), which permits any noncommercial use, sharing, distribution and reproduction in any medium or format, as long as you give appropriate credit to the original author(s) and the source, provide a link to the Creative Commons license and indicate if you modified the licensed material. You do not have permission under this license to share adapted material derived from this chapter or parts of it.

The images or other third party material in this chapter are included in the chapter's Creative Commons license, unless indicated otherwise in a credit line to the material. If material is not included in the chapter's Creative Commons license and your intended use is not permitted by statutory regulation or exceeds the permitted use, you will need to obtain permission directly from the copyright holder.

Research of Hydrodynamic and Energy Capture Characteristics of Wave Energy Device with Eccentric Inner Rotor

Yuxiang Niu[1], Lujun Zhao[2], Yanjun Guan[1], Zifeng Sun[1], Junsheng Zhang[3], and Wanqing Zhang[4(✉)]

[1] Ocean Department, Power China Guiyang Engineering Corporation Limited, Guiyang, China
[2] China Special Vehicle Research Institute, Hubei, China
[3] College of Ocean and Civil Engineering, Dalian Ocean University, Dalian, China
[4] Communications and Industry Department, AVIC Jonhon Optronic Technology Corporation Limited, Henan, China
1763809192@qq.com

Abstract. In view of the characteristics of small wave height and short period in China sea area, a floating eccentric inner rotor wave energy generation device is proposed. Based on frequency-domain and time-domain analysis, the selection and optimization of the numerical calculation model of the eccentric inner rotor wave energy device are carried out, and the comparison with the experimental data of the pool is verified. The results show that the hydrodynamic performance and energy capture efficiency of the ellipse-circular wave energy device with the same length, width and draft depth are superior to those of the ellipse-circular wave energy device with the same length, width and draft depth. Increasing the mass of the inner rotor and the draft depth of the device can improve the hydrodynamic performance and energy capture efficiency of the device, and an optimal PTO damping coefficient can be found to make the energy capture power of the device reach the peak value. After selection optimization, the instantaneous output power surged from 1.16 kW to 2.29 kW, marking a 97.4% improvement, while the average output power ascended from 0.53 kW to 1.08 kW, reflecting a 103.8% increase.

Keywords: Wave Energy · Eccentric Inner Rotor · Hydrodynamic Performance · Energy Harvesting Characteristics · Mooring Systems

1 Introduction

In recent years, China has aligned itself with the global inclination towards the intensification of marine exploration and development. Wave energy converters, innovative devices designed to harness marine energy, are currently in the nascent stages of development, with a paucity of literature addressing their hydrodynamic, power generation, and mooring attributes. These wave energy devices can be categorized into oscillating water column, point absorber, and overtopping types, delineated by their distinct methods of energy capture [1, 2].

For the enhancement of energy capture efficiency in point absorber wave energy devices, contemporary methodologies encompass structural optimization grounded in the hydrodynamic properties of floaters, refinement of multi-stage energy conversion systems, structural optimization rooted in energy capture principles, and optimization of control systems at the load end [3]. In 2008, Widden et al. [4]. scrutinized the energy capture efficiency of a dual-degree-of-freedom PTO mechanism, elucidating a correlation between the elevation of the center of mass of the PTO device and its energy capture efficiency, contingent upon stability prerequisites. Negandari et al. [5]. augmented the energy capture efficiency of a dual-degree-of-freedom point absorber wave energy device through hydrodynamic coefficient and structural shape optimization. Gaspar et al. [3]. modeled the hydraulic system of the WaveStar wave energy device and optimized its energy conversion efficiency employing neural network algorithms, revealing that the maximization of capture bandwidth and energy capture efficiency was attained through velocity control tracking as a reference value.

Given the characteristics of low wave energy flow density, diminutive wave height, and abbreviated period in Chinese seas, traditional wave energy devices are prone to frequent reciprocating impact loads, thereby abbreviating device longevity and yielding suboptimal energy capture efficiency during periods of minimal wave height [1, 8]. The prototype of the wave energy device employed in this discourse is the Penguin, an eccentric internal rotor wave energy device engineered by Wello Company of Finland in 2014 [9]. This device features a fully enclosed floating body structure, with the energy conversion system ensconced within the enclosed floating body, impervious to seawater corrosion and the encrustation of marine organisms, thereby conferring high reliability and extended service life. The primary floating body utilizes an irregular form to amplify motion amplitude and engender a displacement differential with the embedded rotor, thereby facilitating energy capture. As the energy capture efficiency of the device is contingent upon rotor rotation speed, the device exhibits superior energy capture performance under abbreviated wave periods [1, 8]. By streamlining the device, the hydrodynamic performance of the device in Chinese waters is examined, furnishing a referential framework for augmenting the energy capture efficiency of the device in Chinese waters.

2 Model of the Internal Rotor Wave Energy Device

2.1 Structure and Working Principle

This study delves into the Wello Penguin wave energy converter, as illustrated in Fig. 1. A streamlined variant, the eccentric internal rotor wave energy generator, is extrapolated from the Penguin model. This apparatus encompasses a buoyant carrier, an internal energy-harnessing rotor, and a Power Take-Off (PTO) system, as depicted in Fig. 2. The upper and lower shells conjoin to create a hermetic space, shielding the internal components from seawater, thus impeding corrosion. The hydraulic PTO is affixed within a framework and is calibrated to diminish the device's center of gravity. The operational mechanism entails the device's reaction to wave forces, inducing the floating structure to oscillate and the internal rotor to deviate from its nadir potential energy position.

Subjected to gravity, the rotor revolves around its axis, propelling the hydraulic system to engender electricity [10, 11].

Fig. 1. Penguin wave energy converter

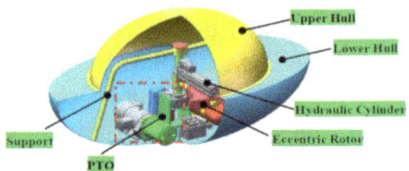

Fig. 2. Internal rotor wave energy device

2.2 Mathematical Model of Motion Response for Wave Energy Devices

Establishing a time-domain mathematical model for a floating internal rotor wave energy device with a Power Take-Off (PTO) system using the impulse response function method [6, 7]:

$$\{M + M_\infty\}\ddot{x}(t) + \int_{-\infty}^{t} C(t - \pi)\dot{x}(\tau)d\tau + \{K\}x(t) + F_{\text{joint}}(t) + F_{\text{PTO}}(t) = F(t) \quad (1)$$

In the equation governing the motion response of wave energy devices, the parameters are delineated as follows:

M_∞ denotes the added mass matrix, which quantifies the inertia augmentation due to the interaction of the device with the surrounding fluid as the wave frequency approaches its zenith.

F_{joint} signifies the vector of resistive forces or moments, which act as the pivotal element in the temporal domain, counteracting the motion of the device and influencing its dynamic response.

F_{PTO} epitomizes the vector of resistive forces or moments engendered by the Power Take-Off (PTO) system in the temporal domain, which are instrumental in converting the mechanical energy of the device's motion into electrical energy.

In this device, the joint function is provided by a unidirectional hinge, with its constraint equation as follows:

$$A_J x(t) = 0, A_J^T f_J(t) = F_{joint}(t) \quad (2)$$

where: f_J represents the column vector of local joint forces or moments in the time domain.

The time-domain velocity potential of the floating body, based on the impulse response function method, is represented as:

$$\phi_i(t) = \dot{x}_i(t)\varphi_j + \int_{-\infty}^{t} \dot{x}_i(\tau)\chi_j(t - \tau)d\tau \quad (3)$$

In the equation elucidating the velocity potential of floating body, the parameters are expounded thus:

ϕ_i embodies the velocity potential of the floating body under the i-th mode, which encapsulates the fluid flow patterns around the body due to its inherent oscillatory motion.

ϕ_j denotes the velocity potential induced by the impulse, reflecting the instantaneous change in momentum imparted to the fluid by the body's motion.

χ_j signifies the velocity potential induced by the impulse after the elapse of time τ, delineating the evolution of the fluid flow patterns consequent to the initial impulse.

In the time domain, the radiation force exerted on the floating body is:

$$F_j(t) = \iint_S \rho \frac{\partial}{\partial t}\left(\frac{\partial \phi_j(t)}{\partial n}\right) ds = m_{ji}\ddot{x}_i(t) + \int_{-\infty}^{t} C_{ji}(t-\tau)\dot{x}_i(\tau)d\tau \tag{4}$$

Assuming the motion of the floating body to be periodic, the following assumptions are made:

$$x_j(t) = \bar{x}_j e^{-i\omega t} \tag{5}$$

The hydrodynamic coefficients in the frequency domain can thus be obtained:

$$\lambda_{ij}(\omega) = m_{ij} - \frac{1}{\omega}\int_0^\infty C_{ij}(t)\cos(\omega t)dt \tag{6}$$

$$\mu_{ij}(\omega) = \int_0^\infty C_{ij}(t)\sin(\omega t)dt \tag{7}$$

The retardation function can be obtained from the hydrostatic restoring coefficient matrix via Fourier transform:

$$C_{ij}(t) = -\frac{2}{\pi}\int_0^\infty \mu_{ij}(\omega)\cos(\omega t)d\omega \tag{8}$$

2.3 Mathematical Model for Wave Energy Conversion

For rotor-type wave energy devices, the primary energy output power can be calculated using the following formula, with units in kW [2, 6]:

$$P = \frac{T \cdot n}{9550} \tag{9}$$

In the formula, n represents the rotor speed. The output torque T can be calculated using the following equation:

$$T = C_{PTO} \cdot \omega \tag{10}$$

In the formula, C_{PTO} is the PTO damping coefficient, and ω is the rotor angular velocity. It is important to ensure uniformity in the units of the damping coefficient and the rotor angular velocity, as well as the conversion relationship between the rotor angular velocity and speed.

2.4 Physical Model and Numerical Model

In adherence to the scale ratio of the real type: model = 5:1, a three-dimensional geometric entity model of the eccentric inner rotor type wave energy device was meticulously crafted. Owing to the intricacies associated with surface machining, the exterior shell of the wave energy device was abstracted into a linear form, as visually represented in Fig. 3. The inner rotor was integrated within the hermetic shell, which was constituted by the upper and lower shells. The floating body, characterized by its dimensions of 1.5 meters in length, 1 meter in breadth, and 0.86 meters in height, was complemented by a physical model with a gross mass of 210 kilograms. The model exhibited a draft depth of 320 mm, with the center of buoyancy positioned 110 mm beneath the waterline, and the center of gravity located 60 mm above the waterline [1, 12]. The mooring arrangement of the device is shown in Fig. 4, and other parameters are shown in Table 1.

Fig. 3. The physical model of the device   **Fig. 4.** Mooring system of the device

Table 1. Other parameters of the device

Name	Value
Center of gravity	(0, 0, -0.06)m
Draft	0.32 m
Depth of water	2 m
Minimum Radius (Lower Shell)	0.25 m
Maximum Radius (Lower Shell)	0.5 m
Mooring radius	2 m
Mooring method	Tensioned type

Numerical simulations of the wave energy device were undertaken with the ubiquitous potential flow software ANSYS AQWA. An equivalent numerical model, congruent in boundary conditions, was constructed within AQWA, as visualized in Fig. 5.

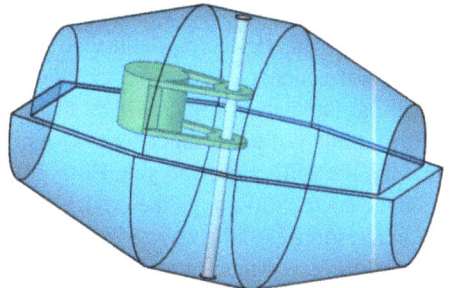

Fig. 5. Numerical model of the device

3 Numerical Simulation Result

3.1 Validation and Comparison

A time-domain numerical model, predicated upon AQWA, was formulated and juxtaposed with the findings from pool experimentation. The investigation centered on the pitch motion response of the device subsequent to the attainment of stable operation (60 s–70 s), as delineated in Fig. 6. The outcomes reveal that the general trajectories of the pool experimental and AQWA numerical simulation results exhibit concordance, albeit with minor disparities in the peak magnitudes of the pitch motion response. Consequently, the employment of AQWA for the numerical simulation of the internal rotor wave energy device is deemed both feasible and precise [12, 13].

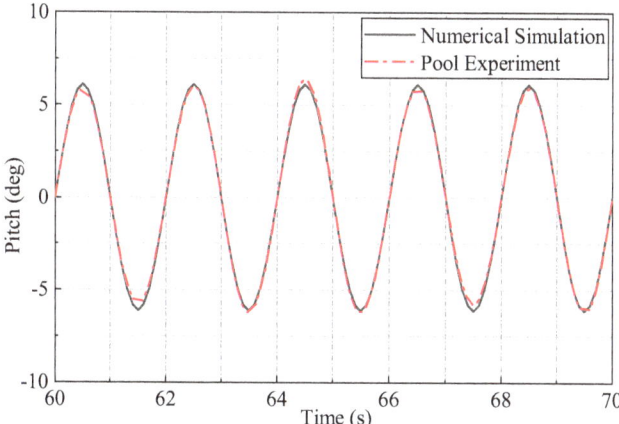

Fig. 6. Comparison of pool experiment and numerical simulation

3.2 Motion Behavior of Floating Bodies in the Frequency Domain

To delve into the motion efficacy of wave energy converters across a diverse array of geometric configurations within the frequency domain, a numerical model of the actual device was meticulously formulated, with a particular emphasis on the scale ratio of the actual model to the physical model being delineated as 5:1. Three distinct wet surface cross-section models were conceived: circular-circular, elliptical-elliptical, and circular-elliptical, as visually articulated in Fig. 7. Given that the device's operational core is predicated upon the rocking motion to engender the rotation of the internal rotor under the influence of gravity, the roll, pitch, and yaw Response Amplitude Operators (RAOs) were employed as the metrics to gauge the device's motion performance. As illustrated in Fig. 8, the motion responses of the wave energy converter were meticulously scrutinized under waves of varying frequencies and a directional incidence of 0°. The findings reveal that the wave energy converter with a circular-elliptical cross-section manifests the most elevated RAO, thereby rendering this model the optimal choice for subsequent computational endeavors.

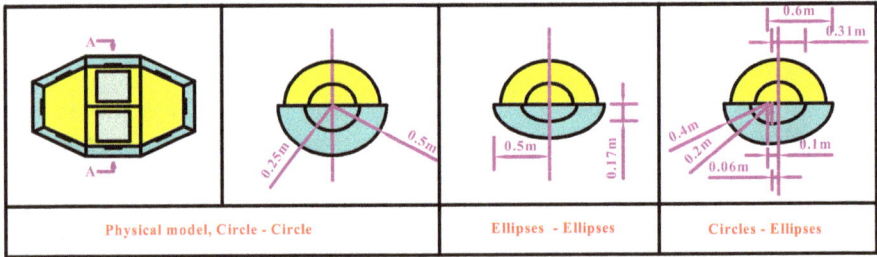

Fig. 7. Wave Energy Converters with Various Cross-Sections

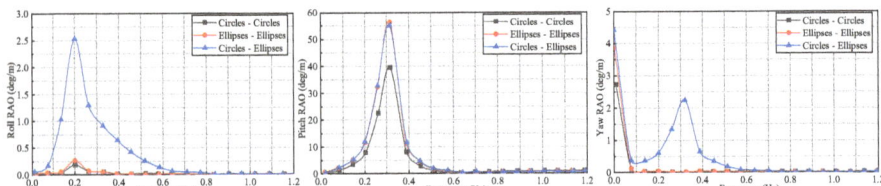

Fig. 8. RAOs of Wave Energy Converters with Various Cross-Sections

Influence of draft and center of gravity position on the motion response of a circular-elliptical cross-section wave energy converter: Fig. 9 delineates the frequency-domain RAOs of the internal rotor-type wave energy converter, subjected to variations in drafts, center of gravity positions, and internal rotor masses, all under the condition of a 0° wave direction. The findings elucidate that more diminutive drafts contribute to amplified motion amplitudes. Furthermore, the identification of optimal center of gravity positions in the pitch and yaw directions can markedly augment the motion response of the wave energy converter, consequentially elevating its power generation efficacy.

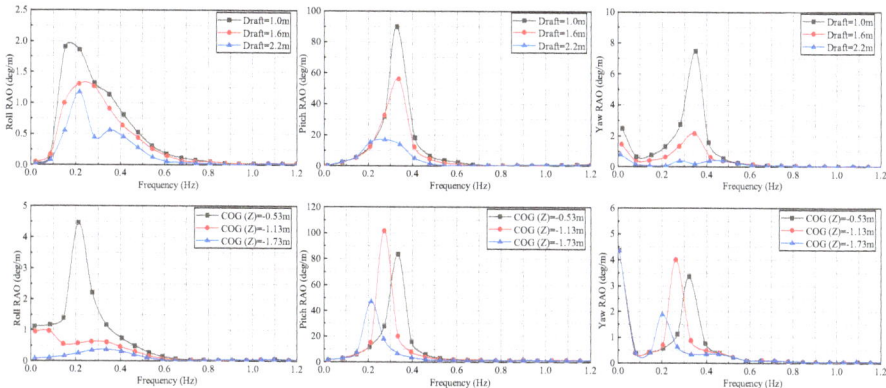

Fig. 9. Roll, pitch, and yaw raos under varying drafts and center of gravity positions

3.3 Research on Energy Capture Characteristics of Device

The energy capture mechanism of the internal rotor-type wave energy converter is predicated on the oscillatory motion of the outer shell induced by wave action, which, in turn, prompts the hinged internal rotor to revolve around its axis under the influence of gravity. This rotational motion actuates the generator, thereby facilitating the generation of electrical energy. Under a constant PTO (Power Take-Off) damping coefficient, the energy capture efficiency of the device is conspicuously mirrored in the rotational velocity of the internal rotor; a more rapid rotor velocity corresponds to a heightened capture efficiency. Employing temporal-domain analysis, this research endeavors to scrutinize the fluctuations in the rotational velocity of the internal rotor and the attendant energy capture characteristics.

For the temporal-domain analysis, regular waves with a height of 2 m and a period of 4.47 s were chosen, alongside a PTO (Power Take-Off) damping coefficient of 200 Nm/(deg/s). The mooring system utilized a taut-line forward three-point configuration, boasting a mooring radius of 10 m. The rotational velocities of the internal rotor, under varying drafts, center of gravity positions, rotor masses, and PTO damping coefficients, are depicted in Fig. 10. This figure reveals that, with a constant PTO damping coefficient, an optimal combination of draft and rotor mass can achieve the maximum rotational speed of the rotor. Additionally, the rotational speed of the rotor increases as the center of gravity of the wave energy device is lowered. Conversely, when the PTO damping coefficient is varied while other parameters remain constant, an increase in the PTO damping coefficient results in a decrease in the rotor's speed.

In accordance with the aforementioned equations, the power generation output of the device is contingent upon the torque and rotor speed, with the output torque being a function of the PTO (Power Take-Off) damping coefficient and rotor speed. Considering the inverse relationship between rotor speed and PTO damping coefficient, there exists an optimal PTO damping coefficient that furnishes a suitable output torque and rotor speed, thereby enhancing the wave energy device's power generation efficiency. Utilizing the primary output power calculation formula, the device's output power under varying PTO damping coefficients can be ascertained, as illustrated in Fig. 11. It becomes apparent

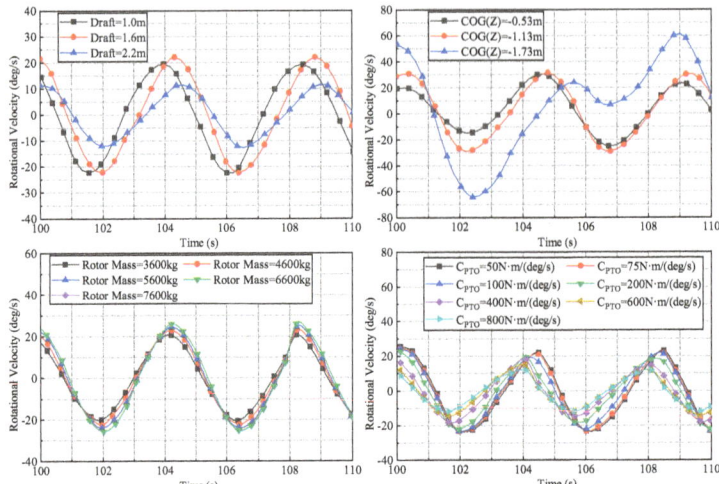

Fig. 10. Rotor speed under different draft, center of gravity, mass and PTO damping coefficient

that the device attains its peak output power when the PTO damping coefficient is 400 Nm/(deg/s).

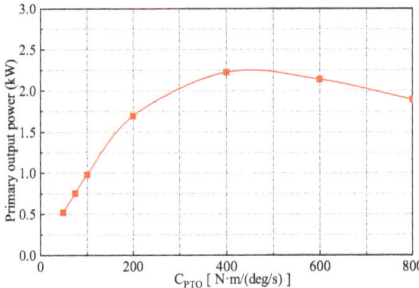

Fig. 11. Output power of wave energy device with different PTO damping coefficient

In light of the aforementioned computational outcomes, optimization of specific parameters for the wave energy device was undertaken, employing a circular-elliptical cross-section, a draft of 1.6 m, a center of gravity position of −1.13 m, an internal rotor mass of 4600 kg, and a PTO (Power Take-Off) damping coefficient of 400 Nm/(deg/s). According to the computational results presented in Fig. 12, the optimized device exhibited substantial enhancements in rotor speed and output power when compared to the pre-optimized device. Specifically, the instantaneous output power surged from 1.16 kW to 2.29 kW, marking a 97.4% improvement, while the average output power ascended from 0.53 kW to 1.08 kW, reflecting a 103.8% increase.

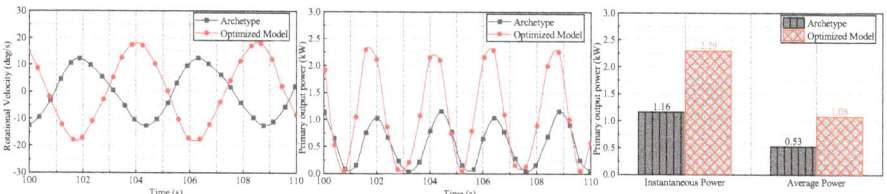

Fig. 12. Output power of wave energy device

4 Conclusion

A groundbreaking floating internal rotor wave energy conversion device has been conceptualized, offering detailed insights into its structural components and operational principles. The fidelity of numerical simulations was verified against tank experiment outcomes, demonstrating strong concordance in both numerical values and observed trends. Frequency domain analysis was employed to scrutinize the roll, pitch, and yaw Response Amplitude Operators (RAOs) of the actual model under various drafts and positions of the center of gravity. The findings revealed that judiciously selected structural parameters could significantly enhance the motion response of the wave energy device. Subsequently, time domain analysis was utilized to examine the rotational speed of the hinged internal rotor, driven by the shell. The study delved into the patterns of rotational speed variations influenced by different drafts, center of gravity positions, rotor masses, and PTO (Power Take-Off) damping coefficients. Leveraging the output power calculation formula, the variations in output power for the optimized device were elucidated. The results indicated that the instantaneous output power surged from 1.16 kW to 2.29 kW, reflecting a 97.4% enhancement, while the average output power ascended from 0.53 kW to 1.08 kW, signifying a 103.8% increase.

References

1. Wanqing, Z.: Research on Hydrodynamic and Energy Capture Characteristics of Floating Inner Rotor Wave Energy Device. Harbin Engineering University (2023) (in Chinese)
2. Yuxiang, N.: Investigation on Hydrodynamic and Energy Conversion Characteristics of a Floating Multi-dimensional Oscillating Pendulum Wave Energy Convertor. Harbin Engineering University (2023) (in Chinese)
3. Zhongliang, M.: Mechanism Research of Wideband Energy Capture for Horizontal Axis Rotor Wave Energy Generator. Shandong University (2021) (in Chinese)
4. Giovanni, B., et al.: A gyroscopic mechanism for wave power exploitation. Mech. Mach. Theory. **46**(10), 1411–1424 (2011)
5. Negandari, M., et al.: Design of A Two-Body Wave Energy Converter by Incorporating the Effect of Hydraulic Power Take-Off Parameters. J. Mar. Sci. Technol.-Taiwan. **26**(4), 496–507 (2018)
6. Zengming, Z.: Research on characteristics and optimization design of an array of cantilever oscillating float wave energy conversion device. Sch. Energy Power Mech. Eng. (2019) (in Chinese)
7. Kim, D.: Multibody analysis of wave energy converters and a floating platform in the time domain. J. Mar. Sci. Eng. **12**(2), 265 (2024)

8. Xinhui, C.: Optimization on the Internal Rotor Wave Energy Converters. Jimei University (2022) (in Chinese)
9. Hongzhou, H., et al.: Design and Optimization on double Screws Internal Rotor Wave Energy Converter. J. Guangzhou. Marit. Univ. **1009-8526**(2022), 03-0045-07 (2022) (in Chinese)
10. Liu, D., et al.: Dynamic response investigation of dynamic cable in wave energy generator under shallow water and severe sea conditions. In: 2021 4th International Conference on Mechatronics and Information Technology., vol. 2021, p. 023 (2021)
11. Liu, D., et al.: Research on the power generation performance of eccentric rotor type wave energy generation device under different wave conditions. In: 2021 4th International Conference on Mechatronics and Information Technology., vol. 2021, p. 022 (2021)
12. Xue, G., et al.: Control parameters optimization of accumulator in hydraulic power take-off system for eccentric rotating wave energy converter. J. Mar. Sci. Eng. **11**(4), 11040792 (2023)
13. Xue, G., et al.: Experimental investigation of mooring performance and energy-harvesting performance of eccentric rotor wave energy converter. J. Mar. Sci. Eng. **10**(11) (2023)

Open Access This chapter is licensed under the terms of the Creative Commons Attribution-NonCommercial-NoDerivatives 4.0 International License (http://creativecommons.org/licenses/by-nc-nd/4.0/), which permits any noncommercial use, sharing, distribution and reproduction in any medium or format, as long as you give appropriate credit to the original author(s) and the source, provide a link to the Creative Commons license and indicate if you modified the licensed material. You do not have permission under this license to share adapted material derived from this chapter or parts of it.

The images or other third party material in this chapter are included in the chapter's Creative Commons license, unless indicated otherwise in a credit line to the material. If material is not included in the chapter's Creative Commons license and your intended use is not permitted by statutory regulation or exceeds the permitted use, you will need to obtain permission directly from the copyright holder.

Design and Analysis of Hexagonal Floating Photovoltaic Platform and Mooring System

Jiayang Sun[✉], Yihou Wang, Tie Ren, and Taocui Yu

Engineering and Technology China Offshore Engineering and Technology Co., Ltd., Shanghai, China
sunjiayang@coffshore.cn

Abstract. Floating photovoltaic platform is recognized as one of the most potential technology development direction in the industry. How to maintain the strength and stability of the structure under harsh marine environmental conditions is of great significance for the development and application of far-reaching marine solar energy. Based on practical engineering, a hexagon truss type floating photovoltaic platform is designed in this paper, which uses buoyancy block to provide buoyancy and adopts a 6 × 1 spread mooring system. The whole platform has the advantages of small waterplane, stable platform desirable motion response and small mooring tension. Firstly, an integrated numerical model of truss-buoyancy block and mooring system is established in commercial software OrcaFlex, subsequently the dynamic coupling calculation is carried out in the full-time domain, and the platform's motion response and mooring tension characteristics under extreme sea conditions with high/low tide levels are investigated, respectively. The results show that the platform has excellent hydrodynamic performance as well as the tension on mooring chains is small, which can maintain stable operation under extreme sea conditions. The analysis and results can provide important reference for practical engineering.

Keywords: Floating photovoltaic platform · Truss structure · Buoyancy block · Spread mooring

1 Introduction

Promoting the development of renewable energy to realizing the replacement of traditional fossil energy is an inevitable choice to ensure national energy security, a great vision and grand blueprint put forward by the "14th Five-Year Plan", and an inevitable requirement to achieve the goal of "carbon peak and carbon neutrality" [1, 2].

In recent years, China has made remarkable achievements in the development and application of solar energy, and the proportion of photovoltaic modules occupying the global market share has exceeded 70%. However, the traditional large-scale land photovoltaic development occupies a lot of land resources, which greatly limits the development of industry and agriculture, so the implementation of offshore photovoltaic can effectively improve the utilization rate of land, and is also an important measure for China to achieve the development strategy of "Marine power" [3].

At present, the offshore photovoltaic project is in the demonstration or pre-commercialization stage, mainly fixed photovoltaic and floating photovoltaic two kinds, fixed photovoltaic often use pile fixed support structure to fix the photovoltaic module, with safe, reliable and low-cost advantages, but it is often used in the coastal shallow waters. Therefore, floating photovoltaic came into being, which is normally composed of photovoltaic module, support structure and mooring anchoring system. At present, there are two common forms, flexible membrane-based platform (Fig. 1(a)) and bowl-based platform (Fig. 1(b)).

(a) Flexible membrane-based photovoltaic platform

(b) Bowl-based photovoltaic platform

Fig. 1. Common floating photovoltaic platform.

Many scholars have conducted a great deal of numerical simulation and experimental studies on floating photovoltaic platforms. Kim [4] conducted numerical simulation calculations on the mooring system of a 1 MW tensioned floating photovoltaic platform, and analyzed the safety and feasibility of engineering construction. Cazzaniga [5] carried out finite element analysis and test verification of the floating offshore photovoltaic structure, and explored the overall motion response of the photovoltaic platform under the action of waves. Han [6] studied a composite photovoltaic platform with a new type of self-floating fiber reinforced polymer as the main body, and explored its hydrodynamic response characteristics under the combined action of wind, wave and current. Degenkamp [7] conducted model test studies on mooring lines of different types of photovoltaic platforms, and measured the tangential and normal soil resistance coefficients of mooring system. Based on the measured wind and wave conditions of the Marine environment, Xu [8] carried out the analysis of the dynamic coupling characteristics of the mooring structure of the deep-water floating photovoltaic platform in the time domain. Zhang [9] used large-scale general software AQWA to conduct hydrodynamic research on three kinds of FPV (Floating Photovoltaic) array systems with floating body structures.

Aiming at a certain sea area in Huang Hai, this paper innovatively proposes a new type of regular hexagon truss type floating photovoltaic platform. Trussed rod elements are selected as the floating foundation and buoyancy blocks are wrapped on the bottom truss to provide buoyancy, which has advantages such as small waterplane, stable platform motion response and small mooring tension. At the same time, according to the symmetry of the platform, a 6 × 1 spread mooring system is used. In this paper, the platform's motion response and mooring tension characteristics under extreme sea conditions with high/low tide levels are investigated, which provides a reliable basis for practical engineering.

2 Basic Theory

2.1 Environment Condition

The fluid motion within the flow field adheres to the continuity equation. Assuming seawater behaves as an ideal, inviscid and incompressible fluid, the continuity equation can be expressed as follows:

$$\nabla \vec{V} = \frac{\partial V_x}{\partial x} + \frac{\partial V_y}{\partial y} + \frac{\partial V_z}{\partial z} = 0 \tag{1}$$

where, V_x, V_y and V_z respectively represent the components of water quality point velocity in the three axis directions; $\nabla^2 = \nabla \cdot \nabla = \frac{\partial^2}{\partial x} + \frac{\partial^2}{\partial y} + \frac{\partial^2}{\partial z}$ represent the Laplacian Operator.

Based on the potential flow theory, there exists a unit velocity potential function $\Phi(x, y, z, t)$, which satisfies:

$$\vec{V} = \nabla \Phi(x, y, z, t) \tag{2}$$

where, t represents the time variable.

Then, the equations of continuity equation can be written as:

$$\frac{\partial^2 \Phi}{\partial x^2} + \frac{\partial^2 \Phi}{\partial y^2} + \frac{\partial^2 \Phi}{\partial z^2} = 0 \tag{3}$$

The surface kinematic boundary condition can be written as:

$$\frac{\partial^2 \Phi}{\partial t^2} + g \frac{\partial \Phi}{\partial z}\bigg|_{z=0} = 0 \tag{4}$$

It can be deduced that the velocity and acceleration components u and a in the x direction of an Airy wave with circular frequency ω at a finite water depth are:

$$u = \omega \zeta_a \frac{\cosh k(z+h)}{\sinh kh} \sin(\omega t - kx) \tag{5}$$

$$a = \omega^2 \zeta_a \frac{\cosh k(z+h)}{\sinh kh} \cos(\omega t - kx) \tag{6}$$

where, k represents the wave number; h represents the water depth; ζ_a represents the wave amplitude.

The rise in the wave front of an irregular wave propagating along the X-axis can be seen as a combination of numbers of Airy waves:

$$\zeta = \sum_{j=1}^{N} A_j \sin(\omega_j t - k_j x + \varepsilon_j) \tag{7}$$

where, A_j, ω_j, k_j and ε_j respectively represents the amplitude, circular frequency, wave number and random phase angle of the jth Airy wave.

Then, the velocity and acceleration components u and a in the x direction of irregular wave are:

$$u = \sum_{j=1}^{N} \omega_j A_j e^{k_j z} \sin(\omega_j t - k_j x + \varepsilon_j) \tag{8}$$

$$a = \sum_{j=1}^{N} \omega_j^2 A_j e^{k_j z} \cos(\omega_j t - k_j x + \varepsilon_j) \tag{9}$$

2.2 Morison Equation

Morison equation is an empirical formula proposed by MORISON to calculate wave load on small-scale structures. It is assumed that the existence of structures has no obvious influence on the movement of waves. The horizontal wave force dF_H of moveable column per unit length dz can be written as follows:

$$dF_H = \frac{1}{2} C_d \rho D (v_x - v_{sx}) |v_x - v_{sx}| dz + \left[(1 + C_m) \rho \frac{\pi D^2}{4} \frac{dv_x}{dt} - C_m \rho \frac{\pi D^2}{4} \frac{dv_{sx}}{dt} \right] dz \tag{10}$$

where, v_x and $\frac{dv_x}{dt}$ are respectively the horizontal velocity and acceleration of the water particle at the height z of the column central axis, including the contribution of wave and current; D is the diameter of the column; C_d is the drag force coefficient perpendicular to the axis of the cylinder, consider taking 0.65 here according to the specification; C_m is the additional mass coefficient, consider taking 1 here; v_{sx} is the horizontal velocity of the moveable column. The first term in formula is the drag force term, and the second term is the inertial force term. The wave force received by the cylinder can be obtained by integrating along the cylinder.

The calculation of the wind load is similar to that of the wave force, which is assumed to be a steady force and does not take into account the inertial force and the motion response of the platform.

3 Model Building and Calculation

3.1 Floating Photovoltaic Platform

The target platform is composed of floating foundation, photovoltaic module and mooring system. The floating foundation is a hexagonal truss structure, and buoyancy is provided by a buoyancy block wrapped on the bottom truss structure. The photovoltaic module is composed of 400 photovoltaic panels, and the power generation can reach 300 kW. Photovoltaic panels are welded to the floating foundation. Considering the symmetrical structure of the floating foundation, the mooring system adopts 6 × 1 multi-point mooring mode. Each mooring line is divided to 2 parts: Upper mooring line (R3S) and lower mooring line (R3S), the former is connected to the floating foundation, and the latter is fixed to anchor. The basic parameters of the platform and it's mooring system are shown in Table 1, and the platform model is shown in Fig. 2.

Table 1. The basic parameters of the platform.

Parameters	Units	Values
Single side of deck	m	28
Depth	m	20
Draft	m	2.5
Total weight	t	400
Height of center of gravity	m	6
Diameter of upper mooring line	mm	80
Length of upper mooring line	m	120
Diameter of lower mooring line	mm	120
Length of lower mooring line	m	145

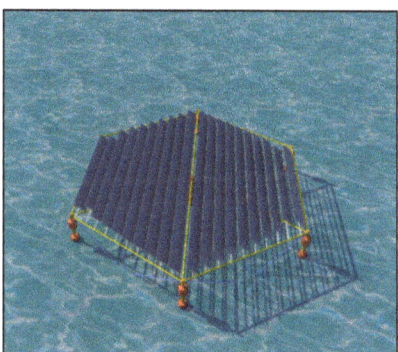

Fig. 2. Schematic Diagram of Floating Photovoltaic Platform.

3.2 Numerical Simulation Model and Coordinate System

Orcaflex was selected for numerical simulation, the photovoltaic panels of the platform have not been considered in numerical simulation, a 6D buoy platform pipe column and buoyancy block model were established in the software, and the hydrodynamic load on the platform was calculated by Morison equation (Fig. 3). At the same time, considering the instantaneous wet surface effect on the platform, the change of the buoyancy block and the wind load caused by the tilt of the photovoltaic panel, the nonlinear time-domain coupling calculation and analysis of the floating foundation and the mooring system are carried out, and the motion response characteristics of the floating foundation and the mooring cable tension characteristics are analyzed.

3.3 Environments and Calculation Conditions

The design or evaluation conditions of marine structure platforms are usually based on environmental conditions. For the platform, 30 m average water depth and environmental

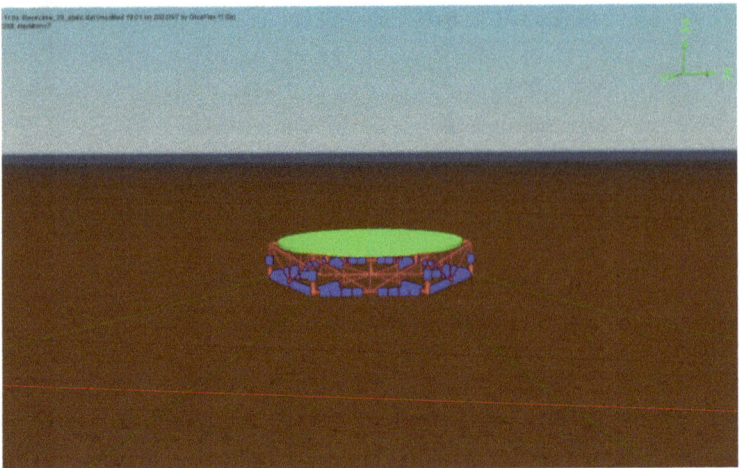

Fig. 3. Numerical Simulation Model of Floating Photovoltaic Platform.

condition with a return period of 50 year as the design basis. Table 2 shows the specific data of wind, wave, current and tidal range. JONSWAP wave spectrum is used, wind and current are simulated as constant.

Table 2. Parameters of environment conditions.

Parameters	Units	Values
Wave Hs	m	6
Wave Tp	s	13.7
Wind velocity	m/s	40
Current velocity	m/s	1.5
High/Low tide level	m	30 ± 1.8

Since the operating water depth of this platform is relatively shallow, it has a great test for the mooring system, so the impact of extreme environmental tidal level difference on the platform needs to be analyzed. Considering the symmetry of platform and mooring cable arrangement, environmental conditions with three directions (0°, 30° and 60°) and three design water depths (average water depth, high water and low water depth) were selected to carry out numerical calculation. Totally 6 working conditions are used in the analysis, as shown in Table 3. Figure 4 shows the arrangement of mooring lines and coordinate system.

Table 3. Working conditions table.

Condition number	Wave direction	Wind direction	Current direction	Water depth
1	0 deg	0 deg	0 deg	30 m
2	0 deg	0 deg	0 deg	31.8 m
3	0 deg	0 deg	0 deg	28.2 m
4	30 deg	30 deg	30 deg	30 m
5	30 deg	30 deg	30 deg	31.8 m
6	30 deg	30 deg	30 deg	28.2 m

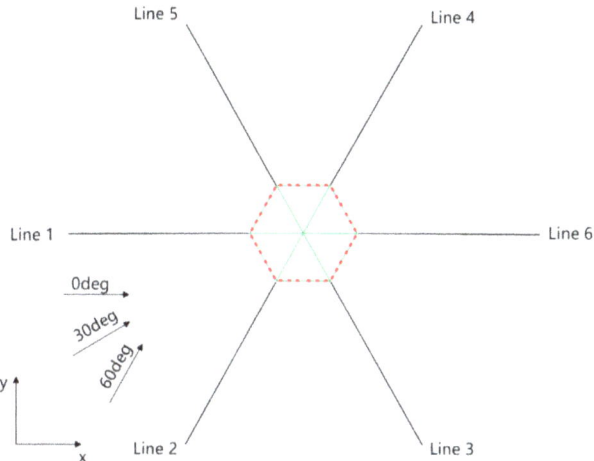

Fig. 4. Arrangement of mooring lines and coordinate system

4 Results and Analysis

4.1 Analysis of FPV Motion Response Characteristics

Table 4 and Fig. 5 show the maximum amplitude of the motion response of the FPV at different tidal ranges under the action of wind, wave and currents. It can be seen that the motion response maxima of the platform is relatively smaller at the high tide level and relatively larger at the low tide level. This is because the pretension of mooring system increases with the increase of tide level, and the recovery stiffness from mooring system also increases, so the platform motion response decreases significantly. At the same time, mooring system provides smaller recovery stiffness in heave direction, thus the difference of heave response under different tidal ranges is not obvious. In addition, the motion response of FPV under 0°. is greater than that under 30°.

Table 4. Maximum response amplitudes of the platform.

Condition number	1	2	3	4	5	6
Surge (m)	2.0	1.8	2.2	1.7	1.6	1.9
Heave (m)	4.8	4.5	4.7	4.8	4.8	4.7
Pitch (deg)	8.2	8.0	8.3	7.2	7.0	7.7

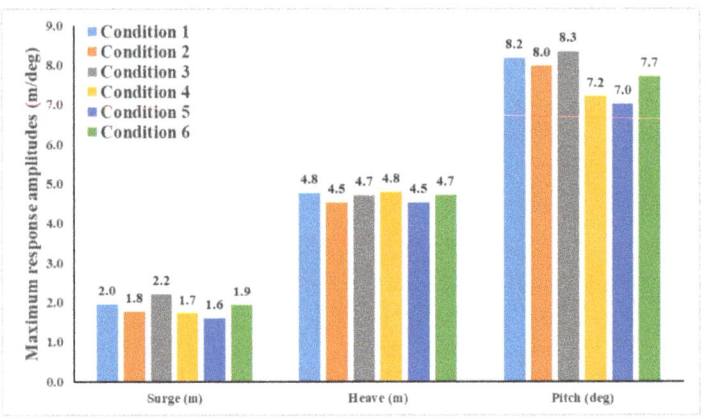

Fig. 5. Maximum response amplitudes of the platform

4.2 Analysis of Mooring Line Tension Characteristics

Table 5 and Fig. 6 shows the maximum mooring tensions of the six mooring cables at different tidal ranges under the action of wind current and waves. It can be seen, higher tension occurs at high tide level as well as smaller tension is produced at lower tide level. This is due to the recovery stiffness provided by the mooring system is larger at higher water depth. At the same time, the mooring tension under 0°. is greater than that under 30°. In all working conditions, the safety factor of mooring cable is greater than 1.67, which meets the rule's requirements (refer to CCS standard).

Table 5. Statistics of mooring tension.

Condition number	1	2	3	4	5	6
Maximum Mooring tension (t)	265.67	272.70	264.36	245.04	258.78	228.35
Safety factor	2.15	2.10	2.16	2.33	2.21	2.50

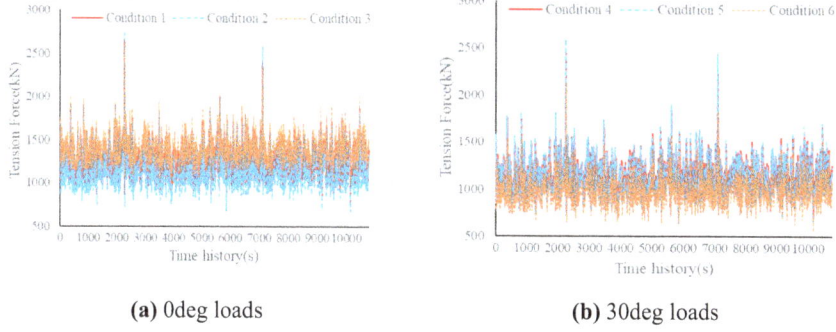

(a) 0deg loads (b) 30deg loads

Fig. 6. Time history of mooring line tensions.

5 Conclusion

In this paper, the platform motion response characteristics and mooring cable tension characteristics of FPV in an engineering project are analyzed under extreme sea conditions of high/low tide in 50 years. The main conclusions are as follows:

1) With the increase of the tide level, the motion response amplitude of FPV decreases significantly, while the maximum mooring tension increases significantly, which is caused by the increase of the pretension of the mooring system.
2) The maxima of motion response amplitude and the maximum mooring tension of FPV occurs at 0 deg. wave/wind direction.
3) In all working conditions, the maximum tension of the mooring cable is 272.70 t, and the safety factor is 2.10, which meets the rule's requirements.
4) It can be seen, the regular hexagonal truss FPV has good motion response characteristics and small mooring tension under extreme sea conditions, which can provide a reliable basis for practical engineering.

There are still some limitations in this paper, such as the lack of comparison between test and numerical analysis, which will be further supported by further research on model tests.

References

1. National Development and Reform Commission: National energy administration. In: Fourteenth Five-Year Plan Modern Energy System Planning. [EB/OL] (2022) (in Chinese)
2. National Development and Reform Commission: National energy administration. In: Implementation Plan for Promoting High Quality Development of New Energy in the New Era: State Office Letter. [EB/OL] (2022) (in Chinese)
3. Chen, J., et al.: Current status and prospect of membrane-based offshore floating photovoltaic technology. South. Energy Constr. **10**(2), 1–10 (2023) (in Chinese)
4. Kim, S.H., et al.: Design and construction of 1 MW class floating PV generation structural system using FRP members. Energies. **10**(8), 1142 (2017)
5. Cazzaniga, R., et al.: Compressed air energy storage integrated with floating photovoltaic plant. J. Energy Storage. **13**, 48–57 (2017)

6. Han, J., et al.: Development of self-floating fiber reinforced polymer composite structures for photovoltaic energy harvesting. Compos. Struct. **253**, 112788 (2020)
7. Degenkamp, G., et al.: Soil resistances to embedded anchor chain in soft clay. J. Geotech. Eng. Div. **115**(10), 1420–1438 (1989)
8. Pu, X., et al.: Dynamic response analysis of mooring structure of deep-water floating photovoltaic platform. Acts Energ. Sol. Sin. **44**(10), 156–164 (2023) (in Chinese)
9. Zhang, J., et al.: Design and optimization of single-module structure of offshore FPV system. Sh. Sci. Technol. **45**(19), 104–110 (2023) (in Chinese)

Open Access This chapter is licensed under the terms of the Creative Commons Attribution-NonCommercial-NoDerivatives 4.0 International License (http://creativecommons.org/licenses/by-nc-nd/4.0/), which permits any noncommercial use, sharing, distribution and reproduction in any medium or format, as long as you give appropriate credit to the original author(s) and the source, provide a link to the Creative Commons license and indicate if you modified the licensed material. You do not have permission under this license to share adapted material derived from this chapter or parts of it.

The images or other third party material in this chapter are included in the chapter's Creative Commons license, unless indicated otherwise in a credit line to the material. If material is not included in the chapter's Creative Commons license and your intended use is not permitted by statutory regulation or exceeds the permitted use, you will need to obtain permission directly from the copyright holder.

Application of Structural Response Prediction Methods for Real Time Structural Health Monitoring of Multi-Linked Floating Offshore Structures

Kichan Sim[1], Byoung Wan Kim[1,2], Sa Young Hong[1,2], Hong Gun Sung[3], and Kangsu Lee[1,2]($\boxtimes$)

[1] Korea Research Institute of Ships and Ocean Engineering School, National University of Science and Technology, Daejeon 34103, Republic of Korea
klee@kriso.re.kr

[2] Eco-friendly Ocean Development Research Division, Korea Research Institute of Ships and Ocean Engineering, Daejeon 34103, Republic of Korea

[3] Deep Ocean Engineering Research Center, Korea Research Institute of Ships and Ocean Engineering, Busan 46729, Republic of Korea

Abstract. Recently, with the emergence of climate change and environmental pollution issues due to global warming, the importance of renewable energy generation has been highlighted. As demand increases, many of these systems are being installed offshore, where space utilization is more feasible. However, various environmental loads such as waves, wind, and currents act on these systems, and access is more difficult compared to onshore installations, resulting in high maintenance costs. Recently, to overcome these challenges, research has been conducted on implementing digital twins. This involves synchronizing sensor data measured from actual operating structures with virtual models in real-time, allowing users to monitor desired responses and make quick maintenance decisions. To prevent accidents, it is crucial for virtual models to compute responses faster using real-time measurement data. In this study, fluid-structure coupled analysis was performed on multi-linked floating offshore structures. And then, transfer functions between virtual sensors and structural responses at locations where significant stresses occur were calculated. These transfer functions were derived to predict structural responses using measured sensor data obtained from model tests, and we validated the application of this prediction method. Furthermore, it was evaluated the performance of the prediction method by comparing its results with measured data. Ultimately, we confirmed predicting structural responses with less than about 3% error compared to measured data, at computational speeds of less than 1 s.

Keywords: Multi-linked Floating Offshore Structures · Transfer Function · Fluid-structure Coupled Analysis · Structural Response Prediction Method

1 Introduction

In the past, marine space utilization technologies have been studied to secure food resources of sea and to support the development of oil and natural gas reserves. However, with the recent emphasis on renewable energy, research on the development of structures for renewable energy generation that can operate at sea has become more active. In offshore environments, where strong environmental loads like wind and waves are present, it is essential to ensure the structural integrity of the structures through proper design, along with regular inspections and maintenance. Although designs are conducted with high safety factors during the design stage of offshore structures, structural damage caused by actual loads on the structure can lead to structural defects. To effectively prevent this, it is necessary to ensure structural integrity by detecting structural damage in real-time and responding quickly with maintenance operations. In order to ensure real-time structural integrity, fast simulations with low computational cost are required. To achieve this, research is being conducted to implement digital twin models through statistical analysis, model order reduction, and machine learning. Vu et al. 2023, [1] created a meta-model based on a support vector regression algorithm to estimate the hydrodynamic coefficients of a ship using its Automatic Identification System (AIS) data. Kim and Kim 2024, [2] performed statistical analysis for each different scenarios to enable faster structural integrity assessments of floating offshore wind turbines. As a result, they were able to estimate the most probable maximum (MPM) equivalent stress with 1/3 of the simulation time. In this study, conversion with distortion base mode was applied and verified to predict the structural response at unmeasured locations using measured data collected from sensors, focusing on the multi-linked floating offshore structures (hereafter, MLFS) in Fig. 1. And the method was validated using measured data from model test.

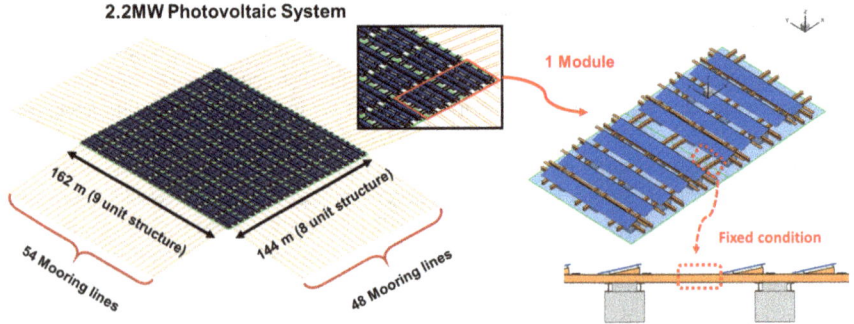

Fig. 1. Configuration of multi-linked floating offshore structures [5].

The MLFS, developed by the Korea Research Institute of Ships and Ocean Engineering (hereafter, KRISO), which can be expanded for various purposes. Kim [3] expanded the MLFS by connecting it with hinged connection to a 2.2 MW photovoltaic system and developed an advanced numerical method capable of time-efficiently calculating the motion and structural response of the structure, verified through model test [4, 5]. In this study, the hinged connection was replaced with fixed connection cause of the nonlinearity at hinged connection, and application of conversion matrix method based on distortion base mode was conducted on MLFS 1 module, consisting of only two unit structures. The distortion base mode refers to the structural deformed mode of the structure under wave loads, and by configuring the conversion matrix with this modes, it effectively reflects the physical characteristics of the structure, enabling accurate structural response prediction.

2 Methodology of Structural Response Prediction

2.1 Principal Particulars

The MLFS 1 module is composed of 18 rectangular floating bodies and 24 I shaped beam connectors, with the two unit structures connected by fixed connections. Additionally, 36 mooring lines connect all the beam connectors at the edges with a taut mooring configuration. Table 1 shows the principal particulars of the MLFS 1 module and the model fabricated in 1/8 scale for model test. The scale ratio for the model test was determined considering the specifications of the ocean engineering basin in KRISO.

Table 1. Specification of scale-down model for model test [6]

Item		Real structure	Model (1/8 scale)
Floating body	Material	Aluminum	
	Breadth × length × Height	1.80 m × 1.80 m × 1.72 m	0.225 m × 0.225 m × 0.215 m
	Draft	1.005 m	0.125 m
Beam Connector	Material	Aluminum	
	Bending stiffness	49,290,720 N-m^2 (I-shaped)	1504 N-m^2 (Rectangular)
Total structure	Breadth × Length × Height	18 m × 36 m × 1.72 m	2.25 m × 4.5 m × 0.215 m
	Draft	1.005 m	0.125 m
	Mass	60,082 kg	117.35 kg

For the numerical model, it was modeled as shown in Fig. 2.(a), based on the real-scale structure, while the model was constructed with scaled dimensions, as depicted in Fig. 2.(b). Since bending stress in the beam connectors is a major structural response for the MLFS, the beam connectors for model test were manufactured with dimensions matching the scaled bending stiffness of rectangular cross-sections, which are easier to produce. For the MLFS, both a panel model for motion analysis and a structural model for structural analysis are needed. The panel model consists of 432 boundary elements, and the structural model comprises 1056 structural beam elements [6].

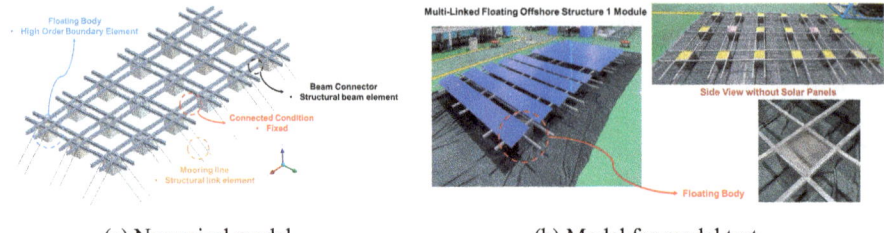

(a) Numerical model (b) Model for model test

Fig. 2. 1 Module structure for numerical analysis and model test

2.2 Fluid-structure Coupled Analysis

The motion and structural analysis of the MLFS 1 module were performed using in-house codes developed by the KRISO. First, the wave loads $f_{wave}(\omega)$ and hydrodynamic characteristics acting on the floating bodies were calculated in the frequency domain using a higher order boundary element method [7]. The motion response of the structure $x(\omega)$ was then calculated using Eq. (1).

$$\left(-\omega^2[M_B + M_{add}] - i\omega[C_B] + [K_B]\right)\{x(\omega)\} = \{f_{wave}(\omega)\} \quad (1)$$

where ω is wave frequency. $[M_B]$, $[M_{add}]$, $[C_B]$, and $[K_B]$ are matrices of mass, added mass, hydrodynamic damping and hydrostatic stiffness, respectively. Subsequently, wave load $f_{wave}(t)$ in time domain were obtained through convolution transformation as described in Eq. (2), [8], and by coupling with the structural model, the coupled motion and structural responses were calculated based on Eq. (3), [9].

$$x(t) = \int_0^t f(t-\tau)h(\tau)d\tau + x_{t=0} \quad (2)$$

$$[M + M_B + M_{add}(\infty)]\{\ddot{u}(t)\} + [C + C_B]\{\dot{u}(t)\} + [K + K_B]\{u(t)\} = \{f_{wave}(t)\} \quad (3)$$

where f(t) is input force and h(t) is the impulse function due to a unit impulse input. In Eq. (3), [M], [C], and [K] represent the mass, structural damping, and stiffness matrix of the MLFS. To reduce computational costs, the floating bodies in the structural model were modelled as lumped bodies. The wave loads acting on the floating bodies were integrated over the body surface and applied as point loads. For the beams transferring wave loads to the beam connectors, they were modeled with very high stiffness beam elements to ensure energy transfer without loss.

2.3 Construction of Conversion Matrix

The conversion matrix method is fundamentally a technique that transforms measured structural responses into predicted structural responses. This method follows the assumption of the mode superposition method, where any structural response S can be expressed as a linear superposition of mode amplitudes ξ and distortion bas modes S^N, as shown in Eq. (4). Both the measured data S_m and unmeasured location data S_p can also be expressed using Eq. (4), and ultimately, the conversion matrix A is calculated according to the relationship in Eq. (5). Where B, M is matrix with distortion base modes for measured and unmeasured [10, 11].

$$S = \xi_1 S^1 + \xi_2 S^2 + \cdots + \xi_N S^N \qquad (4)$$

$$S_p = B \cdot \xi = B \cdot M^+ S_m = A \cdot S_m \because S_m = M \cdot \xi \qquad (5)$$

In the case of the MLFS 1 module, as shown in Fig. 3, the measured and unmeasured locations use an optimized sensor arrangement based on previous studies [12]. A pool of distortion base mode was calculated through fluid-structure coupled analysis under 1000 regular wave conditions, as shown in Table 2. To improve the prediction accuracy for unmeasured locations, the effective distortion base modes must be selected to construct A, and these modes should be orthogonal to each other. To achieve this, modes with the smallest autocorrelation coefficients were extracted through a correlation coefficient analysis. This analysis was performed to select modes that are orthogonal to the distortion base mode with the highest autocorrelation coefficient. By using this approach, the selected modes ensured minimal correlation, thereby allowing for the accurate construction of the conversion matrix and improving the prediction accuracy of unmeasured structural responses.

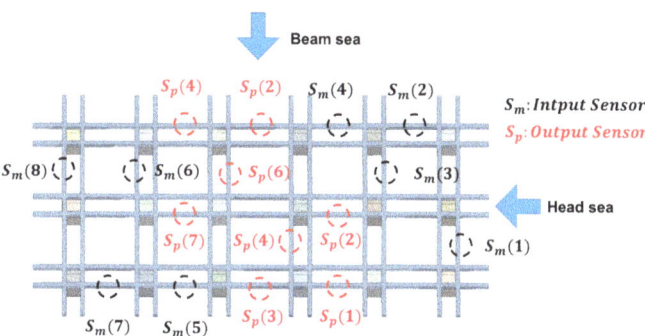

Fig. 3. Sensor placements for conversion matrix method

Table 2. Results of prediction accuracy for irregular wave conditions

Significant wave height	Peak period	Root mean squared error
0.8 m	4.0 s	5.53%
	6.0 s	4.64%
	8.0 s	3.87%
1.0 m	4.0 s	6.20%
	6.0 s	4.75%
	8.0 s	4.26%

3 Results of Structural Response Prediction

The fluid-structure coupled analysis for the MLFS 1 module revealed that, unlike ship structure that large bending stress occur at the midship under head sea conditions, the large bending stress occurs under oblique sea conditions due to the superposition of complex modes. Therefore, in this study, the conversion matrix method was applied and validated using measurement data for irregular wave model test under oblique sea conditions.

The performance of the conversion matrix was evaluated by calculating the root mean squared error (RMSE) between the measured and predicted bending stress, as shown in Table 2. The largest bending stress of 29.4 MPa occurred at sensor Sp(3), located at the connection point of the unit structures corresponding to the midship of the structure, as shown in Fig. 4. Although the conversion matrix tends to predict the bending stress bigger than before in certain regions, the overall trend and phase of response closely match the measured results. Additionally, as shown in Table 2, it was confirmed that as the peak period increases and the wave height decreases, wave nonlinearities such as green water are reduced. This leads to an improvement in the performance of the conversion matrix with the accuracy improving from 6% to 4%, which is based on the linear relationship between fluid and structure. The reduction in nonlinear effects allows the linear-based conversion matrix method to more accurately predict the structural response under these conditions.

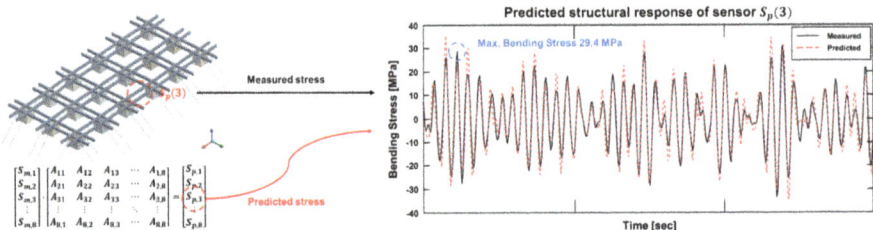

Fig. 4. Predicted bending stress under irregular wave load condition

In the case of Sp(2), since it is located at the midship of the structure like Sp(3), the predicted bending stress has the same error as Sp(3) in head sea condition. For the remaining six sensors, the predicted bending stresses using the conversion matrix were found to be within a maximum error of 8.3%. The increase in error is attributed to the relatively smaller bending stress compared to the Sp(3) sensor, which results in a relatively increased influence of noise.

4 Conclusion

In this study, based on fluid-structure coupled analysis, distortion base mode were generated, and principal modes were selected using orthogonality to construct the conversion matrix. The method were used to accurately predict the structural response, validating using measured data. In actual operation, as corrosion or structural damage occurs, it is necessary to update the numerical model with the changed stiffness or mass to reflect the altered structural characteristics. This requires re-performing numerical simulations for each regular wave load condition, which may incur some computational costs in the offline stage. However, since this method effectively captures the structural characteristics, it enables highly accurate predictions of the structural response. For large-scale photovoltaic systems, maintenance by human can be time-consuming and costly. However, by applying the methods, it is expected to be effective in ensuring structural integrity through real-time monitoring, reducing the need for extensive manual maintenance.

Acknowledgements. This research was supported by the Endowment Project of "Core Technology Development of Hydro-elasticity based Structural Damage Assessment for Offshore Structures considering Uncertainty (5/5)" funded by the Korea Research Institute of Ships and Ocean Engineering (PES5150); and the Korea Institute of Energy Technology Evaluation & Planning (KETEP) and the Ministry of Trade, Industry & Energy (MOTIE) of the Republic of Korea (No. 20223030020240).

References

1. Vu, H.T., Park, J., Yoon, H.K.: Estimating hydrodynamic coefficients of real ships using ais data and support vector regression. J. Ocean Eng. Technol. **37**, 198–204 (2023)
2. Kim, B., Kim, B.: Stress estimation based on stochastic method for strength evaluation of floating wind turbines. J. Ocean Eng. Technol. **38**, 391–401 (2024)
3. Kim, H.S.: Efficient wave load calculation methods for estimating structural responses of highly numerous floating bodies with connection beams. Doctoral dissertation. National University of Science and Technology (2023)
4. Kim, B.W., Kim, H.S., Hong, S.Y.: Coupled and lumped methods in time domain dynamics analysis of floating structure with multi-bodies and connector beams. In: Proceedings of International Conference of Offshore and Polar Engineering (2020)
5. Lee, K., Sim, K.: An experimental study on the motion-structure coupled characteristics of multi-linked floating unit offshore structure with two different connection conditions. Int. J. Nav. Archit. Ocean Eng. **16**, 100614 (2024). https://doi.org/10.1016/j.ijnaoe.2024.100614
6. Sim, K., Lee, K., Kim, B.W.: Structural response analysis for multi-linked floating offshore structure based on fluid-structure coupled analysis. J. Ocean Eng. Technol. **37**, 273–281 (2023)
7. Kim, B.W., Hong, S.Y., Sung, H.G.: Comparison of drift force calculation methods in time domain analysis of moored bodies. Ocean Eng. **126**, 81–91 (2016)
8. Cummins, W.E.: The impulse response function and ship motions. In: Symposium on Ship Theory at the Institute of shipbuilding at the University of Hamburg (1962)
9. Kim, B.W., Sung, H.G., Kim, J.H., Hong, S.Y.: Comparison of linear spring and nonlinear FEM methods in dynamic coupled analysis of floating structure and mooring system. J. Fluids Struct. **42**, 205–227 (2013)
10. Bigot, F., Derbanne, Q., Baudin, E.: A review of strains to internal loads conversion methods in full scale measurements. In: Proceedings of the PRADS 2013 (2013)
11. Bigot, F., Sireta, F.X., Baudin, E., Derbanne, Q., Tiphine, E., Malenica, S.: A novel solution to compute stress time series in nonlinear hydro-structure simulation. In: Proceedings of the ASME 2015 34th International Conference on Ocean, Offshore and Arctic Engineering (2015)

12. Sim, K., Lee, K.: A comparative study on the structural response of multi-linked floating offshore structure between digital model and physical model test for digital twin implementation. J. Mar. Sci. Eng. **12**(2), 262 (2024). https://doi.org/10.3390/jmse12020262

Open Access This chapter is licensed under the terms of the Creative Commons Attribution-NonCommercial-NoDerivatives 4.0 International License (http://creativecommons.org/licenses/by-nc-nd/4.0/), which permits any noncommercial use, sharing, distribution and reproduction in any medium or format, as long as you give appropriate credit to the original author(s) and the source, provide a link to the Creative Commons license and indicate if you modified the licensed material. You do not have permission under this license to share adapted material derived from this chapter or parts of it.

The images or other third party material in this chapter are included in the chapter's Creative Commons license, unless indicated otherwise in a credit line to the material. If material is not included in the chapter's Creative Commons license and your intended use is not permitted by statutory regulation or exceeds the permitted use, you will need to obtain permission directly from the copyright holder.

Two-Dimensional Diffraction Problem of a Floating Rectangular Box Over a Sloping Seabed

Ziqi Li[1] and Bin Teng[2]($\boxtimes$)

[1] Dalian University of Technology, Dalian 116024, China
[2] Ludong University, Yantai 264025, China
bteng@dlut.edu.cn

Abstract. In the article, a hybrid BEM is used to solve the diffraction problem for a box above a sloping topography. An inner and outer domain matching approach is used, with the eigen expansion method in the outer domain and the BIM (Boundary Integral Method) in the inner domain. The wave exciting force applied to the box is calculated. It is found that the wave exciting force curve starts to fluctuate unevenly with the increase of horizontal length and vertical height of the uneven topography. Finally, the formation of the fluctuation is analysed in terms of the reflection coefficient of the waves. There is a phase difference between the reflected waves generated by the slope and the other wave components, and the phase difference varies drastically with wave frequency, causing uneven fluctuations in the wave exciting force curve.

Keywords: Sloping Seabed · Eigenfunction Expansion Method · Boundary Element Method

1 Introduction

In recent years, a large number of scholars have carried out research work on the problem of wave-float interaction on variable topography. Buchner [1] and De Hauteclocque et al. [2] firstly proposed a second-body model, in which the slope topography is independently simplified into a second body, and the size and dimensions of the second object need to be defined. In this case, there is no standardized criterion for the definition of the size and dimensions of the local seabed, and the form of simplification varies for different slope topographies, and no complete theory has been developed. Therefore, the results calculated by this method have low accuracy and change with the different simplified forms of local seabed. And the calculation results show that the selection of local seabed size has a significant effect on the motion response of the floating body.

Later, scholars developed different models in conjunction with the boundary integral equation. Athanassoulis and Belibassakis [3] derived a semi-analytical solution framework-the consistent coupled model for wave transmission over uneven seabed topography that can be accurately computed for arbitrarily variable topography, which implements a modification of the mild-slope equation (Massel [4]; Porter and Staziker

[5]). Then, Belibassakis [6] and Belibassakis [7] based on this foundation, combined the boundary element theory with the inner and outer domain matching method to develop a 3D floating body-variable topography-wave interaction model. In addition, Kim and Kim [8] combined numerical wave tank and time-domain boundary element models to develop a three-dimensional time-domain model for the interaction of wave structures over variable topography. The effects of variable topography on wave exciting force, added mass and radiation damping were investigated. However, in these studies, the computational volume is too large for studying the case of arbitrary large-scale variable topography; and none of them has systematically investigated the effects of individual dimensional parameters of the uneven seabed topography on the wave components and wave forces.

In this paper, a two-dimensional frequency-domain boundary element model is developed for the interaction of wave structures on variable topography. The technique of matching the inner and outer domains is used, with the higher-order boundary element method in the inner domain and the analytical solution theory in the outer domain. Since the velocity potential at the outer boundary element boundary has an analytical solution expression instead, the computational volume is small and the computational efficiency is high, which is especially suitable for arbitrarily varying topography at large scales. Moreover, this paper also systematically investigates the influence of each parameter of the seabed on the hydrodynamic performance of the floating body using the established model.

2 Governing Equation and Boundary Conditions

Assume a seabed with varying bathymetry in an intermediate region with a rectangular floating square box directly above it. The coordinate relation Oxz as shown in Fig. 1(a). And the parameters of the rectangular box are width $2B$ and draft T, and the parameters of the seabed are mean water depth h, height D and length L. Based on the linear potential flow theory, there are the following governing equation and boundary conditions:

$$\Phi(x, z, t) = \text{Re}\left[\phi(x, z)e^{-i\omega t}\right] \quad (1)$$

$$\frac{\partial^2 \phi(x, z)}{\partial x^2} + \frac{\partial^2 \phi(x, z)}{\partial z^2} = 0 \quad (2)$$

$$\frac{\partial \phi(x, z)}{\partial z} = \frac{\omega^2}{g}\phi(x, z), \text{ on } z = 0 \quad (3)$$

$$\frac{\partial \phi(x, z)}{\partial n} = 0, \text{ on } z = -h(x) \quad (4)$$

$$\frac{\partial \phi(x, z)}{\partial n} = 0, \text{ on the body surface} \quad (5)$$

where $\phi(x, z)$ is the time-independent complex velocity potential, t the time, ω the wave frequency, g the gravitational acceleration, n is the unit normal vector outwards on the boundary plane.

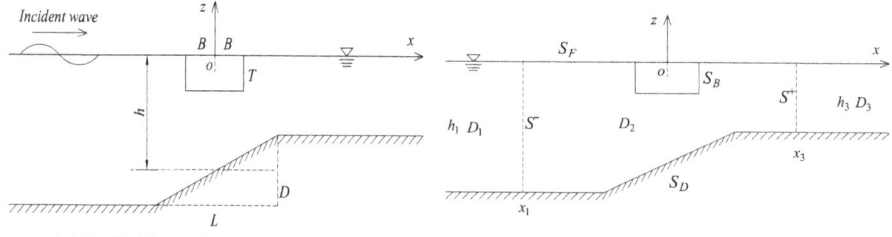

(a) Definition of coordinate system   (b) Decomposition of fluid domain

Fig. 1. The diffraction problem over a two-dimensional variable seabed.

3 The Hybrid BEM Method

Using the technique of matching internal and external domains and applying different methods to the computational domains of different regions can greatly simplify the computation and quickly construct the numerical model.

For the external computational domain with a flat seabed, we use the more common analytical method (see Mei [9]), eigenfunction expansion, to represent the general form of the velocity potential in the external domain:

$$\phi_1(x,z) = -\frac{igA}{\omega}\left(\frac{\cosh k_{10}(z+h_1)}{\cosh k_{10}h_1}\left(e^{ik_{10}x} + R_0^- e^{-ik_{10}x}\right) + \sum_{l=1}^{\infty} R_l^- \frac{\cos k_{1l}(z+h_1)}{\cos k_{1l}h_1} e^{k_{1l}(x-x_1)}\right), x < x_1 \quad (6)$$

$$\phi_3(x,z) = -\frac{igA}{\omega}\left(R_0^+ \frac{\cosh k_{30}(z+h_3)}{\cosh k_{30}h_3}e^{ik_{30}x} + \sum_{l=1}^{\infty} R_l^+ \frac{\cos k_{3l}(z+h_3)}{\cos k_{3l}h_3} e^{-k_{3l}(x-x_3)}\right)x > x_3 \quad (7)$$

where $\phi_1(x,z)$ is the velocity potential on the left, $\phi_3(x,z)$ is the velocity potential on the right, A denotes the wave amplitude of the incident potential, $R_l^\pm$ ($l = 0, 1, 2, \ldots$) the unknown expansion coefficients, $R = R_0^-$ the reflection coefficient. In the velocity potential expression obtained by the eigen-expansion method, the eigenvalues k_{Il} are roots of the dispersion equations

$$\begin{cases} \omega^2 = gk_{I0} \tanh k_{I0}h_I \\ \omega^2 = -gk_{Il} \tan k_{Il}h_I, l = 1, 2, 3, \ldots \end{cases} \quad (8)$$

where $I = 1, 3$ are different exterior computational domains on the left and right hand sides, respectively.

In contrast, for the internal computational domain of the variable seabed, we use the boundary integral method for the solution. In this case, a simple Rankine source is used as the Green's function

$$G(x; x_0) = \frac{1}{2\pi}\ln(r) \quad (9)$$

Substituting the boundary conditions of Sect. 2 into Green's Second Theorem and combining them with a general expression for the velocity potential in the exterior domain yields the boundary integral equation:

$$\alpha\phi(\mathbf{x}_0) - \int_{S_B+S_F+S_D} \frac{\partial G(\mathbf{x};\mathbf{x}_0)}{\partial n}\phi(\mathbf{x})ds + \frac{\omega^2}{g}\int_{S_F} G(\mathbf{x};\mathbf{x}_0)\phi(\mathbf{x})\,d\mathbf{s} - \sum_{l=0}^{L^--1}\int_{S^-} -\frac{igA}{\omega}R_l^-\varphi_{1l}^-(\mathbf{x})F_{1l}(\mathbf{x},\mathbf{x}_0)ds$$
$$-\sum_{l=0}^{L^+-1}\int_{S^+} -\frac{igA}{\omega}R_l^+\varphi_{3l}^+(\mathbf{x})F_{3l}(\mathbf{x},\mathbf{x}_0)ds = -\frac{igA}{\omega}\int_{S^-}\varphi_{10}^+(\mathbf{x})\left[\frac{\partial G(\mathbf{x};\mathbf{x}_0)}{\partial n} + ik_{10}^- G(\mathbf{x};\mathbf{x}_0)\right]ds \quad (10)$$

where α the solid angle, $L^{\mp}$ the truncated terms in the potential expressionn, and the the notation $F_{Il}(x,x_0)$

$$F_{Il}(\mathbf{x},\mathbf{x}_0) = \begin{cases} \frac{\partial G(\mathbf{x},\mathbf{x}_0)}{\partial n} - ik_{I0}G(\mathbf{x},\mathbf{x}_0), & l=0 \\ \frac{\partial G(\mathbf{x},\mathbf{x}_0)}{\partial n} + k_{Il}G(\mathbf{x},\mathbf{x}_0), & l=1,2,\ldots \end{cases} \quad (11)$$

The implementation of the numerical model is carried out using a higher-order element discretization method. The boundary surfaces in the boundary integral equation are discretized, with the free water surface, body surface, seabed, and left/right vertical surfaces divided into N_F, N_B, N_D, and $N^{\pm}$ elements, respectively. Eq. (10) can then be transformed as follows:

$$\alpha\phi(\mathbf{x}_0) - \sum_{n=1}^{N_B+N_F+N_D}\int_{-1}^{1}\frac{\partial G(\mathbf{x};\mathbf{x}_0)}{\partial n}\sum_{k=1}^{K}h^k(\xi)\phi^k|J(\xi)|d\xi + \frac{\omega^2}{g}\sum_{n=1}^{N_F}\int_{-1}^{1}G(\mathbf{x};\mathbf{x}_0)\sum_{k=1}^{K}h^k(\xi)\phi^k|J(\xi)|d\xi$$
$$+\sum_{l=0}^{L^--1}R_l^-\sum_{n=1}^{N^-}\int_{-1}^{1}\frac{igA}{\omega}\varphi_{1l}^-(\mathbf{x})F_{1l}(\mathbf{x},\mathbf{x}_0)|J(\xi)|d\xi + \sum_{l=0}^{L^+-1}R_l^+\sum_{n=1}^{N^+}\int_{-1}^{1}\frac{igA}{\omega}\varphi_{3l}^+(\mathbf{x})F_{3l}(\mathbf{x},\mathbf{x}_0)|J(\xi)|d\xi \quad (12)$$
$$= -\frac{igA}{\omega}\sum_{n=1}^{N^-}\int_{-1}^{1}\varphi_{10}^+(\mathbf{x})\left[\frac{\partial G(\mathbf{x};\mathbf{x}_0)}{\partial n}+ik_{10}^- G(\mathbf{x};\mathbf{x}_0)\right]|J(\xi)|d\xi$$

where $h(\xi)$ the shape function, $|J(\xi)|$ the Jacobian determinant. And by placing the source points at the $L^{\pm}$ nodes of the left and right vertical surface elements and the nodes of the elements on other boundary surfaces, a system of equations can be obtained

$$[A]\begin{Bmatrix}\{\phi\} \\ \{R_l^-\} \\ \{R_l^+\}\end{Bmatrix} = \{B\} \quad (13)$$

Finally, applying the quadratic discretization technique, the boundary integral equations can be solved. The potentials on the boundary surfaces and the reflection coefficients in the eigen-expansion expressions for the velocity potential can be obtained. And also, the wave exciting force of the floating body can be calculated

$$f_i = i\omega\rho\iint_{S_B}\phi n_i\,ds,\quad i=1,2 \quad (14)$$

4 Numerical Results

In this section, the reflection coefficients of waves on the left side of the topography and the wave exciting force of the floating body are investigated utilizing the previously established numerical method. The results under different seabed parameters are compared to summarize and analyze the laws and phenomena of the influence of variable topography on the wave components and wave forces in the wave diffraction.

The first is the reflection coefficients, where $h = 2T$, the box parameter $B = 2T$, and the individual parameters for the variable seabed are $\frac{D}{h} = 0.5, 1.0$, and 1.5 and $\frac{L}{h} = 6.0, 12.0$, and 24.0.

Figure 2 shows the velocity potential reflection coefficient versus wave frequency for the case where there is no box present and only wave propagation. As can be seen from this figure, the amplitude of the reflected waves induced by the variable seabed exhibits significant fluctuations with changes in wave frequency. And as the length of uneven seabed increases, the fluctuation period of the reflection coefficient gradually decreases; as the height of uneven seabed increases, the fluctuation amplitude gradually increases. Figure 3 shows the variation of velocity potential reflection coefficient with wave frequency in the presence of a box. For the case of the sloping topography, the fluctuation patterns of curves are the same as above; however, for flat seabed, the curve has almost no fluctuation.

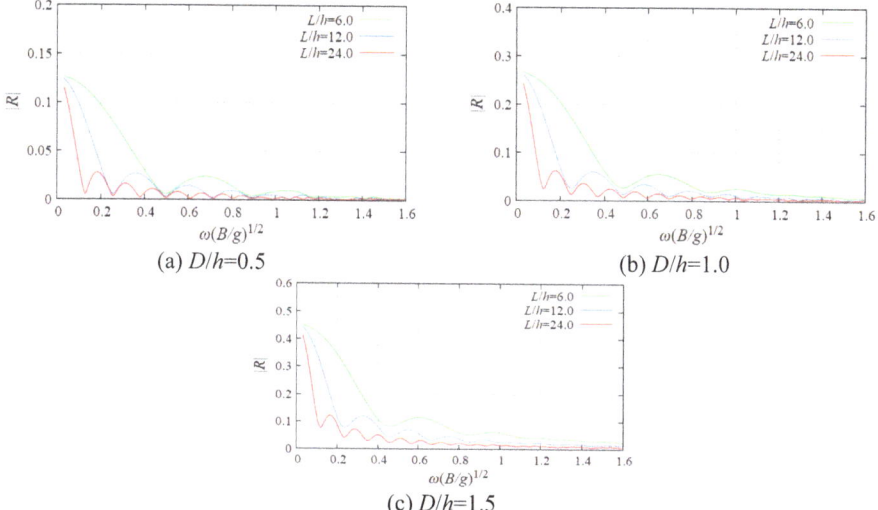

Fig. 2. Wave reflection coefficients for wave propagation conditions.

Subsequently, the wave exciting force of the box was calculated. Based on the parameters of uneven seabed, the exciting forces were studied for three sets of calculations with different mean water depths ($\frac{h}{T} = 2.0, 3.0$, and 4.0), seabed lengths ($\frac{L}{B} = 3.0, 6.0, 12.0$, and 24.0), and seabed heights ($\frac{D}{T} = 1.0, 2.0$, and 3.0).

Figure 4 shows the variation of wave force versus frequency under flat and sloping seabed's with different mean water depths. We can see that the impact of variable seabed on wave exciting force gradually increases with the decrease of mean water depth. It is consistent with the general situation of water wave theory. And the influence on z-direction wave force is more significant.

Figure 5 shows the variation of wave force versus frequency for the box above uneven seabed with different lengths. And the same phenomenon as in Fig. 3 can be seen, as

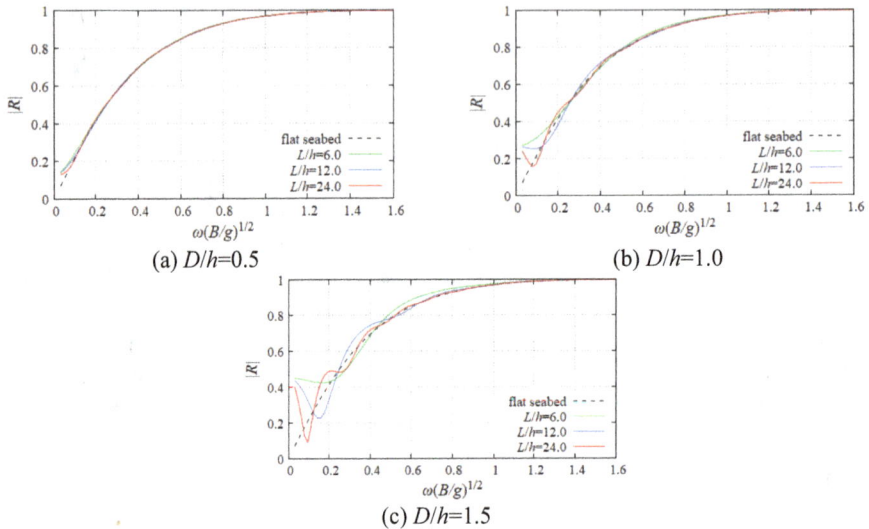

Fig. 3. Wave reflection coefficient in case of wave diffraction.

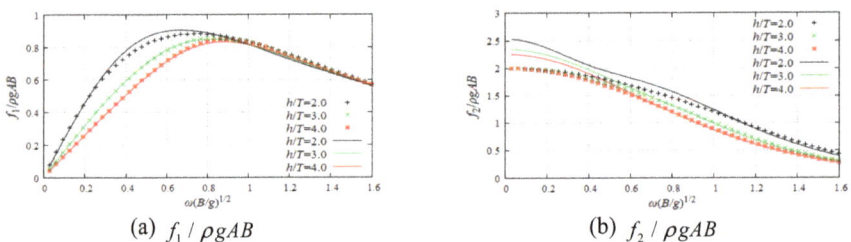

Fig. 4. Wave exciting force on the box at different water depths ($D = 2T$, $L = 6B$).

the length of uneven seabed increases, the wave force starts to fluctuate unevenly with frequency and the period of fluctuation decreases gradually.

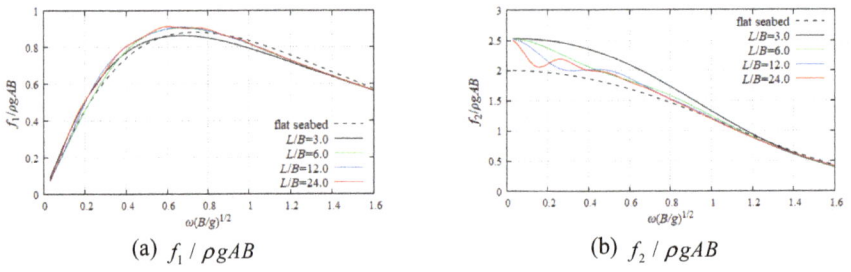

Fig. 5. Comparison of wave exciting force above the uneven seabed at different lengths ($D = 2T$, $h = 2T$).

Figure 6 shows the variation of wave exciting force with frequency for the box above the sloping seabed of different heights. Again, a similar phenomenon to Fig. 3 can be seen, where the amplitude of the wave force fluctuating with frequency increases as the height of uneven seabed increases.

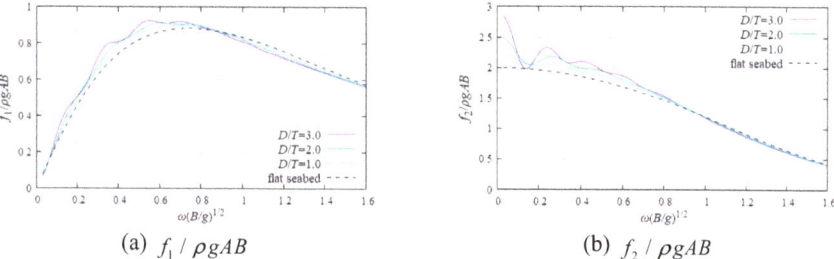

(a) $f_1 / \rho g A B$ (b) $f_2 / \rho g A B$

Fig. 6. Comparison of wave exciting force above the uneven seabed at different heights ($L = 24B$, $h = 2T$).

5 Conclusions

In this paper, a two-dimensional hybrid BEM model of wave-structure interaction on variable topography is developed. The following conclusions can be drawn:

1. The effects of each seabed parameter on the velocity potential reflection coefficient and wave action force are systematically calculated. It is observed that as the length and height of the seabed increase, the curve of wave forces begins to exhibit irregular fluctuations.
2. Combined with the analysis of the reflection coefficient, when the wave propagates on the variable seabed, the seabed produces a reflection wave with sharp fluctuations in amplitude and a small fluctuation period. Therefore, when the floating body hydrodynamic problem is transferred to the variable seabed, a phase difference begins to arise between these wave components, and the phase difference fluctuates unevenly with frequency, leading to the phenomenon of uneven fluctuation of wave action force in this paper.

In this paper, the hydrodynamic problem is simplified to a two-dimensional space for calculation, and certain phenomena and patterns are summarized. However, in practical engineering, due to the complexity of the seabed or the floating body, most cases cannot be simplified to two-dimensional problems. Therefore, it is necessary to extend this research to a three-dimensional hydrodynamic model.

References

1. Buchner, B.: The motions of a ship on a sloped seabed. In: Proceedings of the 25th International Conference on Offshore Mechanics and Arctic Engineering, pp. 339–347 (2006)
2. De Hauteclocque, G., Rezende, F., Giorgiutti, Y., Chen, X.B.: Wave kinematics and seakeeping calculation with varying bathymetry. In: Proceedings of the 28th International Conference on Offshore Mechanics and Arctic Engineering, pp. 515–523 (2009)
3. Athanassoulis, G.A., Belibassakis, K.A.: A consistent coupled-mode theory for the propagation of small-amplitude water waves over variable bathymetry regions. J. Fluid Mech. **389**, 275–301 (1999)
4. Massel, S.: Extended refraction–diffraction equations for surface waves. Coast. Eng. **19**, 97–126 (1993)
5. Porter, D., Staziker, D.J.: Extension of the mild-slope equation. J. Fluid Mech. **300**, 367–382 (1995)
6. Belibassakis, K.A.: Hydrodynamic analysis of floating bodies in general bathymetry. In: Proceedings of the International Conference on Offshore Mechanics and Arctic Engineering, pp. 449–456 (2005)
7. Belibassakis, K.A.: A boundary element method for the hydrodynamic analysis of floating bodies in variable bathymetry regions. Eng. Anal. Bound. Elem. **32**, 796–810 (2008)
8. Kim, T., Kim, Y.: Numerical analysis on floating-body motion responses in arbitrary bathymetry. Ocean Eng. **62**, 123–139 (2013)
9. Mei, C.C.: The applied dynamics of ocean surface waves. Wiley Interscience, New York (1983)

Open Access This chapter is licensed under the terms of the Creative Commons Attribution-NonCommercial-NoDerivatives 4.0 International License (http://creativecommons.org/licenses/by-nc-nd/4.0/), which permits any noncommercial use, sharing, distribution and reproduction in any medium or format, as long as you give appropriate credit to the original author(s) and the source, provide a link to the Creative Commons license and indicate if you modified the licensed material. You do not have permission under this license to share adapted material derived from this chapter or parts of it.

The images or other third party material in this chapter are included in the chapter's Creative Commons license, unless indicated otherwise in a credit line to the material. If material is not included in the chapter's Creative Commons license and your intended use is not permitted by statutory regulation or exceeds the permitted use, you will need to obtain permission directly from the copyright holder.

Modal Decomposition Method for the Dynamic Response of a Submerged Floating Tunnel Under Wave Action

Mei Yu[1] and Bin Teng[2(✉)]

[1] Dalian University of Technology, Dalian 116024, China
[2] Ludong University, Yantai 264025, China
bteng@dlut.edu.cn

Abstract. In this article, a frequency domain modal decomposition method is proposed to determine the hydrodynamic response of a submerged floating tunnel subjected to wave forces. In this method, using the 3D finite element method (FEM) to solve the modal functions and 2D high-order matched boundary element method (HOBEM) is used to solve the hydrodynamic problem of the structure. Using modal decomposition method, the modal functions of the structure are Fourier expanded and combined with 2D hydrodynamic coefficients to construct the 3D generalized hydrodynamic coefficients and generalized wave excitation forces. Among them, the generalized wave exciting force is calculated by considering the ratio of the tunnel's cross-sectional area to the incident wavelength, using either the Morison equation or the potential flow theory. During the research process, the influence of key structural parameters, such as cables diameter and pipe length, on the inherent properties of the system was examined. Additionally, the research compared the effects of various methods for calculating the added mass on SFT displacement. Finally, the research presented that how wave frequencies and incident angles affect the structural dynamic response under wave actions.

Keywords: Submerged Floating Tunnel (SFT) · Hydrodynamic Response · Finite Element Method (FEM) · Hybrid Boundary Element Method (BEM) · Modal Decomposition · Fourier series Expansion

1 Introduction

Submerged floating tunnels (SFTs), an innovative transportation solution for traversing lakes, straits, and oceans, have garnered attention from countries such as Italy [1], Japan [2], China [3] and others. Like other offshore structures, an SFT must withstand various environmental forces, including waves, currents, tsunamis, and seismic activity, as well as dynamic loads from moving traffic and potential impact forces from falling objects. In addition, due to the long length, the dynamic response of the structure due to wave loads exhibits significant nonlinearity. The response of the SFT to these loads is a classic fluid-solid interaction (FSI) problem, which is closely related to the SFT's hydrodynamic properties and requires a coupled approach integrating hydrodynamics and structural mechanics for resolution. Scholars worldwide have conducted substantial research on this issue. Kunisu et al. [4] utilized the BEM and Morison's Equation to calculate the wave force characteristics and dynamic response of SFT under wave loads in a 2D model. Paik et al. [5] investigated the dynamic behavior of an SFT under the influence of external wave loads. In order to investigate the coupled dynamic response of an SFT under wave loads, Jin et al. [6] developed a complete second-order 2D numerical model in the time domain. For a more comprehensive consideration, 3D models have been widely applied. Ge et al. [7] carried out a 3D analytical of the FSI behavior of SFTs in wavefield and performed hydrodynamic calculations based on the SFT of Qiandao Lake. Muhammad et al. [8] estimated the 3D deformations and inner stresses of the SFT subjected to hydrodynamic and earthquake loads, providing a reference for structural optimization design.

This paper employs a joint FEM and BEM. A modal decomposition method is used for modeling the vibration of the SFT. In addition, using Fourier cosine order to expand modal function, and hence the hydrodynamic of the SFT is reduced to 2D, where the matched BEM applies to the computation of hydrodynamic coefficients and exciting forces by Teng et al. [9]. The combination of modal functions and FEM and BEM can obtain the hydrodynamic structural response. The proposed method can be used to simulate the interaction between wave loads and long-span SFT. It does not require the high performance of the computer but can accurately reflect the overall dynamic response of the SFT. At the same time, this calculation model can accurately simulate boundary and anchoring conditions.

2 Mathematical Formulations and Numerical Model

Figure 1 illustrates an SFT submerged T meters beneath the sea surface, within water of depth d. Both ends of the SFT are attached to the shoal, and it is held in place by numerous mooring cables running along its length. A right-hand Cartesian coordinate system $Oxyz$ is established with the zero point at the water plane, the y-axis along the pipeline and z-axis vertically upwards.

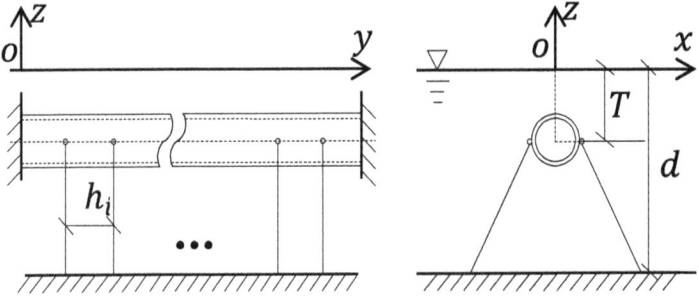

Fig. 1. Structural diagram of an SFT.

2.1 The SFT's Modal Analysis

Because FEM can be applied to the modal analysis of structures with complex geometries. FEM is used here for dry modal analysis and the Euler-Bernoulli theory is used to construct the structural free motion equation without considering structural damping

$$[M]_{4q \times 4q} \{\ddot{X}(t)\}_{4q} + [K]_{4q \times 4q} \{X(t)\}_{4q} = 0 \tag{1}$$

where, q is the number of nodes in the structure. $[M]_{4q \times 4q}$ and $[K]_{4q \times 4q}$ are the global mass matrix and the global stiffness matrix of the SFT system, the displacement at the nodes of the SFT is given by $\{X(t)\}_{4q}$. Axial and torsional displacements are not considered for the structure, meaning that all nodes have 4 degrees of freedom (DOFs).

Based on the oscillation of the structure with an angular frequency ω, the frequency domain free vibration equation can be expressed as

$$\omega^2 [M]_{4q \times 4q} \{\chi\}_{4q} = [K]_{4q \times 4q} \{\chi\}_{4q} \tag{2}$$

where, $\{\chi\}_{4q}$ is the frequency domain SFT nodes deformation.

2.2 The SFT's Generalized Dynamic Equation

For the response of an SFT under the action of linear regular waves, the motion equation of the structure can be constructed as follows:

$$\left[-\omega^2 ([M]_{4q \times 4q} + [a]_{4q \times 4q}) - i\omega ([B]_{4q \times 4q} + [b]_{4q \times 4q}) + [K]_{4q \times 4q} \right] \{\chi\}_{4q} = \{F\}_{4q} \tag{3}$$

where $[a]_{4q \times 4q}$ and $[b]_{4q \times 4q}$ represent the added mass and radiation damping; $[B]_{4q \times 4q}$ is the system damping, encompassing structural and viscous components; $\{\chi\}_{4q}$ and $\{F\}_{4q}$ represent the SFT's response and the applied external loads. Here, the external loads only include wave exciting forces.

Using modal decomposition method to represent modal functions

$$\{\chi\}_{4q} = [\chi_n]_{4q \times J} \{A\}_J \tag{4}$$

where $[\chi_n]_{4q \times J}$ is the eigenmodes, J represents the chosen modal terms number, $\{A\}_J$ denotes modal amplitude and the generalized motion equation is

$$\left[-\omega^2 \left([\hat{M}]_{J \times J} + [\hat{a}]_{J \times J} \right) - i\omega \left([\hat{B}]_{J \times J} + [\hat{b}]_{J \times J} \right) + [\hat{K}]_{J \times J} \right] \{A\}_J = \{\hat{F}\}_J \tag{5}$$

in which

$$\begin{aligned} \left[\hat{M}\right]_{J \times J} &= [\chi_n]_{4q \times J}^T [M]_{4q \times 4q} [\chi_n]_{4q \times J} \\ \left[\hat{K}\right]_{J \times J} &= [\chi_n]_{4q \times J}^T [K]_{4q \times 4q} [\chi_n]_{4q \times J} \\ \left[\hat{B}\right]_{J \times J} &= [\chi_n]_{4q \times J}^T [B]_{4q \times 4q} [\chi_n]_{4q \times J} \end{aligned} \tag{6}$$

are the mass, stiffness and viscous damping matrices in generalized form. $[\hat{a}]_{J \times J}$ and $[\hat{b}]_{J \times J}$ obtained from the 2D hydrodynamic $[a^m]_{2 \times 2}$ and $[b^m]_{2 \times 2}$ corresponding to each eigen value λ_m, and then be transformed into a 3D generalized form using modal functions as

$$a_j(y) = \sum_{m=0}^{M} \chi_j^m [a^m]_{2\times 2} e^{i\lambda_m y}$$
$$b_j(y) = \sum_{m=0}^{M} \chi_j^m [b^m]_{2\times 2} e^{i\lambda_m y} \quad (7)$$

$$[\hat{a}]_{J\times J} = \int [\chi_n]_{4q\times J}^T [N(y)]_{2\times 4q}^T [a(y)]_{2\times J} dy$$
$$[\hat{b}]_{J\times J} = \int [\chi_n]_{4q\times J}^T [N(y)]_{2\times 4q}^T [b(y)]_{2\times J} dy \quad (8)$$

where $\chi_j^m = \left(\chi_{jx}^m, \chi_{jz}^m\right)$ is horizontal and vertical displacement derived from the Fourier analysis. $[a(y)]_{2\times J}$ and $[b(y)]_{2\times J}$ are the modal hydrodynamic coefficients.

By utilizing the node distribution forces and modal functions, we can obtain the generalized wave exciting forces

$$\{\hat{F}\}_J = \int [\chi_n]_{4q\times J}^T [N(y)]_{2\times 4q}^T f_e dy \quad (9)$$

where $[[N(y)]_{2\times 4q}$ is the shape function and $f_e = (f_{ex}, f_{ez})$ is the 2D wave exciting force, which can be calculated using either diffraction theory or Morison equation.

3 Analysis of the Influence of Structural Parameters and Added Mass Calculation Methods

According to the proposed project plan for the SFT, the numerical calculation model shown in Table 1 is selected.

Table 1. Variables of the whole system.

Segment	Physical quantity	Sign	Magnitude	Unit
Piple	Density	ρ	2501.01	kgm^{-3}
	Elastic modulus	E	3.55 × 1010	Pa
	Length	L	4000	m
	Out diameter	D	15	m
	Wall thickness	H	1	m
Cable	Elastic modulus	E_c	1.95 × 1011	Pa
	Length	l_c	99	m
	Diameter	D_c	150	mm
	Inclination angle	α	π/4	rad
	Cable interval	h_i	50	m

3.1 The Influence of Anchor Cables Diameter and the SFT Length

Set the anchor cables diameter to be 50 to 300 mm and the SFT length to be 2000 to 1000 m, and the changes in natural frequencies can be used to show the effect of structural geometry on the stiffness of the system. From Fig. 2., it can be seen that the stiffness of the anchor has a significant impact on the low order natural frequencies of the structure, while the increase in SFT length can significantly reduce the high order natural frequencies of the structure, resulting in a decrease in the difference in natural frequencies between different models.

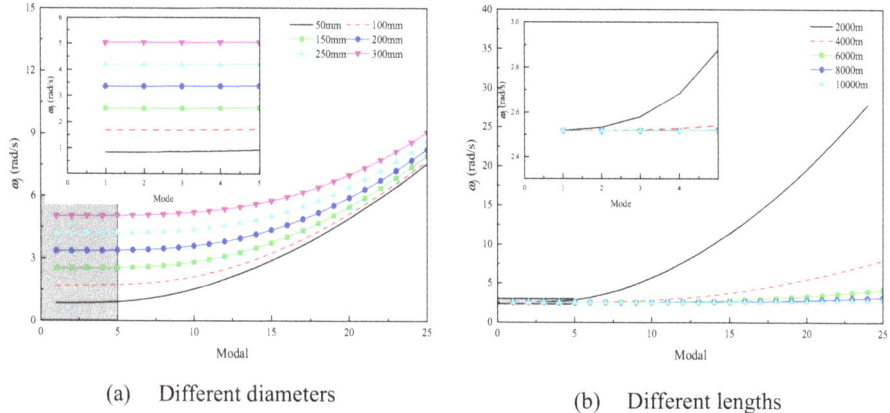

(a) Different diameters　　　　　　　(b) Different lengths

Fig. 2. The variation in natural frequencies.

3.2 The Impact of Added Mass

Using the structural parameters shown in Table 1, select a water depth of 100 m and a submergence depth of $T/d = 0.3$ for SFT. Calculate the dynamic coefficients. The added masses are taken as constants 0 or 1, or the generalized terms calculated by the Fourier expansion method. Following linearization, the structure's response at the $L/2$ position due to a unit wave amplitude is presented. As shown in Fig. 3, the response varies significantly with different added masses as the wave frequency increases. An increase in added mass results in a reduction of the wet intrinsic resonance frequency. The size of the added mass determines the relationship between the natural frequencies of the dry and wet modals.

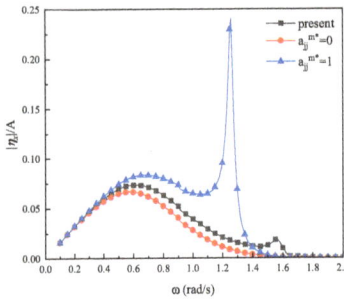 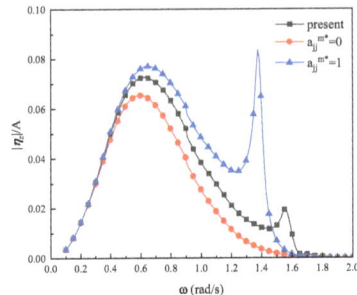

(a) Displacement in the *x* direction at *L*/2   (b) Displacement in the *z* direction at *L*/2

Fig. 3. Comparison of displacement at SFT *L*/2 position based on various added mass calculation techniques.

4 Dynamic Characteristics of an SFT Under Different Wave actions

In this section, the SFT's hydrodynamic response of under the action of forward regular waves at different wave frequencies and different incidence angles were calculated.

4.1 Dynamic Behavior of an SFT Subjected to Normal Regular Wave Loads

Figure 4 illustrates that both displacements initially rise and subsequently fall as the wave frequency increases. The modal analysis reveals that the basic fundamental wet modal frequency equals to 1.56 rad/s, corresponding to a wave period of $T = 4.03$ s. The SFT's response reaches its maximum value at 0.6 rad/s. This is due to the fact that the water particles' velocity and acceleration increase as the wave frequency increases, but decrease sharply at greater depths as the frequency is higher.

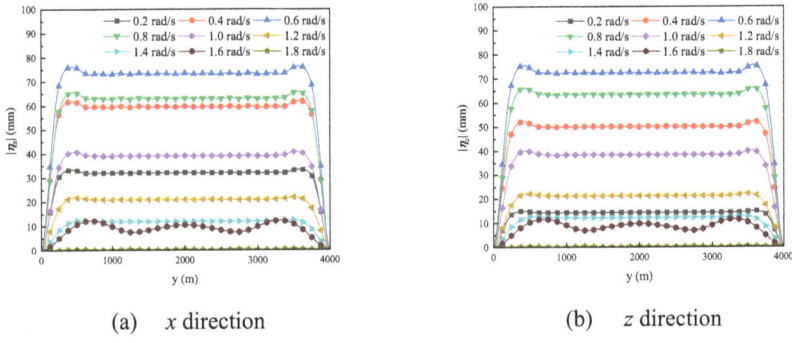

(a) *x* direction   (b) *z* direction

Fig. 4. Analysis of the SFT's displacement along the *x*- and *z*- axes under normal wave conditions.

4.2 Dynamic Behavior of an SFT Subjected to Oblique Regular Waves

Figure 5 presents the amplitude variation of the first six structural modals along the x-axis for different wave conditions. The generalized equations of the structure show that the modal response of the structure is closely related to the generalized exciting force, which is the projection of pressure loading on the modal function. From Fig. 5, demonstrates that for all frequencies, there is a specific direction where the exciting force's longitudinal profile coincides with the modal function, producing a relatively high generalized exciting force and enhanced structural modal response. With increasing wave frequency, the angle of incidence wave required to enhance the hydrodynamic response amplitude becomes smaller.

5 Conclusion

This paper uses a FEM to obtain the vibration functions of the SFT in a 3D model. The mode functions are reconstructed through each Fourier expansion term under each modal. The 2D exciting forces and hydrodynamic coefficients of the SFT are obtained by the HOBEM in the 2D model. By combining with modal functions, construct the 3D hydrodynamic coefficients of the SFT. Finally, by combining the FEM and BEM, a generalized motion equation of the SFT is established to obtain the motion amplitudes of each modal. Using this method, the dynamic displacement of long span SFTs subjected to wave loads can be calculated quickly and efficiently, providing some certain theoretical basis for engineering construction. However, since the hydrodynamic model of the structure is solved in a 2D model, we are unable to obtain the axial deformation of the 3D SFT. This is a problem to be solved in future work.

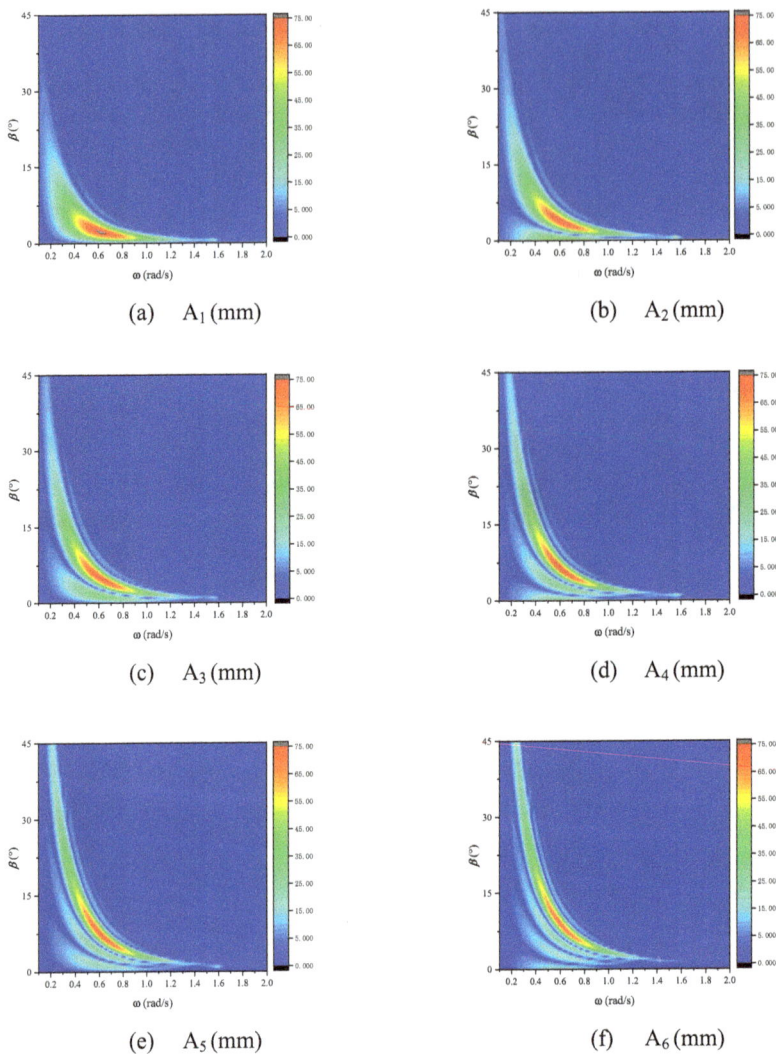

Fig. 5. Modal amplitudes in the x direction for the first six modals at different incident wave frequencies and angles.

References

1. Faggiano, B., Landolfo, R., Mazzolani, F.M.: Design and modeling aspects concerning the submerged floating tunnels: An application to the Messina Strait crossing. In: Strait Crossing, Krobeborg, pp. 511–519 (2001)
2. Kunisu, H., Mizuno, S., Mizuno, Y., Saeki, H.: Numerical analysis of wave force and dynamic response to the submerged floating tunnels. In: Proceedings of the 3rd Symposium on Strait Crossings, pp. 637–644 (1994)
3. Xiang, Y., Yang, Y.: Challenge in design and construction of submerged floating tunnel and state-of-art. Procedia Eng. **166**, 53–60 (2016)

4. Kunisu, H., Mizuno, S., Mizuno, Y.: Study on submerged floating tunnel characteristics under the wave condition. Fourth Int. Offshore Polar Eng. **27-32**, 94–96 (1994)
5. Paik, I.Y., Oh, C.K., Kwon, J.S., Chang, S.P.: Analysis of wave force induced dynamic response of submerged floating tunnel. KSCE J. Civ. Eng. **8**(5), 543–550 (2004)
6. Jin, R., Gou, Y., Geng, B., Zhang, H., Liu, Y.: Coupled dynamic analysis for wave action on a tension leg-type submerged floating tunnel in time domain. Ocean Eng. **212**, 1–14 (2020)
7. Ge, F., Lu, W., Wu, X., Hong, Y.: Fluid-structure interaction of submerged floating tunnel in wave field. Procedia Eng. **4**(4), 263–271 (2010)
8. Muhammad, N., Ullah, Z., Choi, D.H.: Performance evaluation of submerged floating tunnel subjected to hydrodynamic and seismic excitations. Appl. Sci. **7**, 1122 (2017)
9. Teng, B., Yu, M., Gou, Y., Sheng, J.L.: Generalized hydrodynamics coefficients of a circular SFT in finite water depth. Ocean Eng. **273**, 113938 (2023)

Open Access This chapter is licensed under the terms of the Creative Commons Attribution-NonCommercial-NoDerivatives 4.0 International License (http://creativecommons.org/licenses/by-nc-nd/4.0/), which permits any noncommercial use, sharing, distribution and reproduction in any medium or format, as long as you give appropriate credit to the original author(s) and the source, provide a link to the Creative Commons license and indicate if you modified the licensed material. You do not have permission under this license to share adapted material derived from this chapter or parts of it.

The images or other third party material in this chapter are included in the chapter's Creative Commons license, unless indicated otherwise in a credit line to the material. If material is not included in the chapter's Creative Commons license and your intended use is not permitted by statutory regulation or exceeds the permitted use, you will need to obtain permission directly from the copyright holder.

Experimental and Numerical Evaluation of Hydrodynamic Behavior of 10 MW Floating Offshore Wind Turbine

Byoung Wan Kim[✉], Kichan Sim, Kangsu Lee, and Sa Young Hong

Korea Research Institute of Ships & Ocean Engineering (KRISO),
University of Science & Technology (UST), Daejeon, South Korea
kimbw@kriso.re.kr

Abstract. A 10 MW floating offshore wind turbine (FOWT) is developed by a Korea R&D project. The FOWT is composed of three blades, a tower and their supporting floater. Three catenary steel chains are attached to the floater for mooring lines. The design target site is East-South sea of Korea. Since the FOWT is a floating structure in waves, evaluation of hydrodynamic behaviors is a key process in design of FOWT. This paper carried out hydrodynamic analysis for the FOWT to evaluate its responses in waves, winds and currents. Wave forces at floater were analyzed by higher order boundary element method (HOBEM) and the catenary behaviors of mooring lines were analyzed by finite element method (FEM). The time domain body-mooring coupled responses were calculated by convolution method and the floater motions and mooring line tensions were obtained from the coupled analysis. Experimental model test was also carried out in ocean engineering basin in KRISO. Floater motions and mooring line tensions were measured from the model test and all the results were compared with the numerical results.

Keywords: Floating Offshore Wind Turbine · Hydrodynamic Analysis · Model Test

1 Introduction

Wind energy is recently discussed as an alternative to fossil energy and many engineers tried FOWT system as a new type of wind energy converter. A 10 MW FOWT is also in design by Korea R&D project "Development of Disconnectable Mooring System for A MW class Floating Offshore Wind Turbine (DMS project)". The FOWT is a floating body structure with RNA (Rotor Nacelle Assembly), tower, floater and mooring lines. RNA is composed of blade, hub & nacelle and a tower supports the RNA. Floater is composed of deck, column & pontoon and three steel chain lines are attached to FCS (Fairlead Chain Stopper) in the floater as mooring system. KRISO is one of members of DMS project and a model test was carried out in KRISO basin to experimentally evaluate hydrodynamic responses of the FOWT in wave, wind and current. This paper summarized and discussed the model test results. Some figures and photos for the FOWT and model ship are shown in Fig. 1.

(a) Model ship in KRISO basin (scale=1/55)

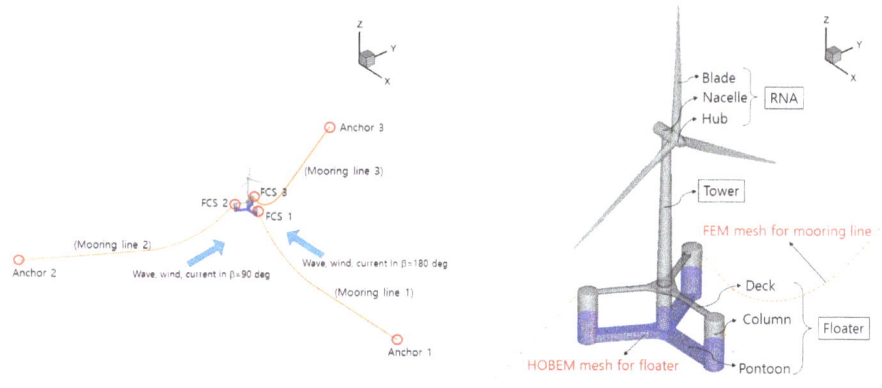

(b) Numerical model

Fig. 1. Experimental and numerical models for a 10 MW FOWT system by DMS project

Pre-tests such as tilt test, free-decay test and pull-out test were done to match GM & natural periods of floating body and excursion curve of mooring lines. Wave calibrations were done to match regular and irregular waves in target condition. The model ship was moved to KRISO basin and model tests were done to measure floating body motions and mooring line tensions in wave, wind and current. All test results were compared with numerical results by hydrodynamic analysis.

2 Design Particulars and Pre-tests

Total weight of the FOWT is 10,728 ton and three steel catenary chains are attached to the floating body as mooring system. The geometric shapes and design particulars are summarized in Fig. 1 and Table 1. They are from [1]. In the table, x_L = horizontal distance from FCS to anchor, z_L = vertical distance from sea-bottom to FCS, s_L = total length of mooring line. SMP is a submersible mooring pulley. C_D, C_M, C_F, C_a are

Table 1. Design particulars [1]

Target site	East-South sea of Korea		Water depth = 150 m	
Floating body (10 MW FOWT)	Length × Breadth × Draft × Freeboard		79.5 × 9.942 × 15.5 × 16.5 m	
	Weight		10,728,000 kg	
	KG/GM		20.080/18.234 m	
	$I_{xx}/I_{yy}/I_{zz}$ (at CoG)		1.967e10/1.961e10/1.298e10 kg-m^2	
	Natural periods w/o mooring ($T_{heave}/T_{roll}/T_{pitch}$)		14.827/22.572/22.544 s	
	Damping ratio in calculation ($\xi_{heave}/\xi_{roll}/\xi_{pitch}$)		10.67/6.14/6.5%	
	C_D	Blade	0.009–0.078 (in parking)	
		Blade cylinder & tower	0.6	
		Column/Pontoon	1.0/1.72	
Mooring line (steel chain with d = 147 mm)	Axial stiffness (EA)		1.845e9 N	
	$x_L/z_L/s_L$		800/163.08/850 m	
	Mass	Chain mass per length (dry/wet)	430.019/373.870 kg/m	
		SMP (dry/wet)	18,664/16,287 kg	
	Marine growth	z = −150 m–−40 m	Thickness = 50 mm	Equiv. C_D = 4.03
		z = −40 m–+2 m	=100 mm	=5.67
		z > +2 m	=0 mm	=0.00
	$C_M/C_F/C_a$		2.0/0.4/0.8	
External forces	Wave forces for RAO	Regular wave	Height (H)	1.912–4 m
			Period (T)	7–40 s
	Design forces	Irregular wave	Sig. height (H_s)	8.34 m
			Peak period (T_p)	13.1 s
			Shape parameter (γ)	1.7
		Wind	Wind speed (V_w)	40.28 m/s
		Current	Current speed (V_c)	1.69 m/s

drag, mass, friction, added mass coefficients. RAO (Response Amplitude Operator) is responses for unit amplitude of wave. Origin of coordinate is x = 0, y = 0 at CoG (Center of Gravity) and $z = 0$ at water-line. Model structures for floating body and mooring lines were fabricated in 1/55 scale and the model tests were done in KRISO basin.

Numerical analyses were also carried out to compare the test results. Hydrodynamic responses in wave, wind & current were calculated by solving the time domain coupled equations as

$$[M_B + M_{add}(\infty)]\{\ddot{x}\} + \int_0^t [R(t-\tau)]\{\dot{x}(\tau)\}d\tau + [K_B]\{x\}$$
$$= \{f_{wave}\} + \{f_{drift}\} + \{f_{damp}\} + \{f_{wind}\} + \{f_{cur}\} + \{f_{moor}\} \quad (1)$$

$$[M]\{\ddot{u}\} + [C]\{\dot{u}\} + [K]\{u\} = \{f\} \quad (2)$$

In (1), $[M_B]$, $[M_{add}]$, $[R]$, $[K_B]$, $\{f_{wave}\}$, $\{f_{drift}\}$, $\{f_{damp}\}$, $\{f_{wind}\}$, $\{f_{cur}\}$, are mass matrix, added mass matrix, retardation matrix, restoring coeff. Matrix, wave force vector, drift force vector, damping force vector, wind force vector, current force vector of floating body and they are formulated by HOBEM [2, 3] and convolution [4]. $\{x\}$ is motion vector of floating body. $\{f_{moor}\}$ is mooring force vector and it is from (2). In (2), $[M]$, $[C]$, $[K]$, $\{f\}$ are mass matrix, damping matrix, stiffness matrix, force vector for mooring lines and they are formulated by catenary elements using FEM [5–7]. $\{u\}$ is displacements vector of mooring lines. The boundary condition of (2) is fairlead motions and they are from (1). Time domain solutions for (1) & (2) were obtained by Hamming method [8] & modified Newmark method [9], respectively. Test set-up and numerical model are shown in Fig. 1. Numerical solution based on HOBEM has been widely applied to evaluating wave-induced responses of floating bodies and it is also shown in recent studies [10–12].

Before main test, tilt test for GM, free-decay test for natural periods and damping, wave calibration, pull-out test for mooring excursion curve were done to match basic similarity between model and real structure. Results of tilt test, free-decay tests w/o mooring and wave calibration were summarized in Tables 2–4. GM, natural periods and wave heights were matched in 97% accuracy. In irregular waves, six repeated waves were generated to check maximum response change with seeds. Results of pull-out test were summarized in Fig. 2. In the figure, 'Cal' is numerical results and 'Exp' is experimental results. Excursion curve of mooring line is also well matched. Figure 3 shows results of free-decay tests w/ mooring lines. The periods well matches but damping shows some difference.

3 Results of Model Tests and Hydrodynamic Analysis in Wave, Wind and Current

Model test results in basin in wave, wind and current are summarized in this section. Numerical results by hydrodynamic analysis were also compared. Figure 4 shows RAO of floating body motions and tension variation RAO of mooring lines in regular waves. In mooring line tensions, it is large at FCS 1 in wave direction $\beta = 180$ deg. It is large at FCS 2 in $\beta = 90$ deg. So, each result is shown in the figure. The trend in RAO matches well between numerical and experimental results.

Table 2. Results of tilt test and free-decay test w/o mooring

Pre-tests	Matched item	Target (real)	Target (model)	Measured (model)	Error
Tilt test	GM	18.234 m	0.332 m	0.329 m	0.802%
Free-decay test w/o mooring	T_{heave}	14.827 s	1.999 s	1.962 s	1.831%
	T_{roll}	22.572 s	3.044 s	2.957 s	2.848%
	T_{pitch}	22.544 s	3.040 s	2.947 s	3.046%

Table 3. Calibration for regular waves

Wave period (T)		Wave height (H)			
Real	Model	Real	Model	Measured (model)	Error
7 s	0.944 s	1.912 m	0.0348 m	0.0342 m	−1.53%
10 s	1.348 s	3.902 m	0.0709 m	0.0717 m	1.04%
15 s	2.023 s	4.000 m	0.0727 m	0.0723 m	−0.57%
17 s	2.292 s	4.000 m	0.0727 m	0.0734 m	0.99%
20 s	2.697 s	4.000 m	0.0727 m	0.0706 m	−2.99%
23 s	3.101 s	4.000 m	0.0727 m	0.0744 m	2.33%
26 s	3.506 s	4.000 m	0.0727 m	0.0730 m	0.40%
30 s	4.045 s	4.000 m	0.0727 m	0.0723 m	−0.63%
35 s	4.719 s	4.000 m	0.0727 m	0.0725 m	−0.30%
40 s	5.394 s	4.000 m	0.0727 m	0.0713 m	−2.02%

Table 4. Calibration for irregular waves

Wave seed no.	Peak period (T_p)		Sig. wave height (H_s)			
	Real	Model	Real	Model	Measured (model)	Error
1	13.1 s	1.766 s	8.34 m	0.152 m	0.1485 m	−2.04%
2	13.1 s	1.766 s	8.34 m	0.152 m	0.1475 m	−2.76%
3	13.1 s	1.766 s	8.34 m	0.152 m	0.1518 m	0.12%
4	13.1 s	1.766 s	8.34 m	0.152 m	0.1500 m	−1.08%
5	13.1 s	1.766 s	8.34 m	0.152 m	0.1475 m	−2.76%
6	13.1 s	1.766 s	8.34 m	0.152 m	0.1491 m	−1.68%

Experimental and Numerical Evaluation of Hydrodynamic Behavior 807

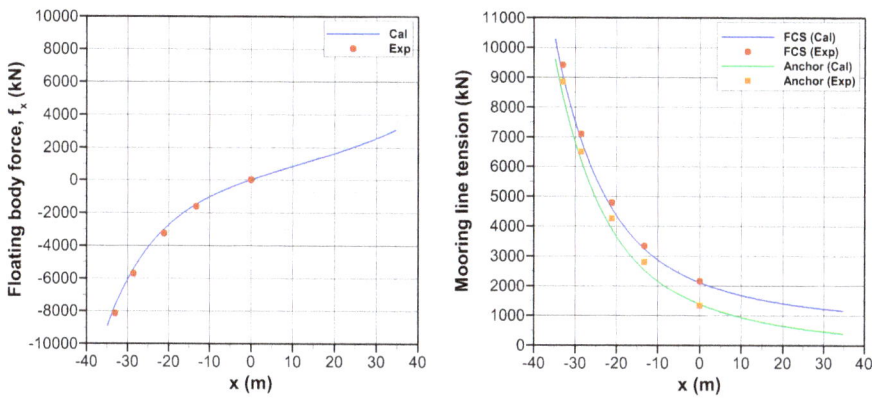

(a) Floating body forces v.s. surge offset (b) Tension of mooring line 1 v.s. surge offset

Fig. 2. Excursion curve from pull-out test

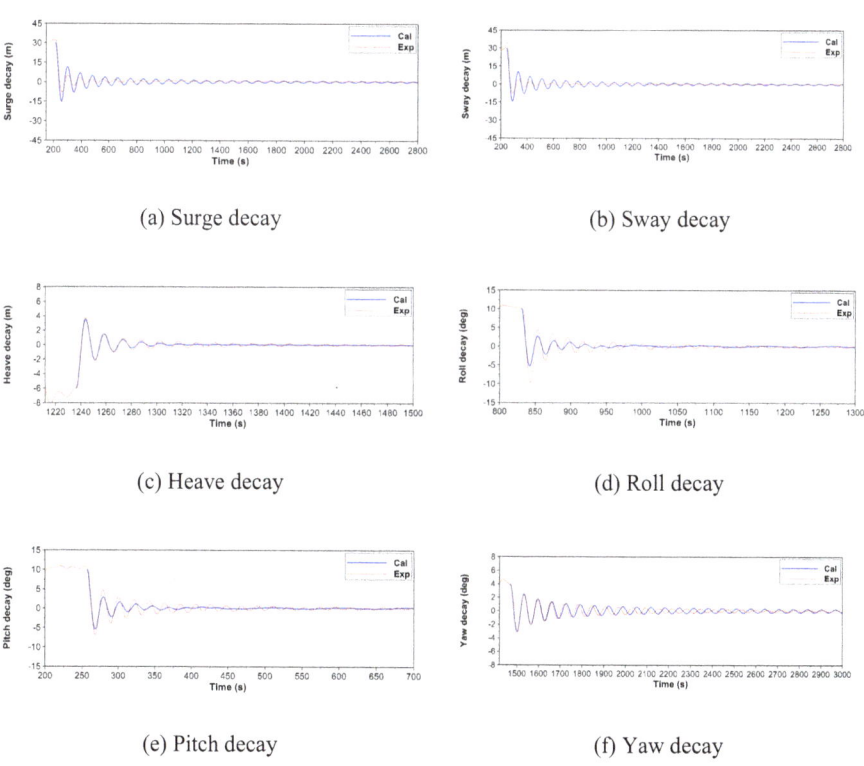

(a) Surge decay (b) Sway decay

(c) Heave decay (d) Roll decay

(e) Pitch decay (f) Yaw decay

Fig. 3. Free-decay motions of floating body w/ mooring

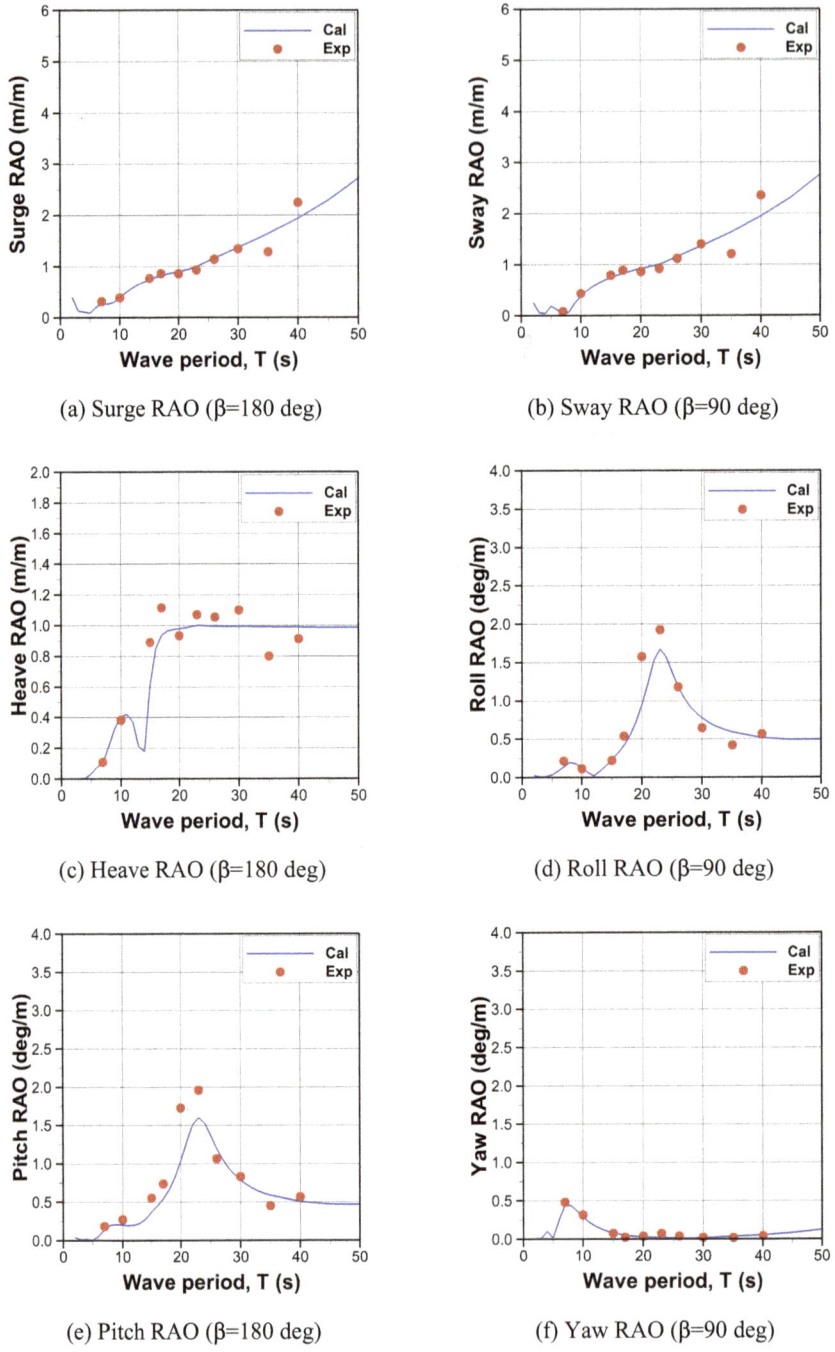

Fig. 4. RAO of floating body motions and tension variation RAO of mooring lines in regular waves

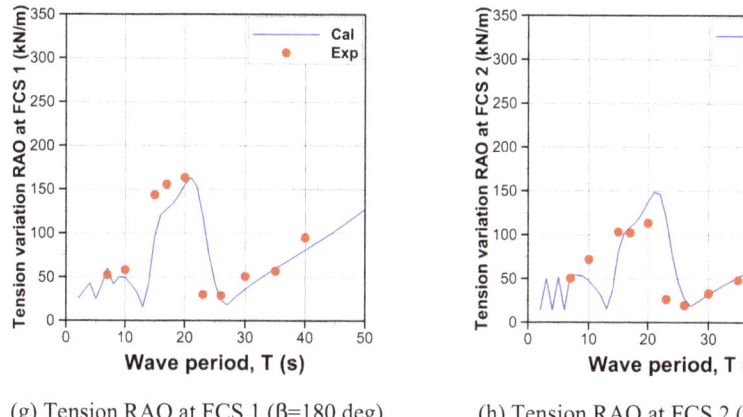

(g) Tension RAO at FCS 1 (β=180 deg)　　(h) Tension RAO at FCS 2 (β=90 deg)

Fig. 4. (*continued*)

Time series of floating body motions and mooring line tensions in design forces such as irregular wave, wind and current are shown in Figs. 5 and 6. Mooring line tension is the largest at FCS 1 in β = 180 deg. So, that case is presented. Six repeated tests were done to check change of maximum tensions versus wave seeds and the results are summarized in Table 5 and Fig. 7. Dynamic variations of tension are similar between numerical and experimental results but maximum values show some difference. MPM (Most Probable Maximum) tension was observed at seed 6. Figures 5 and 6 are the results for the seed 6. Among the results of Figs. 5 and 6, average heights of variation and maximum values are summarized in Table 6. Dynamic variations in surge, sway, yaw, mooring line tensions are similar between numerical and experimental results but maximum responses show some difference. Maximum surge, sway, heave are 33.069 m, 45.629 m, 5.056 m. Maximum roll, pitch, yaw are 7.733 deg., 11.219 deg., 5.842 deg. Maximum mooring line tension is 13,442 kN at anchor and 13,900 kN at FCS. This result for maximum responses will be a guide in selecting offset range of floating body and strength of mooring line for the 10 MW FOWT in DMS project.

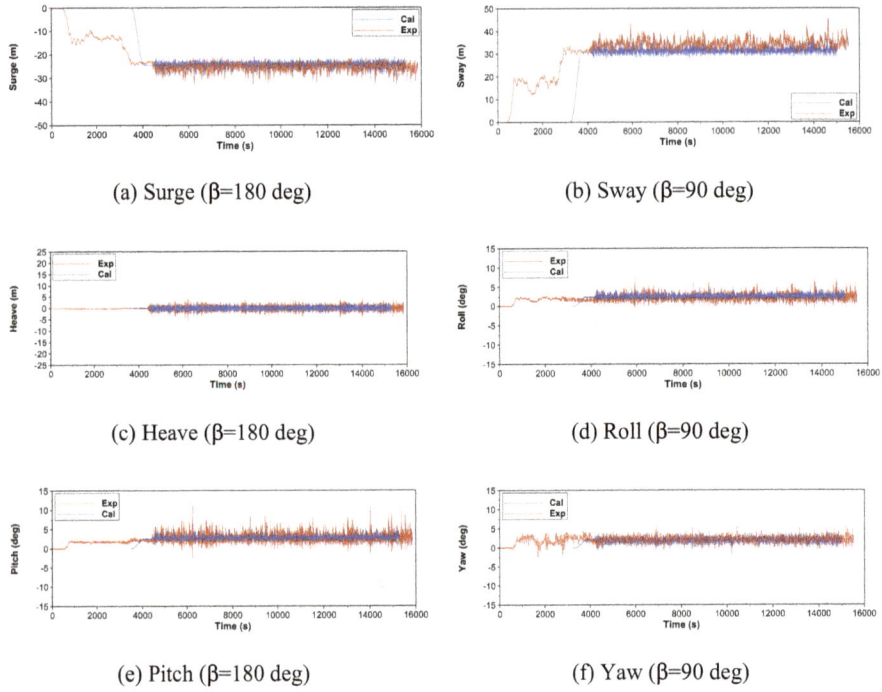

Fig. 5. Time series of floating body motions in design forces (irregular wave, wind & current)

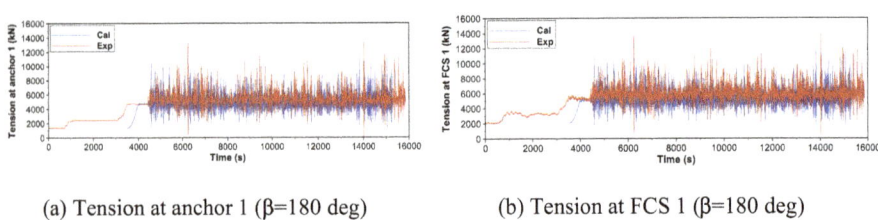

Fig. 6. Time series of mooring line tensions in design forces (irregular wave, wind & current)

Table 5. Tension af FCS 1 v.s. wave seed in design forces ($\beta = 180$ deg)

Wave seed no.	Avg. height of variation		Max.	
	Cal	Exp	Cal	Exp
1	2531 kN	2402 kN	11,455 kN	13,336 kN
2	2595 kN	2424 kN	10,503 kN	15,960 kN (outlier)
3	2580 kN	2493 kN	12,236 kN	14,209 kN
4	2561 kN	2579 kN	12,475 kN	13,575 kN

(*continued*)

Table 5. (*continued*)

Wave seed no.	Avg. height of variation		Max.	
	Cal	Exp	Cal	Exp
5	2583 kN	2447 kN	11,309 kN	13,256 kN
6	2575 kN	2374 kN	11,654 kN	13,900 kN
Mean	2571 kN	2453 kN	11,605 kN	13,655 kN
MPM			11,654 kN	13,900 kN

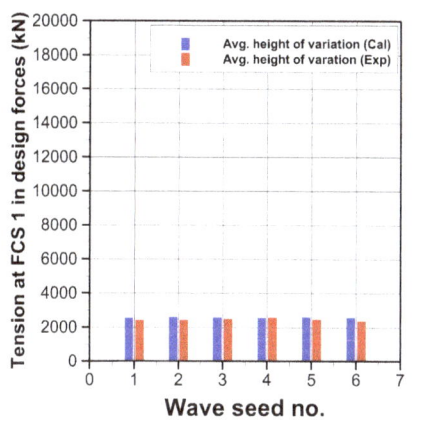
(a) Average height of variation

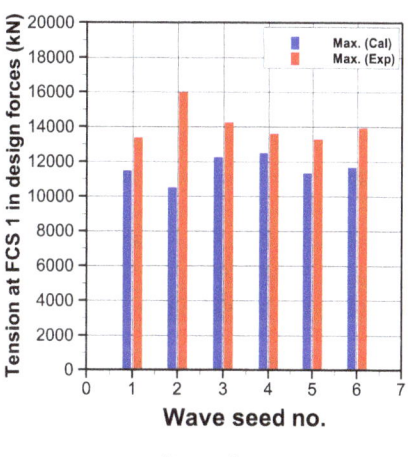
(b) Maximum

Fig. 7. Tension af FCS 1 v.s. wave seed in design forces ($\beta = 180$ deg)

Table 6. Summary of floating body motions and mooring line tensions in design forces

Responses		β (deg)	Avg. height of variation		Max.	
			Cal	Exp	Cal	Exp
Floating body motions	Surge	180	3.020 m	3.102 m	30.156 m	33.069 m
	Sway	90	2.905 m	3.362 m	37.564 m	45.629 m
	Heave	180	1.939 m	2.891 m	3.176 m	5.056 m
	Roll	90	1.104 deg	2.191 deg	5.294 deg	7.733 deg
	Pitch	180	1.127 deg	2.949 deg	5.903 deg	11.219 deg
	Yaw	90	1.455 deg	1.512 deg	4.012 deg	5.842 deg
Mooring line tensions	Anchor 1	180	2533 kN	2488 kN	11,379 kN	13,442 kN
	FCS 1	180	2575 kN	2374 kN	11,654 kN	13,900 kN

4 Conclusions

Model test results for a 10 MW FOWT by DMS project are presented in this paper. GM, natural periods, mooring excursion and wave heights for the test model were matched by tilt, free-decay, pull-out tests and wave calibration in 1/55 scale. After matching tests, the model was moved to KRISO basin and floating body motions and mooring line tensions in regular waves and design forces such as irregular wave, wind and current were measured. The measured responses were also compared with hydrodynamic analysis by HOMEM, convolution and FEM.

Natural periods of floating body motions, excursion curves of mooring line, RAO responses in regular waves and dynamic variations of surge, sway, yaw, mooring tensions in design forces agree well between experimental and numerical results. Damping behaviors and maximum responses show some differences.

Maximum surge, sway, heave are 33.069 m, 45.629 m, 5.056 m. Maximum roll, pitch, yaw are 7.733 deg., 11.219 deg., 5.842 deg. Maximum mooring line tension is 13,442 kN at anchor and 13,900 kN at FCS. Those results will provide a guide to selecting offset range of floating body and strength of mooring line for the 10 MW FOWT in DMS project.

Numerical and test results generally agree in trend and average responses. However, damping behaviors are a little different. Therefore, improvement in damping evaluation is recommended in the future study.

Acknowledgements. This research is a part of Development of Disconnectable Mooring System for A MW class Floating Offshore Wind Turbine (Grant 20213000000030 by Korea Ministry of Trade, Industry and Energy, Grant PNS5090 by KRISO).

References

1. DMS Project Team: Design Basis. Project Management Control Document (2021~2025)
2. Choi, Y.R., Hong, S.Y., Choi, H.S.: An analysis of second-order wave forces on floating bodies by using a higher-order boundary element method. Ocean Eng. **28**, 117–138 (2001)
3. Hong, S.Y., Kim, J.H., Cho, S.K., Choi, Y.R., Kim, Y.S.: Numerical and experimental study on hydrodynamic interaction of side-by-side moored multiple vessels. Ocean Eng. **32**, 783–801 (2005)
4. Cummins, W.E.: The impulse response function and ship motions. Symposium on Ship Theory at the Institut fur Schiffbau der Universitat Hamburg (1962)
5. Garrett, D.L.: Coupled analysis of floating production systems. Ocean Eng. **32**, 802–816 (2005)
6. Kim, B.W., Sung, H.G., Kim, J.H., Hong, S.Y.: Comparison of linear spring and nonlinear FEM methods in dynamic coupled analysis of floating structure and mooring system. J. Fluids Struct. **42**, 205–227 (2013)
7. Kim, B.W., Hong, S.Y., Sung, H.G.: Comparison of drift force calculation methods in time domain analysis of moored bodies. Ocean Eng. **126**, 81–91 (2016)
8. Hamming, R.W.: Stable predictor-correct methods for ordinary differential equations. J. ACM. **6**, 37–47 (1959)
9. Chung, J., Hulbert, G.M.: A time integration algorithm for structural dynamics with improved numerical dissipation: The generalized α-method. J. Appl. Mech. **60**, 371–375 (1993)

10. Kim, H.S., Kim, B.W., Hong, S.Y.: Comparative Study on effect of buoys for floating sunlight generation system with numerous buoys and connection beams. Int. J. Offshore Polar Eng. **30**(2), 209–219 (2020)
11. Kim, H.S., Kim, B.W., Lee, K., Sung, H.G.: Application of average sea-state method for fast estimation of fatigue damage of offshore structure in waves with various distribution types of occurrence probability. Ocean Eng. **246**, 110601 (2022)
12. Kim, B.W., Lee, K.: Hydro-dynamic behavior of multi-body floating solar platform with beam or shell connectors and comparison of design capacity. In: The 34th International Ocean and Polar Engineering Conference, pp. 2357–2366 (2024)

Open Access This chapter is licensed under the terms of the Creative Commons Attribution-NonCommercial-NoDerivatives 4.0 International License (http://creativecommons.org/licenses/by-nc-nd/4.0/), which permits any noncommercial use, sharing, distribution and reproduction in any medium or format, as long as you give appropriate credit to the original author(s) and the source, provide a link to the Creative Commons license and indicate if you modified the licensed material. You do not have permission under this license to share adapted material derived from this chapter or parts of it.

The images or other third party material in this chapter are included in the chapter's Creative Commons license, unless indicated otherwise in a credit line to the material. If material is not included in the chapter's Creative Commons license and your intended use is not permitted by statutory regulation or exceeds the permitted use, you will need to obtain permission directly from the copyright holder.

Numerical Investigation of Coastal Evolution and Protection Measures Under Sea Level Rise

Renqiang Wen[1], Chen Gu[2], Yue Zheng[2], Tianxia Jia[1], Yang Hong[3], and Zhipeng Qu[4(✉)]

[1] China Three Gorges Corporation, Wuhan 430010, China
[2] Shanghai Investigation, Design and Research Institute Co., Ltd., Shanghai 200335, China
[3] Qingdao Huahang Seaglet Environmental Technology Ltd, Qingdao 266041, China
[4] College of Engineering, Ocean University of China, Qingdao 266404, China
hei21guns@163.com

Abstract. Sea level rise accelerates the rate of coastal erosion, threatening the safety of coastal infrastructure. This study uses XBeach to establish a one-dimensional beach profile evolution model, simulating coastal evolution under different engineering measures (submerged breakwaters, berm nourishment, and dune nourishment) and exploring the impact of sea level rise on wave transmission. The research finds that wave height changes little in deep water areas but significantly increases in shallow water areas, with the wave-breaking zone moving closer to the coast after sea level rise. In storm surge events, submerged breakwaters can significantly attenuate wave energy and reduce wave height; however, their protective efficacy diminishes progressively with rising sea levels. Berm nourishment can enhance beach width and effectively mitigate beach erosion, yet sediment loss in nourished areas is exacerbated by sea level rise. Dune nourishment can effectively counteract erosion of land areas resulting from sea level rise, although it is essential to ensure a sustainable supply of sediment. In non-storm wave scenarios, underwater berm nourishment plays a crucial role in beach widening, necessitating an increase in berm height in response to rising water levels. This study provides relevant engineering recommendations to address the impact of sea level rise on the coast, helping to optimize coastal protection projects and promote the sustainable management of coastal zones.

Keywords: sea level rise · storm surge · XBeach · beach protection

1 Introduction

The rising sea levels exacerbate coastal hazards and disrupt coastal ecosystems. The IPCC reports a 0.19 m global sea level rise from 1901 to 2010, with projections exceeding 1 m by 2100 [1]. This elevates nearshore wave heights, exacerbating beach erosion through sediment redistribution and sandbar migration [2]. As shown in Fig. 1, traditional hard structures like submerged breakwaters or seawalls mitigate erosion but often disrupt coastal ecosystems [3]. In contrast, nature-based solutions such as berm nourishment provide sustainable alternatives by restoring sediment balance [4].

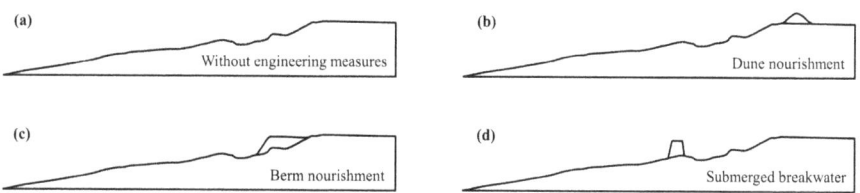

Fig. 1. Conceptual diagram of engineering measures: (a) without engineering measures (b) dune nourishment (c) berm nourishment (d) submerged breakwater.

The Bruun theory remains foundational for predicting equilibrium shoreline response to sea level rise, assuming sediment redistribution within closure depth [5]. Experimental validations demonstrate its utility in estimating beach recession rates [6], though limitations exist in addressing storm impacts and sediment supply [7]. Coastal defense structures critically impair shoreline migration capacity under accelerating sea-level rise [8]. Particularly concerning are seawalls, which although effective for immediate erosion control, induce feedback mechanisms that amplify localized scour processes during marine transgression phases [9]. Compounding factors like storm surges and El Niño events further amplify erosion risks [10,11].

Numerical models enable quantitative assessment of these complex interactions. Process-based tools like XBeach simulate coupled hydrodynamic and morphodynamic responses, including storm impacts and engineering interventions [12,13]. Such models overcome traditional limitations by integrating wave-current interactions and sediment transport physics [14]. This study uses XBeach to analyze three protection strategies (submerged breakwaters, berm nourishment, and dune nourishment; Fig. 1) under sea level rise scenarios. By comparing their hydrodynamic buffering capacities and profile evolution patterns, we aim to inform sustainable coastal management practices.

2 Methods

2.1 XBeach Model

XBeach is an open-source numerical model designed to simulate wave dynamics, sediment transport, and morphological changes in coastal environments [15]. This study uses the non-hydrostatic mode, combining shallow water equations with a pressure correction for accurate wave simulation. It includes dispersive behavior for short wave resolution in intermediate shallow waters [16]. The continuity and momentum equations of the model are as follows:

$$\frac{\partial \eta}{\partial t} + \frac{\partial h u^L}{\partial x} + \frac{\partial h v^L}{\partial y} = 0 \qquad (1)$$

$$\frac{\partial u^L}{\partial t}+u^L\frac{\partial u^L}{\partial x}+v^L\frac{\partial u^L}{\partial y}-fv^L-v_h\left(\frac{\partial^2 u^L}{\partial x^2}+\frac{\partial^2 u^L}{\partial y^2}\right)=\frac{\tau_{sx}}{\rho h}-\frac{\tau_{bx}^E}{\rho h}-g\frac{\partial \eta}{\partial x}+\frac{F_x}{\rho h}+\frac{F_{v,x}}{\rho h} \quad (2)$$

$$\frac{\partial v^L}{\partial t}+u^L\frac{\partial v^L}{\partial x}+v^L\frac{\partial v^L}{\partial y}+fu^L-v_h\left(\frac{\partial^2 v^L}{\partial x^2}+\frac{\partial^2 v^L}{\partial y^2}\right)=\frac{\tau_{sy}}{\rho h}-\frac{\tau_{by}^E}{\rho h}-g\frac{\partial \eta}{\partial y}+\frac{F_y}{\rho h}+\frac{F_{v,y}}{\rho h} \quad (3)$$

where τ_{bx} and τ_{by} represent bed shear stresses, η denotes water surface elevation, F_x and F_y are wave radiation stresses, v_h stands for horizontal viscosity, f is the Coriolis force coefficient, and F_v represents the vegetation term.

The elevation of the bed is calculated according to the following formula:

$$\frac{\partial z_b}{\partial t}+\frac{f_{mor}}{(1-\rho)}\left(\frac{\partial q_x}{\partial x}+\frac{\partial q_y}{\partial y}\right)=0 \quad (4)$$

where ρ represents the porosity, f_{mor} is the factor accelerating terrain changes, and q_x and q_y denote the sediment transport rates in the x and y directions, respectively.

2.2 Experimental Cases and Simulation Work

This study simulates the experimental data of a static water level test conducted by Meng et al. in the wave flume [17]. The experimental setup is shown in Fig. 2, with a sandy area in the flume measuring 14.75 m in length and 0.73 m in width, a static water level of 0.7 m, and a 1/15 initial beach slope. The experiment utilizes natural sand with a median particle size of $D_{50} = 0.284$ mm and a uniformity coefficient of $D_{60}/D_{10} = 1.72$. Nine wave gauges are set up, with the position of the first wave gauge as the reference point, located at 0 m, 1.02 m, 1.73 m, 13.25 m, 13.72 m, 14.47 m, 17.29 m, 19.62 m, and 21.11 m, respectively. Under the long-term combined action of waves, the experimental results have revealed the profile of beach erosion, which is relatively stable with slow changes (Fig. 2b). Based on this experimental profile, this study will further investigate the impact of sea level rise on beaches and engineering measures.

The experiment follows the Froude similarity law with a scale of 1:15. In the simulations, water levels are set at 0.7 m and 0.75 m with a water level difference of 0.05 m, corresponding to a sea level rise of 0.75 m. The study includes the following engineering measures: berm nourishment, dune nourishment, and the placement of submerged breakwaters. Specific experimental and simulated scenarios are shown in Table 1. Additionally, the Iribarren number, defined as $\xi = \tan \beta / \sqrt{H/L}$, is utilized to characterize the type of wave phenomena. In this study, the simulated wave conditions are classified as spilling waves (ξ <0.5) and plunging waves ($0.5 < \xi < 3.3$).

Three typical engineering measures are selected for comparison to address the impact of sea level rise, including the installation of submerged breakwaters, berm nourishment, and dune nourishment, with reference to the situation

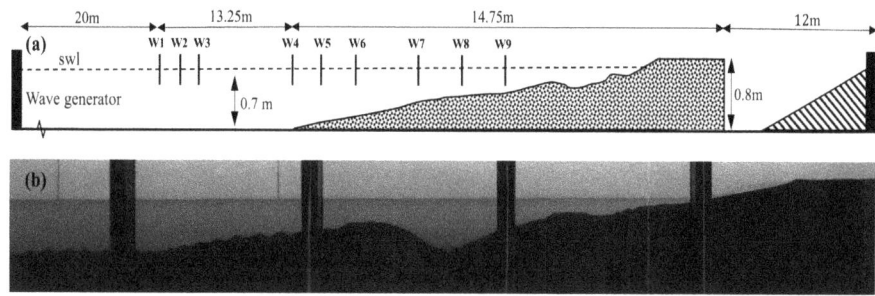

Fig. 2. (a) Schematic diagram of the experimental setup. (b) Initial beach profile captured by a camera.

Table 1. Experimental cases and simulation work. H is wave height, T is wave period, H/L is wave steepness, h is the height of the berm.

	Case	d(m)	H(m)	T(s)	H/L	ξ	Wave type	Measures
Experiment	A1	0.7	0.185	1.6	0.054	0.287	Storm	/
	A2	0.7	0.185	1.6	0.054	0.287	Storm	Berm nourishment ($h = 0.8$ m)
Simulation(XBeach)	A3	0.7	0.185	1.6	0.054	0.287	Storm	Dune nourishment
	A4	0.7	0.185	1.6	0.054	0.287	Storm	Submerged breakwater
	B1	0.75	0.185	1.6	0.053	0.290	Storm	/
	B2	0.75	0.185	1.6	0.053	0.290	Storm	Berm nourishment ($h = 0.8$ m)
	B3	0.75	0.185	1.6	0.053	0.290	Storm	Dune nourishment
	B4	0.75	0.185	1.6	0.053	0.290	Storm	Submerged breakwater
	C1	0.7	0.06	1.6	0.018	0.504	Non-storm	Berm nourishment ($h = 0.68$ m)
	C2	0.75	0.06	1.6	0.017	0.509	Non-storm	Berm nourishment ($h = 0.68$ m)
	C3	0.75	0.06	1.6	0.017	0.509	Non-storm	Berm nourishment ($h = 0.73$ m)

without engineering measures. Specifically, submerged breakwaters, as the most commonly used wave dissipation measure in engineering applications, are placed in the wave breaking zone considering construction difficulty and effectiveness, with a location near x = 20.68 m, a crest elevation of 0.65 m, and a width of 0.5 m. The berm nourishment, as described in the experimental setup by Meng et al. [17], involves extending the leading edge of the beach berm to approximately x = 23.4 m. This extension effectively widens the beach width, with the height of the berm set at 0.8 m. The dune front toe is near x = 25.3 m, and front and back slopes of 1/2 and 1/3 in the dune nourishment.

2.3 Model Validation

As shown in Fig. 3a-d, wave gauges WG1 and WG4 are placed in the deep water area and at the foot of the slope, respectively. WG6 is positioned in the transitional area where the water depth gradually becomes shallower, while WG8 is located near the sandbar. In XBeach simulations, the simulated time series of water surface elevation shows good agreement with the data measured by the wave gauges. Figure 3e shows the profile at t = 15 min during the berm nourishment case, and the model simulation results are consistent with the measured profile.

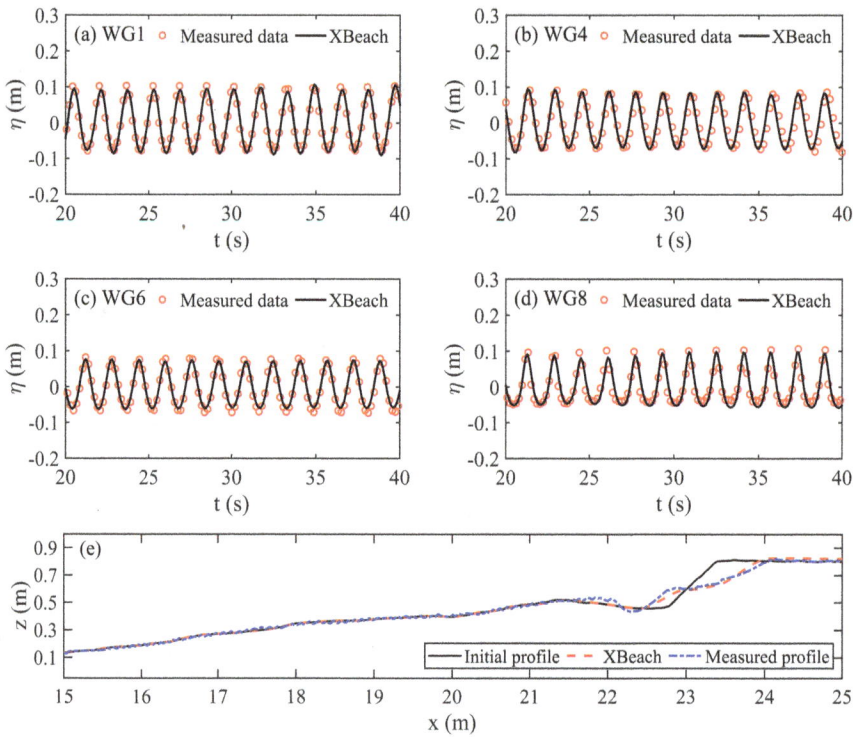

Fig. 3. (a-d) Water level verification. (e) Profile verification.

3 Results

3.1 Hydrodynamic Changes

The effectiveness of wave attenuation by submerged breakwaters is closely related to their length, depth, and structural design [18]. This study presents a typical layout of submerged breakwaters as a reference, with Fig. 4 showing the variation of water surface at t = 10 min. As the water level rises, without engineering measures, the breaking point moves closer to the coastal area (Fig. 4b). The presence of submerged breakwaters can reduce wave heights and amplitudes, especially in the area behind the breakwaters, where the wave height is significantly influenced by the structures (Fig. 4c). Figure 4d illustrates that as the water level increases, the breaking point shifts nearer to the beach berm, leading to diminished wave breaking along the coastline. This shift could weaken the wave attenuation capacity of the breakwaters, potentially allowing more wave energy to reach the shore.

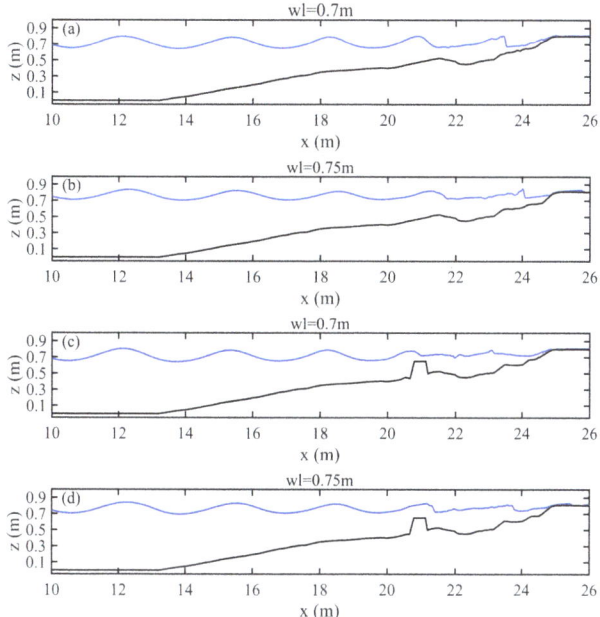

Fig. 4. The variation of water surface: (a) without engineering measures under static water level (b) without engineering measures after sea level rise (c) submerged breakwater arrangement under static water level (d) submerged breakwater arrangement after sea level rise.

Figure 5 compares the variations in average wave height after sea level rise under four different conditions. The results indicate that wave height changes are minor in deep water areas, while shallow water areas are significantly affected by sea level rise. In specific conditions, such as berm nourishment and submerged breakwater layout, the changes in nearshore wave height are complex and require a comprehensive consideration of the combined effects of topography and sea level rise. In the berm nourishment, the sudden shallowing of the topography due to the nourishment location results in strong nonlinear wave behavior, and the impact of sea level rise on the average wave height in shallow water areas is not very pronounced. In the absence of engineering measures and dune nourishment, a significant rise in nearshore average wave height is apparent, particularly within the range of x = 22 − 24.3 m, as illustrated in Fig. 5a and 5d when compared with wave heights in static water level conditions. This phenomenon indicates that sea level rise has a significant amplifying effect on nearshore waves. In the case of submerged breakwater layout, there is a significant decrease in wave height behind the breakwater. However, even with sea level rise, the wave height remains higher than under static water level conditions. This suggests that while submerged breakwaters can effectively reduce wave intensity, they cannot completely offset the impact of sea level rise on nearshore wave conditions.

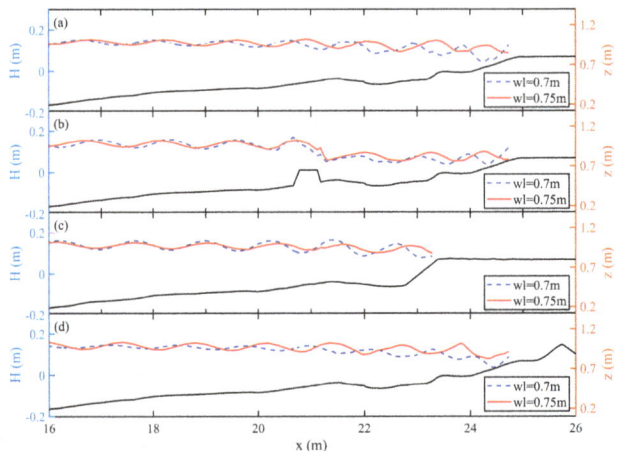

Fig. 5. Average wave height: (a) without engineering measures (b) submerged breakwater arrangement (c) berm nourishment (d) dune nourishment.

As shown in Fig. 6, there are significant differences in wave energy flux changes at x = 22.3 m in the nearshore area. Under static water level conditions, implementing submerged breakwater measures can significantly reduce wave energy flux, effectively maintaining beach stability (Fig. 6b). However, with rising water levels, whether without engineering measures, submerged breakwater placement, berm nourishment, or dune nourishment, there is a noticeable increasing trend in wave energy flux. The purpose of berm nourishment is to attenuate the impact force of waves reaching the shoreline by increasing the width and height of the coastal beach face. However, after sea level rise, the berm nourishment area narrows, underwater sediment transport becomes more frequent, the slope becomes gentler, and nonlinear effects increase, leading to a decrease in wave energy flux (Fig. 6c). The variation in wave energy flux may lead to beach erosion, necessitating ongoing beach maintenance. Dune nourishment typically involves adding dunes near the inland side to prevent erosion of the hinterland. However, with rising water levels, waves are more likely to reach the dune toe, especially in storm surge situations, and increased wave energy flux may exacerbate dune erosion (Fig. 6d).

In conclusion, rising sea levels will increase wave energy flux in the nearshore, exacerbating erosion. Implementing submerged breakwaters can effectively reduce wave propagation, berm nourishment indirectly reduces wave energy flux by expanding the shallow water area, and dune nourishment indirectly affects wave energy flux by enhancing protective capabilities and reflecting wave energy.

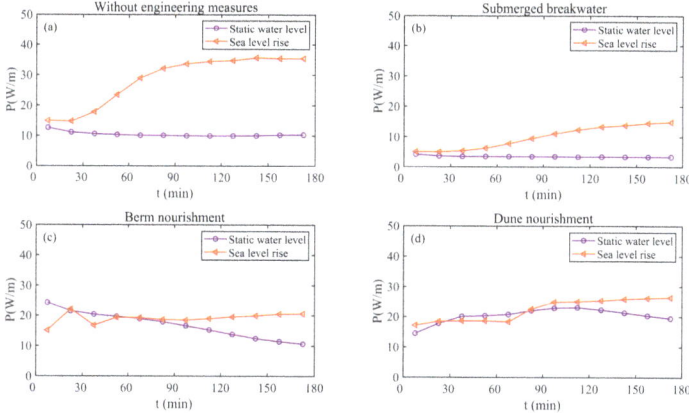

Fig. 6. Wave energy flux changes: (a) without engineering measures (b) submerged breakwater arrangement (c) berm nourishment (d) dune nourishment.

3.2 Morphological Change

As shown in Fig. 7a, under static water levels, there is no significant change in the position of the shoreline. Due to the wave swash, erosion and deposition phenomena occur on the beach berm. Submerged breakwaters mitigate the wave energy in the protected water area, reducing wave energy and making suspended sediments more easily deposited behind the breakwater. However, with the rise in sea level, continuous scouring in the swash zone causes the shoreline to retreat towards the land, consequently leading to the advancement of the erosion zone towards the land (Fig. 7b). The increase in water level results in partial overwash phenomena, transporting sediments to the land area and depositing them.

Under static water levels with the arrangement of submerged breakwaters, the breakwaters weaken the wave energy. Compared to Fig. 7a, the degree of erosion on the beach berm is reduced, and the degree of deposition on the beach is weakened. This is because the presence of submerged breakwaters weakens the waves transmitted to the surf zone, thereby affecting the erosion and deposition status near the shore. In the case of rising sea levels, the deposition phenomenon in the land area weakens, indicating that submerged breakwaters still have a wave attenuation effect (Fig. 7d). However, the existing submerged breakwater structures are unable to cope with the wave breaking and coastal erosion caused by sea level rise closer to the coast. Submerged breakwaters and other hard protective structures require additional optimization to effectively combat the increasing risk of coastal erosion resulting from rising sea levels.

Berm nourishment is a method of directly increasing the width of the coast, which can quickly take effect. The berm stands at a height of 0.8 m, and the shoreline moves from x = 24.5 m to x = 23.5 m towards the sea, effectively preventing erosion in the short term (Fig. 8a). In the early stages of berm nourishment, the beach profile may be steep, but over time, waves will redistribute

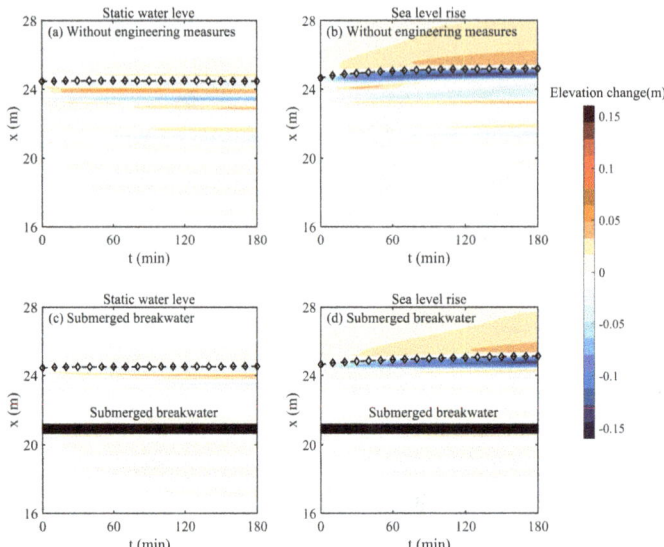

Fig. 7. Morphology changes: (a)(b) without engineering measures and (c)(d) submerged breakwater arrangement conditions, black area indicates submerged breakwater location, dashed line indicates shoreline position.

the sand, forming a flatter and more stable profile. However, the sand replenished will be eroded by waves, and the effectiveness of berm nourishment may weaken over time, requiring regular maintenance and replenishment. As shown in Fig. 8b, sea level rise will lead to significant shoreline retreat, expansion of erosion and deposition areas, and more obvious sediment transport offshore.

Dune nourishment refers to replenishing sand in the land area behind the beach, which can effectively prevent wave erosion and wind erosion, reducing the rate of coastal erosion. As shown in Fig. 8c, its protective effect is not significant in conditions where overtopping or inundation does not occur. However, in scenarios of sea level rise, dunes act as a natural protective barrier, to some extent protecting the beach from erosion and promoting sediment accumulation, thereby changing the beach profile characteristics (Fig. 8d). In other scenarios, due to the blocking effect of dunes, there is no significant accumulation in the land area. However, as seawater advances inland, the erosion of dunes will become more severe, potentially leading to a decrease in the height and stability of the dunes.

The influence of rising sea levels may render shoreline retreat inevitable, particularly in the absence of effective engineering interventions [19]. As illustrated in Fig. 9, the rise in sea level, coupled with wave action, results in a substantial retreat of the shoreline. In scenarios where sand is replenished on the beach berm, although the shoreline experiences the greatest retreat, the width of the beach remains the widest among the four conditions examined. It is important

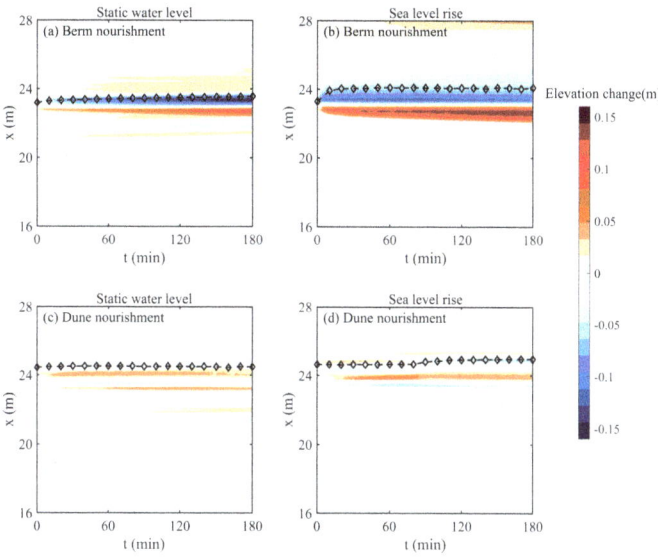

Fig. 8. Morphology changes: (a) (b) berm nourishment and (c) (d) dune nourishment, dashed line indicates shoreline position.

to note that in this study, the position of the dunes is close to the shoreline, thus providing good protection for the shoreline.

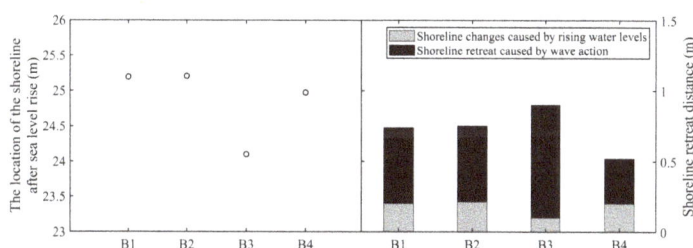

Fig. 9. Shoreline changes.

From the previous sections, it is evident that beach nourishment at the berm significantly contributes to the widening of the coastline. However, due to policy restrictions in certain regions, conducting beach nourishment or artificially widening coastal zones may not be feasible. In engineering applications, implementing underwater nourishment in the original berm area may represent a more viable approach. As illustrated in Fig. 10a, the beach nourishment has been implemented within the berm zone, where the berm heights are set at 0.68 m and 0.73 m, corresponding to Cases C1–C3 in Table 1.

Figure 10b demonstrates that the sand source supplied migrates continuously landward due to wave action, occasionally surfacing above the waterline. This process gradually increases the width of the beach and provides a natural maintenance function under non-storm conditions. However, if the water level rises, the underwater nourishment in the original berm area does not lead to an increase in beach width (Fig. 10c). Therefore, as the water level rises, it is essential to concurrently elevate the height of the underwater sand nourishment at the berm. This adjustment promotes the transport of sediments towards the coast, effectively nourishing the beach (Fig. 10d).

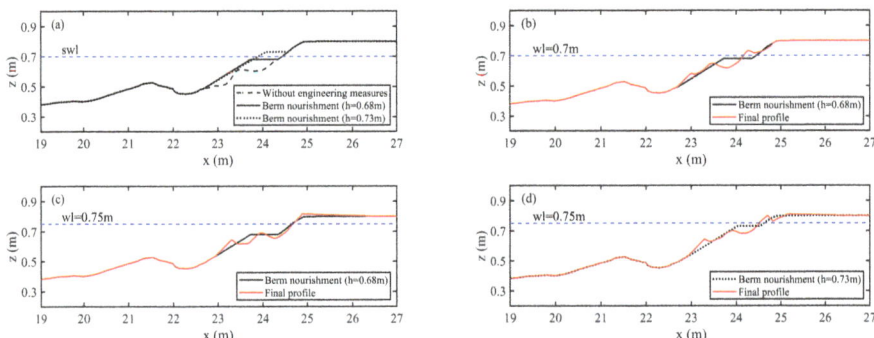

Fig. 10. Non-storm wave conditions: (a) berm nourishment at different heights (b) water level is 0.7 m, the height of the berm is 0.68 m (c) water level is 0.75 m, the height of the berm is 0.68 m (d) water level is 0.75 m, the height of the berm is 0.73 m.

In conclusion, sea level rise will alter the sediment transport patterns near the coast, affecting the sources and distribution of sediment. Waves and tidal currents will redistribute the sediment on beaches, leading to increased accumulation in some areas and decreased in others. The higher water level allows wave energy to act on higher parts of the beach, particularly during storms, exacerbating erosion and reducing the stability of the beach. Under non-storm wave cases, the increase in water level reduces the sediment transport towards the shore, making it impossible to naturally form sediment deposition on the beach surface. Therefore, the impacts of sea level rise need to be evaluated based on specific coastal topography and environmental conditions to assess their effectiveness and long-term stability. Submerged breakwaters need to be raised to maintain their wave reduction effect, while beaches and dunes require additional sand replenishment to combat erosion.

4 Discussion and Conclusion

This study establishes a one-dimensional beach profile evolution model using XBeach to evaluate coastal protection measures under sea level rise. Four scenarios were analyzed: without engineering measures, submerged breakwater protection, berm nourishment, and dune nourishment. Key findings reveal that sea level rise shifts wave-breaking zones landward, amplifying wave energy flux in shallow areas and intensifying coastal erosion. While submerged breakwaters effectively attenuate wave energy, their long-term efficacy diminishes as rising water levels reduce their elevation relative to the sea surface. Proactive elevation adjustments may be required to maintain their protective capacity.

Soft protection strategies demonstrate distinct advantages in adapting to dynamic coastal processes. Berm nourishment directly widens beaches, offering immediate erosion resistance. However, accelerated sediment loss under sea level rise necessitates frequent replenishment, raising concerns about long-term cost-effectiveness. Dune nourishment enhances natural storm buffers but faces challenges from intensified storm surges, which accelerate sand displacement. Sustainable dune management demands scientifically guided sediment sourcing and placement to balance ecological and protective functions.

Notably, underwater berm nourishment emerges as a promising alternative in policy-constrained regions where direct coastline extension is prohibited. While sediment naturally migrates shoreward under normal wave conditions, rising sea levels reduce nearshore accumulation efficiency, suggesting the need for elevated berm designs to align with future water levels.

Coastal protection strategies must reconcile immediate engineering objectives with long-term environmental adaptability. The nonlinearity of wave dynamics, sediment transport complexity, and increasing extreme weather frequency underscore the need for integrated approaches. Future work should prioritize large-scale physical experiments and advanced numerical simulations to refine predictive models, enabling robust, adaptive designs for evolving coastal systems.

Acknowledgments. This research is sponsored by research funding of China Three Gorges Corporation (No. 202003025).

References

1. André, C., Boulet, D., Rey-Valette, H., Rulleau, B.: Protection by hard defence structures or relocation of assets exposed to coastal risks: contributions and drawbacks of cost-benefit analysis for long-term adaptation choices to climate change. Ocean Coastal Manag. **134**, 173–182 (2016)
2. Lomelí-Quintero, V.M., Calderón-Vega, F., Mösso, C., Sánchez-Arcilla, A., García-Soto, A.D.: Impact costs due to climate change along the coasts of Catalonia. J. Marine Sci. Eng. **11**(10), 1939 (2023)
3. Toth, L.T., et al.: The potential for coral reef restoration to mitigate coastal flooding as sea levels rise. Nat. Commun. **14**(1) (2023)
4. Brand, E., Ramaekers, G., Lodder, Q.: Dutch experience with sand nourishments for dynamic coastline conservation–an operational overview. Ocean Coastal Manag. **217**, 106008 (2022)
5. Bruun, P.: The bruun rule of erosion by sea-level rise: a discussion on large-scale two-and three-dimensional usages. J. Coast. Res. 627–648 (1988)
6. Atkinson, A.L., Baldock, T.E.: Laboratory investigation of nourishment options to mitigate sea level rise induced erosion. Coastal Eng. **161**, 103769 (2020)

7. Cooper, J.A.G., et al.: Sandy beaches can survive sea-level rise. Nat. Clim. Chang. **10**(11), 993–995 (2020)
8. Vousdoukas, M.I., et al.: Sandy coastlines under threat of erosion. Nat. Clim. Chang. **10**(3), 260–263 (2020)
9. Beuzen, T., Turner, I., Blenkinsopp, C., Atkinson, A., Flocard, F., Baldock, T.: Physical model study of beach profile evolution by sea level rise in the presence of seawalls. Coast. Eng. **136**, 172–182 (2018)
10. Houston, J.: Beach nourishment versus sea level rise on florida's coasts. Shore Beach 3–13 (2020)
11. Zhang, X., et al.: High winds associated with cold surges and their relevance to climate patterns in the yellow and Bohai seas. Clim. Dyn. (2024)
12. Harter, C., Figlus, J.: Numerical modeling of the morphodynamic response of a low-lying barrier island beach and foredune system inundated during hurricane ike using XBeach and CSHORE. Coast. Eng. **120**, 64–74 (2017)
13. Qu, Z., Meng, Y., Liang, B.: Coastal processes and dune stability: insights from wave transmission and runup modeling. Phys. Fluids **36**(7) (2024)
14. Mendes, J., et al.: Modeling dynamic processes of mondego estuary and óbidos lagoon using delft3d. J. Marine Sci. Eng. **9**(1), 91 (2021)
15. Berard, N.A., Mulligan, R.P., da Silva, A.M.F., Dibajnia, M.: Evaluation of XBeach performance for the erosion of a laboratory sand dune. Coast. Eng. **125**, 70–80 (2017)
16. de Beer, A., McCall, R., Long, J., Tissier, M., Reniers, A.: Simulating wave runup on an intermediate-reflective beach using a wave-resolving and a wave-averaged version of XBeach. Coast. Eng. **163**, 103788 (2021)
17. Meng, Y., Qu, Z., Li, X., Zhu, M., Liang, B.: An experimental study on the evolution of beach profiles under different beach nourishment methods. Front. Marine Sci. **11** (2024)
18. Suh, K., Dalrymple, R.A.: Offshore breakwaters in laboratory and field. J. Waterw. Port Coast. Ocean Eng. **113**(2), 105–121 (1987)
19. Vitousek, S., Vos, K., Splinter, K.D., Erikson, L., Barnard, P.L.: A model integrating satellite-derived shoreline observations for predicting fine-scale shoreline response to waves and sea-level rise across large coastal regions. J. Geophys. Res. Earth Surface **128**(7) (2023)

Open Access This chapter is licensed under the terms of the Creative Commons Attribution-NonCommercial-NoDerivatives 4.0 International License (http://creativecommons.org/licenses/by-nc-nd/4.0/), which permits any noncommercial use, sharing, distribution and reproduction in any medium or format, as long as you give appropriate credit to the original author(s) and the source, provide a link to the Creative Commons license and indicate if you modified the licensed material. You do not have permission under this license to share adapted material derived from this chapter or parts of it.

The images or other third party material in this chapter are included in the chapter's Creative Commons license, unless indicated otherwise in a credit line to the material. If material is not included in the chapter's Creative Commons license and your intended use is not permitted by statutory regulation or exceeds the permitted use, you will need to obtain permission directly from the copyright holder.

Joint Statistical Analysis of Wave Height and Wave Period of Guangdong Sea Area Based on Archimedean Copula

Shiyun Wang(✉), Jiangshan Zheng, Chen Gu, Yingying Li, and Bihong Zhu

Shanghai Investigation, Design and Research Institute Co., Ltd., Shanghai, China
1007934946@qq.com

Abstract. In order to complete the reasonable plan of the design standard of marine engineering in extreme wave events, the joint distribution of annual extreme wave height and annual extreme period was constructed. Based on the SW model hind east wave model from 1996 to 2020 in Guangdong Sea Area, optimal marginal distributions of annual extreme wave height and annual extreme period at each location were obtained through the evaluation of the goodness-of-ft. And the joint distribution function of wave heights and periods was using the Archimedean Copula function to compare the different return periods and risk probably. The results show that there may be similar features in the joint wave height and period distribution in Guangdong Sea Area. Under the same design frequency, the concurrent return period is the largest, the secondary return period is the second and the joint return period is the smallest. The secondary return periods at different design frequency for the four representative sites in Guangdong sea area are 119.49–6.35a, 116.32–8.14a, 115.89–6.56a and 116.39--9.61a. The design values of wave height and period of Guangdong sea area calculated by secondary return period can provide effective guidance for marine engineering construction standards.

Keywords: Extreme wave height · Extreme wave period · Joint distribution · Copula function · Secondary return period

1 Introduction

There is a high risk of damage to coastal and marine structures affected by extreme wind and waves events. The wave process is an organic whole composed of several characteristic elements, such as wave height, wave period, wind speed and duration [1]. Compared to other offshore buildings, offshore wind turbines are subjected to greater wind load contributions and more dynamic responses, which leads to higher demands on fatigue load design. It has been shown that wave is one of the main factors contributing to the fatigue loading of wind turbines by combined effect of wave height and period. In particular, the resonance phenomenon generated by the period of waves close to the self-oscillation period of the structure is a great safety threat to offshore structures [2, 3]. Exploring the characteristics of the multivariate joint distribution of waves is of great significance for the safety and stability of wind turbine structures.

Copula function describing multi-variable correlation structure has been widely used in the field of multi-variable joint distribution and risk probability in the field of ocean engineering. In 2012, Chen Zixun [4] et al. investigated the maximum wave height and the mean period of Shanwei sea area in east Guangdong, and proposed that the design values of concurrent return period and joint return period can be used as upper and lower limits for the design of the wave heights and corresponding periods. In 2016, Vanem [5] studied the multi-dimensional distribution of characteristic wave height and wave period, and the results showed that asymmetric Copula function could be used to simulate the asymmetric correlation. In 2018, Xu [6] et al. applied the Copula function to describe the correlation between three offshore environmental parameters, namely, wave height, wave crest, and mean wind speed. The reliability analysis of marine structures considering long-term fatigue load and extreme response was carried out. In 2022 Song [7] used the Copula method to analyze the joint distribution of wave spectral peak period, which meets the requirement of continuous joint probability distribution of wave elements in risk assessment.

Currently, most of the joint distribution studies focus on joint return period and probability analysis of extreme wave height and corresponding period, while fewer studies on annual extreme wave height and extreme period are conducted. Offshore wind power development in Guangdong Province has great potential and suffers from extreme wind and wave conditions frequently. In this paper, based on hindcast wave data of Guangdong sea area in 1996–2020, the statistical analysis of joint distribution is carried out after sampling extreme wave height and extreme wave period in representative points along the coast. It is significant to deeply understand extreme wave risk, and provide effective guidance for design standards and risk management of offshore wind power in Guangdong Province.

2 Methodology

2.1 Copula Function

In 1959, Sklar [8] proposed the Copula function to construct a multidimensional joint distribution function by connecting the edge distributions of several random variables. Assuming that the marginal distribution function of n-dimensional random variable is, the joint distribution of is:

$$F(x_1, x_2 \ldots x_n) = C[F_1(x_1), F_2(x_2) \ldots F_n(x_n)] \tag{1}$$

The continuous probability distribution of wave elements is difficult to be simply expressed by linear relation. The Copula function is not limited by the type of marginal distribution, which makes it very effective in the representation of nonlinear relationships among complex hydrological variables. The Archimedean Copula function is the most common family of copula functions in extreme hydrological events, as shown in Table 1. The multivariate joint distribution of structural wave element features in the study of wave height and wave period joint distribution characteristics is as follows:

$$F(x, y) = C[F_X(x), \ F_Y(y)] = C[u_1, u_2] \tag{2}$$

where $F_X(x)$ and $F_Y(y)$ are the marginal probability cumulative distribution functions of wave height and wave period; $F(x, y)$ is Copula function of the above two random variables.

Table 1. Copula function definition and parameter range.

Copula Function	Type	$F(x, y)$	Parameter
Archimedean Copula	Clayton	$\left(u_1^{-\theta} + u_2^{-\theta} - 1\right)^{\frac{-1}{\theta}}$	$\theta > 0$
	Gumbel	$\exp\left(-\left((-\ln u_1)^\theta + (-\ln u_2)^\theta\right)^{\frac{1}{\theta}}\right)$	$\theta \geq 1$
	Frank	$-\frac{1}{\theta}\ln\left(\frac{(e^{-\theta u_1}-1)(e^{-\theta u_2}-1)}{e^{-\theta}-1}\right)$	$\theta \neq 0$

2.2 Return Periods and Risk Probability

Return period is an important reference for medium and long term prediction and warning of wave hazard risk, and is widely used in ocean engineering. Joint return period and concurrent return period are the two most common methods to define joint distribution return period, which is called the first repturn period. For the wave extreme event $E(u_1, u_2)$, the critical occurrence probability is defined as event hazard rate P. If one of the multivariable exceeds the hazard threshold, it is recorded as "OR" event; if the multivariable exceeds the hazard threshold at the same time, it is recorded as "AND" event. Joint return period T_{OR} and concurrent return period T_{AND} are given by Eqs. (3) and (4).

$$T_{OR} = \frac{1}{P(X \geq x \cup Y \geq y)} = \frac{1}{1 - C(u_1, u_2)} \quad (3)$$

$$T_{AND} = \frac{1}{P(X \geq x \cap Y \geq y)} = \frac{1}{1 - u_1 - u_2 + C(u_1, u_2)} \quad (4)$$

It is worth noting that in the traditional recurrence period, different u_1, u_2 combinations may produce the same recurrence period as long as they have same cumulative probability. There is no uniqueness in the safety domain defined by Joint return period T_{OR} and concurrent return period T_{AND}. Salvadori et al. [9] use Kendall distribution function to divide three scenarios: Subcritical (security domain), critical (warning event) and supercritical (danger domain). The multi-dimensional extreme events are projected as one-dimensional distributions by finding that the cumulative probability is less than or equal to some critical probability t. Kendall distribution function K_c is given by Eq. (5).

$$K_c(t) = P(C(u_1, u_2) \leq t) = \frac{1}{1 - C(u_1, u_2)} = t - \frac{\varphi(t)}{\varphi'(t)} \quad (5)$$

where $\varphi'(t)$ is the right derivative of the generating element $\varphi(t)$. Secondary return period determined by the Kendall distribution function can be expressed as follows:

$$T_k = \frac{1}{1 - K_c(t)} \tag{6}$$

2.3 Evaluation of Goodness-of-fit

The great likelihood method is applied to parameterize univariate distribution. Six common marginal distribution line patterns for wave height and period were fitted and preferred sequentially based on the K-S test results and root mean square error (RMSE). In order to evaluate the effect of Archimedean-type Copula function fitting, OLS, AIC and BIC ARE used to select the optimal joint distribution model.

3 Case Study

3.1 Data Sources

The MIKE 21 SW wave model is based on the conservation equation of wave action and uses the wave action density $N(\sigma, \theta)$ to describe the generation, growth and decay of waves. SW wave model has been widely used and continuously developed in the design of coastal and ocean engineering. Considering the lack of wave observation data during offshore wind power construction, a wave model covering the whole East China Sea range is established, and the Guangdong sea area is encrypted layer by layer on a grid. Comparing the numerical simulation results of waves in Guangdong sea area with the measured data from buoys, the overall degree of agreement is good. In this paper, the data is obtained from the hour-by-hour hindcast wave element data from 1996 to 2020 at the representative points of important locations in Guangdong waters. As is shown in Fig. 1.

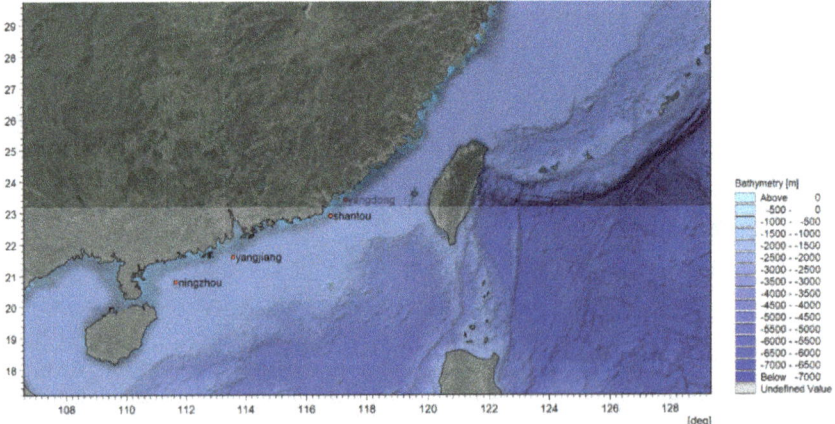

Fig. 1. SW wave model range and representative points in Guangdong sea area.

The annual extreme value sampling method is used to extract the extreme value series of effective wave height (H) and peak period (T) of four representative points (Yangdong, Shantou, Yangjiang and Qingzhou), and the joint distribution of wave height period is studied.

3.2 Univariate Marginal Distribution

The annual extreme wave height (H1–H4) and annual extreme wave period (T1–T4) of the four representative points in Guangdong sea area from 1996 to 2020 are extracted. Six common marginal distribution types (Normal, Lognormal, Gamma, GEV, Rayleigh, Weibull) are used for probability fitting and distribution function optimization. The results are shown in Tables 2 and 3.

Table 2. Evaluation results of goodness-of-fit for univariate Marginal distribution.

Fitness Index	Marginal distribution	H				T			
		H1	H2	H3	H4	T1	T2	T3	T4
K-S value	Normal	0.1416	0.1862	0.2078	0.1990	0.1090	0.1162	0.1067	0.1094
	Lognormal	0.1352	0.1808	0.1979	0.1883	0.1103	0.1183	0.1066	0.1096
	Gamma	0.1367	0.1817	0.2001	0.1909	0.1097	0.1174	0.1066	0.1094
	GEV	0.1366	0.1846	0.2000	0.1894	0.1091	0.1151	0.1073	0.1103
	Rayleigh	0.1868	0.2149	0.2259	0.2216	0.1552	0.1564	0.1587	0.1580
	Weibull	0.1462	0.1884	0.2090	0.2014	0.1080	0.1145	0.1043	0.1094
RMSE	Normal	0.0607	0.0455	0.0644	0.0801	0.0554	0.0597	0.0550	0.0599
	Lognormal	0.0462	0.0364	0.0413	0.0579	0.0612	0.0714	0.0550	0.0518
	Gamma	0.0527	0.0406	0.0501	0.0663	0.0616	0.0688	0.0577	0.0536
	GEV	0.0381	0.0398	0.0318	0.0431	0.0446	0.0450	0.0631	0.0527
	Rayleigh	0.1710	0.1385	0.1236	0.1410	0.2210	0.2107	0.2305	0.2252
	Weibull	0.0627	0.0434	0.0622	0.0793	0.0441	0.0446	0.0506	0.0515

According to K-S test values and RMSE results, the best marginal distribution of wave height of the four representative points in Guangdong sea area is Lognormal distribution, and the best marginal distribution of wave height is Weilbull distribution.

3.3 Multivariate Joint Distribution

Analyzing the correlation between wave height and wave period, the Pearson correlation coefficients at the four representative points are 0.1372, 0.2398, 0.1171, 0.1767, respectively. There is a certain positive correlation between the two variables, which can be expressed as Copula function to characterize the joint distribution of wave elements.

Table 3. Optimal univariate marginal distribution and fitting parameters.

Variable type	Optimal results	variant	fitting parameter
H	Lognormal	H1	$\mu = 1.4324; \sigma = 0.2102$
		H2	$\mu = 1.6696; \sigma = 0.2616$
		H3	$\mu = 1.7444; \sigma = 0.2881$
		H4	$\mu = 1.7214; \sigma = 0.2678$
T	Weibull	T1	$a = 16.2115; b = 9.2353$
		T2	$a = 16.5888; b = 8.3430$
		T3	$a = 17.4124; b = 9.9130$
		T4	$a = 17.1783; b = 9.2332$

Clayton, Frank and Gumbel function commonly used in Archimedean Copula functions are selected for fitting. The parameters θ of three candidate Copula models are obtained by IFM estimation method, which are 0.0319, 0.4614 and 1.0032, respectively. The goodness-of-fit tests are conducted by OLS metrics, AIC criterion, and BIC criterion. The results are shown in Table 4. Taking the representative point Yangdong(H1-T1) as an example, the fitting results between the theoretical and empirical frequencies of the Copula joint distribution model are shown in Fig. 2, and the optimal model is further confirmed by the graphical goodness-of-fit method.

Table 4. Results of parameter estimation and goodness-of-fit tests for Copula function.

Representative point	Copula Function Type	Goodness-of-fit			Copula Function parameter θ
		AIC	BIC	OLS	
H1-T1	Clayton	1.992	3.211	0.062	0.0319
	Frank	1.813	3.032	0.059	0.4614
	Gumbel	1.999	3.218	0.064	1.0032
H2-T2	Clayton	0.430	1.649	0.045	0.3246
	Frank	1.453	2.672	0.052	0.8351
	Gumbel	1.529	2.748	0.053	1.0963
H3-T3	Clayton	−0.119	1.100	0.055	0.5162
	Frank	1.217	2.436	0.067	1.0040
	Gumbel	1.900	3.119	0.075	1.0478
H4-T4	Clayton	1.465	2.684	0.082	0.2972
	Frank	1.532	2.751	0.088	0.7556
	Gumbel	2.000	3.219	0.096	1.0000

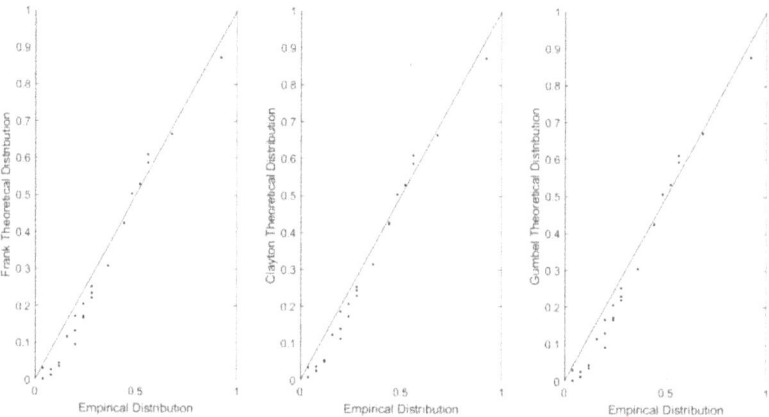

Fig. 2. Fitted plot of theoretical and empirical distribution for 2D Copula function.

According to the results in Table 4, the best fitting function of wave height and period of Yang Dong representative point is Frank Copula function, and the best fitting function of the other three representative points is Clayton Copula function.

3.4 Return Period Calculation and Risk Analysis

According to the optimization results of edge distribution and joint distribution function, the joint distribution model and the contour map combined by wave height and wave period based on Copula function are established under three return periods. It shows that the return period characteristics of four representative points are similar. Figure 3 shows the representative point H1-T1 of Yangdong as an example. As can be seen from Fig. 3, the distribution characteristics of the contour plots for joint and secondary return period are basically consistent, while concurrent return period is quite different. Under the same combination of wave height and period, the value of joint return period must be smaller than that of concurrent return period, and the value of secondary return period is between the first two. Therefore, it can be seen that the risk rate is the highest in joint return period when any element of the wave height and wave period exceeds the standard, and the minimum in concurrent return period when wave height and wave period exceed the standard.

The results of univariate design values, joint return period and secondary return period at different design frequencies for each representative point in Guangdong sea area are shown in Table 5. It indicates that wave height and wave period corresponding to the univariate return period of 100a are largest at Yangjiang representative point 3, followed by Qingzhou representative point4 and Shantou representative point 2, and smallest at Yangdong representative point 1.

The joint return period is about half of the univariate return period, while the secondary return period is slightly larger than the univariate return period. The greater the return period, the lower the corresponding risk rate. Taking Yangdong representative point 1 as an example, the joint return period in 100a is 50.31a, with a hazard rate of 0.0199; the secondary return period in 100a is 119.49a, with a hazard rate of 0.0084.

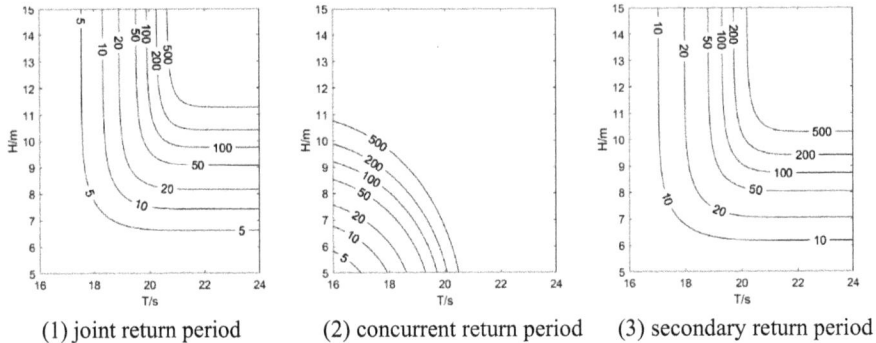

(1) joint return period　　(2) concurrent return period　　(3) secondary return period

Fig. 3. Contour plots of wave height period for different return period.

Table 5. Comparison of return periods and hazard rates for the joint distribution of wave height and period.

Representative Points	Design Frequency	Return Periods	univariate design value		Joint		Secondary	
			H/m	T/s	T_{OR}	P_{OR}	T_k	P_k
H1-T1	0.01	100	6.83	19.13	50.31	0.0199	119.49	0.0084
	0.02	50	6.45	18.79	25.31	0.0395	59.73	0.0167
	0.05	20	5.92	18.26	10.31	0.0969	23.93	0.0418
	0.1	10	5.48	17.74	5.32	0.1880	12.09	0.0827
	0.05	5	5.00	17.07	2.83	0.3539	6.35	0.1575
H2-T2	0.01	100	9.76	19.92	50.37	0.0199	116.32	0.0086
	0.02	50	9.09	19.54	25.37	0.0394	64.46	0.0155
	0.05	20	8.17	18.92	10.37	0.0965	27.90	0.0358
	0.1	10	7.43	18.33	5.37	0.1863	14.73	0.0679
	0.05	5	6.62	17.56	2.87	0.3489	8.14	0.1229
H3-T3	0.01	100	11.19	20.31	50.40	0.0198	115.89	0.0086
	0.02	50	10.34	19.98	25.39	0.0394	58.98	0.0170
	0.05	20	9.19	19.45	10.39	0.0962	24.04	0.0416
	0.1	10	8.28	18.94	5.39	0.1856	12.32	0.0812
	0.05	5	7.29	18.27	2.89	0.3466	6.56	0.1524
H4-T4	0.01	100	10.43	20.27	50.36	0.0199	116.39	0.0086
	0.02	50	9.69	19.91	25.36	0.0394	68.43	0.0146

(*continued*)

Table 5. (*continued*)

Representative Points	Design Frequency	Return Periods	univariate design value		Joint		Secondary	
			H/m	T/s	T_{OR}	P_{OR}	T_k	P_k
	0.05	20	8.69	19.35	10.36	0.0966	31.07	0.0322
	0.1	10	7.88	18.80	5.36	0.1867	16.81	0.0595
	0.05	5	7.01	18.09	2.86	0.3500	9.61	0.1041

During the design of ocean engineering, the univariate return period is usually used as the criterion for estimating wave design elements, ignoring the correlation between wave height and wave period variables, which is not conducive to the accurate assessment of wave hazard events. If joint return period is used, the hazardous domain of the event is enlarged, resulting in a smaller return period of the hazardous event under the same combination of wave height period, and the design standard of wave elements is large under the same return period. On the contrary, concurrent return period of dangerous events is overestimated, and the standard of wave element design is too small. Secondary return period is between the above two and slightly smaller than the univariate design standard, which can reduce the design standard and economic cost of offshore engineering on the basis of meeting the safety requirements. As shown in Table 5, joint return period of 100a, 50a, 20a, 10a and 5a for the four representative points in Guangdong sea area are 50.31–2.83a, 50.37–2.87a, 50.40–2.89a, 50.36–2.86a; secondary return period are 119.49–6.35a, 116.32–8.14a, 115.89–6.56a, and 116.39–9.61a. Considering the safety and economy of the project, it is recommended to adopt the design value of wave height and wave period in Guangdong sea area deduced from secondary return period, as a reference value of design wave elements for offshore engineering construction.

4 Conclusion

In this paper, the wave characteristics of Guangdong sea area in 1996–2020 are analyzed by using hindcast wave data simulated by MIKE 21 SW wave model. The conclusions are as follows:

(1) The optimal marginal distributions of wave height and wave period in Guangdong sea area is Lognormal and Weibull distribution, and the Archimedean Copula function can effectively indicate the joint distribution characteristics between wave height and wave period. The return period characteristics of each representative point in Guangdong sea area are similar.
(2) Joint return period of 100a, 50a, 20a, 10a and 5a for the four representative points in Guangdong sea area are 50.31–2.83a, 50.37–2.87a, 50.40–2.89a, 50.36–2.86a; secondary return period are 119.4––6.35a, 116.32–8.14a, 115.89–6.56a, and 116.39–9.61a. Considering the safety and economy of the project, it is recommended to adopt the design value of wave height and wave period in Guangdong sea area

deduced from secondary return period, as a reference value of design wave elements for offshore engineering construction.

Acknowledgements. This research is sponsored by research funding of China Three Gorges Corporation (No. 202003025).

References

1. Huihua, S., Jian, S., Jiali, X., et al.: Wave forecasting algorithm with stacking ensemble machine learning method. J. Hohai Univ. (Nat. Sci.) **48**(4), 354–358 (2020)
2. Chen, X.B., Li, J., Chen, J.Y.: Numerical calculation of random wind and wave loads time history of offshore wind turbine. Acta Energ. Sol. Sin. **32**(3), 288–295 (2011)
3. Li, H.T., Li, L.B.: Design analysis for supports structure of offshore wind turbine. Ocean Eng. **29**(4), 74–80 (2011)
4. Zishen, C., Shenxi, C.: Long-term joint distribution of wave height and period based on Copula functions. Mar. Sci. Bull. **31**(6):630–635 (2012)
5. Vanem, E.: Joint statistical models for significant wave height and wave period in a changing climate. Mar. Struct. **49**, 180–205 (2016)
6. Xu, X., Lai, F., Li, G.J., et al.: A novel vibration suppression device for floating offshore wind generator. In: Proceedings of ASME 2018 12th International Conference on Energy Sustainability Collocated With the ASME 2018 Power Conference and the ASME 2018 Nuclear Forum. Lake Buena Vista, Florida, USA (2018)
7. Chunyan, S., Pan, L., et al.: Joint distribution of wave period and height based on Copula function. Acta Energ. Sol. Sin. **004**, 043 (2022)
8. Sklar, A.: Fonctions de répartition an dimensions et leurs marges. Public Institute Statistic University, Paris (1959)
9. Salvadori, G.: Bivariate return periods via 2-copulas DJ. Stat. Methodol. **1**(1–2), 129–144 (2004)

Open Access This chapter is licensed under the terms of the Creative Commons Attribution-NonCommercial-NoDerivatives 4.0 International License (http://creativecommons.org/licenses/by-nc-nd/4.0/), which permits any noncommercial use, sharing, distribution and reproduction in any medium or format, as long as you give appropriate credit to the original author(s) and the source, provide a link to the Creative Commons license and indicate if you modified the licensed material. You do not have permission under this license to share adapted material derived from this chapter or parts of it.

The images or other third party material in this chapter are included in the chapter's Creative Commons license, unless indicated otherwise in a credit line to the material. If material is not included in the chapter's Creative Commons license and your intended use is not permitted by statutory regulation or exceeds the permitted use, you will need to obtain permission directly from the copyright holder.

Temporal and Spatial Distribution of Sea Surface Temperature in the East China Sea Based on SODA Data

Yingying Li[1(✉)], Chen Gu[1], Guoqu Cui[2], Shiyun Wang[1], Bihong Zhu[1], and Kan Yi[3]

[1] Shanghai Investigation, Design and Research Institute Co., Ltd., Shanghai 200335, China
li_yingying@ctg.com.cn
[2] Sino-Portuguese Centre for New Energy Technologies(Shanghai) Co., Ltd., Shanghai 200335, China
[3] China Three Gorges Corporation, Beijing 100038, China

Abstract. Sea Surface Temperature (SST) is one of the primary physical parameters of the ocean and a key focus of marine research both domestically and internationally. This study utilizes the Simple Ocean Data Assimilation (SODA) dataset from 1980 to 2022, comprising 43 years of data, to analyze the temporal and spatial evolution characteristics of SST in the East China Sea (ECS). Through a comprehensive analysis of the dataset, this research reveals long-term trends, distribution features, and seasonal variations in SST across ECS. It was found that the SST in ECS has shown a significant upward trend over recent decades, with marked seasonal variations and spatial distribution primarily influenced by solar radiation, ocean currents, and water masses. This study provides valuable insights for understanding the marine environment of ECS under global climate change and offers scientific support for the management of marine resources and ecological protection in the region.

Keywords: SODA data · he East China Sea · SST · Temporal and spatial evolution · Climate change

1 Introduction

ECS is an important marginal sea of the Northwest Pacific Ocean, being the third-largest continental shelf sea globally. It serves as a vital passage for major shipping routes and a habitat for diverse marine species. In recent years, the increasing global climate change has drawn significant attention to the changes in seawater temperature in ECS from both scientists and policymakers. Sea water temperature, as a key factor affecting the dynamics of marine ecosystems, is crucial for predicting and responding to the impacts of climate change on marine ecosystems and environment, with SST being the most extensively studied aspect. The spatial distribution of SST results from the interactions of internal oceanic thermal and dynamic processes and ocean-atmosphere coupling, indirectly reflecting changes in ocean circulation and climate alterations due to heat flux variations. Although previous studies [1–4] have monitored and analyzed sea water temperature through satellite remote sensing and field observations, these data are often limited by short temporal coverage and low spatial resolution.

To enhance the understanding of the changes in seawater temperature in ECS and provide scientific basis for marine resource management, marine ecological protection, and climate change adaptation strategies, this study employs the SODA dataset's monthly average temperature data from 1980 to 2022. It investigates the distribution characteristics and long-term variation patterns of SST in ECS and the PN section (location shown in Fig. 1), presenting the overall, seasonal, and interannual distribution characteristics and trends of SST in the region.

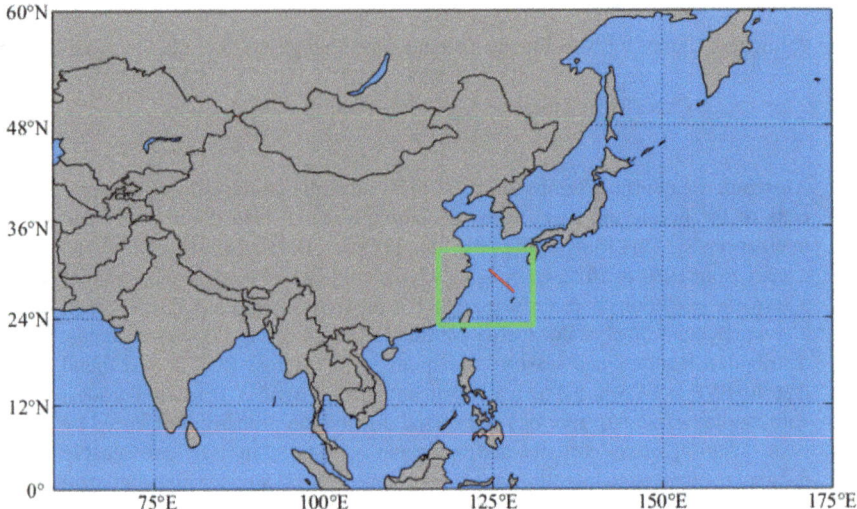

Fig. 1. Schematic Diagram of the Study Area.

2 Data and Methods

2.1 SODA Data

The SODA ocean dataset is a reanalysis product developed collaboratively by the University of Maryland and Texas A&M University using the global Simple Ocean Data Assimilation system. It assimilates a substantial number of temperature and salinity profile data (up to seven million), enabling accurate assessment of global ocean states. This dataset is widely used in marine science research [5–7]. The study uses the SODA dataset's monthly average gridded data with a spatial resolution of $0.5° \times 0.5°$.

2.2 PN Section

The study focuses on the temporal and spatial distribution patterns of SST in ECS, covering the area from 23.00 ° to 33.17 °N and 117.18 ° to 131.00 °E. The PN section, a standard section crossing the main axis of the Kuroshio in the central East China Sea, is richly documented and considered one of the most important sections for studying ECS and the Kuroshio. The PN section stretches northwest-southeast from the Yangtze River estuary in the northwest to the Ryukyu Islands in the southeast, specifically from 30.5 °N, 124.5 °E to 27.5 °N, 128.25 °E, forming a 37 ° angle with the latitude line. Besides, it crosses several major water masses in ECS, making it highly representative for studying the region's hydrological, chemical, and ecological characteristics [8]. Given the large spatial scope of the study, PN section SST data is extracted for analysis.

2.3 Analysis Methods

Univariate linear regression analysis, a statistical method for modeling the relationship between a dependent and an independent variable, is used to explore the linear relationship between variables and make predictions. In this study, a univariate linear regression method based on the least squares method is employed, using the year as the independent variable and SST as the dependent variable, to analyze the annual sequence change trend of SST in ECS from 1980 to 2022. The regression coefficient obtained is the rate of SST change, and the correlation coefficient (r) indicates the trend and significance of SST changes over time.

The Mann-Kendall test, a non-parametric statistical test, is advantageous as it does not require a specific distribution for the sample and is not affected by a few outliers. This study briefly introduces the results of this method, and more details can be found in the relevant literature [9]. The Mann-Kendall test calculates two variables, UF and UB, and their curve charts are used to analyze the time series' trend and mutation. If the values of UF or UB are greater than 0, the sequence shows an upward trend; if less than 0, it shows a downward trend. When UF and UB exceed the confidence interval, the trend is significant; the intersection point of UF and UB within the critical line indicates the time of mutation onset. The Mann-Kendall test is used in this study to examine the SST annual sequence mutations in ECS from 1980 to 2022, which is also a method recommended by the World Meteorological Organization for time series analysis.

3 Results and Analysis

3.1 Spatial Distribution of SST

The average SST distribution in ECS from 1980 to 2022 is shown in Fig. 2, with the red line indicating the location of the PN section. Influenced by solar radiation, SST in ECS generally decreases with increasing latitude, with isotherms trending northeast. The isotherms are denser in the warm current area, indicating clear warm water tongues.

SST along the Zhejiang and Fujian coasts is relatively low due to the passage of cold coastal currents, influenced by seabed topography, making the low-temperature range relatively narrow. Influenced by the Taiwan Warm Current, SST in the western Taiwan Strait is higher, with the warm water tongue extending northwest. The Kuroshio region in the central East China Sea has the highest SST, with isotherms consistent with the Kuroshio flow direction (northeast). SST in the Tsushima Warm Current area in the northeastern East China Sea is slightly lower than in the Kuroshio region, with isotherms aligned with the Tsushima Warm Current flow direction (north-northeast), and the warm

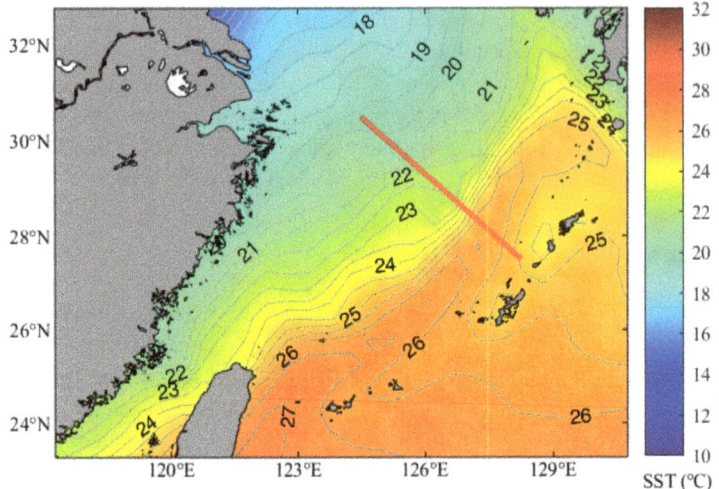

Fig. 2. Distribution of Multi-Year Average SST (Red Line Indicates the PN Section).

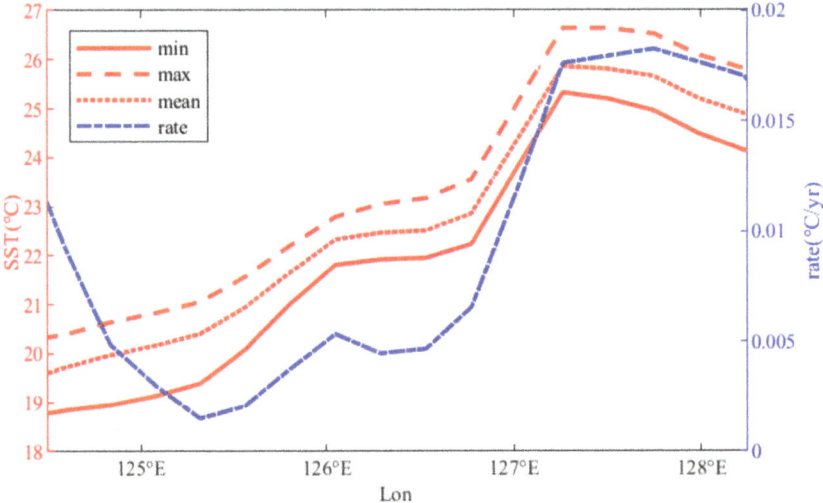

Fig. 3. Annual Sequence Statistics and Interannual Variation Rate of SST on the PN Section.

water tongue extends towards the Korea Strait. SST in the offshore southeastern East China Sea is higher due to strong solar radiation, with isotherms tending to parallel the latitude lines. SST in the northern East China Sea, adjacent to the Yellow Sea, is influenced by the Yellow Sea current, with lower SST on the western side due to the Yellow Sea Cold Water Mass and higher SST on the eastern side influenced by the northwesterly Yellow Sea Warm Current. The annual average SST on the PN section is shown in Fig. 3. As longitude increases, the annual average SST on the PN section increases, with a significant increase near 125 °E due to the influence of the Tsushima Warm Current and a sharp rise near 127 °E due to the Kuroshio.

3.2 Interannual Trend Analysis of SST

After spatially averaging the annual average SST grid data of ECS, the annual SST sequence is obtained, as shown in Fig. 4. The Mann-Kendall test of the annual SST sequence is shown in Fig. 5. From 1980 to 2022, the maximum annual average SST was 23.74 °C in 2020, and the minimum was 22.64 °C in 1984, with a multi-year average of 23.12 °C. Generally, the annual average SST varies slightly, indicating a certain degree of stability. Influenced by climate and ocean currents, the long-term annual average SST shows a fluctuating upward trend, with a regression analysis indicating an average change rate of +0.0138 °C/yr. The correlation coefficient (r) between the SST annual sequence and time is approximately 0.61, indicating a strong correlation and significant upward trend in SST over time. It is predicted that SST will remain at a high level and continue to fluctuate upward in the coming years.

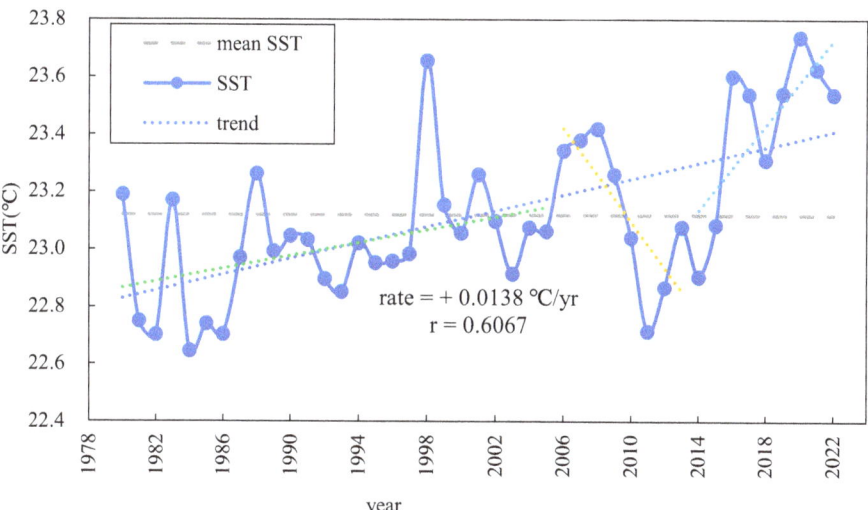

Fig. 4. Trend Changes in SST.

Analyzing the SST annual sequence combined with the Mann-Kendall test curve (Fig. 5), two intersections of the UF and UB curves are observed within the 90% confidence interval, occurring in 2005 and 2013, indicating two significant mutations in the SST annual sequence from 1980 to 2022. Dividing the period into three phases for trend analysis, the results are shown in Fig. 4 and Table 1: From 1980 to 2005, SST showed a stable upward trend with significant fluctuations; from 2006 to 2013, SST initially decreased sharply before recovering, with an overall downward trend and an average change rate of −0.0806 °C/yr.; from 2014 to 2022, SST increased sharply, with an average change rate of +0.0740 °C/yr.

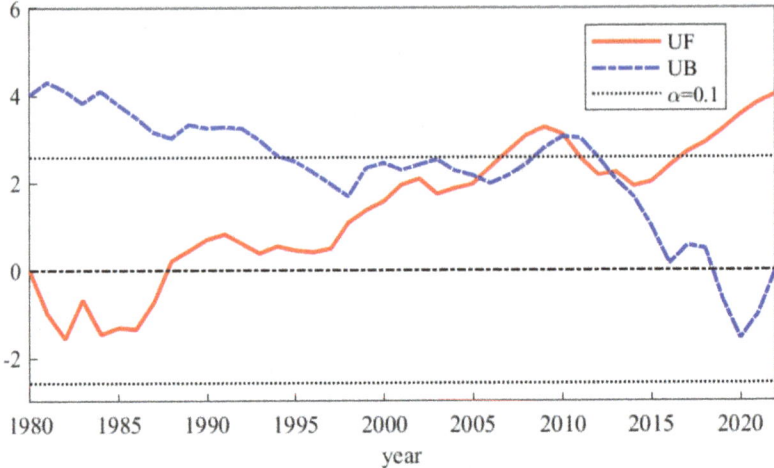

Fig. 5. Mann-Kendall Mutation Test Curve.

Table 1. Trend of SST Changes.

No.	Periods (year)	Rate (°C/yr)	r
1	1980–2005	+0.0112	0.16
2	2006–2013	−0.0806	0.59
3	2014–2022	+0.0740	0.54
4	1980–2022	+0.0138	0.61

The blue dashed line in Fig. 3 shows the annual change rate of SST on the PN section. The SST change rate on the PN section is positive and does not exceed +0.018 °C/yr., indicating a slight warming trend. In the Kuroshio region, the change rate is minimal. Figure 3 also shows the maximum and minimum values of the annual sequence on the PN section, which are generally parallel to the annual average and differ slightly, especially in the Kuroshio region. This is because the Kuroshio continuously brings heat from lower latitudes, maintaining SST stability to a certain extent.

3.3 Analysis of Intra-annual Variation of SST

The monthly SST distribution in ECS is shown in Fig. 6. The study area is divided into four seasons as: Winter (January–March), spring (April–June), summer (July–September), and autumn (October–December). In winter, SST in ECS ranges from 8–24 °C, with significant differences between the south and north, generally trending southwest-northeast, and decreasing from offshore to inland. Isotherms are dense, with clear cold and warm water tongues. In the southern coastal areas of ECS, isotherms are generally parallel to the coastline, influenced by continental cooling and coastal currents. A warm water tongue extending northeast on the eastern side of Taiwan represents the Kuroshio.

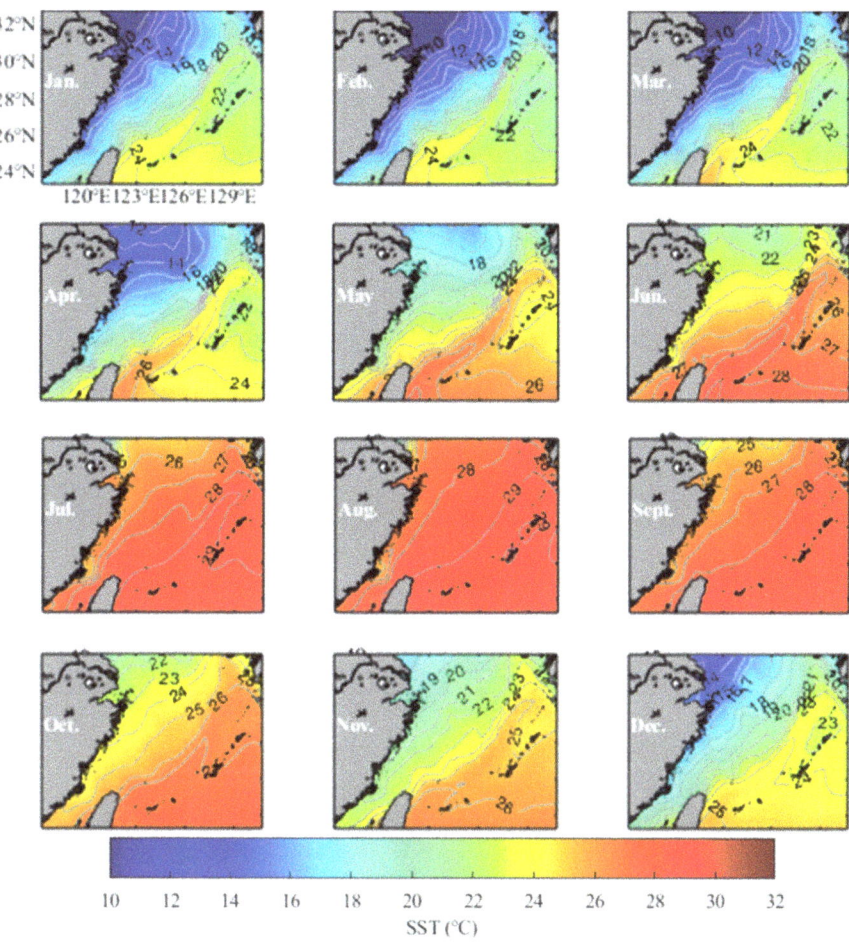

Fig. 6. Monthly Distribution of SST.

In summer, SST in ECS ranges from 23–30 °C, reaching the highest levels of the year, with a relatively uniform overall distribution. There are two cold water tongues in the north and off the coast of Fuzhou, representing the Yellow Sea Cold Water Mass and a small cold water mass caused by topography, respectively. The warm water tongue structure caused by the Kuroshio generally dissipates in summer due to significant stratification of seawater, low surface water density, and insufficient dynamics to drive the Kuroshio's intrusion into surface waters.

In spring and autumn, SST distribution in ECS exhibits transitional characteristics, reflecting the overall SST changes due to solar radiation and the development and dissipation of currents or water masses. Spring SST distribution is similar to that in winter, with overall warming, the formation of the Yellow Sea Cold Water Mass, and a more prominent warm water tongue caused by the Kuroshio. Compared to summer, autumn

SST distribution changes include overall cooling, the gradual retreat of the Yellow Sea Cold Water Mass, and the reappearance of the Kuroshio-induced warm water tongue.

The intra-annual variation curve of SST in ECS is shown in Fig. 7. The curve closely resembles a sine wave, with marked seasonality. The lowest water temperature occurs in February, and the highest in August; the most significant warming occurs from spring and the most significant cooling from autumn.

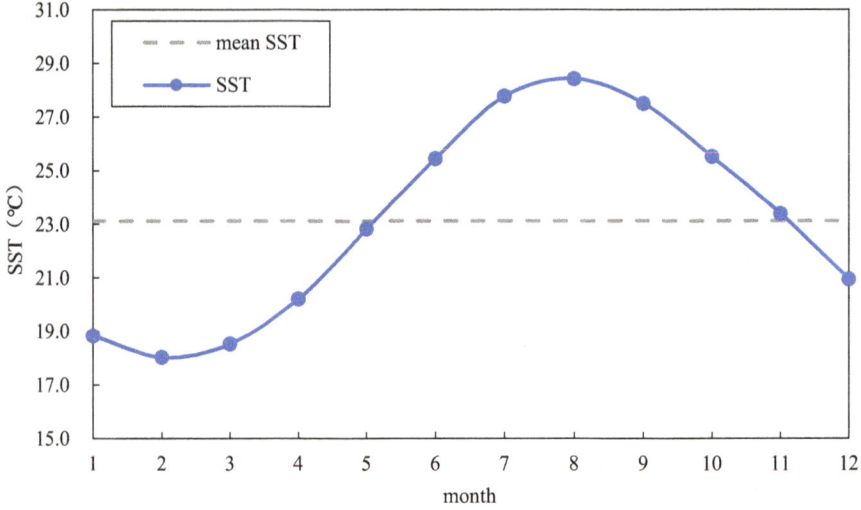

Fig. 7. Intra-Annual Variation of SST.

4 Conclusion

This study analyzed the spatiotemporal evolution characteristics of SST in the ECS using the SODA dataset from 1980 to 2022. The results highlight the significant influence of ocean currents, topography, and solar radiation on SST distribution and variation. The main conclusions are as follows:

(1) The SST distribution in the ECS can be classified into three seasonal types: The winter type, characterized by large spatial differences with distinct warm water tongues; the summer type, which exhibits a relatively uniform distribution with noticeable cold water masses; and the spring-autumn transitional type, representing a gradual shift between winter and summer patterns.
(2) Major ocean currents influencing SST include the Kuroshio, Tsushima Warm Current, Yellow Sea Warm Current, Zhejiang-Fujian Coastal Current, and Yellow Sea Cold Water Mass.
(3) The SST in the ECS exhibits a significant fluctuating upward trend, consistent with global ocean warming. Two notable abrupt changes occurred in 2005 and 2013. From 1980 to 2005, SST remained stable with fluctuations, while from 2006 to 2013, it

experienced a sharp decline before recovering, showing an overall downward trend. From 2014 to 2022, SST increased significantly, and it is predicted that SST will remain high and continue to fluctuate upward in the coming years..

The study relies solely on the SODA dataset, which, despite its reliability, has inherent limitations in resolution and data accuracy. External influences such as extreme weather events, and ecosystem responses were not explicitly considered in the analysis. The study does not incorporate high-resolution regional models or in-situ observational data, which could provide more localized insights.

Future studies should incorporate multiple datasets, such as satellite observations and in-situ measurements, to improve the accuracy and reliability of SST analysis. Additionally, investigating the impact of extreme climate events, including typhoons and heatwaves, on SST variations will provide deeper insights into short-term fluctuations.

This study provides a comprehensive understanding of SST changes in ECS, serving as a foundation for further research on the regional marine environment under the influence of climate change.

Funding. Funded by the Research Project of China Three Gorges Corporation (Contract No.: 202003025).

References

1. Tseung, C.: Temporal and spatial variations of sea surface temperature in the east China sea. Cont. Shelf Res. **20**(4), 373–387 (2000)
2. Shen, J.H., Zhou, F., Dong, Y.L., Cui, X.S.: Surface water temperature distribution in the East China Sea and part of the Yellow Sea in 2004. Res. Oceanogr. **25**(3), 9 (2007)
3. Wang, F., Meng, Q., Tang, X., Hu, D.: The long-term variability of sea surface temperature in the seas east of China in the past 40 a. J. Mar. Sci. **32**, 48–53 (2013)
4. Gao, G., et al.: Drivers of marine heatwaves in the East China Sea and the South Yellow Sea in three consecutive summers during 2016–2018. J. Geophys. Res. Oceans. **125**(8), e2020JC016518 (2020)
5. Danhua, Z., Yuanman, H.U., Miao, L., Yu, C., Lishuang, S.: Geographical variation and influencing factors of spartina alterniflora expansion rate in coastal China. Chin. Geogr. Sci. **30**(1), 15 (2020)
6. Yan, C., Zhu, J.: Evaluation. of an ocean reanalysis system in the Indian and Pacific Oceans. Atmos. **14**(2), 220 (2023)
7. Mishonov, A., Seidov, D., Reagan, J.: Multidecadal variability of ocean climate and circulation of the North Atlantic Ocean. (No. EGU24-2054). Copernicus Meetings (2024)
8. Chen, H.X., Yuan, Y.L., Hua, F.: Multi-core structure of G-PN section in the main section of Kuroshio in East China Sea. Chin. Sci. Bull. **51**(6), 8 (2006)
9. Wei, F.: Modern Climate Statistical Diagnosis and Prediction Techniques, 2nd edn. China Meteorological Press (2007)

Open Access This chapter is licensed under the terms of the Creative Commons Attribution-NonCommercial-NoDerivatives 4.0 International License (http://creativecommons.org/licenses/by-nc-nd/4.0/), which permits any noncommercial use, sharing, distribution and reproduction in any medium or format, as long as you give appropriate credit to the original author(s) and the source, provide a link to the Creative Commons license and indicate if you modified the licensed material. You do not have permission under this license to share adapted material derived from this chapter or parts of it.

The images or other third party material in this chapter are included in the chapter's Creative Commons license, unless indicated otherwise in a credit line to the material. If material is not included in the chapter's Creative Commons license and your intended use is not permitted by statutory regulation or exceeds the permitted use, you will need to obtain permission directly from the copyright holder.

Research on Berth Allocation Problem for Jetty Ro-Ro Terminals Based on Disruption Recovery

Yu Lu[1], Zeyu Shi[1(✉)], Yi Sui[2], Shihao Wu[1], and Jiangang Jin[2]

[1] Marine Design & Research Institute of China, Shanghai, China
1173289264@qq.com
[2] School of Ocean and Civil Engineering, Shanghai Jiao Tong University, Shanghai, China

Abstract. The berth and transshipment road of the ro-ro terminal are the key nodes connecting the ro-ro ship and the land. The terminal improves the operation efficiency of the terminal by formulating the berth allocation and cargo transshipment plan in advance. However, in actual operation, due to the interference of ship delay, equipment failure and extreme weather, the original scheme needs to be adjusted quickly. Based on the disruption management theory, considering the actual constraints such as the tidal time window and the capacity of the jetty channel, the objective function is to minimize the weighted sum of the planned time deviation and the total working time of the ro-ro ship in the wharf, this paper proposes a disturbance recovery model for the berth allocation and vehicle transfer of the jetty ro-ro terminals. By randomly generating three different scale examples, the validity of the model and the accuracy of the heuristic algorithm are verified by comparing the solution results of the heuristic algorithm and the commercial solver CPLEX. Finally, the influence of the adjustment deviation of the interference recovery scheme introduced in the berth allocation and vehicle transshipment scheduling model is analyzed, and the sensitivity analysis of the weight parameters is carried out. The results show that the disturbance recovery model can effectively reduce the time deviation before and after the scheme under the premise of ensuring that the total operation time of the ro-ro ship is basically unchanged.

Keywords: Water Transportation · Berth Allocation · Jetty Ro-Ro Terminals · Operation Efficiency of the Terminal · Simulated Annealing Algorithm

1 Introduction

As an important hub for automobile transportation, ro-ro terminals are mainly used for the loading, unloading and transportation of ro-ro goods such as automobiles. The berth allocation problem of ro-ro terminal is an important part of planning cycle. Berth allocation and scheduling decisions can be foreseen as a challenging problem. There are three main categories of modeling assumptions for berths in the problem of berth allocation, namely, discrete berths, mixed berths and continuous berths which can be seen in Fig. 1.

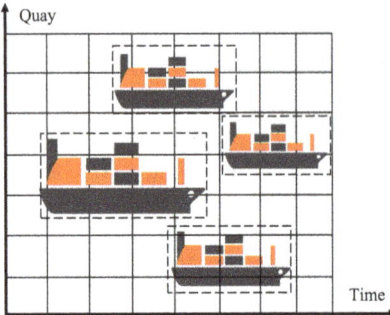

Fig. 1. Continuous berths model of berth allocation problem.

At present, the research on disturbance recovery[1] under uncertainty mainly focuses on container terminals. Yang and Wang [2] proposed the disturbance management decision-making method for the adjustment of berth allocation plan in container terminals. Zeng [3] studied the adjustment of berth plan after container terminal disturbance events, proposed a multi-objective and multi-stage recovery model based on disturbance management. Lin [4] proposed a mixed integer programming model for the integrated allocation of continuous berths and quay cranes and a hierarchical adjustment model based on the adjustment of berths and quay cranes. The simulation software is used to study the optimal adjustment scheme under different ship delay conditions. Meng [5] proposed a reactive recovery strategy to solve the disturbance recovery problem of berth allocation and quay crane allocation in container terminals. The model is solved by a heuristic method based on squeaky wheel optimization. Nitish [6] used the method of minimizing the cost of updating berth plan during the rolling planning period, established data uncertainty through probability distribution, and proposed an optimal recovery algorithm based on set partitioning and intelligent greedy algorithm. Han [7] focused on the periodic internal uncertainty in container transportation, proposed a dynamic decision-making framework based on two-stage approximate optimization, and used a double-layer nested tabu search algorithm to optimize the resource scheduling of container terminals under the uncertainty of ship arrival time and market demand. Jaap-Jan [8] developed a decision model for off-line basic berth planning and on-line recovery. The basic berth plan is formulated by the optimization method, and the periodic ship arrival and tidal window are considered. The real-time disturbance management model adopts a multi-level heuristic method to deal with the deviation of ship arrival and loading and unloading time, and minimizes the impact on the original plan. Lyu [9] introduced the cooperation between terminals into the recovery strategy of container terminals, optimized the terminal cooperation and ship transit connection, and solved large-scale problems by using mixed integer programming and squeaky wheel algorithm. The experimental results show that the cooperation strategy can reduce the terminal operation cost by up to 40 % under various disturbance events. Maxim [10] proposed a new disruption recovery berth allocation scheduling optimization model, taking into account a variety of recovery strategies and terminal resource management methods, focusing on the design of a new DMO optimization algorithm, which promotes the interaction between individuals through the diffusion grid, and uses the problem-specific hybrid technology to verify the superiority of the DMO algorithm in solving the berth scheduling problem.

Based on the operation background of the jetty ro-ro terminal, this paper establishes a disturbance recovery planning model that conforms to the special scene constraints in the jetty ro-ro terminal. Based on the original berth allocation plan, the berth and vehicle transfer recovery plan with minimum system disturbance and as much as possible to maintain the optimality of the original plan is quickly generated. In this paper, the research on the disturbance recovery part is based on the system state when the disturbance occurs, aiming at minimizing the disturbance of the disturbance to the system, that is, adjusting the original plan of the berth allocation and vehicle operation of the jetty ro-ro terminal, and establishing the berth allocation disturbance recovery model with the goal of minimizing the cost of disturbance recovery.

2 Mathematical Model

In the process of berth allocation, it is necessary to arrange the ship to an appropriate berth according to the arrival time of ship, the availability of berth, the type and demand of the ship, and the efficiency and safety of the port operation, so as to optimize the operation process with the goal of maximizing the utilization rate of resources. The whole process needs to consider a number of key factors, including the size and draught of the ship, the arrival and departure time, the type and quantity of loading and unloading goods, as well as the priority and berthing time of the ship. The mathematical model established in this article and the technical route for solving it are shown in Fig. 2.

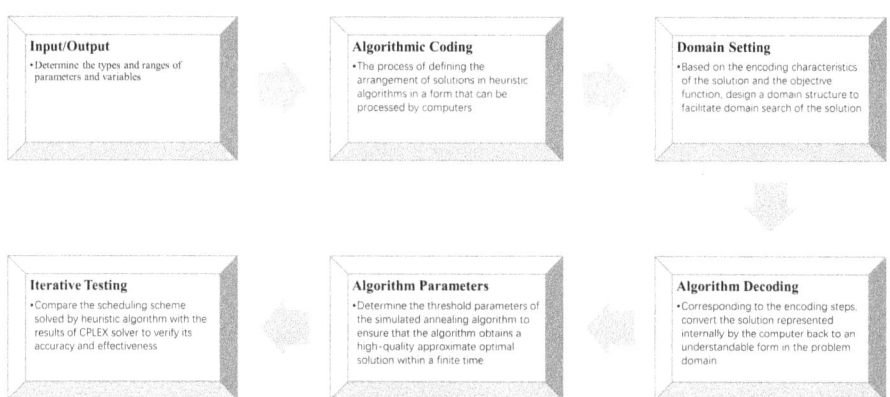

Fig. 2. Technical route for solving berth allocation.

2.1 Parameters and Variables

Based on the above problem description and assumptions, the key variables in the model can be defined as follows. Considering the disturbance recovery scenario and berth allocation problem, the following sets and parameters are defined and shown in Table 1.

Table 1. Parameters definition.

Object sets	Parameters	Definition
V, Ro-Ro Ships Set $V = \{1, 2, ..., |V|\}$	eta_v	Confirmation time of ro-ro ship v
	ton_v	The draught tonnage of ro-ro ship v
	$borad_v$	The springboard type of ro-ro ship v
	bt_v^0	The berthing time of ro-ro ship v in original plan
	dt_v^0	The start unloading / loading time in the original plan of ro-ro ship v
	ubt_v^0	The departure time in the original plan of ro-ro ship v
	Δt_{berth}	The berthing time of ro-ro ship v
	$\Delta t_{unberth}$	The off-shore time of ro-ro ship v
	Δtp_{vb}	Unloading/loading time of ro-ro ship v at berth b
B, Berth Set $B = \{1, 2, ..., |B|\}$	$cton_b$	Maximum loading tonnage of berths
	$jump_b$	The springboard type of ship at berth b
	at_b	The earliest available time of berth b
	$dpos_{b_1 b_2}$	Berth $b_1, b_2 \in B$ is on the opposite side or not
	$spos_{b_1 b_2}$	Berth $b_1, b_2 \in B$ is on the same side or not

Other parameters:

D: Tidal Time Window Set, $D = \{1, 2, ..., |D|\}$. α: The ratio of the total time of the new plan to the weight of the deviation time of the original plan;

M: Positive number large enough.

The decision variables for ro-ro ships are as follows:

x_{vb}: Ro-ro ship v dock in berth b or not;

bt_{vb}: The berthing time of ro-ro ship v in berth b;

dt_{vb}: The start unloading / loading time of ro-ro ship v at berth b;

ubt_{vb}: Departure time of ro-ro ship v at berth b;

$atide_{id}$: Ro-ro vessel v dock at the tidal time window or not;

$dtide_{id}$: Ro-ro vessel v dock at the d tidal time window or not;

z_{vub}: Ro-ro ship u berth after ship v berth b or not;

z_{uv}^1: The berthing of the ro-ro ship u is 1 before the ro-ro ship v departs, and then is 0;

z_{uv}^2: The berthing of the ro-ro ship u is 1 before the ro-ro ship v departs, and then is 0;

z_{uv}^3: The berthing of the ro-ro ship u is 1 before the ro-ro ship v departs, and then is 0;

z_{uv}^4: The berthing of the ro-ro ship u is 1 before the ro-ro ship v departs, and then is 0.

2.2 Model Establishment

The established disturbance recovery model is as follows.

$$\min \alpha \sum_{v \in V} \left(\sum_{b \in B} ubt_{vb} - eta_v \right) + (1-\alpha) \sum_{v \in V} \left[\left(\sum_{b \in B} ubt_{vb} - ubt_v^0 \right)^+ + \left(\sum_{b \in B} bt_{vb} - bt_v^0 \right)^+ \right] \quad (1)$$

Subject to:

$$\sum_{b \in B} x_{vb} = 1, \forall v \in V \quad (2)$$

$$\sum_{u \in V, u \neq v} z_{vub} = x_{vb}, \forall v \in V \quad (3)$$

$$\sum_{u \in V, u \neq v} z_{uvb} = x_{vb}, \forall v \in V \quad (4)$$

$$ubt_{vb} \leq bt_{ub} + (1 - z_{vub})M, \forall v, u \in V, v \neq u, \forall b \in B \quad (5)$$

$$ubt_{vb} \leq x_{vb}M, \forall v, u \in V, v \neq u, \forall b \in B \quad (6)$$

$$bt_{vb} + x_{vb}t_{berth} \leq dt_{vb}, \forall v \in V, \forall b \in B \quad (7)$$

$$dt_{vb} + x_{vb}\Delta tp_{vb} \leq ubt_{vb}, \forall v \in V, \forall b \in B \quad (8)$$

$$eta_v \leq \sum_{b \in B} bt_{vb}, \forall v \in V \quad (9)$$

$$x_{vb}at_b \leq bt_{vb}, \forall v \in V \quad (10)$$

$$\sum_{d \in D} atide_{vd} = 1, \forall v \in V \quad (11)$$

$$\sum_{d \in D} dtide_{vd} = 1, \forall v \in V \quad (12)$$

$$\sum_{b \in B} bt_{vb} \geq \sum_{d \in D} lowt_d \cdot enter_{vd}, \forall v \in V \quad (13)$$

$$\sum_{b \in B} bt_{vb} - t_{berth} \leq \sum_{d \in D} hight_d \cdot enter_{vd}, \forall v \in V \quad (14)$$

$$\sum_{b \in B} ubt_{vb} - t_{unberth} \geq \sum_{d \in D} lowt_d \cdot leave_{vd}, \forall v \in V \quad (15)$$

$$\sum_{b \in B} ubt_{vb} \leq \sum_{d \in D} hight_d \cdot leave_{vd}, \forall v \in V \tag{16}$$

$$\sum_{d \in D} d \cdot enter_{id} \leq \sum_{d \in D} d \cdot leave_{id}, \forall v \in V \tag{17}$$

$$dt_{vb} + \Delta tp_{vb} \leq dt_{up} + (1 - x_{vb})M + (1 - x_{up})M \\ + (1 - dpos_{bp})M + (1 - p_{vu}^o)M, \forall v, u \in V, v > u, \forall b, p \in B, b \neq p \tag{18}$$

$$dt_{up} + \Delta tp_{up} \leq dt_{vb} + (1 - x_{vb})M + (1 - x_{up})M \\ + (1 - dpos_{bp})M + p_{vu}^o M, \forall v, u \in V, v > u, \forall b, p \in B, b \neq p \tag{19}$$

$$cton_b \geq ton_v x_{vb}, \forall v \in V, b \in B \tag{20}$$

$$jump_b = \sum_{\forall v \in V} x_{vb} board_v, \forall b \in B \tag{21}$$

$$bt_{vb} \geq ubt_{up} - M(1 - spos_{bp}) - (1 - x_{vb})M \\ - (1 - x_{up})M - (1 - z_{uv}^1)M, \forall v, u \in V, v > u, \forall b, p \in B, b \neq p \tag{22}$$

$$ubt_{up} - t_{unberth} \geq bt_{vb} + t_{berth} - M(1 - spos_{bp}) - (1 - x_{vb})M - (1 - x_{up})M \\ - z_{uv}^1 M, \forall v, u \in V, v > u, \forall b, p \in B, b \neq p \tag{23}$$

$$ubt_{vb} - t_{unberth} \geq bt_{up} + t_{berth} - M(1 - spos_{bp}) - (1 - x_{vb})M \\ - (1 - x_{up})M - (1 - z_{uv}^4)M, \forall v, u \in V, v > u, \forall b, p \in B, b \neq p \tag{24}$$

$$bt_{up} \geq ubt_{vb} - M(1 - spos_{bp}) - (1 - x_{vb})M \\ - (1 - x_{up})M - z_{uv}^4 M, \forall v, u \in V, v > u, \forall b, p \in B, b \neq p \tag{25}$$

$$bt_{vb} \geq bt_{up} + t_{berth} - M(1 - spos_{bp}) - (1 - x_{vb})M \\ - (1 - x_{up})M - (1 - z_{uv}^2)M, \forall v, u \in V, v > u, \forall b, p \in B, b \neq p \tag{26}$$

$$bt_{up} \geq bt_{vb} + t_{berth} - M(1 - spos_{bp}) - (1 - x_{vb})M \\ - (1 - x_{up})M - z_{uv}^2 M, \forall v, u \in V, v > u, \forall b, p \in B, b \neq p \tag{27}$$

$$ubt_{vb} - t_{unberth} \geq ubt_{up} - M(1 - spos_{bp}) - (1 - x_{vb})M \\ - (1 - x_{up})M - (1 - z_{uv}^3)M, \forall v, u \in V, v > u, \forall b, p \in B, b \neq p \tag{28}$$

$$ubt_{up} - t_{unberth} \geq ubt_{vb} - M(1 - spos_{bp}) - (1 - x_{vb})M \\ - (1 - x_{up})M - z_{uv}^3 M, \forall v, u \in V, v > u, \forall b, p \in B, b \neq p \tag{29}$$

$$x_{vb}, atide_{id}, dtide_{id}, z_{vub}, z_{uv}^1, z_{uv}^2, z_{uv}^3, z_{uv}^4 \in \{0,1\}, bt_{vb}, dt_{vb}, ubt_{vb} \in R^+ \tag{30}$$

Constraint (2)-(6) are the space-time constraints of berth allocation for ro-ro ships. Constraint (2) ensures that each ro-ro ship is allocated with only one berth ; constraints (3)-(5) restrict the berthing of two ro-ro ships in order without time overlap ; constraint

(6) restricts the departure time, when the ro-ro ship v is docked in berth b, the departure time ubt_{vb} is greater than 0, and when it is not docked in berth b, the departure time ubt_{vb} is equal to 0 ; constraint (7) restricts the start unloading/loading time of the ro-ro ship docked in different berths after the ro-ro ship has completed the berthing operation ; constraint (8) restricts the berthing operation of the ro-ro ship docked at different berths after the ro-ro ship completes the unloading/loading operation ; constraint (9) ensures that the berthing time of the ro-ro ship is greater than the earliest available time of the berth ; constraint (10) ensures that the berthing time of the ro-ro ship is greater than the guaranteed time ; constraint (11)-(17) are the berthing and departing time window constraints of the ro-ro ship, which limit the berthing and departing time of the ro-ro window within the tidal time window; constraint (18) and (19) establish the unloading/loading sequence relationship between the two ro-ro ships docked on the opposite side, so as to ensure that the loading time of the opposite ro-ro ship does not overlap; constraint (20) and constraint (21) ensure the tonnage and springboard type matching of ro-ro ships and berths ; constraint (22)-(25) establish the order of berthing and departing between ro-ro ships docked on the same side ; constraint (26) and (27) establish the order of berthing between the same side ro-ro ships ; constraint (28) and (29) establish the order of berthing between the same side ro-ro ships.

3 Simulated Annealing Algorithm

3.1 Model Establishment

The scheduling order of ro-ro ships in the scheduling plan is encoded, and the berth allocation, berthing, transshipment, unloading and departure operations are prioritized in the front position. It is assumed that there are six ro-ro ships that need to make berth allocation scheduling plan. By randomly sorting the six ro-ro ships, the coding order in the following figure is generated, and then the time and berth decision of each ro-ro ship is made according to this order. The encoding diagram is shown in Fig. 3.

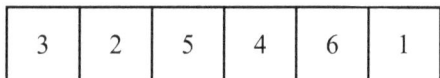

Fig. 3. Encoding structure diagram.

3.2 Neighborhood Structure

By randomly generating the two numbers of the ro-ro ships in the coding, the corresponding positions of the two ro-ro ships in the coding are exchanged to form a new coding structure. The swapping operator diagram is shown in Fig. 4. For example, two ro-ro ships, ro-ro ship 3 and ro-ro ship 6, are randomly generated. Ro-ro ship 3 is in the first position in the coding, and ro-ro ship 6 is in the fifth position in the coding. The position of the two ro-ro ships in the coding is exchanged to form a new coding structure such as {6,2,5,4,3,1}.

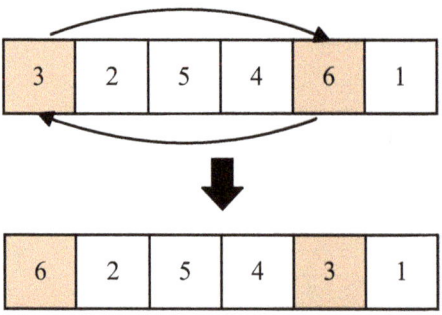

Fig. 4. Swapping operator diagram.

3.3 Algorithm Procedure

In this paper, the heuristic algorithm is used to generate the initial solution, and the key decisions such as berth and time of the ro-ro ship are formed. For example, when there are six ships that need to be scheduled for berthing and unberthing operations, such as '143265', the heuristic algorithm is used to calculate the berthing and unberthing operation plan that is most beneficial to ship 1, and then the berthing and unberthing operation plan that is most beneficial to ship 4 is calculated, and the initial overall scheduling plan is finally obtained by analogy.

The steps of the heuristic algorithm are given as Fig. 5.

Research on Berth Allocation Problem for Jetty Ro-Ro Terminals 855

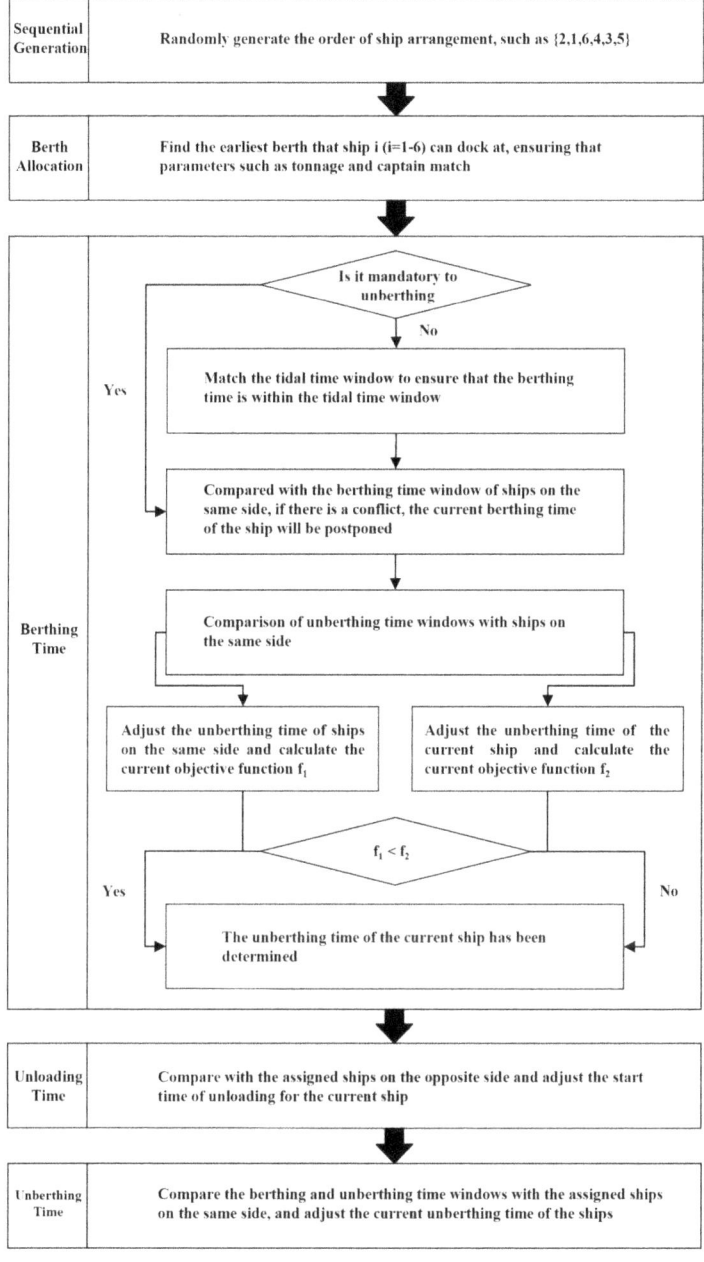

Fig. 5. Swapping operator diagram.

4 Numerical Experiments

In order to verify the rationality of the mathematical model construction and test the solution quality of the improved simulated annealing algorithm, the CPLEX 12.9 commercial solver and the simulated annealing algorithm with parameter setting are used to encode and solve the randomly generated test sequence respectively. The quality and efficiency of the genetic algorithm are evaluated according to the solution results, and the sensitivity analysis of the conflict penalty coefficient is completed. The genetic algorithm is coded by C++. All the test experiments run on AMD NVIDIA GeForce RTX 3070 Ti, Intel (R) Xeon Silver 4215R CPU @ 3.2GHz and RAM 64GB computer.

4.1 Test Example Generation

The numerical experiments in this paper form test cases by random generation. The scale of testing cases is can be seen in Table 2. The main parameters involved are as follows:

(1) Number of berths: According to the number of berths provided by a single ro-ro terminal, the number of berths in the example is determined, and the number of berths is randomly generated within the range of [4, 8].
(2) Number of ro-ro ships: The number of ro-ro ships arriving at the same time period is limited to a random integer between [10,20].
(3) Average loading capacity: Ro-ro ships will load different quantities of ro-ro cargo such as cars according to different load tonnage and draft depth, so the average loading capacity of ro-ro ships is set to [100,300].
(4) Arrival time range: Considering the arrival time range of ro-ro ships, the number of ro-ro ships in the time period affected by the occurrence of disturbance can be reflected side by side, and the expected arrival time range of ships is set to [0.5,2].

Table 2. Scale of testing cases.

Set	Number of ro-ro ships	Average loading capacity	Number of berths	Arrival time range(h)
Set1	[10,20]	[100,300]	4	2
Set2	[10,20]	100	[4, 8]	2
Set3	[10,20]	100	4	[0.5,2]

4.2 Analysis of the Results of the Algorithm

The CPLEX solver is called to solve 15 examples of different sizes. The CPLEX solution time is limited to 1 hour, and the final solution result and solution time of CPLEX are recorded. The simulated annealing algorithm is called to solve each example three times, and the average value and average time of the solution results are recorded, as shown in

Table 3. Set1 Solution result.

Solution group	CPLEX		Simulated Annealing Algorithm			GAP (%)
	required value	Time (s)	mean value	variance	mean time (s)	
Set1-I01	236.43	17.88	236.43	0	0.71	0
Set1-I02	312.59	21.19	312.59	0	0.63	0
Set1-I03	408.29	154.17	408.29	0	0.73	0
Set1-I04	330.95	3600	330.95	0	0.46	0
Set1-I05	465.73	3600	465.95	0	0.96	0.05
Set1-I06	566.68	3600	572.68	0	0.93	1.06
Set1-I07	538.74	3600	537.88	0.1	1.27	−0.16
Set1-I08	710.9	3600	641.05	94.16	1.33	−9.83
Set1-I09	896.1	3600	904.15	0.01	1.48	0.9
Set1-I11	962.41	3600	953.31	0.09	1.46	0.47
Set1-I12	1111.22	3600	1126.15	0.05	2.08	−0.94
Set1-I13	969.35	3600	899	5.77	1.91	1.34
Set1-I14	1164.6	3600	1095.68	0.22	2.43	−7.26
Set1-I15	1374.27	3600	1336.66	0.01	2.05	−5.92

Tables 3 to 5. The GAP in the table represents the relative gap between the simulated annealing algorithm and the CPLEX optimal solution.

GAP = (simulated annealing algorithm target average-CPLEX target value) / CPLEX target value * 100 %

Based on the solution results in the above table, the following conclusions can be obtained:

(1) Accuracy analysis of algorithm results

Under three scale examples, CPLEX can obtain the exact optimal solution of some examples in a fixed time. However, as the number of ro-ro ships increases, CPLEX cannot obtain the optimal solution within the specified time, and can only record the lower bound of the current exact solution when the arrival time limit is reached. At the same time, according to the calculation results of the three test example sets, it can be found that the GAP values are all less than 5%, and the part less than 0 indicates that the approximate optimal solution obtained by the simulated annealing algorithm is better than the optimal lower bound obtained by the CPLEX solver. It shows that the solution accuracy of the simulated annealing algorithm based on model design can propose an effective berth allocation and unloading scheduling scheme on the basis of meeting the scene requirements.

(2) Stability analysis of algorithm results

Table 4. Set2 Solution result.

Solution group	CPLEX		Simulated Annealing Algorithm			GAP (%)
	required value	Time (s)	mean value	variance	mean time(s)	
Set2-I01	228.8	17.88	228.8	0	0.72	0
Set2-I02	147.98	600	150.27	0	1.83	1.54
Set2-I03	108.66	600	108.66	0	0.37	0
Set2-I04	392.17	600	391.99	0	0.94	−0.04
Set2-I05	234.36	600	237.04	0	2.74	1.14
Set2-I06	127.4	600	127.34	0	0.4	−0.05
Set2-I07	487.09	600	486.91	0	1.11	−0.04
Set2-I08	460.83	600	466.11	19.03	3.92	1.15
Set2-I09	227.4	600	227.18	0	0.59	−0.1
Set2-I10	836.75	600	832.93	0	2.12	−0.46
Set2-I11	649.48	600	618.13	1.16	4.2	-4.83
Set2-I12	357.1	600	355.85	0.01	1.51	−0.35
Set2-I13	1034.79	600	992.09	0.18	1.95	−4.13
Set2-I14	779.88	600	806.91	1.77	4.82	3.47
Set2-I15	572.68	600	482.55	0.04	1.84	−15.7

From the average value and variance of the simulated annealing algorithm in Table 2, it can be seen that under different scale examples, even if the number of ships, berths, loaded vehicles in the current scene change, the result value obtained by each solution fluctuates little, and the variance is less than 2. There is almost no gap. Therefore, it is proved that in the process of solving the interference recovery model, although the initial solution generation and solution space search are random, the final output optimal (3) Time analysis of algorithm results

When the CPLEX solver is used to test three scale examples, the maximum solution time is limited to 600s, and the solution time is basically more than 600s. However, the solution time of the simulated annealing algorithm is less than 30s, and it does not increase significantly with the increase of the number of ro-ro ships and berths. Therefore, the optimal berth allocation and offloading resource scheduling scheme can be quickly output. solution remains basically stable.

4.3 Sensitivity Analysis

The weight parameter α in the objective function represents the weight balance between the newly adjusted total working hours and the difference of the previous and subsequent schemes. From the perspective of the global optimal solution, when an emergency occurs, the ro-ro berth, berthing time and departure time of the new scheme must be deviated from the original plan ; however, considering the consistency of the scheme, when

Table 5. Set3 Solution result.

Solution group	CPLEX		Simulated Annealing Algorithm			GAP (%)
	required value	Time (s)	mean value	variance	mean time(s)	
Set3-I01	236.51	600	236.51	0	0.58	0
Set3-I02	234.66	600	237.88	0	0.57	1.37
Set3-I03	224.53	82.26	224.53	0	0.66	0
Set3-I04	339.6	600	339.6	0	0.55	0
Set3-I05	336.12	600	336.12	0	0.85	0
Set3-I06	331.66	600	331.66	0	0.72	0
Set3-I07	549.43	600	549.05	0	1.51	−0.07
Set3-I08	593.4	600	547.04	1.84	0.9	−7.81
Set3-I09	581.47	600	587.17	0.01	0.9	0.98
Set3-I10	853.96	600	846.8	0.02	1.34	−0.84
Set3-I11	779.67	600	776.2	0.39	2.02	−0.44
Set3-I12	733.24	600	736.06	3.71	1.91	0.38
Set3-I13	1101.52	600	1094.26	0.16	1.96	−0.66
Set3-I14	1176.44	600	1079.65	0.07	2.33	−8.23
Set3-I15	1222.29	600	1081.1	1.1	0.58	−11.6

the time deviation and berth deviation between the new plan and the original plan are minimized, there is a certain deviation between the final total working hours of the ro-ro ship and the global optimal solution, and the weight ratio is the key parameter affecting the formulation of the new scheme. Through the objective function, the total working hours and the number of berth changes under different weights, the influence of weight parameters on the overall scheme is obtained.

The result of sensitivity analysis is shown in Fig. 6. When the weight coefficient changes between 0.1 and 1, with the increase of the weight coefficient, the two values of the total working time and the total time change do not change much, and fluctuate in a certain value range, indicating that when the weight coefficient is greater than 0, the weight coefficient has little effect on the overall scheme and the total working time. When the weight coefficient is 0, the total change time increases suddenly, indicating that the presence or absence of weight coefficient will affect the ro-ro ship berthing time. Therefore, the introduction of weight coefficient has little effect on the total working hours of ro-ro ships, and can reduce the difference between the previous and subsequent schemes.

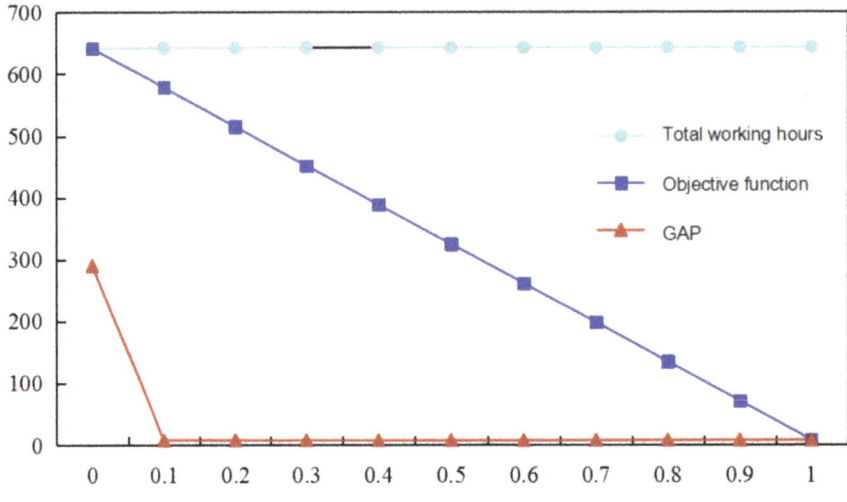

Fig. 6. Sensitivity analysis curve.

5 Conclusion

This paper studies the optimization problem of berth allocation and vehicle transshipment plan under uncertain conditions. Based on the physical constraints and deterministic model of jetty ro-ro terminal, the weighted sum of the total operation time of the ro-ro ship in the jetty ro-ro terminal and the time cost gap of the adjustment scheme is taken as the objective function, and the disturbance recovery model of berth allocation and ship loading and unloading plan is established. Based on the characteristics of the constraint conditions and the problem structure in the model, the simulated annealing algorithm and the heuristic decoding algorithm for calculating the target value are improved. The correctness of the model and the effectiveness of the improved simulated annealing algorithm are verified by examples of different sizes. Through the analysis of the solution results, the time gap between the adjustment scheme and the original scheme is further verified, and the sensitivity analysis of the weight parameters is carried out. The results show that the model after introducing the time gap between the two schemes can significantly reduce the adjustment cost of the interference recovery scheme, and the total working time of the ro-ro ship in the port is basically unchanged.

References

1. Clausen, J., Larsen, J., Larsen, A., Hansen, J.: Disruption management-operations research between planning and execution. Or/Ms Today. **28**(5), 40–43 (2001)
2. Chunxia, Y., Nuo, W.: Research on decision-making method of berth allocation interference management in container terminals. Oper. Res. Manag. **20**(04), 90–95 (2011) (in Chinese)
3. Zeng Qingcheng, H., Xiangpei, Y.Z.: Multi-objective model for disruption recovery of container terminal berth plan. J. Manag. Eng. **27**(02), 154–159 (2013) (in Chinese)
4. Lin Qingfu, H., Zhihua, T.S.: Hierarchical adjustment strategy of container terminal berth-quay crane integrated assignment interference management. J. Chongqing Jiaotong Univ. (Nat. Sci. Ed.). **33**(03), 133–139 (2014) (in Chinese)

5. Li, M.Z., Jin, J.G., Chun Xia, L.: Real-time disruption recovery for integrated berth allocation and crane assignment in container terminals. Transp. Res. Rec.: J. Transp. Res. Board. **2479**, 49–59 (2015)
6. Umang, N., Bierlaire, M., Erera, A.L.: Real-time management of berth allocation with stochastic arrival and handling times. J. Sched. **20**, 67–83 (2016)
7. Xiaole, H., Lina, Q., Zhiqiang, L.: Dynamic interference management of container terminal resource allocation under periodic environment. J. Tongji Univ. (Nat. Sci. Ed.). **46**((02)), 264–272 (2018) (in Chinese)
8. van der Steeg, J.-J., Oudshoorn, M., Yorke-Smith, N.: Berth planning and real-time disruption recovery: a simulation study for a tidal port. Flex. Serv. Manuf. J. **35**, 70–110 (2022)
9. Xiaohuan, L., Rudy-R, N., Xiaoning, S., et al.: A collaborative berth planning approach for disruption recovery. IEEE Open J. Intell. Transp. Syst. **3**, 153–164 (2022)
10. Dulebenets, M.A.: A diffused memetic optimizer for reactive berth allocation and scheduling at marine container terminals in response to disruptions. Swarm Evol. Comput. **80**, 101334 (2023)

Open Access This chapter is licensed under the terms of the Creative Commons Attribution-NonCommercial-NoDerivatives 4.0 International License (http://creativecommons.org/licenses/by-nc-nd/4.0/), which permits any noncommercial use, sharing, distribution and reproduction in any medium or format, as long as you give appropriate credit to the original author(s) and the source, provide a link to the Creative Commons license and indicate if you modified the licensed material. You do not have permission under this license to share adapted material derived from this chapter or parts of it.

The images or other third party material in this chapter are included in the chapter's Creative Commons license, unless indicated otherwise in a credit line to the material. If material is not included in the chapter's Creative Commons license and your intended use is not permitted by statutory regulation or exceeds the permitted use, you will need to obtain permission directly from the copyright holder.

Numerical Study on Local Scour Propagation Below Two Intersecting Pipelines on a Sandy Seabed Under Currents

Zhuo Fang[1], Zhaolong Yin[2], Limeng Zhao[3], and Zhipeng Zang[3(✉)]

[1] Transport Planning and Research Institute, Ministry of Transport, Beijing 100028, China
[2] Daqing Oilfield Engineering Construction Co., Ltd., Daqing 163453, Heilongjiang, China
[3] State Key Laboratory of Hydraulic Engineering Intelligent Construction and Operation, Tianjin University, Tianjin 300350, China
zhipeng.zang@tju.edu.cn

Abstract. With the swift advancement of offshore oil exploration, the intersection of pipelines has become a prevalent occurrence in nearshore areas. The flow dynamics and sediment transport processes around two crossing pipelines exhibit greater complexity compared to those associated with a conventional isolated pipeline. This study investigates the 3D scour development process below two crossing pipelines under steady currents based on numerical simulations. The lower pipeline is positioned normal to the current direction while the upper pipeline is installed above it at an crossing angle. The numerical results suggest that the 3D scour propagation rate on both the left and right sides diminishes with an increase in the crossing angle, and both rates exceed those observed for an single pipeline. Due to the asymmetry of the structure, the propagation rate on the right half is lower than that on the left. When the crossing angle reaches 90°, the results closely resemble those observed in the case of a single pipeline. The current numerical findings offer significant insights for the design of pipelines in practical engineering applications.

Keywords: Propagation rate · Crossing pipelines · Numerical simulation

1 Introduction

The phenomenon of local scour and the ensuing vortex-induced vibrations caused by current loads are critical factors contributing to the failure of submarine pipelines. With the rapid advancement of offshore oil field development, a substantial number of pipelines and cables are being installed on the seabed, particularly in coastal zones, resulting in an increased frequency of submarine pipeline intersections [1]. The flow structures and the resultant sediment transport around these crossing pipelines are more complicated, and the scour process is more pronounced compared to that around a single pipeline. Consequently, investigating the local scour around crossing pipelines holds significant engineering and academic importance.

Initially, investigations into pipeline local scour were approached in a two-dimensional (2D) context, assuming uniform flow and consistent scour along the pipeline's span. Significant research has focused on the equilibrium scour depth [2], the scouring process [3], and associated criteria [4] for pipelines. In practical terms, y, the scour begins at a vulnerable point and subsequently extends in two opposing directions along the pipeline, resulting in the formation of a free span. A few investigators have conducted experiments on the 3D scour propagation rate and their fluid mechanics [5, 6].

Currently, research on local scour predominantly focuses on a single pipeline or two parallel pipelines, with insufficient attention directed towards the local scour characteristics of two crossing pipelines. This study aims to address this research gap by employing physical modeling to investigate the three-dimensional scour phenomena associated with crossing pipelines through numerical simulations.

2 Numerical Model Descriptions and Validations

In this study, the simulation of the incompressible fluids is conducted by solving the Reynolds-Averaged Navier–Stokes (RANS) equations for. To close the equations, the k-ε turbulence closure is employed. The Flow-3D software was utilized to solve the RANS equations and the k-ε turbulence model [7]. The bedload transport is estimated using the Van Rijn's formula, while the convection-diffusion equations of the concentration of sediment is solved to account for the suspended load transport [8].

Figure 1 illustrates the numerical model for the scour process beneath two intersecting pipelines. The lower pipeline is positioned normally to the current direction, with a zero gap between it and seabed. The upper pipe is installed obliquely above the lower one with an angle α. Here, $\alpha = 90°$ presents that the upper pipeline is along the current streamlines, $\alpha = 0°$ denotes that the two pipelines are in tandem configuration. The clearance between two pipelines is fixed at $G/D = 1.0$. In the present study, five crossing angles, i.e., $\alpha = 0°, 30°, 45°, 75°$ and $90°$ are considered to investigated its effect on the 3D scour process. The mesh sensitivity analysis and validation of the model have been carefully addressed in Zhang et al. [9], and are not presented here.

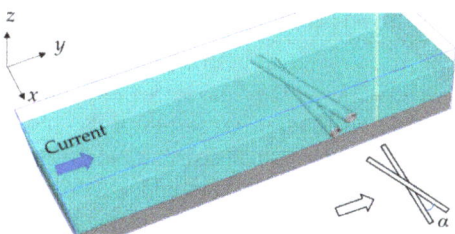

Fig. 1. Mesh system of the numerical model for crossing pipelines.

3 Numerical Results and Discussions

3.1 Development of Scour Hole Below Crossing Pipelines

Figure 2 illustrates the three-dimensional seabed elevations below the two parallel pipelines ($\alpha = 0°$) during the scour propagation. As observed in Fig. 2(a), the scour initiates at the center of the pipeline and progressively extends outward along the axial direction from the initial stage. Under the influence of fluid dynamics, erosion is observed upstream of the pipeline, while sedimentation occurs downstream (Fig. 2b). The extent of scouring upstream typically exceeds the sedimentation range downstream, with both regions exhibiting symmetrical distribution characteristics and pronounced vortices, indicative of turbulent water flow, are formed at the downstream side of the pipeline, as shown in Fig. 2(c) and 2(d).

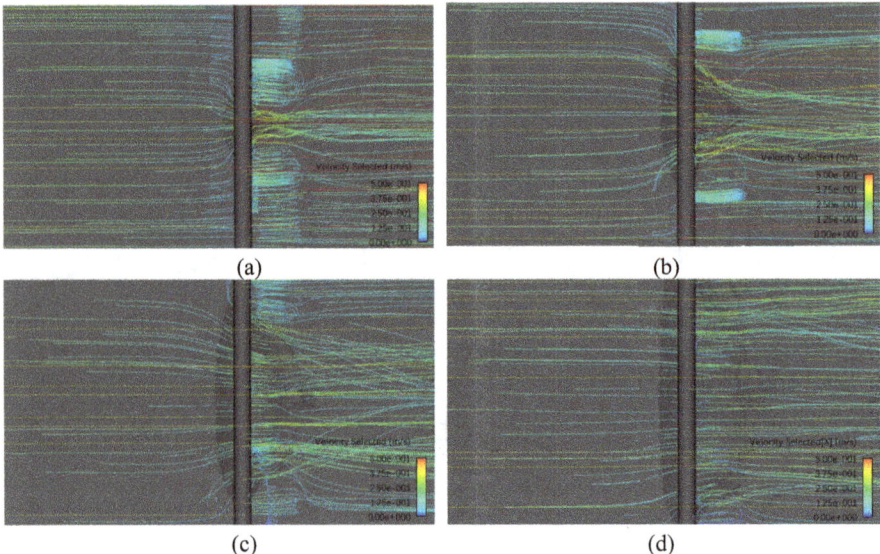

Fig. 2. Scour profiles and flow streamlines around the two crossing pipelines at $\alpha = 0°$, (a) $L = 0.1$ m, (b) $L = 0.3$ m, (c) $L = 0.5$ m, (d) $L = 0.7$ m.

Figure 3 shows the scour holes at the final stages for angles $\alpha = 30°, 45°, 75°$ and $90°$. Owing to the symmetric distribution properties of the pipeline structure, the flow field exhibits a predominantly symmetrical distribution at various intermediate stages of scour. The overall stability of the flow field under scour conditions surpasses that observed at the aforementioned crossing angles. As the boundary of the scour hole progressively contracts toward the sides, the wake vortex dissipates, resulting in a stable overall flow field. Specifically, the flow structures for $\alpha = 90°$ is symmetrical.

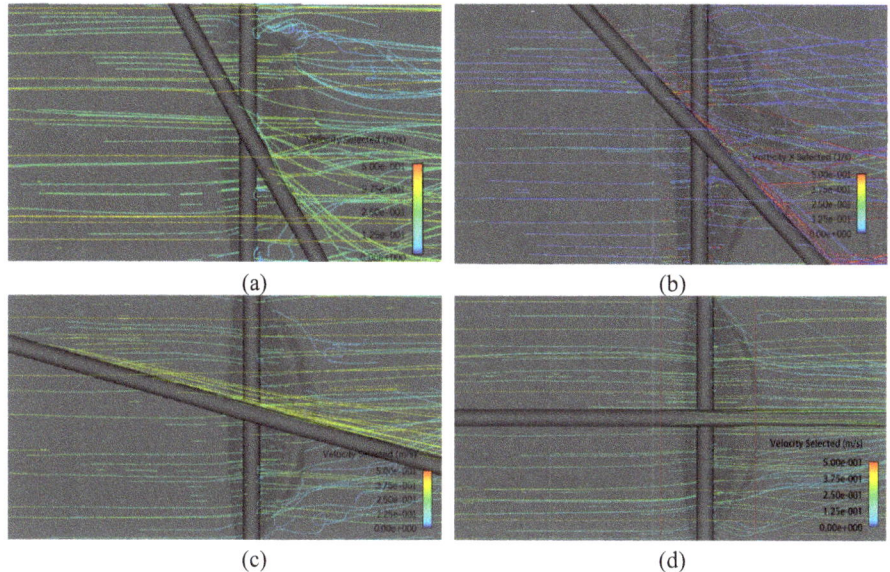

Fig. 3. Scour profiles and flow streamlines around the two crossing pipelines at $\alpha = 30°$, (a) $L = 0.1$ m, (b) $L = 0.3$ m, (c) $L = 0.5$ m, (d) $L = 0.7$ m.

3.2 Propagation Rate of Scour

Based on the analysis of the three-dimensional scour propagation process of crossing pipelines at various angles, the scour expansion lengths on the left and right sides have been systematically analyzed and presented. The asymmetry introduced by the crossing pipelines at an oblique angle results in differential propagation rates of the scour hole on the left and right sides of the pipelines. Figure 4(a) illustrate the time-dependent curve of the free span length as influenced by scour propagation on both sides. The slope of the curve illustrates the rate of scour propagation. Figure 4(b) presents the scour propagation rate on the left and right sides at $D = 0.1$ m and $G/D = 1$, plotted against the crossing angle. The propagation rate is expressed in a nondimensional form, V/v, where V denotes the scour propagation rate and v represents the flow velocity.

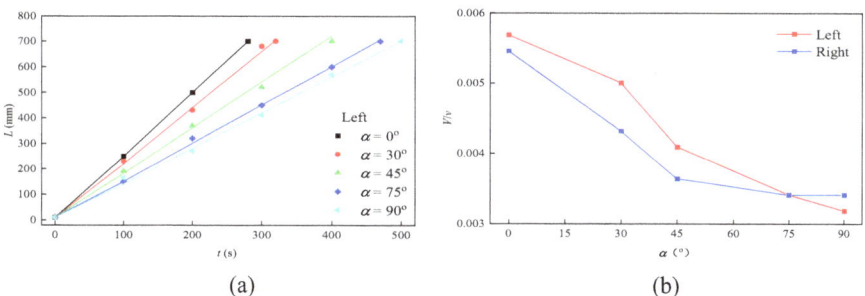

Fig. 4. Propagation of scour below the intersection pipelines: (a) left span length vs. time, (b) propagation rate vs. crossing angle.

Figure 4(b) illustrates that the nondimensional propagation rates on both the left and right sides of the lower pipeline decrease as the crossing angle increases. A comparative analysis of the two sides reveals that the propagation rate on the right half of the pipeline is generally lower than that on the left half; however, the difference between the two rates is minimal. Notably, when the crossing angle surpasses 45°, the erosion expansion rate on the right side remains relatively stable, whereas the propagation rate on the left side exhibits a gradual decline. When the crossing angle exceeds 75°, indicating that the upper pipeline is nearly normal to the lower one and aligned with the current direction, the rates on the left and right sides are essentially equivalent.

4 Concluding Remarks

This paper investigates the variation in erosion extension velocity through numerical simulation. The lower pipe is oriented normal to the current direction, while the crossing angle of the upper pipe relative to the lower pipe ranges from 0 to 90°. The spacing between the pipes is maintained at $G/D = 1$. The numerical results indicate that the scour propagation rate on both the left and right sides decrease as the crossing angle increases. Ultimately, when the crossing angle reaches 90°, the results closely resemble those observed in the case of a single pipeline. During the process of crossing angle variation, the upstream extension of the upper pipe, coupled with the offset pair erosion effect of the water flow, contributes to structural asymmetry. Consequently, the propagation rate on the right half is slower than that on the left half. As the crossing angle increases, the propagation rate on the right diminishes, and the impact on the upper pipe becomes negligible when the crossing angle surpasses 75°. However, a comprehensive analysis should also account for the sedimentation effects induced by the spacing and the presence of the upper pipeline.

Acknowledgements. The authors would like to acknowledge the support from National Natural Science Foundation of China (52371289).

References

1. Yan, T., Meng, X.: Study on detection means of crossed submarine pipelines and cables. Port Eng. Technol. **54**(03), 108–112 (2017)
2. Mao, Y.: The Interaction Between A Pipeline and an Erodible Bed. Technical University of Denmark, Denmark (1987)
3. Zhao, M., Cheng, L.: Numerical investigation of local scour below a vibrating pipeline under steady currents. Coast. Eng. **57**(4), 397–406 (2010)
4. Zang, Z., Tang, G., Chen, Y., Cheng, L., Zhang, J.: Predictions of the equilibrium depth and time scale of local scour below a partially buried pipeline under oblique currents and waves. Coast. Eng. **150**, 94–107 (2019)
5. Cheng, L., Yeow, K., Zang, Z., Li, F.: 3D scour below pipelines under waves and combined waves and currents. Coast. Eng. **83**, 137–149 (2014)
6. Wu, Y., Chiew, Y.: Mechanics of pipeline scour propagation in the spanwise direction. J. Water W. Port Coast. Ocean Eng. **141**(4), 04014045 (2015)

7. Flow-3D: Flow-3D User Manual v9.4. Flow Science, Inc., Santa Fe, NM, USA (2009)
8. Van Rijn, L.C.: Sediment transport, art I: Bedload transport. J. Hydraul. Div. Proc. ASCE. **110**, 1431–1456 (1984)
9. Zhang, F., Zang, Z., Zhao, M., Zhang, J., Xie, B., Zou, X.: Numerical investigations on scour and flow around two crossing pipelines on a sandy seabed. J. Mar. Sci. Eng. **10**, 2019 (2022)

Open Access This chapter is licensed under the terms of the Creative Commons Attribution-NonCommercial-NoDerivatives 4.0 International License (http://creativecommons.org/licenses/by-nc-nd/4.0/), which permits any noncommercial use, sharing, distribution and reproduction in any medium or format, as long as you give appropriate credit to the original author(s) and the source, provide a link to the Creative Commons license and indicate if you modified the licensed material. You do not have permission under this license to share adapted material derived from this chapter or parts of it.

The images or other third party material in this chapter are included in the chapter's Creative Commons license, unless indicated otherwise in a credit line to the material. If material is not included in the chapter's Creative Commons license and your intended use is not permitted by statutory regulation or exceeds the permitted use, you will need to obtain permission directly from the copyright holder.

Numerical Simulation of Cylindrical Gravity Anchor Penetrating into Lakebed

Yong Chen[1], Li Liang[2], Minghui Wang[2], Zan Fan[2], Lei Gao[1(✉)], and Song Liu[1]

[1] COOEC Subsea Technology Co., Ltd., Shenzhen, China
gaolei8@cnooc.com.cn
[2] Kunming Shipborne Equipment Research and Test Center,
China State Shipbuilding Corporation Limited, Kunming, China

Abstract. A three-dimensional finite element numerical model was established to simulate the gravity anchor bottoming out and penetrating into the lakebed, considering different soil conditions. Gravity anchors with different geometric sizes were examined. The penetration process and the flow of soil around the anchor body were analyzed using the fluid-solid coupling and particle tracking finite element technology. The results show that solid cylindrical gravity anchors should be selected with caution, as that in natural lakebed conditions, the full anchor body could penetrate into the lakebed and be buried by soil. If the lakebed soil condition is sand type, the gravity anchor penetration depth will be significantly smaller than that under natural conditions, which coins could be one of the measures to reduce the lifting capacity requirement during anchor recovery. The analysis results presented in this paper can provide reference for the selection, design and use of the anchor body of the test platform in the lake.

Keywords: Gravity anchor · Penetration depth · Soil flow · Finite element

1 Introduction

The main body of the solid cylindrical gravity anchor (referred to as the "gravity anchor" in this paper) is a cylindrical solid steel structure, which has the characteristics of construction convenience, low cost, and reusability. As one key anchoring system component of a test floating platform, installation and recovery of the gravity anchor will directly affect the economic feasibility of the testing facility.

According to the geotechnical data, the lakebed soil in the area where the gravity anchor will be deployed has thick silt. It is inevitable that part or full length of the gravity anchor will penetrate the soil during the deployment process. During the recovery process, the penetration depth of the gravity anchor will greatly increase the load required for recovery, due to its own weight and the suction friction forces between the anchor and the soil [1, 2]. In extreme cases, the anchor cannot be lifted and recovered normally, which will have significant impact on test safety.

The soil will be compressed and flow around during the gravity anchor penetration process, which is highly nonlinear. With the development of computational technology, it has become possible to numerically simulate the penetration process of gravity anchors into soil. The finite element method, as the most widely used numerical method, has been adopted to simulate the penetration process, which not only helps to visually understand the final state of the gravity anchor after the penetration, but also facilitates the optimization design of the gravity anchor. The high cost and uncertainty of experimental research could be reduced in this way.

In this article, the finite element numerical analysis method is utilized to analyze the penetration process of gravity anchors through a three-dimensional FEM model, in order to preliminarily evaluate the gravity anchors recovery feasibility and optimize the design of gravity anchors. ABAQUS, as one of the most advanced large-scale finite element softwares in the world, can analyze complex engineering mechanics problems with its world leading nonlinear mechanics analysis capability [3]. A finite element model of gravity anchor bottom penetration into the soil is established in this article, based on the ABAQUS explicit analysis module. The initial stress state, contact, friction and other coupling effects of the lakebed soil are considered to simulate the dynamic process of gravity anchor penetration. The impacts of anchor mass, soil type and thickness parameters on the penetration depth are examined.

2 Physics Model and Input Data

2.1 Gravity Anchor

The gravity anchor examined in this article can be simplified as a large cylindrical solid steel structure, with a height-to-diameter ratio of about 0.75 and a density of $\rho = 7880$ kg/m3. Typical diameters are selected to be 1 m, 1.5 m, 2 m, 2.5 m, 3 m, 3.5 m, and 4 m respectively. The penetration speed of the gravity anchor is constrained by the wire releasing speed of the anchor winch. Under normal working conditions, the wire releasing speed is very slow (about 0.03 m/s). For simplicity, the bottom penetration of the gravity anchor will assume no initial speed, i.e., a penetration state under the action of self-weight only. Geometric and physical parameters of gravity anchors are shown in Table 1.

Table 1. Gravity Anchor Dimension and Weight Parameters.

ID #	D (m)	Height (m)	Weight in Air (kg)	Bottom Speed (m/s)
1	1	0.75	4640	0
2	1.5	1.125	15,660	0
3	2	1.5	37,120	0
4	2.5	1.875	72,490	0
5	3	2.25	125,260	0
6	3.5	2.625	198,910	0
7	4	3	296,920	0

2.2 Geotechnical Data

According to the geotechnical survey results, the main strata at the site where gravity anchors are deployed are quaternary alluvial/lacustrine layers, which show clear stratification and unevenness. The soil layers from top to bottom are: Silty clay, silt and sandy soil, in which the silty clay layer is in brown-gray color, saturated and flow-plastic, with low dry strength, low toughness and an average thickness of about 2 m. In comparison, the silt layer is gray, saturated, in slightly dense state, with low dry strength, low toughness, and an average thickness of 1 m. The sandy soil layer is brown, saturated and loose, with thickness greater than 5 m. For simplicity, each layer of soil is considered as a uniform cross-section, and the lakebed soil layers can be schematically illustrated in Fig. 1.

Fig. 1. Schematic Diagram of Lakebed Soil Layers.

Mechanical properties of the three soil layers in the lakebed are summarized in are shown in Table 2. In order to simulate the physical conditions, the finite element model will consider the mechanical characteristics of each of the three soil layers respectively.

Table 2. Mechanical Properties for Each Soil Layer of the Lakebed.

ID #	Natural Weight (kN/m^3)	Compression Modulus (MPa)	Shear Strength C Φ (kPa) (°)		Average Thickness (m)
(1)	17	1.5	6	2	2
(2)	18	5.0	7	13	1
(3)	18.5	8.0	9	17	>5

3 Finite Element Model and Verification

3.1 Finite Element Modelling

In the process of the gravity anchor touching the bottom and penetrating into the soil, in order to minimize the influence of boundary conditions, the soil model in the length direction is 5 times the maximum diameter of the gravity anchor [4], that is, 20 m; while the soil thickness modelled is taken as 4 times the anchor height H, which is 12 m. As the lakebed soil has internal stress under its own weight in nature [5], in order to model the in-situ stress distribution of the soil, the finite element model will first calculate the stress in the soil layers without displacement through the geo-stress balance analysis in ABAQUS software. This initial stress distribution output is used as the initial condition for gravity anchor penetration analysis.

3.2 Material Constitutive

The gravity anchor is modelled as an ideal rigid body in this article. Deformation of the gravity anchor during penetration can be considered as negligible, as the strength and elastic modulus of the steel solid cylindrical gravity anchor are much higher than those of the soil. Considering the convergence of finite element modelling calculations which is time-consuming, each soil layer is simplified into an ideal elastic-plastic constitutive model, using the Drucker-Pager yield criterion [6]. The mechanical parameters for each soil layer can be found in Table 2.

3.3 Fluid-Structure Interaction and Particle Tracking Technology

Compared with other engineering problems, the numerical analysis of the gravity anchor bottoming out and penetrating into the soil has many difficulties such as large deformation, changes in soil mechanical properties, and complicated contact issue. In particular, during the flow and deformation of the soil, the meshes can be highly distorted and the analysis can be diverged due to finite volume being compressed to 0, for which general Lagrangian grid and adaptive grid technology are not applicable.

Fluid-structure coupling technology is adopted for numerical simulation in this paper. The coupled Euler-Lagrangian method (CEL) has been widely used to deal with large deformation problems. Based on the symmetry of model boundaries and loading conditions, a three-dimensional finite element model is built in 1/24 sector along the circumference, in order to reduce calculation time and resource consumption.

For the conventional Lagrangian finite element method, the grid nodes and material points always coincide during the entire analysis process. By observing the grid deformation of the calculation results, the displacement of the relevant material points can be studied intuitively. However, in the numerical analysis results from CEL technology, the grid nodes no longer correspond to a fixed material point, while the material points can flow in the grid, and the grid nodes and material points can separate and move from each other. Therefore, the contour plots of the finite element displacement will be meaningless, and the corresponding node variables such as velocity can only represent the variable value of the material point corresponding to the current node, and its variation over time will be also meaningless. To address this problem, ABAQUS provides the function to define the tracer particles in order to track changes in material points, in which way the changing of tracer particle variables over time can be obtained. The particle tracking technology is used in this paper to track material points in each soil layer at different interfaces, and flow state of soil particles in different soil layers during the penetration process can be described.

3.4 Finite Element Model Meshes and Boundary Conditions

In the finite element modelling, the linear reduced integral C3D8R solid element and EC3D8R Euler body element are used to simulate the gravity anchor and soil layer respectively. In order to improve the finite element calculation efficiency, the soil element size is refined along the radial direction of the gravity anchor, with mesh size in the range of 0.05~0.2 m, and the mesh size in the soil depth direction is 0.15 m. Regarding to the boundary conditions, in order to prevent the soil Euler model from flowing out of the finite element boundary, the bottom edge of the soil model is constrained for each degree of freedom in cylindrical coordinates, and the circumferential displacement on both sides of the soil model, and radical external displacement at the outer diameter of the soil model are constrained. For the gravity anchor model, only vertical dropping displacement are allowed. The penetration process of the gravity anchor adopts the self-contact model. At the same time, in order to ensure that the resistance during the penetration is completely determined by the strength of the surrounding soil, the friction coefficient between the gravity anchor and the surrounding soil in this article is set to 1.0(3). After the finite element model boundary conditions are established, the initial gravity load of the gravity anchor (gravity acceleration $g = 10$ m/s^2) is applied, the analysis time is set to 3 s, and the dynamic analysis is performed through the Dynamic/Explicit module.

3.5 Finite Element Model Verification

Taking the gravity anchor with diameter $D = 4$ m as an example: After the finite element model is calculated with the loading conditions set aforementioned, the velocity / displacement curve during the gravity anchor penetration process is shown in Fig. 2.

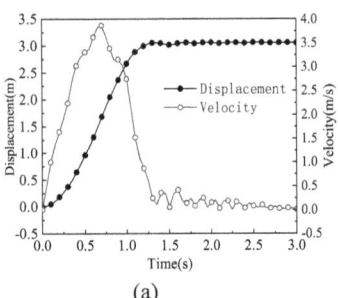
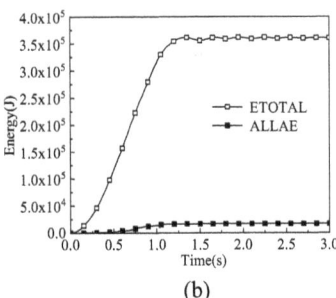

Fig. 2. (a) Velocity / Displacement Curve during Penetration ($D = 4$ m); (b) Energy Response during Penetration (Diameter $D = 4$ m).

As shown in Fig. 2(a), during the penetration process of the gravity anchor with diameter $D = 4$ m, due to the small resistance of the gravity anchor during the initial penetration process, the gravity anchor speed increases initially. With the penetration depth increases, the resistance continues to increase and is equal to the gravity at about 0.7 s, when the gravity anchor speed reaches the maximum value of about 3.8 m/s. The penetration depth is 1.7 m at this moment. After that, as the resistance to the gravity anchor is greater than its own gravity, the penetration speed begins to decrease, and the speed gradually approaches 0 after 1.5 s. The anchor penetration depth also tends to be stable with final depth at about 3.0 m. The entire penetration process of the gravity anchor is in-line with the actual physical test data.

In terms of energy, Fig. 2(b) shows the total energy (ETOTAL) and pseudo strain energy (ALLAE) variation curve during the whole process of bottoming out, for the finite element modelling the gravity anchor with diameter D = 4 m. It is shows that the ratio of pseudo strain energy (ALLAE) to total energy (ETOTAL) (ALLAE/ETOTAL) is always within 5%, which is reasonable. Note that finite element modelling in this article adopts fluid-structure coupling analysis in large deformation in order to control the unrealistic dynamic response due to unit volume penetration or excessive hourglass resistance. The finite element model penetration contour plots at different times of 0.5 s, 1.0 s, and 1.5 s are shown in Fig. 3.

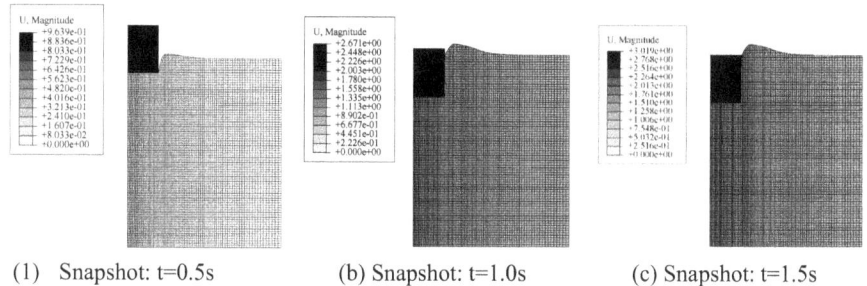

(1) Snapshot: t=0.5s   (b) Snapshot: t=1.0s   (c) Snapshot: t=1.5s

Fig. 3. Penetration Displacement Contour at Different Snapshots (Anchor Diameter D = 4 m).

4 Finite Element Analysis Results

4.1 Analysis of the Penetration Process with Different Anchor Sizes

Figure 4 shows the depth curve of the penetration process of the gravity anchors in different geometric sizes (diameters and height), under their self-weight. At natural lakebed conditions, the penetration depth into the lakebed of gravity anchors in different geometric sizes increases as the mass of the gravity anchor increases, and the penetration time basically reaches a stable state after 1.5 s. The penetration depth of the gravity anchor finally reaches a stable state. All reach or exceed the height of the anchor body itself, that is, the anchor body fully penetrates into the lakebed, which will make subsequent recovery very difficult. Therefore, cylindrical gravity anchors should be cautiously selected for this lakebed.

4.2 Soil Flow Conditions

As mentioned above, the lakebed soil material flows in the grid during the simulation process, so the physical state of the nodes (such as displacement, velocity) obtained

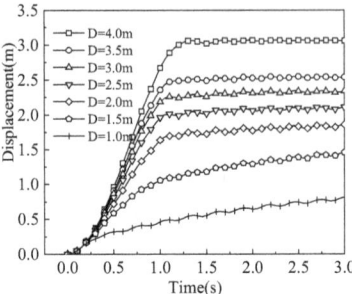

Fig. 4. Penetration depth curve for different gravity anchor geometric sizes.

by finite element calculation only represents the current moment, and time history data cannot be obtained. In order to describe the flow state of soil particles in different soil layers during the penetration process, tracer particles are set at different soil interfaces to obtain the flow conditions of soil around the gravity anchor, as shown in Fig. 5.

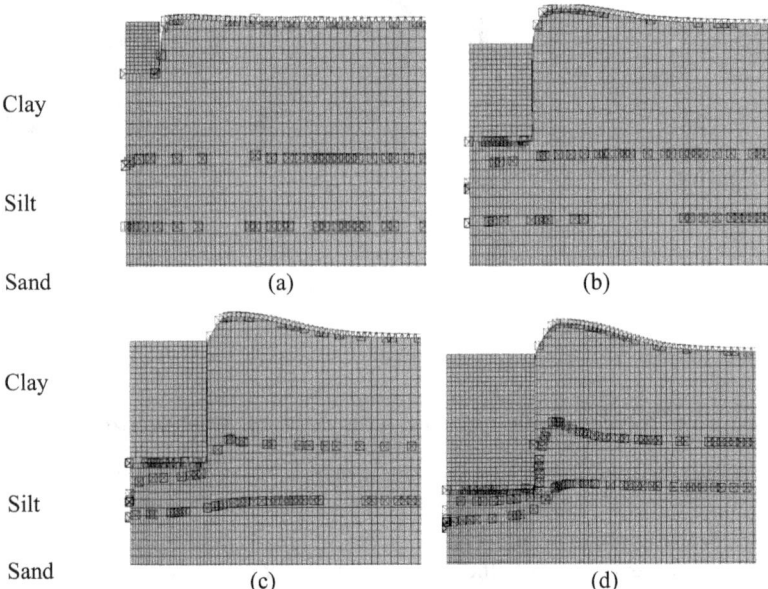

Fig. 5. Soil Flow Situation after Penetration into the Soil for Gravity Anchors in Different Sizes: (a) Diameter = 1.0 m; (b) Diameter = 2.0 m; (c) Diameter = 3.0 m; (d) Diameter = 4.0 m.

For silty clay, the soil around the anchor body is compressed and extruded, resulting in the largest flow deformation and poor bearing capacity. At the end of the gravity penetration, the front end of the anchor is clogged with soil which in general exceeds the upper surface of the anchor body. Under the action of the lake bottom flow, the gravity anchor may be covered and submerged. For silt soil, when the mass of the gravity anchor is small, the flow deformation of the silt layer is not large. However, when the diameter of the gravity anchor is greater than 3 m, the silt layer at the bottom of the anchor body is also compressed and flows out, which shows a certain bearing capacity. For sandy soil, the deformation of the sandy soil layer is small after the penetration of gravity anchors reach stable state, which has good bearing capacity.

4.3 Effect of Lakebed Soil Replacement on Penetration Depth

From the FE analysis, it can be seen that the solid cylindrical gravity anchor selected in this article has poor bearing capacity under natural lakebed conditions which is silty clay in general. The gravity anchor itself completely penetrates into the soil with its own weight. If the testing platform proposed needs to be equipped with a larger size gravity anchor, it is suggested to replace all the lakebed soil with sand soil, in order to avoid great difficulties in recovery and recycling of the anchor piles.

Figure 6 shows the recalculated penetration depth using sandy soil layers. It is shown that after the lakebed soil layers are replaced with sand, the penetration depth of the gravity anchor is greatly reduced compared to natural conditions. The penetration depth of the 4 m diameter gravity anchor is reduced from 3 m to about 1.2 m, for which most of the anchor body can be exposed over the lakebed. In addition, the lifting capacity required for recovery will be reduced. In summary, replacing lakebed soil with sand can reduce the penetration depth of gravity anchors, which is one of the measures to reduce the lifting capacity requirement for recovery.

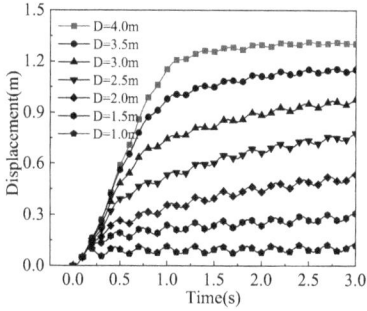

Fig. 6. Gravity anchor penetration depth curve after replacing lakebed with sand.

5 Conclusion

A three-dimensional finite element model is established in this article to simulate the penetration process of a solid cylindrical gravity anchor into the lakebed. Fluid-structure coupling method and particle tracking finite element technology are adopted to model the displacement and soil flow conditions of the gravity anchor in different geometric sizes throughout the entire process of bottoming out and penetrating. Conclusions can be drawn as follows:

(1) At natural lakebed soil conditions, gravity anchors in different geometric sizes will fully penetrate into the lake bed, which has a greater impact on anchor lifting and recovery. Solid cylindrical gravity anchors should be selected with caution.
(2) At natural lakebed soil conditions, the front ends of gravity anchors are clogged with soil that could exceed the upper surface of the anchor body, which may cause the gravity anchor to be covered and submerged by soil.
(3) By replacing the lakebed soil with sand, the penetration depth of the gravity anchor is greatly reduced compared to natural lakebed conditions, which is one of the measures in order to reduce the recovery lifting capacity requirement.

References

1. Hu, Z.M., Chen, W.B., Shi, W.Q.: Research of absorption force of seabed platform applied in estuarine silty waters. J. Ocean Technol. **32**(3), 1–5 (2013)
2. Yang, T.: Study on the Adsorption Capacity and Structure Optimization of Bottom Mounts in Muddy Sea Area, Master's Thesis. Qingdao Univ. of Science and Technology (2017)
3. Fei, K., Zhang, J.W.: Application of ABAQUS in Geotechnical Engineering. China Water & Power Press
4. Lv, Y.: Finite Element Simulation of Suction Cylinder Penetration and Theoretical Research on Inner Wall Friction, Master's Thesis. Dalian University of Technology (2015)
5. Zhang, J.K.: Numerical Simulation of Continuous Penetration of Static Pressure Piles Using ABAQUS Master's Thesis. Qingdao University of Technology (2010)
6. Ma, L.Y., Li, M.Q., Wang, W.: FEM analysis and experimental research of vibration compaction based on Drucker-Prager model. J. Chongqing Jiaotong Univ. (Nat. Sci.). **38**(5), 108–113 (2019)

Open Access This chapter is licensed under the terms of the Creative Commons Attribution-NonCommercial-NoDerivatives 4.0 International License (http://creativecommons.org/licenses/by-nc-nd/4.0/), which permits any noncommercial use, sharing, distribution and reproduction in any medium or format, as long as you give appropriate credit to the original author(s) and the source, provide a link to the Creative Commons license and indicate if you modified the licensed material. You do not have permission under this license to share adapted material derived from this chapter or parts of it.

The images or other third party material in this chapter are included in the chapter's Creative Commons license, unless indicated otherwise in a credit line to the material. If material is not included in the chapter's Creative Commons license and your intended use is not permitted by statutory regulation or exceeds the permitted use, you will need to obtain permission directly from the copyright holder.

Cyclic Behaviour of Completely Overlapped Joint for Offshore Jacket Substructures

Ye Yang[1]($\boxtimes$), Jiewen Wang[1], Jinkun Shi[1], Lei Gao[1], Wie-Min Gho[2], and Song Liu[1]

[1] CNOOC Offshore Engineering Solutions Co. Ltd, Shenzhen, China
ye.yang@qotglobal.com
[2] Maritime Production Research Pte. Ltd, Singapore 915806, Singapore

Abstract. This paper introduces a novel eccentric jacket substructure for offshore wind turbines, engineered to withstand severe environmental forces and provide a robust alternative to conventional X-braced jackets in seismically active regions. The design features fully overlapped joints at all connections, comprising a chord and two braces arranged within a single plane. The cyclic behavior of these joints is analyzed using nonlinear finite element (FE) modeling, calibrated and validated against experimental results. The study demonstrates that while the geometric parameters of the chord member and through brace (β_{CT}, τ_{CT}, γ_C) have insignificant impact, those of the through brace and lap brace (β_{TL}, τ_{TL}, γ_T) significantly influence joint performance under axial cyclic loading of the lap brace. A marked reduction in load-bearing and energy dissipation capacity is observed following lap brace yielding and local buckling. To enhance performance, it is suggested that the dissipative zone for fully overlapped tubular joints subjected to cyclic loading should be located in the short link segment of the through brace connected to the chord member.

Keywords: Completely overlapped joint · Cyclic loading · Jacket substructure

1 Introduction

Fixed offshore wind turbines anchored to the seabed constitute a substantial share of global offshore wind installations. A critical factor in their deployment is the support structures, which must bear turbine loads and endure the harsh marine environment. As offshore wind farms expand into deeper waters, jacket substructures have emerged as a preferred choice for depths between 30 and 50 m [1]. In steel jacket substructures, a key component is the welding joint where the members are joined together. Similar to offshore platform structures, jacket substructures typically feature X-braced vertical frames with welded tubular joints such as K-, T/Y-, and X-joints, as depicted in Fig. 1a. To enhance structural performance and economic viability, Gho and Yang proposed an eccentric jacket substructure designed to replace conventional X-braced jackets in seismically active regions [2]. This innovative configuration incorporates fully overlapped tubular circular hollow section joints at each connection, as illustrated in Fig. 1b, with detailed geometry shown in Fig. 1c and Fig. 2. Compared to traditional designs, the eccentric jacket provides several benefits, including improved load transfer, fewer welded joints, and the use of shorter, thicker wall joint cans. This design enables the short segment to dissipate energy generated by load excitation.

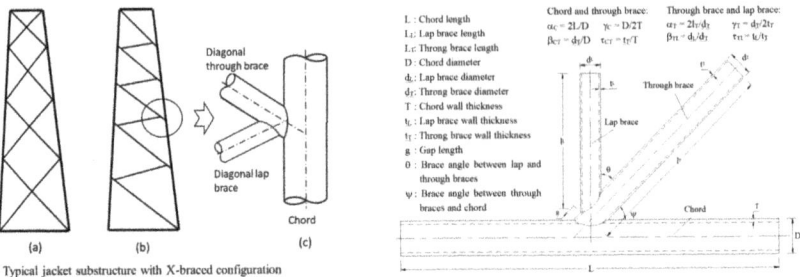

Fig. 1. Offshore Jacket Substructure.

Fig. 2. Geometrical parameters.

Previous research on completely overlapped joints has primarily focused on static strength [3] and stress/strain concentration factors [4]. Gho and Yang's parametric study highlighted the significance of the short diagonal brace segment in controlling joint failure mechanisms under cyclic loading [5]. However, studies on the cyclic behavior of these joints remain limited. A deeper understanding of this behavior is essential for designing seismic-resilient eccentric jackets, especially as Asia is projected to surpass Europe in offshore wind capacity by the 2030s, followed by North America in the 2040s [6]. Wind farm infrastructure in regions like Asia and the U.S. must withstand extreme conditions, including typhoons, earthquakes, and tsunamis. In this study, two tubular joint specimens with complete overlapped braces were tested under monotonic and cyclic loading. Details of the experiments are described in Gho and Yang [2]. The test results were used to validate a finite element (FE) model, which was subsequently employed for an extensive parametric study to investigate the impact of geometric parameters on the cyclic performance of fully overlapped joints.

2 FE Model Verification

The commercial finite element (FE) software MARC MENTAT was utilized to analyze the behavior of fully overlapped tubular joints subjected to axial cyclic loading in the lap brace. The mid-surface of the member walls was modeled using four-node, doubly curved thick shell elements capable of accounting for transverse shear deformation. Exploiting the vertical plane of symmetry, only half of the joint model was simulated, which was sufficient for studying the joint behavior. Axial loading was applied to the end of the lap brace through prescribed displacements, as illustrated in Fig. 3. Fine mesh elements were employed at joint intersections to accurately capture the effects of steep stress gradients, while transition elements were included to optimize the total element count in the computational analysis. A convergence study was performed to confirm that the chosen element size was adequate for reliable results. The joint model accurately replicated the load-displacement and force response-load step behavior of two tubular joint test specimens under monotonic compression and cyclic loading [2].

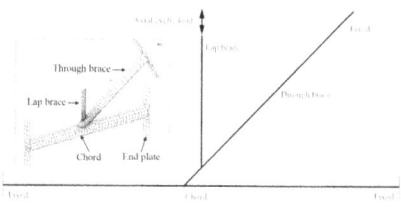

Fig. 3. FE model.

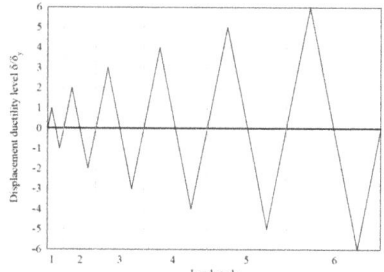

Fig. 4. Loading procedures.

3 Parametric Study

3.1 Scope of Parametric Study

The parametric study of the tubular joint with complete overlapped braces focused on specific geometrical parameters within the following ranges: $0.4 \leq \beta_{CT} \leq 0.8$, $0.4 \leq \beta_{TL} \leq 0.8$, $0.4 \leq \tau_{CT} \leq 1.0$, $0.4 \leq \tau_{TL} \leq 1.0$, $12 \leq \gamma_C \leq 50$. Six load cycles were conducted, with displacement levels incrementally increased for each cycle until the deformation reached six times the yield level, as depicted in Fig. 4.

3.2 Effect of Diameter Ratio β_{CT} Between Through Brace and Chord

Figure 5 illustrates the axial displacement and load relationship at the end of the lap brace. To analyze behavioral differences, the normalized displacement and load relative to the yield displacement (δ_y) and yield load (P_y) were used, respectively. The plot indicates that β_{CT} had minimal influence on the joint's hysteresis behavior.

3.3 Effect of Wall Thickness Ratio τ_{CT} Between Through Brace and Chord

As depicted in Fig. 6a, the hysteresis curves reveal that joints with larger τ_{CT} demonstrate greater stiffness compared to the joint with lower τ_{CT}. Joints with $\tau_{CT} = 0.7$ and 1.0 and a smaller $\gamma_T = 8$ experienced failure due to lap brace local buckling, whereas the joint with $\tau_{CT} = 0.4$ and a larger $\gamma_T = 12$ exhibited through-brace wall plastification. These observations indicate that differences in cyclic behavior are largely influenced by γ_T. Additionally, Fig. 6b shows that joints with $\gamma_T > 12$ also failed through through brace wall plastification, resembling the failure behaviour under axial compression of the lap brace. Consequently, the effect of τ_{CT} appears negligible when the influence of γ_T is accounted for.

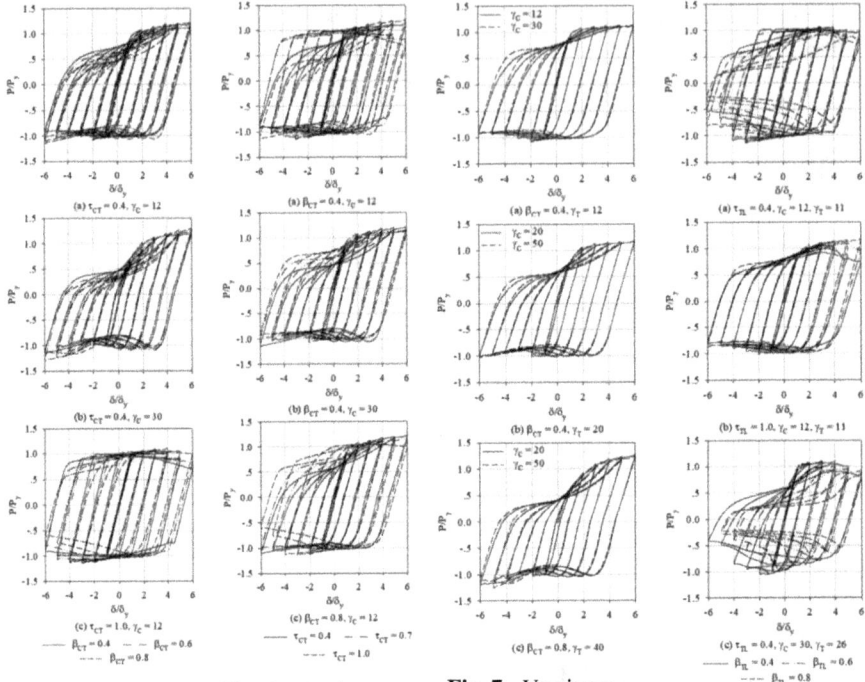

Fig. 5. Varying β_{CT}. **Fig. 6.** Varying τ_{CT}. **Fig. 7.** Varying γ_C.

Fig. 8. Varying β_{TL}.

3.4 Effect of Radius/Wall Thickness Ratio γ_C of Chord

Fig. 7 demonstrates that the hysteresis behaviors for different values of γ_C are generally similar. However, a reduction in strength is observed after the fifth load cycle for joints with $\gamma_C = 50$ in Fig. 7c. This behavior is absent in the joint with smaller γ_T in Fig. 7b, likely due to lap brace local buckling caused by the influence of a large γ_T.

3.5 Effect of Diameter Ratio β_{TL} Between Lap Brace and Through Brace

As depicted in Fig. 8, the analysis for the joint with higher $\beta_{TL} = 0.8$ terminated at the fifth load cycle due to severe FE mesh distortion. The hysteresis behavior at low $\tau_{TL} = 0.4$ remains unaffected by variations in β_{TL}. In Figs. 8a and 8c, joints with higher β_{TL} exhibit poor ductility, with noticeable reductions in stiffness and strength under inelastic cyclic load. Conversely, the hysteresis curves at $\tau_{TL} = 1.0$, shown in Fig. 8b, display stable behavior, indicating that joint capacity is largely unaffected by β_{TL}. Good ductility performance is observed when failure occurs due to through brace wall plastification. However, at medium β_{TL} values (0.4~0.6), significant tensile strength reductions are observed at the fourth load cycle, as illustrated in Fig. 10b. This strength loss is due to failure criteria applied in the model and punching shear stresses on the through brace wall of the joint.

3.6 Effect of Wall Thickness Ratio τ_{TL} Between Lap Brace and Through Brace

The hysteresis behavior of the joint is strongly influenced by τ_{TL}, as illustrated in Fig. 9. At a lower τ_{TL}, the joint demonstrated unstable hysteresis performance, characterized by a sudden loss of stiffness and strength caused by premature yielding of the lap brace.

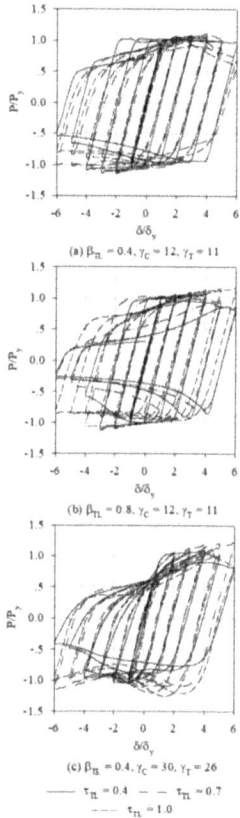

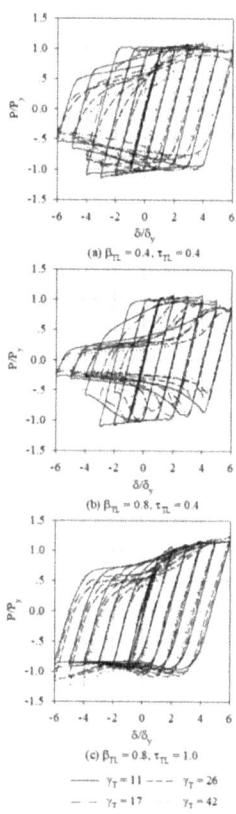

Fig. 9. Varying τ_{TL}.

Fig. 10. Varying γ_T.

3.7 Effect of Radius/Wall Thickness Ratio γ_T of Through Brace

As shown in Fig. 10a, the joint with low β_{TL} (=0.4) and τ_{TL} (=0.4) experienced failure in the lap brace member. The joint stiffness reduced with increasing γ_T, although this effect was minimal for joints with $\gamma_T > 17$. The joint's cyclic load-carrying capacity during the tension phase remained largely unaffected by changes in γ_T. For joints with high β_{TL} and low τ_{TL}, lap brace local buckling occurred, and both stiffness and strength deteriorated with increasing γ_T, as shown in Fig. 10b. However, this effect became negligible for $\gamma_T > 26$. Joints with lower γ_T failed in the compression due to lap brace yielding combined with local buckling, while joints with higher γ_T (=26 and 42) failed due to lap brace local buckling. For the joint with large β_{TL} and τ_{TL}, the failure mode was caused by through brace wall plastification. The joint's hysteresis behavior remained stable under cyclic loading, with stiffness and strength largely unaffected by variations in γ_T (Fig. 10c).

4 Conclusions

The cyclic behavior of the fully overlapped tubular joint was analyzed using nonlinear finite element (FE) simulations, with the model validated by the test data. The findings indicate that the influence of geometrical parameters of the chord and through brace members (β_{CT}, τ_{CT} and γ_C) is less pronounced compared to the parameters of the through brace and lap brace members (β_{TL}, τ_{TL}, γ_T) under axial cyclic loading of the lap brace. The study highlights that the joint's load-carrying and energy dissipation capacities significantly degrade if the lap brace fails before the joint collapses. Therefore, considering the effects of local buckling is crucial when designing fully overlapped tubular joints with low τ_{TL}, high β_{TL} and high γ_T. Joints that fail due to through brace wall plastification exhibit more stable and ductile cyclic behavior. It is suggested that the dissipative zone for such joints be positioned at the short link segment of the through brace where it connects to the chord. The conclusion primarily focuses on the strength performance of the joints under cyclic loading. Future work should include an assessment of the fatigue performance of the eccentric jacket.

Acknowledgement. This paper is part of the doctoral research conducted by Dr. Ye Yang, who was sponsored by the Nanyang Technological University (NTU) of Singapore.

References

1. He, B., Lv, N., Shi, R.: Design optimization of jacket substructure of offshore wind turbine. In: Feng, G. (ed.) 10th International Conference on Civil Engineering ICCE, vol. 526, pp. 699–706. Springer, Singapore (2024)
2. Gho, W.M., Yang, Y.: Ultimate strength of completely overlapped joint for fixed offshore wind turbine jacket substructures. J. Mar. Sci. Appl. **18**(1), 99–113 (2019)
3. Gho, W.M., Yang, Y.: Parametric equations for static strength of tubular circular hollow section joints with complete overlap of braces. J. Struct. Eng. **134**(3), 393–401 (2008)
4. Gao, F., Gho, W.M.: Parametric equations to predict SCF of axially loaded completely overlapped tubular circular hollow section joints. J. Struct. Eng. ASCE. **134**(3), 412–420 (2008)
5. Gho, W.M., Yang, Y.: Cyclic performance of completely overlapped tubular joints. In: Proceedings of The 15th International Offshore and Polar Engineering Conference, ISOPE, pp. 320–324, Seoul, Korea (2005)
6. DNV Homepage., https://www.dnv.com/news/dnv-launches-new-joint-industry-project-to-tackle-earthquake-challenges-for-wind-farms-242844/ (2024). Accessed 1 Oct 2024

Open Access This chapter is licensed under the terms of the Creative Commons Attribution-NonCommercial-NoDerivatives 4.0 International License (http://creativecommons.org/licenses/by-nc-nd/4.0/), which permits any noncommercial use, sharing, distribution and reproduction in any medium or format, as long as you give appropriate credit to the original author(s) and the source, provide a link to the Creative Commons license and indicate if you modified the licensed material. You do not have permission under this license to share adapted material derived from this chapter or parts of it.

The images or other third party material in this chapter are included in the chapter's Creative Commons license, unless indicated otherwise in a credit line to the material. If material is not included in the chapter's Creative Commons license and your intended use is not permitted by statutory regulation or exceeds the permitted use, you will need to obtain permission directly from the copyright holder.

Dynamic Anti-rolling Mechanism of Floating Wind Turbine Based on Liquid-Sloshing Principles

Chongwei Zhang[✉], Donghai Li, Shenghui Huang, Xunhao Zhu, and Dezhi Ning

State Key Laboratory of Coastal and Offshore Engineering, Dalian University of Technology, Dalian 116024, China
chongweizhang@dlut.edu.cn

Abstract. The floating wind turbine has high structure and low stiffness, and it is easy to produce rolling motion under the combined action of wind and waves, affecting their normal operation and structural safety. This study proposes a scientific approach to broadly suppress different modes resonate simultaneously in floating wind turbine structures using a multi-tuned liquid column damper (multi-TLCD) system. By simplifying the floating wind turbine structure through the concentrated mass method, we establish a mechanical model of the multi-degree-of-freedom motion system of the floating wind turbine-TLCD. Vibrational characteristic analysis of the floating wind turbine structure is conducted to identify natural frequencies and vibration modes. The resonance characteristics of various modes of the floating wind turbine structure under TLCD are studied to elucidate the suppression mechanism of the multi-TLCD on different modes resonate simultaneously. Results indicate that the TLCD's position significantly affects the suppression effect on different resonance modes of the floating wind turbine. By properly arranging multi-TLCD, effective overall suppression of different modes resonate simultaneously in the floating wind turbine structure can be achieved.

Keywords: Floating Wind Turbine · Liquid-Sloshing · Dynamic Anti-rolling · Multi-TLCD · Multi-degree-of-freedom

1 Introduction

Offshore wind energy is vital for optimizing energy structures and combating climate change. Fixed-foundation turbines installation in deeper waters is challenging due to limited resources and ecological constraints. Therefore, floating offshore wind turbines become a more feasible solution when the water depth exceeds 60 m [1]. Floating offshore wind turbines are subject to harsh environmental loads, such as waves, wind, and currents, which can induce structural vibrations and multi-mode resonance. Multi-mode resonance not only reduces energy conversion efficiency but also poses significant safety risks, affecting turbine operation and structural integrity [2]. Thus, it is crucial to mitigate these vibrations to ensure the structural safety and longevity of wind turbines.

The researchers have conducted extensive research on vibration control of wind turbines under load using passive devices such as TMD, TLD and TLCD. The TLCD consists of a U-shaped tube. Its internal structure comprises only oscillating liquid. Colwell and Basu [3] first introduced TLCDs for offshore wind turbines, highlighting their effectiveness in reducing displacement and improving fatigue life. Zhang and Høeg [4] and Park et al. [5] further demonstrated that well-designed TLCDs significantly mitigate lateral bending and roll motions in monopile and TLP structures. However, previous studies have not focused on the suppression of multi-mode resonance in floating wind turbines.

This study presents a method utilizing multiple TLCDs to suppress multi-mode resonance in offshore floating wind turbine structures. A multi-degree-of-freedom motion model of the wind turbine structure is developed, allowing for analysis of its dynamic vibration characteristics and resonance behaviors under varying TLCD physical parameters.

2 Mathematical Model

Figure 1 illustrates the floating wind turbine, featuring a nacelle, turbine blades, and a flexible tower above, with a floating foundation below. The turbine experiences wind and wave loads. To simplify the structure, the concentrated mass method is employed. The floating foundation is treated as a single mass point with mass m_{N+1} and horizontal displacement q_{N+1}. Mooring lines are modeled as springs with stiffness k_{N+1} and damping c_{N+1}. The tower consists of $N-1$ mass points with masses m_i (for $(i < N)$), while the nacelle and blades form a single mass point m_N. The horizontal displacements are $q_i + q_{N+1}$, with stiffness and damping between mass points as k_i and c_i, respectively. Wind and wave loads are modeled as concentrated forces F_{wind} and F_{wave} on the nacelle and foundation. In the two-dimensional model, two TLCDs are positioned at both the nacelle and foundation, with liquid height variations represented as u_1 and u_2. The kinetic energy T of the entire system is expressed accordingly

$$\begin{aligned} T = & \frac{1}{2}m_1(\dot{q}_1 + \dot{q}_{N+1})^2 + \cdots + \frac{1}{2}m_N(\dot{q}_N + \dot{q}_{N+1})^2 + \frac{1}{2}m_{N+1}\dot{q}_{N+1}^2 \\ & + \rho A_1(h_1(\dot{u}_1^2 + \dot{q}_{N+1}^2) + B_1(\dot{u}_1 + \dot{q}_{N+1})^2) \\ & + \rho A_2(h_2(\dot{u}_1^2 + \dot{q}_{N+1}^2) + B_2(\dot{u}_2 + \dot{q}_N + \dot{q}_{N+1})^2). \end{aligned} \quad (1)$$

In the equation, $\dot{q}_i = \mathrm{d}q_i/\mathrm{d}t$, ρ is the density of the liquid filled in the TLCD, A_{bi} and A_{hi} represent the cross-sectional areas of the horizontal and vertical segments of the TLCD liquid column, respectively. h_i is the height of the TLCD vertical liquid column, B_i is the length of the TLCD horizontal segment, and $\dot{u}_i$ is the velocity of the liquid relative to the external structure of the TLCD. The potential energy V of the entire system can be expressed as

$$V = \frac{1}{2}k_1 q_1^2 + \frac{1}{2}k_2(q_1 - q_2)^2 + \cdots + \frac{1}{2}k_N(q_N - q_{N+1})^2 + \frac{1}{2}k_{N+1}q_{N+1}^2 \qquad (2)$$
$$+ \rho g(A_1 u_1^2 + A_2 u_2^2).$$

The dissipated energy E_d of the entire system can be expressed as

$$E_d = \frac{1}{2}c_1 |\dot{q}_1| \dot{q}_1 + \frac{1}{2}c_2 |\dot{q}_1 - \dot{q}_2| (\dot{q}_2 - \dot{q}_1) + \cdots \qquad (3)$$
$$+ \frac{1}{2}c_N |\dot{q}_N - \dot{q}_{N-1}| (\dot{q}_N - \dot{q}_{N-1}) + \frac{1}{2}c_{N+1} |\dot{q}_{N+1}| \dot{q}_{N+1}.$$

The non-conservative forces acting on this system include wind load, wave load, and the damping effect of the orifice plate on the fluid in the TLCD, expressed as

$$Q = \frac{1}{2}\rho A_1 \xi_1 |\dot{u}_1| \dot{u}_1 + \frac{1}{2}\rho A_2 \xi_2 |\dot{u}_2| \dot{u}_2 + F_{\text{wind}} + F_{\text{wave}}. \qquad (4)$$

ξ_i is the damping coefficient. The Lagrange equation is

$$\frac{d}{dt}\left(\frac{\partial T}{\partial \dot{q}_i}\right) - \frac{\partial T}{\partial q_i} + \frac{\partial V}{\partial q_i} + \frac{\partial E_d}{\partial q_i} = Q \qquad (5)$$

Substituting Eqs. (1) - (4) into Eq. (5) yields the equation of motion for the entire structure

$$M\ddot{q} + C\dot{q} + Kq = Q. \qquad (6)$$

where $\ddot{q} = d^2q/dt^2$, M is the mass matrix, C is the damping matrix, and K is the stiffness matrix.

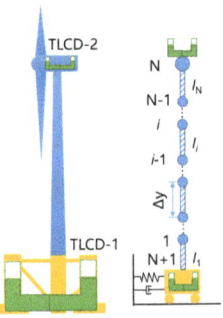

Fig. 1. Floating wind turbine structure diagram and physical model diagram.

3 Results and Discussion

3.1 Wind Turbine Structural Parameters

The upper structure of the wind turbine is a 5 MW offshore model [6], with the following parameters: tower height is 87.60 m, and the interval between concentrated mass points is $\Delta y = 5$ m. The tower's material density is $\rho_s = 8500\,\mathrm{kg/m^3}$ and elastic modulus $E = 2.10 \times 10^{11}$ Pa. The nacelle and blades weigh is 3.50×10^6 kg. The tower's outer diameters are 3.87 m (top) and 6.00 m (bottom), with inner diameters of 3.83 m and 5.95 m, respectively. The floating foundation mass is $m_{N+1} = 1.89 \times 10^6$ kg, and the structural damping ratio is 0.01. Each TLCD mass is 0.01 of the total structure mass, with 0.7 of the liquid length in the horizontal section. The TLCD head loss coefficient is $\xi_1 = \xi_2 = 3.97$. TLCD-1's natural frequency matches the offshore platform's swaying resonance, while TLCD-2's frequency aligns with the tower's lateral bending resonance.

3.2 Identification of Resonant Modes and Frequencies

By solving the undamped vibration equations, the resonant modes and frequencies of the floating wind turbine were determined. Figure 2 shows that mode one is the resonant sway of the floating foundation, while mode two is the lateral bending of the tower. This paper focuses on the vibration suppression of the first two modes of floating wind turbine.

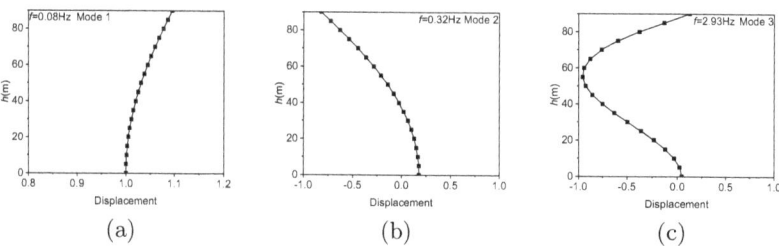

Fig. 2. Each resonant mode diagram of floating wind turbine. (a) Mode 1. (b) Mode 2. (c) Mode 3.

3.3 Suppression of Platform Mode

When a periodic load is applied to the floating platform, the suppression effect of TLCD-1 placed at different positions of the wind turbine on the platform mode vibration is compared. TLCD-1 is placed at four evenly spaced positions from the top of the tower to the floating platform. The results are shown in Fig. 3. The Fig. 3(a) displays the maximum displacement of the nacelle and the floating platform as a function of the load frequency. It can be seen that when

TLCD-1 is placed on the floating platform, the suppression effect on the floating platform mode vibration is optimal. The maximum displacement at the floating platform and nacelle is significantly reduced near the resonance frequency of the floating platform. In contrast, placing TLCD-1 at other positions does not provide an ideal vibration suppression effect on the wind turbine structure. The TLCD-1 is fixed at the floating platform and periodic loads are applied at different locations on the wind turbine. The suppression effect of TLCD-1 on platform vibration under different position loads is shown in Fig. 3 (b). The graphs illustrate the variation of maximum displacement at the floating platform with load frequency, comparing scenarios without TLCD-1 and with TLCD-1. F_1, F_2, and F_3 represent loads applied at the nacelle, mid-tower, and floating platform positions, respectively. Although the vibrations at the floating platform caused by loads at different positions vary, the vibrations at the floating platform are significantly reduced with the addition of TLCD-1. TLCD-1 effectively suppresses vibrations from loads at different positions and frequencies near the floating foundation's resonance frequency. Therefore, optimal suppression can be attained by positioning the TLCD at the point of maximum displacement under resonance conditions, while ensuring that the natural frequency of the TLCD aligns with the resonance frequency.

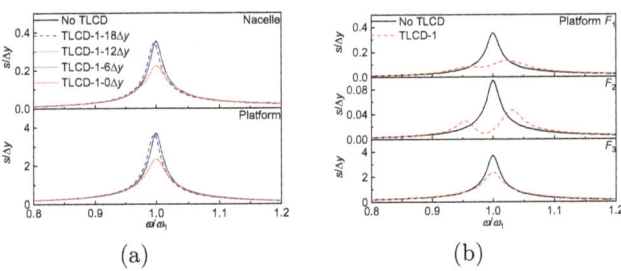

Fig. 3. Inhibition effect of TLCD-1 on platform vibration. (a) The TLCD-1 is placed in different locations, (b) The load is placed in different locations.

3.4 Random Load Action

The wind load is calculated using Kaimal's empirical wind spectrum expression, and the wind load F_{wind} acting on the nacelle is derived from the blade thrust coefficient. Wave load is computed based on the Morison equation. TLCD-1 is placed at the floating foundation, matching its resonance frequency with the sway mode, while TLCD-2 is positioned at the nacelle, aligning with the tower mode. Under simultaneous wind and wave loading, displacement time histories with and without TLCDs at nacelle and floating foundation locations are shown in Fig. 4(a), demonstrating effective structural vibration suppression. Fourier analysis in Figs. 4(b) and (c) compares vibration amplitudes with and without

TLCDs in the frequency domain, confirming TLCD-1 and TLCD-2 effectively suppress vibrations in both platform and tower modes.

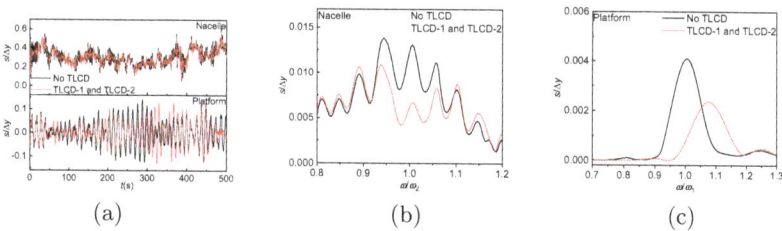

Fig. 4. Comparison of structural motion with and without TLCDs. (a) Time history curve. (b) Platform mode suppression effect. (c) The mode suppression effect of the tower.

4 Conclusions

This study proposes a method utilizing multi-TLCDs to suppress multi-mode resonance in floating wind turbines. A multi-degree-of-freedom motion model of wind turbine was established, and its dynamic vibration characteristics and resonance behavior under different TLCD physical parameters were analyzed. The vibration modes of wind turbine structure are identified. The results show that TLCD-1, which maintains the natural frequency of the platform mode under the action of the load acting on the platform, has a good suppression effect on the vibration of the wind turbine platform and the nacelle when placed on the floating platform. It also effectively suppresses platform vibrations induced by loads at various positions. TLCD-1 and TLCD-2 are simultaneously placed in the floating platform and nacelle, respectively. Under the combined action of wind and wave, the multi-TLCDs can effectively suppress the vibration of the wind turbine in the platform and tower modes.

This study focuses solely on calculations for the two-dimensional horizontal motion of the wind turbine system, without mathematical modeling for a more realistic three-dimensional model. Future work should explore the vibration suppression effects of multiple TLCDs on wind turbines under a three-dimensional model considering the six-degree-of-freedom motion of the platform.

References

1. Rehman, S., Alhems, L.M., Alam, M.M., Wang, L., Toor, Z.: A review of energy extraction from wind and ocean: technologies, merits, efficiencies, and cost. Ocean Eng. **267**, 113192 (2023)
2. Chou, J.S., Tu, W.T.: Failure analysis and risk management of a collapsed large wind turbine tower. Eng. Fail. Anal. **18**(1), 295–313 (2011)

3. Colwell, S., Basu, B.: Tuned liquid column dampers in offshore wind turbines for structural control. Eng. Struct. **31**(2), 358–368 (2009)
4. Zhang, Z., Høeg, C.: Dynamics and control of spar-type floating offshore wind turbines with tuned liquid column dampers. Struct. Control. Health Monit. **27**(6), e2532 (2020)
5. Park, S., Glade, M., Lackner, M.A.: Multi-objective optimization of orthogonal TLCDs for reducing fatigue and extreme loads of a floating offshore wind turbine. Eng. Struct. **209**, 110260 (2020)
6. Jonkman, J.: Definition of a 5-mw reference wind turbine for offshore system development. National Renewable Energy Laboratory (2009)

Open Access This chapter is licensed under the terms of the Creative Commons Attribution-NonCommercial-NoDerivatives 4.0 International License (http://creativecommons.org/licenses/by-nc-nd/4.0/), which permits any noncommercial use, sharing, distribution and reproduction in any medium or format, as long as you give appropriate credit to the original author(s) and the source, provide a link to the Creative Commons license and indicate if you modified the licensed material. You do not have permission under this license to share adapted material derived from this chapter or parts of it.

The images or other third party material in this chapter are included in the chapter's Creative Commons license, unless indicated otherwise in a credit line to the material. If material is not included in the chapter's Creative Commons license and your intended use is not permitted by statutory regulation or exceeds the permitted use, you will need to obtain permission directly from the copyright holder.

Nonlinear Dynamics of Sloshing Liquid with Variable Mass

Donghai Li, Chongwei Zhang(✉), Xunhao Zhu, Shenghui Huang, and Dezhi Ning

State Key Laboratory of Coastal and Offshore Engineering, Dalian University of Technology, Dalian 116024, China
chongweizhang@dlut.edu.cn

Abstract. The floating storage and regasification unit (FSRU) has several large LNG tanks in the body. Under unloading conditions, the mass of liquid in the cabin changes dynamically, and the time-varying mass of liquid can have a transient sloshing phenomenon when excited by waves. In this paper, a physical model experiment was carried out to study the transient sloshing characteristics of time-varying mass liquid in a rectangular tank, and the influence of the mass change rate of liquid and the initial water depth in the tank on the sloshing time-history characteristics of liquid under different excitation conditions was analyzed. It is found that the sloshing of time-varying mass liquid has significant nonlinear characteristics. Under the same excitation conditions, the maximum relative slosh amplitude of the liquid in the tank increases with the increase of the initial water depth, but decreases with the increase of the liquid mass change rate and the excitation amplitude.

Keywords: FSRU · Sloshing · Nonlinear Dynamics · Time-Variable Mass · Transient Resonance

1 Introduction

In recent years, the global demand for floating storage and regasification units (FSRUs) has grown rapidly. FSRUs function as offshore terminals that directly receive liquefied natural gas (LNG), store it in insulated tanks, and subsequently regasify the LNG before transferring the natural gas to onshore stations. Due to prolonged wave excitation, FSRUs inevitably experience oscillations, leading to sloshing of the liquid in partially filled tanks, which poses significant safety risks due to the nature of the cargo [1]. Unlike LNG transport tanks, the liquid level in FSRU tanks changes during loading and unloading, causing transient shifts in the natural sloshing frequency. When the filling level during drainage or loading reaches a point where the natural sloshing frequency coincides with the external excitation frequency, the previously non-resonant liquid may experience temporary resonance. This transient resonance phenomenon is rarely mentioned in the literature.

Researchers have extensively studied sloshing under fixed water depths through theoretical [2], experimental [3], and numerical [4] methods. Faltinsen et al. [5] derived a discrete infinite-dimensional modal system for the nonlinear sloshing of incompressible, irrotational fluids and applied multi-dimensional modal theory to analyze the sloshing problems of various tank geometries. Gandara et al. [3] combined experiments and simulations to study sloshing in partially filled tanks. Qiu et al. [4] conducted numerical research on sloshing in cryogenic liquid hydrogen. Jiang et al.[6] conducted numerical simulations of variable mass tanks and analyzed the effects of filling rates, sloshing periods, and baffles on sloshing behavior. However, they only considered non-resonant conditions and did not address transient resonance behavior.

In this study, the transient resonance of sloshing liquid with time-variable mass is investigated through physical experiments. A rectangular water tank with drain pipe is excited harmonically by a shaking table. Ultrasonic wave gauges are used to measure the free-surface elevation histories in the tank. Effects of the initial water depth, the drainage rate, and the excitation conditions on the sloshing behavior are examined.

2 Experimental Setup

The experiment was conducted at the State Key Laboratory of Coastal and Offshore Engineering, Dalian University of Technology. Figure 1 shows the experimental setup for horizontal excitation of the liquid tank with a drainage pipe. The main equipment includes a linear motor, a shaker table, and a liquid tank. The upper part of the tank measures 0.5 m in length, 0.1 m in width, and 0.6 m in height. A drainage pipe with an inner diameter of 0.006 m is installed at the center of the tank's bottom and is connected to a graduated cylinder. The flow rate is controlled by adjusting a control valve. Non-contact ultrasonic wave gauges are mounted on both sides to measure the water surface height. The linear motor provides horizontal excitation of 0–20 Hz with a maximum displacement of ± 0.3 m, measured by a laser displacement sensor.

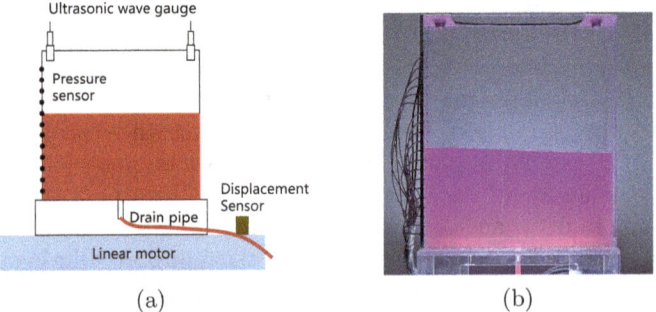

Fig. 1. Experimental setup of liquid tank undergoing horizontal excitation: (a) concept design; (b) physical model.

The depth of the tank at the beginning of the test was $h = H_s$. The linear motor starts to shake the tank with horizontal vibration with excitation amplitude A and excitation frequency f. Open the control valve when the free surface amplitude tends to be constant. The liquid in the tank drains at a constant rate. Close the control valve when the liquid depth reaches $h = H_e$. Until the free surface amplitude is again constant, the linear motor stops vibrating. The experimental conditions are listed in Table 1.

Table 1. Physical parameters of tank

Physical parameter	Parameter value
Initial depth H_s	0.15 m, 0.20 m, 0.25 m
End depth H_e	0.10 m
Excitation frequency f	0.80, 0.93, 0.96, 0.99, 1.02, 1.05, 1.08, 1.11, 1.14, 1.17, 1.20, 1.30 Hz
Excitation amplitude A	0.001 m, 0.003 m
Drainage rate v	$v_1 = 4.5 \times 10^{-4}$ m/s, $v_2 = 4.9 \times 10^{-4}$ m/s, $v_3 = 5.6 \times 10^{-4}$ m/s

3 Results and Discussion

3.1 Analysis of the Whole Process of Sloshing

Figure 2 shows the experimental results under drainage rate v_2, initial water depth $H_s = 0.25$ m, excitation amplitude $A = 0.001$ m, and excitation frequency $f = 1.11$ Hz. The upper part is the free water elevation η_0 and the average water depth $h(t)$. The lower part shows the oscillatory component η, which is the difference between η_0 and $h(t)$. The critical water depth h_c is defined when the resonance frequency equals the excitation frequency, and the critical time t_c is when the water level reaches h_c.

The wave height time-history curve shows three main stages: the initial, drainage, and final stages. In the initial stage, non-resonant sloshing occurs in the tank with a fixed water depth, starting with a transient phase marked by a wave envelope, where the period is related to the difference between the excitation and natural frequencies. Over time, viscous dissipation stabilizes the free surface amplitude. During the drainage stage, the slope of the liquid depth represents the drainage rate $h' = v_2$, and as the average water level nears the critical depth h_c, the oscillation amplitude increases rapidly but peaks after the critical time t_c. As the water level drops further, the oscillation amplitude decreases sharply. At $t = t_e$, the valve is closed, and the water level stays constant. In the final stage, the oscillation period shortens, and the amplitude stabilizes, eventually ceasing as the liquid comes to rest.

3.2 Effect of Drainage Rate on Liquid Sloshing

The maximum relative sloshing amplitude $\eta_{\max}/\eta_s$ for drainage rates v_1, v_2, and v_3 is illustrated in Fig. 3. The η_s represents the wave amplitude at the end of the initial stage. The $\eta_{\max}$ represents the maximum wave amplitude during the drainage stage. The experiments were conducted with an excitation amplitude of $A = 0.001$ m and an initial water depth of $H_s = 0.25$ m. Although the trends of $\eta_{\max}/\eta_s$ with varying excitation frequency are consistent across different drainage rates, a higher drainage rate results in a smaller $\eta_{\max}/\eta_s$. At an excitation frequency of 0.99 Hz, $\eta_{\max}/\eta_s$ reaches its maximum values of 27, 24, and 23 for v_1, v_2, and v_3, respectively. When the excitation frequency exceeds 1.17 Hz, $\eta_{\max}/\eta_s$ approaches 1, indicating no transient resonance, and the oscillation amplitude gradually decreases throughout the drainage stage.

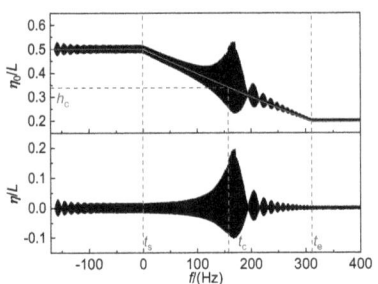

Fig. 2. Measured free-surface elevation history, instantaneous average water level, and oscillatory component of wave elevation in liquid tank

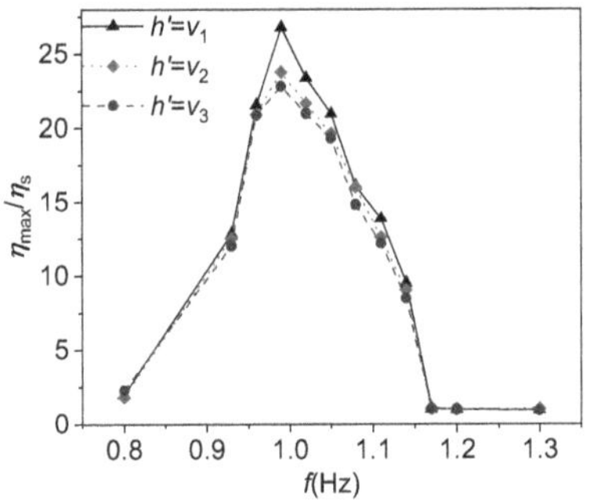

Fig. 3. $\eta_{\max}/\eta_s$ at different excitation frequencies, where the drainage rates are v_1, v_2 and v_3, respectively

3.3 Effect of Initial Water Depth on Liquid Sloshing

Figure 4 shows the peak amplitude of η under different excitation frequencies, with the initial water depth of $H_s = 0.25, 0.20, 0.15$ m, respectively. For cases with $H_s = 0.25, 0.20, 0.15$ m, the maximum η_s occurs at $f = 1.17$ Hz, 1.14 Hz, and 1.11 Hz (i.e., the fundamental natural sloshing frequencies determined by three water depths), respectively. The maximum $\eta_{\max}$ occurs at $f = 1.14$ Hz, 1.11 Hz, and 1.08 Hz, respectively. In the same excitation condition, the ratio $\eta_{\max}/\eta_s$ decreases for the case with a smaller initial water depth H_s. The maximum value of $\eta_{\max}$ can reach over 23 times the η_s. With the excitation frequency larger than 1.17 Hz, the ratio $\eta_{\max}/\eta_s$ tends to be 1.

3.4 Effect of Excitation Amplitude on Liquid Sloshing

The maximum amplitude of η under different excitation frequencies, with excitation amplitudes of $A = 0.001$ m and 0.003 m, is presented in Fig. 5. Under the same excitation frequency, both η_s and η_{max} in the case of $A = 0.003$ m are always greater than those of $A = 0.001$ m. However, the maximum η_{max} appears at $f=1.05$ Hz and 1.14 Hz for the case of A=0.001 m and 0.003 m, respectively. At the same frequency, the ratio η_{max}/η_s under $A =0.001$ m is larger than that under $A =0.003$ m.

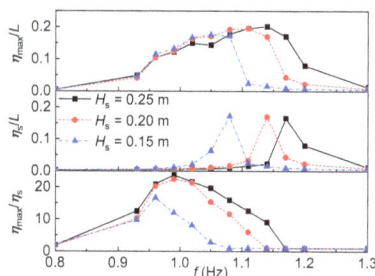

Fig. 4. Maximum amplitude of η under different excitation frequencies, with initial water depths of $H_s = 0.25$ m, 0.20 m and 0.15 m.

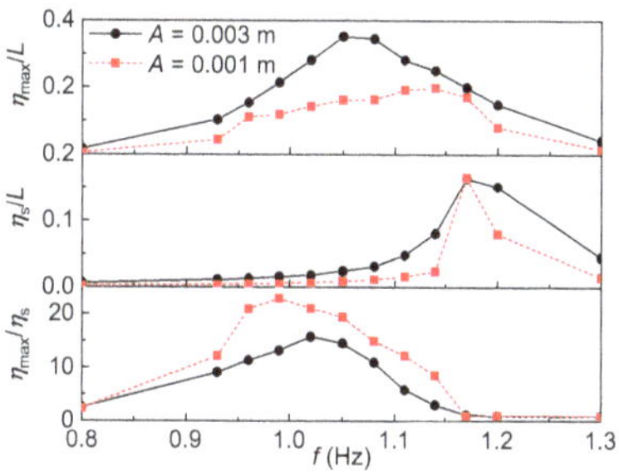

Fig. 5. Maximum amplitude of η under different excitation frequencies, with excitation amplitude of $A = 0.001$ m and 0.003 m.

4 Conclusions

This study investigated transient resonance caused by time-varying liquid mass through physical model experiments. During the experiment, as the liquid level in the tank decreases, the liquid transitions from a non-resonant state to a resonant state when the natural frequency of the average water level matches the excitation frequency. As the liquid level continues to decrease, the liquid transitions back to a non-resonant state. The results showed that the maximum sloshing amplitude during transient resonance can reach up to 27 times the initial stable phase. The smaller the drainage rate and excitation amplitude, the larger the maximum relative sloshing amplitude. Additionally, the larger the initial water depth, the greater the maximum relative sloshing amplitude under transient resonance conditions.

The external excitation in this study is the forced motion of a linear motor. Future research should focus on the sloshing behavior of variable-mass liquid inside the floating platform.

References

1. Martins, M.R., Pestana, M.A., Souza, G.F.M.d., Schleder, A.M.: Quantitative risk analysis of loading and offloading liquefied natural gas (lng) on a floating storage and regasification unit (fsru). J. Loss Prevent. Process Ind. **43**, 629–653 (2016)
2. Xue, M.A., Cao, Z., Yuan, X., Zheng, J., Lin, P.: A two-dimensional semi-analytic solution on two-layered liquid sloshing in a rectangular tank with a horizontal elastic baffle. Phys. Fluids **35**(6) (2023)
3. Gándara, T., Barrio, E.C.D., Cruchaga, M., Baiges, J.: Experimental and numerical modeling of a sloshing problem in a stepped based rectangular tank. Phys. Fluids **33**(3) (2021)
4. Qiu, Y., Bai, M., Liu, Y., Lei, G., Liu, Z.: Effect of liquid filling level on sloshing hydrodynamic characteristic under the first natural frequency. J. Energy Storage **55**, 105452 (2022)
5. Faltinsen, O.M., Timokha, A.N.: A multimodal method for liquid sloshing in a two-dimensional circular tank. J. Fluid Mech. **665**, 457–479 (2010)
6. Jiang, Z., Shi, Z., Jiang, H., Huang, Z., Huang, L.: Investigation of the load and flow characteristics of variable mass forced sloshing. Phys. Fluids **35**(3) (2023)

Open Access This chapter is licensed under the terms of the Creative Commons Attribution-NonCommercial-NoDerivatives 4.0 International License (http://creativecommons.org/licenses/by-nc-nd/4.0/), which permits any noncommercial use, sharing, distribution and reproduction in any medium or format, as long as you give appropriate credit to the original author(s) and the source, provide a link to the Creative Commons license and indicate if you modified the licensed material. You do not have permission under this license to share adapted material derived from this chapter or parts of it.

The images or other third party material in this chapter are included in the chapter's Creative Commons license, unless indicated otherwise in a credit line to the material. If material is not included in the chapter's Creative Commons license and your intended use is not permitted by statutory regulation or exceeds the permitted use, you will need to obtain permission directly from the copyright holder.

Numerical Study of the Reeling Installation on China's First Reel-Lay Vessel

Aijin Xu[1], Junhan Yan[2], Weiping Xu[1], Lin Yuan[2(✉)], and Yan Chen[3]

[1] Shanghai Salvage Company, Ministry of Transport, Shanghai 200090, China
[2] State Key Laboratory of Hydraulic Engineering Intelligent Construction and Operation, Tianjin University, Tianjin 300350, China
lin_yuan@tju.edu.cn
[3] Tianjin Pipe Corporation, Tianjin 300300, China

Abstract. During the reel-lay installation, the pipeline undergoes multiple cycles of plastic bending, resulting in a highly complex stress state due to interactions with the reel, aligner, and straighteners. Accurate prediction and control of the pipeline's structural response are crucial to preventing local buckling and instability. In this study, uniaxial tensile and cyclic loading tests were conducted on materials used in reel-lay installation to determine their parameters. The material response was then fitted using both the isotropic hardening model and the Chaboche model. Based on the actual parameters of China's first reel-lay vessel, "Shenda," a full-process finite element model was developed in ABAQUS to simulate the pipeline's winding, unwinding, and straightening stages. Time-history curves of key parameters, such as curvature and axial strain, were obtained, and the pipeline's deformation characteristics were analyzed throughout the process. The effects of the isotropic and Chaboche models on the pipeline's bending moment, stress, and accumulative plastic strain were also investigated. The results indicate that the Chaboche model generally yields lower bending moment, stress, and plastic strain throughout the process, while the isotropic model produces higher values. Designing pipeline parameters based on the isotropic model may lead to overly conservative outcomes.

Keywords: Subsea pipeline · Reel-lay · Cyclic loading · Plastic deformation · Finite element

1 Introduction

Subsea pipelines are critical infrastructures for offshore oil and gas transportation. Among existing installation methods, the Reel-lay technique has gained prominence in deepwater projects due to its efficiency and adaptability. However, the pipeline undergoes multiple plastic bending cycles during reeling, unreeling, and straightening (Fig. 1), leading to complex stress states and potential local buckling [1]. Accurate prediction of structural responses is thus essential to ensure pipeline integrity while avoiding over-conservative designs.

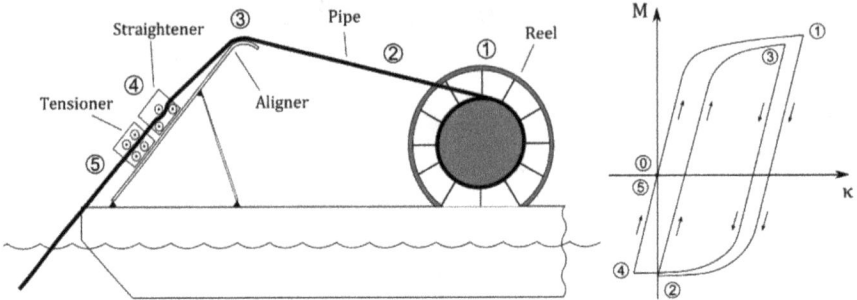

Fig. 1. Schematic diagram of the reel-lay process and pipeline cyclic bending loading [11].

Many scholars have conducted studies on the critical issues related to Reel-lay installation. Martinez and Brown [2] investigated the impact of Reel-lay installation on pipeline geometry and mechanical performance through experiments and numerical simulations. Meissner et al. [3] conducted full-scale and small-scale reeling installation experiments, evaluating the impact of repeated plastic bending on the strength of pipeline materials. Manouchehri [4] discussed the reeling mechanisms involved, from the manufacturing of the reel drum base to the reeling process, offshore operations, unspooling, and subsea installation. Yuan and Kyriakides [5–7] uncovered the evolution of local wrinkling and liner collapse under pure bending and also examined the influence of various parameters on liner instability. Roy et al. [8] utilized finite element analysis to adjust and optimize the straightening settings of reel-lay vessels, thereby implementing the residual curvature method in reel-lay installation, which was applied in real projects. Chatzopoulou et al. [9] studied the effect of cyclic loading during reeling on the ovalization of pipeline cross-sections, material anisotropy, and its impact on external pressure and pressurized bending performance. Yuan and Kyriakides [10] developed numerical models to analyze the deformation history of lined pipes and the causes of liner wrinkling. They conducted a series of sensitivity analyses to explore methods for delaying liner buckling.

Despite these advancements, the isotropic hardening model remains widely adopted in engineering practice, potentially overestimating bending moments and residual stresses due to neglecting cyclic hardening effects. In contrast, the Chaboche model, incorporating nonlinear kinematic hardening rules, offers a more accurate representation of Bauschinger effects under cyclic loading. Nevertheless, its applicability in full-scale Reel-lay simulations, particularly for China's first reel-lay vessel "Shenda," (Fig. 2) has not been systematically investigated.

This paper studied the cyclic plastic bending behavior of pipelines during the reeling and unreeling procedure based on the "Shenda" reel-lay vessel. The variation of local curvature and axial strain responses were closely captured. In addition, the effects of isotropic and Chaboche hardening models on the pipeline's bending moment, stress, and accumulative plastic strain were also investigated.

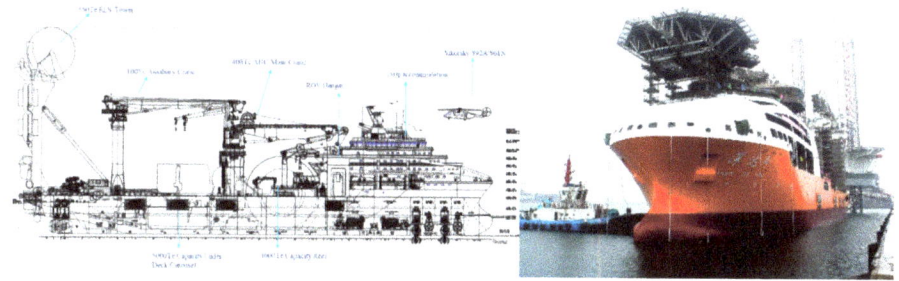

Fig. 2. China's first reel-lay vessel, "Shenda".

2 Material Test

2.1 Uniaxial Tensile Test

In this work, X65 is selected for the first sea trial of pipeline installation due to its exceptional strength and durability. The results of the chemical composition analysis of the pipeline material are shown in Table 1, highlighting the specific elements and their concentrations that contribute to its mechanical properties.

Table 1. Results of the chemical composition analysis of the pipeline material (unit: %).

C	Si	Mn	P	S	Ni	Cr	Mo	Cu	Al	Ca	V
0.09	0.28	1.30	0.008	0.001	0.03	0.11	0.01	0.06	0.022	3E-4	0.05
Nb	Ti	B	Pb	Sn	As	Sb	Bi	N	H	CEpcm	
0.026	<0.01	<5E-4	<0.001	0.0044	0.0048	0.0021	<0.001	0.0054	1E-5	0.18	

Longitudinal specimens were cut from the pipe for uniaxial tensile tests, as shown in Fig. 3. L_t represents the total length of the specimen, L_c the parallel length, L_o the original gauge length, b_o the original width of the parallel length, B the width of the gripping section, R the transition arc radius, and t the original wall thickness. The specific parameters are shown in Table 2.

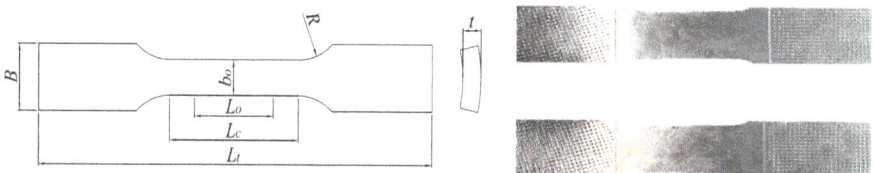

Fig. 3. X65 uniaxial tensile specimen.

The stress-strain curves obtained from the two sets of tensile tests are shown in Fig. 4. These curves exhibit a distinct plastic plateau, which is characteristic of low-carbon steels

Table 2. Tensile specimen parameters.

Specimen No.	L_t (mm)	L_c (mm)	L_o (mm)	b_o (mm)	B (mm)	R (mm)	t (mm)
1	219.10	70.00	50.00	38.20	50.00	25.40	15.89
2	219.10	70.00	50.00	38.20	50.00	25.40	14.85

like X65. Additionally, the curves showcase a well-defined yield point and demonstrate large plastic strain ability, which is essential for applications involving reeling. The ultimate tensile strength is also evident, reflecting the maximum stress the material can withstand before failure. Table 3 presents the detailed parameters, including elastic modulus (E), yield strength ($\sigma_{0.5}$), ultimate tensile strength (σ_u), and total extension at maximum force (A_{gt}).

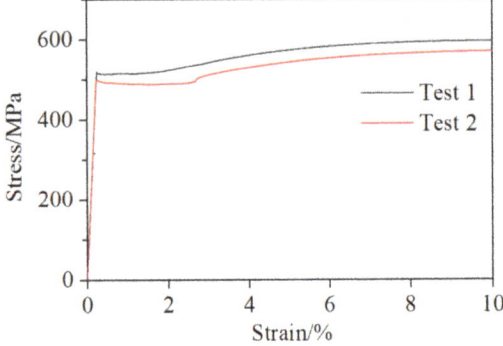

Fig. 4. Stress-strain curves from uniaxial tensile test.

Table 3. Critical material parameters.

Specimen No.	E (GPa)	$\sigma_{0.5}$ (MPa)	σ_u (MPa)	A_{gt} (%)
1	208.8	513.9	595.0	10.3
2	211.6	492.4	570.3	11.5

2.2 Cyclic Loading Test

Rod specimens were cut from the pipe material for cyclic loading tests, as shown in Fig. 5. As shown in Fig. 6, the stress-strain curve exhibits linear elastic behavior, followed by a well-defined yield plateau. When the tensile strain reaches 3%, the specimen begins to unload and is then subjected to compression up to 3%. As cyclic loading progresses, a pronounced hysteresis loop develops, reflecting cyclic plasticity and material hardening. The stress-strain curve exhibits some difference between the maximum tensile and compressive stresses, with the compressive stress reaching a higher magnitude than the tensile stress. Additionally, the hardening behavior is more pronounced in the compressive segment, where the curve shows a greater deviation compared to the tensile side. The characteristics mentioned above are crucial for accurately modeling its behavior in engineering applications.

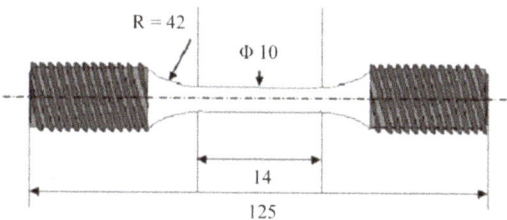

Fig. 5. X65 uniaxial cyclic loading specimen (unit: mm).

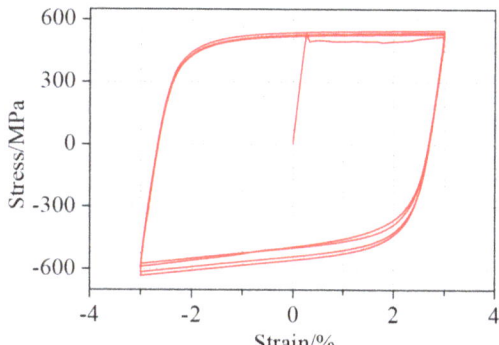

Fig. 6. X65 uniaxial cyclic loading test curve.

3 Finite Element Analysis of Reel-Lay Installation

This chapter begins with a discussion of the material constitutive models used in finite element simulations, focusing on the distinctions and advantages of the Chaboche model compared to the isotropic hardening model. Subsequently, a finite element model was established based on the actual parameters of "Shenda" to simulate the stress, strain, and bending moment responses of the pipeline. Since the pipeline undergoes repeated plastic bending during the reel-lay installation process, the Chaboche model incorporates a kinematic hardening term to simulate the Bauschinger effect, thereby providing a more accurate description of the material's mechanical behavior.

3.1 Chaboche Model

The Chaboche model is a widely used nonlinear elastoplastic constitutive model in engineering, designed to describe the complex behavior of materials under cyclic loading. The Chaboche model is based on a nonlinear kinematic hardening rule and can accurately simulate the material's hysteresis curves and stress-strain behavior by introducing multiple kinematic hardening terms. In contrast, the isotropic hardening model typically considers only the hardening process of materials and cannot capture these complex behaviors. Therefore, in scenarios such as reel-lay installation involving repeated bending, the Chaboche model, which accurately describes the nonlinear cyclic behavior of materials, is more suitable than the isotropic hardening model.

The Von Mises yield criterion is employed, which can be expressed as:

$$F = \sqrt{\frac{3}{2}(s-\alpha):(s-\alpha)} - \sigma_0 = 0 \tag{1}$$

where F represents the yield function, s is the deviatoric stress tensor, α is the back stress tensor, and σ_0 is the initial yield stress.

The associated flow rule adopted by the Chaboche model can be expressed as:

$$\frac{\partial Q}{\partial \sigma} = \frac{\partial F}{\partial \sigma} = \frac{3}{2}\frac{s-\alpha}{\sigma_e} \tag{2}$$

where Q is the plastic potential function, σ is the stress tensor, and σ_e is the effective stress.

The Chaboche kinematic hardening rule [12] can be expressed as:

$$\alpha = \sum_{1}^{3} \alpha_i \tag{3}$$

$$d\alpha_i = \frac{2}{3}C_i d\varepsilon_p - \gamma_i \alpha_i dp \tag{4}$$

where C_i and γ_i are material parameters, $d\varepsilon_p$ is the increment of plastic strain, and dp is the accumulative plastic strain rate.

Figure 7 illustrates the stress-plastic strain curve after the superposition of three back stress components. The first back stress curve α_1 exhibits a very high plastic modulus and quickly stabilizes, indicating that C_1 and γ_1 are large. The second back stress curve α_2 describes the transient nonlinear portion of the response. The third back stress curve α_3 represents the linear hardening behavior at larger strains.

The Chaboche model parameters obtained by fitting the experimental curve in Fig. 6 from Sect. 2.2 are shown in Table 4.

3.2 Methodology for Determining Critical Parameters

During the reeling process, the pipeline is subjected to a combination of bending moment and axial load, posing a risk of local buckling. For reel-lay installation, before submerging the pipeline, environmental loads and accidental loads are not considered, and there is no

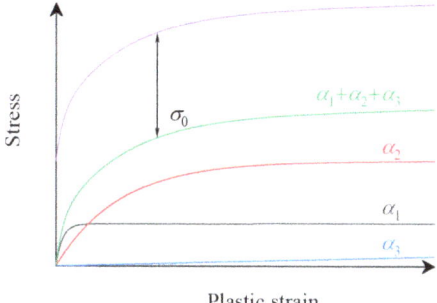

Fig. 7. Superposition of three back stresses of Chaboche model.

Table 4. Chaboche model fitting parameters.

σ_0 (MPa)	C_1 (MPa)	γ_1	C_2 (MPa)	γ_2	C_3 (MPa)	γ_3	Q_∞ (MPa)	b
465	2000	1000	16,000	300	1400	0.01	10	10

pressure difference inside the pipeline. According to DNV standards, the displacement control criterion for local buckling of the pipeline under the combined effect of bending moment and axial force is as follows [13]:

$$\varepsilon_{Sd} = \varepsilon_F \cdot \gamma_F \cdot \gamma_c \leq \varepsilon_{Rd} = \frac{\varepsilon_c}{\gamma_\varepsilon} \quad (5)$$

$$\varepsilon_c = 0.78 \left(\frac{t}{D} - 0.01 \right) \alpha_h^{-1.5} \alpha_{gw} \quad (6)$$

where ε_{Sd} is the design compressive strain, ε_F is the functional strain, γ_F is the load effect factor for functional loads, γ_c is the condition load effect factor, ε_{Rd} is the design compressive strain resistance, γ_ε is the resistance factor, ε_c is the characteristic bending strain resistance, α_h is the hardening factor, α_{gw} is the girth weld factor.

The pipeline is considered reelable if its wall thickness exceeds the minimum thickness required to prevent buckling on the reel. The minimum reelable wall thickness refers to the thickness at which the buckling strain equals the nominal bending strain. By transforming Eqs. (5) and (6), the minimum wall thickness can be derived from Eqs. (7) and (8) [14]:

$$t \geq D \cdot \left(\varepsilon_F \cdot \frac{\gamma_F \cdot \gamma_c \cdot \gamma_\varepsilon}{0.78 \cdot \alpha_h^{-1.5} \cdot \alpha_{gw}} + 0.01 \right) \quad (7)$$

$$\varepsilon_F = \frac{D}{2R_{reel} + D} \quad (8)$$

where t is the wall thickness, D is the pipeline diameter, and R_{reel} is the reel radius. Eq. (8) represents the bending strain of the pipeline during the reel-laying process under pure displacement control.

To prevent the pipeline from slack on the reel, DNV provides a minimum back tension based on industry experience:

$$T \geq 1.5 \cdot \frac{f_y \cdot (D-t)^2 \cdot t}{R_{reel}} \qquad (9)$$

where T is the back tension during reeling, and f_y is the yield stress.

3.3 Finite Element Model

This study utilizes ABAQUS to establish a finite element model to simulate the reeling, unreeling, and straightening process (Fig. 8). The model assumes that the sectional properties and material characteristics of the pipeline are uniformly distributed along its length, neglecting discontinuities such as welds.

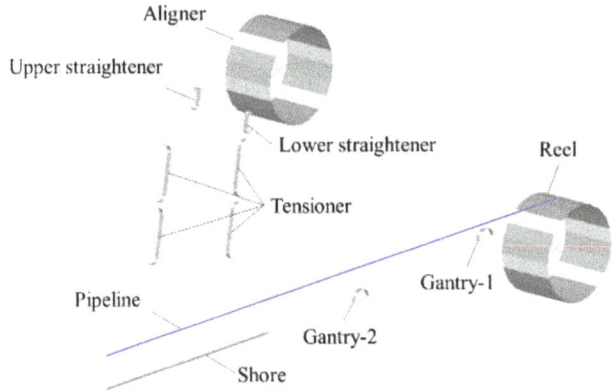

Fig. 8. Finite element model layout.

The shore, gantries, reel, aligner, straighteners, and tensioners are all modeled using analytical rigid surfaces. The pipeline is modeled using beam elements (PIPE31). The contact properties between the pipeline and rigid surfaces are defined as soft contact in the normal direction and penalty friction in the tangential direction, with a friction coefficient of 0.01. Due to the significant curvature changes that the pipeline experiences throughout the process, mesh size control is necessary. After convergence analysis, a final mesh size of 0.5 m was adopted.

The analysis process of the model consists of eight steps and can be divided into the following stages:

0: At the initial moment, the pipeline remains parallel to the ground, and its right end is coupled to the reel

0–2: Back-tension is applied to the left end of the pipeline, and gravity is applied to the entire pipeline, causing it to eventually fall onto the shore and gantries.

2–4: Rotation is applied to the reel, driving the pipeline to reel on.

4–6: The left end of the pipeline moves towards the aligner, then bypasses it and eventually reaches the vicinity of the predetermined position near the upper straightener.

6–7: The straighteners and tensioners move into contact with the pipeline and reach their predetermined positions.

7–8: The pipeline end passes sequentially through the upper straightener, lower straightener, and tensioner, straightening the subsequent pipeline. The pipeline's displacement is set to 40 m. Figure 9 shows the illustration of the completed finite element simulation.

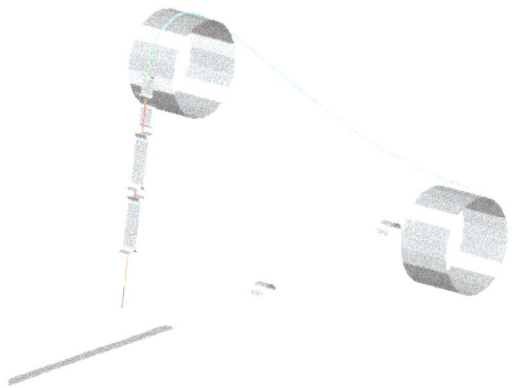

Fig. 9. Model view after the unreeling process.

4 Results and Discussion

In the base case, the pipeline has a total length of 160 m, an outer diameter of 323.9 mm (12″), and a wall thickness of 15.9 mm, which meets the requirements of Eq. (7). The reel has a diameter of 18 m, and the back-tension is set at 129.0 kN, complying with Eq. (9). The steel grade is X65, with a density of 7850 kg/m^3. A series of data was extracted from a node located 20 m from the left end of the pipeline, as this section experienced the entire process of reeling, unreeling, and straightening.

Figure 10 shows the curvature-time history curve of the isotropic model, with the different stages of the pipeline process marked (corresponding to the time settings in Sect. 3.3). When passing over the gantries, although there were some fluctuations in curvature, the pipeline remained in the elastic deformation stage and did not experience plastic bending. The pipeline underwent a total of five cycles of plastic bending: reel-on, reel-off, on to aligner, on to upper straightener, and on to the lower straightener. Reel-off and on to upper straightener involve reverse bending, compared to reel-on, on to aligner, and on to the lower straightener. After straightening, the pipeline's curvature stabilized at a relatively low level, approximately 0.004 m^{-1}.

Figure 11 shows the time-history curves of tensile strain at the top (12 o'clock position) and compressive strain at the bottom (6 o'clock position) of the isotropic pipe material. The tensile strain time-history curve at the top exhibits good consistency with the trend of the curvature strain curve. The maximum strain during the reeling process occurs immediately after reeling is completed, with a tensile strain of approximately

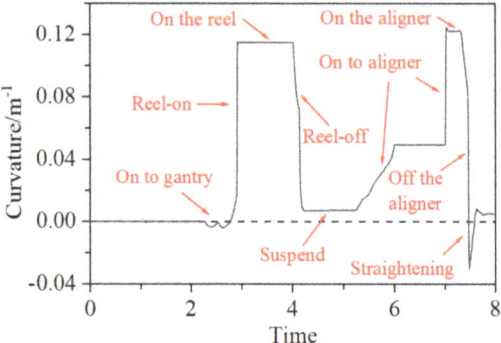

Fig. 10. Curvature time history curve.

1.79% and a compressive strain of about 1.76%. The pipe's configuration at this point is shown in Fig. 12(a). Although the tensile strain is slightly larger than the compressive strain due to the axial back tension, the difference between the two is minimal. As reeling continues, the pipe rotates as a whole with the reel, causing a slight reduction in axial strain, which then stabilizes at around 1.78%, closely matching the theoretical strain (1.77%) predicted by Eq. (8). During the unreeling process, the axial strain of the pipe decreases to a lower level and remains stable in the suspended section. Subsequently, when the pipe passes through the aligner, the axial strain increases rapidly again, with the tensile strain reaching a maximum of about 1.95% and the compressive strain at 1.89%. The pipe's configuration is shown in Fig. 12(b). Notably, the horizontal curve during time 6–7 indicates that this analysis step is for the straightener to fit against the pipe, and the pipe itself does not move. Subsequently, when the pipe is located on the aligner, the strain level remains stable. As the pipe continues to move away from the aligner and enters the upper straightener, the strain curve shows a noticeable decline, indicating that the pipe experiences significant reverse bending. The pipe then passes through the lower straightener and undergoes reverse bending again, with the axial strain ultimately reaching a lower level, with tensile strain at about 0.12% and compressive strain at about 0.02%.

Figure 13 shows the moment-time history curves for the isotropic model and the Chaboche model throughout the entire process. The trends are generally consistent with the curvature time history curve and the axial strain time history curve. It can be observed that, except for the stage when the pipe is on the reel, the moment levels of the Chaboche model are lower than those of the isotropic model at all other times. After the straightening is completed, the residual moment of the Chaboche model is smaller, being approximately 33% lower than that of the isotropic model.

Figure 14(a) shows the stress time history curves for the isotropic model and the Chaboche model throughout the entire process. During the stable segments on the reel, the stress of the Chaboche model is relatively high. In the suspended section, the stress level of the isotropic model is higher. To better illustrate the stress changes during the straightening process, the enlarged view in Fig. 14(b) shows the stress time history curve specifically for the period of 6.0–8.0. When the pipe is on the aligner, on to the upper straightener, and on to the lower straightener, the stress of the Chaboche pipe material is

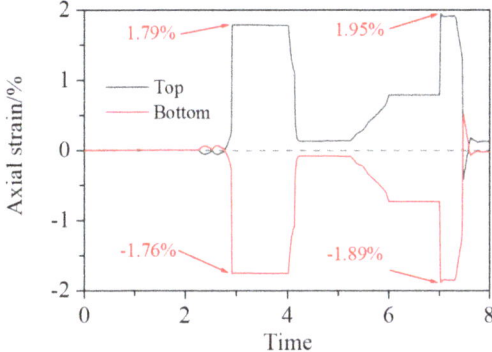

Fig. 11. Axial strain time history curve.

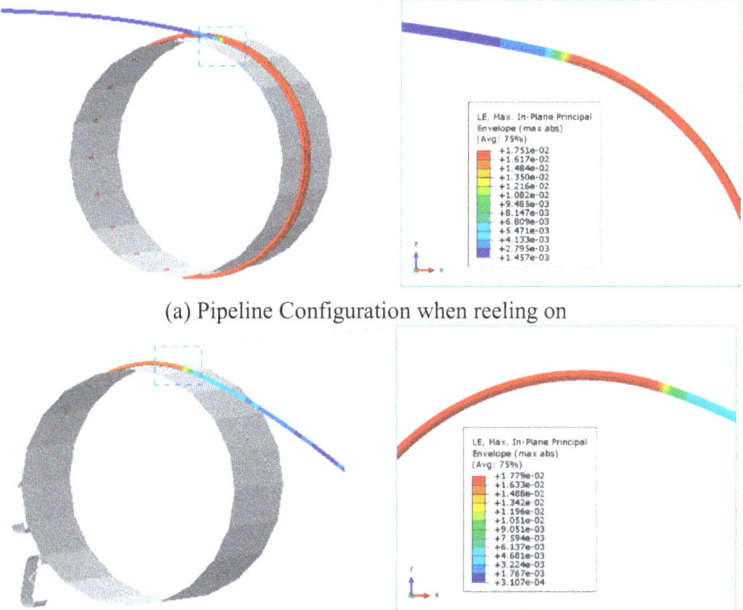

(a) Pipeline Configuration when reeling on

(b) Pipeline configuration when passing the aligner

Fig. 12. Pipeline Configuration at different time.

lower than that of the isotropic model. After the straightening is completed, the residual stress of the Chaboche model is approximately 333 MPa, which is about 31% lower than that of the isotropic model. This discrepancy arises from the Chaboche model's ability to capture the Bauschinger effect and nonlinear kinematic hardening, which are critical under cyclic loading conditions.

Figure 15 shows the accumulative plastic strain time history curves for the isotropic model and the Chaboche model. Since plastic strain occurs during significant bending of the pipe, its growth rate is rapid, resulting in a stair-step appearance of the curve,

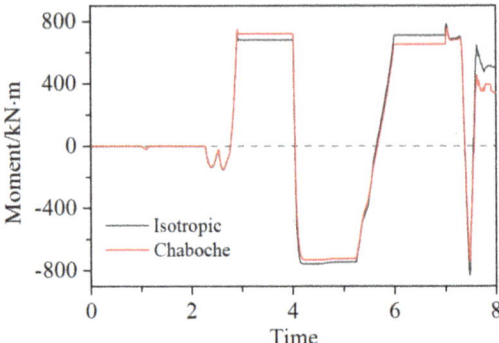

Fig. 13. Moment time history curves.

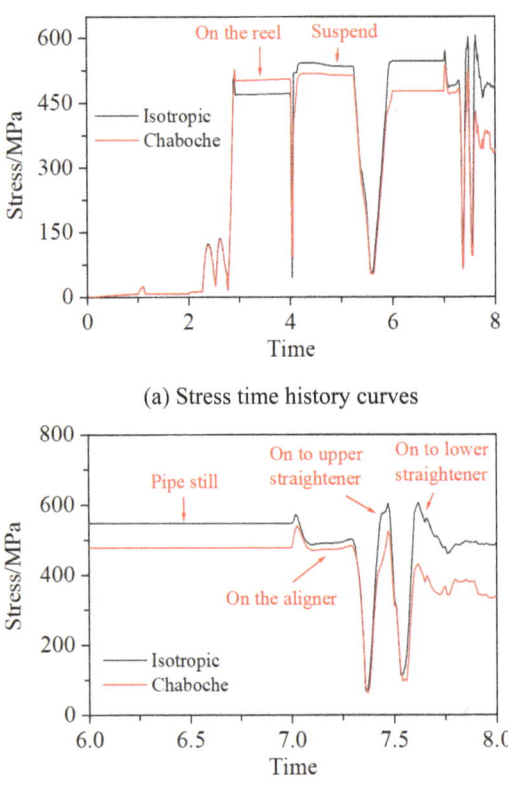

(a) Stress time history curves

(b) Enlarged curves for the time of 6.0-8.0

Fig. 14. Stress time history curves and local magnification. (a) Stress time history curves. (b) Enlarged curves for the time of 6.0–8.0.

which stabilizes after bending is completed. It can be seen from the figure that at each stage, the accumulative plastic strain of the isotropic model is higher than that of the

Chaboche model, with the difference reaching a maximum of approximately 0.37% after straightening is completed. The isotropic model's overestimation of plastic strain accumulation may lead to conservative design decisions, such as unnecessary wall thickness increments.

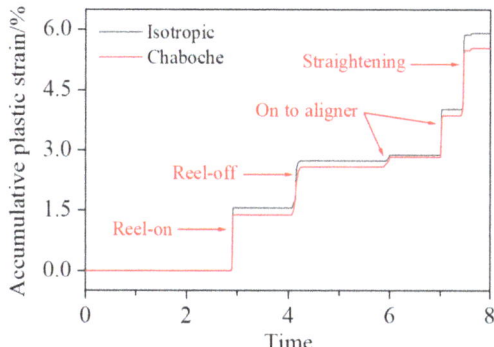

Fig. 15. Accumulative plastic strain time history curves.

5 Conclusion

This study initially conducted uniaxial tensile and cyclic loading tests on the pipe material to obtain key material parameters. Next, based on the actual parameters of China's first reel-lay vessel "Shenda," a finite element model was developed to simulate the entire pipe reeling and laying process, focusing on comparing the performance of the isotropic hardening and Chaboche models. The main conclusions are as follows:

The Chaboche model outperforms the isotropic hardening model in predicting pipeline behavior under cyclic plastic bending, particularly in capturing residual stress and moment reduction (31% and 33%, respectively). Using isotropic hardening models may lead to overly conservative designs, emphasizing the necessity of adopting advanced constitutive models for reel-lay engineering.

Future work should focus on parametric sensitivity analysis of the Chaboche model to examine its adaptability to various working conditions and material properties.

Acknowledgments. This work was supported by the National Key Research and Development Program of China "Key Deep-sea Technology and Equipment" Key Project "Physiological Research on Safety Labor Intensity of Deep Saturation Diving and Tour Diving Ability" (2018YFC0309500) and National Natural Science Foundation of China (Grant number 52271288). The support is acknowledged with thanks.

References

1. Xu, J., Vishnubhotla, S., Aamlid, O., Collberg, L.: Reeling Analysis and Limit State Criteria, Pipelines, Risers, and Subsea Systems, American Society of Mechanical Engineers, Busan, South Korea, 5, V005T04A044 (2016)
2. Martinez, M., Brown, G.: Evolution of Pipe Properties during Reel-Lay Process: Experimental Characterisation and Finite Element Modelling. In: Omae2005, 24th International Conference on Offshore Mechanics and Arctic Engineering, vol. 3, pp. 419–429 (2005)
3. Meissner, A., Erdelen-Peppler, M., Tanja S.: Impact of reel-laying on mechanical pipeline properties investigated by full- and small-scale reeling simulations. In: Paper presented at The Nineteenth International Offshore and Polar Engineering Conference, Osaka, Japan (2009)
4. Manouchehri, S.: A discussion of practical aspects of reeled flowline installation. In: Volume 3: Pipeline and Riser Technology, pp. 531–542. American Society of Mechanical Engineers, Rio de Janeiro, Brazil (2012)
5. Yuan, L., Kyriakides, S.: Wrinkling Failure of Lined Pipe under Bending, Volume 4B: Pipeline and Riser Technology, American Society of Mechanical Engineers, Nantes, France, p. V04BT04A029 (2013)
6. Yuan, L., Kyriakides, S.: Liner wrinkling and collapse of bi-material pipe under bending. Int. J. Solids Struct. **51**(3–4), 599–611 (2014)
7. Yuan, L., Kyriakides, S.: Plastic bifurcation buckling of lined pipe under bending. Eur. J. Mech.—ASolids. **47**, 288–297 (2014)
8. Roy, A., Rao, V., Charnaux, C., Ragupathy, P., Sriskandarajah, T.: Straightener settings for under-straight residual curvature of reel laid pipeline. In: Proceedings of the ASME 2014 33rd International Conference on Ocean, Offshore and Arctic Engineering, vol. 6B: Pipeline and Riser Technology. San Francisco, California, USA. June 8–13, V06BT04A046. ASME (2014). https://doi.org/10.1115/OMAE2014-24513
9. Chatzopoulou, G., Karamanos, S.A., Varelis, G.E.: Finite element analysis of cyclically-loaded steel pipes during deep water reeling installation. Ocean Eng. **124**, 113–124 (2016)
10. Yuan, L., Kyriakides, S.: Liner buckling during reeling of lined pipe. Int. J. Solids Struct. **185–186**, 1–13 (2020)
11. Gavriilidis, I., Karamanos, S.A.: Structural response of steel lined pipes under cyclic bending. Int. J. Solids Struct. **234**, 111245 (2022)
12. Chaboche, J.L.: Time-independent constitutive theories for cyclic plasticity. Int. J. plast. **2**(2), 149–188 (1986)
13. Submarine pipeline systems. Det Norske Veritas (2013)
14. Guideline for Installation of Rigid Pipeline—Rigid Limit States Rev 3. DNV GL (2015)

Open Access This chapter is licensed under the terms of the Creative Commons Attribution-NonCommercial-NoDerivatives 4.0 International License (http://creativecommons.org/licenses/by-nc-nd/4.0/), which permits any noncommercial use, sharing, distribution and reproduction in any medium or format, as long as you give appropriate credit to the original author(s) and the source, provide a link to the Creative Commons license and indicate if you modified the licensed material. You do not have permission under this license to share adapted material derived from this chapter or parts of it.

The images or other third party material in this chapter are included in the chapter's Creative Commons license, unless indicated otherwise in a credit line to the material. If material is not included in the chapter's Creative Commons license and your intended use is not permitted by statutory regulation or exceeds the permitted use, you will need to obtain permission directly from the copyright holder.

Study on the Influence of Tension in Deepwater REEL-Lay Process Based on Finite Element Method

Liwei Li[1(✉)], Zhibing Yu[2], Jing Hou[1], and Rundong Zhang[2]

[1] CNOOC Research Institute Ltd., Beijing 100010, China
weiweidaolai1986@126.com

[2] Offshore Oil Engineering CO., LTD., Tianjin 300450, People's Republic of China

Abstract. The reel-lay method, developed in the late twentieth century, is a novel pipe-laying technique with advantages such as high laying efficiency and excellent welding quality. This paper utilizes Abaqus's nonlinear processing capabilities to simulate the process of a pipe being reeled onto a drum and the deepwater laying process under hydrodynamic loads. The rotation of the drum is simulated by constraining its rotational displacement, while the software's contact analysis feature is employed to model the interaction between the pipe and the drum. A parametric analysis of the effect of tension on the plastic deformation of the pipe is conducted. Results show that during the increase of the reeling traction, the main strain of the pipe is concentrated on the drum and remains stable. However, when the traction force becomes too large, the axial strain of the unreeled portion of the pipeline increases. Appropriate laying tension can effectively reduce the interference of deepwater wave and current loads on the laying curvature. The positions of the straightener and calibrator need to be adjusted according to the actual conditions, laying angle, and tension.

Keywords: REEL-LAY · Laying Tension · Plastic Strain

1 Introduction

Reel-lay is a technique commonly used in subsea pipeline construction, where pipe materials are pre-wound into reels onshore or onboard, and then deployed to the seabed at the target location using specialized equipment, as shown in Fig. 1. Compared to traditional S-lay and J-lay methods [1], reel-lay offers higher efficiency and economic benefits. In recent years, R-lay technology has been widely applied in marine engineering, but many challenges and unresolved issues remain. Reel-lay technology is limited by factors such as seabed topography, water depth, and sea conditions in practical applications, imposing higher requirements on equipment and processes. Additionally, with the continued exploration of marine resources, the demand for pipeline installation in deeper waters has increased. Compared to shallow waters, the impact of hydrodynamic loads on pipelines is more pronounced, imposing greater challenges for the reel-lay technique.

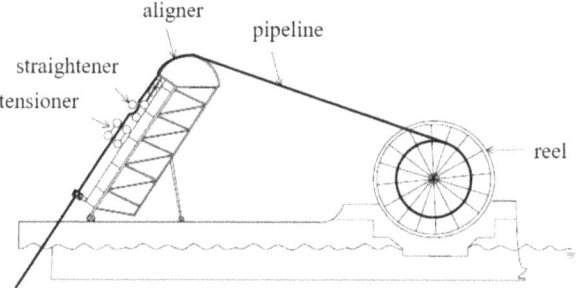

Fig. 1. Schematic of REEL-LAY Method.

Table 1. Pipeline Property Parameters.

Material Grade	ASTM A694 F65
Outer Diameter	323.9 mm
Wall Thickness	27 mm
Material Standard	API 5 L PSL2 X65
Manufacturing Type	SMLS
Steel Density	7850 kg/m^3
Young's Modulus	207,000 MPa
Poisson's Ratio	0.3
Minimum Yield Strength	450 MPa
Ultimate Tensile Strength	530 MPa

This paper primarily focuses on the stress variation during the reeling process of steel pipes and studies the impact of tension. A finite element simulation is also conducted to investigate the effect of laying tension on the pipeline installation process under deepwater conditions.

2 Research Status

Since the 1970s, reel-lay has been applied in countries like the United States, Brazil, and Europe. The DNV-OS-F101 standard [2] designs the pipeline laying system based on experience but does not consider the residual stress of the pipeline before immersion and its impact on subsequent laying. Szczotka M [3] conducted a dynamic simulation of the reeling and unreeling processes, accounting for the passage through tensioners and other equipment, and obtained the plastic deformation characteristics of the pipeline. In 2001, Daly et al. [4] carried out a global finite element analysis of pipe-in-pipe installations, simulating the reeling process and the static configuration of the submerged portion after installation, studying the internal force distribution and stress of the subsea pipeline. M. Martinez et al. [5] performed full-scale reeling and straightening experiments, comparing the results with corresponding numerical simulations to investigate the impact of cyclic reeling and straightening operations on the pipe's ovality. Liu [6] found through finite element analysis that the discontinuity in wall thickness and yield stress leads to abrupt local changes in curvature, and the degree of disturbance is determined by the ratio of the pipe to drum diameter and the bending strain applied. Ruan et al. [7], based on finite element analysis, investigated the coupling effect between flexible pipes and drums during reeling, discovering that friction has a significant impact on pipe stability.

Bai Yong et al. [8] used theoretical analysis and finite element methods to study the stability of flexible pipes under different conditions, loads, and environmental factors during reel-lay installation. Zhang Jiuju et al. [9] used theoretical analysis and finite element methods to identify the main parameters affecting the ovality of marine pipelines and their variation Patterns. Yu Yang et al. [10], based on elastoplastic theory, derived analytical solutions for the residual axial stress during unreeling. Wang Yanhui et al. [11] numerically simulated the dynamic process of pipe reeling and unreeling, obtaining the history of axial strain, and curvature changes.

3 Numerical Calculation Model

For the calculation analysis, a 12″ pipeline was selected. The numerical simulation was conducted using an elasto-plastic constitutive model.

The analysis model primarily includes two components: the pipe and the reel. The vessel Shen Da Hao was used to represent the reel-lay installation ship, with a drum diameter of 18 m and a total pipe length of 2400 m. The reel was modeled as a rigid analytical body, and the pipeline was simulated using B31 beam elements with a mesh size of 1 meter. The reference point of the analytical rigid body model was coupled to the start of the reel, and frictionless contact was established between the pipe and the reel. A steady-state analysis step was employed for the simulation.In addition to the pipe and reel, a calibrator and two straighteners (upper and lower) were incorporated into the reel-lay installation model. The tension provided by the tensioner was simulated by applying tensile force to the free end of the pipeline.

Considering the deepwater installation conditions, the water depth was set to 1500 m, and wave and current loads were defined using the AQUA solver. The environmental parameters are provided in Table 2. The hydrodynamic characteristics of the pipeline were defined using the CLOAD keyword, and the drag coefficient, added mass coefficient, inertia coefficient, and lift coefficient are listed in Table 2.

Table 2. Environmental and Hydrodynamic Parameters.

Significant Wave Height Hs	8.3 m
Significant Period Ts	10.7 s
Surface Current Velocity	100.5 cm/s

(*continued*)

Table 2. (continued)

Significant Wave Height Hs	8.3 m
Mid-layer Current Velocity	90.5 cm/s
Bottom Current Velocity	66.4 cm/s
Drag Coefficient Cd	1.2
Added Mass Coefficient CA	1.0
Inertia Coefficient CM	2.0
Lift Coefficient Cf	0.7

4 Results and Analysis

4.1 Reel-in Analysis

During the reeling process, initial traction is applied to ensure smooth reeling of the pipe. The motion and strain of the pipe vary under different initial traction forces. To investigate the effect of initial traction on the reeling process, three initial traction forces were set: 100 kN, 300 kN, and 500 kN.

Figure 2 shows the stress distribution when the pipe is fully reeled. It can be observed that the high-stress regions are primarily concentrated on the portion of the pipe in contact with the reel, with relatively uniform stress distribution. The peak stress reaches 464 MPa, indicating some plastic deformation in the pipe.

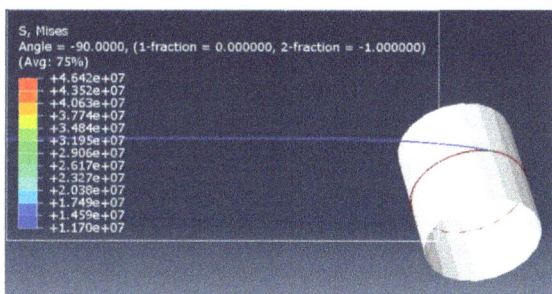

Fig. 2. Stress distribution during reeling.

Figure 3 illustrates the configuration of the pipe during reeling under different initial traction forces. It shows that the pipe undergoes vertical displacement during reeling, and after a certain displacement, the pipe configuration stabilizes. Under fixed traction conditions, the reeling process becomes steady. Comparing the conditions at 0.5 rad of reel rotation, it is observed that when the initial traction is 100 kN, the pipe still tends to experience vertical displacement.

Figure 3(d) presents a comparison of the pipe configurations at 8 rad of reel rotation under different initial traction forces. It can be seen that under low traction forces, the

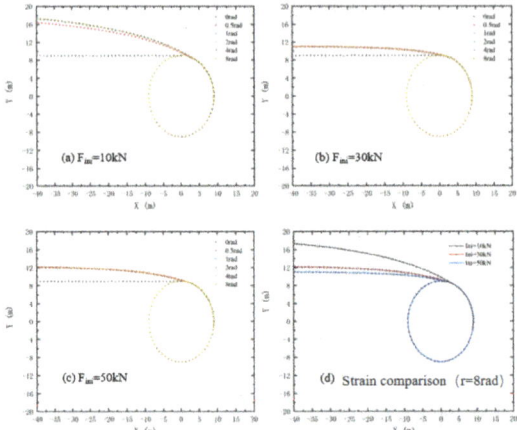

Fig. 3. Comparison of pipe configurations during reeling under different initial tractions.

pipe exhibits larger vertical displacements. As the traction increases, the pipe maintains the reeling curvature more effectively. However, when the initial traction reaches 300 kN and 500 kN, there is little difference in the pipe configuration, indicating that once the initial traction exceeds a certain threshold, the pipe configuration becomes stable. Excessive traction leads to the pipe bearing more axial loads, which could negatively affect the reeling performance.

Figure 4 shows the strain distribution at different locations along the pipe under varying initial traction forces. At L = 0 m, the pipe is initially coupled to the reel. It can be observed that the strain variation during reeling under different traction forces follows a similar trend. The strain of the pipe already wound on the reel fluctuates with the reel's movement but remains relatively stable, around 0.017, indicating that the coupling between the pipe and the reel plays a dominant role in determining the strain during stable reeling. As shown in Fig. 4(d), under smaller traction forces, the strain in the unreeled section is higher. In contrast, the strain in the reeled section shows the opposite trend.

4.2 Pipe Unreeling Analysis

REEL-Lay Procedure. To investigate the effect of tension from the tensioner during the unreeling process, this section analyzes three different unreeling tensions: 120 kN, 240 kN, and 360 kN. Figure 5 shows the strain curve of a pipeline element at L = 100 m from the reel to the seabed during the entire process. The curve reflects the following stages

Curve O-A: The pipeline begins to reel in slowly under the initial pulling force.

Curve A-B: The pipeline element continues to reel and remains stable on the reel.

Point B: Gravity and flow loads are introduced into the calculation model, causing a slight increase in the pipeline element's strain.

Curve B-C: The pipeline is on the reel, not yet being laid, with axial strain around 0.017.

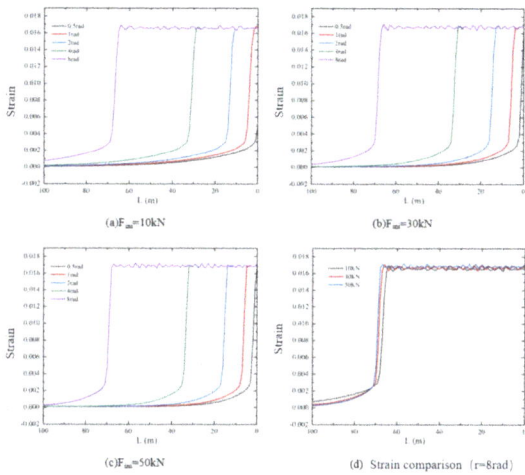

Fig. 4. Comparison of pipe strains under different initial traction forces.

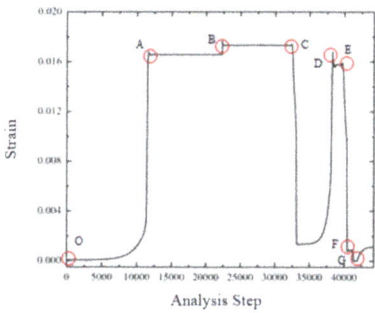

Fig. 5. Strain variation curve during the pipe laying process.

Curve C-D: The unreeling process begins, and as the pipeline gradually unwinds from the reel, the axial strain drops sharply to near zero and remains constant for a period before increasing sharply near the aligner.

Curve D-E: The pipeline enters the aligner and bends over it. As the pipeline moves entirely into the aligner, the strain fluctuates slightly but generally remains around 0.016, indicating some dependence on the aligner's position.

Curve E-F: The pipeline moves from the aligner to the upper straightener track. During this process, the pipeline shifts from a positive to a negative bend, causing the axial strain to drop rapidly.

Curve F-G: The pipeline moves from the upper straightener track to the lower one.

Laying Tension and Flow Loads. In deep-water environments, due to the length of the suspended section of the pipeline, the pipeline is significantly affected by flow loads underwater. Figure 6 compares the pipeline configuration under different laying tensions, with and without flow load. Under low tension (F = 120 kN), the pipeline is more influenced by the flow load. Due to the insufficient tension, the pipeline is poorly constrained at the seabed, leading to a reduced laying angle and increased curvature. When tension increases, the pipeline near the seabed becomes more constrained, while the part closer to the surface experiences lateral displacement due to incoming flow. Buoyancy causes the pipeline at the seabed to lift, reducing the bending curvature. However, when the laying tension is too high, the lifting effect is more pronounced. Therefore, an appropriate laying tension can effectively reduce the interference of deep-water wave and current conditions on the pipeline, optimizing the laying curvature

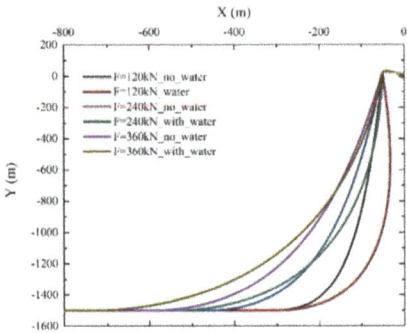

Fig. 6. Comparison of pipeline configurations under different laying tensions with and without flow load.

Analysis of Laying Tension Effects. Figure 7 compares the stress changes of pipeline elements at L = 100 m as they pass through the reel, aligner, and straightener under different tensions. It can be observed that during the process of unreeling from the reel to the aligner, the strain differences under various tensions are minimal. This is because, at this stage, the strain is still primarily governed by the coupling between the pipeline and the reel and aligner. However, as the pipeline enters the aligner and bends, noticeable strain differences emerge. It is worth noting that this change is not directly proportional to the tension. Excessive tension can cause stress concentration due to interference between the pipeline and the straightener, reducing the deformation at the rear of the aligner

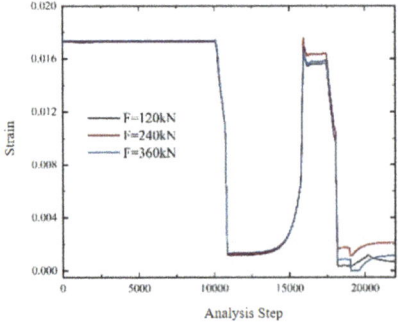

Fig. 7. Strain comparison of pipelines under different laying tensions.

Figure 8 shows the stress cloud diagram near the straightener when the reel has rotated 10 radians during the laying process. A comparison shows that as the tension increases, the coupling between the pipeline and the upper straightener intensifies, while the coupling with the lower straightener weakens, leading to stress concentration in the pipeline. In practical engineering, it is essential to adjust the position of the straightener according to different laying tensions to ensure optimal stress distribution along the pipeline.

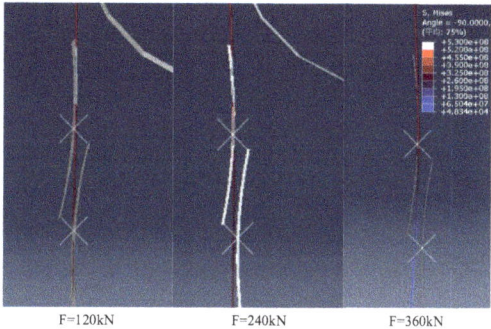

Fig. 8. Stress comparison near the straightener under different laying tensions.

5 Conclusions

The deep-water REEL-lay pipeline installation is a complex nonlinear process influenced by various factors during each stage of reeling and unreeling. Using Abaqus' nonlinear analysis capabilities, this study numerically simulates the pipeline reeling process and conducts a parametric analysis of the effects of initial pulling force and laying tension on pipeline plastic deformation, leading to the following conclusions:

(1) When the initial pulling force increases beyond a certain threshold, the pipeline's reeling configuration stabilizes. However, excessive pulling force causes the pipeline to bear more axial loads, negatively affecting reeling performance. For the unreeled pipeline, lower pulling force results in greater strain, while the opposite trend is observed for the reeled pipeline.
(2) Appropriate laying tension can effectively reduce the interference of deep-water wave and current conditions on the pipeline. Within a certain range, increased laying tension mitigates the impact of flow loads, but excessive tension may compromise pipeline stability.
(3) Laying tension affects the coupling strength between the pipeline and components such as the aligner and straightener. The selection of laying tension should be coordinated with the position of the aligner and straightener; otherwise, the pipeline elements may experience extreme strain values when passing through different components, potentially causing damage to the pipeline.

References

1. Guomin, S., Wang, H., Chunhong, H., et al.: Key equipment and technology development and prospects of deep-water REEL-lay installation [J]. Mar. Eng. Equip. Technol. **6**(6), 6 (2019)
2. DNV-OS-F101, Submarine Pipeline Systems. Det Norske Veritas, Norway, 2010
3. Szczotka, M.: Pipe laying simulation with an active reel drive[J]. Ocean Eng. **37**(7), 539–548 (2010)
4. Daly, R., Bell, M.: Reeled pipe in pipe steel catenary riser[C]. In: Proceedings of OMAE'01 20th International Conference on Offshore Mechanics and Arctic Engineering. Rio de Janeiro , Brazil, vol. 1, pp. 139–147 (2001)
5. Martinez, M., Brown, G.: Evolution of pipe properties during reel-lay process: Experimental characterization and finite element modelling[C], ASME 2005 24th International Conference on Offshore Mechanics and Arctic Engineering, pp. 419–429 (2005)
6. Liu, Y., Kyriakides, S.: Effect of Geometric and Material Discontinuities on the Reeling of Pipelines[C], Asme International Conference on Ocean. American Society of Mechanical Engineers (2015). https://doi.org/10.1115/OMAE2014-24474
7. Ruan, W., Qi, K., Sun, B., et al.: Multi-scale coupling numerical modeling of metallic strip flexible pipe during reel-lay process[J]. Ocean Eng., (Sep.1), 283 (2023)
8. Yong, B., Tang, J., Binbin, Y., et al.: Stability analysis of flexible pipe REEL-lay installation [J]. Mar. Eng. **31**(6), 22–29 (2013)
9. Jiuju, Z., Duan, M., Xianwei, H.: Curvature model and residual stress calculation of reel-lay pipeline installation [J]. Mar. Eng. **33**(3), 19–28 (2015)
10. Yang, Y., Liu, Z., Jianxing, Y., et al.: Sensitivity analysis of pipeline deformation in REEL-lay method based on ABAQUS [J]. Mar. Eng. **002**, 040 (2022)
11. Yanhui, W., Peng, X., Zhiqin, C., et al.: Time-domain dynamic analysis of REEL-lay installation process [J]. Ship Eng. **37**(7), 5 (2015) CNKI:SUN.0.2015-07-022

Open Access This chapter is licensed under the terms of the Creative Commons Attribution-NonCommercial-NoDerivatives 4.0 International License (http://creativecommons.org/licenses/by-nc-nd/4.0/), which permits any noncommercial use, sharing, distribution and reproduction in any medium or format, as long as you give appropriate credit to the original author(s) and the source, provide a link to the Creative Commons license and indicate if you modified the licensed material. You do not have permission under this license to share adapted material derived from this chapter or parts of it.

The images or other third party material in this chapter are included in the chapter's Creative Commons license, unless indicated otherwise in a credit line to the material. If material is not included in the chapter's Creative Commons license and your intended use is not permitted by statutory regulation or exceeds the permitted use, you will need to obtain permission directly from the copyright holder.

A New 6-DOF Maneuvering Model for Open Frame ROVs Considered the Flow Memory Effect

Ruinan Guo[1], Duanfeng Han[1], Yingfei Zan[1(✉)], Nan Sun[2], Binggang Yin[3], Fuxiang Huang[3], and Yaogang Sun[2]

[1] College of Shipbuilding Engineering, Harbin Engineering University, Harbin, China
zanyingfei@hrbeu.edu.cn
[2] Ship Standardization Research Center China Institute of Marine Technology and Economy, Beijing, China
[3] Offshore Oil Engineering Co. Ltd., Tianjin, China

Abstract. The trajectory of ROV is complex, making it difficult for the surrounding turbulence to develop into a stable state. The motion and turbulence memory have an unsteady impact on subsequent loads. However, existing ROV maneuvering models are based on steady assumptions and cannot express unsteady phenomena. A new 6 degrees of freedom (DOF) mathematical model of viscous hydrodynamic loads on the Remotely Operated Vehicle (ROV) is established to consider the flow memory effect. The new model builds the relationship between the unsteady viscous force and the velocity via response functions and convolutions. When considering the flow memory effect, the motion simulation results show that the hydrodynamic amplitude levels off after about 4 to 5 periods. The flow memory effect delays the phases of the unsteady roll moments. The flow memory effect increases the swing diameters about 3 times.

Keywords: Maneuvering model · ROV · Flow memory effect · Motion simulation · Unsteady hydrodynamic loads

1 Introduction

In steady theory, a ship or submarine's hydrodynamic force is determined by its instantaneous steady motion disturbance. By defining hydrodynamic derivatives, these forces are expressed via linear or nonlinear equations with constant coefficients. Assuming all forces and moments depend on instantaneous velocity and acceleration, they are expanded into the Taylor series relative to motion speed. For slow speeds, only linear terms are retained. Various steady derivatives are defined by linearizing motion equations, and experimental methods determine coefficients to predict the spacecraft's turning and heading retention.

Fossen initiated studies on the nonlinear dynamic model of ROVs, deriving a six-DOF rigid body kinematic equation from momentum principles and Newton's second law [1, 2]. This model encompasses mass, inertia, velocity, acceleration, and external forces. He also established a nonlinear dynamic equation incorporating wave force, restoring force, added mass inertial force, and viscous damping. Building on Fossen's work, numerous scholars have researched ROV maneuverability and control, including robust submarine control [3], adaptive control optimization [4], and ROV uncertainty impacts [5]. Soylu, et al. [6] proposed a multi-input multi-output control law for a 4-DOF model, while Korytskyi [7] developed a mathematical model for underwater vehicle control system design. Dong, et al. [8] implemented a fuzzy PID depth control strategy, and Kadiyam, et al. [9] discussed actuator configurations. Zhao, et al. [10] studied hydrodynamic coefficients, Cardaillac, et al. [11] proposed hull inspection methods, Herrero, et al. [12] designed model basin test methodologies, and Luo, et al. [13] presented a double-closed-loop sliding mode tracking control.

Research on unsteady hydrodynamics and fluid memory effects for underwater vehicles is limited. Submerged objects exhibit fluid memory similar to wings. When a cylinder starts impulsively in water, resistance rapidly increases to the inertial force magnitude, then decreases, slowly increases, and stabilizes. Studies at low Reynolds numbers show resistance becomes constant over ten diameters of travel [14]. Doyle, et al. [15] used URANS to study submarine unsteady motion, discovering fluid memory from vortex shedding. Paifelman, et al. [16] applied the Wagner function and vertical velocity convolution for a towed submarine model with hydrofoils. Tinker [17] used functional analysis to establish an impulse response relation for submarine horizontal hydrodynamic forces, creating a linear unsteady model. Z-type steering simulations showed reduced displacement peaks with the unsteady model compared to quasi-steady models. Impulse response models are more accurate for motion prediction and load response calculation [18]. Javanmard, et al. [19] presented a method to extract translational added mass coefficients of underwater vehicles based on the impulse response relation. Figure 1 shows the influence of fluid memory effect on hydrodynamic and motion.

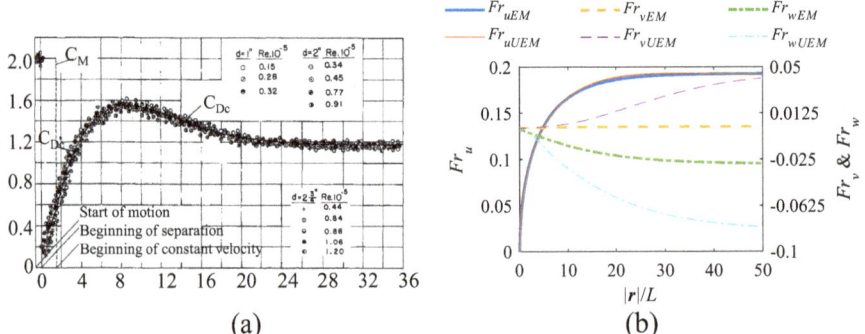

Fig. 1. The (a) drag coefficients for a circular cylinder in impulsively started laminar flow [14] and (b) the difference between motions results from the models with (UEM) and without (EM) the flow memory effect [20]

The motion form of ROV is different from that of submarines or ships, as its operating range is limited and often in a dynamic positioning state within a certain range. The flow around ROVs cannot be fully developed like when submarines and ships are sailing.

However, based on the steady theory, it is mostly used to calculate the hydrodynamic coefficients after the flow is fully developed, so the established steady model is not entirely applicable to ROVs. The hydrodynamic modeling of ROV is more suitable for using the impulse response relation, as underwater operations of ROVs have the characteristics of cross degrees of freedom, unsteady, and highly susceptible to high-frequency motion caused by ocean currents and umbilical cables. Avila and Adamowski [21, 22] found that the damping and added mass of ROV vary with the amplitude and frequency of the oscillation when studying the underwater oscillation motion based on the Morrison equation. This phenomenon is difficult to explain with steady-state theory, but the impulse response relationship is completely suitable for solving this problem. This study aims to establish a set of unsteady ROV hydrodynamic models based on pulse response relationships and study the flow memory effect on the hydrodynamics and motions of ROV.

2 Mathematical Model

Motion can be described by the two coordinate systems, one is the Earth-fixed coordinate system O-$x_0 y_0 z_0$, and the other is the body-fixed coordinate system G-xyz, as shown in Fig. 2. The origin of G-xyz is the center of gravity. The linear and angular velocities are defined. The Euler angles are used to translate the velocities and positions between O-$x_0 y_0 z_0$ and G-xyz [23]. The notations of the ROV are shown in the appendix.

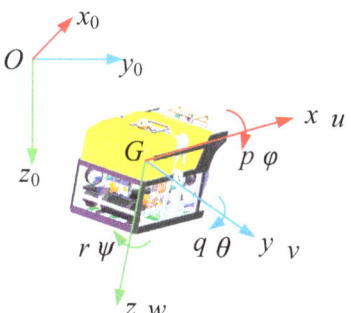

Fig. 2. The Earth-fixed and the body-fixed coordinate system

When propeller performance is not considered, the new unsteady hydrodynamic models of the ROV are as follows:

$$m(\dot{u} - vr + wq) = X_{\dot{u}}\dot{u} + X_{\dot{v}}\dot{v} + X_{\dot{w}}\dot{w} + X_{\dot{p}}\dot{p} + X_{\dot{q}}\dot{q} + X_{\dot{r}}\dot{r} + \int_0^{\tau_u} X_{uu} \Phi_X(\tau_u - \tau_0) \frac{du^2}{d\tau_0}(\tau_0) d\tau_0 \quad (1)$$

$$m(\dot{v} - wp + ur) = Y_{\dot{u}}\dot{u} + Y_{\dot{v}}\dot{v} + Y_{\dot{w}}\dot{w} + Y_{\dot{p}}\dot{p} + Y_{\dot{q}}\dot{q} + Y_{\dot{r}}\dot{r} + \int_0^{\tau_v} Y_{vv} \Phi_Y(\tau_v - \tau_0) \frac{dv^2}{d\tau_0}(\tau_0) d\tau_0 \quad (2)$$

$$m(\dot{w} - uq + vp) = Z_{\dot{u}}\dot{u} + Z_{\dot{v}}\dot{v} + Z_{\dot{w}}\dot{w} + Z_{\dot{p}}\dot{p} + Z_{\dot{q}}\dot{q} + Z_{\dot{r}}\dot{r} + \int_0^{\tau_w} Z_{ww} \Phi_Z(\tau_w - \tau_0) \frac{dw^2}{d\tau_0}(\tau_0) d\tau_0 \quad (3)$$

$$I_x \dot{p} + (I_z - I_y)qr = K_{\dot{u}}\dot{u} + K_{\dot{v}}\dot{v} + K_{\dot{w}}\dot{w} + K_{\dot{p}}\dot{p} + K_{\dot{q}}\dot{q} + K_{\dot{r}}\dot{r} + \int_0^{\tau_u} K_p \Phi_K (\tau_u - \tau_0) \frac{dp}{d\tau_0}(\tau_0) d\tau_0 \quad (4)$$
$$+ \Delta BG \cos(\theta) \sin(\varphi)$$

$$I_y \dot{q} + (I_x - I_z)rp = M_{\dot{u}}\dot{u} + M_{\dot{v}}\dot{v} + M_{\dot{w}}\dot{w} + M_{\dot{p}}\dot{p} + M_{\dot{q}}\dot{q} + M_{\dot{r}}\dot{r} + \int_0^{\tau_u} M_q \Phi_M (\tau_u - \tau_0) \frac{dq}{d\tau_0}(\tau_0) d\tau_0 \quad (5)$$
$$+ \Delta BG \sin(\theta)$$

$$I_z \dot{r} + (I_y - I_x)pq = N_{\dot{u}}\dot{u} + N_{\dot{v}}\dot{v} + N_{\dot{w}}\dot{w} + N_{\dot{p}}\dot{p} + N_{\dot{q}}\dot{q} + N_{\dot{r}}\dot{r} + \int_0^{\tau_u} N_r \Phi_N (\tau_u - \tau_0) \frac{dr}{d\tau_0}(\tau_0) d\tau_0 \quad (6)$$

where the viscous coefficients and the response functions are defined as follows:

$$\begin{cases} X_{uu} = X_{uu}^+, \Phi_X = \Phi_X^+, \tau_u > 0 \\ X_{uu} = X_{uu}^-, \Phi_X = \Phi_X^-, \tau_u < 0 \end{cases} \begin{cases} Y_{vv} = Y_{vv}^+, \Phi_Y = \Phi_Y^+, \tau_v > 0 \\ Y_{vv} = Y_{vv}^-, \Phi_Y = \Phi_Y^-, \tau_v < 0 \end{cases} \begin{cases} Z_{ww} = Z_{ww}^+, \Phi_Z = \Phi_Z^+, \tau_w > 0 \\ Z_{ww} = Z_{ww}^-, \Phi_Z = \Phi_Z^-, \tau_w < 0 \end{cases}$$
(7)

The non-dimensional trajectories are defined as follows: $\tau_u = \frac{2ut}{L}$, $\tau_v = \frac{2vt}{L}$, and $\tau_w = \frac{2wt}{L}$.

To compare the steady and unsteady hydrodynamic and motion characteristics, the steady models are also given as follows:

$$m(\dot{u} - vr + wq) = X_{\dot{u}}\dot{u} + X_{\dot{v}}\dot{v} + X_{\dot{w}}\dot{w} + X_{\dot{p}}\dot{p} + X_{\dot{q}}\dot{q} + X_{\dot{r}}\dot{r} + X_{uu}u^2 + X_{u|u|}u|u| + X_u u + X_{|u|}|u| \quad (8)$$

$$m(\dot{v} - wp + ur) = Y_{\dot{u}}\dot{u} + Y_{\dot{v}}\dot{v} + Y_{\dot{w}}\dot{w} + Y_{\dot{p}}\dot{p} + Y_{\dot{q}}\dot{q} + Y_{\dot{r}}\dot{r} + Y_{vv}v^2 + Y_{v|v|}v|v| + Y_v v + Y_{|v|}|v| \quad (9)$$

$$m(\dot{w} - uq + vp) = Z_{\dot{u}}\dot{u} + Z_{\dot{v}}\dot{v} + Z_{\dot{w}}\dot{w} + Z_{\dot{p}}\dot{p} + Z_{\dot{q}}\dot{q} + Z_{\dot{r}}\dot{r} + Z_{ww}w^2 + Z_{w|w|}w|w| + Z_w w + Z_{|w|}|w| \quad (10)$$

$$I_x \dot{p} + (I_z - I_y)qr = K_{\dot{u}}\dot{u} + K_{\dot{v}}\dot{v} + K_{\dot{w}}\dot{w} + K_{\dot{p}}\dot{p} + K_{\dot{q}}\dot{q} + K_{\dot{r}}\dot{r} + K_p p + \Delta BG \cos(\theta) \sin(\varphi) \quad (11)$$

$$I_y \dot{q} + (I_x - I_z)rp = M_{\dot{u}}\dot{u} + M_{\dot{v}}\dot{v} + M_{\dot{w}}\dot{w} + M_{\dot{p}}\dot{p} + M_{\dot{q}}\dot{q} + M_{\dot{r}}\dot{r} + M_q q + \Delta BG \sin(\theta) \quad (12)$$

$$I_z \dot{r} + (I_y - I_x)pq = N_{\dot{u}}\dot{u} + N_{\dot{v}}\dot{v} + N_{\dot{w}}\dot{w} + N_{\dot{p}}\dot{p} + N_{\dot{q}}\dot{q} + N_{\dot{r}}\dot{r} + N_r r \quad (13)$$

The nomenclatures are shown in the appendix. The coefficients and response functions were obtained from the computational fluid dynamics (CFD) simulations. The CFD simulations were verified by the model experiments in [24]. The hydrodynamic loads from Eq. (1) were verified by the CFD results in our other research on the ROV.

3 Hydrodynamic Loads in Captive Motions

The convolution in the unsteady model indicates the hydrodynamic loads will depend on motion speeds, accelerations, and time. Figure 3 shows the longitudinal forces from the steady and unsteady models with the longitudinal speed of $u = Uc + 0.1 \sin(2\pi f_u t)$ m/s and Uc = 0.3 m/s. The longitudinal forces from the steady model vary with a constant amplitude and frequency in one case. The amplitude increases with the motion frequency. However, under the effect of convolutions, the amplitude of the longitudinal forces from the unsteady model changes with time in one case. The amplitude levels off after about 4–5 periods. The force amplitudes from the unsteady model are larger than those from the steady model. The occurrence of the maximum and minimum longitudinal forces from the unsteady model is lagging as the frequency increases.

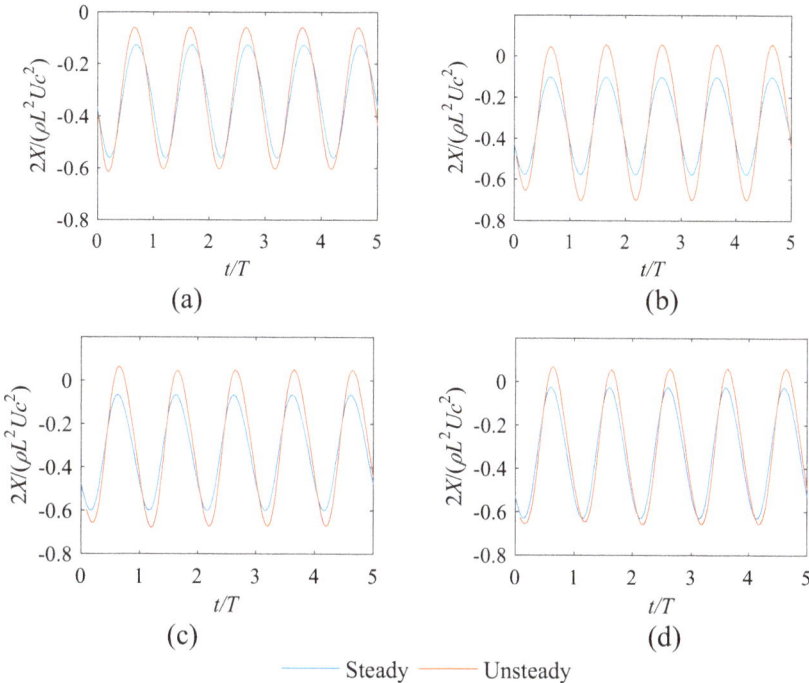

Fig. 3. The longitudinal force results of the steady and unsteady models under the motion frequencies f_u are (a) 0.1 Hz, (b) 0.2 Hz, (c) 0.3 Hz, and (d) 0.4 Hz

Figure 4 shows the roll moment with the rolling angle velocity $p = 0.05 \sin(2\pi f_p t)$ rad/s and longitudinal speed $u = 0.3$ m/s. The amplitudes of both steady and unsteady roll moments increase with the rotating frequency but the increasing ratio of the unsteady moment is larger. The unsteady moment amplitude is less than the steady one when the rotating frequency is lower than 0.3 Hz and the unsteady moment amplitude is 1.2 times the steady one as the frequency of 0.4 Hz. The sizes of the steady roll moments with the same magnitudes of the positive and negative rotating angles are equal because the roll moments in the steady and unsteady model are linear odd functions. However, the phases of the unsteady roll moments lag behind the steady model results.

4 Maneuverability Analysis

The motions of the ROV with constant longitudinal speeds are studied in this section. The speeds of other DOFs are under no extra control except the longitudinal one. Figure 5 illustrates the normalized motions from the steady model (index of st) and unsteady model (index of unst) with longitudinal speeds of 0.1, 0.3, and 0.5 m/s. The ROV tends to rotate to the left and rise for both steady and unsteady results. The unsteady results show the ROV rotates faster. The difference between the steady and unsteady motion results increases with time and longitudinal velocity in vertical. The roll and pitch angles fluctuate in the initial stage of the motions. The amplitudes of the angles decrease and

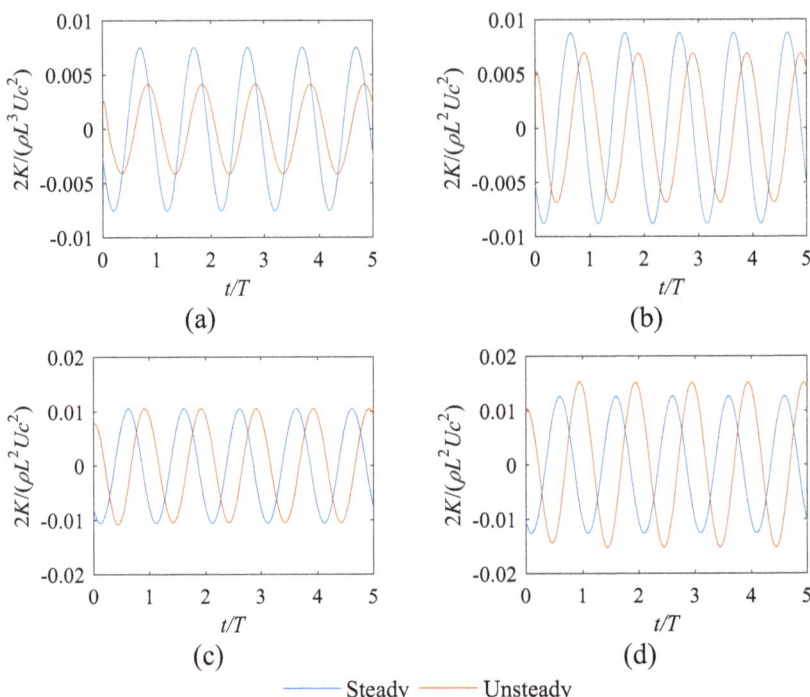

Fig. 4. The roll moment results from the steady and unsteady model under the motion frequencies f_p are (a) 0.1 Hz, (b) 0.2 Hz, (c) 0.3 Hz, and (d) 0.4 Hz

stay at nonzero constant values gradually as time increases. The fluctuation of the roll and pitch angles are induced by the asymmetric added mass matrix and Coriolis forces, the motions are damped by the drag moments, and the restoring moments decide the magnitudes of the nonzero constant values. However, under the initial fluctuation of the roll and pitch motions, the unsteady roll and pitch motions keep fluctuating when considering the flow memory effect. The amplitude increases with the longitudinal speed and it varies with time.

Figure 6 shows the trajectory and motion angle of the ROV with non-dimensional positive and negative yaw moments $Nt' = 2Nt/(\rho L^3 Uc^2)$. The ROV has a constant longitudinal velocity of 0.3 m/s and the simulations stop when the yaw angle is 720°. The ROV has a smaller swing diameter with a negative yaw moment. The unsteady results show that the ROV rotates faster when considering the flow memory effect. The swing diameters of the positive rotations are 10.94 L and 3.24 L m for the steady and unsteady motions. The swing diameters of the negative rotations are 10.63 L and 2.27 L respectively. The ROV goes downward in all rotations. The ROV moves slower when considering the flow memory effect. The roll angles with positive yaw moments are

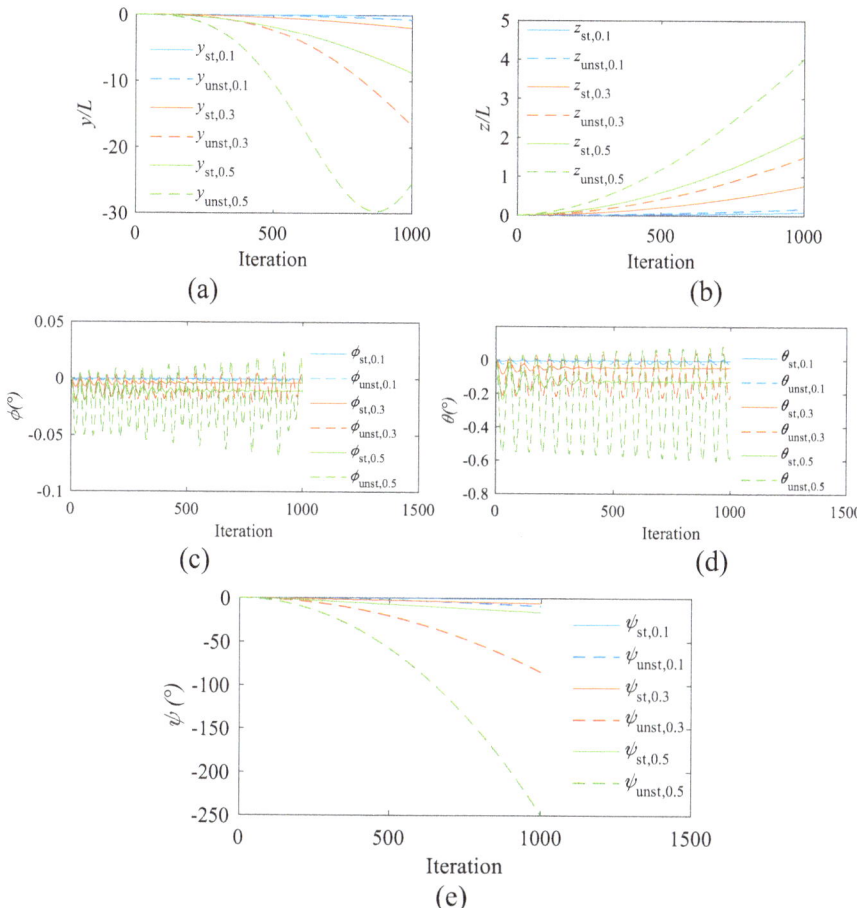

Fig. 5. The steady and unsteady motions of the ROV with constant longitudinal speeds. (a) The lateral motions, (b) the vertical motions, (c) the roll angles, (d) the pitch angles, and (e) the yaw angles

larger. The roll angles are unsteady under the flow memory effect. The yaw moment has a limited effect on pitch angles. The pitch angles from the unsteady model are larger than the steady results and the damping of amplitude is not obvious.

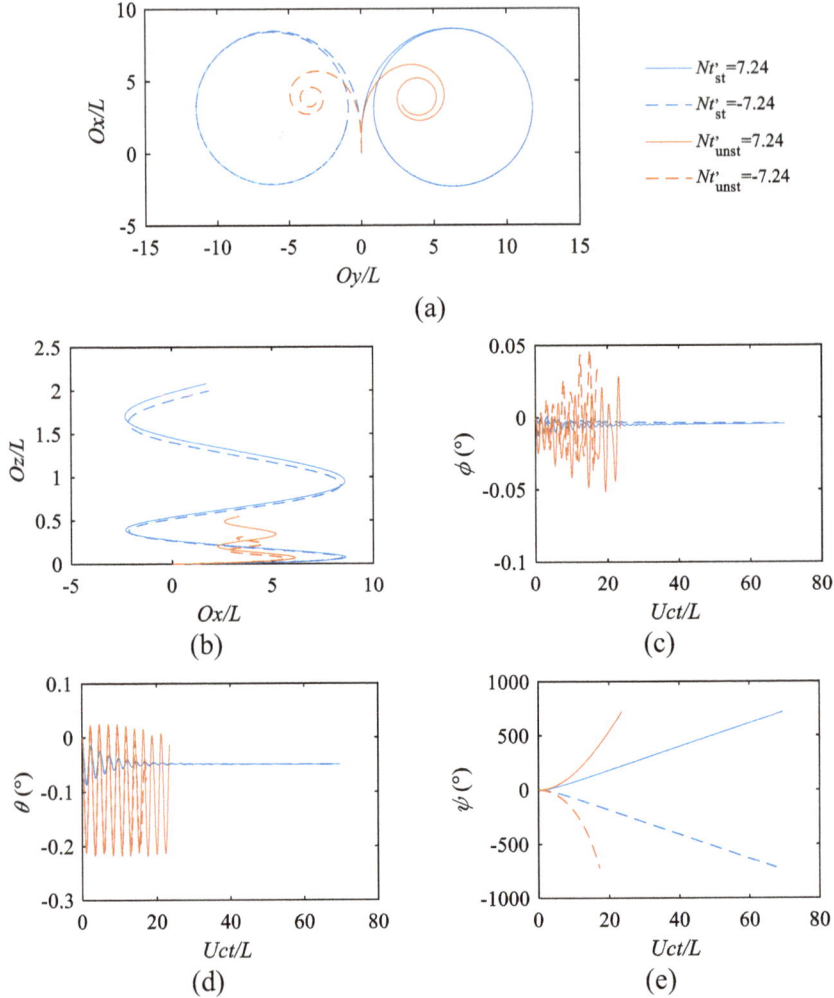

Fig. 6. The motions of the ROV in rotations. (a) The horizontal motions, (b) the vertical motions, (c) the roll angles, (d) the pitch angles, and (e) the yaw angles

5 Conclusion

A novel unsteady hydrodynamic model of an open-frame ROV was presented. The unsteady model considers the flow memory effect on the hydrodynamic loads and motions of the ROV. The flow memory effect is caused by the vortex separation and the influence of the circulation in the wake. The flow memory effect induces multiple magnitudes of the hydrodynamic loads even when the moving speeds of the ROV are the same. The unsteady model is based on the convolutions that relate to the speeds and trajectories of the ROV. The captive motion and free-running motion simulation results are compared with the steady model. The significant effects of the flow memory and motion history were discovered. The main findings are as follows:

(1) The flow memory effect increases the amplitude of the longitudinal hydrodynamic loads and delays the occurrence of the maximum and minimum loads of the ROV.
(2) The flow memory effect increases the sensitivity of the roll hydrodynamic moments related to the roll frequency of the ROV.
(3) The ROV moves farther in translational motions with a constant longitudinal speed when considering the flow memory effect.
(4) The flow memory effect makes the roll and pitch angles more unstable.
(5) The flow memory effect improves the maneuverability of the ROV. The swing diameters from the unsteady model are smaller than the steady results.

The model proposed in this study only includes the viscous hydrodynamic force on the diagonal of the damping matrix. To establish complete damping considering the fluid memory effect, it is necessary to study the unsteady characteristics of the hydrodynamic force in the non-moving direction. If the impulse response motion with drift angle or pitch angle is carried out, the unsteady viscous hydrodynamic force of yaw motion can be established. This study does not consider the influence of current, but referring to the existing research, it is a feasible method to calculate the influence of current on ROV velocity by using the method of relative velocity.

Appendix

Table 1. The nomenclature of unsteady and steady model

Notations	Meanings	Notations	Meanings
$u\ \dot{u}$	longitudinal velocity and acceleration	$v\ \dot{v}$	lateral velocity and acceleration
$w\ \dot{w}$	vertical velocity and acceleration	$p\ \dot{p}$	roll velocity and acceleration
$q\ \dot{q}$	pitch velocity and acceleration	$r\ \dot{r}$	yaw velocity and acceleration
φ	roll angle	θ	pitch angle
ψ	yaw angle	m	mass
$I_x\ I_y\ I_z$	longitudinal, lateral, and vertical moment of inertia	$X_{\dot{u}}\ X_{\dot{v}}\ X_{\dot{w}}\ X_{\dot{p}}\ X_{\dot{q}}\ X_{\dot{r}}$	longitudinal inertial hydrodynamic coefficients
$Y_{\dot{u}}\ Y_{\dot{v}}\ Y_{\dot{w}}\ Y_{\dot{p}}\ Y_{\dot{q}}\ Y_{\dot{r}}$	lateral inertial hydrodynamic coefficients	$Z_{\dot{u}}\ Z_{\dot{v}}\ Z_{\dot{w}}\ Z_{\dot{p}}\ Z_{\dot{q}}\ Z_{\dot{r}}$	vertical inertial hydrodynamic coefficients

(*continued*)

Table 1. (*continued*)

Notations	Meanings	Notations	Meanings
$K_{\dot{u}}\ K_{\dot{v}}\ K_{\dot{w}}\ K_{\dot{p}}\ K_{\dot{q}}\ K_{\dot{r}}$	roll inertial hydrodynamic coefficients	$M_{\dot{u}}\ M_{\dot{v}}\ M_{\dot{w}}\ M_{\dot{p}}\ M_{\dot{q}}$	pitch inertial hydrodynamic coefficients
$N_{\dot{u}}\ N_{\dot{v}}\ N_{\dot{w}}\ N_{\dot{p}}\ N_{\dot{q}}\ N_{\dot{r}}$	yaw inertial hydrodynamic coefficients	$X_{uu}\ Y_{vv}\ Z_{ww}\ K_p\ M_q\ N_r$	viscous hydrodynamic coefficients
$\Phi_X\ \Phi_Y\ \Phi_Z\ \Phi_K\ \Phi_M\ \Phi_N$	hydrodynamic response functions	$\tau_u\ \tau_v\ \tau_w$	longitudinal, lateral, and vertical non-dimensional trajectories
L	ROV length	t	time
BG	distance between the center of the buoyancy and gravity	Δ	displacement
ρ	water density	Uc	sailing velocity

Acknowledgments. This research was supported by the Research on Supervision, Inspection and Evaluation of Deepwater Underwater Facilities and Emergency Repair Technology (WBS: E-0822 J014).

References

1. Fossen, T.I.: Nonlinear Modeling and Control of Underwater Vehicles. Norwegian University of Science and Technology. Trondheim, Norway (1987)
2. Fossen, T.I.: Adaptive macro-micro control of nonlinear underwater robotic systems. In: Fifth International Conference on Advanced Robotics 'Robots in Unstructured Environments, vol.2, pp. 1569–1572. IEEE, Pisa, Italy (1991). https://doi.org/10.1109/ICAR.1991.240508
3. White, B.A.: Robust control of an unmanned underwater vehicle. In: Proceedings of the 37th IEEE Conference on Decision and Control, vol.3, pp. 2533–2534. IEEE, Tampa, FL, USA (1998). https://doi.org/10.1109/CDC.1998.757830
4. Wang, F., Li, Y., Wan, L., Xu, Y.: Modeling and motion control strategy for autonomous underwater vehicles. In: 2009 International Conference on Mechatronics and Automation, pp. 4851–4856. IEEE, Changchun, China (2009). https://doi.org/10.1109/ICMA.2009.5246392
5. Ding, N., Li, Z., Yang, C., Ge, T.: Robust adaptive motion control for Remotely Operated Vehicles with velocity constraints. In: 2010 IEEE International Conference on Robotics and Biomimetics, pp. 932–937. IEEE, Tianjin, China (2010). https://doi.org/10.1109/ROBIO.2010.5723451
6. Soylu, S., et al.: Precise trajectory control for an inspection class ROV. Ocean Eng. **111**, 508–523 (2016)

7. Korytskyi, V.I.: Improvement of the mathematical model for spatial motion of a remotely operated underwater vehicle with technological equipment. Ship Build. Mar. Infrastruct. **1**(9), 44–52 (2018)
8. Dong, M., et al.: Depth control of ROV in nuclear power plant based on fuzzy PID and dynamics compensation. Microsyst. Technol. **26**(3), 811–821 (2019)
9. Kadiyam, J., et al.: Simulation-based semi-empirical comparative study of fixed and vectored thruster configurations for an underwater vehicle. Ocean Eng. **234**, 109231 (2021)
10. Zhao, B., et al.: Hydrodynamic coefficients of the DARPA SUBOFF AFF-8 in rotating arm maneuver: Part I: Test technology and validation. Ocean Eng. **266**, 113148 (2022)
11. Cardaillac, A., et al.: ROV-Based Autonomous Maneuvering for Ship Hull Inspection with Coverage Monitoring. J. Intell. Robot. Syst. **110**, 59 (2024)
12. Herrero, E.R., et al.: Experiment design for model basin tests with a remotely operated vehicle. Ocean Eng. **307**, 118215 (2024)
13. Luo, G., et al.: ROV trajectory tracking control based on disturbance observer and combinatorial reaching law of sliding mode. Ocean Eng. **304**, 117744 (2024)
14. Sarpkaya, T.: Separated flow about lifting bodies and impulsive flow about cylinders. AIAA J. **4**(3), 414–420 (1966)
15. Doyle, R., et al.: Hydrodynamic impulse generated by slender bodies undergoing unsteady motion in viscous flows. Ocean Eng. **236**, 109532 (2021)
16. Paifelman, E., et al.: An optimal indirect control of underwater vehicle. Int. J. Control. **94**(2), 312–326 (2019)
17. Tinker, S.J.: Fluid memory effects on the trajectory of a submersible. Int. Shipbuild. Prog. **25**(290), 261–269 (1978)
18. Scragg, C.: Memory Effects in Deepwater Maneuvering. J. Ship Res. **23**(3), 175–187 (1979)
19. Javanmard, E., et al.: A new CFD method for determination of translational added mass coefficients of an underwater vehicle. Ocean Eng. **215**, 107857 (2020)
20. Guo, R., et al.: Flow Memory Effect on Open-Frame Remotely Operated Vehicle Motion. J. Field Robot. **42**(6), 2611–2638 (2025)
21. Avila, J.P.J., Adamowski, J.C.: Experimental evaluation of the hydrodynamic coefficients of a ROV through Morison's equation. Ocean Eng. **38**(18), 2162–2170 (2011)
22. Avila, J.P.J., et al.: Experimental model identification of open-frame underwater vehicles. Ocean Eng. **60**, 81–94 (2013)
23. Fossen, T.I.: Handbook of marine craft hydrodynamics and motion control. Wiley, Trondheim, Norway (2021)
24. Guo, R., et al.: Evaluation of hydrodynamic coefficients and sensitivity analysis of a work-class remotely operated vehicle using planar motion mechanism tests. Ocean Eng. **312**, 119037 (2024)

Open Access This chapter is licensed under the terms of the Creative Commons Attribution-NonCommercial-NoDerivatives 4.0 International License (http://creativecommons.org/licenses/by-nc-nd/4.0/), which permits any noncommercial use, sharing, distribution and reproduction in any medium or format, as long as you give appropriate credit to the original author(s) and the source, provide a link to the Creative Commons license and indicate if you modified the licensed material. You do not have permission under this license to share adapted material derived from this chapter or parts of it.

The images or other third party material in this chapter are included in the chapter's Creative Commons license, unless indicated otherwise in a credit line to the material. If material is not included in the chapter's Creative Commons license and your intended use is not permitted by statutory regulation or exceeds the permitted use, you will need to obtain permission directly from the copyright holder.

Modeling and FEM Analysis of the IEA15MW Wind Turbine Blade

Ya Luo[1] and Yan Gao[1,2(✉)]

[1] Qindao Innovation and Development Centre of Harbin Engineering University, Qingdao 266400, China
yan.gao@hrbeu.edu.cn
[2] College of Shipbuilding Engineering, Harbin Engineering University, Harbin 150001, China

Abstract. The structural performance of wind turbine blades is of paramount importance as it directly influences the overall performance of the entire floating offshore wind turbine system. In this study, the deformation and stress distribution of the IEA15MW wind turbine blade are investigated. The research commences by precisely fitting the blade geometry in accordance with the distribution of airfoil parameters. Subsequently, a sophisticated 3D finite element model is developed using Abaqus. Under wind loads with both uniform and linear distributions, the stress and displacement distributions of the blades are analyzed in exhaustive detail. By carefully analyzing the maximum stress, the most dangerous region of the wind turbine blade can be accurately identified. The article also investigates the effect of different load directions on the magnitude of blade stresses. This research not only provides highly valuable insights for enhancing the design and safety of wind turbine systems but also holds significant implications for the burgeoning renewable energy industry. As wind power continues to play an increasingly crucial role in meeting global energy demands, understanding the structural performance of wind turbine blades becomes even more essential.

Keywords: Wind turbine blade · IEA 15 MW · Stress distribution · Finite element model

1 Introduction

Wind energy is currently one of the most promising clean energy sources [1]. As the core component of the wind turbine, the wind turbine blade converts wind energy into mechanical energy during operation, hence the structural strength and stability of the blade play a very important role in the reliability of the wind turbine [2].

The finite element method is a reliable and simple method to analyze large-sized blades considering the relatively acceptable computation expense. For example, R. Gukendran et al. [3] used ANSYS to analyze the composite blade and investigate the effect of different materials on the blade response. According to the static analysis of the result, the Epoxy carbon material had the smallest deformation; Himayat Ullah et al. [4] did a structural analysis of a large composite wind turbine blade under extreme loads, results demonstrated that the tip deflection was comparable to that of other similar systems and the blade will not hit the tower during extreme gust conditions.

Therefore, in this paper, the dynamic response of the blade is analyzed by finite element software to figure out the most dangerous region of the blade under different loads, and stress variations were also studied by applying loads in different directions to the blades. Which will provide a reference for the structural design of large wind turbine blades.

2 Finite Element Model of Blade

The aerodynamic loads on offshore floating wind turbine blades are alternating and dynamic. Cracks often propagate in areas with high stresses. Therefore, identifying the locations of maximum stress is crucial for the safety of wind turbine blades. This paper establishes a finite element model of a single blade using Abaqus to obtain the deformation response, and then determine the locations of maximum stress of the blade under wind loads. Abaqus is utilized in this paper to model the IEA 15 MW reference wind turbine blade. The modeling steps are as follows:

Blade Coordinate Conversion. According to the data of blade (Fig. 1), the coordinates of the standard airfoil can be translated, scaled, and rotated through Eqs. (1) and (2) to obtain the three-dimensional coordinate data of each cross-section airfoil.

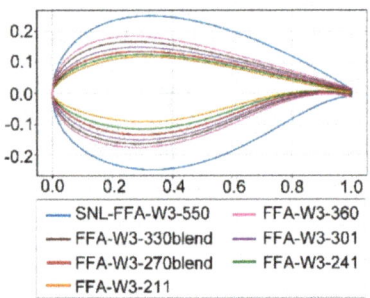

Fig. 1. Airfoils of the 15 MW wind turbine blade [5].

$$x' = c \frac{|x - x_{pitch}|}{x - x_{pitch}} \sqrt{(x - x_{pitch})^2 + y^2} \cos\left(arc \frac{y}{x - x_{pitch}} + \beta\right) \quad (1)$$

$$y' = c \frac{|x - x_{pitch}|}{x - x_{pitch}} \sqrt{(x - x_{pitch})^2 + y^2} \sin\left(arc \frac{y}{x - x_{pitch}} + \beta\right) \quad (2)$$

where x and y are the normalized airfoil coordinates; x' and y' are the transformed 3D coordinates; x_{pitch} is the pitch axis; c is the chord; β is the twist.

Finite Element Model of Blade Shell. The coordinates of each airfoil obtained in the previous step will be used to generate closed spline curves in Abaqus through the spline curve command, and then the closed spline curves will be connected to generate the blade shell model through the command of the curve group (Fig. 2 (a)).

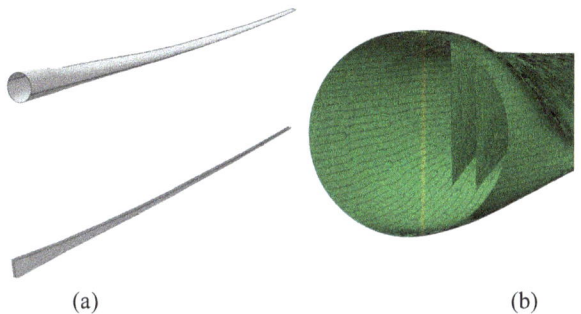

(a)   (b)

Fig. 2. Geometric models. (a) Shell and webs, (b) Blade mesh generation.

Finite Element Model of Blade Webs. For the modeling of blade webs, according to the distribution of the webs at different cross-sections, the upper and lower spline curves of the webs are generated by spline interpolation, and the webs are similarly generated by the command of the curve group (Fig. 2 (a)).

The blade mainly consists of a blade shell, spar cap, and webs (Fig. 3). In this paper, the material of the blade shell is simplified according to [5], consisting of Medium Density Foam and Glass Triax. The blade spar cap material consists of CarbonUD, and the web material consists of Medium Density Foam and Glass Biax. The specific material properties are shown in Table 1. In the selection of laminate lay-up angle, considering that the blade shell is mainly subjected to bending load, the lay-up angle of 0° and 90° is adopted; in contrast, the webs are mainly subjected to shear load, the lay-up angle of 0° and ± 45° is adopted. Both the blade shell and the webs were modeled using S4R shell elements, reduced integration together with hourglass control is employed to enhance computational efficiency and accuracy. The blade shell is meshed into 28,172 elements, and the webs are meshed into 6659 elements (Fig. 2 (b)).

3 Dynamic Responses of Blade

3.1 Validation for the FE Model

In order to verify the finite element model developed in this paper, the modal frequency of the blade is calculated and compared with the one in [5], as the modal frequency of the first eight orders of the blade shown in Fig. 4. Good match between the two simulation results is obtained, which although has an error because the finite element model of this paper is simplified, which will be refined in the future study.

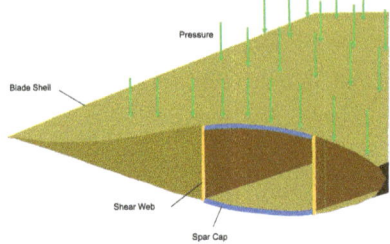

Fig. 3. Schematic diagram of blade model section setting.

Table 1. Material properties.

Material	E_1(MPa)	E_2(MPa)	G_{12}(MPa)	ν_{12}	ρ(t/mm^3)
Medium Density Foam	1.292×10^3	1.292×10^3	4.895×10^3	0.32	1.300×10^{-10}
Carbon UD	1.145×10^5	8.390×10^3	5.990×10^3	0.27	1.220×10^{-9}
Glass Triax	2.870×10^4	1660×10^4	8.400×10^3	0.50	1.940×10^{-9}
Glass Biax	1.110×10^4	1.110×10^4	1.353×10^3	0.50	1.940×10^{-9}

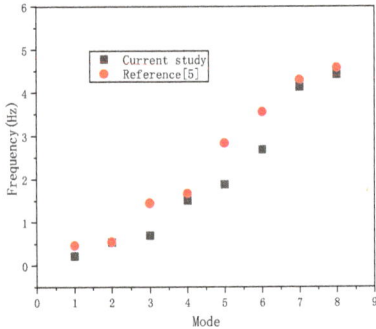

Fig. 4. Modal frequencies of the blade.

3.2 Boundary Conditions and Load Application

Regarding the boundary condition, the blade root is rigidly fixed to represent the connection to the hub. The other parts are free to deform. Two loading conditions, i.e. uniform load and linear distributed load, are considered in the simulation, as illustrated in Fig. 5.

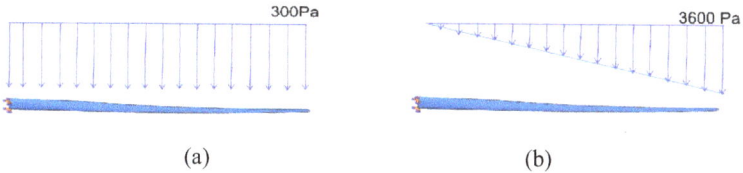

Fig. 5. Loading diagram. (a) Uniform load, (b) Linear load.

3.3 Analysis of Blades With Different Load Forms

The response of the blade during 0–10 s is simulated using the dynamic implicit analysis method. The time step is set as 0.025 s. Firstly, a uniform load with a pressure of 300 Pa is applied to the blade. The stress distribution of the blade is analyzed. It is observed that the maximum stress occurs at 0.2R from the root of the blade, this result is consistent with the simulation of Zhang [5], which has a maximum value of 33.47 MPa. The stresses are higher at the blade spar cap and lower at the leading and trailing edges of the blade. This area is most prone to failure when the blade is actually in operation, so it can be thickened to improve the strength of the blade. The closer to the tip of the blade, the lower the stress. The maximum stress in webs is 12.68 MPa, occurring where webs

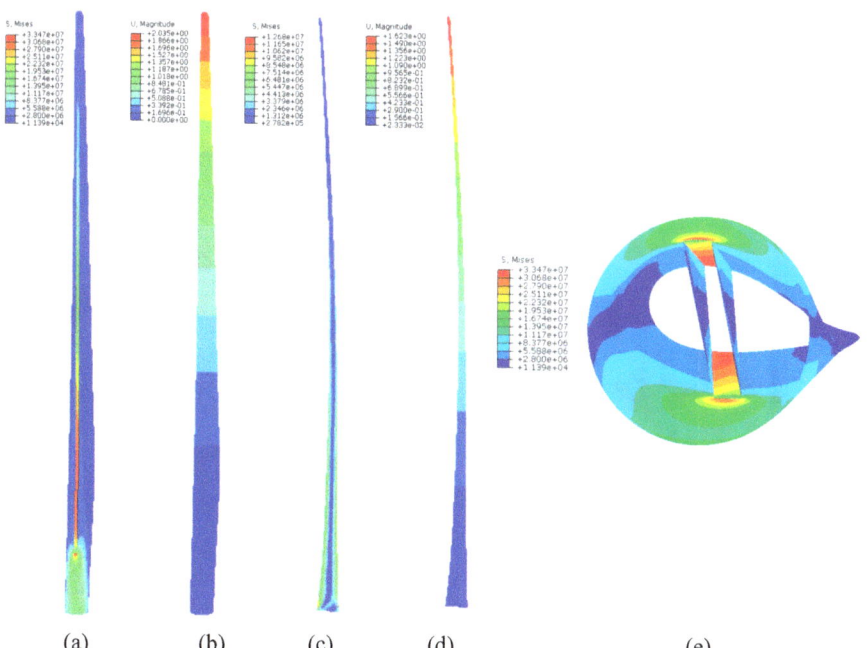

Fig. 6. Dynamic response of blade under uniform load. (a) Blade shell stress distribution, (b) Blade shell displacement distribution, (c) Blade web stress distribution, (d) Blade web displacement distribution, (e) Stress distribution near the positions of maxima.

contact the blade and close to the blade root. From Fig. 6 (e), it can be seen that the stress reaches its maximum at the spar cap. The stress decreases sharply in the region without spar cap, so the higher stress is borne at the junction of the web and the shell, which is very easy to fracture. Therefore, connection strengthening is crucial to ensure the safety of the blade. Regarding the blade displacement distribution, the displacement of the blade and webs is roughly linearly distributed with maximum values of 2.035 m and 1.623 m, respectively (Fig. 6). The deformation at the tip of the blade is the most severe, since the thickness and chord length of the airfoil at the tip are decreasing and there is no web support near the tip of the blade.

The load is transformed into a linear distribution and reaches its maximum at the tip of the blade with a value of 3600 Pa, with other settings keeping the same. It is observed that the maximum stress occurs at 0.3R from the root of the blade, which has a maximum value of 278.0 MPa. The location of the maximum stress in webs is similar to the uniform loading results, which has a maximum value of 90.78 MPa. The stress values in this condition are much higher compared to the uniform condition. Similarly, the stress at the blade spar cap is higher than the one in other regions. Due to the sharp variation of stress distribution at the junction of the web and shell, this region needs to be strengthened in order to ensure the structural stability of the blade. The deformation at the tip of the blade is still more pronounced compared to the root of the blade under this condition, and the displacement distribution shows a linear variation as well. The maximum values of blade and web displacements are 18.58 m and 14.45 m, respectively (Fig. 7).

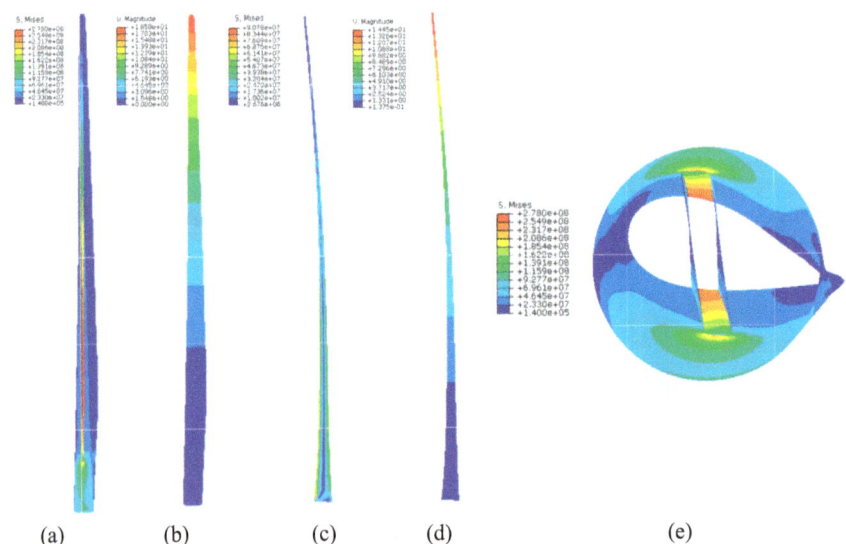

Fig. 7. Dynamic response of blade under linear load. (a) Blade shell stress distribution, (b) Blade shell displacement distribution, (c) Blade web stress distribution, (d) Blade web displacement distribution, (e) Stress distribution near the positions of maxima.

3.4 Analysis of Blades With Different Directional Loads

To study the trend of blade stresses under different loading directions, five load directions, i.e. 30°, 45°, 90°, 135°, and 150°, are considered. The angle is the direction of the load to the direction of the blade leading edge, pointing towards the trailing edge. From Fig. 8, it can be seen that the maximum stress of the blade under both uniform and linear loads shows a trend of increasing and then decreasing with angle variation. Figure 9 shows the stress distribution of the blade under linear loading in different directions. It can be found that all the stresses at the spar caps of the blade are relatively high, therefore, the design of this region should be paid attention.

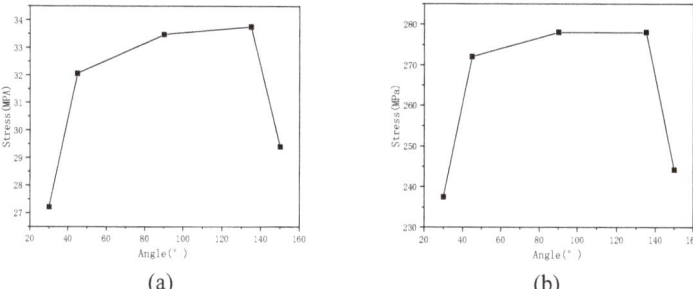

Fig. 8. Trends in blade stress. (a) Uniform load, (b) Linear load.

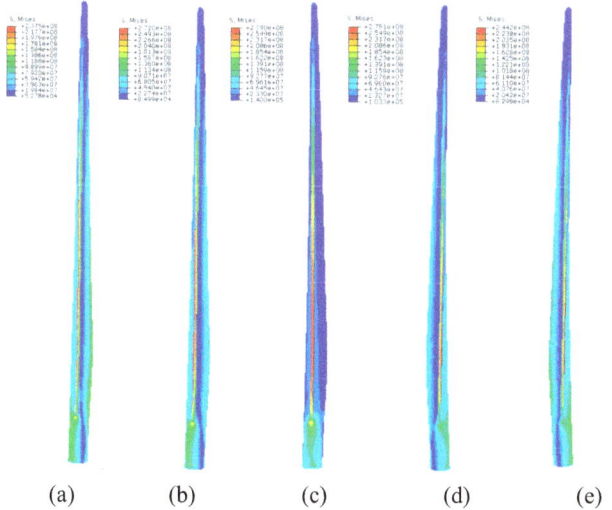

Fig. 9. Blade stress distribution under linear load. (a) 30°, (b) 45°, (c) 90°, (d) 135°, (e) 150°.

4 Conclusions

This paper develops a finite element model through Abaqus to investigate the IEA 15 MW wind turbine blade dynamic response under both uniformly and linearly distributed loading conditions. Besides, the effect of load direction is also discussed via comparing the deflection and stress of the blade under five load directions. The main conclusions of this paper are as follows:

(1) It is found that the maximum stress of the blade occurs at a distance of 0.2R from the blade root under uniform loads and 0.3R from the blade root under linear loads. The stresses are relatively higher at the spar cap of the blade than the ones in other region. And stress concentration is observed at the web-to-shell joint. Therefore, extra attention needs to be paid to this region in the blade structural design.
(2) The deformation at the tip of the blade is greater than that near the root of the blade. And the blade deformation under linear load is more serious than the one under uniform load. Consequently, during the design stage of the wind turbine blade, more loading conditions rather than simple uniform loads should be considered to ensure the blade safety.
(3) The stresses analyses with different load directions show that the maximum stress value occurs when the load is perpendicular to the blade shell. However, no matter how the load direction changes, the stress at the blade spar cap is greater than that in other regions. Particular attention needs to be paid to the strengthening of this region in future studies.

Acknowledgments. This research is funded by Shandong Provincial Natural Science Foundation (Grant No. 2024HWYQ-085), the Shandong Higher Education Young Science and Technology Support Program (Grant No. 2023KJ082), and National Natural Science Foundation of China (Grant No. 52401317).

References

1. Roga, S., Kumar, Y., Dubey, S.K., et al.: Recent technology and challenges of wind energy generation: A review. Sustain. Energy Technol. Assess. **52**, 1–17 (2022)
2. Gaertner, E., Rinker, J., Sethuraman, L., et al.: Definition of the IEA 15-megawatt offshore reference wind turbine (2020)
3. Gukendran, R., Sambathkumar, M., Sabari, C., et al.: Structural analysis of composite wind turbine blade using ANSYS. Mater. Today: Proc. **50**, 1011–1016 (2022)
4. Ullah, H., Ullah, B., Riaz, M., et al.: Structural Analysis of a Large Composite Wind Turbine Blade under Extreme Loading. In: 2018 International Conference on Power Generation Systems and Renewable Energy Technologies (PGSRET), pp. 1–6. IEEE (2018)
5. Zhang, Y., Song, Y., Shen, C., et al.: Aerodynamic and structural analysis for blades of a 15MW floating offshore wind turbine. Ocean Eng. **287** (2023)

Open Access This chapter is licensed under the terms of the Creative Commons Attribution-NonCommercial-NoDerivatives 4.0 International License (http://creativecommons.org/licenses/by-nc-nd/4.0/), which permits any noncommercial use, sharing, distribution and reproduction in any medium or format, as long as you give appropriate credit to the original author(s) and the source, provide a link to the Creative Commons license and indicate if you modified the licensed material. You do not have permission under this license to share adapted material derived from this chapter or parts of it.

The images or other third party material in this chapter are included in the chapter's Creative Commons license, unless indicated otherwise in a credit line to the material. If material is not included in the chapter's Creative Commons license and your intended use is not permitted by statutory regulation or exceeds the permitted use, you will need to obtain permission directly from the copyright holder.

Peridynamic Analysis of Wind Turbine Tower Crack Propagation Under Axial Force

Yingying Chen[2] and Yan Gao[1,2(✉)]

[1] College of Shipbuilding Engineering, Harbin Engineering University, Harbin 150001, China
yan.gao@hrbeu.edu.cn
[2] Qindao Innovation and Development Centre of Harbin Engineering University, Qingdao 266400, China

Abstract. With wind energy being more and more popular, the study of crack propagation in wind turbine tower is of great significance and necessity for the safety of the whole floating wind turbine. Peridynamics is a non-local theory and does not rely on the continuum assumption, making it suitable for failure analysis since no extra failure criteria are required. In this paper, a peridynamic model is constructed to represent a section of the IEA 15 MW wind turbine tower. The geometry varying of the diameter along the tower height direction is considered in the model while the thickness keeps same. The deformations of the tower under axial force predicted by the peridynamic model and ABAQUS simulation are compared to validate the developed model. Subsequently, the peridynamic model is employed to simulate and analyze the crack growth behavior with two pre-existing cracks located in the middle of the tower section. The crack is observed to initiate growth and propagate along the cylindrical circumference direction over time, which may lead to the potential safety hazard of the tower.

Keywords: Wind turbine tower · IEA15MW · Peridynamics · Crack propagation

1 Introduction

As a crucial connecting structure between the rotor—nacelle assembly (RNA) and the floating platform, the offshore wind turbine tower is subject to not only environmental loads but also internal forces generated by the operation of the wind turbine. Once the deformation exceeds its critical value, any local or micro crack can start to propagation at very high speed, which is hard to notice promptly in the early state. Therefore, it is of great significance to predict the crack propagation behavior of the wind turbine tower.

Regarding the wind turbines safety analysis, many scholars have analyzed the fatigue failure mode of towers under dynamic loads, but few on the brittle crack propagation prediction. M. Capaldo et al. [1] analyze the influence of cracks on the plastic strain and critical buckling load of wind turbine tower by using finite element method. Through several analyses with different crack position and length, it can conclude that the crack affects more when it is located on the compressive side than on tension side. Polyzois, D. J. et al. [2] conduct a experiment to study the behavior of a glass fiber-reinforced polymer (GFRP) wind turbine tower under static transverse loading. It is observed that specimens fail in shear rupture or local buckling mode near the support location. Lin et al. [3] analyze the plastic tensile damage between different transition segments of wind turbine tower. Results show that although the failure mechanisms differ for the different segments, the plastic tensile damage evolution trend is similar. Although extensive research has been conducted on the dynamic responses of wind turbine tower, its crack propagation prediction still remains a challenging topic due to its complex failure mechanism. Consequently, this paper utilizes peridynamic theory, which is suitable for failure analysis, to analyze the dynamic response including the crack propagation of the wind turbine tower.

2 Peridynamics Theory

2.1 Concept of Peridynamics

Being different from the classical continuum mechanics (CCM) using partial differential equations of motion, Peridynamics (PD) proposed by Silling [4] employs integral-differential equations, which remain valid at discontinuities. Since each PD material point is free to interact with all material points within its neighborhood, PD theory is also a nonlocal theory. As illustrated in Fig. 1, the neighborhood of material point x is denoted by H_x with a radius of δ. All material points within H_x are called family members of **x**. The coordinates of the point in the undeformed and deformed configurations are denoted as **x** and **y**, respectively.

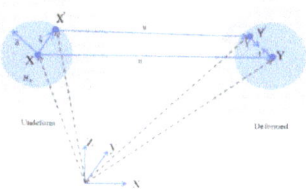

Fig. 1. Domain of interaction for a material point in PD theory.

2.2 Equation of Motion

In bond-based PD theory, the paired force density vectors are equal in values and parallel to the relative vector of the two points in deformed configuration. The equation of motion can be written as [4].

$$\rho(\mathbf{x})\ddot{\mathbf{u}}(\mathbf{x},t) = \int_{H_x} \mathbf{f}(\mathbf{u}(\mathbf{x}',t) - \mathbf{u}(\mathbf{x},t), \mathbf{x}' - \mathbf{x})dV + \mathbf{b}(\mathbf{x},t) \tag{1}$$

where ρ is the mass density, V is the volume, **f** is the pairwise response function, **b** (**x**, t) is the volumetric body force. The expression of PD force density is expressed as [4].

$$\mathbf{f}(\mathbf{u}' - \mathbf{u}, \mathbf{x}' - \mathbf{x}) = cs\frac{\mathbf{y}' - \mathbf{y}}{|\mathbf{y}' - \mathbf{y}|} \tag{2}$$

where c is the peridynamic material parameter, s is the bond stretch and it can be defined as [4].

$$s = \frac{|\mathbf{y}' - \mathbf{y}| - |\mathbf{x}' - \mathbf{x}|}{|\mathbf{x}' - \mathbf{x}|} \tag{3}$$

3 Peridynamic Model of Wind Turbine Tower

As a structural linking the nacelle and the floating platform, the base of the tower is typically in a fixed condition, while its top is connected to the nacelle and the wind turbine blades, with the latter end being essentially unconstrained. Due to this configuration, the deflection of the wind turbine blades causes the root of the tower to experience internal tensile forces. These forces usually lead to the stretching of the tower in turn and the growth of cracks potentially. In this study a peridynamic model is developed to simulation the deformation and crack propagation of the tower root under tensile loading condition.

3.1 Peridynamic Model Description

The IEA15MW wind turbine tower section from 15 m to 28 m is constructed according to information provided in [5]. As shown in Fig. 2, the outer diameter of the tower decreases linearly with the increasing height, while the thickness stays the same. The geometry parameters are shown in Table 1. The tower is made of steel, with its material properties as Young Modulus $E = 200\mathrm{e}11$ Pa, Shear Modulus $G = 793\mathrm{e}10$ Pa, and density $\rho = 785\mathrm{e}3$ kg/m^3.

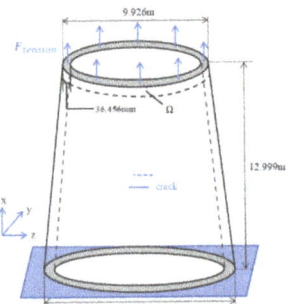

Fig. 2. Problem Description of the tower section under tensile loads.

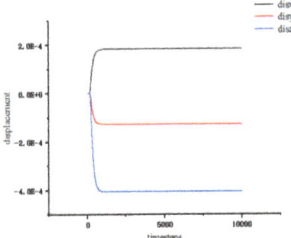

Fig. 3. Convergence Analysis of the Tower.

Table 1. Tower Dimensions as a Function of Height.

Height[m]	Outer Diameter[m]	Thickness[mm]
28.001	10.000	36.456
41.000	9.926	36.456

The PD discretization of the tower is set non-uniformly. The grid size in X and Y directions is $\Delta x_1 = 1.0 \times 10^{-1}$ m while in Z direction is $\Delta x_2 = 9.114 \times 10^{-3}$ m. The horizon size is chosen as $\delta_1 = 3.015\Delta x_1$ and $\delta_2 = 3.015\Delta x_2$, respectively. Since the energy release rate of the material is not clearly, the critical stretch parameter s_c is supposed to be 4.61e-3. The timestep is set as $\Delta t = $ 1e-6 s after convergence study.

Given that the tower is fixedly to the floating platform, the bottom of the tower model is fully constrained, while the remaining parts of the model are allowed to deform freely. Besides, the top surface of the tower is subjected to uniaxial tensile load $F_{tension}$ = 20 GPa, as shown in Fig. 2. In peridynamics, the above-mentioned external loads should be applied as body force density through the region Ω, which is actually made of the volume of one real layer of material points. Firstly, the accuracy of the model is verified by comparing the deformation both predicted by PD and ABAQUS without allowing failure, the analysis is conducted as a static one with the technique adaptive dynamic relaxation (ADR) [6] being used. Convergence analysis is considered prior to the crack propagation simulation. The displacements of a specific point in all directions in the peridynamic model are traced and presented in Fig. 3. Results show that the displacements end being stable after 1500 steps.

4 Numerical Results

4.1 Displacement and Crack Propagation Analysis

Secondly, initial central cracks are added to the model on both sides. The cracks are aligned with the *y*- axis and have a length of 2 m. Then the deformation and crack propagation are simulated. As presented in Table. 2, considering that the principal strain is oriented along the *x*- direction, the response of the tower in the *x*- direction is compared with that predicted by an identical model configured in ABAQUS. The results indicate that the average displacement is 10e-3 m and the maximum displacement occurs at the top of the tower model, with its value being 1.29e-2 m. The displacement distributes in a linear fashion along the tower axial direction. The good match of the displacement between the PD predictions and ABAQUS results validates the accuracy of the developed model, as shown in Fig. 4.

Table 2. Displacement distribution of the tower.

	Peridynamics	ABAQUS
Displacement distribution in X direction		

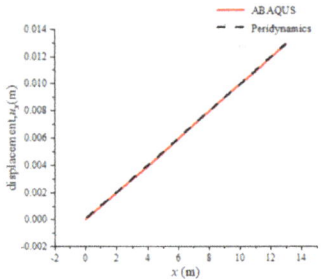

Fig. 4. Displacement in x- direction along the height. ($y = 0, z = 0$).

Regarding the crack propagation prediction, as shown in Table. 3, the crack at three time points are monitored in this case and the crack propagation path evolves as time increases. The crack initiates its propagation in the y- direction from the tip of the initial crack at 0.0028 s, and then the crack length steadily increases along the circular direction, which means that under the tension load, the crack exhibits an open mode appearance.

Table 3. Bond damage of the tower.

time	0.0028s	0.0048s	0.0060s
Peridynamics			

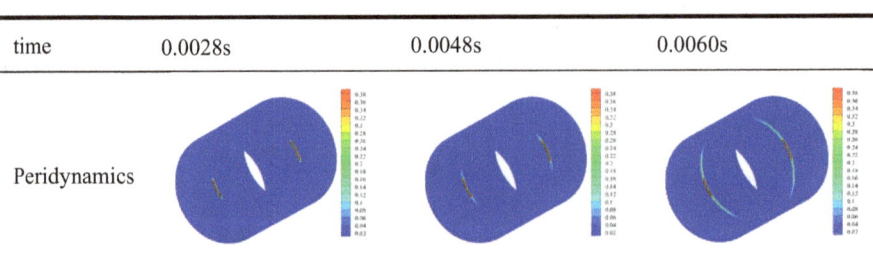

5 Conclusions

This paper develops a peridynamic model for the IEA 15 MW wind turbine tower root to investigate the crack propagation behavior under tensile loading condition. Firstly, the PD model is validated by matching the predicted deformation with the one obtained from ABAQUS. The analysis shows that under the axial loading condition, deformation experiences a linear distribution although the diameter of the tower changes. Then the crack propagation behavior is investigated to analyze the safety of the tower root, the results show that the crack propagates along the circular direction at a very high speed, which may lead to the collapse of the tower. This study output is helpful for the optimization of the tower structural design and safety analysis of the wind turbine tower. It should be noted that due to the limitation of the adopted bond-based peridynamic theory, complex loads such as bending moment cannot be directly imposed on the tower model. Also, only section not the whole tower is modelled and analysis due to the computational expenses. Therefore, further peridynamic analysis with high computational efficiency will be performed in the future, especially for the complex loading conditions such as out-of-plane loadings.

Acknowledgement. This research is funded by National Natural Science Foundation of China (Grant No. 52401317), the Shandong Higher Education Young Science and Technology Support Program (Grant No. 2023KJ082), and Shandong Provincial Natural Science Foundation (Grant No. 2024HWYQ-085).

References

1. Capaldo, M., et al.: Influence of cracks on the buckling of wind turbine towers. J. Phys: Conf. Ser. **1618** (2020)
2. Polyzois, D.J., Raftoyiannis, I.G., Ungkurapina, N.: Static and dynamic characteristics of multi-cell jointed gfrp wind turbine towers. Compos. Struct. **90**(1), 34–42 (2009)
3. Lin, L., Zhang, X., Zhang, D., Wu, X., Liu, Y., Wang, X., et al.: Damage evolution and failure analysis of the advanced transition segment behavior of wind turbine tower. Eng. Fail. Anal. **152** (2023)
4. Silling, S.A., Askari, E.: A meshfree method based on the peridynamic model of solid mechanics. Comput. Struct. **83**(17/18), 1526–1535 (2005)
5. Gaertner, Evan, et al.: Definition of the IEA 15-megawatt offshore reference wind turbine(2020)
6. Kilic, B., Madenci, E.: An adaptive dynamic relaxation method for quasi-static simulations using the peridynamic theory. Theor. Appl. Fract. Mech. **53**(3), 194–204 (2010)

Open Access This chapter is licensed under the terms of the Creative Commons Attribution-NonCommercial-NoDerivatives 4.0 International License (http://creativecommons.org/licenses/by-nc-nd/4.0/), which permits any noncommercial use, sharing, distribution and reproduction in any medium or format, as long as you give appropriate credit to the original author(s) and the source, provide a link to the Creative Commons license and indicate if you modified the licensed material. You do not have permission under this license to share adapted material derived from this chapter or parts of it.

The images or other third party material in this chapter are included in the chapter's Creative Commons license, unless indicated otherwise in a credit line to the material. If material is not included in the chapter's Creative Commons license and your intended use is not permitted by statutory regulation or exceeds the permitted use, you will need to obtain permission directly from the copyright holder.

Author Index

A
An, Zhong-chang 244

B
Bai, Xu 330, 339
Bai, Yichao 1
Bi, Cheng 62
Bi, Jingqiang 666
Bi, Yuantao 41, 74, 711

C
Cai, Feng 224
Calheiros-Cabral, Tomás 118
Cao, Jun 85
Chang, Ming 613
Chen, Chen 275
Chen, Haoyu 275
Chen, Jianjun 62
Chen, Jianting 318
Chen, Pengfei 95
Chen, Shagu 633
Chen, Shude 74, 724
Chen, Weimin 309, 650
Chen, Wenjin 666
Chen, Xueli 623
Chen, Yan 897
Chen, Yikang 460
Chen, Yingying 942
Chen, Yong 868
Cheng, Hong 266
Cui, Can 734
Cui, Guoqu 837

D
Dai, Min 143
Dai, Shaoshi 694
Deng, Wei 74
Deng, Zhipeng 460
Ding, Weiqi 565
Ding, Xin 559
Du, Xiaoyu 254

F
Fan, Cunlong 602
Fan, Zan 868
Fang, Chao 108
Fang, Cuiwen 254
Fang, Zhen 297
Fang, Zhuo 862

G
Gao, Kaixian 402
Gao, Lei 868, 877
Gao, Yan 933, 942
Gao, Yuan 633
Gho, Wie-Min 877
Giannini, Gianmaria 118
Gu, Chen 814, 827, 837
Gu, Dongliang 266
Gu, Haoyun 565
Guan, YanJun 742
Guan, Yanjun 756
Guo, Ruinan 921
Guo, Tao 572
Guo, Yu 108

H
Han, Bing 587, 602
Han, Duanfeng 921
Han, Leilei 215
Han, Peng 527
Han, Yijian 74, 724
Han, Yundong 394
Hao, Hao 127, 135, 650
Hong, Sa Young 777, 802
Hong, Yang 814
Hou, Jianjun 224, 476, 515
Hou, Jing 912
Hou, Lingyi 515
Hu, Dinghua 676
Hu, Hao 62
Hu, Hewen 62
Hu, Jianlong 551
Hu, Qinyou 623

Hu, Shenping 254, 411, 602
Huang, Fenglai 496
Huang, Fuxiang 921
Huang, Haibo 152
Huang, Jun 95
Huang, Qian 394
Huang, Qianwen 171
Huang, Shenghui 884, 891

J

Ji, Hongtao 85
Ji, Renwei 330
Jia, Tianxia 814
Jiang, Haoran 402
Jiang, Xiaogang 237
Jiang, Xiao-gang 244
Jiang, Xiaogang 537
Jiang, Yongxin 421
Jiao, Tingyu 565
Jin, Fang 587
Jin, Jiangang 847
Jin, Wei 143
Jun, Song 207

K

Ke, Wu 687
Kim, Byoung Wan 777, 802
Kong, Yulong 430, 488

L

Lan, Guohui 394
Lee, Kangsu 777, 802
Lei, Guoqiang 330
Lei, Yu 62
Li, Bibo 95
Li, Bin 108
Li, Chunxiao 62
Li, Dawei 443, 452, 515
Li, Di 1
Li, Donghai 884, 891
Li, Hongqiang 303
Li, Jinxing 15
Li, Jun 613
Li, Liwei 912
Li, Minghai 215
Li, MingKai 742
Li, Song 143
Li, Tao 572
Li, Wei 189

Li, Weifeng 369, 377, 384
Li, Yingjie 41, 711
Li, Yingying 827, 837
Li, Yongyue 309
Li, Yuhao 360
Li, Zhuang 402
Li, Ziqi 785
Liang, Li 868
Libo, Yao 207
Lin, Lin 551, 559
Lin, Pengfei 613
Lin, Yu 95
Liu, Jing 303
Liu, JingJing 15
Liu, Jingjing 694, 724
Liu, Peng 152
Liu, Ruichao 62
Liu, Sihan 496
Liu, Song 868, 877
Liu, Xiao 587
Liu, Xin 62
Liu, Yajun 687
Liu, Yun 659
Lu, Nan 29
Lu, Yu 847
Luo, Jiaxuan 339
Luo, Pingping 347
Luo, Xiaofang 339
Luo, Ya 933
Lv, Mingdong 237
Lv, Ming-dong 244
Lv, Mingdong 537

M

Ma, Chunhui 496
Ma, Hairui 215, 394
Ma, Yichen 496
Mei, Bin 369, 377, 384
Meng, Fanjun 394
Meng, Xiangkun 297, 303
Meng, Xiangqian 411

N

Nan, Mingjun 62
Nie, Yan 95
Ning, Dezhi 884, 891
Niu, Jianjie 330
Niu, YuXiang 742
Niu, Yuxiang 756

Author Index

O
Ogooluwa, Idwimoh Daniel 347

P
Pan, Weibo 108
Patlaichuk, Oleksii V. 53

Q
Qi, Yingchun 711
Qiao, Xizhen 666
Qin, Tingrong 347, 623
Qin, Zhengchuan 360
Qu, Wenxin 85
Qu, Zhipeng 814

R
Ramos, Victor 118
Ren, Haikui 318
Ren, Tie 767
Rosa-Santos, Paulo 118

S
Serbin, Serhiy I. 53
Sheng, Minghui 171
Shi, Guoyou 369, 377, 384
Shi, Jinkun 877
Shi, Zeyu 847
Sim, Kichan 777, 802
Song, Chengguo 572
Su, Lei 189
Su, Shipeng 505
Su, Yi 152
Sui, Yi 847
Sun, Hang 443, 515
Sun, Jiayang 767
Sun, Nan 921
Sun, Yaogang 921
Sun, Yindong 687
Sun, ZiFeng 742
Sun, Zifeng 756
Sung, Hong Gun 777

T
Tang, Ruitao 496
Taveira-Pinto, Francisco 118

Teng, Bin 785, 793
Tu, Tianyu 676

W
Wang, Fei 15, 724
Wang, Hanjia 527
Wang, Hongbo 402
Wang, Jiewen 877
Wang, Lihao 734
Wang, Lucai 179
Wang, Maquan 62
Wang, Min 602
Wang, Minghui 868
Wang, Shiyun 827, 837
Wang, Xiao 285, 452, 476
Wang, Xinxin 74
Wang, Xi-wei 244
Wang, Xu 285
Wang, Yihou 767
Wang, Yongtao 587
Wei, Zhiyuan 476
Wen, Renqiang 814
Wu, Hao 572
Wu, Jianjun 411
Wu, Ming 488
Wu, Qihang 29
Wu, Shihao 847
Wu, Xianjun 171
Wu, Yong 572, 587
Wu, Zicheng 1

X
Xi, Yongtao 152, 602
Xia, Zhaowang 551, 559
Xie, Tianhua 179
Xie, Xiaozhong 633
Xing, Dan 85
Xing, Lei 309
Xing, Sihao 711
Xiong, Junting 198
Xu, Aijin 897
Xu, Lei 613
Xu, Weiping 897
Xu, Weizhe 85
Xu, Wenbin 687
Xu, ZiKai 15, 41

Xu, Zikai 694
Xue, Jiahui 734

Y

Yan, Haoxuan 602
Yan, Haoyu 452
Yan, Junhan 897
Yan, Linglong 602
Yang, Bo 285
Yang, Haoyu 443
Yang, He 659
Yang, Jinglei 460
Yang, Qi 198
Yang, Xue 297, 303
Yang, Ye 877
Yang, Yilin 127, 135
Yang, Yuepeng 505, 527
Yang, Zhenbang 330
Ye, Jiayun 369, 377, 384
Yi, Kan 837
Yin, Binggang 921
Yin, Zhaolong 862
Yu, Fu 421
Yu, Guanghui 189
Yu, Kaibo 224
Yu, Mei 793
Yu, Peng-yao 244
Yu, Qing 360
Yu, Taocui 767
Yu, Zhibing 912
Yuan, Lin 897
Yuan, Zhijang 237
Yuan, Zhi-jiang 244
Yuan, Zhijiang 537
Yuan, Zhuoping 143
Yuehua, Peng 207

Z

Zan, Yingfei 339, 921
Zang, Zhipeng 862
Zavvar, Esmaeil 118
Zeng, Ji 1, 29, 565
Zhang, Chongwei 884, 891
Zhang, Chunyu 659
Zhang, Heng 687
Zhang, JunSheng 742
Zhang, Junsheng 756
Zhang, Li 309, 318
Zhang, Liang 676
Zhang, Rundong 912
Zhang, Wanqing 756
Zhang, Wei 15, 41, 74, 694, 724
Zhang, Wen 330
Zhang, Wenjun 297, 303
Zhang, Xiao 95
Zhang, Xiaojing 347
Zhang, Xinxin 254
Zhang, XinYu 742
Zhang, Yong 62, 198, 266
Zhang, Yongjie 443, 452
Zhang, Yu 565
Zhang, Yuchen 207
Zhang, Yujia 369, 377, 384
Zhang, Zhiyou 421, 430
Zhao, Congcong 369, 377, 384
Zhao, Heda 613
Zhao, Jianjian 62
Zhao, Lei 676
Zhao, Limeng 862
Zhao, Lujun 756
Zhao, Ruipu 496
Zhao, Xianrui 53
Zheng, Jiangshan 827
Zheng, Mao 572
Zheng, Minmin 127, 135, 650
Zheng, Yue 814
Zheng, Zhenyu 430, 488
Zheng, Zhilin 179
Zhou, Guohong 15
Zhou, Xiangxing 360
Zhou, Xiangyu 297
Zhou, Xiang-Yu 303
Zhu, Bihong 827, 837
Zhu, Liu 411
Zhu, Xiaoming 402
Zhu, Xunhao 884, 891
Zuo, Bin 505